AMERICAN DEFENSE POLICY

AMERICAN DEFENSE POLICY

Sixth Edition

Edited by

SCHUYLER FOERSTER and EDWARD N. WRIGHT

The Johns Hopkins University Press, Baltimore and London

The Johns Hopkins University Press, 701 West 40th Street, Baltimore, Maryland 21211
The Johns Hopkins Press Ltd., London

The views and conclusions expressed by the military personnel
in articles contained in this volume are those of the authors and
not necessarily those of the USAF Academy, the Department
of Defense, or any particular Service thereof.

∞

The paper used in this publication meets the minimum requirements
of American National Standard for Information Sciences—
Permanence of Paper for Printed Library Materials,
ANSI Z39.48-1984.

Library of Congress Cataloging-in-Publication Data

American defense policy / edited by Schuyler Foerster and Edward N. Wright
6th ed.
p.   cm.
Includes bibliographical references.
ISBN 0-8018-3867-3 (alk. paper)
ISBN 0-8018-3868-1 (pbk. : alk. paper)
1. United States—Military policy.   I. Foerster, Schuyler.   II. Wright, Edward N.
UA23.A626   1990
355′.033573—dc20   89-24687   CIP

CONTENTS

FOREWORD
Teaching American Defense Policy

ERVIN J. ROKKE, BRIGADIER GENERAL, USAF

The birth of the United States Air Force Academy coincided with the genesis of a major defense debate in this country about the role of nuclear weapons and the increasingly complex nexus of strategy, force procurement, and arms control. It seemed to many that the unprecedented availability of destructive military power had produced with it an unprecedented vulnerability of the nation-state. Recognition of this "security dilemma" coincided in turn with the realization that the spectrum of conflict had widened dramatically. America's national security interests were not to be as easily solved by "the Bomb" as some had believed in the lengthy shadows of Alamogordo, Hiroshima, and Nagasaki.

One charter of the new Air Force Academy, in the words of the 1950 Stearns-Eisenhower Report, was for graduates "to anticipate the dynamism of the nuclear age, providing direction to the profession based on an intellectual understanding of the future and an operational experience with the past." Teaching defense policy at the Air Force Academy is thus both an academic and a professional endeavor. Students are not principally scholars who digest the subject as detached observers; rather, they are enjoined to examine both the content and the process of defense policy as soon-to-be practitioners, implementers, and eventually, policymakers themselves.

The professional imperative of teaching defense policy at the Air Force Academy took shape in that early defense debate, and this volume represents the sixth edition of a textbook produced by the Academy's Department of Political Science to meet that imperative. The first edition of this book, published in 1965, was an early attempt to pull together in a single volume much of the literature of that defense debate. Edited by Colonel Wesley W. Posvar, first Permanent Professor and Head of the Department of Political Science at the Academy, and Captain John C. Ries, the first edition broke new ground in defining the parameters of a fledgling field of study called "defense policy."

Subsequent editions of this text have empha-

sized different theoretical and practical issues as the focus of defense policy analysts changed over the years. The second edition, published in 1968 and edited by Major Mark E. Smith III and Lieutenant Colonel Claude J. Johns, Jr., was essentially a new reader, indicative of the growth of the literature in the field. A substantial portion of that volume dealt with the theory of limited war and the problems of conflict and escalation control.

By the early 1970s, the US had reached a new milestone in its own strategic thinking, both at the strategic nuclear level and in its understanding of the perils of protracted yet limited war. The third edition of this book, which I coedited with Major (now Brigadier General) Richard G. Head, appeared in 1973. That volume included a retrospective look at the evolution of strategy, the early Strategic Arms Limitation (SALT) negotiations, and insurgency. Two relatively new areas were also included: a detailed examination of bureaucratic politics in defense policymaking, with special attention to weapons procurement, and the problems of a professional military in a democratic society.

Emphasis on defense decisionmaking and the identity of the military profession within a democracy dominated the fourth edition, edited by Colonels John E. Endicott and Roy W. Stafford, Jr., and published in 1977. By the time the fifth edition was published in 1982, the variety and complexity of the defense debates in this country had grown to tremendous proportions. Editors John F. Reichart and Steven R. Sturm, both now Lieutenant Colonels, found themselves dealing with fundamental debates on nuclear strategy, arms control, regional interests, defense decisionmaking, and the sociological and moral dimensions of the military profession.

For several years now, this country has been embroiled in a series of debates reminiscent of the early defense debates which animated the editors and contributors of the first edition. The questions raised by these debates go to the very heart of the profession. What is the utility of nuclear weapons? What are the conditions of strategic stability? What is the impact of new technology on strategy? Is there, indeed, the prospect of defense against nuclear weapons, and what are the implications of such a defense? What is the role of space as a new dimension of international and even military competition? How does a democracy sustain its military prowess in a world in which threats range from the invisible individual terrorist to the prospect of "nuclear winter"? How relevant and enduring are 40-year-old alliance systems? How capable is an increasingly pluralistic policy process in meeting these new demands?

Now, suddenly, there are new questions as well, occasioned by the revolutionary changes that gripped the Communist world as the 1980s drew to a close. What are the prospects for transforming the East-West confrontation that determined much of American postwar thinking on defense and security? What are the long-term political and military effects of arms control agreements that substantially change not only the threats to the West, but also the West's ability to defend? What impact will these developments have on our alliances and, indeed, on our own defense outlook?

One of the unique contributions of the early editions of this book was to bring together in a single volume the essential core of that defense literature and provide a framework to make that literature intelligible to the undergraduate student. The quantity of that defense literature has grown exponentially over the years, however, and no single volume can do justice to its breadth, variety, and nuance. Moreover, we have found that students of defense policy, at this and other undergraduate institutions, respond best to a return to basics. Today's undergraduates were only being born when the first edition was published and the second was being prepared. The Cuban missile crisis seems like ancient history, Vietnam a crucible of an-

other generation, and "détente" a word that has no real historical context.

This sixth edition, then, departs from the format of previous editions. The editors, Lieutenant Colonels Schuyler Foerster and Edward N. Wright, have taken on the task of writing a textbook rather than assembling an updated compendium of readings. Both are Tenure Associate Professors in the Department of Political Science at the Air Force Academy. Colonel Foerster (D. Phil., Oxford) is Director of International and Defense Studies, and Colonel Wright (Ph.D., Georgetown) is Director of American and Policy Studies. Contributors to this volume are predominantly Air Force officers and alumni of the Political Science Department who are specialists in their respective areas. Their work represents a continuing tradition in the Department where teaching defense policy to future Air Force officers assumes prominent importance.

It is often said that defense policy is "Janus-like," involving both international and domestic dimensions. This dichotomous theme has characterized all previous editions and is continued in this volume. The student of defense policy, whether from an academic or a professional orientation, will always have to reconcile those two dimensions. In teaching defense policy, we hope to convey to students that such a confluence of domestic and international factors makes strategic choices less than clear. Whatever the strategic questions, they will always have to be confronted in the context of a pluralistic political system where rationality is elusive and variously defined. We hope this volume will assist those who will assume positions of professional leadership and all of us who participate as citizens in the definition of national security in understanding the complexity of defense in a nuclear age fast approaching its half-century.

PREFACE

Reminiscing on the purposes of undergraduate education, former British Prime Minister Sir Harold Macmillan recalled one of his professors at Oxford who had noted, "nothing that you will learn in the course of your studies will be of the slightest possible use to you—save only this—that if you work hard and intelligently, you should be able to detect when a man is talking rot." This volume is an attempt to convey the context, complexity, and character of American defense policy so that students, citizens, and professionals may be better able to detect when someone "is talking rot."

In our combined 13 years on the faculty of the Air Force Academy, we have discovered that getting students to know when someone "is talking rot" is a difficult task indeed. Nowhere is this more true than in the realm of defense policy. Perhaps it is the nature of the subject. Issues of peace and war are no trivial matters. Emotional and ideological predispositions as well as the urgency of the subject often entice one to simple answers. For better or for worse, we have found no simple answers in this business.

Our task is certainly not to convey simplicity. The pursuit of national security has always been riddled with dilemmas and contradictions. A century and a half ago, Clausewitz offered the seductively simple proposition that "war is the continuation of politics by other means." He eschewed, however, any litany of "principles of war" such as those outlined by many of his contemporaries. A century later, E. H. Carr, in his classic *The Twenty Years Crisis, 1919–1939,* analyzed the dialectical relationship between idealism and realism. Though he clearly leaned toward the latter, Carr could not articulate any clear synthesis of the two. In recent decades, the nuclear age and the continuing revolutions in both international politics and military technology have only exacerbated these contradictions, while exaggerating their consequences.

NATURE OF THE BOOK

This edition of *American Defense Policy* takes a different approach than its predecessors. Previous editions took the form of readers, compendia of published articles designed to capture the highlights of contemporary defense debates. In initiating this series of textbooks in the early 1960s, the Air Force Academy Political Science Department had sought to bring together in a single volume much of the growing defense policy literature that was not always readily accessible to students or professionals. Over the years, however, this literature has grown to tremendous proportions, both in breadth and depth. Excellent compendia have been published that focus on discrete aspects of defense policy, whether it be nuclear strategy, conventional force posture, theater defense, maritime issues, arms control, defense decisionmaking, the weapons acquisition process, manpower and personnel, or broader civil-military relationships. Nonetheless, we doubt that a single defense policy reader can do justice to the richness of that literature.

In our own efforts to teach defense policy at the undergraduate level, we have also observed that, as the literature grew more diverse and the theoretical and policy debates became more complex, it was harder to find pieces that laid out the essential first principles upon which an understanding of those debates rested. Neither were there many good summary pieces that surveyed the evolution of American defense policy or the broad scope of defense decisionmaking, even though many excellent detailed studies in both areas have appeared.

In producing this sixth edition, then, we have sought to provide a basis for teaching both undergraduates and graduate students, a survey of the field for both the professional and the lay citizen. We have tried to introduce the literature of defense policy through the inclusion of both bibliographic essays and selected readings. By incorporating some of the more timeless pieces in the defense policy literature, we hope to have preserved the best of the compendium tradition. Our principal purpose, however, has been to lay out the essential historical and theoretical foundations for understanding contemporary defense policy issues.

A formidable challenge in producing such a textbook is to maintain a balance between the need to teach the fundamentals of defense policy and the desire to offer an up-to-date treatment of contemporary policy issues. The result has been a protracted process of writing, editing, and updating in an attempt to keep pace with the quickening drumbeat of change. We have discovered, not surprisingly, that a change in presidential administrations offers a convenient plateau for shifting from editing to production, even if that plateau is anything but flat. We expect that future editions will appear every four years, following the inaugural year of a new presidential term of office, when its goal will be to present a foundation for understanding the issues faced by the president in the new administration.

The editorial burden in a compendium typically revolves around the selection, placement, and introduction of written work. In a textbook such as this, however, the editorial function includes authorship itself. Over the years, the Air Force Academy faculty has been privileged to include within its ranks a steady stream of officers who distinguished themselves both as professionals and as scholars. We have been fortunate to be able to draw upon this talent and teaching experience in soliciting authors for each of the chapters and for most of the bibliographic essays. Similarly, we have been fortunate to have within our classrooms some especially bright and talented cadets, some of whom we are glad to introduce in this volume as authors of bibliographic essays.

Between the two of us, we were able to divide the editorial labor according to our respective academic and professional specialties, consistent with our view that defense policy has both domestic and international dimensions. As director of International and Defense Studies within the Political Science Department, Sky Foerster assumed responsibility for those parts of the book dealing with the content of American defense policy. As director of American and Policy Studies, Ed Wright took charge of those parts dealing with defense decisionmaking. We each authored or coauthored the first chapters of each part of the book to preserve thematic continuity. As the senior editor, Ed also took on the burden of managing the administration of the project.

FORMAT OF THE BOOK

This volume is divided into four parts. Part I explains the context in which American defense policy is made. Opening chapters discuss the international and domestic political environments of defense policymaking. The next chapter introduces the nature of Soviet power, since the Soviet Union remains the principal focus of US defense policy. Finally, we include an overview of the theoretical foundations of defense policy, with particular attention to the problems of defense and deterrence in the nuclear age.

Part II spans both various global regions and the spectrum of conflict that US policy must address. We begin by considering the role of the US as a world power, empirically and historically, and then discuss US policy toward the Soviet Union since World War II. The next two chapters address, in turn, the evolution of commitments and strategies for defense in Western Europe and Asia, where the US retains deployed forces as part of formal alliances. This part concludes with a look at a growing concern that is global rather than regional in nature—the need to project power in defense of US security interests outside of standing alliance structures—and the problems associated with the use of military force as an instrument of American policy.

After this broad survey, Part III focuses on the interaction of diplomacy and defense policy in the pursuit of arms control. The opening chapter discusses the theory of arms control and the scope of contemporary arms control efforts. Subsequent chapters provide an overview of the US experience with arms control negotiations, both at the strategic nuclear level and as they apply to the European nuclear and conventional balance.

The last part of the book examines the defense policymaking process, the domestic political context in which strategic choices are made and the instruments of strategy are developed. The first two chapters of this part survey the essential elements of defense policymaking, the institutional structures of government that play in that process, and the role of military officers in particular. The next chapter introduces force planning, which is receiving greater attention as the inexorable decline in resources increasingly constrains weapons procurement decisions. Force planning, a process that takes place largely within the Department of Defense, is an attempt to translate doctrinal choices into force structures. That planning process ultimately interacts with a larger policy process that involves all aspects of the American political structure. Given the events surrounding the Iran-contra affair and the debate, especially in the military profession, concerning the proper role of military officers in the policy arena, it is especially fitting that the final readings in this part pertain to this issue, which goes to the heart of the military's place in American society.

The chapters included in each part of the book represent our attempt to define the contours of American defense policy—the nature of its challenges, the scope of the policy instruments available, and the systemic and institutional pressures on the decisionmaking process. We have tried to inject a historical perspective in understanding these issues, but we must not allow ourselves to become prisoners of the past. Too often we view the past as prologue to the present, ignoring the revolutionary ways in which the future will bring change. The last chapter, therefore, takes us into that future, where the dynamics of politics, economics, and technology will confront us with new challenges as we attempt to create a more secure world.

ACKNOWLEDGMENTS

A volume such as this can only be produced with the aid and counsel of a substantial number of people. As with the previous editions, this book reflects the insights and lessons learned from teaching defense policy to successive classes of cadets at the Air Force Academy. It is through teaching that we have been able to crystallize and refine our own thinking on this vital subject. In that sense, publication of successive editions of a textbook on American defense policy involves the entire Department of Political Science.

This book has benefited from the comments of hundreds of cadets who were subjected to successive drafts of the manuscript in our courses at the Air Force Academy and from the critiques of our colleagues in the department. While serving as the head of the department, Colonel Robert P. Haffa, Jr., played an instrumental role in the early decisions about the content and the focus of the book. The continued support provided by Colonels Douglas J. Murray and Paul R. Viotti, now the department head and deputy head respectively, has been essential to whatever success we have had in achieving what we set out to accomplish in this book. Others who merit special attention are former editors—especially Brigadier Generals Ervin J. Rokke and Richard G. Head—and Dr. Curtis Cook, former acting head of the department, whose review and encouragement were most valuable in the early decisions about the book's approach. Editorial and research assistance provided by Lieutenants Andrew Carlson, Richard Klumpp, and especially Diane Provost was invaluable in the final completion of the text. They are outstanding officers and future leaders in our Air Force.

We would especially like to thank the staff of Johns Hopkins University Press for their ever-present patience and understanding. Henry Tom has trained a succession of editors of this volume as well as its companion, *The Defense Policies of Nations*. Both personally and as a department, we remain in his debt. The able assistance of Denise Dankert and the editorial staff at the press have been nothing less than outstanding. Similarly, Therese Boyd and the production staff were superb. In addition, we were blessed with the support of a truly remarkable copy editor, Anne Schwartz, whose expert assistance and dedication to consistency have enormously improved the quality of this work.

Finally, we are indebted to our wives, Jan and June, and our families for their support; for their understanding on the too many nights when the book, rather than their needs, required our attention; and for their patience with the distractions that an undertaking such as writing and editing a book always produces. May we always be deserving of the love they have shown throughout this endeavor.

CHAPTER 1

THE TWIN FACES OF DEFENSE POLICY INTERNATIONAL AND DOMESTIC

SCHUYLER FOERSTER and EDWARD N. WRIGHT

This is a book about American defense policy. As such, it addresses the various dimensions of America's attempt to "provide for the common defense" as mandated by the Constitution. Understanding America's approach to its national security, however, requires a sensitivity to both the international and domestic contexts in which defense policy evolves. That is the purpose of this first chapter. The next chapter addresses more specifically the nature of Soviet power, since the Soviet Union has been the principal competing power for the United States and the central—though hardly the sole—object of American defense policy. Finally, the last chapter of this part of the book surveys in some detail the theoretical foundations of deterrence in the nuclear age. The nuclear age poses some unique problems for those charged with providing for the common defense, for which there are rarely clear solutions.

The issues addressed in this chapter are not unique to the nuclear age. Violent international conflict has always been a feature of human history. The causes of war have been debated for centuries. Some have suggested that the tendency for conflict and violent behavior lies in human nature itself, that state aggression is merely an extension of human aggression. Others have argued that conflict results from the way societies are ordered. Groups discover that there is power in their collective numbers and may act with a boldness that individuals might otherwise not possess. In groups, individuals can be anonymous instruments of violence while avoiding personal responsibility for their actions. Likewise, organizations pursue their interests jealously. Organizations articulate values which can often be more powerful than personal values, especially if those values are alleged to apply to all. Classes—social strata based on economic or other status criteria—may assume positions of dominance or subordination within society. Marx, for example, viewed conflict as the natural outgrowth of classes struggling either to preserve or to overthrow such a hierarchical structure.

These explanations of violent conflict exist essentially on two different levels, one having to do with the nature of the individual, or human nature, and one having to do with the nature of the groups, organizations, classes or other collective structures in which societies are ordered.[1] In a way, states are an extension of this second "level of analysis." A form of political organization which has dominated modern political life since the seventeenth century, states share four key characteristics: a population, fixed territory, a government, and sovereignty, the ultimate authority possessed by government to maintain order, provide security, and fulfill other functions within a state.

For many, the causes of war lie in the very nature of states. States of course differ in their forms of government, the scope and source of their governments' sovereignty, and the purposes which their governments are expected to fulfill. The simplest distinction is perhaps that between "authoritarian" and "democratic" governments, in which the latter's sovereignty derives from the consent of its citizens and is limited by constitutional or other institutional means. The eighteenth-century philosopher Immanuel Kant, for example, argued in his *Perpetual Peace* that the causes of war lay in the ambitions of government, but the victims of wars were the citizens.[2] If governments were accountable to their citizens, therefore, warfare could be rendered obsolete because governments would be prohibited by their citizens from waging it.

Contemporary states do not fall neatly into these twin categories of authoritarian and democratic. What's more, we know that contemporary nationalism, as other forms of political fervor, is quite capable of motivating citizens to violence on the state level. An additional element for understanding modern warfare, then, is the role of ideology. Ideologies are "world views," or sets of values which both explain the way the world is and outline the world as one would like it to be. Wars may result because pursuit of a world "as one would like it to be" may be viewed as justifying the violence of war. Indeed, in modern times, states have often gone to war either to impose their view of what the world ought to be or to defend this against the challenge posed by those with a different view. Thus, "making the world safe for democracy" is no less ideological in its purpose than "making the world safe for socialism" or the pursuit of "living space" for the Aryan "superman," however preferable the former may be. Western notions of a just war in fact pivot on the question of whether the values to be achieved by war are the "right" ones.

Values, then, are very much a part of the nature of warfare and, by implication, of the defense policies which nations adopt to protect themselves and their ways of life. If we accept the definition of politics as "the authoritative allocation of values in society," then war and defense policy are very much political phenomena. Defense policies are shaped by the politics of both the international system in which states exist and the domestic system of government in which defense policy must be made.

For some international relations theorists, the structure of the international political system—a third level of analysis—is itself a cause of war.[3] The modern state system is often viewed as one of sovereign states who are autonomous actors without any overarching authority. States thus compete with each other in the pursuit of their own values. In the absence of any world government authority, that competition often takes violent forms as states rely on self-help to preserve their integrity and advance their interests. Countries must therefore deal with this reality of an anarchic world in providing for their defense. *Defense* is indeed the proper word. It is an elastic word, allowing propositions such as "the best defense is a good offense," equally applicable to the stadium and to the battlefield. The implications of such an international system for defense policy will be surveyed in greater detail below.

A state's defense policy is more than a response to international political factors, however. States have goals other than national defense. In the Preamble to the US Constitution, government is established not only to "provide for the common defense" but also to "form a more perfect Union, establish Justice, insure domestic Tranquility, . . . promote the general Welfare, and secure the Blessings of Liberty to ourselves and our Posterity." In some respects, these charters are complementary. Certainly "the common defense" facilitates the fulfillment of other goals. Yet, securing the means for providing the common defense is not an open-ended charter either. Americans are traditionally wary not only of governmental authority but also of elevating too high the position of the military profession in society. Authoritarian regimes often possess an efficiency in policy-making and a latitude in allocating national resources that enable the effective mobilization of military power. A democracy, on the other

hand, must always confront the tension between being able to mobilize military power to serve its defense requirements and preserving the very fabric of its political institutions. In the United States, for example, the structures and processes of governmental decisionmaking reflect the democratic foundations of American political values. Defense policy decisions—on strategy and on the creation, structure, and use of military force—occur in the context of this pluralistic political environment. This, too, is addressed in some detail below.

THE INTERNATIONAL CONTEXT: SECURITY IN THE MODERN WORLD

The nineteenth-century military thinker, Carl von Clausewitz, whose work *On War* remains a masterpiece of enduring relevance, stressed that "war is the continuation of politics by other means."[4] Politics is about conflict—the conflict between competing values and interests—and war is its violent manifestation at the state level. There are, in effect, only three ways to resolve conflict—consensus, compromise, or coercion—and societies' principal tasks involve the management and, if possible, peaceful resolution of conflict on whatever level it exists. If all parties agree on a particular issue, interest, or value, then resolution of conflict is clear. If all parties do not agree, but are willing to cooperate in accommodating each others' interests, then there is at least the potential for compromise. If all parties do not agree, and prefer to compete using the resources available to them rather than find a compromise, then the resolution of conflict will often result from one side's successful coercion of others. In practice, societies engage in a mix of these three alternatives in managing both domestic conflict and conflict between states.

One need not conclude that states are, by their nature, inclined to violent conflict and coercive aggression to accept the need for military force as an instrument of defense. Michael Howard, for one, has argued "that force is an ineluctable element in international relations, not because of any inherent tendency on the part of man to use it, but because the *possibility* of its use exists."[5] In many respects, the possibility of the use of force—and, hence, at least one cause of war—is attributable to the nature of the international political system itself.

The origins of the modern state system, in which sovereign states are the principal actors, are generally traced to the Peace of Westphalia (1648), which ended the Thirty Years War. Sovereign states, which by definition are free from any higher authority, were recognized as free and equal members of an international society, thus marking the end of an era in which a unified Church had provided a relatively cohesive and overarching authority. At about the same time, Thomas Hobbes wrote the *Leviathan*, an explanation of the growth of government in society, which continues as an analogy for understanding the nature of the international political system as well.[6]

Hobbes began with the premise that man, in a hypothetical "state of nature," was motivated principally by the "fear of violent death." This state of nature was an anarchic condition, where life was "nasty, brutish, and short." Conflict was endemic, and coercion by the threat or use of force was the normal way of resolving conflict. As an analogy for international politics, this Hobbesian view holds that sovereign states—like Hobbes's sovereign individuals—exist in a condition of perpetual conflict, either potentially or in fact. Lacking any externally imposed moral or legal restrictions, states are free to resort to violent means to achieve their ends, constrained only by existing power relations. International law is only consensual; states are bound to restrictions to which they consent, but there is no higher enforcement mechanism to compel compliance with those restrictions. In such a world, security derives fundamentally from the ability to pose sufficient countervailing power to insure against attack. Maintenance of whatever order states seek within the system is a self-help enterprise.

The logic of this Hobbesian view does not rule out a search for order. For Hobbes, man "freed" himself from this anarchic existence by voluntarily surrendering sovereignty to a higher authority, the "Leviathan" (government), whose principal charter was the maintenance of law and order and the preservation of security. Thus, the notion of the "social contract" was born, which others like John Locke and Jean-Jacques Rousseau—with more optimistic assumptions about human nature—developed, providing in turn the philosophical foundations of Western democracy. As an analogy for international relations, this notion of a social contract is suggestive. In the same way that man, out of self-preservation, emerged from the state of nature and "contracted" to form a higher governmental authority, many have hoped that states would, in similar fashion, create a world government to which they could surrender some or all of their authority, rather than continue the competitive and often violent existence which has characterized the modern state system.

While that goal of world government has remained elusive, states have at various times, and with varying degrees of success, sought to regulate their interaction to their mutual benefit and security. Two mechanisms in particular have dominated this search for order: international law and balance of power. Since the genesis of the modern state system, states have

devised "rules of coexistence" to provide both order and predictability to their affairs. Immunity for diplomatic envoys, which was systematically recognized as early as the sixteenth century, is but one such rule; the inviolability of treaties, still a central tenet of international law, is another.[7] International law remains constrained by its consensual nature, and a private commercial contractual relationship provides a more precise analogy than does civil law. Nonetheless, international legal regimes offer vehicles for maintaining order in which at least some conflicts are managed by compromise and bargaining rather than by coercion.

While states have attempted to apply international law to the problems of war, this is perhaps the area where its effectiveness has been the most limited. International law requires that states consent to be bound by it; hence, some commonality of interest is necessary to enable compromise. Such conditions rarely exist when states have resorted to force to resolve disputes. There are examples of laws which have effectively regulated the conduct of war (such as the Geneva Convention on prisoners of war), but laws regarding the initiation of conflict have often been ignored out of expediency.

Traditionally, the more effective instrument for restraining the initiation of violent conflict between states has been "balance of power." A diplomatic rather than legal institution of international politics, balance of power considerations have been at the heart of most European attempts to preserve peace, from the Treaty of Westphalia in 1648, to the Treaty of Utrecht in 1713, to the Congress of Vienna in 1815. The notion of balance of power has many connotations, having been used historically to mean a "policy," a "condition" of the international system, a justification both for peace and for war, and an argument both for "parity" in a military balance and for military superiority. Most commonly, however, balance of power is viewed as a mechanism for preserving a stable order among states, principally by insuring that no state or coalition of states possesses sufficient power to overturn the established order.[8]

The goal of preserving order in a balance of power system, however, is ironically not necessarily synonymous with the goal of preserving peace. Indeed, going to war to preserve the balance has been viewed as a corollary to a state's inherent right of self-defense and therefore justified within international law. "Collective security" arrangements, such as those envisioned under the League of Nations and later the United Nations, reflect this principle. These arrangements differed from classical balance of power traditions only in that the League of Nations and the UN represented an international community of nations rather than a self-made coalition or alliance of states. Nonetheless, the willingness of states to go to war to preserve the balance has been an important premise of balance of power. Winston Churchill, in advocating British rearmament in the 1930s to deal with the rising power of the Third Reich, highlighted this argument quite eloquently:

For four hundred years the foreign policy of England has been to oppose the strongest, most aggressive, most dominating Power on the Continent . . . Here is the wonderful unconscious tradition of British foreign policy . . . Observe that the policy of England takes no account of which nation seeks the overlordship of Europe . . . It is a law of public policy which we are following, and not a mere expedient dictated by accidental circumstances, or likes or dislikes, or any other sentiment.[9]

One can debate whether the tradition of balance of power persists in the nuclear age. Churchill himself later coined the phrase "balance of terror," suggesting that atomic weapons might even contribute to a lasting, albeit fragile, peace. When a potential attacker must reckon with the reasonable certainty that he will be destroyed in a retaliatory blow, the likelihood that he will take the risk of initiating war is lowered dramatically. Indeed, this "Crystal Ball effect"—the ability to envision the horrendous outcome of a nuclear war without experiencing it—is often cited as the fundamental reason why the tensions between East and West have so far produced, at worst, cold war and not erupted into violent military confrontation.[10]

Contemporary notions of nuclear deterrence are based on the premise that states must possess both the capability to use this destructive power in retaliation, and the credible will to do so. In this respect, nuclear deterrence invokes the fundamental premise of balance of power. On the other hand, in a world in which nuclear war could mean the destruction of civilization as we know it, the corollary of balance of power—which asks states to employ force to preserve the balance—entails risks which are qualitatively different from those which confronted states in the pre-nuclear age. While we will return to this dilemma in subsequent chapters, the immediate conclusion this suggests is that the nature of power itself has changed.

POWER IN TODAY'S INTERNATIONAL SYSTEM

While power can have many different definitions, we view power here as simply "the capacity to influence." Military force is an instrument of power, but the capacity of the military instrument to influence other actors differs according to circumstances. It depends, for instance, whether one seeks to "deter"—to prevent a change in another's behavior—or to "compel"—to cause a change in another's behavior. Success in the latter is easier to demonstrate, but harder to accomplish, because compellence involves a change in the status quo. Success in the former is easier to accomplish, but harder to demonstrate, because deterrence merely reinforces the status quo.[11]

Power is thus relative, rather than absolute; contextual, rather than universally applicable. The real question in assessing a state's power is whether the instruments of power are usable in a particular context. Traditionally, state power tended to be viewed largely as an aggregate measure, a sum of its component dimensions. Those dimensions included, among others, military force, economic strength, social cohesion, diplomatic reputation, and political will. A state's characteristics—its geographic position, its population, and its raw materials—all contributed to an aggregate measure of its power. While today we may view these attributes as part of a state's power base, it is increasingly necessary to ask how and against whom those instruments of power are to be used before we can assess how powerful a state really is.

We commonly speak of a hierarchy of states, implying that superpowers are the most powerful states in the international system. Coined by William T. R. Fox toward the end of World War II, the term *superpower* referred to those states that would be indisputably the most powerful because they would emerge from that war undefeated.[12] Today, we typically include the US and the Soviet Union in that exclusive club, largely by virtue of their possession of substantial nuclear arsenals. Yet superpowers do not always possess inherent advantages in all that they do. In the first place, their instruments of power may not be usable. Moscow is undoubtedly more impressed by US nuclear power than Hanoi ever was, although the US and the Soviet Union have not fought each other, while the US committed substantial military forces in combat in Vietnam. Secondly, the various dimensions of state power are not necessarily interchangeable. Superior military strength, for example, does not always translate into political influence or economic leverage. While America's trading partners have to deal with the US in terms of economic power, American military force has little, if any, effect on the outcome of those bargaining processes. Likewise, the Soviet Union is a superpower by virtue of its military might, although its economic reach is still largely confined to its empire and dramatically in decline.

This "paradox of power"—that possessing superior instruments of military power does not necessarily mean that one's effective power is superior—relates to a contemporary "security dilemma." While nuclear weapons may have arguably kept the peace between East and West, they have just as arguably not increased the security of the states which possess them. As long as deterrence rests on threats of offensive retaliation, more retaliatory power can only affect the force of one's retaliatory threats, it cannot alter the fact that the holder of those weapons is vulnerable to destruction. As long as societies can be held hostage to destruction by another, then, it may be more precise to talk in terms of a "balance of vulnerability" among states than a balance of power. Debates on how one should resolve this dilemma are often colored by two opposing perspectives, "realism" and "idealism."

PRESCRIPTIONS FOR SECURITY: REALISM AND IDEALISM

While the most significant change in the postwar world has clearly been the advent of the tremendously destructive power of nuclear weapons, the debate on how to maintain security has deeper historical roots. Even before the atom had been split, this century had been dubbed "the century of total war." The wars of 1914–1918 and 1939–1945 had demonstrated that warfare was no longer principally confined to the armed forces of the opposing sides. Rather, whole societies were profoundly affected by warfare, were vulnerable to the effects of warfare, and had to be mobilized on a grand scale to wage war. The seemingly fruitless carnage in and between the trenches of World War I had a revolutionary impact on the way nations viewed warfare. Not unlike the conclusions one drew from the awesome imagery of an atomic detonation, the conclusions many drew from World War I were that war (and therefore armaments) no longer served any useful purpose, and that it (and they) somehow had to be abolished.

E. H. Carr's *The Twenty Years' Crisis, 1919–1939* remains a classic exposition of two opposing schools of thought on how to deal with the problems of warfare in this century of total war. The twin poles of *realism* and *idealism* reflect many of the same views attributed earlier in this chapter to Thomas Hobbes and Immanuel Kant. Generally taken as a critique of idealism (what Carr called utopianism) and an apology for realism, Carr's essay actually offers a subtle balancing of these antithetical poles. Carr's thesis is that World War I stimulated a "utopian" attempt to eradicate nineteenth-century "realist" balance of power traditions which were viewed as an actual cause of war. In their place, new international political institutions, animated by moral vision and faith in the power of human reason to resolve conflict, needed to be created. This utopian alternative, Carr argued, succeeded only in inviting a second major war. Neglecting the realities of power in the world, noted Carr, can be fatal, just as neglecting mo-

rality deprives one of a needed vision and purpose.[13]

The difference between realism and idealism is, in essence, the difference between stressing what is and stressing what ought to be. The latter poses a vision of change (with a clear value preference), while the former emphasizes the limits to that change. The idealist states a moral imperative and insists on overcoming whatever "technical" obstacles impede progress toward that goal. The realist, on the other hand, accedes to Bismarck's counsel that "politics is the art of the possible" and retorts that obstacles to achieving any alleged moral imperative are hardly "technical," but rather inherent to human nature.

The debate between the realist and the idealist is fundamentally a debate about human nature and the relationship between man and his environment. The realist is a pessimist: human nature is basically one of self-interest. However statesmen wrap their policies in moral trappings; their policies reflect self-interest, and indeed, there is no greater self-interest than survival. For Machiavelli—whom Carr called "the first important political realist"—"morality is the product of power." There is no universal moral imperative, no "absolute good" to be achieved. Besides, the world in which man finds himself imposes its own sometimes deterministic constraints. Even if one could envision some ultimate moral imperative, attempting to achieve it would be futile. Thus the realist views the idealist's exhortations of a moral vision both with skepticism about the practicability of achieving it and with distrust about its desirability.

The idealist, on the other hand, is significantly more optimistic about the capacity of human reason, not only to discover the truth about what ought to be, but also to devise ways to get there. Thus, the application of reason can overcome even the sources of human conflict. In the tradition of the Enlightenment, one assumed that there were universal laws which could be discovered by the power of human reason. Physicists' reason enabled the discovery of physical laws of motion, laws which bound mankind together in a single universe. In the same way, this age of reason also offered lessons for human affairs. Reasonable men would begin to discover that they too were bound together in a common universe and shared common rather than antagonistic interests. Not only could human reason discover ways to resolve conflict, it was merely a question of the will to implement, once those means of conflict resolution were discovered.

One can argue that, at least in the West, the realist tradition typically found expression in the realm of foreign affairs and national security policy, while the idealist tradition had its roots in theories about government, specifically in the foundations of democratic institutions. The clash between these two traditions erupted in the aftermath of World War I where, in the West, the trauma of that war stimulated a search for ways to ensure that the Great War was, in fact, "the war to end all wars." Carr argued that the utopians fashioned notions of "collective security" which relied too little on power and too much on public opinion galvanized by the horrors of the trenches as a way of ensuring the end of war. With respect to the League of Nations, as a leading British advocate put it in 1919, "there is no attempt to rely on anything like a superstate; no attempt to rely upon force to carry out a decision. What we rely upon is public opinion."[14]

World War II shattered many illusions about the power of human reason to deter war and the capacity of states to resolve conflict through reasonable compromise and a recognition of common interests. "Appeasement"—once the formal label for a state policy which assumed that war was unnecessary for the resolution of conflict—took on derogatory connotations. "Munich" became a metaphor for how one might indeed invite aggression. With the onset of the cold war, such lessons seemed all the more applicable to this broader global antagonism.

REALISM AND IDEALISM IN THE NUCLEAR AGE

In many ways, the nuclear age has witnessed the reaffirmation of both realist and idealist traditions. The realist tradition accepts the inevitability of state conflict, even if it is not manifested in war, and asserts the need to balance the power of potential adversaries as a necessary means to avoid war. By the same token, the advent of nuclear weapons seemed to offer both an "absolute weapon" which made war unthinkable and, more fundamentally, an absolute moral imperative that war be avoided.

This reaffirmation of both realist and idealist traditions stems, in part, from the two total wars which this century has already produced. In one, war arguably came because aggressive powers were encouraged in their desire to restructure the geopolitical map, by the failure of others to check those ambitions with countervailing power. The "1914 analogy," on the other hand, suggests to many a fundamentally different set of lessons.[15] World War I was "the war that nobody wanted," but it occurred nonetheless because of miscalculation, anxiety, uncontrolled arms competition, and rigid mobilization requirements. One's defensive actions were viewed by others as provocative and threatening, stimulating a spiral of countering actions on an inexorable path to war.

As the East-West confrontation in Europe appears to dissipate, the need to balance the prescriptions of realism and idealism becomes greater. In the short term, the threat of aggression decreases, but the potential for miscalculation in a rapidly changing environment increases. In the long term, the fragmentation of Eastern Europe can lead to a resurgence not just of democracy, but also of nationalism, which has been the culprit behind many wars throughout history.

In the nuclear arena, states are more vulnerable both to the failure to create a credible deterrent posture and to the failure to manage conflict and weaponry so as to avoid accident, miscalculation, and provocation. Labeling different periods of the postwar era as the cold war, détente, the post-détente era, or the age of *glasnost*, obscures the reality that the two superpowers and their allies have pursued competitive and cooperative strategies simultaneously, although in different proportions over the years.

Even as the Soviet empire in Eastern Europe crumbles and the Soviet leadership struggles with managing its domestic problems, while seeking a "breathing spell" in its competition with the West, we are each conscious of the need to preserve our security and to exercise caution. Conversely, the more antagonistic the relationship between East and West became at times, the greater was the sense that cooperative instruments had to be found to insure that war did not result from political competition. As we will see in later chapters, concepts of arms control—as a practical alternative to more utopian disarmament—developed out of the heightened tensions of the late 1950s and early 1960s. Quite apart from attempts to negotiate limitations on armaments, arms control has an important and often less-noticed component— the attempt to create mechanisms, such as the Hot Line and other "confidence and security building measures," to reduce the chances of miscalculation in a crisis.

Such a blending of competitive and cooperative strategies in superpower relations reflects the very nature of the nuclear age and the problems posed by nuclear weapons. States remain independent sovereign actors, unconstrained by any higher authority in the pursuit of their interests, and states still seek to influence the actions of their opponents by threatening unacceptable punishment. This requires projection of political power, no less than in the prenuclear age. Yet the way states employ this power is in many ways more subtle. Instead of manipulating actual forces, it is more precise to say that they manipulate an opponent's calculation of risks.[16] Until states are able to defend against nuclear attack, they will remain vulnerable to total destruction, suggesting that a balance of vulnerability is at least as important as a balance of power in preserving stability in a nuclear relationship. To some, the stakes involved are too high to rely on such a nuclear stalemate to preserve peace. The current debate about the desirability of such a framework is reminiscent of the debate which Carr described, with popular sentiment and moral issues intermingled with the problems of applying power in the nuclear age.

DEFENSE POLICY IN THE NUCLEAR AGE

While the splitting of the atom has clearly changed the international system, it has not eliminated warfare as an instrument of state policy. Military policy may be constrained in its usability, but no substitute has yet been found for military force as a means of defending one's interests. Nuclear states continue to rely on nuclear weapons as a means of exerting political influence in their relations with nuclear adversaries. Judging by the incidence of war among nonnuclear rival states in the Middle East, Africa, Asia, and Latin America, military force remains an active instrument of policy for a large number of states in the world. Although nuclear states may be constrained from resorting to the nuclear instrument in conflicts with nonnuclear states, conflicts nevertheless persist in terms that appear more traditional than novel.

For American defense policy in particular, military force remains a significant component of American power. Both the nature of the international system and the nature of the American interests require that American defense policy deal with three different forms of political conflict. One is of course the bilateral superpower relationship, essentially a nuclear relationship. The second form is an alliance system which confronts an opposing alliance system in an East-West context. This has both nuclear and nonnuclear dimensions. Finally, the US has global interests which it defends through military and nonmilitary means in areas commonly known as the Third World. Militarily, the nuclear dimension is considerably less relevant in the Third World, while the risk of violent conflict there is commensurately greater. Such conflicts are largely low intensity, involving insurgent and terrorist challenges to American interests where much of American military might is unusable.

The number of sovereign states in the world has tripled since the end of 1945. The political conflict between East and West—a legacy of World War II—persists, although its shape is rapidly changing. Even though it might return to more traditional forms, it more likely will evolve into a new security framework that could bear a number of yet unpredictable consequences: the unity of Europe, or its nationalist fragmentation; the disengagement of the super-powers, or their affirmation of a joint responsibility to manage the new structure. Alongside the East-West conflict has grown a conflict between "North" and "South"—between developed industrialized states, many of which are moving into a postindustrial stage of development, and developing states which lag increasingly behind their richer neighbors to the north. While this conflict exists largely in an economic domain, poverty provides a powerful tool for violence and for ideas which perpetuate that violence, contributing to greater regional conflict.

The international system is likewise increasingly interdependent. Few states—and perhaps least of all the US—are immune from the effects of world events, even if those events are beyond control. Regional conflicts, often historical rivalries which are ignorant of the profound changes of the last few decades, often involve American interests and sometimes threaten to embroil both superpowers in a broader conflict. Thus the US faces numerous challenges in fashioning a defense policy which is capable of securing American interests across a broad spectrum of conflict.

Subsequent chapters in this book examine in more detail how the US seeks to deal with these defense challenges. There is a common denominator to these challenges, however. Despite its isolationist roots, the US can no longer remove itself from global affairs. The US is vulnerable to an array of challenges to its national interests, including the threat of destruction. Thus, the US must manage conflict to ensure that crises—which to the realist are inevitable—do not threaten American security. The tradition of idealism which exists in American society demands that substitutes for military force be found, and the US is as capable as any society in advancing that prospect. The world remains, however, a potentially dangerous place, and it is the task of defense policy—as opposed to other elements of state policy—to deal with those dangers. How the US does that is conditioned both by the nature of the international political system and by the nature of its own domestic political system. It is the latter to which this discussion now turns.

THE DOMESTIC CONTEXT: DEFENSE POLICY IN A PLURALISTIC SOCIETY

American foreign and defense policies are decisions of government intended to achieve objectives in relation to the international system. Specifically, policy in the defense arena revolves around the creation of forces and their use. In fact, the decisions of government called defense policy are the determinants of strategy when viewed as the link between ends and means.[17] Although Part IV of this text provides an in-depth examination of defense decisionmaking, it is important at the outset to examine the domestic context in which defense policy is made.

Although differences exist between domestic and defense decisionmaking, defense policy formulation is subject to the same processes, influences, and constraints as most other forms of public policy. Policy in the defense arena is affected by the nature of the American political system as much as it is by the anarchic international environment. Consequently, the study of defense policy requires an understanding of the way the American political system comprehends its objectives and roles, internationally and at home, and then formulates a course of action for their accomplishment. In that regard the American political system is easily criticized as lethargic, atomistic, compromise-laden, and inefficient. Because criticism comes easily, it is important to examine first the founding principles or virtues from which the aforementioned vices derive. American democracy is ordered by the concept of limited, or constitutional, government with an emphasis on *how* members of the polity are treated and treat each other and *how* decisions are made. As a result, the powers of government are articulated in, and are limited by, a written constitution structured to protect individual liberty and to insure that government decisions or policies result from a pluralistic and competitive process. The cumbersome and often difficult process created by the framers of the Constitution was intended to insure that broad compromises would be necessary in order to govern. The Founders believed that forcing compromise would assure that a "greater good" for a larger proportion of the polity was achieved. Likewise, multiple actors would prevent long-term dominance by any singular force in the process.

Two principal characteristics of the American polity have perpetuated its competitive and compromise-laden nature. First, the constitutional principles of separation of powers, and checks and balances among the branches of government require that a common ground between the branches must be found. All must act in concert to accomplish the function of government. James Madison argued eloquently in the *Federalist Papers* that this system of interlocking powers distributed among separate branches of the government was essential to the preservation of liberty. Among the branches "ambition must be made to counteract ambition" so as to prevent arbitrary acts of the government.[18] As a result,

the magistrate in whom the whole executive power resides cannot of himself make a law, though he can put a negative on every law; nor administer justice in person, though he has the appointment of those who do administer it. The judges can exercise no executive prerogative, though they are shoots from the executive stock; nor any legislative function, though they may be advised with by the legislative councils. The entire legislature can perform no judiciary act, though by the joint act of two of its branches the judges may be removed from their offices, and though one of its branches is possessed of the judicial power in the last resort. The entire legislature, again, can exercise no executive prerogative, though one of its branches constitutes the supreme executive magistery, and another, on the impeachment of a third, can try and condemn all the subordinate officers in the executive department.[19]

Most constitutional scholars have concluded that the system of government described by James Madison and others in the *Federalist Papers* reflects not a true separation of powers as much as separate institutions sharing power. This fragmented authority and the dependent nature of each branch on the others has a significant impact on how policy, domestic or defense, is made.

Pluralism, or the competition in the policy process among multiple elites or groups holding particularistic interests, is the strongest characteristic of the American political system. In its purest theoretical form, David Truman would define "public interest" as whatever results from the competition of interest groups in the decisions of government.[20] Pluralism and the competition of particularistic interests in the policy process are spawned by the fragmented authority found in the structure of the national government, the enormous expansion of the government, and the multiple points of access which have been created. More important in domestic policy than in defense, the interrelationship between the rational government and state and local governments further multiplies the competition at work in the policy process.

THE AMERICAN POLITICAL SYSTEM

One approach to the study of the American political system views it as just that, a system.[21] The government consists of institutions and processes established by the Constitution and modified over time in response to contemporary needs. Americans elect members of Congress and the president who function (ideally) within the parameters established by the Constitution and the laws of the United States. In the context of finite resources, the government authoritatively allocates available resources through the policy process. The output of government, or public policy, includes regulations about what a fast-food chain can serve you and call it hamburger, laws prescribing what constitutes a criminal offense against society, a court decision protecting free speech, or an appropriation to deploy a weapon system such as the B-1 bomber or to fund Strategic Defense Initiative research. Figure 1.1 diagrams the flow of policy in a schematic of the American political system. The institutions of government react to input and demands which influence the content of its output or policy. That input is delivered through channels of communication which have developed based on the decentralized and pluralistic nature of American society and its government.

SYSTEM INPUTS

Inputs to the decisionmaking apparatus of government, the institutions and associated bureaucracy, come in various forms. Needs and wants may result from a drought in the Southeast or a series of natural disasters in the West. Demands may take the form of a protest march on the Capitol calling for congressional action to establish a constitutional amendment to outlaw abortion or an action to ban nuclear weapons. Political socialization generates support for the American constitutional system. Supports are important inputs to the system and include attitudes and behavior which help to legitimize the political system and its policies at all levels. By obeying the speed limit on a stretch of interstate highway, paying one's taxes, or voting in a referendum, individuals demonstrate their support for and acceptance of the legitimacy of the government and its policies, an important element of political stability.

Apathy, too, is an input to the political system. A large portion of the American public chooses not to vote, or at least does not vote, in elections. Public-opinion research has demonstrated that a large proportion of the population is indifferent to a broad range of policy issues. The consequences of apathy on government policy decisions are important because, as the proportion of apathetic citizens increases, the freedom of government to act without concern for the public also increases.

As indicated in figure 1.1, inputs may be delivered by individual acts or through collective channels. Clearly, individual contact with government officials is encouraged. For example, members of Congress allocate a substantial portion of their resources to responding to constituents, both in constituent service and in activities designed to ascertain constituent views on current issues. Logically, the more political resources an individual holds, the more access and influence that individual will have in the political process. As Charles O. Jones reminds us, in "getting problems to government," the degree to which interests are aggregated, organized, and represented to the government has enormous influence on how government acts.[22]

Collective channels, which provide for the aggregation, organization, and representation of needs, wants, demands, expectations, and supports, are of two basic types: interest groups and political parties.

Political Parties

Historically, American political parties have been loose confederations of state and local parties which unite every four years to elect a president of the United States. In the American two-party electoral system, maintained and nurtured by federal and state laws, political parties have several important functions. These functions historically have included the fostering of consensus through the reconciliation of competing interests, educating the public about issues in the political system, managing the electoral system and structuring its organization, and recruiting and grooming political talent, among others.

The most important function of parties is to provide labels which the public uses to simplify voter choices. Additionally, the label facilitates the task of organizing the government, especially the Congress, because no other means of identification can accommodate the broad spectrum of political, ideological, and social values found in each of the parties. Both major parties consist of members from across the political spectrum and both have conflicting factions on nearly all social and political issues. American parties are centrist parties and parties of accommodation, which over the long term have been more easily motivated by electoral prospects than by ideological issues.

The role of parties in organizing and conducting political campaigns has increasingly deteriorated. Candidates run independent campaigns, so as not to be tied to their party, and primary nomination campaigns are fought in the media with the winner decided by direct primary election, rather than by party convention.

Kenneth Dolbeare has concluded that in making public policy, "parties are not effective means of translating voter preferences into government action. In some respects the parties seem to have abdicated policy and program functions they may have once performed. In others, conditions have changed so drastically (the rise of the electronic media, primary laws) that there seems little prospect that they could come to play a stronger role."[23] In making policy, the natural rival of political parties is interest groups. As observed by Jeffrey Berry, "Interest groups have inherent advantages over American type parties that make them highly appealing to the politically active."[24]

Interest Groups

The fragmented authority structure and the federal system of government created by the Constitution, coupled with the expansive growth of government, have created a multitude of access points to make inputs to government. Similarly, they have allowed interest groups to flourish. Political interest groups, commonly referred to

Figure 1.1
Schematic of the American Political System

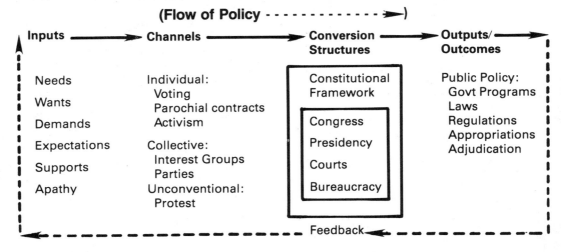

as lobbies or pressure groups, seek to influence the decisions of government in a much more narrowly focused policy area than is possible for political parties. An interest group's success is directly related to its political resources. The size and cohesiveness of its membership and the degree to which its cause is viewed as legitimate in the public eye all influence the interest group's ability to raise money, which in turn affects its ability to organize its political resources. The prestige or status of its members also influences the degree to which the interest group gains access to officials in government. Finally, the political skills of its leaders frequently can result in greater or lesser effectiveness than might be expected from its other characteristics.

Tactics used by interest groups are varied and diverse. They range from direct lobbying, visiting officials and their staffs where they frequently provide much-needed information about an important issue, to indirect lobbying at the grassroots level. Grassroots lobbying occurs when an interest group attempts to influence government officials through their constituency, rather than directly.

A particularly important role of interest groups is found in their association with other interested parties in a particular policy area. Close ties between executive agencies and lobbyists from the affected policy area frequently result in cooperative relationships called subgovernments. These triangular relationships among bureaus, congressional members and staff, and interest groups have come to dominate some policy areas almost to the exclusion of other outside interests. Policy agreements

reached by subgovernments are frequently accepted by Congress or quasi-legislative bodies without significant modification.

CONVERSION STRUCTURES

The institutions of government and their associated bureaucracies comprise the conversion structures of the political system. These institutions receive the varied, and frequently conflicting, system inputs and convert them into decisions of government. The outputs of government consist of government programs addressing needs, wants, and demands; laws and regulations governing new or modified programs; appropriations to provide durable goods or services; and adjudication of conflicts in society. The process by which the institutions act on system inputs reflects the division of authority among the branches and creates in each its own area of authority. Yet each branch is not fully independent with respect to that authority. As a result there are "turf battles" within and among the institutions of government.

In the policy process, the primary function of government is to formulate proposals or plans responsive to the inputs received. Formulation is primarily the responsibility of the bureaucracy, with extensive assistance from others both inside and outside of government. Participation by subgovernment representatives of congressional interests and lobbyists is common in the development of proposals. Legitimation or the selection and approval of the best alternative must then be accomplished. Important here is the realization that public problems involve the allocation of available values and judgments about what ought to be done. As a result, their resolution is as much, or more, political than technical. Selecting the best alternative as the solution to a problem requires value judgments and the compromise of conflicting interests as well.

Once a proposal is selected as the best alternative, it must receive the approval and support of those in authority. In the case of major government actions, this requires approval through authorization and appropriations by the legislative body. The legislative process in the US Congress is labyrinthine and difficult. Numerous obstacles exist in the path of a proposed program before it can reach final form. A series of successive majorities must be built from the subcommittee to the full committee, leadership conferences and councils, and finally to the floor of both the House and Senate. At each stage the proposal is likely to be compromised to make it more politically attractive to the membership. A central part of the approval process is the requirement that policy proposals be budgeted. Here, the competition among different policy priorities intensifies. Particularly when resource questions are involved, Congress is more likely to approve policy proposals that are broadly defined and appeal to the constituency-based decisionmaking of individual congressmen.

Once Congress approves a program and allocates the resources necessary to bring it into being, the government must turn to the task of implementing the program. Primary responsibility for implementation rests with the executive branch. Interpretation of the intentions of the program's authors is the job of the bureaus. Accordingly, their role is critical to the way in which any program is implemented. Where bureau goals and objectives coincide with the approved solution, action is more likely to proceed. However, when bureau goals and objectives are not furthered by the approved solution, or when the desires of the political leadership conflict with bureau goals and objectives, then policy conflict and bureaucratic inertia are likely to arise.

SYSTEM OUTPUTS AND DEFENSE POLICY

Once programs are created and resources are allocated the program's effectiveness must be assessed. The solution which finally makes its way through government may be very successful in addressing the problem as intended. However, it may miss the mark. Results of the policy must be measured and adjustments made where needed to increase the policy's effectiveness or efficiency or to terminate the program. As is the case with policy formulation, policy evaluation is at least as political as it is technical. Value choices must be made as to what constitutes

progress or accomplishment. Once the effectiveness of a particular program is assessed, then adjustments can be made. The path that these adjustments take is depicted in figure 1 as *feedback* to the input side of the political system. As is the case with initial inputs, individuals and groups participate in the identification of needed change in existing programs. Importantly, the government also influences how its policies are perceived and evaluated outside the government.

Randall Ripley and Grace Franklin have observed that most studies of defense policy, particularly with respect to the interactions between the Congress, the president, military and executive bureaucracies, and interest groups, have tended to treat defense policy as a "single, undifferentiated area."[25] Based on their study of foreign and defense policy, in comparison with the politics involved in different types of domestic policy, Ripley and Franklin concluded that there are three different types of defense policies: structural, strategic, and crisis.

Structural policies are those which involve the development and deployment of military forces—both weapons systems and personnel. Included here are decisions about military basing, procurement of new weapons systems, and the selection of specific weapons systems to be sold to foreign countries. Because structural decisions affect enormous constituency-based resources, Ripley and Franklin found that they are characterized by a high degree of congressional interest and conflict.

Strategic policies are more frequently viewed as foreign policy decisions because they deal with strategies of how to deal with other nations. The decision to offer an arms sale to a foreign country, the nature of American obligations arising out of treaties, and the basic mix of weapons systems and forces needed to achieve general and specific objectives are all examples of strategic policy decisions. Ripley and Franklin conclude that, generally, the Congress has tended to support presidential requests in this area or to offer a "competing point of view in such a way as not to remove all administration flexibility."[26]

Crisis policies are characterized by congressional legitimation of presidential actions already taken. Typically, in the case of international crises, the president and his closest advisors maintain tight control of information and make decisions centrally. In such cases, the tendency of the Congress and the American people at large is to support the president and to "rally around the flag." Classic examples of crisis politics were the US invasions of Grenada and Panama. In the case of Grenada, immediate criticism of President Reagan by some leaders of the Congress quickly turned to support after it became clear that the president's actions were viewed favorably in public opinion.

Each of these types of defense policy is characterized by the differing roles of the various actors involved, the degree to which any one actor plays a dominant role, and the amount of conflict among those actors. The role of the Congress, the president, and executive branch departments and agencies in the defense decisionmaking process is the focus of Part IV.

CONFLICT IN THE FORMULATION OF DEFENSE POLICY

The formulation of defense policy within the executive branch alone is characterized by the conflict among priorities and roles of the various departments and agencies involved. Organizational mechanisms such as the National Security Council exist to facilitate the development of a coordinated policy, sometimes called horizontal coordination, across the spectrum of departmental and agency positions involved. The policy environment outside the executive branch is also characterized by conflict which results from competing interests—now of different branches of government. There is the constitutionally imposed tension between the executive and congressional branches in the formulation and execution of national security policy and, further, the competition for resources in the national budgetary process. In an era of unprecedented deficits, each dollar spent on defense is viewed as coming at the expense of needed social programs or as adding to a mounting deficit. However, the president's proposed budget is only the entering argument. It is the Congress which must finally reconcile the executive's request and competing claims on resources and interests.

EXECUTIVE-CONGRESSIONAL RELATIONS

It has been said that the Constitution provides "an invitation to struggle" between the executive and congressional branches in the formulation of foreign and defense policy.[27] In accordance with the principle of separation of power, the president is the commander-in-chief of the armed forces, yet only the Congress can declare war, authorize manpower levels and their organization, and appropriate funds to maintain and equip the military services. The president recognizes foreign governments by receiving and sending ambassadors, but the Senate must confirm all major diplomatic, political, and military appointments and must give its advice and consent to treaties by a two-thirds majority. The president can articulate a policy such as support for "the democratic resistance" in Nicaragua, while the Congress can prohibit the expenditure of appropriations for that declared purpose. As is manifestly clear in this example, the Congress can investigate the executive branch's conduct of foreign policy. Ultimately the Congress can hold the executive accountable through constitutionally established impeachment proceedings.

In the area of foreign and defense policy, the interests of the Congress revolve around the political element of public support, the relative balance among spending priorities, and its power relationship with respect to the presidency. The president's interests center on a need to act with speed, secrecy, and continuity of purpose.

Following World War II, which many believed resulted in part from a lack of support for President Wilson, the League of Nations, and the Treaty of Versailles, an era of bipartisan support of presidential foreign policy initiatives occurred.[28] Both political parties were committed to supporting the national interest as articulated by the president. Partisan political conflict over foreign policy objectives was viewed as potentially weakening the stability of US policy as well as its credibility. Consequently, American public opinion and the Congress were essentially supportive of presidential leadership in foreign and defense policy. The constraints common to domestic policy issues, such as active opposition by involved interest groups, vocal public opinion, and an assertive Congress, were absent.

The era of bipartisan support for presidential leadership in foreign affairs came to an end in the Vietnam war. As observed by Richard Haass

The war in Indochina, the issue which represented to many the essence of the "imperial presidency," also became the instrument through which Congress more seriously challenged presidential prerogative. Beginning only a few years after the expression of overwhelming support for the Gulf of Tonkin Resolution, and continuing through 1975, Congressional opposition to both the expansion and continuation of American military involvement in Southeast Asia, had a major impact on the policies of all the countries involved. Televised hearings of the Senate Foreign Relations Committee raised widespread public doubts about the wisdom of the Administration's official policy.[29]

President Nixon's invasion of Cambodia in 1970, pursued without consultation with Congress, precipitated a growing unrest among the American people and increasing congressional resistance to the war. Senators Cooper and Church attempted to cut off funding for all American military activities in Cambodia. Though it failed, "the effort limited the length and scale of the military initiative and forced the Administration to reassess its political support at home."[30] Congress's opposition to the war in Vietnam fueled an increasing sense of mistrust of the presidency and marked the beginning of what Frans Bax describes as a period of acrimony (1970–76) characterized by disputes over Vietnam and nearly every other aspect of foreign policy.[31]

The acrimonious relationship between the Nixon administration and Congress did not pertain to foreign and defense policy alone. In the domestic arena, major clashes occurred over social spending and President Nixon's refusal to spend monies appropriated by the Congress. As a result major restrictions of presidential prerogative with respect to both domestic and foreign policy were enacted by the Congress. Two especially clear reassertions of congressional authority were the War Powers Resolution of 1973 and the Budget and Impoundment Control Act of 1974. The ultimate manifestation of the conflict between the presidency and the Congress resulted in President Richard Nixon's resignation in 1974 in the midst of impeachment proceedings associated with Watergate.

In the post-Vietnam era, the Congress has been reluctant to play a less assertive role in foreign and defense policy. The Strategic Arms Limitation Treaty negotiated by President Carter faced sure defeat until he withdrew it from the Senate in 1980. Similarly, his Panama Canal Treaty was severely challenged. President Reagan enjoyed enormous success early in his term. He initiated a long-term buildup in defense expenditures, launched the Strategic Defense Initiative, and essentially viewed the Soviet Union in a way reminiscent of the earlier US policy of containment. President Reagan's policy to support the "freedom fighters" in Nicaragua, however, was rebuked by Congress when it passed legislation denying US military aid to the contra forces. This refutation of Reagan policies and the subsequent actions by members of the Reagan administration resulted in the most serious challenge to executive-congressional relations since Watergate. Congressional reaction will be to constrain further the president's power to conduct foreign affairs, especially covert operations.

Constitutional separation of powers and division of authority for foreign and defense policy were designed to permit a president to act in an emergency but yet be restrained from the pursuit of ill-founded policy or policies which lack the support of the American people. The tension which results is systemic and was intended to produce a balanced exercise of government. Another source of conflict is the competition for federal resources among the multitude of programs which are generally categorized as social or defense programs. In the last years of the Reagan administration, a growing deficit, mounting challenges in the Congress to increases in defense spending, and political pressures to protect social programs resulted in a declining defense budget in real terms in the last three years of the administration. The magnitude of the federal deficit and the competing interests represented in the federal budget will be enduring problems for American presidents into the next century.

How Much Is Enough?

Few people disagree that the United States must maintain an adequate defense to assure its national interests. In contrast to many other policy areas, no one disagrees that security is the responsibility of the national government. Beyond that, however, the consensus breaks down. What constitutes an adequate defense, the relative priority defense spending should enjoy over other types of spending, and at what cost are questions on which there is little agreement, either between the executive and congressional branches or within the general public at large.

In any discussion of defense spending, it is important to recognize that an objective measure of what constitutes national security exists only in the extreme. Defense spending is intended to provide forces—weapons systems, personnel, and facilities—to insure the effective attainment of national objectives. Ultimately defense decisions concerning the mix of forces, the modernization and replacement of weapons systems, and the capability they are intended to achieve are subject to widely differing perceptions of the nature of the threat to US security and the best means of responding to it. These decisions are driven as much by political considerations as they are by analytical ones.

One characterization of the complexity of America's commitment to this endeavor is provided by Leonard Sullivan.

The primary purpose of the DOD is to provide ready operational forces of sufficient stature to deter any potential adversary from resorting to war against the United States, its allies, or its friends as a means of attaining national objectives. To accomplish this fundamental objective, the US military has evolved into a vast, complex conglomerate of active and reserve forces totaling over 3.25 million military and 1.1 million civilian personnel, operating weapons and other equipment costing over $2.8 trillion to replace (in FY 86 dollars) from facilities with a replacement cost of over $400 billion, excluding land.[32]

As can be seen in table 1.1, a significant portion of available resources is required to maintain this investment.

On assuming office in 1981, President Reagan pledged to increase dramatically the defense budget while reducing federal spending in nondefense, or social, programs. He was equally committed to cutting federal taxes. The Reagan administration argued that huge increases in defense spending, especially in the modernization of weapons systems, were necessary to over-

Table 1.1

Selected Measures of Defense Spending

Fiscal Year(s)	Percentage of GNP	Percentage of Federal Budget
1955	9.1	51.5
1960	8.2	45.0
1965	6.8	38.8
1970–76	6.1	30.7
1977–81	4.8	22.8
1982–87	6.2	25.9

Source: DoD Annual Report, 1982–88, FY 1990.

come debilitating underfunding in the 1970s.

President Reagan's commitment to increases in defense spending prevailed in the Congress and in public opinion as well. According to Secretary of Defense Carlucci, who inherited the defense buildup engineered by Caspar Weinberger,

To appreciate fully the difficult challenges we faced in adjusting our current budget, it is important to understand clearly the resources and effort that were required to overcome defense underfunding in the 1970s. Then, economic considerations played the major role in setting the level of defense expenditures, as the federal government tried to cope with what many thought to be long-term runaway inflation. Annual double-digit price increases were driving up the cost of government, especially in entitlement programs, and fiscal restraint was decreed for all areas of federal spending. Unfortunately, reducing the defense budget became the primary means of lowering federal expenditures and, consequently, real defense spending declined by over 20 percent from 1970 to 1980.[33]

Consequently, annual real growth in defense spending averaged 8.9 percent in the first term, after which the growing federal deficit and political reverses in the Congress resulted in significant cuts in Reagan's defense budget requests. Budget levels approved by the Congress for fiscal years 1986–88 resulted in a cumulative decrease to 10 percent in real growth.[34] As federal deficits grew to the $180–200 billion range in fiscal years 1983–87, and with the loss of congressional support, the Reagan administration was forced to accept significant decreases in its defense budget requests. In late 1987 President Reagan and the Congress reached a stalemate. In an attempt to reach deficit reduction objectives mandated by previous legislation, the Congress refused to fund the president's defense spending request without increasing taxes. President Reagan refused to support tax increases and demanded further cuts in social programs. At the height of the budget standoff crisis, both sides agreed to an executive-congressional summit to break the logjam. After eight days of marathon negotia-

tions ending in October, a compromise two-year deficit reduction package acceptable to President Reagan and the congressional leadership was fashioned.

The net effect of the budget compromise on the defense budget for FY 1989 was a reduction of $32.5 billion which resurrected Defense complaints that, "We are now again witnessing the consequences of defense underfunding" and a situation where economic considerations displace analytical responses to the national security threat facing the nation.[35] This significant reduction in resources was translated into Department of Defense plans to deactivate two Air Force tactical fighter wings, one Army brigade, one Navy carrier airwing, and the air-launched miniature vehicle antisatellite (ASAT) program. Accordingly, DoD argued that these reductions would result in "significantly greater risks to our ability to achieve our strategic goals."[36]

The Reagan administration left office suffering significant cuts in its defense budget. Into the early 1990s, defense spending will compete with mounting pressures to decrease the deficit and maintain social spending.

The prospect of automatic budget cuts imposed by the Gramm-Rudman-Hollings deficit reduction law motivated the Bush administration to consider cutting as much as $180 billion from its original defense plan for fiscal years 1992 through 1994. In addition, the drastic changes in Eastern Europe and Soviet willingness to reduce its defense posture, both unilaterally and through arms control, led many in Washington to contemplate even more radical defense reductions. Independent of any strategic analysis about future defense requirements, the debate quickly shifted focus from *whether* defense spending would be cut to *how deeply* it would be cut. Moreover, the competing demands for spending such a so-called peace bonus included not just deficit reduction and domestic programs, but also the rebuilding of a reviving democratic Eastern Europe. The debate about how much is enough and the trade-offs associated with the need for defense spending persists. In this regard defense spending is like the weather. Observers of all political and ideological persuasions comment on its size and past spending levels and prognosticate as to what ought to be spent in the future. On rare occasions, for short periods of time, the Congress and the executive reach a consensus or agreement on the need for more or less spending for defense. These periods of agreement are rare and short-lived. In its early years the Reagan administration enjoyed considerable success in increasing defense spending, but later it had to contend with substantial disagreement over policy priorities and, ultimately, the complete collapse of consensus in its final years. The challenge of the 1990s will be to forge a new consensus regarding the needs of defense and its priority relative to other needs.

CONCLUSION

This book is dedicated to an examination of defense policy, both from the perspective of the international environment from which arises the security dilemma and the evolution of American policy in response to that dilemma. Clearly, there are no easy answers to be found, and neither does this text offer any.

If anything, the business of defense policy is becoming increasingly complex. The debate that will dominate defense policymaking in the years ahead will continue to revolve around familiar themes—the relevance of balance of power traditions in the nuclear age in the face of significant changes in East-West relations, West-West relations, and North-South relations; the relationship between diplomatic and military instruments in maintaining security and creating a more stable security environment; the effectiveness of pluralist government in fashioning an effective defense policy; and the resource tradeoffs that dominate the national agenda. There will be new dynamics as well, such as the opportunities and challenges afforded by new technology, and the need to adapt traditional international and domestic political institutions to changes in the way power can be effectively wielded in both systems.

In approaching these challenges, we should recognize that there is a tendency to view history as a process leading up to the present and to view the present as somehow permanent and impervious to further change. But historical experience suggests that such myopia bears within it the seeds of crisis—unanticipated at the time but, for future historians, clearly predictable to those sensitive enough to attend to the winds of change. Thus, this book is dedicated to understanding the questions, for only then can the answers be relevant to our changing needs.

NOTES

1. Thucydides' *Peloponnesian War* is a graphic portrayal of the roots of conflict, seen from the vantage point of ancient Greece.

2. Immanuel Kant, *Perpetual Peace: A Philosophical Proposal*.

3. See Kenneth N. Waltz, *Man, the State, and War* (New York: Columbia University Press, 1959).

4. Carl von Clausewitz, *On War*, ed. and trans. Michael Howard and Peter Paret (Princeton: Princeton University Press, 1976).

5. Michael Howard, *Studies in War and Peace* (New York: Viking Press, 1970), p. 13.

6. Thomas Hobbes, *Leviathan*.

7. Hedley Bull, *The Anarchical Society* (New York: Columbia University Press, 1977), p. 32.

8. See Bull's discussion, ibid., pp. 101–12.

9. Quoted from Winston S. Churchill, *The Second World War*, Vol. 1, *The Gathering Storm*, in Hans J. Morgenthau, *Politics among Nations: The Struggle for Power and Peace*, 6th ed., rev. Kenneth W. Thompson (New York: Alfred A. Knopf, 1985), pp. 213–14.

10. The Harvard Nuclear Study Group (Albert Carnesale et al.), *Living with Nuclear Weapons* (New York: Bantam Books, 1983), pp. 43–44.

11. Robert J. Art, "To What Ends Military Power?" *International Security* 4, no. 4 (1980): 3–35.

12. William T. R. Fox, *The Super Powers* (New York: Harcourt Brace, 1944).

13. E. H. Carr, *The Twenty Years' Crisis, 1919–1939* (New York: Harper and Row, 1946), pp. 11–101.

14. See ibid., p. 35.

15. See, for example, Miles Kahlor, "Rumors of War: The 1914 Analogy," in *Foreign Affairs* 58, no. 2 (1979–80): 374–97.

16. Michael Howard, "The Relevance of Modern Strategy," *Foreign Affairs* 51, no. 2 (1973): 264.

17. Michael Howard, "The Forgotten Dimensions of Strategy," *Foreign Affairs* 57, no. 5 (1979): 975–86.

18. James Madison, *Federalist*, no. 51.

19. Madison, *Federalist*, no. 47.

20. David Truman, *The Governmental Process* (New York: A. A. Knopf, 1971).

21. See David Easton, *The Political System* (New York: A. A. Knopf, 1960).

22. C. O. Jones, *An Introduction to the Study of Public Policy*, 3d ed. (Monterey, Calif.: Brooks/Cole, 1982), p. 46.

23. Kenneth M. Dolbeare, *American Public Policy: A Citizen's Guide* (New York: McGraw-Hill, 1982), p. 40.

24. Jeffrey M. Berry, *The Interest Group Society* (Boston: Little, Brown, 1984), p. 53.

25. Randall B. Ripley and Grace A. Franklin, *Congress: The Bureaucracy and Public Policy*, 4th ed. (Chicago: Dorsey Press, 1987), p. 177.

26. See ibid., p. 191.

27. Edward S. Corwin, *The President: Office and Powers, 1787–1957* (New York: New York University Press, 1940), p. 200.

28. Cecil V. Crabb, Jr., *Bipartisan Foreign Policy* (New York: Row, Peterson, 1957).

29. Richard Haass, *Congressional Power: Implications for American Security Policy* (London: International Institute for Strategic Studies, 1979). As found in John F. Reichart and Steven R. Sturm, eds., *American Defense Policy*, 5th ed. (Baltimore: Johns Hopkins University Press, 1982), p. 549.

30. Ibid.

31. Frans R. Bax, "The Legislative-Executive Relationship in Foreign Policy: New Relationship or New Competition?" *Orbis* (Winter 1977): 881–904.

32. Leonard J. Sullivan, Jr., "The Defense Budget," *American Defense Annual: 1985–1986* (Lexington: Lexington Books, 1985), p. 53.

33. Frank C. Carlucci, *Annual Report to the Congress: FY 1989*, pp. 121–22.

34. Ibid., p. 123.

35. Ibid., p. 122.

36. Ibid., p. 21.

CHAPTER 2

THE COMPETITION
The Nature and Purpose of Soviet Power

CARL W. REDDEL

"How many divisions does the Pope have?" Such was Stalin's alleged reply during World War II to a question about the extent of the Pope's power. The tenor of this retort confirms Stalin's oft-quoted statement concerning the backwardness of "old Russia," which he believed led to Russia's being beaten again and again.[1] The special role of military force in the Soviet view of power during Stalin's rule, and since, is undeniable. Indeed, Stalin's special contribution to the Soviet period of Russian history was to lead Russia to successful industrialization for war and to sustain and to expand the territory of the Soviet Union under conditions of modern warfare.[2]

This fact is a special achievement of the Communist leadership of the Soviet Union, a matter which its opponents have found difficult to digest, because the final success of the Soviet Union is predicated upon the demise and ultimate collapse of its enemies. The nature of Soviet military power is thus tied to larger purposes in a manner which makes many non-Communist observers especially suspicious of Soviet military might. This suspicion of Soviet purposes complicates the task of assessing Soviet power. Soviet power, especially its military power, is strikingly evident throughout the world, and including a chapter on Soviet power in a textbook on American defense policy speaks for itself as testimony to the global nature of Soviet power and its ability to threaten the national security and the international interests of the United States, the world's most powerful state.

The reality of Soviet power appears indisputable. However, the controversial nature of the origins and development of the Soviet state; the resulting lack of widely accepted legitimacy, both within and outside the Soviet Union; and the complex issues surrounding the subsequent development of the power of that state have made it a subject beset by ambiguities, claims, and counterclaims. In an essay on the evolution of contemporary values, Jacob Bronowski once noted, "I do not think that truth becomes more primitive if we pursue it to simpler facts."[3] Accordingly, the significance of Soviet power requires careful elaboration and reduction to its simplest elements.

CONTRASTS BETWEEN US AND SOVIET DEFINITIONS OF POWER

One might argue that military power, the capability of the state to apply force in its own interests, has always been significant in matters of national security, and it is possible to analyze meaningfully the role of the Soviet Union in international affairs from the vantage point of *Realpolitik* and the primacy of force in its relationships with other nations.[4] However, accepting this approach does not by itself carry one's understanding of Soviet power beyond a certain point; it certainly does not elucidate the distinctive nature and peculiar elements of Soviet power. Therefore, it is useful to begin with some geopolitical facts which partially determine the nature and extent of Soviet power.

The Russian state has been the largest on this planet since the mid-seventeenth century.[5] Its size and resulting dominance of the Eurasian continent, among other factors, destined it for a global role in world affairs. Handicapped by its extreme northerly location and the lack of access to warm-water ports, it also possesses an exceptional range and amount of natural resources, rendering it more self-sufficient than any other major industrial nation. Its size, location, and long-term Mongol occupation isolated it from Western European nations, whose success at modernization enabled centuries of large-scale European intervention in global affairs. But Imperial Russia, and later the Soviet Union, chose to overcome its isolation and to cast its lot with the general thrust of events occurring in Western Europe. Subsequently, the Russians also found themselves the objects of European designs, an inescapable development when leaders such as Napoleon and Hitler sought world domination. Sitting astride the Eurasian continent, in isolation from the major centers of historical civilizations, Russians tended to think of themselves as possessing a special purpose and role in history, a persuasion expressed in the twentieth century in the form of communist ideology. The global cast of this view was supported by the gathering of an extraordinary number of ethnic groups and nationalities within the Russian empire as it steadily expanded through the centuries. These nationalities have remained within the Soviet Union in the twentieth century, thus making it the most ethnically diverse major state in the world today.[6] This fact complicated its relationships with the states on its borders and at the same time promoted the Soviet Union's view of itself as international in its basic makeup and in its legitimate global concerns. The Soviet Union's great size; large, diverse, and literate population; immense natural resources; and domination of the Eurasian continent constitute unusual potential for a central role, not only in Asian and European affairs, but also in global affairs.

Such geopolitical considerations are universal in their relevance to the evolution of the power of any state in the twentieth century, but they also specifically influence the historical role which the various national actors have come to play on the planetary stage in modern times. For example, the United States today possesses many of the same universal characteristics of power as the Soviet Union: large, continental size; a sizeable, diverse population; great natural resources; and domination of strategic areas of the earth's surface. But history and geopolitics have treated the United States differently. Its isolation, due to the immense expanses of the Atlantic and Pacific Oceans, was not as formidable or difficult to penetrate, and the oceans provided zones of "free security" for much of the US early history.[7] Its expansion over a continent left it with fewer, weaker, and friendlier neighbors than the Soviet Union, and it has never faced on its frontiers and at its borders land-based powers seeking continental or world domination at its expense. In short, geographical determinants were a major factor in enabling the United States to be more selective than the Russian empire and the Soviet Union in its relationship to European and Asian powers.

Out of its shorter, more recent, and less security-conscious history, America has arrived at its own understanding of power. On the other hand, the Russian empire, and later the Soviet Union, learned through historical experience that an essential dimension of power is a military one. Moscow's power derived first and foremost from its ability either to protect itself from invasion and occupation or to extend its interests overland by military force. The United States developed a largely economic view of national power and influence. Mastering its continental resources in a relatively short period of time, aided by its more southerly location and the year-round navigable oceans on its coasts, the United States came to play an increasingly larger role in world economic affairs and eventually to dominate them, partially as a result of two world wars which caused immense devastation to the homelands of other nations, including the Soviet Union, but not the United States. Only in recent times have the oceans bordering the United States come to harbor immediate and mortal nuclear threat to its security, having constituted for most of US history, and even today, a means of ready access to the world's markets and peoples.

This global access, American entrepreneurial skills, and its impressive industrial productivity have given the United States extraordinary economic power throughout the world, a dimension of influence in which its achievements are largely unchallenged by the Soviet Union. In contrast to the global economic successes of the United States, the Soviet Union has nurtured its self-sufficiency, facing an extremely hostile world in the 1920s and a bankrupt one in the 1930s. Preoccupied with its own rehabilitation in the post–World War II years, the Soviet Union only became significant in world trade in recent times, and much of that trade is with other socialist nations and in support of its political purposes.[8] The relative autarky pursued by the Soviet Union provides obvious military advantages, but it has not consistently enhanced the national and international economic base of the Soviet Union, on which Marxism-Leninism says the military power of the Soviet Union must ultimately rest.

In spite of the great difference in their economic influence, both the Soviet Union and the United States share great difficulty in dealing with the cultural diversity of the world, thereby limiting their political effectiveness.[9] However, the United States retains the advantage of exploiting to its benefit European experience and expertise in world affairs and with few exceptions has never believed itself essentially inferior to Europe. This posture stands in striking contrast to the Russian empire and the Soviet Union, which were not only directly threatened by European military power but also considered themselves culturally inferior. A profound sense of insecurity is thus combined with deep-rooted feelings of inferiority, the latter illustrated in part by the manner in which the Soviet Union takes its own measure by Western standards.

In contrast, the US ties with Europe have been a natural outgrowth of the European discovery and settlement of North America and the subsequent emigration of Europeans to American shores. Following independence and the early years of the nineteenth century, American suspicion of European motives and purposes was not buttressed by the experience or serious prospect of large-scale European invasion. More significantly, the United States was founded and shaped in the modern period of world history and directly benefited from the European political, economic, and industrial revolutions. In parallel with the general process of modernization which determined the nature of modern warfare, the United States acquired an economically based advantage in waging modern war which it sustains to this day. By the same token, the loss of that economic dominance would cause the United States enormous difficulties in maintaining its national security.

Partially as a result of its largely entrepreneurial approach to economic development and its extraordinary success in projecting its influence economically in the markets of the world, the United States has lagged in its mastery of the intellectual basis of modern warfare. By and large it has not had to outthink its enemies in two world wars; rather, it has outproduced them in the means to wage war. To date, the pragmatism of American military leaders has both confounded and mastered the enemies in two world wars, but it did so less convincingly in the Korean conflict, and it suffered outright defeat in South Vietnam. The strikingly different Soviet view of military power is a major element in this contrast between the two superpowers.

In contrast to the eclectic American view of power and war, the Soviet view is codified.[10] Criticized by Westerners for being hidebound and inflexible, it nonetheless possesses certain advantages. It is systematic, comprehensive, and internally logical. It is also historically based and regards the influence of history upon contemporary events as profound and basic. The Soviet view considers the role of human beings as crucial, but gives more weight to what it terms objective factors, foremost among which are the economic forces which it holds are the major determinants of political and social outcomes, a fact which engenders enormous Soviet respect for the economic strength and vitality of the United States. The use and application of military force are in turn seen as a direct extension of politics. This general view is specified in an analytical concept, or paradigm, termed the "correlation of forces" (*sootnoshenie sil*), which is used to evaluate the relative power of different nations as it relates to the international competition between socialism and capitalism.[11]

The correlation of forces broadly encompasses all factors which the Soviet Union believes determine the relative weight and influence of the forces of capitalism and socialism at any given time. Not designed to identify broad trends, being instead a review of status of forces, it nonetheless includes such general categories as political, economic, military, and international considerations. Within the category of military concerns, Soviet calculations are designed to assess those factors which will determine success or failure in military conflict. Viewed as a scientific undertaking, ideally leading to an accurate assessment of military and combat power, the methodology of the correlation of forces is also considered imperfect, subject to many variables and part of the larger dynamic processes of history. Compared to the way Americans tend to analyze the preparation and waging of war, however, the Soviets consider their methodology superior, partially because they believe it is consistent with their more progressive sociopolitical system and ideology.[12] The American approach to planning for modern warfare is viewed by the Soviets as ad hoc and reflective of the divisiveness within American society and politics. They also see it as focused on a narrower definition of the requirements for success in modern warfare.[13]

Differing American and Soviet definitions of power emerge from the discussion above. The Soviet definition is more open to general agreement by outside observers, partially because of its codification in the publications used in the professional education of its military officers and elsewhere.[14] Neither American military professionals nor their political leaders have agreed on such matters. Apart from the differences deriving from distinctive, unique historical experiences which have resulted in cultures alien to each other, the ideologies of the two nations also emphasize differences. However, the shared struggle for global power and influence provides a commonality and a touchstone of reality, occasionally piercing the political rhetoric of both nations and clarifying the essential nature of their differences.

What do the two nations share? The study of the evolution of the pursuit of power in modern history suggests that there is indeed an "ineluctable geometry of power," to which major nations are drawn as they calculate and pursue their respective destinies.[15] These calculations as well as traditional balance of power considerations have been greatly complicated by the introduction of nuclear weapons, but they continue, albeit modified by ever newer weapons and the influence of forces resulting from the dynamics of a modernizing world. Both nations are influenced by the "geometry of power," and they consider the impact of geopolitical factors, but neither follows a purely geopolitical strategy in its pursuit of power.

More important for illustrating the differences between the two nations are the grammar and syntax of power, epitomized in Lenin's question of *kto-kogo*, "Who does what to whom?" This question focuses on the struggle for power and is especially pertinent to understanding the distinguishing elements of the Soviet view. As important as the Soviet definition of power is, the key element has been raw experience in the struggle itself. Out of this struggle have come distinctive Soviet values, purposes, and approaches to power which are at great variance with those of the United States. For example, historical experience has reinforced the Soviet view that the struggle for global power is, and will be, indefinite and continuing at multiple levels below that of nuclear war and outright military conflict. The United States possesses a shorter-term, less historical view which encourages it to seek meaningful restraints on conflict through negotiations, and magnifies their importance in resolving conflict. The Soviet view, on the other hand, is based on an understanding that the basic differences with the United States are substantive and irreconcilable in the long term. This suggests that, if the Soviet Union successfully avoids military conflict, especially the nuclear war which it dreads, it still anticipates a long-term struggle leading to eventual political and social victory.

Central to the Soviet view of power, therefore, are the conditions of struggle out of which the Soviet Union originated and developed. This historical experience has reinforced Soviet confidence in its ideology and in its overall assessment of world affairs. Surveying this history will illustrate how military power has come to play such a central role in Soviet considerations, an outcome which appears to Soviet eyes not only a natural outgrowth of the events which have occurred in the twentieth century, but also inevitable and necessary.

THE DEVELOPMENT OF SOVIET POWER: 1917–1945

Exploring the evolution of Soviet power historically is a deeply Russian and Soviet approach to the subject, for the Russians' discovery and fascination with their own history in the nineteenth century was matched in the twentieth century by the profoundly historical orientation of Marxism-Leninism and the immeasurable respect of Communists for the processes of history. Although from the Soviet viewpoint the October Revolution constituted a major turning point in world history, Soviet analysts examining the evolution of Soviet power find it inappropriate to ignore completely the imperial period of Russian history. Western scholars similarly identify elements of change and continuity on both sides of the October Revolution of 1917.

The continuing growth of the Russian empire and then the Soviet state is fascinating. Of all the great empires of the nineteenth century, the only empire to survive and to expand into the last quarter of the twentieth century is the Russian empire. This may have been due to its land-based nature which rendered the frontier ever-present and inescapable. Russian frontiers were either lines of defense or attack, but rarely something from which Russians could sail to the refuge of their homeland or *rodina*, as did on occasion the British, French, Dutch, Spanish, and Portuguese. To retreat from the frontier was to bring the frontier along, leaving it close to the heart of the empire. The Russians' frontier presence during the imperial period was sometimes superficial relative to the more numerous native populations, but it also was enduring. Russian expansion might also be explained in the context of the experience of other historical "march states" which expanded on the margins of power and eventually came to dominate more advanced, smaller states at one-time centers of wealth and skill. [16] Whatever the accurate, objective explanation of the Russian empire before 1917 may be, Communist ideology provides Soviet leaders a rationale and justification for the continuing expansion of the Soviet state, as it tests the outreaches of its power in the margins of the Eurasian continent.

Although the growing influence of Communism and the continuing expansion of the Soviet state following the revolution were anticipated by the Bolsheviks in a vague and undefined form, the objective basis for such expansion did not exist on several counts in October 1917. The resources and institutions of the Russian empire were inadequate to meet the challenges of modern war manifested in World War I. The transition to a Communist leadership would gradually bring major changes in the governmental, economic, and military institutions. In the meantime, however, revolutionary fervor was insufficient to make up deficiencies in industrial power and military effectiveness, and Lenin took the Soviet Union out of the war with the Treaty of Brest-Litovsk in March 1918. Later, the newly developed Soviet institutions would enable an exploitation of Soviet resources sufficient to meet the challenges of modern war; meanwhile, Lenin and his fellow Bolsheviks concentrated on eliminating competitors for Soviet power within the boundaries of the former empire.

The birthing experience of the civil war, out of the womb of a world war made possible by the results of the industrialization of war on a global scale, proved a traumatic and indelible experience and left an international and military cast of mind subsequently reflected in the Soviet regime's language, organization, and view of the world. [17] The civil war brought troops and support for anti-Bolshevik Russians to Soviet soil from fourteen nations, including Great Britain, Japan, France, and the United States. The desperate struggle for Bolshevik survival in the civil war became known as War Communism and demonstrated, among other things, the efficacy of the Communist party's military style of organization, discipline, and ruthlessness. But success in this conflict, aided greatly by Trotsky's skill as a military commander, left the

Soviet Union prostrate, and Lenin sought to revive the moribund economy with a New Economic Policy (NEP). It served this short-term purpose and even contributed to specific military purposes, but at the time of Lenin's death in 1924 the Soviet Union remained beleaguered, isolated, uncertain of its future, and lacking the means to wage modern war successfully against more advanced industrial states.

Following the emergence of a former Georgian, seminarian, and fellow-revolutionary as Lenin's successor, the future direction of Soviet economic and military policies was clarified. Stalin subordinated the international thrust of the early Bolshevik leadership to the needs and dictates of "building socialism in one country." It was a prerequisite for both the survival of the Soviet Union as a nation-state and as a base for eventual success in pursuing international revolution. He succeeded in collectivizing the Soviet economy through extraordinary means, which for want of a better word came to be called totalitarian, because he used modern technology and political organization to involve the whole of the Soviet population in forced measures of unparalleled sacrifice.[18] The centralized, authoritarian direction embodied in the Five Year Plan which Stalin used to set and to achieve his goals required complete political control of all sectors of Soviet society. He achieved this through censorship, purges, terror, and the elimination of all overt opposition, crushing and demoralizing the peasantry and destroying any possible nuclei for the leadership of an opposition in both the army and the Communist party.

The impact of these changes upon Soviet society was so dramatic and destructive that the outside world largely underestimated Stalin's contribution to the substantive increase in Soviet power which went beyond the rehabilitation of the NEP period. This was partially because of the extreme political rhetoric employed by the Soviets, their careful manipulation of foreign visitors,[19] and the intense anti-communism of Western observers who minimized the Soviets' achievements and magnified their difficulties. Many Europeans, and certainly Adolf Hitler, would not have believed possible the eventual destruction wrought on Germany by the Soviet Union, given the domestic problems of the Soviet Union during the interwar years and the poor performance of the Soviet army in the war with Finland before World War II.[20]

Nonetheless, the Communist development of a modern, industrialized state was undeniable. Indeed, the successful reemergence of a powerful Russian state, surpassing in its growing political power and social control the fantasies of any autocrat of Tsarist Russia, proved one of the great continuities and central facts of modern Russian history. Some Marxian theorists thought the state would increasingly wither away as socialism was progressively realized, but the opposite has occurred and the state assumed a life and vitality of its own, thereby superseding the imperatives of history envisioned by some socialists. Instead, it proved a major, overarching element of connectedness between the tsars and the commissars as they wrestled with the problem of governing the diverse peoples of the world's largest nation.

Under Stalin, the Soviet state developed into the basic form which persists today, embodying elements of the Russian past clothed in the garb of the Soviet present. The lack of a highly developed economic and social infrastructure, so characteristic of modern Europe, reflected an economic and social lag in the development of the Russian empire. Moreover, the small intelligentsia and nascent middle class which developed in the nineteenth and early twentieth centuries were either destroyed or reshaped in the throes of Soviet development. This gap in Russian development persisted into the Soviet period and reinforced the continuing operation of the "command principle," characteristic of the tsarist autocracy and unhindered by competing institutions or social and economic classes which might successfully deflect or divert its power. Under Stalin's leadership, this "command principle" remained in force and was used to mobilize Russian resources for industrialization, just like it had for the tsars in the 1890s.[21] However, the new Soviet state was distinctively energized by the Communist party, which brought the bureaucratic dimension of state power in Russian history to new heights of development and refinement, far beyond that enjoyed by the supporting bureaucracy of the defunct Tsarist government. Moreover, whereas the tsars had eventually freed both the aristocracy and the serfs from direct service obligations (the former in 1762 and the latter in 1862), Stalin was determined to bring all citizens back into the service of the state. Collectivization broke peasant resistance and initiative; draconian discipline kept workers at their lathes; and purges kept in line the army leadership, the Communist party, and the remnants of the intelligentsia.

The interwar period was enormously important for the Soviet Union, constituting a Stalinist "revolution from above" which brought the Soviet Union fully into the ranks of industrialized nations and which provided the means to wage modern war.[22] Disguising the fact of the achievement was that the Soviet Union was at war with itself as Stalin faced stubborn resistance in establishing the political controls he believed necessary to industrialize peasant Russia, resulting in an unparalleled bureaucratization of organized violence within the society at large.

The enormous increase in state power, operating on the "command principle" for the purpose of modernizing Russia through the Five Year Plans, provided the Soviet Union with the institutionalized, managed economy characteristic of the belligerents in World War I.[23] The Bolshevik revolution, occurring in the cauldron of global war and popular dissatisfaction, was succeeded by Stalin's revolution from above which demonstrated that the Communist leadership learned well from World War I about the needs and demands of modern warfare.

Soviet power, however, was not great enough to warrant self-confidence in security from capitalist and, increasingly, fascist forces. With the long-term goals of modernizing the Soviet Union and supporting revolution in the capitalist nations, the Soviet Union from its position of relative weakness required stability in relations with the outside world, at least until the day arrived when it could securely play a larger role in world affairs. A double-tiered approach to international relations was reflected in the effort to secure and maintain conventional state-to-state relations while simultaneously directing the subversive and revolutionary activities of the Communist International (Comintern) against those same states. As the power of German fascism grew, Stalin's effort to establish security against that immediate threat through alliances and collective action grew more desperate, eventually resulting in a pragmatic compromise with Hitler himself. Stalin appears to have correctly assessed, earlier and more accurately than Western leaders of the time, the extent and depth of the Hitlerian menace as a destabilizing threat in world affairs. Again, as in the immediate post–World War I period, Soviet survival and security came first, and, as a result of Stalin's compromise with Hitler, enormous damage was inflicted on the international revolutionary networks sponsored and developed by the Soviet Union. But the lessons learned by Bolshevik leaders in the revolutions of 1917 and the civil war growing out of World War I were lessons about the lack of power, survival, and the potentially revolutionary results of defeat in modern war for any nation.

World War II was both a tremendous catastrophe and an immense opportunity for the Soviet Union, attesting to the correctness of the Communist understanding of full-scale modern war as a seminal historical event and as a desperate struggle for survival. Although the commitment to forced modernization while remaining in beleaguered isolation during the interwar period was extraordinarily stressful and costly for the Soviet population, its demands were surpassed by the suffering and sacrifice wrought in World War II. It is difficult to find words adequate to describe the losses undergone by the Soviet people. It is also beyond statistics. The catalog of economic losses was summarized by one economist as "equal to, and possibly even somewhat greater than, the total wealth created during the industrialization drive of the 1930s."[24] The human losses proved numbing and beyond meaning to those of ordinary experience. Officially the Soviet Union eventually reported a loss of 20 million lives, and competent Western sources estimate as many as 30 million total casualties, approximately 15 percent of the 1940 population.[25] Adding the 10 million unborn babies of the 1940s, which would have been part of the normal demographic projection, compounds the enormity of the war's destructiveness.[26]

The German invasion was an extraordinary test of the Soviet Union in all dimensions of its existence and organization, yet the Soviet Union survived.[27] Unfortunately, scholars have only begun to examine in detail the total Soviet experience of the war, and we are not sure why the Soviet Union survived. From the Western viewpoint it could have collapsed. Maybe it almost did, but the Communist leadership's successful methods for industrializing the country during the 1930s, with development of the associate economic infrastructure of centralized planning and direction, proved as capable of mobilizing the country for war as it had in forced draft modernization before the war. Ironically, the traumatic experience of the Soviet Union having been at war with itself proved to be preparation for war with the Germans.[28] The forced requisitions and movements of population, the development of industrial enterprises from the ground up, and the ruthless exactions of the political police and the Communist party which characterized the interwar period proved useful training for effective action against the Soviet Union's foreign enemies. Stalin also exhibited a new level of organizational flexibility and versatility in creating the State Defense Committee, which was superimposed as a wartime measure on the existing state structure and provided efficient and effective direction of the total war effort.

SOVIET POWER IN THE POSTWAR PERIOD

Successfully waging modern war in World War II, in contrast to the Russian failure in World War I, raised new opportunities and challenges for the Soviet Union. In the simplest sense, it reestablished the Soviet state roughly at the limits of the tsarist empire at the beginning of World War I.[29] The defeat of Hitler was both more and less rewarding for the Russians than the defeat of Napoleon, with Alexander I having victory parades in Paris, whereas Stalin's troops marched in Berlin. However, the Soviet presence behind the lines of its soldiers' farthest advance was comprehensive and enduring, and indefinitely changed the map of central and eastern Europe. This expansion of Soviet influence was based on the military power developed out of the fight for survival in World War II, which provided the military means for realizing some of the advanced ideas of Soviet military theorists, a matter to be dealt with in greater detail below. Having defeated the German army, the most professionally capable modern military force of the twentieth century, the Soviet army in victory constituted an extraordinary new powerful arm of the Soviet state. And it was out of the continuing development of this military power that Soviet influence was eventually to move in the postwar period from the continental to the global stage of political destiny. But realization of this influence was not to come quickly or easily.

In spite of its large, successful military forces, the Soviet Union was a weakened giant, demographically as well as economically.[30] However, it held on to the territories it occupied and even tenaciously pursued additional gains, such as an expansion of its influence in Greece and Iran. At home, Stalin instituted new draconian measures of recovery and rehabilitation, offering his subjects little respite for their immense wartime sacrifices. From the viewpoint of its own security and survival, the Soviet leadership may have had little choice, given the central role of the economic base which Marxist-Leninists see as the source of the military power. The US long-term purposes, originally developed in conjunction with its World War II allies, were less clear, but became increasingly focused on what it believed to be the aggressiveness and the revolutionary aims of the Soviet Union, with the conflict between the two states reaching such intensity that it was called a cold war. For most of the four decades since World War II, thinking about the relationship between the two nations has been dominated by this cold war concept, and it is used to measure greater or lesser hostility in their dealings with each other.

Ideological and technological influences upon both nations provided catalytic impulses which repeatedly varied the temperature of the cold war relationship between them and which continue to have this effect today. Both influences have origins and supporting contexts which go beyond the individual nations themselves, being rooted in the broader developments of modern history. They are also related to the basic power of both states, being both part of their respective levels of power and at the same time stimuli to the competition for power. Neither, however, is as essential in its impact as the relative economic strength of the two nations. In this regard, the United States' economic dominance has been unquestioned throughout the postwar period and has provided both advantage and flexibility in its relationship with the Soviet Union. The lack of comparable economic power on the part of the Soviet Union has severely constrained the achievement of both its domestic and international goals.

The ideological differences between the two nations are basic and manifold. Both nations are highly ideological in their makeup, the Soviet Union more consciously so.[31] Without reviewing the full catalog of ideological differences, claims, and counterclaims, it might be said that the ideological element which serves as the bedrock of their disputes, and which is also a fundamental dimension of political power, is the question of legitimacy. From its inception the Soviet Union has challenged and rejected the long-term legitimacy of states it deems capitalist, although it grants de jure recognition for its own purposes as part of the double-tiered relationship described earlier. Marxist-Leninist historical analysis finds capitalist states unduly exploitive and doomed to eventual failure. The United States, for its part, has never accepted what it believes are the fundamentally undemocratic origins of the Soviet state, although it grudgingly granted de jure recognition to the Soviet Union in 1933. The US view of the future of the Soviet Union is less clear than the Soviet view of the United States future. Americans sustain perpetual optimism that the Soviet Union is capable of future transformation into a political and social entity acceptable to the United States and simultaneously view it as the stereotypical antithesis of everything which the United States represents. Both nations have universalized their political values and believe themselves indissolubly involved with the general fate of humanity, each sustaining a view of themselves as truly exceptional in the course of human history. This vagueness means that the frontiers of their national interest are roving frontiers and may be found in parts of the world far from their respective continental limits.[32] For the Soviet Union during the postwar period, this has meant supporting the cause of national liberation wherever it may be. For the United States, the frontier of national interest has retreated somewhat from the Rhine River in World War II, but it is difficult to understand the impulse, motivation, and purpose of American efforts in the Korean and Vietnam wars without the ideological dimension. Both nations continue to challenge each other on the battlefield of legitimacy, considering it an essential element of power and seeking support for their respective positions in the national and international arenas.

The influence of technology has had an enormous impact upon the power of the Soviet

Union. Having shown its mastery of modern land warfare in World War II, the Soviet Union was confronted by the stark reality of the development and use of nuclear weapons by the world's leading capitalist nation. To use Soviet terminology, it caused a postwar "revolution" in the course of military affairs and has greatly complicated the assessment and use of Soviet military power.[33] However, the Soviet Union has demonstrated its own mastery of nuclear weapons development and of the means to use them. In spite of both nations' mastery of these weapons of unimaginable destructiveness, they are prisoners to the momentum and pace of the speeding technological development which grows out of their huge industrial and military establishments. Each new technological military development poses the need for countermeasures and for a new weapon to outreach the means of the opponent. Both nations now possess enormous military power, but at what expense? The impact on both nations is unclear, and economists cannot definitely show cause and effect. Are the domestic economic difficulties of the Soviet Union today related to its defense expenditures? Is the historically unprecedented American deficit related to the Vietnam war, to the modernization of its military forces, and to the development of new weapons?

What is clear is the unprecedented extent of Soviet military power in modern Russian history. This power has made Western Europe something of a bridgehead on a Eurasian continent dominated by the Soviet Union and another Communist power, the People's Republic of China. Fortunately, from the viewpoint of the United States, historical and cultural differences between China and the Soviet Union have proven more powerful in recent years than ideological ties and political similarities. This complicates the use and application of Soviet military power, as does the restive nature of some of its Warsaw Pact allies, requiring military forces in place not only to oppose capitalist forces but also to secure the loyalty of Communist states. The increasingly pluralistic nature of international communism since Stalin's death renders the unprecedented Soviet military power subject to an extraordinary range of demands. Indeed, the alliances of the Soviet Union and their attendant complications may work to the advantage of the United States and actually isolate the Soviet Union politically and strategically.[34]

Given these complications, the maintenance of Soviet military power is a difficult task facing numerous challenges, but Soviet leaders must maintain it because it is the one dimension of Soviet power which entitles the Soviet Union to superpower status and which finally guarantees its security in a hostile world. Possessing highly professional and powerful military forces is a great equalizer in the Soviet Union's relationship with the United States, but it cannot diminish or compensate for the long-term significance of the United States' much greater economic strength and the technological advantages which accrue to the United States in its open competition with other capitalist states. Historically, the Soviet Union has turned to the intangibles of power to exert influence when it has not been able to compete effectively in the substantive elements of power. Today, the Soviet Union compensates by exploiting the "geopolitics of information," using information and misinformation systematically to control its own citizenry, to mask its weaknesses, to magnify its strengths, to diminish its enemies in the world's eyes, and to promote its own view of events and history.[35] The scale of this particular effort is unprecedented in history, and its final outcome remains to be seen, but the United States may underestimate the significance of this Soviet endeavor to its own peril and detriment. More measurable perhaps is the success of the Soviet effort to compensate for the technological advances of the West by developing intelligence resources to capture advanced Western technology for its own military purposes. The United States has decided that losses in this respect are significant enough to warrant serious countermeasures. In spite of many obstacles and difficulties in maintaining its military power, the Soviet Union has not only sustained its former levels of power and the territorial acquisitions of the Russian empire, but has gone beyond them.

The physical expansion of the Soviet state has continued into the late twentieth century. Having secured in 1975 at Helsinki a measure of legitimacy for its World War II conquests, the Soviet Union invaded Afghanistan in 1979, seeking subordination of its Asiatic, Moslem population. The appropriate historical case study for understanding the possible long-range results and impact of such intervention is not the comparison with the American experience in the Vietnam war gleefully employed by American observers shortly after the invasion. A more useful analogy is the decades-long experience of the Russian empire in conquest of Caucasus,[36] or the struggle to incorporate Central Asia into the Soviet Union in the 1920s. The rationale for the invasion of Afghanistan and the continuing expansion of the Soviet state and its military power cannot be fully understood by solely using Western criteria for efficiency and effectiveness. They are usefully complemented by an examination of the overarching Soviet political-military doctrine which provides the paradigm within which Soviet decisions are made.

SOVIET POLITICAL-MILITARY DOCTRINE

Two major instruments of power for the Soviet state are Marxism-Leninism, the state's ideological arm, and its military forces, which are guided by the military doctrine growing out of Marxism-Leninism. Marxism-Leninism is in and of itself viewed as an element of power in the correlation of forces mentioned at the outset. As an ideological and propaganda tool in the projection of power, Marxism-Leninism specifies and defines the enemies of the Soviet state. Out of this identification naturally grows the military doctrine which provides guidance on how military force is to be used for political and military purposes against these enemies. Within both Marxism-Leninism and its attendant military doctrine are to be found the basic Soviet understanding of war and international conflict, as manifested politically, socially, and economically.

This understanding is based on a holistic approach to the problem of conflict in the modern world; the separation of Communist doctrine into political and military components is artificial and not in keeping with the Soviet view.[37] This may appear strange and inexact to Western observers, but the Soviet Union represents a very different culture from our own and poses peculiar obstacles to understanding. As a unique blend of a distinctive Eurasian history with Leninist modifications of Marxism, Soviet political culture has to be approached on its own merits if we are to have a chance of understanding it outside the bounds of our own ethnocentrism.[38] For this reason Soviet political-military doctrine must be viewed as grand strategy in the largest sense, with military doctrine being an essential but subordinate part of a larger view in which political strategy is paramount. From this political context flows a military doctrine which in turn leads to military science, art, and strategy. Last in this structural hierarchy are operational art and tactics, which is the level at which US military doctrine is usually developed.[39] For this reason, it is important to follow the Soviet thrust, which suggests that military doctrine begins politically with the identification of enemies in their social and economic context.

The opponents of the Soviet Union are numerous and diverse, but they all share an unwillingness to subordinate themselves to either the Communist party of the Soviet Union or the Soviet state. They range from the epitome of imperialist powers, the United States, which has evolved to the highest stage of capitalist development, to the People's Republic of China, a fellow Communist state which is considered a serious enough threat to warrant the stationing of large, combat-ready Soviet forces on its borders. Additional threats to Soviet security include other nations that constitute greater or lesser obstacles to the achievement of Soviet aims, ranging, for example, from members of the North Atlantic Treaty Organization, which is specifically directed against Soviet military forces, to fellow members of the Warsaw Pact such as Poland, which are unreliable enough to cause serious concern to the Soviet leadership. Whatever the relative gravity of the difficulties posed by these states to the Soviet Union, they are frequently considered by the Soviets to have at least some of their origin and support in the United States, the Soviet Union's chief enemy.

The view of the United States as the chief enemy of the Soviet Union is systemic, not racial or personal. With complete candor and honesty, Soviet Communists can say they hold nothing against the American people collectively or against individuals personally, because their view is that both the American people in general and any given individual are subject to historical forces largely beyond their control, as a result of which they are manipulated and exploited for the purposes of the ruling class which controls them. To the Soviets the economic system of state monopoly capitalism and international imperialism, of which the United States is the major support, is a regrettable but inevitable stage in the unfolding of the economically based historical process which governs us all.[40] The class contradictions in the United States will eventually cause the demise of domestic capitalism and thereby end the international, imperialist role of the United States. Meanwhile, the United States continues to exploit its own people and the peoples of the world, with deleterious effect and a resultant denial of the realization of individual and national destinies. As the foremost leader of those progressive forces in the world which are at a higher stage of historical evolution, the Soviet Union has no choice but to assume the responsibility for accelerating the demise of the economic and political system of the United States and for limiting its ability to influence negatively the affairs of the world. The risk entailed in this enterprise is enormous. Although Soviet leaders believe progressive forces will eventually win out, the specific winner of any particular competition is not foreordained, and the episodic nature of the competition may in the short term bring suffering to many. Indeed, until the mutual destructiveness of nuclear weapons was accepted by Stalin's successors, war between the opposing economic systems of capitalism and socialism was considered inevitable.[41]

The evolution of the Bolshevik understanding of its enemies began before 1914, when revolution was the primary goal, and when neither World War I nor the revolutions of 1917 were specifically anticipated. These events pointed toward major deficiencies in socialist ideology, and Lenin played the crucial role in adjusting Marxist thought to the fact of modern war. His insight and pragmatism showed how it was possible to pursue revolutionary goals within the context of modern warfare and resulted in the melding of political and military thought which is the hallmark of the Soviet military profession. Lenin placed war into the center of Marxist thought, and, building upon the thinking of Clausewitz and the experience of fighting for survival during the civil war, contributed to the "militarization of Marxism" and the eventual militarization of the Soviet state.[42]

Central to Lenin's thinking about military affairs was the integration of the results of practical experience into a theoretical exposition which would not only explain but also predict with some certainty the eventual course of events. For Lenin, the formulation of doctrine was a dynamic process requiring continual review and input from direct experience. World War I demonstrated to Lenin the irrelevance of the militia-style army envisioned by socialists before the war and advocated by some afterwards. In its place he saw the need for an army possessing some of the same elements of professionalism he expected of revolutionaries. The fears of Trotsky and other Bolsheviks concerning the threat of Bonapartism were superseded in Lenin's thinking by the needs of survival and the changing nature of modern warfare. Having recognized the central role of modern war in history, Lenin affirmed the necessity for a professional army to sustain the Soviet state in a condition of long-term conflict with economically superior capitalist nations. War in Lenin's eyes was essentially political and the result of class conflict, a view which partially explained the revolution of 1917 and which remains at the core of Soviet military thinking today. He pursued the development of a professional standing army with supporting staff and officer education organizations. With this army and with state socialism he waged total war against Bolshevik enemies during the civil war, and elements of the mode of struggle developed for survival of the Soviet state during the civil war were eventually projected into the larger global struggle between capitalist and socialist systems.

The necessity of armed force in the international arena was demonstrated early when much of the Soviet state had to be surrendered to German invaders in World War I and foreign anti-Bolsheviks had to be ejected during the civil war. The limits of revolutionary influence became clear in unsuccessful Hungarian and German revolutions following the war; the limits of Soviet military power were demonstrated when Soviet armed forces were stopped in Poland in 1920. In light of the Soviet state's limited resources, continuing to field revolutionaries

under the guidance of the Comintern remained easier than fielding and equipping armies, and military thought and doctrine ran far ahead of the resources necessary for the Soviets to conduct modern war as they understood it. Nonetheless, the Bolshevik revolution had laid the basis for a revolution in how to think about war and how to organize for it.[43] The changes underway in Soviet society and the economy following the revolution of 1917 were later reflected in Soviet military thought and eventually in the conduct of war itself.

Certain advantages accrued to Soviet military thinkers. Although more affected and influenced by Russian military history and tradition than they perhaps realized, they believed themselves unencumbered by the immediate weight of the past, if for no other reason than so much had been destroyed and so little had been created to replace it. They possessed a Marxist-Leninist world view which seemed to be vindicated by events, an analysis of history born in the throes of the Industrial Revolution which dictated the nature of modern war as assuredly as it would provide Soviet military leaders with the means to win such wars once Russia itself was industrialized. This was a heady and inspiring time, when the future appeared limited only by imagination and correctly understanding the ineluctable processes of history. Their self-confidence in the face of enormous obstacles and difficulties derived from their position as Marxist-Leninists in the vanguard of revolutionary and progressive forces. They believed themselves better suited than anyone else to understand the significance of historical developments.

This optimism was reinforced by Soviet belief in the epistemological and methodological advantages of Marxism-Leninism when applied to explaining modern war and its causes.[44] Marxism-Leninism does not explain well all it purports to explain, but it does well with war. Why? Perhaps because it originated in the bowels of the Industrial Revolution which shaped the modern world and because it is rooted in a view which emphasizes conflict in a dynamic environment during the passage of time. Marxism-Leninism is also systematic, comprehensive, and internally logical. It views modern war as a total experience, operating at all dimensions of human existence and organization—political, economic, social, psychological, and cultural—neither ignoring nor excluding anything that is part of the human historical experience. It emphasizes human resources in the conduct of modern war,[45] the result of both military theory and the Soviet need to compensate with human resources for technological deficiencies.

Whatever the intellectual advantages of Marxism-Leninism for accurate analysis of modern war, its leading practitioners were both aided and constrained by history itself. Soviet leaders debated policy alternatives while continuing the process of modernization which the tsars had begun. Soviet military thinkers were directly affected by the distinctive features of Stalin's leadership, with his purges of the military, and by the lack of Soviet success in the Spanish civil war, the Soviet-Finnish war and the early phases of World War II. Soviet military thinking also had to adapt to the superior military technology of the capitalist West at the end of World War II, when nuclear weaponry was temporarily the monopoly of the United States, its leading capitalist opponent. The doctrinal reaction to this development of unparalleled destructive power was bound up with the broader Soviet reaction to Stalin's death, the end of the man whose individual influence in Russian and global history had no historical parallel. Although largely successful in rehabilitating the Soviet Union from the enormous destruction of World War II, the Communist hierarchy under Stalin's leadership did not succeed in developing the economic momentum which would enable the Soviet Union to keep pace with the speed and scope of Western development in the broadest sense, and it became necessary for Soviet military thinkers to adjust to this lag. One of the results may have been that Soviet military responses to American strength have been designed in part to take advantage of what it perceives as American military weaknesses while moving away from confronting America in areas of great strength.[46] Meanwhile, the Soviet leadership gave priority to resource allocations for military development and expanded intelligence activities to capture the best of Western military technology, while at the same time seeking ways to extend economic improvement to the whole of Soviet society.

Given the broader, global historical context of the development of the Soviet state, and the violent changes occurring within Soviet society itself, the irregular nature of Soviet doctrinal development following Lenin's death in 1924 is not surprising. For all of the doctrinal unevenness, however, certain elements of Soviet military thinking have remained more or less constant, although Westerners have often misread postwar Soviet doctrine just as they underestimated Soviet military developments in the 1930s. The blinders of ethnocentricity which lead to misunderstanding are universal and are not unique to Americans. In this case, more significant are the particular differences in the approaches of the respective military establishments of the two nations to doctrine, wherein the Soviet Union considers doctrine a central feature in the evolution of grand strategy and professional military education, and the American military does not. Soviet military professionals retain a dominant role in the formulation of doctrine, a subject greatly influenced by civilian specialists in the United States. Accordingly, Soviet doctrine has by and large hewed to military fundamentals as it evolved. Specifically, the warfighting nature of Soviet military forces has never been in dispute, whether in conditions of nuclear or conventional war.[47]

Two other elements have often escaped Western observers of Soviet military doctrine, one pointed toward the past and the other toward the future. Given the basic nature of historical processes in Marxist-Leninist thinking, military history plays an exceptionally large and central role in the development of Soviet military thought. The Soviet *Military History Journal* has no equivalent and is probably the best publication of its kind, a peculiar synthesis of professional military and historical concerns.[48] Soviet experience in past wars, more accurately *Russian* experience in past wars including the Imperial period, is viewed as the source of relevant criteria for analyzing problems in command and control, operations, and strategy. For the Soviet military professional, the study of military history is operational research.[49] Military history in the Soviet Union is not only an exercise for obtaining accurate information about the past, but also part of the method for developing theory and for teaching professionals about war. This does not lead to an obsession with the past for its own sake. Rather, in keeping with Lenin's view that doctrine must be predictive, it means that Soviet doctrine is not only concerned with the past and present, but also heavily oriented toward the future. The failure to assess correctly this Soviet orientation toward the future may help explain why the West failed to recognize that Soviet strategic nuclear force development predated the Cuban missile crisis and that later Soviet combined-arms applications were a natural outgrowth of prior doctrinal developments and Russian historical experience.

THE EVOLUTION OF SOVIET MILITARY DOCTRINE

In tracing doctrinal developments after the death of Lenin, the following evolutionary stages can be identified:

1917–41: Early development of Soviet military thought

1941–53: The Great Patriotic War and the last years of the Stalin era

1953–59: The "Revolution in Military Affairs"

1960–68: The strategic nuclear buildup

1969–73: Development of a controlled-conflict capability

1974–80: Opening era of power projection[50]

The early development of Soviet military thought included not only the central ideas of Lenin discussed above, but also the special contribution of M.V. Frunze (1885–1925), the most notable Soviet military thinker during the time. His thought included assessments of World War I, the civil war experience, and the special opportunities created by the establishment of the world's first socialist state to wage modern war successfully. He viewed modern war as total, requiring a unity of effort between state and people that was possible under socialist leadership at a new and higher level than ever previously achieved. His writings remain central to Soviet military thinking today, and the highest Soviet award for excellence in military publications on military theory or history is the Frunze prize. Other notable contributions from military figures of the time included the ideas of M. N. Tukhachevskii (1893–1937) concerning offensive operations and deep penetration which were to achieve realization in World War II, and A. A. Svechin's (1878–1938) views on strategy which are studied by Soviet officers today. All three of these officers developed and promoted ideas which exceeded the means of the Soviet state to support them, and all three were to experience untimely deaths at the hands of Stalin who was to dominate the direction of Soviet military development until his own death in 1953.[51]

Stalin's indisputable military significance is difficult to assess, because of its sheer breadth and paramountcy. As maker of the industrial means to wage modern war and as supreme commander of the Soviet military forces which defeated the Germans in World War II, Stalin achieved a level of success which overshadowed his numerous mistakes and failures. Stalin achieved much with state socialism and the professional military forces he developed, but he also destroyed, stifled, and constrained creative and innovative influences in Soviet military thinking and in Soviet society in general. He codified the principles for the success of the Soviet Union in the Great Patriotic War in his statement of the "permanent operating factors" in modern warfare.[52] His insistence on complete political control meant that these principles were not only dominant, but also unquestionable. This prevented creative discussion and debate among Soviet military thinkers about the impact of nuclear weapons on modern warfare until after Stalin died. However, Stalin had in the meantime promoted an effective intelligence effort to secure information on Western development of nuclear weapons, supported the work of captured German scientists in rocket development, and essentially laid the groundwork for subsequent Soviet achievements in developing nuclear weapons, with Soviet testing of an atomic bomb in August 1949 and a hydrogen bomb in August 1953.

Following Stalin's death it was dramatically clear that Soviet military doctrine needed revision in view of nuclear weapons development, but the Soviet Union is above all a conservative society. Political control of all facets of Soviet life is an absolute requirement for the Communist leadership, and the nature of this control results in cautious, careful, and even subterranean approaches to major changes in matters of public policy. Gradually, recognition of the need to reassess the impact of nuclear weapons upon Soviet military doctrine outside the limits of Stalin's "permanent operating factors" emerged in the military press and was given a heightened sense of urgency by the Soviet belief that a revolution had occurred in military affairs.[53] This belief was based on the combination of nuclear weapons with ballistic missiles, the Soviet Union having tested the first intercontinental ballistic missile in 1957. Institutional recognition of the decisive nature of nuclear weapons in Soviet military doctrine was embodied in the formation, in December 1959, of a new Soviet military service, the Strategic Rocket Forces, which was acknowledged as the primary service among the other Soviet military services—the Air Forces, the Troops of National Air Defense, the Ground Forces, and the Navy.

The results of this doctrinal change were emphasis on missile development, improvement of Soviet nuclear weapons, and increased overall strategic nuclear capability. These developments took place during the relaxation of controls under Khruschev's leadership; with the ebullient Soviet leader also came recognition at the 20th Party Congress in 1956 that "war is not fatalistically inevitable."[54] Further recognition by Khruschev of the decisive nature of nuclear weapons came with his reduction of conventional forces, from about 5.7 million men in uniform in 1950 to approximately 3 million in 1964.[55] The most striking and comprehensive strategic analysis of nuclear war publicly available was V. S. Sokolovskii's *Military Strategy*, first published in 1962 and reappearing in 1963 and 1968. This work made the unparalleled destructive results of nuclear war strikingly clear not only for professional military forces but also for the nation at large.[56] Nonetheless, being able to fight nuclear war successfully was the key to Soviet survival and security, and the continuing modernization and development of Soviet strategic nuclear forces were paramount concerns in this analysis.

Soviet modification of its emphasis on strategic nuclear weapons came with its increasing strength in nuclear weapons and the changing context of international developments, including the 1967 Middle East war and the increasing involvement of the United States in Southeast Asia. NATO also engaged in discussions of the advantages of "flexible response," which it officially adopted in 1967. It was characteristic of Soviet military thinkers to consider the full spectrum of weapons usage, from conventional to thermonuclear weapons, as integral to the problem of war between the superpowers, without the either/or dichotomy stressed by Western civilian strategic specialists who focused on nuclear weapons and deterrence doctrine in their strategic studies.[57] In any case, the Soviet belief in the need for warfighting capability remained constant, whatever the weaponry envisioned, and a definite continuum existed in the Soviet view of conventional and nuclear weapons. Eventually, by the early 1970s, Soviet combined-arms doctrine emerged, synthesizing both conventional and nuclear weapons capabilities, a natural outgrowth of preceding doctrinal discussions when viewed retrospectively.

These broad military capabilities, which grew out of the strategic nuclear buildup of the 1960s and the subsequent development of a controlled-conflict capability in the early 1970s, set the stage for new doctrinal developments. US recognition of burgeoning Soviet defense capability was implicit in the conditions of détente between the military superpowers and explicit in their various arms limitations agreements. The confidence produced by these developments led to an expanded role in Soviet doctrine for Soviet military forces. Based on a greater capability to project its military power, the Soviet minister of defense, Marshal A. A. Grechko, announced in 1974 that Soviet military forces were no longer restricted to defensive functions, but were to be actively involved in supporting progressive forces and resisting imperialism through the planet,[58] a fact dramatically expressed both in the Soviet invasion of Afghanistan in 1979 and in Soviet support for "wars of national liberation" on virtually every continent.

The recently developed Soviet capability to project its military power should be understood both for its potential and for its limitations. The sense of potential is heightened by the global and holistic approach of Soviet military doctrine. By definition it excludes nothing that has a bearing on the exercise of military force in the political interests of the Soviet state. This means that all the media available for the projection of power—land, sea, air, and space—remain obstacles until they are conquered in Soviet interests. Until that time they have the potential of being exploited against Soviet interests. Geography and limited industrialization had long dictated the land-bound nature of the Russian state, and the attempt to mitigate this status was tied to the larger historical experience of modernization.

The mastery of modern weapons technology which freed the Soviet military forces from their land-bound status was consistent with the continuing effort of all the rulers of Russia in this century to master technology in the interest of national security. Just as naval development preceded the Soviet period of Russian history, so also had Russian interest in the use of airplanes for military purposes begun under the tsars.[59] However, it took the industrial base initiated under the tsars but fully established by the Communist leadership to make the Russian weaponry capable of defeating the Germans in World War II and then, in the postwar period, of enabling the projection of power beyond the vast expanse of the Soviet Union itself through mastery of the media of water, air, and space.

The Soviet capability to project its military power was greatly enhanced by the expansion of Soviet naval and air forces. Following the building of the Strategic Rocket Forces, major resources were devoted to the navy's expansion and modernization. Building upon the great advantages accruing to the largely land-bound Soviet Union from nuclear power plants and the revolution in weaponry caused by the union of nuclear weapons and ballistic missiles, the navy became part of the Soviet strategic nuclear forces with nuclear armed missiles in its wide-ranging submarines. The chief sponsor of Soviet naval development, Admiral of the Fleet S. G. Gorshkov, consistently advocated a unified military strategy, wherein the Soviet navy's ocean-going fleet with its multipurpose weapons and capabilities constituted part of the Soviet Union's single military strategy.[60]

The Soviet navy is only one of the military instruments available to the Soviet state; another is the Soviet air force.[61] During the first two decades of the postwar period the Soviet air force remained a largely defensive force with aircraft limited by their short range. Since then, Soviet capabilities have increased dramatically across a broad front in support of Marshal Grechko's statement that Soviet armed forces were no longer restricted to historic defensive functions.[62] The improved aircraft capability was seen in the Czechoslovakian invasion of 1968, military assistance to Egypt and Syria in 1973, and the 1979 invasion of Afghanistan. Other improvements included large increases in payload capacity and the ranges of attack aircraft. A broad and expanded use of helicopters was also underway and clearly evident in the fighting in Afghanistan. This greatly enlarged offensive capability troubled Western observers and raised questions about the purposes and intentions of

Soviet leaders. Is examination of Soviet doctrine a possible key to answering these questions?

The limits of assessing these unknowns in Soviet doctrinal statements are seen in attempting to understand the purposes of the Soviets in space. No publicly available explicit Soviet military doctrine for the use of space exists, in spite of the large, systematic, and long-term Soviet space program which appears to be run and maintained largely under Soviet military direction. The Soviet Union consistently denies military purposes and uses of space and accuses the United States of militarizing space, contrary to the 1967 Outer Space Treaty and the 1977 Soviet–US Agreement on Cooperation in the Exploration and Use of Space for Peaceful Purposes. From what is known about Soviet military doctrine in the broadest sense, there is nothing in the Soviet space program which suggests that it is not available for military purposes in a manner which meets the generalized Soviet doctrinal requirements for aiding in the achievement of Soviet objectives through superior military force used offensively in a combined-arms approach. Moreover, given the achievements of the United States in space, it would be irresponsible on the part of Soviet leaders not to consider the potential use of space by its enemies against Soviet security interests. Not to prepare for its own military use of space would appear illogical in the Soviet construct.

Clearly, the limits to the projection of Soviet power do not stem from ideas and values. Rather, those limits are resources; the inadequacies of the Soviet political, social, and economic system; and the pressures of remaining competitive within the dynamics of global history in the twentieth century. These are severe constraints which probably make the maintenance of Soviet military power an even greater priority for the Soviet leadership than it might otherwise be. Within Soviet military doctrine, the rationale for both defensive and offensive postures exists. Has the Soviet Union sufficient military forces and power for its own defense? How much is enough? And what military forces beyond that unknown constitute an unacceptable threat to the security of the United States?

Michael McGwire, a leading British specialist on the Soviet military, has analyzed Soviet military objectives as determined by doctrine and as supported by a wide range of secondary evidence. He concluded, in part, that "the top three levels of national objectives" are the following:

—Promote the well-being of the Soviet state.

—Avoid world war.

—If war is inescapable do not lose.[63]

Maintaining the sufficiency of Soviet forces necessary to support these objectives in a hostile world, in which the Soviet Union possesses superpower status only in terms of its military forces, must be a deeply troubling problem to Soviet leaders. How can they pursue to their maximum advantage the objectives cited above? It probably cannot be done by accepting the definitions of deterrence and security proposed by their enemies; it may be to the Soviets' advantage to exploit the "essential asymmetry of the arms race," as argued by James M. McConnell, wherein the Soviets lack any comparative advantage in superior military technology and therefore "compete by taking up options vacated or neglected by the West."[64]

What does this mean in terms of the nature, organization, and structure of Soviet military forces in being today? What are their capabilities and by implication their purposes?

SOVIET MILITARY FORCES

How does one measure the substance of power in the last quarter of the twentieth century? From the foregoing, it would appear a complex matter and rightly so. In the broadest sense, military power cannot be viewed in isolation from other elements of national power, such as the economic, social, and political dimensions of a nation's strength. This is true from the viewpoint of both Western analysts and the Soviets' own thought on the matter. Both the Soviet Union and the United States are caught up in the dynamic of forces much larger than themselves, such as the economics and politics of varying energy supplies, forces to which they themselves have contributed but which they do not always understand and therefore do not control or anticipate. Both nations are also tied to international political and economic systems which both constrain them and at the same time give them influence.

A true measure of Soviet power, therefore, cannot be taken by a review of Soviet military forces apart from other factors. However, the destructive capability of modern weapons systems, both conventional and nuclear, is so large, and the speed with which such destruction can be unleashed is so great, that military forces must be dealt with on their own merits before placing them in a larger strategic context. The potential destructive impact of these military forces is also heightened by the increasingly complex nature of modern, urban societies, wherein large population centers often comprise critical transportation and communications nodes and contain considerable industrial resources. The vulnerability of such centers and the difficulty of defending them with confidence makes the task of defense planners enormously complex, but they must begin somewhere, and a logical point of departure is the analysis of Soviet military forces themselves.

Given modern technical means of reconnaissance and surveillance, the gross numbers of Soviet military forces are probably not a matter of serious dispute or difference, and they appear with mind-numbing frequency in newspapers, journals, and magazines. One of the immediate and obvious criticisms of the accountant's approach to the question of who has more of what is both its static nature and a lack of discrimination with regard to the critical military criteria of efficiency and effectiveness.[65] Partly for this reason, understanding the structure and organization of Soviet military forces is more useful than knowledge of the force levels themselves in appreciating their long-term significance, especially if their evolution is put into some historical context. The numbers and specific capabilities will continue to change, but the meaning of the changes in terms of efficiency and effectiveness will probably be found when they are analyzed in the context of Soviet military structure and organization. This is also the dimension of Soviet military forces which gives substance and meaning to Soviet ideas about war (discussed in the preceding section on Soviet political-military doctrine). The organizational nexus between thought and action is critical in judging Soviet statements and supported intentions. In the final analysis, organization governs the employment of Soviet military forces and their resulting efficiency and effectiveness.

How much change has taken place in Soviet military structure and organization in this century? Soviet writers describe the October Revolution as an epoch-making event which began a new phase in global history, but in military affairs the revolution transcended neither geography nor basic problems in national security. The destruction of World War I, the innovations and adaptations of Soviet leaders during the civil war, and the creation of the new Red Army were part of an unsettling period of transition suggesting great departures from the past, but when the debris settled, much of it fell back into the molds of previous historical forms and shapes. When the enormous stresses and strains of World War II occurred, they called forth other traditional elements of the Russian military past. Russian historical influences of the imperial military past remained strong, if for no other reason than geographical constraints. Peter the Great had sought to conquer Russian geography for reasons of national security, relocating the capital and bringing about major social, political, and administrative reorganization. It is not without reason that Peter the Great has remained a powerful image for his successors in both the imperial and Soviet periods.[66] He faced the problems of enormous space and distance, of hostile neighbors, and of mobilizing an obdurate citizenry for civic and military purposes—problems with which both a tsarist autocrat and a Communist dictator could sympathize.

History and geography provide continuities between the imperial and the Soviet periods of Russian military history. Moreover, war continued as the central fact of modern Russian history, and when Russian military successes of the late 1700s and early 1800s could no longer be sustained in the face of a rapidly modernizing West and East, notably with Russian defeat in the Crimean War of 1854–56 and the Russo-Japanese War of 1904–5, Imperial Russia undertook military reforms which left their mark in the Soviet period. One of the most important organizational measures was the military district system which survives in altered form into the present.[67] It furnished a means of both mobilizing manpower for war and of providing a command structure capable of organizing resources appropriate to the threats and circumstances of the widely varying regions of the Russian empire. The military district is the Russian organizational answer to the question of how one mobilizes, moves, and concentrates large military forces effectively over such large spaces and distances in relatively short periods of time, and its validity is demonstrated by its survival into the Soviet period of Russian history.

The persistence of the military district organization from the imperial into the Soviet period of Russian history attests to the resilience of certain factors in national security regardless of the history period. The factors of space, distance, and size continue as major concerns for the Russian military professional no matter who sits in the Kremlin, because Russian domination of the Eurasian landmass remains the central issue in territorial security. Threats to that security since the middle of the nineteenth century have largely been based on the advantages of technology which have accrued to Western enemies who had successfully exploited the Industrial Revolution. The needs of national security have called for military industrialization, specifically the mastery of the Industrial Revolution, a process which has in turn transformed original Russian purposes and Russia itself.[68] The dynamics of these forces have also contributed to the continuing evolution of threats to the security of the Soviet Union, as a result of which Soviet military professionals must now prepare for the prospect of nuclear weapons launched from the depths of the oceans, from continents away, and perhaps ultimately from space itself.

Whence the Threat to Peace is a Soviet publication devoted to a description of US military power viewed as a threat to the Soviet Union.[69] It depicts a global military power with the most advanced military technology, both conventional and nuclear, based throughout the world and posing a threat to both the periphery and the center of the Soviet Union. Land, air, the seas and oceans, and space itself are described as harboring actual and potential threats to the Soviet Union. When complemented by Soviet fears of China and the US NATO allies, the Soviet Union would appear to have need of an organization and structure of military forces capable of meeting threats from all directions and in all the media of land, air, water, and space.

To accommodate Soviet security needs, such a military organization would need a number of characteristics. It would be derived from the land-based nature of Russian warfare which resulted from territorial threats to the Russian state over the centuries. Given the growing capacity of the seas and oceans of the world to harbor weapons which can strike the Soviet Union, it would have to be an organizational structure enabling the extension of Soviet defensive perimeters out to sea and into the land areas bordering the Soviet Union. Recognizing the capabilities of enemies to strike the Soviet Union from air and space, it would also have to provide for an integration of air and space defenses into the land- and sea-based forces. Under conditions of modern warfare, it would have to accommodate the element of surprise and preemptive attack by its enemies, enabling its military forces to move from a defensive to an offensive posture once adaptation to the initial effects of the attack was made. Such an organization would have to be flexible, constituting a military readiness and a capacity for mobilization sufficient to deter enemies from attacking in the first place; it would be an autonomous unit, complete enough in its organizational structure to deal with a regional threat on its own in the event that supporting elements could not be developed over vast distances from elsewhere in the Soviet Union. Finally, this organizational entity would have to be open to combination with other organizational elements so that it could move as part of a system beyond regional threats to broadly strategic concerns.[70]

For these reasons, no map is more important for understanding Soviet military forces than one showing the military districts, the theaters of military operations, and theaters of war—the Soviet organization for war.[71] This map would have been essentially recognizable to tsarist military commanders, because it grew out of military reforms in the middle of the nineteenth century and has remained the basic skeleton of Russian military organization ever since. Today it epitomizes the distinctively Russian approach to mobilizing, organizing, and directing military forces over the huge spaces, great distances, varying terrain, and climate of the Soviet Union. It provides a complete wartime integration of the civilian and military sectors; it bonds civil resources and military organizations for tactical

and strategic applications; it is an administrative and institutional structure for manpower management in peacetime which can become an operational entity and major level of command in war; it serves as the organizational device which enables a changeover from defense to the counteroffensive in the trans-attack situation; it is designed for survivability and postattack recovery in both conventional and nuclear warfare, and in both short and protracted wars; it integrates civil defense into the military district command structure; it is the means by which combined-arms applications and the packaging of military forces are organized to achieve the Soviet strategic objective of penetrating deeply the rear of the enemy through the redeployment of the military forces generated by the military district; it provides for transition from peace to war and transition to peace after war; and, it integrates both tactical and strategic forces into a command and control net which accommodates direction by the Soviet general staff. The Soviet military district is both a microcosm and macrocosm of Soviet military ecology which accommodates a globally directed strategic view and makes Soviet space and size manageable.[72]

Russian military professionals began moving toward this organizational solution after 1862. It derived from an analysis of Russian military power, strength, and weakness by a staff member of the Saint Cyr Military Academy, Leon Dussieux, published in Paris in 1854. He concluded that Russian security derived from the capacity of the Russian empire to maintain "a 'military presence' in each of the 'theatres of war' into which the vast frontiers of the Empire could be divided." From 1862 to 1885, the Russian empire was first divided into military districts. The subsequent evolution of the military district led to its becoming a significant factor in military planning and organization as a "theater of war" in European Russia during the two decades spanning the nineteenth and twentieth centuries. As the role of the military district changed, especially with regard to its development within the organization of military command, it resulted by 1914 in a strategic field-command structure embracing the military districts under a Supreme Commander-in-Chief.[73]

Soviet military professionals and historians studied this organizational evolution in the Imperial period, particularly its effectiveness in World War I, and contributed to the continuing development of the military district in the Soviet period. Frunze served as a key figure in this transition, viewing the military district as vital in uniting the front and the rear. When located on the frontier, the military district was an administrative body providing a territorial base enabling effective peacetime deployment

of military forces and effective wartime mobilization. His concept of the military district as
a theater of war which mobilized civil manpower
and resources together with military forces was
a new Soviet dimension to the military district.
During the 1920s and 1930s, various military
districts were formed, reflecting traditional Russian ideas about war, approaches to the problems of size, varying continental geography and
climate, and prior experience with the military
district system of organization. During World
War II these military districts were converted
into fronts and were significant in the defeat of
Germany.[74]

Postwar Soviet analysis of the preparations
and planning for World War II suggests the following three conclusions: (1) significant measures to strengthen the defense of the country
were successfully completed before the war; (2)
Soviet military administration remained effective, in spite of Stalin's purges; (3) the military
district proved itself, "both as a senior organ of
administration and as an embryo senior wartime
field command."[75] These analyses also give special attention to the successful regrouping of
large field formations to attain political goals or
strategic objectives, as well as the major redeployment of Soviet military forces in the Far
East which led to the defeat of the Japanese
Kwantung Army in 1945.[76] Given the heavy Soviet emphasis on learning from military history,
the importance of these historical studies in the
development of Soviet military science could be
easily underestimated by Westerners.

Building upon the experience of World War
II, with the historically unprecedented movement and strategic coordination of such large
land forces over enormous distances, the Soviet
military wrestled with the meaning of the historical experience of World War II in light of
the traditional Russian concern for buffers as
elements in territorial security, the continuing
role of space and distance in recovery from attack, and the impact of modern war upon these
problems, especially from the viewpoint of time
and the speed of military operations. As with so
much work in the open source literature on Soviet military forces, as one draws closer to the
present the accuracy of the descriptions and the
significance of the data become less certain. It
does appear, however, that the theater concept
first identified by Leon Dussieux in the mid-
nineteenth century, and as modified by the
experiences of World Wars I and II, remains
central to the operational planning of Soviet military forces. It is compatible with the Soviet
view of war which is necessarily global, requiring coordinated operations on a global scale.
Given the existence of international socialism,
the East European security buffer of socialist
states, and the Warsaw Pact, such planning

must also occur within the framework of coalition warfare.

How is this pertinent to understanding the
nature of Soviet military power today? Its relevance derives from the value of understanding
how the national security problems of the Soviet
state can become regional and then global in
their professional military definition, apart from
the international aspirations of communism. Issues of national security transcending the Imperial and Soviet periods do exist. The planning
and use of military force within an organizational
framework so different from that of the United
States should cause us to question analyses of
Soviet military efficiency and effectiveness
which are based on assumptions derived from
Western military structure and organization. Is
it possible for the Soviets to do more with less
than their Western counterparts? Is there a resulting military efficiency and effectiveness
which is greater than the sum of the constituent
parts, a greater total effect than would otherwise
be possible and which is a significant compensation for the economic, social, and technological deficiencies of the Soviet Union? Answers
to such questions should help us understand the
nature of Soviet power in a dynamic international environment, within which the Soviet
leadership must continually react to changing
and sometimes unforeseen developments. The
key elements in Western understanding of this
process may be those constants from the Soviet
side which govern both their initiatives and
their reactions.

The evolution of the Russian landscape of war
is continually affected by its enemies' skills and
capabilities, the traditional elements of Russian
geography and politics, and the larger dictates
of the requirements of modern warfare. Space
and distance remain significant dimensions of
Soviet military operations, as does the concept
of "theaters" of war originally employed by Dussieux to define the effective employment of military forces. Persistent factors not open to final,
or short-term resolution are political instability
in Eastern Europe, Asian antipathy, international Communist rivalry, and Moslem fundamentalism. The potential threat which these
factors pose may be acceptable when taken individually, but, when taken together against the
backdrop of a globally active and technologically
advanced enemy, they represent an overall
threat scenario of enormous complexity for Soviet political and military planners.

Part of the Soviet reaction to this threat,
growing out of its military district structure, is
the Soviet theater organization for war. The
challenges of theater warfare relate to the traditional problems of territorial defense in which
the frontier remained a source of threat and
insecurity.[77] Given the wide regional variations

in geographical, ethnographical, climatic, and political conditions, the composition and purpose of theater military forces can vary greatly. In Eastern Europe and the Far East they probably share the objectives of preventing or stemming invasion and of mounting offensives to push back an enemy and to increase the distance between enemy forces and the homeland. The industrialization of war has made theater organization both possible and necessary; it has both created its major problems and proved possible solutions. Traditional factors, such as space and distance, remain important in Russian recovery from invasion. However, they have been diminished in their significance by the impact of modern military operations upon time, and the theater organization for war makes it possible to extend defensive perimeters outward, while providing for the offensive postures which the Soviet Union considers the best defense. Soviet military planners and staff face a number of problems as they wrestle with the formation and use of groupings of military forces which vary in their composition and purpose, but perhaps no challenge is greater than that of command and control, wherein the Soviet leadership seeks flexibility and autonomy at lower levels of command while sustaining central control over strategic operations.

The general result of these developments has been to shape and reshape organizationally those elements which contribute to more effective theater employment of military forces.[78] Within this organization significantly different results can occur with changes in command and with variations in the combination of organizational elements such as air defense forces, strategic air strike assets, and theater field forces. The geographical extension of the theater concept to seas and oceans, and an integration with land and air forces, would appear a natural evolutionary result of both Soviet military thought and organization. Building upon the operational concept of a combined-arms force in a theater battlefield, John Erickson and other Soviet specialists believe that by "the early 1970s this requirement was maturing into planning and preparation for coordinated operations on a *global scale* and with[in] the framework of *coalition warfare*."[79]

The weapons to be used in such military operations are both conventional and nuclear, and the Soviet Union has accumulated them in extraordinary numbers and at ever higher levels of quality and performance. John M. Collins, senior specialist in national defense of the Library of Congress, emphasizes certain continuing elements of Soviet military power, especially when put into the context of liabilities and assets on both the American and Soviet sides.[80] At the base of Soviet military power is the enduring, long-term commitment to national security, which enables a continuing dedication of resources to steady improvement of Soviet military forces. Collins concludes that, between 1960 and 1980, "few trends related to the U.S.–Soviet balance favored America" and that, although the US made "impressive" absolute improvements, its "capabilities declined in relative terms, because quantitatively superior Soviet forces modernized faster."[81] After reviewing Soviet and US advancement, advantages, and disadvantages, he points out that, in spite of changing problems for Soviet decision-makers, "their pattern of competition, however, remains remarkably consistent, when compared with sweeping U.S. change."[82] Continuing modernization and improvement in the organization and deployment of Soviet military forces are trends supported by two decades of data and which appear unlikely to change, especially in view of the open-ended, formal and informal, global commitments in which the Soviet Union parallels the United States. But what about the prospects for a stable balance between Soviet and US military forces? Collins concludes, "The U.S.–Soviet military balance has been relatively stable for more than 20 years, and likely will remain so for some lengthy period."[83]

IMPLICATIONS OF SOVIET POWER

The foregoing pages have shown why military power has emerged as the most important dimension of Soviet power in the international competition with the United States. They have also shown why and how the Soviet Union continues to organize for war, but they do not address the question of appropriate responses from the United States. However, partial outlines of the policy problems facing the United States do emerge.

Basic to the long-term concern of the American side of the relationship must be the continuing effort to arrive at a distinctively American definition of the competition. Criteria for success in the competition are debatable, but if the United States were somehow a winner at the cost of nuclear destruction or the loss of distinctive civilizational values, the success would, at best, be questionable. The increasingly long-term nature of the competition and the fact that for some time both the United States and the Soviet Union have possessed sufficient nuclear weaponry to destroy each other without doing so suggest that nuclear destruction should not be the sole concern. The military focus of the competition, in which the Soviet Union has the greatest leverage, is not in the US long-term interest, but broadening the base of the competition is not easily achieved and brings its own problems. Reacting to the Soviet definition of the international scene and remaining rooted in the more defensive postures of anti-communism which have been at the core of much of US policy since World War II are tendencies with considerable inertia.

The long-term implications of Soviet power are in conflict with the basic purposes of the United States, which have historically been an effort to balance power, morals, and liberty to enhance the personal destiny of each human being. The failure to promote and support a "program of strategic ideals" growing out of our own history may indicate that we have not studied our own history well enough,[84] nor respected sufficiently the influence of history upon the Soviet Union. The significance of these historical ideals remains muted in competition with the Soviet Union as long as American economic and technological dominance continues, because the United States may be in the greatest danger vis-à-vis the Soviet Union when it is economically weak. At such a time the advantages of a command society may come to the fore: when it appears that human energies must be controlled and directed; when society at large may not willingly sacrifice to sustain its defense; when in the interest of national security the United States might assume some of the salient characteristics of its Soviet opponent. Ironically, the United States may be weakest relative to the military strength of the Soviet Union when it is not true to the peculiar sources of its vitality and potency, the largely intangible, nonmilitary factors in its history.

Articulating an American definition of the competition must take into account both American and Russian historical traditions. Such a definition should recognize both the transcendental nature of security problems facing all masters of the vast Eurasian landmass and the culturally distinctive approaches of Russian military professionals to those problems. One should not assume, for example, that the overwhelming logic and reasonableness of Western notions of deterrence are necessarily acceptable to the Soviet Union. Moreover, for the leaders of the Soviet military, the weight of the experience of wars suffered on Russian territory has proved equally, if not more, influential than the power of revolutionary inspiration and logic in determining how to organize military forces for national security. The weight of the historical past for the Soviet leadership as it approaches military matters appears greater than for American leaders. At first glance this appears paradoxical in light of the relative recency of the Bolshevik revolution. However, the more distant and aged American Revolution was less founded in the challenges of national security, modern warfare, and territorial aggression. These historical differences are basic in under-

standing both how the United States and the Soviet Union differ today and what they share in facing similar problems and issues.

The most important element of asymmetry between the two nations is the difference in the breadth of the two countries' respective power bases. The broad economic, technological, and social advances and advantages of the United States over the Soviet Union outside the military realm seem to be increasing rather than decreasing. This is probably the single most important strategic fact facing the Soviet leadership today. It also carries special irony for Soviet military leaders, because they have developed a more systematic and profound professional military analysis of modern warfare than American military professionals, especially in showing how success in modern warfare depends on the total resources of a country and their effective mobilization. In certain respects the problem of national security has trapped Soviet leaders and kept them within military confines. Moreover, the primacy of the national security problem has made the Soviet military establishment the leading edge of science, technology, and information about advances in the rest of the world, rendering it the spearpoint of the modernization process. Given the political requirement for control and secrecy, the result is that the huge bulk of a society which has only recently undergone modernization is restricted from free access to and participation in its leading developments. This contrasts sharply with the United States, where the professional military is carried along by the momentum of civilian-based scientific and technological developments, for which there is frequently no military doctrine in the Soviet meaning of the word.

This large asymmetry in the substantive makeup of the two countries is matched by an equally great difference in the general perception and understanding of modern warfare. The Marxist-Leninist view of the world blurs the distinction between war and peace and sees war as less of a historical anomaly than the Western view. The continuing fact and prospect of war in Soviet history have contributed to a command organization of society and an approach to war which is intellectually holistic, nationally survivalistic, global in its orientation, and organizationally mobilizational and integrational—an approach which in each respect is in almost direct contrast to that of the United States.

Perhaps no better touchstone exists for the differences in the two approaches and views than the differing attitudes and policies concerning civil defense. Americans familiar with the facts of Soviet population losses during World War II stand in awe of that nation's capacity for suffering and sacrifice and the ability of the Soviet leadership to exact resources from the nation at large for military purposes. In view of the horrors of modern war, especially from the use of nuclear weapons, the Soviet Union continues to propose and to develop civil defense measures, whereas the United States does not. Equally significant are the possible alternative interpretations which the Soviet mindset could draw from this American position on civil defense. Such Soviet interpretations could include the following: that preemptive strikes by American forces would render American civil defense less essential; that American analysis of the nature and conditions of modern warfare may be inadequate or invalid; that the American political or military leadership is irresponsible with regard to the survival of its citizenry in modern war; and that lack of civil defense measures is evidence that the American leadership does not actually believe the United States is under true threat of military attack. This difference in Soviet and American approaches to defending civilian populations grows out of the shared, general, and overarching problem of national security in the context of enormous advances in weapons technology.

With advanced conventional weapons both countries have explored the limits of their military power since World War II: the United States in Vietnam, where it found that the political will and organization of an indigenous Asiatic population could negate the usefulness of an extraordinary range and number of modern weapons; the Soviet Union in Afghanistan, where it found that the religious identity and will of an indigenous Asiatic population without strong political identity and organization frustrated its designs in the twentieth century as it did in the nineteenth century, when the Russian Empire competed with the British Empire in the same region. The problem of national security is indeed international when both the United States and the Soviet Union see their national interests threatened by peoples and cultures possessing clearly insufficient political or military power to destroy or threaten either the American or Soviet homelands.

Both the United States and the Soviet Union face the increasingly difficult problem of translating into political utility the enormous destructive power of nuclear weapons. A major frustration for both nations is that nuclear weaponry continues to grow in its power and refinement but not in its political efficacy. Nuclear weapons were used in neither Vietnam nor in Afghanistan, and both American and Soviet military policymakers have come to recognize their lack of utility. In a letter to the historian Theodore Draper, Secretary of Defense Caspar W. Weinberger noted that "we regard the enormous damage and devastation created by the use of nuclear weapons as having rendered the concept of victory (as it has classically and his-

torically been defined in a political-military context) meaningless."[85] Marshal Ogarkov described the continuing growth of nuclear arsenals as "senseless" and the world stores of nuclear weapons "absurd from a military point of view."[86] The stores of both nuclear and conventional weapons have been acquired at great direct cost to the United States and the Soviet Union, apart from whatever societal benefits could have been realized with the expenditure of these funds elsewhere. However, nations will not ignore their national security.

Neither nation has consciously sought national security positions fraught with so many dilemmas and complex problems. Their capacity to judge the wisdom of various policy decisions has been complicated by the changing nature of their own societies and their relative positions internationally.[87] After entering a war it did not seek, and emerging from World War II at an unparalleled pinnacle of world power with a monopoly of nuclear weapons, the United States has had to accommodate itself to a decline in its relative influence in the world and to the emergence of its major competitor as a military superpower. The Soviet Union also emerged from World War II with extraordinary influence, but unable to match the political, economic, and military power of the United States. Nonetheless, it possessed the confidence of military victory in a global war and the knowledge of being able to sustain exceptional economic growth. It did not foresee that, following rehabilitation from World War II, it would have created a society particularly unsuited to meeting the special challenges of continuing modernization and poor at adapting to the economic and social changes occurring in the rest of the world.

The decline in growth of its overall power base is a devastating reality for the Soviet Union and renders the problem of national security all the more acute, because by its own definition it can only arrive at a satisfactory level of security at the expense of other nations and ultimately with the failure of nonsocialist systems. While national well-being and security are not demonstrably a zero-sum game, that remains the Soviet view of the world. US losses therefore are implicit or explicit gains for the Soviet Union. Soviet gains in the global context will be difficult to achieve until it secures the clear-cut dominance necessary to sustain and implement its unitary view of the world. In the meantime, burgeoning economies such as Japan's will continue to develop and to expand, with results affecting the broader distribution of power in the world. Without basic systemic change enabling the Soviet Union to engage fruitfully with the global forces changing the world, or the arrival of a powerful leader in keeping with the tradition of forceful change that has been part of Russia's past, the possibility of the Soviet Union's benefiting as much as the United States from broader global changes remains unlikely, except insofar as it obtains advantage from the instability caused by these changes.

NOTES

1. For example, Theodore H. Von Laue, *Why Lenin? Why Stalin?: A Reappraisal of the Russian Revolution, 1900–1930*, 2d ed. (Philadelphia: J. B. Lippincott, 1971), pp. 195–96. More recently by Mark Harrison, *Soviet Planning in Peace and War, 1938–1945* (Cambridge: Cambridge University Press, 1985), pp. 46–47.

2. A summary of Russian expansion resulting from World War II is found in Allen F. Chew, *An Atlas of Russian History* (New Haven: Yale University Press, 1967), pp. 104–7.

3. J. Bronowski, *Science and Human Values*, rev. ed. (New York: Harper and Row, 1975), p. 52.

4. Early argued by Samuel L. Sharp, "National Interest: Key to Soviet Politics," *Problems of Communism* 7 (Mar.–Apr. 1958): 15–21.

5. Richard Pipes, *Russia Under the Old Regime* (New York: Charles Scribner's Sons, 1974), p. 83.

6. The implications of this are far reaching, for as Richard Pipes states, "The Soviet Union is the only major power where the dominant nationality barely has a majority," in "Introduction: The Nationality Problem," *Handbook of Major Soviet Nationalities*, ed. Zev Katz (New York: Free Press, 1975), p. 1.

7. C. Vann Woodward, "The Age of Reinterpretation," *American Historical Review* 66 (Oct. 1960): 1–19.

8. Christer Jonsson, *Superpower: Comparing American and Soviet Foreign Policy* (New York: St. Martin's Press, 1984), p. 95.

9. Michael Howard, "The Bewildered American Raj: Reflections on a Democracy's Foreign Policy," *Harper's Magazine*, Mar. 1985, pp. 59–60.

10. The best study of the Soviet view is by P. H. Vigor, *The Soviet View of War, Peace and Neutrality* (London: Routledge and Kegan Paul, 1975), esp. pp. 25–58. For a review of the evolution of the American view of strategy see the introduction to Russell F. Weigley, *The American Way of War* (New York: Macmillan, 1973), pp. xvii–xxiii. Weigley states: "It is true that during 1941–45 and throughout American history until that time, the United States usually possessed no national strategy for the employment of force or the threat of force to attain political ends, except as the nation used force in wartime openly and directly in pursuit of military victories as complete as was desired or possible" (p. xix).

11. For a detailed discussion see Julian Lider, *Correlation of Forces: An Analysis of Marxist-Leninist Concepts* (New York: St. Martin's Press, 1986).

12. Col. S. A. Tyushkevich, "The Methodology for the Correlation of Forces in War," *Voennaya mysl'* 6 (June 1969), FPD 0008/70 (30 Jan. 1970) in *Selected Readings from "Military Thought," 1963–1973*, Studies in Communist Affairs (Washington, D.C.: G.P.O., 1982), vol. 5, Part 2, pp. 57–70.

13. Gen. Albert C. Wedemeyer, USA (Ret.), a principal planner for US victory in World War II, saw the American understanding of strategy as too narrow in 1986 and called for formation of a National Strategy Council in "Memorandum on a National Strategy Council," *Military Planning in the Twentieth Century*, ed. H. R. Borowski (Washington, D.C.: G.P.O., 1986), pp. 409–15. Benjamin S. Lambeth writes of the Soviets' "general contempt for U.S. strategic doctrine and concepts" (p. 149), Soviet puzzlement over US military behavior in Vietnam (pp. 143–44) and American unpredictability (pp. 152–54) in "Uncertainties for the Soviet War Planner," *International Security* (Winter 1982–83).

14. A representative sample of such publications is found in the volumes translated and published under the auspices of the United States Air Force in a series entitled "Soviet Military Thought" and available from the US Government Printing Office. The Soviet armed forces have had several "Officer's Library" series, including a series for the 1980s. For a description of the series, see Harriet Fast Scott and William F. Scott, "Soviet Bibliographies and Their Use as Research Aids" (Defense Nuclear Agency Report, DNA, 6175T, 31 Dec. 1981), pp. 35–36.

15. A phrase used by William H. McNeill in his masterful study of *The Pursuit of Power: Technology, Armed Force, and Society since A.D. 1000* (Chicago: University of Chicago Press, 1982), p. 309.

16. McNeill, *Pursuit of Power*, p. 148n.

17. An example of this impact is described in the article by Bertram D. Wolfe, "The Influence of Early Military Decisions Upon the National Structure of the Soviet Union," *American Slavic and East European Review* 3 (1950): 169–79.

18. Robert C. Tucker, *The Soviet Political Mind: Stalinism and Post-Stalin Change*, rev. ed. (New York: W. W. Norton, 1971), pp. 20–46.

19. Sylvia R. Margulies, *The Pilgrimage to Russia: The Soviet Union and the Treatment of Foreigners, 1924–1937* (Madison: University of Wisconsin Press, 1968), pp. 185–93.

20. Professional military assessment of the Red Army by the Allies is reviewed by Ronald R. Rader in "Anglo-French Estimates of the Red Army, 1936–1937," *Soviet Armed Forces Review Annual*, ed. D. R. Jones (Gulf Breeze, Fla.: Academic International Press, 1979), vol. 3, pp. 265–80. Rader concludes, "By the end of 1937, as the war clouds gathered over Europe, the Red Army's value had sunk to practically zero in western calculations" (p. 274). He adds that "the Allies' basic failure was that they underestimated the Russians' capacity to analyze the weaknesses and their determination, once urgency was necessary, to make corrections in a forceful and decisive fashion" (p. 275).

21. In *Pursuit of Power*, William H. McNeill traced the impact of command upon societies, armies and technology from premodern times "in Asia, when goods and services were needed to put an army in the field, the rulers' commands sufficed to mobilize whatever was, or could be, mobilized" (p. 111) to the post–World War II period. William L. Blackwell summarized the tradition of "military mobilization" and its historical impact in *The Industrialization of Russia: An Historical Perspective* (Arlington Heights, Ill.: Harlan Davidson, 1970), pp. 173–74. I use the phrase *command principle* to incorporate total mobilization for war under autocratic or dictatorial leadership as described by these two authors.

22. Leonard Schapiro comments on the significance of Stalin's " 'Third Revolution' of enforced collectivization and rapid industrialization" in "Epilogue: Some Reflections on Lenin, Stalin and Russia,"

Stalinism: Its Impact on Russia and the World, ed. G. R. Urban (Cambridge: Harvard University Press, 1986), pp. 421–27.

23. McNeill, *Pursuit of Power*, p. 345.

24. James R. Millar, "Conclusion: Impact and Aftermath of World War II," *The Impact of World War II on the Soviet Union*, ed. S. J. Linz (Towota, N.J.: Rowman and Allanheld, 1985), p. 283.

25. Millar, "Conclusion," p. 284.

26. Barbara A. Anderson and Brian D. Silver, "Demographic Consequences of World War II on the Non-Russian Nationalities of the USSR," *The Impact of World War II on the Soviet Union*, ed. S. J. Linz (Towota, N.J.: Rowman and Allanheld, 1985), p. 207.

27. A significant factor in its survival must be the performance of the Soviet "rear" (*tyl*)—both its military rear services and the organization of the home front in general. Because of both the enormous scope of the subject and the paucity of Western scholarship, planners chose to exclude this subject for the Tenth Military History Symposium (USAF Academy, 20–22 Oct. 1982) on *The Homefront and War in the Twentieth Century*, ed. James Titus (Washington, D.C.: G.P.O., 1984). A series of ground-breaking lectures on this subject was given by John Erickson at Yale University, 1987.

28. There may be more profound, systemic reasons for the success of the Soviet Union in the war against Germany. According to Harrison, Stalin believed with some justification in the "superiority of the Soviet system at both peaceful and warlike tasks" (*Soviet Planning*, pp. 222, 243).

29. For specific exceptions and variations from this broad generalization compare information and maps on pp. 92–93, with that on pp. 100–102 and 104–7 in Chew, *Atlas of Russian History*.

30. Anderson and Silver note that "in 1941, the total population of the Soviet Union was 200 million. This number was not achieved again until 1956," "Demographic Consequences of World War II," p. 208.

31. George C. Lodge, *The New American Ideology* (New York: Alfred A. Knopf, 1975), esp. chap. 1, "The Importance of Ideological Analysis," pp. 3–43.

32. As George Boas stated, "In 1941 the Rhine was said to be the American frontier. In 1964 the frontier moved to the Gulf of Tonkin." In *The History of Ideas* (New York: Charles Scribner's Sons, 1969), p. 74.

33. Col. Gen. N. A. Lomov, ed., *Nauchno-tekhnicheskii progress i revoliutsiia v voennom dele* (Moscow, 1973), trans. and published under the auspices of the US Air Force as *Scientific-Technical Progress and the Revolution in Military Affairs* (Washington, D.C.: G.P.O., 1974).

34. Stephen M. Walt, "Alliance Formation and the Balance of World Power," *International Security* (Spring 1985), p. 41.

35. The term "geopolitics of information" is taken from the title of a book devoted to Third World concern with media practices and the dissemination of news, by Anthony Smith, *The Geopolitics of Information: How Western Culture Dominates the World* (New York: Oxford University Press, 1980). A survey of the Soviet "array of overt and covert techniques for influencing events and behavior in, and the actions of, foreign countries" through disinformation is contained in the book by Richard H. Shultz and Roy Godson, *Dezinformatsia: Active Measures in Soviet Strategy* (Washington, D.C.: Pergamon-Brassey's International Defense 1984). The long-term changes and the institutionalization of the changes taking place under the policy of *glasnost'* remain to be seen.

36. For a review of Russian and Soviet military interest in Afghanistan, which had declined as a military priority by 1900, see the article by Ronald R. Rader, "The Russian Military and Afghanistan: An Historical Perspective," *Soviet Armed Forces Review Annual*, ed. D. R. Jones (Gulf Breeze, Fla.: Academic International Press, 1980), Vol. 4, pp. 308–28. For early Russian efforts to conquer the Caucasus, see the article by E. Willis Brooks, "Nicholas I as Reformer: Russian Attempts to Conquer the Caucasus," *Nation and Ideology: Essays in Honor of Wayne S. Vucinich*, ed. I. Banac et al. (Boulder, Colo.: East European Monographs, 1981), pp. 227–63.

37. Condoleezza Rice also uses the term "holistic" to describe Marxism in the beginning of her article on "The Making of Soviet Strategy," in *Makers of Modern Strategy*, ed. Peter Paret (Princeton: Princeton University Press, 1986), p. 648.

38. Argued by Ken Booth, *Strategy and Ethnocentrism* (New York: Holmes and Meier, 1979), esp. pp. 100–104.

39. John J. Dziak, *Soviet Perceptions of Military Doctrine and Military Power: The Interaction of Theory and Practice* (New York: Crane, Russak, 1981), p. 31.

40. O. V. Kuusinen, et al., *Fundamentals of Marxism-Leninism* (Moscow, 1963), pp. 292–95.

41. As stated by Khrushchev in February 1956, "war is not fatalistically inevitable." In "The Twentieth CPSU Congress and the Doctrine of the 'Inevitability of War,' " Intelligence Report No. 7284 (Washington, D.C., 1956) in *Readings in Russian History*, ed. W. B. Walsh, 3d ed. (Syracuse, N.Y.: Syracuse University Press, 1959), p. 700.

42. The discussion of Lenin's role in this and the following paragraph is based on the superb article by Jacob W. Kipp, "Lenin and Clausewitz: The Militarization of Marxism, 1914–1921," *Military Affairs* 4 (Oct. 1985): 184–91.

43. No revolution is without its antecedents. A survey of the broad elements of continuity is found in the article by David R. Jones, "Russian Military Traditions and the Soviet Military Establishment," *The Soviet Union: What Lies Ahead?* ed. K. M. Carrie and G. Varhall (Washington, D.C.: G.P.O., 1985), pp. 21–47. Specific exploration of the evolution of early Soviet military thought and its Tsarist antecedents is found in a forthcoming article by Jacob W. Kipp, "Mass, Mobility and the Origins of Soviet Operational Art, 1918–1936," in the Proceedings of the Twelfth Military History Symposium, 1–3 Oct. 1986, USAF Academy, to be published by the Office of Air Force History under the title, *Transformation in Russian and Soviet Military History*.

44. The advantage and the disadvantage of using Marxism-Leninism to explain modern war are its "relatively narrow but more cohesive" interpretation. Thomas W. Wolfe pointed out that "communist thought characteristically has sought to explain war in terms of a single body of theory," in "The Communist Theory of War," *Marxism, Communism and Western Society*, vol. 8, ed. C. D. Kernig (New York: Herder and Herder, 1973), p. 308. The application of Marxism-Leninism in the social sciences to military problems has had extraordinary development in the Soviet Union. See the ground-breaking study by Jack

L. Cross and James T. Westwood, *Social Sciences in the Soviet Armed Forces*, SAFRA Paper no. 2 (Gulf Breeze, Fla.: Academic International Press, 1988).

45. Carl W. Reddel, "The Soviet View of Human Resources in War," *The Soviet Union: What Lies Ahead?* ed. K.M. Carrie and G. Varhall (Washington, D.C.: G.P.O., 1985), pp. 415–31.

46. James M. McConnell states that "the American exploitation of asymmetry is usually based on superior technology. The Soviets have no such comparative advantage; they compete by taking up options vacated or neglected by the West," in "Shifts in Soviet Views on the Proper Focus of Military Development," *World Politics* (Apr. 1985): 341.

47. Ibid., pp. 318, 326–27.

48. For American historians the journal may be too purposefully military and not scholarly enough, with its heavy Marxist-Leninist emphasis on "lessons to be learned"; for American military professionals the journal is probably too historical, occasionally reaching into the past before World War II for its subject matter. For a Soviet description see V. A. Matsulenko, "Voenno-istoricheskii zhurnal" [*Military History Journal*], *Sovetskaia voennaia entsiklopediia* (Moscow, 1976), II, p. 224.

49. A view expressed to the author by John Erickson, author of the best two-volume study of the Soviet army in World War II, *The Road to Stalingrad* (New York: Harper and Row, 1975) and *The Road to Berlin* (Boulder, Colo.: Westview Press, 1983). For Erickson's specific views on the scope and complexity of Soviet operational research, see his article, "Soviet Military Operational Research," *Strategic Review* (Spring 1977): 63–73.

50. Harriet Fast Scott and William F. Scott, eds., *The Soviet Art of War: Doctrine, Strategy, and Tactics* (Boulder, Colo.: Westview Press, 1982), p. 11. This periodization is useful for the student because it is supported by translated selections from Soviet military thinkers. A broader, more interpretive periodization based on "revolutions" in Soviet military affairs is found in William E. Odom's "Soviet Force Posture: Dilemmas and Directions," *Problems of Communism* 4 (July-Aug. 1985): 1–14.

51. The richness of Soviet military thought during this period, to include examination of the contributions of V. K. Triandafilov (1894–1931) and others, is found in Kipp, "Mass, Mobility and the Origins of Soviet Operational Art, 1918–1936," passim.

52. As stated by Stalin on Red Army Day following the halting of German advances on Moscow and their subsequent retreat, "Now the fate of the war will be decided not by such transitory aspects as the aspect of surprise, but by the permanently operating factors: the stability of the rear, the morale of the troops, the quantity and quality of divisions, the armaments of the army, and the organizational ability of the command personnel of the army," in "Order of the People's Commissar of Defense, 23 February 1942, No. 55," in Scott, *Soviet Art of War*, p. 80.

53. Described as the "second military revolution" by Odom, "Soviet Force Posture," pp. 4–6.

54. See note 41.

55. Thomas W. Wolfe, *Soviet Power and Europe, 1945–1970* (Baltimore: Johns Hopkins University Press, 1970), p. 166.

56. Published in three editions (1962, 1963, and 1968), the total number of copies printed was 90,000. The third edition was one of five books nominated

for the 1969 Frunze Prize. See the "Editor's Introduction," pp. xv–xvii, by Harriet Fast Scott in V. D. Sokolovskiy, *Soviet Military Strategy*, 3d ed. (New York: Crane, Russak, 1975).

57. Michael MccGwire, "Deterrence: The Problem—Not the Solution," *SAIS Review* (summer/fall 1985): 105–6; Jonathan S. Lockwood, *The Soviet View of U.S. Strategic Doctrine: Implications for Decision Making* (New Brunswick, N.J.: Transaction Books, 1983), p. 32; Benjamin S. Lambeth, "The State of Western Research on Soviet Military Strategy and Policy" (RAND N-2230-AF, Oct. 1984): 21–25; General-Lieutenant I. G. Zav'yalov wrote in *Red Star* (*Krasnaia zvezda*) on 30 Oct. 1970: "The art of conducting military actions involving the use of nuclear weapons and the art of conducting combat actions with conventional weapons have many fundamental differences. But they are not in opposition to one another and are not naturally exclusive or isolated one from the other. On the contrary, they are closely interrelated and are developed as an integrated whole," as trans. and published under the auspices of the US Air Force in *Selected Soviet Military Writings, 1970–1975*, Soviet Military Thought Series, No. 11 (Washington, D.C.: G.P.O., 1977), pp. 210–11.

58. A. A. Grechko, "The Leading Role of the CPSU in Building the Army of a Developed Socialist Society," *Problems of the History of the CPSU*, May 1974 (trans. FBIS, May 1974), as quoted by Harriet Fast Scott and William F. Scott, *The Armed Forces of the USSR*, 2d rev. ed. (Boulder, Colo.: Westview Press, 1981), p. 57.

59. D. R. Jones, "The Birth of the Russian Air Weapon, 1909–1914," *Aerospace Historian* 3 (1974): 169–71; "The Beginnings of Russian Air Power, 1907–1922," in R. Higham and J. Kipp, eds., *Soviet Aviation and Air Power: A Historical View* (Boulder, Colo.: Westview Press, 1977), pp. 15–25; Von Hardesty, "Aeronautics Comes to Russia: The Early Years, 1908–1918," *National Air and Space Museum Research Report 1985* (Washington, D.C., 1985), pp. 23–44.

60. To be sure, Soviet naval officers' views on the employment of the Soviet navy are not always in agreement with the army-dominated Soviet high command, as described by Michael McGwire in his article "Naval Power and Soviet Global Strategy," *International Security* (Spring 1979): 134–89. But in the final analysis, as MccGwire writes, "In the event of war with the West, the primary purpose of maritime operations is to contribute to the success of the battle on land. The Soviet Union has a unified strategy for the conduct of such a war, whereby each branch of the armed forces makes a specific contribution, so that the effect of the whole is greater than the sum of its parts" (p. 187).

61. These comments on the Soviet air force in the context of Soviet strategy overall are from the article by R. A. Mason, "Military Strategy," in *Soviet Strategy Toward Western Europe*, ed. E. Moreton and G. Segal (London: George Allen and Unwin, 1984), pp. 174, 178–80. For an overview of Soviet air force capabilities, see John W. R. Taylor and R. A. Mason, *Aircraft, Strategy, and Operations of the Soviet Air Force* (London: Jane's, 1986).

62. See note 58.

63. Michael McGwire, *Military Objectives in Soviet Foreign Policy* (Washington, D.C.: Brookings Institution, 1987), pp. 39–40.

64. McConnell, "Shifts in Soviet Views," p. 341.

65. In a classic article, John Erickson reviewed the major elements of this problem by answering the question, "How, then, does the Soviet system operate?" (p. 20), in "The Soviet Military System: Doctrine, Technology and Style," in *Soviet Military Power and Performance*, ed. John Erickson and E. J. Feuchtwanger (Hamden, Conn.: Archon Books, 1979), pp. 18–43. Also see John Erickson, "The Soviets: More Isn't Always Better," *Military Logistics Forum* 1(2) (Sept./Oct.): 58–61, 64.

66. Nicholas V. Riasanovsky, *The Image of Peter the Great in Russian History and Thought* (New York: Oxford University Press, 1985).

67. The most significant studies of the military district have taken place over several years under the leadership of John Erickson. Most recently, see John Erickson et al., *Organizing for War: The Soviet Military Establishment Viewed Through the Prism of the Military District*, College Station Paper 2 (College Station, Tex.: Center for Strategic Technology, Texas A & M University System, 1983). The discussion of the military district which follows is largely based on this and earlier studies.

68. Ironically for the Russians, neither the separate, anti-Western path suggested by the Slavophiles of the mid-nineteenth century, who were deeply opposed to following the path of industrialization, nor the antibourgeois, socialist path charted by Marx and Engels out of the bowels of hostility to the Industrial Revolution have been realizable options, partially because the needs of Russian national security have called for military industrialization.

69. *Whence the Threat to Peace*, 3d ed. (Moscow, 1984).

70. Erickson et al., *Organizing for War*, pp. xli, xliii–xlvii.

71. For maps of the individual military districts, see Erickson et al., *Organizing*, pp. 314–29. For a map including some theaters of military operations (TVDs) as well as the military districts, see "Soviet Theater Estimates: Combat Organization," *Air Force Magazine* (Mar. 1989): 82–83. The terminology and the figures used in the latter are speculative. For a discussion of the Soviet concept of theater and theaters of military operations (TVD) *teatr voennykh deistvii*) and the variations from Western-ascribed meanings, see the article and accompanying maps by Viktor Suvorov, "Strategic Command and Control: The Soviet Approach," *International Defense Review* 12 (1984): 1813–20.

72. Erickson et al., *Organizing*, pp. xliii–xlvii, 1–9, 31–42, 47–61.

73. Ibid., pp. 31–32, 48–50.

74. Ibid., pp. 50–55. As pointed out by Suvorov, "Although there are military-geographical regions in the Soviet Union, the military districts, they do not take part in the war. In wartime the military district exists only as long as there is no conflict in that particular area. When the enemy appears, the military district is transformed into a Front which is a purely military unit. During the second world war, many Fronts moved forward hundreds and even thousands of kilometres and fought on territories which had no connection at all with their names: the 1st Ukrainian Front stormed Berlin, while the 3rd Ukrainian Front fought in Romania, Bulgaria, Yugoslavia, Hungary and Austria," in "Strategic Command and Control," p. 1815.

74. Erickson et al., *Organizing*, p. 56.

76. Lt. Col. David M. Glantz made the point that "Scope, magnitude, complexity, timing, and marked success have made the Manchurian offensive a continuing topic of study for the Soviets, who see it as a textbook case of how to begin war and quickly bring it to a successful conclusion," *August Storm: The Soviet 1945 Strategic Offensive in Manchuria*, Leavenworth Papers, no. 7 (Feb. 1983), p. xiv.

77. Erickson et al., *Organizing*, pp. 149–50. See also the forthcoming article by Bruce W. Menning, "Army and Frontier in Russia," in the Proceedings of the Twelfth Military History Symposium, 1–3 Oct. 1986, USAF Academy, to be published by the Office of Air Force History under the title, *Transformation in Russian and Soviet Military History*.

78. See note 71.

79. Erickson et al., *Organizing*, p. 2.

80. John M. Collins, *U.S.-Soviet Military Balance, 1980–1985* (Washington, D.C.: Pergamon-Brassey's International Defense, 1985), p. xix.

81. Ibid., p. 3.

82. Ibid., p. 10.

83. Ibid., p. 156.

84. Adrienne Koch, *Power, Morals and the Founding Fathers* (Ithaca, N.Y.: Cornell University Press, 1961), p. 138.

85. Theodore Draper, *Present History* (New York: Vintage Books, 1984), p. 53.

86. As quoted by Mary C. Fitzgerald, "Marshal Ogarkov on the Modern Theater Operation," *Naval War College Review* (Autumn 1986): 22.

87. Thirty-five experts in seven disciplines produced an extraordinary, pessimistic review of the Soviet Union's strengths and weaknesses, in *After Brezhnev: Sources of Soviet Conduct in the 1980s*, ed. Robert F. Byrnes (Bloomington: Indiana University Press, 1983). See esp. the conclusion by Byrnes, "Critical Choices in the 1980s," pp. 423–40, and variations on the same theme in "Change in the Soviet Political System: Limits and Likelihoods," *Review of Politics* 4 (Oct. 1984): 502–15.

CHAPTER 3

THEORETICAL FOUNDATIONS
Deterrence in the Nuclear Age

SCHUYLER FOERSTER

Deterrence has virtually become a household word. One often speaks of deterrence with no apparent need to define the concept for the audience, suggesting that its meaning is simple and universally well understood. But deterrence is neither simple nor generally understood. This chapter sets out a primer on deterrence in the nuclear age. It is a primer in that it seeks to highlight first principles, that is, the basic elements that make up a complex concept.

American military power is designed to deter potential adversaries and to defend American interests in the event deterrence fails. While the latter objective is clearly vital, the former is the more immediate objective since, if deterrence is successful, defense is unnecessary. To be sure, defense and deterrence are closely interrelated, but they are not identical. Neither, for that matter, are the forces which contribute to each interchangeable. It is a paradox of the nuclear age that forces which contribute to deterrence may not contribute to defense, and forces which contribute to defense may not contribute to deterrence.

This is but one of the many dilemmas that plague the business of defense policy. This chapter lays out the basic elements of deterrence theory and highlights those contradictions. Subsequent chapters will build on this theoretical foundation in discussing the application of American defense policy in its various contexts. The bibliographic essay on deterrence theory following this chapter serves as a more detailed guide to the literature in which that body of theory has evolved. Understanding that foundation, however, requires a framework—one based on first principles.

WHAT IS DETERRENCE?

In the first chapter, we defined power as "the capacity to influence" and distinguished two forms of influence—*deterrence* and *compellence.* Compellence involves the attempt to cause another political actor to change behavior—to do something not necessarily done otherwise, or to stop doing something currently being done. If we assume that actors are rational, we assume that behavior is a product of some calculation of costs and benefits and that a particular course of action is based on a projection of relatively lower costs and risks and/or relatively higher benefits. Compellence typically involves altering those cost-benefit calculations and demonstrating that the costs involved in a certain course of action are actually higher than had been anticipated. Compellence thus seeks to cause an actor to change a course of action by suggesting that a different course of action would bring higher benefits at lower cost.

In deterrence, on the other hand, one seeks to reinforce existing behavior. Specifically, in the case of deterring war, one seeks to reinforce another state's apparent preference not to resort to military force. Again assuming a rational actor, one deters by reinforcing the benefits of peace relative to war and by highlighting the costs that would be involved in war. This seems simple enough, but there are four general problems which plague our ability to define what an adequate deterrent is.

The first problem is that one cannot say with any certainty whether deterrence is working. If State A has no intention of starting a war, then peace results not from State B's deterrent influence but from State A's desire to stay at peace. What State B presumes is a successful deterrent force might actually have been an ineffective deterrent had State A in reality been determined to go to war. The requirements for deterrence thus hinge on assessments about a political adversary's intentions. In fact, the contemporary debate about the efficacy of the deterrent power of the United States often revolves around a debate about Soviet intentions. These intentions can never truly be known, however, and they can change over time. The effectiveness of deterrence, therefore, can never be proven; only its ineffectiveness can.

A second difficulty is that deterrence involves the calculations of the state which is to be deterred, not those of the state which is doing the deterring. As a result, there is a built-in uncertainty as to whether deterrence is sufficient. In the jargon of game theory, notions of risk and payoff are probabilistic rather than absolute. Even assuming a "rational" decisionmaker, states project future risks and potential payoffs. Yet decisionmakers also suffer from the human inability to be clairvoyant: they can easily be wrong regardless of how rational the process is by which they attempt to deal with ambiguity. States which seek to deter must influence the probabilistic calculations of the other state and provide for a sufficient margin of error in the other state's projections. Moreover, "rationality" is itself not an absolute. What seems rational to one actor may be irrational to another, because the values being pursued may differ. Chapter 2 showed how the Soviet Union approaches the problems of defense, deterrence, and security; that approach does not, for a variety of reasons, mirror that of the United States. Thus, what might reasonably deter the United States may not deter the Soviet Union.

While the foregoing assumes that decisionmakers are rational, we also know that states do not always behave rationally. State policies often reflect human error and miscalculation. Wars are often lost by the state that initiated hostilities, and we cannot necessarily conclude that the attacking states committed an "operations research error" in making a presumably rational decision to go to war. Wars typically erupt out of times of political tension and crisis, when fear and anxiety prevail and the ingredients of rational decisionmaking are often absent. Deterrence must be most effective in circumstances when rationality can least be assumed.

The final reason why the effectiveness of deterrence cannot be precisely measured is that the requirements of deterrence are not static. Rather, technology provides a dynamic variable which affects both the deterrer and the state to be deterred. A state seeks to deter a potential adversary from using the military capabilities it presently possesses. But in planning and building its deterrent capability, a state must do so with a view to what capabilities a potential adversary will have in the future. Fielding military forces requires a long lead time—perhaps 10 to 15 years including research and development. Thus rival military powers like the United States and the Soviet Union are actually engaged in a dynamic rather than a static deterrent relationship. Assessing the long-term threat is an inherently uncertain process, one in which the risks of underestimating future requirements of deterrence are unacceptable. For this reason, there is, understandably, an inherent tendency on the part of defense planners to conduct "worst-case analyses." Having charted force requirements based on such pessimistic assumptions, however, defense planners must still optimize force development within the political,

economic, and strategic constraints that—just as inevitably—will limit that force development.

Much of the debate about American defense policy, in this or any other period, concerns the adequacy of America's ability to deter challenges to its vital interests. We often judge specific defense policies and programs according to how they contribute to deterrence. But deterrence is a complex phenomenon, not only because of the factors discussed above which render it imprecise, but also because of the varying contexts in which deterrence applies. In short, when we speak of deterrence, we must be specific about *whom* and *what* we want to deter before we can grapple with the question of *how*.

THE OBJECT OF DETERRENCE: WHOM AND WHAT?

The evolution of nuclear deterrence thinking in the West has been largely driven by a desire to find a mechanism which would prevent war on the scale witnessed twice before in this century. Each of those wars spawned attempts to point out, often in graphic terms, the horror which man's warfighting technologies were demonstrably capable of inflicting, suggesting that warfare on any scale had ceased to be a rational instrument of political policy. Consider, for example, the following observation:

The very magnitude of the disaster that is possible may prove to be a restraining influence. Because the *riposte* is certain, because it cannot be parried, a belligerent will think twice and again before he initiates a mode of warfare, the *final outcome of which is incalculable*. The deterrent influence may, indeed, be greater than that. *It may tend to prevent not only raids on cities but resort to war in any form.* No one can tell what will happen if war does come. Its momentum may carry it to lengths not intended before it began to gather speed. *Wars have a way of deteriorating in their course.* [1]

This observation includes many of the elements of contemporary thinking on deterrence—the prospect of unprecedented destruction of cities, the certainty of retaliation, the incalculability of risk, the inherent uncontrollability of war, the potential for deterring all forms of warfare. Yet, the words are of a British writer in 1938, J. M. Spaight, building on the notions of Giulio Douhet and others who advanced early theories of strategic bombing and the new dimension of warfare represented by airpower. [2]

Such theories were more visionary than accurate, since the fear of new dimensions of warfare and destruction did not deter a second world war, and the application of strategic bombing in that war was largely inconclusive. The advent of atomic weapons at the end of World War II, however, stimulated renewed emphasis on deterrence and the revolutionary impact which these new weapons would have on warfare and on the fundamental purposes of military power. In a now classic analysis of the impact of the new atomic age, written in 1946, Bernard Brodie wrote that we no longer need be "concerned about who will *win* the next war in which atomic weapons are used. Thus far the chief purpose of our military establishment has been to *win* wars. From now on its chief purpose must be to *avert* them. It can have almost no other useful purpose."[3] The primacy of the nuclear dimension of deterrence derives from the tremendous destructiveness of such weapons, hence Brodie's reference to them as the "absolute weapon." Implicit in this early American view of the atomic age is the recognition that whole societies were becoming vulnerable to utter devastation. For the United States in particular, this has been an unprecedented vulnerability, one which took on even more profound significance when thermonuclear weapons were combined with the delivery means of the ballistic missile for which there has been no defense.

The great irony of the nuclear age, however, is that the "absolute weapon" has not offered a means to eliminate war, even if one concedes that it has succeeded in deterring the kinds of total wars which this century has already witnessed. The very destructiveness of nuclear weapons has confined their utility to certain kinds of conflicts, most clearly to those conflicts in which nuclear powers (and their allies) would be pitted against other nuclear powers (and their allies). Nonetheless, the postwar period has not been without conflict, and one might even argue that the incidence of warfare among nonnuclear states has actually increased in recent decades.

One can hardly even argue that nuclear weapons have given states which possess them an unchallenged deterrent. The United States is a global power, with far-reaching global interests, and those interests have often been successfully challenged by nonnuclear states. To be sure, the survival of the United States has not been at stake in those conflicts, and the interests being challenged were not so vital as to warrant nuclear use or even to make the threat of nuclear use particularly credible. Yet, clearly, to be a nuclear superpower is not to be invincible, or even unassailable. With or without nuclear weapons, the interests being challenged and the contexts in which conflicts take place are important variables in determining whether or not military force will be used in defense of those interests and what form that military response will take.

It is therefore difficult for any nation, and particularly a global power such as the United States, to speak of deterrence in a generic sense.

Rather, the US has found that it has to tailor its deterrence to meet a variety of threats to a range of vital and nonvital interests. Obviously, the Soviet Union poses the most direct danger to the US because only the Soviet Union possesses a nuclear arsenal capable of threatening national survival. Yet the most likely threats are the most indirect ones, and such threats do not always emanate from the Soviet Union. Thus, in protecting its global interests, the United States must be capable of dealing with threats which range across the entire continuum of conflict—from nuclear war to conventional conflict, to low-intensity conflict such as insurgency, to terrorist acts by individuals and groups.

If nuclear weapons have not provided a deterrent to all conflict, then what kinds of conflicts are they capable of deterring, and under what conditions? On the one hand, few would doubt the proposition that nuclear weapons can effectively deter nuclear use by an adversary. On the other hand, few, if any, would credit nuclear weapons with the capability of deterring low-intensity conflict. Two questions remain, however. First, are nuclear weapons a *symmetrical deterrent*—that is, capable only of deterring nuclear use by an adversary—or can they also be an *asymmetrical deterrent*, capable of deterring nonnuclear conflict as well? Second, to the extent that nuclear weapons are an asymmetrical deterrent, what are the conditions which determine the limits of nuclear deterrence? For example, can nuclear weapons deter conventional war but not unconventional or low-intensity conflict? Can nuclear weapons deter attacks on vital national interests regardless of the level of conflict involved?

These questions can have no definitive answers for the same reasons, noted above, that the effectiveness of deterrence cannot be precisely measured. Nonetheless, these questions have important policy implications. Nuclear weapons tend to be substantially less expensive than conventional forces, since conventional forces require the maintenance of high levels of military manpower trained in the employment of a diverse array of weapons and equipment and capable of being deployed to disparate theaters of operations. It is not surprising, therefore, that the United States has at times tended to rely more on nuclear deterrence because the costs of meeting every threat with like forces were deemed excessive. Yet, to the extent that the US relies more on nuclear deterrence, it incurs greater risks because of its vulnerability to nuclear retaliation. Thus, it is also not surprising that the US has at other times sought to reduce its reliance on nuclear deterrence, particularly when periods of crisis stimulated concerns both that the nuclear threat might not be universally credible and that the US lacked sufficient response options which were less risky.

The issue of whether nuclear weapons are an effective deterrent to nonnuclear conflict begs the further question of whether using nuclear weapons first in a conflict is appropriate. On the one hand, one can argue that such escalatory threats are little more than irresponsible bluff, for two reasons. First, such threats are arguably incredible, since they imply that the US is willing to place its survival at risk in a conflict which is otherwise geographically remote. As long as the likely theaters of conflict do not involve the territories of the nuclear powers, these powers remain a sanctuary in nonnuclear conflict even though other vital interests may be at stake. Second, some argue that nuclear weapons would have little or no military utility, particularly since the devastation resulting from a nuclear exchange on the battlefield would likely be focused on the territory of the defender. On the other hand, one can also argue that such escalatory threats enhance deterrence. First, because massed offensive formations of conventional armies are especially vulnerable to nuclear attack by the defender, the requirements for mounting an offensive operation are compounded considerably (although such problems are not insoluble for the attacker). Second, the threat to escalate a conventional conflict to nuclear war puts the attacker's homeland at risk, thereby denying the attacker a sanctuary and making the risks of attack "incalculable."[4]

While it is commonplace to note that nuclear weapons have contributed—perhaps decisively—to the avoidance of major war since 1945, it is not possible to demonstrate exactly why or to what extent this is true. Nuclear powers have not relied on them exclusively and, over time, have tended to stress the need for complementary forms of military power which are more credible, less risky, and more appropriate to defending national interests in the event deterrence failed. Nonetheless, nuclear weapons—and the escalatory threats they pose—remain integral to US defense policy.

Thus, the question of whom and what kinds of conflict the United States wants to deter leads to two further questions. Since the US seeks to defend at least some of its interests and deter at least some forms of conflict by the threat of nuclear retaliation, can it extend its nuclear deterrent on behalf of friends and allies? Secondly, what form of threatened retaliation would enhance the deterrent effect of nuclear weapons? Examination of these two questions will highlight further the dilemmas which continue to affect US defense policy.

THE SCOPE OF DETERRENCE: ON BEHALF OF WHOM?

The United States generally operates within the context of a coalition strategy which depends fundamentally on the cooperative efforts of friends and allies. The US does not seek to maintain its global interests alone. Beginning in the late 1940s, most notably with the formation of the North Atlantic Treaty Organization (NATO), the US established a series of alliance relationships as part of a global containment strategy directed against the Soviet Union and other communist countries. That immediately raised the question of what deterrence obligations the US assumed with respect to those alliances.

During the Eisenhower administration in the 1950s, US policy placed primary emphasis on the threat of nuclear relationship—"massive retaliation"—as the most important guarantee of its allies' as well as its own security. Allies, particularly those in Asia who were increasingly vulnerable to insurgent threats to their security, found nuclear retaliatory threat less relevant. Even in Western Europe, which faced a significant conventional and nuclear threat, the increasing vulnerability of the United States to nuclear attack suggested greater emphasis on complementary, nonnuclear defense capabilities. In the early 1960s, the Kennedy administration affirmed the obligation "to pay any price, bear any burden" in support of its friends and allies. But the United States also realized that it had to strengthen its deterrent capabilities across the broader range of unconventional, conventional, and nuclear military instruments to fulfill those obligations within a reaonable degree of risk.

The resource burden associated with such a broad-based deterrent posture was formidable indeed, ultimately leading the Nixon administration, for example, to qualify its alliance commitments. The Nixon Doctrine, articulated in 1970, suggested that the United States would rely increasingly on allied self-defense and provide materiel assistance but not necessarily US troops in local conflicts, particularly conflicts such as that which had drained US resources in Southeast Asia.

The question of whether the United States would necessarily send troops all over the world to maintain alliance commitments existed alongside the question of whether the US nuclear umbrella extended over all of those commitments. The nature of the US nuclear guarantee involves Western Europe most of all, since it is against Western Europe that the most potent Soviet conventional military threat exists. The US nuclear guarantee is arguably the most contentious aspect of American defense policy, precisely because that guarantee forms the basis of US global military power. For the US, that guarantee involves the risk of destruction in the event deterrence fails. For the allies, that guarantee requires faith that the US will accept that risk.

Deterrence requires both the *capability* to

carry out a threat and the *credibility* that the threat will be carried out. For the most part, deterrent threats involving nonnuclear forms of modern warfare involve questions of capability. The real issue in deterring challenges to US interests in most areas around the globe is whether the US is able to muster and sustain the right kind of nonnuclear military force to defeat those nonnuclear and often low-intensity challenges to its interests. For the US nuclear deterrent, on the other hand, the capability is not in doubt; the pertinent question is one of credibility. In assessing the credibility of the US nuclear deterrent, it is helpful to distinguish between types of deterrence according to who the intended beneficiary is.

Basic deterrence is deterring an attack on ourselves. Because it can largely be assumed, basic deterrence can also be called "passive" deterrence. To be sure, one can question the credibility of any threat of nuclear retaliation on the grounds that a threat to obliterate an attacker is not only morally tendentious but also ineffective in that retaliation does not alter the fact of one's own destruction. Yet, "thinking the unthinkable" has provided the foundation of nuclear-deterrence thinking since the early postwar years. If the use of nuclear weapons is credible at all, such use is most credible when one's own survival is threatened. States possessing nuclear weapons therefore need not constantly reiterate their commitment to use such weapons in their own defense, since that commitment is implicit in the very possession of those weapons.

Extended deterrence, on the other hand, is a threat of nuclear use by one state on behalf of one or more allied states, as when the United States insists that an attack on its NATO allies may invoke a US nuclear retaliatory strike. Such a threat, however, also implies that the US is willing to risk its own destruction in the interest of deterring an attack on an ally which does not possess a basic deterrent of its own. What makes extended deterrence a persistently contentious issue in alliance relationships is that the credibility of such a threat is subject to question, perhaps more so from allies than from adversaries. Michael Howard had rightly pointed out that there are in fact two challenges for a country such as the US which extends its deterrent guarantee to its allies.[5] On the one hand, the US must deter the Soviet Union as the principal threat to the integrity of NATO allies. On the other hand, the US must reassure its allies of the continuing viability of that guarantee.

Clearly, no one can say categorically that there is a 100 percent probability that, in the event of a Warsaw Pact attack on NATO, the US will employ its strategic nuclear arsenal against the USSR and risk its own survival in the process. Yet, deterrence is not a question of certainty: it is a question of marginal probability. It may well be that the Soviet Union would be deterred by *any* possibility that the US would retaliate, just because the risks to the Soviet Union from such retaliation would be unacceptable. Since the effectiveness of deterrence can only be hypothesized but never precisely measured in the absence of war, the existence of a nuclear retaliatory capability and the affirmed threat to use it—however questionable the credibility of that threat may be to some—may well be sufficient to render the risk in the eyes of the Soviet leadership as "incalculable."

The problem of extended deterrence for an alliance, however, is not fundamentally a military issue. It is political in nature, since the problem plagues peacetime allied assessments as to whether allied security is adequately served. Allies who are the beneficiaries of the US extended-deterrent guarantee—the credibility of which is inherently imperfect—require reassurances that the nuclear threat is operative. Indeed, reassurance of allies may demand a greater degree of certainty that nuclear retaliation will be forthcoming in war than the degree of certainty that is sufficient to deter an adversary in a crisis. Thus, when a former secretary of state, Henry Kissinger, declared to an audience in Brussels in September 1979 that NATO should not unduly rely on a US nuclear threat which was suicidal in its implications, the political repercussions in Western Europe were substantial.[6] The Soviet Union, on the other hand, seemed indifferent to such an ostensibly authoritative statement.

Theoretically, one prescription for reducing the political anxieties surrounding extended deterrence is for allies to seek the marginally more credible alternative—or complement—of developing their own basic deterrent. France, in fact, pursued precisely this course of action when, under former President de Gaulle, it developed its own *force de frappe*, demanded the withdrawal from its soil of all nuclear weapons not under French control, and, in 1966, opted out of NATO's integrated military command structure (albeit remaining a member of NATO). Quite apart from the political motives underlining France's refusal to rely exclusively on the US nuclear umbrella, the strategic rationale was clear enough: it was fundamentally irrational for a sovereign state to threaten suicide to preserve the integrity of another state, and equally irrational for a sovereign state to surrender wittingly its vital security interests to another state.[7] Britain, having earlier developed its independent national nuclear force in concert with the United States, likewise found occasions to justify that nuclear force in part as a hedge against the possibility that the US nuclear guarantee might not be permanent.[8] Nonetheless, both of these powers remain beneficiaries of the US extended nuclear deterrent guarantee. Moreover, despite US skepticism in the 1960s about the reliability and credibility of independent national nuclear deterrents, NATO has come to accept that the combination of these nuclear forces adds to the incalculability of Soviet risk assessments.

Most of the NATO allies, however, and most significantly those allies who border on the Warsaw Pact, do not possess a basic nuclear deterrent and, as signatories to the Nuclear Nonproliferation Treaty, expect to remain dependent on the US extended-deterrent guarantee within an alliance framework. Thus the political issues associated with extended deterrence are likely to remain. Countries which do possess nuclear weapons must consider the form that deterrent may take and be clear about whether those weapons are to serve as a basic or extended deterrent.

THE MEANS OF DETERRENCE: HOW?

Brodie's admonition in 1946 that deterrence of war had supplanted the pursuit of victory in war as the principal purpose of military force was not accompanied by a lengthy analysis of the means by which such a deterrent would be achieved. The means of deterrence were inherent in the unprecedented and largely indiscriminate destructiveness of atomic and, later, thermonuclear weapons.

Technology has since opened various possibilities for nuclear employment and inexorably fueled debates over how different modes of nuclear weapons employment contribute to deterrence. As long as nuclear weapons were exclusively large in yield, limited in numbers, unable to be delivered with anything but the grossest accuracies against targets, and constrained by marginally reliable command and control systems, they could have only one wartime purpose: destruction of an enemy's high-value socioeconomic targets. One could draw the theoretical distinction between such "countervalue" targeting and the targeting of an adversary's military capability ("counterforce" targeting), but in the early phases of the nuclear age, the two were indistinguishable from the standpoint of operational effect.

Complementing this reality was the fact that there was no real defense against nuclear weapons, a feature of the nuclear age which theoretically contributed to deterrence. Before the advent of a ballistic missile delivery system, even an unusually robust air defense would only be partially effective. The damage that could be inflicted by surviving bombers was still viewed as an unacceptable cost, not only because of the tremendous death and destruction expected in the immediate target area but also, and perhaps more fearfully, because of the poisoning shadow of radioactivity that magnified this cost over both distance and time.

The initial advances in nuclear weapons technology seemed only to affirm this apocalyptic image of the nuclear age. Fusion-based thermonuclear weapons promised even greater and more indiscriminate destructiveness. The development in the early 1950s of so-called tactical nuclear weapons—with yields comparable to those of the Hiroshima and Nagasaki bombs—offered some theoretical selectivity to nuclear use, but only insofar as they enabled employment by battlefield commanders in specific theaters of operation with comparably apocalyptic results.[9] Ballistic missile delivery systems ultimately afforded greater accuracy against designated targets, but initially they rendered the delivery systems invulnerable to anything short of system malfunction. Defense was out of the question. Along with the increase in the nuclear stockpile, technology had afforded the means of assuring the destruction of an enemy's society. It seemed that the vision of Douhet and others who argued the primacy of strategic bombing in the 1920s had become achievable.

The next major technological development reaffirmed further the seeming inevitability of nuclear retaliation. As the missile age dawned, symbolized by the Soviet launch of *Sputnik* in 1957, the prospect of a Soviet intercontinental ballistic missile (ICBM) capability prompted deep concerns in the United States about the survivability of the US deterrent posture. The Gaither Committee subsequently recommended several measures to President Eisenhower to ensure that US retaliatory assets would be protected against a Soviet first strike. By the end of the Kennedy administration, the US was well on its way to developing an invulnerable retaliatory capability in its strategic triad of ICBMs in hardened underground silos, submarine-launched ballistic missiles (SLBMs) patrolling invisibly under the seas, and strategic bombers which could be launched from "quick-alert" postures and dispersed while remaining under positive control. These developments provided the basis for a deterrent concept, formalized in the 1960s under the Defense Secretary McNamara, known as *mutual assured destruction (MAD)*.

DETERRENCE BY THE THREAT OF PUNITIVE DESTRUCTION

The concept of *mutual assured destruction* stemmed from a conviction, first articulated after *Sputnik* by Albert Wohlstetter, one of the key authors of the Gaither Committee Report, that deterrence could not be considered "automatic," but that it required deliberate attention to the elements of strategic "stability."[10] The mere presence of a nuclear retaliatory capability was by itself inadequate for stable deterrence. If that retaliatory capability were vulnerable, it only invited preemption by an adversary and forced the defender into the dangerous position of having to launch a retaliatory strike upon warning of attack in order to keep from losing that capability. An invulnerable retaliatory capability, on the other hand, theoretically enables the defender to ride out the first strike, secure in the knowledge that the retaliatory capability will remain intact. More importantly, the potential attacker's knowledge that retaliation is virtually inevitable serves as a powerful deterrent to attack in the first place.

Besides an invulnerable retaliatory—or "second strike"—capability, MAD requires the capability of assuring the destruction of the adversary's socioeconomic base. That is not necessarily a difficult task since such "soft" targets are especially vulnerable to the devastating effects of nuclear weapons. Having at various times operationally defined "assured destruction" to mean the destruction of 20–33 percent of the Soviet population and 50–75 percent of the Soviet industrial base, McNamara's planners could then calculate the number and types of warheads needed to inflict that degree of damage, factor in the projected reliability of our own weapons systems, and arrive at a desired force posture to achieve that level of destruction.[11] To be sure, other factors—many of them having more to do with bureaucratic politics and cost considerations than with strategic calculations—contributed to the design of the US strategic force structure in the 1960s.[12] Nonetheless, an important strategic logic was involved. It was theoretically possible to devise a strategic nuclear force structure quite independent of the force structure of the adversary. Tanks might fight tanks, but missiles did not fight missiles. As long as one's own forces were invulnerable and the desired targets were vulnerable, the number of adversary missiles remained theoretically—if not politically—irrelevant.

That logic, however, involved two important assumptions—the invulnerability of retaliatory systems and the vulnerabililty of societies—the foundations upon which MAD was based and premises which have since been called increasingly into question. The technological dynamic which ushered in the era of MAD also offered possibilities that threatened the concept's viability. MAD was a concept of convenience in many respects. It exploited the characteristics of the nuclear age which had evolved to that point, characteristics which structured thinking about nuclear weapons as instruments of punishment. By this logic, the more awesome the prospect of punitive destruction, the more "un-

thinkable" war became. The more invulnerable retaliatory assets were, the more assured the punitive blow. Thus, the attacker would never choose—regardless of his frame of rationality—to initiate a conflict whose outcome would be suicidal.

Such logic, moreover, offered a potential recipe for strategic stability from the point of view of both adversaries. If *both* superpowers possessed invulnerable retaliatory capabilities *and* vulnerable societies, their relationship would resemble two scorpions in a bottle: even though deadly enemies, and even though each possesses the undeniable capability to destroy the other, neither can escape the other's lethal retaliatory sting. Without the means of disarming the other, there is no incentive to attack first, since suicide is inevitable. With each society in such a "mutual hostage relationship," the absence of preemptive incentives offers strategic stability between superpowers. Neither need fear the other, therefore, and each could pay more attention to insuring that the safety catch was on, while relaxing the trigger finger.

Nuclear powers other than superpowers have developed much smaller nuclear arsenals—variously called *finite*, *minimal*, or *proportional* deterrents—which were also designed to inflict punishment on an attacker's society. Beyond their limited bomber threats to the Soviet Union, the French nuclear arsenal, for example, includes only 210 land- and sea-based missile warheads capable of striking the Soviet Union, while the British SLBM threat consists of 64 launchers, each with three warheads that cannot be independently targeted. (By comparison, the US has approximately 8000 and the Soviet Union over 10,000 strategic ballistic missile warheads, not counting cruise missiles or bombers.)[13] While both France and Britain are working to modernize their nuclear forces, their arsenals do not pose an assured destruction threat to the Soviet Union. They do, however, threaten high-value Soviet socioeconomic targets and pose to the Soviet Union the likelihood that such targets would be struck if French or British territorial integrity were threatened. Such a "finite" or "minimal" deterrent is also "proportional" since, it is argued, even the limited costs inflicted on the Soviet Union would be at least proportional to whatever gain the Soviets might seek to achieve by targeting France or Britain. In an apt French metaphor, such limited forces could "tear the arm off the Russian bear," an arm worth more to the Soviets than a French or British head.

While the concept of deterrence by punishment derives from the very nature of nuclear weapons, so do the problems associated with that concept. Notions of nuclear deterrence suffer from a fundamental *usability paradox*.[14] On the one hand, the deterrent value of nuclear weapons stems from their ability to inflict tremendous destruction, leading one to conclude that the more destructive they are, the greater their deterrent value. On the other hand, the more destructive they are, the more "unthinkable" their use, thereby casting doubt on the credibility of the deterrent threat. Reversing this logic, the more one attempts to bolster the credibility of one's deterrent threat, by making nuclear use more "thinkable," the more likely nuclear weapons might be used, thus potentially undermining the deterrent.

Unraveling this paradox recalls the earlier discussion of what one wants to deter and on behalf of whom. The most credible threat of nuclear use is as a basic deterrent against nuclear attack on one's own country. That is the context in which MAD evolved into a construct for stability in the bilateral nuclear relationship between two superpowers separated by intercontinental distances. Likewise, that is the case for the more "finite" deterrents of medium nuclear powers. Yet both these forms of punitive deterrence are still plagued by questions of credibility. If deterrence fails, executing such a deterrent threat does not in any way negate one's own destruction since the threat is only one of retaliatory destruction. One is also justified in questioning the morality of targeting noncombatants as disproportionate and misguided punishment for the deliberate or perhaps even miscalculated aggression of a society's political leadership. The same questions apply, arguably with greater urgency, in the case of punitive nuclear use in response to a nuclear or even conventional attack on an ally, particularly when such use may eventually invite nuclear devastation of the country that fulfilled that extended deterrent guarantee.

Theoretically, one can reinforce the credibility of one's punitive nuclear deterrent threats by making retaliation more an automatic than a discretionary act.[15] French notions in the 1960s of a "plate glass window," American notions in the 1950s of a "trip wire" in Europe which would trigger "massive retaliation," and occasional Soviet references to "launch on warning" all reflect attempts to make the "unthinkable" more credible by emphasizing that the defender would have little choice but to retaliate promptly with the full force of his nuclear arsenal. Such postures are inherently destabilizing, however, since they focus attention on the trigger rather than the safety catch, render the entire strategic relationship extremely vulnerable to miscalculation, and deprive political leadership of necessary latitude in a crisis.

The credibility problem becomes particularly poignant when one seeks to deter any attack—conventional or nuclear—against an ally. Barring a purely bilateral nuclear conflict betwen the superpowers, this is, however, the context in which US deterrent strategies have developed. Even as the US adopted *rhetorical* postures of "massive retaliation" and later "mutual assured destruction," US political leaders and military planners have sought ways to increase response options which did not preordain national suicide and which reduced collateral damage, increased the usability of nuclear weapons to achieve identifiable political and military objectives, reinforced the credibility of extended deterrent threats, and enhanced strategic stability. The technology of nuclear weapons employment which had enabled the concept of *deterrence by punishment* in time offered a complementary prospect, *deterrence by denial*.[16] Maximizing the advantages of each, while reducing the dangers of each, is a continuing challenge of the nuclear age.

DETERRENCE BY THE THREAT OF DENYING VICTORY

Notions of purely punitive deterrence share the common assumption that defense is not possible in the nuclear realm. In the sense that Brodie talked of an "absolute weapon," deterrence is based on the threat of offensive retaliation, rather than the more traditional way in which military forces sought to deter war—by threatening to defeat the attacker. Thus, punitive deterrence focuses on the prewar deployment of nuclear weapons and the threat these deployments imply. It does not address specifically the employment of nuclear weapons *in war*, since nuclear weapons are assumed to have little if any rational operational military utility. As to the question of whether such punitive threats are moral or credible, the answer is that their purpose is deterrence, and, as long as that is successful, consideration of how such weapons would be *employed* if deterrence fails is not only moot but perhaps even provocative and destabilizing.

Such theoretical solace is of little help to decisionmakers who must face the prospect that deterrence might fail. Over time, aided by technological advancements in targeting accuracy, new delivery means, and improved command and control mechanisms, competing notions of deterrence have evolved which are more traditional in their roots. These notions—sometimes called "warfighting" strategies—have focused on how a defender could respond if deterrence were to fail. The term *warfighting* has become a value-laden term in a strategic debate that is often emotional. A more precise term is *deterrence by denial*, because it emphasizes the traditional objective of military defense—threatening to deny the attacker success in the achievement of military and political objectives, thereby deterring an attempt that would be not only costly but, more to the point, unsuccessful.

Deterrence by denial, in contrast to *deterrence by punishment*, therefore, stresses defense. An integral part of such defensive orientation is *damage limitation*, the attempt to limit the damage the defender will suffer so as to secure survival, to deny the attacker success, and to enable retaliatory attacks to terminate the war on the defender's terms. In nuclear strategy, damage limitation can take several forms, only one of which is compatible with the conceptual or doctrinal framework represented by MAD, and that is to reduce the vulnerability of one's own retaliatory assets.

Theoretically, there are four ways in which potential targets can be protected in the interest of damage limitation. One is by passive protection measures, such as hardening or mobility. This can protect retaliatory assets, such as missiles and bombers, as has been the norm for decades. Protection of socioeconomic assets, however, requires civil defense programs on a scale that has been unacceptable in the West. A second mechanism for damage limitation is by having more targets than an attacker can possibly cover, thus insuring that a certain proportion of assets survive an attack. Given the destructive capacity of contemporary nuclear arsenals, however, this is an unlikely means for protecting a society. One can proliferate retaliatory assets so that there are more potential launch points which an attacker must target. To be effective, this would require some form of constraint on the adversary's ability to increase its own offensive capability and target coverage. Neither arms control nor the costs of armaments are likely to provide such an enduring constraint.

These first two damage-limitation options are designed to render an attacking force ineffective and, for the most part, either represent accepted practice (hardening of ICBM silos, mobility and dispersal of SLBMs and bombers) or have been judged implausible. The other two damage limitation options, on the other hand, involve the actual destruction of attacking forces—either by destroying the adversary's missiles before they launch or by intercepting them in flight. The former reflects a capability which has been evolving in the US strategic posture for many years; the latter refers to a form of strategic defense—represented by the Reagan administration's Strategic Defense Initiative (SDI)—which is under investigation but has not yet been realized. In both cases, however, the distinctions between offensive and defensive capabilities are obscure. Both options continue to be a source of intense debate which centers on their effect on strategic stability and, hence, whether they ultimately contribute to or detract from the goal of deterrence.

Since notions of punitive deterrence offer few if any realistic response options in the event

deterrence fails, deterrence by denial reflects a search for a coherent nuclear *strategy*. As one critic of MAD has put it:

Theories of pre-war deterrence, however sophisticated, cannot *guarantee* that the United States will never slip into an acute crisis wherein a President has to initiate strategic nuclear employment or, *de facto*, surrender. In such a situation, a President would need realistic war plans that carried a vision of the war as a whole and embodied a theory of how military action should produce desired political ends. *In short, he would be in need of strategy.*[17]

Such advocacy for a clear strategy is reasonable and, were it not for the dilemmas of the nuclear age, beyond controversy. In the tradition of Clausewitz, weapons are merely means to a political end, and sound strategy requires defining the political ends before fashioning the military means to achieve them. While deterrence is arguably a political end, a strategy designed to deter war is incomplete, since deterrence can fail. By necessity, one's deterrent strategy ought to be consistent with a wartime strategy designed to achieve identifiable political ends in the event deterrence fails.

In an ideal form, a nuclear strategy designed to deter by denial requires the capability to limit damage to oneself while targeting the warmaking capabilities and, ultimately, the political cohesion of the adversary. Defensively, it requires not only invulnerable retaliatory assets—as with punitive deterrence—but also the capability to protect one's own political, economic, and societal values. Offensively, it requires the ability to target the adversary's hardened political and military assets to the extent necessary to destroy the adversary's capability and/or will to continue the war. Fundamentally, it implies that nuclear weapons—like other more traditional instruments of warfare—can be employed selectively against their targets, and that such wartime employment can be controlled and executed in a protracted conflict. Advocates of deterrence by denial argue that such a strategy is more credible and more consistent with traditional norms of warfare because, unlike punitive deterrent threats, it does not prescribe the wholesale and indiscriminate destruction of socioeconomic targets. Moreover, it is arguably a more effective deterrent because it threatens what the adversary presumably fears the most—defeat of the state.[18]

Strategy involves, among other things, articulation of ends which are compatible with means. Warfighting strategies, ironically, are subject to criticism precisely because the ends appear out of reach, the means excessive, or the assumptions questionable. While one can theoretically construct a simple formula for defining the proper size of an assured destruction or finite deterrent capability, determining the resources necessary to achieve an ideal denial capability is an open-ended exercise. Moreover, the ability to limit nuclear conflict and to control its escalation over time is only hypothetical and cannot be demonstrated. Some argue that not only is such an assumption erroneous, it is also dangerous since it encourages an arguably illusory view that nuclear weapons use can be controlled. As such, it undermines rather than reinforces deterrence.

Clearly, a coherent strategy of nuclear war remains elusive, largely due to the very nature of nuclear weapons and their destructiveness. Nonetheless, the realities of the nuclear world require that states devise a strategy of deterrence that can accommodate the dilemmas which are inherent to the age in which we live. Political leaders must make deliberate choices about doctrine and weapons, despite the uncertainties that inevitably plague that decision process. While there is obviously no clear and simple formula for deterrence in the nuclear age, there are some identifiable criteria to guide the necessary choices.

THE SEARCH FOR A STABLE DETERRENT

In translating notions of deterrence into strategic policy, the essential criterion is whether strategic postures contribute to the stability of the deterrence relationship. The notion of stability is used here in a broad sense, encompassing conditions which both limit the incentive to resort to violence and, if conflict starts, enhance the capability to terminate that conflict. Strategic stability thus involves a dynamic relationship between the protagonists and the postures which each has to threaten the other. It also implies that deterrence must be operative not only in peacetime, but also in times of crisis and heightened tension, and even after conflict has been initiated, so that conflict can be terminated or at least managed within certain proportions.

Crisis stability requires that, during periods of tension, adversaries do not perceive an advantage in preemptive attack. In the nuclear relationship, this demands that the weapons of all powers are sufficiently invulnerable so as to deny an adversary the opportunity of disarming one's retaliatory capability. This could also suggest that weapons should not be capable of disarming the adversary, lest the adversary feel compelled to launch first to avoid being disarmed. If *either* side feels that it could be deprived of a retaliatory capability, then there is a powerful incentive for *both* sides to strike first.[19]

Maximizing crisis stability was, as we have seen, one of the central aims of MAD, since it focused on keeping retaliatory assets invulnerable while stressing the capability to target socioeconomic values rather than hardened military targets. It is well documented that, while MAD was an enduring *rhetorical* concept, US strategic-targeting doctrine had in fact attempted, within the limits of technology, to focus on military targets and the selective application of nuclear weapons on more limited scales.[20] The reasons are not hard to discern.

Strategic stability rests fundamentally on the credibility of the retaliatory threat since an incredible threat is itself an incentive to attack. This reality assumed greater force as the superpower relationship evolved into one of strategic parity. In an extended-deterrent context, mutual superpower vulnerability suggested—at least theoretically—that the Soviets could conclude that they could initiate conventional attack on US allies while deterring the nuclear response that is inherent to NATO's doctrine of flexible response.

Moreover, it was clear by at least the mid-1970s that—in the age of the multiple independently targetable reentry vehicle (MIRV)—the proliferation of accurate Soviet land-based warheads posed a threat to US land-based ICBMs. While the US might still possess, after a first strike, sufficient SLBMs to pose an assured destruction threat, launching those missiles against soft Soviet socioeconomic targets might only invite the assured destruction of the US by the residual Soviet forces. In short, the logic of MAD seemed—at least in theory—to undermine US retaliatory threats as long as the Soviets retained substantial "warfighting" (or "denial") options at the nuclear or conventional level.

Since there is no guarantee that deterrence will not fail, it is also important to think of deterrence in the context of *conflict management* and, ultimately, *conflict termination*. Here again, MAD as a policy choice is not necessarily helpful. If conflict begins at any level there are two ways to manage it. Preferably, one can defend successfully and defeat the attack at the lowest possible level of violence. This is not always possible, and in the case of NATO, not necessarily desirable. A NATO doctrine which prescribes only conventional defense in the face of conventional attack may not be sufficient to deter the Soviets from launching an attack which, regardless of the outcome, leaves Western Europe weakened (if not devastated) by the destruction of modern conventional warfare, while leaving the Soviet Union a virtual sanctuary. The alternative is to threaten the prospect of nuclear escalation in the hope that the aggressor will reconsider the risks involved. This requires sufficiently flexible options so that the demonstration of escalatory potential does not itself trigger a massive and uncontrolled nuclear exchange or a Soviet attempt to test US nuclear resolve.

This criticism of MAD, however, does not suggest that strategies of deterrence by denial necessarily offer a clear solution to the search for a stable deterrent. Damage limitation policies can enhance crisis stability insofar as they enhance the invulnerabilty of one's own retaliatory assets, but they can be destabilizing if they focus on the ability to threaten the adversary's retaliatory capability. Nuclear weapons are both offensive and defensive; while there is a difference between retaliatory intent and offensive intent, the same weapon serves both purposes. Thus, weapons which are capable of destroying an adversary's hardened military targets can be properly justified as contributing to strategic stability because they offer a credible retaliatory option. Nonetheless, those same weapons remain physically indistinguishable from weapons which are designed for a disarming first strike. The issue, then, is how one side perceives the other's intent, a judgment based both on *objective* factors (such as the size of the adversary's hard-target-kill capability, hardened command and control links, and defensive capability) and, significantly, on *subjective* assessments of political intent.

Denial strategies can enhance deterrence once a conflict starts by offering the flexibility that is needed both to demonstrate the potential for escalation and, at least theoretically, to limit the scope of the conflict. Nonetheless, there is no certainty that nuclear conflict can be controlled, and the effects of using nuclear weapons on the battlefield can at best be judged as unpredictable.

Just as there are clear limits to the ability of punitive deterrence to contribute to crisis stability—for which it seems best suited—there are also clear limits to the ability of denial strategies to contribute to conflict termination while fulfilling their avowed purposes of defeating the adversary. In attempting to defeat the Soviet state, for example, destroying the military and political capacity of the adversary to prosecute the war may also render impossible any attempt to achieve a termination of the conflict through political means.

It is perhaps impossible to reconcile the competing demands of maximizing crisis stability and conflict management, much less the broader theoretical contest between punitive and denial strategies. Overshadowing this problem, however, is the persistent reality of the destructiveness of nuclear weapons and the unpredictability which surrounds consideration of their use. In arguing for what was earlier described as an "ideal" denial strategy, some have stressed that political leaders have, on more than one occasion, demonstrated their willingness to risk high casualties and damage to their nation's economic base in the pursuit of political gain. If the answer is to threaten the defeat of the state, one must ask if there is any practical difference between targeting the political and military apparatus of a state and targeting the socioeconomic foundations of that state. In the former, one destroys the ability of the political leadership to govern; in the latter, one renders the society ungovernable. In reality, given the destructiveness of nuclear weapons, the difference is substantially greater in theory than in practice.

It is in this sense that some have argued that, while MAD does not represent a viable policy choice, the reality of mutual superpower vulnerability is inescapable. Deterrence of global nuclear war may well rest indefinitely on the inability to find any political end which justifies such a conflagration. Critics of MAD have consistently argued that the Soviet Union does not accept the desirability of societal vulnerability as a foundation for stable mutual deterrence. Critics of deterrence by denial strategies have, in response, argued that mutual vulnerability is a fact of the nuclear age. The superpowers are locked into a mutual hostage relationship regardless of the dictates of US or Soviet doctrine.[21] The debate over the Strategic Defense Initiative is, in large measure, about whether that mutual hostage relationship can be changed, and whether one ought to try and change the foundations of nuclear deterrence which have been, by default, based on the threats of offensive retaliation. That debate will be with us for some time to come.

BEYOND NUCLEAR DETERRENCE?

Since the advent of the nuclear age, the United States has presumed that nuclear weapons offer a deterrent to war which is historically unprecedented. We know only that there are limits to their deterrent effect, but not the precise nature of those limits. We theorize that, by and large, basic deterrence is a simpler proposition than extended deterrence, but we do not have the luxury of ignoring the requirements for the latter. We assume that nuclear weapons are most capable of deterring nuclear use by an adversary, but we rely on them to do more even though we cannot be sure of how much more they will deter. We pose the incalculable risks of punitive nuclear destruction as an ultimate deterrent, yet we consistently seek to enhance the credibility of that deterrent by developing options which are arguably more usable and not so unthinkable. In short, we at times embrace nuclear weapons as an "absolute weapon," while we cling to the traditional purposes of military force and the assumption that weapons—including nuclear weapons—can be employed in limited ways.

Such contradictions are the legacy of the nuclear age. Just as previous wars in this century provided a catalyst for ideas on how mankind could protect itself from its capacity for self-destruction, the growing recognition that nuclear deterrence has its more fragile characteristics has prompted thinking about whether and how the foundations of nuclear deterrence might be changed. One such stimulus was the articulation of the concept of *nuclear winter*, the notion that even a limited exchange of nuclear weapons could trigger a global climatic change sufficient to undermine the very foundations of human sustenance.[22] The contours of that debate are instructive.

On the one hand, the theoretical prospect of nuclear winter reinforced the notion that nuclear weapons were indeed as Brodie had initially characterized them—the absolute weapon. Lacking military utility themselves, nuclear weapons promised not only cataclysmic destruction but ecological disaster on a global scale as well, thereby imposing an undeniable deterrent to the kind of military adventurism that could lead to their use. On the other hand, the possibility of human extinction compelled many to advocate dramatic reductions in nuclear arsenals so that, if deterrence failed, such limited arsenals would not be sufficient to trigger the hypothesized apocalypse.

The debate about nuclear winter resides more in scientific than in political circles. Many questions remain about the validity of the theory and, in particular, about whether there is any identifiable threshold of nuclear effects which would trigger such far-reaching climatic change. Politically, it raises questions which cannot be answered. If there is a threshold below which nuclear weapons could be used and the envisioned climatic effects avoided, should that threshold be identified? If nuclear weapons were to be reduced below such a threshold, is the intent to make the world safe for nuclear conflict, accidental or otherwise? Or is it preferable to avoid specificity and reinforce the uncertainty about nuclear destructiveness which supports at least in part the notion of an absolute weapon?

The difficulty in answering such questions highlights not only the complexity of deterrence but also the political context in which deterrence exists. Fundamentally, deterrence involves political perceptions of adversaries who compete over interests and issues of power and ideology. It is worth recalling that weapons are merely instruments of policy; opposing arsenals are symptomatic of political disputes, not their cause. Reducing weapons or even eliminating categories of weapons may alter the framework of that political and military competition, but it will likely not change the basis of the political conflict. It may make the conflict more stable,

but it may also make it unstable by removing constraints which had previously deterred the resort to violence.

In reality, military force—including both nuclear and conventional weapons—is designed to deter not just nuclear war but war itself, while at the same time providing a state with the means to defend its interests if deterrence fails. A stable deterrent thus limits incentives to resort to violence of any kind while offering the ability to terminate that conflict within reasonable bounds. As with the questions posed by the nuclear winter thesis, the debate over "no first use" of nuclear weapons and the controversy surrounding the prospect that the superpowers might have been willing at Reykjavik to consider the phased elimination of strategic nuclear weapons posed fundamental questions about the desirability of removing the nuclear shadow from certain domains of international conflict. The essential qustion is whether such steps would increase the stability of an adversarial political relationship, even if one assumed that they successfully reduced the dangers of *nuclear* war.

In the absence of alternative safeguards, there is little indication that states are as yet willing to make the world safe either for nuclear conflict or for conventional aggression against their vital interests. Technology is an important dynamic, however. Advances in conventional weapons technologies suggest that future wars will be characterized by dramatically increased intensity, in terms of both firepower and scope of destruction. That in itself may supplant nuclear weapons as an ultimate deterrent and conceivably render nuclear weapons less relevant to the strategic relationship. Technology may also provide for a defense against nuclear attacks, altering the very foundation of postwar theories of deterrence based on the threat of nuclear retaliation. Such prospects remain futuristic, and the problems of transitioning between deterrence frameworks are monumental. These prospects do not, however, remove the need for deterrence. Barring a fundamental change in the political character of the international system and the conflicts which persist within that system, we shall continue to need to maintain some structure for deterrence, be it in the nuclear age or in some post-nuclear age.

NOTES

1. J. M. Spaight, *Air Power in the Next War* (London: Geoffrey Bles, 1938), p. 126, quoted in George H. Quester, *The Future of Nuclear Deterrence* (Lexington, Mass.: D.C. Heath, 1986), p. 64 (emphasis added).

2. See Quester's chapter, "The Newness of Deterrence," in Quester, *Future of Nuclear Deterrence*, pp. 59–72.

3. Bernard Brodie et al., eds., *The Absolute*

Weapon (New York: Harcourt, Brace, 1946), p. 76.

4. See McGeorge Bundy et al., "Nuclear Weapons and the Atlantic Alliance," *Foreign Affairs* 60, no. 4 (1982) and the rejoinder by Karl Kaiser et al., "Nuclear Weapons and the Preservation of Peace: A German Response to No First Use," *Foreign Affairs* 60, no. 5 (1982). See also Robert S. McNamara, "The Military Role of Nuclear Weapons," *Foreign Affairs* 62, no. 1 (1983).

5. Michael E. Howard, "Reassurance and Deterrence," *Foreign Affairs* 61, no. 2 (1982–83).

6. See Kissinger's speech, "NATO: The Next Thirty Years," in *Survival* 21 (Nov./Dec. 1979).

7. See Lawrence Freedman, *The Evolution of Nuclear Strategy* (London: Macmillan, 1981), pp. 313–24.

8. See Lawrence Freedman, *Britain and Nuclear Weapons* (London: Macmillan, 1980).

9. The CARTE BLANCHE NATO exercise in June 1955, for example, simulated a limited NATO employment of tactical nuclear weapons. Even without Soviet use of nuclear weapons, the hypothetical results involved over five million Germans killed or incapacitated. See James L. Richardson, *Germany and the Atlantic Alliance* (Cambridge: Harvard University Press, 1966), p. 40.

10. Albert Wohlstetter, "The Delicate Balance of Terror," *Foreign Affairs* 37, no. 2 (1958).

11. See Alain C. Enthoven and K. Wayne Smith, *How Much Is Enough? Shaping the Defense Program, 1961–1969* (New York: Harper and Row, 1971), esp. chap. 5, "Nuclear Strategy and Forces."

12. See, for example, Desmond J. Ball, *Politics and Force Levels: The Strategic Missile Program of the Kennedy Administration, 1961–1963* (Berkeley and Los Angeles: University of California Press, 1980).

13. Data adapted from the International Institute for Strategic Studies (IISS), *The Military Balance, 1989–1990* (London: IISS, 1989).

14. See Albert Carnesale et al., The Harvard Nuclear Study Group, *Living with Nuclear Weapons* (New York: Bantam Books, 1983), pp. 34–35.

15. For an analysis, see Graeme P. Auton, "Nuclear Deterrence and the Medium Power: A Proposal for Doctrinal Change in the British and French Cases," *Orbis* 20, no. 2 (1976), esp. pp. 373–76. For advocacy of such automaticity in the French case, see Pierre Gallois, *The Balance of Terror: Strategy for the Nuclear Age* (Boston: Houghton Mifflin, 1961).

16. Glenn Snyder's *Deterrence and Defense* (Princeton: Princeton University Press, 1961) is the classic exposition of the distinction between these two concepts.

17. Colin Gray, "Nuclear Strategy: The Case for a Theory of Victory," in *International Security* 4, no. 1 (1979): 87.

18. See, for example, Gray's "War Fighting for Deterrence," in *Journal of Strategic Studies* 7, no. 2 (1984).

19. On the nature of "preventive war" and the implications of pursuing a strategy that gives up that option, see Bernard Brodie, *Strategy in the Missile Age* (Santa Monica: RAND Corporation, 1959), chap. 7, "The Wish for Total Solutions," and pp. 393–400.

20. For an excellent summary, see Jeffrey Richelson, "PD-59, NSDD-13, and the Reagan Strategic Modernization Program," *Journal of Strategic Studies* 6, no. 2 (1983).

21. See Spurgeon M. Keeny and Wolfgang K.H. Panofsky, "MAD Versus NUTS: Can Doctrine or Weaponry Remedy the Mutual Hostage Relationship of the Superpowers?" *Foreign Affairs* 60, no. 2 (1981–82).

22. For a summary of the TTAPS thesis on "nuclear winter," see Carl Sagan, "Nuclear War and Climatic Catastrophe," *Foreign Affairs* 62, no. 2 (1983–84).

BIBLIOGRAPHIC ESSAYS

SOVIET DEFENSE POLICY

DAVID L. GIDDENS

INTRODUCTION

The serious student of the social phenomenon which has existed in the Soviet Union since 1917 frequently encounters, in research and readings, an introductory qualification. This qualification is intended to enhance the reader's understanding of the material and to point out common errors in perception. So it is with this essay, for there exists in every student a strong tendency to apply personal values, beliefs, and attitudes to the Soviet experience. This, in turn, leads to "mirror-imaging," misperception, and finally to misunderstanding of the subject under study. Students frequently attempt to analyze a subject such as Soviet defense policy without even understanding their own country's defense policy. However, being armed with detailed knowledge about one's own country and its defense policy will not assure success. The tendency to simplify Soviet defense policy and make it fit the parameters of one's own is frequently misleading and causes false perceptions which linger and are hard to change. All of this is further complicated by two related phenomena. First, the Soviet Union is a closed society and the resulting lack of hard data about the institutions and processes of Soviet decision-making severely complicates our attempt at true understanding. Second, the integrity of Soviet writings on defense policy has been the focus of continuing controversy. Here we encounter such problems as the rewriting of history, ideological propaganda, and disinformation. As a result, the student must be extremely careful to aaply research methods which allow for critical examination and multiple sources.

A word of caution is also in order concerning non-Soviet publications. In the United States, for example, there exists in government, business, and academia a relatively large body of influential experts. They have arrived at their position as a result of years of education, work, and research. They are widely respected and widely published. However, they are not of one mind when it comes to Soviet defense policy. They are divided into hawks and doves, conservatives and liberals, left and right, or whatever adjectives one chooses to employ. Further, they often represent their own personal biases as well as their own institutions. They represent one of the greatest strengths of Western democracies: free and open discussion of public issues. All of them should be read, again most critically. However, not one of them will be totally right on any given issue. At best, each will merely provide another viewpoint, another piece of the puzzle.

Finally, our ability to understand Soviet defense policy is hampered by the language barrier. Some have noted that there are more *teachers* of English in the Soviet Union than there are *students* of Russian in the United States. As this situation continues, the problems both of accurate translation and full communication will put us at a disadvantage.

HISTORICAL SOURCES AND OTHER REFERENCES

Soviet defense policy has been a subject of high interest throughout the West since the end of World War II. Prior to this time, the entire Soviet experience was of only marginal interest. The Soviet Union had been defeated by Germany in World War I, and the Bolshevik revolution prompted a negative reaction on the part of most Western nations. This early period in Soviet history is critical to one's understanding of modern Soviet defense policy.

LENIN AND THE BOLSHEVIK REVOLUTION

Lenin (Vladimir I'lich Ulyanov) was the undisputed leader of the November 1917 revolution. A beautifully written biography of his life can be found in Edmund Wilson's *To the Finland Station* (New York: Doubleday, 1940). Another excellent book, with somewhat more detail, is by the noted historian Louis Fischer, *The Life of Lenin* (New York: Harper and Row, 1964). Lenin was, of course, a Marxist. Although Karl Marx's *Das Kapital* (*Capital*) is an enormous and difficult work, *The Communist Manifesto*, coauthored by Marx and Friedrich Engels, is a short, easily read work. (One excellent paperback version is by Simon and Schuster, New York, 1964.) Also, a helpful booklet is *An Introduction to Marxist Economic Theory* by Ernest Mandel (New York: Pathfinder Press, 1983).

The Bolshevik revolution today serves as the ultimate source of legitimacy for the Soviet leadership. It has been variously interpreted by Western observers as largely an accident, as by R. V. Daniels in *Red October—The Bolshevik Revolution of 1917* (New York: Scribner, 1967), by Adam Ulam as a minority-led coup d'etat in *The Bolsheviks* (New York: Macmillan, 1965), and as a genuine, full-fledged revolution by Alexander Rabinowitch in *The Bolsheviks Come to Power* (New York: W. W. Norton, 1978). All three of the books are excellent references for this epoch event in early Soviet history. More importantly, they help to explain the political developments which became the foundation for Soviet defense policy. Another outstanding source for information on the early Soviet period is George Kennan. Kennan was a young diplomat who observed the development of Soviet Russia firsthand. His books provide invaluable insight into the events of the time as well as their relevance to the international arena. Of particular note are *Soviet-American Relations, 1917–1920; Russia Leaves the War*, vol. 1; *The Decision to Intervene*, vol. 2 (Princeton: Princeton University Press, 1956–58) and *American Diplomacy 1900–1950* (Chicago: University of Chicago Press, 1951).

THE CIVIL WAR AND THE INTERWAR PERIOD

A little-known and often neglected historical fact of this period is the Allied intervention in Russia. Shortly after the disastrous Treaty of Brest-Litovsk in March 1918, the country erupted into civil war. In a desperate attempt to bring Russia back into the Alliance, several countries, including the United States, sent troops to occupy Russian soil. Kennan's book, *The Decision to Intervene* (Princeton: Princeton University Press, 1956), is rich in the detail of this intervention as well as original souce material. E. H. Carr, the noted historian, produced a three-volume collection on this period, entitled *The Bolshevik Revolution, 1917–1923* (New York: Macmillan, 1951–53). Also, William H. Chamberlin contributed another classic of the times with *The Russian Revolution* in two volumes (New York: Macmillan, 1935). Isaac Deutscher presents a somewhat more personalized accounting of these times with his three books, *The Prophet Unarmed: Trotsky 1921–1929* (Oxford: Oxford University Press, 1959); *The Prophet Outcast: Trotsky 1929–1940* (Oxford: Oxford University Press, 1963); and *Stalin: A Political Biography*, 2d ed. (Oxford: Oxford University Press, 1967). Trotsky himself made a major contribution to Western understanding in *The History of the Russian Revolution 1917; A Personal Record* (New York: Oxford University Press, 1932).

Stalin eventually succeeded Lenin as the national leader. His path to power is marked by the infamous purges and his establishment of a new authoritarian political system—totalitarianism. The totalitarianism model of the Soviet system is one that has gathered an impressive following, even into the mid-1980s. Hannah Arendt's book. *The Origins of Totalitarianism* (New York: Harcourt Brace Jovanovich, 1973), which originally appeared in 1951, asserted that totalitarianism was a new and unique form of political rule, and Carl Friedrichs' famous "six point syndrome" of 1954 asserted that the characteristics of totalitarianism were common to both the fascist (Nazi) and communist types. These developments had a profound impact on American perceptions of the Soviet Union, especially in the post–World War II period. A full exposition of this type of regime is found in Carl J. Friedrich and Zbigniew Brzezinski, *Totalitarian Dictatorship and Autocracy* (New York: Praeger, 1956).

Stalin's collectivization of Soviet agriculture in the late 1920s and his introduction of the First Five Year Plan in 1928 returned the country to economic socialism after the respite provided by Lenin's New Economic Policy (NEP). Alec Nove covers these developments in a concise, yet thorough, manner in *An Economic History of the USSR* (New York: Harmondworth/Penquin, 1984). This "Revolution from Above" and the resistance to it, set the stage for the great purges and Robert Conquest's book, *The Great Terror* (New York: Macmillan, 1968). This terrorism, in turn, swelled the ranks of the prisoners in Stalin's forced labor camps, which are so meticulously described in Alexander Solzhenitsyn's works, especially *The Gulag Archipelago*, 3 vols. (New York: Harper and Row, 1974, 1975, and 1977).

As Stalin consolidated his rule in the Soviet Union, he established an enormous bureaucracy with himself alone at its head. Jerry Hough and Merle Fainsod produced an updated version of a classic in *How the Soviet Union Is Governed* (Cambridge: Harvard University Press, 1979), which explains the new Soviet government and its functioning. Robert Tucker's book, *Stalinism* (New York: W. W. Norton, 1977), is another excellent source for understanding this Soviet dictator.

The years before World War II are discussed in such works as Kennan's *Russia and The West Under Lenin and Stalin* (Boston: Little, Brown, 1961) and Adam Ulam's *Expansion and Coexistence: The History of Soviet Foreign Policy, 1917–1967* (New York: Praeger, 1968). Also, in the military area, John Erickson's *The Soviet High Command* (New York: St. Martin's Press, 1962) provides continuity from the time of the 1917 Revolution up to World War II. Erickson explains the relatively minor role played by the Soviet military during these years as it grappled with the dual problems of a need for military expertise coupled with a highly disloyal cadre of military personnel who came initially from the tsar's former army. Erickson's subsequent books, *The Road to Stalingrad* (New York: Harper and Row, 1975) and *The Road to Berlin* (Boulder, Colo.: Westview Press, 1983), have brought him international recognition even from Soviet military leaders.

In addition to Erickson's prodigious works on World War II (or the Great Patriotic War, as the Soviets call it), another interesting source is that of Albert Seaton, *The Russo-German War, 1941–1945* (New York: Praeger, 1970). Max Jakobson's *The Diplomacy of the Winter War: An Account of the Russo-Finnish War, 1939–1940* (Cambridge: Harvard University Press, 1961) provides marvelous detail of the short war which embarrassed Soviet military power before the entire world. Yet, the Soviet Union would, within the next six years, endure an invasion which would have defeated any other country and emerge victorious after suffering unimaginable losses. For example, the Soviets suffered higher losses in the single battle for Leningrad than the United States suffered in both theaters of the entire war. Both Leon Goure, with *The Siege of Leningrad* (Stanford: Stanford University Press, 1962), and Harrison Salisbury with *The 900 Days: The Siege of Leningrad* (New York: Harper and Row, 1969), convey this point extremely well. Seweryn Bialer also made a major contribution to Western understanding of the Soviet war effort in *Stalin and His Generals: Soviet Military Memories of World War II* (New York: Praeger, 1969).

The Postwar Period

The immediate postwar period was marked by rising hostility and the beginning of the cold war. Daniel Yergin, in his book, *Shattered Peace* (Boston: Houghton Mifflin, 1977), points out that the United States had still not satisfactorily answered the question of whether the Soviet Union was merely another nation-state trying to defend its national interests or rather some evil force or conspiracy driven by a hostile ideology to conquer the world. Yergin contrasts Roosevelt and the Yalta "axioms," which emphasized accommodation, with Truman (at Potsdam and thereafter) and his Riga "axioms," which tended toward confrontation. Diane Shaver Clemens' book, *Yalta* (New York: Oxford University Press, 1970) and Herbert Feis's *Between War and Peace: The Potsdam Conference* (Princeton: Princeton University Press, 1960), provide all the relevant detail to permit the reader to define one's own position on this critical period.

This was indeed a critical period because, by the spring of 1947, events in Poland, Greece, Turkey, Iran, and elsewhere had effectively destroyed the wartime alliance. On 12 March 1947, President Truman proclaimed, "It must be the policy of the United States to support free peoples who are resisting attempted subjugation by armed minorities or outside pressures" and thus codified the Truman Doctrine. Five months later, Kennan wrote, in his famous "X" article in the July 1947 issue of *Foreign Affairs*, that "the main element of any United States policy toward the Soviet Union must be that of a long-term, patient but firm and vigilant containment of Russian [sic] expansionist tendencies." In September a year earlier, presidential aide Clark Clifford had completed a lengthy synthesis of advice from within the government on relations with Russia as requested by Truman. This document advocated a global policy of containing the Soviet Union through the use of propaganda, economic aid, and even military force, not excluding atomic or biological warfare, if necessary (see Clifford's September 1946 report in Arthur Krock's memoirs, *Sixty Years on the Firing Line* [New York: Funk and Wagnalls, 1968]). The cold war was on, and containment became the national strategy of the United States.

World War II, and the cold war which followed, marked the beginning of the modern Soviet-American superpower relationship. To the Soviets, the Great Patriotic War confirmed the superiority of their socialist system over the evil forces of Nazi Germany. A country which had been humiliated on the field of battle only one generation earlier and which had suffered the enormous costs of the Treaty of Brest-Litovsk, 25 years later stood astride Europe, in Berlin, Prague, and Vienna. To the United States, victory in World War II meant something different. America emerged as the most powerful nation in the world with a monopoly on atomic weapons. Once again, a European war had forced US participation. This time, however, the United States would not be allowed to retreat into isolationism. This time, the US would champion the cause of world government (the United Nations) and eventually become the defender of freedom and western democracy in the face of the Soviet threat. John L. Gaddis, in his book, *Strategies of Containment* (New York: Oxford University Press, 1982), is one of several scholars who are currently reviewing this period with special emphasis on the Chinese revolution and the Korean War.

Through the Truman and Eisenhower years,

the United States generally followed a hard line in its relations with the Soviets. Fred Kaplan's book, *The Wizards of Armageddon* (New York: Simon and Schuster, 1983), provides a revealing insight into US strategic thinking during this period and continuing to the present.

Stalin died in 1953 and was eventually succeeded by Khrushchev and the famous 20th Party Congress of 1956, during which Khrushchev made his initial denunciation of the excesses of Stalin. Myron Rush's *Political Succession in the USSR* (New York: Columbia University Press, 1965) highlights one of the more controversial phenomena in the Soviet system, as does Michel Tatu's *Power in the Kremlin: from Khrushchev to Kosygin* (New York: Viking, 1968).

The 15 years following Stalin's death included three major foreign policy crises for the Soviets: The Hungarian revolution in 1956, the Cuban missile crisis in 1962, and the Soviet invasion of Czechoslovakia. While literature abounds on these pivotal events, the following three works are excellent surveys and offer superb analyses of Soviet policy perspectives: Zbigniew Brzezinski's *The Soviet Bloc: Unity and Conflict* (Cambridge: Harvard University Press, 1961), Adam Ulam's *The Rivals: America and Russia Since World War II* (New York: Viking, 1971), and Juri Valenta's *Soviet Intervention in Czechoslovakia, 1968* (Baltimore: Johns Hopkins University Press, 1979). For discussions on the challenges and performance of post-Stalin Soviet leaders, see George Breslauer's *Khrushchev and Brezhnev as Leaders* (London: Allen and Unwin, 1982) and Seweryn Bialer's *Stalin's Successors* (London: Cambridge University Press, 1980).

The Soviet invasion of Czechoslovakia represents, for many observers, an assertion of Soviet control over its European empire which, for the Soviet leadership, was a necessary precondition to engaging in "détente" with the US and Western Europe. This process eventually led to major agreements not only in arms control, but also in an attempt to expand political, economic, and social interaction between East and West. Yet, if the 1960s and 1970s were marked by major Soviet-American agreements and improving relations, the 1980s have been described by some as a worsening of relations and even a return to the cold war.

The development of nuclear weapons has had a profound effect on each of the two superpowers as well as on their bilateral relationship. This complex process called "détente" is discussed by Richard Pipes in his book, *U.S.-Soviet Relations in the Era of Détente* (Boulder: Westview Press, 1981). Dan Caldwell's *Soviet International Behavior and US Policy Options* (Lexington, Mass: D. C. Heath, 1985) is an especially good recent overview of Soviet policy

in this era of nuclear statemate. Contributors to this edited work offer excellent bibliographies in their respective areas of emphasis. Finally, R. Judson Mitchell provides a valuable analysis of how Soviet ideology has come to terms with the new global role of the Soviet state in his *Ideology of a Superpower: Contemporary Soviet Doctrine on International Relations* (Stanford: Hoover Institution Press, 1982).

It is perhaps easier for a student of Soviet policy to delve into the history of specific events or periods than it is to stand back and try and understand Soviet policy with a longer view. It is appropriate, then, to round out this survey of historical sources by noting two books in particular which deserve special mention for their contribution to Western understanding of Soviet historical experience. These are D. W. Treadgold's *Twentieth Century Russia* (Chicago: University of Chicago Press, 1981) and Nicholas Riasanovsky's *A History of Russia* (New York: Oxford University Press, 1984).

UNITED STATES PERIODICALS

One of the greatest strengths of Western democracies is the continuing public debate of national and international issues using such vehicles as recurring periodicals in order to reach as many readers as possible. In the national security area, articles on the Soviet-American relationship can be found in journals ranging from *Reader's Digest* to *Playboy*. Certain academic journals publish articles on the Soviet Union on a regular basis. They include a standard list with which the student should become familiar. *Orbis* is published quarterly by the Foreign Policy Research Institute in association with the International Relations Graduate Group of the University of Pennsylvania. *International Security* is also published quarterly by the Program for Science and International Affairs at Harvard University. *Armed Forces and Society* is published quarterly by the Inter-University Seminar on the Armed Forces and Society. *Comparative Strategy* is also published quarterly for the Strategic Studies Center by Crane, Russak. *Strategic Review* is published quarterly by the United States Strategic Institute. *Defense Monitor* is published monthly by the Center for Defense Information.

Soviet military power and defense policy are also discussed somewhat less frequently and generally more indirectly in such journals as *Foreign Affairs*, which is published five times a year by the Council on Foreign Relations. *Foreign Policy* is published quarterly by the Carnegie Endowment for International Peace. *International Studies Quarterly* is published by SAGE Publications, and *World Politics* is published quarterly by Princeton University Press. The International Communications Agency publishes the highly informative *Problems of Communism* six times annually, and it can be ordered through the Government Printing Office. The government has become more and more active in disseminating literature on Soviet military power in recent years. A particularly useful series is entitled "Soviet Military Power" in four editions (1981–1985, available through the US Government Printing Office) and presented by the secretary of defense. Additionally, the Joint Economic Committee of Congress publishes a report entitled, "Annual Economic Indicators for the U.S.S.R.," through the US Government Printing Office. The Central Intelligence Agency is also but one of several agencies which periodically issue unclassified reports on the Soviet Union, particularly in such areas as biographical data on the current Soviet leadership and economic forecasting.

Also, each of the senior service schools of the Army, Navy and Air Force publishes a journal on defense policy issues. They are *Military Review* (monthly), *Naval War College Review* (bimonthly) and the *Air University Review* (bimonthly). This last journal has devoted whole issues to the Soviet experience (see for example, the Mar./Apr. 1985 issue) and includes valuable articles on understanding the Soviet military. One such article, by Carl Reddel, "Future of the Soviet Empire" (July/Aug. 1982), and another by Jonathan Adelman, "The Evolution of Soviet Military Doctrine, 1945–1984" (Mar./Apr. 1985) are excellent examples.

Additionally, the March issue of *Air Force Magazine* annually carries a special section called the "Soviet Defense Almanac," which discusses military-related issues and newly acquired Soviet weapons systems. Finally, *Aviation Week and Space Technology* frequently is the first "public" source of new Soviet weapons developments and technological innovations.

SOVIET SOURCES

Working with Soviet sources presents the student with some frustrating problems. First of all, there is the irritating requirement for all Soviet publications to follow the proper ideological and propagandistic "party line." Second, as the number of qualified translators decreases, the quality and quantity of translations will surely suffer. Third, the lack of an "open and free" discussion of the issues in the Soviet media tempts one to abandon them for the ease and convenience of researching on Western sources. In spite of such problems, research aids do exist which allow Western scholars to utilize Soviet writings.

The most familiar and the most general aid is the *Current Digest of the Soviet Press*, pub-

lished weekly by the American Association for the Advancement of Slavic Studies. This document is widely available and comes with a useful quarterly index. A similar document is the US government's *Foreign Broadcast Information Service* (FBIS), published as a daily report with translations covering not only the press but also radio and television broadcasting. Also, the government's *Joint Publications Research Service* (JPRS) translates entire books as well as articles and is available through the National Technical Information Service of the Department of Commerce.

The above translations work directly from the Soviet press. Daily newspapers include *Pravda* ("*Truth*"), the organ of the Central Committee of the Communist Party of the Soviet Union, which has a daily circulation of 10.7 million copies (1983) and which is reportedly available in 150 countries. *Izvestia* ("*News*") is the organ of the Council of Peoples' Deputies of the Soviet Union and is published by the Presidium of the Supreme Soviet. It has a circulation of over six million (1982). *Krasnaya Zvezda* ("*Red Star*") is the organ of the Ministry of Defense of the USSR and is published six times a week with a circulation of 2.5 million.

To this list of the three most important newspapers should be added a similar list of recurring periodicals. The most important journal is *Kommunist Vooruzhennykh Sil'* ("*Communist of the Armed Forces*"), which is published twice monthly by the Main Political Administration of the Soviet Armed Forces. As such, it deals with the political education, training, and conduct of the Soviet military. *Voyennyy Vestnik* ("*Military Herald*") is published monthly by the Ministry of Defense of the Soviet Union and is, on occasion, the vehicle for public discussions of new weapons systems and their impact on tactics. *Voenno-Istoricheskiy Zhurnal* ("*Military Historical Magazine*") is published monthly by the Ministry of Defense.

William F. Scott, former US Air Attache to Moscow, has written a compendium entitled *Soviet Sources of Military Doctrine and Strategy* (New York: Crane, Russak, 1975), which is an invaluable aid because of its comprehensive coverage of Soviet military writings since 1960. Scott has a personal library of Soviet military writings which is truly extensive and to which he is continually adding new sources. This book is particularly valuable because of its annotated bibliography.

The classic work on Soviet military strategy is that compiled under the editorship of Marshal V. D. Sokolovsky, entitled *Military Strategy*. Originally published in 1962, with revised editions appearing in 1963 and 1968, this book was the first definitive work in the area of military science since before the death of Stalin (1953). The translation by Harriet F. Scott (New York:

Crane, Russak, 1975), which draws comparisons to the earlier versions, is most useful.

The *Soviet Military Thought Series*, available from the US Government Printing Office, is another valuable source for Soviet military writings. This series began in 1975 under the auspices of the United States Air Force with the publication of A. A. Sidorenko's 1970 classic, *The Offensive* (vol. 1) and includes a useful *Dictionary of Basic Military Terms* (vol. 9), the late Minister of Defense Marshal Grechko's *Armed Forces of the Soviet State* (vol. 12) and the informative *Officers' Handbook* (vol. 13). The latest volumes in this continuing series include *The Command and Staff of the Soviet Army Air Force in the Great Patriotic War, 1941–1945* by M. N. Kozhevnikov (vol. 17), *Fundamentals of Tactical Command and Control* by three authors, D. A. Ivanov, Z. P. Savel'yev and P. V. Shemansky (vol. 18), *Soviet Armed Forces: A History of Their Organizational Development* by S. A. Tyushkevich (vol. 19), and *The Initial Period of War* by S. P. Ivanov (vol. 20). These volumes can be ordered from the US Government Printing Office. The great value of this series is that the books offer a Soviet view of the issues—a view that is often substantively different from that of the West.

Another publications series by the Air Force is "Soviet Press Selected Translations," issued monthly. Articles are chosen from the Soviet press and translated with emphasis on dissemination to English readers and minimal commentary. In addition, the same office publishes "Studies in Communist Affairs," a six-volume series which began in 1976 with Joseph Douglass, Jr.'s *Soviet Theater Nuclear Offensive*. The second volume by Paul Murphy was entitled, *Naval Power in Soviet Policy*. The third volume by Phil Peterson was *Soviet Air Power and the Pursuit of New Military Options*. The fourth, also by Peterson, was *Soviet Policy in the Post-Tito Balkans*. The fifth was *Selected Readings from Soviet "Military Thought" 1963–1973* in two volumes and is drawn, in part, from the limited circulation Soviet military journal, *Voyennaya Mysl'* ("Military Thought"). The final volume is *The Soviet Union—What Lies Ahead*, edited by Ken Currie and Gregory Varhall.

The Soviet Aerospace Handbook published in 1978 by the Government Printing Office remains a particularly useful reference for those elements of the Soviet Armed Forces engaged in aerospace operations. It is a comprehensive unclassified review of the Soviet Air Force, Strategic Missile Forces, National Air Defense Forces, Naval Aviation and space programs.

The Ministry of Defense of the Soviet Union has recently published two documents which will be of enormous value to students of the Soviet military. The first of these is a growing multivolume publication of the *Soviet Military Encyclopedia*. This work has brought more appreciation for the Soviet view than any other and will greatly expand Western understanding of the Soviet military as more and more articles are translated. It was, unfortunately, first published with a small number of copies and is, therefore, extremely difficult to find, even in the Soviet Union. It remains, nonetheless, a uniquely valuable source. The second document is the *Soviet Military Encyclopedic Dictionary*, published by Military Press in Moscow in 1983, under the editorship of then Chief of the General Staff Marshal N. V. Ogarkov to assist in the understanding and research of military problems. It is interesting and useful to pause and note the flavor of the articles of this dictionary. For example, the article on defense policy ("Voennaya Politika") includes a discussion on the differences between the defense policies of capitalist and socialist countries:

defense policy of capitalist states serves the interests of monopoly capital and is directed against socialist countries toward preparation for predatory, aggressive wars and the suppression of national liberation and revolutionary movements. The defense policy of countries of the socialist community are directed at the defense of the revolutionary accomplishments of the workers, of social progress as a whole, at the strengthening of peace and the avoidance of nuclear war. (p. 137.)

Memoirs are a particularly rich source of information about the Soviet military. Among an endless list of published memoirs, the following are representative of the best: *The Memoirs of Marshal Zhukov* (New York: Delacorte, 1971) and General S. M. Shtemenko's *General'niy Shtab v Gody Voiny* (*The General Staff During the War Years*) (Moscow: Voenizdat, 1968 and 1973).

EUROPEAN SOURCES

When one thinks of European sources of material dealing with the Soviet Union, London's International Institute of Strategic Studies (IISS) immediately comes to mind. This prestigious institute publishes several relevant documents. First, there is an often-cited unclassified source of national and alliance military capabilities in the annual publication, *Military Balance*. Also published annually is *Strategic Survey*, which reviews major international events from the previous year. *Adelphi Papers* are published periodically on a variety of high-interest security-related subjects. The *Military Balance* is particularly useful when coupled with the Secretary of Defense's annual report to Congress with its expanded addendum, *Soviet Military Power* (noted above) and even more so when weapons capabilities are methodically explained in the series of *Jane's All the World's Weapons Sys-*

tems, published annually by Jane's Publishing Co., London.

Other European sources include the *Journal of Strategic Studies*, published quarterly by Frank Cass, London, *NATO Review*, published bimonthly by the NATO Information Service in Brussels, and the *RUSI Journal*, published quarterly by the Royal United Services Institute in London. The *International Defense Review* is published nine times annually by Interavia in Geneva and deals more with weapons development and current defense issues in general rather than specifically Soviet-related ones. An interesting article by David C. Isby in IDR, which described the Soviet combat experience in Afghanistan, appeared in 1982 with Soviet order-of-battle information and commentary on Soviet combat performance.

A SELECTIVE REVIEW OF CURRENT LITERATURE

Published books, periodicals, and articles on Soviet military power are innumerable, and the volume grows on a daily basis. The available material is so voluminous that it might overwhelm the newly interested student. Yet, a systematic and continuous effort will not only permit enhanced understanding but will also explain the associated controversy and eventually prepare the student to engage more actively in the continuing discussion.

The following section is intended to highlight some of the more interesting documents appearing recently, as well as to guide the student through a representative sampling of the contemporary issues in the Soviet-American relationship and the discussions surrounding them.

Any military establishment can be described as being merely a reflection of its society. The Soviet military is no exception and, as such, reflects the goals and intentions of the Soviet leadership, if not directly those of the Soviet people. An expanded appreciation of modern Soviet society will strengthen the student's understanding of the similarities and differences of Soviet society as compared to American society. US journalists stationed in the Soviet Union have frequently returned to document their experiences. Two such books have become familiar to the student of the Soviet Union and have, indeed, even been used as textbooks for courses on modern Soviet society. They are Hedrick Smith's *The Russians* (New York: Ballantine Books, 1976) and Robert G. Kaiser's *Russia: The Power and the People* (New York: Pocket Books, 1976). Two similar, more recent books are David Shipler's *Russia: Broken Idols, Solemn Dreams* (New York: Times Books, 1983) and Kevin Klose's *Russia and the Russians: Inside the Closed Society* (New York: W. W. Norton, 1984). Another book that approaches this topic from a somewhat different perspective is *MIG Pilot: The Escape of Lieutenant Belenko*, edited by John Barron (New York: Reader's Digest Press, 1980), which provides the personal observations of Victor Belenko on life in the Soviet Union. Barron is, of course, the noted author of the now classic *KGB: The Secret Work of Soviet Secret Agents* (New York: Bantam Books, 1974), which he has recently followed with *The KGB Today: The Hidden Hand* (New York: Reader's Digest Press, 1983) dealing with the problems of technology transfer and industrial espionage.

Soviet military power, including its strengths and its weaknesses, is the subject of several relatively new books. Victor Suvorov's *Inside the Soviet Army* (New York: Macmillan, 1982) is another excellent example of documentation presented from the perspective of a defector who was a high-ranking Soviet officer. Andrew Cockran's *The Threat: Inside the Soviet Military Machine* (New York: Random House, 1983) is extremely critical of the Soviet military, which he describes as a "paper tiger." Another book based largely on emigre insights into the Soviet military is Richard Gabriel's *The New Red Legions* (Westport, Conn.: Greenwood Press, 1980). Although based on a small sampling, it includes interesting comments on morale and combat effectiveness. John Erickson has published *Organizing for War: The Soviet Military Establishment Viewed Through the Prism of the Military District* (College Station: Texas A&M University Press, 1983) which takes an in-depth look at the role of the military district in Soviet military planning. Peter H. Vigor, in *Soviet Blitzkrieg Theory* (New York: St. Martin's Press, 1983), uses case studies of Czechoslovakia and Afghanistan to highlight Soviet theory of how to wage war using a rapid offensive and exploiting the element of surprise. An excellent summary of Soviet strategic policy, emphasizing intentions as well as capabilities, can be found in *Soviet Strategic Forces: Requirements and Responses* (Washington, D.C.: Brookings Institution, 1982) by Robert Berman and John Baker.

For a European interpretation of Soviet strategy, one should review *Soviet Strategy toward Western Europe* (Winchester: Allen and Unwin, 1984), edited by Edwina Moreton and Gerard Sega, as well as Julian Critchley's, *The North Atlantic Alliance and the Soviet Union in the 1980s* (London: Macmillan, 1982), which summarizes strengths and weaknesses of the Alliance's northern, central, and southern fronts. In the recent two-volume work compiled by Uwe Nerlich, *Soviet Power and Western Negotiating Policies* (Cambridge: Ballinger, 1983), 25 Western military experts set the agenda for modernizing NATO's military strategy.

Stephen Kaplan's *Diplomacy of Power* (Washington, D.C.: Brookings Institution, 1981), examines 190 incidents since World War II using excellent selected case studies, especially in Eastern Europe, which often highlight limits to Soviet power. For an analysis of Soviet military thought since 1964, Mark Katz's *The Third World in Soviet Military Thought* (Baltimore: Johns Hopkins University Press, 1982), examines the Soviet view of the nature of war. It is an excellent reference and contains a valuable bibliography of both Soviet and Western sources.

Moving to the political arena, David Holloway has produced a balanced and thoughtful review of Soviet strategic doctrine in *The Soviet Union and the Arms Race* (New Haven: Yale University Press, 1983). Additionally, Adam Ulam's new book, *Dangerous Politics: The Soviet Union in World Politics 1970–1982* (New York: Oxford University Press, 1983), again offers the author's keen insight into the Soviet mentality with specifics on Soviet military forces and force buildup. In addition, Robert Byrnes has assembled a well-balanced and stimulating compilation of inputs from a variety of respected American specialists under the title *After Brezhnev* (Bloomington: Indiana University Press, 1983). Finally, Stephen F. Cohen's book, *Rethinking the Soviet Experience* (New York: Oxford University Press, 1985), questions the traditional totalitarian model applied to the Soviet Union. Cohen feels the intellectual consensus surrounding this practice is too simplistic and, therefore, unhealthy. It is an excellent and thought-provoking book which will, doubtless, cause considerable controversy.

The periodical literature has likewise been rich and substantive in recent years. According to David Holloway in "War, Militarism and the Soviet State" (*Alternatives* 6, no. 1 [1980]), there are valid obstacles to disarmament in the USSR. Holloway feels societal sources of militarism are strong but not absolute and that the West should not foreclose disarmament as an option. The growing role of the Soviet military in the decisionmaking process is the subject of Roman Kolkowicz's article, "The Military and Soviet Foreign Policy" (*Journal of Strategic Studies* 4, no. 4 [1981]).

In addition, two RAND reports by Benjamin Lambeth are excellent critiques of the differences in strategic perceptions of the US and the USSR when analyzing military policy. They are "Trends in Soviet Military Policy" (RAND Report P-6819, Oct. 1982) and "On Thresholds of Soviet Military Thought" (RAND Report P-6860, Mar. 1983).

Two important articles dealing with the Soviets' expanding naval capability are former Secretary General of NATO, Joseph M. A. H. Luns' "Political-Military Implications of Soviet Naval Expansion" (*NATO Review*, Feb. 1982) and Admiral (Ret.) Harry D. Train's "Challenge at Sea: Naval Strategy for the 1980's" (*NATO's Fifteen Nations*, Special 2, 1982). These articles highlight the far-ranging aspect of Soviet naval power, such as Angola (1975), Vietnam (1979), Lebanon (1981) and the Soviet-Syrian naval maneuver of July 1981).

Colin Gray argues, in "Understanding Soviet Military Power" (*Problems of Communism* 30, no. 2 [1981], that the USSR is more powerful today than ever and that its drive for power is necessitated by the regime's need to impress internal factions as well as external adversaries. These enhanced military capabilities come only at great cost, obviously. This is the subject of Franklyn D. Holzman's article, "Soviet Military Spending: Assessing the Numbers Game" (*International Security* 6, no. 4 [1982]). In RAND Report P6908, Abraham S. Becker writes "Sitting on Bayonets? The Soviet Defense Burden and Moscow's Economic Dilemma" (Sept. 1983), an excellent analysis of Soviet cost accounting and economic predictions.

The Soviets are, of course, confronted by a large variety of problems. Two useful articles which deal with Soviet problems in maintaining their global influence include Malcolm MacIntosh's "Political Problems in Eastern Europe" (*RUSI Journal* 1981) as well as Andrew K. Semmel's "Security Assistance: US and Soviet Patterns" (*SAGE International Yearbook of Foreign Policy Studies* 7, 1982), which contrasts the two approaches over 30 years in a variety of dimensions.

CONCLUSION

This essay is intended to assist the student in gaining an enhanced awareness and understanding of the Soviet Union and of the enormous volume of documentation now in existence and growing each day. Soviet General Secretary Gorbachev's rhetoric on "new Soviet thinking" has stimulated considerable speculation about the future of Soviet defense policy. As the literature continues to grow, there will surely be competing assessments of the impact of Gorbachev's initiatives. As students grapple with this literature, we should remember that no author, no analyst, has the full picture. Each argument will have to be approached critically.

DETERRENCE IN THE NUCLEAR AGE

BRENT D. BRANDON

The concept of nuclear deterrence has not evolved much since scholars first began writing about it in the 1940s. While the ebb and flow of strategic thought have laid bare the many facets of deterrence, and theorists have wielded their scalpels in different manners, the fundamental concept has remained unchanged. Thus, a review of the state of the literature yields an overview that is reassuringly coherent. The reader will encounter a relatively small family of prolific thinkers who draw on each other's works and contribute to an increasingly interrelated web of knowledge. Perspectives vary, of course, and debates have raged over targeting and basing applications; creative phrases describing strategy have come and gone; and "schools" or "camps" have gathered and disbanded. Still, the "balance of terror"—delicate or not—has remained a profoundly powerful element of strategic thought. The reader will conclude that the strategic thinkers of our era tend to agree on what deterrence is, but offer radically different perspectives on how best to achieve it.

JOURNALS

Nowhere is this relationship more evident than in journals of international security where points are presented and rebutted with each publication. *International Security*, published quarterly by the Center for Science and International Affairs at Harvard University, for example, is an excellent source which has served as a vehicle for the vigorous exchange of ideas among scholars. Similarly, *Foreign Affairs*, published by the Council on Foreign Relations, is an extremely valuable source, both for keeping abreast of current issues, and for tracing the thoughts of the early strategic theorists. It began publication in 1922 and is issued five times a year.

Orbis, of the Foreign Policy Research Institute, is published in association with the Graduate Program in International Relations of the University of Pennsylvania and frequently yields timely and stimulating articles on deterrence. *Foreign Policy*, too, must be noted as a source of considerable value. Founded in 1970, it is published quarterly by the Carnegie Endowment for International Peace and regularly includes articles by authorities in the field.

Other journals which often address deterrence theory and application are *World Politics*, published quarterly by Princeton University Press; *International Studies Quarterly* of SAGE

Publications; and *Strategic Review*, a product of the United States Strategic Institute.

The service schools of the Army, Navy, and Air force also publish journals which include articles on deterrence: *Military Review* (monthly), *Naval War College Review* (quarterly), and *Air University Review* (bimonthly).

FRAMEWORK FOR REVIEW

Literature on deterrence in the nuclear age has spanned four decades and can be viewed in terms of four broad waves of strategic thought. The first wave, which laid the foundation for later writing, was inaugurated in 1946 by Bernard Brodie's *The Absolute Weapon* (New York: Harcourt, Brace, 1946). The theorists of this wave outlined the fundamental issues of deterrence in the nuclear age and wrote throughout the 1950s and into the early 1960s.

Second-wave thinkers refined deterrence theory, often quantifying its complex dimensions. These writers constructed models of deterrence and added rigor and structure to the doctrine. Deterrence theory grew in complexity throughout the decade of the 1960s as derivative components of general deterrence were developed and analyzed.

The growth in deterrence theory led to divergence and debate in the 1970s and early 80s. The third wave of strategic rationalists often clashed, presenting and supporting doctrine and challenging the assumptions of others. The vocabulary of the debate grew accordingly: skirmishes over "countervalue" versus "counterforce," "warfighting" versus "deterrence," and "flexible response" versus "assured destruction" all swirled around the fundamental concept of deterrence.

A fourth wave is now emerging which seeks to abandon many of the traditional views on deterrence in the nuclear age. While the forum for debate is expanding, and scientists and laymen have joined the strategic theorists in the 1980s in an attempt to provide new approaches, much of the focus has been on the potential transformation of deterrence implied by strategic defense. This, in turn, has produced a contemporary debate at least as significant and fundamental as any previous debates.

THE FOUNDATION

The first wave of thinkers was quick to grasp the enormous implications of the atomic bomb and generated books and articles which today appear refreshingly straightforward. Bernard Brodie's *Absolute Weapon* epitomizes the approach of the early writers and reflects the thoughts of an age of American atomic monopoly. Written in the transitional period of the late 1940s, Brodie's book includes essays which address the fundamental question: "Is war more or less likely in a world which contains atomic bombs?" Brodie, Arnold Wolfers, and others answer the question in understandable terms.

Wolfers' essay on "Political Consequences," included in *The Absolute Weapon*, anticipated the "early end of our monopoly" and noted, "Once the Russians have the bomb, we shall depend for the very existence of our civilization on the success of our efforts to maintain friendly relations with the Russians, or, failing that, on our ability to deter them from undertaking actions which we would feel obliged to oppose with armed might." Brodie laid the groundwork for later writers by noting the significance of rationality in the atomic age and emphasizing the awesome destructive power of "the weapon" with practical and detailed scenarios which included figures on refueling, bomb loads, and scales of destruction.

"The Requirements of Deterrence" by William W. Kaufmann is found in *Military Policy and National Security* (Princeton: Princeton University Press, 1956), an early collection of works on strategic thought. After outlining the requirements of deterrence, Kaufmann applied them to the conditions which then existed and critiqued the doctrine of "massive retaliation." He noted that "essentially deterrence means preventing certain types of contingencies from arising" and included communication, credibility, operational capacity, and public opinion as four key elements.

The importance of communication was argued by Henry A. Kissinger in *Nuclear Weapons and Foreign Policy* (New York: Harper and Bros, 1957). In this classic analysis of the complexities and dangers of international relations in the nuclear era, Kissinger argued that intentions need to be communicated explicitly, or the asset of a deterrent is lost. In countering the "Peace at any Price" movement, he stressed the importance of communicating our objective to end aggression by eliminating military targets and thereby raised the issue of counterforce targeting. Kissinger likewise emphasized credibility, especially in striking the balance between "the desire for posing the maximum threat and the need for a strategy which does not paralyze the will." It was in this context that Kissinger made one of the most provocative statements in the work: "In this task of posing the maximum *credible* threat, limited nuclear war seems a more suitable deterrent than conventional war."

Bernard Brodie's "The Anatomy of Deterrence" in *Strategy in the Missile Age* (Princeton: Princeton University Press, 1959) is considerably more refined—and provocative—than *The Absolute Weapon*. Brodie fired the opening salvo of the "Deterrence versus Warfighting Debate" by stating that total nuclear war "is too all consuming to permit survival of final values." Clearly wishing to avoid a doctrine of "warfighting," Brodie concluded, "a plan and policy which offers a good promise of deterring war is therefore by orders of magnitude better in every way than one which depreciates the objective of deterrence in order to improve somewhat the chances of winning." This classic work also addressed credibility and distinguished "basic deterrence" from other varieties, while noting that deterrence "does not depend on superiority." While, in *The Absolute Weapon*, Brodie submitted that "governments are of course ruled by considerations not wholly different from those which affect even enlightened individuals," in *Strategy in the Missile Age* he modified that position and asserted that "governments do not think like ordinary human beings."

In Albert Wohlstetter's "The Delicate Balance of Terror" (*Foreign Affairs* 37, no. 2 [1959]), one finds a rigorous examination of the mechanics of deterrence. In this classic article, Wohlstetter specified the "successive obstacles to be hurdled by any system providing a capability to strike second." In addition to the ability to make and communicate the decision to retaliate, he noted the importance of survivability and the capacity to penetrate both active and passive defenses. Deterrence, he stressed, was not an automatic condition. Wohlstetter's insightful work addressed the need to integrate deterrence theory into the broader context of foreign policy and concluded, "though deterrence is not enough in itself, it is vital."

Another work by Henry Kissinger which is relevant to this period is *The Necessity for Choice* (New York: Harper and Bros., 1960). In it, he analyzed the psychological dimension of deterrence—"an intangible quality"—and stressed the importance of "the state of mind of the potential aggressor." Indeed, "deterrence," according to Kissinger, "requires a combination of power, the will to use it, and the assessment of these by the potential aggressor." In an excellent analysis of targeting and finite deterrence, Kissinger noted the difficulty in reconciling counterforce targeting and the strategic defensive. His rigorous examination yielded three valuable conclusions. First, the prerequisite of deterrence is an invulnerable retaliatory force. Second, if the goal is stability, invulnerability should be sought through measures which, in so far as possible, convey a defensive intent. Third, in maintaining deterrence, the strategic relationship must not be viewed as static, and present readiness must not be subordinated to long-term balance in procurement.

Another early theorist of the first wave is Glenn H. Snyder, author of *Deterrence and Defense* (Princeton: Princeton University Press, 1961). From the theoretical introduction contrasting deterrence and defense through sections on the logic of deterrence and "denial versus punishment," Snyder's analysis is superb. One especially interesting piece deals with the differences and interactions between the balances of terror and power. Snyder's book is thoughtful and well organized and includes a helpful "note on terminology," providing definitions of key terms such as "first strike," "strike-back," "punitive," and "counterforce." After reviewing the range of applications for deterrence and defense, Snyder offered an excellent section on "Declaratory Policy and Force Demonstrations," and examined a variety of issues relevant to threats, including clarity versus ambiguity, reiteration, and publicity versus secrecy. The book concludes with a chapter on the reconciliation of deterrence and defense.

REFINEMENT

While the strategic theorists of the first wave laid the broad foundation of nuclear deterrence, some of the writers of the 1960s sought to cast the subject in more rigorous, analytic terms. Thomas Schelling and Herman Kahn inaugurated this era of strategic thought with their innovative works, *The Strategy of Conflict* (New York: Oxford University Press, 1960) and *On Thermonuclear War* (Princeton: Princeton University Press, 1961).

Schelling's *Strategy of Conflict* provided a general framework for his work on game theory. In it he noted that "there are enlightening similarities between, say, maneuvering in limited war and jockeying in a traffic jam, between deterring the Russians and deterring one's own children, and between the modern balance of terror and the ancient institution of hostages." In applying the theory of conflict to the "retarded science of international strategy," Schelling focused on deterrence as "a typical strategic concept" which he wove throughout the remainder of the book. His stimulating handling of deterrence incorporated game matrices and the "prisoner's dilemma," as well as notions of payoffs, threats, randomization, fear, and communication.

"Dedicated to the goal of anticipating, avoiding, and alleviating crises," Kahn's *On Thermonuclear War* employed systems analysis in examining "the military side of what may be the major problem that faces civilization." Kahn reviewed alternative national strategies and noted that the notion of "minimum deterrence" is "dramatic": "No nation whose decision making

ers are sane would attack another nation which was armed with a sufficiently large number of thermonuclear bombs." In developing a national strategy, Kahn started with minimum deterrence and added "insurance" for reliability (finite deterrence); against unreliability (counterforce targeting); and against a change in policy (pre-attack mobilization base). His inclusion of a credible first-strike capability was based primarily on the need to fulfill treaty obligations.

Deterrence, Arms Control, and Disarmament (Columbus: Ohio State University Press, 1962) by J. David Singer continued in the flavor of the second wave by employing a probability-utility model to examine a range of relevant issues: destruction levels, retaliator's nuclear stockpiles, acceptable destruction, and passive defenses. One interesting component of this empirical treatment is Singer's analysis of delivery systems. Another is his discussion of elements which add to the "invulnerability" of retaliatory systems: numbers, hardening, dispersal, distance, concealment, and mobility. In his conclusion, Singer called for a retaliatory force which is "finite in size, relatively impregnable to attack, and extremely patient and slow to react." Finally, Singer stressed the importance of one "variable" in particular: the adversary's policy—"his preference, predictions, intentions, and capabilities."

Glenn Snyder addressed an irony of nuclear deterrence in "The Balance of Terror and the Balance of Power" in *The Balance of Power* (San Francisco: Chandler, 1965). He outlined the "Stability-Instability Paradox" and asserted that mutual second-strike capabilities could lead to conventional war. This insight was a harbinger of later debates within the Western alliance as to whether or not superpower parity and a doctrine of mutual assured destruction could provide a basis for strategic stability in Europe.

Thomas Schelling continued to capture the essence of nuclear deterrence in his *Arms and Influence* (New Haven: Yale University Press, 1966). In this stimulating analysis, Schelling integrated examples from the Persian expedition of Xenophon the Greek and the Athenian siege of Acropolis to John F. Kennedy and the Cuban missile crisis. In doing so, he enriched the concept of deterrence within the broader framework of the principles which underlie the diplomacy of violence.

Deterrence and Strategy (New York: Praeger, 1966) is a superbly constructed work by Andre Beaufre which provides a solid foundation for embarking on an analysis of deterrence. Dedicated to Liddell Hart, a strategic thinker of a different era, Beaufre's work stressed two major themes: "The Laws of Deterrence" and "The Consequences." "The object of deterrence is to prevent an enemy power from making the decision to use armed force." The psychological result, said Beaufre, is the product of the combined effect of a calculation of the risk incurred compared to the issue at stake and of the fear engendered by the risks and uncertainties of the conflict. Having built a general framework, Beaufre examined the concept from several perspectives, distinguishing, for example, defensive deterrence (preventing the enemy from initiating an action) from offensive deterrence (preventing the enemy from resisting some action).

Philip Green captured the outlook of the second-wave theorists in *Deadly Logic: The Theory of Nuclear Deterrence* (Columbus: Ohio State University Press, 1966). Especially worthwhile for the undergraduate student is his brief introductory chapter, "This Idea of Nuclear Deterrence," in which Green reviewed "a coherent body of doctrine—a theory of war and peace in the nuclear age." Drawing on Schelling and Snyder, Green employed the process of definition by exclusion: deterrence theorists are those who reject both a disarmament strategy and a win-the-war nuclear strategy. He stated five requisite conditions of nuclear deterrence: (1) some portion of one's own nuclear striking force must be capable of surviving an enemy's first strike (generally referred to as "invulnerability"); (2) the survivable force must be capable of retaliating with a certain effect; (3) the retaliatory force "must not be trigger happy"; (4) mutual vulnerability (much like Schelling's "exchange of hostages"); and (5) a doctrine for response to limited provocations. Green's *Deadly Logic* also includes three valuable chapters which underscore the core ideas of second-wave thinkers: "Systems Analysis and National Policy"; "Game Theory and Deterrence"; and "Deterrence Rationality and Ethical Choice." His concluding piece, "Social Science, Nuclear Deterrence, and the Democratic Process," provides his overall perspective.

DIVERGENCE AND DEBATE

While strategic thinkers of the first and second waves shared some common ground and were noticeably similar in their basic outlook on deterrence, the third wave is marked by significant differences of opinion. The more notable debates centered on the merits of flexible response, the distinction between assured destruction and "warfighting," and the potential consequences of a first-use option, although broader critiques contributed to the field.

Deterrence in American Foreign Policy (New York: Columbia University Press, 1974) by Alexander George and Richard Smoke is a broad critique of deterrence theory that focuses on the effort of the United States to deter limited conflicts. Part 1 of this considerable work addresses

the nature of contemporary deterrence and offers a critical assessment: "it has been markedly less useful to policy-makers than it might have been." Because the authors believe that many of the premises on which deterrence theory has been based are "oversimplified" and "often erroneous," Part 2 presents 11 case histories in which the United States applied (or considered applying) a deterrence strategy. As such, the theoretical assumptions are compared with the complex variables and processes of "real-life" cases. Given their belief that systematic study of the historical experience can assist in developing a better theory of deterrence, George and Smoke began a reformation of deterrence theory in Part 3, deriving conclusions from the empirical material in Part 2. Emphasis is given to the development of a differentiated theory, one that discriminates among varieties and patterns of potential deterrent situations.

Deterrence is addressed in the broader context of global relations in *Power and Community in World Politics* (San Francisco: W. H. Freeman, 1974), by Bruce M. Russett. Part 2, "International Violence: Deterrence and Restraint," integrates within the book's framework both conventional and nuclear deterrence. While the first two chapters of Part 2 discuss relevant issues ("Cause, Surprise, and No Escape" and "The Complexities of Ballistic Missile Defense"), it is Russett's "The Calculus of Deterrence" which is the most valuable. In it he traces theoreticians' answers to the persistent question of how to defend "third areas." Herman Kahn, for example, is seen as placing emphasis on the overall strategic balance as the determining factor in defending the "periphery," while Thomas Schelling focuses on credibility and notes that it can be enhanced by increasing the defender's potential loss.

Though the in-depth assessments by George, Smoke, and Russett show the diversity of the third-wave thinkers, the controversial nature of deterrent strategy is captured in the journals noted earlier. Paul Nitze's "Deterring Our Deterrent" (*Foreign Policy*, no. 25 [1976–77]) raises Henry Kissinger's earlier view that any war between the US and USSR would be nuclear and would inevitably result in millions dead. Such a view implies that it makes no difference whether (1) the USSR has more or bigger offensive warheads; (2) the USSR improves technologically; (3) Soviet defenses are active or passive; or (4) either side strikes first. "No more serious question faces us than whether these propositions are true or false," wrote Nitze. He then describes the Clausewitzean nature of Soviet doctrine by first paraphrasing an issue of *Communist of the Armed Forces*, "a war involving nuclear missiles should and can be an extension of policy," and then asserting that the

Soviets are attempting to meet all the force requirements necessary to make war an extension of policy.

Political Science Quarterly contained a robust article by Robert Jervis which argued that flexible response is costly, ineffective, and dangerous. In "Why Nuclear Superiority Doesn't Matter" (*Political Science Quarterly* 94, no. 4 [1979–80]), Jervis defended the policy of assured destruction, noting that the vulnerability of population centers has transformed strategy. Flexible response, he contended, "misunderstands the nature of nuclear deterrence." Thus, a military advantage no longer assures a decisive victory, and traditional views of warfighting are no longer appropriate. In this sense, Jervis directly challenges the logic of flexible response, at least as it applies to the nuclear level.

Related to the assured destruction–flexible response debate, the third wave of deterrent theorists also included skirmishes on "warfighting." In what may be considered a response to the article by Jervis, Colin Gray wrote "Nuclear Strategy: The Case for a Theory of Victory" (*International Security* 4, no. 1 [1979]). Gray's article bluntly attacks the acute deficiency of strategic thinking inherent in assured destruction: "Indeed, MAD is the antithesis of strategy." Assured destruction asks military power, he noted, to punish a society for the sins or misjudgments of its rulers. Instead, Gray called for a deterrent doctrine which threatens political defeat of the Soviet Union—"the forcible demise of the Soviet state." He questioned the utility of the nuclear doctrine espoused by Bernard Brodie and noted "the world, perhaps fortunately, is not ruled by strategic rationalists."

In response to Gray and others, Michael E. Howard's "On Fighting a Nuclear War," (*International Security* 5, no. 4 [1981]) opens in defense of Bernard Brodie's view. Quoting from Brodie's *Absolute Weapon*, he related the view that the chief purpose of the American security program was (and is) to *avert* wars. While potentially usable on a limited scale in the European theater, nuclear weapons, Howard asserted, have no utility as instruments of policy or strategic tools in a general war. Howard clearly sides with Brodie on the fundamental issue of nuclear warfighting and notes that criticism of the deterrent posture comes from two linked sources: the continuous inventiveness of the scientific community (which "has made the pursuit of a stable nuclear balance of mutually assured deterrence seem to be the chase for . . . a will o' the wisp") and the widespread belief that the Soviet Union does not share in holding the original concept of deterrence. He noted the Clausewitzian framework of Soviet doctrine and submitted that it would be best not to imitate it, but rather to convey that it

will not work. Placing both Gray and Nitze as members of the "maximalist school," Howard provided a challenging response. The subsequent volume (*International Security* 6, no. 1, [1981]) included a lively exchange of correspondence between Gray and Howard on this issue.

Colin Gray's article cited above distinguishes between a "first school"—essentially advocates of deterrence through the threat of assured destruction—and a "second school"—those who prefer, as he does, deterrence through a warfighting posture which threatens victory in case of aggression. An opposing preference, using the same dichotomy, comes from Spurgeon M. Keeny, Jr., and Wolfgang K. H. Panofsky in their "MAD Versus NUTS: Can Doctrine or Weaponry Remedy the Mutual Hostage Relationship of the Superpowers?" (*Foreign Affairs* 60, no. 2 [1981–82]). The authors contrast mutual assured destruction with the tongue-in-cheek doctrine of "Nuclear Utilization Target Selection," the latter corresponding to the warfighting school. The authors' preference for MAD is less a prescription for a doctrine which one can "choose" from other alternatives, but more an acknowledgment that societal vulnerability is inescapable in the nuclear age. The superpowers exist in a "mutual hostage relationship" which, the authors argue, cannot be overcome either by doctrine or weaponry, either offensive or defensive in nature. The argument by advocates of a warfighting strategy— that the USSR does not accept mutual vulnerability as desirable, so neither should the US— is countered by stressing that vulnerability is a fact of life which has no prospect of changing, regardless of the doctrinal desires of any nuclear power. The authors emphasize, moreover, that the pursuit of a NUTS strategy is not only futile but also destabilizing to the overall relationship. NUTS assumes that the use of nuclear weapons can be both limited and controlled, a premise which the authors assert is dangerously illusory.

Foreign Affairs was the forum for a debate over the "first-use" doctrine of the NATO Alliance. McGeorge Bundy, George F. Kennan, Robert S. McNamara, and Gerard Smith produced a controversial article (*Foreign Affairs* 60, no. 4 [1982]) entitled "Nuclear Weapons and the Atlantic Alliance." In it, they asserted that the willingness of the United States to use nuclear weapons first to defend against aggression in Europe is weakening the coherence of the Alliance and threatening the safety of the world while its deterrent credibility declines. Any use of nuclear weapons in Europe, they claimed, "carries with it a high and inescapable risk of escalation into general nuclear war which would bring ruin to all and victory to none." They advocated development of NATO conventional forces capable of defending against conventional attack and a corresponding adoption of a policy of no-first-use of nuclear weapons. McNamara's later article, "The Military Use of Nuclear Weapons: Perceptions and Misperceptions" (*Foreign Affairs* 62, no. 1 [1983]), continues the same argument.

The remarks by Bundy, Kennan, McNamara, and Smith gave rise to a vigorous debate on "first-use" and prompted a rebuttal by four West German strategists, Karl Kaiser, George Leber, Alois Mertes, and Franz-Josef Schulze. "Nuclear Weapons and the Preservation of Peace: A Response to an American Proposal for Renouncing the First Use of Nuclear Weapons" (*Foreign Affairs* 60, no. 5 [1982]) touched on a number of deterrent issues, including the distinction between "first-use" (the first use of a nuclear weapon regardless of its yield and place) and "first strike" (a preemptive disarming nuclear strike aimed at eliminating as completely as possible the entire strategic potential of the adversary). Before addressing the "profoundly disturbing" conclusions and "skeptical arguments" forwarded by the four Americans, the four Germans outlined the NATO strategy of flexible response and the pillars on which the doctrine rests. Their rebuttal rests fundamentally on the political necessity of having a demonstrable linkage between the US nuclear arsenal and the defense of Western Europe, rather than a military argument for "warfighting." Any alternative foundation for NATO would suggest to Moscow that the USSR would be a sanctuary in war: "Wherever nuclear weapons are present, war loses its earlier function as a continuation of politics by other means . . . the coupling of conventional and nuclear weapons has rendered war between East and West unwageable and unwinnable up to now."

While acknowledging "that the present strategic system is dangerous and that nuclear deterrence is peculiarly so," Laurence Martin does not find any of the "radical alternatives plausible in terms of either strategic efficacy or political attainability." Thus, Martin delivers a series of six 30-minute lectures on the "centerpiece of the contemporary strategic scene": nuclear deterrence. In *The Two-Edged Sword* (New York: W. W. Norton, 1982) he outlined warfighting strategies by analyzing John Newhouse's statement that "killing people is good; killing weapons is bad." Noting that *threatening* to kill people is the essence of deterrence because the idea of mutual mass killing is "unthinkable," Martin submitted that attacking weapons is not unthinkable since doing so would not be "immediately and utterly catastrophic." His initial analysis of assured destruction is also painted in fundamental terms; "city busting," the classical notion of deterrence, is seen as having certain advantages: there are a finite number of cities, these cities are not mobile, and there is no need

to multiply and refine existing nuclear weapons. Thus, Martin concluded that there is no incentive to launch first, or engage in an arms race. His lecture series is well balanced and covers the European balance, the Third World, and arms control.

The final third-wave theorist to be noted is Louis Rene Benes, author of *Mimicking Sisyphus: America's Countervailing Nuclear Strategy* (Lexington, Mass.: D. C. Heath, 1983). This highly critical analysis of the Reagan administration's strategy challenges the feasibility of limited nuclear war and the likelihood of national survival. Benes also addresses the "curious assumption" that the USSR is more likely to be deterred by the threat of limited American counterforce reprisals than by the threat of overwhelming nuclear retaliation. American nuclear strategy, he concluded, has become an incremental policy of strategic response. In developing a series of cogent responses to the "five major assumptions of the current American strategy," Benes highlighted several of the tradeoffs involved in deterrence. The provocative final chapter underscores elements of "a rapid and far-reaching disengagement" from the trends of counterforce targeting and nuclear warfighting. In doing so, Benes comments on issues such as the Comprehensive Test Ban, a no-first-use policy, the nuclear-freeze movement, and the creation of nuclear-free zones. All of these are issues very much at the heart of the fourth wave.

FUTURE APPROACHES

The fundamental orientation of deterrence, grounded in the first wave of thinkers, refined in the second, and debated in the third, is profoundly challenged by the strategic theorists of the fourth wave. While most of the fourth wave thinkers mentioned here have come to prominence in the 1980s, Fred C. Iklé inaugurated a "right wing revisionism" (in the words of Leon Wieseltier). In Ikle's "Can Nuclear Deterrence Last Out the Century?" (*Foreign Affairs* 51, no. 2 [1973]), he claims that deterrence is really "mutually assured genocide." "There exists no rational basis for deterrence," he writes, "because those calculated decisions which our deterrent seeks to prevent are not the sole processes that could lead to nuclear war . . ." Indeed, he submits that deterrence makes survival dependent on the rationality of all future leaders in all major wars. This early disenchantment with mutual vulnerability as the basis for the deterrence later gave rise to calls for a fundamental transformation epitomized by the Strategic Defense Initiative.

Nuclear Weapons and World Politics: Alternatives for the Future (New York: McGraw-Hill, 1977) offered four nuclear regimes for the future. Written by David C. Gompert, Michael Mandelbaum, Richard L. Garwin, and John H. Barton, it is based on the "widely held recognition" that many of the assumptions, policies, and institutions that have characterized international relations during the past 30 years are inadequate to the demands of today and the foreseeable demands into the 1990s. The first regime, presented by Mandelbaum, is essentially the current regime projected into the future and rests on the premise that nuclear weapons have fostered, if not forced, moderation and stability in international policies. The second regime, "an ensemble of arms control prescriptions," in Garwin's words, is derived from the belief that nuclear weapons are an inescapable burden and that our efforts should be devoted to reducing dependence on them in the conduct of world politics and to maintaining international security through unilateral and bilateral measures. John H. Barton's denuclearized regime, the third, develops the concept of proscribing nuclear weapons. It sees nuclear weapons as an intolerable menace and therefore seeks to ban them. The possibility of "collapse and calamity" and the belief that "it is morally corrosive for peace and stable world politics to depend in perpetuity upon the capacity and expressed willingness of leaders to destroy one another's societies" demand the creation of "workable and enforceable arrangements for the abolition of nationally held nuclear weapons." Finally, the fourth regime, "strategic deterioration," anticipates a number of plausible developments in technology and politics over the next decade that could undermine strategic stability. An excellent appendix by Franklin C. Miller completes this work, along with a glossary that defines terms admirably.

The Future of Strategic Deterrence (Hamden, Conn.: Archon Books, 1981), edited by Christopher Bertram, is a collection of thoughtful papers presented at the International Institute for Strategic Studies. It is based on premises similar to those of *Nuclear Weapons and World Politics*: technological developments and changes in the political relations of the nuclear powers have made it difficult to reach an agreement restraining nuclear competition and have undermined most of the conditions for stable mutual deterrence. The varying perspectives on the future of deterrence include papers by exceptional theorists. McGeorge Bundy, for example, addresses the fundamental requirements for strategic forces (they must be able to inflict "totally unacceptable retaliatory damage" even after the strongest foreseeable first strike by the adversary). His paper includes a valuable section on "extended deterrence" in which he asserts that, while deployment of nuclear weapons in Europe is not necessary, the "American strategic guarantee" is critical. Hedley Bull's

"Future Conditions of Strategic Guarantee" is also superb. After noting that "deterrence has been the leading strategic idea of our times," Bull proceeds to outline the limitations of the doctrine of peace through mutual nuclear deterrence. Bull usefully distinguished among strategic, general, and extended forms of deterrence. A third paper worth noting is Edward Luttwak's "The Problem of Extending Deterrence," which is more narrow in focus than the other two. It addresses both the theory and application of extended deterrence and deterrence by counterforce.

E. P. Thompson's *Beyond the Cold War* (New York: Pantheon Books, 1982) advocates a policy much like Barton's denuclearized regime. In calling for the "zero option" of a nuclear-free Europe, Thompson disdains deterrence theory—"It is in truth a pitiful light-weight theory"—and writes of "Deterrence and Addiction" and "Exterminism." His work shows the emotion inherent in the issue and displays an unorthodox logic which is useful in gaining a sense of the broad spectrum of writers of the '80s.

A book which addresses deterrence in refreshing and understandable terms (in contrast to E. P. Thompson's claim that "It is espoused, in its pristine purity, only by a handful of monkish celibates, retired within the walls of strategic studies") is *Nuclear War, Nuclear Peace* (New York: Holt, Rinehart and Winston, 1983), by Leon Wieseltier. "Only a citizen who writes," Wieseltier provides well-balanced and readable critiques of both the "party of peace" and the "party of war." Noting that "the great nuclear debate has consisted mainly in the thrashing of deterrence," Wieseltier is not given to one extreme or the other. His piece is a superb attempt to validate the concept of deterrence and to place it in a contemporary context.

An interesting contribution to the nuclear debate, and one which underscores the spectrum of fourth-wave writers, is "Nuclear War and Climatic Catastrophe: Some Policy Implications" by Carl Sagan (*Foreign Affairs* 62, no. 2 [1983–84]). After predicting the climatic and biological consequences of nuclear war, Sagan explores the possible strategic policy implications of such a "nuclear winter." These implications, writes Sagan, "point to one apparently inescapable conclusion: the necessity of moving as rapidly as possible to reduce global nuclear arsenals below levels that could conceivably cause the kind of climatic catastrophe and cascading biological devastation predicted by these studies." In examining a spectrum of nuclear exchanges ranging from 0.8 to 75 percent of the world's strategic arsenals, Sagan's prognosis for the resultant effects is dismal and includes a litany of blights ranging from epidemics and psychiatric disorders to disruption of the global ecosystem.

His implications are as startling: in the event of a first strike, prevailing winds would carry the "nuclear winter" to the *aggressor*. In addition to calling for a ban on all warheads greater than 300 to 400 kilotons, Sagan notes the need for treaties on targeting. The last three of his eight prescriptions call for a nuclear freeze, "build-down" arms control schemes, and deep cuts in the strategic arsenals of the world.

Although the outlook and background of fourth-wave writers are considerably broader than those of earlier eras, as illustrated by the writings of Thompson and Sagan, a central debate has emerged which has focused the attention of the strategic thinkers of the 1980s. The debate over the potential transformation of deterrence from offense to defense is at least as significant as "the great debate" of the late 1950s and early 1960s. *Foreign Affairs* has been a principal forum for this debate, and a review of recent exchanges captures the essence of the debate and provides the context for contemporary readers.

Under the journal's heading "Star Wars Debate," Keith Payne and Colin Gray develop a controversial argument in favor of President Reagan's call for a "potentially radical departure in US strategic policy," noting that the president deemed a policy of deterrence through the threat of strategic nuclear retaliation "inadequate." In "Nuclear Policy and the Defensive Transition" (*Foreign Affairs* 62, no. 4 [1984]: 820–42), Payne and Gray expand on this inadequacy, focusing on the basis for the current offensive concept of deterrence: mutual vulnerability.

After reviewing the frenzied intergovernmental activity sparked by the mandate to examine those technologies that could eliminate a ballistic missile threat, they move to the pivotal issue: the likely effect of strategic defense on stability. In concluding that a transition to strategic defense "could reduce both the probability and the consequences of nuclear war," the authors fire an explicit volley in this great debate of the fourth wave.

A distinctly different conclusion is reached by William Burrows in the same issue, however. His "Ballistic Missile Defense: The Illusion of Security" provides a scathing critique of the ballistic missile defense concept. Labeling the Strategic Defense Initiative "a dangerous hoax and a cruel and potentially expensive exercise in self-deception," he submits that offense and defense "have forevermore become indistinguishable." Burrows then reviews the history of missile defense systems and emphasizes the formidable technical challenges faced by the designers of the proposed system. His argument is well organized and draws frequently on historical examples, especially from the Nike-Zeus

system of the Kennedy era, to underscore the likelihood of a "leak" in any defensive umbrella.

Robert Tucker joins the fray in "The Nuclear Debate," an article in the next volume (*Foreign Affairs* 63, no. 1 [1984]). He approaches the issue along theoretical lines, noting that "to the deterrent faithful," deterrence is not only an "inherent property," but a "self-sufficient property." He interprets the controversy over "the necessary and sufficient conditions of deterrence" as a debate over politics and perceptions, not weapons. Thus, Tucker concludes, the real indictment of the Reagan administration is not of its military strategy but of its politics, which have resulted in a lapse of faith in deterrence. The logical implication, for Tucker, is a renewal of the spirit of détente as a means of restoring faith in deterrence, rather than the traditional options of regaining strategic superiority, or a policy of withdrawal.

The "four Americans" of the no-first-use debate clearly reinforce the arguments of both William Burrows and Robert Tucker. In "The President's Choice: Star Wars or Arms Control" (*Foreign Affairs* 63, no. 2 [1984–85]) McGeorge Bundy, George F. Kennan, Robert S. McNamara, and Gerard Smith reiterate the claim that strategic defense cannot be "100 percent leakproof": "What is centrally wrong with the President's objective is that it cannot be achieved." They then enumerate the consequences of pursuing such a strategy, noting the three causes of instability which would result from "Star Wars." First, it would destroy the Anti-Ballistic Missile (ABM) Treaty ("the most important arms control agreement" in effect). Secondly, it would directly stimulate both offensive and defensive systems of the USSR. And thirdly, it would sharpen the anxieties of the superpower relationship, a point with which Robert Tucker would concur.

In the next volume familiar writers show the degree to which strategic thinkers are divided. Fred Iklé and Leon Wieseltier present starkly opposing perspectives in Iklé's "Nuclear Strategy: Can There Be A Happy Ending?" and Wieseltier's gloomily titled "When Deterrence Fails" (*Foreign Affairs* 64, no. 4, [1985]: 810–47).

Ikle, like Payne and Gray, castigates "the theory" which equates mutual vulnerability with strategic stability. He discredits the oversimplification of issues and states that the stable equilibrium of "two sides" has never existed. Implying that foreign policy and strategic doctrine suffer from "mirror imagery," Ikle emphasizes their dynamic and complex nature and urges that "we disenthrall ourselves of the dogma of consensual vulnerability." In calling for a transformation, Ikle concludes with an interesting and upbeat analogy, likening the potential technology of the SDI with the "new" economic factors which negated the iron law of Thomas Malthus and rejuvenated the "dismal science" of economics.

As the title of Leon Wieseltier's article implies, he is concerned with the essential fragility of nuclear deterrence, flatly stating, "Failure is a plain possibility." While only briefly mentioning Ronald Reagan's defensive "dream," he faults Kenneth Waltz and the school of liberal strategists who plead for nuclear proliferation. In the terse style of his earlier work, Wieseltier counters, "Nuclear weapons are not there to create deterrence. Deterrence is there to cope with nuclear weapons." He then turns to the main thread of his argument, tracing "the conventional fallacy" from Curtis LeMay through Schlesinger's "limited nuclear options," Harold Brown's "countervailing strategy" and Caspar Weinberger's "prevailing strategy," all of which represent the "war fighting or classical strategy approach to deterrence." Instead, Wieseltier presents war termination as the proper objective of nuclear strategy, a form of "intra-war" deterrence.

CONCLUSION: DETERRENCE IN PERSPECTIVE

With 40 years, four broad waves of strategic thinkers, and numerous intellectual debates and skirmishes as a backdrop, some of the better surveys presently available warrant review. These are often broad enough in scope to draw on many of the classic works cited, yet are current enough to incorporate the monumental debate of the fourth wave.

The Evolution of Nuclear Strategy (New York: St. Martin's Press, 1981) by Lawrence Freedman is just such a book. Well organized and highly readable, Freedman's work spans the nuclear age and more, citing H. G. Wells's prophetic *The War in the Air*, which spoke of "airborne atomic warfare" in 1908.

As impressive as the scope of his work is Freedman's insight into the cyclical nature of debates about strategy. Candidly pointing out that "evolution is somewhat misleading" when used in the title, Freedman emphasizes that "much of what is offered today as a profound and new insight was said yesterday, and usually in a more concise and literate manner." That observation has two important results. First, Freedman carefully reviews "what was said yesterday" and thus provides a rich and thoughtful perspective on deterrence which is extremely well documented. Secondly, in a larger sense, the cyclical nature of strategic debates puts the contemporary debate in context. Thus, Freedman's section on "The Possibility of Defense," which quotes President Truman's confidence

that "every new weapon will eventually bring some counter defense to it," sheds some historical light on the SDI controversy.

Michael Mandelbaum has also been a prolific contributor to the field, and his trilogy of strategic works frames contemporary issues. *The Nuclear Question: The United States and Nuclear Weapons, 1946–1976* (Cambridge: Cambridge University Press, 1979), while a history of American nuclear weapons policy, is more than a mere chronology. Mandelbaum interprets the development of American policy "in light of the history of international politics." It is a broad book and provides a suitable foundation for his second, *The Nuclear Revolution: International Politics Before and After Hiroshima* (Cambridge: Cambridge University Press, 1981), which dramatically captures the profound impact of the use of atomic weapons in World War II on the process and substance of international relations. *The Nuclear Future* (Ithaca: Cornell University Press, 1983) draws on the historical framework of his first two works while providing a future orientation.

Having assimilated the material of his earlier works, Mandelbaum writes the third more lucidly. He attempts to diffuse the emotional aspects of the issue by noting that the "nuclear priesthood" recognizes the bomb as an "incurable but not fatal" disease, and thus views these weapons with the "dispassion of the doctor, not the anxiety and horror of the patient." Together, Mandelbaum's trilogy offers different approaches (and styles) to the nuclear debate and some unstartling predictions: the nuclear future will follow a middle path between nuclear war and nuclear disarmament.

Living with Nuclear Weapons (New York: Bantam Books, 1983) is a highly readable work which asks—and answers—important questions. Written by the Harvard Nuclear Study Group (Albert Carnesale, Paul Doty, Stanley Hoffman, Samuel P. Huntington, Joseph S. Nye, Jr., and Scott D. Sagan), it is based on the premise that "nuclear weapons are too important to be left just to classrooms and specialized journals." As such, it introduces complex facets in understandable terms, and provides frequent and well-chosen examples. It is also well organized. The book first discusses the nature of the nuclear predicament, provides a general overview of the current nuclear inventory, and surveys the terrain of contemporary strategic issues, before asking "What Can Be Done?" After sobering the reader by likening our current predicament to that of the Greek mythological character Sisyphus (recall Louis Benes's book) the authors provide a comforting assessment that is much like Mandelbaum's: "the future is not limited to a choice between nuclear holocaust and universal disarmament." The ultimate conclusion of this "populist" book is an exhortation to American citizens: "Avoid atomic escapism . . . and honestly address the true dilemmas of the Nuclear Age," they urge, for "living with nuclear weapons is our only hope."

A work which squarely addresses those dilemmas is *The Nuclear Reader: Strategy, Weapons, War* (New York: St. Martin's Press, 1985), edited by Charles W. Kegley, Jr., and Eugene R. Wittkopf. It is designed to expose the range of "opinion and prescription" relating to "the maintenance of stable deterrence, the utility of nuclear weapons as instruments of foreign policy, the means of limiting their quantity (and perhaps abolishing them altogether), and to the horrors that might befall humanity should nuclear deterrence fail." Without doubt, the articles selected by Kegley and Wittkopf successfully accomplish what *The Nuclear Reader* was designed to, and taken as a whole yield one of the finest works yet published on deterrence in the nuclear age.

Each of the three parts, "Strategy, Weapons, and War," thoroughly addresses pivotal issues. In all, 24 interrelated themes are presented: from the (im)morality of mutual assured destruction, to the (ir)revelance of a nuclear freeze, and the practical problems of civil defense. The authors are as diverse as the topics and include the National Conference of Catholic Bishops, the Union of Concerned Scientists, Jonathan Schell, and Theodore Draper, as well as many whose works have been cited earlier: Keith Payne and Colin Gray, the Harvard Nuclear Study Group, and Carl Sagan. In addition to a useful glossary, "Nuclear Nomenclature: A Selective Dictionary of Terms," the book opens with a thoughtful introductory essay by the editors which develops the three organizing concepts and illustrates how interconnected they are. With essentially the same conclusion as Mandelbaum and the Harvard Study Group,

that a nuclear-free world is not probable, Kegley and Wittkopf declare that the purpose of *The Nuclear Reader* "is thus to describe this reality of structural terrorism" and help the reader understand the developments of the past four decades which have led to it.

Finally, a book that lacks breadth and rigor, but warrants mention nonetheless, is *Nuclear Strategy and Common Sense* (Moscow: Progress Publishers, 1981), ostensibly written by Nikolai Luzin, a Soviet. The English translation of Luzin's book underscores two points which help put 40 years of deterrence literature in perspective. The first is that the literature reviewed is almost exclusively American and is completely Western in orientation. The diligent reader will thus be left with a distinctly American outlook on strategic deterrence. This is not necessarily a liability: nuclear deterrence is "mostly" an American invention. Thus, Luzin's book, and the more systematic and widely read work by Marshal V. D. Sokolovski, *Soviet Military Strategy* (London: Prentice Hall, 1963, translation), rarely use the term "deterrence" and devote no chapter or section of their works to it. This is not to say, of course, that the Soviets are not deterred, or are unaware of deterrence theory. Luzin cites Brodie, Kahn, and Kissinger frequently—and out of context. That is the second point. Although an extensive review of American literature encompasses all facets of deterrent theory, the concept *is* studied on the other side of the Atlantic, and on the other side of Churchill's "iron curtain." With that in mind, four broad waves of deterrent theory as a backdrop, and the current debate as a focal point, it is perhaps fitting to recall Churchill's assessment of it all: "It may be that we shall, by a process of sublime irony, have reached a stage in history where safety will be the sturdy child of terror, and survival the twin brother of annihilation."

SELECTED READINGS I

THE FORGOTTEN DIMENSIONS OF STRATEGY
MICHAEL E. HOWARD

I

The term "strategy" needs continual definition. For most people, Clausewitz's formulation "the use of engagements for the object of the war," or, as Liddell Hart paraphrased it, "the art of distributing and applying military means to fulfill the ends of policy," is clear enough. Strategy concerns the deployment and use of armed forces to attain a given political objective. Histories of strategy, including Liddell Hart's own *Strategy of Indirect Approach*, usually consist of case studies, from Alexander the Great to MacArthur, of the way in which this was done. Nevertheless, the experience of the past century has shown this approach to be inadequate to the point of triviality. In the West the concept of "grand strategy" was introduced to cover those industrial, financial, demographic, and societal aspects of war that have become so salient in the twentieth century; in communist states all strategic thought has to be validated by the holistic doctrines of Marxism-Leninism. Without discarding such established concepts, I shall offer here a somewhat different and perhaps slightly simpler framework for analysis, based on a study of the way in which both strategic doctrine and warfare itself have developed over the past 200 years. I shall also say something about the implications of this mode of analysis for the present strategic posture of the West.

II

Clausewitz's definition of strategy was deliberately and defiantly simplistic. It swept away virtually everything that had been written about war (which was a very great deal) over the previous 300 years. Earlier writers had concerned themselves almost exclusively with the enormous problems of raising, arming, equipping, moving, and maintaining armed forces in the field—an approach which Clausewitz dismissed as being as relevant to fighting as the skills of the swordmaker were to the art of fencing.

None of this, he insisted, was significant for the actual conduct of war, and the inability of all previous writers to formulate an adequate theory had been due to their failure to distinguish between the *maintenance* of armed forces and their *use*.

By making this distinction between what I shall term the *logistical* and the *operational* dimensions in warfare, Clausewitz performed a major service to strategic thinking; but the conclusions he drew from that distinction were questionable, and the consequences of those conclusions have been unfortunate. In the first place, even in his own day, the commanders he so much admired—Napoleon, Frederick the Great—could never have achieved their operational triumphs if they had not had a profound understanding of the whole range of military activities that Clausewitz excluded from consideration. In the second place, no campaign can be understood, and no valid conclusions drawn from it, unless its logistical problems are studied as thoroughly as the course of operations; and as Dr. Martin van Creveld has recently pointed out in his book *Supplying War*, logistical factors have been ignored by 99 military historians out of 100—an omission which has warped their judgments and made their conclusions in many cases wildly misleading.

Clausewitz's dogmatic assertion of priorities—his subordination of the logistical element in war to the operational—may have owed something to a prejudice common to all fighting soldiers in all eras. It certainly owed much to his reaction against the super-cautious "scientific" generals whose operational ineptitude had led Prussia to defeat in 1806. But it cannot be denied that in the Napoleonic era it *was* operational skill rather than sound logistical planning that proved decisive in campaign after campaign. And since Napoleon's campaigns provided the basis for all strategic writings and thinking throughout the nineteenth century, "strategy" became generally equated in the public mind with *operational* strategy.

Reprinted with minor revisions and by permission from Foreign Affairs, Summer 1979. Copyright © 1979 by the Council on Foreign Relations, Inc. Originally titled "The Forgotten Dimensions of Strategy."

But the inadequacy of this concept was made very clear, to those who studied it, by the course of the American Civil War. There the masters of operational strategy were to be found, not in the victorious armies of the North, but among the leaders of the South. Lee and Jackson handled their forces with a flexibility and an imaginativeness worthy of a Napoleon or a Frederick; nevertheless they lost. Their defeat was attributed by Liddell Hart, whose analyses seldom extended beyond the operational plane, primarily to operational factors, in particular, to the "indirect approach" adopted by Sherman. But fundamentally, the victory of the North was due not to the operational capabilities of its generals, but to its capacity to mobilize its superior industrial strength and manpower into armies which such leaders as Grant were able, thanks largely to road and river transport, to deploy in such strength that the operational skills of their adversaries were rendered almost irrelevant. Ultimately the latter were ground down in a conflict of attrition in which the *logistical* dimension of strategy proved more significant than the operational. What proved to be of the greatest importance was the capacity to bring the largest and best-equipped forces into the operational theater and to maintain them there. It was an experience that has shaped the strategic doctrine of the US armed forces from that day to this.

But this capacity depended upon a third dimension of strategy, and one to which Clausewitz was the first major thinker to draw attention: the *social*, the attitude of the people upon whose commitment and readiness for self-denial this logistical power ultimately depended. Clausewitz had described war as "a remarkable trinity," composed of its political objective, of its operational instruments, and of the popular passions, the social forces it expressed. It was the latter, he pointed out, that made the wars of the French Revolution so different in kind from those of Frederick the Great, and which would probably so distinguish any wars in the future. In this he was right.

With the end of the age of absolutism, limited wars of pure policy fought by dispassionate professionals became increasingly rare. Growing popular participation in government meant popular involvement in war, and so did the increasing size of the armed forces which nineteenth-century technology was making possible and therefore necessary. Management of, or compliance with, public opinion became an essential element in the conduct of war. Had the population of the North been as indifferent to the outcome of the Civil War as the leaders of the Confederacy had initially hoped, the operational victories of the South in the early years might have decisively tipped the scales. The logistical potential of the North would have

been of negligible value without the determination to use it. But given equal resolution on both sides, the capacity of the North to mobilize superior forces ultimately became the decisive factor in the struggle. Again Clausewitz was proved right: *all other factors being equal*, numbers ultimately proved decisive.

III

In one respect, in particular, other factors were equal. The Civil War was fought with comparable if not identical weapons on both sides, as had been the revolutionary wars in Europe. The possibility of decisive *technological* superiority on one side or the other was so inconceivable that Clausewitz and his contemporaries had discounted it. But within a year of the conclusion of the American Civil War, just such a superiority made itself apparent in the realm of small arms, when the Prussian armies equipped with breech-loading rifles defeated Austrian armies which were not so equipped. Four years later, in 1870, the Prussians revealed an even more crushing superiority over their French adversaries thanks to their steel breech-loading artillery. This superiority was far from decisive: the Franco-Prussian War in particular was won, like the American Civil War, by superior logistical capability based upon a firm popular commitment. But technology, as an independent and significant dimension, could no longer be left out of account.

In naval warfare, the crucial importance of technological parity had been apparent since the dawn of the age of steam, and in colonial warfare the technological element was to prove quite decisive. During the latter part of the nineteenth century, the superiority of European weapons turned what had previously been a marginal technological advantage over indigenous forces, often counterbalanced by numerical inferiority, into a crushing military ascendancy, which made it possible for European forces to establish a new imperial dominance throughout the world over cultures incapable of responding in kind. As Hilaire Belloc's Captain Blood succinctly put it: "Whatever happens, we have got The Maxim gun, and they have not." Military planners have been terrified of being caught without the contemporary equivalent of the Maxim gun from that day to this.

So by the beginning of this century, war was conducted in these four dimensions: the *operational*, the *logistical*, the *social*, and the *technological*. No successful strategy could be formulated that did not take account of them all, but under different circumstances, one or another of these dimensions might dominate. When, in 1914–15, the operational strategy of the Schlieffen Plan, for the one side, and of the

Gallipoli campaign, for the other, failed to achieve the decisive results expected of them, then the logistical aspects of the war, and with them the social basis on which they depended, assumed even greater importance as the opposing armies tried to bleed each other to death. As in the American Civil War, victory was to go, not to the side with the most skillful generals and the most courageous troops, but to that which could mobilize the greatest mass of manpower and firepower and sustain it with the strongest popular support.

The inadequacy of mere numbers without social cohesion behind them was demonstrated by the collapse of the Russian empire in 1917. But the vulnerability even of logistical and social power if the adversary could secure a decisive technological advantage was equally demonstrated by the success of the German submarine campaign in the spring of 1917, when the Allies came within measurable distance of defeat. The German empire decided to gamble on a technological advantage to counter the logistical superiority which American participation gave to their enemies. But they lost.

IV

From the experiences of the First World War, different strategic thinkers derived different strategic lessons. In Western Europe, the most adventurous theorists considered that the technological dimension of war would predominate in the future. The protagonists of armored warfare in particular believed that it might restore an operational decisiveness unknown since the days of Napoleon himself—the first two years of the Second World War were to prove them right. Skillfully led and well-trained armed forces operating against opponents who were both militarily and morally incapable of resisting them achieved spectacular results.

But another school of thinkers who placed their faith in technology fared less well; this school included those who believed that the development of air power would enable them to eliminate the operational dimension altogether and to strike directly at the roots of the enemy's *social* strength, at the will and capacity of the opposing society to carry on the war. Instead of wearing down the morale of the enemy civilians through the attrition of surface operations, air power, its protagonists believed, would be able to attack and pulverize it directly.

The events of the war were to disprove this theory. Technology was not yet sufficiently advanced to be able to eliminate the traditional requirements of operational and logistical strategy in this manner. Neither the morale of the British nor that of the German people was to be destroyed by air attack; indeed, such attack was found to demand an operational strategy of

a new and complex kind in order to defeat the opposing air forces and to destroy their logistical support. But operational success in air warfare, aided by new technological developments, did eventually enable the Allied air forces to destroy the entire logistical framework that supported the German and Japanese war effort, and rendered the operational skills, in which the Germans excelled until the very end, as ineffective as those of Jackson and Lee.

Technology had not in fact transformed the nature of strategy. It, of course, remained of vital importance to keep abreast of one's adversary in all major aspects of military technology, but given that this was possible, the lessons of the Second World War seemed little different from those of the First. The social base had to be strong enough to resist the psychological impact of operational setbacks and to support the largest possible logistical buildup by land, sea, and air. The forces thus raised had then to be used progressively to eliminate the operational options open to the enemy and ultimately to destroy his capacity to carry on the war.

V

The same conclusions, set out in somewhat more turgid prose, were reached by the strategic analysts of the Soviet Union—not least those who in the late 1940s and early 1950s were writing under the pen name of J. V. Stalin. But Marxist military thinkers, without differing in essentials from their contemporaries in the West, naturally devoted greater attention to the social dimension of strategy—the structure and cohesiveness of the belligerent societies. For Soviet writers this involved, and still involves, little more than the imposition of a rigid stereotype on the societies they study. Their picture of a world in which oppressed peoples are kept in a state of backward subjection by a small group of exploitative imperialist powers, themselves domestically vulnerable to the revolutionary aspirations of a desperate proletariat, bears little resemblance to the complex reality, whatever its incontestable value as a propagandistic myth. As a result their analysis is often hilariously inaccurate, and their strategic prescriptions either erroneous or banal.

But the West is in no position to criticize. The stereotypes which we have imposed, consciously or unconsciously, on the political structures that surround us, have in the past been no less misleading. The cold war image of a world which would evolve peacefully, if gradually, toward an Anglo-Saxon style of democracy under Western tutelage if only the global Soviet-directed Marxist conspiracy could be eradicated was at least as naive and ill-informed as that of the Russian dogmatists. It was the inadequacy of the sociopolitical analysis of the societies with which we were dealing that lay at the root of the failure of the Western powers to cope more effectively with the revolutionary and insurgency movements that characterized the postwar era, from China in the 1940s to Vietnam in the 1960s. For in these, more perhaps than in any previous conflicts, war really was the continuation of political activity with an admixture of other means; and that political activity was itself the result of a huge social upheaval throughout the former colonial world which had been given an irresistible impetus by the events of the Second World War. Of the four dimensions of strategy, the social was here incomparably the most significant; and it was the perception of this that gave the work of Mao Zedong and his followers its abiding historical importance.

Military thinkers in the West, extrapolating from their experience of warfare between industrial states, naturally tended to seek a solution to what was essentially a conflict on the social plane either by developing operational techniques of "counterinsurgency," or in the technological advantages provided by such developments as helicopters, sensors, or "smart" bombs. When these techniques failed to produce victory, military leaders, both French and American, complained, as had the German military leaders in 1918, that the war had been "won" militarily but "lost" politically—as if these dimensions were not totally interdependent.

In fact, these operational techniques and technological tools were now as ancillary to the main sociopolitical conflict as the tools of psychological warfare had been to the central operational and logistical struggle in the two world wars. In those conflicts, fought between remarkably cohesive societies, the issue was decided by logistic attrition. Propaganda and subversion had played a marginal role, and such successes as they achieved were strictly geared to those of the armed forces themselves. Conversely, in the conflicts of decolonization which culminated in Vietnam, operational and technological factors were subordinate to the sociopolitical struggle. If that was not conducted with skill and based on a realistic analysis of the societal situation, no amount of operational expertise, logistical backup or technical know-how could possibly help.

VI

If the social dimension of strategy has become dominant in one form of conflict since 1945, in another it has, if one is to believe the strategic analysts, vanished completely. Works about nuclear war and deterrence normally treat their topic as an activity taking place almost entirely in the technological dimension. From their writings not only the sociopolitical but the operational elements have quite disappeared. The technological capabilities of nuclear arsenals are treated as being decisive in themselves, involving a calculation of risk and outcome so complete and discrete that neither the political motivation for the conflict nor the social factors involved in its conduct—nor indeed the military activity of fighting—are taken into account at all. In their models, governments are treated as being as absolute in their capacity to make and implement decisions, and the reactions of their societies are taken as little into account as were those of the subjects of the princes who conducted warfare in Europe in the eighteenth century. Professor Anatole Rapoport, in a rather idiosyncratic introduction to a truncated edition of Clausewitz's *On War*, called these thinkers "Neo-Clausewitzians." It is not easy to see why. Every one of the three elements that Clausewitz defined as being intrinsic to war—political motivation, operational activity, and social participation—are completely absent from their calculations. Drained of political, social, and operational content, such works resemble rather the studies of the eighteenth-century theorists whom Clausewitz was writing to confute and whose influence he considered, with good reason, to have been so disastrous for his own times.

But the question insistently obtrudes itself: In the terrible eventuality of deterrence failing and hostilities breaking out between states armed with nuclear weapons, how will the peoples concerned react, and how will their reactions affect the will and the capacity of their governments to make decisions? And what form will military operations take? What, in short, will be the social and the operational dimensions of a nuclear war?

It is not, I think, simply an obsession with traditional problems that makes a European thinker seek an answer to these questions. If nuclear war breaks out at all, it is quite likely to break out here. And in Europe such a conflict would involve not simply an exchange of nuclear missiles at intercontinental range, but a struggle between armed forces for the control of territory, and rather thickly populated territory. The interest displayed by Soviet writers in the conduct of such a war, which some writers in the West find so sinister, seems to me no more than common sense. If such a war does occur, the operational and logistical problems it will pose will need to have been thoroughly thought through. It is not good enough to say that the strategy of the West is one of deterrence, or even of crisis management. It is the business of the strategist to think what to do if deterrence fails, and if Soviet strategists are doing their job and those in the West are not, it is not for us to complain about them.

But it is not only the operational and logistical dimensions that have to be taken into account; so also must the societal. Here the attention devoted by Soviet writers to the importance of the stability of the social structure of any state engaged in nuclear war also appears to me to be entirely justifiable, even if their conclusions about contemporary societies, both their own and ours, are ignorant caricatures.

About the operational dimension in nuclear war, Western analysts have until recently been both confused and defeatist. In spite of the activities of Defense Secretary Robert McNamara and his colleagues nearly 20 years ago, and in spite of the lip-service paid to the concept of "flexible response," the military forces in Western Europe are still not regarded as a body of professionals, backed up where necessary by citizen-soldiers, whose task it will be to repel any attack upon their own territories and those of their allies. Rather they are considered as an expendable element in a complex mechanism for enhancing the credibility of nuclear response. Indeed, attempts to increase their operational effectiveness are still sometimes opposed on the grounds that to do so would be to reduce the credibility of nuclear retaliation.

But such credibility depends not simply on a perceived balance, or imbalance, of weapons systems, but on perceptions of the nature of the society whose leaders are threatening such retaliation. Peoples who are not prepared to make the effort necessary for operational defense are even less likely to support a decision to initiate a nuclear exchange from which they will themselves suffer almost inconceivable destruction, even if that decision is made at the lowest possible level of nuclear escalation. And if such a decision were made over their heads, they would be unlikely to remain sufficiently resolute and united to continue to function as a cohesive political and military entity in the aftermath. The maintenance of adequate armed forces in peacetime, and the will to deploy and support them operationally in war, are in fact symbols of that social unity and political resolve which are as essential an element in nuclear deterrence as any invulnerable second-strike capability.

So although the technological dimension of strategy has certainly become of predominant importance in armed conflict between advanced societies in the second half of the twentieth century—as predominant as the logistical dimension was during the first half—the growing political self-awareness of those societies and, in the West at least, their insistence on political participation have made the social dimension too significant to be ignored. There can be little doubt that societies, such as those of the Soviet Union and the People's Republic of China, which have developed powerful mechanisms of social control, enjoy an apparent initial advantage over those of the West, which operate by a consensus reached by tolerating internal disagreements and conflicts; though how great that advantage would actually prove under pressure remains to be seen.

Whatever one's assessment of their strength, these are factors that cannot be left out of account in any strategic calculations. If we do take account of the social dimensions of strategy in the nuclear age, we are likely to conclude that Western leaders might find it much more difficult to initiate nuclear war than would their Soviet counterparts—and, more important, would be perceived by their adversaries as finding it more difficult. If this is the case, and if on their side the conventional strength of the Soviet armed forces makes it unnecessary for their leaders to take such an initiative, the operational effectiveness of the armed forces of the West once more becomes a matter of major strategic importance, both in deterrence and in defense.

Most strategic scenarios today are based on the least probable of political circumstances— a totally unprovoked military assault by the Soviet Union, with no shadow of political justification, on Western Europe. But Providence is unlikely to provide us with anything so straightforward. Such an attack, if it occurred at all, would be likely to arise out of a political crisis in central Europe over the rights and wrongs of which Western public opinion would be deeply and perhaps justifiably divided. Soviet military objectives would probably extend no farther than the Rhine, if indeed that far. Under such conditions, the political will of the West to initiate nuclear war might have to be discounted entirely, and the defense of West Germany would depend not on our nuclear arsenals but on the operational capabilities of our armed forces, fighting as best they could and for as long as they could without recourse to nuclear weapons of any kind. And it need hardly be said that hostilities breaking out elsewhere in the world are likely, as they did in Vietnam, to arise out of political situations involving an even greater degree of political ambiguity, in which our readiness to initiate nuclear war would appear even less credible.

The belief that technology has somehow eliminated the need for operational effectiveness is, in short, no more likely to be valid in the nuclear age than it was in the Second World War. Rather, as in that war, technology is likely to make its greatest contribution to strategy by improving operational weapons systems and the logistical framework that makes their deployment possible. The transformation in weapons technology which is occurring under our eyes with the development of precision-guided munitions suggests that this is exactly what is now happening. The new weapons systems hold out the possibility that operational skills will once more be enabled, as they were in 1940–41, to achieve decisive results, either positive in the attack or negative in the defense. But whether these initial operational decisions are then accepted as definitive by the societies concerned will depend, as they did in 1940–41 and in all previous wars, on the two other elements in Clausewitz's trinity: the importance of the political objective, and the readiness of the belligerent communities to endure the sacrifices involved in prolonging the war.

These sacrifices might or might not include the experience, on whatever scale, of nuclear war, but they would certainly involve living with the day-to-day, even the hour-to-hour, possibility that the war might "go nuclear" at any moment. It is not easy to visualize a greater test of social cohesion than having to endure such a strain for a period of months, if not years, especially if no serious measures had been taken for the protection of the civil population.

Such measures were projected in the United States two decades ago, and they were abandoned for a mixture of motives. There was, on the one hand, the appreciation that not even the most far-reaching of preparations could prevent damage being inflicted on a scale unacceptable to the peoples of the West. On the other was the reluctance of those peoples to accept, in peacetime, the kind of social disruption and the diversion of resources which such measures would involve. The abandonment of these programs was then rationalized by the doctrine of mutual assured destruction. And any attempt by strategic thinkers to consider what protective measures might have to be taken if the war which everyone hoped to avoid actually came about was frowned on as a weakening of deterrence. But here again, there seem to have been no such inhibitions in the Soviet Union; and their civil defense program, which some Western thinkers find so threatening, like that of the Chinese, seems to me no more than common sense. It is hard not to envy governments which have the capacity to carry through such measures, however marginally they might enhance the survivability of their societies in the event of nuclear war.

The Western position, on the other hand, appears both paradoxical and, quite literally, indefensible, so long as our operational strategy quite explicitly envisages the initiation of a nuclear exchange. The use of theater nuclear weapons within Western Europe, on any scale, will involve agonizing self-inflicted wounds for which our societies are ill-prepared, while their extension to Eastern European territory will invite retaliation against such legitimate military targets as the ports of Hamburg, Antwerp, or Portsmouth, for which we have made no preparations at all. The planned emplacement of nuclear weapons in Western Europe capable of matching in range, throw-weight, and accuracy those which the Russians have targeted onto that area may be necessary to deter the Soviet Union from initiating such an exchange. But it will not solve the problem so long as the Russians are in a position to secure an operational victory without recourse to nuclear weapons at all. Deterrence works both ways.

VII

It cannot be denied that the strategic calculus I have outlined in the above pages has disquieting implications for the defense of the West. We appear to be depending on the technological dimension of strategy to the detriment of its operational requirements, while we ignore its societal implications altogether—something which our potential adversaries, very wisely, show no indication of doing. But the prospect of nuclear war is so appalling that we no less than our adversaries are likely, if war comes, to rely on "conventional" operational skills and the logistical capacity to support them for as long as possible, no less than we have in the past.

Hostilities in Europe would almost certainly begin with the engagement of armed forces seeking to obtain or to frustrate an operational decision. But as in the past—as in 1862, or in 1914, or in 1940–41—social factors will determine whether the outcome of these initial operations is accepted as decisive, or whether the resolution of the belligerent societies must be further tested by logistical attrition, or whether governments will feel sufficiently confident in the stability and cohesion of their own peoples, and the instability of their adversaries, to initiate a nuclear exchange. All of this gives us overwhelming reasons for praying that the great nuclear powers can continue successfully to avoid war. It gives us none for deluding ourselves as to the strategic problems such a war would present to those who would have to conduct it.

THE SOURCES AND PROSPECTS OF GORBACHEV'S NEW POLITICAL THINKING ON SECURITY

STEPHEN M. MEYER

In his short time as general secretary, Mikhail Gorbachev has beyond a doubt revitalized the discourse on Soviet military doctrine in both the Soviet Union and the West. Under the banner of "new political thinking" on security, Soviet academics, party, military, and other government officials are reexamining many long-standing tenets of Soviet security policy.[1] Some of the ideas being articulated are doctrinally and ideologically revolutionary in the context of traditional Soviet security policy. Indeed, the ongoing doctrinal dialogue, if carried to its logical extreme, could imply even greater changes in Soviet military policy than those associated with Khrushchev's doctrinal initiatives of the late 1950s and early 1960s.

It is also possible that this new political thinking on security could remain just that—thinking—with little long-term impact on the basic forces, capabilities, and approaches that underlie Soviet military policy. Or, following a short-term shift in military policy, there could be a subsequent regression to more standard Soviet security formulations, as happened after Khrushchev's ouster.

Therefore, in considering the potential impact of this new security thinking on Soviet defense policy in the years ahead, it is essential to assess both the likelihood of realization and the durability of the new thinking over time. Can these ideas serve as the foundation for a long-term framework for Soviet military policy in the decades to come? Are they credible principles to which future Soviet political and military leaders could easily subscribe, or are they merely a temporary doctrinal structure with little staying power?[2]

I will argue that conceptual elements of Gorbachev's new thinking on security are first and foremost tools for gaining control of the Soviet defense agenda. The new thinking is highly specific to Gorbachev's approach to rebuilding the Soviet economic-industrial base, and is decidedly not the result of deterministic forces driving Soviet defense or foreign policy. While some of his notions are easily incorporated into more traditional Soviet security frameworks, and thus readily adopted by future Soviet general secretaries, other aspects are quite antithetical to conventional Soviet thinking, and threaten important institutional interests. As is true for all of Gorbachev's programs, there are powerful political and institutional forces that must be coerced or co-opted into implementing policy. Thus, the durability of the entire new thinking framework—which would have far more radical implications for Soviet defense policy—depends on Gorbachev's ability to institutionalize its conceptual elements in political and military decisionmaking.

However, many of the more radical embellishments on the new thinking offered by Soviet academics are not presently part of Gorbachev's framework. Confusion among Western observers over what the Soviet leader has, and has not, advocated has created some unrealistic expectations about future directions in Soviet defense policy.

The discussion that follows is divided into three major sections. First, the purpose and the process behind Gorbachev's effort to stimulate doctrinal discussion are reviewed. Without this context, any effort to separate substance from noise in the new thinking would be futile. Next, the major elements of Gorbachev's new political thinking on security are examined, along with the critiques of the "old thinkers." This article concludes with an assessment of the implications of both the substance and process of the new political thinking for future Soviet security policy.

ENGINES OF DOCTRINAL CHANGE

Mikhail Gorbachev became general secretary of the Communist Party of the Soviet Union in March 1985. The first unambiguous hint of his effort to recapture the defense agenda appeared in the preparatory work for the draft program of the 27th Party Congress. Gorbachev and his associates in the Central Committee Secretariat proposed a cryptic, though significant, change in the standard formulation of the Party's commitment to defense resource allocation.[3] The 27th Congress became a kind of watershed, after which one doctrinal principle after another was subjected to review, discussion, and revision.[4] From that point on, Gorbachev's new political thinking on security grew by accretion.

Reprinted with minor revisions from International Security 13, no. 2 (1988): 124–63, by permission of the MIT Press, Cambridge, Mass. © 1988 by the President and Fellows of Harvard College and of the Massachusetts Institute of Technology.

What would compel the new general secretary to assume the arduous task of revising Soviet military doctrine while still in the throes of consolidating his leadership position? Previous general secretaries had waited until after they had firmly established themselves to take on the national security establishment. For answers, some Western analysts look to rational military-strategic analysis: In recognition of the awesome power of nuclear weapons and the futility of nuclear war, they say, the Soviet leadership since late 1966 has systematically been altering doctrine to reduce the likelihood of nuclear war. By this analysis, Gorbachev's new political thinking on security is merely the logical extension of Soviet military-strategic trends spanning the last twenty years.[5]

Others have suggested that deterministic forces within Soviet society produced an inevitable and irreversible shift in doctrine. In particular, a shift from extensive economic growth to intensive growth, and the expansion of the Soviet "intellectual class"—for whom defense holds a lower priority—are seen as important engines of change.[6]

Curiously, both of these explanations put little or no weight on the personae of the Soviet leadership—and in particular, on the man who is general secretary. Rather, they hold that Soviet leaders are compelled to change policy by deterministic forces beyond their control: by the military-technological revolution, or by social-economic evolution.

These putatively new "engines" for the new political thinking on security have, however, long been present. Soviet leaders have been aware of, and have even publicly warned, of the destructiveness of nuclear war for over 30 years. Similarly, efforts to shift the relative contributions of extensive and intensive growth in the Soviet economy have been underway for over a decade, and despite assertions to the contrary, there is no obvious indication that the size, the composition, or the character of the policy-relevant "intellectual elite" in the Soviet Union has changed significantly in 15 years.[7] Indeed, most of the Soviets identified as "new thinkers" by Western scholars have been around since the early 1970s. While these underlying "forces" may quite plausibly influence policy trends in general, they fail to answer the most important question: Why are significant changes to Soviet military doctrine suddenly being proposed *now*? Why weren't they raised 5, 10, or 15 years ago; or 5, 10, 15 years in the future?

If one looks carefully at the history of military doctrinal change in the Soviet Union, one inescapable conclusion is clear: individual general secretaries do matter. Individual perceptions, impressions, recollections, and biases affect significantly the articulation and elaboration of the Soviet defense agenda.[8] It is not so much a matter of "personalities" as it is personal policy agendas, priorities, and images of what has gone before and what needs to be done now. In this respect, there is no reason to believe that Andropov or Chernenko (had they lived), or Romanov (had he been selected instead of Gorbachev) would have chosen to travel Gorbachev's path of doctrinal reform.

When Gorbachev came to power his only obvious aim was economic revitalization. There is no evidence—or reason to believe—that he was contemplating a grand scheme for defining a new defense agenda. He quickly learned, however, that there were at least two forces that constrained economic change. One was the entrenched attitude of the bureaucracy and labor force toward responsibility, accountability, and authority. Thus, *glasnost* and democratization were born out of the need to create accountability and responsibility by basing rewards and sanctions on performance, not position.[9]

The other constraining force was the defense agenda—not simply the existing defense budget. The problem was not the resources that were already going to the military, so much as the future resource commitments implied by threat assessments and requirements derived from traditional thinking. More elbow room for economic *perestroika* (restructuring) required lifting the shadow of a further Soviet military buildup in the 1990s and beyond, which in turn meant that Gorbachev had to gain control over and restructure the Soviet defense agenda.

At the same time, global political and economic trends seemed to have raised doubts among the Gorbachev coterie about the long-term foundation of Soviet superpower status. By the time Gorbachev became general secretary, Soviet superpower status was precariously balanced on a single leg: military power. The building of Soviet military power had, indeed, been a noteworthy achievement of previous regimes, but it was now threatened with rapid depreciation by the Western technological challenge. Meanwhile, as Soviet military power was growing, Soviet communism has declined significantly as a political-economic and social model for the world: Soviet political and economic influence was spotty at best, and Soviet technological power was falling further behind world standards. From Gorbachev's perspective, rebuilding the political, economic, and social bases underlying Soviet superpower status was intimately tied to economic reform that, in turn, had important implications for defense.[10]

Seen in this light, Gorbachev's agitation for new political thinking on security is more a product of instrumental necessity than of military-strategic enlightenment. Given the problems facing the Soviet economy as he (and his advisers) perceived them, his impressions of what had been attempted in the past and the reasons for their failure, and the goals he set, gaining control of the defense agenda was necessary. At the same time, Gorbachev and his allies do not appear to be affected by the traditional Soviet philosophical-ideological blinders that restricted the flexibility of some of his predecessors and many of his current colleagues.[11] Thus, a broader array of approaches has found its way onto the Gorbachev defense agenda.

CHANGING THE PROCESS

The skepticism that has greeted Gorbachev's doctrinal agenda in Western policy circles is easily understood. Some of the items on Gorbachev's list, such as complete denuclearization by 2000, smack more of propaganda than serious policy.

Compounding doubts is the deadening effect on Western thinkers of the 18-year rule of the Brezhnev regime. Comfortable with the glacial nature of its institutional-consensus style of decisionmaking, long-time Western students of Soviet military policy now have trouble imagining how the kinds of novel statements emanating from the Gorbachev regime could be anything but propaganda. Certainly there is little reason to expect radical shifts in thinking from the long-dominant bureaucracies of the Ministries of Defense and Foreign Affairs.

What this line of thinking fails to recognize, however, is that Gorbachev has been altering the Soviet process of national security policymaking. In the Soviet system, substantive change almost always requires process change and organizational restructuring. This is due in large part to the natural inclinations of institutional actors to resist changes in their organizational roles, missions, and resource allocations.[12]

Gorbachev has brought policy initiation out into the open; under his predecessors public doctrinal discussions reflected decisions already taken.[13] Most importantly, Gorbachev—adopting a policy Andropov appears to have set in motion—has moved to take back the defense agenda-setting function from the military bureaucracy and to redistribute it between the general secretary's office (as Khrushchev did at the very end of the 1950s) and the Central Committee staff.[14] Having become well acquainted with the institutional inclinations and the performance of the traditional national security structure during both the Andropov and Chernenko regimes, Gorbachev has turned instead for new ideas—at least for now—to nonmilitary professionals in the Ministry of Foreign Affairs, the Central Committee staff, and the Academy of Sciences.[15]

As a result, these new actors find themselves in a kind of competition—both among themselves and with the more traditional national security bureaucracy—to catch the general secretary's ear and to provide him with new ways of thinking about old problems. They appear to be aware that their influence and impact on Soviet security policy will in the long term be determined in large part by their near-term success in supplying the general secretary with new ideas that have visible and immediate payoffs.[16] A string of flops could see Gorbachev turn back, willingly or otherwise, to more traditional centers of national security advice.[17]

Of course, the Ministry of Defense continues to be an essential actor in defense policy by sheer virtue of its expertise and its responsibilities for implementing defense policy. With the reduction and limitation of its agenda-setting responsibilities, some of its long-standing avenues of influence have been narrowed. However, while the Ministry of Defense may now share agenda-setting functions with other institutional actors, it remains the primary source of detailed analysis and is still the primary implementer of defense policy. Thus, the Ministry of Defense may find that it still has a potent ability to affect defense policy through its continuing roles in policy option formulation and implementation.[18]

ANALYTIC CAVEATS

Like Khrushchev, Gorbachev's effort to recast Soviet military doctrine has created a great deal of confusion and uncertainty in what was a very stable policy environment. Yet things are even more tentative today because Gorbachev and the new thinkers are still engaged in agenda-setting and option formulation.

Meanwhile, the burgeoning of *glasnost* has resulted in a significant increase in the "noise" found in Soviet discussions of military policy. Many of the ideas being tossed about by Soviet "new thinkers" represent little more than personal views. The competition for influence among the new thinkers further reinforces this tendency, as each attempts to get his own ideas on the national agenda.[19] Thus, it would be a mistake to assume that the entire ensemble of arguments articulated under the banner of new political thinking—even those things presented by Gorbachev—reflects accepted decisions on Soviet military doctrine.

Noteworthy counterarguments to aspects of the new political thinking on defense are already beginning to appear. For the most part, these counterarguments are being voiced by professional military officers; but such exchanges do not reflect a purely civil-military rift. First, these criticisms would not appear in such a diversity of party, government, and media fora without significant high-level political support. The civilians who control these information organs are allowing—if not encouraging—probing questions about the new political thinking on security. Second, the professional military's specialized expertise, its institutional role in the national security bureaucracy, and its defense policy implementation functions make it the logical authoritative countervoice against significant departures from previous defense policies. However, the military lacks the political power to launch an independent campaign against the new political thinking on security. As in war, the men in uniform may be on the front lines, but elements of the political leadership are in command.

Moreover, the professional military is not a homogeneous body; service agendas and priorities differ, as do the views of individual military leaders. There are parts of Gorbachev's doctrinal agenda that segments of the Soviet military could find quite acceptable. For example, the Air Defense Forces might be quite pleased with greater emphasis on defensive operations in military planning. Other segments would have reason to be less pleased—e.g., the Strategic Rocket Forces could not find much comfort in the reduction in emphasis on nuclear forces.

The discussion that follows identifies each of the major elements of the new thinking. Each core concept is examined and placed in context, significant variants are described, and the instrumental utility of the concept is discussed. The traditionalists' or old thinkers' critiques are then presented, followed by an examination of the possible points of convergence between the new and old thinking on security.

THE GORBACHEV DOCTRINAL FRAMEWORK

Given its instrumental origins, it is not surprising that Gorbachev's defense agenda has emerged piecemeal, rather than as a single master blueprint. Nevertheless, by mid-1987 the basic threads of Gorbachev's new political thinking on security had been articulated:

—War prevention is a fundamental component of Soviet military doctrine;

—No war—including nuclear war—can be considered a rational continuation of politics; and inadvertent paths to nuclear war are as likely, if not more likely, than deliberate paths;

—Political means of enhancing security are more effective than military-technical means;

—Security is mutual: Soviet security cannot be enhanced by increasing other states' insecurity;

—Reasonable sufficiency should be the basis for the future development of the combat capabilities of the Soviet armed forces;

—Soviet military strategy should be based on "defensive" (nonprovocative) defense, not offensive capabilities and operations.

WAR PREVENTION IN SOVIET MILITARY DOCTRINE

War prevention has always been a keystone of Soviet national security policy. It was certainly a fundamental part of Stalin's policy vis-à-vis Nazi Germany in the late 1930s and early 1940s. Though Stalinist military doctrine assumed that a cataclysmic war between capitalism and socialism was inevitable, Stalin's goal was to postpone that eventuality as long as possible.

One of Khrushchev's first moves toward de-Stalinization was the revocation of the principle of the "inevitability of East-West war." As first secretary he demonstrated that—crisis-provoking policies notwithstanding—he was not actually eager to go to war and would do what was necessary to avoid it.

Brezhnev proved to be even more cautious, shying away from actions that posed a risk of direct confrontation. Yet, despite his strong and enduring de facto policy of avoiding war, war avoidance never attained a formal place in Soviet military doctrine.[20]

New Thinking

One of the hallmarks of Gorbachev's new political thinking on security was the formal shift in emphasis in military doctrine to prevent war. Authoritative Soviet political and military sources now unambiguously state that preventing war is the fundamental goal of Soviet military doctrine.[21] Gorbachev has also chosen to reemphasize that the Party does not consider nuclear war between East and West to be inevitable.[22] This places a stamp of ideological correctness on the redirection of military doctrine toward avoiding war.

The formal inclusion of war prevention in Soviet military doctrine provides Gorbachev with a legitimate doctrinal basis for trading off current forces and future increments of military power (which might be useful should war occur) for increments of politico-military stability (which might reduce the probability of war). This lends greater weight generally to politico-military considerations (such as strategic stability), and particularly places politico-diplomatic efforts, such as arms control and actions that reduce regional and international tensions, on a par with military technical efforts. It offers a doctrinal basis for attempting to reduce tensions and engage in truly cooperative behavior.

Counterpoint and Convergence

The incorporation of war prevention into Soviet military doctrine is perhaps the least controversial component of the new thinking. Indeed, it is not really a substantive change at all, but more of a political highlighting. An inability to develop the high-confidence capabilities that would provide for meaningful victory in strategic nuclear war has long tempered Soviet military doctrine and strategy, notwithstanding propaganda to the contrary.[23] There do not appear to be any overt disagreements with the thesis that war between socialism and capitalism can be avoided indefinitely.

There are differences of opinion, however, over the best strategy for preventing nuclear war. In particular, skeptics take a much more pessimistic view of intentions of the West. As would be expected, as policy moves from concept to implementation, disputes arise, as described in detail below.

INADVERTENT VERSUS DELIBERATE PATHS TO NUCLEAR WAR

Closely related to the theme of nuclear war prevention is the new emphasis on inadvertent paths to nuclear war. In the past, the problem of accidental nuclear war received cursory treatment in Soviet military doctrine. Deliberate paths to nuclear war were the dominant scenarios.[24] The new thinkers, however, have given the threat of inadvertent nuclear war equal, if not greater, prominence.

New Thinking
The new thinking argues that in the contemporary era, war, especially nuclear war, cannot be considered a rational continuation of politics; and that inadvertent paths to nuclear war are as likely, if not more likely, than deliberate paths.

For the new thinkers, the deliberate path to nuclear war—the cold rational decision to initiate a nuclear strike for political gain—is an increasingly remote possibility.[25] The roots of this view were discernible in Brezhnev's 1977 Tula speech, where he argued that the aggressor could not expect to fight and win a nuclear war.[26] But Brezhnev only went half-way, leaving unresolved the question of the defender's—presumably the Soviet Union's—prospects of winning a nuclear war.

Gorbachev and the new thinkers have gone much further. Nuclear war is not war in the traditional sense and cannot, therefore, be considered a rational means for seeking political goals—even for capitalist states.[27] The risks of nuclear escalation inherent in any East-West military engagement imply that war in general can no longer be considered a means for achieving political goals.[28]

Instead, the new thinkers perceive a growing threat of inadvertent nuclear war. They see an obvious tradeoff here: while ever greater quantities of nuclear arms make deliberate nuclear war unthinkable, they simultaneously raise the likelihood of nuclear war by accident or miscalculation. Higher levels of nuclear armaments, they argue, could even undermine the deterrent value of strategic nuclear parity.[29] Hence, they see parity *at lower levels* as enhancing Soviet security with respect to both deliberate and inadvertent causes of nuclear war.

In the new thinking, therefore, nuclear war is portrayed as a threat in its own right—irrespective of its political content. The great danger, as one Central Committee consultant put it (in most un-Marxist terms), is that "nuclear war could begin and end without political decisions."[30] In other words, technology, not politics, might be the cause of nuclear war.

The new thinkers—Gorbachev included—have also expanded this discussion to the ideological front, arguing that, contrary to traditional Marxist-Leninist interpretations, capitalism need not be inherently militaristic.[31] Thus, the argument that the threat of nuclear war is technological rather than political becomes ideologically acceptable.

What is the instrumental utility of making inadvertent nuclear war a serious threat to Soviet security while simultaneously downplaying the threat of deliberate attack? This notion establishes the ideological and doctrinal context for reducing the role of military power in Soviet foreign policy in general, and for undertaking major force reductions in particular. It is the basis for arguing that Soviet security can be increased by reducing strategic forces in step with corresponding Western reductions. This is not a repudiation of past Soviet policies. Rather, it is an ideological justification: the East-West security dilemma has reached a qualitatively new stage, driven by new technologies and capabilities for destruction, while at the same time capitalism is maturing to a less militaristic stage. Thus, new approaches to security become possible, if not required.

Counterpoint
While it initially scoffed at the dangers of its own accidental or "unsanctioned" nuclear use in the 1960s, in subsequent years the Soviet military has implemented measures designed to limit the possibility of such events.[32] Since the early 1980s, technological and procedural changes have been underway to decrease further the likelihood of unsanctioned or unauthorized Soviet nuclear use.[33] Thus, it is clear that the military's appreciation of the dangers of inadvertent nuclear war has grown over time.

Nevertheless, military analysts have attempted to restore some "balance" to the consideration of accidental versus deliberate paths to nuclear war in doctrinal discussions. While acknowledging the dangers of accidental nuclear war, they insist that—contrary to the new thinkers' hopes—the West continues to believe quite firmly that war is a continuation of politics, and thus the threat of deliberate attack to the Soviet Union is real. For these Soviet observers the physical evidence is obvious: the West continues to buy the forces it needs to start either a nuclear war or a conventional war.[34] Accordingly, as Minister of Defense Dmitri Yazov emphasizes, the Soviet armed forces must be prepared to fight all forms of war—nuclear or conventional.[35]

The issue here seems to be one of the perception of relative risk. How far can the Soviet Union go in implementing measures to reduce the risk of accidental nuclear war before impinging on militarily important capabilities, and thereby raising the risk of deliberate nuclear war? The suggestion by some new thinkers that unilateral Soviet moves may be desirable is seen by more traditional ideological elements in the Soviet leadership as increasing the danger of aggressive American behavior. This would in turn, they believe, increase the likelihood of a US-Soviet confrontation and, ultimately, nuclear war.

Convergence

While most Western discussions treat the threat of accidental nuclear war as a technical problem, it is first and foremost a political-ideological issue in Soviet discussions. This means that in Soviet policymaking the relative emphasis placed on inadvertent war versus deliberate war scenarios is more a matter of political-ideological interpretation than military-technical analysis. Thus, it is unlikely that there will ever be convergence on this issue. Gorbachev and the new thinkers must highlight the dangers of inadvertent nuclear war; the ideological conservatives and the military establishment must emphasize the dangers of deliberate nuclear war.

The relative dominance of scenarios of inadvertent war over those of deliberate war will continue as long as it is useful to the Gorbachev program and there is no consensus against it in the political leadership. In other words, Gorbachev or a like-minded future general secretary does not need a consensus to support this view; the absence of a consensus against it is sufficient.[36]

POLITICAL MEANS OF GUARANTEEING SECURITY

Given the new thinking on war avoidance and accidental war, an increased emphasis on political means of guaranteeing security begins to look attractive. Of course, political means have always figured prominently in Soviet security policy. Treaties, agreements, and political-economic relationships have long been used to create a better security environment.[37] However, few if any of the Soviet Union's international partners have been seen as reliable or benign. Therefore, the Soviet political leadership viewed military power as the fundamental basis for maintaining Soviet security. Political means were merely an adjunct to military-technical means.

New Thinking

Here again Gorbachev established the initial direction of the discussion. In concert with the new emphasis on war avoidance, Gorbachev argued that "new realities" meant that political means, rather than military-technical means, had become the primary tools for guaranteeing the security of the Soviet Union. He posited that political means may be the only means for ultimately solving Soviet security problems.[38]

This evolution toward political solutions is portrayed as a consequence of the military technologies of the "nuclear-space" era: modern weapons make it impossible for states to defend themselves by military-technical means alone. War means catastrophe. Negotiations and diplomacy, therefore, can buy the Soviet Union more security than could allocation of additional defense rubles.[39] Reducing the threat facing the Soviet Union accomplishes more than countering it. An example of this approach was offered by Chief of the General Staff Sergei Akhromeyev when he was asked why the Soviet Union was giving up more weapons than the United States in the INF (intermediate-range nuclear forces) Treaty. He replied that it was worth the trade since the removal of the American Pershing IIs and GLCMs (ground-launched cruise missiles) completely eliminated the threat of limited nuclear war to the European USSR.[40] From this perspective, similarly, halting American Strategic Defense Initiative (SDI) deployments via negotiation would be seen as far more efficient and effective than developing and deploying countermeasures.

Counterpoint

Not surprisingly, Gorbachev's emphasis on political means for guaranteeing Soviet security—and the consequent subordination of military-technical means—provoked a reaction from more traditional national security elements. For example, Minister of Defense Yazov and First Deputy Chief of the General Staff Varennikov felt compelled to point out that while, from the Soviet perspective, providing for security may seem to be increasingly a political task, the United States and NATO do not appear to see it that way.[41] These critics point out that the West remains committed to attaining military-technical superiority over the Soviet Union and the Warsaw Pact, adducing as evidence alleged Western failures to respond to unilateral Soviet initiatives and US duplicity in arms control.[42]

Politburo member Ligachev, in his statement before the Supreme Soviet Foreign Affairs Commission reviewing the INF Treaty, noted that there are many questions about the US commitment to comply with arms control in general, and the INF Treaty in particular.[43]

As a consequence, more traditional thinkers argue, such thinking may be correct, but implementation of such policies is premature: Political means cannot work without the equal balance of adequate military-technical means. Military writers warn that if the adversary perceives weakness (i.e., an absence of military-technical preparedness), it will not be restrained from aggression.[44]

Until mid-1987 Marshal Akhromeyev—the military man most closely identified with Gorbachev's new political thinking on security— appeared to give unflinching support to Gorbachev's views on the dominance of political means over military-technical means of providing security. This is interesting, since Akhromeyev's primary responsibility as chief of the general staff is to guarantee the military-technical security of the Soviet state. However, by fall 1987 he had shifted the tone of his argument toward a more cautious view, arguing that military-technical solutions make equally important contributions to avoiding war and to restraining the aggressive impulses of probable enemies.[45]

Convergence

There is general appreciation in the Soviet Union that peacetime military activities have important political dimensions, and Soviet military procurement and deployment decisions do take into account a reading of the political environment. It follows that political efforts— whether improving the international environment, improving bilateral relations, or engaging in arms control—can affect military requirements. To the extent that defense resources are constrained in the USSR—i.e., that the defense constituency cannot do everything it would like at once—then political actions can help sort out military-technical priorities.[46] For example, if political accommodation and arms control could at least constrain, if not eliminate, the need to counter SDI, then R&D (research and development) and procurement priorities could continue to emphasize improving systems for conventional war. Obviously, this is one part of a politico-military strategy that both new and old thinkers can agree on.

In the West, Gorbachev's call for increased reliance on political means is always interpreted in the most benign fashion. There is, however, a side to this approach that should make it more worrisome to the West, while also making it more acceptable to Soviet Union traditionalists. That is: political means also include active measures to divide and weaken NATO politically. They also imply enhancing Soviet political authority around the world as a counterweight to American political presence. This is, of course, a direct copy of Khrushchev's strategy. To the extent that enhanced use of political means can increase Soviet political "presence" while weakening the political bonds of the Western Alliance, then the new thinkers and the traditionalists will find common ground.

SECURITY IS MUTUAL

Since the end of World War II, Soviet military doctrine seems to have viewed the Soviet security dilemma as a zero-sum game: the USSR's own security could best be guaranteed by posing an overwhelming threat to its neighbors, whether putative adversaries or friends. For a prime example of the application of this approach, one need only recall the massive Soviet buildup of nuclear and conventional forces facing China beginning in 1964. It is impossible to determine whether the underlying intent was eventual conquest, a somewhat less ambitious goal of political-economic dominance, merely deterrence by defense, or some combination of these. But the basic fact remained that Soviet leaders felt that Soviet security was enhanced by increasing the insecurity of the Soviet Union's neighbors.

New Thinking

Gorbachev's new political thinking turns this zero-sum view on its head. Now, say the new thinkers, Soviet security must be viewed as inevitably intertwined with American and, indeed, global security.[47] Security is a mutual problem. Once again this is presented as a direct consequence of the nature of nuclear war. Since no country can defend itself by military-technical means alone, national and international security are indivisible; the other side's security concerns must be taken into account.[48]

Obviously, this has important implications for military policy. First, it means that decisions about new weapons and deployments must be based on more than just military-technical and military-operational criteria. How other countries' perceptions of threat will be affected must be assessed; the military power of one state must not even *appear* to threaten other states.[49] Second, it means that political solutions, rather than military-technical ones, come to the forefront of policy.[50] This, in turn, presupposes a willingness to take into account the interests of adversaries and, above all, to make concessions.[51] There can be no denying that, since the beginning of 1987, traces of this principle have been reflected in Soviet political-military behavior.

Counterpoint

Disagreements over the thesis of interdependent security have a distinctly philosophical and ideological flavor. Critics of the idea of the mutuality of security take refuge in Marxist-Leninist orthodoxy: It is the political-social essence of a state, not its force posture, that determines whether its doctrine is threatening (offensive) or nonthreatening (defensive).[52] The Soviet Union and the Warsaw Pact states cannot be perceived as a threat by anyone, they argue, because they are classless socialist societies. Thus, Soviet military preparations for the legitimate defense of socialism cannot threaten anyone no matter what the scale of those efforts.

Skeptical Soviet commentators observe that, despite Soviet peaceful initiatives and numerous concessions, the West continues to build raw military-technical power with the aim of acquiring military superiority.[53] The failure to induce the United States to join the unilateral Soviet nuclear test moratorium is one often-cited example. At the same time, they detail what they describe as past US duplicity in arms control. Indeed, they note that even as the INF agreement, with numerous Soviet concessions, was being signed, the West was already planning ways to circumvent the agreements by deploying "compensating" nuclear forces.[54] In short, there appears to be a substantial body of opinion—political and military—that rejects the premise that anything the USSR might do to enhance its security could be interpreted as threatening to truly peaceful states.

Convergence

It is not obvious where there are significant points of convergence here. The suspension of the historically dominant "us-them" dichotomy in Soviet ideology and foreign policy would seem to be heavily dependent on the context (for example, the World War II alliance against the Nazis), the political power of the general secretary, and personalities. This perspective is likely to affect Soviet military doctrine only as long as the general secretary has the political capabilities to pursue policies along this line. There is no reason to assume that this view of the mutuality of security has any independent durability over time.

REASONABLE SUFFICIENCY

Of all the ideas falling under the umbrella of new political thinking on security, none is more enigmatic than the notion of "reasonable sufficiency." There is general agreement among Soviet new and old thinkers that the Soviet defense posture should be based on sufficiency.[55] Differences emerge when the issues of "what is sufficient?" and "how is sufficiency determined?" are addressed.

Moreover, there is a curious—some might say comedic—disagreement among new thinkers over the meaning of the term and how it differs from past Soviet policy. For some Soviet commentators, it connotes a shift away from the action-reaction policy of the past to one of internal determination of defense requirements. Soviet weapons programs and deployments in the past were largely the product of reactions to Western programs, it is explained. Now—under reasonable sufficiency—the Soviet Union will procure only those weapons that are necessary to fulfill the defense goals it sets for itself. Correspondingly, the characteristics of those weapons would be a function of Soviet-determined requirements, not emulatory or reactive behavior.

Other Soviet spokesmen describe reasonable sufficiency in precisely opposite terms. Unlike the past, they say, where Soviet weapons programs and deployments were determined by internally set defense requirements and goals, now the Soviet Union will procure new weapons only in strict reaction to Western military deployments. Reasonable sufficiency implies the strict maintenance of parity by matching Western military efforts.

It is quite clear that these are two completely opposite notions of what Soviet defense policy was, and what reasonable sufficiency implies. This division of views does not break down along obvious institutional lines, nor does it divide among new and old thinkers. It suggests confusion, not political or institutional alignment. Ironically, Gorbachev's best interests are served by this confusion, making reasonable sufficiency a wild card that can be applied as the situation suits him. Reasonable sufficiency can be invoked to call into question any defense program that the general secretary perceives as superfluous to Soviet security.

New Thinking

The notion of reasonable sufficiency first appears in Gorbachev's political report to the 27th Party Congress, where he said that the Soviet Union would limit its nuclear potential to levels of reasonable sufficiency.[56] But he qualified this by stating that the character and level of reasonable sufficiency would be determined by the actions of the United States and NATO. In other words, Gorbachev's invocation of reasonable sufficiency was along the lines of the second interpretation noted above: an action-reaction formulation. Later, reasonable sufficiency was generalized to the entire military posture, and was purportedly adopted as the basis for Warsaw Pact military planning in May 1987.[57]

One of the clearest elaborations of the conceptual underpinnings of reasonable sufficiency was written by Lev Semeyko, a senior researcher at the Institute for the Study of USA and Canada, in the government newspaper *Izvestia* in August of 1987.[58] The old political thinking, he argued, assumed that "more is better" in security affairs; acquiring military superiority over an opponent was a necessary and sufficient condition for victory in war. In contrast, the new political thinking—as captured by the notion of reasonable sufficiency—recognizes that this is no longer the case.

Reasonable sufficiency, according to Semeyko, has three dimensions: political, military-technical, and economic. The political aspect captures the nonaggressive orientation of Soviet military doctrine. Priority is placed on political solutions to security issues (through arms control, for example). The military dimension of reasonable sufficiency ties together several other principles of the new political thinking on security. The military power of the Soviet Union must be sufficient to absorb any attack under the worst imaginable conditions, and to rebuff the enemy, while at the same time not be so great as to threaten the security of other states. There should be no unreasonable surplus of military potential.[59] Manifestations of reasonable sufficiency should be evident in the structure, character, composition, and deployment of Soviet forces.

Lastly, Semeyko argues that the economic aspect of reasonable sufficiency recognizes that there is a law of diminishing returns in the arms race: marginal improvements of military power come at ever greater cost. However, those increments of military power add less and less to the net military capability of the state.

Confusion and disagreement among the new thinkers reared its head once they attempted to flesh out the policy implications of reasonable sufficiency. Consonant with Gorbachev's economic agenda, some of the discussions accented the political and economic implications of reasonable sufficiency. They posited that the underlying purpose of current US military policy was to manipulate the arms race to bleed the Soviet economy, in part by compelling the Soviet Union to buy inappropriate military hardware (a thesis echoed by Gorbachev.)[60] Reasonable sufficiency could neutralize this Western strategy via resort to *asymmetric* responses: reacting to US and NATO actions with offsetting rather than emulating responses. Here the goal is to maintain strict parity of potentials and capabilities, not parity of categories of weapons. Thus, for example, the Soviet Union would respond to SDI with an offensive buildup, not an "SDI-skiy" program.[61]

Others have pushed the bounds of asymmetry further, into the domain of unilateral Soviet actions. They argue for a unilateral restructuring of Soviet forces, consistent with reasonable sufficiency and with other tenets of the new thinking on security.[62] That is, the Soviet Union would not wait for the West to agree to mutual force changes. Rather, the Soviet Union could go forward in restructuring its military as it desired—keeping in mind that all such changes would represent a net increase in Soviet security.[63]

Negotiated deep reductions in strategic forces is a concept that has been embraced by all the new thinkers. Mutual reductions of 50 percent are seen as the first logical step of the process, and as such is the official Soviet negotiating position in START (Strategic Arms Reduction Talks).[64] The new thinkers argue, however, that the ultimate near-term goal should be mutual cuts of 95 percent in strategic forces.[65]

Even the possibility of unilateral Soviet force reductions has been suggested.[66] Past unilateral measures such as the Soviet nuclear test moratorium did not undermine the Soviet Union's security, or so they argue. Unilateral reductions (restructuring), in combination with bilateral and multilateral actions, could contribute significantly to Soviet security by reducing the overall level of the East-West military confrontation.[67]

Counterpoint

In contrast to the new thinkers, those associated with the more traditional national security establishment focus attention on the military dimension of sufficiency. Acknowledging that the "old way of thinking"—that security was proportional to military forces deployed—can no longer be considered valid in the strategic nuclear arena, they center their concern on the importance of maintaining strict parity.[68] This involves careful analysis, forecasting of Western military trends, and timely procurement of forces needed to compensate for any deviations from strict parity.

The centrality of strategic nuclear parity is captured by the refusal of the skeptics to use the term "reasonable sufficiency." Instead they prefer to speak in terms of "defensive sufficiency," or just "sufficiency." While acknowledging the foolhardiness of matching every Western military development with a duplicate Soviet effort, they nonetheless emphasize the importance of possessing every type of weapon and maintaining strict strategic parity.[69]

Sufficiency in general-purpose forces is allowed more latitude by the skeptics, perhaps because "strict" parity is impossible to define for highly complex conventional force structures. Here the criteria are given as: forces and means sufficient to ensure the peaceful work of socialist labor, and sufficient to defend the socialist alliance reliably.[70] Sufficiency is not necessarily pegged to parity or even equivalence; it can mean superiority as perceived by the other side.

While the more traditional thinkers agree that

asymmetries in the European theater need to be corrected, they marshal considerable evidence demonstrating that all but a few of the meaningful asymmetries favor NATO.[71] In particular, they include naval forces in their assessments, arguing that to count only land forces biases the picture against the Warsaw Pact. Thus, one cannot talk about alleged Soviet advantages in land forces absent consideration of US naval power when theater balances are calculated.[72] When all is considered, they assert, the Warsaw Pact military posture today is at the level of sufficiency.

In this respect, all military references to *perestroika* (restructuring) within the last year have been explicitly restricted to discipline, morale, and attitude issues.[73] They have summarily excluded any notion that *perestroika* is related to force structure questions. For the military, *perestroika* means improving the work habits of officers and soldiers, not changing the basic configuration of Soviet forces. These commentators correspondingly reject the suggestion that unilateral reductions could improve Soviet security; indeed, they argue that such actions would magnify the threat.[74] Unilateral restructuring might be permissible, they suggest, but only as a means of improving the Soviet military posture, not of reducing force capabilities.

Convergence

There is good reason to believe that reasonable sufficiency will remain a political concept that will avoid explicit operational definitions. Otherwise, the idea would lose its political utility as a wild card for Gorbachev. However, as a consequence, some areas of convergence are likely.

The concept of strategic nuclear sufficiency has been on the professional military's agenda at least as far back as since at latest the early 1980s. Then Chief of the General Staff Nikolai Ogarkov very explicitly called into question the military utility of procuring additional nuclear weapons.[75] Given present-day military-technical and economic realities, strategic parity appears to be the endpoint of Soviet nuclear planning for the foreseeable future for both new and old thinkers alike. Strategic parity at lower levels is a corollary that is also acceptable, though divergence is likely to grow in proportion to the size of the proposed reduction in forces.[76]

The convergence points with respect to conventional forces are harder to detect. The traditionalist view—quite strongly expressed in a number of documents—is that sufficiency exists *now* on the Warsaw Pact side of central Europe.[77] Thus, sufficiency at lower levels in the theater is possible, but only with strictly matched compensating reductions on the NATO side.[78]

Another apparent convergence point may be found on the question of reducing asymmetric advantages in the theater. Given the skeptics' portrayal of the situation in Europe as one favoring NATO, it is not hard for them to agree with the general principle of reducing advantages even through unilateral action. Of course, it is NATO asymmetric advantages that interest them, and NATO unilateral reductions. More usefully, the traditional thinkers seem prepared to consider the mutual reduction of asymmetries as part of formal agreements on the balance in Europe.[79]

DEFENSIVE DEFENSE

The Soviets have always maintained that their military doctrine was "defensive." This meant that the Soviet Union would only resort to military force when the "gains of socialism" were threatened. Its only goal in war would be defense of the socialist world. To support that defensive doctrine, however, the Soviet Union had a decidely offensive military strategy.[80] Its forces were configured and deployed to carry any conflict onto the territory of its opponent— to defeat and occupy its adversaries in theaters of war adjacent to the Soviet Union.

The new concept of "defensive defense" represents a significant departure from this formula; with it, Gorbachev's new political thinking crosses from the realm of doctrine to the realm of military strategy.

New Thinking

Defensive defense connotes a force posture and military strategy sufficient to repel a conventional attack, but incapable of conducting a surprise attack with massive offensive operations against the territory of the other side.[81] It is a military strategy of offensive self-denial: Soviet theater forces would be configured and deployed to stop an attack, hold against further penetration of Warsaw Pact territory, and then push enemy forces back to the border. This defensive force would not possess the capabilities to threaten the other side's territory.[82]

Proponents of defensive defense argue that the mutual fear of surprise attack in Europe can be eliminated by restructuring existing forces— that is, replacing offensive weapons with defensive weapons.[83] Thus, their motivation for this shift is partially military-political. They also point out that conventional weapons represent the largest part of a state's military spending. At least superficially then, the combination of reasonable sufficiency and defensive defense suggests an opportunity for significant reductions of the defense burden.[84]

Counterpoint

Among all the activities of the new thinkers, none strikes more at a traditional preserve of Soviet military professionalism than this intrusion into the realm of military strategy for theater war. As the strongest proponents of defensive defense point out, moreover, nothing is as much anathema to traditional Soviet military thinking as is a defense-dominant theater strategy and force posture.[85]

The opposing voices have reacted with according vigor. First, they strongly assert that individual weapons systems cannot be isolated from a complex conventional force structure and characterized as offensive or defensive. How a weapon is employed—the strategy that dictates its use and the purposes to which it is put—determines whether it is offensive or defensive.[86]

Second, one cannot successfully defend one's territory or allies without ultimately carrying the attack to the enemy (the 8-year Iran-Iraq war is a good example). Pure defense is not a viable military strategy. Regardless of whether a military force is operating at the tactical, operational, or strategic level, the opposing force must ultimately be routed and expelled—it cannot merely be slowed and exhausted.[87] From the perspective of the old thinkers, defensive operations do have an important role, in this respect, by providing time for Soviets forces to regroup and, if it has not been possible to begin offensive operations at the start of the war, to move to the counteroffensive.[88] The successful defensive operation is one that allows for a quick and powerful shift to the offensive.

Third, military history demonstrates the inferiority of purely defensive strategies. The reliable defense of the homeland cannot be based on ahistorical concepts.

Fourth, once again the ideological argument is made: that the Soviet military posture could be considered provocative, irrespective of its technical attributes, flies in the face of classical Marxist-Leninist teachings. Socialist states cannot, by definition, have military doctrines or postures that are threatening.[89] Thus, there are serious ideological issues to consider.

Convergence

Many Western students of Soviet military affairs were pleasantly surprised when, following the new thinkers' discussions of "defensive defense," they detected increased attention to defensive actions in Soviet military writings. Unfortunately, appearances in this instance are deceiving.[90]

In fact the Soviet military already had a burgeoning interest in defensive operations prior to Gorbachev's rise to power. By the early 1980s a number of studies on defensive operations were underway, conducted under the auspices of the General Staff, and a series of high-level military conferences explored the role of defensive operations in combat.[91] There is little doubt that one of the most visible military commentators on defensive defense—General-Colonel

Gareyev, believed to have headed the Military-Science Directorate of the General Staff at that time—directed much of the work.[92] What the military had in mind, however, was something quite different from the new thinkers' ideas on defensive defense.

The primary impetus for the military's interest in defensive operations was the realization that the Soviet theater posture was premised on a number of questionable assumptions: (1) that the USSR would know when war was imminent; (2) that Soviet political leaders would be able to make timely decisions to enable the armed forces to seize the initiative on the battlefield and begin the war with offensive operations; (3) that the war would be waged on NATO territory; and (4) that NATO would not have the forces or means to wage offensive war against the Pact in Eastern Europe. Evolving American concepts such as "Airland Battle" and "Follow-On-Forces Attack," which advocated deep strikes against second-echelon forces, as well as more radical NATO proposals for counteroffensive strikes against Pact territory, evidently started ringing alarm bells in the Soviet Ministry of Defense.

The need for increased emphasis on defensive operations, then, grew out of military pessimism (or realism). Changes had already occurred in military planning and training at the tactical and operational levels by the time Gorbachev came to power.[93] Some evidence suggests that there have even been corresponding procurement and force structure changes within the Soviet and Warsaw Pact forces.[94]

What distinguishes the military's interest in defensive operations from the new thinkers' ideas about nonprovocative defense is the relative role of defensive operations in war. The military sees defensive operations as merely a component of a larger offensive-oriented strategy. Defensive operations hold the enemy and buy time until the appropriate forces and means can be assembled for transition to the counteroffensive. Defensive operations, then, are a worst-case option in theater war. At the tactical level there may be defensive actions as Soviet divisions off the main axis of attack attempt to contain NATO counterstrikes. At the operational level there may also be a considerable number of defensive operations along certain sectors of the front. There may even be strategic defensive operations at the beginning of the war, especially if the Warsaw Pact is caught by surprise. But this defensive phase of the war would be temporary, a transition phase, in the view of traditional thinkers.

It seems, therefore, that there is an "unstable" convergence point between the views of the new thinkers and those of the military leadership towards increased emphasis on defense in Soviet theater planning. Her political weight is unlikely to be of much use to the new thinkers. Conventional forces are complex organisms, and are not as easily disassembled as are strategic nuclear forces. In this respect, the military still has no analytic rivals comparable to the civilian expertise on nuclear forces that grew up in the Foreign Ministry and the Academy of Sciences during 20 years of strategic arms negotiations.[95] While there are indications of efforts to develop civilian expertise in conventional warfare, the substantive and institutional hurdles suggest a long, uphill process.[96] However, no matter how strong a cadre of civilian conventional force analysts eventually emerges in the Soviet Union, the military will remain the dominant actor in planning and implementing changes in the basic structure and organization of Soviet theater forces.[97]

Moreover, conventional force planning—more than any other aspect of defense policy—cuts to the heart of defensive military professionalism. It would seem that this is one battle that would ignite the institutional ire of the armed forces.[98] This, in turn, would be a fairly potent force for Gorbachev's political rivals on the Politburo to employ if they could organize opposition.

Over the long term, any movement in Soviet theater strategy toward the new thinkers' notions of defensive defense will most likely fit within the military's rubric for a more balanced theater strategy. The new thinkers may have to be content with declaring "victory" as the military sets about the task of improving the capabilities of the armed forces to undertake defensive actions.

CONCLUSIONS

The new political thinking on security is neither a methodology for force development nor a system of criteria for defining security requirements. It is a set of notions that, *ex post facto*, can be used by the general secretary to justify, doctrinally and ideologically, his efforts to manipulate the Soviet defense agenda. The various elements of the new thinking could be used to rationalize almost any force posture and defense policy. Thus, it would be fruitless to use this framework to attempt to predict specifically what Soviet forces might look like in 10 years, since neither Soviet Minister of Defense Yazov, Chief of the General Staff Akhromeyev, nor Gorbachev himself could do this objectively. It is not surprising in this respect that Yazov found it difficult to spell out in concrete terms the thrust of "reasonable sufficiency" in his March 1988 conversations with US Secretary of Defense Frank Carlucci.

Gorbachev's new thinking does, however, reflect his regime's liberation from the conceptual and ideological baggage that has so heavily influenced Soviet defense policy in the past. It also provides clues about the general directions in which Gorbachev would prefer to steer Soviet defense policy in the 1990s.

Some of the notions now hailed under the banner of Gorbachev's new political thinking on security are likely to become enduring elements of Soviet defense policy. The emphasis on avoiding nuclear war, attention to the dangers of inadvertent nuclear war, and acknowledgment of strategic and theater nuclear sufficiency all fit well within established trends in Soviet political and military thinking.

However, the extent to which the Soviet Union will pursue solutions to its security problems that represent significant departures from past policy will depend on the acceptability and durability of the more controversial ideas of the new political thinking: the mutuality of security, the centrality of political means for guaranteeing security, reasonable sufficiency in general-purpose forces, and defensive defense. The thesis of this article is that this has much more to do with leadership politics than with strategic thinking.

While Gorbachev remains in power, for example, the defense policy impact of his new thinking will be sensitive to the performance of his other programs, most notably economic and social *perestroika*. Should Gorbachev's much-heralded economic and social reforms fail to produce the significant improvements that were promised, or should they spawn new problems, then more traditional elements in the leadership would have an opening to attack the full spectrum of new political thinking—including new political thinking on security. The turmoil in Azerbaijan during 1987–88 is one case in point. Equally ominous—and perhaps more directly connected with external society—would be another burst of political self-assertion, tinged with anti-Soviet behavior, in Eastern Europe. Powerful members of the Politburo seeking to contain the power and authority of the general secretary could use such events to create a consensus against the new political thinking, whether the links were real or trumped-up.

Similarly, Western behavior, and American behavior in particular, can affect the long-term impact of the new thinking. This is because the most outstanding difference between the new thinkers and the old thinkers is their divergent perceptions of the threat posed by the United States and NATO. Western actions that can be perceived as invalidating the assumptions of the new thinking would be potent tools for those opposed to Gorbachev's security framework. For example, NATO's contemplation of compensating nuclear deployments in the wake of the INF Treaty—whether appropriate or not—undermines the credibility of the new thinking: the threat to the Soviet Union has not been controlled or reduced by the treaty, merely redirected.

In the short term, then, Gorbachev does not

need a political consensus supporting the new thinking on security. He only needs to prevent the formation of a political consensus against it. He must prevent the emergence of an issue that will galvanize opposition.

Looking beyond Gorbachev, there is no reason to expect that future general secretaries—or other members of the Politburo—will be "new thinkers." In the absence of truly noteworthy achievements in the near term, especially in economic-industrial performance, subsequent leaders could simply write off the new thinking as a desperate measure that, in retrospect, was ill-conceived. Indeed, if one traces the policy history of the Soviet leadership from Lenin to Stalin, Stalin to Khrushchev, Khrushchev to Brezhnev, and Brezhnev to Gorbachev, such behavior is the rule for transitions. The same political mechanisms that enabled Gorbachev to push aside many of the "Brezhnevite" members of the Politburo could be used by a wily future general secretary to oust any remnants of the Gorbachev coterie.

For the longer term, then, Gorbachev must institutionalize the new thinking within the Soviet national security decisionmaking structure—a lesson he must have learned from Khrushchev's failures. Indeed, Gorbachev's entire agenda-setting strategy for the new thinking seems explicitly designed to build a multi-institutional constituency.[99] But agenda-setting is only the first stage of the process: The history of defensive defense policymaking shows quite clearly how the intended course of policy can be severely altered in the transition from agenda-setting and option formulation to decision to implementation.

Barring resort to Stalin's preferred strategy of killing off and replacing everyone occupying Party and government posts, Gorbachev will have to play off institutions against one another to prevent a consensus against the new thinking from forming, while attempting to convince the institutions that they will ultimately benefit from his programs. While Gorbachev may be able to depend heavily on the Foreign Ministry, the Central Committee Staff, and the Academy of Sciences for new ideas during the agenda-setting phase of his restructuring of defense, ultimately he will have to turn to the Ministry of Defense for implementation. In particular, Gorbachev will need the military's support if he is to tackle successfully the largest and, from a resource-saving perspective, potentially most fruitful element of restructuring the Soviet military posture: general-purpose forces. There is some evidence that Gorbachev recognizes this fact and that the military has been told that their willingness to support the new thinking and forthcoming limitations on defense spending will yield technologically superior weapons in the longer run.[100]

With these caveats about the durability of the new thinking in mind, let us assume that Gorbachev remains general secretary over the next five years or more, that his political clout remains about as strong as it appears today, and that his other programs proceed without major reversals. What might we expect of Soviet defense policy and behavior?

THE ROLE OF NUCLEAR WEAPONS

If one considers the major convergence points described above, then the relative de-emphasis of nuclear weaponry in Soviet military policy is one of the strongest parallels between the new thinking and trends in the old thinking. I stress the word "relative" because this de-emphasis is in terms of the Soviet military's past preferences for strategy and operations, not of any putative NATO or US use of nuclear weapons.[101]

Strategic Nuclear Forces

As noted earlier, the top Soviet military leadership (the General Staff and the Collegium of the Ministry of Defense) has come to see strategic parity as the best strategic nuclear posture that the Soviet Union can hope to attain in the foreseeable future. This is primarily because very high levels of nuclear armaments are already in the stockpiles, and it is implausible that either side could prevent a retaliatory second strike by the other.[102]

Given that there is a ceiling on the total amount of resources going to Soviet defense procurement and given the relative scarcity of some precious resources (e.g., microelectronics), the Soviet military faces a number of trade-offs and opportunity costs as it pieces together its procurement program. What is put into one system is not available for others. The Soviet defense constituency recognizes that the United States can easily match any additional strategic nuclear buildup by the Soviet Union. At the same time, the obvious and complete vulnerability of modern industrial societies to the immense destructiveness of nuclear weapons produces a fairly sharp decline in their marginal utility. Thus, the marginal increment in usable military power from each defense ruble allocated to more strategic nuclear weapons is quite small, while the opportunity costs may be high. Strategic parity, then, becomes a reasonable goal.

In this respect, strategic parity at lower levels of forces might not necessarily appear to detract from Soviet security. Although the military establishment might not have proposed 50 percent reductions in strategic nuclear forces, it is not obvious why it should necessarily oppose such cuts initiated by others—if at least some of the freed high-technology resources could be channeled to general-purpose force modernization programs.[103] Indications are that this is one of the few concrete parts of Gorbachev's defense agenda with which most of the military establishment does agree.[104] In this respect, the litmus test of Gorbachev's new thinking should be the manner in which the Soviet Union implements its strategic force cuts. Specifically, if the new thinking is in force, then the Soviet Union should be willing to make disproportionately large cuts in its most potent counterforce capabilities.

Quite naturally, then, we arrive at the issue of complete denuclearization. The idea of a nuclear-free world was put on the Soviet foreign policy agenda by Gorbachev in January 1986.[105] There is no serious enthusiasm for this notion within the Soviet national security establishment, and it is highly doubtful that the majority of the Soviet political leadership shares this vision. (In fact, the degree of political opposition is likely to increase in direct proportion to the size of the cuts contemplated.)[106]

Some Western analysts have argued that the Soviet military would be in favor of a denuclearized world because it would leave the Soviets with overwhelming conventional military superiority. The image of this superiority is, however, a Western perception not readily shared by those in Moscow. Without strategic nuclear weapons, the Soviet Union would instantly lose its global military power status. It would be unable to project significant and decisive military force beyond the Eurasian landmass, and most importantly, it would not be capable of posing any significant military threat to the US homeland. For the foreseeable future—well beyond 2000—Soviet superpower status requires strategic nuclear weapons.

Theater Nuclear Forces

The situation is different for theater denuclearization. The Soviet military began more than 20 years ago to move away from heavy reliance on nuclear fire support in theater warfare planning. The strong preference for fighting an all-conventional war—should war arise at all—is well documented.[107] It would not be surprising, therefore, that most of the General Staff, the commands of the Ground Forces, the Air Forces, and the Air Defense Forces would perceive the *mutual* large-scale denuclearization of theater forces to be a gain.[108]

CONVENTIONAL FORCES

The major implications for general-purpose forces can be readily deduced from the discussion of defensive defense. There can be little doubt that Soviet military strategy (as it pertains to theater operations) was already moving toward increased attention to defensive actions and operations. The intent was to raise the status of defensive operations, not, as the new thinkers would have it, to lower the status of offensive operations.

Soviet military studies of defensive operations point unambiguously to the forces and means required for a stable defensive posture. Mobile tank forces, mobile and deployed artillery, extensive fighter and fighter-bomber air support are the most prominent.[109] While the basic weapons types will remain the same, new generations with improved capabilities—currently in testing—will enter service. It is quite possible that we will also witness important changes in the structure and composition of ground combat units in the years ahead. Interestingly, this used to be called "force modernization"; now it seems less threatening to refer to it as "unilateral restructuring." There also appears to be renewed interest in fortifications and other forms of defensive engineering work, but only as an adjunct to help stabilize the overall theater position.[110] In particular, an increase in defensive engineering work throughout Eastern Europe is likely.

At the same time, there is increasing interest in high-technology conventional weaponry where quality dominates quantity. There is widespread recognition that conventional weapons technologies are on the verge of a major revolution in capabilities. Nothing better characterizes the Soviet perception than the often-encountered description that these new weapons will have capabilities "equivalent to battlefield nuclear weapons," not in explosive firepower, but in lethality to forces in the field. Indeed, some in the General Staff appear to believe that a new phase in the scientific-technical revolution in military affairs is on the horizon.[111] Nevertheless, the Soviet military continues to believe that more traditional systems of conventional war will dominate the battlefield in the near term—especially given the great quantities of these systems in both sides' arsenals. At best, then, the new and old technologies of theater war will exist side by side for decades to come.

Any major changes likely to occur in the next decade will most closely fit the traditional approach, not the new thinking.[112] In fact, the wholesale transition from offense-dominance to defense-dominance in the theater, preferred by the new thinkers, would still require passing through the kinds of changes currently desired by the more traditional thinkers. Given the sheer mass of Soviet theater forces, any such change would take more than a decade; the Soviet weapons procurement process would require at least that to fill orders. Retraining of commanders and troops is a very slow process, and thus a more rapid change could pose serious risks to the Soviet military posture in Europe and Asia. Therefore, it might not be possible to distinguish the two approaches for another decade.

In sum, one should expect to see noticeable changes in the Soviet theater force posture that reflect increased emphasis on defense, but not at the expense of offense. For the most part, however, Soviet military thinking on defensive operations is characterized by changes in the "software" of war, not the hardware.

ARMS CONTROL

It is fair to say that defensive arms control behavior in the last year has been uncharacteristically flexible, even accommodating. But it would be a mistake to take such behavior for granted. First, if one steps back to view the larger picture of Gorbachev's arms control initiatives of the past two years, one gets the image of a juggler dashing madly to keep a dozen plates spinning in the air. Why Gorbachev has felt compelled to offer up so much so quickly is unclear. Some of his compulsion can be explained by the fact that while the new political thinking is supposed to validate the new approach to arms control, the "success" of arms control in moderating the military threat is in turn supposed to validate the new political thinking. Again, threat perception is at the heart of the old thinkers' critique, and Gorbachev must demonstrate an ability to reduce that threat via the tools of the new thinking. Thus, in the early stages of the new political thinking Gorbachev needs to keep arms control at the forefront of defensive policy. It is hard to imagine, however, how he can continue to keep all his plates in the air through the whole show.

Second, there appears to be considerable sentiment among the political and military leadership that some of Soviet arms control actions and "concessions" have been ill-advised. The Soviet unilateral nuclear test moratorium was interpreted by many—including initial supporters—to have been a humiliating failure because the United States flaunted its continued testing while Soviet nuclear test sites were dormant.[113] The asymmetry of the INF cuts has been implicitly criticized in many corners of the Soviet Union, as has the agreement's failure to control British and French nuclear forces.[114]

Third, as already mentioned, professed Western interest in compensating actions to shore up NATO capabilities in the wake of the withdrawal of US INF missiles from Europe has tarnished some of the new thinking's luster. Even with significant Soviet concessions, it is argued, the West is finding ways to expand the threat to the Soviet Union. In short, it may be that the period of Gorbachev's greatest freedom of action in arms control is passing.

NOTES

The author would like to thank Judyth Twigg and Matthew Partan for their comments on earlier drafts of this article.

1. Gorbachev's new political thinking on security is concerned primarily with the tenets of Soviet military doctrine. Formal definitions aside, Soviet military doctrine broadly addresses the questions: (1) What are the greatest external dangers facing the Soviet Union? (2) What are Soviet options for coping with those dangers? (3) What kinds of military capabilities are required? (4) How much is enough, and how should it be allocated? More formal discussions of the content and scope of military doctrine can be found in: S. F. Akhromeyev, "Doktrina predotvrashcheniya voyny, zashchity mira i sotsializma," *Problemy mira i sotsializma*, no. 12 (Dec. 1987): 26–27; G. Kostev, "Nasha voyennaya doktrina v svete novogo politicheskogo myshleniya," *Kommunist vooruzhennykh sil*, no. 17 (Sept. 1987): 10–13; L. Semeyko, "Vmesto gor oruzhiya . . . O printsipe razumnoy dostatochnosti," *Izvestia*, 13 Aug. 1987, p. 5. Kostev is a vice admiral in the Soviet Navy, and a professor. Semeyko is a senior researcher at the Institute of USA and Canada. (Translations and transliterations from the Russian are by the author, except where otherwise noted.)

2. I reject immediately the argument that the ongoing doctrinal dialogue in the Soviet Union is mere propaganda (disinformation), though there is no doubt it is being propagandized. The breadth and scope of the discussion and the variety of fora where it is occurring are such that one cannot dismiss the seriousness of the substance under consideration. Careers have been staked on some of these issues. The subsequent disruptions to the domestic political body have been significant; some rather obvious policy actions, consistent with the principles of the new thinking, have been taken.

3. The old formulation was an open and absolute framework: "The party will do [provide] everything necessary to reliably defend the homeland." The new form is a closed and relative framework: "The party will do everything to ensure that the imperialist countries do not attain military superiority." There was some internal debate over this change in wording when the draft appeared, apparently initiated by those who felt this new formulation represented a weakening of the Party's commitment to defense. The new wording was retained in the final program. See *Programma kommunisticheskoy partii Sovetskogo Soyuza, Novaya redaktsiya* (Moscow: Politizdat, 1986), p. 115. Information on the internal debate is based on interviews with Soviet officials and academics in Moscow during fall 1987.

4. Gorbachev claims that the philosophical-political basis for the Washington summit—thus also, by implication, for the new political thinking on security—was the 27th Party Congress. See "Vystupleniye M. S. Gorbachev po sovetskomu televideniyu," *Krasnaya zvezda*, 15 Dec. 1987, p. 1.

5. For some imaginative speculation on this point see Michael McGwire, *Military Objectives in Soviet Foreign Policy* (Washington, D.C.: Brookings Institution, 1987).

6. See Jack Snyder, "The Gorbachev Revolution: A Waning of Soviet Expansionism?" *International Security* 12, no. 3 (1987–88): 93–131. Extensive growth refers to the use of additional capital and labor to increase economic output. Intensive growth refers to increasing the productivity of existing capital and labor to increase output.

7. Snyder argues that the growth of the intellectual elite can be gauged by the increase in the number of individuals with higher education. Ibid., p. 112. There is no reason to assume, however, that

higher education is in any way correlated with the rise and expansion of an "intelligentsia" class in the USSR. Many of the people who are naturally part of this "class"—artists, writers, poets, etc.—often do not have higher education degrees. At the same time, it is most unlikely that Party bureaucrats, engineers, scientists, literary figures, and industrial managers— all with higher education—share similar sociophilosophical and intellectual perspectives.

8. I am not suggesting an explanation based on personality types or psychological profiles, though these may be important variables. I merely wish to argue that individuals—even individuals from a fairly homogeneous cohort—make a difference in the Soviet system. A complete examination of the evidence is presented in my *Defending the USSR: Building Soviet Military Power* (forthcoming). For studies outside the national security area see the very interesting work of Valerie Bunce, "Leadership Succession and Policy Innovation in Soviet Republics," *Comparative Politics* 11, no. 4 (1979):379–402; Bunce, "The Succession Connection: Policy Cycles and Political Change in the Soviet Union and Eastern Europe," *American Political Science Review* 74, no. 4 (1980): 966–77; and Philip Roeder, "Do New Soviet Leaders Really Make a Difference? Rethinking the 'Succession Connection,'" *American Political Science Review* 79, no. 4 (1985): 958–77.

9. This derivation for *glasnost* and *perestroika* was described in interviews in Moscow and Boston with two of Gorbachev's advisers. See also Gorbachev's comments in "Otvety M.S. Gorbacheva na voprosy gazety 'Vashington post' i zhurnala 'Nyusuik,'" *Krasnaya zvezda*, 24 May 1988, p. 2.

10. "'Rech' tovarishcha Gorbacheva," *Krasnaya zvezda*, 2 Oct. 1987, p. 2.

11. This includes a view of an inimical and hostile world that engenders an "us versus them" mentality, a basic distrust of foreign ideas, and a need for autocracy.

12. When Khrushchev attempted to institute radical changes in doctrine and strategy, he also found it necessary to create a new armed service that would be institutionally committed to his approach. So the Strategic Rocket Forces were born. Even within the Ministry of Defense, changes in strategy and force concepts almost always conclude intriguingly with force reorganizations. The multiple reorganizations of the Air Force and Air Defense Forces over the last decade are cases in point.

13. Soviet Chief of the General Staff Marshal Sergei Akhromeyev claims that Soviet military doctrine is *in the process* of reformulation. S. F. Akhromeyev, "Slava i gordost' sovetskogo naroda," *Sovetskaya rossiya*, 21 Feb. 1987, p. 1. Of course styles differ. Khrushchev set the agenda, along with his staff devised policy options, and then announced decisions. Often, only a handful of Politburo members knew what to expect. The professional military, for the most part, was an institutional bystander, left to implement policies with which it often disagreed. Public doctrinal discussions, therefore, were part of the implementation phase of policymaking (including postdecision drum beating) when the political leadership attempted to inform the elite, justify the action, and cajole the skeptical. Brezhnev, in contrast, cultivated an institutional-consensus approach to defense policymaking. Ideas and policy options were put forward by the responsible organizations, heavily dominated by the Ministry of Defense. This dominance was rein-

forced by the fact that the General Staff served as the staff for the Defense Council, and the chief of the General Staff acted as its secretary. While decision selection remained the prerogative of the Politburo (and Defense Council), control of the agenda-setting and option-development functions could be potent instruments for steering political leaders to desired decisions. Public doctrinal discussions during Brezhnev's tenure were little more than postdecision elaborations of policy, as all the key individuals and institutions were already well aware of the outcome. I suspect that a noteworthy exception was Brezhnev's 1977 Tula speech, which marked the beginning of his effort to recapture the defense agenda-setting function. See "'Rech' tovarishcha L.I. Brezhneva," *Krasnaya zvezda*, Jan. 1977, pp. 1–3.

14. The establishment of the Committee of Soviet Scientists for Peace and against Nuclear War (hereafter called the Committee of Soviet Scientists) under General Secretary Yuri Andropov appears to have begun largely as a propaganda effort. At the same time, Andropov is alleged to have been searching for an institutional alternative to the General Staff—a Soviet analogue to the US National Security Council staff—as a source of defense policy advice and analysis. Andropov died before this plan could be implemented. (This information derives from interviews with Soviet officials and academicians in Moscow, Washington, New York, and Boston in late 1987 and 1988, hereafter cited as "interviews.") There are indications that Soviet academics, acting through the Committee of Soviet Scientists, have subsequently begun to evolve into a new institutional source of advice on security issues in general, and nuclear weapons issues in particular. (Based on interviews.) As a result of this process, Soviet academics have also begun to have access to high-level policy information. See A. Arbatov, "Glubokoye sokrashcheniye strategicheskikh vooruzheniy" (Part 1) *Mirovaya ekonomika i mezhdunarodnyye otnosheniya (MEMO)*, no. 4 (1988):21–22. Aleksy Arbatov is head of the disarmament and security department at the Institute of World Economy and International Relations (IMEMO). His department is becoming the locus for serious academic research on general-purpose forces.

15. The Gorbachev regime's use of personnel from the Academy of Sciences and the Central Committee staff was confirmed by Central Committee Secretary Anatoly Dobrynin in a recent article: A. Dobrynin, "Za bez'yadernyy mir, navstrechu 21 veku," *Kommunist*, no. 9 (1986): 21, 26. Foreign ministry personnel are also more heavily involved than in the past in the early stages of security-related agenda-setting and policy option formulation. (Based on interviews.)

16. This competition was described by several Soviet institutional "competitors" in interviews.

17. Where Snyder sees these "intellectuals" as being the engines of new political thinking, I see the current prominence of nonmilitary national security "intellectuals" in the Soviet Union as a byproduct— a consequence—of Gorbachev's quest for new political thinking. Where Snyder sees Gorbachev worrying about this constituency, I see them worrying about his continued willingness that let them play in defense politics. See Snyder, "Gorbachev Revolution," pp. 110–15.

18. To be sure, military-technical considerations remain important to policy deliberations. Gorbachev has noted that all preparations for the Washington

summit were discussed in the Politburo, and that everything was considered from a military-technical viewpoint. "Vystupleniye M. S. Gorbachev po sovetskomu televideniyu."

19. Interviews. Some of the more provocative arguments by individuals such as Yevgeniy Primakov, Vitaly Zhurkin, and Deputy Chief of the General Staff General-Colonel Gareyev were emphatically characterized as their personal views, not government policy.

20. This curious fact is also noted by Akhromeyev, "Doktrina predotvrashcheniya voyny," p. 25.

21. D. T. Yazov, "Voyennaya doktrina Varshavskogo Dogovora—doktrina zashchity mira i sotsializma," Krasnaya zvezda, 28 July 1987, p. 2; Akhromeyev, "Slava i gordost'," p. 1.

22. Programma kommunisticheskoy partii, p. 50.

23. See Stephen M. Meyer, "Soviet Nuclear Operations," in Ashton B. Carter, John D. Steinbruner, and Charles A. Zraket, eds., Managing Nuclear Operations (Washington, D.C.: Brookings Institution, 1987), chap. 15.

24. See Stephen M. Meyer, "Soviet Perspectives on the Paths to Nuclear War," in Graham T. Allison, Joseph S. Nye, Jr., and Albert Carnesale, eds., Hawks, Doves, and Owls (New York: Norton, 1985), chap. 7, pp. 167–205.

25. See Vitaly Zhurkin, Sergei Karaganov, and Andrei Kortunov, "Vyzovy bezopasnosti—staryye i novyye," Kommunist, no. 1 (1988): 43, 46. In Gorbachev's speech in Murmansk in fall 1987 he noted that if one gauged the threat facing the Soviet Union by the rhetoric and statements of Western leaders, then the threat would not appear to be declining. However, he pointedly observed, such words are always forgotten in a few days. See " 'Rech' tovarishcha Gorbacheva," p. 2.

26. " 'Rech' tovarishcha L. I. Brezhneva," p. 2.

27. See Yu. Zhilin, "Faktor vremeni v yardernyy vek," Kommunist, no. 11 (1986):115. This is a serious revision of a long-standing Leninist principle of military doctrine.

28. Gorbachev is quoted to this effect by Akhromeyev, "Doktrina predotvrashcheniya voyny," p. 25. Reinforcing this image, Soviet discussions have begun to emphasize that the presence of nuclear power plants, chemical industry facilities, and hydroelectric complexes in the likely theater of war means that the destructiveness of conventional war will be unprecedented. See comments by Akhromeyev in "Dogovor po RSD-RMD: Protsess ratifikatsii," Vestnik ministerstva inostrannykh del SSSR, no. 5 (15 Mar. 1988): 23; also Zhurkin, Karaganov, and Kortunov, "Vyzovy bezopasnosti—staryye i novyye," p. 43. There is, of course, a strong propagandistic element at work here, but the issue is nonetheless a valid one.

29. This notion was sanctioned by Gorbachev in "Politicheskiy doklad tsentral'nogo komiteta KPSS XXVII s'yezdu KPSS," 27th S'yezd KPSS, Stenograficheskiy otchet #1 (Moscow: Politizdat, 1986), p. 88.

30. Yu. Zhilin, "Faktor vremeni," p. 120. It is important to realize how heretical this notion is from the standpoint of traditional Marxist-Leninist ideology, which holds that all international conflict is class-based. War is seen as an act of political and political-economic purpose and therefore must begin with political decisions. In contrast, the new thinkers seem to be particularly concerned with the way in which

military technology is eroding decisionmaking time. Ibid., pp. 119–20; Dobrynin, "Za bez'yadernyy mir," p. 19. Ye. Velikhov, "Prizyv k peremenam," Kommunist, no. 1 (1988): 50–51.

31. M. S. Gorbachev, "Oktyabr' i perestroyka; Revolyutsiya prodolzhayetsya," Kommunist, no. 17 (1987): 31–32; Ye. Primakov, "Leninskiy analiz imperializma i sovremennost'," Kommunist, no. 9 (1986): 109; G. Arbatov, "Militarizm i sovremennoye obshchestvo," Kommunist, no. 2 (1987): 113–14.

32. See Meyer, "Soviet Nuclear Operations."

33. These include mechanical and electronic control systems, personnel reliability programs, and new handling and deployment methods. (Interviews.) The issue was first broached publicly by former Minister of Defense Dmitri Ustinov in 1982; (see D. Ustinov, "Otvesti ugrozu yadernoy voyny," Pravda, 12 July 1982, p. 2,) and remains an important consideration. See D. T. Yazov, Na strazhe sotsializma i mira (Moscow: Voyenizdat, 1987), p. 32. General of the Army Yazov is Minister of Defense.

34. D. T. Yazov, Na strazhe sotsializma i mira (Moscow: Voyenizdat, 1987), pp. 30–32; V. Serebryannikov, "S uchtom real'nostey yadernogo veka," Kommunist voorezhennykh sil, no. 3 (1987): 10, 13–15; D. Volkogonov, "Imperativy yadernogo veka," Krasnaya zvezda, 22 May 1987, pp. 2–3. Serebryannikov is a General-Lieutenant of Aviation, Doctor of Philosophy, and professor; Volkogonov is a General-Colonel, Doctor of Philosophical Sciences, and professor. He is now the head of the Ministry of Defense's Military Historical Institute.

35. Yazov, Na strazhe sotsializma i mira, p. 31.

36. Khrushchev was able to implement many of his most controversial programs because the opposition was unable to organize against him. He rarely had a majority of the political leadership supporting him.

37. For a Soviet history of these efforts, see A. A. Gromyko and B. N. Ponomarev, Istoriya vneshney politiki SSSR, vols. 1 and 2 (Moscow: Nauka, 1986).

38. "Politicheskiy doklad," p. 86.

39. Ye. Primakov, "Novaya filosofiya vneshney politiki," Pravda, 10 July 1987, p. 4.

40. See A. Gorokhov, "Na puti yadernomu razoruzheniyu," Pravda, 16 Dec. 1987, p. 4.

41. Yazov, Na strazhe sotsializma i mira, pp. 5–20; V. Varennikov, "Na strazhe mira i bezopasnosti," Partiynaya zhizn, no. 5 (1987): 11–12. General Varennikov is the First Deputy Chief of the General Staff. There is a curious impression that General Yazov has been a supporter of Gorbachev's new political thinking on security. This is clearly not the case. Even a cursory reading of Yazov's comments reveals that while he "mouths the words," he summarily criticizes the ideas and alters their meaning.

42. See Yazov, Na strazhe sotsializma i mira, p. 19; Marshal V. G. Kulikov, "Strazh mira i sotsializma," Krasnaya zvezda, 21 Feb. 1988, p. 2. Marshal Kulikov is commander in chief of the Warsaw Pact and a first deputy Minister of Defense.

43. Of course, Ligachev never said that these were his questions, but the implications were obvious. See "Protsess ratifikatsii Dogovora po RSD-RMD nachalsya V Prezidiume Verkhovnogo Soveta SSSR," Vestnik ministerstva inostrannykh del SSSR, no. 4 (1 Mar. 1988): 12–19.

44. Varennikov, "Na strazhe mira i bezopasnosti," p. 12.

45. Compare and contrast Akhromeyev, "Slava i

gordost'," and Akhromeyev, "Doktrina predotvrashcheniya voyny." It is as though two different individuals wrote those articles. Perhaps this change in tone
is due to: (1) his failure to be named Minister of Defense after Marshal Sokolov's ouster; (2) peer pressure
from his more traditional colleagues; or (3) concern
over the growing influence of the nonmilitary "new
thinkers" on defense policy. For whatever reason,
Akhromeyev has changed his tone toward the Gorbachev defense agenda.

46. There are two forms of resource constraints
that must be considered: explicit constraints imposed
by decisions to limit resource flows to defense, and
implicit constraints that reflect scarcity of resources
in the Soviet economy (e.g., microprocessors).

47. Gorbachev, "Politicheskiy doklad," p. 87;
Dobrynin, "Za bez'yadernyy mir," p. 24. Interestingly, this has been a theme in the Soviet academic-
institute literature for a number of years.

48. Dobrynin, "Za bez'yadernyy mir," p. 24; Primakov, "Novaya filosofiya."

49. Semeyko, "Vmesto gor oruzhiya."

50. Vitaly Zhurkin, Sergei Karaganov, and Andrei Kortunov argue that relying exclusively on military-technical means is, by definition, setting one's
own security against the security of others. "Reasonable Sufficiency—Or How to Break the Vicious Circle," New Times, no. 40 (12 Oct. 1987): 13–15; and
"O razumnoy dostatochnosti," SShA: Ekonomika,
politika, ideologiya, no. 12 (1987): 14. Zhurkin is the
Director of the Institute for West European Studies.
Karaganov and Kortunov are senior researchers at the
Institute for the Study of the United States and Canada.

51. Zhurkin, Karaganov, and Kortunov, "Rea-
sonable Sufficiency," p. 13; Zhurkin, Karaganov, and
Kortunov, "O razumnoy dostatochnosti," p. 14; and
Dobrynin, "Za bez'yadernyy mir," p. 24.

52. Yazov, Na strazhe sotsializma i mira, p. 29;
Akhromeyev, "Doktrina predotvrashcheniya voyny";
and Kostev, "Nasha voyennaya doktrina," p. 10.

53. Varennikov, "Na strazhe mira i bezopasnosti," p. 11; Yazov, Na strazhe sotsializma i mira,
pp. 16, 19; Kostev, "Nasha voyennaya doktrina," pp.
9–10.

54. Yazov, Na strazhe sotsializma i mira, p. 19;
Akhromeyev, "Doktrina predotvrashcheniya voyny,"
pp. 23–24. American and British suggestions that
NATO must build up its theater nuclear firepower to
compensate for the removal of the INF systems has
been a major theme in Soviet political and military
discussions.

55. Brezhnev first raised the notion of sufficiency
for defense in his Tula speech in 1977. See " 'Rech'
tovarishcha L. I. Brezhneva," p. 2.

56. "Politicheskiy doklad," p. 90.

57. "O voyennoy doktrine gosudarstv uchastnikov Varshavskogo Dogovora," Krasnaya zvezda, 30
May 1987, pp. 1–2.

58. Semeyko, "Vmesto gor oruzhiya," p. 5.

59. On this point others have been far more explicit, arguing that the notion of accumulating enough
military power to balance all potential adversaries simultaneously is unrealistic and dangerous. See Zhurkin, Karaganov, and Kortunov, "Reasonable
Sufficiency," p. 14; and, by the same authors, "O
razumnoy dostatochnosti," p. 20.

60. Primakov, "Novaya filosofiya"; Zhurkin, Karaganov, and Kortunov, "Vyzovy bezopasnosti," pp.
47–48; and "Rech' tovarishcha Gorbacheva," p. 3.

61. Zhurkin, Karaganov, and Kortunov, "Reasonable Sufficiency," p. 14. These authors elaborate
further in "O razumnoy dostatochnosti," pp. 16–17.
It is important to understand that some Soviet writers
are using the word "asymmetry" primarily to connote
differences between the technical forms of the threat
and the response. They mean non-emulating responses, which presumes that equivalent military potential is retained and that resource requirements
would be more manageable (by Soviet standards).

62. Zhurkin, Karaganov and Kortunov, "O razumnoy dostatochnosti," pp. 17–20; Semeyko,
"Vmesto gor oruzhiya." Only one military officer,
General-Colonel Gareyev, a deputy chief of the General Staff, initially seemed to endorse the unilateral
restructuring of Soviet forces. See "Voyennaya doktrina organizatsii Varshavskogo Dogovora i eye pre-
lomleniye v mezhdunarodnoy politika," Vestnik
ministerstva inostrannykh del SSSR, no. 1 (22 June
1987):52–62. However, it is reported that Gareyev
was disappointed with that particular press conference, feeling that his views were not properly reported. Irrespective of Gareyev's actual views, other
Soviet military officials asserted that this idea was not
the position of the Ministry of Defense (Interviews).
It should be kept in mind that the Soviet forces have
been undergoing unilateral restructuring—what military analysts have traditionally referred to as force
modernization—on an almost continual basis since
1946. Gareyev may have been referring to ongoing
changes in Soviet forces.

63. Zhurkin, Karaganov and Kortunov, "O razumnoy dostatochnosti," p. 18. See also Gareyev's
comments in "Voyennaya doktrina organizatsii Varshavskogo Dogovora." Some involved in the development of this new political thinking have argued that
if the USSR waits for mutual agreements, nothing
will ever happen. They fear political inertia and stress
the belief that there is a three-year window to redirect
Soviet military policy. Soviet military officials disagreed with Gareyev's emphasis on unilateral restructuring (Interviews.)

64. A. Arbatov, "Glubokoye sokrashcheniye"
(Part 1), pp. 10–22; and by the same author, "Glubokoye sokrashcheniye strategicheskikh vooruzhe-
niye," MEMO, no. 5 (1988): 18–30.

65. Strategic Stabilnost' v usloviyakh radika-
l'nykh sokrashcheniye yaedernykh vooruzheniy (Moscow: Komitet sovetskkh uchenykh v zashchitu mira,
protiv yadernoy ugrozy, 1987).

66. Zhurkin, Karaganov, and Kortunov, "O razumnoy dostatochnosti," pp. 17–20. I believe that
this pushing of the policy boundaries is a direct consequence of the competition for influence among the
new thinkers, and between this group and the more
traditional elements in Soviet national security policy.
As such, I fail to find any evidence that such notions
reflect policy at the present time.

67. This is not a widely shared perspective among
new thinkers, and some openly disagree with their
colleagues who advocate unilateral Soviet reductions
in the name of reasonable sufficiency.

68. Yazov, "Voyennaya doktrina"; Varennikov,
"Na strazhe mira i bezopasnosti"; Serebryannikov, "S
uchetom real'nostey yadernogo veka," p. 10.

69. Kostev, "Nasha voyennaya doktrina," p. 12;
Yazov, Na strazhe sotsializma i mira, pp. 33–34.

70. See Yazov, "Voyennaya doktrina"; Yazov, Na
strazhe sotsializma i mira, pp. 33–34, defines sufficiency as having forces sufficient to carry out the tasks

of collective defense. Of course, military commentators proclaim that the Pact does not possess superiority and does not seek it, but they never dismiss its utility from a military standpoint (as they do dismiss the utility of strategic nuclear superiority). Akhromeyev, "Doktrina predotvrascheniya voyny," p. 26, defines sufficiency as having just enough forces to attain their objectives!

71. See, for example, V. Chernyshev, "O disbalansakh i asimmetrii," *Krasnaya zvezda*, 28 Jan. 1988, p. 3.

72. Interviews. See S. F. Akhromeyev, "Chto kroyetsya z bryussel'skim zayavleniyem NATO," *Krasnaya zvezda*, 20 Mar. 1988, p. 3. Soviet military spokesmen have added a new wrinkle to this issue by discussing NATO naval forces as "mobile groups" capable of devastating offensive actions against the rear areas of the Warsaw Pact. This is an explicit effort to develop a counter to Western concerns about Soviet tank army mobile groups (so-called operational maneuver groups).

73. See Yazov, *Na strazhe sotsializma i mira*, pp. 44–56; Kulikov, "Strazh mira i sotsializma," p. 2. Gareyev's early statement suggesting otherwise is an anomaly. See note 62.

74. Akhromeyev, "Doktrina predotvrashcheniya voyny," p. 26; statements by General-Colonel Chervov in "Voyennaya doktrina organizatsii Varshavskogo Dogovora," p. 56. Chervov is the head of the General Staff directorate concerned with treaties and arms control. In contrast to the new thinkers' positive view of Khrushchev's unilateral conventional force cuts in the late 1950s and early 1960s, authoritative Soviet military spokesmen have described those episodes as "sorry experiences" that hurt Soviet security while providing only temporary economic benefits. See the interview with General Tret'yak, in Yuri Teplya, "Reliable Defense First and Foremost," *Moscow News*, no. 8 (1988): 12. Tret'yak, a general of the Army, was named commander-in-chief of the Air Defense Forces following the Matthias Rust affair. Interestingly, Tret'yak was approved for that post by Gorbachev just eight months prior to this interview.

75. N. V. Ogarkov, "Zashchita sotsializma: opyt istorii i sovremennost'," *Krasnaya zvezda*, 9 May 1984, p. 2. Ogarkov suggested that some elements in the military did not subscribe to his notion that additional nuclear weapons had little marginal military utility.

76. Chervov, "Voyennaya doktrina organizatsii Varshavskogo Dogovora," p. 60; Yazov, "Voyennaya doktrina Varshavskogo Dogovora."

77. Noting that Warsaw Pact forces already are sufficient for defense, one military writer points out that in the future sufficiency can be maintained by: (1) keeping the present force levels on both sides, (2) mutual reductions to lower levels, or (3) increases to higher levels. He observes that, of course, the Soviet Union prefers the second alternative. P. Skorodenko, "Voyennyy paritet i printsip razumnoy dostatochnosti," *Kommunist vooruzhennykh sil*, no. 10 (1987): 17. See also Yazov, *Na strazhe sotsializma i mira*, pp. 33–41.

78. On this issue Gorbachev seems to be more in line with the old thinkers than he is with the new thinkers. See his interview in "Otvety M. S. Gorbacheva," p. 1.

79. V. Tatarnikov, "Do uroveney razumnoy dostatochnosti," *Krasnaya zvezda*, 5 Jan. 1988, p. 3. Here again they insist strongly on including the naval balance as a major element of the asymmetric force balance. (Interviews.) See also Akhromeyev, "Chto kroyetsya za bryussel'skim zayavleniyem NATO."

80. Strategy refers to the operational principles for employing forces in war. See "Strategiya voyennaya," *Voyennyy entsiklopedicheskiy slovar'* (Moscow: Voyenizdat, 1986), pp. 711–12; P. A. Zhilin, "Istoriya voyennogo iskutsstva" (Moscow: Voyenizdat, 1986), p. 406. General-Lieutenant Zhilin, recently deceased, was head of the Military Historical Institute of the Ministry of Defense. See note 1 on doctrine.

81. A. Kokoshin, "Razvitiye voyennogo dela i sokrashcheniye vooruzhennykh sil i obychnykh vooruzheniy," *MEMO*, no. 1 (1988). See also the Warsaw Pact declaration: "O voyennoy doktrine gosudarstv—uchastnikov Varshavskogo Dogovora," *Za novoye politicheskoye myshleniye v mezhdunarodnykh otnosheniyakh* (Moscow: Politizdat, 1987), p. 564.

82. A. Kokoshin and V. Larionov, "Kurskaya bitva v svete sovremennoy oboronitel'noy doktriny," *MEMO*, no. 8 (1987): 32–40, at p. 37, argue that the idea of needing to carry the war to an enemy's territory from the very beginning of the conflict is unscientific and nonanalytic, and that it precludes proper thinking about defense! All Soviet participants in these discussions—pro and con—are adamant that the notion of "defensive defense" applies only to theater warfare. In their view, it cannot apply to strategic or theater nuclear war. (Interviews.) In particular, while agitating for movement towards mutual defensive postures in the theater, Soviet "new thinkers" reject the idea that the US and the USSR could move to mutual strategic defensive defense along the road suggested by President Reagan in his SDI speech of March 1983. This philosophical and doctrinal inconsistency is explained by the vast destructiveness of nuclear weapons, which they claim invalidates any notion of defensive postures at the strategic nuclear level. The implications of this curious—but not surprising—perspective are discussed in greater detail below.

83. Zhurkin, Karaganov, and Kortunov, "Vyzovy bezopasnosti," p. 45; Kokoshin and Larionov, "Kurskaya bitva"; Kokoshin, "Razvitye voyennogo dela."

84. In fact, they ignore the possibility that a defensive posture could be more expensive than an offensive posture. The costs of establishing defense production lines based on new weaponry, new output requirements, new product mixes, etc., would be staggering.

85. Kokoshin and Larionov, "Kurskaya vitva," p. 33.

86. For example, Marshal of the Armor Troops Losik makes the case that tanks are a vital system for defensive operations. See O. Losik, "Issledovaniye opyta tankovykh armiy," *Krasnaya zvezda*, 8 Mar. 1988, p. 2. General-Lieutenant Zhilin offers the idea that some forms of nuclear weapons give defensive capabilities to the troops to stop enemy offensives; P. A. Zhilin, *Istoriya voyennogo iskusstva*, p. 416. The view that weapons could not be neatly categorized as offensive or defensive was vehemently repeated in interviews with Soviet military officials in Moscow.

87. Yazov, *Na strazhe sotsializma i mira*, p. 33, points out that a combination of offensive and defensive capabilities is necessary for security. In his view, counteroffensive capabilities do not contradict a defensive doctrine. This fundamental point actually predates the new thinking; see P. A. Zhilin, *O voyne i*

voyennoy istorii (Moscow: Voyenizdat, 1984), p. 437. See General Tret'yak's comments regarding the need for attack capabilities to defeat the enemy, in Teplya, "Reliable Defense," p. 12.

88. This is confirmed in both military writings and exercises. See P.A. Zhilin, *Istoriya voyennogo iskusstva*, p. 414; by the editors of the *Voyenno-istoricheskiy zhurnal*, "Itogi diskussii o strategicheskikh operatsiyakh Velikoy Otechestvennoy voyny 1941–1945," *Voyenno-istoricheskiy zhurnal*, no. 10 (1987): 8–24. For a typical contemporary tactical exercise report, see V. Efanov and V. Pryadkin, "Stoykost' oborony—Reportazh s ucheniya 'Druzhba-88,'" *Krasnaya zvezda*, 6 Feb. 1988, p. 1.

89. It is reported that in US Secretary of Defense Frank Carlucci's March 1988 meeting with Soviet Minister of Defense Yazov, Yazov insisted that the Soviet military posture in Europe is already defensive. See George Wilson, "Soviet Military Chief Calls Forces Defensive," *Washington Post*, 17 Mar. 1988, p. 38.

90. If one simply "dropped in" on the Soviet military literature during the past year it would be easy to find an article or two that appeared to support the view that the military is busy turning defensive defense into operational concepts. One would not understand, however, the context of those articles nor would one realize, for example, that *Voyenno-istoricheskiy zhurnal*, (the *Military Historical Journal*) had been running articles on strategic defensive operations since the end of the 1970s (see note 91), or that *Voyennyy vestnik (Military Herald)* had devoted a special issue to defensive operations each year since 1983 (see note 93).

91. Yu. Maksimov, "Razvitiye vzglyadov na oboronu," *Voyenno-istoricheskiy zhurnal*, no. 10 (1979): 10–16. M.M. Kozlov, "Organizatsiya i vedeniye strategicheskoy oborony po opytu Velikoy Otechestvennoy voyny," ibid., no. 12 (1980): 9–17; M. M. Kozlov, "Osobennosti strategicheskoy oborony i kontranastupleniya i ikh znacheniye dlya razvitiya sovetskogo voyennogo iskusstva," ibid., no. 10 (1981): 28–35. The culminating conference is reported in M. M. Kozlov, *Akademiya generalnogo shtaba* (Moscow: Voyenizdat, 1987), p. 185. General of the Army Maksimov was appointed Commander in Chief of the Strategic Rocket Forces in 1985. General of the Army Kozlov was Head of the General Staff Academy from 1979 until 1987.

92. Many Western observers have pointed to Gareyev as an example of an influential figure who appears to be "on board" the train of defensive defense. In fact Gareyev's latest book clearly shows that this perception of his views is wrong. See M. A. Gareyev, *Sovetskaya voyennaya nauka* (Moscow: Znaniye, 1987). I suspect that some tendentious editing of the general's comments by the Soviet news agency TASS may have misled some readers. See "Doktrina predotvrashcheniya voyny," *Krasnaya zvezda*, 23 June 1987, p. 3, and compare it with a more complete transcript of the press conference in "Voyennaya doktrina organizatsii Varshavskogo Dogovora." See also note 62.

93. See, for example, G. Ionin, "Sovremennaya oborona," *Voyennyy vestnik*, no. 4 (1981): 15–18; V. Galkin, "Oboronyaetsya batal'on," *Voyennyy vestnik*, no. 3 (1982): 16–19; V. Kravchenko and Yu. Upeniyek, "Vzvod PTUR v oborone," *Voyennyy vestnik*, no. 5 (1984): 67–68.

94. Soviet air defenses were reorganized several times between 1978 and 1985. These changes were intended to improve the integration of strategic and tactical air defense capabilities in general, and the defensive capabilities of Soviet ground forces in particular. See International Institute of Strategic Studies (IISS), *The Military Balance 1984–1985* (London: IISS, 1984), p. 14; US Department of Defense, *Soviet Military Power—1985* (Washington, D.C.: G.P.O., 1985), pp. 48–49. During the early 1980s the Soviet Ground Forces began experimenting with corps structures, the equivalent of two divisions but with a smaller logistical support requirement. IISS, *The Military Balance 1987–1988*, p. 27; *Soviet Military Power—1986*, pp. 65–66. If this corps structure is adopted, then it would imply a reduction in Soviet military manpower without a reduction in military capability. It may be that General Gareyev was thinking of these types of changes when he claimed that defensive concepts are affecting Soviet force structure already. See Gareyev's statement in "Voyennaya doktrina organizatsii Varshavskogo Dogovora."

95. Interviews.

96. Within the Academy of Sciences there is some interest in establishing international security as a recognized field of scientific inquiry. See Kokoshin, "Razvitiye voyennogo dela," p. 20. This would enable the Academy to set up research and training departments under its auspices to create a generation of civilian defense intellectuals. (Interviews.) General Gareyev, however, seems to have suggested that civilian scientists do not have the proper training to be involved in such issues. See Gareyev, *Sovetskaya voyennaya nauka*, p. 32.

97. This does not preclude a political decision to cut troop strength, move to lower readiness levels, or rebase certain forces. However, such changes are not in any way comparable to the kinds of radical restructuring of theater forces proposed by the new thinkers. In fact, such activities have occurred in both NATO and Warsaw Pact forces in the past.

98. This might explain the lessening of Marshal Akhromeyev's support for the new political thinking. As it moved from the abstract to the concrete and began encroaching on the professional domain of the military, he may have found himself drawn to defend his institution.

99. In their zeal, however, some of the new thinkers are having just the opposite effect. In particular, the writings of Zhurkin, Karaganov, and Kortunov, and of Kokoshin, are probably alienating the military. More ideological elements in the political leadership are being unnerved by the sudden reversal of formerly sacrosanct ideological concepts appearing in the works of Yu. Zhilin and others.

100. "27th s'ezd KPSS o dal'neyshem ukreplenii oboronosposobnosti stray i povyshenii boyevoy gotovnosti Vooruzhennykh Sil," *Voyenno-istoricheskiy zhurnal*, no. 4 (1986): 3–12; B. A. Zubkov, "Zabota KPSS ob ukreplenii ekonomicheskikh osnov voyennoy moshchi sotsializsticheskogo gosudarstva," *Voyenno-istoricheskiy zhurnal*, no. 3 (1986): 3–8. At least one military official seems to accept this line of argument. See V. Shabanov, "Shchit rodiny," *Ekonomicheskaya gazeta*, no. 8 (Feb. 1988): 18. General of the Army Shabanov is the Deputy Minister of Defense for Armaments.

101. Meyer, "Soviet Perspectives on the Paths to

Nuclear War"; John G. Hines, Phillip A. Petersen, and Notra Trulock III, "Soviet Military Theory from 1945–2000: Implications for NATO," *Washington Quarterly* 9, no. 4 (1986): 117–37.

102. Ogarkov, "Zashchita sotsializma"; Yazov's statements in "K miru bez yadernogo oruzhiya," *Pravda*, 10 Feb. 1988, p. 2; P. A. Zhilin, *Istoriya voyennogo iskusstva*, p. 407.

103. For a discussion of this problem see Stephen M. Meyer, "The Role of Economic Constraints in Soviet Defense," paper presented at the Conference on the Soviet Defense Economy, Hoover Institution, Stanford University, 23–24 March 1988.

104. See Akhromeyev's comments in "Za bez'yadernyy boleye bezopasnyy mir," *Vestnik Ministerstva inostrannykh del SSSR*, no. 3 (15 Feb. 1986): 42–49; Gareyev, *Sovetskaya voyennaya nauka*, p. 15.

105. Gorbachev's 15 Jan. 1986 statement can be found in "Zayavleniys general'nogo sekretarya Tsk KPSS M. S. Gorbacheva," *Za novoya politicheskoye myshleniye v mezhdunarodnykh otnosheniyakh* (Moscow: Politizdat, 1987), pp. 220–33.

106. There has already been a rebuttal of the new thinkers' recommendation for 95 percent cuts in strategic forces: Sergei Vybornnov, Andreo Gusenkov, and Vladimir Leontiyev, "Nothing Is Simple in Europe," *New Times*, Mar. 1988, pp. 37–39.

107. Stephen M. Meyer, "Soviet Theater Nuclear Forces," *Adelphi Papers* no. 187 and 188 (London: IISS, Winter 1983–84); Hines, Peterson, and Trulock, "Soviet Military Theory."

108. At the same time, one should also not be surprised that the Soviet military continues to try to work out credible applications of nuclear weapons on the battlefield. In particular, nuclear weapons have been the subject of increased interest in *defensive* operations. See P. A. Zhilin, *Istoriya voyennogo iskusstva*, pp. 412–17; G. Reznichenko, ed., *Taktika* (Moscow: Voyenizdat, 1984), pp. 152–73, on offensive actions, and pp. 174–221 on defensive actions.

109. See sources in notes 91 and 93, as well as A. I. Yevseyev, "Opyt osushchestvleniy manyovra s tsel'yu sosredotocheniya usiliy protiv udarnoy gruppirovki protivnika v khode oboronitel'noy operatsii fronta," *Voyenno-istoricheskiy zhurnal*, no. 9 (1986): 12–23; L. Zaytsev, "Ognevoye porazheniye protivnika v oborone," *Voyenno-istoricheskiy zhurnal*, no. 9 (1983): 20–24.

110. See, for example, V. I. Pevykin, *Fortifikatsiya: proshloye i sovremennost'* (Moscow: Voyenizdat, 1987).

111. Marshal Ogarkov is the Soviet military man most closely identified with this view, due to his numerous pronouncements on the subject while he was chief of the General Staff. In fact many in the Ministry of Defense shared his view, but Ogarkov was the only one who spoke out publicly on the subject. (Interviews.) Ogarkov did not, however, advocate scrapping current weapons systems and moving rapidly to new technologies—a position often ascribed to him and used to explain his removal as chief of the General Staff. Better than anyone, Ogarkov understood that new technologies—qualitative improvements in weaponry—have a minimal initial impact on military capabilities. The accumulation of quantities of weapons based on new technologies sufficient to make a difference takes a decade or more. See Ogarkov, *Vsegda v gotovnosti k zashchite otechestva* (Moscow: Voyenizdat, 1982), pp. 37–41.

112. One can never exclude the possibility that Gorbachev will order the withdrawal of several combat divisions, or even an entire tank army, from Eastern Europe. This, however, is a token gesture that would not seriously affect the military balance.

113. Interviews.

114. The issue surfaces repeatedly in Soviet official and public settings. See, for example, the interview with General-Colonel Chervov, "Dogovor, kotoryy povyshayet unroven' bezopasnosti," *Krasnaya zvezda*, 19 Dec. 1987, p. 1; and the letter to the editor of *Pravda* by H. Prozhogin, "Dogovor po RSD-RMD: za ili protiv," *Pravda*, 20 Feb. 1988, p. 4.

THE ABSOLUTE WEAPON

BERNARD BRODIE

WAR IN THE ATOMIC AGE

Most of those who have held the public ear on the subject of the atomic bomb have been content to assume that war and obliteration are now completely synonymous, and that modern man must therefore be either obsolete or fully ripe for the millennium. No doubt the state of obliteration—if that should indeed be the future fate of nations which cannot resolve their disputes—provides little scope for analysis. A few degrees difference in nearness to totality is of relatively small account. But in view of man's historically tested resistance to drastic changes in behavior, especially in a benign direction, one may be pardoned for wishing to examine the various possibilities inherent in the situation before taking any one of them for granted.

Excerpted and reprinted with minor revisions from Bernard Brodie, ed., The Absolute Weapon: Atomic Power and World Order (New York: Harcourt, Brace, 1946), pp. 21–24, 70–77. With permission.

It is already known to us all that a war with atomic bombs would be immeasurably more destructive and horrible than any the world has yet known. That fact is indeed portentous, and to many it is overwhelming. But as a datum for the formulation of policy it is in itself of strictly limited utility. It underlines the urgency of our reaching correct decisions, but it does not help us to discover which decisions are in fact correct.

Men have in fact been converted to religion at the point of the sword, but the process generally required actual use of the sword against recalcitrant individuals. The atomic bomb does not lend itself to that kind of discriminate use. The wholesale conversion of mankind away from those parochial attitudes bound up in nationalism is a consummation devoutly to be wished and, where possible, to be actively promoted. But the mere existence of the bomb does not promise to accomplish it at an early enough time to be of any use. The careful handling required to assure long and fruitful life to the Age of Atomic Energy will in the first instance be a function of distinct national governments, not all of which, incidentally, reflect in their behavior the will of the popular majority.

Governments are of course ruled by considerations not wholly different from those which affect even enlightened individuals. That the atomic bomb is a weapon of incalculable horror will no doubt impress most of them deeply. But they have never yet responded to the horrific implications of war in a uniform way. Even those governments which feel impelled to the most drastic self-denying proposals will have to grapple not merely with the suspicions of other governments but with the indisputable fact that great nations have very recently been ruled by men who were supremely indifferent to horror, especially horror inflicted by them on people other than their own.

Statesmen have hitherto felt themselves obliged to base their policies on the assumption that the situation might again arise where to one or more great powers war looked less dangerous or less undesirable than the prevailing conditions of peace. They will want to know how the atomic bomb affects that assumption. They must realize at the outset that a weapon so terrible cannot but influence the degree of probability of war for any given period in the future. But the degree of that influence or the direction in which it operates is by no means obvious. It has, for example, been stated over and over again that the atomic bomb is *par excellence* the weapon of aggression, that it weights the scales overwhelmingly in favor of a surprise attack. That if true would indicate that world peace is even more precarious than it was before, despite the greater horrors of war. But is it inevitably true? If not, then the effort to make the

reverse true would deserve a high priority among the measures to be pursued.

Thus a series of questions present themselves. Is war more or less likely in a world which contains atomic bombs? If the latter, is it *sufficiently* unlikely—sufficiently, that is, to give society the opportunity it desperately needs to adjust its politics to its physics? What are the procedures for effecting that adjustment within the limits of our opportunities? And how can we enlarge our opportunities? Can we transmute what appears to be an immediate crisis into a long-term problem, which presumably would permit the application of more varied and better considered correctives than the pitifully few and inadequate measures which seem available at the moment?

It is precisely in order to answer such questions that we turn our attention to the effect of the bomb on the character of war. We know in advance that war, if it occurs, will be very different from what it was in the past, but what we want to know is: How different, and in what ways? A study of those questions should help us to discover the conditions which will govern the pursuit of security in the future and the feasibility of proposed measures for furthering that pursuit. At any rate, we know it is not the mere existence of the weapon but rather its effects on the traditional pattern of war which will govern the adjustments which states will make in their relations with each other.

The Truman-Attlee-King statement of 15 November 1945 epitomized in its first paragraph a few specific conclusions concerning the bomb which had evolved as of that date: "We recognize that the application of recent scientific discoveries to the methods and practice of war has placed at the disposal of mankind means of destruction hitherto unknown, against which there can be no adequate military defense, and in the employment of which no single nation can in fact have a monopoly."

This observation, it would seem, is one upon which all reasonable people would now be agreed. But it should be noted that of the three propositions presented in it the first is either a gross understatement or meaningless, the second has in fact been challenged by persons in high military authority, and the third, while generally admitted to be true, has nevertheless been the subject of violently clashing interpretations. In any case, the statement does not furnish a sufficient array of postulates for the kind of analysis we wish to pursue.

It is therefore necessary to start out afresh and examine the various features of the bomb, its production, and its use which are of military importance. Presented below are a number of conclusions concerning the character of the bomb which seem to this writer to be inescapable. Some of the eight points listed already enjoy fairly universal acceptance; most do not . . .

1. The power of the present bomb is such that any city in the world can be effectively destroyed by one to ten bombs.

2. No adequate defense against the bomb exists, and the possibilities of its existence in the future are exceedingly remote.

3. The atomic bomb not only places an extraordinary military premium upon the development of new types of carriers but also greatly extends the destructive range of existing carriers.

4. Superiority in air forces, though a more effective safeguard in itself than superiority in naval or land forces, nevertheless fails to guarantee security.

5. Superiority in numbers of bombs is not in itself a guarantee of strategic superiority in atomic bomb warfare.

6. The new potentialities which the atomic bomb gives to sabotage must not be overrated.

7. In relation to the destructive powers of the bomb, world resources in raw materials for its production must be considered abundant.

8. Regardless of American decisions concerning retention of its present secrets, other powers besides Britain and Canada will possess the ability to produce the bombs in quantity within a period of five to ten years hence.

IMPLICATIONS FOR MILITARY POLICY

Under conditions existing before the atomic bomb, it was possible to contemplate methods of air defense keeping pace with and perhaps even outdistancing the means of offense. Long-range rockets baffled the defense, but they were extremely expensive per unit for accurate, single-blow weapons. Against bombing aircraft, on the other hand, fighter planes and antiaircraft guns could be extremely effective. Progress in speed and altitude performance of all types of aircraft, which on the whole tends to favor the attacker, was more or less offset by technological progress in other fields where the net result tends to favor the defender (e.g., radar search and tracking, proximity-fused projectiles, etc.).

At any rate, a future war between great powers could be visualized as one in which the decisive effects of strategic bombing would be contingent upon the *cumulative effect of prolonged bombardment efforts*, which would in turn be governed by aerial battles and even whole campaigns for mastery of the air. Meanwhile—if the recent war can serve as a pattern—the older forms of warfare on land and sea would exercise a telling effect not only on the ultimate decision but on the effectiveness of the strategic bombing itself. Conversely, the strategic bombing would, as was certainly true against Germany, influence or determine the decision mainly through its effects on the ground campaigns.

The atomic bomb seems, however, to erase the pattern described above, first of all because its enormous destructive potency is bound vastly to reduce the time necessary to achieve the results which accrue from strategic bombing—and there can no longer be any dispute about the decisiveness of strategic bombing. In fact, the essential change introduced by the atomic bomb is not primarily that it will make war more violent—a city can be effectively destroyed with TNT and incendiaries—but that it will concentrate the violence in terms of time. A world accustomed to thinking it horrible that wars should last four or five years is now appalled at the prospect that future wars may only last a few days.

One of the results of such a change would be that a far greater proportion of human lives would be lost even in relation to the greater physical damage done. The problem of alerting the population of a great city and permitting resort to air raid shelters is one thing when the destruction of that city requires the concentrated efforts of a great enemy air force; it is quite another when the job can be done by a few aircraft flying at extreme altitudes. Moreover, the feasibility of building adequate air raid shelters against the atomic bomb is more than dubious when one considers that the New Mexico bomb, which was detonated over 100 feet above the ground, caused powerful earth tremors of an unprecedented type lasting over 20 seconds.[1] The problem merely of ventilating deep shelters, which would require the shutting out of dangerously radioactive gases, is considered by some scientists to be practically insuperable. It would appear that the only way of safeguarding the lives of city dwellers is to evacuate them from their cities entirely in periods of crisis. But such a project too entails some nearly insuperable problems.

What do the facts presented in the previous add up to for our military policy? Is it worthwhile even to consider military policy as having any consequence at all in an age of atomic bombs? A good many intelligent people think not. The passionate and *exclusive* preoccupation of some scientists and laymen with proposals for "world government" and the like—in which the arguments are posed on an "or else" basis that permits no question of feasibility—argues a profound conviction that the safeguards to security formerly provided by military might are no longer of any use.

Indeed, the postulates set forth would seem to admit of no other conclusion. If our cities can be wiped out in a day, if there is no good reason to expect the development of specific defenses against the bomb, if all the great powers are already within striking range of each other, if even substantial superiority in numbers of aircraft and bombs offers no real security, of what possible avail can large armies and navies be? Unless we can strike first and eliminate a threat before it is realized in action—something which our national Constitution apparently forbids— we are bound to perish under attack without even an opportunity to mobilize resistance. Such at least seems to be the prevailing conception among those who, if they give any thought at all to the military implications of the bomb, content themselves with stressing its character as a weapon of aggression.

The conviction that the bomb represents the apotheosis of aggressive instruments is especially marked among the scientists who developed it. They know the bomb and its power. They also know their own limitations as producers of miracles. They are therefore much less sanguine than many laymen or military officers of their capacity to provide the instrument which will rob the bomb of its terrors. One of the most outstanding among them, Professor J. Robert Oppenheimer, has expressed himself quite forcibly on the subject:

"The pattern of the use of atomic weapons was set at Hiroshima. They are weapons of aggression, of surprise, and of terror. If they are ever used again it may well be by the thousands, or perhaps tens of thousands; their method of delivery may well be different, and may reflect new possibilities of interception, and the strategy of their use may well be different from what it was against an essentially defeated enemy. But it is a weapon for aggressors, and the elements of surprise and terror are as intrinsic to it as are the fissionable nuclei."[2]

The truth of Professor Oppenheimer's statement depends on one vital but unexpressed assumption: that the nation which proposed to launch the attack will not need to fear retaliation. If it must fear retaliation, the fact that it destroys its opponent's cities some hours or even days before its own are destroyed may avail it little. It may indeed commence the evacuation of its own cities at the same moment it is hitting the enemy's cities (to do so earlier would provoke a like move on the opponent's part) and thus present to retaliation cities which are empty. But the success even of such a move would depend on the time interval between hitting and being hit. It certainly would not save the enormous physical plant which is contained in the cities and which over any length of time is indispensable to the life of the national community. Thus the element of surprise may be less important than is generally assumed.[3]

If the aggressor must fear retaliation, it will know that even if it is the victor it will suffer a degree of physical destruction incomparably greater than that suffered by any defeated nation in history, incomparably greater, that is, than that suffered by Germany in the recent war. Under those circumstances no victory, even if guaranteed in advance—which it never is—would be worth the price. The threat of retaliation does not have to be 100 percent certain; it is sufficient if there is a good chance of it, or if there is belief that there is a good chance of it. The prediction is more important than the fact.

The argument that the victim of an attack might not know where the bombs are coming from is almost too preposterous to be worth answering, but it has been made so often by otherwise responsible persons that it cannot be wholly ignored. That the geographical location of the launching sites of long-range rockets may remain for a time unknown is conceivable, though unlikely, but that the identity of the attacker should remain unknown is not in modern times conceivable. The fear that one's country might suddenly be attacked in the midst of apparently profound peace has often been voiced, but, at least in the last century and a half, it has never been realized. As advancing technology makes war more horrible, it also makes the decision to resort to it more dependent on an elaborate psychological preparation. In international politics today few things are more certain than that an attack must have an antecedent hostility of obvious grave character. Especially today, when there are only two or three powers of the first rank, the identity of the major rival would be unambiguous. In fact, as Professor Jacob Viner has pointed out, it is the lack of ambiguity concerning the major rival which makes the bipolar power system so dangerous.

There is happily little disposition to believe that the atomic bomb by its mere existence and by the horror implicit in it "makes war impossible." In the sense that war is not something to be endured if any reasonable alternative remains, it has long been "impossible." But for that very reason we cannot hope that the bomb makes war impossible in the narrower sense of the word. Even without it the conditions of modern war should have been a sufficient deterrent but proved not to be such. If the atomic bomb can be used without fear of substantial retaliation in kind, it will clearly encourage aggression. So much more the reason, therefore, to take all possible steps to assure that multilateral possession of the bomb, should that prove inevitable, be attended by arrangements to make as nearly certain as possible that the aggressor who uses the bomb will have it used against him.

If such arrangements are made, the bomb cannot but prove in the net a powerful inhibition to aggression. It would make relatively little difference if one power had more bombs and was better prepared to resist them than its opponent. It would in any case undergo incalculable destruction of life and property. It is clear that there existed in the thirties a deeper and probably more generalized revulsion against war than in any other era of history. Under those circumstances the breeding of a new war required a situation combining dictators of singular irresponsibility with a notion among them and their general staffs that aggression would be both successful and cheap. The possibility of irresponsible or desperate men again becoming rulers of powerful states cannot under the prevailing system of international politics be ruled out in the future. But it does seem possible to erase the idea—if not among madmen rulers then at least among their military supporters—that aggression will be cheap.

Thus, the first and most vital step in any American security program for the age of atomic bombs is to take measures to guarantee to ourselves in case of attack the possibility of retaliation in kind. The writer in making that statement is not for the moment concerned about who will *win* the next war in which atomic bombs are used. Thus far the chief purpose of our military establishment has been to win wars. From now on its chief purpose must be to avert them. It can have almost no other useful purpose.

Neither is the writer especially concerned with whether the guarantee of retaliation is based on national or international power. However, one cannot be unmindful of one obvious fact: for the period immediately ahead, we must evolve our plans with the knowledge that there is a vast difference between what a nation can do domestically of its own volition and on its own initiative and what it can do with respect to programs which depend on achieving agreement with other nations. Naturally, our domestic policies concerning the atomic bomb and the national defense generally should not be such as to prejudice real opportunities for achieving world security agreements of a worthwhile sort. That is an important proviso and may become a markedly restraining one.

Some means of international protection for those states which cannot protect themselves will remain as necessary in the future as it has been in the past. Upon the security of such states our own security must ultimately depend. But only a great state which has taken the necessary steps to reduce its own direct vulnerability to atomic bomb attack is in a position to offer the necessary support. Reducing vulnerability is at least one way of reducing temptation to potential aggressors. And if the technological realities make reduction of vulnerability largely synonymous with preservation of striking power, that is a fact that must be faced. Under those circumstances any domestic measures which effectively guaranteed such preservation of striking power under attack would contribute to a more solid basis for the operation of an international security system . . .

NOTES

1. *Time*, 28 Jan. 1946, p. 75.
2. "Atomic Weapons and the Crisis in Science," *Saturday Review of Literature*, 24 Nov. 1945, p. 10.
3. This idea was first suggested and elaborated by Professor Jacob Viner. See his paper: "The Implications of the Atomic Bomb for International Relations," *Proceedings of the American Philosophical Society* 90, no. 1 (1946): 53ff. The present writer desires at this point to express his indebtedness to Professor Viner for numerous other suggestions and ideas gained during the course of several personal conversations.

NUCLEAR STRATEGY
The Case for a Theory of Victory
COLIN S. GRAY

For good or ill, or even perhaps for some of both, 1979 is almost certain to see the most intensive debate over strategic postural and doctrinal issues since the days of the misprojected missile gap back in 1959–60. SALT II is bringing it all together: the state of the balance, predictions of trends, the relevance (or otherwise) of strategic forces to superpower diplomacy, developments in high technology, Soviet intentions and Soviet performance, and the character of a desirable strategic doctrine.

Reprinted with minor revisions from International Security 4, no. 1 (1979): 54–87, *by permission of the MIT Press, Cambridge, Mass.* © 1979 by the President and Fellows of Harvard College and of the Massachusetts Institute of Technology.

The great SALT II debate, when finally joined, will probably cast as much shadow as light because much of the argumentation will avoid reference to truly fundamental issues. Indeed, a similar problem besets the quality of debate over individual weapon and related program questions (i.e., does the United States need a follow-on [to Minuteman III] ICBM, and if so of what kind?—or, does the United States need a civil defense program?—and so forth). Much of the earnest and even occasionally rather vitriolic debate over SALT, the MX-ICBM, cruise missiles, and the like, is almost purely symptomatic of disagreement over basic strategy—indeed, so much so that if attention were to be focused on the latter, then the generic, though not detailed, solutions to the former problems should follow fairly logically. As a somewhat inelegant axiom, this author will argue that a defense community which has not really decided what its strategic force posture is for has no business either engaging in strategic arms control negotiations, or in passing judgment on the merits of individual weapon systems.

A NEED FOR STRATEGY

Notwithstanding the popular, and indeed official, nomenclature which classifies our centrally based nuclear launch systems as *strategic*, the fact remains that there is an acute deficiency of strategic thinking pertaining to those forces. To many people, apparently, it is not at all self-evident that there are any issues of operational strategy relevant to the so-called strategic nuclear forces. Strategic nuclear war, presumably, is deterred by the prospect of the employment of those forces; while, should such a war actually occur, again presumably, each side executes its largely preplanned sequence of more and more punishing strike options in its Single Integrated Operational Plan (SIOP) and then dies with the best grace it can muster. This author has difficulty seeing merit (let alone moral justification) in executing the posthumous punishment of an adversary's society, possibly to a genocidal level of catastrophic damage, and hence has some difficulty discerning the value of such an option brandished as an intended prewar deterrent.[1] Of course, the US government has not been planning to *execute* even a rough facsimile of genocide for many years. But official, and even presidential, language (and perhaps thinking),[2] and war planning, have long been recognized to be somewhat different activities. This author is not confusing post–NSDM 242 nuclear weapon employment policy (NUWEP) guidance with assured destruction thinking,[3] although he believes that both would prove fatal to the US prospect of success in the event of war. In addition, this author does not accept the argument that US war plans are in good order: the real deficiency lies in the strategic forces that have been acquired to attempt to implement them (though there is considerable merit in that argument).

Absurd and murderous though mutual assured destruction (MAD) reasoning is to a strategic rationalist, one has to admit that the world, perhaps fortunately, is not ruled by strategic rationalists. Readers should be warned that this author does believe that there is a role for strategy—that is, for the sensible, politically directed application of military power in thermonuclear war. However, it is entirely possible that politicians of all creeds and cultures are, and will be, deterred solely by the undifferentiated prospect of nuclear war—which may be translated as meaning the fear of suffering societal punishment at an unacceptable level. Even if one suspects that the politician, a rank amateur in strategic analysis, will be deterred where a professional strategic analyst would advise that he should not be, there remain good reasons for listening to the cautionary words of the professional.

First, however unlikely the possibility from the perspective of American political culture, there could come to power in the Soviet Union a leader, or a group of collegial leaders, who would take an instrumental view of nuclear war. Whether or not such a group already is in place is very much a moot point. It could be profoundly imprudent simply to assume that strategic analysis has no bearing on the likelihood of occurrence of nuclear war. In a political context where a decision to act or not to act was finely balanced, military confidence and promises, or the lack thereof, could have a large influence on the political decision. *One of the essential tasks of the American defense community is to help ensure that in moments of acute crisis the Soviet general staff cannot brief the Politburo with a plausible theory of military victory.*

Second, it should not be forgotten that an important role for strategic analysis is the underpinning of a strategic doctrine which makes for the orderly management of, and choice between, defense programs.[4] If sufficient deterrent effect is believed to repose simply in the undifferentiated threat of nuclear war (of doing a lot of damage in a short space of time), on what basis does one choose what to buy? The essentially arbitrary guidelines for the "required" levels of assured destruction make some sense of this reasoning, (i.e., nobody claims that some "magic fraction" of threatened damage is needed for deterrence—even if annual Posture Statements do lend themselves to being misread in that fashion—but, some doctrinal guidance, beyond simply doing a lot of damage, is required for the provision of rules of thumb and for the suggestion of appropriate measures of merit).[5] Unfortunately, arbitrary doctrinal guidance for force sizing (and even quality) devised for the convenience of orderly administration tends to acquire an aura of strategic authority that was not originally intended and which it cannot bear.[6]

Third, and most important of all, it is sometimes easy to forget that a central nuclear war really could occur. Whatever the prewar feelings, thinking, and even instincts of a politician may have been, in the event of war it is safe to predict that he would demand a realistic war plan. The promise of imposing catastrophic levels of damage on Soviet society may, or may not, have merit as a prewar declaratory stance, but the politician would find that his learning curve on nuclear strategy rose very rapidly indeed following a deterrence failure.[7] Killing people and blowing down buildings, on any scale, cannot constitute a strategy—unless, that is, one has some well-developed theory which specifies the relationship between societal damage, actual and threatened, and the achievement of (political) war aims. Unless one is willing to endorse the proposition that nuclear deterrence is all bluff, there can be no evading the requirement that the defense community has to design nuclear employment options that a reasonable political leader would not be self-deterred from ever executing, however reluctantly.

Several years ago, Kenneth Hunt argued that among the NATO's more important duties was the need to guarantee to the Soviet Union that it could not avoid having to initiate a "major attack" should it move westwards in Europe.[8] NATO's function, on this theory, is not to defend Western Europe (at least, not directly); rather, it is to impose a high threshold for military-political adventure—to compel Soviet leaders, of any degree of intelligence or rabid hostility, that there are no relatively cheap or risk-constrained military options available. A similar logic underlies the policy positions of a major school of thought on strategic nuclear issues: all that we should, or need to, ask of the US strategic nuclear posture is that it be capable of inflicting a lot of punishment on Soviet society. Precisely how much punishment, and of what kinds, we need to promise must be a matter for conjecture, but fortunately for the robustness of a nuclear deterrent regime, precision is not required. Essential to this thesis are the beliefs that nuclear war cannot be won, that there is no way in which damage in such a war can be held down to tolerable levels in the face of an adversary determined to impose major damage, and that notwithstanding the many differences between the superpowers in strategic culture, each side should have no difficulty identifying, threatening, and, if necessary, ef-

fecting a level of damage that the other would find unacceptable.

Strategic debate of recent years on SALT and strategic forces' issues has become so polarized and has involved such a high level of polemical "noise" that the transmission of signals between contending camps has been difficult. It is a considerable oversimplification to assert that there are two schools of thought on nuclear deterrence—there are not: there are many. However, while admitting the many nuances that separate the exact philosophies and policy prognoses of individuals, it is useful to recognize that the impending major debate over SALT II is being nurtured by what amounts to a fundamental dispute over the requirements, and even the place, of a deterrent policy. It is argued that the premises of the two loose coalitions of policy contenders drive the debate that surfaces all too often with reference to specific defense and arms control issues[9]—at a level of detail where the policy action (or inaction) advocated can have integrity only if it is related to basic assumptions and explicit *desiderata* (it is analogous to discussing strategy without reference to war aims).

ASSURED DESTRUCTION AND ITS DESCENDANTS: A SICKLY BREED

The first school of thought,[10] which currently holds policy-authoritative sway in Washington (though it is not unchallenged within the government), may be thought of as the heir to the assured destruction ideas of the mid- to late 1960s. In 1964–65, the US defense community substantially abandoned the concept of damage limitation. It was believed that strategic stability (*the* magic concept—far more often advanced and cited than defined),[11] largely by virtue of a logic in technology (a truly American theme), could and should repose in what would amount to a strategic competitive stalemate. Each side could wreak unacceptable damage on the other's society, and neither could limit such prospective damage through counterforce operations or through active or passive defenses. Ballistic missile defense (BMD), in principle if not in contemporary technical realization, did of course pose a potentially fatal threat to this concept. A good part of the anti-ABM fervor of the late 1960s, which extended from Secretary of Defense Robert McNamara to local church and women's groups, can be traced to the strange belief that the goals of peace, security, arms control, stability, and reduced resources devoted to defense preparation, could all flow from a context wherein societies were nearly totally vulnerable and strategic weapons were nearly totally invulnerable. It is important to note both that US operational planning never reflected any close approximation to the assured destruction concept,[12] and that the legatees of MAD reasoning in the late 1970s have made some adjustments to the doctrine for its better fit with contemporary reality.

The adjective "strange" was applied to MAD reasoning in the paragraph above from the perspectives of the historian of strategic ideas and of the sociologist of strategic culture. A detached observer might well ask and observe as follows:

— Is MAD a matter of logic in technology (wherein offense-dominance is a physical law), or is it more in the nature of a self-fulfilling prophecy?

— If one side to the competition pursues the assured destruction path, how great a risk is it taking should the other side, for whatever blend of reasons, choose differently?

— MAD and its variants assume a noticeable measure of functional convergence of strategic ideas. But, strategic sociology tells us that each security community tends to design unique solutions to uniquely defined problems.[13]

— History may not tell us much with assurance, but certainly it suggests that technology cannot be frozen through arms control regimes: some qualitative boundaries upon its inventory expansion may be accomplished, but the slender historical arms control record suggests that politicians are as likely to freeze the wrong, as the right (i.e., stabilizing!), developments, and that prohibitions in one area serve to encourage the energetic pursuit of capabilities in other areas.[14]

— Although the existence of nuclear weapons encourages nuclear-armed states to be extremely careful in their mutual dealings, the fact of nuclear weapons has yet to be transcribed into some absolute injunction against war. The late Bernard Brodie has offered the thought that "if it is not yet an established fact it is at minimum a strong possibility that, at least between the great powers who possess nuclear weapons, the whole character of war as a means of settling differences has been transformed beyond all recognition."[15] Notwithstanding some recent claims by Raymond Garthoff,[16] it is not at all certain that Brodie was correct with respect to the Soviet Union, while—even if he were correct—it remains the case that a nuclear war could occur. Mutual assured destruction, whatever its (highly dubious) merits as a prewar deterrent declaratory stance, clearly has no appeal as operational guidance. Indeed, MAD is the antithesis of strategy—it relates military power to what?—to the punishment of a society for the sins or misjudgments of its rulers.

— That the prewar deterrent focus of MAD reasoning is appealing, but the historian is distressed by the realization that the advent of nuclear weapons has affected, but has not transformed the character of, international politics. Insane, drugged, or drunk American chief executives might seek to punish Soviet society, but a more responsible leadership has to be presumed to be likely to wish to adhere to an ethic of consequences (rather than revenge).

PREWAR DETERRENCE: A MISLEADING FOCUS?

The second school of thought embraces a coalition of people who are convinced: that Soviet strategic-nuclear behavior is difficult to equate even with a very rough facsimile of MAD reasoning; that the technical-postural basis for the American MAD thesis of the late 1960s has been eroded severely;[17] and that the theory of mutual assured destruction, even as amended officially in the 1970s in favor of greater flexibility, appears to have little of merit to offer as an operational doctrine. To state the central concern of this article, US official thinking and planning do not embrace the idea that it is necessary to try to effect the *defeat* of the Soviet Union. First and foremost, the Soviet leadership fears *defeat*, not the suffering of damage—and *defeat*, as is developed below, has to entail the forcible demise of the Soviet state. The second school of thought is edging somewhat tentatively towards the radical thesis that the theory of nuclear deterrence espoused, for example, by Bernard Brodie from 1946 until 1978, a theory which stressed the "utility in nonuse [of nuclear weapons],"[18] has had extremely deleterious effects upon the quality of Western strategic thinking and hence upon Western security. Above all else, our attention has been directed towards the effecting of prewar deterrence, at the cost of the neglect of operational strategy.[19] Incredible though it may seem, it has taken the United States' defense community nearly 25 years to ask the two most basic questions of all pertaining to nuclear deterrence issues: these are, first, what kinds of threats should have the most deterring effect upon the leadership of the Soviet state?—and, second, should prewar deterrence fail, what nuclear employment strategy would it be in the United States' interest actually to implement?

The debate has yet fully to be joined, but this revisionist school of strategic theorists sees little merit in contemporary official US deterrent policy (though the trend, as is discussed below, is mildly encouraging). The argument launched in public in late 1973 by then Secretary of Defense James Schlesinger concerning selective nuclear targeting[20] served, in retrospect at least, more to encourage persuasive fallacies than it did to focus attention upon the real problem. Our real problem, according to this view, is that the United States (and NATO-Europe) lacks a theory of victory in war (or satisfactory war termination). If, basically, one has no war aims (one has no image of enforced and favorable war termination, or of how the balance of power may be structured in a postwar world), on what grounds does one select a strategic nuclear employment policy, and how does one know how to choose an appropriate strategic posture? The answer, in this perspective, is that one does not know.[21] The answer provided by the first school of thought is that one chooses an employment policy (at least at the declaratory level), with roughly matching equipment, that has little if anything to do with the intelligent conduct of war. By definition it is assumed that nuclear war cannot be waged intelligently for rational political ends: the overriding function of nuclear weapons is the deterrence, not the waging, of war.

The second school of thought objects to the above reasoning on several grounds. First, the heavy focus upon nuclear threat, as opposed to nuclear execution, has encouraged a basic lack of seriousness about the actual conduct of a nuclear war—which feeds back into an impoverished deterrent posture and doctrine. Second, although peace may be its profession, one day— arising out of political circumstances that no one could foresee with any confidence—SAC might discover that war is its business, and it would be better for our future if, in that event, SAC were guided by some theory of how it should wage the war to a tolerable outcome. As noted earlier, the somewhat irresponsible ideas that pass for orthodox nuclear deterrent wisdom, with their bottom-line focus upon damaging Soviet economic assets, would (as a prediction) evaporate in their official appeal in the event of a deterrence failure. Fundamentally, they are not serious. SAC does, of course, have plans to wage a central war in a fairly serious way. But, and it is a very large but, (1) US strategic weapon acquisition policy (under four presidents) has failed to provide SAC with the means to prosecute the counter-military war very effectively, and (2) our counter-military planning (however well or poorly it could be executed) continues to be deprived of the overarching political guidance that it needs—a definition and a concept of victory.

Superficially at least, Schlesinger's strategic flexibility, as reflected in NSDM-242 and eventually in actual nuclear employment plans,[22] marked a noteworthy improvement in the quality of US deterrent policy. A richer menu of attack options, small and large, would provide a president with less-than-cataclysmic nuclear initiatives, should disaster threaten, or occur, in Europe or elsewhere.[23] Selectivity of scale and kind of attack, it was and is still argued,

enhances deterrence because it promotes the vital quality of credibility. As far as it goes, that line of thought has much to recommend it. Few would deny that a president should feel less inhibited over the prospective dispatch of (say) 30 (or even 130) reentry vehicles, than over the dispatch of 1000 to 3000—particularly when the targets for those 30 to 130 reentry vehicles had been chosen very carefully with a view to inflicting the minimum possible population loss on the Soviet Union. This theme of restraint, selectivity, and usability—all in the interest of enhancing credibility for the improvement of deterrence—attracted predictable negative commentaries from quarters prone to argue that a more usable nuclear deterrent was a nuclear deterrent more likely to be used[24] (similar arguments surfaced in connection with the protracted debate over enhanced radiation weapons). As Herbert Scoville explained: "[a] flexible strategic capability only makes it easier to pull the nuclear trigger."[25]

The second school of thought has no quarrel whatsoever with the ideas of flexibility, restraint, selectivity, minimal collateral damage, and the rest. But, it does have some sizable quarrel with strategic selectivity ideas that are bereft of a superordinate framework for the conduct and favorable termination of the war. Against the background of a fairly steadily deteriorating strategic nuclear balance,[26] the selectivity thesis simply adds what could amount to bigger and slightly more effective (i.e., the Soviets pay a higher military price) ways of losing the war. With a healthy strategic (im)balance in favor of the United States on the scale of, say, 1957 or 1962, one can see some logic to strategic flexibility reasoning. However, in the late 1970s and the 1980s, there are many reasons why a Soviet leadership might be less than fully impressed by constrained US strategic execution, and might well respond with a constrained nuclear reply that would (and indeed should) most likely impose a noteworthy measure of escalation discipline upon the United States.[27] Selective nuclear options, even if of a very heavily counter-military character, make sense, and would have full deterrent value only if the Soviet Union discerned behind them an American ability and will to prosecute a war to the point of Soviet political defeat.

TARGETING THE RECOVERY ECONOMY

Of very recent times, much of the nuclear strategy debate has narrowed down to a dispute over the validity of the thesis that *the real* (and *ultimate*) *deterrent* to Soviet risk-taking/adventure is the threat that our strategic nuclear forces pose to the Soviet recovery economy. Orthodox assured destruction thinking has evolved since the late 1960s. Notwithstanding the worthy deterrent motives of their authors, it is a fact that the last several annual Posture Statements of Robert McNamara endorsed a mass murder theory of nuclear "war" (to stretch a term). In the event of a central nuclear war, our declaratory policy was to kill tens of millions of Soviet citizens and destroy Soviet industry on a heroic scale.[28] Fortunately, under Presidents Nixon and Ford, killing people and blowing down buildings per se ceased to be strategic objectives (though, to repeat, this is not to impugn the motives of the MAD theorists of the 1960s—they wished to deter war: a highly ethical objective—it is only their judgment that is challenged here). Instead, it was noticed (belatedly—though welcome for all that) that recovery from war was an integral part of the Soviet concept of victory[29]—ergo, the United States should threaten the postwar recovery of the Soviet Union.[30]

The counter-recovery theory was not a bad one, but in practice several difficulties soon emerged. First, and most prosaically, American understanding of the likely dynamics of the Soviet postwar economy was (and remains) far short of impressive. In the same way that arms controllers have been hindered in their endeavors to control the superpower strategic arms competition by their lack of understanding of how the competition "worked," so our strategic employment planning community has found itself in the position of being required to be able to do that which nobody apparently is competent to advise it how to do. To damage the Soviet recovery economy would be a fairly elementary task, but to damage it in a calculable (even a roughly calculable) way is a different matter. Furthermore, the discovery, year by year through the mid- to late 1970s, of more and more Soviet civil defense preparation, threw into increasing doubt the "damage" expectancy against a very wide range of Soviet economic targets.

Second, it appears that the counter-economic recovery theme is yet another attempt to evade the most important strategic question. Should war occur, would the United States actually be interested in setting the Soviet economy back to 1959, or even 1929? Such an imposed retardation might make sense if it were married to a scheme for ensuring that damage to the American economy were severely limited. However, no such marriage has yet been mooted in policy-responsible circles.[31] Third, it is possible that the posing (even credible posing) of major economic recovery problems to the Soviet Union might be insufficiently deterring a prospect if Soviet arms could acquire Western Europe in a largely undamaged condition to serve as a recovery base; if the stakes in a war were deemed by Moscow to be high enough; and if the Soviet

Union were able, in the course of the war, to drive the United States back to an agrarian economy.[32] It is difficult to disagree with Henry Kissinger's comment on massive counterpopulation strikes.

> Every calculation with which I am familiar indicates that a general nuclear war in which civilian populations are the primary target will produce casualties exceeding 100m. Such a degree of devastation is not a strategic doctrine: it is an abdication of moral and political responsibility. No political structure could survive it.[33]

TARGETING THE SOVIET STATE

Nonetheless, the counter-recovery theme of the 1970s has prompted an interesting line of speculation. Namely, perhaps the recovery that should be threatened is not economic in character, but rather political.[34] Some revisionists of the second school of deterrence theorists argue that any kind of counter-economic strategy is fundamentally flawed because it leads into Soviet strength. The Soviet Union, like Czarist Russia, knows that it can absorb an enormous amount of punishment (loss of life, industry, productive agricultural land, and even territory), recover, and endure until final victory—provided the *essential assets of the state* remain intact. The principal assets are the political control structure of the highly centralized CPSU and government bureaucracy; the transmission belts of communication from the center to the regions; the instruments of central official coercion (the KGB and the armed forces); and the reputation of the Soviet state in the eyes of its citizens. Counter-economic targeting should have a place in intelligent US war planning, but only to the extent to which such targeting would impair the functioning of the Soviet state.

The practical difficulties that would attend an endeavor to wage war against the Soviet state, as opposed to Soviet society, have to be judged to be formidable. However, one would at least have established an unambiguous and politically meaningful war aim (the dissolution of the Soviet political system) that could be related to a postwar world that would have some desirable features in Western (and Chinese) perspective. More to the point perhaps, identification of the demise of the Soviet state as the maximum ambition for our military activity encourages us to attempt to seek out points of high leverage within that system. For examples, we begin to take serious policy note of the facts that:

—The Soviet peoples as a whole have no self-evident affection for, as opposed to toleration of, their political system or their individual political leaders.

—The Soviet Union, quite literally, is a colonial

empire—loved by none of its non–Great Russian minority peoples.[35]

—The Soviet state has to be enormously careful of its domestic respect and reputation, so fragile is the system deemed to be (evidence of Soviet official estimates of this fragility is located in the very character of the police state apparat that is maintained, and in the extreme sensitivity historically displayed in response to threats to Soviet authority in Eastern Europe).

—The entire Soviet political and economic system is critically dependent upon central direction from Moscow. If the brain of the Soviet system were destroyed, degraded, or—at a minimum—isolated from those at lower levels of political command who traditionally have been discouraged from showing initiative, what happens to the cohesion, or pace of recovery, of the whole?

—The peoples of Eastern Europe, and the minority republics in the Soviet Union itself, respect the success and power of the Soviet state. What happens in terms of the acquiescence of these peoples in Soviet (and Great Russian) hegemony if Soviet arms either are defeated, or are compelled to wage a long and indecisive struggle?[36]

Improbable though it may seem to many, this discussion is beginning to point towards a not-implausible theory of victory for the West. The alternative theory of deterrence/war waging proposed by some people within the second school, by way of contrast to the mass murder, punishment theme of the first school, comprises, essentially, the idea that the Soviet system be encouraged to dissolve itself. We resist the external military pressure of the Soviet Union, and effect carefully selected kinds of damage against the capacity of the Soviet state to function with authority at home. Soviet leaders can reason as well as Western defense analysts that large-scale counter-economic strikes would not serve Western interests (if only because of the retaliation that they would invite), whereas a war plan directed at the destruction of *Soviet* power would have inherent plausibility in Soviet estimation.

THE DECLINE (BUT NOT FALL) OF ASSURED DESTRUCTION

The essential backcloth to this counter-political control strategy has to be the ability to deny the Soviet Union any outcome approximating military victory in a short war. No matter how intelligent our ultimate goals may be for World War III, if the Soviet Union can (or believes that it can) win a rapid campaign against NATO-Europe[37] and, if need be, could escalate to do unmatchable damage to US strategic forces, while holding virtually all American economic assets at nuclear risk, then the second school would have failed to think through the totality of the deterrence problem. Needless to say, scarcely less significant a weakness in orthodox deterrence thinking than the fact that it focuses upon the threat of effecting the kinds of damage to the Soviet Union that should not be of interest to American policy makers actually to execute is the fact that it discounts totally the intrawar self-deterrent implications of the vulnerability of American assets. Foreign policy, in good part, is about freedom of action. Mutual assured destruction thinking, which still lurks in our declaratory policy and, presumably, in our war plans, virtually ensures self-deterrence and denies us the freedom of strategic-nuclear action that is a premise of NATO's strategy of flexible response.[38]

It is no exaggeration to claim that still-orthodox punishment-oriented deterrence thinking stemmed to a notable degree from a group of theorists who tended to think of the superpowers as though they were two missile farms: the attainment of an assured destruction capability by both sides would encourage the establishment and endurance of a technologically imposed peace.[39] The idea was fundamentally apolitical, astrategic, and was contrary to what the Soviet Union discerned, very sensibly, as its self-interest. Overall, as John Erickson has observed, American thinking on mutual deterrence, with its technological premises, reflects a "management" approach by way of contrast to "the Soviet 'military' inclination."[40] This author has difficulty understanding how a country like the United States, which has accepted obligations to project power at great distances in support of forward-located allies, could have seen any noteworthy attractions in the mutual hostage theory of deterrence. Of all countries, the United States needs a credible strategic force posture married to a theory of feasible employment. The catastrophic retaliation thesis, whether or not preceded by very selective nuclear employment options, is an idea it would be hard to improve upon were one seeking to minimize the relevance of (American) strategic weapons to world politics. It is probably appropriate largely to dismiss the deterrence-through-punishment ideas of the 1960s (or, at least, as formalized and codified in the 1960s) as the products of a defense community that was neither trained nor inclined to think strategically.[41] After all, the codification of the mutual punishment theory of deterrence as explicit policy—between 1964 and 1968—coincided exactly, and scarcely totally by chance, with gross *strategic* mismanagement in, and concerning, Vietnam. The same Department of Defense policymaking hierarchy that could not (or would not) design a theory of victory for Vietnam,[42] similarly abandoned such an apparently extravagant notion in the realm of strategic nuclear policy.

Until the mid-1960s, it is probably true to say that the quality of American strategic thinking concerning central war execution was a matter of relatively little importance: defeat for the Soviet Union was virtually implicit in the sheer scale of the strategic imbalance (i.e., even if the United States, in the event of war, had executed a foolish strategy, it would have done so on so massive a relative basis that the Soviet Union could not possibly have emerged from such a conflict with any net profit). But, as the capability of the two sides approached rough equivalence in the early to mid-1970s, the quality of strategic thinking, as reflected in actual plans, could easily make the difference between victory and defeat, or recovery and no recovery.[43]

The strategic debate referred to repeatedly in this article thus far is in a curious condition—with neither side quite sure of which positions are really worth defending. Notwithstanding the high polemical noise level, there has been a very notable narrowing of real differences of opinion over the past three to four years. For prominent examples:

—There is now widespread endorsement of the thesis that Soviet strategic thinking differs markedly from American. Indeed, recognition of what in the West we term a "warfighting" focus (on the part of the Soviet Union) has helped greatly to promote insecurity in the minds of many over either the inherent wisdom, or the practical advisability (or both), of a punishment-oriented theory of nuclear deterrence. It is many years since commentators in the United States have written about "raising the Russian learning curve."[44]

—Today, there is virtually universal agreement that, notwithstanding the many and accelerating weaknesses in the Soviet system, most of the major military balances have been moving to the disadvantage of the West. There is no consensus over whether or not those trends will continue into and through the 1980s, nor as to whether or not those adverse trends constitute cause for alarm as opposed merely to concern.[45]

—To the knowledge of this author, in the United States' defense and arms control community today there are no strong adherents to anything approximating the pure theory of mutual assured destruction. But, those who have disengaged from the arguments of Robert McNamara's *Posture Statements* for 1968 and 1969 seem to be uncertain as to what other doctrinal haven there might be available, while many of those who have rejected

MAD reasoning outright are less than confident that they have identified any superior alternative.

A CATALOGUE OF CONFUSION

The admittedly unsatisfactory designations, "first school" of thought and "second school" of thought, have been employed here because there is a considerable danger of unintended misrepresentation and undue simplification should any less neutral titles have been chosen. At this stage in the article, it may be safe to introduce the claim that the "first school" corresponds roughly to a focus upon "deterrence through punishment," while the "second school" tends to focus upon deterrence through the expectation of a militarily effective prosecution of war. Alas for neatness of description, neither group closely approaches its ideal type.[46]

The first school has recognized the immorality, inflexibility, and plain incredibility of having a strategic force posture preprogrammed to deliver only massive strikes against Soviet economic assets per se. However, because it rejects any thoroughgoing "warfighting" alternative as being certain to stimulate the arms competition, perhaps to render war more likely (through the believed consequent increase in strategic instability), and to make the prospects for negotiated measures of arms control far less encouraging, it has endeavored to design what might be termed "assured destruction with a human face." In place of the grisly (though superficially anodyne) prose of 1967–68 vintage McNamara, we are told about the deterrent virtues of strategic flexibility and the ultimately dissuasive merits of impairing Soviet economic recovery to a catastrophic degree. However, as observed above, the first school has yet to cope adequately with the rather obvious critical point that strategic flexibility and counter-recovery targeting are options that two can exercise. An intelligent strategy, if feasible, would be to design nuclear threats and employment options that the adversary either cannot or dare not match (or overmatch). Also, the first school has been increasingly overtaken by developments both in American weapon laboratories and, above all else, in the force posture that the Soviet Union is deploying. There is no logical reason why one should shift from a selective punishment thesis as a consequence of observing the Soviet strategic developments of recent years (if one endorses the punishment thesis); but it does appear that many commentators have been uneasy and defensive in a context where the Soviet Union is apparently challenging every major tenet of the American theory of strategic stability.[47]

First-school adherents are obliged by the contemporary climate of opinion in the United States to endorse the proposition that there should be an "essential equivalence" in strategic prowess between the superpowers, but what can this mean when there are very large asymmetries in strategic doctrine? Does it mean that the United States should invest in strategic capabilities that it deems to be destabilizing (say, hard-target counterforce and civil defense), solely in order to provide a perceptual match with Soviet capabilities?[48] In practice, first-school theorists are finding it very difficult to resist venturing into program regions which really have no place in their philosophy. The result, as may be seen in the curious mix of half-heartedly promoted programs and ill-assorted ideas that constitute current strategic policy, is something for nearly everybody. Because the official theory of nuclear deterrence is so uncertain, one sees the following:

—A new commitment to civil defense, qualified near-instantly by assurances that the new commitment will be neither very expensive nor so serious as to pose a threat to strategic stability.

—A commitment to preserve a survivable ICBM leg to the strategic forces triad, but one that will pose as little (and as late) a threat to fixed-site Soviet assets as the domestic SALT-related traffic will permit.

—A commitment to the devising of a new strategic nuclear employment doctrine, but not one which challenges any of the basic premises of the deterrence through punishment thesis.

—A commitment to second-strike, hard-target, counterforce prowess, on a scale which should fuel little first-strike anxiety in Moscow.[49]

—A commitment to a SALT process, and to a SALT II outcome, that has no reference to a stable strategic doctrine that has political integrity.[50]

As the period of intense debate over SALT II begins, it is fair to note that the US government sees merit in strategic flexibility, in some counterforce, in some degree of direct protection for the American public (though not much), and in the ability, in the last resort, to blow down large sections of the urban Soviet Union. This may be sufficient for deterrence, but a defense community should be capable of providing strategic direction that has more political meaning.

REVISIONIST CLAIMS: MYTHS AND REALITY

The second school of deterrence theory waxes eloquent on the absence of strategy in official policy, and indeed on the rarity of strategic thinking within the defense community,[51] but remains slightly abashed at the boldness, and even apparent archaism, of the logic of its own position. Today's revisionists are challenging the mature judgment of the finest flowering of American strategic thought. In policy terms at least, "The Golden Age" of American strategic thought extended roughly from 1956 until (at the outside) 1965.[52] The author of probably the single most highly regarded book to appear in this period[53] has written as follows in the pages of this journal:

The main war goal upon the beginning of a strategic nuclear exchange should surely be to terminate it as quickly as possible and with the least amount of damage possible—on both sides.[54]

Of course, the best prospect of all for minimizing (prompt) damage lies in surrendering preemptively. If Bernard Brodie's advice were accepted, the West would be totally at the mercy of a Soviet Union, which viewed war in a rather traditional perspective. The second school of nuclear deterrence is concerned lest its debating adversaries, neglecting the degree to which their ideas rest upon an unacknowledged measure of US firepower (if not strategy) superiority, which no longer exists, may mislead American policymaking into ignoring the possibility that nuclear age crises and wars can be lost, in a quite unambiguous fashion.

The various arguments of the second school (really a loose coalition) of strategic theory do, it must be admitted, lend themselves fairly easily to grotesque misrepresentation. For example, responsible theorists of this persuasion do not claim:

—That the Soviet Union believes that it will win a thermonuclear war (instead, it is claimed that there is an impressive apparent consensus among Soviet authorities to the effect that victory [and defeat] is possible).[55]

—That the Soviet Union either wants or expects to have to wage a central war with the United States. Military power can be most useful, and cost-effective, when the mere promise of its exercise achieves desired deterrent and compellent outcomes. It is very likely indeed that the Soviet Union sees its strategic forces largely in a counterdeterrent role—functioning to seal off local conflicts from influence by US strategic forces. However, any Soviet skepticism over the likelihood of central war does not (to the best of our knowledge) spill over into defense programs and doctrine in the form of weapons and ideas that make little or no military sense. Because war is possible, one prepares sensibly for it.

—That the Soviet Union anticipates achieving ultimate victory in war at little cost (much,

—though by no means all, of the argument of recent years in the United States concerning Soviet civil defense is really beside the point). Cautious committeemen in the Politburo could not afford to assume that T. K. Jones's optimistic studies (in Soviet perspective) were even close to the mark.[56] Second-school theorists, by and large, anticipate Soviet expectation of the necessity of accepting human and industrial loss on a catastrophic scale. However, catastrophic loss need not be intolerable loss—and may indeed be loss of a kind that the Soviet Union is willing to absorb, if the political stakes in the conflict are high enough (and if the alternatives to extreme measures of military action are very unattractive in their likely returns). It is fundamental to the Soviet theory of victory that the essential (and as much else in the) homeland be preserved. It is a sobering thought that the loss of 30 or 40 million people might well be compatible with a context defined by a Soviet leadership as victory: it would depend very much on who was among that 30 or 40 million.[57]

—Any certain knowledge concerning the requirements of deterrence or the proper conduct of thermonuclear war for a politically acceptable outcome. What is claimed is that the ideas of the 1960s (the assured destruction of people and industry) and the 1970s (small- and large-scale attack options of a carefully constrained nature, counter-economic recovery targeting, and the currently increasing interest in even more counter-military options [than in the past—which was fairly extensive]) cannot withstand critical examination, given the adverse evolution of the major East-West military balances, and the more mature Western understanding of the Soviet approach to the waging of war.

COUNTER-MILITARY TARGETING

Newspaper reports in late 1978 and early 1979 suggested that the Department of Defense was attracted to the idea of a substantially counter-military targeting doctrine, in contrast to the counter-economic recovery theme.[58] However, intra-governmental opposition to this idea is substantial, in part for reason of its budgetary implications, and in part because it offends some still fairly authoritative notions pertaining to the sacrosanct concept of stability. In 1978, a State Department publication claimed that "it is our policy not to deploy forces which so threaten the Soviet retaliatory capability that they would have an incentive to strike first to avoid losing their deterrent force."[59] Counter-military targeting is not, of course, even close to being a novel idea in US war planning. Indeed, one may

speculate to the effect that counter-military targeting already comprises the lion's share of strategic resource allocations in SIOP planning—a thought supported amply by the bevy of official commentaries offered in 1974–75 in support of the "Schlesinger shift" in targeting, largely following the guidance provided by NSDM-242 (of 1974). If there is a shift impending in favor of (still more) large counter-military strike options in the SIOP, one can speculate that such a shift might imply the paying of heavier attention to Soviet projection forces or, at a more basic level, a commitment to purchasing the ability to neutralize a far higher fraction of really hard Soviet military (and political) targets than is the case at present.

For the US government to endorse a full-fledged warfighting doctrine in the strategic realm would constitute a doctrinal revolution. Such a doctrine would deny the validity of the stability theory that has informed US defense and arms control policy since the mid-1960s.[60] Strategic stability, in the standard formula, requires that societal assets (people and industry) be nearly totally vulnerable, while strategic weapons be invulnerable. The Soviet Union has always believed in the value of the assured destruction option vis-à-vis the United States, but not in *mutual* assured destruction. It is too early to be certain, but even if the United States under President Carter *might* be willing to shift its declaratory focus (and eventually its actual targeting plans)—and to invest in actual military capability—more toward military targets than is the case at present, it is unlikely that it will be able to overcome its fundamental skepticism over the wisdom of approaching a central nuclear war as one should approach (or did approach, in pre-nuclear days) nonnuclear war. Pending the occurrence (and resolution in favor of change) of a sophisticated debate over the worth of still-fashionable ideas concerning crisis and arms race instability, American strategic policy will be shifted at the margin rather than rewritten. Also, a particular strategic posture, even one as large as that to be maintained in the 1980s by the United States (with the blessing of SALT II), is not omni-competent.

At the present time the United States does not have a strategic posture capable of seeking a military outcome to a war in which Western political authorities could place any confidence. Moreover, on the record extant, the interest of the Carter administration in purchasing such a posture rapidly has to be judged to be distinctly lukewarm. Carter's record on the character and timing of the MX ICBM program, on the B-1 bomber (aborted), and on civil defense, would have to fuel incredulity over the likely postural matching for any proclaimed new strategic doctrine with a war-waging, as opposed to a prewar deterrent, orientation. If, as (almost certainly over-) reported, the US government should inch towards a very heavily counter-military strategic nuclear employment doctrine, it will need to understand the requirements and limitations of such an approach. A nondefense professional might be somewhat puzzled by this discussion. As a general rule, one might observe, US war planning surely has *always* been oriented most heavily towards Soviet military targets (strategic forces, projection forces, command and control targets, and war-supporting industry and transport networks)—so what is new? The answer lies in the scope of the military targeting, in its ability to cope with a much harder target set than before, and in its design for separation from civil society. Anybody who sought to argue that the United States suddenly had discovered counter-military targeting as an interesting option would of course be guilty of misleading his audience. For example, Richard Burt of the *New York Times* wrote recently that:

The Carter Administration has revised the United States strategy for deterring nuclear war by adopting a concept that would require strategic forces to be capable of large-scale precision attacks against Soviet military targets as well as all-out retaliatory blows against cities.

The new strategy, which has emerged after months of debate in the Pentagon, represents a significant departure from the long-held concept that the United States needs only to threaten all-out retaliation against Soviet cities to deter Moscow from launching a nuclear attack.[61]

Clausewitz wrote of war that "[i]ts grammar, indeed, may be its own [i.e., war should be waged in a way that makes military sense, given its unique dynamics], but not its logic."[62] A US SIOP oriented towards different kinds of military targets should be guided by a political logic—what are our war aims? A rewriting and recomputing of the SIOP in an even more heavily counter-military direction than is the case at present could place the United States in a somewhat worse position than that occupied by the (major) allied politicians of World War I—there could be a determination to do *military* damage to the enemy (which is very sensible), but a lack of commitment to the idea of prosecuting the war to the point where the enemy is defeated militarily (unlike World War I). The question of just how the military damage to be wrought is to be translated into political advantage could easily be evaded.

It may be ungenerous to proffer such a negative (or, at least, skeptical) verdict, but it does seem that the official (at least in the Department of Defense) redirection of US strategic nuclear targeting preferences continues to neglect factors that bear upon the issue of desired war outcomes. As noted earlier, a counter-political control strategy cannot succeed unless a Soviet military offensive, at the theater and/or inter-

continental levels, is thwarted. However, it would be foolish to wage the military war without taking proper prior account of the Soviet perspective upon Soviet vulnerabilities. There is considerable danger that the United States, looking to the damage promise of an inventory of cruise missiles and (much later) MX ICBMs, will neglect the very important political criteria for strategic targeting. A theory of victory over the Soviet Union can be only partially military in character—the more important part is political. The United States and its allies probably should not aim at achieving the military defeat of the Soviet Union, considered as a unified whole; instead, it should seek to impose such military stalemate and defeat as is needed to persuade disaffected Warsaw Pact allies and ethnic minorities inside the Soviet Union that they can assert their own values in very active political ways. It is possible that a heavily counter-military focused SIOP might have the same insensitivity to Soviet domestic fragilities as may be found in the counter-economic recovery orientation of the 1970s.

In important respects, a heavily counter-military SIOP would be the kind of war plan that the Soviet Union is well equipped to counter. Notwithstanding its apparent war-waging focus, the American authors and executors of such a doctrine would be unlikely to have considered the conduct of war as a whole: really they would still be seeking, very substantially, to be responsive to prewar deterrence needs. With a clear political war aim—to encourage the dissolution of the Soviet state—much of the military war might not need to be fought at all. The apparently resolute determination of the American defense community not to think through its deterrence needs, which would involve addressing the question of war aims, promises to produce yet another marginal improvement in doctrine (after all, US strategic forces have always been targeted against Soviet military power—whatever annual Posture Statements may have said).[63] It may be worth reminding American policymakers in 1979 that the United States had a counterforce doctrine in Vietnam. A focus upon counter-military action, bereft of an overarching political intent, save of the vaguest kind, is unlikely to serve American interests well, except by unmerited luck.

The warfighting theme which now has limited, though important, official support in Washington, comprises no more than half of the change in thinking that is needed. It is essential for prewar deterrent effect that Soviet leaders not believe they could wage a successful short war. But, for reasons that none could predict in advance, war might occur regardless of the prewar theories and the postures of the two sides. In that event, it will be essential that the United States has a theory of war responsive to its political interests.[64] Because a counter-military focus in the SIOP is not informed by a clear goal of political victory against the Soviet state, the United States is unlikely to be able to wage an intercontinental nuclear war in a very intelligent fashion. In World War II, American wartime leaders declined to attempt to look beyond the battlefield, so long as the war was still in progress, with results of impressive negative educational value for succeeding generations. How much more intelligent it would be to have explicit war aims that should, in and of themselves, have considerable pre- and intrawar deterrent value.

One hesitates to criticize the reported current trend in official thinking, so healthy a change is it in its war-waging focus from the murderous and impolitic counter-economic themes of the 1960s and (most of the) 1970s. Nonetheless, the point has to be made that there continues to be an absence of political judgment overseeing US strategic nuclear employment policy and, ergo, there is a neglect of strategy. A possible change in the 1980s in strategic employment orientation towards the counter-military is fully compatible with a US defense community which would not be able to bring itself to think of thermonuclear war in terms of victory or defeat. The US defense community, substantially coerced in its thinking by the adverse trends in the major East-West military balances, has progressed from a counter-economic to a counter-military focus in its nuclear employment reasoning (although the mechanical details of war planning may well have focused more upon Soviet military assets than the US defense community generally understood to be the case), but it has yet to accept a *strategic* focus and advance to a counter-political control thesis. Unlike Soviet defense analysts, Western commentators continue to be bemused by the reality-numbing concept of "war termination." Wars are indeed terminated, but they are also won or lost. Moreover, if the US defense community envisages (as it must, realistically) the sacrifice (presumably unwilling) of tens of millions of Americans in a thermonuclear war, that sacrifice should be undertaken only in a *very* worthwhile cause. If there is no theory of political victory in the US SIOP, then there can be little justification for nuclear planning at all.

STABILITY AND THE NEED FOR DEFENSE

The principal intellectual culprit in our pantheon of false strategic gods is the concept of stability. For more than 15 years, influential members of the US defense and arms control community have believed that it is useful, or even essential, that the Soviet Union have guaranteed unrestricted strategic access to American societal assets. Such unrestricted access *was* believed to have a number of stabilizing consequences. In and of itself it should limit arms competitive activity (such activity as remained would stem from "normal" modernization and from efforts to offset counterforce-relevant developments on the other side),[65] while—more basically—it should promote some relaxation of tension, in that the Soviet Union would, belatedly, be assured of its ability to deter (through punishment) the United States.[66] (This theory has some features in common with the view that the four-fold rise in oil prices in 1973–74 was "good for us"—compelling us to confront the implications of our own profligacy in the energy consumption field.)

Analysts of all (or perhaps most) doctrinal persuasions have come at last to accept the view that the Soviet Union does not relax as a consequence of its achieving a very high quality assured destruction capability: the excellent reason for such continued effort is that the assured destruction of American societal assets plays no known role in Soviet deterrent or wartime planning—save as a threat to deter American counter-economic strikes. In addition, Soviet planners probably see considerable political coercive value for a postwar world in a very large counter-societal threat. Backward though it has seemed to some, the Soviet Union has provided unmistakable evidence of believing that wars, even large nuclear wars, can be won or lost. The mass murder of Americans makes a great deal of sense in terms of the authority structure of a postwar world (since the Soviet Union cannot consummate a victory properly through the physical occupation of North America), but such a grisly exercise has little or nothing to do with the prosecution of a war (save as a counterdeterrent threat).

American strategic (and arms control) policy, since the mid-1960s, has been misinformed by stability criteria which rested (and rest) upon a near-total misreading of Soviet phenomena. Soviet leaders are opportunists with a war-waging doctrine as their strategic leitmotiv. Supposedly sophisticated self-restraint in American arms competitive activity, designed so as not to stimulate "destabilizing" Soviet responses, has simply presented the Soviet Union with an upcoming period of strategic superiority of uncertain duration. The American stability theorem held only for so long as both sides endorsed it. There is legitimate dispute today over the quality of Soviet strategic programs, but no one, to the knowledge of this author, disputes the contention that the Soviet Union is seeking both to protect its societal assets (assured survival, not mutual assured destruction) and to pose the maximum threat to American strategic forces compatible with the adequate manipulation of Western hopes and fears for the future for the purpose of discouraging a strong American competitive response. Unlike the Soviet Union, the United States has declined to recognize (courtesy of its still-authoritative stability theory) that an adequate strategic deterrent posture requires the striking of a balance between offensive and defensive elements. There is a painful irony of several dimensions in this American intellectual failing.

First, among the more pertinent asymmetries that separate the US from the Soviet political systems, is the acute sensitivity of the former to the *personal* well-being of its human charges. It is little short of bizarre to discover that it is the Soviet Union, and not the United States, that has a serious civil defense program.[67] Second, potentially the strongest element in the overall Western stance vis-à-vis the Soviet Union is its industrial mobilization capacity. Reasonably good American BMD carries healthily terminal implications for Soviet opportunism or adventure. A BMD system that works well enables the United States to wage a long war and to mass produce the military means for eventual victory. So great is American mobilization potential, vis-à-vis the extant strategic posture, that US defense policy, logically, should endorse a defensive emphasis. Such an emphasis is the guarantor of strategic forces in overwhelming numbers *tomorrow*.

Third, if US strategic nuclear forces are to be politically relevant in future crises, the American homeland has to be physically defended. It is unreasonable to ask an American president to wage an acute crisis, or the early stages of a central war, while he is fearful of being responsible for the loss of more than 100 million Americans. If escalation discipline is to be imposed upon the Soviet Union, even in the direst of situations, potential damage to North America has to be limited. Damage-limitation has to involve both counterforce action and active and passive defenses. The claim that actually to protect (even very imperfectly) Americans and their industry would be destabilizing is a doctrinal cliché whose shallowness merits uncompromising exposure. Since virtually all Western commentators recognize that the Soviet Union is not moved in its strategic policy by assured destruction criteria, and since no one can deny that an American president could not threaten, or implement, even highly intelligent strikes against the Soviet body politic if American society is totally open to Soviet retaliation, the stability concept is in need of fundamental redefinition. As long as American society is essentially unprotected by BMD, air defense, and civil defense, the United States will have to lose any process of competitive escalation against the Soviet force posture anticipated for the 1980s.

Fourth, even if the arms controllers' argument were correct, that a defensive emphasis would stimulate the Soviet Union into working harder so as to be able to overcome it through offensive force improvement,[68] so what? Generically, the claim that this or that American initiative will catalyze Soviet reactions tends to be accorded far too respectful a hearing. Certainly it is sensible to consider adversary reactions and to take a full systemic look at possibilities, but a country as wealthy (and as responsible for international security) as the United States should not be deterred by the mere prospect of competition from undertaking necessary programs. (For example, an MX ICBM, deployed in a multiple protective structure [MPS] mode, will certainly have some noteworthy impact upon Soviet arms competitive activity, but such recognition does not constitute proof of the folly of deploying MX/MPS.)[69] Crude though it may sound, the United States would probably achieve more in the field of arms control if it decided to achieve and sustain a politically useful measure of strategic superiority,[70] than if it continues its endorsement of the elusive quality known as essential equivalence.

SUPERIORITY FOR STABILITY

If it is true, or at least probable, that a central war could be won or lost, then it has to follow that the concept of strategic superiority should be revived in popularity in the West. Superiority has a variety of possible meanings, ranging from the ability to dissuade a putative adversary from offering resistance (i.e., deterring a crisis), through the imposition of severe escalation discipline on opponents, to a context wherein one could prosecute actual armed conflict to a successful conclusion. There is certainly no consensus within the United States defense community today over the issue of whether or not any central war outcome is possible which would warrant description as victory. However, a consensus is emerging to the effect that the Soviet Union appears to believe in the possibility of victory, and that the time is long overdue for a basic overhaul of our intellectual capital in the nuclear deterrence field.[71] At the very least, most defense analysts would endorse the proposition that it is important for the United States to be able to deny the Soviet Union victory on its own terms.

There is a need for Western strategic thinkers to address and overcome the emerging tension between the (probable) requirements of high-quality deterrence and the still-authoritative and inhibiting ideas of crisis and arms race instability which have directed the US defense community away from programs that speak to Soviet reality. A false choice has misinformed the structure of our thinking. The historical record of the arms competition since the mid-

1960s shows that the choice has not been between, on the one hand, a measure of US restraint which would facilitate Soviet acquisition
of an assured destruction capability—an
achievement that would promote the prospects
of arms control negotiations intended to codify
a more stable strategic environment—and, on
the other hand, an absence of American restraint which would serve to stimulate Soviet
countervailing programs and which would be
doomed to failure anyway. Instead, the choice
has been between a measure of American restraint which facilitated the Soviet drive to
achieve a not implausible war-winning capability, and a relative absence of restraint which
would greatly complicate the life of Soviet defense planners. Overall, the evidence suggests
that the Soviet Union has not been seeking a
deterrent, as that concept and capability has
been (mis?)understood in the West for nearly
twenty years. The choice the United States confronts today is whether or not it will tolerate
Soviet acquisition, unmatched, of an emerging
warfighting capability which might, with some
good judgment and some luck, produce success
in crisis diplomacy and in war.

The instability arguments that are leveled
against those who urge an American response
(functionally) in kind are somewhat fragile. For
example, there is good reason to believe that
the Soviet Union would be profoundly discouraged by the prospect of having to wage an arms
competition against an American opponent no
longer severely inhibited by its long-familiar stability theory. In addition, an American warfight-
ing-oriented strategic posture, if well designed,
should not contribute to crisis instability. The
fact that the United States might pose a theoretical first-strike threat to much of the Soviet
strategic posture should not give aid and comfort to the "use them or lose them" argument.
A central purpose informing US strategic posture would be its denial of any plausible Soviet
theory of victory. Why the Soviet Union would
be interested in starting a war that it would
stand little, if any, prospect of winning is, to say
the least, obscure.

The contemporary debate over strategic doctrine, whatever its eventual effect may be upon
US war planning and declaratory policy, has
registered a qualitative advance over most of
the strategic thinking of the past 15 years. The
debate has focused upon what might be needed
to deter *Soviet* leaders, *qua* Soviet leaders, and
some (still unduly limited) attempt has been
made to consider operational, as opposed to prewar declaratory, strategy. Theories of prewar
deterrence, however sophisticated, cannot
guarantee that the United States will never slip
into an acute crisis wherein a president has to
initiate strategic nuclear employment or, de
facto, surrender. In such a situation, a president
would need realistic war plans that carried a

vision of the war as a whole and embodied a
theory of how military action should produce
desired political ends. In short, he would be in
need of strategy. Fortunately, still orthodox wisdom notwithstanding, there is no necessary tension between a realistic wartime strategy (and
the posture to match) and the prewar deterrence of undesired Soviet behavior.

NOTES

1. The actual execution of SIOP-level attacks
upon Soviet population and economic targets, on the
canonical scale advertised in the late 1960s, would be
either an act of revenge (and without political purpose), or—as initiative—would likely trigger a Soviet
response in kind. Assured destruction would leave an
adversary's (presumably surviving) political leaders
with nothing left to lose. Prominent among the political weaknesses of assured destruction reasoning is
the consideration that just as not all credible threats
need deter (if the threat is insufficiently awesome),
so not all awesome threats need deter (if they are
insufficiently credible).

2. President Carter, in his State of the Union
Message for 1979, advertised the "overwhelming" deterrent influence that reposed in only one Poseidon
SSBN (nominally bearing 160 reentry vehicles of 40
kt: 16×10). The president neglected to mention that
although 40 kt warheads could destroy a lot of buildings, it was not obvious that one Poseidon SSBN could
accomplish anything very useful by way of forwarding
the accomplishment of US war aims. "Transcript of
President's State of Union Address to a Joint Session
of Congress," *New York Times*, 24 Jan. 1979, p. A.13.

3. See US Congress, House Committee on
Armed Services, *Full Committee Consideration of
Overall National Security Programs and Related
Budget Requirements, Hearings*, 94th Cong., 1st
sess. (Washington, D.C.: G.P.O., 1975), testimony
of Edward Aldridge.

4. See Benjamin S. Lambeth, *How to Think
About Soviet Military Doctrine*, P-5939 (Santa Monica, Calif.: RAND, Feb. 1978), pp. 15–16.

5. On the rationales for the "magic fractions" of
damage that permeated assured destruction reason-
ing in the 1960s, see Alain C. Enthoven and K.
Wayne Smith, *How Much Is Enough? Shaping the
Defense Program, 1961–1969* (New York: Harper and
Row, 1971), chaps. 5–6.

6. NATO's 23/30 guideline is a case in point. For
planning convenience, a baseline "threat" had to be
identified in order to ensure that NATO did not underestimate its possible operational problems. It was
assumed, as a guideline only—*not* as a strategic prediction—that the Warsaw Pact would take 30 days to
mobilize for war in Europe and that NATO would
identify the character of the threat only 7 days into
the Pact mobilization, thereby granting 23 days for
countermobilization. The 30-day assumption was
never intended to stand as a judgment that the Soviet
Union would attack *only* after such a lengthy period
of mobilization; rather it was intended to generate a
large, as opposed to a more modest, theater threat.
Almost needless to say, 23/30 came to assume doctrinal significance.

7. As James Schlesinger once said: "[b]ut I might
also emphasize, Mr. Chairman, that doctrines control
the minds of men only in periods of non-emergency.

They do not necessarily control the minds of men during periods of emergency. In the moment of truth, when the possibility of major devastation occurs, one is likely to discover sudden changes in doctrine." Testimony in US Congress, Senate Committee on Foreign Relations, *Nuclear Weapons and Foreign Policy, Hearings*, 93d Cong., 2d sess. (Washington, D.C.: G.P.O., 1974), p. 160.

8. In *The Alliance and Europe: Part II: Defence with Fewer Men, Adephi Paper* 98 (London: IISS, 1973), p. 20 and passim.

9. The "two camps" premise is not defended in detail in the text because (a) it is very close to being a self-evident truth, and (b) such an exercise in description would divert the discussion away from ideas and towards a summary of debate—with details required that are really of secondary importance, at most, to the theme of the article. Opinion, of course, exists on a spectrum. However, this author predicts that if one designed a simple questionnaire containing say, ten "litmus-paper type" test questions of an either/or character, and submitted this questionnaire to 100 members of the US national security community, inside and outside of government, there would be little cross-voting by individuals between "liberal" and "conservative" replies. Moreover, if one knew what an individual's final judgment was on SALT II, yea or nay, that fact would be extremely helpful in predicting his/her position on a wide range of other security issues.

10. In some important respects, it is more accurate and more satisfactory, at least for the limited purposes of this article, to talk of two schools, really loose coalitions of functional allies, of thought, than it would be to attempt to design a sophisticated multidimensional categorization of attitude and opinion. The latter implies a commitment to an accuracy in personal detail that verges upon the trivial and yet which could never really be complete. Probably the most satisfactory attempt at the categorization and analysis of strategic attitudes was Robert A. Levine, *The Arms Debate* (Cambridge: Harvard University Press, 1963). However, even this excellent book suffered from the vices of its virtues. The very comprehensiveness of its coverage compelled the author to take at least semiserious note of opinions that are of no policy relevance.

11. See Edward Luttwak's contribution to "The Great SALT Debate," *Washington Quarterly* 2, no. 1 (1979): esp. 84–85.

12. US strategic nuclear planning was essentially unrevised from the Kennedy years until the early 1970s. See Desmond J. Ball, *The Strategic Missile Program of the Kennedy Administration, 1961–1963.* Unpublished manuscript (n.d.), part 3, chap. 2. For a definitive judgment we will have to await the eventual publication of the war plans (SIOPs) of the 1960s under the auspices of the Freedom of Information Act.

13. The concept of strategic culture is a fascinating one and is as obvious as it has been neglected. For a brief and interesting introduction to the subject, see Jack L. Snyder, *The Soviet Strategic Culture: Implications for Limited Nuclear Operatons*, R-2154-AF (Santa Monica, Calif.: RAND, Sept. 1977). The protracted SALT history has served to diminish enthusiasm for the strategic intellectual convergence thesis, but the US government is only at the beginning of attaining a due appreciation of the policy implications of the distinctive strategic culture thesis. This is one of those cases of rediscovery of the wheel. Most American strategic thinkers have always *known* that there was a uniquely "Soviet way" in military affairs, but somehow that realization was never translated from insight into constituting a serious and enduring factor influencing analysis, policy recommendation, and war planning.

14. Naval arms limitation by treaty in the 1920s and 1930s (with its heavy focus upon battleships and, eventually, cruisers) should stand as a classic lesson for all time. Also, it is worth recalling Bernard Brodie's judgment on the complex naval competition of the last decades of the nineteenth century. "It is very likely that a more costly and politically more dangerous competition was avoided because the Powers permitted the building to go on steadily, subject only to self-imposed restraints, which in a period of such rapid obsolescence of new material were certain to be very real." *Sea Power in the Machine Age* (Princeton: Princeton University Press, 1941), p. 254. A brilliant contemporary analysis of the unintended damage that can be wrought through the (mis-)control of technology is Edward N. Luttwak, "SALT and the Meaning of Strategy," *Washington Review of Strategic and International Studies* 1, no. 2 (1978): 16–28.

15. "The Continuing Relevance of *On War*," in Carl von Clausewitz, *On War*, ed. Michael Howard and Peter Paret (Princeton: Princeton University Press, 1978), p. 49.

16. "Mutual Deterrence and Strategic Arms Limitation in Soviet Policy," *International Security* 3, no. 1 (1978): 112–47. For a preliminary reply to Garthoff, see the commentary by Donald G. Brennan in *International Security* 3, no. 3 (1978–79).

17. Assured destruction may have residual merit today in the strict context of deterring a Soviet counter-societal assault, but US strategic forces have the same formal extended-deterrent duties that they have always had. As the Soviet Union has cancelled the more obvious US strategic nuclear advantages, and as the US continues to decline to seek to secure some measure of strategic superiority, so the attempt has been made to design "strategy offsets" for the adverse trend in the basic weapons balance. Very selective nuclear strike options, counter-economic recovery targeting, selective counter-military (and perhaps, in the 1980s, counter-political control) targeting, are all—to some degree—endeavors to effect an end run around the logical implications of an eroding military balance. This problem is well described with reference to the probable needs of NATO-Europe in Lawrence Martin's contribution to "The Great SALT Debate," pp. 29–37.

18. The title of chap. 9 of his *magnum opus*, *War and Politics* (New York: Macmillan, 1973). The historical boundaries of Bernard Brodie's career as a theorist of nuclear affairs were marked by *The Absolute Weapon* (New York: Harcourt, Brace, 1946), pp. 21–110; and "The Development of Nuclear Strategy," *International Security* 2, no. 4 (1978): 65–83.

19. Note the scorn which Brodie pours upon the idea of "war-winning strategies" in "The Development of Nuclear Strategy," p. 74. Nonetheless, a little earlier Brodie did observe that civilian scholars have "almost totally neglected" the question of "how do we fight a nuclear war and for what objectives?"—if deterrence fails (ibid., p. 66).

20. Probably the most powerful single exposition of "the Schlesinger doctrine" was Schlesinger's testimony in US Congress, Senate Committee on For-

eign Relations, Subcommittee on Arms Control, International Law and Organization, *U.S.-U.S.S.R. Strategic Policies, Hearing*, 93d Cong., 2d sess. (Washington, D.C.: G.P.O., 4 Mar. 1974).

21. The strategic flexibility theme was much criticized by representatives of the first school of deterrence theory (see Herbert Scoville, "Flexible MADness," *Foreign Policy*, no. 14 [1974]: 164–77, and Barry Carter, "Nuclear Strategy and Nuclear Weapons," *Scientific American* 230, no. 5 [1974]: 20–31), but those representatives—reasonably enough, from their perspective—did not offer the most telling line of criticism: namely, that strategic flexibility, however desirable in and of itself (a view which Scoville and Carter did not share), does not constitute, or even approximate, a strategy.

22. See Lynn E. Davis, *Limited Nuclear Options: Deterrence and the New American Doctrine*, Adelphi Paper 121 (London: IISS, 1975–6); and Desmond Ball, *Déjà Vu: The Return to Counterforce in the Nixon Administration* (Los Angeles: California Seminar on Arms Control and Foreign Policy, December 1974). The Davis characterization, with its focus upon limited nuclear options (LNOs), is very substantially misleading as to the basic thrust of NSDM-242.

23. See Martin's contribution to "The Great SALT Debate," passim.

24. For examples, see George W. Rathjens, "Flexible Response Options," *Orbis* 18, no. 3 (1974): 677–88, and Herbert Scoville, "First Use of Nuclear Weapons," *Arms Control Today* 6, no. 7/8 (1975): 1–3.

25. Ibid, p. 2.

26. Assessment of trends in the strategic balance tend to be driven to a nonmarginal degree by the doctrinal preferences of the assessor: somehow, people manage to find their beliefs supported by statistics. Nonetheless, it is difficult to analyze the contemporary strategic balance and find much comfort therein (almost regardless of one's doctrinal preferences). A group of analysts in ACDA has succeeded in achieving this quite remarkable goal. See *U.S. and Soviet Strategic Capability through the Mid-1980s: A Comparative Analysis* (Washington, D.C.: US Arms Control and Disarmament Agency, Aug. 1979). Needless to say, there are several important premises that one has to grant for the analysis to turn out as it does. Gloomier prognoses for the United States include: Santa Fe Corporation, *Measures and Trends: U.S. and U.S.S.R. Strategic Force Effectiveness*, DNA 4602Z (Washington, D.C.: Defense Nuclear Agency, Mar. 1978); John Collins, *American and Soviet Military Trends since the Cuban Missile Crisis* (Washington, D.C.: Center for Strategic and International Studies, Georgetown University, 1978); and John Collins and Anthony Cordesman, *Imbalance of Power: An Analysis of Shifting U.S.-Soviet Military Strengths* (San Rafael, Calif.: Presidio, 1978).

27. This thesis is argued forcefully in Paul H. Nitze, "Deterring Our Deterrent," *Foreign Policy*, no. 25 (1976–77): 195–210. On the subject of possible Soviet responses to American selective strike options see Benjamin S. Lambeth, *Selective Nuclear Options in American and Soviet Strategic Policy*, R-2043-DDRE (Santa Monica, Calif.: RAND, Dec. 1976). Also note the very brief discussion in Harold Brown, *Department of Defense Annual Report, FY 1979* (Washington, D.C.: G.P.O., 2 Feb. 1978), pp. 55–56, 62. Rather more interesting is Harold Brown, *Department of Defense Annual Report, FY 1980*

(Washington, D.C.: G.P.O., 25 Jan. 1979), pp. 77–78. These two paragraphs, weak though they are, constitute the strongest Posture Statement language in favor of (second-strike) hard-target counterforce that has been seen for well over a decade.

28. *Statement of Secretary of Defense Robert S. McNamara on the FY 1968–72 Defense Program and 1968 Defense Budget* (Washington, D.C.: G.P.O., 23 Jan. 1967), chapter 2; and *Statement by Secretary of Defense Robert S. McNamara on the FY 1969–73 Defense Program and the 1969 Defense Budget* (Washington, D.C.: G.P.O., 22 Jan. 1968), chap. 2. Also see Jerome H. Kahan, *Security in the Nuclear Age: Developing U.S. Strategic Arms Policy* (Washington, D.C.: Brookings Institution, 1975), pp. 94–106.

29. Counter-recovery targeting was not, of course, invented in the 1970s. In 1967, Robert McNamara said that "it seems reasonable to assume that in the case of the Soviet Union, the destruction of, say, one-fifth to one-fourth of its population and one-half to two-thirds of its industrial capacity *would mean its elimination as a major power for many years*." (Emphasis added.) *Statement on the FY 1968–72 Defense Program*, p. 39. Counter-recovery targeting has come, in the 1970s, to imply attacks on a more discrete character than those suggested in McNamara's words.

30. General George Brown, then chairman of the Joint Chiefs of Staff, was very explicit on this subject. "We do not target population per se any longer. What we are doing now is targeting a war recovery capability." Quoted in *The Defense Monitor* 6, no. 6 (1977): 2.

31. On the contrary, the current secretary of defense has written as follows: "I am not persuaded that the right way to deal with a major Soviet damage-limiting program would be by imitating it. Our efforts would almost certainly be self-defeating, as would theirs. We can make certain that we have enough warheads—including those held in reserve—targeted in such a way that the Soviets could have no expectation of escaping unacceptable damage." *Department of Defense Annual Report, FY 1979*, p. 65. Of course the United States *could* impose unacceptable damage upon the Soviet Union, but there is no good reason to believe that the current administration (1) knows what unacceptable damage means in Soviet terms; (2) would be willing to fund a US strategic posture capable of imposing truly unacceptable damage; or (3) would be capable of understanding that *our* offensive strategy will avail us very little if *our* domestic assets are totally at risk.

32. This author sees some merit in Bernard Brodie's comment (on the targeting of war recovery capability) that "[w]hatever else may be said about this idea, one would have to go back almost to the fate of Carthage to find an historical precedent." "The Development of Nuclear Strategy," p. 79.

33. "Kissinger's Critique [of SALT II]," *Economist*, 3 Feb. 1979, p. 18. Kissinger watchers should note that their subject traditionally has been as poor a strategic theoretician as he has been a strong foreign policy analyst.

34. This idea has had some US official status for at least five years, but its detailed meaning has never been probed rigorously.

35. The Soviet imperial thesis has been advanced strongly by Richard Pipes. See: *Russia Under the Old Regime* (New York: Scribner's, 1974); and "Détente:

Moscow's View," in Pipes, ed., *Soviet Strategy in Europe* (New York: Crane, Russak, 1976), pp. 3–44.

36. It is one thing if the Soviet state is able, as in The Great Patriotic War, to assume the mantle of defender of Mother Russia and if the general populace discerns no reasonable political alternative to Soviet power. It is quite another if the external enemy is being combatted militarily far from home, if "that enemy" seeks intelligently to exploit the latent fragilities of the Soviet system, and if the military damage suffered on Soviet soil is very substantially confined to Soviet-state type targets.

37. A very persuasive recent discussion of this area is C. N. Donnelly, "Tactical Problems Facing the Soviet Army: Recent Debates in the Soviet Military Press," *International Defense Review* 2, no. 9 (1978): 1405–12. Also see Peter Vigor, *The Soviet View of War, Peace, and Neutrality* (London: Routledge and Kegan Paul, 1975), pp. 14–15 and passim.

38. For an expansion upon this argument see Colin S. Gray, "The Strategic Forces Triad: End of the Road?" *Foreign Affairs* 56, no. 4 (1978): 771–78.

39. A brief clear statement of this thesis permeated Robert McNamara's statement introducing the supposedly China-oriented *Sentinel* ABM system. "McNamara Explanation of 'Thin' Missile Defense System," *Washington Post*, 19 Sept. 1967, p. A.10. Also see Wolfgang K. H. Panofsky, "The Mutual-Hostage Relationship between America and Russia," *Foreign Affairs* 52, no. 1 (1973): 109–18. Interestingly enough, some of the more intense denials of Mc-Namara-Panofsky reasoning seem to focus unduly upon the tactical merits of particular weapon systems.

40. "The Chimera of Mutual Deterrence," *Strategic Review* 4, no. 2 (1978): 16. Also very useful is the discussion in Fritz Ermarth, "Contrasts in American and Soviet Strategic Thought," *International Security* 3, no. 2 (1978): 138–55.

41. Policymakers in Washington might profit from frequent reminders of Clausewitz' definition of strategy. Strategy teaches "*the use of engagements for the object of the war*" (emphasis in the original). *On War*, p. 128.

42. As Alexis de Tocqueville and many lesser commentators have observed, the conduct of foreign policy is not, and (given its political structure) cannot be, an American forte. For a sense of perspective, it is worth noting that very few countries can wage long, *losing* (or perpetually inconclusive) wars and emerge with little, if any, domestic damage. If Americans feel ashamed, in different ways, over their Vietnam record, they should consider what the war in Algeria did to France. Admittedly, Algeria was a true *colonial* war, but still it was a case of a democracy attempting to cope with the consequences of military success and political defeat. Any American president should know that the only kind of war his country can fight, and fight very well, is one where there is a clear concept of victory—analogically, the marines raising the flag on Mt. Suribachi is the way in which a president should think of American wars being terminated. The more distant the Mt. Suribachi analogue from the case at hand, the more doubtful a president should be over committing US forces to action. The US public could have understood, and almost certainly would have approved, the US Marine Corps seizing Hanoi (intact or rubble—no matter) in 1965 or 1966, and *compelling* Ho Chi Minh (or a successor—again, no matter) to sign a peace treaty. That would be victory. American academic theorists of "limited" (and "sublimited"

in the late 1950s and early 1960s) war simply failed to understand their own country. Most Americans believe that if wars are not worth winning (in fairly classical terms), they are not worth fighting.

43. See T. K. Jones and W. Scott Thompson, "Central War and Civil Defense," *Orbis* 22, no. 3 (1978): 681–712; and Director of Central Intelligence, *Soviet Civil Defense* (Washington, D.C.: CIA, July, 1978). This latter study claims that Soviet casualties (only half of which would be fatalities) could be held to "the low tens of millions," though only under the most favorable conditions for the Soviet Union (p. 4). Some Boeing civil defense studies have suggested, by way of contrast, that under the most favorable conditions Soviet population fatalities would be less than ten million. Most commentators agree that a proper mix of offensive and defensive programs should make a dramatic difference to the prospects of early postwar recovery.

44. John Newhouse, *Cold Dawn: The History of SALT* (New York: Holt, Rinehart, and Winston, 1973), p. 4. Also of interest is Raymond L. Garthoff, "Salt 1: An Evaluation," *World Politics* 31, no. 1 (Oct., 1978): esp. pp. 3, 24.

45. It is beginning to be fashionable to concede that the West will have to endure several years of unusual peril in the early 1980s, in terms of military balances considered narrowly, but that that condition will be transformed in the latter half of the decade as the US strategic force posture accommodates cruise missiles, a follow-on ICBM and, eventually, the Trident 2 SLBM. A similar phenomenon is claimed for the trend in the theater balance in Europe: NATO's long-term defense program should have a very noticeable cumulative impact by the mid- to late 1980s. This author grants the *possible* validity of this theory, but is disturbed by the fragility of almost all of its premises. Henry Kissinger has commented persuasively on the early to mid-1980s being "a period of maximum peril" in "Kissinger's Critique," p. 20.

46. The US Department of Defense, in its declaratory policy, and even more in its actual operational planning (though *not* in its force acquisition), stands squarely between the two schools. DoD planning *looks* as though it is about the serious prosecution of war, but (a) the proper means are lacking, and (b) (to repeat a now familiar refrain) there is no theory of Soviet defeat to be discerned.

47. Soviet offensive-force development will, on current trends, pose an unacceptably high threat to the prelaunch survivability of the US ICBM force by the early 1980s (for contrasting analyses of this problem see John D. Steinbruner and Thomas M. Garwin, "Strategic Vulnerability: The Balance between Prudence and Paranoia," *International Security* I, no. 1 [1976]: 138–81, and Colin S. Gray, "The Future of Land-Based Missile Forces," *Adelphi Paper* 140 [1977]), while the Soviet Union continues to invest very heavily in active and passive defense of its homeland—thereby rejecting the thesis that it is desirable for Soviet assets to be at nuclear risk as hostages to the prudent behavior of Soviet (and American and Chinese) leaders.

48. On this subject see Paul C. Warnke "Apes on a Treadmill," *Foreign Policy*, no. 18 (1975): 12–29.

49. Brown, *Department of Defense Annual Report, FY 1980*, pp. 77–78.

50. It is difficult to tell a convincing story in support of SALT II, when the strategic doctrine that provides the political meaning in the strategic forces

posture is very uncertain. On what basis can one assess adequacy?

51. For examples, see Daniel O. Graham, "The Decline of U.S. Strategic Thought," *Air Force Magazine* 60, no. 8 (1977): 24–29; and Luttwak, "SALT and the Meaning of Strategy." The scope of *strategic* thinking may, of course, be reduced if one discerns no, or hardly any, political value in military action at the level of *nuclear* operations. In the words of Fritz Ermarth: "For many years the prevailing US concept of nuclear war's consequences has been such as to preclude belief in any military or politically meaningful form of victory." "Contrasts in American and Soviet Strategic Thought," p. 144. One might reformulate Clausewitz's definition of strategy so as to read "the use of [*the threat*] of engagements for the object of the war" (*On War*, p. 128, my addition in brackets) in order to accommodate the *strategy* of deterrence and compellence, but there is grave danger in the judgment offered by Bernard Brodie in 1946: "Thus far the chief purpose of our military establishment has been to win wars. From now on its chief purpose must be to avert them. *It can have almost no other useful purpose*" (*The Absolute Weapon*, p. 76). This is a prime example of a good idea becoming a poor idea when it is taken too far: at worst, it is a doctrinal formula for losing wars.

52. Two as yet unpublished manuscripts discuss the rise of (civilian) nuclear-age strategic theorizing in great detail. These are James King, *The New Strategy*, and my own *Strategic Studies and Public Policy: The American Experience*.

53. Bernard Brodie, *Strategy in the Missile Age* (Princeton: Princeton University Press, 1959).

54. Brodie, "The Development of Nuclear Strategy," p. 79.

55. See Richard Pipes, "Why the Soviet Union Thinks It Could Fight and Win a Nuclear War," *Commentary* 64, no. 1 (1977): 21–34.

56. See US Congress, House of Representatives, Committee on Armed Services, *Civil Defense Review, Hearings*, 94th Cong., 2d sess. (Washington, D.C.: G.P.O., 1976), pp. 206–67.

57. For an unsympathetic but useful review of revisionist arguments on Soviet civil defense, see William H. Kincade, "Repeating History: The Civil Defense Debate Reviewed," *International Security* 2, no. 3 (1978): 99–120.

58. Richard Burt, "U.S. Moving Toward Vast Revision of Its Strategy on Nuclear War," *New York Times*, 30 Nov. 1978, pp. A1, A7; Bernard Weinraub, "Pentagon Seeking Shift in Nuclear Deterrent Policy," *New York Times*, 5 January 1979, p. A5; and Richard Burt, "Carter Shifts U.S. Strategy for Deterring Nuclear War," *New York Times*, 10 Feb. 1979, p. 5. As Mark Twain said of the story of his alleged death: these reports are highly exaggerated.

59. The next sentence is intriguing. "However, this policy is contingent on similar Soviet restraint." *The Strategic Arms Limitation Talks, Special Report* no. 46 (Washington, D.C.: Department of State, July, 1978), p. 3. This official logic fails, even on its own terms, if US strategic forces are not vulnerable to preemptive attack.

60. But it had its intellectual genesis in the late 1950s. For examples, see Thomas C. Schelling: "Surprise Attack and Disarmament," in Klaus Knorr, ed., *NATO and American Security* (Princeton: Princeton University Press, 1959), chap. 8; "Reciprocal Measures for Arms Stabilization," in Donald G. Brennan, ed., *Arms Control, Disarmament, and National Security* (New York: Braziller, 1961), chap. 9.

61. "Carter Shifts U.S. Strategy for Deterring Nuclear War."

62. *On War*, p. 605.

63. In his valuable study of the counterforce debate of the early 1970s, Desmond Ball quotes an Air Force general as claiming (in Feb. 1973) that the SIOP was "never reworked under (President) Johnson. It is still basically the same as 1962." *Déjà Vu: The Return to Counterforce*, p. 17.

64. Such a war might not be tripped by a military accident that related to political intention on either side. It might pertain to matters of vital interest to both sides. In short, the US defense community might discover that it did have political goals that far transcended Brodie's prediction that the earliest possible war termination would likely be the superordinate objective (see note 54).

65. Classic "period-piece" statements of the arms-race stability thesis were George W. Rathjens: *The Future of the Strategic Arms Race: Options for the 1970s* (New York: Carnegie Endowment for International Peace, 1969); and "The Dynamics of the Arms Race," *Scientific American* 220, no. 4 (1969): 15–25.

66. Strange to note, the theory of arms race dynamics that featured as its centerpiece the proposition that each side acts and reacts in a fairly mechanistic fashion in pursuit of a secure assured destruction capability has now been discredited pretty well definitively by the historical facts, but the strategic policy premises that flow from that flawed theory have not been overhauled thoroughly. Since virtually all US commentators agree that the Soviet Union is not attracted to MAD reasoning, the long-familiar "instability" case against urban-area BMD and nonmarginal civil defense provision is simply wrong. We are still in search of an adequate explanatory model for the strategic arms competition. See Colin S. Gray, *The Soviet-American Arms Race* (Lexington, Mass.: Lexington, 1976).

67. To have a serious civil defense program does not mean that a country is preparing for war, any more than equipping a ship with lifeboats means that the shipping line is preparing to operate the ship in a dangerous manner.

68. An argument central to the case against urban-area ABM defense was that its banning by treaty would break the action-reaction cycle of the arms race: the Soviet Union would not need to develop and deploy offensive forces to overcome such an American deployment (in order to preserve their assured destruction capability). It is a matter of history that the ABM treaty banned the ABM defense of US cities, but Soviet offensive force improvements have marched steadily onward. The action-reaction thesis was logical and reasonable; it just happened to be wrong (it neglected the local color, the domestic engines of the arms competition).

69. If US MX/MPS should induce the Soviet Union to proceed down a similar path, then stability (by anyone's definition) would be promoted. Rubles spent on MPSs are rubles not spent on missiles and warheads. It is true that an MPS system in place might attract the Soviet Union to producing large numbers of missiles, undetected, to be surge-deployed in a period of acute need. However, the Soviet

government can produce ICBMs secretly now—in the absence of an MPS system they could be fired from presurveyed "soft" sites. The verification argument against MX/MPS is not a telling one, but—as a hedge—deployment of a fairly thin (preferentially assigned) ballistic missile defense system around the MPS could purchase an extraordinary degree of leverage vis-à-vis any secretly (or suddenly) deployed Soviet missiles. See Colin S. Gray, *The MX ICBM: Multiple Protective Structure (MPS) Basing and Arms Control*, H1-2977-P (Croton-on-Hudson, N.Y.: Hudson Institute, 1979).

70. On the political meaning of strategic power, see Edward N. Luttwak, *Strategic Power: Military Capabilities and Political Utility, The Washington Papers*, vol. 4 (Beverly Hills, Calif.: SAGE, 1976).

71. Noteworthy endeavors since the mid-1970s include: Robert Jervis, "Deterrence Theory Revisited," *World Politics* 31, no. 2 (1979): 289–324; Patrick Morgan, *Deterrence: A Conceptual Analysis* (Beverly Hills, Calif.: SAGE, 1977), and Richard Rosecrance, *Strategic Deterrence Reconsidered, Adelphi Paper* 116 (London: IISS, 1975). Even well-considered judgments published as recently as 1975 can look a little fragile in 1979. Consider these words of Professor Rosecrance: "Thus it is possible to say that although the deterrent requirements that were deemed necessary to protect Europe in the 1950s are probably not currently being met, they may not have to be met. An improvement in the Soviet-Western relationship makes them less necessary now than they were then." Ibid., p. 36.

ON FIGHTING A NUCLEAR WAR

MICHAEL E. HOWARD

Thirty-five years have passed since Bernard Brodie, in the first book that he or indeed anyone else had written about nuclear war, set down these words:

The first and most vital step in any American security program for the age of atomic bombs is to take measures to guarantee to ourselves in case of attack the possibility of retaliation in kind. The writer in making this statement is not for the moment concerned about who will *win* the next war in which atomic bombs have been used. Thus far the chief purpose of our military establishment has been to win wars. From now on its chief purpose must be to avert them. It can have almost no other useful purpose.[1]

For most of those 35 years, the truth of this revolutionary doctrine was accepted in the Western world as self-evident, and our whole military posture became based on the concept of deterrence that Brodie had defined so presciently and so soon. The fear of nuclear war *as such* was considered sufficient deterrent against the initiation of large-scale violence on the scale of the Second World War, and the policy of the United States, its allies, and perhaps also its adversaries was to create a strategic framework that made it not only certain that a nuclear attack would provoke a nuclear response, but likely that an attack with conventional weapons would do so as well. It was agreed almost without a dissenting voice that nuclear wars were "unwinnable." A nuclear exchange on any scale would cause damage of a kind that would make a mockery of the whole concept of "victory."

When Brodie died, that consensus was beginning to disintegrate. In the last article he

published, on "The Development of Nuclear Strategy," he reprinted the passage and defended it. While accepting the changes that had occurred both in weapons technology and in the structure of the military balance over the past 30 years, he saw no reason to alter his view. It was necessary to develop and to deploy nuclear weapons in order to deter their use by others. Such weapons, he believed, might perhaps be utilizable on a limited scale in the European theater. But as instruments of policy, as strategic tools in a general war, they could have no utility. Nuclear war was unfightable, unwinnable.

Was he wrong? Let me offer my own answer to that question.

Bernard Brodie's belief that a nuclear war could not be "won" did not of course mean that he did not hold strong views about the optimal deployment and targeting of nuclear weapons, a matter on which his opinion was greatly valued by successive chiefs of the Air Staff. The maintenance of a credible capacity for nuclear retaliation, and its structuring so as to create maximum political and psychological effect, was a problem with which he concerned himself deeply throughout his career. But if deterrence failed and these weapons had in fact to be used, what then? What should the political objective of the war be, and how would nuclear devastation help to attain it? How and with whom was a peace to be negotiated? Above all, in what shape would the United States be, after suffering substantial nuclear devastation herself, to negotiate any peace?

Reprinted with minor revisions from International Security 5, no. 4 (1981): 3–17, *by permission of the MIT Press, Cambridge, Mass.*

Brodie could have asked further questions, and no doubt did. What would be the relations between the Soviet Union and the United States after such a war? What would their position be in an international community that could hardly have emerged intact from a nuclear battle on so global a scale? In what way could the post-nuclear world environment be seen as an improvement on the prewar situation and one which it was worth enduring—and inflicting—such unimaginable suffering in order to attain? It is not surprising that, in his 1978 article, Brodie should have defined "the main war goal upon the beginning of a strategic nuclear exchange" as being "to terminate it as quickly as possible and with the least amount of damage possible—on both sides."

This phrase was taken up a year later by one of Brodie's keenest critics, Mr. Colin Gray, who countered: "Of course the best prospect of all for minimizing (prompt) damage lies in surrendering preemptively. If Bernard Brodie's advice were accepted, the West would be totally at the mercy of a Soviet Union, which viewed war in a rather more traditional perspective."[2] By the time that article appeared, Bernard Brodie had died, but I can well imagine his sardonic response: "Would the Soviet Union or anyone else view *anything* in a 'traditional perspective' *after* a nuclear exchange?"

Colin Gray's question does however go to the heart of the problem that has led many people in the United States to reject the conventional wisdom of Brodie's position as no longer relevant in a new and harsher age. Today, there is widespread doubt that a posture of nuclear deterrence, however structured, will be enough to prevent a Soviet Union that accepts nuclear war as an instrument of policy and has built up a formidable nuclear arsenal from thinking the unthinkable, from not only initiating but fighting through a nuclear conflict in the expectation of victory, whether the United States wishes to do so or not. And if such a conflict is forced upon the United States, how can she conduct it effectively unless she also has a positive objective to guide her strategy, other than the mass annihilation of Soviet civilians? Should she not also regard nuclear war "in a rather more traditional perspective"?

The controversy over this question will be familiar to every reader of foreign policy and military affairs. But it is worthwhile to at least briefly review it, in an effort to clarify current perceptions on the matter.

Let me make it clear from the outset that on the fundamental issue of nuclear warfighting, I side with Bernard Brodie rather than with his critics. As I understand it, the criticism of the deterrent posture arises from two linked sources. The first is the development of missile technology: the growing accuracy of guidance

systems, the miniaturization of warheads, the increasing capability of target acquisition processes, the whole astounding panoply of scientific development in which the United States has, to my knowledge, in every instance taken the lead, with the Soviet Union, at great cost to its economy, keeping up as best it can. Incidentally, I find it curious that a scientific community that was so anguished over its moral responsibility for the development of the first crude nuclear bombs should have ceased to trouble itself over its continuing involvement with weapons systems whose lethality and effectiveness make the weapons that destroyed Hiroshima and Nagasaki look like clumsy toys. Be that as it may, it is the continuous inventiveness of the scientific community, and I am afraid primarily the Western scientific community, that has made the pursuit of a stable nuclear balance, of mutual assured *deterrence* (which seems to me the correct explication of that much abused acronym MAD) seem to be the chase for an *ignis fatuus*, a will o' the wisp.

The second ground for criticizing the original concept of deterrence is the widespread belief that the Soviet Union does not share it, and never has shared it. The absence from Soviet textbooks of any distinction between a "deterrent" and a "warfighting" capability; the reiterated statements that nuclear weapons cannot be exempted from the Clausewitzian imperative that military forces have no rationale save as instruments of state policy; the confident Marxist-Leninist predictions of socialism ultimately prevailing over the capitalist adversary whatever weapons systems or policies it might adopt: does this not make it clear that American attempts to indoctrinate the Soviet Union in strategic concepts quite alien to their ideology and culture have failed? And this perception of the Soviet Union as a society prepared to coolly contemplate the prospect of fighting a nuclear war as an instrument of policy is enhanced by a worst-case analysis of its capacity to do so; its first-strike capacity against US land-based ICBMs; its much discussed civil defense program to reduce its own casualties to an "acceptable" level; and an historical experience of suffering which, according to some authorities, enables their leaders to contemplate without flinching the prospect of frightful damage and casualties running into scores of millions if it enables them to achieve their global objectives.

Now in dealing with those who hold this view of the Soviet Union I am conscious, only too often, that I am arguing with people whose attitude, like that of committed pacifists, is rooted in a visceral conviction beyond the reach of any discourse that I can command. It was a realization that was borne in on me by their deployment of two arguments in particular. The first was that since the Soviet Union had suffered some 20 million dead in the Second World War, they might equally contemplate further comparable losses in pursuit of a political objective sufficiently grandiose to warrant such a sacrifice. Now the United States has never suffered such losses, and I hope that it never will. But it is a matter of historical record to what shifts and maneuvers Stalin was reduced in his attempt to *avoid* having to fight that war; and speaking as a representative of a people who sustained nearly a million war dead between 1914 and 1918, I suggest that the record also shows that readiness to risk heavy losses in another war does not necessarily increase in direct ratio with the sacrifices one endured in the last.

The second argument that I encountered with even greater astonishment was that which maintained that Soviet civil defense measures provided incontrovertible evidence of Soviet intentions to launch a first strike. These arguments were all the more curious in that they were advanced almost simultaneously, and often from the same sources, as the very well-reasoned advocacy of a United States civil defense program of an almost identical kind, a program adopted by President Kennedy's administration and abandoned only in face of the kind of popular and congressional resistance with which the Soviet leadership does not have to contend. The difference between the development of civil defense in the two countries thus tells one rather more about their respective political structures than about their strategic intentions. Until recently indeed it would not have occurred to anyone outside a tiny group of strategic analysts in this country that civil defense preparations were anything except prudent and proper precautions for a remote but horribly finite possibility. Unfortunately their view is now more widely spread; and those of us in Europe who have been urging the advisability of taking even minimal precautions for civil defense now find ourselves accused by true believers on the left, as the Russians are accused by true believers on the right, of planning to precipitate the very catastrophe against which we seek to insure.

The debate about Soviet intentions has been conducted by people so far more expert than myself that I shall not seek to add to it beyond underlining and reinforcing Bernard Brodie's gently understated comment on those who see the buildup of Soviet forces over the past two decades as incontrovertible proof of their aggressive intentions:

Where the Committee on the Present Danger in one of its brochures speaks of 'the brutal momentum of the massive Soviet strategic arms build-up—a build-up without precedent in history,' it is speaking of something which no student of the American strategic arms build-up in the sixties could possibly consider unprecedented.[3]

In fact one of the oldest "lessons of history" is that the armaments of an adversary always seem "brutal" and threatening, adjectives that appear tendentious and absurd when applied to one's own.

The sad conclusion that I draw from this debate is that no amount of argument or evidence to the contrary will convince a large number of sincere, well-informed, highly intelligent and now very influential people that the Soviet Union is not an implacably aggressive power quite prepared to use nuclear weapons as an instrument of its policy. My own firmly held belief, however, is that the leadership of the Soviet Union, and any successors they may have within the immediately foreseeable future, are cautious and rather fearful men, increasingly worried about their almost insoluble internal problems, increasingly aware of their isolation in a world in which the growth of Marxian socialism does little to enhance their political power, deeply torn between gratification at the problems which beset the capitalist world economy and alarm at the difficulties which those problems are creating within their own empire, above all conscious of the inadequacy of the simplistic doctrines of Marxism-Leninism on which they were nurtured to explain a world far more complex and diverse than either Marx or Lenin ever conceived. Their *Staatspolitik*, that complex web of interests, perceptions and ideas which Clausewitz believed should determine the use of military power, thus gives them no clearer guidance as to how to use their armed forces than ours gives to us.

The evidence for this view of Soviet intentions seems to me at least as conclusive as that for the beliefs of, for example, the Committee on the Presdent Danger, who will no doubt consider me to be as visceral, emotive, and irrational in my beliefs as I have found all too many of their publications. I would only say, in defense of my own views, first that they take rather more account of the complexities of the historic, political, and economic problems of the Soviet Union than do theirs; and secondly, for what it is worth, that they correspond more closely with those held by most of the Europeans among whom I move than do those of the Committee on the Present Danger. Naturally in Europe as elsewhere there is a diversity of views, and there can be few more enthusiastic supporters of Mr. Paule Nitze than my own prime minister. Nonetheless, I have found in Europe a far more relaxed attitude towards the Russians than I have ever encountered in the United States because, paradoxically, we are not more frightened of them than are Americans, but rather less. I think we find it easier to see them as real people, with real, and alarming, problems of their own: people of whom we must be constantly wary and whose military power and propensity to use it when they perceive they can safely do so is certainly formidable; but with whom it is possible to do business, in every sense of the word, and certainly not as people who have any interest in, or intention of, deliberately unleashing a nuclear war as an instrument of policy.

And here again I should like to underline, if possible with even greater emphasis, what Bernard Brodie had to say about the Soviet dedication to Clausewitz's theory of the relationship of war to policy, if only as an act of personal homage to both Brodie and Clausewitz. Clausewitz was a subtle, profound, and versatile thinker, and his teaching about the relationship between war and policy was only one of the many insights he provided into the whole phenomenon of war. "War," he wrote, "is only a branch of political activity; it is in no sense autonomous . . . [It] cannot be divorced from political life—and whenever this occurs in our thinking about war, the many links that connect the two elements are destroyed, and we are left with something that is pointless and devoid of sense."[4]

Insofar as the Russians believe this and hammer it into the heads of successive generations of soldiers and politicians, we should admire and imitate them. When they castigate us for ignoring it and for discussing nuclear war, as we almost invariably do, *in vacuo*, they are absolutely right, and we should be grateful for their criticism. When I read the flood of scenarios in strategic journals about first-strike capabilities, counterforce or countervailing strategies, flexible response, escalation dominance and the rest of the postulates of nuclear theology, I ask myself in bewilderment: this war they are describing, *what is it about*? The defense of Western Europe? Access to the Gulf? The protection of Japan? If so, why is this goal not mentioned, and why is the strategy not related to the progress of the conflict in these regions? But if it is not related to this kind of specific object, what are we talking about? Has not the bulk of American thinking been exactly what Clausewitz described—something that, because it is divorced from any political context, is "pointless and devoid of sense"?

When I made these comments in a now much quoted article in *Foreign Affairs*, I was gratified but slightly alarmed by the response. I certainly did not expect my arguments to be cited in support of the thesis that a nuclear war is fightable and winnable, and that we should base our strategy on that assumption; and I am grateful for this opportunity to distance myself from that school of thought and explain why I do so.

I do not deny that Soviet theoreticians attempt to fit nuclear weapons into their Clausewitzean framework, and maintain that, should

nuclear war occur, nuclear weapons should be used in order to forward the overall goals of policy and ensure a victory for the armed forces of the Soviet Union. But one only has to state the opposite of this doctrine to accept that it is in theory unexceptionable, and that it would be difficult for them to say anything else. The ideological and bureaucratic framework within which this Soviet teaching has evolved has been convincingly described in recent articles by such experts as Ambassador Raymond Garthoff and Mr. Benjamin Lambeth,[5] and I find convincing the formulation propounded by the latter:

> [The Russians] approach their strategic planning with the thoroughly traditional conviction that despite the revolutionary advances in destructive power brought about by modern weapons and delivery systems, the threat of nuclear war persists as a fundamental feature of the international system and obliges the Soviet Union to take every practical measure to prepare for its eventuality . . . They appear persuaded that in the nuclear age no less than before, the most reliable way to prevent war is to maintain the appropriate wherewithal to fight and win it should it occur.

Logical as this doctrine may appear, and no doubt necessary for the maintenance both of ideological consistency and of military morale in the Soviet Union, the West's response, it seems to me, should *not* be to imitate it but to make it clear to the Russians, within their own Clausewitzean framework, that it simply will not work: that there is no way in which the use of strategic nuclear weapons could be a rational instrument of state policy, for them or for anyone else.

This view commands a satisfyingly wide consensus—or did until recently. Mr. Paul Nitze himself, in his famous plea for a maximalist US defense posture, emphasized that the object of the measures he proposed "would not be to give the United States a war-fighting capability: it would be to deny to the Soviet Union the possibility of a successful war-fighting capability"[6]; and another leading thinker of the maximalist school, Mr. Colin Gray, also accepts that "one of the essential tasks of the American defense community is to help ensure that in moments of crisis the Soviet general staff cannot brief the Politburo with a plausible theory of military victory."[7] But Mr. Gray believes that another task of the defense community is to brief the White House with a plausible theory of military victory, and that is surely a very different matter.

Mr. Gray is a Clausewitzean and believes that US strategy should be geared to a positive political object. "Washington," he suggests, "should identify war aims that in the last resort would contemplate the destruction of Soviet political authority and the emergence of a postwar world order compatible with Western values."[8]

For this it would be necessary to destroy, not the peoples of the Soviet Union in genocidal attacks on cities, but the apparatus of the Soviet state. The principal assets of the latter, he identifies as "the political control structure of the highly centralized Communist Party of the Soviet Union and government bureaucracy; the transmission belts of communication from the center to the regions; the instruments of central official coercion (the KGB and armed forces); and the reputation of the Soviet state in the eyes of its citizens. . . . The entire Soviet political and economic system," he writes, "is critically dependent upon central direction from Moscow. If the brain of the Soviet system were destroyed, degraded, or at minimum isolated . . . what happens to the cohesion, or pace of recovery, of the whole?"[9]

Now about this scenario there are several things to be said, and I am only sorry that Bernard Brodie is not still around to say them. The first problem is one that Mr. Gray quite frankly admits himself. "Is it sensible," he asks, "to destroy the government of the enemy, thus eliminating the option of negotiating an end to the war?"[10] The answer is no, it is not; unless we believe that out of the midst of this holocaust an alternative organized government would somehow emerge, capable, in spite of the destruction of all internal communications networks, of taking over the affairs of state. The alternative is, presumably, the conquest, occupation, and the reeducation of the Soviet peoples in "Western values"—an interesting but ambitious project which might be said to require further study.

Secondly, it is quite unrealistic to assume that such strikes against centers of government and communications could be carried out without massive casualties, numbering scores of millions, among the peoples we would be attempting to "liberate." And if historical experience is any guide at all, such sufferings, inflicted by an alien power, serve only to strengthen social cohesion and make support for the regime, however unpopular it might be, a question literally of physical survival. We now know that the strategic bombardment of Germany only intensified the control exercised by the Nazi regime over that unhappy nation. The sufferings inflicted on the Russian peoples during the Second World War—those 20 million casualties which, we are asked to believe, only whetted their appetite for starting a Third—not only strengthened Stalin's tyranny; it went far to legitimize it. The prospect of any regime in the least compatible with what Mr. Gray calls "Western values" emerging from a bloodbath on yet more horrific scale is, to put it mildly, pretty remote.

Finally, what would be going on here while the strategic strike forces of the United States were conducting their carefully calibrated and controlled nuclear war? I shall leave out of account the problems of command and control, of maintaining fine-tuning and selective targeting under the kind of nuclear retaliation that is to be expected during such an attack. This is the famous "C³I factor" addressed in President Carter's recent Directive No. 59, and some of the United States' finest technologists are no doubt working on the problem. Even if they do come up with plausible solutions, however, nobody can possibly tell whether in practice they will work; and for strategic planners to prepare to fight a nuclear war on the firm assumption that they would work would be criminally irresponsible. Nor do I address the question, more interesting to the allies than perhaps it is to the superpowers, of what would be happening in Western Europe during such an exchange. Dr. Kissinger, in that very curious speech he delivered in Brussels in September 1979, informed his audience that "the secret dream of every European was . . . if there had to be a nuclear war, to have it conducted over their heads by the strategic forces of the United States and the Soviet Union."[11] In fact I have yet to meet an intelligent European who thinks that anything of the kind would be possible, or that Western Europe would under any circumstances be omitted from the Soviet targeting plan. Few of us believe that there would be much left of our highly urbanized, economically tightly integrated and desperately vulnerable societies after even the most controlled and limited strategic nuclear exchange.

But it is the implications for the United States itself that I want to consider. Mr. Gray and his colleagues admit that there is a problem here, but they assert that "strategists can claim that an intelligent United States offensive strategy, wedded to homeland defenses, should reduce US casualties to approximately 20 million, which should render US strategic threats more credible."[12] Well, perhaps they can claim it, in the same way that Glendower could claim, in Shakespeare's *Henry IV*, that he could "call spirits from out the vasty deep"; they should be asked in return though, "but will they come when you do call for them?" How valid is such a claim—especially since a Soviet leadership in its death-throes would have no possible incentive, even if it had the C³I capability, to limit the damage to 20 million or to any other figure?

But even if it *is* valid, and granted that 20 million is a preferable figure to 180 million, it is not clear to me that Mr. Gray has thought through all the implications of his suggestion. Those twenty million *immediate* casualties— and we leave out those dying later from residual radiation—are only the visible tip of an iceberg of destruction and suffering of literally incalculable size. Most readers will be familiar with the very careful and sober report by the Congressional Office of Technology Assessment on "The Effects of Nuclear War,"[13] which came to the conclusion that

The effects of nuclear war that cannot be calculated are at least as important as those for which calculations are attempted. Moreover even these limited calculations are subject to very large uncertainties . . . This is particularly true for indirect effects such as deaths resulting from injuries and the unavailability of medical care, or for economic damage resulting from disruption and disorganization rather than direct destruction.

As for a small or "limited" attack, the impact of this, points out the Report, would be "enormous . . . [and] the uncertainties are such that no government could predict with any confidence what the results . . . would be, even if there was no further escalation." Certainly the situation in which the survivors of a nuclear attack would find themselves would be unprecedented. "Natural resources would be destroyed; surviving equipment would be designed to use materials and skills that might no longer exist; and indeed some regions might be almost uninhabitable. Furthermore, prewar patterns of behavior would surely change, though in unpredictable ways."

As for the outcome of the conflict in which these sufferings were incurred, I can only quote from a memorandum that Bernard Brodie wrote for RAND over 20 years ago but which has lost none of its relevance: "Whether the survivors be many or few, in the midst of a land scarred and ruined beyond all present comprehension, they should not be expected to show much concern for the further pursuit of political-military objectives."[14]

Under such circumstances, the prime concern of everyone, American, Russian, European—to say nothing of the rest of the world, which is always left out of these scenarios—would be simply to survive, and in an unimaginably hostile environment. As to what would become of "Western values" in such a world, your guess is as good as mine. It is my own belief that the political, cultural, and ideological distinctions that separate the West from the Soviet Union today would be seen, in comparison with the literally inconceivable contrasts between *any* pre-atomic and *any* post-atomic society, as almost insignificant. Indeed I am afraid that the United States would probably emerge from a nuclear war with a regime which, in its inescapable authoritarianism, looked much more like that which governs the Soviet Union today than that of the Soviet Union would in any way resemble the government of the United

States; and this would almost certainly be the case in Western Europe.

Admittedly, this is all guesswork. But what is absolutely clear is that to engage in nuclear war, to attempt to use strategic nuclear weapons for "warfighting" would be to enter the realm of the unknown and the unknowable, and what little we do know about it is appalling. Those who believe otherwise, whether they do so, like the Soviet writers, because of the constraints imposed by their ideology and cultural traditions, or, as do some Americans, out of technological hubris, are likely to be proved equally and dreadfully wrong, as wrong as those European strategists who in 1914 promised their political masters decisive victory before Christmas.

I take issue with Mr. Colin Gray in particular not because I do not admire his work, but simply because he has had the courage to make explicit certain views that are now circulating widely in some circles in the United States and which, unless publicly and firmly countered, might become influential, with catastrophic consequences. I also believe that if a thinker as intelligent as Mr. Gray is unable to provide nuclear strategy with a positive political object, no one else is likely to succeed any better. But this does not mean that Clausewitz's theory has to be abandoned, and that nuclear weapons can serve no political purpose. Clausewitz accepted that strategy might have a *negative* object and pointed out that, historically, this had more often than not been the case. This negative object he defined as being to make clear to the other side "the improbability of victory . . . [and] its unacceptable cost."[15] In a word, deterrence; or to reiterate Mr. Gray's own admirable words, to "ensure that in moments of acute crisis the Soviet general staff cannot brief the Politburo with a plausible theory of military victory."

This takes us back to where we began, to Bernard Brodie's warning that: "The first and most vital step in any American security program . . . is to take measures to guarantee to ourselves in case of attack the possibility of retaliation in kind." In principle nothing has changed since then, even though in practice the problem has become enormously more difficult. In particular, Bernard's phrase "in kind" has acquired a significance that he could not possibly have anticipated. With the diversification of nuclear delivery systems, deterrence becomes an ever more complex business; and prudent account has to be taken of the contingency that deterrence might fail, so as to provide feasible alternatives between holocaust and surrender. But the object of such "intra-war deterrence" would still be, as Mr. Brodie put it, "to terminate the strategic exchange as quickly as possible, with the least amount of damage possible—on both sides," in the interests not just of the United States but of mankind as a whole. Can one doubt that in 1914, rational European statesmen would have cut their losses and made peace at the end of the year if they had not been driven on by popular pressures and delusive expectations of victory? Or that if they had done so, the world would be a rather better place than it is today?

What about Bernard Brodie's other pronouncement, that "thus far the chief purpose of our military establishment has been to win wars. From now on its chief purpose must be to avert them"; does this remain valid? Well yes, it does; but with respect to Bernard's shade, there is nothing new about this. It has always been the role of military establishments in peacetime to dissuade their opponents from using force as an instrument of policy—even if needs be of *defensive* policy—by making it clear that any such action on their part would be counterproductive, either because they would *lose* in such a war, or because they could gain victory only at an unacceptably high cost. That still seems to me to be true. And it also seems to me to be true that such a deterrent posture lacks conviction if one does not have the evident capacity to fight such a war—in particular, to defend the territory which our opponent may wish to occupy.

This is where a "warfighting capability" comes in; a capability, not to fight through a war to an impossible, mutually destructive "victory"— and let us remember Clausewitz's epigram "in strategy there is no such thing as victory"—but to set on victory for our opponent a price that cannot possibly be afforded. And for this we must have the evident will and readiness to defend ourselves and one another: something that can only be made clear by the presence or availability of armed forces capable of fighting for territory—adequately armed, adequately trained, adequately supported, and, in our market economies, adequately *paid*.

This is the warfighting capability that acts as the true deterrent to aggression, and the only one that is convertible into political influence. There is as little reason to suppose that Soviet nuclear superiority will give them political advantages in the 1980s as that American nuclear superiority lent weight to the foreign policy of the United States in the 1960s. The neighbors of the Soviet Union are primarily impressed by the warfighting capacity of its *conventional* forces, and the rest of the world by its growing capacity to project that force beyond its frontiers. Within Europe, the "theater nuclear balance" concerns only a tiny group of specialists. The presence and fighting capability of the United States army and air forces are seen as

the real, and highly effective, deterrent against Soviet attack. I have expressed elsewhere my regret that the British government should have decided to spend five billion pounds out of our very restricted defense budget on a strategic nuclear strike force, which can only be at the expense of the conventional forces we can contribute to the Alliance. And however much the new administration of the United States may feel it necessary to spend on new strategic nuclear weapons systems to match or overmatch those of the Soviet Union, the effect on America's influence within the international community is likely to be negligible if it is not matched by a comparable and evident capability to defend American interests on the oceans and on the ground with forces capable of *fighting*.

For the best part of a century the peoples of industrial societies have been applying technological expertise and industrial power, initially to assist but increasingly to replace the traditional military skills and virtues on which they formerly relied for the protection of their political integrity. As a result they have been able to attain their objects in war only at the cost of enormous and increasingly disproportionate destruction. With the advent of nuclear weapons, the disproportion becomes insensate. It is politically so much easier, so much less of a social strain, to produce nuclear missiles rather than trained, effective military manpower and to believe that a valid tradeoff has somehow been made between the two. It has not. And the more deeply we become committed to this belief, on both sides of the Atlantic, the greater will be the danger that we are trying to avoid: on the one hand the impossibility of defending the specific areas and interests that are seriously threatened by a potential adversary, and, on the other, the possibility that, in a lethal mixture of hubris and despair, we might one day feel ourselves compelled to initiate a nuclear war. Such a war might or might not achieve its object, but I doubt whether the survivors on either side would very greatly care.

NOTES

1. Bernard Brodie, ed., *The Absolute Weapon* (New York: Harcourt Brace, 1946), p. 76.

2. Colin Gray, "Nuclear Strategy and the Case for a Theory of Victory," *International Security* 4, no. 1 (1979): 54–87.

3. Bernard Brodie, "Development of Nuclear Strategy," *International Security* 2, no. 4 (1978): 65–83.

4. Carl von Clausewitz, *On War*, Book 8, chap. 6B, "War Is an Instrument of Policy" (Princeton: Princeton University Press, 1976).

5. Raymond L. Garthoff, "Mutual Deterrence and Stategic Arms Limitation in Soviet Policy," *International Security* 3, no. 1 (1978): 112–47; and Ben-

jamin Lambeth, "The Political Potential of Soviet Equivalence," *International Security* 4, no. 2 (1979): 22–39.

6. Paul Nitze, "Deterring Our Deterrent," *Foreign Policy*, no. 25 (1976–77).

7. Colin Gray, "Nuclear Strategy: The Case for a Theory of Victory," *International Security* 4, no. 1 (1979): 54–87.

8. Colin Gray and Keith Payne, "Victory Is Possible," *Foreign Policy*, no. 39 (1980).

9. Gray, *Nuclear Strategy*.

10. Gray and Payne, "Victory Is Possible."

11. *Survival*, Nov.-Dec. 1979, p. 266.

12. Gray and Payne, "Victory Is Possible."

13. US Congress, Office of Technology Assessment, "The Effects of Nuclear War," 1979.

14. "Implications of Nuclear Weapons on Total War," RAND Memorandum, p. 1118, July 1957.

15. Clausewitz, *On War*, Book 1, chap. 2, "Purpose and Means in War."

MAD VERSUS NUTS
Can Doctrine or Weaponry Remedy the Mutual Hostage Relationship of the Superpowers?

SPURGEON M. KEENY, JR., and WOLFGANG K. H. PANOFSKY

Since World War II there has been a continuing debate on military doctrine concerning the actual utility of nuclear weapons in war. This debate, irrespective of the merits of the divergent points of view, tends to create the perception that the outcome and scale of a nuclear conflict could be controlled by the doctrine or the types of nuclear weapons employed. Is this the case?

We believe not. In reality, the unprecedented risks of nuclear conflict are largely independent of doctrine or its application. The principal danger of doctrines that are directed at limiting nuclear conflicts is that they might be believed and form the basis for action without appreciation of the physical facts and uncertainties of nuclear conflict. The failure of policymakers to understand the truly revolutionary nature of nuclear weapons as instruments of war and the staggering size of the nuclear stockpiles of the United States and the Soviet Union could have catastrophic consequences for the entire world.

Military planners and strategic thinkers for 35 years have sought ways to apply the tremendous power of nuclear weapons against target systems that might contribute to the winning of a future war. In fact, as long as the United States held a virtual nuclear monopoly, the targeting of atomic weapons was looked upon essentially as more effective extension of the strategic bombing concepts of World War II. With the advent in the mid-1950s of a substantial Soviet nuclear capability, including multimegaton thermonuclear weapons, it was soon apparent that the populations and societies of both the United States and the Soviet Union were mutual hostages. A portion of the nuclear stockpile of either side could inflict on the other as many as 100 million fatalities and destroy it as a functioning society. Thus, although the rhetoric of declaratory strategic doctrine has changed over the years, mutual deterrence has in fact remained the central fact of the strategic relationship of the two superpowers and of the NATO and Warsaw Pact alliances.

Most observers would agree that a major conflict between the two hostile blocs on a worldwide scale during this period may well have been prevented by the specter of catastrophic nuclear war. At the same time, few would argue that this stage of mutual deterrence is a very reassuring foundation on which to build world peace. In the 1960s the perception of the basic strategic relationship of mutual deterrence came to be characterized as "Mutual Assured Destruction," which critics were quick to note had the acronym of MAD. The notion of MAD has been frequently attacked not only as militarily unacceptable but also as immoral since it holds the entire civilian populations of both countries as hostages.[1]

As an alternative to MAD, critics and strategic innovators have over the years sought to develop various warfighting targeting doctrines that would somehow retain the use of nuclear weapons on the battlefield or even in controlled strategic war scenarios, while sparing the general civilian population from the devastating consequences of nuclear war. Other critics have found an alternative in a defense-oriented military posture designed to defend the civilian population against the consequences of nuclear war.

These concepts are clearly interrelated since such a defense-oriented strategy would also make a nuclear warfighting doctrine more credible. But both alternatives depend on the solution of staggering technical problems. A defense-oriented military posture requires a nearly impenetrable air and missile defense over a large portion of the population. And any attempt to have a controlled warfighting capability during a nuclear exchange places tremendous requirements not only on decisions made under incredible pressure by men in senior positions of responsibility but on the technical performance of command, control, communications and intelligence functions—called in professional circles "C³I" and which for the sake of simplicity we shall hereafter describe as "control mechanisms." It is not sufficient as the basis for defense policy to assert that science will "somehow" find solutions to critical technical problems on which the policy is dependent, when technical solutions are nowhere in sight.

In considering these doctrinal issues, it should be recognized that there tends to be a very minor gap between the declaratory policy and actual implementation expressed as targeting doctrine. Whatever the declaratory policy might be, those responsible for the strategic forces must generate real target lists and develop procedures under which various combinations of targets could be attacked. In consequence, the perceived need to attack every listed target, even after absorbing the worst imaginable first strike from the adversary, creates procurement "requirements," even though the military or economic importance of many of the targets is small.

In fact, it is not at all clear in the real world of war planning whether declaratory doctrine has generated requirements or whether the availability of weapons for targeting has created doctrine. With an estimated 30,000 warheads at the disposal of the United States, including more than 10,000 avowed to be strategic in character, it is necessary to target redundantly all urban areas and economic targets and to cover a wide range of military targets in order to frame uses for the stockpile. And, once one tries to deal with elusive mobile and secondary military targets, one can always make a case for requirements for more weapons and for more specialized weapon designs.

These doctrinal considerations, combined with the superabundance of nuclear weapons, have led to a conceptual approach to nuclear war which can be described as Nuclear Utilization Target Selection. For convenience, and not in any spirit of trading epithets, we have chosen the acronym of NUTS to characterize the various doctrines that seek to utilize nuclear weapons against specific targets in a complex of nuclear warfighting situations intended to be limited, as well as the management over an extended period of a general nuclear war between the superpowers.[2]

While some elements of NUTS may be involved in extending the credibility of our nuclear deterrent, this consideration in no way changes the fact that mutual assured destruction, or MAD, is inherent in the existence of large numbers of nuclear weapons in the real world. In promulgating the doctrine of "countervailing strategy" in the summer of 1980, President Carter's Secretary of Defense Harold Brown called for a buildup of nuclear warfighting capability in order to provide greater deterrence by demonstrating the ability of the United States to respond in a credible fashion without having to escalate immediately to all-out nuclear war. He was very careful, however, to note that he thought that it was "very likely" that the use of nuclear weapons by the superpowers at any level would escalate into general nuclear war.[3] This situation is not peculiar to present force structures or technologies, and, regardless of future technical developments, it will persist as long as substantial nuclear weapon stockpiles remain.

Despite its possible contribution to the deterrence of nuclear war, the NUTS approach to military doctrine and planning can very easily become a serious danger in itself. The availability of increasing numbers of nuclear weapons in a variety of designs and delivery packages at all levels of the military establishment inevitably encourages the illusion that somehow nuclear weapons can be applied in selected circumstances without unleashing a catastrophic series of consequences. As we shall see in more detail below, the recent uninformed debate on the virtue of the so-called neutron bomb as a selective device to deal with tank attacks is a depressing case in point. NUTS creates its own endless pressure for expanded nuclear stockpiles with increasing danger of accidents, accidental use, diversions to terrorists, etc. But more fundamentally, it tends to obscure the fact that the nuclear world is in fact MAD.

The NUTS approach to nuclear warfighting will not eliminate the essential MAD character of nuclear war for two basic reasons, which are rooted in the nature of nuclear weapons and the practical limits of technology. First, the destructive power of nuclear weapons, individually and most certainly in the large numbers discussed for even specialized application, is so great that the collateral effects on persons and property would be enormous and, in scenarios which are seriously discussed, would be hard to distinguish from the onset of general nuclear war. But more fundamentally, it does not seem possible, even in the most specialized utilization of nuclear weapons, to envisage any situation where escalation to general nuclear war would probably not occur given the dynamics of the situation and the limits of the control mechanisms that could be made available to manage a limited nuclear war. In the case of a protracted general nuclear war, the control problem becomes completely unmanageable. Finally, there does not appear to be any prospect for the foreseeable future that technology will provide a secure shield behind which the citizens of the two superpowers can safely observe the course of a limited nuclear war on other people's territory.

II

So much has been said and written about the terrible consequences of nuclear war that any brief characterization of the problem seems strangely banal. Yet it is not clear how deeply the horror of such an event has penetrated the public consciousness or even the thinking of knowledgeable policymakers who in theory have access to the relevant information. The lack of public response to authoritative estimates that general nuclear war could result in 100 million fatalities in the United States suggests a general denial psychosis when the public is confronted with the prospect of such an unimaginable catastrophe. It is interesting, however, that there has been a considerable reaction to the campaign by medical doctors in several countries (including the United States and the Soviet Union), which calls attention to the hopeless plight of the tens of millions of casualties who would die over an extended period due to the total inability of surviving medical personnel and facilities to cope with the situation. One can stoically ignore the inevitability of death, but the haunting image of being among the injured survivors who would eventually die unattended is a prospect that few can easily accept fatalistically.

It is worth repeating the oft stated, but little comprehended, fact that a *single* modern strategic nuclear weapon could have a million times the yield of the high explosive strategic bombs of World War II, or one hundred to a thousand times the yield of the atomic bombs that destroyed Hiroshima and Nagasaki, killing 250,000 people. The blast from a single one-megaton weapon detonated over the White House in Washington, D.C., would destroy multistory concrete buildings out to a distance of about three miles (ten pounds per square inch overpressure with winds of 300 miles per hour)—a circle of almost complete destruction reaching the National Cathedral to the northwest, Kennedy Stadium to the east, and across National Airport to the south. Most people in this area would be killed immediately. The thermal radiation from the same weapon would cause spontaneous ignition of clothing and household combustibles to a distance of about five miles (25 calories per centimeter squared)—a circle of raging fires reaching out to the District line. Out to a distance of almost

nine miles there would be severe damage to ordinary frame buildings and second-degree burns to exposed individuals. Beyond these immediate effects the innumerable separate fires that had been ignited would either merge into an outward-moving conflagration or more likely create a giant fire storm of the type Hamburg and Tokyo experienced on a much smaller scale in World War II. While the inrushing winds would tend to limit the spread of the fire storm, the area within five to six miles of the explosion would be totally burned out, killing most of the people who might have escaped initial injury in shelters.

The point has been forcefully made recently by members of the medical community that the vast numbers of injured who escape death at the margin of this holocaust could expect little medical help. But beyond this, if the fireball of the explosion touched ground, the resulting radioactive debris would produce fallout with lethal effects far beyond the site of the explosion. Assuming the prevailing westerly wind conditions, a typical fallout pattern would indicate that there would be levels of fallout greater than 1000 rems (450 rems produce 50 percent fatalities) over an area of some 500 square miles, and more than 100 rems (the level above which there will be significant health effects) over some 4000 square miles reaching all the way to the Atlantic Ocean. In the case of a single explosion the impact of the fallout would be secondary to the immediate weapons effects, but when there are many explosions the fallout becomes a major component of the threat, since the fallout effects from each weapon are additives and the overlapping fallout patterns would soon cover large portions of the country with lethal levels of radiation.

Such levels of human and physical destruction are difficult for anyone, layman or specialist, to comprehend even for a single city, but when extended to an attack on an entire country they become a dehumanized maze of statistics. Comparison with past natural disasters is of little value. Such events as dam breaks and earthquakes result in an island of destruction surrounded by sources of help and reconstruction. Nuclear war involving many weapons would deny the possibility of relief by others.

When General David Jones, chairman of the Joint Chiefs of Staff, was asked at a hearing of the Senate Foreign Relations Committee on 3 November 1981 what would be the consequences in the northern hemisphere of an all-out nuclear exchange, he had the following stark response:

We have examined that over many, many years. There are many assumptions that you have as to where the weapons are targeted. Clearly, the casualties in the northern hemisphere could be, under the worst conditions, into the hundreds of millions of fatalities. It is not to the extent that there would be no life in the northern hemisphere, but if all weapons were targeted in such a way as to give maximum damage to urban and industrial areas, you are talking about the greatest catastrophe in history by many orders of magnitude.

A devastating attack on the urban societies of the United States and Soviet Union would in fact require only a very small fraction of the more than 50,000 nuclear weapons currently in the arsenals of the two superpowers. The United States is commonly credited with having some 30,000 nuclear warheads of which well over 10,000 are carried by strategic systems capable of hitting the Soviet Union. It is estimated that the Soviet Union will soon have some 10,000 warheads in its strategic forces capable of hitting the United States. An exchange of a few thousand of these weapons could kill most of the urban population and destroy most of the industry of both sides.

But such figures are in themselves misleading because they are already high on a curve of diminishing returns, and much smaller attacks could have very severe consequences. A *single* Poseidon submarine captain could fire some 160 independently targetable nuclear warheads (each with a yield several times larger than those of the weapons that destroyed Hiroshima and Nagasaki) against as many Soviet cities. If optimally targeted against the Soviet population, this alone could inflict some 30 million fatalities. One clear fact of the present strategic relationship is that the urban societies of both the United States and the Soviet Union are completely vulnerable to even a small fraction of the other side's accumulated stockpile of nuclear weapons.

III

The theme that nuclear weapons can be successfully employed in warfighting roles somehow shielded from the MAD world appears to be recurring with increasing frequency and seriousness.[4] Support for Nuclear Utilization Target Selection—NUTS—comes from diverse sources: those who believe that nuclear weapons should be used selectively in anticipated hostilities; those who believe that such capabilities deter a wider range of aggressive Soviet acts; those who assert that we must duplicate an alleged Soviet interest in warfighting; and those who are simply trying to carry out their military responsibilities in a more "rational" or cost-effective manner. The net effect of this increasing, publicized interest in NUTS is to obscure the almost inevitable link between any use of nuclear weapons and the grim "mutual hostage" realities of the MAD world. The two forces generating this link are the collateral damage associated with the use of nuclear weapons against selected targets and the pressures for escalation of the level of nuclear force once it is used in conflict. Collateral effects and pressures for escalation are themselves closely linked.

To appreciate the significance of the collateral effects of nuclear weapons and the pressure for escalation, one must look at actual warfighting scenarios that have been seriously proposed. The two scenarios that are most often considered are Soviet attempts to carry out a disarming, or partially disarming, attack against US strategic forces in order to force the surrender of the United States without war, and the selective use of nuclear weapons by the United States in Western Europe to prevent the collapse of NATO forces in the face of an overwhelming Soviet conventional attack. One can expect to hear more about the selective use of nuclear weapons by the United States in the Middle East in the face of an overwhelming Soviet conventional attack on that area.

The much discussed "window of vulnerability" is based on the fear that the Soviets might launch a "surgical" attack against vulnerable Minuteman ICBM silos—the land-based component of the US strategic triad—to partially disarm the US retaliatory forces, confident that the United States would not retaliate. The scenario then calls for the United States to capitulate to Soviet-dictated peace terms.

Simple arithmetic based on intelligence assessments of the accuracy and yields of the warheads on Soviet missiles and the estimated hardness of Minuteman silos does indeed show that a Soviet attack leaving only a relatively small number of surviving Minuteman ICBMs is mathematically possible in the near future. There is much valid controversy about whether such an attack is in fact operationally feasible with the confidence that a rational decision-maker would require. But what is significant here is the question of whether the vulnerability of Minuteman, real or perceptual, could in fact be exploited by the Soviets without risking general nuclear war. Would a US president react any differently in response to an attack against the Minuteman force than to an attack of comparable weight against other targets?

Despite the relatively isolated location of the Minuteman ICBM fields, there would be tremendous collateral damage from such an attack, which under the mathematical scenario would involve at least 2000 weapons with megaton yields. It has been estimated by the Congressional Office of Technology Assessment that such an attack would result in from 2 to 20 million American fatalities, primarily from fallout, since at least half the weapons would probably be ground burst to maximize the effect of the attack on the silos. The range of estimated fatalities reflects the inherent uncertainties in fallout calculations due to different assumptions on such factors as meteorological conditions, weapon yield and design, height of burst and amount of protection available and used. Estimates of fatalities below 8 to 10 million require quite optimistic assumptions.

It seems incredible that any Soviet leader would count on any president suing for peace in circumstances where some ten million American citizens were doomed to a slow and cruel death but the United States still retained 75 percent of the strategic forces and its entire economic base. Instead, Soviet leadership would perceive a president, confronted with an incoming missile attack of at least 2000 warheads and possibly many more to follow in minutes, and with the action options of retaliating on warning with his vulnerable land-based forces or riding out the attack and retaliating at a level and manner of his own choosing with substantial surviving air- and sea-based strategic forces.

It is hard to imagine that this scenario would give the Soviets much confidence in their ability to control escalation of the conflict. If the Soviets did not choose to attack US command, control, communications and intelligence (C^3I) capabilities, the United States would clearly be in a position to retaliate massively or to launch a more selective initial response. If vulnerable control assets were concurrently attacked, selective responses might be jeopardized, but the possibility of an automatic massive response would be increased since the nature of the attack would be unclear. But even if these control assets were initially untouched, the Soviets could not be so overly confident of their own control mechanisms or so overly impressed with those of the United States as to imagine that either system could long control such massive levels of violence, with increasing collateral damage, without the situation very rapidly degenerating into general nuclear war.

The question of nuclear warfighting in Europe has a long and esoteric history. Tactical nuclear weapons have been considered an additional deterrent to a massive Soviet conventional attack by threatening escalation to general nuclear war involving strategic forces—the so-called coupling effect. At the same time, tactical nuclear forces have been looked on as a necessary counterbalance to Soviet conventional forces in a limited warfighting situation. To this end, the United States is said to have some 6000 to 7000 tactical United States weapons in Europe.[5] The existence of this stockpile has been public knowledge so long that it is largely taken for granted, and the power of the weapons, which range in yield from around a kiloton to around a megaton, is not appreciated. It is interesting to note that we have in Europe one nuclear weapon (with an average yield probably comparable to the weapon that destroyed Hiroshima) for every 50 American soldiers stationed there including support troops. Tactical nuclear weapons are, of course, no longer a US monopoly. The Soviets are building up comparable forces and have had for some time long-range theater nuclear missiles, earlier the SS-4 and SS-5, and now the SS-20, for which the United States does not have a strict counterpart. In this regard, it must be remembered that it is always feasible for the United States or the Soviet Union to employ some of their long-range strategic missiles against targets in Europe.

There is now a great debate, particularly in Europe, about the proposed deployment on European soil of US-controlled long-range Pershing II and ground-based cruise missiles capable of reaching the territory of the Soviet Union, in response to the growing deployment of Soviet SS-20 mobile medium-range ballistic missiles. This discussion tends to consider the SS-20s and the proposed new forces as a separate issue from the short- and medium-range nuclear weapons already deployed in Europe. There is indeed a technical difference: the proposed Pershing II missile is of sufficient range to reach Soviet territory in only a few minutes, and the SS-20 is a much more accurate and flexible weapons system than earlier Soviet intermediate-range nuclear systems. Yet, the overriding issue which tends to be submerged in the current debate is the fact that *any* use of nuclear weapons in theater warfare in Europe would almost certainly lead to massive civilian casualties even in the unlikely event the conflict did not escalate to involve the homelands of the two superpowers.

Calculations of collateral casualties accompanying nuclear warfare in Europe tend to be simplistic in the extreme. First, the likely proximity of highly populated areas to the combat zone must be taken into account. One simply cannot assume that invading enemy columns will position themselves so that they offer the most favorable isolated target to nuclear attack. Populated areas could not remain isolated from the battle. Cities would have to be defended, or they would become a safe stepping-stone for the enemy's advance. In either case, it is difficult to imagine cities and populated areas remaining sanctuaries in the midst of a tactical nuclear war raging around them. Then one must remember that during past wars in Europe as much as one-half of the population was on the road in the form of masses of refugees. Above all, in the confusion of battle, there is no control system that could assure that weapons would not inadvertently strike populated areas. Beyond immediate effects, nuclear fallout would not recognize restrictions based on population density.

The common feature of the above examples is that specialized use of nuclear weapons will as a practical matter be difficult to distinguish from unselective use in the chaos of tactical warfare. A case in point is the much-publicized neutron bomb, which has been promoted as a specialized antitank weapon since neutrons can penetrate tank armor and kill the crew. It is frequently overlooked that the neutron bomb is in fact a nuclear weapon with significant yield. While it does emit some ten times as many neutrons as a comparable "ordinary" small nuclear weapon, it also kills by blast, heat, and prompt radiation. For instance, one of the proposed neutron warheads for the Lance missile has a one-kiloton yield, which would produce the same levels of blast damage experienced at Hiroshima at a little less than one-half the distance from the point of detonation.

An attack on tanks near a populated area or a targeting error in the heat of battle would clearly have a far-reaching effect on civilians and structures in the vicinity. Moreover, the lethal effects of the neutrons are not sharply defined. There would be attenuation by intervening structures or earth prominences, and there is a wide gap (from 500 to 10,000 rems) between a dose which would eventually be fatal and that which would immediately prevent a soldier from continuing combat. Under actual war conditions no local commander, much less a national decisionmaker, could readily tell whether a neutron weapon or some other kind of nuclear weapon had been employed by the enemy. Thus, the threat of escalation from local to all-out conflict, the problems of collateral damage of nuclear weapons, and the disastrous consequences of errors in targeting, are not changed by the nature of the nuclear weapons.

In short, whatever the utility of the neutron bomb or any other "tactical" nuclear weapon in *deterring* Soviet conventional or nuclear attack, any actual use of such weapons is extremely unlikely to remain limited. We come back to the fundamental point that the only meaningful "firebreak" in modern warfare, be it strategic or tactical, is between nuclear and conventional weapons, not between self-proclaimed categories of nuclear weapons.

IV

The thesis that we live in an inherently MAD world rests ultimately on the technical conclusion that effective protection of the population against large-scale nuclear attack is not possible. This pessimistic technical assessment, which follows inexorably from the devastating power of nuclear weapons, is dramatically illustrated by the fundamental difference between air defense against conventional and nuclear attack. Against bombers carrying conventional bombs, an air defense system destroying only 10 percent of the incoming bombers per sortie would,

as a practical matter, defeat sustained air raids such as the ones during World War II. After ten attacks against such a defense, the bomber force would be reduced to less than one-third of its initial size, a very high price to pay given the limited damage from conventional weapons even when over 90 percent of the bombers penetrate. In contrast, against a bomber attack with nuclear bombs, an air defense capable of destroying even 90 percent of the incoming bombers on each sortie would be totally inadequate since the damage produced by the penetrating 10 percent of the bombers would be devastating against urban targets.

When one extends this air defense analogy to ballistic missile defenses intended to protect population and industry against large numbers of nuclear missiles, it becomes clear that such a defense would have to be almost leakproof since the penetration of even a single warhead would cause great destruction to a soft target. In fact, such a ballistic missile defense would have to be not only almost leakproof but also nationwide in coverage since the attacker could always choose the centers of population or industry to target. The attacker has the further advantage of not only choosing targets but also deciding what fraction of the total resources to expend against any particular target. Thus, an effective defense would have to be extremely heavy across the entire defended territory, not at just a few priority targets. The technical problem of providing an almost leakproof missile defense is further compounded by the many technical measures the attacking force can employ to interfere with the defense by blinding or confusing its radars or other sensors and overwhelming the system's traffic-handling capacity with decoys.

When these general arguments are reduced to specific analysis, the conclusion is inescapable that effective protection of the population or industry of either of the superpowers against missile attack by the other is unattainable with present ABM (anti-ballistic missile) defense technology, since even the most elaborate systems could be penetrated by the other side at far less cost. This conclusion is not altered by prospective improvements in the components of present systems or by the introduction of new concepts such as lasers or particle beams into system design.

These conclusions, which address the inability of ballistic missile defenses to eliminate the MAD character of the strategic relationship, do not necessarily apply to defense of very hard point targets, such as missile silos or shelters for mobile missiles. The defense of these hardened military targets does offer a more attractive technical opportunity since only the immediate vicinity of the hardened site needs to be defended, and the survival of only a fraction of the defended silos is necessary to serve as a deterrent. Thus, the technical requirements for the system are much less stringent than for population or industrial defense, and a much higher leakage rate can be tolerated. When these general remarks are translated into specific analysis which takes into account the many options available to the offense, hard site defense still does not look particularly attractive. Moreover, such a defense, even if partially successful, would not prevent the serious collateral fallout effects from the attack on the population discussed above. Nevertheless, the fact that these systems are technically feasible, and are advocated by some as effective, tends to confuse the public on the broader issue of the feasibility of urban defense against ballistic missiles.

The United States has a substantial research and development effort on ballistic missile defenses of land-based ICBMs as a possible approach to increase survivability of this leg of the strategic triad. The only program under serious consideration that could be deployed in this decade is the so-called LOAD (Low Altitude Defense) system. This system, which would utilize very small hardened radars and small missiles with small nuclear warheads, is designed to intercept at very close range those attacking missiles that might detonate close enough to the defended ICBM to destroy it. This last-ditch defense is possible with nuclear weapons since the defended target is extremely hard and can tolerate nuclear detonations if they are not too close. While such a system for the defense of hard sites is technically feasible, there has been serious question as to whether it would be cost-effective in defending the MX in fixed Titan or Minuteman silos since the system could be overwhelmed relatively easily. In the case of the defense of a mobile MX in a multiple shelter system, the economics of the exchange ratios are substantially improved if the location of the mobile MX and mobile defense system are in fact unknown to the attacker; however, there are serious questions whether the presence of radiating radar systems might not actually compromise the location of the MX during an attack.

Looking further into the future, the US research program is considering a much more sophisticated "layered" system for hard site defense. The outer layer would involve an extremely complex system using infrared sensors that would be launched on warning of a Soviet attack to identify and track incoming warheads. Based on this information, many interceptors, each carrying multiple, infrared-homing rockets with nonnuclear warheads, would be launched against the cloud of incoming warheads and attack them well outside the atmosphere. The warheads that leaked through this outer exoatmospheric layer would then be engaged by a close-in layer along the lines of the LOAD last-ditch system described above.

It has been suggested that the outer layer exoatmospheric system might evolve into an effective area defense for population and industry. Actually, there are many rather fundamental technical questions that will take some time to answer about the ability of such a system to work at all against a determined adversary in the time frame needed to deploy it. For example, such a system would probably be defeated by properly designed decoys or blinded by nuclear explosions and, above all, may well be far too complex for even prospective control capabilities to operate. Whatever the value of these types of systems for hard site defense to support the MAD role of the deterrent, it is clear that the system holds no promise for population or industry defense and simply illustrates the technical difficulty of dealing with that problem.

While the government struggles with the much less demanding problem of whether it is possible to design a plausible, cost-effective defense of hardened ICBM silos, the public is bombarded with recurring reports that some new technological "breakthrough" will suddenly generate an "impenetrable umbrella" which would obviate the MAD strategic relationship. Such irresponsible reports usually rehash claims for "directed energy" weapons which are based on the propagation of extremely energetic beams of either light (lasers) or atomic particles propagated at the speed of light to the target. Some of the proposals are technically infeasible, but in all cases one must remember that for urban defense only a system with country-wide coverage and extraordinarily effective performance would have an impact on the MAD condition. To constitute a ballistic missile defense system, directed energy devices would have to be integrated with detection and tracking devices for the incoming warheads, an extremely effective and fast data-handling system, the necessary power supplies for the extraordinarily high demand of energy to feed the directed energy weapons, and would have to be very precisely oriented to score a direct hit to destroy the target—as opposed to nuclear warheads that would only have to get in the general vicinity to destroy the target.

There are fundamental considerations that severely limit the application of directed energy weapons to ballistic missile defense. Particle beams do not penetrate the atmosphere. Thus, if such a system were ground-based, it would have to bore a hole through the atmosphere and then the beam would have to be focused through that hole in a subsequent pulse. All analyses have indicated that it is physically impossible to accomplish this feat stably. Among other things, laser systems suffer from the fact that they can only operate in good weather since clouds interfere with the beam.

These problems involving the atmosphere could be avoided by basing the system in space. Moreover, a space-based system has the desirable feature of potentially being able to attack missiles during the vulnerable launch phase before the reentry vehicles are dispersed. However, space-based systems involve putting a very complex system with a large power requirement into orbit. Analysis indicates that a comprehensive defensive system of this type would require over a hundred satellites, which in turn would need literally thousands of space shuttle sorties to assemble. It has been estimated that such a system would cost several hundred billion dollars. Even if the control mechanisms were available to operate such a system, there are serious questions as to the vulnerability of the satellites to physical attack and to various measures that would interfere with the system's operation. In short, no responsible analysis has indicated that for at least the next two decades such "death ray weapons" have any bearing on the ABM problem or that there is any prospect that they would subsequently change the MAD character of our world.

Defense against aircraft further illustrates the inherently MAD nature of today's world. Although the Soviets have made enormous investments in air defense, the airborne component of the US strategic triad has not had its damage potential substantially reduced. Most analyses indicate that a large fraction of the "aging" B-52 fleet would penetrate present Soviet defenses, with the aid of electronic countermeasures and defense suppression by missiles. It is true that the ability of B-52s to penetrate will gradually be impaired as the Soviets deploy "look down" radar planes similar to the much publicized AWACS (Airborne Warning and Control System). However, these systems will not be effective against the air-launched cruise missiles whose deployment on B-52s will begin shortly; their ability to penetrate will not be endangered until a totally new generation of Soviet air defenses enters the picture. At that time, one can foresee major improvements in the ability of both bombers and cruise missiles to penetrate through a number of techniques, in particular the so-called "stealth" technology which will reduce by a large factor the visibility of both airplanes and cruise missiles to radar.

In short, there is little question that in the defense-offense race between air defenses and the airborne leg of the triad, the offense will retain its enormous damage potential. For its part, the United States does not now have a significant air defense, and the limited buildup proposed in President Reagan's program would have little effect on the ability of the Soviets to deliver nuclear weapons by aircraft against this country. Consequently, the "mutual hostage" relationship between the two countries will continue, even if only the airborne component of the triad is considered.

It is sometimes asserted that civil defense could provide an escape from the consequences of the MAD world and make even a general nuclear war between the superpowers winnable. This assertion is coupled with a continuing controversy as to the actual effectiveness of civil defense and the scope of the present Soviet civil defense program. Much of this debate reflects the complete failure of some civil defense advocates to comprehend the actual consequences of nuclear war. There is no question that civil defense could save lives and that the Soviet effort in this field is substantially greater than that of the United States. Yet all analyses have made it abundantly clear that to have a significant impact in a general nuclear war, civil defense would have to involve a much greater effort than now practiced on either side and that no amount of effort would protect a large portion of the population or the ability of either nation to continue as a functioning society.

There is evidence that the Soviets have carried out a shelter program which could provide fallout and some blast protection for about 10 percent of the urban population. The only way even to attempt to protect the bulk of the population would be complete evacuation of the entire urban population to the countryside. Although to our knowledge there has never been an actual urban evacuation exercise in the Soviet Union, true believers in the effectiveness of Soviet civil defense point to the alleged existence of detailed evacuation plans for all Soviet cities. Yet, when examined in detail, there are major questions as to the practicality of such evacuation plans.

The US Arms Control and Disarmament Agency has calculated, using a reasonable model and assuming normal targeting practices, that even with the general evacuation of all citizens and full use of shelters, in a general war there would still be at least 25 million Soviet fatalities. Such estimates obviously depend on the model chosen: some have been lower but others by the Defense Department have been considerably higher. The time for such an all-out evacuation would be at least a week. This action would guarantee unambiguous strategic warning and provide ample time for the other side to generate its strategic forces to full alert, which would result in a substantially greater retaliatory strike than would be expected from normal day-to-day alert. If the retaliatory strike were ground burst to maximize fallout, fatalities could rise to 40 to 50 million; and if part of the reserve of nuclear weapons were targeted against the evacuated population, some 70 to 85 million could be killed. Until recently little has been said about the hopeless fate of the vast number of fallout casualties in the absence of organized medical care or what would become of the survivors with the almost complete destruction of the economic base and urban housing.

Finally, there is no evidence that the Soviets are carrying out industrial hardening or are decentralizing their industry, which remains more centralized than US industry. This is not surprising since there is nothing they can do that would materially change the inherent vulnerability of urban society in a MAD world.

V

In sum, we are fated to live in a MAD world. This is inherent in the tremendous power of nuclear weapons, the size of nuclear stockpiles, the collateral damage associated with the use of nuclear weapons against military targets, the technical limitations on strategic area defense, and the uncertainties involved in efforts to control the escalation of nuclear war. There is no reason to believe that this situation will change for the foreseeable future since the problem is far too profound and the pace of technical military development far too slow to overcome the fundamental technical considerations that underlie the mutual hostage relationship of the superpowers.

What is clear above all is that the profusion of proposed NUTS approaches has not offered an escape from the MAD world, but rather constitutes a major danger in encouraging the illusion that limited or controlled nuclear war can be waged free from the grim realities of a MAD world. The principal hope at this time will not be found in seeking NUTS doctrines that ignore the MAD realities but rather in recognizing the nuclear world for what it is and seeking to make it more stable and less dangerous.

NOTES

1. See, for example, Fred Charles Iklé, "Can Nuclear Deterrence Last Out the Century?" *Foreign Affairs* 51, no. 2 (1973): 267–85.

2. The acronym NUT for Nuclear Utilization Theory was used by Howard Margolis and Jack Ruina, "SALT II: Notes on Shadow and Substance," *Technology Review*, Oct. 1979, pp. 31–41. We prefer Nuclear Utilization Target Selection, which relates the line of thinking more closely to the operational problem of target selection. Readers not familiar with colloquial American usage may need to be told that "nuts" is an adjective meaning "crazy or demented." For everyday purposes it is a synonym for "mad."

3. See Harold Brown, Speech at the Naval War College, 20 August 1980, the most authoritative public statement on the significance of Presidential Directive 59, which had been approved by President Carter shortly before.

4. For a particularly clear statement of this view, see Colin S. Gray and Keith Payne, "Victory Is Possible," *Foreign Policy* (Summer 1980): 14–27. For opposing arguments, see Michael E. Howard, "On Fighting Nuclear War," *International Security* (Spring 1981): 3–17, and a further exchange between Messrs. Gray and Howard in *International Security* (Summer 1981): 185–87.

5. For a discussion of the usefulness of theater nuclear forces in NATO as of that date, see Alain C. Enthoven, "U.S. Forces in Europe: How Many? Doing What?" *Foreign Affairs* (Apr. 1975): 523–31.

NUCLEAR POLICY AND THE DEFENSIVE TRANSITION

KEITH B. PAYNE and COLIN S. GRAY

On 23 March 1983, President Reagan delivered a televised speech to the nation in which he initiated a potentially radical departure in US strategic policy. The president suggested that the policy of transition deterrence through the threat of strategic nuclear retaliation is inadequate, and called upon the vast American technological community to examine the potential for effective defense against ballistic missiles:

Would it not be better to save lives than to avenge them? Are we not capable of demonstrating our peaceful intentions by applying all our abilities and our ingenuity to achieving a truly lasting stability? I think we are—indeed we must.

After careful consultation with my advisers, including the Joint Chiefs of Staff, I believe there is a way . . . It is that we embark on a program to counter the awesome Soviet missile threat with measures that are defensive. Let us turn to the very strengths in technology that spawned our great industrial base . . . I know this is a formidable technical task, one that may not be accomplished before the end of the century. Yet, current technology has attained a level of sophistication where it is reasonable for us to begin this effort.[1]

Reprinted with minor revisions and by permission from Foreign Affairs 62, no. 4 (1984): 820–42. Copyright © 1984 *by the Council on Foreign Relations, Inc.*

The central problem of nuclear deterrence is that no offensive deterrent, no matter how fearsome, is likely to work forever, and the consequences of its failure would be intolerable for civilization. The president's call was a direct challenge to the offensive concept of deterrence that has dominated US strategic policy for decades. That concept is based upon the simple and still widely accepted argument that neither the United States nor the Soviet Union will launch a nuclear first strike or engage in other highly provocative actions if both sides are vulnerable to nuclear retaliation. The president's speech suggested that vulnerability to a Soviet nuclear attack is not an acceptable condition in the long term, and that the United States would examine avenues to counter the threat of nuclear missiles.

Following the president's speech, National Security Study Directive 6–83 mandated an examination of the technology that could eliminate the threat posed by nuclear ballistic missiles to the security of the United States and its allies. Accordingly, between June and October 1983, two studies assessed the technical and policy issues of a national commitment to ballistic missile defense (BMD). James C. Fletcher, former administrator of the National Aeronautics and Space Administration, headed a Defensive Technologies Study Team, and Fred S. Hoffman, director of Pan Heuristics (a policy analysis organization based in Los Angeles), led an extragovernmental Future Security Strategy Study. A senior interagency group integrated the two studies, and on behalf of Secretary of Defense Caspar Weinberger and then National Security Adviser William P. Clark, recommended a technology development plan to the president. The interagency group reportedly advised a vigorous research program to support an early decision concerning BMD development and deployment options, and recommended funding estimates for Fiscal Year 1985 through FY 1989 ranging from $18 billion to $27 billion. The interagency group apparently estimated the total cost, through deployment, of a multilayered defensive system to the year 2000 at approximately $95 billion; a recent Department of Defense study, however, has placed the cost of a laser defense for US cities much higher.[2]

On 6 January 1984, President Reagan reportedly signed National Security Decision Directive 119, which authorized an expansive research program to demonstrate the technical feasibility of intercepting attacking enemy missiles. The president requested $1.99 billion for the "strategic defense initiative" for FY 1985, a level of support that is $250 million above the budget figure for FY 1984.

Although a small core of strategic defense enthusiasts has always been present within the US defense community, this level of officially expressed interest in strategic defense is an unprecedented development in recent US strategic policy. The goal of actively defending the American homeland in the event of nuclear conflict has not received serious official endorsement since the 1960s. Also unprecedented is the fact that the president has set policy in front of technology. If the United States does, in fact, deploy a multilayered system for defense against ballistic missiles, it will be the result of policy leading technology, not the more familiar "technology creep" generating enthusiasm and a constituency for a weapons system which then "finds" a policy rationale as it is developed.

In essence, what would be involved would be a new direction in US nuclear policy, a transition period of possibly two decades, involving a new and serious commitment to strategic defensive forces. Of course, such a commitment could not limit itself to countering the threat from ballistic missiles, but would also call for greatly improved capabilities to defend against strategic bomber and cruise missile threats.

Strategic defense has not been debated seriously since the antiballistic missile (ABM) debate of 1968–1971. Many of the technological, political and strategic factors pertinent to the old debate have changed considerably since then, but the key issues remain unchanged. They are:

—What role is ballistic missile defense expected to play;

—What effect will a commitment to strategic defense have on stability;

—What is the role of strategic offensive forces during and after a defensive transition;

—How are the European allies likely to react;

—What is Soviet policy concerning ballistic missile defense, and how is the Soviet Union likely to respond to an American initiative; and finally,

—What are the arms control implications of a defensive transition?

II

The role that strategic defense might play in US national security policy is now a contentious issue. Should it be expected to provide an "astrodome" covering American military forces and cities comprehensively, or is a more limited objective acceptable, such as only defending US retaliatory weapons? The criteria for defense effectiveness have profound implications for the research programs to be pursued and the types of defensive systems to be deployed.

For example, if a comprehensive defense for the American homeland is the only objective deemed worthy of the cost, then "exotic" defensive technology such as space-based, directed-energy beam systems or hypervelocity guns will be essential—technology that may take many years to mature. If, however, a more limited defensive mission is endorsed, such as the defense of US retaliatory forces against a Soviet first strike, then more conventional ground-based systems incorporating radars, infrared sensors, and rocket interceptors would be appropriate. Proponents of the "astrodome" approach fear that the diversion of attention toward less sophisticated and less effective systems could harm the chances for the deployment of any defensive forces. They feel that the promise of a comprehensive defense of cities by exotic systems and the transcending of offensive-oriented deterrence is a goal that will capture the imagination and support of the American people—support that should not be jeopardized by discussion of limited defenses for US retaliatory forces.

A limited defense for US retaliatory forces, however, need not be inconsistent with a future exotic defense of cities. Indeed, the two roles and systems would be highly compatible, perhaps essential to a stable defensive transition. A comprehensive BMD system would require multiple defensive layers, including conventional earth-based rocket interceptors as well as exotic beam technology. Such a system would use multiple tiers of defensive protection, intended to intercept Soviet missiles during different phases of flight: the boost and post-boost phase, early and late mid-course, and the terminal phase of flight. More or fewer defensive tiers could exist depending upon the number and types of systems deployed. The layering of defensive forces into multiple tiers of interceptors could provide an extremely capable system for defense. For example, five tiers of defensive interceptors achieving 85 percent effectiveness in each layer would reduce the overall attack to less than 0.01 percent of the original number of attacking weapons. If 10,000 nuclear warheads were launched at the United States with such a multilayered defensive system in place, at most a single weapon would be likely to penetrate to its target.

Current, more conventional earth-based BMD technology is relevant to the interception of warheads at various points in the mid-course and terminal phases of missile flight, but it does not promise effectiveness in the boost phase, wherein an attacking missile would be intercepted prior to releasing its host of individually targetable warheads (MIRVs). Intercept during this stage exerts great defensive leverage over the attacking force because each missile destroyed eliminates all the warheads carried by that missile.

Nevertheless, near-term BMD technology could provide the means for the important lower tiers of conventional defense designed to defend US intercontinental ballistic missiles (ICBMs), strategic bomber bases, and selected critical command, control, and communication facilities. These ground-based defensive systems designed to intercept Soviet warheads in their mid-course and terminal phases of flight would likely be nonnuclear, i.e., they would not use nuclear-tipped interceptors, and could be available by late in this decade. The deployment of such near-term defenses for limited coverage would actually help to facilitate a subsequent commitment to a more comprehensive defensive system involving exotic space-based systems.

The period of transition to a comprehensive defense incorporating additional layers of more advanced defensive technology could require two decades or longer for full deployment of the systems. It must be recognized that this transition period could be dangerous if precautions were not taken to ensure political and strategic stability during that period. The Soviet Union, for example, is likely to achieve an initial temporary advantage in defensive capability given its existing extensive radar network and rapidly deployable ground-based BMD. A unilateral Soviet BMD system of even limited effectiveness could be highly destabilizing in the context of existing Soviet offensive first-strike capabilities and extensive air defense and civil defense preparations: the US deterrent threat could be severely degraded by the combination of the Soviet first-strike potential to destroy American strategic nuclear forces and a Soviet defense against surviving American forces.

This combination of Soviet offensive and defensive capabilities could increase first-strike incentives during a crisis if Soviet leaders were persuaded that the USSR's defenses might be capable of largely absorbing the much-diminished US retaliatory capability. Admittedly, Soviet strategic defenses on a modest scale could do little to limit damage against a coordinated and large-scale attack by undamaged US offensive forces; but alerted Soviet defenses could be very effective in defending against a US force sharply reduced in size and impaired by a Soviet first strike.

Thus, early deployment by the United States of a ballistic missile defense for its retaliatory forces would not only help to ensure the survival of US strategic weapons, but would also assist in preserving the credibility of the US offensive deterrent during an otherwise potentially unstable transition period. Even such limited conventional defensive coverage for US retaliatory forces would create enormous uncertainties for Soviet planners considering the effectiveness of a strategic first strike, in addition to those long-standing doubts pertaining to the calculated effectiveness of their offensive forces.[3] In the context of Soviet defensive deployments, enhancing the survivability and potential effectiveness of US retaliatory forces by means of both strategic defense and more immediate special "penetration aids" for offensive weapons would help to ensure that a transition to a comprehensive defensive capability could be pursued safely.

The crucial role played by US retaliatory forces to safeguard stability during the initial phase of a defensive transition should also be clear. MX-Peacekeeper intercontinental ballistic missiles, a new small ICBM, B-1B bombers, cruise missiles, and Trident submarines are essential for deterrence stability during the decades of transition. New defensive weapons should not be considered a substitute for, or alternative to, the current modernization of strategic offensive forces; indeed, a strategic defensive initiative necessitates modernization of offensive forces to help sustain stability during the defensive transition. The leaders of the Soviet Union are likely to share this view.

Finally, a transition to a comprehensive defense for the American homeland would require the support of many administrations and necessitate major technological advances. Completion of such a transition could be waylaid for technological or political reasons over the course of the decades that any such transition will require. As a result, each phase of a defensive transition must be valuable in and of itself, and should complement any subsequent investment in strategic defense. Early deployment of BMD coverage for US retaliatory forces would constitute the first phase of a comprehensive defensive transition; it would safeguard the process of transition, and would be strategically valuable on its own merits, even if subsequent phases were delayed or terminated for political or technical reasons.

An important by-product of a ballistic missile defense system, even of limited effectiveness, is protection of the United States (and Soviet Union) from the accidental launch of a missile. For both sides to be protected against an accident would be far preferable to the current condition which carries a very high risk that disastrous consequences would follow from any such mishap. Further, limited BMD could help protect the United States against lightly armed nuclear powers of the future.

There are dozens of technical possibilities for future strategic defense systems; the debate over the proposed defense transition is not about the prospective effectiveness of one or two systems. Faced with the very rich menu of technical options for defense, it is simply untenable to assert that "they" (ground-based conventional interceptors and beam weapons, and space-based or deployable directed- and kinetic-energy weapons) will not work. Very near-

term technology almost certainly could effectively provide limited but important defensive coverage for US strategic forces.[4]

The more sophisticated exotic technology necessary for a comprehensive defense, although still in its infancy in some regards, shows tremendous potential. Commenting upon the recent classified governmental studies reviewing BMD, the president's science adviser, Dr. George Keyworth, noted:

We can now project the technology—even though it hasn't been demonstrated yet—to develop a defense system that could drastically reduce the threat of attack by nuclear weapons, not only today, but those that could reasonably be expected to be developed to counter such a defense system.[5]

All of recorded history has shown swings in the pendulum of technical advantage between offense and defense. For the strategic defense to achieve a very marked superiority over the offense over the next several decades would be an extraordinary trend in the light of the last 30 years, but not of the last hundred or thousand years. Military history is replete with examples of defensive technology and tactics dominating the offense.

In sum, different levels of defensive technology can make important, and distinct, contributions to strategic stability. The near-term role for limited ballistic missile defense suitable to the technology likely to be available soon would be to provide protection for the US nuclear deterrent and perhaps protection of the nation against small or accidental attacks. As the initial phase in a transition to comprehensive BMD coverage, defense of US retaliatory forces would help provide the stability necessary for the long-term development of the more effective systems needed for a comprehensive defense of US cities. If the technology for effective city defense proves to be attainable, then strategic defense can expand in scope and depth to provide coverage for urban and industrial America. If so, limited coverage provided by earlier conventional ballistic missile defense will still be valuable, indeed essential, as a backstop against "leakage" through space-based tiers of defense.

III

What is the likely effect of strategic defense on stability? If the Reagan administration (or its successors) cannot answer critics who charge that strategic defense would be "destabilizing," it will have little hope of generating or sustaining the necessary congressional and public support. During the initial phase of a defensive transition, stability could be safeguarded by defending US offensive forces and by enhancing their potential to penetrate Soviet ballistic missile defenses.

A more difficult question concerns how stability would be maintained during the later phases of a transition wherein, theoretically, neither side could pose a credible threat to inflict very widespread nuclear destruction upon the other's homeland. Deterrence in the nuclear age has come to be understood in terms of mutual threats of nuclear devastation varying only in kinds of targets, i.e., counter-military, counter-industrial, counter-city, or all of these. The nuclear missile age has so far been an age of defenselessness against these threats. The question now is: could stability be maintained if the current condition of mutual homeland vulnerability were to be altered drastically?

Before addressing this question, it should be noted that neither the United States nor the Soviet Union could have complete confidence in their defensive systems. Exotic defensive systems could fail catastrophically under actual operational conditions. Very sophisticated systems, which must integrate numerous and complex component functions in the face of countermeasures, entail a degree of uncertainty. This uncertainty concerning the operational effectiveness of the total system could persuade leaders of both superpowers that an "unacceptable" level of destruction could still result from a clash of arms. The prospect of a catastrophic failure of the defense, or substantial leakage, should serve as a residual offensive deterrent. It certainly should preclude the possibility that any political leader would believe that the world had been rendered completely "safe" for large-scale nonnuclear war.

If, however, one assumes a very high degree of confidence by US and Soviet leaders in their defensive systems, deterrence should still function, but it would no longer be the long-familiar deterrence from mutual vulnerability. The US deterrent would rest on defensive capability to deny plausibility to any Soviet "theory of victory." That is, US defenses would thwart Soviet strategy and deny the Soviet Union its requirements for military and political success. These include: the destruction of US military potential such that the Soviet Union, though not escaping damage in nuclear war, would survive, recover, and continue to function; the destruction of opposing forces in Europe; and the seizure of critical strategic assets worldwide.[6] Soviet military writers caution against any nuclear "adventurism" in the absence of a capability to meet these requirements for success.

This type of defensive deterrent is not totally removed from the current offensive-oriented deterrent. Currently the United States seeks to deny the Soviet Union its theory of victory by promising a devastating nuclear retaliation. In contrast, a defensive deterrent would deny the Soviet Union its theory of victory by ensuring its inability to defeat the United States—promising a long and potentially unwinnable war

which would allow the vastly superior US and US-allied, military-industrial potential to come into play. Soviet leaders are acutely sensitive to the probable negative political consequences of such wars and are highly respectful of US military-industrial potential. The prospect of waging a protracted war would be a deterring prospect for Soviet leaders. Perhaps most important, unlike the current condition, a defensive deterrent would combine the prospect for denial of Soviet victory with the avoidance of US defeat and destruction.

It is difficult to compare the relative efficacy of offensive and defensive approaches to deterrence. The current offensive-oriented deterrent threatens dire short-term consequences, but its credibility is subject to grave doubt. Would an American president actually invite national self-destruction by unleashing US nuclear forces in response to a limited Soviet attack (against US allies perhaps), thereby triggering massive Soviet nuclear retaliation? Some reply with a cautious "yes," others with a skeptical "no." The point is that no matter how the United States refines its offensive-oriented nuclear strategy, it is apparent that American society could not withstand Soviet nuclear retaliation. Consequently, a solely offensive-oriented deterrent must lack credibility vis-à-vis most threats; its effectiveness is extremely constrained.

A defensive-oriented deterrent would not impose comparably dire short-term punishment, but there would be no doubt concerning the credibility of a defensive deterrent being used. Just as no one questions whether NATO's conventional forces would fight if attacked, so there should be no question that strategic defenses would be used, particularly if they were of a nonnuclear kind. It is a matter of no small importance to have forces whose credibility of employment is 100 percent. While it is difficult to compare the relative efficacy of these different types of deterrents, there is one critically important distinction. In the event deterrence fails, extremely effective defenses could enable the United States (and perhaps the Soviet Union) to avoid a nuclear holocaust, while a purely offensive approach to deterrence virtually ensures a holocaust.[7]

In short, a transition to strategic defense would not be inconsistent with deterrence. Rather it would introduce a different approach to deterrence, an approach that could reduce both the probability and the consequences of nuclear war.

IV

A defensive deterrent would thus present powerful disincentives against a Soviet nuclear first strike. It is likely, however, to be less appropriate for the current policy of extending deterrence coverage to allies and global interests.

The US strategic nuclear threat, which is integral to NATO's "flexible response" doctrine, would be less deterring in the presence of Soviet strategic defenses. Moreover, the Soviet Union might believe that the potential benefits of conventional conquest in Europe or the Persian Gulf would be worth the risk if there were a strategic defensive stalemate with the United States. Indeed, control and exploitation of the industrial and energy resources of Western Europe and the Gulf may be seen by Soviet leaders as the way of overcoming their otherwise long-term structural economic disadvantages in the global competition with the United States, notably in the production of military high-technology items.

A possible solution to this potential problem is the same as that suggested for the solution to NATO's current over-reliance upon the threat of nuclear retaliation, namely the enhancement of NATO's conventional forces. That solution has been understood and advocated by every US administration for the past two decades. The European allies, however, have long resisted incurring the social and economic costs associated with providing NATO with conventional forces sufficiently large, well equipped, and intelligently deployed to compel the Soviet Union to think in terms of a high-risk, high-cost nuclear attack. Instead, NATO Europe has preferred for years to rely heavily on a largely US-provided nuclear deterrent. Should, however, the Soviets develop an effective strategic defense, the US "nuclear umbrella" would appear much less fearsome to the USSR, compelling Western Europeans to seek an alternate means of preserving their security.

Thus, it is doubtful that America's European allies will ever be enthusiastic about a defensive transition in US national security policy. They are likely to see it as a weakening of the US commitment to provide a nuclear umbrella over Western Europe. Moreover, European countries confront a wider spectrum of threat than does the United States. A BMD system that effectively protected the United States *and* its European allies from strategic nuclear attack would still leave the Europeans vulnerable to conventional and some kinds of tactical nuclear attack. This asymmetry in vulnerability, and hence the perception of an asymmetry in American and European interests, could be exacerbated by a new US defensive deterrence policy.

A defensive transition by both superpowers would also degrade, perhaps nullify, the British and French independent strategic deterrents. It was clear during the SALT I negotiations that the British and French wanted BMD limited to very low levels so that their relatively small independent nuclear forces would retain effectiveness. There is little to indicate that the British or French have a different perspective today.[8]

During a defensive transition, however, some of the allies might seek to parallel US and Soviet efforts and acquire their own strategic defenses. French President François Mitterrand apparently indicated interest in strategic defense when he recently stated that a "European space community" with the capability "to fire projectiles that would travel at the speed of light" would "be the most appropriate answer to the military realities of the future."[9] The British and French may also adopt a more aggressive "penetration aids" program for their offensive nuclear forces to enhance their effectiveness against Soviet defenses. In the near- to mid-term, however, it would seem unlikely that independent European offensive or defensive capabilities could compete successfully with those of Soviet forces.

On an objective analysis, there are some respects in which a defensive transition could enhance security in Europe. First, US BMD technologies could help protect the Western Europeans from Soviet long-range theater nuclear weapons (such as SS-4s, SS-5s, SS-20s, and variable-range ICBMs and submarine-launched ballistic missiles or SLBMs). Given the relative ease of Soviet military access to Western Europe, such a defense would not be comprehensive, but it certainly could reduce NATO Europe's vulnerability significantly and provide a politically effective counter to the much discussed SS-20.

Second, if America is defended, the president is likely to see a lower level of risk involved in responding to a Soviet invasion of Europe than if America were naked to Soviet nuclear attack. This fact alone should significantly reduce any Soviet inclination to attack NATO Europe. Even though the United States could not pose an "assured destruction" level of threat against the Soviet Union in response to an invasion of NATO Europe,[10] the prospect of waging a long war for control of Europe with a defended and mobilized America could not help but be a highly deterring prospect. Thus, the US "extended deterrent" over Europe could still be effective to an important degree.

Nonetheless, the reaction of major NATO countries to a US BMD program would be likely to reflect the concerns already noted, as well as a fundamental difference between American and European perspectives on security. NATO allies have long criticized what they identify as the US penchant for a technological rather than a political solution to security concerns. No matter how sound the strategic case for a defensive transition, many Europeans will be more impressed by the effect a defensive transition might have on the political foundations of the familiar East-West security system.

In sum, many Europeans are likely to prefer what they judge to be the political benefits of détente and an intact ABM Treaty to the strategic benefits that would follow a transition to effective strategic defenses. Ultimately the United States may have to decide whether concessions to the concerns of its allies about strategic defense are worth the price of forgoing the potential for a comprehensive defense and leaving the American homeland vulnerable.

V

How would the Soviet Union respond to an American defensive transition? Would the Soviet Union cooperate, tacitly or explicitly, with a defensive transition, by negotiated reductions in offensive weapons to ease the defense burden, or would it choose to compete comprehensively in an offense-defense race?

First, it should be noted that the Soviet Union has exhibited much more enthusiasm for strategic defense over the past two decades than has the United States. Soviet strategic defensive activities were roughly five times US outlays in 1970 and increased to 25 times US outlays in 1979.[11] While the United States drastically reduced its number of interceptor aircraft and deactivated its air defense, surface-to-air missile (SAM) batteries during the 1960s (on the argument that if defense against ballistic missiles was infeasible, there was little point in bomber defense), the Soviet Union modernized and increased its air defenses. While the United States reduced its commitment to civil defense to a marginal level, the Soviet Union expanded its civil defense efforts. Additionally, Soviet offensive-force modernization over the last decade has been directed toward achieving the capability to destroy US retaliatory forces before they could be used—"active defense" in Soviet military parlance. It is apparent that the notion of defending the homeland is central to Soviet strategic thinking.

The Soviet Union has approached ballistic missile defense through two avenues. First, the Soviet Union has maintained, and is now modernizing, the world's only operational ballistic missile defense site around Moscow. Moreover, it has continued to upgrade its extensive network of air defense radars and interceptors (such as the SA-5 and SA-12 interceptors), giving them some capability against strategic ballistic missiles and intermediate-range theater nuclear missiles such as the Pershing II. The United States tried to address this possibility by including articles in the 1972 ABM Treaty prohibiting the testing of air defense components "in an ABM mode." This is a key area where the Soviet Union is alleged to have violated the letter and the spirit of the ABM Treaty by testing air defense radars and interceptors against ICBMs. Recently, US photo reconnaissance satellites discovered a new, gigantic radar under construction near Abalakova that is difficult to explain except as an ABM battle-management

facility. The location (far from the Soviet periphery) and the direction of this radar appear to be in direct violation of the ABM Treaty.

Second, the evolving Soviet ballistic missile defense research and development program accommodates directed-energy beam systems and a more conventional BMD system involving transportable radars and high- and low-altitude interceptors that could be deployed rapidly.[12] The combination of an existing infrastructure of large battle-management radars, rapidly deployable BMD interceptors, and transportable missile site radars has led some in the intelligence community to conclude either that the Soviet Union is preparing to "break out" of ABM Treaty constraints and initiate a defensive transition, or is engaged in a "creeping" break-out, to be followed by rapid deployment of BMD.

Yet there are several reasons why it is unlikely that the Soviet Union would be the first to withdraw formally from the ABM Treaty and initiate an overt transition to ballistic missile defense. While the treaty is in effect, the Soviet Union can continue to pursue its gradual upgrading of air defense to achieve greater ballistic missile defense capability, with less likelihood that the United States will react strongly. This sub-rosa avenue to greater strategic defense would permit the Soviet Union to avoid the political fallout and frantic US response that would probably result from outright Soviet renunciation of the ABM Treaty.

Second, the Soviet Union obviously is aware that for the first time in two decades the United States is making a very serious commitment to explore the technical promise of strategic defense. Soviet leaders may suspect that the United States will petition for revision or withdrawal from the ABM Treaty within the decade. A prudent tactic for the Soviet Union would be to wait for such a US initiative, and then to insist on significant US arms control concessions in return for Soviet endorsement of the revisions sought by the United States. In this case the Soviet Union could achieve a major propaganda success with the charge that it was the United States that had sought to weaken or terminate this important symbol of détente, indeed, this perceived monument to the mutual commitment to prevent nuclear war. Given the enduring West European commitment to 1970s-vintage détente, the Soviet Union could further its traditional objective of dividing NATO by presenting itself as the defender of the ABM Treaty.

Finally, if the Soviet Union should "break out" of the ABM Treaty in the near term, it is almost certain that the United States would respond with deployment of a system based upon current defensive technology. US ballistic missile defense research and development have focused upon ICBM defense for almost two decades, and it is clear that the defense of ICBM silos is within our grasp. The Soviet Union, however, has spent billions of rubles deploying its fourth-generation ICBMs (particularly the SS-18s and SS-19s) for the purpose of putting US ICBM silos at risk. It also is quite clear that the Soviet Union signed the ABM Treaty substantially in order to forestall the potential of the US Safeguard missile defense program, then under deployment to protect US ICBMs from attack. In effect, the ABM Treaty provided the Soviet SS-18s and SS-19s unimpeded access to US ICBM silos. With MX-Peacekeeper ICBMs scheduled to go into silos in 1986, it seems unlikely that the Soviet Union would choose to surrender its ability to threaten US ICBMs by abrogating the ABM Treaty. Such a course of action would be reasonable only if the Soviet Union were very confident in the ability of its defenses to intercept US ICBM warheads. In that case, the Soviet Union would not have to rely on its offensive forces to destroy US retaliatory forces and would need to be far less concerned about a US capability to defend its ICBMs.

Given all of these considerations, it would appear unwise for the Soviet Union overtly to break out of the ABM Treaty regime. What then would be the likely Soviet responses to a US defensive initiative? Our judgment is that the Soviet Union would be most likely to pursue a dual-track response—combining arms control and diplomatic initiatives with strong military programs. Such behavior would be in keeping with the traditional Soviet proclivity for pursuing arms control negotiations and a dynamic arms buildup simultaneously.

Specifically (assuming that the United States had gone ahead), the Soviet Union is likely to attempt to deploy its own BMD systems and to provide effective countermeasures for its offensive forces in order to thwart US defensive systems. If the Soviet Union seeks to maintain its nuclear threat against the United States and to provide strategic missile defense for its homeland, then an effort to provide offensive countermeasures to US defenses is inevitable. If the Soviet Union were able to achieve the capability to nullify the US retaliatory deterrent while maintaining its own offensive threat against the United States, it would have achieved a strategic condition of major military advantage. The United States obviously must guard against such a condition.

Some of the potential Soviet countermeasures include: an attempt to prevent the United States from deploying defensive weapons in space through direct military interdiction; an increase in the number of offensive weapons sufficient to saturate the US defense; and passive protection of Soviet offensive forces to ensure penetration of US defenses.

The first of these options is the least likely, given the risks involved in initiating a war of attrition in space, particularly if the Soviet Union hopes to deploy its own space-based defenses. The Soviet Union historically has not taken direct action in response to US deployment of a new type of military system. A basic precept of Soviet policy is that war is a political phenomenon; it is not a proper response to an American peacetime weapons deployment.

Among offensive and defensive military countermeasures, the Soviet Union might attempt to increase its number of ICBM and SLBM warheads, intending thereby to saturate the US defensive systems. A second tactic would be to attempt to sidestep US defenses against ballistic missiles by increasing the bomber and cruise missile threat. The Soviet Union could also attempt to avoid all US strategic defenses by emphasizing "unconventional" operations (e.g., crisis and wartime sabotage) involving nuclear, chemical, or biological weapons.

Finally, the Soviet Union could attempt to provide "passive" protection for its offensive forces against US defensive beam weapons. This could be done, for example, by hardening the missiles, increasing their speed of ascent and thereby reducing their time at maximum risk, adding ablative material to missiles so as to absorb the heat transferred by a high-energy laser beam, or even by spinning the missile during its ascent in order to reduce the dwell time of the laser beam.

If the Soviet Union lacked confidence that it could succeed in this high-technology competition, however, it would have greater incentives to pursue constraints on the United States through arms control. The ABM Treaty is an informative precedent. During the early 1970s the Soviet Union chose to limit US superiority in ABM technology through arms control rather than by relying upon offensive countermeasures alone.

VI

This brings us to the question of the impact on arms control of a determined US BMD program and of the likely Soviet response to such a program. Here the first question is the impact on existing arms control agreements, notably the ABM Treaty.

The United States and the Soviet Union would have to revise the ABM Treaty to permit deployment of BMD systems of even limited effectiveness. A comprehensive defense would likely necessitate withdrawal from the treaty, which Article XV permits on six months' notice.

The United States certainly has sound strategic and arms control reasons to reconsider its continued endorsement of the ABM Treaty. At the time the treaty was signed, the United States established a clear linkage between offensive and defensive arms control limitations. Such a linkage made good sense; the United

States could accept severe constraints on BMD, which might defend US ICBMs and strategic bomber bases, if the Soviet offensive threat to US retaliatory forces could be constrained and reduced on a long-term basis through arms control.

Thus, US Unilateral Statement A, accompanying the ABM Treaty, stated specifically that a failure to achieve agreement within five years, providing for more comprehensive limitations on offensive forces than those contained in SALT I, could be grounds for withdrawal from the ABM Treaty. Unilateral Statement A also said: "The U.S. Delegation believes that an objective of the follow-on negotiations should be to constrain and reduce on a long-term basis threats to the survivability of our respective strategic retaliatory forces."[13]

Unfortunately the standard thus set by the United States for continued support of the ABM Treaty has not been met. Indeed, the Soviet offensive threat to US retaliatory forces has increased dramatically since the signing of SALT I, and the signed but unratified SALT II agreement would not have eased the problem of strategic force vulnerability given the types and numbers of weapons permitted.

Yet the strict prohibitions on BMD systems that could defend retaliatory forces remain intact in the form of the ABM Treaty. Since the signing of SALT I, the United States has, to a large degree, dismissed the sensible linkage between offensive and defensive limitations established at those negotiations.

Thus, the United States and the Soviet Union do have the legal right to withdraw from the ABM Treaty, given proper notice, and the United States does have a strategic and an arms control rationale for reconsidering the ABM Treaty. The treaty should not be considered sacrosanct.

If the United States does decide to seek revisions to, or withdrawal from, the ABM Treaty to permit extensive deployment of BMD, when would a change in the status of the treaty be necessary? It is unlikely that revision or withdrawal would be required until relatively late in the 1980s. This is because the ABM Treaty constrains deployment much more severely than it does research and development.

The potential near-term BMD system that could provide limited but important defensive coverage during the initial phase of transition could run into trouble with the treaty prohibition against testing, development or deployment of mobile BMD components (Article V); certainly actual deployment of any effective system would run afoul of the limitations of Article III, which restrict the number of permitted sites to two (amended to one site in the 1974 ABM Treaty Protocol) and the number of permitted interceptors and launchers to 100.

Although mobility would probably be sought in a near-term BMD system, to ensure survivability, the initial development and testing of subsystems need not be in a mobile mode, and the actual mating of these subsystems to constitute a mobile component or system need not occur until relatively late in the development and testing process. Thus there is no great urgency to revise the treaty in the very near future. If the United States decides to proceed with the defensive transition, however, it should begin to examine the types of treaty revisions that are necessary, and the appropriate timing for those revisions.

The treaty does allow great freedom for development and testing (but not deployment) of exotic beam defenses, particularly as long as these are tested from fixed, ground-based installations. Nevertheless, in the long run, if the United States decides to pursue comprehensive BMD coverage using multiple layers of defensive interceptors, the ABM Treaty will be unlikely to survive in anything resembling its current form.

On the other hand, neither the prospective near-term defensive systems considered here, nor later more sophisticated defenses, would require atmospheric testing of nuclear weapons.[14] Most of the technologies under consideration are nonnuclear and would not involve space-based "weapons of mass destruction" (i.e., nuclear, chemical, or biological). Hence a defensive transition need not run afoul of either the Limited Test Ban Treaty of 1963 or the Outer Space Treaty of 1967, banning such weapons.

In sum, a defensive transition would compel initial revision and in all probability later withdrawal from the ABM Treaty. This could be done legally, and would not be inconsistent with the US arms control position established at SALT I. But the most basic question is its broader impact on arms control efforts, particularly to limit offensive weapons on both sides.

In this broader context, there are several reasons why defense and arms control could be mutually beneficial. A defensive transition could establish a necessary basis for deep offensive force level reductions. First, even limited near-term defense systems designed to protect retaliatory forces could alleviate the verification difficulties associated with deep force level reductions. At current relatively high force levels a degree of ambiguity in the ability to verify an agreement is considered acceptable because only large-scale violations would have a "significant" impact upon the strategic balance, and such violations are likely to be noticed. Reducing US and Soviet strategic arsenals to the level initially proposed by the United States at START (e.g., reportedly, a ceiling of 850 for deployed ballistic missiles) or lower, however, would place a higher premium on each delivery system. At much lower force levels even a relatively small level of noncompliance could have a significant impact upon the strategic balance, and thus be of great concern.

In the context of small numbers of US retaliatory forces, the covert retention or deployment of even a few MIRVed ICBMs could be a threat to the survivability of an important fraction of US deterrent forces, and thus provide the Soviet Union with an increased incentive to strike first. And the importance of very strict verification for deep reductions is incongruous with the increasing difficulty of totally monitoring the deployment of new types of nuclear weapons such as cruise missiles and mobile ballistic missiles. Consequently, in the absence of a defensive transition the US insistence upon the ability to verify an agreement with very high confidence could reduce the chance for any deep arms reductions. A transition to strategic defense, however, would reestablish the condition wherein deception on a very large scale would be necessary before the strategic balance could be jeopardized by cheating. US retaliatory forces that were defended would be less vulnerable to deceptively deployed Soviet forces— hence the need for very rigorous verification standards could be somewhat relaxed.

Second, it is clear that one reason for the Soviet commitment to large numbers of strategic weapons is to achieve a damage-limiting effect through offensive "counterforce" capabilities, i.e., the ability to disrupt US command channels and to destroy US retaliatory forces before they could be launched against the Soviet Union. It is likely that the Soviet Union has been so reluctant to agree to US proposals for deep reductions in heavy MIRVed ICBMs because these weapons are the primary counterforce instruments in the Soviet arsenal. As noted above, damage limitation for national survival is a key objective of Soviet strategic doctrine, and that objective currently is pursued primarily through offensive counterforce preparations.

A defensive transition could provide the damage-limiting capability mandated by Soviet doctrine, with strategic defense replacing offensive forces as the principal means for limiting damage. Strategic defense could, in effect, take over the damage-limitation mission now given to offensive counterforce weapons. Such a development should reduce the long-noted reluctance of the Soviet Union to accept negotiated cuts in its large ICBM force.

Finally, if defensive technology proves to be extremely effective, it could reduce the incentives for the offensive arms competition by rendering it futile. The Soviet ballistic missile defense program of the 1960s probably was truncated because it became apparent that US advances in MIRV technology would easily counter the Soviet defensive system. Thus, US technological advances in offensive systems probably discouraged the Soviets from continuing to deploy what had become obsolete defensive technology. The development of highly effective defensive technology could, and logically should, have a similar impact upon offensive weapons programs.

Of course, a transition to defense could lead to a competition in the development and deployment of increasingly advanced defensive technologies. Nevertheless, restructuring the arms competition toward a "defense race" would have a benign impact upon the catastrophic potential of nuclear war, and would be far preferable to an indefinite continuation of the competition in offensive nuclear arms.

VII

In one other aspect, related to arms control, a defensive transition could serve the interest of all mankind in a critically important fashion. A recent study on the "Global Consequences of Nuclear War" indicates that a relatively "small" nuclear war, involving between 500 and 2000 detonations, could result in climatic changes that would trigger a global catastrophe.[15] That number of detonations would reflect the use of only a small fraction of the nuclear weapons in US and Soviet arsenals. Strategic defense is the only candidate answer to this potential threat to humanity. Suggested alternatives simply are ineffective.

The most obvious and effective solution to this danger would be for nuclear weapons never to be used. However, it simply is not within the power of a US president to determine whether a nuclear war will occur. Despite the best efforts of the United States to avoid nuclear war, the Soviet Union or another nuclear-armed country could employ nuclear weapons against the United States or a local foe. The first step in understanding this issue is to recognize that whatever the United States does, or does not do, cannot ensure the prevention of nuclear war. It is beyond reason to believe that all nuclear-armed powers would agree to ban the use of nuclear weapons and abide by that agreement under all conditions.

Equally incredible is the prospect for arms control to reduce the global arsenal of over 50,000 nuclear weapons to numbers below the threshold reportedly necessary to cause a climatic catastrophe. This does not mean that the arms control process has no value for the US pursuit of strategic stability. Rather, what is suggested is that this avenue can hardly be relied on to prevent a climatic catastrophe that might stem from even a "small" nuclear exchange.

In sum, a defensive emphasis and nuclear disarmament are essential allies. Advocates of a radical scale of nuclear disarmament need to appreciate that truly deep reductions in offensive nuclear arsenals would be feasible only in the event of heavy deployment of strategic defensive systems. The United States could never verify strict Soviet compliance with a possible START regime that mandated reductions in offensive forces down to the low hundreds of weapons. But, with strategic defenses deployed, the superpowers could be confident that cheating would have to be conducted on a massive scale before it could provide a capability sufficient to yield important military or political advantage. If we assume that the United States and the Soviet Union will be political rivals for many years into the future, strategic defenses offer the *only* path to a nuclear disarmament agreement with which both parties could live.

A defensive transition—particularly one including the global coverage that space-based systems or components could provide—would reduce the risk of a global climatic catastrophe by intercepting nuclear weapons after they are launched. Given the fact that nuclear deterrence can never provide the certainty of stability, and given the reported possibility of climatic disaster resulting even from the limited use of nuclear weapons, transition to a defense-dominant strategic policy should be seen as a moral imperative. At the very least, given the perils of the arms competition in offensive systems, the United States is obligated to seek to alleviate those perils through defense. Success in this venture is not guaranteed, but there would seem to be no excuse for the United States not making the attempt.

VIII

Whether defensive technology will be sufficiently robust to defeat potential active and passive countermeasures is an important issue, one which must be examined before critical decisions concerning development and deployment of potential defensive systems can be made. The preliminary conclusions in this regard, drawn from the recent official reviews of BMD, are quite optimistic about the potential for a robust defense against ballistic missiles. Ultimately, only time and further advanced investigation into the potential for defensive technology and offensive countermeasures will enable us to conclude whether the defense will become again the stronger form of warfare.

Unfortunately, though not unexpectedly, the debate that is shaping up promises to be yet another stale and unimaginative confrontation between those who judge homeland defense to be destabilizing and those who do not. Neither the government nor private commentators are well equipped at present with an understanding of how offense and defense can proceed together in complementary, synergistic fashion for the benefit of more stable deterrence. It is very desirable that those who are strongly committed to President Reagan's vision of an America defended against ballistic missile threats think constructively about the positive roles that US offensive and near-term defensive forces can play both to safeguard the defensive transition and, perhaps, to help stabilize deterrence beyond the transition.

Neither superpower, at least in the early stages of an essentially competitive defense transition, is going to cooperate tacitly in assisting the defenses of the other side to achieve high effectiveness. The United States and Soviet Union have, after all, been involved in an arms competition expressing deep-rooted political rivalry. No matter how great the technical success of US and Soviet defense transitions, the competition between offense and defense will not stop. Neither superpower is likely to abandon permanently all hope of gaining a major advantage by developing both effective offensive and defensive weapons.

One must assume that both the Soviet Union and the United States prefer a condition wherein *both* their offensive *and* their defensive capabilities are effective to a condition wherein only their defensive weapons can perform as intended. Neither side, however, is likely to anticipate an enduring advantage in the strategic offensive and defensive systems; both will be constrained to accept much more limited offensive targeting capabilities than now exist. Future missions for US strategic offensive forces may include the following: guarding the defense transition; holding at risk so many high-value assets of the Soviet state that the Soviet leaders perceive a substantial net advantage in negotiating a major bilateral drawdown in offensive forces (thereby assisting the US defense transition); providing an enduring hedge against sudden revelation of weaknesses in defensive systems; and providing some deterrent effect in order to help discourage gross misbehavior by third parties.

The public debate over the orientation of future US strategic policy that was triggered by President Reagan's defense initiative proposal of 23 March 1983 has revealed all too plainly that there are more and less sensible ways to think about defense.

Strategic defense should not be viewed in terms of an all-or-nothing "astrodome." "Star wars" defenses, no matter how great their promise, will not constitute the last move in high-technology arms competition, and strategic defensive technology will not solve the fundamental problems of political rivalry. But strategic defense, embracing a wide range of near-term and far-term weaponry, promises to strengthen the stability of deterrence by imposing major new uncertainties upon any potential attack. In the long run, it holds out the possibility of transforming, though not transcending, the Soviet-American deterrence relationship.

NOTES

1. The full text of President Reagan's speech is in the *New York Times*, 24 Mar. 1983, p. 20.

2. Reported in *Defense Daily*, 29 Nov. 1983, p. 137.

3. For example, a "light" area defense should be capable of defeating Soviet precursor attacks against the US strategic command and control system and of denying the Soviet Union the option of pinning down US ICBMs in their silos so that they could not launch under attack. On the technical uncertainties for the offense, even in the absence of ballistic missile defenses, see Matthew Bunn and Kosta Tsipis, "The Uncertainties of a Preemptive Nuclear Attack," *Scientific American*, Nov. 1983, pp. 38–47.

4. For an excellent technical discussion that supports this claim, see Dr. Patrick Friel, "Status of U.S. BMD Technology," a paper presented to the annual conference of the American Association for the Advancement of Science, Washington, D.C., 6 January 1982. See also William Davis, "Ballistic Missile Defense Will Work," *National Defense* 66, no. 373 (1981): 16.

5. Quoted in "Keyworth: Space-Based Defense Possible," *Air Force Times*, 31 Oct. 1983, p. 29.

6. For an examination of Soviet victory requirements, see John Dziak, *Soviet Perceptions of Military Doctrine and Military Power: The Interaction of Theory and Practice* (New York: National Strategy Information Center, 1981), p. 28.

7. The qualifier, "virtually," acknowledges the presence of the existing theory of damage control and limitation. To be specific, damage to the American homeland in war might be limited if the Soviet government chose to exercise restraint in its nuclear targeting.

8. See for example the discussion in Lawrence Freedman, "The Small Nuclear Powers," in *Ballistic Missile Defense*, ed. Ashton B. Carter and David N. Schwartz (Washington, D.C.: Brookings Institution, 1984), pp. 151–74.

9. Remarks by President François Mitterrand at a luncheon offered by the Council of Ministers of the Kingdom of The Netherlands, The Hague, 7 Feb. 1984. Embassy of France, Press and Information Service, Memo 84/6, p. 8.

10. This assertion rests on the assumption that Soviet technical prowess in BMD is roughly equivalent to postulated US prowess. Given the respective strengths of the superpowers as competitors in this area of military high technology, it is very possible that US BMD will be significantly better than will Soviet BMD. Our argument does *not* require for its validity that the US achieve and sustain a lead in defensive systems. But such a lead would have beneficial consequences for strategic stability.

11. Central Intelligence Agency, National Foreign Assessment Center, *Soviet and U.S. Defense Activities, 1970–1979: A Dollar Cost Comparison*, SR–80–10005, Jan. 1980, p. 9.

12. For an excellent review of Soviet BMD activities see Sayre Stevens, "The Soviet BMD Program," in *Ballistic Missile Defense*, pp. 182–220. See also Department of Defense, *Soviet Military Power* (Washington, D.C.: G.P.O., 1983), pp. 27–28, 68.

13. See US Arms Control and Disarmament Agency, *Arms Control and Disarmament Agreements: Texts and Histories of Negotiations* (Washington, D.C.: G.P.O., 1980), p. 146.

14. Presidential Science Adviser George Keyworth recently observed that there does not appear to be an important role for nuclear weapons involved in the transition to strategic defense. This comment was made during Dr. Keyworth's presentation at the Forum on The Future of Ballistic Missile Defense, Brookings Institution, 29 Feb. 1984. His statement appears to be an authoritative refutation of the notion that a nuclear-pumped X-ray laser system would be a critical component.

15. See Carl Sagan, "Nuclear War and Climatic Catastrophe," *Foreign Affairs* (Winter 1983–84): 257–92.

PART II

US DEFENSE COMMITMENTS AND STRATEGIES FOR DETERRENCE

CHAPTER 4

THE UNITED STATES AS A WORLD POWER
An Overview

SCHUYLER FOERSTER

Earlier chapters in this volume have focused on the context in which American defense policy is formulated. Those chapters stressed the characteristics of the international political system in which the US seeks to define and protect its interests, as well as the role of domestic politics in the development and execution of policy; examined the nature of Soviet power which dominates American calculations of its security requirements; and highlighted the unique features of the nuclear age and the associated dilemmas that confront considerations of deterrence and defense.

Part II of this text turns to the development of US defense policy to deal with this security environment. The chapters which follow examine, in turn, different facets of a global defense policy designed to contain Soviet power; to protect and manage alliances in Europe and Asia; and to defend global interests against an array of threats requiring an equally diverse array of military and nonmilitary responses.

Chapter 5 focuses on the evolution of the US-Soviet adversarial relationship and the strategies employed by the United States to deal with that relationship. In doing so, it examines four recurring themes that have guided the United States as it has sought to manage Soviet power. The first is *containment*, the foundation of US security policy since the advent of the cold war, which has been variously applied by successive administrations to reinforce the global character of US defense policy. The second is *strategy*—"the calculated relationship between ends and means"—which continues to confront policy planners with the challenge of reconciling finite means with the tendency to define expansive ends. A third characteristic of US defense policy is the fact that the expansion of US power and the definition of US global interests have been accompanied by a persistent sense of *vulnerability* which is, at least for the United States, without historical precedent. Finally, the recurring theme of *legalism/moralism* reflects the fact that the definition of US security interests and the means to protect those interests go beyond considerations of power alone. In becoming a "world power," the United States has done so with a view—sometimes explicit, sometimes implicit—to the legal and moral purposes which that power is to serve.

Chapters 6 and 7 take a more regional approach by examining the evolution of alliance commitments and strategies in Europe and Asia, respectively. Although the nature of US formal alliance commitments varies significantly—and includes many countries throughout the world not included in these chapters' discussions—the structure of US military forces is in many ways determined by the permanent military presence in Western Europe and in the Pacific region in support of multilateral and bilateral treaty commitments. These discussions also reflect the problems of alliance management, the constraints as well as the benefits afforded by a structure of global power built on coalition pillars.

The final chapter in this part of the book recognizes that US global interests go far beyond direct alliance commitments, and that the threats to those interests do not emanate exclusively from the Soviet Union. While the international political order that grew out of the rubble of World War II has endured in the continuing East-West conflict and the confrontation of two alliance systems opposite each other in Europe, the rest of the world has witnessed a fundamental transformation that is revolutionary in its implications. The volatility that characterizes the less-developed regions of the planet cries out for solutions which the East-West framework is ill-equipped to provide. The persistent political, economic, and social problems in those regions threaten US security interests directly while challenging traditional thinking about the way in whichmilitary force can be used to defend those interests.

Understanding these disparate challenges to US interests and the difficult defense policy choices they pose requires a clear sense of what it means for the United States to be a world power and of the requirements for sustaining that power over time.

WHAT IT MEANS TO BE A WORLD POWER

The tendency for individual powers to play dominant roles in a political system predates the birth of the modern system of sovereign states. Within the post-Westphalian state system, political order has been maintained largely by dominant "great powers" who tended to interact in a system of balance of power and shifting alliances. Those great powers—European until the emergence of the United States as a global power—assumed certain privileges by virtue of their dominant role. At the same time, the responsibility for maintaining the stability of the system fell to the great powers, a role generally accepted out of self-interest since, by preserving the status quo, the great powers could perpetuate their dominant role in the system.

Even before this century, some European states were "world powers" in that they were able to project their power to far corners of the globe and to establish and maintain vast colonial structures.[1] The eighteenth and nineteenth centuries witnessed the growing globalization of international politics as well as the growing interrelationship among continents heretofore isolated from each other. Great-power rivalries were projected into diverse global regions, at times playing a role in the formation of new nations, including the United States. Once established, the United States heeded its founding fathers' counsel against engaging in "entangling alliances" and enjoyed the luxury of expanding its own national scope with relatively little interference from or even regard for events elsewhere in the world.

The advent of total war in the twentieth century removed that luxury. The United States entered World War I with a recognition that its vital interests were involved even though, from a geographical standpoint, the United States itself remained invulnerable. "Making the world safe for democracy" reflected an ideological interest that was nothing less than global in scope. Although precipitated by the Japanese attack on Pearl Harbor, the entry of the United States into World War II was probably inevitable because of the Roosevelt administration's conviction that US interests were inextricably linked with the survival of Britain and the defeat of the Third Reich. Even though the interwar years had witnessed a return to isolation, they had also witnessed the evolution of an international political and economic role for the United States which had strong security implications. Sponsorship of naval armaments conferences involving both Asian and European maritime powers and underwriting European economic recovery in the 1920s are but two examples of that role. The international impact of the Great Depression was evidence of the extent to which the United States had already become a "world power."

World War II reaffirmed that the United States could not pretend to be immune from global turmoil and gave birth to the notion of the United States as a "superpower." Along with Britain and the USSR, the United States was expected to assume the responsibility for reconstructing and preserving a new world order to avoid the horrors of yet another "total war." The Truman Doctrine and the Marshall Plan in 1947, the articulation of "containment" and the beginning of peacetime alliances such as NATO, and the US response to the attack in Korea in 1950 all represented major milestones in the emergence of the United States as *the* Western superpower that could lead to a coalition of states in containing the opposing superpower and its bloc of satellites, which had a quite different ideological and political vision of what that postwar world order should be. While the Rio Pact (1947) represented a long-standing US concern for hemispheric security, rooted in the 1823 Monroe Doctrine, the formation of NATO (1949) reflected an unprecedented peacetime commitment. Additional collective security treaties included the Australia, New Zealand, US Pact (ANZUS, 1951); the Southeast Asia Treaty Organization (SEATO, 1954); the Central Treaty Organization (CENTO, 1955); and bilateral defense agreements with the Philippines (1951), South Korea (1953), Taiwan (1954), and Japan (1960). These were complemented by a variety of bilateral defense cooperation agreements with friendly states around the world in forming the structure for both global military commitments and a global military presence.

That global structure has changed over the years. NATO, of course, remains the premier US alliance in terms of the scale of US defense investment dedicated to fulfill that commitment. CENTO and SEATO have been disbanded as organizations, although the Southeast Asia Treaty remains in force; the defense commitment to Taiwan was terminated when the United States recognized the People's Republic of China. In the wake of Vietnam, moreover, the US military presence in Asia contracted, with Korea providing the only remaining permanent US military presence on the Asian mainland. Political changes in North Africa and the Middle East have likewise altered and, to some extent, restricted US military presence and access in those regions. Such restrictions in its global military presence also reflect US domestic, political, and economic considerations. Greater circumspection about US military intervention in regional conflicts, plus a decline in the resources allocated to foreign security assistance and economic aid, have constrained American investment in an overseas military infrastructure.

Nonetheless, by any measure, the United States is a global power, not only in military terms but also in political and economic terms. Defending its global interests—fundamentally an interest in the nature and shape of a global world order—requires a global defense strategy based on a clear vision of the objectives to be sought and the means available for achieving them.

US SECURITY OBJECTIVES AS A WORLD POWER

In Defense Secretary Caspar Weinberger's final *Annual Report to the Congress*, US national interests were framed as follows:

US national interests encompass both broad ideals and specific security assets. America's paramount national interests are peace, freedom, and prosperity for ourselves and for our allies and our friends, and for others around the world. We seek an international order that encourages self-determination, democratic institutions, economic development, and human rights. We endorse the open exchange of ideas and other measures to encourage understanding between peoples.

More specifically, we maintain our steadfast concern for the security and well-being of our allies and other nations friendly to our interests. We oppose the expansion of influence, control, or territory by nations opposed to freedom and other fundamental ideals shared by America and its allies.

The peaceful existence and prosperity of democracies is the core US interest.[2]

What is noteworthy about this statement is that US national interests are defined principally in terms of the defense of ideals and of those who share those ideals. It is not just a question of territorial defense, but a basic recognition that the "American way of life," however defined, is bound up in the existence of a particular international order that must be defended and nurtured.

Such a definition of national interests—the kind of "core" interests that merit defense, if necessary, by force of arms—leads to a number of specific security objectives. Such objectives can be variously phrased but generally include the following:[3]

—Deterring aggression against the United States, its forces, and its allies.

—Should deterrence fail, defeating armed aggression and terminating conflict on favorable terms at the lowest possible level of conflict.

—Securing the United States and its allies from coercion or intimidation by hostile powers.

—Facilitating the ability of allies and friends to defend themselves against aggression, coercion, subversion, insurgency, and terrorism.

—Ensuring US access to critical resources, markets, the oceans, and space.

—Where possible, reducing the dominant threat of Soviet military power by a variety of means short of the direct use of military force (political influence; restrictions on access to military technology; and agreements to control, limit, or reduce armaments).

These security objectives are global in scope—indeed, the inclusion of "space" suggests that they are more properly characterized as potentially universal in scope. As a guide to shaping a defense establishment to fulfill these objectives, however, these objectives are only part—and not necessarily the determining part—of the equation. As one analyst has noted and as subsequent chapters affirm, US strategy in the postwar period has alternated "between the belief that limited means require differentiated interests, on the one hand, and the belief that undifferentiated interests require unlimited means on the other."[4] Unless the United States is prepared to distinguish between vital and nonvital interests, the defense requirements to maintain that global position are potentially open-ended.

There are, to be sure, significant constraints on the ability of the United States to defend all of its interests. One constraint, of course, is *resources*. While the United States has seen periods of sustained growth in defense spending, those periods have tended to oscillate with periods of retrenchment and decline, not because the security environment changed but because the domestic political and economic environment demanded it. Another constraint is *geography*. It is often forgotten that, at least in a geostrategic sense, the United States is an island. Its interests are far-flung; its defense requirements are substantially more extraterritorial than territorial. The US global position is thus subject to changing political climates in global regions in which events are often beyond its control.

These two constraints—resources and geography—account to a large extent for the fact that the United States has pursued a *coalition strategy* in the defense of its interests. This reality, however, has its own paradoxical features that shape many of the defense issues that continue to confront the United States.

IMPLICATIONS OF A COALITION STRATEGY

US security rests in many ways on a structure of alliances and bilateral defense relationships which, on the one hand, *enables* the United States to remain a global power. Those relationships provide the physical infrastructure for a global military presence. The pursuit, with allies, of a collective defense rather than uni-lateral national defense potentially enables the United States to maintain its own security at reduced costs, provided the others share the burdens of that collective defense. At the same time, however, those defense commitments be-come *objects* themselves to be defended, at times generating their own independent ration-ale for military strength. Thus, coalitions pro-vide a means to defend global interests, while the security of coalition partners becomes a de-fense objective in its own right.

In theory, this need not be contradictory, since one might assume that the defense of those nations with which the United States has a de-fense commitment would be in the US interest regardless of whether that defense commitment exists. This is perhaps most evident with NATO, since the United States has already twice dem-onstrated its willingness to defend Western Eu-rope even though no prior commitment existed. Yet it is also true that the definition of vital interests changes over time. Commitments made to defend a country can take on a political force of their own, to be defended because they are US commitments, independent of who the beneficiary of that commitment is. For example, many have argued that, in the case of Vietnam, the United States justified the scale of its in-volvement in terms of upholding US commit-ments in general, quite apart from whether the defense of South Vietnam itself was a vital in-terest.

Alteration of commitments, by the same token, can take on acute political significance, even though one might decide that it is no longer in the US interest to be committed to the defense of a particular country. Taiwan is a case in point, where the political dynamics in Asia changed, and the decades-old commitment to defend Taiwan was terminated. Even altering the *form* in which the United States chooses to defend its standing commitments takes on con-siderable political importance. A persistent fea-ture of the US relationship with its NATO allies is that even discussing the possibility of signif-icant modifications in the military presence in Europe creates anxieties among allies who de-mand reassurance that the commitment remains in effect. The same is also true of South Korea, as evidenced by the Carter administration's abortive attempt to withdraw US forward-deployed ground forces from that country.

There is an additional paradox that suggests that coalitions can be both a means and an end in shaping defense policy. In its simplest sense, an alliance is a voluntary association of sovereign states so that a collective benefit will derive from collective action. As a coalition of states oper-ating from a common interest, an alliance thus affords greater resources and capabilities to achieve a desired end. In practice, however,

alliances also place constraints on states. Typi-cally, alliance members surrender some free-dom of action to facilitate collective defense, perhaps giving up some options in return for some guarantees. Common interests do not imply identical interests, however, and, in a coalition of sovereign states, attempts to har-monize collective actions also involve compro-mises among competing interests.

When the United States embarked on its global coalition strategy in the first decade fol-lowing World War II, it was the unchallenged, dominant Western power. US alliance relation-ships tended to reflect a one-way dependency more characteristic of a patron-client relation-ship than of a coalition of equal partners. Today, those relationships have changed. It is still the case that the United States plays a unique role in each of its alliance relationships, owing fun-damentally to its continuing position as *the* Western nuclear superpower, the irreplaceable importance of its nuclear deterrent on which its allies ultimately depend, and the sheer force of its military and industrial might. But many US allies in both Europe and Asia have themselves matured into significant political, economic, and—in some cases—even military powers. Al-liance political relationships no longer reflect the model of a patron and its clients, but instead are indicative of a growing interdependence that, in the peacetime business of alliance man-agement, often overshadows the still-singular nuclear and military role of the United States. US leadership within its alliance structures per-sists—indeed, the United States is criticized most of all when it is seen as faltering in that leadership role—but the nature of that lead-ership is more consensual and conditional than it once was.

The dilemmas of a coalition strategy are par-ticularly evident in the case of the US decision, in 1987, to increase substantially its naval pres-ence in and around the Persian Gulf and to solicit allied cooperation in that effort. The prox-imate causes of that increased role were the intensification of the Iran-Iraq war, the growing threat from Iran to neutral shipping in the Gulf, as well as US determination to defend the long-standing principle of freedom of navigation. US naval presence in the Gulf region was not new, and the assertion that vital Western interests were involved—maintaining the flow of oil through international waterways—was neither novel nor contested. What was controversial was, first, US agreement to reflag Kuwaiti ships to make them eligible for US protection as US-flag carriers; and second, the US decision to rein-force the Navy's small, long-standing Mid-East Force in the Gulf and its decade-old battle group presence in the Indian Ocean with a size-able force including carrier and battleship battle groups, a contingent of Marines on amphibious landing ships, and minesweepers. To its allies, the United States appeared to have raised both the stakes of involvement and, commensurately, the likelihood of conflict.

Many allies did in fact respond. The British and French—both of whom had also traditionally maintained a nominal presence in the region—increased their naval contingents as well, the latter with a carrier battle group. As the threat of Iranian-laid mines and air and surface attacks increased, Belgium, the Netherlands, and Italy dispatched, along with surface combatants, mine-countermeasure ships to complement those sent by the United States, Britain, and France, providing a capability in which they were, in many ways, better equipped than the United States. The Federal Republic of Germany (FRG) and Luxembourg—the only Western European Union (WEU) members not participating in Gulf operations—contributed indirectly: the FRG sent a surface contingent to the Mediterranean to compensate for NATO-committed assets deployed to the Gulf, and Luxembourg provided financial resources to Belgium and the Netherlands. Japan agreed to set up a Gulf navigation system.

US *allies* ultimately responded to what they viewed as a challenge to their own, not just US, interests. They did so, however, conspicuously outside of an *alliance* framework involving the United States. Given that the NATO "Treaty Area" (defined by Article 6) does not include the Persian Gulf region, it should not have been surprising that NATO did not address this threat in terms of a corporate response. But NATO had conceded that its interests were affected by "out of area" events—particularly in the Gulf—and several allies' merchant ships had themselves been threatened in that region. Regardless of Article 6, there was nothing to preclude cooperation among *allies* in the defense of common interests even if that cooperation took place outside the context of a formal and corporate *alliance* response. Yet, following US overtures for such cooperation, those allies which did respond either cast their initial responses in terms of defense of their own *national* interests or wrapped their participation in the context of an independent *European* (WEU) response. To many, association with the United States in such an effort seemed a political liability. No country justified its response in terms of solidarity with the United States. Still others preferred the aura of noninvolvement in what they viewed as largely a US anti-Iranian adventure, in some cases even denying port access to allied ships en route to the Gulf even though they privately acknowledged that those ships were defending a shared interest.

The political dynamics of Western cooperation in the defense of freedom of navigation in the Gulf are indicative of the paradox of US power and alliance leadership. On the one hand, the reluctance of allies to respond to US overtures and their clear preference for seizing alternative rationales for their own involvement suggest that they might have been more forthcoming in defending their own interests in the Gulf if the evolution had not, in fact, assumed such a leading, substantial, and politically visible role. On the other hand, one can also argue that, if the United States had not assumed the lead in highlighting the threats to common Western interests, had not responded to those threats in such a demonstrably decisive way, and had not challenged the allies to do likewise, then those allies that did respond might not have invested the political, military, and financial capital that they have in the Gulf.

The fact remains that the United States is a superpower with global interests which is allied with small and medium powers whose interests are more normally regional in scope. To the extent that US coalition partners share an interest in preserving a world order—certainly a "global" interest—it is often left to the United States to bear the principal responsibility for the defense of that order. (This is most evident in the case of Japan, whose economic power *and* economic dependency make it a state whose interests in many respects transcend the Pacific region but which is deliberately restricted in what it can contribute to the military defense of regional as well as global interests.) Moreover, the US global interests are more than the sum total of its alliance interests, although the United States must rely on its formal allies and others to help defend those global interests.

This disparity in both power and interests between the United States and its allies around the world creates the potential for tension, a tension which manifests itself in two paradoxical fears. The first fear is that the United States may not actually come to the defense of its allies, particularly if it means risking its own survival and well-being. As a result, a vital part of the process of alliance management for the United States is the investment of political, military, and financial capital to reassure allies of the efficacy of the US commitment quite apart from that effort necessary to deter a potential adversary from challenging that commitment. From the allies' point of view, the United States must remain *entangled*.

The second fear is that the United States, in the pursuit of global interests and in its broader competition with the Soviet Union, may embroil its allies in a conflict which allies view as unnecessary and undesirable. This fear—one might call it *reverse entanglement*—underpins allied anxieties about too close an association with the United States in conflicts such as the Gulf, in US-Soviet confrontation in any Third World region in which allied security is not directly threatened, or even in perceived US attempts to challenge Soviet domination in Eastern Europe (as in the Polish crisis in 1981).

Managing the sometimes competing and sometimes complementary requirements of being, on the one hand, a global power defending global interests and, on the other hand, a coalition leader defending commitments to far-flung allies is arguably the preeminent challenge to US security policy. It is a problem that involves more than just the exercise of military force but rather the exercise of the entire spectrum of political, economic, and social instruments of national power. Within the realm of defense policy, however, managing this problem places particular burdens on decisions regarding the deployment and employment of military force, in which resource and geographic constraints play a significant role. After a period of sustained real growth in defense spending, the United States, in the late 1980s, entered into a period of fiscal retrenchment in which the trade-offs between competing interests became more acute. This does not suggest that the United States will be unable to sustain its commitments and its position as a world power. As indicated by earlier periods of restrained defense spending, however, it is likely that the means by which the United States fulfills its global responsibilities may well change over the coming decade.

FORCE POSTURE FOR A GLOBAL DEFENSE POLICY

By any measure, the United States has maintained a substantial investment in defense. With annual defense appropriations holding at approximately $300 billion in the late 1980s, US defense spending increased in the first half of that decade by over 60 percent after inflation, from 5 percent of gross national product (GNP) in 1980 to over 6 percent.[5] Without a frame of reference, however, that defense effort can be exaggerated, distorted, and misunderstood. Soviet defense spending, for example, has consistently been in an estimated range of 15 to 17 percent of GNP.[6] Similarly, with over 2.1 million men and women on active duty in the armed forces, the United States has the largest military force in the West. By comparison, however, Soviet active-duty strength is estimated at over 5.2 million—the largest in the world—with the Chinese second, at 3.2 million. Including reserves, moreover, the US military comprises only 1.4 percent of the total population, a percentage smaller than that of every continental European nation—East or West—save Luxembourg.[7]

The United States and the West in general have tended to rely more on the qualitative advantages offered by a vibrant educational, technological, and industrial base rather than on massive amounts of military manpower and weaponry. A professional, all-volunteer force in a prosperous society is an expensive proposition, consuming a large portion of the defense dollar. (By contrast, most US allies continue to conscript.) As table 4.1 shows, military personnel plus operations and maintenance (O&M) costs—necessary to sustain the readiness and combat effectiveness of the force, particularly as weapons systems become more sophisticated—consistently claim the most of the defense budget. Research, development, testing, and evaluation (RDT&E) plus procurement of new weapons systems—often the most politically visible and controversial elements of the defense budget—constitute a somewhat smaller share of that budget authority.

US force structure is shaped in part by the fact that the United States is, in effect, an island nation with geographically dispersed interests and commitments. As such, US forces prepare to defend against three distinct categories of threats. One is the direct threat to territorial integrity and security which emanates exclusively from Soviet strategic nuclear capabilities. Second is the Soviet threat to the territorial integrity and security of allies in Europe, Asia, and Latin America. Third is the threat to US and Western interests in the Third World, typically lower in the spectrum of conflict intensity and not always emanating from the Soviet Union. Of these three threats, only in the first—the threat of Soviet nuclear attack on the United States—is the United States a "frontline" state concerned with the defense of its own territory. Strategic nuclear deterrent forces are an absolutely vital element of the US force posture, not only for deterrence of attack on the United States but also as an extended deterrent on behalf of allies. Yet, those forces comprise a relatively small portion of the entire defense budget—less than 25 percent—because they are relatively few in number, less manpower-intensive, and thus cheaper to procure and maintain than general-purpose forces.

A substantial majority of the US defense efforts, therefore, derives not from the requirement to maintain a viable nuclear deterrent force, but from the requirement to procure and maintain conventional or general-purpose forces—air, land, and maritime—capable of dealing with threats which, although not *immediately* directed against US territory, nonetheless involve vital US security interests. These forces are designed to meet direct military threats to US allies which, in many cases, are "frontline" states themselves; to protect friends and allies that are vulnerable to subversion and other forms of low-intensity conflict; and to preserve US access to strategically vital resources and lines of communication. To meet

Table 4.1
Percentage Allocation of Total Budget Authority

Category	FY86	FY87	FY88	FY89
Military Personnel	24.1	26.2	26.9	27.0
O&M	26.6	27.9	28.5	29.4
Procurement	32.8	30.2	28.6	27.5
RDT&E	11.9	12.6	12.9	13.1
Other	4.6	3.1	3.1	3.0
Total ($ million)	281.4	281.7	283.8	290.2

Source: US Department of Defense, *Annual Report to the Congress, FY 1990*, p. 83; *Annual Report, FY 1989*, p. 127; *Annual Report: Fy 87*, p. 82; and *Defense Almanac*, p. 29.

these threats, the United States has developed a resource-intensive defense posture designed to project military power—both actively and passively—as a means of deterring and, if necessary, defending against challenges to its vital interests. In particular, a substantial portion of US forces is already forward-deployed in peacetime to likely theaters of conflict in which US alliance commitments or other vital interests are involved. Moreover, those forces based in the United States are for the most part designed for rapid mobilization, reinforcement, and deployment to potential theaters around the world.

Most US coalition partners, whether in Europe or Asia, are "frontline" states in that they are contiguous to or near the periphery of the Soviet Union and its allies and therefore face a direct threat to their territorial integrity. As such, the military postures of most US allies are designed principally for territorial self-defense or—notably in the case of NATO—to aid in the defense of proximate neighbors as a forward defense of their own territory. The requirement for the United States to respond quickly in crises and to deploy forward to potential or actual theaters of conflict, on the other hand, places a greater premium on force readiness. Thus, while continental powers tend to have conscript forces with even larger reserve forces, the United States—as indicated in table 4.2—has a relatively large proportion of its total forces on active duty. (The Soviets, for example, are estimated to have 6.2 million reserves in addition to their 5.2 million standing force.)[8] Moreover, the US "Ready Reserve"—consisting of Selected Reserve, Individual Ready Reserve, and National Guard units—is maintained in a relatively high degree of readiness and training. In a national emergency, the United States will likely not be able to afford the relatively extended mobilization, training, and deployment pattern that characterized US involvement in the two previous world wars in this century.

Not only is the US active-duty military presence such a significant portion of the overall military posture, but the same requirement to

defend far-flung commitments and interests imposes an additional defense burden. As table 4.3 highlights, almost one-fourth of those active-duty forces are stationed outside the United States—almost one-third including those afloat. Given the economic, political, and military constraints associated with stationing ground forces in potential theaters of conflict, the bulk of US Army combat units are stationed in the United States. Nonetheless, four divisions, three brigades, and two regiments are stationed in Germany; two divisions are stationed in Korea and Okinawa; and one light brigade is in Panama—meaning that about one-third of the Army is stationed outside the United States. In the case of land-based tactical aircraft, approximately half of the Air Force's tactical fighter squadrons are already stationed in Europe and the Pacific to enable rapid response in a crisis.[9] In addition, approximately one-third of the Navy's *operational force* is typically forward-deployed at any one time.[10]

This substantial overseas deployment reflects the fact that the United States, together with its allies, is committed to a strategy of *forward defense*. For the United States, the logistical difficulties of transporting large units of personnel, heavy equipment, and support structure are substantial. Moreover, the politically sensitive decisions to reinforce potential theaters of conflict in a crisis would—if those reinforcements were to be effective—have to be made in sufficient time before conflict broke out, but under circumstances in which warning indicators would likely be ambiguous. Thus, to facilitate more timely reinforcement, the United States has pre-positioned substantial amounts of equipment and munitions, both in theater and afloat, so that initial reinforcing units could engage within one to two weeks rather than one to two months. In the case of NATO, where this problem demands most attention, even with pre-positioned materiel enabling some combat units to be deployed by air, over 90 percent of NATO's equipment, supplies, munitions, and petroleum, oil, and lubricants (POL) needed in the first three months of a conflict would have

Table 4.2
US Active and Reserve Force Levels (in thousands)

Category	Army	Navy	Marines	Air Force	Total
Total Active Duty	771.3	583.5	198.9	611.5	2168.2
National Guard	462.2			114.2	576.4
Selected Reserve	310.8	146.4	41.4	79.3	577.9
Individual Ready Reserve	305.6	70.4	47.8	46.8	470.6
Total Reserve	1079.0	228.1	90.9	265.2	1063.2
Total Military	1853.5	811.6	289.8	876.7	3831.4
On Active Duty (%)	41.8%	71.9%	68.6%	70.0%	56.6%

Source: *Defense Almanac*, pp. 29, 34–35. Data as of 31 Mar. 1987.

to traverse the Atlantic where ships would be subject to interdiction.[11]

If the defense of forward allied commitments in Western Europe and Northeast Asia were the only requirements driving US conventional force posture, defense planning would be a much more straightforward—although hardly simple—affair. The United States is not, however, solely a hemispheric power with two extrahemispheric regional commitments. A major conflict in Western Europe or Northeast Asia would probably not be confined to those theaters, but would more likely take a more global dimension. Moreover, the more likely regional and low-intensity conflicts that might occur in the years ahead (as noted in chapter 8) would probably occur in diverse regions where the United States has vital interests at stake but does not have an established or substantial permanent military presence. Thus, for example, it is not surprising that US military deployments outside Western Europe and Northeast Asia are predominantly maritime (see table 3), since naval and marine forces provide flexible instruments of power projection that are designed to be less dependent on substantial territorial-based support infrastructure.

In meeting the requirements of a global defense policy, therefore, the United States has developed a force structure that has a relatively high proportion of those forces outside the United States. Those forces remaining in the United States—still three-quarters of the active strength, plus reserves—must be trained and maintained in a sufficiently high state of readiness to mobilize and deploy forward to meet contingencies that can range from support of low-intensity conflict to full engagement in a major war encompassing multiple theaters of operation.

To support that forward-deployment requirement, the United States must also maintain a base structure that sustains not only the permanently stationed forces but also those forces that would be deployed in a crisis. Prepositioning stores of equipment, munitions, and supplies both on land and at sea enables greater flexibility in the rapid deployment of forces. In

addition, access to bases and port facilities, transit rights, and other forms of host-nation support assume particular significance if forces are to be able to deploy and remain on station. Thus, the United States maintains, in addition to its formal alliance commitments, an array of defense cooperation agreements with countries around the world to maintain access in times of crisis. At times, these relationships can become politically sensitive, especially in those countries in which the internal political climate is volatile. Particularly in areas such as the Middle East, where vital interests are clearly involved, the United States may find itself responding to contingencies without having full access to a land-based support infrastructure. As in the case of the Persian Gulf, the burden falls principally on maritime forces.

The US role in the Third World has, particularly since Vietnam, been one of the most controversial aspects of US policy. In dealing with regional conflicts and various forms of low-intensity conflict, the United States has, with obvious exceptions, tended to apply military force in an indirect and passive way rather than to engage actively. In part this reflects the fact that, while US interests might be at stake, those interests are often not vital and do not justify the direct use of military force. It also reflects the fact that, in many cases, the direct application of US military force is not an appropriate instrument for dealing with many internal conflict situations. Security assistance, military support, and advice and training to indigenous forces are more common instruments, recognizing that, in the long run, the ability of a government to sustain its own legitimacy and to defend itself against internal and external threats is the more enduring means to security and stability.[12] Nonetheless, besides checking Soviet power, there may be other US interests involved, such as the protection of US citizens, commercial and economic interests, and military facilities. Thus, the United States must demonstrate its willingness and capability to respond to threats against those interests. The fact that the United States has deployed military force, principally naval forces, over two

Table 4.3
Geographic Disposition of US Active Duty Forces (in thousands)

Location	Army	Navy	USMC	USAF	Total	% Afloat	% Total
US Territories	512.9	499.8	169.2	478.4	1660.3	11.5%	76.6%
West/So. Europe	217.7	30.9	3.2	91.6	343.4	5.5%	15.8%
E. Asia/Pacific	33.5	34.8	24.6	38.2	131.1	17.2%	6.0%
Afr/ME/So. Asia	1.8	11.6	0.6	0.4	14.4	72.2%	0.7%
Other W. Hemi.	8.3	6.1	1.2	2.9	18.5	11.4%	0.9%
Total	774.4	583.5	198.9	611.5	2168.2	11.3%	100.0%
Outside US (%)	33.8%	14.4%	14.9%	21.8%	24.4%	—	—

Source: Defense Almanac, pp. 28–29. Data as of 31 Mar. 1987. US and US territories also include unspecified special locations.

hundred times since 1945 reflects the important role that projecting a military presence has in the exercise of global power.[13]

Notwithstanding the requirement to be able to project military power in crises around the world, the principal determinant of the scope of the defense effort is the need to deter Soviet challenges to its vital interests and, if necessary, to defend against Soviet aggression. That in itself is a global endeavor, involving not only strategic nuclear forces and a substantial investment in general-purpose forces but also the forward deployment of a significant portion of those forces throughout the world. In any major conflict with the Soviet Union, the Soviets will possess the initiative in determining when, where, and how conflict will start. The burden on the United States and its allies will be to respond in a timely fashion with defenses that are of sufficient readiness and to sustain those defense efforts over time and with long sea and air lines of communication. The principle of forward defense suggests not only defense at the points of attack but also securing flanks and ensuring that the aggressor is at risk and does not perceive itself as a sanctuary. Given the geographical position of the Soviet Union and its allies, that principle can only be applied on a global scale.

STRATEGY, RESOURCES, AND RISKS: DILEMMAS OF A SUPERPOWER

Throughout the postwar period, the structure of US military force has varied to accommodate changes in strategy and fluctuations in the availability of resources to meet the requirements of strategy. The persistent issue in fashioning a military strategy and designing the forces to match is whether the costs associated with a particular strategy and force structure can be tolerated. In the final analysis, a strategy which exceeds the means available is a hollow strategy, creating expectations that cannot be met. Yet, a strategy whose ends, albeit affordable, do not secure vital national interests is blatantly insufficient and dangerous if it does not convey the

necessary expectations to deter a potential adversary. The problem for defense policy is to develop a strategy which is neither insufficient to the task nor a hollow shell. Neither serves as a deterrent. Neither is capable of defending national interests if deterrence fails.

The "costs" associated with defining a particular strategy are not just economic but also political and military. Whereas one must examine the economic costs of what one does, one must also examine the political and military costs associated with what one does not do. In 1950, in National Security Council document NSC-68, the United States charted its military requirements to meet the needs of global power. Yet three years later, recoiling against the fiscal implications of a global force posture based on conventional land, sea, and air forces and the experience of the Korean War, the new Eisenhower administration opted for a strategy based principally on the threat of "massive retaliation" and the relatively inexpensive power offered by nuclear weapons.

Reliance on nuclear weapons might enable an economically affordable strategy, but it brings with it commensurate military risk of nuclear war and the political risk that the threat may not be altogether credible. Following the 1954 speech in which Secretary of State John Foster Dulles articulated the new strategy of "massive retaliation," for example, his predecessor, Dean Acheson, dissented in an editorial in the New York Times in a way that illustrates this point: "Strategic bombing is not our first but our last resort . . . If it is said . . . that we cannot afford another war like Korea, the answer is that such a war is the only kind of war we or anyone else can afford.[14] Acheson's critique was based on the premise, shared to varying degrees by subsequent administrations, that nuclear weapons were clearly an instrument to deter nuclear war; they could contribute to deterrence of major conventional war, but they could not substitute for an effective conventional defense posture. Conventional forces had to be capable not only of dealing with limited contingencies around the world but also of bolstering the credibility of the defense of allies in Europe and Asia even though nuclear retaliation might be part of that defense.

The Kennedy administration subsequently adopted a strategy of *flexible response* based on similar premises. While particularly controversial in NATO (as discussed in chapter 6) flexible response was also a global strategy designed to provide the United States with the capability of responding with conventional and even unconventional forces to global contingencies of varying levels of intensity. While the Kennedy administration launched a strategic modernization program based on the "triad" of intercontinental ballistic missiles (ICBMs), submarine-launched ballistic missiles (SLBMs), and intercontinental bombers, it also pursued a major expansion of general-purpose forces designed to be able to fight two major wars and one limited contingency simultaneously (the "two-and-one-half" strategy). As one participant in that force-planning process recalled, the force structure planned in 1962 anticipated that the United States would have some thirty days' warning (thus enabling a degree of reliance on reserves over active forces and on active strategic reserves over forward-deployed forces); that certain forces (like the two existing divisions in Korea) could be withdrawn to the United States as part of the active strength reserve; and that the United States would develop the necessary maritime force projection capability and requisite airlift and sealift capacity to move forces quickly to where they were needed.[15] Over the ensuing years, the latter two assumptions were not met, while the former has become optimistic.

After a period of drawdown in conventional force posture and overall defense investment beginning in the early 1970s—both as a consequence of the US withdrawal from Southeast Asia and as a reflection of greater fiscal conservatism—the United States found itself, in many respects, with a force structure whose components were less able to respond quickly to crises and generally in a lower state of readiness. Improvements during the Carter administration included a greater emphasis on prepositioned stocks of munitions, equipment, and supplies in Europe, owing in part to a revised estimate that the United States would have only 14 days' warning of a Soviet attack there. A focus on improving the readiness of forces already forward deployed, particularly in Europe, suggested a greater reliance on coalition partners as an adjunct to an attempt to avoid undue reliance on nuclear weapons. Despite the fact that the Nixon administration had, in 1970, replaced the "two-and-one-half" strategy with a "one-and-one-half" strategy—and withdrawn one division from Korea—the actual force requirements to sustain US interests had changed little.

Despite the conclusion that China did not pose a direct threat, the Korean contingency remained, alongside the major commitment to NATO, and Carter ran into severe opposition in his attempt to withdraw the remaining division from Korea. In addition, the "Carter Doctrine," articulating a commitment to the Persian Gulf region, led to plans for a Rapid Deployment Joint Task Force (RDJTF) which had the additional requirement to be flexible and versatile enough to respond to crises elsewhere as well.[16]

The Reagan administration's initial defense programs reflected a dramatic reversal in the decade-long decline in US defense spending and were designed to rectify what the administration and others called a "strategy-force mismatch." On the one hand, the Reagan administration continued the Carter administration's preference for a "countervailing" nuclear strategy capable not just of inflicting punitive destruction on the Soviet Union but of being able to wage a protracted nuclear war if necessary. Such an approach to nuclear deterrence spawned a major strategic modernization program, accelerated and expanded over the Carter program, and, with it, a substantial investment in strategic command, control, communications, and intelligence.[17] Subsequently, that strategic investment encompassed as well the administration's Strategic Defense Initiative (SDI) which developed alongside increasing efforts to achieve reductions of strategic offensive weapons through arms control.

The Reagan administration's defense programs also substantially increased investment in conventional defense programs. Like its predecessors, the Reagan administration sought to avoid an excessive reliance on nuclear weapons even though it disagreed with those who argued for a declared policy of "no first use" of nuclear weapons. In addition to introducing major new equipment items such as the "Abrams" tank, "Bradley" infantry fighting vehicle, and "Apache" attack helicopter, the Army, by 1986, had also expanded from sixteen to eighteen active divisions and from eight to ten reserve divisions. A new generation of tactical fighter aircraft for the Air Force, Navy, and Marine Corps was coupled with an almost 10 percent increase in the number of active fighter and attack squadrons and a 17 percent increase in the number of reserve squadrons between 1980 and 1987. Similarly, the Navy's "total deployable battle force" increased from 479 ships in 1980 to 569 in 1987, with a "600-ship Navy" projected by 1989, including fifteen rather than thirteen carrier battle groups. In addition, the Reagan administration's program has resulted in substantial expansion in airlift and sealift capacity. Finally, in terms of manpower, active-duty military strength has grown 6 percent since

1980, and selected reserves increased by almost one-third, while the number of military personnel overseas has increased almost 7 percent.[18]

The Reagan administration's defense program has been variously criticized by many observers as an excessive investment in a strategy that is open-ended in its requirements, does not differentiate between vital and peripheral interests, and assumes an unnecessarily offensive character in both its continental and maritime dimensions.[19] The substantial debate in the early years of the Reagan administration that accompanied the dramatic initial increases in defense spending was colored in part by a variety of notions advanced by private commentators—such as "warfighting" and "horizontal escalation"—that were not actually reflected in administration defense policy in the manner they were often presumed to be. Nonetheless, as suggested in the earlier discussion on national security objectives, the Reagan administration's strategy has been global in its context; forces have been substantially expanded and modernized; and the United States has developed a greater capability to project power globally, to defend allied commitments, and, in response to Soviet aggression, to mobilize forces to defend not only at the points of attack but to threaten the Soviets directly. Given the US geostrategic position, it is not a question of preferring offense over defense, but a question of how, given the potential theaters of conflict and the exposure of allies, one defends and replies to aggression.[20]

During the final year of the Reagan administration—faced with continuing high budget deficits compounded by a staggering trade deficit, increased foreign indebtedness, a declining dollar, and uncertainty about the strength of the economy following an unprecedented fall of the stock market—major cuts in the defense budget were necessary. To some extent, the capitalization of the force structure through procurement of major weapons systems in the 1980s was already a sunk cost and therefore an improvement that would endure. More likely options included manpower cuts, curtailment of future procurement and modernization plans, cutbacks on operations and maintenance, and some reductions in force structure. Indeed, Defense Secretary Carlucci's revised FY 1989 budget, submitted to Congress in February 1988, incorporated significant reductions in the projected force structure for all three services. There are, however, no easy choices. Historically, major reductions in defense spending had been accommodated by a deliberate decision to rely more on nuclear weapons, to make incremental reductions in the size and readiness of the force structure, and to rely more on allies and coalition partners. All of these remain problematic.

The deliberate decision to rely more on nuclear weapons was arguably an option only in the early 1950s, at a time when the United States enjoyed substantial strategic superiority and before a time when the United States became vulnerable to a Soviet assured destruction capability. As discussed in the previous chapter, extended deterrence on behalf of allies has always been politically complex, while the threat of nuclear use in defense of Third World interests is not entirely credible. In fact, the direction of the US-Soviet strategic relationship, as reflected in efforts to negotiate reductions in nuclear weapons, seems to be to rely less on strategic nuclear weapons as an instrument beyond the deterrence of nuclear war while continuing to pose the risk of escalation to nuclear war as a deterrent to major conventional aggression.

Incremental reductions to force structure and readiness may lead to a hollow strategy which cannot be executed, and may only provide short-term solutions since the long-term effects are expensive to correct. Technology may offer some compensating benefits by making some military functions less manpower intensive than they used to be. The Navy, for example, boasts that its new *Ticonderoga*-class cruiser requires a crew that is little more than one-fourth the size required on the *Chicago*-class cruiser which it replaces.[21] Similarly, improved ground force equipment enables increased firepower and military capability out of proportion to the manpower needed to operate and maintain it, while greater maintainability of aircraft may allow for some manpower savings. Nonetheless, there are clear limits to these savings. Unless one cuts the force structure itself, the costs of training and maintaining the readiness of such technology-intensive forces may consume some of the potential savings.

In addition, increasingly sophisticated weaponry will likely be more expensive to develop and procure, adding to the tradeoff between technology and manpower. There are limits as well to the extent to which more effective weapons can compensate for the fact that there are fewer of them. In the past, procurement savings on an annual basis have been made by stretching out production, but the result has often been a smaller force at the same cost as unit costs increase. Similarly, incremental savings by reducing investment in support equipment and spare parts may simply yield a less ready force.

Reliance on allies to do more is principally a function of their political will and economic capability. European allies will likely experience more rather than fewer constraints in their ability to shoulder a greater burden of NATO defense, while Japan will continue to have unique political constraints on what it can contribute to collective defense. Moreover, the bulk of US overseas military deployments remains tied to commitments to allies which remain acutely sensitive to indications of a possible US disengagement, a persistent allied fear that assumed greater prominence in the context of the 1987 US-Soviet agreement to eliminate land-based intermediate-range nuclear missiles. To the extent that the United States was to reduce its peacetime forward deployment of military force in support of its allies and to rely more on reinforcement potential, the global base infrastructure as well as the means to transport men and materiel and to protect vital sea and air lines of communication would assume commensurately greater importance. Moreover, unless forces redeployed to the United States were to be demobilized, the savings would likely be relatively marginal.

Within these parameters, US force structure and capabilities must remain consistent with the ends that strategy is to achieve. To the extent that military capabilities decline in a more fiscally constrained environment, the United States must adjust its strategy to find other means to achieve its security objectives and may well have to reassess those objectives. There are clear limits here as well. US vital interests are unlikely to change significantly. The United States will have little choice but to rely more heavily on allies, while seeking ways to use limited resources more effectively and more collaboratively with them. Nonetheless, maintaining the cohesion of its alliance structure will require that the US commitment to allied defense remain credible, even though the form of that commitment may be forced to change somewhat.

Even before the Reagan administration ended, such a strategic reassessment began to gather momentum in the public debate. In late 1986, the administration itself chartered a "Commission on Integrated Long-Term Strategy." Cochaired by Fred Iklé—who, as under secretary of defense for policy, fulfilled the congressional mandate that the commission include an administration official—and Albert Wohlstetter, the commission included former cabinet officials, national security advisers, retired military leaders, and distinguished scholars. The commission's report, released in January 1988, advocated a strategy of "discriminate deterrence" and enumerated the principles of that strategy, "some calling for radical adjustment, some reaffirming key elements in the current defense effort."[22]

One of the commission's more controversial judgments was that the United States should reduce the extent to which its forces are specifically dedicated to individual threats and should increase its flexibility in employing its forces in global contingencies. This would imply forces which are less forward-deployed, more mobile, and more versatile to accommodate anticipated reductions in force structure and allegedly deteriorating US access to bases throughout the world. In addition, the commission rejected notions of punitive deterrence that relied on threats of nuclear retaliation that would invite annihilation of the United States and its allies if carried out. Instead, the commission suggested that the United States should emphasize conventional weapons technologies that could reduce reliance on nuclear weapons and emphasize nuclear weapons technologies that could effectively defeat aggression but would avoid "deterioration into an apocalypse."[23] In arguing for both conventional and nuclear capabilities for controlled and selective offensive operations, the report seemed to imply a marked shift to a "warfighting" strategy, although the resource implications of such a strategy were less clear.

The release of the commission's report on "Discriminate Deterrence" demonstrated the difficulties associated with any attempt to contemplate the possibility of long-term change in the US strategic posture. Many in Europe saw within the report evidence of US nuclear decoupling and the specter of a US desire to shift the burdens of both conventional defense and, to some extent, nuclear deterrence, to the allies. Despite administration assurances about the constancy of the US commitment and of US confidence regarding the vitality of standing NATO strategy, allied anxieties remained focused on the longer-term implications of this debate, a debate which coincided with both the onset of an election year and Senate hearings on the INF Treaty and its implications for NATO.

Arms control, a subject reserved for Part III of this text, offers to many the prospect of a changed strategic environment in which US military posture may be reduced with less risk. Although often touted as a source of savings in defense spending, arms control cannot, based on historical experience, be expected to reap much economic gain. In fact, arms control may be expensive. In one sense, the reduction of weaponry—as in the case of the INF Agreement—involves short-term costs of dismantling and destruction. Moreover, to the extent that reductions, either in nuclear or conventional forces, require, as they would, highly intrusive, on-site monitoring by human as well as technical sensors, that in itself is a resource-intensive proposition. Just as with maintaining military preparedness, monitoring military reductions, redeployments, and disengagements is a process in which economic short-cuts entail higher security risks.

In another sense, arms control may alter the security environment by its long-term impact on the relevant force structures to be maintained. Increasingly, arms control has the potential of restructuring the military forces of its participants by altering the choices and calculations on which those choices are based. It is in this light that the growing emphasis on reducing nuclear weapons—if not eliminating them altogether—must be assessed. Indeed, some have suggested that, with or without arms control, the relevance of nuclear weapons to global security may be diminishing as conventional weapons technologies improve, offering more usable military power at lower risk.[24]

While there are clear risks in basing one's global power on one's nuclear arsenal—a risk noted by Acheson in his 1954 retort to Dulles— there are also clear risks in abandoning that uniquely important element of being a superpower. In a world without nuclear weapons— to take an extreme case—the Soviet Union might not be a global superpower, but the regions it would dominate would likely encompass the entire Eurasian landmass by virtue of the political influence that would derive from a preponderance of military power. The United States, while essentially invulnerable to military attack, would nonetheless find it increasingly difficult to maintain its global position. Unable to pose to the Soviet Union the absolutely unacceptable risks of aggression, the United States might well find it more difficult to maintain the alliance structure on which an important element of the US position rests. The defense requirements to secure global national security interests would be substantially greater than they are now and, conceivably, could not be sustained in peacetime. Regardless of whether conventional deterrence could ever substitute for nuclear deterrence, it does not appear that the requirements would be easily achieved.

CONCLUSION

American defense policy is designed to secure the United States in its role as a world power, a role which the United States inherited as the clear legacy of World War II. All of the elements that shape that global role, however, are changing, whether in superpower relations, the US role in Europe, the US military presence in Asia, the growing challenges of nonterritorial security threats, or the resources likely to be able to sustain that role. Yet, in focusing on the elements of a global defense posture and the problems associated with sustaining the forces necessary to secure US interests and commitments, it is easily forgotten that the United States is a world power not simply because of its military prowess or because of the unique role derived from being a nuclear superpower. That global position is, fundamentally, derived from the political, economic, social, and philosophical influence that the United States has, an influence that has, over the past four decades

and more, tolerated a degree of elasticity in the structure of US military forces and the form in which the United States has met its commitments. The United States and its allies have been able to accommodate the fluctuations in US defense posture also because the overriding purpose of that defense posture is deterrence, itself a perceptual phenomenon whose dimensions are more than just military.

Nonetheless, credible military power is vital both to deterrence, and, fundamentally, to the interests and commitments of a world power. There have (as the next chapters describe) always been limits to that military power, both in creating it and in exercising it. The United States does not enjoy the luxury of choosing between being a maritime power and a coalition leader. As a geostrategic island with global interests and far-flung allies, it must be both, finding a way for them to complement each other and choosing among their competing demands. As the following chapters suggest, there are clear limits to unilateralist policies along with a requirement to be able to take unilateral action; there are clear constraints as well as benefits associated with the maintenance and management of alliance structures around the world.

The United States enjoys a unique position in the world, but with it come unique responsibilities. Chapter 5's discussion of the evolution of the US-Soviet relationship highlights the fact that the United States has a singular role in engaging the Soviet Union either in pursuing greater cooperation or in coping with the risks of confrontation. Chapters 6 and 7 stress the problems of sustaining alliance commitments where the risks, responsibilities, and benefits of coalition partnership are inherently asymmetrical. Finally, chapter 8 focuses on those contingencies in which the potential for low-intensity conflict imposes perhaps the hardest choices for defining the US interests and how those interests are to be defended. The issues described in the following chapters provide the context in which choices are made and risks are assessed in structuring a sustainable and effective global defense posture.

NOTES

1. A classic work on this subject is Martin Wight's *Power Politics*, ed. Hedley Bull and Carsten Holbraad (London: Pelican Books, 1979), esp. pp. 1–68.

2. US Department of Defense, *Annual Report to the Congress, Fiscal Year 1988*, Jan. 1987, p. 41.

3. Paraphrased from *Annual Report: FY 88*, p. 42.

4. John Lewis Gaddis, "The Rise, Fall, and Future of Détente," *Foreign Affairs* 62, no. 2 (1983–84): 356.

5. Data from the Department of Defense as contained in the annual "Defense Almanac," *Defense 87*, Sept.–Oct. 1987, p. 21. Also Gordon Adams and Ste-

phen Alexis Cain, "The Defense Budget in the 1990s," in Joseph Kruzel, ed., *American Defense Annual 1989–1990* (Lexington, Mass.: D. C. Heath, 1989), p. 34.

6. See the Defense Department's annual *Soviet Military Power 1987*, 6th ed., Mar. 1987, p. 5, plus the additional comparative and analytical data in the International Institute for Strategic Studies annual *Military Balance, 1987–88* (London: IISS, Oct. 1987), pp. 29–33 and 215.

7. *Military Balance, 1987–88*, pp. 212–14. NATO and the US have comparable percentages of *active duty* forces as a percentage of population— approximately 1 percent. See also US Department of Defense, *Report to the Congress on Allied Contributions to the Common Defense*, Apr. 1987, pp. 28–29.

8. *Military Balance, 1987–88*, p. 212.

9. *Defense Almanac*, pp. 44–45.

10. Chief of Naval Operations, Adm. James D. Watkins, before the Seapower and Strategic and Critical Materials Subcommittee, House Armed Services Committee, *The 600-Ship Navy and the Maritime Strategy*, 99th Cong., 1st sess., 24 June 1985, p. 19.

11. Adm. Wesley L. McDonald, "Anti-Ship Cruise Missiles," *Signal* 38, no. 4 (1983): 48–55.

12. See *Annual Report: FY 88*, p. 59, for Defense Secretary Weinberger's comments on restraints on the use of military force.

13. The classic study remains that was conducted by Barry M. Blechman and Stephen S. Kaplan and summarized in their "US Military Forces as a Political Instrument since World War II," in *Political Science Quarterly* 94, no. 2 (1979): 193–209.

14. *New York Times*, 29 Mar. 1954.

15. William W. Kaufmann, *Planning Conventional Forces, 1950–1980* (Washington, D.C.: Brookings Institution, 1982), pp. 4–17.

16. Ibid., pp. 14–24. See also Robert W. Komer, "Maritime Strategy vs. Coalition Defense," in *Foreign Affairs* 60, no. 5 (1982): 1126–28.

17. See Jeffrey Richelson, "PD-59, NSDD-13, and the Reagan Strategic Modernization Program," *Journal of Strategic Studies* 6, no. 2 (1983): 125–46.

18. *Annual Report: FY 88*, app. B, p. 334; and app. C, tables 2 and 3, pp. 338–39.

19. Including Gaddis, "Rise, Fall, and Future"; Komer, "Maritime Strategy" and his book by a similar title (Cambridge: Abt Books, 1984); also, Barry Posen and Stephen Van Evera, "Reagan Administration Defense Policy: Departure from Containment," in *International Security* 7, no. 1 (1983): 3–45. For a comprehensive bibliography on the maritime dimension of the debate, see "The Maritime Strategy Debates: A Guide to the Renaissance of US Naval Strategic Thinking in the 1980s" by Captain Peter M. Swartz, USN, OPNAV P-60-3-81.

20. See, for example, Scott D. Sagan's "1914 Revisited: Allies, Offense, and Instability," *International Security* 11, no. 2 (1986): 151–75. I am grateful to Captain Peter Swartz, cited above, for pointing this source out to me.

21. Secretary of the Navy John Lehman's testimony to the Seapower and Strategic and Critical Materials Subcommittee, House Armed Services Committee, *The 600-Ship Navy*, 99th Cong., 1st sess., 24 June 1985, p. 86.

22. *Discriminate Deterrence: Report of the Commission on Integrated Long-Term Strategy*, 11 Jan. 1988 (Washington, D.C.: G.P.O., 1988), p. 2.

23. Ibid., esp. pp. 26–30, 33–37.

24. For example, Edward N. Luttwak, "An Emerging Postnuclear Era?" *Washington Quarterly* 11, no. 1 (1988): 5–15.

CHAPTER 5

DEALING WITH THE SOVIETS

RONALD J. SULLIVAN

The management of Soviet power has been the central task of US policy since World War II. Within the framework of this comprehensive foreign policy objective, US defense policy has played a central role. This chapter examines the evolution of the US security relationship with the Soviet Union and considers some recurring themes which have guided that evolution.

The last 45 years of US-Soviet relations is a remarkable study in both consistency and variation. The fundamental thrust of US policy has remained the containment of Soviet power. The particular policies and strategies selected to support this basic principle, however, have fluctuated considerably. The analysis of this fluctuation lies at the heart of the study of postwar US security policy. In examining this evolution of security policy from a historical perspective, it is possible to exploit the advantages of hindsight and focus on certain themes, sometimes imperceptible even to the actors involved at the time, that simplify our inquiry. These themes represent various perceptions, policies, and philosophies that have consistently recurred in US security policy since World War II and have framed each administration's vision of the security dilemma it faced. We can generally categorize the most influential and persistent themes as containment, strategy, vulnerability and legalism/moralism.

CONTAINMENT

Containment—the bottling up of Soviet expansionism—has been the foundation of US security policy since 1946. While the details and policy history of containment will be addressed shortly, its broader intellectual elements are of interest here. Although often described as a policy, that label is misleading. Containment is a desired condition in US-Soviet relations, rather than a policy prescription for achieving that condition. It is an intermediate security objective or goal which is supported by specific military, political, and economic policies. Its ultimate objective is a stable and peaceful international system. Containment, then, is neither a policy nor an ultimate goal. The original formulation of containment in the 1940s was designed to

circumscribe Soviet expansionism in order to (1) save the international system from a revolutionary state, and (2) force internal changes in the USSR which, over time, would produce an accepting, peaceful, and cooperative participant in the international system. An important US policy directive of the late 1940s summarized the objectives toward the Soviet Union this way:

A. To reduce the power and influence of the USSR to limits which no longer constitute a threat to the peace, national independence, and stability of the world family of nations.
B. To bring about a basic change in the conduct of international relations by the government in power in Russia, to conform with the purposes and principles set forth in the U.N. Charter.[1]

Containment was the key vehicle aimed at achieving these larger goals. The value of reviewing the original formulation of containment lies in the comparison to later formulations when containment was treated more as a goal in itself. In those later instances, strategies and policies that focused too narrowly on containment as an intrinsic value lost sight of the larger systemic objectives that characterized containment in its original incarnation. This subtle but critical shift of emphasis from the means of policy to the ends of policy, when it has occurred, has had a major impact on US security policy.

STRATEGY

A second thematic element of great value in analyzing the evolution of US security policy is the concept of strategy. In its purest definition, strategy is simply the plan or process that relates the resources available for foreign policy to the objectives of that policy. It is the game plan which presses a nation's means into the service of its ends. The simplicity of this definition of strategy vastly understates the complexity of fashioning a strategy in the United States. Differing perceptions of what and how many resources are available, debates over the suitability of objectives, and constant challenges regarding the feasibility of any given strategy are but a few of the dilemmas that confront the US strategic planner. A foremost analyst of US post-war strategy, John L. Gaddis, reduces the strategic equation to two basic but vexing options: Shall interests be restricted to keep them in line with available resources, or shall resources be expanded to bring them into line with proclaimed interests?"[2] An assessment of how each policymaking regime has resolved this strategic dilemma is important in understanding US containment policy since World War II.

VULNERABILITY

A third concept that is central to our analysis of US security policy is national vulnerability. This vulnerability goes beyond the normal sense of insecurity that any nation has in the international system, encompassing in the nuclear age the fundamental question of national and societal survival. Vulnerability was an alien concept to the US national psychology. For generations, geography and the prevailing European power constellations had allowed the US to enjoy isolation and a genuine, but basically unearned, security. From the early 1950s, however, the US recognized a susceptibility to nuclear attack from an adversary which seemed authentically capable of such an attack.

This unprecedented sense of vulnerability was composed of two identifiable elements. The first was the technological component of the Soviet threat. Since World War II, the United States has chosen to rely on its advanced technological competence as the foundation of its defense. When the Soviet Union, which possessed advantages in manpower and sizeable conventional forces, began closing the technology gap in nuclear weapons and delivery systems, it was perceived to be driving for superiority across the board. Technological advancements by the Soviet Union were seen not as routine or predictable force modernizations, but as challenges to the stability of the international strategic balance, as well as positive proof of Soviet hostile intentions. Various "gaps" developed over the course of the postwar period—a bomber gap, a missile gap, a window of vulnerability for ICBMs—all of which were tremendously disturbing for a nation whose principal security advantage over a totalitarian regime was perceived to be the advanced technology produced by its vibrant free enterprise system. It is ironic, of course, that such feelings of technological vulnerability in the US have been stimulated by a state which itself possesses such a profound sense of technological inferiority.

The second element in the US sense of vulnerability was a social vulnerability. A free and open democratic state, forced by its nature to respond to the demands of an open consumer society, was seen to be at a disadvantage in dealing with a secretive police state. The Soviet command economy could simply direct large resources into the military without troublesome congressional committees, public opinion, or media attention. Moreover, Soviet subversion and espionage, dramatized by a few spectacular postwar spy cases, added to this sense of social vulnerability that was manifested in part by Senator McCarthy's zealous search for communists within the US policy establishment. The Soviet leadership helped to sustain this sense of insecurity. When Khrushchev declared to the US in the 1950s that "Your grandchildren will be communists," or "We will bury you," most Americans understandably saw a direct threat to their national survival. One important result of this was the sense that the US was uniquely vulnerable to a first strike planned and executed by a closed and totalitarian society. Although such an option for the US technically existed in US war plans, it was really not an acceptable policy for an open and democratic society.

This sense of disadvantage and vulnerability to advancing Soviet technology and to the intrigues of a secretive police state has been a major influence in much of US defense policy since World War II, particularly its inclination toward worst-case defense planning. This perceived Soviet proclivity to use any means, even surprise attack, to subdue its capitalist adversary obviously presumed total Soviet disregard for traditional concepts of law and morality.

LEGALISM/MORALISM

Legalism/moralism, a term coined by George Kennan, represents a major and recurring theme in American foreign and security policy. This multifaceted theme is deeply rooted in the American experience, but is not an ideology. Its key component is a Wilsonian desire for a recognized standard of international behavior, usually embodied in an international organization, based on peace, order, democracy, and justice. This idea implies an American moral superiority, certainly a superiority of key American values like democracy, individual rights, and free enterprise. An important corollary is that capitalist/democratic states are peaceful in their approach to international politics, while totalitarian states are inherently warlike. Legalism/moralism allows for the "carrying over into the affairs of states of concepts of right and wrong, the assumption that state behavior is a fit subject for moral judgment." Where there is no question that US policy should be moral by its own standards, the discussion over legalism/moralism has revolved around the extent and vigor with which the US applies its own morality to the conduct of international affairs. This concept has introduced, to varying degrees, a juridical and ideological element into US foreign policy that, according to many observers, has on occasion prevented the US from "taking the awkward conflicts of national interest and dealing with them on their own merits."[3]

This idea has enjoyed cyclical periods of prominence and decline. When legalism/moralism assumed a powerful force in an administration, its approach to the Soviet Union tended to be more militant, less conciliatory, and more committed to a firmer containment policy. Never in the postwar era, however, has it been completely absent from US policy aimed at the management of Soviet power.

The practical policy dilemma caused by legalism/moralism is the difficulty it creates in dealing with an adversary perceived to be immoral. The problem has been summed up this way: "It seems contradictory to recognize that the Soviet Union is repressive and expansionist and also to believe that we should seek to manage rationally the fundamentally competitive Soviet-American relationship."[4] The more powerful the legalistic/moralistic tradition is in any given administration, the more difficult this basic contradiction has been to deal with.

An important subset of the concept of legalism/moralism is what might be called the "Munich" analogy. This is a powerful image, applied by some policy framers to project strong parallels between the Soviet Union and Nazi Germany and to predict the "appeasement" of Soviet power will only render it more aggressive and dangerous. It views the conciliatory policies of the Western democracies toward Hitler, epitomized at Munich, to be self-defeating. Concession anywhere, even in remote corners of the globe where the US has no direct interest, could start a chain of aggression, with each episode strengthening the aggressor and making the ultimate confrontation more dangerous. The Munich analogy tacitly postulates that the internal totalitarianism of the Soviet State will translate into unbridled external expansion. "Munich," as a particularly acute observer has noted, "came to be not merely an analogy, but an iron law and a moral principle."[5] While it is often risky to discuss psychological images and their impact upon policy, the strength and endurance of the Munich analogy give it a remarkable and identifiable substance. Regardless of the validity of such "lessons," the danger of appeasement has become one of the most forceful historical "lessons" of this century.

While the importance of these key themes should not be overstated, the notions of containment, strategy, vulnerability, and legalism/moralism have been powerful and consistent considerations in US postwar security policy. Throughout the following discussion, these themes will serve as a framework for examining the complex history of US security policy toward the Soviet Union.

THE DECLINE OF THE "GRAND ALLIANCE" AND THE ORIGINS OF THE COLD WAR

Assessing the origins of the so-called cold war has been a major preoccupation of US diplomatic historians over the last 40 years. There are three broad approaches to this issue. Two of these are normative, while the other is rooted in realism, or *Realpolitik*.

The first two schools of interpretive history seek to apportion blame for the cold war. Focusing on the breakdown of the anti-Nazi coalition and the drift into the East-West confrontation, these schools see the dangerous and costly clash between the US and the USSR as an avoidable tragedy, a product of ignorant or reprehensible policies on one side or the other. They are distinguished, of course, by their conclusions about which side bears the onus of guilt. The first, or traditional approach, views the Soviet Union as an aggressive and ideological stage, determined to expand relentlessly into the global chaos left in the wake of World War II and openly manipulative of Western good intentions. According to one noted scholar, "Leninism and totalitarianism created a structure of thought and behavior which made postwar collaboration between Russia and America . . . inherently impossible."[6] More epigrammatic are the words of a former US national security adviser: "In the clash between Western naivete and Stalinist power—power prevailed."[7]

The second school of thought has more variations, but all essentially shift the burden of guilt to the West, particularly the Truman administration. This school explains Soviet behavior as an understandable and defensive reaction to its history and sees US policies in the cold war as provocative attempts to ward off threats to the Western political-economic system and to roll back legitimate Soviet security gains in Eastern Europe. This revisionist view believes that American "exhilaration" with its monopoly of atomic weapons led to a provocative diplomacy that "irretrievably embittered" Soviet-American relations.[8]

Ultimately, the question of who is to blame for the cold war is unresolvable. After 40 years, the relevance of that debate lies more in its ability to highlight the normative dimensions of American foreign policy toward the Soviet Union which still retain currency. The third approach to the origins of the cold war is perhaps more illuminating about the continuing struggle of the two superpowers, as it is rooted in the dynamics of geopolitics. With two great power centers, based on antithetical ideologies, facing one another astride a prostrate and powerless Europe, the chance that the US and USSR would remain allies was probably minuscule. Add to this the "inability of the US and the USSR to communicate meaningfully in the absence of a common enemy,"[9] and the stage is set for a dispute. There were blunders, misperceptions, and ill-will on both sides, rendering inescapable a political confrontation of some unpredictable magnitude. The reality of the cold war—whatever its origins—thus sets the stage for the more germane question: not whether the US-USSR confrontation could have been avoided, but rather whether it could have been managed and channeled into a more benign, less militarized encounter.

The alliance between the US and the Soviet Union during World War II was a genuine but uneasy partnership. The relations between the two nations had never been good. Problems with recognition, intervention, debts, subversion, and human rights had dominated the relationship since 1917. The US harbored a fundamental distrust of the new Soviet state on ideological grounds, and Stalin's geopolitical maneuvering with Hitler and his attacks on his neighbors further embittered Americans. For his part, Stalin saw Western policy of the late 1930s as clearly designed to channel Hitler's aggression toward the Soviet Union, an impression reinforced by his perception of Western delay in opening a second front during the war. Only the vision of the Nazi war machine victorious throughout Europe could have overcome such a history of poor relations and bitterness to bring the two nations into an alliance. During the war, the alliance was at its healthiest when the struggle against Hitler was most desperate. As the tide of the war began to turn, antagonism became more evident.

President Roosevelt's policy toward the management of Soviet power was twofold. The first element, reflecting his grasp of *Realpolitik*, was to establish a postwar balance of power based on four powers—the US, Soviet Union, Great Britain, and less realistically, China. The Soviet Union would be integrated into the power balance by conciliation and creation of trust. In other words, the traditional Soviet insecurity was to be addressed forthrightly and overcome by clear understanding, aid, and, if necessary, Western geopolitical concessions to the Soviet sense of insecurity.

The second part of Roosevelt's policy appealed to the strong skein of legalism/moralism at home. To ensure public support for a postwar order, simple balance of power calculations were complemented by a somewhat idealistic formula of UN collective security. Roosevelt was well aware, however, that collective security, without agreement between the great powers, was unrealistic. His two-tiered approach was a delicate policy, one that required a certain amount of forthrightness with the American people.[10] Roosevelt's untimely death, however, meant that a new leadership had to manage this policy dilemma even as the war continued.

The immediate roots of US-Soviet alienation were in Eastern Europe. Once it was clear that the tide of battle had been turned irreversibly in the Soviets' favor, Stalin's military strategy became more clearly political. Enormous resources were spent in the Balkans and Hungary in a clear effort to stake out a postwar sphere of influence and, particularly in Poland, to preserve Soviet dominance. That the Soviets should become attached to the geopolitical advances that the Red Army had won at such a staggering cost is not remarkable. In fact, recognizing that Soviet occupation would determine the matter in any case, Roosevelt and Churchill had tacitly conceded Eastern Europe as a sphere of influence to Stalin in 1943 and 1944.[11] The problem was reconciling this fact with the popular vision of a postwar world rooted in the Atlantic Charter's noble commitment to self-government, free international trade, and a renunciation of territorial aggrandizement.

The US leadership change in April 1945 left Roosevelt's policy without an advocate. Harry Truman was selected for the vice presidential nomination based on his influence with the Senate which would be necessary for peace treaty ratification. He had been in office less than three months when he assumed the presidency. Truman had never even been invited to the White House for dinner, let alone informed about the Manhattan Project or the subtleties of the president's Soviet policy. Continuity in policy was therefore very difficult, and the period from April 1945 to early 1947 was characterized by heavy-handed, reactive decisions in Washington, and, for that matter, in Moscow. Stalin puzzled even those US leaders who expected the worst with his callous, provocative, and even insulting policies. Beyond style, the catalogue of events that contributed to the US-Soviet split is well known. The Soviet consolidation of Eastern Europe, its intransigence in the new UN, and Stalin's territorial demands—stretching from North Africa to Southwest Asia to the Far East—troubled US leaders. For its part, US policy on aid, lend lease, occupation, and reparations created hostility and suspicion in the Kremlin.

By early 1946, the deterioration had become public knowledge. Frustration was beginning to replace the euphoria and self-satisfaction that had marked the immediate postwar period. The Soviet Union seemed unresponsive to any US demarche, whether conciliatory or threatening. In February 1946, a leading Soviet expert—a State Department officer stationed in Moscow named George Kennan—offered an analysis of Soviet behavior that shaped his nation's foreign policy for the next generation.

THE ADVENT OF CONTAINMENT

In his "Long Telegram" of 1946 and in his *Foreign Affairs* article of 1947, Kennan "came as close to authoring the diplomatic doctrine of his era as any diplomat in our history."[12] In Kennan's view, American security required a balance of power in the world. US policy was to be aimed at maintaining the independence of the key global industrial areas. Soviet policy, however, based on historical and ideological factors, was antagonistic to such a balanced international system. Immune to most external influences, Soviet policy focused on advancing the "relative strength for the USSR as a factor in international society" while missing no opportunity to reduce the strength of, and exploit the differences between, the "capitalist powers." Kennan was glum as he summarized the USSR's attitude toward its relationship with the US: "with the US there can be no permanent *modus vivendi* . . . it is desirable and necessary that the internal harmony of our society be disrupted, our traditional way of life be destroyed, the international authority of our state broken, if Soviet power is to be secure."[13] Although Kennan never intended his ideas on containment to be construed primarily in military terms, it is not surprising that many interpreted these early writings as a call to arms. In any case, the policy to counter such Soviet hostility was to be "a long-term, patient but firm and vigilant containment of Russia's expansive tendencies."[14]

Beyond simply bottling up the Soviets to preserve the balance of power, Kennan saw containment as eventually changing the internal Soviet system, forcing on the Kremlin "moderation and circumspection" that would result in either "the breakup or the gradual mellowing of Soviet power."[15]

These were dramatic descriptions and recommendations, but ones that had the ring of truth to the US policy establishment. They seemed to explain so much of Soviet behavior—the hostility, the rhetoric, their truculence with the "capitalist" powers at the UN. The official embrace of containment was so complete that certain subtleties were overlooked and certain details presumed.

Kennan was not advocating a global military response to Soviet expansion; US resources were far too limited for that. Rather, the US should stabilize those areas critical to the balance of power against Soviet opportunism and then patiently await Soviet acceptance of the international system based on this balance. It was Soviet political opportunism, not military aggression, that Kennan saw as the primary threat. Furthermore, Kennan was anti-Soviet rather than anti-Communist in his thinking. He felt that fissures in the Communist bloc could be exploited—as they eventually were.[16] Containment was a geopolitical rather than an ideological or military strategy. These "fine points," however, represented substantial subtleties for an American public accustomed to the absolutes of isolation or total war.

In spite of differing interpretations about implementation, containment became the overarching framework for US security policy. Its uniqueness lay in providing a rationale for the first US peacetime balance of power policy. Ten years before, it would have been inconceivable to convince the American people of their vital interests or responsibility in maintaining a peacetime balance of power in Europe or the Far East. Kennan's clear formulation of policy helped to accomplish precisely that. However, to Kennan's disappointment, this was not a wholesale conversion of US public opinion to *Realpolitik*. The policy that came to dominate US thinking was a hybrid that combined realism and traditional American idealism. The moral lubricant that made containment work in the domestic American context was a self-righteous and increasingly militant anti-Communism.

The Truman administration accepted Kennan's ideas of containment and used them to add rigor and direction to its own emerging ideas of dealing with the Soviets. The first clear policy manifestation was the Truman Doctrine, announced in March 1947. In a dramatic departure from traditional US foreign policy, the Truman Doctrine extended economic and military aid to Greece and Turkey, both of whom were struggling against communist insurgencies. Although it seemed to make a sweeping offer to help all "free peoples," the Truman Doctrine was specifically limited in scope and means.

Limited means characterized the containment policy until the Korean War. The Truman administration was very sensitive about the economic health of the US and notoriously frugal in its estimation of resources available for security policy. This, of course, imposed a certain strategic rigor and forced a prioritization of policy goals.[17]

In June 1947, the Marshall Plan for the economic recovery of Europe was announced, signaling the point of no return in the US-Soviet split. It was clear that the US was committed to remain in Western Europe and deny it as a sphere of influence to the Soviets. By early 1948, the Czechoslovakian crisis led to the establishment of a Soviet puppet regime in the very nation whose sacrifice on the altar of appeasement in 1938 had become such a powerful moral symbol for the West. In 1948 Soviet blockage of Berlin, followed by the Allied airlift, produced a new symbol of Western resolve to defend its isolated outpost behind the "Iron Curtain." In April 1949, the US signed the North Atlantic Treaty with the Allies, agreeing that "an armed attack against one or more of them in Europe or North America shall be considered an attack against them all."

SECURITY POLICY IN THE AGE OF ATOMIC MONOPOLY

By the end of World War II, the new term *national security* was finding increasing acceptance in policy circles in Washington. It was a much more sweeping concept than the ideas of diplomacy and defense which it replaced, reflecting the technological, political, and geographic forces that had emerged during World War II. It implied that the safety of the nation was dependent on a wide, interrelated range of foreign policy issues much more diverse and complex than anything the nation had previously faced. Military force and preparedness became more important elements in the national sense of well-being. Geographically remote areas and issues took on more importance than before the war, and military and foreign policy became much more closely intertwined.

This national security mind-set has been criticized for militarizing and radicalizing US foreign policy. While there may be truth in this, in a larger sense it reflected the increasing interdependence of actors and issues in the shrinking international arena. It also reflected the US perception of vulnerability to a well-orchestrated totalitarian threat. In 1946, Truman's Special Counsel, Clark Clifford, urged a more unified approach upon the president. Citing a police state's advantages in security matters, Clifford warned, "There must be such effective coordination within the government that our military and civil policies concerning the USSR, her satellites, and our Allies are consistent and forceful."[18] There was a strong sense, one that continued far into the future, that the Soviets had a clear, unified, centrally controlled plan to their foreign policy. Later the State Department pointed out that, in dealing with such a conspiratorial and centralized state, the US must maintain a state of "unvacillating mental . . . and military preparedness."[19] Nonetheless, this rising national security consciousness did not prevent a massive demobilization after the war and high optimism that the collective security apparatus of the UN would solve the nation's security dilemma.

By 1947, a formal national security mechanism was commissioned. The National Security Act created a unified Department of Defense with an autonomous Air Force, a Joint Chiefs of Staff system, the Central Intelligence Agency, and the National Security Council. For better or worse, a coordinated national security apparatus had been put in place.

There was little such coordination in the days immediately after hostilities ended. The demobilization reduced military manpower from 12 to 1.6 million in less than two years. It was less the amount of reduction (after all, the war was over) than the precipitate, almost frenzied manner of it. This gave the impression to many observers, almost certainly Stalin among them, that the US was preparing to return to more or less isolated splendor in the Western Hemisphere. The Soviets were demobilizing as well, but more slowly, particularly in the highly visible European theater. Once demobilization had run its course, the Truman administration was not willing (before Korea) to pay the high price of substantially reversing it. Clinging to the theme that national security was threatened as much by budget deficits as by Red Army divisions, defense spending was capped at about $12 billion per year through the late 1940s (approximately $42 billion in FY 85 dollars).

These budgets dramatically reduced conventional forces and contributed to the increasingly important role of the atomic weapon in US policy. Early attempts to parlay atomic weapons into diplomatic gains failed. In fact, the entire period of US atomic monopoly was one of unprecedented Soviet adventurism. Nonetheless, the one group pleased about the increasing reliance on atomic weapons was their custodian, the new United States Air Force. The theories of strategic bombardment—that the enemy could be defeated by the aerial destruction of its socioeconomic structure and popular will—had not proved very accurate during World War II. Only with the arrival of atomic weapons and their apparent successes against Japan did strategic bombardment regain its credibility. An emphasis on atomic strategic bombardment was in the Air Force's best institutional interest, helping to ensure its autonomy as a separate service and later protecting its portion of the small postwar defense budgets. In a larger sense, the atomic strategy reflected a well-established US preference to substitute technology for manpower in combat. By 1948, cultural, institutional, political, and budgetary pressures converged to create a US reliance on an atomic strategy.

The only serious arms control attempt of the forties, the Baruch Plan, was a sincere, but ultimately nonnegotiable US offer to internalize atomic power. The Soviets rejected its inspection and punishment provisions and it was soon buried in the deteriorating US-Soviet relationship.

The fundamentals of deterrence were understood in both the military and diplomatic communities, but it fell to an academic, Bernard Brodie, to present the first systematic analysis of the impact of atomic weapons on strategy. Brodie pointed out that deterrence rather than military victory in war would become the central mission of military establishments. Superiority, at least in terms of quantitative superiority, had lost most of its significance in the era of "atomic bomb warfare." Guaranteed retaliation became the key functional element in deterrence. Important issues of force structure, targeting policy, and timing remained to be answered, but the fundamentals governing the strategic debate for the next 40 years had been laid down.

By 1948, deterrence had been formally incorporated into US defense planning. Yet, through the end of the decade, the US atomic capability was hardly impressive. In 1947, the head of the Atomic Energy Commission, David Lilienthal, noted that the US had one bomb "that had good chances of being operable" and that "there were really no bombs in a military sense."[20] Things began improving that year, but it was clear that there was no hope of an atomic "knockout blow." All US war plans virtually conceded Western Europe to the Soviets, even after a US atomic strike on Soviet urban industrial concentrations. The war was then expected to be a replay of World War II lasting "not less than five years."[21] In spite of these doubts about US atomic capability, doctrinal reliance on these weapons increased. In 1948, at the height of the Berlin crisis, President Truman dispatched some 60 US B-29 "atomic" bombers to Europe as an expression of US resolve. In spite of the fact that these bombers were not equipped to carry atomic weapons, the message to the Soviets was clear. Atomic deterrence had arrived in American diplomacy.

By 1949 several new influences were operating on US national security policy. The surprisingly early Soviet detonation of an atomic device closed the door on the unproven luxury of unilateral nuclear policy. In Asia, the Chinese Communists drove the Nationalists from the Mainland, and the accusations about who "lost China" created a powerful and virulent anti-Communist backlash within Congress. The State Department's realistic assessment that the turn of events in China was beyond US influence was met with charges of disloyalty and even treason against the highest department officials. The storm over "losing" China was so violent and enduring that few future US leaders would be willing to accept the political damage of "losing" another nation to Communism.

Early in 1950, Truman made an important decision to begin development of thermonuclear weapons. To add some perspective to this decision amidst the prevailing atmosphere of uncertainty, he asked as well for a thorough "reexamination" of US objectives and strategic plans.

NSC-68: TRUMAN'S REAPPRAISAL OF US SECURITY POLICY

The chief architect of this remarkable document was Paul Nitze, director of the State Department's Policy Planning Staff. The paper, designed as an instrument of advocacy more than analysis, made a strong, sometimes emotional plea for a buildup in US military forces opposing Soviet expansion. It was the first comprehensive approach to US strategy and foreign policy. NSC-68 agreed with Kennan's original formulation of containment as a policy vehicle aimed at maintaining a balance of power. Its definition of that balance was, however, much more comprehensive and delicate than Kennan's. Not only were military aggression and economic collapse in key industrial areas likely to cause a shift in the balance, but even perceptions of US weakness or humiliation could be destabilizing. For example, the "destruction of Czechoslovakia" was a shock: "Not in the measure of Czechoslovakia's material importance to us . . . it was in the intangible scale of values that we registered a loss more damaging than the material loss."[22] Placing such importance on intangible values was a formula for a major extension of containment. NSC-68 was also much more explicit about the means of containment than Kennan had been, with the military element being key: "Without superior aggregate military strength . . . containment . . . is no more than a policy of bluff."[23] Being strategic on its own terms, NSC-68 felt that US society could easily provide much greater resources to implement this vision of containment.

As might be expected, coming so soon after the Soviet atomic test, vulnerability played an important role in this document. NSC-68 responded to US vulnerability on two levels. First, Soviet nuclear capability accented the dangers of US conventional weakness, since a confrontation with the Soviet Union presented the possibility of either humiliation or atomic war in which, Nitze estimated, the Soviet Union would be able by 1954 to inflict "unacceptable damage" on the US. Secondly, NSC-68 again expressed great concern with respect to US social vulnerability. Dictatorial governments could act in secrecy and with speed while "democratic processes" were vulnerable. "A democracy can compensate for its natural vulnerability only if it maintains clearly superior overall power."[24]

Finally, NSC-68 did not hesitate to cast its descriptions into emotional and moralistic terms that clearly equated the Soviets with the Nazis: "The concentration camp is the prototype of the society . . . in which the personality of the individual is so broken and perverted that he participates affirmatively in his own degradation." This moral opposition applied not just to the Soviet Union, but to the "international communist movement"—a more encompassing definition of the threat.

It is much easier to criticize NSC-68 after almost four decades than it was in 1950. In essence it recommended expanding the scope of containment and the instruments available to effect containment. It foresaw mutual deterrence and the implications that it had for conventional defense against Soviet expansion. It was not a militarization of containment, since the general response to Soviet expansion had already begun to take on military dimensions during the Berlin crisis and with the formation of NATO. It did represent, however, the institutionalization of a national security mind-set.[25] Beyond this, perhaps the most important element in NSC-68 was its timing. It was presented to a skeptical and still frugal President Truman in early April 1950. Less than three months later the North Koreans launched an invasion of South Korea. Many of the NSC-68 projections seemed to come true overnight. By September, the president had, in effect, approved NSC-68, and the defense budget began to soar to a fourfold increase over its prewar level. Nuclear weapons funding was increased, military assistance shot up, combat troops were dispatched to Europe, and the US built a chain of overseas military facilities that virtually surrounded the Eurasian landmass. With the Korean invasion there is little doubt that US military spending and activity would have increased even in the absence of NSC-68. Like Kennan's original formulation of containment, however, NSC-68 provided a rationale, a direction, and a justification to US policymakers in a period of uncertainty and confusion.

The first five years of the nuclear age wrought unprecedented changes in the global situation and in US policy. Containment and national security had come to dominate the country's thinking about foreign affairs. The fundamental goals that would determine policy over the next 40 years were in place. Soviet power must be contained. Deterrence of general war must succeed. The Soviet system must somehow be transformed and pacified. These goals have remained stable, although the different strategies and policies that various administrations have selected in their pursuit have varied widely. Much of this variation has been due to the interpretation of the four themes discussed earlier—containment, strategy, vulnerability, and legalism/moralism. Differences in interpretation of these elements were evident even by 1950. Containment could be seen as a stable policy aimed at a general balance of power in the world; or it could be seen as an end in itself, a fragile structure in which any disturbance was harmful to US security. Strategy could be rigorously based on limited resources and aimed at prioritized objectives; or it could be much more expansive, based on a generous estimate of resources and aimed at a wide range of global objectives. Vulnerability could be seen as an acute problem demanding a margin of superiority to make up for US weaknesses; or as a simple fact of international life shared with other states. Finally, the amount of free reign and expression given to the traditional US sense of legalism/moralism varied dramatically.

THE EISENHOWER ERA

THE PERSPECTIVE

The spectacle of stalemate in the Korean War accentuated the rising public frustration with both the passivity of containment and its expense. During the 1952 campaign, the Eisenhower/Nixon ticket promised to end the war in Korea and to take a more assertive policy toward the Soviet Union. Containment was to be augmented into an active policy of "liberation." In fact, the new administration never did go beyond rhetoric in its policy of "liberation" and maintained a substantial continuity with the basic policy goals of the Truman administration. Eisenhower did, however, overhaul the means for accomplishing those goals.

An NSC policy statement early in 1953 described the "two principal threats" to the survival of the US to be "the formidable power and aggressive policy of the communist world," and "the serious weakening of the economy of the US that may result from the cost of opposing the Soviet threat."[26] Similar to the Truman administration before Korea, the new administration perceived US resources to be limited and saw financial instability as nearly as dangerous as the Soviet threat. The idea was security and solvency—to regain American initiative in foreign policy without bankrupting the nation.

THE POLICY

The policy aimed at achieving this goal had two elements. The first was the "New Look" defense policy. In essence, this also resembled early Truman policy in that it emphasized a reliance on nuclear weapons rather than conventional forces. The nuclear element of this policy has come to be known as massive retaliation. This explicitly considered the possibility of US nuclear retaliation for any Soviet provocation—even those beyond an attack on the US or Western Europe. It was designed to reintroduce the initiative into US policy by using its large and growing strategic superiority. The Strategic Air Command's expanding bomber force, the burgeoning overseas base structure, and technical advances in nuclear production had all put the US far ahead in strategic power. Massive retaliation played to this strength by ambiguously suggesting that piecemeal aggression might in the future be met with nuclear strikes rather than conventional forces. Tactical nuclear weapons were soon introduced in quantity in Europe, while US conventional strength and manpower were drawn down. Massive retaliation was an attempt to extract an advantage from US nuclear superiority, although its architects recognized that this era of nuclear superiority would only be temporary. The NSC document that summarized this "New Look" warned that, as the Soviets increased their nuclear capability, "the entire policy of the United States toward the USSR will have to be radically reexamined."[27]

The second element in Eisenhower's national security policy was the formation of a global alliance system. Besides containing Soviet expansion by tacitly extending the nuclear umbrella, the US hoped to develop indigenous manpower pools so that future conventional operations would not necessarily involve US troops. On the other hand, such an alliance system substantially increased the scope of US interests and created a much larger number of clients that could be "lost" to communism. The mechanics of this "loss," in the eyes of the administration, would very likely be a chain reaction of regional defeats. A premium was therefore put on a very scrupulous and global definition of containment.

Eisenhower basically kept his promise to restore some initiative to US policy and to lower costs. There was virtually no real increase in defense spending during his tenure. At the same time he successfully fashioned a forthright containment policy, and peacefully overcame crises in East Asia, the Middle East, and Berlin.

Ultimately, however, the rise of Soviet nuclear power—first to a retaliatory capability, and, later, to a perceived first strike threat—undermined a policy rooted so deeply in atomic superiority. By mid-1954, the National Security Council warned that the emerging Soviet power meant that a nuclear war "would bring about such extensive destruction as to threaten the survival of Western civilization." The implications for massive retaliation were clear. Could the US continue to suggest credibly that the offshore islands in the Formosa Straights or even Berlin for that matter, were worth the demise of "Western civilization"?[28] By 1956, the credibility of massive retaliation had become a public topic of debate in the growing national security community where ideas of limited war and nuclear options and issues of force survivability came under serious consideration. The emerging wisdom was that, if containment were to be enforced without facing the absolutes of humiliation or nuclear war, then a policy more realistic and flexible would have to be found.

In fact, massive retaliation was never so inflexible. It preserved a range of options, both nuclear and conventional. In 1958, in part to counter criticisms of this defense policy, and in part to bolster his Middle Eastern containment policy known as the Eisenhower Doctrine, the president ordered a major conventional intervention in Lebanon. A division-sized force, complete with tactical nuclear weapons, spent some four months successfully stabilizing that nation. It was a limited military operation that accomplished its objectives.

This did not, however, serve to stem the criticism of massive retaliation which increasingly came to be based not just on the Soviets' ability to retaliate, but on their ability to threaten the US strategic force. By early 1955, a presidentially appointed panel of experts warned of the increasing Soviet capability for a Pearl Harbor–style attack that could destroy SAC on the ground and leave the nation defenseless. Later that year a surprising Soviet display of nuclear-capable aircraft suggested a "bomber gap." In 1957 the Soviets tested an ICBM and orbited an artificial satellite. Suddenly a "missile gap" had opened, thrusting strategic issues directly into the popular consciousness. Media and congressional attention was enormous. A genuine sense of vulnerability was aroused as the Soviets seemed to have bested the US in its own strong suit—high technology. Another prestigious panel, the Gaither Commission, was convened, and not unlike NSC-68, called for a major defensive buildup irrespective of the cost involved. US strategic forces were felt to be in danger of a surprise attack. From the late 1950s on, survivability of the strategic forces would be a primary consideration in both force structure and arms control decisions.

Despite these alarms, Eisenhower remained unperturbed. Quite possibly his background as a major field commander made him skeptical of "expert" predictions about a Soviet first strike. He had seen and endured Clausewitz's "fog of war," and he personally knew and understood the conservatism of many Soviet military leaders. With 2000 US bombers at dozens of continental and overseas bases, a first strike would not be a routine undertaking for a Russian military inexperienced at strategic operations outside the Eurasian landmass. More fundamentally, however, Eisenhower worked on judgments about Soviet intentions rather than analysis of their capabilities. Why, given all the historical and political factors that determine their outlook, would the Soviets want to risk so much on a first strike at the US? He fended off pressure to increase strategic spending dramatically and insisted on maintaining what was officially termed an "adequate" force structure.

By 1958, in spite of the reliance on nuclear weapons that he himself had authored, Eisenhower thought that the US atomic arsenal was becoming unnecessarily large. Resisting requests for more funding, he noted that the US already had enough "to destroy every conceivable target all over the world plus a three fold reserve." Speculating on a topic that would gain much attention in the 1980s, he feared so much radiation would be released that "there just might be nothing left of the Northern Hemisphere."[29]

Arms control in the Eisenhower years was based on the "Open Skies" proposal for mutual aerial reconnaissance, and the 1958 conference on surprise attack. Neither bore any fruit, but both were rooted in the increasing US sense of vulnerability to a Soviet first strike. Later in the second term, Eisenhower had access to U-2 aerial reconnaissance which, although inconclusive, showed little evidence of a massive Soviet strategic buildup. As the fifties drew to a close, the "weak and dangerous" strategic policy of the Eisenhower administration was under pressure from an assortment of presidential hopefuls in the Democratic party who, despite Eisenhower's assurances to the contrary, publicly warned about the growing "missile gap."

THE KENNEDY/JOHNSON YEARS

THE PERSPECTIVE

The Kennedy campaign stressed the inadequacies it saw in the Eisenhower administration's security policy. Eisenhower's ideas of "adequacy" had apparently allowed a missile gap to develop, while the New Look's de-emphasis on conventional forces had seriously reduced US options in dealing with limited aggression. The new administration perceived itself to be in a position so aptly described in NSC-68 eleven years before of having no better choice than to capitulate or precipitate a global war. The pedigree here was quite clear. Paul Nitze was the chief of Kennedy's preinaugural task force on national security.

THE POLICY

The policy initiated to increase options was known as "flexible response." This was a major shift away from reliance on nuclear weapons. Conventional and even antiguerrilla forces were to be substantially increased. The goal was to be able to execute a "two-and-one-half" war strategy, requiring the capability to face a Soviet attack in Europe, a Chinese attack in Northeast or Southeast Asia, and a minor contingency elsewhere, such as Cuba. This conventional emphasis, however, did not imply a cutback in the strategic force structure. In fact, the Kennedy administration doubled the already substantial Eisenhower program for Minuteman missiles and Polaris submarines and slowed the decommissioning of the B-47 bomber fleet. By late 1961, reliable intelligence had shown the missile gap to be a myth. The Soviets then had only a handful of highly vulnerable ICBMs.

The concept of flexibility was extended into nuclear targeting policy as well. The new secretary of defense, Robert McNamara, was particularly disturbed by the inflexibility and destructiveness inherent in the war plans he had inherited. Two thousand targets in the USSR, China, and Eastern Europe were to be struck with some eighteen thousand warheads. The expected casualties were estimated to be 400 million.[30] McNamara opted instead for a policy of "controlled response" that avoided cities and focused on Soviet military targets. Hopefully, the Soviets in a nuclear exchange would then have the incentive to avoid US cities. It offered the possibility of some damage limitation by destroying Soviet military assets and also served to increase the credibility of the US guarantee to NATO. Controlled response was short-lived as US policy. It was abandoned in the face of criticism that it implied a US first strike in order to limit damage to the nation. It also relied on the premise that the Soviets would—and, given their technology, could—limit themselves to counterforce targets in the exchange. This policy also stimulated demands by the services for enormous US strategic forces.

By the mid-sixties the declaratory US strategic policy had shifted to assured destruction. This was more a shift in emphasis than a drastic policy revision. Most of the counterforce options in the strategic war plans remained. Assured destruction was, at least from a psychological point of view, more oriented toward deterrence than was controlled response which dwelt on options after deterrence had failed. The logic of assured destruction had been foreseen by Bernard Brodie in 1946: in the event of an attack, secure second-strike forces would inflict unacceptable damage on the Soviet Union. The "assurance" in assured destruction was formally rooted in the triad of strategic forces. Together these forces were highly survivable and presented an unmanageable first-strike targeting problem to the Soviets. Offensively, each leg of the triad had unique operational capabilities and presented unique and expensive defensive problems to Soviet planners.

The "destruction" in assured destruction was basically a product of the analytic approach that McNamara brought to all aspects of his management of the Defense Department. In an interesting interplay of economic analysis and military strategy, the level of assured destruction was found to be 400 megatons of explosive power delivered on Soviet targets. This was the point on the curve of diminishing marginal returns where large increases in delivered megatonnage produced only small increases in destruction. The levels of destruction at this point were roughly 30 percent of Soviet population and one-half to three-quarters of Soviet industry. This was important to McNamara in sizing the US strategic force. Beyond 400 megatons, even assuming the worse-case scenario of a successful surprise attack, there was simply no rationale for more forces.[31] McNamara had come to the conclusion that nuclear superiority had lost any operational meaning. The Soviets were clearly developing an assured destruction capability of their own, and the period of mutual assured destruction was at hand. McNamara did not see technological fixes such as ballistic missile defense or multiple warheads as altering that basic fact.

In the conventional area, flexible response seemed to get off to an auspicious start during the Cuban missile crisis. In the summer of 1962, the Soviets attempted to install some 75 intermediate-range nuclear missiles in Cuba. In effect, this would have tripled their number of missile warheads targeted at the US in a very short period of time. In spite of US forward bases surrounding the USSR, the Kennedy administration refused to accept such an alteration in the balance of power in the Western Hemisphere. The confrontation escalated, and Khrushchev finally withdrew the missiles. US conventional options, notably the naval blockade, had contributed materially to resolving the issue without recourse to general war.

The results of flexible response in Southeast Asia were less positive. Observers have noted that flexible response was much more reactive and less assertive than massive retaliation. It remained wedded to a global vision of US interests, however, and, like the Eisenhower and the late Truman administrations, the Kennedy/Johnson administrations felt that they could not afford the political costs of "losing" a nation to communism.

This fear of loss led to a vision of containment that was herculean in breadth. Containment became, in effect, an end in itself in which any loss would presumably set off a chain of defeats. Any communist advance was a serious problem capable of upsetting the delicate balance. As NSC-68 had warned in 1950, "the assault on free institutions is now worldwide . . . a defeat for free institutions anywhere is a defeat everywhere." Fifteen years later President Johnson, with the Munich analogy clearly in mind, used a similar rationale concerning our commitment to Vietnam: "If we ran out on Southeast Asia, I could see trouble ahead in every part of the globe—not just in Asia but in the Middle East and in Europe, in Africa and in Latin America. I was convinced that our retreat from this challenge would open the path to World War III."[32] Containment had become so important in itself that a communist advance even in a geographically distant, culturally alien, and politically unstable nation appeared dangerous. Any strategy, flexible response or some other, would have been hard pressed to deal with such an expansive perception of the threat.

THE NIXON-FORD-KISSINGER PERIOD

THE PERSPECTIVE

The arrival of Richard Nixon and his national security adviser, Henry Kissinger, at the White House in 1969 ushered in a major transformation in US foreign and security policy. For the first time in the nuclear era, the Soviet Union was on the verge of genuine strategic parity with the US. All of the theories and strategies developed during the days of US superiority now had to submit to the stern test of Soviet equivalence. Of equal importance, the arrival of Soviet parity coincided with the evaporation of the US domestic policy consensus that had attended containment since the late 1940s. With perverse timing, the ordeal of Vietnam had destroyed the consensus behind conventional containment just as the geopolitical and technological environment was accenting its importance. The essence of Nixon/Ford/Kissinger security policy was to maintain containment during a period of external challenge and internal discord.

The Nixon administration brought diplomatic skill and historical vision to US policy. Kissinger in particular shared a vision of containment that was very similar to the original thoughts of George Kennan. Kissinger's studies of the classical European power balance convinced him that a multipolar equilibrium was the key US interest. He saw Soviet expansion, not an undifferentiated fear of communism, as the central threat to this balance. Communist advances at the periphery would have only a minor impact on the overall picture of US security. Time was seen to be on the side of the West. If the Soviets could be prevented from projecting their internal tensions into international crises, then their own system would eventually change. Once again, containment had become an instrument of policy rather than an end in itself. This approach was deeply rooted in *Realpolitik*. The Soviets were not considered zealous ideologues, but rather "ruthless" opportunists. The "essence of the West's responsibility" was to "foreclose Soviet opportunities."[33]

The leaders of the Nixon and Ford administrations were soberly grounded in the realization that resources for any national strategy would be severely circumscribed by the public mood. They responded to this limitation by curbing the policy goals inherited from the previous administration. As Melvin Laird, Nixon's new secretary of defense, noted, "The fundamental weakness in our national security policy during the 1960s was that it was over-ambitious and over-militarized."[34] The new administration reduced both the global sensitivity of the US to communist expansion and the direct role of the US in containing it. This retreat from the "bear any burden" vision of previous administrations to a more constrained approach reflected the new realism and implied as well a de-emphasis of moralism/legalism in US policy. Nonetheless, subjective perceptions of the US by the rest of the world remained important. For example, the Vietnam commitment was still crucial, not because of any tangible geopolitical advantage, but because the "confidence of free peoples" in American commitments had to be maintained.[35] This was an elusive and judgmental element in an otherwise realistic policy.

The Vietnam experience limited the Nixon administration in other more subtle ways. The war had deprived the US of its sense of moral superiority. The moral standards that the US had routinely applied to other nations now seemed sharply lacking in its own foreign and security policy. US overseas commitments and forces were seen by many to be the cause of instability, not a deterrent to it. Likewise, the arms race was seen in many influential circles as an inherent product of the US military-industrial complex. Functionally related to this sense of moral uncertainty was the apparent recession of the Soviet threat. The USSR seemed preoccupied with concerns closer to home—domestic problems, the opening to West Germany, reasserting control over its Eastern European empire, and the deterioration of its relationship with China. Even the "pacification" of Czechoslovakia in 1968 seemed less threatening than previous such interventions. In reality, since the Cuban missile crisis, there had been opposite tensions in US-USSR security policy. While the US froze its strategic strength and began a turn toward détente and arms control, the Soviets had temporarily harnessed their overt adventurism and begun a dramatic military buildup.

Nixon and Kissinger faced the dilemma of an unprecedented and growing Soviet military power aimed at an American society and body politic which, perhaps justifiably, had lost its taste to compete. The administration had little choice but to demilitarize containment, define it more realistically, and seek other instruments for maintaining the global equilibrium. US policy found itself limited by the effects of Vietnam.

This limitation would only increase as Watergate intruded, and as Congress began a concerted effort to reduce presidential power in foreign and security policy.

THE POLICY

The keystone of the Nixon administration's policy toward the Soviet Union was détente. Simply put, détente was an effort to carry on the containment of Soviet expansionism by means that were less confrontational and more realistic. Incentives and pressures were to be combined in a carefully orchestrated policy that created motivations within the Soviet system to accept a stable world order. Neither the Nixon administration nor the Soviets saw détente as a neutralization of differences. Rather, it was an attempt to create a network of interests so that the competition would take place within more benign rules of engagement. Détente required extensive negotiations with the Soviet Union. It expected a measure of restraint on the part of the Soviets, not because of any US military threat, but rather because the Soviets would recognize that it was in their self-interest to maintain a network of profitable contact.

The key mechanism in détente was linkage—the idea that any one area of US-Soviet relations was contingent upon acceptable behavior in other areas of the relationship. Hence, trade and arms control, or technology transfer and regional disputes, would be "linked" so that benefits in one area required Soviet restraint in the other. Linkage was designed to exploit Soviet self-interest. The Soviets were interested in trade with the West, gaining access to Western technology, cooperating in the European security process, diverting the US from reapproachment with China, and restraining US defense modernization. For Nixon and Kissinger, détente was a vehicle for submerging the increasing US military disadvantage into a multidimensional policy that could still divert Soviet expansion. Kissinger noted that Vietnam led to a retraction of US foreign involvement that opened the door for Soviet expansion. Later, Watergate weakened the executive power necessary to resist that expansion. Détente, as Kissinger later recalled, did not "cause these conditions, but was one of the necessities for mastering them."[36] Ultimately, détente in this original formulation proved too difficult to orchestrate. Its finely tuned dynamic of carrots and sticks could not withstand the erosion of executive power associated with Watergate and the general post-Vietnam climate of mistrust and anti-militarism.

The nuclear policy inherited from the Johnson administration was mutual assured destruction (MAD). Impending Soviet parity, however, had begun to cast this doctrine into some doubt. The logic and stability of this formulation were, and remain subject to, debate on several grounds. Politically, morally, and from an alliance perspective, concern had already been expressed about the policy, but within an overall context of US superiority such criticisms seemed peripheral. More serious doubts about MAD arose in the context of the burgeoning Soviet threat in the early 1970s—after the SALT agreements. It was already clear that the Soviet ICBM force, with easily predictable improvements, would soon pose a serious threat for most of the US land-based missiles. With congressional resistance mounting against either an improved ICBM or an antiballistic missile system for defense, opportunities to reduce this serious US vulnerability seemed elusive. With characteristic realism, the administration accommodated the Soviet buildup by espousing a doctrine called strategic sufficiency, and later, essential equivalence. These ideas accepted Soviet equivalence so long as the US retained power enough for stable deterrence and freedom from political coercion. The fundamental question that remained, however, was, what do we do if deterrence fails?

The theoretical specter raised by this strategic situation was that of a US president, having lost his land-based counterforce assets in a Soviet first strike, being forced into the position of doing nothing or launching an SLBM strike against Soviet urban-industrial targets, thereby inviting a second Soviet attack against US cities. In other words, the credibility of the assured destruction response was brought into question. Another problem with the assured destruction strategy was the difficulty it caused as a force-sizing device. As designed by McNamara, assured destruction specified certain calculable levels of destruction that are deemed necessary. Once having achieved the requisite force levels, however, further force enhancements cannot be justified. In the anti-defense atmosphere of the early 1970s this logic was expanded to work as well against prudent force modernization. Any request for modernized weapons was likely to be met with charges of "overkill."

The administration gradually developed a new targeting policy designed to increase the president's strategic options and to offer a better justification for force modernization. Early in 1974, President Nixon signed National Security Decision Memorandum (NSDM)-242. This represented a shift of emphasis away from the mutual assured destruction strike options in the strategic war plans toward more limited and flexible options designed to control escalation and neutralize any Soviet advantage. The logic here was unremarkable. As the Soviet nuclear forces became large and more diversified, they

clearly offered a larger repertoire of Soviet nuclear options. NSDM-242 emphasized, in declaratory policy at least, that the US also had options. Defense Secretary Schlesinger pointed out that "We are simply ensuring that in our doctrine, our plans, and our command and control we have—and are seen to have—selectivity and flexibility."[37] In fact, the US force structure was not that well designed for limited nuclear options. Overall, the accuracy, command, control, and distribution of warheads were much better suited for large-scale retaliation against unhardened targets. It would be well into the Carter administration before the hardware necessary for credible limited options would begin to take shape. Critics of this new policy on the left nonetheless felt that it lowered the nuclear threshold and was in fact a "warfighting" rather than a deterrent strategy. They argued that such discussions of escalation control were fantasy and that, once the nuclear threshold was crossed, escalation would be inevitable. Critics on the right had no theoretical problem with increased flexibility but, without major force improvements, it was a strategy without substance. They felt it merely placed more pressure on already overstressed and underfunded strategic forces.

SALT

The strategic arms limitation agreements of the early 1970s played a central role in the Nixon/Kissinger détente policy. The accords themselves were neither as good as their proponents claimed nor as bad as their critics claimed. They represented an intelligently negotiated first step in what was to be a continuing process. Unfortunately, the follow-up agreements needed to address the serious problems deferred by SALT I were never effectively negotiated. Because of US political weakness, and the inability of the Soviet leadership to curb its own opportunism and desire for strategic advantages, the early negotiations to follow up on the progress in SALT I were unable to produce results. Hence, the imperfect and limited SALT I agreement on offensive weapons was extended into the late 1970s into a political and technological environment that negated much of its usefulness.

SALT I was composed of two separate agreements. The first was a formal treaty that limited ABM deployments on both sides to only two sites. From the Soviet point of view, this capped the dangerous US lead in ABM technology. For the US, this agreement seemed to codify Soviet acceptance of the idea that a mutual hostage relationship between the superpowers offered stability in their strategic relationship.

The second document was an Interim Agreement on Offensive Arms, freezing the number of offensive weapons until 1977. This preserved several force inequalities. The Soviets had more than 300 "heavy" missiles for which the US had no counterpart. The Soviets were also allowed higher ICBM/SLBM ceilings—roughly 2400 Soviet to 1700 US. The justification for such inequality was that the formidable US bomber force was completely excluded as were US forward-based systems in Europe, and the US had already begun to deploy MIRVed systems, thus enhancing its warhead advantage. Additionally, US technology, reliability, and accuracy were all far superior to the Soviet force. The US negotiators felt that, by the time the Soviets could exploit their advantage in quantitative measures, either a new SALT agreement could be hammered out, or unilateral US force modernizations could neutralize the disparity.

That these hopes never materialized in the 1970s should not diminish the accomplishments of SALT I for the era in which it was negotiated. The agreement arguably provided a modest cap on Soviet expansion at a time when they were building and the US was not. In fact it had very little impact on US force structure. Every defense program was under severe attack in Congress. Kissinger, although not an impartial observer, was on target when he described SALT I as a "snapshot of the strategic relationship that had evolved over the preceding decade."[38]

In spite of the public fanfare that greeted SALT I, the administration knew that it had merely delayed and not ended the threat posed by the Soviet buildup. The serious problem of MIRVs was deferred until SALT II because the Soviets refused to negotiate on a system the US was deploying and they had not yet tested. It was this inability to constrain the MIRVing of the huge Soviet ICBM force that later in the decade led to a new era of vulnerablity known as the "window of vulnerability."

It soon became clear that the Soviets had more in mind with the SALT process than simply capping the US ABM. By late 1973, it was clear that the Soviets "were using the quantitative freeze to engage in a qualitative race."[39] They began flight-testing an entirely new generation of missiles including a new "heavy" missile, leading many US observers to doubt their commitment to a concept of strategic stability grounded in arms control agreements. As Secretary Schlesinger put it, "Apparently they had deliberately held up such activity (introducing new missiles) until the American signature was dry on the agreements." Kissinger, not an innocent in dealing with the Soviets, defended SALT I by noting that a bilateral negotiation with the Soviets could not do things for US security that it was unwilling to do for itself. He was referring, of course, to the continued popular and congressional pressure on defense spending. Although the elements of an extensive force modernization were begun in the Nixon administration, the clear legacy of Vietnam was that US strategic forces had been neglected while the Soviets had concentrated on theirs.

By the mid 70s, the US negotiating position was very weak due to the erosion of executive authority and continued pressure on defense programs. The Soviets, in contrast, had, with surprising speed, tested and begun deployment of MIRVs and had stepped up their regional intrusions. It was then clear that the critical follow-on negotiations, necessary to transform the SALT I "first step" into lasting stability, were impossible.

During the 1972 congressional discussion of SALT, Senator Henry Jackson successfully backed an amendment requiring all future SALT agreements to be based on equal aggregates. This was manifested in the Vladivostok arms accord signed in late 1974 between President Ford and General Secretary Brezhnev. Both sides agreed to 2400 total strategic launch vehicles, and to a subceiling of 1320 MIRVed vehicles. These were equal aggregates on paper only. The US was well below both the total ceiling and the MIRV subceiling, even though US intercontinental bombers were now included. More importantly, the Soviet heavy missile program remained largely unconstrained, and the MIRV ceiling was so high that the mating of Soviet MIRVs with missiles of large throw-weight would continue to threaten the US ICBM force. Moreover, a new Soviet bomber, the Backfire, with at least some potential for intercontinental strikes, was excluded from the ceilings. Few in the US were satisfied by the Vladivostok accords. The arms control community saw the ceilings as unconscionably high, while the more conservative defense analysts saw the agreement as inviting genuine Soviet strategic superiority. Far from equal aggregates, the SALT process seemed to be locking in dangerous inequalities. Even the deeply analytical Secretary of Defense James Schlesinger noted ominously that opponents may feel tempted to exploit imbalances "as Hitler did so successfully in the 1930s, particularly with Neville Chamberlain."[40] The Munich analogy had returned as an explicit element in the US arms control calculations, and the politicization of the arms debate deepened.

THE NIXON DOCTRINE

If the Nixon administration's moves in the area of arms control were dramatic and controversial, its policy on conventional force issues was much more dramatic and much less controversial.

During the early years of the Nixon administration, the public demand for a drawdown in
US overseas commitments was overwhelming.
Recalling the early days of his vice presidential
tenure in the Eisenhower administration, President Nixon responded to pressure with two related policy decisions—the Nixon doctrine and
the "one-and-one-half" war strategy.

The Nixon doctrine was a decisive geographical and functional retrenchment of US containment policy. While the US would continue to
honor all treaty commitments, military and economic assistance in areas beyond formal treaties
would be considered "as appropriate." In any
case the "nation directly threatened" would assume the "primary responsibility for providing
the manpower for its defense." But "where our
interests or treaty commitments are not involved, our role will be limited." Although
downplayed by its originators, the contrast with
President Kennedy's commitment to "bear any
burden, pay any price, support any friend" is
stark. The first manifestation was the Vietnamization program, followed by a conventional cutback in Korea and a general drawdown in active-
duty military personnel of some one million.

Closely connected to these adjustments was
the shift from a "two-and-one-half" war strategy
to a "one-and-one-half" war strategy, calling for
the capability to handle simultaneously a major
conflict with the Soviet Union (presumably in
NATO) and a minor contingency. The US never
had the physical capability for a "two-and-one-
half" strategy, so this was a reasonable alignment of declaratory policy and force structure.
Significantly, this realignment signaled to the
People's Republic of China that the US saw it
as separable from the Soviet threat and not constituting a threat of its own.

By the end of the Ford administration, détente, a hopeful element in US-Soviet relations
just a few years earlier, was under serious attack. SALT had not produced the benefits desired, Soviet Third World adventurism was on
the rise, and Soviet domestic policies were becoming more oppressive. To some extent the
architects of the détente policy were to blame
for failing to point out to the American public
that détente was a modest change in the terms
of the East-West competition and not a solution
to that competition. On the other hand, the
delicate powers of control necessary to link up
a network of incentives and punishments were
denied to the executive as the aftermath of Vietnam and Watergate eroded its power. A period
of US military decline and social confusion had,
however, been weathered without a major loss
in the nation's geopolitical position. The fundamental global equilibrium had been maintained,
and the emergence of an amicable People's Republic of China perhaps strengthened.

THE CARTER ADMINISTRATION

THE PERSPECTIVE

The team that took charge of US foreign and
security policy in January 1977 effected a genuine if only temporary change in US geopolitical
focus.

The nation temporarily discarded its obsession with communism and, with it, the centrality of the US-Soviet relationship that had
characterized the Nixon-Ford years. The Carter
administration emphasized a more global
agenda, concentrating on regional issues, the
North-South relationship, the economic interdependence of the industrial democracies, and
human rights. The clear implication of this was
a major de-emphasis of traditional containment.
Soviet interference in Third World disputes
brought much milder reactions than in previous
administrations. The heavy burden of Vietnam
was a constant reminder that an overcommit-
ment of the US to containment could precipitate
national tragedy. There was little popular sense
of vulnerability vis-à-vis the USSR in the early
days of the Carter administration. The worst
setbacks which the US had encountered in the
early 1970s, many argued, had occurred because of its own overextension in Asia and the
Arab oil embargo, neither of which was directly
connected with overt Soviet expansion.

Another important departure for the Carter
administration was a renewed emphasis on moralism in US policy. The *Realpolitik* of the Kissinger years was scorned during the presidential
election campaign as an immoral and alien interlude in US foreign policy. Human rights became the new rallying cry, the vehicle for
reasserting US moral superiority that had been
frittered away in recent years by administrations
more interested in a global balance of power
than a global moral order. Like all new administrations, as the complexities of governing and
negotiating became more apparent, these policy
initiatives eventually became less pronounced,
and in some cases were discarded altogether.

THE POLICY

Jimmy Carter, during his inaugural address,
spoke of his hope that "nuclear weapons would
be rid from the face of the earth." His original
moves in the strategic arena reflected this hope.
His first approach to the Soviets was an attempt
to move beyond the Vladivostok accords and to
garner deep cuts in Soviet forces. Basically this
early 1977 offer sought a cut of about 25 percent
in Soviet strategic forces, including 50 percent
of the worrisome "heavy" missiles. In exchange,
the US offered a ban on the undeveloped MX
missile and an exemption for the Soviet's new

Backfire bomber. This was a bold first step for the administration, but it provided little incentive for the Soviets. As a national security aide later described it, "We would be giving up future draft choices in exchange for cuts in their starting lineup."[41] The Soviets, already incensed by some Carter administration moves on the human rights front, publicly and insultingly rejected this proposal. SALT II then retreated into two years of painful negotiations which built incrementally on the Vladivostok accords.

US ICBM vulnerability bothered the early Carter administration less than the Ford administration. In the view of Carter's advisers, the enormous operational complexity of such a strike reduced its likelihood to virtually zero. The new secretary of defense, Harold Brown, doubted that the conservative Soviet leadership was interested in such a "cosmic roll of the dice."

The remainder of the Carter strategic policy, while substantial in its strategic spending and prodigious in its efforts to achieve an arms control agreement, reflected a certain confusion, or at least a lack of vision. In June 1977, for example, the B-1 bomber was cancelled—too late to help sweeten the deep cuts proposal for the Soviets, but too early to extract any concessions in the SALT II negotiations. The Trident submarine was delayed, and great difficulty attended decisionmaking concerning the MX and its basing mode.

By June of 1979, SALT II was complete and signed in Vienna. It was never ratified by the Senate. It was, at best, a modest agreement that did not substantially limit Soviet heavy missiles and again accepted high subceilings on MIRVed missiles. Still, it represented some cap on Soviet expansion while limiting very little in US programs. The chances of Senate approval of SALT II were slim at best, but the Soviet invasion of Afghanistan in December 1979 led to its withdrawal from consideration.

From an arms control perspective, SALT II was criticized because it allowed virtually all planned force programs to be built, and hence it represented only a questionable attempt at genuine arms control. Conservative criticism was more determined. From this perspective, SALT II represented something worse than shoddy negotiations. It codified major US inequalities and vulnerabilities. With moral fervor, Senator Jackson compared the administration's Soviet policy to that of Britain in the 1930s when the British public was assured that "Hitler's Germany" would never achieve military equality—let alone superiority." SALT II, he continued, represented "appeasement in its purest form."[42]

A side effect of the high ceilings in strategic offensive systems allowed in SALT II was revival of interest in strategic defense. The ABM Treaty

was clearly contingent upon "more complete" offensive force reductions which would reduce the threat to the survivability of US "retaliatory forces." In a unilateral statement attached to the treaty, the US had warned that if such limitations were not achieved, "US supreme interests could be jeopardized" thus providing a basis for withdrawal from the ABM Treaty. The conservative defense community began to rally to the point that, since SALT II did not reduce the threat, ABM limits were no longer in the US interest.

Outside the SALT framework, the administration's nuclear policy continued along the lines set down in NSDM-242 and the Schlesinger policy of increased flexibility. In July 1980, President Carter signed Presidential Directive 59 (PD-59) directing preparation for a more-prolonged nuclear engagement and a further increase in US targeting options to include Soviet leadership and control apparatus. This was, again, an evolutionary development in US policy, rather than a dramatic break. It occasioned, however, the same controversy that had met earlier statements of limited nuclear war including charges that it lowered the nuclear threshold, made war more likely, and was fantastically optimistic in expecting escalation to be controlled.

THEATER FORCES

The one area where options were desperately needed was in Europe. Extended deterrence has always placed greater demands on credibility because it was a political commitment to defend allies, rather than the fundamental commitment to defend oneself. In addition, the majority of both theoretical and governmental attention had been paid to the central nuclear arena. The European and other regional arenas, where a genuine superpower confrontation was perhaps more likely to occur, had been largely neglected, and the conceptual framework for employing nuclear weapons in a European or regional framework was less developed. While the prospect of Soviet parity and mutual deterrence had stimulated the US and NATO's strategy of flexible response, Soviet deployment of an array of capable new theater weapons threatened to neutralize that strategy. Far from being flexible in its response, NATO was beginning to appear to have no credible response at all. In reaction to the Soviet SS-20 missile and the Backfire bomber deployments, the West Germans requested a NATO reply that would reinvigorate strategy and reintroduce danger and uncertainty into Soviet calculations.

The Carter administration worked hard to address the theater deterrent problem, and its push for increased allied conventional forces in particular was effective. In the nuclear arena, the best of intentions were unable to avert the difficulties faced by the Alliance late in the 1970s. After an embarrassing episode in which the US abruptly cancelled the enhanced radiation warhead program it had earlier urged on its allies, the administration focused on the modernization of Intermediate Nuclear Forces (INF). In December 1979, the US and its allies agreed to a modernization program based on the "dual track" of deployment and arms control. This NATO decision gave the Soviets an irresistable entree into both an allied arms deployment decision and into allied domestic politics.

In its approach to conventional force requirements, the early Carter administration, legitimately reflecting the public mood, took the approach of "no more Vietnams." The exercise of America's power, outside of the clearest threats to basic American alliance commitments, was seen as dangerous and possibly immoral. This is not to say that the Carter administration neglected the American defense effort. Outlays went up every year of Carter's presidency, reversing the substantive declines of the Nixon-Ford years. But the trend was to avoid the use of American military force as a vehicle of containment. Carter began plans to withdraw the US forces from Korea and deferred opportunities to challenge Soviet advances in the Third World. Gradually, over the course of his tenure, and finally with dramatic suddenness after the Soviet invasion of Afghanistan and the seizure of American hostages in Iran, the Carter administration returned to a more global and military vision of containment. The Soviet advances in Angola, its encouragement of Vietnamese imperialism, its intrusion into the Horn of Africa and South Yemen, its tacit and later overt support for the Iranian militants, as well as its continued military buildup and obstructiveness at the SALT talks, finally convinced the Carter administration to move US-USSR relations back to center stage.

The first manifestation of this recommitment to containment was the dramatic restatement of US vital interest in the Persian Gulf and the initiative to form a Rapid Deployment Force. The Carter Doctrine, like the Eisenhower Doctrine of a quarter century before, was aimed at reminding the Soviet Union that there were regional interests that the US would protect with force. With it came a revival—in part rhetorical, in part genuine—of the pre-Vietnam concept of containment.

The Soviet invasion of Afghanistan finally closed the door on the policy experiment known as détente. Its modest gains in arms control were being overwhelmed by the continuing Soviet buildup and the increasing technological complexity of nuclear issues. The Soviets were actively pursuing advantages in many areas of the Third World, and their human rights policy was as repressive as anytime since Stalin. Conversely, their adventurism had driven the US to potentially the most momentous geopolitical event of the postwar era—the rapprochement with China. The advantages of trade and technology transfer that the Soviets had originally envisioned never came about. The Soviets also remained frozen out of the Middle East peace process. Finally, by the end of the Carter administration, the US, in large part due to Soviet activities, was on the verge of a domestic conservative backlash that was intent on turning the "correlation of forces" against the Soviet Union for the entire decade of the 1980s and beyond.

Détente, like most US schemes for managing Soviet power, was rooted more in domestic considerations than in objective analysis of Soviet policy. While it was true that there were many procedural problems with the implementation of détente—the weakened executive, congressional manipulation, inflated public expectations—its basic problem was its incompatibility with the US public view of international affairs. This moralistic public view was "Manichean": states were either good or bad, relations were either hostile or amicable. As the author of détente later pointed out, "The proposition that to some extent we had to collaborate with our adversary while resisting him" proved unconvincing. Americans clung to the "verities of an earlier age."[43]

THE REAGAN ADMINISTRATION

THE PERSPECTIVE

The Reagan administration assumed office after a solid victory over Carter, the first victory over an elected incumbent president since Hoover fell to Roosevelt in 1932. Carter's difficulties in foreign policy were instrumental in his defeat. The public vision, spurred by the hostages in Iran and the Soviet Army in Afghanistan, was one of America in retreat. Inconsequential oil sheikdoms seemed to be dominating the US economy, and anti-Americanism was making visible gains in Central America.

The new administration initiated a broad anti-Soviet program in public diplomacy. The keys to the Reagan foreign policy were to be: military and economic revitalization, revival of alliances, stable progress in the Third World, and a firm Soviet policy based on Russian reciprocity and restraint. The arms control process which had resulted in the "fatally flawed" SALT II treaty was abruptly terminated to await a US buildup that would make negotiations more fruitful. The new secretary of state, Alexander Haig, was clear in his determination to return arms control to its rightful and modest position as an adjunct to national security, rather than its centerpiece. Likewise, Secretary of Defense Caspar Weinberger pointed out that between the signing of the SALT I and SALT II agreements the Soviets had increased their strategic warheads by 60 percent. Moreover, the quality of the Soviet effort in counterforce capability and in strategic/ civil defense led many observers to doubt whether the Soviets accepted the doctrine of mutual assured destruction. Soviet forces seemed designed to undercut strategic stability and to reduce the credible retaliatory options of the US.

The administration's vision of containment was similar to the expansive visions from the Eisenhower and early Kennedy years. Noting that the threat had grown since containment was developed in the Truman administration, the administration's second secretary of state, George Shultz, testified before the Senate that the US must do more than simply foreclose Soviet opportunities. "Now our goal must be to advance our own objectives, where possible foreclosing and when necessary actively countering Soviet challenges wherever they threaten our interests." The familiar theme of avoiding a loss anywhere to prevent a chain reaction of defeats was also evident. Referring to Central America, President Reagan warned in 1983, "if we cannot defend ourselves there, we cannot expect to prevail elsewhere. Our credibility would collapse, our alliances would crumble, and the safety of our homeland would be put in jeopardy."[44]

The Reagan administration's strategy for supporting this broader definition of containment was based on the general objective of increasing US strength, both military and economic. The administration did not feel it had the time for a finely crafted strategy in the classic sense. As Secretary Weinberger later put it, "In 1981, we could not delay rebuilding American military strength while we conducted a lengthy conceptual debate."[45] The Soviet challenge was seen most clearly in military terms. By 1981, it was obvious that the Soviet system had developed little ideological, economic, or cultural influence in world affairs. Soviet military strength and its political side effects were the key threats. US strategy, then, was to counter this strength by reversing the across-the-board disadvantages that the new administration saw in American forces. Like the Kennedy administration, this strategy would provide the strength for any contingency. With the anticipated US economic recovery, resources were felt to be easily within reach for such a buildup.

The early Reagan administration felt an acute sense of vulnerability. The SALT-induced "window of vulnerability" of US ICBMs was the most visible public manifestation of this issue. Not unlike the Kennedy team twenty years before,

the administration felt it had witnessed the unnecessary decline of US power. The erosion of US defense potential—in part due to SALT, in part due to lack of leadership—had created both strategic and conventional vulnerability. In fact, many important players in the Reagan administration mistrusted the entire process of arms control as a sort of dangerous exercise that exploited social vulnerability and lulled the public away from a realistic assessment of the nation's defense needs. Likewise, Soviet noncompliance with arms control agreements was seen, at least in part, as an exploitation of US social vulnerability by a centralized and secretive regime. Conventional helplessness, epitomized by the disastrous 1980 desert rescue attempt in Iran, was equally disturbing.

The Reagan administration brought a substantial element of traditional moralism back into US policy. Due to rhetorical flourish and the religious character of some of President Reagan's political support, this element was often overstated. Nonetheless, the Manichean idea of a moral struggle with the Soviet Union, so prominent in the early years of containment, had indeed regained considerable influence.

THE POLICY

The "revitalization of containment"[46] undertaken by the Reagan administration was energized by a major increase in defense spending. Real spending in FY 85 was over 50 percent greater than in 1980. The fruits of this buildup were distributed across the spectrum of military missions.

Strategic forces were major beneficiaries. Adopting and amplifying the major elements of Carter's PD-59 into National Security Decision Directive (NSDD)-13, the Reagan administration began assembling an appropriate force structure. Believing that the Soviets were best deterred by a US ability to fight and prevail in a possibly protracted nuclear conflict, substantial resources were devoted to counterforce capability and, particularly, to command and control improvements. Following a trend traceable back to the Kennedy administration, Secretary Weinberger believed that more options offered better deterrence. Also, if deterrence failed, a flexible nuclear strategy could help limit damage and terminate the conflict sooner. Requests were submitted for 100 MX missiles, the B-1 bomber was reinstated, and funding for strategic command and control burgeoned.

In March 1983, in a potentially dramatic departure from prevailing nuclear strategy, President Reagan embarked on the Strategic Defense Initiative (SDI), designed to replace the mutually assured destruction relationship with mutually assured survival. This research initiative soon became one of the largest single defense programs and created one of the largest defense debates since Vietnam. Strategic, fiscal, technical, arms control, and alliance questions generated a vortex of controversy but did not move the president from his commitment.

Although Congress initially accepted the high Reagan defense budgets, resistance grew over specific programs. The "warfighting" capability was particularly disturbing to many in Congress, as was the administration's policy of benign neglect towards arms control. This discontent culminated with the debate over the MX missile. After discarding the Carter administration's painfully crafted basing mode, the Reagan administration had great difficulty finding an acceptable replacement. All seemed awkward, and none seemed to address adequately the original issue of ICBM survivability. In late 1982, Congress cut MX missile funding pending the formulation of an acceptable survivable basing mode. President Reagan, in response, appointed former National Security Adviser Brent Scowcroft to chair a panel that eventually reviewed the entire US strategic program including its relationship to arms control.

The comprehensive and influential Scowcroft Commission Report had several results. First, it reported that the vulnerability problem was vastly overstated and oversimplified. In elaborating the immense difficulty and risk of Soviet attack on the US triad, it seemed to close the "window of vulnerability." This finding cleared the way for a recommendation that the MX be deployed in fixed silos, justified by arguments that ranged from arms control bargaining to perceptions of national will, as well as operational capability. The report also stressed the importance of arms control and set the stage for the idea of a "build-down"—that is, as strategic forces are modernized, incentives would be provided for both sides to construct smaller, more stable systems. Along these lines, the Commission recommended development of the so-called Midgetman mobile, single-warhead ICBM. This missile would be survivable because of its mobility and stabilizing because its single-warhead design reduced the preemptive incentive to Soviet decisionmakers.

Building on a carefully crafted bipartisan consensus, the commission's recommendations on ICBMs were accepted by both Congress and the executive. From a bureaucratic perspective, the Scowcroft commission, essentially an honest broker between Congress and the administration, signaled the unprecedented and apparently permanent entry of Congress into specific issues of arms control and strategic force structure. For the first three decades of the postwar period, the executive determined the character and size of the strategic force while Congress generally set only broad budgetary constraints. By the mid-1980s, however, key decisions concerning issues such as the number of MX missiles or the configuration of the Midgetman were being decided primarily in Congress rather than by the executive.

By 1984, the Reagan administration was ready to begin serious arms control negotiations. Its previous moves in this area seemed to many to be designed more to satisfy various Western audiences than to engage the Soviets. By late 1985, with budgetary difficulties clearly ahead, the tempo of arms control, at least on the public relations level, had accelerated dramatically. More importantly, after years of faltering Soviet leadership, Mikhail Gorbachev took control in the Soviet Union in early 1985 determined to reinvigorate both the country's economy and its foreign policy. Gorbachev's skills in public diplomacy led to a series of well-publicized offers on arms control and nuclear testing which, although somewhat unrealistic, began to seize the policy initiative that the Reagan administration had enjoyed since coming to office. More substantively, the American SDI program, like the ABM of 15 years earlier, had created anxieties in the Soviets about a costly new arms race in an area of US technological advantage. Several additional new US systems—missiles, bombers, and antisatellite weapons—were also looming on the horizon, further darkening the Soviet strategic outlook.

Increasing Soviet interest in arms control was not immediately matched by the Reagan administration. In June 1986, citing Soviet infractions, the administration discarded its policy of observing SALT II limits, and it continued to exclude the SDI from the negotiating process. The administration remained committed to the difficult vision of both strategic defense deployment and a drastic offensive weapon drawdown. The odd and inconclusive Reykjavik summit in October, in spite of some unrealistic and disturbing proposals about the elimination of all missiles and nuclear weapons, did eventually produce a broad agreement to reduce strategic weapons by 50 percent and to eliminate all land-based INF missiles in Europe.

As progress continued in 1987 toward eliminating INF forces, the European allies found themselves hesitant about a situation in which nuclear capability was drastically cut, while Soviet conventional superiority remained intact. Through 1987 and into 1988, as the elimination of land-based theater-range nuclear weapons looked to be becoming a reality, Europeans began to question the reliability of the American nuclear guarantee. Indeed, many argued that the INF deal was a net Soviet advantage since reducing European-based nuclear weapons capable of striking the Soviet Union appeared to place greater emphasis on the need for a force balance, an area of clear Soviet advantage. On both sides of the Atlantic, many discovered the logic which had inspired Eisenhower's original strategy of reliance on nuclear weapons for deterrence in Europe. In general, however, Gorbachev's skillful public diplomacy was successful in reducing perceptions of the Soviet threat amongst Western European publics. It became more and more difficult to convince parliaments of the need for increased defense spending in the conventional area. Instead, many looked to conventional arms control to remedy the unstable conventional relationship, in spite of the obvious difficulty of attempting to negotiate with the Soviets from a position of such inferiority.

Nonetheless, at the Washington summit in December 1987, the US and USSR signed the INF Treaty and formalized their commitment to a 50 percent reduction in strategic offensive arms. The Soviets had come to accept an impressive measure of the Reagan administration's agenda in strategic and theater arms control. The total elimination of a weapons category, on-site inspection, and asymmetrical Soviet reductions were important and unprecedented elements of the INF agreement, an agreement that vindicated, at least in principle, the administration's original logic of negotiating from strength.

The conventional force policy since 1981 had been to refurbish general capability for action without limiting planning to specific goals such as "two-and-one-half" or "one-and-one-half" war strategy. Hoping to play to US strength, early administration policy called for geographic escalation of any conflict. Rather than respond to Soviet aggression on its own terms, the US warned it might use its mobile naval striking power to attack Soviet assets far removed from the original conflict. The force structure to support this policy included a commitment to a 600-ship Navy outfitted with 15 carrier battle groups. Always a source of some controversy, particularly as defense budgets declined after 1985, this maritime strategy became a major source of debate both inside and outside the administration. In 1983, a new joint service command—CENTCOM—was established to deal specifically with contingencies in Southwest Asia, while in 1987, under congressional pressure, a new special operations command was instituted. Just as the Reagan administration sought to improve conventional force posture, it also demonstrated a clear willingness to use measured amounts of conventional force when circumstances called for it. Examples include the Grenada invasion, the air strikes on Libya, and the Persian Gulf escort operation.

By early 1986, a new element of strategy informally known as the "Reagan Doctrine" had quietly made an appearance. This policy, made possible in part by a revitalization of US intelligence services, sought to roll back Soviet and Cuban gains in the Third World by active support of liberation movements in areas such as Nicaragua, Angola, and Afghanistan. This doctrine sought to turn Soviet overextension into costly and finally demoralizing adventures beyond its capacity to sustain. US troops were not to be used, but financial, material, and intelligence support would be directed at reversing communist gains in the Third World. In many respects, the Reagan doctrine represented a reversion to the early 1950s vision of global containment. It was strongly moralistic in its tone and was strategic in the sense of extracting maximum cost from the Soviets at minimum expense for the US. Although the momentum of the Reagan doctrine was stalled by the Iran-contra affair, by early 1987 aid was flowing to anti-Soviet insurgents in Africa, Asia, and Central America. In the spring of 1988 the Soviets formally agreed to withdraw from Afghanistan. It is difficult to assess how much US aid to the resistance contributed to this decision. Gorbachev may have simply calculated that the war's costs had come to exceed any potential gains. Nevertheless, the withdrawal of the Soviets from a country where their forces were actively engaged was unprecedented in the postwar era and must be considered a victory in some measure for the Reagan administration's approach.

EPILOGUE: CHALLENGES FOR THE BUSH ADMINISTRATION

In spite of the Reagan administration's more active approach to security policy, it is clear that its vision of containment was not the expansive, crusading vision that its earlier declaratory policy had implied. Although the Reagan rearmament program did increase spending over the Carter projections in terms of real defense outlays, it was still modest in comparison with earlier spending programs of the 1950s and 1960s. On specific issues Reagan consistently employed a conservative, and therefore traditional approach to containment. The ABM Treaty was extended even as the superpowers debated the prospects for strategic defense; offensive arms control remained a key agenda item; grain trade with the Soviets was resumed; and US reactions to both the Polish crisis and the Korean airliner attack were restrained. As for the employment of force, it would seem that the apparent lessons of over-extension, particularly in Lebanon, had been well studied. In November 1984, during a period when many anticipated direct US military intervention in Central America, Secretary Weinberger outlined a very cautious program that called for force to be used only in the highest priority situations, aimed at clear objectives, and with unqualified public and congressional support. Nevertheless, the decision to deploy substantial naval forces to the Persian Gulf in 1987 in defense of "freedom of navigation," and to actively engage Iranian targets when provoked, reflected an abiding belief that American military power played a central role in US global interests.

The formation of a coherent strategy was a controversial issue for the Reagan administration. Critics claimed the administration never carefully matched resources to objectives because it never defined its objectives. Some observers had noted that "deciding how much is enough depends on first deciding enough to do what?"[47] The administration responded that all US national interests are important and should not be discarded because of resource contraints. As defense budgets declined in real terms in the late Reagan years, the strategic approach became a more flexible one of reducing risks to national interests, rather than attempting to secure the interests outright. The key strategic question became not "how much is enough to secure an interest" but rather "how much are we willing to pay to reduce the risk to this interest to an acceptable level?"[48]

The administration overcame the abiding sense of vulnerability that characterized its campaign and early tenure in office, although there was little change in the concrete aspects of the US-Soviet strategic relationship. Like earlier gaps, the window of vulnerability was more a symbolic than operational problem for American policy. The traditional sense of social vulnerability felt by the US was not eased by the series of espionage scandals that plagued the US in the mid-'80s. Balancing that, however, the new Soviet policy of openness began to expose in detail some of the enormous inadequacies of the Soviet system, while Gorbachev's style and apparent cooperation in arms control also tempered US perceptions of the Soviet threat.

Finally, while the American experience demands that morality be an element in foreign policy, and while appealing to moralism can be very effective in rallying short-term domestic support for a given policy, its shortcomings have consistently been recognized by the leaders who must deal over the long term with a powerful and realistic adversary. If the series of Reagan-Gorbachev summit meetings that began in 1985 served no other purpose, they at least highlighted the necessity for the pragmatic moderation of antagonism and for defusing overbearing moralism by demonstrating one can indeed deal with the devil.

Such pragmatism and a tendency to deemphasize "that vision thing" characterized the new Bush administration as it took office in January 1989. President Bush's National Security Advisor, Brent Scowcroft, immediately launched a major review of national strategy, later concluding in a televised interview in the spring that "the future looks a lot like the present, on a straight line of projection."

By the end of the president's first year in office, however, the world—and, in particular, the US-Soviet relationship—hardly looked at all like it had before. Beginning with the election of a noncommunist Solidarity government in Poland, one border after another in Eastern Europe opened for free travel to the West. In November, the Berlin Wall opened, and the long-dormant "German Question" suddenly reappeared on the agenda. By the end of 1989, every Warsaw Pact Communist leader in Eastern Europe had been replaced, in most cases by a noncommunist government and with the overt encouragement and support of Soviet General Secretary Gorbachev himself. Within the Soviet Union, nationalistic pressures grew, challenging the primacy of the Communist Party itself and raising speculation about Gorbachev's ability to endure politically as he continued his controversial *perestroika* ("restructuring") of Soviet society. At least outwardly, Gorbachev's rhetoric about "new defense thinking" began to become visible. The unilateral withdrawals of Soviet forces from Eastern Europe announced by Gorbachev at the United Nations in December 1988 had begun, as forecast, and the US intelligence community reportedly agreed that Soviet defense spending was declining and that, at least in some cases, Soviet weapons procurement had leveled off or been reduced.

The Bush administration's initial response to these changes was cautious and, in some corners, skeptical. Following a meeting between Secretary of State James Baker and Soviet Foreign Minister Schevardnadze in Wyoming in September, however, Secretary Baker suggested that the "era of containment" had perhaps come to an end, a theme subsequently advanced by President Bush when he met with Gorbachev off the coast of Malta in early December. Both superpowers appeared to recognize common interests in maintaining stability in the midst of revolutionary political changes and were even explicit about accepting each other's legitimate security interests and role in preserving European security. At the same time, the Bush administration was careful to warn against a euphoric abandonment of traditional security concerns, pointing to continuing Soviet modernization of strategic weapons and cautioning that Gorbachev might not himself endure. Thus, containment was declared to be on the verge of success, but hardly irrelevant. What was important, the administration stressed, was to make that victory concrete and irreversible while the opportunity presented itself, but also to be prudent in the process.

By the end of its first year, the Bush administration seemed to have fashioned a broad strategy to deal with these tumultuous changes. The president had taken the initiative in advancing proposals in START, the new Conventional Armed Forces in Europe (CFE) negotiations, and in chemical weapons. In the so-called Baker doctrine announced in a speech in Berlin in December, the US recognized that it had an enduring interest in reshaping a Europe that appeared bent on reshaping itself with or without the US. However much containment might be declining as a driving force in US strategy, creating a new security framework would increase commensurately as a major objective. This commitment to continued engagement in world affairs extended to other regions as well. In Asia, the administration initiated quiet overtures to China to preserve that geostrategic relationship, despite Beijing's ruthless suppression of democratic protesters in the spring. Just prior to meeting Gorbachev near Malta, President Bush also agreed to the limited employment of US forces to help put down an attempted coup against Philippine President Corazon Aquino. Closer to home, President Bush launched a pre-Christmas invasion of Panama to oust Manuel Noriega, to establish the duly elected Endara government, and to bring Noriega to the United States for trial on drug charges. Hence, the Bush administration showed itself willing to use military force to secure its interests even as it was willing to de-emphasize conflict and confrontation in its new relationship with the Soviet Union.

This willingness to employ force in Panama perhaps reflected a new sense of vulnerability on the part of the United States. Instead of fearing a growing Soviet threat, many Americans were already celebrating victory in the cold war, even if they were somewhat oblivious to the dangers that might still lie beyond. Topping a list of threats to the general welfare, Americans agreed, were nontraditional security concerns, such as drugs and a decaying environment. On a broader level, lingering controversy about whether America was a declining world power provided an incentive to assert continued American will in defending its interests and, if necessary, punishing those who openly challenged them.

Despite the Bush administration's emphasis on prudence and pragmatism, a moral dimension of American security policy persisted, most visibly in the notion that democracy was everywhere in triumph. To some, this even heralded the "end of history" in the sense that, with such an inexorable victory of popular self-determination, traditional conflicts would become a thing of the past. The greater challenge, however, has become one of redefining a moral purpose for American security policy that does not have at its center a simplistic Manichean struggle against a singular national and ideological

evil. As the decades-old confrontation between East and West dissipates, international conflict may be less clear-cut but will by no means become obsolete.

The future of US-Soviet relations is not necessarily a cause for great optimism even if the Reagan and Bush administrations go down in history as the architects of unprecedented agreements to reduce nuclear, conventional, and chemical weapons. A look at the history of relations since World War II indicates that both powers have been surprisingly introspective in dealing with each other, often responding to cliches, ideological images, and cultural stereotypes. On the other hand, as one particularly acute observer has written, "Given all the conceivable reasons for having had a major war in the past four decades—reasons that in any other age would have provided ample justification for such a war—it seems worthy of comment that there has not in fact been one; that despite the unjust and wholly artificial character of the post–World War II settlement, it has persisted for the better part of half a century."[49] As laudable as this achievement is, the US-Soviet relationship remains dangerous, despite the justifiable but fragile hope that the character of that relationship may change. If there is a formula for progress it probably resides in the superpowers' developing a more realistic, less introspective assessment of each other. Realism does not imply capitulation. Rather it implies striving for an environment in which the altogether noble and necessary containment of Soviet power can take place in an arena less dangerous for all concerned.

NOTES

1. *Papers Relating to the Foreign Relations of the United States* [hereafter *FRUS*] *1948*, Pt. 2, NSC-20/4, p. 667.

2. John L. Gaddis, "The Rise, Fall, and Future of Détente," *Foreign Affairs* 62, no. 2 (1983–84): 356.

3. George Kennan, *American Diplomacy, 1900–1950* (Chicago: University of Chicago Press, 1951), pp. 96–100.

4. Marshall Shulman, "What the Russians Really Want," *Harpers*, Apr. 1984, p. 64.

5. Daniel Yergin, *Shattered Peace* (Boston: Houghton Mifflin, 1977), p. 198.

6. Arthur M. Schlesinger, Jr., "Origins of the Cold War," *Foreign Affairs* 46, no. 1 (1967): 50.

7. Zbigniew Brzezinski, "The Future of Yalta," *Foreign Affairs* 63, no. 2 (1984–85): 279.

8. Charles Lash, "The Cold War Revisited and Revisioned," in Erik P. Hoffman and Frederic J. Fleron, eds., *The Conduct of Soviet Foreign Policy* (New York: Aldine, 1980), p. 271.

9. Adam Ulam, *Expansion and Coexistence* (New York: Praeger, 1968), p. 437.

10. Yergin, *Shattered Peace*, p. 48.

11. Ulam, *Expansion and Coexistence*, p. 364.

12. Henry Kissinger, *The White House Years* (Boston: Little, Brown, 1979), p. 135.

13. *FRUS, 1946*, 1:292–95.
14. Ibid.
15. Ibid.
16. John L. Gaddis, *Strategies of Containment* (New York: Oxford University Press, 1982), p. 47.
17. Ibid., pp. 58–59.
18. Thomas H. Etzold and John L. Gaddis, *Containment: Documents of American Policy and Strategy, 1945–50* (New York: Columbia University Press, 1976), p. 73.
19. Ibid.
20. Gregg Herken, *The Winning Weapon* (New York: Vintage Books, 1981), p. 197.
21. Etzold and Gaddis, *Containment*, p. 313.
22. *FRUS, 1950*, 1:234–92.
23. Ibid.
24. Ibid.
25. Steven L. Rearden, *The Evolution of American Strategic Doctrine: Paul H. Nitze and the Soviet Challenge*, SAIS Papers no. 4 (Boulder, Colo.: Westview Press, 1984), p. 34.
26. *FRUS, 1952–54*, "National Security Affairs," 2:379.
27. Ibid., p. 514.
28. Ibid., p. 716.
29. Stephen Ambrose, *Eisenhower, the President* (New York: Simon and Schuster, 1984), pp. 493–94.
30. David A. Rosenberg, "The Origins of Overkill: Nuclear Weapons and American Strategy, 1945–1960," *International Security* 7, no. 4 (1983): 3–71.
31. Fred Kaplan, *Wizards of Armageddon* (New York: Simon and Schuster, 1983), p. 319.
32. Lyndon Johnson, *The Vantage Point* (New York: Popular Library, 1971), p. 147.
33. Kissinger, *White House Years*, p. 119.
34. Melvin Laird, "Strong Start in a Difficult Decade: Defense Policy in the Nixon-Ford Years," *International Security* 10, no. 2 (1985): 24.
35. Kissinger, *White House Years*, p. 227.
36. Henry Kissinger, *Years of Upheaval* (Boston: Little, Brown, 1982), p. 235.
37. James Schlesinger, *Report of the Secretary of Defense to Congress, FY 1975*, p. 8.
38. Kissinger, *Years of Upheaval*, p. 256.
39. Ibid., p. 1011.
40. James Schlesinger, *Report of the Secretary of Defense to Congress, FY 1976 and Transition Budgets*, Part 2, p. 7.
41. Strobe Talbott, *Endgame* (New York: Harper and Row, 1979), p. 61.
42. Ibid., p. 5.
43. Kissinger, *Years of Upheaval*, p. 981.
44. Ronald Reagan, "Central America: Defending Our Vital Interests," *Department of State Bulletin* 83, no. 2075 (1983): 5.
45. Caspar Weinberger, "US Defense Strategy," *Foreign Affairs* 64, no. 4 (1986): 676.
46. Robert Osgood, "The Revitalization of Containment", *Foreign Affairs: America and the World* 60, no. 3 (1981): 465–502.
47. Barry Rosen and Steven Van Evera, "Defense Policy of the Reagan Administration," *International Security* 8, no. 1 (1983): 4.
48. Weinberger, "US Defense Strategy" p. 678.
49. John L. Gaddis, "The Long Peace: Elements of Stability in the Postwar International System," *International Security* 10, no. 4 (1986): 100.

CHAPTER 6

ALLIANCE COMMITMENTS AND STRATEGIES
Europe

SCHUYLER FOERSTER

On 4 April 1949, twelve nations—Belgium, Canada, Denmark, France, Iceland, Italy, Luxembourg, Netherlands, Norway, Portugal, the United Kingdom and the United States—signed the North Atlantic Treaty in Washington, thereby establishing the North Atlantic Treaty Organization (NATO). Two years later, Greece and Turkey acceded to the treaty, followed by the Federal Republic of Germany (FRG) in 1955, and, in 1982, by Spain. This peacetime commitment of the United States to the defense of Western Europe was unprecedented and remains, almost four decades later, a unique and vital element of US national security policy as well as a source of continuing controversy.

The core of the Washington treaty is in Article 5, which states "that an armed attack against one or more of them in Europe or North America shall be considered an attack against them all," and commits each NATO member to "assist the Party or Parties so attacked by taking forthwith, individually and in concert with the other Parties, such action as it deems necessary, including the use of armed force, to restore and maintain the security of the North Atlantic area." (See the appendix to this chapter for the full text of the treaty.)

In terms of defense spending, the United States currently provides roughly two-thirds of the Alliance's investment in defense, with approximately 325,000 military personnel stationed in Western Europe in support of NATO's forward defense and as a symbol of the much broader US commitment of reserves, reinforcements, maritime forces, and strategic nuclear power.[1] The magnitude and seeming permanence of this tangible commitment suggest that it is part of some "natural order"—that it has always been, that it was meant to be, and that it will endure indefinitely. Yet an understanding of the US commitment to NATO must begin with the realization that it is neither historically inevitable nor even universally viewed as legitimate at the time it was adopted; certainly the scale and duration of that commitment were not anticipated at the outset. Thus, an appreciation of NATO's origins is instructive of NATO's current problems and the future character of the US commitment.

THE ORIGINS OF NATO

In 1917, the United States entered World War I and broke the stalemate of trench warfare that had already persisted for three years. At the end of that war, President Wilson sponsored the formation of the League of Nations as a vehicle of "collective security" to ensure that the Great War would indeed be "the war to end all wars." The failure of the US ultimately to participate in the League has been, for many, the principal reason why the League itself failed. German and Italian aggression in the 1930s demonstrated the impotence of the League to deal with the challenges to peace and security in a Europe which remained traumatized by the legacy of the Great War. World War II began in 1939. Within a year, virtually all of Western and Central Europe was under Axis control, while Britain held on by virtue of a combination of geography, heroism, and largely clandestine American assistance. By the end of 1941, the Third Reich's conquest ranged from the Atlantic to the outskirts of Moscow. In the wake of the Japanese attack on Pearl Harbor, Hitler declared war on the US as well. Despite the involvement of the US in the war, however, allied forces did not land on the continent until they did so in Italy in 1943, and the massive allied landing at Normandy to roll back Nazi control in Western Europe did not occur until June 1944.

Viewed from the perspective of those in Western Europe who had twice been the victims of German aggression in the space of thirty years, the United States was clearly a necessary actor in maintaining a balance of power in Europe, yet it could not necessarily be relied on to assume that responsibility in a timely fashion. As French Premier Henri Queuille pointedly noted in 1949, the pattern of US involvement in the two world wars was not one that could acceptably be repeated: next time, the United States would "liberate a corpse."[2]

The attempt after World War II to forge a new system of collective security based on the United Nations was not in itself a sufficient guarantee of peace. The new "superpowers," on whom the system depended, were themselves ideological and political rivals. Hopes that the US and Soviet Union would collaborate on both the maintenance of peace and the restoration of Europe were quickly dashed on the rocks of the cold war.

William T. R. Fox, who coined the term "superpower" in 1944, had already warned that "to divide Europe between East and West not only would give a wholly undesirable inflexibility to the postwar political system but would in addition outrage the sensibilities of Europeans of all kinds."[3] The problem, Fox continued, was that the only foundation for a united Europe was a united Germany whose power would have to be contained more effectively than it had been in the past. Indeed, for the two years immediately following World War II, US policy in Europe was focused largely on the problems of allied occupation, in the search for formulas that could enable reunification of Germany as a self-sufficient and viable economic and political entity. In late 1945, and again in 1946 and 1947, the US offered Stalin a Four Power Treaty to guarantee German demilitarization. This initial US political commitment to European security was designed to allay Soviet (and French) fears of a future German resurgence, thereby undermining Stalin's rationale for control over Eastern Europe (especially Poland).

In the absence of Soviet agreement, however, the US, Great Britain, and France agreed in 1947 to the unification of the three Western zones of occupation in Germany, leading eventually to the creation of the Federal Republic of Germany (FRG) in 1949. The US also initiated the Marshall Plan to rebuild the war-shattered economies of Western Europe. None of the occupation zones in Germany was by itself economically viable, and rebuilding the German economy was essential to a revitalization of Western Europe's economic base. The US, for its part, was anxious to remove what was quickly becoming an indefinite commitment of American resources to support an impotent German economy. Secretary of State George Marshall complained that "the patient is sinking while the doctors deliberate."[4] The result—unification of the three Western zones of Germany and the concomitant focus on consolidating Western Europe—was the de facto partition of Germany and the accompanying recognition of a divided Europe.

These initiatives were, in a fashion, integral to the evolving US security policy known as "containment," as was the earlier enunciation of the Truman Doctrine. Announced in March 1947, the Truman Doctrine was specifically aimed at providing substantial economic assistance to Greece and Turkey ($400 million) in response to a British plea that it could no longer continue such assistance. President Truman went further, however, declaring that "it must be the policy of the United States to support free peoples who are resisting attempted aggression."[5] In a similarly Manichean vein, Under Secretary of State Dean Acheson had told congressional leaders two months before, "Only two great powers remain in the world . . . For the United States to take steps to strengthen countries threatened with Communist subversion, was not to pull British chestnuts out of the fire; it was to protect the security of the United States—it was to protect freedom itself."[6]

The notion of "containment" of Soviet expansionism later became the touchstone of US global involvement, and the strength with which this concept was articulated in the immediate postwar years was undoubtedly due in part to the clear need to overcome latent isolationist sentiments, particularly in Congress.[7] Arguments for assistance to Europe, however, could be made with reference to both internationalist and isolationist sentiments. The restoration and consolidation of Western Europe as a viable political, economic, and military force in international affairs presumed, on the one hand, the intimate involvement and *commitment* of the US *to* Western Europe's future. On the other hand, investing in the security and reconstruction of Europe in the short term was also the prerequisite for an ultimate *disengagement* of the US *from* Europe, for the eventual termination of the US role as occupier and subsidizer, and for the return of the US to its more familiar posture of "splendid isolation." Such paradoxical pressures have persisted in various forms for the past four decades.

Once the US turned its focus to the consolidation of Western Europe to contain Soviet power, French rather than Soviet cooperation became a vital ingredient of US policy. The French were much more concerned about a future German threat than they were about any existing or projected Soviet threat. French cooperation required, therefore, a British and ultimately an American security guarantee. In 1947, Britain and France signed the Treaty of Dunkirk—a pointed reference to the evacuation of British forces in 1940 prior to the fall of France—which committed each to mutual military assistance in the event of future German aggression. In 1948, Britain, France, and the Benelux countries (Belgium, the Netherlands, and Luxembourg) signed the Brussels Treaty, establishing the "Western Union." (When the FRG and Italy acceded to the Brussels Treaty in October 1954, the "Western Union" became the "Western European Union" or WEU.) Citing Article 51 of the UN Charter, the Brussels Treaty was modeled after the 1947 Rio Treaty which the US had already signed and ratified to establish an Inter-American regional security agreement. The Brussels Pact specified no particular threat (references to future German aggression were excised after the FRG acceded to the treaty in 1954) and committed signatories to provide "all the military and other aid and assistance in their power" in the event any signatory became "the object of an armed attack in Europe."

British Foreign Secretary Ernest Bevin had conceived the idea of the Western Union in late 1947 as a precursor to an explicit American commitment to European security, a commitment which, as Bevin told Marshall, was necessary to allay French doubts about rebuilding a strong Western Germany and to provide a secure base for the eventual integration of Western Europe.[8] In parallel to the Brussels Treaty, Bevin proposed an "Atlantic Approaches Pact" to include the US and Canada. Bevin's concept bore fruit in 1949 with the signing of the North Atlantic Treaty.

In many respects, NATO represents a "transatlantic bargain"[9] between the US—as the principal guarantor of West European security—and the European allies whose territory and way of life were most directly threatened. From the beginning the US urged concerted European defense efforts as necessary to justify the American commitment. Europeans, on the other hand, demanded continued demonstration of the American commitment as necessary to stimulate European contributions to collective defense. The Alliance's chronic debates about "burden sharing" over the years have reflected this theme, already evident in the struggle to craft the initial bargain.

In making his proposal for a parallel Atlantic Approaches Pact, Bevin argued that, in the absence of formal assurances by the US, the Western Union might not materialize. Many in the US—including George Kennan, head of the State Department's Policy Planning Staff and author of "containment"—argued, however, that US participation in European deliberations on a regional security pact was inappropriate.[10] US economic assistance through the Marshall Plan, plus general political backing, would be sufficient. Hesitation within the Truman administration also reflected concerns that congressional isolationist sentiments might ultimately block any US commitment, dashing expectations in Europe such as had been the case with Wilson's League of Nations after World War I. Soliciting congressional approval for military assistance might undermine the bipartisan consensus that had developed in support of the formidable economic commitments represented by the Truman Doctrine and the Marshall Plan.

In the midst of this debate a series of cold war crises stimulated greater US willingness to associate itself with an Atlantic Approaches Pact. The Soviet takeover of Czechoslovakia in February 1948 highlighted the fragility of the European security environment. Following ominous warnings from General Lucius Clay, US military governor in Germany, the CIA could only assess that war "was not probable in sixty days."[11] Later in March, the Soviets began to pressure Norway about a nonaggression pact involving Soviet rights of access, similar to the treaty which the USSR had just signed with Finland. Combined with anxieties about a possible Communist electoral victory in Italy, these pressures stimulated discussions—later reinforced by the start of the Berlin Blockade in June—not only about an Atlantic Approaches Pact that included the Northern Flank but also about a security agreement to encompass the Mediterranean.

On the day the Brussels Treaty was signed, March 17, Truman told a joint session of Congress that the treaty deserved the "full support" of the US, "by appropriate means." The critical element of bipartisan congressional support came in the form of the Vandenberg Resolution, passed by the Senate (64 to 3) on June 11. Arthur Vandenberg, chairman of the Senate Foreign Relations Committee—a Republican with strong isolationist credentials who had supported administration attempts to bolster Western Europe—sponsored a Senate resolution that encouraged "association of the US, by constitutional process, with such regional and other collective arrangements as are based on continuous and effective self-help and mutual aid."[12]

The Vandenberg Resolution thus invited the administration to pursue a security relationship with Western Europe, although within important bounds. In fact, treaty consultations had already begun in March among the US, Great Britain and Canada. What remained, however, was the critical issue of articulating the nature and scope of that "transatlantic bargain" to determine the basis on which this unprecedented peacetime US commitment would be implemented.

DEFINING THE NATURE OF THE AMERICAN COMMITMENT

A key issue for the treaty drafters was how to describe the specific allied commitments. Europeans—most of all the French—preferred an explicit and virtually automatic guarantee of military assistance in the event of attack, as was the case in the Dunkirk and Brussels agreements. The US, on the other hand, strongly resisted such an obligation, arguing that it was unnecessary. Since the very existence of the treaty would represent an unprecedented statement of a US peacetime commitment to Western Europe, it would already serve as a deterrent to aggression. Thus, the US, as a signatory, was committed (in Article 5) only to take "such action as it deems necessary, including the use of armed force," while Article 11 further qualified that commitment to acknowledge Congress's independent right to declare war by providing for implementation of the treaty "in accordance with their respective constitutional processes."

For the US, the treaty represented a fundamentally *political* commitment that did not necessarily require a long-term *physical* commitment. Article 3 of the treaty bore the imprint of Senator Vandenberg in its reference to "continuous self-help and mutual aid" as the basis for achieving the treaty's objectives. Vandenberg himself conceded that "a man can vote for this treaty and not vote a nickel to implement it."[13] During treaty ratification hearings in the Senate, Iowa Republican Senator Bourke Hickenlooper asked Secretary of State Acheson whether the US "was going to be expected to send substantial numbers of troops over there as a more or less permanent contribution to the development of these countries' capacity to resist." Acheson replied, "The answer to that question, Senator, is a clear and absolute 'No.' "[14]

As the US began to assume the responsibility of *global* security in the atomic age, Western Europe was clearly expected to contribute substantially to its own *regional* defense. Within hours of the treaty's ratification by the Senate in July 1949, the administration submitted its request for a $1.4 billion foreign military aid authorization. Vandenberg insisted that, as a precondition, NATO had first to implement the provisions of Article 9 which called for the establishment of a council and a defense committee and then to develop a "strategic concept" for its defense. The new North Atlantic Council met on 17 September 1949, and established its skeletal structure to fulfill Article 9. Six days later, the Soviet Union announced that it had detonated an atomic bomb, effectively terminating congressional debate on the Mutual Defense Assistance Program which became law on October 6. In January 1950, Truman approved NATO's initial "strategic concept," reflecting a vague division of labor in which the US was principally responsible for strategic bombing, and European allies were to provide "the hard core" of ground forces. On that basis, military assistance flowed to Europe.[15]

Ultimately, military assistance was not enough. Following the Soviet entry into the atomic age, and Truman's January 1950 decision to proceed with the "super" or hydrogen fusion weapon, the administration began to reassess, in NSC-68, its national security policy. Concluding that the Soviet Union would, by 1954, be able to "seriously damage" the US in an atomic attack, NSC-68 declared a preference for "a war policy restricted to retaliation against

prior use by an enemy" of atomic weapons.[16] In contrast, however, the existing US defense posture was such that General Omar Bradley, chairman of the Joint Chiefs of Staff (JCS), had told Congress that the atomic bomb was "our principal initial weapon in any war."[17] To make containment "more than a policy of bluff," NSC-68 called for "superior aggregate military strength," including a major mobilization of US ground forces both as a strategic reserve and for forward deployment in Europe and elsewhere.

Although Truman did not formally approve NSC-68 until 30 September 1950, the sudden onset of the Korean War in June had again injected urgency into US security policy deliberations. On 9 September, Truman announced his decision to send four additional US combat divisions to Europe (there had only been one US division there, largely engaged in occupation functions), noting that "a basic element in the implementation of the decision is the degree to which our friends match our action in this regard."[18] Implicit in this general reference to the need for European burden sharing was the view that, for Western Europe to be credibly defended, the Alliance had to be expanded further to incorporate a rearmed West Germany. In effect, the four additional US combat divisions represented the ante to entice Europe—particularly France—to accept the unacceptable. For the French, NATO and the American security guarantee had offered a way to keep Germany disarmed and militarily impotent. For the US—and to some extent the British who likewise preferred to avoid a permanent commitment of ground forces to the European continent—German rearmament was the logical outgrowth of a commitment to the defense of continental Western Europe.

GERMAN REARMAMENT

Although Bevin's initial concept had been a "North Atlantic Approaches" pact, the practical requirements of securing US access to Europe had already dictated an expanded scope for Alliance membership. Iceland's geostrategic location required its adherence, despite the fact that Iceland possessed no armed forces. Likewise, the importance of the Azores and Greenland as refueling bases necessitated the inclusion of Portugal and Denmark. Norway was important for its geostrategic location on the Northern Flank (as was Sweden which, however, insisted on retaining its historical position as an armed neutral). Moreover, the inclusion of Norway was critical as a way of rebuffing Soviet attempts to entice Norway into an arrangement similar to that of Finland. France, for its part, insisted on the inclusion of Algeria (then considered a part of "Metropolitan France") and Italy, and the latter offered a vital base for the

US Sixth Fleet in the Mediterranean. Greece and Turkey, the original beneficiaries of the Truman Doctrine, filled out the "Mediterranean contingent" in 1951, prompted in large measure by the Korean War crisis and the hardening of the cold war.

The issue of German membership, however, posed special problems. As early as the initial US-UK-Canada working-level consultations in 1948, delegates had already envisioned the possible inclusion of Germany, Austria, and Spain, although the State Department's draft position paper noted that "this objective . . . should not be publicly disclosed."[19] The military logic behind German participation in NATO's defense coalition was clear: the Brussels Treaty Military Commission had already committed itself "to fight as far east in Germany as possible . . . that sufficient time for American military power to intervene decisively can be assured."[20] Without German participation, however, there were hardly sufficient forces to make such a forward defense realistic, and a substantial portion of the French army was involved in trying to maintain French colonial presence in North Africa and Southeast Asia. Yet, French ratification of the North Atlantic Treaty reflected the clear premise that Germany would remain disarmed and excluded from the Alliance.[21]

There were also those, like George Kennan, who objected on broader political grounds to a process of expanding the Alliance to incorporate virtually all of Western Europe. This, Kennan wrote in 1948, "would amount to a final militarization of the present line through Europe . . . Such a development would be particularly unfortunate, for it would create a situation in which no alteration or obliteration of that line could take place without having an accentuated military sigificance."[22] Kennan's concern reflected the larger dilemma that had plagued US security policy since the end of World War II. The logic of the cold war had demanded the consolidation of Western Europe into a strong and cohesive political and military alliance to contain Soviet power, a process that recognized that Europe (and Germany) had already been divided in fact. By reinforcing that division, it complicated any longer-term desires to transform the postwar international order to enable a mutual disengagement of the superpowers from Europe.

Such concerns about the longer-term possibilities for a different international constellation in Europe conflicted with other concerns about the future of Germany outside a Western security framework. As Secretary of State Dean Acheson recalled:

The probability was that we would lose Germany politically and militarily, without hope of getting it back

if we did not find means for that country to fight in event of an emergency . . . If there were to be any defense at all, it had to be based on a forward strategy, Germany's role must not be secondary but primary—not only through military formations, but through emotional and political involvement.[23]

Thus, barring an East-West rapprochement which might make the US security guarantee superfluous, or the creation of a unified Western Europe (including West Germany) which might make the US military presence in Europe unnecessary, German rearmament was in many respects the price for securing the American commitment to NATO's defense. The Korean War had merely provided a catalyst for acting on a logic which already existed.

In September 1950, Secretary of State Acheson declared that US agreement to provide an American Supreme Allied Commander, Europe (SACEUR) to an integrated NATO military structure and to increase both financial assistance and the number of US troops in Europe was contingent on allied agreement to West German membership in NATO and the mobilization of ten German divisions under NATO command. France dissented and offered the Pleven Plan as a counterproposal. The plan would allow establishment of small German "combat teams" to be integrated into a multinational European army, subordinate to a European Minister of Defense. Since the plan also envisioned a supranational European Council of Ministers, a European Parliament, and a common European defense budget, it, in effect, would postpone German rearmament until the political integration of Western Europe was achieved.[24]

By the time of the NATO ministerial meeting in Lisbon in February 1952, the elements of a workable compromise seemed to have been established. The US had already appointed General Dwight Eisenhower as the first SACEUR and deployed four additional divisions to Europe. Senate approval of these actions in April 1951, however, included a nonbinding "sense of the Senate" resolution that "no ground troops in addition to such divisions should be sent to Europe in implementation of Article 3 of the North Atlantic Treaty without further Congressional approval."[25] Against this backdrop of US pressure for a substantial European defense contribution, NATO ministers approved in Lisbon the elements of a European Defense Community (EDC) Treaty and agreed to a set of force goals for NATO's conventional defense—50 divisions by the end of 1952, and 96 allied divisions, with over one-third fully combat ready, to be available by 1954. German divisions, integrated into a European army, were eventually to comprise 12 of that total active number, with German manpower available for NATO defense "under strong provisional controls" while the permanent system of integrating European defense was being developed.[26]

The EDC Treaty was signed in May 1952, and the US and the UK—neither of which was included in the integrated European defense force—promptly ratified it. The FRG ratified it in March 1953, following an acrimonious debate in Bonn in which the broader "German question" was the major issue. FRG Chancellor Adenauer argued that West German integration into Western European defense was the necessary price for allied support for eventual German reunification and, more immediately, for allied agreement to full sovereignty for the FRG and an end to occupation. The Social Democratic (SPD) opposition, however, argued that integration with the West hindered reunification (which required Soviet approval as well).[27] Ironically, it was the French, in August 1954, who killed the EDC by failing to ratify it. In the wake of the French defeat in Indochina, a weak coalition government was unable to overcome objections, particularly from the French right, that, in subordinating the German military to an integrated European defense force, the French army would also be subordinated. Fearing that the FRG would ultimately dominate the EDC and doubting that the US presence in Europe would endure indefinitely, the French chose maximum sovereignty for itself over European integration.

By October 1954, a final compromise was reached, by which the allies invited the FRG to join NATO and (with Italy) the WEU, agreed to German rearmament with 12 divisions under NATO command, and ended the occupation of Germany. Inclusion of a rearmed Germany in NATO marked a watershed development for the Alliance and the role of the US in Europe. A German contribution to Western defense had become a key element in justifying congressional support for the US military commitment to Europe. The EDC had been a means to that end, and the new secretary of state, John Foster Dulles, had threatened an "agonizing reappraisal" of US policy if the EDC failed. The resulting compromise did not, in the end, require a reappraisal of the US commitment to Europe. Nevertheless, a different form of reappraisal had already been initiated in the Eisenhower administration that fundamentally altered the form of that American commitment. Having succeeded in achieving German rearmament in NATO, the US was also stressing nuclear weapons as the basis for NATO defense. While a nuclear emphasis justified a less costly US investment in NATO's defense, it ultimately increased the potential risk to the US and the political costs associated with further modifying the form of its commitment.

THE ADVENT OF NATO'S NUCLEAR STRATEGY

During the period in which the NATO treaty was being developed, the US possessed a monopoly of atomic weapons. Nonetheless, the US tended to view its commitment to NATO less in military terms—as the bearer of an atomic retaliatory sword—but more in political terms. The very existence of a US guarantee was presumed to serve as a deterrent to Soviet aggression, since it ostensibly removed any doubt that an attack on Western Europe would engage the full military and industrial power that the US had twice before demonstrated on behalf of its European allies.

That more sanguine US approach to European security evaporated once the atomic monopoly disappeared. As expressed in NSC-68, the US felt itself potentially vulnerable to a Soviet atomic attack which could render "unacceptable damage" to the US. Out of this sense of vulnerability came a desire to restore US military power to a point where atomic weapons would be "restricted to retaliation against prior use by an enemy."[28] As expressed in NSC-68, the US did not envision atomic weapons as an effective instrument of warfighting since they could not prevent a Soviet occupation of Western Europe or inflict sufficient damage on the Soviet Union to cause the Kremlin to sue for peace.[29] Hence, pressure for German rearmament and the articulation, at Lisbon, of substantial force goals for NATO defense reflected a US desire to mount a substantial conventional defense of Europe. To be sure, Europeans would have to bear the burden of initial forward defense, since the US peacetime military presence in Europe was limited by congressional mandate to five divisions. Nonetheless, by 1953, the US had another nine active divisions in its strategic reserve, and total US ground forces had nearly tripled in size from 1950.[30]

Driven largely by fiscal constraints, however, the newly elected Eisenhower administration revised the emphasis on Truman's defense policy. In an article in *Life* magazine written before the 1952 presidential election, Dulles warned against a persistently unbalanced budget and vain attempts to match the quantitative strength of Communist armed forces which "would mean real strength nowhere and bankruptcy everywhere."[31] The Eisenhower administration's "New Look" policy, articulated in October 1953 in NSC-162/2, called for a reduction both in US defense expenditures and in military assistance to allies, declaring that "the major deterrent to aggression against Western Europe is the manifest determination of the US to use its atomic capability and massive retaliatory striking power if the area is attacked."[32] Subsequently, Dulles presented this new policy of "massive retaliation" to the public in terms almost identical to those he had used in his *Life* article almost two years before: "The basic decision was to depend primarily upon a great capacity to retaliate, instantly, by means and at places of our own choosing."[33]

Eisenhower's "New Look" occurred in the context of two technological developments which invited greater reliance on threats of massive retaliation. In November 1952, the US successfully tested its first thermonuclear device (the Soviets followed in August 1953), which offered a quantum increase in destructive capacity over atomic weapons. Then, in April 1953, the US successfully tested an atomic device for use on the battlefield and began deployment of atomic artillery in October 1953. Besides these technological developments, the administration assessed, in NSC-162/2, that a deliberate Soviet attack on the US and NATO was not likely,[34] a judgment reinforced by the death of Stalin in March 1953 and the appearance of a general "thaw" in East-West relations that continued through 1955 and the Geneva summit.

In December 1954, NATO approved MC-48 which authorized NATO commanders to plan on the basis of the availability and probable use of nuclear weapons. While it did not exclude the possibility that nuclear weapons might be used only on the battlefield, its central premise was that tactical nuclear weapons would pose to the Soviet Union the virtual certainty that aggression would lead to massive strikes by the substantially superior US strategic arsenal.[35] The Lisbon Force Goals had been supplanted by a judgment that battlefield nuclear weapons would substantially reduce the number of divisions necessary for forward defense. In 1956, NATO adopted a five-year plan (MC-70) to deploy thirty active frontline divisions equipped with tactical nuclear weapons under US control. A year later, NATO endorsed MC-14/2—its fundamental "strategic concept"—which further articulated NATO's reliance on nuclear weapons both as a strategic deterrent and as its principal means of defense on the battlefield.[36]

There were complementary allied developments as well. Having become a nuclear power in its own right, Britain issued its 1957 White Paper on defense, which, while emphasizing ballistic missile technology, announced the end to conscript military service and a 45 percent cut in the overall size of its armed forces. By 1957, France likewise had embarked on its nuclear weapons program, a process which was accelerated a year later by de Gaulle but which already felt the influence of those like Pierre Gallois and Andre Beaufre who argued that France could not ultimately count on the US nuclear guarantee.[37]

These developments had a profound impact on the Germans who were prohibited from becoming a nuclear power. No sooner had the FRG weathered its debate over the EDC, rearmament, and entry into NATO, than it was confronted with a debate on whether the FRG would allow nuclear weapons to be stationed with the Federal German Armed Forces (*Bundeswehr*) albeit under US control. In June 1955, NATO's exercise CARTE BLANCHE simulated a limited employment of tactical nuclear weapons, with 268 of the total 335 detonations on German soil (no Soviet detonations were considered). The publicized results of these hypothetical attacks included over 5 million German casualties, undermining Chancellor Adenauer's claims that membership in NATO would prevent the FRG from being the battlefield in a future war.[38] A year later, press reports of a plan by Admiral Radford, chairman of the US Joint Chiefs of Staff, to cut up to half of US forces in Europe similarly caught Adenauer off guard in the midst of a vociferous German debate on conscription.[39] The image of German troops serving as the Alliance's "cannon fodder" on an atomic battlefield assumed a powerful force in the German debate, forcing the Eisenhower administration to make special efforts to reassure the Germans about the strength of the US commitment, not only to German security but also to other German interests such as reunification.

In part because of the need to reassure the West Germans and to keep them firmly anchored in the Western Alliance, the Eisenhower administration supported Adenauer's government in rejecting the proliferation of proposals beginning in 1956 for the "disengagement" of US and Soviet forces from Europe and the creation of "nuclear free zones" in Central Europe. While some of these proposals were designed to enable the reunification of Germany, their principal focus was the creation of a new European security framework which was less reliant on opposing blocs confronting each other across the inter-German border. Proposals such as the Polish Rapacki Plan and those from within the British Labour party and FRG Social Democratic party differed in many respects. They shared, however, a hope that, by creating a "neutral belt" in Europe, the "final militarization of the present line through Europe" to which Kennan had referred a decade before could be overcome.[40] Kennan himself, in his BBC Reith Lectures in late 1957, argued that the continued division of Germany, with an indefinite American presence, "expects too much, and for too long a time, of the United States, which is not a European power."[41]

An essential element of such "disengagement" proposals, however, was that the United States, with its strategic nuclear deterrent, could help guarantee the security of this new European order. Within the FRG, however, the continuing presence of US forces constituted the tangible link between German security and the US strategic guarantee. Likewise, tactical nuclear weapons in the FRG, integrated into the *Bundeswehr* force structure as well as with US forces, were to insure that the nuclear threshold was low, so that an aggressor would have to anticipate a NATO nuclear response and, eventually, the employment of US strategic nuclear forces. Without that tangible evidence of "coupling," the credibility of the US extended deterrent guarantee was not self-evident to its beneficiaries. In the wake of the Soviet launch of *Sputnik* in October 1957, symbolizing the imminent vulnerability of the US homeland to Soviet strategic missile attack, the need for such tangible demonstration of the US commitment increased all the more.

In addition to the persistent fear that the US nuclear guarantee might not be entirely credible, several other questions plagued the evolution of NATO's nuclear strategy in the 1950s. First, if the US strategic nuclear forces constituted the principal deterrent, what was the purpose of battlefield nuclear weapons and the 30 allied divisions which were organized to deploy them? Second, how would nuclear weapons be employed, if at all, for defense against local aggression, particularly in the case of Berlin? Finally, how should the NATO machinery be designed to manage its nuclear power both to enhance deterrence and to ensure crisis stability? These questions dominated the wide-ranging defense debate that began in the late 1950s, and attempts to grapple with them shaped the evolution of NATO's strategy to its eventual incarnation known as "flexible response."

PRELUDE TO "FLEXIBLE RESPONSE": THE IMPETUS FOR A STRATEGY OF LIMITED WAR

Although NATO did not formally adopt (in MC-14/3) its current "strategic concept" of flexible response until 1967, its roots go back to the 1950s. The impetus for a new NATO strategy came from the US—in particular, the Kennedy administration—but its final form reflected a compromise between competing perspectives on the role of nuclear weapons both as a deterrent and as a means of defense. In essence, the US sought to raise the nuclear threshold so that the Alliance would not automatically find itself having to resort to nuclear weapons in the face of a Soviet conventional attack in Europe. Many West Europeans, on the other hand, tended to fear the worst: that the United States was searching for a means of disengaging from its commitment to European security and of ensuring that conflict would be confined to the European theater. Even in the late 1950s, a consensus was emerging on both sides of the Atlantic that massive retaliation was not an enduringly viable foundation for NATO strategy. Nonetheless, the differing geostrategic and political perspectives inherent to an alliance such as NATO complicated the search for an alternative.

The search for alternatives to massive retaliation can be divided into two basic schools of thought. One argued for an essentially "symmetrical" defense position, so that, in the event of a Warsaw Pact attack using only conventional weapons, NATO would respond symmetrically, that is, at the conventional level. Likewise, if the Warsaw Pact attacked with tactical nuclear weapons, NATO would respond in kind. Only if the US were faced with general strategic attack by the Soviet Union would the retaliatory force of the US strategic arsenal come into play.

This "symmetrical" strategy became the preferred alternative of those in the Kennedy administration who, after 1961, pushed the Alliance in the direction of a new strategy called "flexible response." Many of its advocates had advised the Truman administration and were critical of the Eisenhower administration's emphasis on nuclear weapons. The other alternative to massive retaliation, focusing greater attention on the role of battlefield nuclear weapons in Europe, stressed the need for a graduated *nuclear* response to Warsaw Pact aggression. In 1957, for example, the US SACEUR, General Lauris Norstad, developed the concept of a NATO "shield"—using the 30 divisions envisioned by MC-70 equipped with tactical nuclear weapons—which would "give pause" to Soviet leaders who might be inclined to initiate aggression on a limited scale:

In an era of nuclear plenty and of delivery means adequate in number and in effectiveness, the NATO Shield provides us with an option more useful than the simple choice between all or nothing. Should we fail to maintain reasonable Shield strength on the NATO frontier, then massive retaliation could be our only response to an aggression, regardless of its nature. There is real danger that inability to deal decisively with limited or local attacks could lead to our piecemeal defeat or bring on general war. If, on the other hand, we have means to meet less-than-ultimate threats with a decisive, but less-than-ultimate response, the very possession of this ability would discourage the threat, and would thereby provide us with essential political and military maneuverability.[42]

The notion of graduated nuclear response stemmed from a growing conviction that the threat of strategic massive retaliation made general war the inevitable outcome of any aggression, and from a parallel recognition that threats to the West—in Europe and elsewhere—were more likely to be regional and limited. For US forces in Europe simply to be a "trip wire" for retaliation by the US strategic "sword" was unacceptable because it invited limited aggression against the West for which there was no appropriate response. The US strategic nuclear capability served as a deterrent to general war, but it was otherwise unusable as an instrument of defense. For the US to be able to deter limited war, therefore, the US needed a limited war capability that, in Europe, was to be provided by NATO's shield of tactical nuclear weapons.

Dulles himself, the author of massive retaliation, urged this view in late 1957, citing technological advances in nuclear weapons design which would allow blast and radiation to be "confined substantially to predetermined targets":

The United States has not been content to rely upon a peace which could be preserved only by a capacity to destroy vast segments of the human race. Such a concept is acceptable only as a last alternative . . . In the future it may thus be feasible to place less reliance upon deterrence of vast retaliatory power. It may be possible to defend countries by nuclear weapons so mobile, or so placed, as to make military invasion with conventional forces a hazardous attempt.[43]

This view reflected as well a substantial body of argument in the US—as old as the doctrine of massive retaliation—that feared that the US might itself be deterred from initiating general nuclear war in defense of its interests and felt that the US should complement its deterrent power with "usable" weapons.[44] The essential element of the argument was that the Soviet Union, rather than the US, should bear the burden of choosing between nuclear war and defeat on the battlefield. Only with such unacceptable choices could deterrence in a dangerous nuclear world be assured.

Arguments in the US for a less exclusive reliance on strategic retaliatory capability assumed greater public force in the wake of the Soviet launch of *Sputnik* in October 1957. A month later, the Gaither Commission projected an American strategic vulnerability based on an anticipated Soviet force of one hundred intercontinental ballistic missiles (ICBMs) by 1959, each with megaton-sized warheads. The commission urged greater emphasis on developing a survivable retaliatory capability and increased forces for limited war. The US Atlas ICBM and Polaris submarine-launched ballistic missile (SLBM) programs were still in the research and development stage. The Eisenhower administration rejected proposals for a buildup of conventional forces and a major investment in civil defense, but did respond with proposals to disperse the US nuclear retaliatory capability. At the December 1957 NATO conference, Dulles offered to station Thor and Jupiter intermediate-

range ballistic missiles (IRBMs) under US control on allied territory. The UK agreed, followed by Italy and Turkey in 1959. The French refused, and Adenauer asked that the FRG not be offered any such missiles.

Fears in Europe over nuclear weapons took on two seemingly contradictory facets. On the one hand, as one German editorial noted, *Sputnik* opened "the justifiable question whether the Americans, in their susceptibility to Soviet missile attacks, will be ready to continue threatening American atomic reprisals."[45] At the same time, there was the converse fear that, while the US would not use its strategic nuclear deterrent to repel Soviet attack, the US could initiate nuclear conflict confined to Europe while remaining a sanctuary by not resorting to strategic nuclear weapons. Britain, for example, having already accepted the possibility of basing IRBMs in March 1957, could justify its acceptance of the American offer citing both fears. Strengthening its nuclear relationship with the US, Britain could proceed to develop an economical national nuclear deterrent with US help which reduced Britain's reliance on the US nuclear guarantee. (Later, in the 1962 Nassau Agreement, the US agreed to provide the UK with Polaris missiles, minus warheads, for British submarines.) At the same time, as Prime Minister MacMillan pointed out, the UK would have a veto over the use of US nuclear weapons in Britain, by virtue of its "dual-key" control arrangement in which both countries had to agree to launch the missiles.[46]

French refusal to accept IRBMs, on the other hand, reflected its frustration over US restrictions on sharing nuclear technology and US unwillingness to assist France in the development of its national nuclear deterrent. The Eisenhower administration had succeeded in persuading Congress to amend the McMahon Act to allow Britain greater access to nuclear weapons information, reflecting an Anglo-US "special relationship" in the nuclear weapons field that dated back to British participation in the Manhattan Project during World War II. France therefore felt discriminated against in its efforts to restore its national status to one of the great powers. In June 1958, de Gaulle became president of the new Fifth French Republic, accelerated France's independent nuclear weapons program, and subsequently ordered the removal of all nuclear weapons on French soil which were not under French control—in effect all nuclear weapons, including those to be delivered by tactical aircraft. Lacking the option of developing its own nuclear deterrent, the FRG vigorously stressed the need for the US to demonstrate its willingness to use nuclear weapons in the forward defense of Germany and viewed the nuclearization of the *Bundeswehr* as the necessary price for maintaining the US commitment. Yet, the Germans faced the prospect that its principal allies—the US, the UK, and France—were pursuing nuclear policies designed increasingly to protect their respective national self-interests. In war, Germany's allies might be willing to see the battlefield confined to Germany. In peace, with growing public pressures in the West for some kind of disarmament initiative, Germany's allies might be willing to engage the Soviet Union in arms control negotiations which limited the risk of war but which, in effect, ratified the division of Germany.

The growing anxieties over the viability of NATO's nuclear strategy erupted into crisis during the series of disputes involving Berlin, beginning with Khrushchev's ultimatum of November 1958 in which he threatened to terminate Four Power control over Berlin and turn all responsibility for Berlin and access thereto over to East Germany, unless the West agreed to end its occupation and make West Berlin a "free city." While NATO formally agreed to defend allied rights in and access to Berlin, there was substantial pressure, both publicly and privately, to make concessions to avert conflict. In the US, Senate Deputy Majority Leader Mike Mansfield and Foreign Relations Committee Chairman J. William Fullbright joined Kennan in renewing proposals for a "disengagement" of US and Soviet forces in Europe in exchange for Soviet acceptance of a stable situation in Berlin.[47]

Even though the Eisenhower administration held firm on Berlin, and Khrushchev's successive deadlines passed without incident, the Berlin crisis heightened the debate about NATO's strategy to deal with such a conflict. Eisenhower had refused to mobilize reserves in the crisis and declared in a press conference that he would not fight a ground war in Europe over Berlin. Furthermore, when questioned if he would therefore use nuclear weapons, Eisenhower retorted that nuclear war was "self-defeating."[48] Subsequently, Christian Herter, in his nomination hearings to become Secretary of State after Dulles's death, told a Senate committee that he could not conceive that any president would initiate a nuclear war unless the US was itself threatened with devastation.[49]

By the time John Kennedy entered the White House, the US posture for the defense of NATO had evolved to a heavy emphasis on the use of tactical nuclear weapons, with IRBMs deployed in three allied countries while the Strategic Air Command represented the strategic nuclear deterrent. Eisenhower's reluctance to provoke conflict over Berlin was understandable. The IRBMs in the UK, Italy, and Turkey were vulnerable targets themselves, unsuitable for limited war but likely to be used early in a conflict lest they be destroyed in a preemptive strike. NATO's 30-division shield was, at best, a hair trigger for nuclear use. In fact, SACEUR only had sixteen active divisions available in theater by 1961, and those forces were structured not for conventional defense but for nuclear release.[50] Moreover, nuclear control arrangements were inadequate. In inspecting those control arrangements in 1960, the Joint Congressional Committee on Atomic Energy had found, in the words of one observer, "fighter aircraft with nuclear bombs on the edge of runways with German pilots inside the cockpits and starter plugs inserted. The embodiment of control was an American officer somewhere in the vicinity with a revolver."[51] In bequeathing this legacy to its successors, the Eisenhower administration had also begun the development of the US strategic missile force, and the first Polaris SLBM became operational in December 1960. In marking that event, the outgoing administration also put forward the idea of a NATO Multilateral Force (MLF), in which NATO allies would participate in the shared custody and control of sea-based MRBMs to complement the US strategic deterrent. That idea had its roots even before *Sputnik*, when Dulles had offered to examine possibilities for developing a "NATO stockpile of weapons" so that allies would not "be in a position of supplicants . . . for the use of atomic weapons."[52]

In an attempt to accommodate allied desires that they have "a finger on the trigger as well as the safety catch" of NATO's nuclear forces, General Norstad had proposed the development of a NATO land-mobile MRBM force under SACEUR control. Norstad's proposal never got beyond the consultation stage, and the fundamental problem of how nuclear control would be shared in such an arrangement was never resolved.[53] Herter's MLF proposal represented an extension of the Norstad proposal, but it also sought to delay any real progress on creating a shared nuclear role for NATO. Its principal purpose was political: to divert allied desires for independent nuclear arsenals by conveying a modicum of control over nuclear weapons committed to alliance defense. In this respect, the MLF was designed to allay German as well as French concerns. But the US MLF proposal also required that allies be forthcoming on conventional force improvements, foreshadowing a theme that was to be central to the Kennedy administration's approach to NATO defense.

THE MCNAMARA SHIFT: NONNUCLEAR OPTIONS AND NUCLEAR CONTROL

NATO's uncertain attempt in the 1950s to fashion a strategy of forward defense based on graduated nuclear response represented, on the one hand, a growing uneasiness over the credibility of threats of massive retaliation and, on the other hand, the perception that an effective conventional defense of Europe was neither economically sustainable nor politically desirable. Already evident were the persistent differences in geostrategic perspectives on either side of the Atlantic. The United States, facing the prospect if not the reality of strategic vulnerability to nuclear weapons, was searching for ways to deal with challenges to its European interests which did not automatically risk its own survival. European allies, by and large, viewed the US as attempting to restrict its commitment to European defense. The more the US focused on the threats of limited conflict—threats that did not challenge fundamental US survival—the more the allies questioned the US commitment to European defense.

Ironically, steps the US took to bolster the credibility of its strategic guarantee involved an increasing emphasis on local forces to deal with localized threats. If one accepted that the US strategic nuclear threat was insufficient by itself to deter likely Soviet challenges to European security, the remedy of the late 1950s had been to move more in the direction of posing a limited nuclear response to aggression. This placed the burden of defense—and focused the risk more—on the allies, particularly those like the FRG which were most directly vulnerable to attack and which did not possess the option of developing a national nuclear deterrent. The result was increased anxiety among those, again like the FRG, which were both most dependent and most vulnerable.

To the extent that tactical nuclear weapons provided a means of defending Central Europe from an attack, they reinforced the European fear that the US intended to keep the conflict limited to Europe as long as the Soviet Union did not threaten the US directly with nuclear attack. To the extent that tactical nuclear weapons were not a vehicle for limited war, but a more visible and tangible insurance that war would escalate sooner rather than later to general nuclear war, they were arguably an effective deterrent. This, however, posed the question of whether the US would willingly trigger such escalation and risk general war. As long as nuclear use seemed virtually automatic—since control procedures were fragile and forces were organized principally to effect nuclear release—the question of credibility was more theoretical than real. Yet, the reality was an unstable nuclear posture that was acutely dangerous, especially when viewed in the climate of crisis that dominated the late 1950s and early 1960s.

Kennedy's defense secretary, Robert McNamara, viewed this state of affairs as unacceptable. In language that was reminiscent of the "symmetrical" strategy reflected in NSC-68 and that subsequently underpinned his later advocacy of "no first use" of nuclear weapons in the 1980s, McNamara told a Senate Subcommittee in April 1961, "the decision to employ tactical nuclear weapons in limited conflicts should not be forced on us simply because we have no other means to cope with them."[54] While the Kennedy administration stressed the development of an invulnerable strategic triad of silo-based ICBMs, SLBMs, and strategic bombers as the basis of a more secure and more reliable US strategic guarantee, it likewise sought to disentangle NATO's forward defense from its predominantly nuclear character. McNamara pressed the allies to meet their force goals, urged the FRG to accelerate its plans to bring the *Bundeswehr* up to its full strength of 12 divisions, and restructured US forces in Europe to make them more oriented to fighting a conventional rather than nuclear forward defense.

Whereas NATO strategy had moved in the late 1950s to encompass at least the possibility of limited nuclear war, McNamara wanted the option of limiting conflict to the conventional level in the event of a Warsaw Pact conventional attack. While he did not preclude the use of tactical nuclear weapons in limited war situations, McNamara did not share Dulles's earlier faith in the ability to limit conflict and collateral damage once the nuclear threshold was crossed. As he cautioned the House Armed Services Committee in 1963,

Nuclear weapons, even in the lower kiloton ranges, are extremely destructive devices, and hardly the preferred weapons to defend such heavily populated areas as Europe. Furthermore, while it does not necessarily follow that the use of tactical nuclear weapons must inevitably escalate into global nuclear war, it does present a *very definite threshold, beyond which we enter a vast unknown.*[55]

As nuclear weapons became more numerous, were deployed more widely, and assumed a greater variety of forms, the Kennedy administration also focused on the element of control. New weapons-control technologies, such as permissive action links, reduced the probability of accidental release, and new procedures were instituted to segregate warheads under US control from the delivery and launch vehicles which were often under allied control. In a broader sense, the Kennedy administration hoped to stem the proliferation of nuclear weapons, increasingly a common interest of both superpowers. As Kennedy noted before the election, "The bulk of the job of deterring Soviet nuclear capabilities must continue to lie with the United States . . . The *instabilities* that might result from this diffusion of nuclear weapons are equally dangerous to Russian and American interests."[56] In his desire to pursue a nuclear nonproliferation agreement with the Soviet Union,

Kennedy was moving at cross purposes with the inclinations of major allies. Such a policy was anathema to the French who viewed it as an attempt to relegate France to perpetually second-rate status. It also conflicted with the German hope that, through the vehicle of a NATO nuclear force, the FRG could assume a role in NATO's nuclear defense and shed some of its absolute dependency.

From a strategic perspective, McNamara's interest in centralizing control of Alliance nuclear weapons was aimed, first, at reversing the trend toward automatic and uncontrolled escalation that had characterized NATO strategy and, second, at rebutting the argument (expressed mostly by the French) that independent national nuclear powers contributed to deterrence. Both at the NATO Council meeting in Athens in May 1962, and subsequently in public at a commencement speech in Ann Arbor, Michigan, McNamara argued forcefully for "unity of planning, executive authority, and central direction in the management of the Alliance deterrence system," noting that, "in a nuclear war, there are no theaters, or rather, the theater is world-wide." In direct criticism of the French *force de frappe*, McNamara concluded, "in short, then, weak nuclear capabilities, operating independently are expensive, prone to obsolescence, and lacking in credibility as a deterrent."[57]

Despite these arguments McNamara did affirm Kennedy's offer, in March 1961, "of eventually establishing a NATO sea-borne missile force, which would be truly multilateral in ownership and control."[58] In seeming to endorse the MLF, however, Kennedy had initially placed three crucial conditions. First, it had to be "desired and found feasible by our allies." Second, such a force could only be established "once NATO's non-nuclear goals have been achieved." Finally, the MLF would have to be "subject to any agreed NATO guidelines on their control and use." None of these conditions was likely to be achieved. In the wake of the Cuban missile crisis, the US withdrew the last of the Thor and Jupiter missiles which had been deployed in the late 1950s, having offered to commit five Polaris submarines to NATO as a more survivable and therefore more stabilizing substitute. Consistent with his theme of "no theaters in a nuclear war," however, McNamara told the allies in Athens that the employment of the committed Polaris would "not be limited to the support of any single theater or major command."[59]

The anguished story of the MLF represents one of the great crises of confidence in the Alliance, a crisis that proceeded in parallel with the related and equally anguished debate over the strategy of flexible reponse (adopted in 1967), as well as with developments in US arms control policy that resulted, first, in the Partial Test Ban Treaty (signed in 1963), second, the Nuclear Non-Proliferation Treaty (signed in 1968), and the subsequent initiation of discussions concerning a strategic arms limitation agreement with the Soviet Union. Rather than an MLF, the Alliance ultimately settled on the formation, in 1966, of the Nuclear Planning Group (NPG) as a forum for allied political consultation on Alliance nuclear matters. The MLF had been, in many respects, a hardware solution to a fundamentally political problem. Allies wanted a genuine sense of participation in nuclear planning despite the fact that they could never have, and likely did not want, ultimate responsibility for nuclear defense. Even though the US retained the ultimate veto on the use of its nuclear weapons, there was benefit as well for the US in engaging the allies in the complexities of nuclear planning, inviting the allies to share the responsibility for designing NATO's nuclear deterrent posture and, in a fashion, reinforcing the need for effective conventional defense by exposing allies more directly to both the risks and the limitations of nuclear weapons.

In the end the French decision to withdraw from NATO's integrated military command structure in 1966 facilitated the resolution of NATO's nuclear control issue. In so doing, France also opted out of NATO's Nuclear Planning Group (NPG), leaving NATO with an incomplete consultative forum but one that—because French objections no longer blocked consensus—has enabled that forum to proceed with its business. Similarly, by removing itself from NATO's Military Committee, France thereby surrendered its veto over NATO's new "strategic concept" of flexible response, approved by the Military Committee and adopted by the Council in 1967 as MC-14/3. While its origins lie with McNamara's earlier overtures to the Alliance as articulated in Athens, MC-14/3 continues to reflect a fundamental compromise between the McNamara preference for a "symmetrical" strategy and the European insistence that NATO's deterrent remain based on the threat of escalation to nuclear war.

A STRATEGIC AND POLITICAL COMPROMISE: NATO'S STRATEGY OF FLEXIBLE RESPONSE

Fundamental to the US perspective in the 1960s—and not only then—was the view that the requirements of global defense again demanded that regional allies assume the appropriate burdens of regional defense. Besides the primary responsibility to defend the US, the US had, in 1961, nine formal treaty commitments to more than 40 countries, half of which were in the Western Hemisphere. In addition, there were informal commitments to several others. In 1962, McNamara's military planners identified 16 potential theaters of conflict, 11 of which would require the deployment of ground and tactical air forces. From this analysis, McNamara adopted the "two-and-one-half" war concept, which assumed that the US would not have to fight more than two major nonnuclear contingencies and one lesser contingency simultaneously.[60]

From an American perspective—increasingly preoccupied with regional threats around the globe for which the nuclear threat was ill-suited—"flexible response" implied not only the need for NATO to have multiple options in confronting the Warsaw Pact threat, but also the need for the US to maintain sufficient flexibility in its own force deployments to meet all of its global commitments. McNamara's 1962 force study called for the US to have fifteen active and reserve divisions for NATO's Central Region and one division each to reinforce the Northern and Southern flanks. The US would keep its five divisions stationed in Germany, with three others pledged to NATO.[61] McNamara had also explored the possibility of increasing the *Bundeswehr* ceiling from 500,000 to 750,000 men to displace two US divisions in the FRG, although NATO was not sympathetic.[62] In 1964, McNamara announced the withdrawal of the 7500 troops that had been deployed during the 1961–62 Berlin crisis, a decision that had been deferred by Kennedy the year before out of concern for German political sensitivities.[63]

McNamara also argued that the Warsaw Pact conventional threat was considerably more manageable than had been believed in the 1950s. Earlier estimates that the Soviet threat to NATO amounted to 175 divisions had discounted the smaller size of Soviet divisions which were often grossly undermanned, the relative inferiority of Soviet equipment, the potential unreliability of Warsaw Pact forces, and the growing threat which China posed to the Soviet Union. Thus, McNamara sought to demonstrate that a conventional defense of NATO was possible. Moreover, NATO would likely have 30 days warning of a Warsaw Pact attack, with 23 days to reinforce Europe once the decision to mobilize was made.[64] The US commitment to forward defense did not, therefore, require that the bulk of US forces be stationed in Europe. Given sufficient airlift and sealift capabilities (which, however, have never fully materialized) plus pre-positioning of major equipment and munitions stocks, NATO's defense needs could be met while affording the United States the necessary flexibility in its force structure.

Viewed in this light, the US advocacy of a strategic flexible response—both for itself and for NATO—implied three troubling conclusions for the allies. First, the Alliance was to be less reliant on the escalatory threats of nuclear weapons. Second, allies were expected to assume an increasing proportion of NATO's conventional defense burdens. Third, the United States would not only not increase its troop presence in Europe beyond the five divisions, but it would also retain the flexibility to deploy its forces from NATO to meet global contingencies (as indeed the US later did during the conflict in Southeast Asia). In combination with an increased American enthusiasm for engaging the Soviet Union in arms control negotiations to check the proliferation of nuclear weapons and ultimately to limit strategic arsenals, these aspects of flexible response suggested to allies that the United States was again seeking to qualify its commitment to European defense.

Kennedy himself seemed to imply as much in early 1963:

> The day may come when Europe will not need the United States and its guarantee. I don't think that day has come yet, and we would welcome that. We have no desire to stay in Europe except to participate in the defense of Europe. Once Europe is secure and feels secure, then the United States has 400,000 troops there, and we would, of course, want to bring them home.[65]

Kennedy's remark came in response to French President de Gaulle's press conference of the day before in which de Gaulle had vetoed British entry into the Common Market, rejected the US offer of Polaris missiles as part of a NATO MLF, and pressed for a Franco-German Friendship Treaty as the core of a European "Third Force" that would be "emancipated" from the US. Still trying to cope with the implications of the Berlin Wall crisis in August 1961, Bonn was confronted with two competing "Grand Designs"—Kennedy's image of an "Atlanticist" bond between the US and a united Europe (represented, in part, by the MLF) and de Gaulle's vision of a Europe "from the Atlantic to the Urals" which would define its destiny independent of the superpowers.[66] Bonn needed to cling to the US nuclear guarantee—hence the German endorsement of the MLF—but it also needed to keep France from pursuing an independent course, lest France become alienated from the Alliance and pursue a "traditional" relationship with Moscow.

Yet, having decided to cling all the more to the US, Bonn grew increasingly suspicious of US motives. In response to McNamara's argument that US troops in Germany could be based in the US and be airlifted back in time of crisis, FRG Defense Minister von Hassel openly wrote that the demonstration of a rapid transatlantic lift capability suggested as well that US divisions stationed in Germany could be pulled *out* of the theater rapidly in a crisis.[67] When President Johnson allowed the MLF concept to die by neglect in late 1964 in the interest of pursuing a nuclear nonproliferation treaty, it only fueled German suspicions. The growing prospect of détente in the mid-1960s—symbolized by movement on arms control, by US as well as domestic German pressures on the West German government to seek some accommodation with the East to defuse the "German question," and by de Gaulle's visit to Moscow in June 1966—suggested the possibility that the US might conclude, sooner rather than later, that Europe was secure and that, as Kennedy had predicted, the US would bring its troops home.[68]

The culmination of this crisis of confidence in the Alliance came when de Gaulle announced, in March 1966, that France would withdraw from NATO's integrated military command structure effective 1 July 1966. The French announcement came at the height of a burden-sharing dispute among the allies in which Bonn faced demands for increased "offset payments" from both the US and the UK to maintain troop levels. In April 1966, McNamara announced the "temporary withdrawal" of 15,000 troops from the FRG. One month later, von Hassel advised McNamara of Bonn's inability to meet fully its pledge to offset US troop costs by military purchases, and McNamara threatened further troop cuts in proportion to the offset deficit. Ultimately, the FRG agreed to accommodate limited troop reductions for both the US and the UK and to take steps to support the dollar, reflecting as well the economic dimension to Alliance security relationships.[69]

These developments, taken in conjunction with approval of MC-14/3 by NATO's Military Committee in May 1967, reflected a West German willingness to accept fewer allied troops and, as one German put it, to make the US commitment "less expensive, mutually acceptable, and thus more permanent."[70] Despite France's withdrawal from NATO's integrated command structure, French forces in Germany remained there under a bilateral agreement.

NATO's new "strategic concept," however, had changed substantially from the earlier McNamara vision. It did not demand substantially increased force goals to create a "symmetrical" conventional defense capability. Indeed, McNamara himself deemphasized the importance of troop levels to a Senate subcommittee in June, arguing that the priority lay with the creation of an adequate infrastructure in NATO to sustain NATO's combat capability over time.[71] In addition, the US had, during the McNamara period, continued to deploy tactical nuclear weapons, increasing the number from approximately 1000 to more than 6000 warheads. Given McNamara's initial strategic orientation, this increase in nuclear weapons on European soil is ironic, but it reflected the overriding political requirement of reassuring the allies at a time in which strategic preference, economic constraints, and broader political dynamics did not match. Thus, while MC-14/3 provided explicitly for the "direct defense" of Western Europe—including conventional defense against conventional attack—this principle was set alongside the need for NATO to possess the capability for "deliberate escalation" as well as a "general nuclear response."

NATO's new strategy thus continued to envision the clear possibility of a nuclear response to conventional attack, including both tactical and strategic nuclear weapons. The ambiguity enshrined in this concept was designed to serve both strategic and political interests. Militarily, NATO possessed more response options than had previously been the case; in one sense, therefore, MC-14/3 ratified the manner in which force structure had already evolved. Nonetheless, NATO continued to pose to the Soviet Union the likelihood that aggression would lead, if not to Soviet defeat, at least to general war. Ultimately, as the US ambassador to NATO noted at the time, the "real deterrent is uncertainty."[72] Politically—and perhaps more fundamentally—MC-14/3 represented a compromise between irreconcilable strategic perspectives that allowed the Alliance to remain intact.

NATO AND THE ERA OF DÉTENTE

The adoption of NATO's strategy of flexible response marked a major accommodation to the disparate strategic perspectives that have been inherent to the Alliance since its beginnings. Significantly, the endorsement of MC-14/3 came at a time when, in anticipation of a major East-West détente, the political integrity of the Alliance assumed substantially greater importance than the articulation of a precise military strategy. The imminent threats of the 1950s and early 1960s had passed, and the justification for maintaining this long-term peacetime defense investment had to be cast in a broader political context. The allies' reaffirmation of the North Atlantic Treaty on its twentieth birthday in 1969—at which time, according to Article 13, any party could denounce the treaty—came in the wake of a major review of the purpose of the Alliance and, in effect, a redefinition of that purpose.

It was no accident, therefore, that the North Atlantic Council endorsed, in conjunction with MC-14/3, the Harmel Report on the Future Tasks of the Alliance in December 1967. Specifically, the report declared:

The Atlantic Alliance has two main functions. The *first* function is to maintain adequate military strength and political solidarity to deter aggression and other forms of pressure and to defend the territory of member countries if aggression should occur. Since its inception, the Alliance has successfully fulfilled this task. But the possibility of a crisis cannot be excluded as long as the central political issues in Europe, first and foremost the German Question, remain unsolved. Moreover, the situation of instability and uncertainty still precludes a balanced reduction of military forces. Under these conditions, the Allies will maintain as necessary a suitable military capability to assure the balance of forces, thereby creating a climate of stability, security, and confidence.

In this climate the Alliance can carry out its *second* function to pursue the search for progress towards a more stable relationship in which the underlying political issues can be solved. *Military security and a policy of détente are not contradictory but complementary.* Collective defense is a stabilizing factor in world politics. It is the necessary condition for effective policies directed towards a greater relaxation of tensions. The way to peace and stability in Europe rests in particular on the use of the Alliance constructively in the interests of détente. *The participation of the USSR and the USA will be necessary to achieve a settlement of the political problems of Europe.*[73]

The Harmel Report gave primacy to the "ultimate political purpose of the Alliance . . . to achieve a just and lasting peaceful order in Europe accompanied by appropriate security guarantees." Central to this was a "solution to the German Question which lies at the heart of present tensions in Europe." Faced with the prospect that the Alliance might unravel as each ally pursued its independent détente strategy, NATO sought to embrace détente as an *Alliance* policy, in which the US played a vital role not only in the pursuit of specific agreements but also in maintaining its strategic guarantee on behalf of whatever future European security framework might evolve. Some feared that, in the absence of such an alliance context, the US might seek arms control agreements which enabled a US disengagement, the FRG might pursue an *Ostpolitik* that reawakened neutralist tendencies in Germany, and France—out of fear of Germany and suspicion of US reliability—might renew its "traditional" relationship with Moscow.

In addition to pursuing its own interests bilaterally with the Soviet Union—in particular in strategic arms control—the US endorsed détente not only as a vehicle for stabilizing the East-West relationship but also for maintaining alliance discipline in the broader process. As Henry Kissinger, Nixon's National Security Adviser, later reflected:

In previous decades, American rigidity had become a target for leftist criticism in Europe . . . forcing Europe's leaders to shift to a détente line, purporting to act as a "bridge" between East and West. The stark fact was that if America was intransigent, *we risked being isolated within the Alliance and pushing Europe towards neutralism . . .* We came to the conclusion that *we could best hold the Alliance together by ac-* *cepting the principle of détente and by establishing clear criteria to determine its course.*[74]

In the "era of negotiations" that followed, the principal Western actors were the US and the FRG. Each, in fact, harbored suspicions that the other would sacrifice allied interests for the pursuit of national interests, thus leading each to attempt to link together the various negotiating efforts. The result was a complex relationship encompassing the Strategic Arms Limitation Talks (SALT); Mutual and Balanced Force Reductions (MBFR); Bonn's "Eastern Treaties" with Moscow, Warsaw, Prague, and the German Democratic Republic (GDR); the Quadripartite Agreement on Berlin; and the Conference on Security and Cooperation in Europe (CSCE).[75]

The common denominator among these various negotiating efforts was to maintain the integrity of the Alliance. In previous decades, détente had long been associated with the potential for US disengagement and, indeed, US force levels in Western Europe had, by 1970, declined by more than 25 percent from the postwar high during the 1962 Berlin crisis. Given the material and psychological pressures of Vietnam, the domestic US pressures for unilateral troop withdrawals had increased substantially. In 1967, Senator Mansfield had introduced the first of a series of congressional resolutions calling for substantial reductions in US forces in Europe. By 1973, Mansfield almost succeeded in securing Senate approval for legislation that would have mandated a 40 percent reduction.[76] The Nixon administration, in 1970, likewise called for greater allied self-reliance to allow the US to play a "balanced and realistic American role . . . over the long pull."[77] Nonetheless, the Nixon administration feared that domestic "neoisolationist" pressures would undermine European willingness to sustain their defense efforts. Thus Nixon assured the allies in December 1970 that there would be no unilateral withdrawals and, one year later, announced the return of 20,000 US troops to Europe. Similarly, ten European allies, with strong urging from the FRG and concerned with demonstrating their capacity for "self-help" to maintain the US commitment, agreed, in December 1970, to a five-year, $1 billion European Defense Improvement Program (EDIP).[78]

The pursuit of arms control negotiations in the early 1970s by the US and its NATO allies manifested concerns that were often more political than strategic. The US, for example, viewed the prospect of negotiations on conventional force reductions, in MBFR, as a useful counter to Congressional pressures for unilateral troop withdrawals. Even though NATO's May 1970 Declaration on MBFR included the criterion that reductions should include both "indigenous" (read: FRG) and "stationed" (read: allied) forces,[79] US interest in MBFR focused on the prospect, at least initially, of negotiating reductions in US and Soviet forces in Europe. Bonn ultimately agreed, having received assurances that indigenous forces would be included in a second phase of reductions.[80]

Despite general allied skepticism that MBFR would lead to agreement, Bonn's enthusiasm for MBFR increased in the wake of the SALT I agreement, signed by the US and Soviet Union in May 1972. While the allies generally welcomed SALT I as a necessary ingredient to a broader and fundamentally political East-West détente, SALT raised substantial strategic concerns.[81] Britain and France in particular welcomed the ABM Treaty, since, in limiting Soviet ballistic missile defense, it preserved the viability of the limited British and French strategic arsenals. Bonn, however, was ambivalent. German leaders from both the conservative Christian Democratic Union (CDU) and the Social Democratic party (SPD) warned that Europe would be worse off than before if SALT did not address Soviet medium-range ballistic missiles (SS-4s and SS-5s) that were targeted on Europe.[82] Yet, Bonn had to abandon this demand to keep US nuclear-capable aircraft stationed in Europe from also being included in negotiations. Such so-called US forward-based systems (FBS) assumed increasing importance as an escalatory mechanism in NATO's response capability, both because of the prospect of US troop withdrawals and because of the possibility that tactical nuclear weapons would be reduced in the context of an MBFR agreement.

After 1969, the new German defense minister, Helmut Schmidt, increasingly argued that the advent of superpower parity—which SALT reflected—required greater emphasis on achieving a balance in conventional forces. Assuming that the US would likely decrease rather than increase its military presence in Europe, and arguing that the allies would not tolerate a substantial increase in German military capability, Schmidt's emphasis turned to conventional arms control. Schmidt likewise argued that tactical nuclear weapons offered a useful bargaining tool for achieving Soviet troop reductions, consistent with his view that NATO ought to de-emphasize tactical nuclear weapons in its own strategy. The US, Schmidt argued, was more inclined to view tactical nuclear weapons as an instrument of defense rather than escalation in response to a Warsaw Pact attack. Thus, they would more likely be used later rather than earlier, with devastating effect on the territory of both the FRG and the GDR.[83] US forward-based systems, on the other hand, not only served the escalatory function with greater flexibility, but they also put the Soviet Union at risk.

Despite earlier expectations, SALT II again deferred the knotty issues of nuclear systems targeted on—and from—Western Europe. SALT II negotiations also increased European fears that the US was willing to pursue its own strategic interests at the expense of European interests. The Carter administration's agreement to exclude the Soviet Backfire bomber and the SS-20 from SALT II ceilings, while conceding a three-year protocol limiting the range of cruise missiles, offered further evidence of the limits of a bilateral process. Despite the fact that the cruise missile protocol had no effect on deployment plans for cruise missiles, it nonetheless appeared to many Europeans as an advance concession for SALT III which was expected, finally, to deal with systems based in or targeted on Europe.

By 1976, the "era of negotiations" had largely run its course. The FRG had successfully weathered its own internal crisis and redefined its relationship to the East in a way that reinforced rather than undermined its position within the Western Alliance and facilitated Four Power agreement in Berlin. The CSCE process, culminating in the August 1975 Helsinki Final Act, offered a forum for discussing confidence- and US security-building measures (CSBMs) with the East, while symbolically reaffirming the US as a "European" power. In MBFR, NATO had offered proposals for the reduction of both US troops and nuclear weapons in Europe, but the Soviets rejected the notion of common force ceilings. While negotiations proceeded, ultimately to be mired in disputes over data, prospects for near-term agreement evaporated. In the US, a surplus balance of payments situation and fading congressional support for unilateral troop withdrawals after the fall of Saigon removed much of the impetus for reaching agreement in the face of Soviet intransigence. The SALT II process looked to be a drawn-out affair, increasingly complicated by evolving weapons technologies that had not featured in SALT I.

On a broader political level, the "era of negotiations" had witnessed substantial accomplishments in East-West relations, but it had not altered their fundamentally adversarial nature. By the mid-1970s, there were clear limits to détente. Soviet challenges to Western and particularly US interests in other regions of the world had demonstrated, most acutely in the 1973 Middle East war, that crises could erupt in ways that threatened war. East-West trade—long viewed by many on both sides of the Atlantic as a vehicle both for enticing the Soviet Union into an interdependent relationship with the West and for loosening Moscow's grip over Eastern Europe[84]—likewise revealed its limitations, both because of the limits to what Western creditors could allow their Eastern European trading partners and because of the growing fear that such trade was offering Moscow excessive access to militarily significant technology.

Militarily, the Soviet Union had made substantial improvements such that, in many ways, the threat to NATO had increased. At the strategic level, Soviet emphasis on heavy ICBMs with multiple, independently targetable reentry vehicles (MIRVs) had fundamentally altered the nature of the Soviet strategic threat to the US, leading critics to suggest that the US ICBM force was vulnerable to a preemptive first strike.[85] On the theater level, the advent of the Backfire and SS-20 represented a more potent and more invulnerable nuclear threat to Western Europe, suggesting a Soviet capability to preempt critical NATO targets which, in itself, might deter NATO from executing its flexible response strategy. Moreover, substantial improvements in the Warsaw Pact's conventional forces seemed designed to offer the Soviet Union an offensive capability that did not require the use of nuclear weapons for its success.

Thus, in the wake of the "era of negotiations," NATO found itself faced with a significantly improved military threat from the East and the realization that the fundamentally adversarial relationship between East and West had not changed. In many respects, NATO had adopted MC-14/3 fully cognizant of the inherent ambiguities which it contained. Many had hoped that initiatives in détente and arms control would provide for a stable security framework in which those ambiguities would remain tolerable. The perceived instabilities that had resulted from these arms control efforts, however, suggested an agonizing process designed to accomplish multiple objectives: increase the credibility of the US strategic guarantee, improve NATO's theater defenses, and keep open the possibility for further arms control agreements to stabilize the balance.

MANAGING NATO'S DEFENSE DILEMMAS

For most of the 1970s, US defense spending declined dramatically in real terms, reflecting not only the US disengagement from Southeast Asia but also a general disaffection for maintaining past levels of defense investment. By contrast, the major European allies had continued to increase their real defense spending through those years.[86] In 1975, the US began to increase its total active division strength and, in 1978, President Carter moved to pre-position equipment in Germany to support six reinforcing divisions which were to be deployable to NATO within ten days. In May 1977, NATO agreed to a Long Term Defense Program (LTDP) focusing on conventional defense and readiness "to withstand the initial phases of attack." In addition, the allies agreed to an annual real increase "in the region of 3 percent" in defense spending, a goal endorsed not only by the US but also by allies who were intent on arresting the downward trend in US defense spending and securing an increased US commitment.[87]

In agreeing to these new commitments, the allies cited the improvements in Soviet conventional warfighting ability and the possibility that the Soviets were stressing a short-war strategy which would enable them to secure a conventional victory before NATO could mobilize. Underneath this view of a growing Soviet conventional threat was the sense that NATO's nuclear retaliatory and escalatory capacity was inadequate both as a deterrent and as a defense. As FRG Chancellor Helmut Schmidt noted at the May 1977 NATO summit:

Strategic parity . . . will make it necessary during the coming years . . . to reduce the political and military role of strategic nuclear weapons as a normal component of defense and deterrence; the strategic nuclear component will become increasingly regarded as an instrument of last resort, to save the *national* interest and *protect the survival of those who possess these strategic weapons of last resort.*[88]

To Schmidt, SALT was an understandable and useful bilateral process between the superpowers. Yet, even if SALT were to stabilize the strategic nuclear balance—and many argued that it would not succeed in doing so—it "magnifies the disparities between East and West in nuclear . . . and conventional weapons."

In essence, Schmidt argued that strategic nuclear parity made it less likely that the US would cross the nuclear threshold to deal with a conventional attack by the Warsaw Pact that caught NATO unprepared. If NATO had to rely on tactical nuclear weapons to compensate for conventional imbalances, the result would be a deterrent that was both incredible—because they might not be used—and undesirable—since, if they were used, Germany would be destroyed. To the extent that tactical nuclear weapons were to be used, they should be designed to produce as little collateral damage as possible. More importantly, however, NATO needed to develop a capability to hold the Soviet Union at risk in ways that did not rely exclusively on the US strategic nuclear triad. These were the essential complements to a NATO effort to improve its conventional force posture since it remained out of the question for NATO to mount a conventional defense that excluded nuclear retaliation. First, the economic and political costs could likely not be sustained. Second, it was strategically undesirable since a conventional defense would be similarly destructive for Central Europe while leaving the Soviet Union a sanctuary.

In conjunction with its conventional force improvements program, the Carter administration was also pursuing the development of the "enhanced radiation warhead," or "neutron bomb." Designed to maximize radiation effects while minimizing blast and heat, the neutron weapon offered the possibility of reducing substantially the collateral damage associated with tactical nuclear weapons. Because it seemed to blur the traditional distinction between "conventional" and "nuclear" weapons and to lower the nuclear threshold by making such weapons ostensibly "more usable," the neutron warhead program was instantly surrounded in controversy. Having asked for an allied commitment on deployment before proceeding with production, Carter put the burden of decision on his allies. Despite fierce public opposition in the FRG, Schmidt gave support for deployment in the spring of 1977; nonetheless, Carter vacillated, ultimately deciding, in April 1978, to defer the production decision indefinitely.[89]

The neutron bomb debacle was a political blow to the Alliance because it lent a sense of indecision and lack of resolve in grappling with the knotty issues of nuclear deterrence. The principal challenge to NATO's defense posture, in the allies' view, was the need to deal with what Schmidt had referred to as the imbalance in "Eurostrategic" weapons. In anticipation of arms control negotiations that would ultimately encompass Soviet nuclear weapons targeted on Europe, NATO had little to trade. Outside of a small number of aging British and US aircraft, there were no dedicated nuclear weapons in theater capable of striking Soviet territory. To be sure, the British and French possessed a limited nuclear capability, but NATO has continued to insist that these are incomparable to Soviet systems. While Soviet SS-20s posed a credible hard-target kill capability, French and British systems remained essentially a minimum deterrent to nuclear attack on their respective territories, and neither country has been willing to surrender its nuclear status in an arms control forum.

In its two-track decision of December 1979, therefore, NATO foreign and defense ministers (less France) announced their intention to deploy, beginning in 1983, 108 Pershing II missiles in the FRG and 464 Ground-Launched Cruise Missiles (GLCMs) in the FRG, the UK, Italy, and, later, Belgium and the Netherlands—unless an arms control agreement with the Soviet Union on long-range intermediate nuclear forces (LRINF) altered that requirement.[90] NATO justified the LRINF deployments on the grounds that Soviet SS-20 deployments had "highlighted the gap in the spectrum of NATO's available nuclear response to aggression" and that NATO's LRINF deployments would therefore provide the currency for negotiating a reduction in Soviet

LRINF. Indeed, NATO subsequently endorsed, in December 1981, President Reagan's proposed "zero option" in which NATO's LRINF deployments would be cancelled in exchange for Soviet dismantling of all SS-4, SS-5, and SS-20 missiles, including those deployed in the Ural Mountains and the Soviet Far East.[91]

The strategic arguments for LRINF focused less on the specific threat by Soviet SS-20s than on the broader threat which Soviet nuclear and conventional improvements posed to the credibility of flexible response. Some argued that LRINF based on West European soil were more likely to be used to strike the Soviet Union than would US central strategic systems, since they would be "nonstrategic" and specifically dedicated to SACEUR's theater requirements for selective nuclear employment. Moreover, since NATO's LRINF would be under exclusively US control, the Soviet Union would view them as a US strategic asset and view their use as indicative of US determination to escalate to general war if necessary. Thus, LRINF seemed to enhance coupling at a time when many in the US were speaking of a "window of vulnerability" regarding the US strategic triad.[92]

Yet coupling is essentially a political phenomenon and is, like beauty, often in the eyes of the beholder. As the debate on LRINF deployment raged in the early years of the Reagan administration, there were also those in Europe who suspected that LRINF might actually be decoupling. Antinuclear movements in the prospective basing countries focused on the notion that deployments would merely invite preemptive Soviet attack. Fueled by comments in Washington that suggested the US was willing to fight a protracted nuclear war in Europe, many suspected that the US would view LRINF as an alternative means to striking the Soviet Union while hoping to remain a sanctuary in such a conflict. Despite assurances from both Washington *and* Moscow that striking the Soviet Union from Europe would undoubtedly lead to strategic nuclear war, Europeans increasingly feared that the Reagan administration was becoming dangerously enamored with "nuclear warfighting" strategies.[93]

The four-year hiatus between the LRINF two-track decision and the beginning of LRINF deployments invited a major debate not only about NATO's arms control policy but also about the fundamental precepts of NATO's strategy. A substantial wave of antinuclear sentiment—more widespread than in the late 1950s—expressed itself in massive "antimissile" demonstrations throughout Western Europe, particularly in the prospective LRINF basing countries. In the United States, the "nuclear freeze" movement broadened its public base as the National Conference of Catholic Bishops issued its pastoral letter questioning the foundations of nuclear deterrence, while popular commentary on the dangers of nuclear war received a scientific underpinning through notions of "nuclear winter."[94] The effect of this broad-based sentiment was to reinforce the argument by particularly prominent individuals on both sides of the Atlantic that NATO's strategy of flexible response—in particular its reliance on the prospect of first use of nuclear weapons—was strategically bankrupt if not also morally wrong.[95]

The most notable critique of NATO's strategy came in the form of an article in *Foreign Affairs* by McGeorge Bundy, George Kennan, Robert McNamara, and Gerard Smith, all of whom had held senior positions in earlier Democratic administrations.[96] In essence, the four argued for a "symmetrical" strategy reminiscent of the views that underlay NSC-68 in the Truman administration and the initial US arguments for flexible response in the Kennedy administration. First the threat of nuclear use in response to a conventional attack was fundamentally incredible as a deterrent since it involved too high a risk for the US. Second, the use of nuclear weapons as a battlefield defense was excessively destructive, could not be controlled, and was inherently escalatory. Third, conventional defense of Europe *was* possible: it was not at all out of the question for NATO to sustain a conventional defense that would deter the Soviet Union from believing it could win such a conflict. The authors concluded that NATO should commit itself to developing its conventional defense posture to a point that would allow NATO to embrace a policy of "no first use of nuclear weapons."

Not surprisingly, the German response was an immediate rebuttal, cast again in the pages of *Foreign Affairs* and authored by a comparably stellar group of four scholars, retired senior military officers, and former senior government officials.[97] While acknowledging the strategic importance of a strong conventional deterrent, they (like Helmut Schmidt) focused their rebuttal on the central role which nuclear weapons played in keeping the risks to the Soviet Union "incalculable." To rely solely on conventional defense against a Warsaw Pact attack made forward defense—the essential element of Germany's "bargain" with NATO—implausible. Even if such a defense succeeded, the associated destruction in Germany would be unacceptable. Fundamentally, they argued, a policy of "no first use" undermined deterrence of *war* (as opposed to *nuclear* war), because it invited the Soviet Union to initiate aggression, secure in the belief that the Soviet Union would remain a sanctuary in such a conflict.

Combined with the general antinuclear sentiment, arguments for a no-first-use policy stimulated NATO to address improvements in conventional defense even as it reaffirmed the necessity of keeping the Soviet Union at risk with nuclear weapons (the essential strategic argument for LRINF). For example, General Bernard Rogers, NATO's SACEUR, energetically advocated a policy of "no *early* first use" of nuclear weapons in an effort to translate antinuclear sentiment into a greater willingness to increase defense spending aimed at improving NATO's conventional defense posture.[98] Rogers's argument seemed to strike an acceptable balance between, on the one hand, the need to follow through with LRINF deployments to prevent "nuclear blackmail" by the Soviet Union and, on the other hand, the need to make the necessary defense investment to give NATO the flexibility to respond to Soviet attack without "undue" recourse to nuclear weapons. The question was simply whether NATO was willing to pay the bill.

The debate over *whether* flexible response should rely on nuclear weapons generated as well a substantial literature on *how* NATO should employ its resources for a viable conventional defense.[99] Some argued that NATO should abandon forward defense and adopt a strategy of "maneuver warfare" to absorb and defeat a Soviet attack by striking its flanks. Others argued for an offensive strategy that envisioned counterattacks into Eastern Europe. Still others suggested a form of "defensive defense," relying not on standard armored formations to confront an attack, but on small units armed with "purely defensive" antitank missiles and a pervasive partisan force to frustrate the aspirations of an occupying power. Many of these alternatives have required too much of NATO's conventional force structure, expected too much of particular classes of weapons systems, or involved political compromises, such as abandoning the very notion of forward defense, that are blatantly unacceptable to NATO.

Yet, there was a sense that improvements in conventional weapons technologies, both munitions and delivery systems, could enable a vastly improved conventional defense posture in an environment in which allied economic and demographic resources remained highly constrained. General Rogers, for example, argued that NATO could not be content with blunting the spearhead of a Warsaw Pact attack, since subsequent attacking echelons would ultimately overwhelm NATO's defenses. NATO needed, therefore, the capability to interdict targets deep in Warsaw Pact territory, thus disrupting the Warsaw Pact's ability to exploit breakthroughs and maintain the momentum of attack that was so crucial to the Soviets' offensive doctrine. To be sure, the requirements for an effective Follow-on-Forces Attack (FOFA) capability are substantial. Whether NATO will be able to fulfill these requirements depends not only on resources but also on the effective development of the weapons systems and integrated battlefield management technologies on which it depends.[100] In the context of the defense debate that emerged in the early 1980s, however, the concept seemed to many to be a reasonable alternative to abandoning forward defense.

Much of the anxiety surrounding these debates about the relationship between nuclear weapons and conventional defense in NATO strategy derives from public concerns about the future of arms control. Through the first half of the decade, neither START nor LRINF negotiations seemed to be progressing. In retrospect, much of the lack of progress can be attributed to a virtual paralysis in the Soviet political apparatus as successive leaders assumed office and died, in the transition from Brezhnev to Gorbachev. Ironically, given later anxieties about the implications of eliminating LRINF, the Reagan administration's 1981 proposal of a "zero option" in LRINF was criticized by many on both sides of the Atlantic as demanding too much from the Soviets, impeding progress in arms control, and allowing the Soviets to continue to deploy SS-20s. When President Reagan announced the Strategic Defense Initiative (SDI) in March 1983, many countered that it posed yet a further obstacle to arms control. When the Soviets walked out of the START and LRINF negotiations in December 1983, those criticisms mounted.

When the Geneva negotiations reopened in March 1985, with Gorbachev as the new Soviet leader and after Reagan's landslide reelection, the arms control agenda included START, LRINF, and "Defense and Space Talks" (DST), with Gorbachev holding progress on the first two hostage to US agreement to limit severely the US SDI program. By this time, a broad-based debate over the strategic implications of SDI had developed, reinforcing criticisms that SDI stood in the way of progress in arms control. Allies were wary of SDI because it seemed to suggest an American willingness to retreat behind a defensive shield, leaving NATO exposed. Given the persistent doubts that Europeans harbored about whether the US, vulnerable to nuclear attack, would in fact risk nuclear war, one might have thought that the allies would have welcomed an effort designed to reduce US strategic vulnerability. The more dominant allied view, however, was that SDI might enable the US to pursue a limited war-fighting strategy employing, in part, the very missiles which were being deployed in Europe to enhance coupling. The prospect of both superpowers ultimately protected by strategic defenses, moreover, suggested that Europe might remain vulnerable to a conflict, either nuclear or conventional, confined to the theater. Given the vast resource investment which SDI was likely to entail, the burden of defense would fall increasingly on the allies who might conceivably also have to shoulder substantially higher risks.

These allied fears of being left vulnerable to Soviet military power while the US disengaged its strategic deterrent, though exaggerated, gained greater intensity as the notion of a "denuclearized" world assumed considerable prominence in superpower rhetoric.[101] In early 1986, both Gorbachev and Reagan offered not only various formulations of a zero LRINF agreement, but also the prospect of removing all nuclear weapons by the end of the century. Subsequently, in October, both leaders met in Reykjavik. Both seemed to agree to a zero LRINF in Europe and an approximate 50 percent reduction in each side's strategic warheads to a ceiling of 6000. Gorbachev further proposed the elimination of all "strategic nuclear *weapons*" in ten years; Reagan countered with a proposal to eliminate all "strategic ballistic *missiles*" in ten years. Gorbachev, however, made any deal contingent on US abandonment of SDI, which Reagan refused.[102]

Among the allies, initial disappointment that a potential arms control breakthrough had aborted at Reykjavik soon faded into relief that no agreement had been reached. The possibility that the two superpowers might conceivably have agreed, in a fateful instant, to such a dramatic reduction of nuclear weapons was enough to give the allies pause. The possibility that the US might have agreed to such a fundamental change in the structure of its deterrent posture without consulting the allies sent tremors through the Alliance, almost overshadowing the fact that, indeed, no agreement had been made. In the wake of Reykjavik, the sudden prospect that nuclear arms control might take on substantial and unpredictable momentum led allies to raise their concerns more directly.

First, the notion of eliminating all strategic ballistic missiles threatened to undermine French and British plans to modernize their independent nuclear forces. Both governments rejected outright the prospect that European security could be preserved in the foreseeable future by the removal of the West's strategic nuclear retaliatory threat. In Britain, the Opposition Labour party had long criticized Prime Minister Thatcher's decision to replace Polaris with the US Trident SLBM. Arguing for a nonnuclear Britain as an electoral platform, the Labour party even cited Reagan administration statements as evidence that Britain's Conservative government was out of step with historical trends. With French support, Thatcher traveled to Washington on the eve of NATO's fall 1986 Nuclear Planning Group ministerial meeting, and, while endorsing the shorter-term goal of a 50 percent reduction in strategic weapons in

START, urged the Reagan administration to de-emphasize the longer-term vision of eliminating ballistic missiles entirely.[103]

Second, the sudden prospect that a zero LRINF agreement might actually become a reality caused many to consider the implications of what had in fact been NATO policy since 1981. In the FRG, Chancellor Kohl highlighted the growing danger of Soviet forward-deployed shorter-range INF (SRINF) missiles which could threaten many of the targets covered by the SS-20s, arguing that an LRINF agreement should also incorporate constraints on SRINF. When the Soviets, in April 1987, offered as well a zero SRINF agreement encompassing land-based missiles with ranges of 500 to 1000 kilometers, Bonn again hesitated. If the only nuclear missiles left in Europe were to have ranges below 500 kilometers, Bonn argued, then the FRG and GDR would find themselves with a "singular" risk as the only countries threatened with total devastation in a European war.[104]

In consulting with its allies on the prospects of a "double zero" agreement, the US found, on the one hand, general political sympathy for such an agreement based largely on domestic political sentiment. On the other hand, there was a nagging fear that Gorbachev had seized the initiative. Bonn's call for further negotiations on short-range nuclear forces (SNF) below 500-kilometer range to remove its presumed (and exaggerated) "singular" nuclear risk contrasted with the prevailing allied view that further negotiations on theater nuclear systems in the short term would invite a dangerous "denuclearization" of Europe. It was necessary to "draw the line" for the moment to avoid "successive zeros" that would undermine NATO's flexible response strategy. Further serious discussion of eliminating ballistic missiles entirely or on SNF systems in Europe would have to wait for substantial progress in redressing the imbalances in conventional and chemical weapons which compelled NATO to rely as it did on its nuclear threat.[105]

Both politically and militarily, the prospects of arms control agreements to remove all US land-based missiles in Europe above 500-kilometer range and to reduce substantially the US strategic missile threat to the Soviet Union have highlighted the central issues of NATO strategy. The fact that such agreements would also have a substantial impact on Soviet nuclear force structures has been virtually overshadowed by the political implications for the West in altering the structure of its nuclear deterrent. Politically, the Alliance could not retreat on its long-standing endorsement of the "zero option" for LRINF. Moreover, zero SRINF was clearly appealing because, in fact, the US had no weapons deployed in that category, and the FRG's 72 Pershing Ia missiles (with US warheads) were

aging and their modernization was in doubt.[106] The alternative to accepting "double zero"—a separate SRINF agreement providing for equal ceilings—would be for NATO to deploy up to that ceiling, but the parallels with the anguished debates over LRINF deployments were too disturbing.

NATO's response in accepting a "double zero" agreement has been to affirm the principles and vitality of flexible response, to stress the importance of nuclear weapons to deterrence, and to emphasize the need for Alliance cohesion. Not only would the search for an alternative strategy be politically devastating, an alternative strategy is not necessary. Flexible response does not rely on any particular type or class of weapons system. It does require, however, not only a credible nuclear deterrent that keeps risks incalculable to the Soviets, but also an adequate conventional defense that upholds the principle of forward defense and preserves NATO's response options.

The signing of an INF agreement and the prospect for further substantial negotiated reductions in both nuclear and conventional arms have highlighted issues about how NATO should implement that flexible response strategy. Clearly, the need for an effective conventional deterrent is paramount. Yet, the combination of arms control possibilities and economic imperatives, in both the East and West, will likely mean reduced force levels and greater reliance on mobilization and reinforcement in a crisis. NATO's nuclear deterrent will likely assume different forms, inviting sometimes vociferous debate on what kinds of nuclear forces should be deployed where, who would control them, and how those weapons would be employed in war. Central to this debate will be the political issues of whether nuclear risks are shared appropriately throughout the Alliance. NATO had already agreed, in 1983, to reduce the number of tactical nuclear weapons by 1400, following an earlier agreement, in 1979, to reduce that number by 1000.[107] That 1983 Montebello decision had been predicated, however, on modernization of the remaining stockpile and on deployment of LRINF. Yet, even before the dramatic political changes in Eastern Europe, NATO had already agreed at its May 1989 summit to accelerate the timetable for negotiating the reduction of shorter-range nuclear weapons in Europe.[108] As those political changes unfolded, any prospect for modernizing nuclear weapons, especially on German soil, dissolved.

Some observers have noted that nuclear weapons will, in any event, become increasingly irrelevant as a means of deterring war. Others have suggested that political change within the Soviet Union and Eastern Europe has *already* led to a fundamental change in the magnitude of the Soviet threat to European security. Still others predict that the transatlantic security guarantee will not endure and that the evolution of European defense cooperation will eventually supplant NATO's existing institutions. It is worth recalling that each of these notions has historical antecedents. The role of nuclear weapons, the nature of the US extended deterrent guarantee, the possibility of an enduring East-West rapprochement, and alternative structures for European security have all been contemplated at various stages since the Alliance was founded. NATO's defense dilemmas assume new and various forms, but they are, in essence, historical, familiar, and by no means unmanageable.

CONCLUSION: NATO AT A CROSSROADS

Throughout the literature on the evolution of NATO and its successive strategic debates, one is struck by how often NATO has been considered to be "at a crossroads." This can be attributed to NATO's inherent tensions which have a number of sources. First, despite the fact that NATO endures as an unprecedented peacetime alliance of sovereign nations, there remain fundamental tensions in the strategic outlook of its members and, indeed, regarding its strategy for dealing with the continuing antagonisms between East and West. Second, more so than with any other region, NATO's tensions reflect the underlying tensions of the nuclear age in which the role of nuclear weapons has eluded precise definition and remains a subject of high public consciousness and political sensitivity. Third, NATO is more than simply a military alliance; rather, it reflects a broader political and economic relationship that itself is replete with tensions as transatlantic interdependence increases.

The persistence of these tensions has always provided the potential for revolutionary change in the way the postwar European order has been structured and in the way the US relates to the structure. Nonetheless, the tendency has been toward evolution rather than revolution, largely because alternative political structures have never been viewed as preferable to the admittedly imperfect "transatlantic bargain" that was struck almost four decades ago. Notwithstanding the growing importance in the last decade of Japan and the Pacific, Central America, and the Middle East in the US outward orientation, the US remains tied more closely to Western Europe than to any other global region, not only in terms of defense policy but also in terms of its broader political and economic outlook.

But the transatlantic security relationship is not static. The prospect of an INF agreement triggered renewed fears that the US commitment to Europe might not be indefinite, stim-

ulating thinking about alternative, or at least complementary, forms of defense cooperation and about how Europe might tend to its defense needs in the event that US commitment is lessened. Increased Franco-German defense coordination on a bilateral basis in parallel to the NATO integrated military command structure, contemplation of greater Anglo-French nuclear cooperation, reemphasis of the Western European Union as a "European pillar" of defense, and recognition of the need to revitalize European defense industries—all reflect attempts to reduce, albeit on the margins, European security dependency on the US.

The prospect of agreement in the negotiations on Conventional Armed Forces in Europe (CFE) that began in March 1989, pressures for unilateral defense cuts in the West, and dramatic political changes in the East all increase speculation that the United States might in fact disengage from Western Europe.

Yet the demise of NATO is hardly imminent. The US has always hoped that Europe would increase its self-sufficiency in meeting its defense requirements. Arguably, however, NATO—and for that matter the Warsaw Pact—retains important functions in managing a fundamental transition to a new and less confrontational security relationship and maintaining a degree of stability amid radical and potentially dangerous political change. Thus, it was important for US Secretary of State Baker to reaffirm, in December 1989, the continued commitment of the US to the defense of Europe and the continued engagement of the US in helping to shape a new Europe.[109]

The United States remains by far the most powerful member of the Alliance and continues to bear responsibilities of leadership that have not, in essence, changed since the founding of the Alliance. Yet there are limits on the exercise of that leadership. The Alliance is a coalition of sixteen sovereign and very different states. Some are medium powers in their own right. Others are minuscule in terms of the resources they can, even under the best of circumstances, bring to bear on the problems of collective defense. None can substitute for the US. Yet all have a voice in shaping the direction that the Alliance takes, and each can set clear boundaries to unilateral US actions.

The frustrations often encountered by the US in exercising its leadership role in the Alliance do not, however, diminish the fact that the Alliance and the US commitment thereto remain very much in the US interest. Despite occasional political sentiment to the contrary, there is no alternative to a coalition framework for the pursuit of US security interests, especially in Europe where the legacy of the cold war has been most visible and where changing that decades-old security framework will be a delicate proposition. The forthcoming challenges for the US, therefore, involve managing that Alliance relationship in such a way so to preserve its essential institutions while discarding those that are obsolete, adapting to changes, and finding mechanisms for greater cooperation and more effective use of resources. The centrifugal forces of disengagement, protectionism, and national prerogatives remain powerful corrosive forces in any alliance. They cannot, by their nature, be overcome by complacency. In that respect, one can argue that the recurring cries of "Alliance crisis" and suggestions that NATO is "at a crossroads" are useful reminders that the inherent tensions of the "transatlantic bargain" require continuous attention.

NOTES

1. US Department of Defense, *Annual Report to the Congress, FY 1988*, p. 334, and *Report on Allied Contributions to the Common Defense*, Apr. 1987, p. 7.
2. See Claude Delmas, "A Change of Heart: The Discussions in France." In Andre de Staercke et al., *NATO's Anxious Birth: The Prophetic Vision of the 1940s* (London: C. Hurst, 1985), p. 64.
3. William T. R. Fox, *The Super-Powers* (New York: Harcourt, Brown, 1944), p. 112.
4. Quoted in John Gimbel, *The American Occupation of Germany: Politics and the Military, 1945–1949* (Stanford: Stanford University Press, 1968), p. 5.
5. See Clark Clifford's essay on the Truman Doctrine in de Staercke, *NATO's Anxious Birth*, pp. 1–10.
6. See Joseph M. Jones, *The Fifteen Weeks* (New York: Viking, 1955), pp. 138–41.
7. Refer to the previous chapter by Ronald J. Sullivan for a discussion of "containment." Kennan's *Foreign Affairs* article, "The Sources of Soviet Conduct," is reprinted at the end of Part II of this volume.
8. See, for example, Bevin's correspondence with Marshall, in US Department of State, *Papers Relating to the Foreign Relations of the United States* [hereafter *FRUS*] *1948, Western Europe* (Washington, D.C.: G.P.O., 1974), 3:32–33, 46–48, 79–80, 122–23, and 138. Also, Harry S. Truman, *Volume II. Years of Trial and Hope, 1946–1953* (London: Hoddes and Stoughton, 1956), pp. 254–76.
9. This theme of a "transatlantic bargain" was noted by Harlan Cleveland, former US ambassador to NATO. See Stanley R. Sloan, *NATO's Future: Toward a New Transatlantic Bargain* (Washington, D.C.: National Defense University Press, 1985), for an elaboration of that theme.
10. See Alan K. Henrikson, "The Creation of the North Atlantic Alliance," in John F. Reichart and Steven R. Sturm, eds., *American Defense Policy*, 5th ed. (Baltimore: Johns Hopkins University Press, 1982), pp. 299–302.
11. Ibid., p. 301.
12. See Theodore C. Achilles, "The Omaha Milkman: The Role of the United States in the Negotiations," in de Staercke, *NATO's Anxious Birth*, pp. 30–34.
13. Quoted in Henrikson, "Creation," p. 308.

14. See Dean Acheson, *Present at the Creation: My Years at the State Department* (New York: Norton, 1969), pp. 285–90.
15. See Henrikson, "Creation," pp. 311–12.
16. The full text of NSC-68, sent to the president on 7 Apr. 1950 and approved on 30 Sept. 1950, is in *FRUS, 1950, National Security Affairs; Foreign Economic Policy* 1:235–92. Excerpts are included in the selected readings at the end of Part II of this volume.
17. *FRUS, 1950* 1:16.
18. Quoted in Sloan, *NATO's Future*, p. 11.
19. *FRUS, 1948* 3:72–75.
20. *FRUS, 1948* 3:123–26.
21. French Foreign Minister Schuman's assurances to the French National Assembly during the ratification debate are quoted in Alfred Grosser, *Germany in Our Time: A Political History of the Postwar Years* (London: Pall Mall, 1971), p. 304.
22. Kennan's memo to Secretary of State Marshall, 24 Nov. 1948, in *FRUS, 1948* 3:283–89, here p. 287.
23. Acheson, *Present at the Creation*, pp. 436–37.
24. See Robert McGeehan, *The German Rearmament Question: American Diplomacy and European Defense after World War II* (Urbana: University of Illinois Press, 1971), for a full discussion of this issue.
25. Quoted in Coral Bell, *Negotiation from Strength: A Study in the Politics of Power* (London: Chatto and Windus, 1962), p. 54.
26. *North Atlantic Treaty Organization: Facts and Figures* (Brussels: NATO Information Service, 1983), pp. 30–31. See also Sloan, *NATO's Future*, pp. 15–18.
27. For a detailed discussion, see the author's *Détente and Alliance Politics in the Postwar Era: Strategic Dilemmas in United States–West German Relations*, unpublished D.Phil. thesis, Oxford University, 1982, pp. 130–48.
28. See Nitze's memo to Acheson, 17 Jan. 1950, in *FRUS, 1950* 1:13–17.
29. *FRUS, 150* 1:264–68.
30. William P. Mako, *U.S. Ground Forces and the Defense of Central Europe* (Washington, D.C.: Brookings Institution, 1983), pp. 11–13.
31. John Foster Dulles, "A Policy of Boldness," *Life*, 19 May 1952.
32. From NSC 162/2, quoted in David N. Schwartz, *NATO's Nuclear Dilemmas* (Washington, D.C.: Brookings Institution, 1983), pp. 23–24.
33. Dulles, "The Evolution of Foreign Policy," *Department of State Bulletin* 30, 25 Jan. 1954, p. 107. Excerpts are included in the supplemental readings for this part of this volume.
34. Cited in Schwartz, *NATO's Nuclear Dilemmas*, p. 23.
35. Sloan, *NATO's Future*, pp. 39–40.
36. For a detailed discussion, see Robert E. Osgood, *NATO: The Entangling Alliance* (Chicago: University of Chicago Press, 1962), chap. 5.
37. For detailed discussions, see Andrew J. Pierre, *Nuclear Politics: The British Experiences with an Independent Strategic Force, 1939–1970* (London: Oxford University Press, 1972), and Wilfred L. Kohl, *French Nuclear Politics* (Princeton: Princeton University Press, 1971).
38. See Catherine M. Kelleher, *Germany and the Politics of Nuclear Weapons* (New York: Columbia University Press, 1975), chap. 2.

39. The plan did not materialize. See Robert Drummond and Gaston Coblentz, *Duel at the Brink: John Foster Dulles' Command of American Power* (London: Weidenfeld and Nicholson, 1960), pp. 45ff.

40. For a general review of the disengagement proposals based on a Chatham House study group, see Michael E. Howard, *Disengagement* (London: Penguin, 1958).

41. George F. Kennan, *Memoirs, 1950–1963* (London: Hutchinson, 1972).

42. Norstad's Nov. 1957 Cincinnati speech is quoted in Schwartz, *NATO's Nuclear Dilemmas*, p. 58.

43. Dulles, "Challenge and Response in United States Foreign Policy," *Foreign Affairs* 35, no. 1 (1957): 24–43, here p. 31.

44. See, for example, Morton H. Halperin's compendium, *Limited War in the Nuclear Age* (New York: John Wiley, 1963). A particularly strong argument came from Henry Kissinger in "Force and Diplomacy in the Nuclear Age" in *Foreign Affairs* 34, no. 3 (1956), and developed further in his *Nuclear Weapons and Foreign Policy* (New York: Houghton Mifflin, 1957). Kissinger retreated from his advocacy of employing tactical nuclear weapons as usable instruments of policy in *The Necessity for Choice: Prospects of American Foreign Policy* (New York: Harper, 1960). See also the works of William Kaufmann, Bernard Brodie, and Thomas Schelling in the bibliographic essay on "Deterrence Theory" in this volume.

45. Bonn's *General-Anzeiger*, 13 Nov. 1957.

46. *New York Times*, 3 Jan. 1958, cited in Schwartz, *NATO's Nuclear Dilemmas*, p. 67.

47. *New York Times*, 5 Jan., 5 Feb., 17 Mar., and 3 Apr. 1959. See also Kennan, "Disengagement Revisited," *Foreign Affairs* 37, no. 2 (1959): 187–210.

48. *New York Times*, 12 Mar. 1959.

49. Quoted in Dean Acheson, "The Practice of Partnership," *Foreign Affairs* 41, no. 2 (1963): 352–53.

50. William Kaufmann, *The McNamara Strategy* (New York: Harper and Row, 1964), p. 106.

51. Quoted in John D. Steinbruner, *The Cybernetic Theory of Decision* (Princeton: Princeton University Press, 1974), p. 182.

52. Dulles's 16 July 1957 press conference, *New York Times*, 17 July 1957.

53. See Kelleher, *Germany*, pp. 128–30, Osgood, *NATO*, pp. 160–63, and Steinbruner, *Cybernetic Theory*, pp. 174–89.

54. Testimony before the Senate Subcommittee on Defense Appropriations, Apr. 1961, quoted in Kaufmann, *McNamara Strategy*, pp. 59–60.

55. Testimony before the House Armed Services Committee, *Hearings on Military Posture*, 88th Cong., 1st sess., Feb. 1963, pp. 299–300.

56. See Kennedy's review of B. H. Liddell Hart's *Deterrence or Defense*, in *Saturday Review of Literature*, 30 Sept. 1960, quoted in Walter Stuetzle, *Kennedy und Adenauer in der Berlin-krise, 1961–1962* (Bonn-Bad Godesberg: Neue Gesellschaft, 1973), p. 48.

57. See Schwartz, *NATO's Nuclear Dilemma*, pp. 156–65 for extensive excerpts from McNamara's remarks in Restricted Session in Athens. Subsequent to the Ann Arbor speech McNamara qualified his criticism of independent nuclear forces to exclude Britain (leaving only France), citing the nominal integration of Britain's nuclear forces in NATO. Six months later,

this special relationship with Britain was reinforced by the Nassau Agreement. See Pierre, *Nuclear Politics*, pp. 235ff.

58. Kennedy's speech to the Canadian Parliament in Ottawa, Mar. 1961, cited in Kaufmann, *McNamara Strategy*, p. 107. See also Schwartz, *NATO's Nuclear Dilemma*, pp. 82–135, Kelleher, *Germany*, chaps. 6 and 7, and Steinbruner, *Cybernetic Theory*, passim, for detailed discussion of the MLF.

59. Quoted in Schwartz, *NATO's Nuclear Dilemma*, p. 164.

60. William W. Kaufmann, *Planning Conventional Forces, 1950–1980* (Washington, D.C.: Brookings Institution, 1982), pp. 4–5.

61. See ibid., pp. 4–13 for a brief recapitulation of the 1962 General Purpose Force Study.

62. *Frankfurter Allgemeine Zeitung*, 3 Aug. 1962, *Baltimore Sun*, 11 Aug. 1962, and *New York Times*, 18 Dec. 1962.

63. *Oral History Interview, Roswell L. Gilpatrick*, John F. Kennedy Library, Harvard University, pp. 83–84. Also press reports, *New York Times*, 22 Oct. 1963 and 11 Apr. 1964.

64. See Alain C. Enthoven and K. Wayne Smith, *How Much Is Enough? Shaping the Defense Program, 1961–1969* (New York: Harper and Row, 1971), pp. 132–42.

65. *New York Times*, 15 Jan. 1963.

66. See F. Roy Willis, *France, Germany, and the New Europe: 1945–1967*, 2d ed., (London: Oxford University Press, 1968), pp. 310–18.

67. See Kai-Uwe von Hassel, "Détente through Firmness," *Foreign Affairs* 42, no. 2 (1964): 184–94.

68. For a detailed discussion of conflicting pressures for an East-West détente in the 1960s and the political implications for US-FRG relations, see Foerster, *Détente and Alliance Politics*, pp. 282–358.

69. *Washington Post*, 13 Apr. 1966, 17 May 1966; *New York Times*, 6 June 1966.

70. Wilhelm Cornides, "The Power Balance and Germany," *Survey*, no. 58 (1966): 159.

71. See Henry M. Jackson, ed., *The Atlantic Alliance: Jackson Subcommittee Hearings and Findings*, Subcommittee on National Security and International Operations, US Senate Committee on Government Operations (New York: Praeger, 1967), p. 259.

72. Harlan Cleveland, "The Real Deterrent," *Survival* 9, no. 12 (1967).

73. *NATO-Basic Documents* (Brussels: NATO Information Service, 1981), pp. 98–99. Emphasis added.

74. Henry A. Kissinger, *The White House Years* (London: Weidenfeld and Nicolson, Michael Joseph, 1979), p. 403. Emphasis added.

75. For a detailed discussion, see Foerster, *Détente and Alliance Politics*, pp. 359–427.

76. See Phil Williams, "Whatever Happened to the Mansfield Amendment?" in *Survival* 18, no. 4 (1976): 146–53.

77. *United States Foreign Policy for the 1970s: A New Strategy for Peace* (The White House, 18 Feb. 1970), pp. 4–5.

78. *New York Times*, 5 Dec. 1970; *Washington Post*, 16 Dec. 1971.

79. See *NATO: Final Communiques, 1949–1974* (Brussels: NATO Information Service, 1975), pp. 237–38.

80. See FRG Defense Minister Leber's statement, *New York Times*, 30 July 1973.

81. For example, FRG Defense Minister Schmidt's article in *Washington Post*, 2 Apr. 1970.

82. See FRG Chancellor Kiesinger's Aug. 1969 interview with *Suedwestfunk*, in Boris Meissner, ed., *Die deutsche Ostpolitik, 1961–1970: Kontinuitaet und Wandel (Dokumentation)* (Cologne: Wissenschaft und Politik, 1970), pp. 374–76. Also, Helmut Schmidt's interview in *Die Welt*, 16 Feb. 1970.

83. This view, contained in the FRG's 1970 Defense White Paper, is consistent with Schmidt's writing almost a decade before. Cf. Schmidt's *Defense or Retaliation: A German Contribution to the Consideration of NATO's Strategic Problems* (Edinburgh: Oliver and Boyd, 1962) and his *The Balance of Power: Germany's Peace Policy and the Superpowers* (London: William Kimber, 1971).

84. See, for example, Kissinger's argument to the Senate Foreign Relations Committee, 19 Sept. 1974, in Kissinger, *American Foreign Policy*, 3d ed. (New York: W. W. Norton, 1977), pp. 149–50.

85. Paul H. Nitze, "Assuring Strategic Stability in an Era of Détente," *Foreign Affairs* 54, no. 2 (1976): 207–32.

86. Data based on indices of defense spending, in constant dollars, from the annual *Military Balance* (London: International Institute for Strategic Studies).

87. For a summary of the Defense Planning Committee's 1977 Ministerial Guidance, see *NATO: Final Communiques, 1975–1980* (Brussels: NATO Information Service, 1981), pp. 71–74.

88. Schmidt's speech at the 10 May 1977 NATO summit, in *Survival* 19, no. 4 (1977): 177–78. Schmidt elaborated these concerns five months later at the IISS Alastair Buchan Memorial Lecture, 28 Oct. 1977, in *Survival* 20, no. 1 (1978): 2–10.

89. For an excellent summary, see Hans J. Neuman, *Nuclear Forces in Europe: A Handbook for the Debate* (London: International Institute for Strategic Studies, 1982), pp. 45–46. See also Gregory Treverton, *Nuclear Weapons in Europe*, Adelphi Paper 168 (London: IISS, 1981).

90. See *NATO: Final Communiques, 1975–1980*, pp. 121–23.

91. For President Reagan's speech of 18 Nov. 1981, in which he proposed not only the LRINF "zero option" but also his proposals for the Strategic Arms Reduction Talks (START), see *Survival* 24, no. 2, (1982): 87–89. For NATO's endorsement of the "zero option," see the 11 Dec. 1981 Ministerial Declaration in *NATO: Final Communiques, 1981–1985* (Brussels: NATO Information Service, 1986), p. 43.

92. See Leon Sigal, *Nuclear Forces in Europe: Enduring Dilemmas, Present Prospects* (Washington, D.C.: Brookings Institution, 1984), pp. 37–48.

93. For an excellent overview of the antinuclear movements in Europe during these debates, see Sigal, *Nuclear Forces*, pp. 70–103.

94. Charles W. Kegley, Jr. and Eugene R. Wittkopf, eds., *The Nuclear Reader: Strategy, Weapons, War* (New York: St. Martin's Press, 1985), incorporates an excellent sampling of this literature, including the Catholic Bishop's Letter, a critique by Albert Wohlstetter, plus excerpts from Jonathon Schell's *Fate of the Earth* (New York: Alfred A. Knopf, 1982), and Carl Sagan's "Nuclear War and Climatic Catastrophe" from *Foreign Affairs* 62, no. 2 (1983–84): 257–92.

95. Robert S. McNamara, "The Military Role of Nuclear Weapons: Perceptions and Misperceptions," *Foreign Affairs* 62, no. 1 (1983): 59–80, and those whom he cites in the opening paragraphs of the article.

96. Bundy et al., "Nuclear Weapons and the Atlantic Alliance," *Foreign Affairs* 60, no. 4 (1982): 753–68.

97. Karl Kaiser, Georg Leber, Alois Mertes, and Gen. Franz-Josef Schulze, "Nuclear Weapons and the Preservation of Peace," *Foreign Affairs* 60, no. 5 (1982): 1157–70.

98. Bernard W. Rogers, "The Atlantic Alliance: Prescriptions for a Difficult Decade," *Foreign Affairs* 60, no. 5 (1982): 1145–56.

99. See, for example, the Report of the European Security Study (ESECS), *Strengthening Conventional Defense in Europe: Proposals for the 1980s* (London: Macmillan Press, 1983). See also Keith A. Dunn and William O. Staudenmaier, eds. *Military Strategy in Transition* (Boulder, Colo.: Westview Press, 1984) and John D. Steinbruner and Leon V. Sigal, eds. *Alliance Security and the No First Use Question* (Washington, D.C.: Brookings Institution, 1983).

100. For a comprehensive review and critique of FOFA, see Jeffrey Record, "NATO's Forward Defense and Striking Deep," *Armed Forces Journal International* 121, no. 4 (1983): 42–48.

101. See, for example, Paul Nitze's articulation of the Reagan administration's "Strategic concept" in his "Alastair Buchan Memorial Lecture," in *Survival* 27, no. 3 (1985): 104.

102. See Secretary of State George Shultz's speech to the *Chicago Sun-Times* Forum, USIS Wireless File, 17 Nov. 1986.

103. For example, Henry A. Kissinger, "The Reykjavik Revolution: Putting Deterrence in Question," in *Washington Post*, 18 Nov. 1986. The NPG communique explicitly endorsed the maintenance of the US and UK nuclear deterrents (NATO Press Release M-NPG-2[86]32, 22 Oct. 1986).

104. See CDU/SCU Bundestag Floor Leader Volker Ruehe's comments in Bonn (27 Apr. 1987 *Koelner Stadt-Anzeiger*) and similar comments in London reported 14 May 1987 by the Associated Press. Also, "NATO's Tight Deadlines for Tough Missile Choices," in the *Independent* (UK), 21 Apr. 1987.

105. See the June 1987 North Atlantic Council (NAC) communique, NATO Press Information Service, for the expression of this arms control agenda. For President Reagan's own assurances that an INF accord would not lead to the "denuclearization" of Europe, see his 28 Apr. 1987 interview, in USIS Wireless File, 28 Apr. 1987. This "Reykjavik formula" continued to be repeated in subsequent Ministerial statements and in the declaration issued by NATO Heads of State and Government in Mar. 1988.

106. In a 26 Aug. 1987 press conference, Chancellor Kohl offered unilaterally to dismantle the Pershing Ias once the superpowers had carried out the reductions in a "double zero" INF accord. This effectively kept the FRG's systems outside of a bilateral superpower agreement while probably enabling that agreement to go forward.

107. For the text of the Montebello decision, annexed to the fall 1983 NPG Ministerial communique, see *NATO: Final Communiques, 1981–1985* (Brussels: NATO Press Information Service, 1986), p. 106.

108. See "Declaration of the Heads of State and

Goverment," NATO Press Communique M-1(89)21 (Brussels: NATO Press Information Office, 30 May 1989.)

109. This reaffirmation was particularly strong in Baker's speech before the Berlin Press Club, 12 De-

cember 1989—dubbed the "Baker Doctrine"—following the Malta summit and a Four Power meeting on Berlin and just prior to a meeting of NATO foreign ministers that endorsed a draft treaty for the CFE negotiations.

THE NORTH ATLANTIC TREATY

The full text of the North Atlantic Treaty is presented here.

WASHINGTON D.C., 4 APRIL 1949[1]

The Parties to this Treaty reaffirm their faith in the purposes and principles of the Charter of the United Nations and their desire to live in peace with all peoples and all Governments.

They are determined to safeguard the freedom, common heritage and civilization of their peoples, founded on the principles of democracy, individual liberty and the rule of law.

They seek to promote stability and well-being in the North Atlantic area.

They are resolved to unite their efforts for collective defence and for the preservation of peace and security.

They therefore agree to this North Atlantic Treaty:

ARTICLE 1

The Parties undertake, as set forth in the Charter of the United Nations, to settle any international dispute in which they may be involved by peaceful means in such a manner that international peace and security and justice are not endangered, and to refrain in their international relations from the threat or use of force in any manner inconsistent with the purposes of the United Nations.

ARTICLE 2

The Parties will contribute toward the further development of peaceful and friendly international relations by strengthening their free institutions, by bringing about a better understanding of the principles upon which these institutions are founded, and by promoting conditions of stability and well-being. They will seek to eliminate conflict in their international economic policies and will encourage economic collaboration between any or all of them.

ARTICLE 3

In order more effectively to achieve the objectives of this Treaty, the Parties, separately and jointly, by means of continuous and effective self-help and mutual aid, will maintain and develop their individual and collective capacity to resist armed attack.

ARTICLE 4

The parties will consult together whenever, in the opinion of any of them, the territorial integrity, political independence or security of any of the Parties is threatened.

ARTICLE 5

The Parties agree that an armed attack against one or more of them in Europe or North America shall be considered an attack against them all, and consequently they agree that, if such an armed attack occurs, each of them, in exercise of the right of individual or collective self-defence recognized by Article 51 of the Charter of the United Nations, will assist the Party or Parties so attacked by taking forthwith, individually, and in concert with the other Parties, such action as it deems necessary, including the use of armed force, to restore and maintain the security of the North Atlantic area.

Any such armed attack and all measures taken as a result thereof shall immediately be reported to the Security Council. Such measures shall be terminated when the Security Council has taken the measures necessary to restore and maintain international peace and security.

ARTICLE 6[2]

For the purpose of Article 5, an armed attack on one or more of the Parties is deemed to include an armed attack
—on the territory of any of the Parties in Europe or North America, on the Algerian De-

Reprinted from the NATO Handbook, *May 1986.*

partments of France,[3] on the territory of Turkey or on the islands under the jurisdiction of any of the Parties in the North Atlantic area north of the Tropic of Cancer;
—on the forces, vessels, or aircraft of any of the Parties, when in or over these territories or any area in Europe in which occupation forces of any of the Parties were stationed on the date when the Treaty entered into force or the Mediterranean Sea or the North Atlantic area north of the Tropic of Cancer.

ARTICLE 7

This Treaty does not affect, and shall not be interpreted as affecting, in any way the rights and obligations under the Charter of the Parties which are members of the United Nations, or the primary responsibility of the Security Council for the maintenance of international peace and security.

ARTICLE 8

Each Party declares that none of the international engagements now in force between it and any other of the Parties or any third State is in conflict with the provisions of this Treaty, and undertakes not to enter into any international engagement in conflict with this Treaty.

ARTICLE 9

The Parties hereby establish a Council, on which each of them shall be represented to consider matters concerning the implementation of this Treaty. The Council shall be so organized as to be able to meet promptly at any time. The Council shall set up such subsidiary bodies as may be necessary; in particular it shall establish immediately a defence committee which shall recommend measures for the implementation of Articles 3 and 5.

ARTICLE 10

The Parties may, by unanimous agreement, invite any other European State in a position to further the principles of this Treaty and to contribute to the security of the North Atlantic area to accede to this Treaty. Any State so invited may become a party to the Treaty by depositing its instrument of accession with the Government of the United States of America. The Government of the United States of America will inform each of the Parties of the deposit of each such instrument of accession.

ARTICLE 11

This Treaty shall be ratified and its provisions carried out by the Parties in accordance with their respective constitutional processes. The instruments of ratification shall be deposited as soon as possible with the Government of the United States of America, which will notify all the other signatories of each deposit. The Treaty shall enter into force between the States which have ratified it as soon as the ratification of the majority of the signatories, including the ratifications of Belgium, Canada, France, Luxembourg, the Netherlands, the United Kingdom and the United States, have been deposited and shall come into effect with respect to other States on the date of the deposit of their ratifications.

ARTICLE 12

After the Treaty has been in force for ten years, or at any time thereafter, the Parties shall, if any of them so requests, consult together for the purpose of reviewing the Treaty, having regard for the factors then affecting peace and security in the North Atlantic area including the development of universal as well as regional arrangements under the Charter of the United Nations for the maintenance of international peace and security.

ARTICLE 13

After the Treaty has been in force for twenty years, any Party may cease to be a Party one year after its notice of denunciation has been given to the Government of the United States of America, which will inform the Governments of the other Parties of the deposit of each notice of denunciation.

ARTICLE 14

This Treaty, of which the English and French texts are equally authentic, shall be deposited in the archives of the Government of the United States of America. Duly certified copies will be transmitted by the Government to the Governments of the other signatories.

NOTES

1. The Treaty came into force on August 24, 1949, after the deposition of the ratifications of all signatory states.
2. As amended by Article 2 of the Protocol to the North Atlantic Treaty on the accession of Greece and Turkey.
3. On January 16, 1963, the French Representative made a statement to the North Atlantic Council on the effects of the independence of Algeria on certain aspects of the North Atlantic Treaty. The Council noted that insofar as the former Algerian Departments of France were concerned the relevant clauses of this Treaty had become inapplicable as from July 3, 1962.

CHAPTER 7

ALLIANCE COMMITMENTS AND STRATEGIES
Asia

WILLIAM E. BERRY, JR.

In the post–World War II era, a constant goal of US foreign policy has been to contain the Soviet Union and its allies, first in Europe and then later in Asia. Nonetheless, the means to achieve this goal have changed from time to time. President Franklin Roosevelt hoped that by integrating the Soviet Union into the international system, the Soviets could be contained without more overt economic or military pressure. However, a series of events in Europe, the Middle East, and Asia occurred after the war which called into question whether this approach would be successful.

John Lewis Gaddis has noted that two distinct strategies of containment have evolved in the postwar years.[1] The first of these is asymmetrical containment and is most closely associated with George Kennan. The second is symmetrical containment, most closely associated with Paul Nitze. The primary distinction between these two strategies involves what variable drives the policy formulation process. In asymmetrical containment, US economic resources are viewed as limited, requiring that national interests be differentiated into vital and peripheral categories. Vital interests must be protected, but peripheral interests do not necessarily require an American military response when challenged. The economic means available play an important role in determining the policy goals to be pursued. Symmetrical containment, on the other hand, requires that the threat be met on the same level as it is issued. Threat, rather than resources, then becomes the independent variable. Since there is a more universal view of the threat, the distinction between vital and peripheral interests is blurred. Responding to that threat requires that resources must be expanded to meet it.

In this chapter on American security interests in Asia, the Asian policies of each administration from Harry Truman to Ronald Reagan are examined to determine which style of containment was used to meet the Soviet threat in Asia. US relations with the Philippines, Japan, and the Republic of Korea are emphasized because the

United States has mutual defense treaties with each of these countries and military forces stationed in all three. Both the treaty obligations and the forward-basing systems have been and remain critical to the attainment of US foreign policy goals in Asia. The American relationship with the People's Republic of China, the Republic of China (Taiwan), the countries of Indochina, and the members of the Association of Southeast Asian Nations (ASEAN) also are discussed although in less detail.[2]

THE TRUMAN ADMINISTRATION, 1947–1950

The two different containment strategies noted above have generally alternated between administrations, and even within a particular administration such changes have occurred on occasion. Such was the case with the Truman administration where asymmetrical containment more closely identifies US Asian policy between 1947 and early 1950 and symmetrical containment from mid-1950 until January 1953. In his famous March 1947 speech in which he articulated the "Truman Doctrine," President Truman pledged that "it must be the policy of the United States to support free peoples who are resisting attempted subjugation by armed minorities or by outside pressures."[3] Such a pledge appeared to commit the United States to assist countries throughout the world who were threatened either from internal or external sources. In actual practice, however, Truman's policies in Asia between 1947 and 1950 were much more limited in scope.[4]

In evaluating Truman policy initiatives in Asia during this period, it appears that three underlying assumptions were at work. First, while it was important that Asia not be dominated by any one hostile country, it was more important for the United States to maintain stability in Europe. In other words, the "Europe first" strategy that dominated military planning during the war continued to dominate the postwar period. Second, because of limited resources, the US should avoid military involvement on the Asian mainland. Third, the United States should attempt to align itself with the forces of Asian nationalism as the most effective means to counter the Soviet Union in Asia.[5]

These assumptions are closely associated with asymmetrical containment because resources were recognized as limited which in turn required a prioritization of interests. The Joint Chiefs of Staff in a 1947 report voiced similar views in their prioritization of US interests in Asia.[6] Japan and the Philippines were given particular attention as being important to US national security in this report.[7] These two countries plus Korea and China were all ranked in a listing of sixteen countries, but none of them was higher on the list than a European or Latin American country.[8]

In December 1949, the National Security Council published a Top Secret document, NSC-48/1, which attempted to evaluate the US position in Asia as the decade of the 1940s concluded.[9] The NSC identified the Soviet Union as the primary threat to US interests in Asia and recommended that the United States must contain Soviet expansion in the region. While cautioning that Europe remained more important than Asia, the NSC suggested that the US maintain its defenses in Asia along a line running from Japan in the north to the Philippines in the south. Conflict on the Asian mainland should be avoided because of the strain such involvement would place on limited resources.[10] Similarly, in his famous speech before the National Press Club in January 1950, Secretary of State Dean Acheson also described a defensive perimeter from the Aleutian Islands through Japan, the Ryukyu Islands, and the Philippines. He indicated the US was determined to protect and defend this perimeter.[11]

As the US began to define its postwar commitments to Asia, four points stand out. While European recovery clearly took precedence, the threat to US interests posed by the USSR and, after 1949, China remained significant in the US view. As the global power position of the US expanded, the need to preserve allied commitments grew. Yet, economic constraints and public demands for demobilization precluded any increase in military forces. Offshore island bases thus assumed preeminent importance. The following reviews of Truman policies toward the Philippines, China, Korea, and Japan highlight these points.

THE PHILIPPINES

As a former American colony, the US military presence in the Philippines dated to 1898, interrupted only by the Japanese defeat of US forces at Bataan and Corregidor in 1942. Shortly after the Japanese attack, President Roosevelt promised the Philippine people that their freedom would be restored and their independence granted and protected.[12] In response to a presidential request, the Congress passed Joint Resolution 93 in June 1944 which authorized the president to acquire land in the Philippines (in consultation with Philippine leaders) for US military bases to protect mutual US and Philippine interests.[13] From the American perspective in 1943–44, military bases in the Philippines were deemed important because of their possible use to launch attacks on the Japanese home islands. By mid-1945, however, this view changed as islands closer to Japan, such as Taiwan and Okinawa, were captured. Consequently, the Philippine bases became more important as a deterrent against future aggression in the Pacific.[14]

By late 1946 and early 1947, both the US Army and Navy were somewhat ambivalent about the retention of bases in the Philippines. In the case of the Navy, there was concern that the Congress would not appropriate the necessary funds to retain its bases in the Philippines and still build the facilities the Navy wanted on Okinawa and Guam.[15] The Army also was influenced by economic considerations. Since Philippine President Manuel Roxas was adamant that US bases not be located in Manila, expensive bases would have to be built elsewhere, especially port facilities. Like their Navy colleagues, Army leaders feared the Congress would be reluctant to fund these new bases, particularly when new facilities were required in Europe.[16]

Influenced by these arguments and others, President Truman directed in December 1946 that all US Army forces were to be withdrawn from the Philippines.[17] President Roxas opposed this planned withdrawal in the strongest possible terms and recalled past pledges of the American commitment to Philippine security. Without the US military presence, he argued, the American showcase of democracy in Asia could be seriously threatened.[18] Truman subsequently rescinded his withdrawal order but withdrew most of the troops, leaving only a skeletal force.[19]

In March 1947, negotiators representing the two countries concluded the Military Bases Agreement (MBA).[20] An involved document, the MBA provided for the retention of US military bases in the Philippines as well as the addition of necessary facilities in the future. The two major installations, Clark Air Base and Subic Bay Naval Base, were included in the list of retained bases. The MBA was to remain in effect for 99 years.

American ambivalence on the retention of bases in the Philippines was influenced by economic, strategic, and political considerations. The rapid and extensive demobilization of US military forces coupled with increased tension in Europe and the Middle East dictated that the number of forces in the Philippines be reduced significantly. The Truman administration's stringent fiscal policies also were important in reducing these forces. From the strategic perspective, there were other locations (such as Guam and Okinawa) which were more attractive to the military than were the facilities in the Philippines. However, there were also political factors involved. The United States had a special relationship with the Philippines because of the colonial history. If the US did not make the effort to provide for Philippine security until the Philippines acquired the necessary capabilities, American prestige would suffer in the international community. The retention of a reduced presence served as a compromise among these competing considerations.

By the latter part of 1949, US security policy involving the Philippines shifted somewhat as the Truman administration focused more attention on Asia. First, the international situation in Asia had changed with Mao's victory in China and the establishment of the People's Republic of China (PRC) in October 1949. Second, Philippine leaders constantly expressed their concern that the United States did not have sufficient forces and military hardware in the Philippines to deter or defend against an attack.[21]

In August, President Elpidio Quirino made a state visit to Washington to confer with President Truman and other administration officials and to request additional US military assistance.[22] While promising as much military and economic assistance as posssible, Truman cautioned that his most immediate concern was overcoming congressional resistance to funding the European Recovery Program (ERP or the Marshall Plan). Any other aid programs would have to wait until the ERP was passed.[23]

Fiscal constraints also were foremost in the minds of the US military establishment. In a letter written in September 1949, Secretary of Defense Louis Johnson recognized the increased strategic importance of US bases in the Philippines as developments in China became more ominous. But he believed the major problem confronting the US military in the Philippines was economic. The military budget simply did not provide sufficient funding during peacetime to keep all of the military forces at the levels required if war occurred. In Johnson's view, the US should maintain the most essential Philippine bases so that force levels and facilities could be increased if a crisis occurred.[24]

Johnson's concerns about the military budget were valid. The fiscal year 1949 military allocation was $14.4 billion, but Truman had indicated that maintenance of a balanced budget would require reductions in military spending by as much as $5 billion.[25] By the end of 1948, the postwar demobilization had largely been accomplished with the total US military force reduced from 10 million at the end of the war to approximately 1.3 million.[26]

The Department of State also attempted to assuage Philippine doubts about US resolve to defend the Philippines. In a draft letter to the National Security Council in June 1950, just a few days before the beginning of the Korean war, the State Department affirmed the Philippines' political and strategic importance. Politically, the "Philippine experiment," as it was described, was an attempt to foster democracy in Asia based on the American model and had to be preserved. The Philippines also had an important strategic mission: "To contain the tide of Communism in its present limits and eventually to reverse its present expansionist trend."[27]

In order to forestall the collapse of the government, the State Department advocated economic assistance so that the Philippines could increase its exports and reduce its balance of payments deficit, and recommended internal reforms to reduce graft and corruption.[28] There was no recommendation to increase US forces stationed in the Philippines or to upgrade base facilities.

The Philippine-US security relationship between 1947 and 1950 illustrates a broader dilemma which continues to confront US policymakers. As it assumed a truly international role, the US not only had to adjust to changing conditions in the international system but also had to weigh the relative merits of political-economic initiatives versus military involvement in sustaining the American presence in the region. In the case of the Philippines, most experts believed the former were more important, but the verdict was not unanimous, and the debate would continue.

The combination of a "Europe-first" prioritization of interest and the paramount concern for limited resources was typical of asymmetrical containment, making it difficult for the US to convince Asian allies of its commitment to peace and security in the region. American prestige and reputation as the dominant international actor remained a major factor in maintaining the status quo in Asia. This tension between fiscal constraints on the one hand and US prestige and reputation as a dependable ally is an enduring feature of American policy.

THE CHINA ISSUE

Controversy still surrounds the Chinese civil war (1945–49) and whether the United States could have done more to save Chiang Kai-shek and his Kuomintang regime. The purpose here is not to examine in detail the "loss of China," but rather to survey briefly some of the political, economic, and military factors which influenced the Truman administration policies involving China from 1947 to 1950.[29] In December 1945, President Truman dispatched General George Marshall to China in an attempt to find a political solution to the Chinese civil war. Marshall remained in China until January 1947, where he unsuccessfully tried to influence Mao Tse-tung and Chiang Kai-shek to form some sort of coalition government.

Upon his return to the US, Marshall submitted a report to the president which indicated that urgent political and economic reforms were necessary in China if Chiang Kai-shek were to succeed. He recommended continued economic support for China but cautioned against US military involvement.[30] Marshall's advice and recommendations were particularly important because of Truman's respect for Marshall and because Truman had serious doubts about how much more the US could or should do to save Chiang.[31]

For his part, Chiang perhaps overestimated his political leverage as he tried to influence the US to increase American assistance and place less emphasis on reforms. There also was a dispute between the State Department and the Defense Department as to the proper US policy.[32] The Defense Department tended to view the Chinese civil war in the context of US-Soviet competition. Consequently, the JCS, for example, urged increased military assistance programs and placed less emphasis on domestic reforms in China than did the State Department which feared that increased military assistance could lead to disastrous American military involvement.[33]

Because of these differences concerning what the correct policy should be, the president sent Lieutenant General Albert Wedemeyer to China in July 1947 where he remained until September. He recommended increased economic and military assistance for Chiang as being the only means to prevent a communist victory, and he advocated sending both economic and military advisers to China.[34] Marshall, now secretary of state, opposed sending military advisers and increasing military support. The secretary of state, for whom Europe remained a higher priority, was concerned that the presence of advisers would lead to greater military intervention and involvement in a land war in Asia which he wanted to prevent.[35] Ultimately, the administration recommended a China aid program of $570 million, but Congress provided only $275 million in economic assistance and $125 million for military equipment.[36]

Conditions in China continued to deteriorate during 1948 both on the battlefield and in the economy, leading to a further loss of confidence in the Chiang regime.[37] In NSC-22, published in July 1948, the NSC warned that further increases in economic assistance for China would jeopardize the ERP.[38] NSC-34, published in October 1948, even raised the question of how reliable a China under Mao's leadership would be as an ally of the Soviet Union, citing both Chinese nationalism and the Tito example in Yugoslavia. Fundamentally, the NSC recognized that there were definite limitations on what the US could do to influence the course of events in China.[39] In the final analysis, events in China determined the outcome, and on 1 October 1949, Mao established the People's Republic of China, and Chiang Kai-shek retreated to the island of Taiwan which became the Republic of China (ROC).

After the communist victory, the Truman administration had to address three issues. Should the US recognize the PRC? Which country should have the Chinese seat in the United Nations? What should be US policy toward the ROC?

The recognition question involved both domestic and international issues. The "China Lobby" in the US argued strongly that the Truman administration had not done enough to "save" China, and therefore recognition should not be extended.[40] After the PRC seized the US consular premises and property in Peking in January 1950, public support for recognition declined further.[41] Moreover, the Sino-Soviet Treaty of Friendship and Alliance, signed in 1950, convinced many in the US that China would be a Soviet puppet.[42] Great Britain, on the other hand, argued that communist control of the mainland was a fact that had to be accepted. Despite US objections, diplomatic relations between Britain and the PRC were established in January 1950.[43]

The question of the United Nations seat was closely associated with the diplomatic recognition issue, but the US tried a slightly different approach. Rather than directly opposing the PRC assumption of the UN seat, the US representative to the UN tried to drive a wedge between the USSR and PRC by criticizing Soviet actions in Manchuria. When the Soviets presented a motion before the Security Council to seat the PRC, the American delegate indicated that the US would not use its veto and would accept the decision for admission if seven members of the Council so voted. On 13 January 1950, the vote was taken and PRC admission was rejected by a six to three vote.[44] Interestingly enough, it was after this vote that the Soviet delegate walked out of the security Council and would not return until after the UN decision to send troops into Korea.

The US relationship with the ROC was perhaps the most difficult of these three issues to resolve, primarily because of the China Lobby. Again a dispute developed between the departments of Defense and State. Here the dispute was not over recognition, since the US still recognized the Nationalist government even though its seat of power had changed, but rather whether the US should provide military assistance, and if so, what kind? General Douglas MacArthur viewed Taiwan as an "unsinkable aircraft carrier" and of extreme importance to US security interest in Asia. As such he recommended that the US regard an attack on Taiwan as an act of war and urged an increase in military assistance programs.[45] Secretary of State Dean Acheson, on the other hand, advised against any military efforts to assist in the defense of Taiwan.[46] Interestingly, when Acheson outlined his defensive perimeter in his January 1950 National Press Club speech, he did not mention Taiwan as part of that perimeter.

US policy toward China from 1947 to 1950 reflected many of the same factors evident in US-Philippine relations, specifically economic constraints and a realization that there were limits to what the US could accomplish. Other countries were viewed as more important to US national security interests, and economic and military assistance programs for China reflected this prioritization. However, US prestige was involved, making it more difficult to reduce support for Chiang Kai-shek without serious domestic and international implications.

KOREA

During World War II, Korea did not figure prominently in the conferences and discussions among allied leaders. At the Cairo conference in late 1943, Chiang and Churchill agreed with the Roosevelt initiative that Japan should lose its colonial possessions after the war. In reference to Korea, which Japan annexed in 1909, the Cairo Communique indicated that "in due course Korea shall become free and independent."[47] However, Stalin was not a participant at the Cairo conference, so Soviet views on the future of Korea at that time were unclear.

At the Yalta conference, Roosevelt suggested that a Four-Power trusteeship be established until Korea could become independent. The four powers were to be the US, Great Britain, China, and the Soviet Union. Reportedly, Stalin at least verbally approved this suggestion.[48] After the Soviets entered the war against Japan in early August 1945, Soviet troops were dispatched to the northern part of the Korean peninsula. The quick collapse of the Japanese forced the United States to make a decision concerning the future American military presence in Korea. The US recommended that the country be tem-

porarily divided at the 38th parallel with the Soviets north of this line and the Americans in the southern half.[49] US military forces actually did not arrive until September 1945, and they were ill-prepared as an occupation force in Korea since they originally were to participate in the occupation of Japan.[50]

As the occupation of Korea continued, US leaders confronted many of the same conflicting goals which were evident in Asia in general. Once the Amerian presence was established, US prestige became involved which made a withdrawal more difficult without jeopardizing US commitments to other countries. Even within the military, there was a difference of opinion. One faction argued that if the US withdrew its military forces, the Soviets would gain control of the entire peninsula.[51] By the end of 1947, however, the Joint Chiefs of Staff formally agreed with the opposite view that Korea had little strategic value, and the 45,000 US troops stationed there could be deployed elsewhere more effectively. US air power from Japan could be used to offset any Soviet gains in the southern part of Korea.[52] The National Security Council recommended in April 1948 that the US support its allies in the south within "practical and feasible limits" while at the same time limiting its military presence.[53]

Since efforts to unify the peninsula were unsuccessful, and because the US military decided that Korea was not vital to US security interests, the State Department urged that the United Nations should be encouraged to play a more important role in Korea. Accordingly the UN established the United Nations Temporary Commission on Korea (UNTCOK) which arrived in Korea during January 1948. UNTCOK's major function was to supervise the nationwide elections which the General Assembly mandated in 1947. However, because UNTCOK representatives were not able to proceed north of the 38th parallel, these representatives decided that elections were possible only in the south.[54] In May 1948, National Assembly elections were held, and in July the Republic of Korea (ROK) came into existence through the adoption of a constitution by the newly elected National Assembly. This legislative body then elected Syngman Rhee as the first ROK president. In September 1948, the Democratic People's Republic of Korea (DPRK) was formed in the North under the leadership of Kim Il Sung.

In the effort to reduce tensions on the peninsula after the creation of two independent states, the UN General Assembly passed a resolution in December 1948 that all occupation forces should be removed from Korea "as early as practicable."[55] The Soviet Union indicated that it intended to comply with the UN resolution by the end of 1948. Similarly, in March

1949, the National Security Council issued a report recommending that all US forces, with the exception of a small advisory unit, be removed by June 1949.[56] By the end of June 1949, the last US combat units were withdrawn.

The decision to withdraw military forces from South Korea was influenced by both economic and security considerations. The Joints Chiefs of Staff believed that the US was overextended and that the forces assigned in Korea could be better used elsewhere. An opposing and less persuasive argument was that, with the demise of Chiang Kai-shek's forces in China, Korea provided the last foothold on the Asian mainland, and was, therefore, of increased importance to the success of the occupation in Japan.[57] The administration decided that there was no acceptable choice other than to reduce its presence in Korea because of the importance of other commitments. Economic considerations dominated political and security interests, and this would remain true until June 1950. Therefore, Acheson's January 1950 National Press Club speech reflected the culmination of a policy debate which had been ongoing for some time and was not a new initiative on the part of the Secretary of State as some of his critics claimed.

JAPAN

Whereas US occupation planning for Korea was largely nonexistent before American troops arrived, such was not the case with the occupation of Japan between 1945 and 1952. Also, the United States dominated the occupation of Japan to a far greater extent than was true in either Germany or Korea.[58] A State Department document entitled "United States Initial Post-Surrender Policy for Japan," dated 6 September 1945 and approved by President Truman, indicated that two of the most important occupation goals were the demilitarization and democratization of Japan.[59]

Demilitarization involved disarming and demobilizing Japanese military forces. This was relatively easy to accomplish since many Japanese blamed the military for leading the country into war. A more difficult task was to try to prevent the reversion to militarism in the future. Occupation authorities, particularly General Douglas MacArthur, the Supreme Commander Allied Powers (SCAP), determined that a new Japanese constitution should be written which would address these long-term concerns. MacArthur's staff drafted a new constitution in 1946 which was subsequently accepted by the Japanese government in May 1947 as an amendment to the old 1889 Meiji Constitution. Article 9 of the new constitution addressed demilitarization directly and stated in part:

The Japanese people forever renounce war as a sovereign right of the nation and the threat of the use of force as means of settling international disputes.

In order to accomplish the aim of the preceding paragraph, land, sea, and air forces, as well as other war potential, will never be maintained. The right of belligerency of the state will not be recognized.[60]

This article is important not only because of the obvious restriction on the sovereign rights of Japan, but also because of the effects it would have on future American-Japanese security issues.

Occupation officials and US policymakers in Washington realized that these demilitarization measures must be supplemented by fundamental democratic reforms if Japan were to be less likely to go to war in the future. Accordingly, the 1947 constitution provided for numerous political, economic, and social reforms. Some of the most important political reforms included reducing the role of the Emperor and increasing power for the legislature (the Diet); extending the franchise to women and reducing the voting age to twenty; and decentralizing the educational system.[61] Economic reforms included an attempt to break up the *Zaibatsu*, the family-dominated holding companies which exercised tremendous economic and political power in prewar Japan; labor liberalization including the right to organize and enter into collective bargaining; and a land reform program which attempted to create a new class of independent farmers. Social reforms provided for equal rights for women and equal distribution of property to children regardless of their sex.[62]

As the occupation continued, Japan became more important to the United States as a potential major Asian ally particularly as the situation in China deteriorated during the civil war and as US-Soviet relations soured. Occupation policies reflected these changing conditions, particularly in the emphasis placed upon Japanese economic recovery in the 1947–48 period. Toward this goal of economic recovery and rehabilitation, the US Congress appropriated $150 million in May 1948, and occupation officials concentrated on controlling inflation, ending reparation payments, and slowing some of the earlier efforts to decentralize the economy.[63]

As important as these changes in occupation policies were, however, a larger issue developed concerning the negotiation of the peace treaty with Japan. The State and Defense departments had different views on when a peace treaty should be negotiated. NSC-49, dated 15 June 1945, represented the latter's view that a peace treaty should be delayed because of the high strategic value of Japan to the United States. The offshore island chain of bases was viewed by military planners as being extremely important and, if there were no peace treaty, the US would be assured of continued access to

Japanese bases. For the Defense Department the primary threat to Japan was external in nature, and the US military presence served as a deterrent to aggression.[64]

The State Department agreed that Japan was of increasing importance to the US. The difference, however, was over an evaluation of the threat and how the US could best confront this threat. Whereas the Defense Department emphasized the external threat, the State Department stressed the internal threat and stated that a continuation of the occupation over a long period jeopardized future US-Japanese security relations because the Japanese people were becoming disillusioned with the occupation. American security interests could best be protected by an early peace treaty which included provisions "for essential U.S. military needs in Japan."[65]

By late 1949, the Truman administration, with MacArthur's support, approved the State Department's position. There was no disagreement over whether US military forces should remain in Japan; the issue was how best to assure the presence of these forces.[66]

SUMMARY: 1947–1950

This review of Truman administration security policies in Asia between 1947 and early 1950 is important because it highlights some of the contradictions in US policy during this period of time and how policymakers attempted to resolve them. The administration understood that the international system had changed as a result of the war and that the United States had a global role to play. However, because resources were limited, priorities had to be established, and the administration saw Europe as being more important to US interests than Asia. Although the regional situation in Asia changed, particularly with Mao's victory in China and the development of closer Chinese-Soviet relations, the United States limited its reponses to these events. The adminstration wanted to contain the Soviet threat but chose to do so by concentrating on a series of offshore bases rather than by continuing or increasing its military presence on the Asian mainland. The president understood that American prestige as a superpower was affected adversely by events in China and Korea, but economic factors precluded the US from taking stronger measures to protect its interests. This application of asymmetrical containment would change dramatically with the beginning of the Korean war in June 1950.

THE TRUMAN ADMINSTRATION, 1950–1953

Even before the Korean War began, President Truman became convinced that a reevaluation of American security policy was needed, based on the victory of Mao's forces in China, the Soviet Union's explosion of a nuclear device in 1949, and the American decision to build a fusion bomb. Consequently, a special State and Defense Department study group, under the direction of Paul Nitze, began its examination of US security policy in February and March 1950. In April, this group submitted its report, NSC-68, to President Truman.[67]

The authors of NSC-68, which one authority has called "the first comprehensive statement of a national strategy,"[68] viewed the international system as a zero-sum game in which a gain for the Soviet Union was a loss for the United States. This interpretation tended to blur the distinction between vital and peripheral interests and required an American response to Soviet aggression wherever it occurred. As such, NSC-68 refuted many of the earlier views on containment associated with George Kennan, which had influenced the early years of the Truman administration. Whereas Kennan believed emphasis should be placed upon Soviet intentions in attempting to predict Russian behavior, Nitze believed it was more important to examine Soviet capabilities. As he and his associates saw increasing Soviet capabilities, they recommended strongly that the US must build up its own capabilities to meet this challenge.[69]

Prior to 1950, economic considerations, particularly the threat of rampant inflation, served to constrain defense budgets. NSC-68 challenged this view by referring to the World War II example when defense expenditures tripled without debilitating inflation. Although no cost figure was included in NSC-68, estimates suggested that the required military buildup would cost approximately $50 billion annually, or nearly 3½ times the defense budget of $13.5 billion for fiscal 1951.[70] In the final analysis, NSC-68 was an attempt to convince the bureaucracy, Congress, and the public that the Soviet threat was real, and that the US must respond to this challenge by increasing its military capabilities. Although President Truman indicated his support for most of the NSC-68 recommendations, congressional approval of the required funds was in doubt. After the North Korean attack on the Republic of Korea on 25 June 1950, however, Congress appropriated $48.2 billion for fiscal 1951.[71]

The Korean war confirmed the sentiments of those who believed the US must be prepared to meet Soviet aggression even in areas such as Korea which previously had been identified as not vital to American security interests. The Truman administration responded on both political and military levels. On the former, the US delegate to the Security Council introduced a resolution to that body calling upon North

Korea to cease hostilities and withdraw from the ROK. Since the Soviets were still boycotting the Security Council over the presence of the Nationalist Chinese, the resolution was approved by a 9 to 0 vote.[72] On 27 June, a second resolution was introduced, again with a Soviet boycott, calling upon UN members to "furnish such assistance to the Republic of Korea as may be necessary to repel the armed attack and to restore international peace and security in the area." This resolution also passed with only Yugoslavia voting in opposition and was the justification for the UN police force action in Korea.[73]

On the military front, the president proposed several courses of action which he announced in a statement delivered on 27 June. The president ordered General MacArthur to provide Korean military forces with arms and equipment beyond what was authorized in the Military Assistance Program. Also, American military advisers were to remain with their Korean units rather than be withdrawn. For Truman, the North Korean attack was clear evidence that communism had become a more overt threat in Asia and required American action beyond the Korean peninsula. Accordingly, he ordered parts of the Seventh Fleet to sail from the Philippines and assume a position between Taiwan and the mainland to prevent a PRC attack on Taiwan or vice versa. He directed that US forces in the Philippines be strengthened and that military assistance be increased. Finally, Truman decided that the United States should increase its military assistance to French forces fighting in Indochina and dispatched a US military mission to the region "to provide close working relations with those forces."[74] Since there was concern that the attack in Korea might only be a diversion for a Soviet move against Western Europe, the administration also decided in September 1950 to send between four and six divisions to Europe as part of a NATO defense force.[75]

The importance of the Korean war to the American alliance system in Asia cannot be overstated. One expert has identified this as "the decisive event in the evolution of American alliance policy."[76] The changes in policies toward Taiwan and Korea are evidence of this rather rapid evolution since neither country had previously been viewed as vital to American security interests. The increased assistance to the French and the stationing of US military advisers in Indochina risked the possibility of American involvement in another war on the Asian mainland, a possibility which only a short time before in 1949 influenced the decision to withdraw US combat forces from Korea. President Truman believed the Soviets were responsible for the North Korean attack and, after the Chinese became actively involved in the fighting, Truman applied US containment policy to the PRC. The Soviet-Chinese relationship was perceived as a monolith which had to be confronted. The president was convinced that aggression could not be appeased if a repetition of World War II was to be avoided.[77]

In addition to the immediate military steps which were taken to support the war in Korea and the increased military assistance programs for the Philippines and the French in Indochina, the US alliance system in Asia was strengthened by a series of bilateral and multilateral defense treaties during the remainder of the Truman administration. Initially, Truman and other administration officials were not enthusiastic about a mutual defense treaty with the Philippines because they considered the American bases and military units assigned there to be sufficient to deter aggression. However this view began to change as progress was made toward the conclusion of the peace treaty with Japan. Truman was anxious to end the occupation because he feared its continuation would adversely affect Japanese-American relations, particularly important during the conduct of the Korean war. To conclude this treaty with Japan, Truman understood he would need the support of a number of countries in Asia including the Philippines, Australia, and New Zealand.

Truman selected John Foster Dulles to be the primary negotiator with the Japanese for the peace treaty and to secure the support of US allies in Asia. To obtain this support, Dulles had to overcome both economic and security concerns which these countries had relating to Japan. The security problem was resolved by a series of three interlocking pacts.[78] The first of these was the US-Philippine Mutual Defense Treaty (MDT), signed on 30 August 1951; the second was a tripartite agreement among the United States, Australia, and New Zealand, known as the ANZUS Pact; the third was the US-Japanese Security Treaty which provided for Japanese defense through the stationing of American military forces there. The last two treaties were signed on 8 September 1951.[79] Ten days later, the peace treaty with Japan was signed in San Francisco, officially ending World War II.[80]

The Truman administration underwent a major transformation in its security policies after 1950, particularly in Asia. Whereas economic considerations were major determining factors prior to 1950, the communist threat became more important after the Korean war. As a result, fewer distinctions were made between vital and peripheral interests. The containment of the Soviet Union and its Chinese ally became much more important regardless of where this challenge occurred. The United States not only increased its own military capabilities in Asia, but entered into a series of security treaties with several of its allies and increased military assistance programs to others. Consequently, American prestige became more closely identified with protecting the status quo in Asia. The change from the asymmetrical to the symmetrical style of containment had occurred with far-reaching results.

THE EISENHOWER ADMINISTRATION

When Dwight Eisenhower became president in January 1953, he was confronted with a dilemma concerning America defense policies, particularly in Asia. The Korean war had become unpopular in the United States: it was a limited war in the sense of American military goals but expensive to the US Treasury. There was strong public sentiment to bring the troops home and reduce defense spending. On the other hand, American commitments in Asia had expanded during the Truman administration, and the new president found it difficult to reduce these obligations without risking the loss of American prestige in the international community. Achieving "security with solvency" became a major goal of the Eisenhower administration.[81]

The administration's answer to this dilemma was the "New Look" or massive retaliation strategy included in NSC-162/2 and approved by the president in October 1953. Secretary of State John Foster Dulles presented one of the clearest statements on the economic and security factors involved with the formulation of the massive retaliation strategy in a speech delivered in January 1954. In this speech, Dulles criticized the economic costs of the Truman defense efforts, arguing that the resultant budgetary deficits could not continue without grave economic and social consequences.

Dulles's answer was to depend "primarily upon a great capacity to retaliate, instantly, by means and at places of our own choosing . . . As a result, it is now possible to get, and share, more basic security at less cost."[82] The Eisenhower administration had been critical of President Truman for not effectively incorporating nuclear weapons into US security planning. For Eisenhower, both strategic and tactical nuclear weapons would play a more predominant role in such plans.[83]

President Eisenhower envisioned nuclear weapons as deterring an attack not only on the United States but on American allies as well. He also believed it necessary to expand the American alliance system in Asia to confront what he perceived to be a more threatening Sino-Soviet challenge. Like Truman, Eisenhower feared appeasing an aggressor, and this fear found expression in what became known as Eisenhower's "domino theory."[84]

In order to keep the dominoes from falling in Asia, President Eisenhower and Secretary

Dulles continued the alliance-building begun by President Truman. On 1 October 1953, the US and ROK signed a mutual defense treaty which committed each side to come to the aid of the other "in accordance with its constitutional processes." Article IV of this treaty also provided for US land, air, and sea forces to be assigned in the ROK.[85] In December 1954, a similar mutual defense treaty with the Republic of China (Taiwan) was signed which again committed each side to assist the other "in accordance with its constitutional processes."[86] Article IV stipulated that not only Taiwan but also the nearby Pescadores Islands were included, and Article VII allowed US land, air, and sea forces to be stationed in the Republic of China. With this treaty, the offshore chain of bases, envisioned by Acheson and MacArthur earlier, was completed so that the US had basing rights from Japan in the north to the Philippines in the south, plus similar rights on the Asian mainland in Korea.

In September 1954, primarily at the urging of the United States, representatives of eight countries met in Manila to sign the Manila Pact and create the Southeast Asia Treaty Organization (SEATO).[87] Article 4 provided that armed aggression in the treaty area should be resisted by the signatories to meet the common danger "in accordance with its constitutional processes." The term "treaty area" was defined in Article 8 as the general area of Southeast Asia. In an attached understanding, the United States stipulated that the aggression mentioned in Article 4 referred only to "Communist aggression."

The pact also included a protocol which contained a very important provision. Cambodia, Laos, and the "free territory under the jurisdiction of the State of Vietnam" were designated as states covered by the provisions of Article 4. In other words, SEATO members could "act to meet the common danger" if communist aggression occurred in these three states. The eight signatories of the Manila Pact and thereby SEATO members were the United States, Australia, France, New Zealand, Pakistan, the Philippines, Thailand, and Great Britain. Interestingly, only the Philippines and Thailand were Southeast Asian countries. The US Senate ratifed the SEATO treaty in February 1955 by a vote of 82 to 1.[88]

What, if any, were the successes of the massive retaliation strategy as it was applied to Asia during the Eisenhower years? The evidence is somewhat ambiguous in response to this question. During the negotiations to end the Korean war in May 1953, American spokesmen apparently suggested that President Eisenhower was at least considering using nuclear weapons against Chinese military bases if the negotiations to end the war were not successful.[89] The possibility of using nuclear weapons was also considered in April 1954 as the French position at Dienbienphu became increasingly tenuous. Admiral Arthur Radford, chairman of the Joint Chiefs of Staff, recommended that US air power, apparently including nuclear weapons, be used against Viet Minh positions to save the French. Eisenhower, however, insisted that, before the US would commit such forces, he would need the cooperation and support of US allies, particularly Britain, as well as the support of the Congress. When this cooperation and support were not forthcoming, Eisenhower decided not to intervene.[90]

In late 1954 and early 1955, the PRC conducted military operations against the islands of Quemoy and Matsu, between the mainland and Taiwan. Chiang Kai-shek had military forces stationed on both these islands and feared that if the PRC took them, Taiwan and the Pescadores Islands would be in jeopardy. Dulles advised Eisenhower that nuclear weapons would have to be used against Chinese airfields if the US chose to defend these two small islands. In March, Dulles made a speech that stressed that the US had strategic striking power available. Shortly afterwards, the president explained that he saw no reason why nuclear weapons couldn't be used on military targets. Subsequently in May 1955, a cease-fire was established and the crisis over Matsu and Quemoy passed for the time being.[91] Based on this review, the answer to the question about the effectiveness of the threat of massive retaliation in Asia is mixed. What is clear, however, is that the Eisenhower administration at least considered using nuclear weapons and conveyed this threat more than the previous administration had.

INDOCHINA

Although references have already been made to Indochina, more detail is required to understand how President Eisenhower enlarged the American commitment there, particularly in Vietnam. President Truman had increased American military support to the French in Indochina after the beginning of the Korean war and had also sent American advisers. What began as relatively modest aid ($10 million in 1950) increased by fiscal 1954 to $1.063 billion or approximately 78 percent of the French military burden.[92] Despite this level of assistance, the French were unable to defeat the Viet Minh. In fact, with the defeat at Dienbienphu in May 1954, the French military efforts in Vietnam came to an end.

The Geneva Conference, beginning in July 1954, attempted to resolve this conflict. The Geneva Agreements provided that the Democratic Republic of Vietnam (DRV) would control the territory north of the 17th parallel. French control over the area to the south of this parallel

was to continue for a period of time, but administrative control was ceded to Emperor Bao Dai and his "state of Vietnam." General elections were to be held in July 1956 to reunite the country. Cambodia and Laos were to be demilitarized, neutral, and free of foreign intervention. However, since no general elections were conducted, two states occupied the Vietnamese peninsula after 1954, the DRV above the 17th parallel and the Republic of Vietnam (RVN) below that line.[93]

The RVN government from June 1954 was headed by Ngo Dinh Diem who quickly gained the support of the Eisenhower administration in the form of economic and military assistance. Initially, Diem performed relatively well and introduced some reforms which contributed to his popularity and effectiveness. Over time, however, his autocratic and dictatorial practices plus the abuses of some of his relatives in high positions began to work against him. In addition, the DRV under the leadership of Ho Chi Minh was slowly rebuilding its economy after the war against France, and the goal of reuniting the country under DRV control remained very important to Ho and his supporters.[94] By the end of the Eisenhower administration, the early successes Diem achieved were beginning to wane and popular discontent increased. In 1960 some of those in the south opposed to the Diem regime formed the National Liberation Front which became the successor to the Viet Minh. The United States was becoming more deeply involved in Vietnam and tried to persuade Diem to initiate reforms, but with little success.

CHINA

Few efforts were made to improve US-PRC relations during the Eisenhower years. Chinese participation in the Korean war, closer US-ROC relations as evidenced by their mutual defense treaty and support for the ROC's position on Quemoy and Matsu, and the PRC–North Vietnamese relationship all made progress very difficult. In addition, Eisenhower and Dulles became convinced that the best way to play upon differences between China and the Soviet Union was to force the former to make increasing demands upon the latter for economic and military assistance. Therefore, the US attempted to put pressure on the PRC during the Quemoy and Matsu crises in both 1954–55 and again in 1958 in part to force the PRC to make requests which the Soviets couldn't meet and thereby cause dissatisfaction in the alliance.[95] Arguably, the administration was at least partially successful in this effort.

JAPAN

The Eisenhower administration's security policies with Japan revolved around two related issues: the creation of the Japanese Self Defense Force (SDF) and the renegotiation of the 1951 security treaty. Secretary of State Dulles strongly believed that Japan must be prepared to do more to provide for its own defense, however, the Japanese government under Prime Minister Yoshida resisted this pressure because of constitutional restrictions under Article 9 and memories of the military role in prewar Japan.

Nonetheless, the two countries entered into a Mutual Defense Assistance Agreement in March 1954 which included the provision that Japan would fulfill its military obligations under the security treaty in the context of its own defense.[96] Later in the summer of 1954, Yoshida was able to convince the Diet to pass the Defense Agency Establishment Law and the Self-Defense Forces Law after lengthy debate. These laws established the National Defense Agency and the Ground, Air, and Maritime Self Defense Forces. The initial SDF authorization was for 152,100 men, but this number increased to 214,182 in 1956 partly in response to the reduction in US forces stationed in Japan from approximately 200,000 in December 1954 to 90,000 men in December 1956.[97]

The Japanese had found several provisions in the 1951 Security Treaty onerous as the decade of the 1950s progressed, and the government pressed the United States to renegotiate this treaty.[98] Article I of the 1951 treaty not only allowed the US to station its military forces in Japan, but also provided that these forces could be used to suppress internal disturbances if the Japanese government so requested. This affront to Japanese sovereignty was obvious and became a point of contention between the two countries. The treaty also made no provision for consultations between the two countries concerning the use of US bases on Japanese soil, and there was no expiration date for the treaty.

After long negotiations, the two countries signed the Treaty of Mutual Cooperation and Security Between the United States and Japan on 19 January 1960.[99] In this version, there was no mention of a domestic role for US military forces to play in Japan, provisions were established for consultations between the two countries, and the treaty was to be in force for 10 years with the option for extension. Article V stipulates that the two countries recognize that an "armed attack against either Party in the territories under the administration of Japan" would jeopardize international peace and should be responded to in accordance with constitutional provisions. The significance of this article is that Japan is not obliged to come to the assistance of the United States if the US is attacked elsewhere in the Pacific outside of Japanese territory. In part this article was representative of the defensive nature of Japanese military forces and the prohibitions associated with Article 9. However, the Japanese respon-

sibilities remain less than those of other Asian countries with whom the US has been aligned and has become a source of contention between the two countries over the years.[100]

SUMMARY: THE EISENHOWER YEARS

The Eisenhower administration extended the containment strategy in Asia, primarily because of the perceived Chinese threat, by increasing the American commitment to the region. The style of containment reverted to the asymmetrical form associated with the early Truman administration mainly because economic factors were so important. When applicable, the threat of nuclear weapons was used, and the American alliance network was expanded significantly through bilateral treaties with Korea and Taiwan and the long-term basing rights established with each country. SEATO was created, and the American involvement in and commitment to Indochina increased as evidenced by larger economic and military assistance programs and political support for the Diem regime in Vietnam.

Despite these commitments, US budgetary policies and military manpower programs raised questions concerning whether these commitments could be met. For example, during the Eisenhower years, expenditures for nuclear weapons increased appreciably while the expenditures for conventional forces declined.[101] While this expenditure trend is consistent with the massive retaliation doctrine of the New Look strategy, the reduction of conventional forces in Asia was not consistent with expanded commitments. As mentioned, US troop levels in Japan were cut by more than half between 1954 and 1956. This can be only partly attributed to the overall reduction in force after the Korean war. In the Philippines, US military forces were reduced even more significantly during the Eisenhower second term.

By 1960, it was not self-evident that the United States would use nuclear weapons to counter an insurgency in Indochina with all of the attendant risks involved, nor was it clear that conventional forces were available for the task. Containment of Sino-Soviet expansion remained the preeminent US goal in Asia, but as his administration came to a close, there were doubts as to whether Eisenhower's policies and the means to enforce them could be continued successfully into the next decade.

THE KENNEDY ADMINISTRATION

Seldom have the contrasts between administrations been more distinct than when John Kennedy succeeded Dwight Eisenhower in January 1961. This contrast was particularly evident in defense policy. Kennedy criticized massive retaliation both because it lacked credibility and because it did not provide the president with satisfactory options in the event of a crisis. "Flexible response" replaced massive retaliation, and Kennedy sought to increase substantially military forces across the entire spectrum of strategic weapons, tactical nuclear weapons, and conventional forces.[102] Such a buildup would require extensive military spending, but Kennedy and his advisers believed it was possible to expand overall government spending to accommodate defense increases without reducing other budgetary allocations. Economic factors were not to impede the rebuilding of American defense capabilities to respond to larger commitments abroad.[103] The switch from the asymmetrical to the symmetrical style of containment began once again.

The new administration requested a supplemental appropriation of $1.2 billion to begin the modernization of the strategic triad consisting of nuclear submarines, long-range bombers, and intercontinental ballistic missiles. In addition, Kennedy increased the number of Army divisions from 11 to 16, and total military manpower grew by more than 200,000 even before the buildup for the Vietnam war began in 1965. With this buildup, the Kennedy administration developed the "two-and-one-half" war strategy whereby the US could fight a major war both in Europe and Asia as well as a smaller war someplace else.[104]

In addition to increasing defense budgets and expanding the force, Kennedy believed that American foreign policy in general needed to be rejuvenated. In his oft-quoted inaugural address, he emphasized the need for the US to "pay any price, bear any burden, meet any hardship, support any friend, oppose any foe to assure the survival and the success of liberty."[105] In the late 1950s and early 1960s, a large number of former colonies gained their independence, and the superpower competition for influence in these countries was strong.

The president's actions were influenced somewhat by the commitments previous adminstrations had made in Asia. He won the 1960 election by a very small margin over Richard Nixon and did not believe his political position was strong enough to make major breaks from previous policies of containment without appearing too weak. Internationally, if the United States did not meet its commitments in Asia, the developing countries of the region would take note as would China and the Soviet Union. Kennedy believed the US needed to channel Third World nationalism in directions that would serve long-term American interests.[106] Since he also was convinced that the United States had the necessary means to compete with the Soviets and Chinese politically, militarily, and economically in the developing world, he embarked upon a policy in Asia which would

quickly increase the American commitment and involvement in this region.

INDOCHINA

Few events in American history have been more contentious than the US involvement in Indochina, particularly during the 1960s and early 1970s. It is not the purpose here to make a normative judgment about this involvement, but rather to provide a partial explanation of how this involvement escalated and the effects of this escalation on the broader American security policy in Asia at the time and during subsequent administrations.

In reading the accounts written by close Kennedy associates concerning the administration's decisionmaking process on Indochina, one is struck by the general feeling that the president did not believe he had many options available because of the commitments made by previous presidents.[107] Kennedy took seriously the threat he perceived from both the Soviet Union and China in Southeast Asia. In a speech given in April 1961, Kennedy noted Khrushchev's pledge to support wars of national liberation and pointed out the "relentless pressure" being exerted by the Chinese on Southeast Asia.[108] These comments implied that Kennedy feared the falling dominoes just as Eisenhower before him, and he was determined to prevent this from happening.

In Laos, a coalition government was formed in 1957 which was composed of three factions: a rightist group including the Royal Lao army; the Pathet Lao, a communist force; and a neutralist group let by Prince Souvanna Phouma. The Eisenhower administration had opposed the formation of this coalition government and supported the rightist forces.[109] When the coalition government was unsuccessful in governing the country, a civil war erupted in May 1960. After assuming office, Kennedy eschewed the idea of a military solution and instead decided to pursue a political settlement. The Geneva conference was reconvened and met over a period of 15 months until an agreement was reached in July 1962 which provided that Laos was to be a neutral country, and all foreign military forces were to be withdrawn within 75 days. Souvanna Phouma formed a new government of national union.[110]

The Kennedy policy in Laos influenced the adminstration's decisions on Vietnam. Since the president opted for a political solution in Laos, a decision criticized by many, there was more pressure to use military means in support of the Diem regime in South Vietnam. In May 1961, Kennedy approved National Security Action Memorandum 52 which stated that the US objective was "to prevent communist domination of South Vietnam."[111] Also in May, Vice President Lyndon Johnson traveled to Asia and toured Vietnam. Upon his return, he submitted a report to the president which contained a thesis quite similar to Eisenhower's domino theory. If the United States was serious about combating communism in Asia, the line had to be drawn in Vietnam. Although Johnson did not recommend sending American military forces to the RVN, he did suggest that military assistance to Diem be increased.[112]

In late 1961, Kennedy dispatched General Maxwell Taylor, chairman of the JCS, and Walt Rostow, a political adviser, to Vietnam to evaluate both the political and military situations. The Taylor-Rostow report was very important because it recommended that American ground units be introduced to support South Vietnamese forces, raise Vietnamese morale, and indicate American commitment and resolve to resist communist aggression. Taylor and Rostow also recommended that Kennedy consider conventional bombing against the DRV (North Vietnam).[113] At about the same time, Secretary of State Dean Rusk and Secretary of Defense Robert McNamara also recommended that American ground forces be introduced into South Vietnam, but warned that it might require far more troops than suggested in the Taylor-Rostow report.[114]

Despite these recommendations, Kennedy refused to authorize the introduction of ground forces into Vietnam in other than an advisory role. Nonetheless, the American military presence was increasing significantly. When Kennedy assumed the presidency, there were 875 Americans assigned to the Military Advisory Group (MAAG). The number of advisers increased to 3164 in 1961; 11,326 in 1962; and 16,263 by 1963, and the MAAG was replaced by the Military Assistance Command Vietnam (MACV) in 1962.[115]

Kennedy resisted a larger combat role because he believed this would limit his flexibility, particularly in convincing Diem to institute the reforms necessary for popular support in the RVN. However, Diem resisted such reforms in a way similar to Chiang Kai-shek many years before in China. He believed that the US had no choice but to support him because of the lack of alternatives.[116] By late 1963, the military and political situations had deteriorated to such a point that the administration proved Diem wrong. The Vietnamese military conducted a coup and killed Diem and his brother on 1 November 1963.[117]

Unfortunately, Kennedy was himself assassinated in Dallas later that month. Whether Kennedy would have changed American policy in Vietnam if he had lived remains a topic of debate for which no definitive answer is possible.[118] However, there are some conclusions that can be drawn concerning Kennedy's security policies in Asia. First, Kennedy, like Truman and Eisenhower, believed that containment of the Soviet Union and China was a primary policy goal in Asia. He shared to a certain extent Eisenhower's concept of falling dominoes and was willing to increase the US military commitment in Vietnam to meet the challenge. The question most frequently addressed was how to save Vietnam, not whether Vietnam could or should be saved. In this manner, Kennedy was responding to cold war tensions.

Second, the president and most of his advisers were convinced that massive retaliation was not effective in combating insurgencies or wars of national liberation. These conflicts required increases in conventional forces and the development of nonconventional units such as the Green Berets and others. Consequently, Kennedy increased defense spending and expanded the military capabilities appreciably so that flexible response could give him more options. The style of containment once more changed from asymmetrical to symmetrical as the means were expanded to meet what were perceived to be increased threats. As American prestige grew more involved in support of these commitments, it grew more difficult to change policy directions.

Third, Kennedy's Asian security policies were very narrowly focused on Indochina. The other countries in the region did not receive much attention outside of their relationship to the struggle in Vietnam. Little progress was made in improving relations with China because of Chinese support for North Vietnam, even though some members of the administration began to observe strains in the Chinese relationship with the Soviet Union.

THE JOHNSON ADMINISTRATION

The report which Lyndon Johnson authored after his visit to Vietnam in 1961 is instructive in understanding his decisions when he later became president in November 1963. In that earlier report he stressed the importance of defending Vietnam to prevent the spread of aggression elsewhere in Asia and why it was imperative that the United States meet its commitments to South Vietnam.[119] In his first address to Congress on 27 November 1963, Johnson reiterated this position: "We will keep our commitments from South Vietnam to West Berlin."[120] For the new president, the continuation of the American course of action in Vietnam was consistent with his interpretation of US policy in Southeast Asia since the end of World War II, and he was not about to change this policy.

In his memoirs, Johnson presented several arguments to support his decision to hold firm in Vietnam. First, he believed the future of Southeast Asia hinged on US success in Vietnam, and he criticized those who did not believe the domino theory to be valid in the mid-1960s. Second, Johnson was concerned about what the domestic reaction would be in the United States if Vietnam fell. He drew an analogy with China in 1949, except that the fall of Vietnam would be worse because of US treaty commitments in Southeast Asia through the Manila Pact and SEATO. Any debate over "who lost Vietnam" would divide the country and invite isolationist sentiments which would jeopardize American commitments to other countries, in his view. Third, American allies and friends throughout the world would lose confidence in the United States if it failed to protect an ally. Fourth, the Soviet Union and China would move to exploit any perceived weakness in US policies.[121]

In March 1964, Johnson sent Secretary McNamara and General Taylor to Vietnam to evaluate the situation. They reported that the Viet Cong (National Liberation Front founded in 1960) was gaining in strength throughout the country, and large groups of the South Vietnamese population were becoming disillusioned. McNamara and Taylor recommended substantial increases in military assistance to the Army of Vietnam (ARVN) and that the US consider the possibility of air attacks over North Vietnam.[122] Shortly thereafter, National Security Action Memorandum 288, entitled "US Objectives in South Vietnam," was written and again emphasized that the stakes were high in Vietnam, but warned that overt US military action against North Vietnam would possibly be counterproductive. However, it did authorize planning studies for possible strikes against the north.[123] The president resisted the arguments to initiate overt attacks against North Vietnam at this time primarily because he wasn't certain the political and military base in South Vietnam was strong enough to sustain probable retaliatory strikes and because he was unsure of the likely Chinese and/or Soviet responses.[124]

Covert actions against the north had been occurring since September 1963. These actions were designed to convince the North Vietnamese leaders to stop supporting aggression in Laos and the RVN. In August 1964, two US Navy ships supporting these covert actions were allegedly fired upon by North Vietnamese patrol boats. In response, President Johnson authorized an air strike against North Vietnamese naval bases. He then decided to request from the Congress wide-ranging powers to support his policies in Southeast Asia. A joint resolution was introduced in the Congress on 6 August 1964 and passed the next day. What came to be known as the Gulf of Tonkin resolution was passed unanimously in the House of Representatives and by a vote of 88 to 2 in the Senate. It contained the following provisions which later would become extremely controversial:

That the Congress approves and supports the determination of the President, as Commander in Chief, to take all necessary measures to repel any armed attack against the forces of the United States and to prevent further aggression.

Consonant with the Constitution of the United States and the Charter of the United Nations in accordance with its obligations under the Southeast Asia Collective Defense Treaty (SEATO), the United States is, therefore, prepared, as the President determines, to take all necessary steps, including the use of armed force, to assist any member or protocol state of the Southeast Asia Collective Defense Treaty requesting assistance in defense of its freedom.[125]

For all intents and purposes, the Congress had thus given President Johnson a blank check to conduct the war in Indochina as he deemed appropriate.

Partly because 1964 was a presidential election year, US escalation did not begin until early in 1965. Walt Rostow, one of Johnson's closest advisers, supported a program known as "graduated escalation." This program called for the systematic bombing of North Vietnam beginning at a relatively low level of violence and increasing in intensity as needed. Its purposes were to punish the DRV and indicate that the US was serious in its support for the south, while increasing the morale of the political leaders in Saigon and the South Vietnamese population. Rostow also hoped the north's supply lines and infiltration routes would be disrupted so that the DRV would be forced to negotiate an end to the conflict.[126]

When the Viet Cong launched a series of raids on American military advisers' bases in February 1965, President Johnson decided to initiate ROLLING THUNDER bombing operations against the north.[127] The subsequent decision to send American ground forces into Vietnam did not take as long as the decision to bomb north of the 17th parallel, since it was in many ways just another part of the graduated escalation of the conflict which the Johnson administration hoped would lead to a negotiated settlement and provide for South Vietnam's survival. Johnson made this decision to commit ground forces himself with little consultation with others, and, in March 1965, the first division of US soldiers in a combat role arrived in Vietnam.[128]

One of the major problems with Rostow's plan for graduated escalation was that there were no limits placed on these operations, whether in the air or on the ground. Another problem was that such a proposal did not take into account how high a price the North Vietnamese were willing to pay to achieve their goal of national unification. As a result, over 550,000 American military personnel were in Vietnam by the end of 1967, and the war had become largely an American conflict. The Tet offensive at the end of January 1968 proved to be a fatal blow to the Johnson presidency. Clearly the war was not going as well as the administration was saying, and it would be a long time before the war could be brought to a successful conclusion, if ever.

When General William Westmoreland, the MACV Commander, requested 206,000 more troops for Vietnam duty, the president balked. To meet this request Johnson would have been forced to call up reserve units and quite possibly raise taxes, neither of which was attractive in an election year.[129] The political situation had become so untenable for Johnson that in March 1968, he announced that he would not seek re-election to the presidency.[130]

The increased American involvement in Vietnam during the Kennedy-Johnson years influenced US security relations with other countries in Asia, particularly the Philippines and Republic of Korea. Johnson was interested in convincing allies to send at least token military forces to Vietnam to legitimize the expanding American role. Both the Philippines and the ROK complied with Johnson's request. During a visit to Washington in October 1964, Johnson convinced Philippine President Macapagal to send two medical/civic action teams to Vietnam which were in place by the end of 1964. In late 1965, Ferdinand Marcos was elected president, and Johnson requested that the Philippine contingent be increased. Consequently, Marcos formed the Philippine Civic Action Group (PHILCAG) and sent it to Vietnam in August 1966 with a full engineering battalion of 2048 men following in October 1966.[131] The Republic of Korea stationed its forces in Vietnam between 1965 and 1973 with approximately fifty thousand troops involved during this period.[132]

Both countries received political and economic rewards for the services provided. The United States funded the total costs of each contingent while in Vietnam. In the Philippine example, the total US cost for PHILCAG in 1969 was estimated at almost $39 million.[133] A comparison of US military aid to the ROK before and after the decision to send troops is instructive as to the economic benefits. In 1964, the US provided $124 million, which increased to $173 million in 1965, the year of the introduction of Korean troops, and eventually reached $556 million in 1971.[134]

The political rewards were important too for Marcos and for President Park Chung Hee of Korea. Marcos visited the US in September 1966 and addressed a joint session of Congress which was viewed as very important in increasing Marcos's popularity at home because of the traditional US-Philippine relationship. During this visit, Johnson also pledged to assist the Philippines in obtaining loans from international institutions such as the World Bank.[135] Finally the use of the Philippine military bases to support the US role in Vietnam was critical to this effort, providing Marcos with additional leverage for future discussions with the United States on the retention of these bases.

President Park Chung Hee was motivated by several considerations to contribute to the Vietnam war. Two of the most important concerned his own consolidation of power and the continuation of the American security guarantee to Korea. Park came to power in 1961 through a military coup which did not particularly endear him to the Kennedy administration. By sending troops to Vietnam, Park was able to improve his relations with Washington which had domestic political benefits as well.[136]

Even more important, however, was Park's concern about the US troop presence in Korea. As the US increased its military commitment to Vietnam, Park feared some of the US forces in his country would be redeployed to Vietnam, thereby reducing the deterrent function of these forces. By sending Korean troops, he was not only showing he was a loyal ally, but hopefully limiting the possibility of US troop reductions in the ROK.[137] When Vice President Humphrey visited the ROK in 1966, he strongly endorsed the continued US military presence in Korea, and Johnson affirmed this commitment when the two presidents met in April 1968.[138] Park looked upon Johnson's pledge as being especially important because North Korean acts of belligerency had increased with an attack on Park's residence in 1967 and the seizure of the USS *Pueblo* early in 1968. As the Johnson administration came to a close, US relations with the Philippines and Korea had improved over what they had been earlier in the decade.

SUMMARY: THE KENNEDY-JOHNSON YEARS

The Kennedy-Johnson years were examples of the symmetrical style of containment with the overwhelming focus of attention in Asia being on the war in Vietnam. The United States became bogged down in a war it could not win under acceptable political and economic costs. It became so unpopular with the American public and with the Congress that it not only forced an incumbent president to decide not to run, but also influenced the options available to his successor. The trauma resulting from the US war in Vietnam contributed to a major change in American security policy in Asia during the administration of Richard Nixon and beyond. Once again, the symmetrical style of containment would give way to the asymmetrical style as economic factors, and public dissatisfaction with the war became predominant.

THE NIXON ADMINISTRATION

President Nixon was faced with significant domestic and international constraints when he assumed office in January 1969. The unpopularity of the war in Vietnam had grown substantially since the Tet offensive. Within Congress, demands for reduced military spending could not be ignored. During the Nixon and Ford administrations, a tremendous reallocation of resources from military to domestic programs occurred. As a percentage of the national budget, defense spending decreased from 44 percent when Nixon became president to 24 percent when Gerald Ford left office in January 1977. As a percentage of the gross national product, the reduction was from 8.7 to 5.2 percent over the same period of time.[139] In addition, costs due to inflation were increasing so that the conventional forces lost effectiveness in comparison with those available in the Kennedy-Johnson years. The "two-and-one-half" war strategy which the Kennedy and Johnson administrations had supported no longer was a credible strategy.[140]

Another domestic constraint on the president was the reassertive role of the Congress in the security policy arena. The days when the Gulf of Tonkin Resolution could be passed were over. The Congress also was willing to use its appropriation powers to limit the actions of the president. The Cooper-Church amendment in 1970 prohibited the spending of any more funds in support of military operations in Cambodia. The State Department Authorization Act of 1973 effectively terminated the war by prohibiting funds for US military forces after 15 August in hostilities "in or over or from off the shores of North Vietnam, South Vietnam, Laos, or Cambodia unless specifically authorized hereafter by the Congress."[141]

Even more restrictive was the War Powers Resolution passed by the Congress in October 1973, vetoed by President Nixon, and then passed over his veto by both houses in November 1973. This legislation restricted the president's authority as commander in chief to commit US military forces into a hostile environment for extended periods without the concurrence of the Congress.[142] Although the Congress had repealed the Gulf of Tonkin Resolution in 1971, the War Powers Resolution was designed to prevent a future president from assuming the sweeping authority provided to President Johnson in 1964.

The changing international system also influenced Nixon's Asian security policy. Stanley Hoffmann described this as a switch from tight bipolarity after World War II, with the United States and the Soviet Union being the dominant powers, to polycentrism by the late 1960s as additional states began to influence the system.[143] Japan emerged as the dominant economic power in the region, and many in the US thought that Japan should assume more responsibility for its own defense rather than relying almost totally on the US. The Soviet Union, through its assistance program to North Vietnam, began to exert more influence in Southeast Asia. As a result, Washington's Asian policy began to focus more on Northeast Asia and the interrelationship of the Soviet Union, Japan, the People's Republic of China, and the US.[144]

THE NIXON DOCTRINE

The Nixon Doctrine, enunciated by the president during a brief visit to Guam on 24 July 1969, reflected the influence of these domestic and international factors.[145] While noting that the US would remain involved in the region, he stressed that some of these allies, particularly Japan, were able to assume more of the burden to ensure "the peaceful progress of the area."[146] Concerning security relationships, there were three major elements of the new strategy outlined by Nixon.

1. The United States will keep its treaty commitments.

2. We (the United States) shall provide a shield if a nuclear power threatens the freedom of a nation whose survival we consider vital to our security and the security of the region as a whole.

3. In cases involving other types of aggression we shall furnish military and economic assistance when requested and as appropriate. But we shall look to the nation directly threatened to assume the primary responsibility of providing the manpower for its defense.[147]

Through this policy, President Nixon hoped to create what he described as a "careful balance" which would enable the allied states to rely on American economic and military assistance to build up their forces so that they could provide for their own defense from external aggression and domestic dissident activity.[148] Clearly, a security policy based on the three principles outlined above represented a shift from Kennedy and Johnson's flexible response to a security policy more like the asymmetrical style of containment during the Eisenhower years. The emphasis switched from the reliance upon US ground forces to more dependence on nuclear weapons and "self-help" to deter aggression.[149]

The "two-and-one-half" war strategy of the previous two administrations was changed so that only one major war and one minor conflict could be fought simultaneously.[150] The president reduced the numbers of US military personnel in several Asian countries during this first term. In Vietnam, the force level dropped from 500,000 when he assumed office to less than 3000 by early 1973. In Korea, the number dropped from 60,000 to 40,000; from 39,000 to 27,000 in Japan; from 48,000 to 43,000 in Okinawa; and most of the 16,000 troops in Thailand were removed. Smaller numbers of military personnel were reassigned from the Philippines and Taiwan during the same period.[151]

VIETNAM

US relations with Vietnam during the Nixon years had widespread effects not only on the bilateral relationship, but, more important, with US relations with other countries in Asia. During the 1968 presidential campaign, candidate Nixon indicated he had a "secret plan" to end the American involvement in the Vietnam war. Once elected, however, the process of this withdrawal proved to be quite slow, and the Paris cease-fire agreement and final US military withdrawal did not occur until January 1973.[152] The "secret plan" actually turned out to be a policy of Vietnamization of the war whereby the American role was reduced, and the Vietnamese assumed more and more responsibility until the cease-fire agreements went into effect.[153]

KOREA

Nixon's policies toward the Republic of Korea were considerably cooler than those of President Johnson. When Nixon met with President Park Chung Hee in August 1969, the Joint Communique issued at the conclusion of this meeting was significantly different from that issued after the Park-Johnson meeting in April 1968.[154] Rather than the pledge to offer "prompt and effective assistance to repel armed attacks," which his predecessor made, Nixon agreed only "to meet armed attack against the Republic of Korea and the United States."

In 1971, President Nixon began to reduce the US ground presence in Korea by sending the 7th Infantry Division home. After completion of this withdrawal, only the 2d Infantry Division remained.[155] Despite the earlier articulation of the Nixon Doctrine, ROK leaders objected that the deterrent value of the American presence was reduced and that the Nixon administration had not consulted with them sufficiently beforehand.

In the early 1970s President Park launched a major effort to increase the industrial capabilities of his country so that the ROK would become as self-sufficient as possible in the production of military hardware. Quite ob-

viously, Korean confidence in American reliability was shaken by the events in Southeast Asia and by the Nixon troop withdrawal decision. The latent fears of "decoupling," not unlike European fears, were exacerbated by this withdrawal.

CHINA

In February 1972, President Nixon and a delegation of other American officials including National Security Adviser Henry Kissinger visited the PRC and held discussions with Mao Tsetung and Premier Chou En-lai, a visit which Nixon described as "the week that changed the world."[156] While perhaps slightly overstated, it is true that the Nixon visit and the evolution of US-PRC relations during the remainder of the 1970s to the present have influenced significantly US security policy in Asia. During previous administrations, some efforts had been made to improve relations with the PRC, but the Korean war experience, Chinese support for North Vietnam, and the US relationship with Taiwan precluded these efforts from being successful.

Further, the Great Proletarian Cultural Revolution which engulfed China in turmoil during the mid to late 1960s made diplomatic contacts difficult. The president's well-established anticommunist credentials gave him a unique opportunity to approach the PRC without being subjected to charges of being "soft" on communism. In addition, both Nixon and Kissinger believed that the PRC should be reintegrated into the international community to moderate Chinese foreign policy and to increase American options in the US-Soviet relationship.[157]

China also had substantial reasons for desiring to improve relations with the US. The UN seat issue had been settled in 1971 when the PRC took Taiwan's place. US involvement in Vietnam was decreasing. Chinese leaders also saw economic incentives in improving the relationship. Moreover, tensions between the Soviet Union and China along parts of their common border had escalated to the extent that fighting had broken out in 1969 along the Ussuri River. The Soviet invasion of Czechoslovakia in 1968 had caused concern because of the obvious implications for a socialist state which deviated from the Soviet line.[158]

At the conclusion of the Nixon visit, he and Chou En-lai issued what became known as the Shanghai Communique, in which Nixon made two very important concessions.[159] First, he agreed with Chou that each country "is opposed to efforts by any other country or group of countries to establish hegemony" in the Asia-Pacific region. The reference to "hegemony" was important because it is the code word used by the Chinese to refer to Soviet designs in the region.

Therefore, the US was pledging to oppose Soviet expansion in Asia in concert with the PRC. Second, Nixon agreed that there was only one China, and that Taiwan was part of China. Nixon reaffirmed his hope that a peaceful settlement could be reached to resolve the question of Taiwan if the tension between the PRC and Taiwan decreased.

For Nixon and Kissinger, improving relations with China made sense because of economic and geopolitical considerations. The old "two-and-one-half" war strategy of the Kennedy-Johnson years envisioned one of the major wars occurring in Asia with the PRC as the likely protagonist. As this strategy was modified to a "one-and-one-half" war concept in reaction to the Vietnam war and fiscal constraints, the possibilities of a major war in Asia needed to be reduced, and rapprochement with the PRC was a means to this end. Since both Nixon and Kissinger saw the international system evolving to a multipolar arrangement, China would become an important part of the balance of power in Asia, and the US would be well served to improve the bilateral relationship and use this improvement as a tool in the developing détente between the US and Soviet Union.

JAPAN

The direction of US-Japanese relations also began to change. By 1967, Japan had developed the world's third strongest economy behind those of the United States and Soviet Union.[160] Its export industries were booming, and Japan was becoming a major economic competitor of the US, particularly in Southeast Asian markets. In addition, Japanese exports to the US increased so that, by 1972, the US trade deficit with Japan grew to $4 billion, approximately two-thirds of the total US balance of trade deficit for that year.[161] Because of domestic economic problems in the early 1970s and also because of Nixon's belief that Japan and the European community were not cooperating in reducing their imports to the US, he announced his New Economic Policy in August 1971 which imposed a ten percent surcharge on dutiable imports.[162] The Japanese viewed this action as being directed primarily at them and objected strenuously.

One key aspect to the success of the Nixon Doctrine in Asia was to convince Japan to provide more for its own defense.[163] In response to this, Japan projected the Self Defense Forces to increase from 180,000 to 271,000 between 1972 and 1976 and at least began to recognize the need for assuming a larger role in conjunction with the United States for the security of the region. The Japanese also started to assume more of the economic costs in support of American military presence in Japan. Significantly,

after Nixon met with Prime Minister Sato in November 1969, Sato conceded that the "security of the Republic of Korea was essential to Japan's own security." Sato also indicated that, after consultation with the Japanese government, Japan would respond "positively" to requests to use the bases in Japan to support a contingency on the Korean peninsula.[164] In view of the defensive orientation of Article 9 in the Japanese Constitution, this was a major concession.

These economic and security issues plagued the US-Japanese relationship during the Nixon administration. The Japanese referred to both the Nixon decision to visit the PRC without prior consultation with Japan and the New Economic Policy as the "Nixon shocks." One expert has suggested that Nixon intended these "shocks" to convince the Japanese that more had to be done on Japan's part to reduce the economic issues and assume a more dominant political role in Asia.[165] There was a growing sentiment in the US that the US would no longer be willing to provide for Japan's defense at the same time Japan was becoming an economic competitor.

SUMMARY: THE NIXON YEARS

Containment of the Soviet Union and its allies remained a predominant goal as had been the case for each of his postwar predecessors. However, the means which Nixon and Kissinger devised to pursue this end were both dramatic and innovative. The unpopularity of the Vietnam war, severe economic constraints, and the reemergence of the Congress in the foreign policy process were important domestic factors. The same was true for changing international conditions such as the Sino-Soviet split, the growth of the Japanese economy, and increased stability in parts of Southeast Asia through the development of the regional association, ASEAN.

The Nixon Doctrine reflected these domestic and international influences. The unpopularity of the war and economic constraints made it impossible to retain the US military presence in Asia at the level which had developed in the 1960s, particularly as the Congress became more assertive. Japan's economic development convinced the president that Japan could assume a larger role in maintaining stability in Asia, allowing the US to reduce its overt military presence in Asia. Equally important, the Sino-Soviet split made the old fear of a major war in Asia involving the Soviets and Chinese as allies against the United States seem highly unlikely, thereby enabling a change from a "two-and-one-half" war strategy to a "one-and-one-half" war strategy. The US rapprochement with the PRC not only took advantage of the Chinese fears of the Soviet Union, but also attempted to moderate Chinese foreign policy by drawing the PRC into the international community.

Nixon and Kissinger also pursued détente with the Soviet Union. Détente with the Soviets was not an abandonment of containment. Rather, détente was another means of influencing Soviet behavior by linking improved relations with the US to a less aggressive Soviet foreign policy.[166] Achievement of the Nixon administration's foreign policy goals was hindered by the Watergate crisis which weakened President Nixon until he was forced to resign in the middle of his second term.

THE FORD ADMINISTRATION

Gerald Ford became president in August 1974 upon the resignation of President Nixon. By this time, the US had withdrawn all of its military forces from the Republic of Vietnam, and serious questions were being raised by American allies in Asia concerning America's commitments to remain an Asian power and fulfill its security obligations. These concerns were accentuated in the spring of 1975 with the communist victories in Vietnam, Cambodia, and Laos. The major task for the new administration was to convince other countries that the US intended to remain an Asian power and meet its commitments. This task was complicated because Watergate had increased tensions between the legislative and executive branches, and the trauma of Vietnam had raised great doubts within the US about the proper American role abroad, particularly in Asia.

Philippine President Marcos expressed the concerns of many Asian leaders in April 1975, when he outlined his view of the situation in Southeast Asia and his anxiety over American bases in his country and the Mutual Defense Treaty.[167] Marcos mentioned an emerging quadrilateral power system in Asia which included the Soviet Union, Japan, the People's Republic of China, and the United States. If this system were to remain, it was imperative, in his view, that the US maintain its presence in Southeast Asia. He applauded President Ford's stated intention that the US would remain an Asian power, but expressed doubts about Ford's efforts to gain congressional approval to support this intention in the face of widespread American disillusionment after Vietnam.

President Ford attempted to respond to these concerns by traveling to Asia and affirming the continued US role in the region. In late 1974, after meeting with Soviet leader Leonid Brezhnev in Vladivostok, Ford stopped in Seoul to consult with President Park. In the joint communique issued at the end of this meeting, Ford reverted to the language used by President Johnson in 1968 when he pledged "prompt and effective assistance to repel armed attack against the Republic of Korea." He went on to state directly that "the United States had no plan to reduce the present level of United States forces in Korea."[168] This statement was considerably stronger than President Nixon's in 1969 and was welcomed in Korea particularly after the withdrawal of the 7th Infantry Division in 1971.

At the conclusion of a subsequent Asian trip in December 1975, President Ford attempted to clarify further his Asian policy. In a speech delivered in Hawaii, he outlined his "Pacific Doctrine."[169] The United States would remain an active Asian actor because the regional balance of power required the US presence. He identified the US-Japanese partnership as "a pillar of our strategy" and expressed confidence that economic differences could be resolved between the two countries. President Ford also promised to move toward normalization of relations with the People's Republic of China based on the provisions in the 1972 Shanghai Communique. In stating that the US and PRC shared "opposition to any form of hegemony in Asia," Ford reflected growing concern about the Soviet Union and the effectiveness of détente. With respect to Southeast Asia, Ford pledged continued US involvement and military assistance. US relations with Vietnam, Laos, and Cambodia would be determined by their conduct towards the US, particularly the return of the remains of Americans killed during the Vietnam war.

Secretary of State Kissinger made a major policy address in Seattle on 22 July 1976 which contained even more specific policy references to Asia than Ford's "Pacific Doctrine."[170] He pointed out that previously, American commitments in Asia sometimes determined US interests rather than vice versa, and he recommended bringing these commitments into balance with interests and the means to fulfill them. His comments about the Philippines and the ROK were particularly interesting because of the security policy implications. He agreed to renegotiate the 1947 Military Bases Agreement (MBA) because retention of these bases was tangible evidence of the American commitment to maintain its military presence in Asia. Kissinger reiterated Ford's pledge to the ROK and stated emphatically that the US would not unilaterally withdraw any more of its military forces for fear of contributing to increased tension on the Korean peninsula.

Philippine-American base negotiations began in earnest during the summer of 1976 and rather quickly became bogged down as Philippine demands for increased sovereignty conflicted with American desires to retain unhampered military control over the bases.[171] In December 1976, the US announced that a tentative agreement had been reached.[172] The settlement reportedly called for the US to provide $1 billion to the

Philippines over a five-year period to be divided equally between military and economic assistance programs. In return, the US was to retain the use of the bases. The following day, however, President Marcos denied that an agreement had been reached and stated that the reported compensation was not sufficient and that several issues remained to be resolved.[173] The negotiations conducted in 1976 ended in failure, and the possibilities of an early resolution of the differences seemed even more remote as Gerald Ford was replaced by Jimmy Carter, who, as a candidate, had indicated that respect for human rights would be a major determinant concerning which countries would receive military and economic assistance from his administration.[174]

President Ford attempted to retain US credibility in Asia in the aftermath of the Vietnam war and the collapse of US allies in Indochina by pledging that the United States would remain an Asian power and meet its commitments. As with the Nixon Doctrine, there was a clear recognition that American resources were limited and Asian allies were going to have to assume an increased share of the security burden. Although the president indicated he wanted to proceed with normalizing relations with the PRC, he was restricted in these efforts by resistance from the Congress. Supporters of Taiwan argued that it would be a major mistake for the US to establish normal diplomatic relations with the PRC because this would require the US to abrogate its defense treaty with Taiwan, remove its military forces, and break diplomatic relations with Taiwan. These actions would further exacerbate the fears among Asian allies that the United States was intent on withdrawing from Asia according to this argument. Accordingly, President Ford was unable to make much progress on the normalization goal. He continued the asymmetrical style of containment within the constraints imposed upon him by both international and domestic factors.

THE CARTER ADMINISTRATION

President Jimmy Carter was forced to confront many of the same problems in Asia as was his predecessor concerning the perception of American willingness to remain an Asian power and meet its commitments. During confirmation hearings in March 1977, Richard C. Holbrooke, Carter's designee for assistant secretary of state for East Asia and Pacific affairs, outlined the administration's broad policy guidelines before a House subcommittee.[175] The US would remain an Asian power in part by preserving "a balanced and flexible" military policy in the western Pacific. Close ties with Japan were to be continued, and Holbrooke indicated that efforts to normalize relations with the People's Republic of China would be increased, while preserving the security of Taiwan would remain a Carter administration goal.

Holbrooke also hoped that normalization of relations with Vietnam would be possible. Concerning the Republic of Korea, he indicated that the remaining US ground forces would be withdrawn "while ensuring that the security of Korea is in no way threatened." Holbrooke promised that close consultation with the ROK and Japan would be carried out as the withdrawal progressed. He attempted to reassure Southeast Asia of the American commitment and was optimistic that the US and Philippines would successfully renegotiate a new agreement to allow the US to retain its bases.

While several of these policy guidelines were continuations of efforts during previous administrations, others, such as normalizing relations with Vietnam and withdrawing all ground forces from the ROK were new. It would prove difficult if not impossible to normalize relations with the PRC without breaking relations with Taiwan and removing the US miliary forces there. To abrogate the security relationship with Taiwan would likely be counterproductive to convincing other countries that the US was a dependable ally. On top of these problems, President Carter's strong stand on human rights as a cornerstone of his foreign policy raised other questions, particularly in the Philippines and the ROK.[176]

KOREA

The planned troop withdrawal from Korea would prove to be one of the most controversial proposals during Carter's tenure in the White House. As early as 1975, candidate Jimmy Carter indicated he would withdraw American ground forces from Korea if he became president.[177] The Korean reaction was more muted than might have been expected both because in 1975 it didn't appear Carter would win the Democratic nomination, let alone the presidency, and because campaign rhetoric in the US is frequently not translated into policy.

At a press conference in March 1977, however, Carter announced that he was going to follow through on his campaign pledge, arguing that the 32,000 US troops could be withdrawn over a four- to five-year span so that the ROK could continue to prepare for its own defense. In addition, he anticipated that American air support would remain in the ROK for years to come.[178] In May 1977, Presidential Decision 12 was sent to the Departments of State and Defense ordering them to implement Carter's planned withdrawal.[179]

The President justified his troop withdrawal order primarily because he thought US troops were not required to maintain stability. He stressed that the Mutual Defense Treaty remained in effect and that US air and naval forces would be sufficient for deterrence purposes. What he termed "strategic considerations" had changed from the late 1940s and early 1950s in the relationships between the US and the PRC, between the US and USSR, and between the USSR and PRC.[180] These changing "strategic considerations" would reduce the possibility of a repetition of the North Korean invasion in June 1950. Moreover, the ROK had developed a strong economy and could provide for its own defense within the five-year transition period.

Experts outside the administration have suggested additional reasons for Carter's decision. One was that he desired additional flexibility in how, or if, the US would respond to an attack against the ROK. If American forces remained deployed along the major invasion routes, his choice of options was limited. Closely associated with this rationale was that Carter wanted to use the 2nd Infantry Division as a reserve force stationed in the US which could be dispatched to trouble spots elsewhere in the world.[181] Others argued that Carter wanted to cut the defense budget as he had promised in the campaign, and he saw withdrawing ground forces from the ROK as a means to this end. Finally, because of his stress on human rights, Carter found the regime of President Park objectionable and wanted to distance himself from it.[182]

For whatever reasons, the president's decision drew immediate criticism from domestic and foreign sources. For the purposes here, the responses from the ROK and Japan are the most significant.[183] The Korean reaction from both political and military officials was overwhelmingly negative based on how the American forces were viewed in Korea. Even opposition political leaders supported the continuation of the American presence. A further criticism was that the Carter administration had not even consulted with the Park government before making the announcement.[184]

Initially, Prime Minister Fukuda of Japan questioned the decision not only because of the possible effects in Korea, but also for the stability of Northeast Asia. Subsequently, Fukuda moderated his opposition by indicating the decision was really a bilateral one between the ROK and US. It has been suggested by at least two authorities that Fukuda changed his public position on the troop issue because he feared Carter would pressure Japan to accept more of the responsibilities for the ROK defense.[185] Nonetheless, official government views in both Seoul and Tokyo remained critical of the Carter plan.

An intelligence reassessment of the comparative military capabilities of the ROK and Democratic People's Republic of Korea (DPRK–North Korea) began in the summer of 1978 involving several US intelligence agencies. The results of this reassessment, announced in June 1979, cast grave doubts over the troop withdrawal schedule. Not only were the DPRK's military capabilities far greater than estimated in 1977, but its military forces were deployed in a manner which could enable a lightning attack across the 38th parallel.[186] The domestic and foreign criticisms plus this intelligence reassessment forced Carter to rescind his withdrawal order. In a visit to the ROK in the summer of 1979, Carter emphasized the importance of the ROK-US relationship and stressed that the American military commitment to Korea's security was "strong, unshakable, and enduring."[187] Finally in July 1979, President Carter announced that the withdrawal plan was being held in abeyance until at least 1981.[188]

The troop withdrawal plan is instructive concerning Carter's security policy goals in Asia. As with Nixon and Ford before him, financial constraints influenced him to adopt an asymmetrical style of containment at least in the first years of his administration. However, the withdrawal plan caused further fears in Asia concerning the American commitment. Because of these concerns, the base negotiations with the Philippine government became more important.

PHILIPPINES

Negotiations on US bases began in February 1977 and dragged on until an agreement was finally signed in January 1979. It proved difficult to find a mutually satisfactory compromise between Philippine demands for increased sovereignty over the bases and the US position that unhampered military control over operations had to be maintained.[189] There were also problems determining how much compensation the US was willing to pay. The Vietnamese invasion of Cambodia in December 1978 and its subsequent military occupation underscored the importance of retaining the bases as an affirmation of American commitment in Southeast Asia.[190] The 1979 amendment to the 1947 MBA and accompanying letters allowed the US to retain the bases, although some concessions to Philippine sovereignty were made.[191] The US agreed to review the MBA every five years until its scheduled expiration date in 1991 and also agreed to pay the Philippines $500 million over this five-year period.

VIETNAM

Efforts to normalize relations with Vietnam began in 1977.[192] Progress was slow, however, as the Vietnamese drew closer to the Soviet Union first by joining COMECON (the Soviet–

East European economic group) in June 1978 and then by entering into the Vietnam-Soviet Friendship Treaty in November.[193] President Carter also recognized the difficulty of trying to normalize relations with both Vietnam and the PRC at the same time. The Vietnamese invasion of Cambodia in late December 1978, leading to the overthrow of the Pol Pot regime and the establishment of the puppet Heng Samrin government, made any further efforts at normalization futile.[194]

Partly as a result of this failure to establish normal diplomatic relations, Vietnam drew even closer to the Soviets. Several countries in Southeast Asia, particularly Thailand, viewed the closer Soviet-Vietnamese relationship as increasing the possibility of further Soviet involvement in the region. To meet the threat to Thailand, Carter met with Prime Minister Kriangsak in Washington during February 1979 and assured him that America's commitment to Thailand under the 1954 Manila Pact remained valid. In addition he promised to increase military assistance programs to this frontline state.[195]

CHINA

President Carter's efforts to normalize diplomatic relations with the PRC was eventually successful, but the strongest impediment was the US relationship with Taiwan. The US had recognized in the 1972 Shanghai Communique that Taiwan was part of China. However, the PRC insisted that the US fulfill three preconditions for establishing normal diplomatic relations: abrogate the US-Taiwan Mutual Defense Treaty, sever diplomatic relations with Taiwan, and withdraw all US military forces from Taiwan.[196]

These demands presented serious problems for the Carter administration. Not only did the president feel a moral obligation for Taiwan's protection and security, he feared that breaking diplomatic relations and abrogating the security treaty would send the wrong signals about American commitment, particularly since his ROK troop withdrawal proposal had been so controversial. Additionally, Carter knew that he would face opposition from segments of the Congress and the population at large over this issue.

Nonetheless, there were specific benefits to normalizing relations with the PRC. In the administration's view, trade and technological links could be established, and the US could exercise greater diplomatic flexibility in East Asia.[197] Specifically in Northeast Asia, President Carter hoped that the PRC would exercise greater influence over North Korea to reduce tensions on the Korean peninsula.[198] Although President Carter in his memoirs denies that

there was any anti-Soviet intent involved in the decision to normalize relations with the PRC, the administration clearly recognized that up to fifty Soviet divisions encompassing as many as one million troops were deployed along the Sino-Soviet border because of the Soviet fear of a war with the PRC. As long as these forces were so deployed, the US and its NATO allies enjoyed greater flexibility in Europe.[199]

On 15 December 1978, the two governments issued a joint communiqué announcing formal normalization of diplomatic relations, effective 1 January 1979. Although President Carter in an accompanying message stated he hoped there would be a "peaceful resolution" of the Taiwan problem, there was no such provision in the communique itself.[200] In fact, a unilateral statement issued by the PRC clearly indicated that the reunification of China was "entirely China's internal affair."[201] Concurrent with normalization, the US severed diplomatic relations with Taiwan, abrogated the security treaty, and began withdrawing the remaining US military forces on Taiwan.[202] President Carter also announced that Chinese Vice Premier Deng Xiaoping would visit the US in January 1979. When the PRC invaded Vietnam in February 1979, many inferred that the administration had agreed to this Chinese plan. In his memoirs, however, President Carter states that he advised Deng not to attack Vietnam.[203]

Carter rightly expected opposition from Congress. Initially, Senator Barry Goldwater threatened to challenge the president's authority to abrogate a treaty which had been ratified by the Senate.[204] When this proved unsuccessful, members of Congress supporting Taiwan turned to the legislative process. The Taiwan Relations Act was passed by the Congress and signed by the president on 10 April 1979.[205] This act authorized the continuation of commercial, cultural, and other relations between the US and Taiwan and also stipulated the American expectation that the "future of Taiwian would be determined by peaceful means." To ensure this peaceful transition, the US was to provide defensive weapons "in such quantity as may be necessary to enable Taiwan to maintain a sufficient self-defense capability." The US willingness to supply defensive weapons to Taiwan has continued to present difficulties in the development of the US-PRC relationship. During the latter stages of the Carter administration, the US agreed to sell first nonlethal military equipment and then "dual use" systems to the PRC, but these offers did not offset Chinese disillusionment over weapon transfers to Taiwan.

JAPAN

The Carter security relationships with Japan focused on two related issues: economic strains caused by the increasing US trade deficit with Japan and questions over Japan's alliance responsibilities and defense spending. As with the Nixon years, economic differences increased objections in the US to Japan's alleged "free ride" on defense. In 1976 the cabinet of Prime Minister Miki established a limit of 1 percent of Gross National Product (GNP) for defense spending. The Carter administration objected to this decision on the grounds that it was unreasonable to establish such a ceiling when it was uncertain whether the threat would increase.[206] The Japanese were concerned over the decision to remove US ground forces from Korea and objected to Carter's pressure for increased defense spending. In his memoirs, however, Carter makes no references to these problems.[207]

SUMMARY: THE CARTER YEARS

During the administration's first two to three years, it seems clear the president subscribed to the asymmetrical style of containment. The Korean troop withdrawal plan, the normalization of relations with the PRC, and the efforts to gain Japanese concurrence for a larger defense role reflect this awareness of limited resources and limited ends. However, a series of events including the Vietnamese invasion of Cambodia, the Iranian seizure of the American Embassy in Teheran, and the Soviet invasion of Afghanistan all convinced the president that stronger measures were required. Somewhat similar to the Truman administration before and after the North Korean invasion in 1950, President Carter changed policies toward the end of his administration. As the 1980 election approached, the president began to take a harder line, particularly on Soviet aggression, and his defense budgets increased. Nonetheless, the perceived weakness and vacillation in Carter defense policies were major issues in the campaign. His opponent, Ronald Reagan, was able to capitalize on these problems and win the presidential race by an overwhelming margin.

THE REAGAN ADMINISTRATION

The primary security goal of the Reagan administration in Asia during his presidency remained the containment of Soviet influence.[208] There is no question that Reagan and his closest foreign policy advisers viewed the international political system through globalist lenses. The Soviet-American competition dominated much of US foreign policy, reminiscent in many respects of the zero-sum calculations of the Kennedy-Johnson years. One expert has referred to this effort to limit Soviet influence as the "revitalization of the containment of Soviet expansion," particularly as Soviet military power has increased and

opportunities for intervention have become more numerous in various Third World countries.[209] President Carter had become concerned about Soviet military expansion in Asia, but this concern developed relatively late in his administration. President Reagan, on the other hand, placed the highest priority on limiting the effects of this expansion early in his tenure.[210]

The Soviets clearly increased their military capabilities in Asia during the past decade. The High Command of the Far Eastern Theater of Military Operations was established in 1979 and, shortly thereafter, SS-20 intermediate-range ballistic missiles and the Backfire bomber were deployed in the Siberian and Transbaikal Military Districts. These weapon systems not only threatened China, but also American forces stationed in Japan, Korea, and the Philippines.[211] The Soviet Navy also was improved, and the Pacific Fleet incorporated approximately 30 percent of the total Soviet naval strength, including submarines, surface ships, and *Kiev*-class aircraft carriers. The substantial Soviet air arm was improved, incorporating bombers and modern fighters such as Mig 25 interceptors.[212]

After Moscow and Hanoi signed a treaty of friendship and cooperation in November 1978, the Soviets gained access to Vietnamese port and airfield facilities, particularly the former US base at Cam Ranh Bay. Since that time, the Soviet military presence increased, and, as one administration official stated in 1984, "On any given day 20-25 Soviet surface ships, and 4-6 submarines are in that strategically located port."[213] This presence provided the Soviet Union with far greater flexibility and maneuverability and increases the threat of Soviet interdiction of vital sea lanes which pass through Southeast Asia.

To meet this broad threat, the Reagan administration devised a three-part strategy which included retaining US forward-deployed military forces in allied countries in Asia, strengthening bilateral and multilateral security relations, and providing security assistance programs to friends and allies so that they can contribute more effectively to security and stability in the region.[214]

JAPAN

Japanese-US security relations during the Reagan presidency offered one of the most interesting examples of how the president attempted to implement his strategy and the problems which developed. Early on, administration officials, particularly Secretary of Defense Weinberger, encouraged the Japanese to assume more responsibility for their own defense.[215] Contrary to the Carter administration, the new president and his advisers focused more on the roles and missions that Japan's military forces should perform than on the percentage of GNP allocated to defense.[216] However, there were definite limits as to how far President Reagan could push Prime Minister Suzuki in this regard as evidenced by Suzuki's visit to Washington in May 1981. In the final communiqué after this visit, the two leaders referred to the "alliance" that existed between the two countries and spoke of a need for a division of roles in the military field. These references caused such a furor in Japan that Prime Minister Suzuki was forced to ask for the resignation of Foreign Minister Ito because Ito was responsible for the wording of the communiqué.[217] Nonetheless, it was during this meeting that Suzuki pledged to protect Japanese sea lanes out to 1000 miles from the four main islands, equally controversial in his country.[218]

Resistance to increased Japanese military roles and missions remains primarily based on domestic and regional factors. The 1976 government decision not to spend more than 1 percent of GNP for defense and Article 9 of the Constitution are two important domestic factors.[219] Regionally, Japan's political and economic leaders fear that an expanded military presence, particularly in southeast Asia, could revive fears of another Japanese "Great East Asia Coprosperity Sphere" reminiscent of the 1930s and 1940s.[220]

Despite these constraints, Japanese leaders in the early 1980s believed that Soviet actions justified increased military spending. The Soviet invasion of Afghanistan; the deployment of Soviet troops and sophisticated weapons on Etorofu, Kunashiri, Shikotan, and Habomai Islands (which Japan considers its own territory); the stationing of SS-20 missiles and Backfire bombers within range of Japan; the buildup of the Soviet Pacific Fleet; and the closer Soviet-Vietnamese relationship were some of the more important of these Soviet actions.[221] Accordingly, Japan and the United States agreed in September 1982 to station up to fifty US F-16 fighter aircraft at Misawa Air Base, approximately 375 miles form the Soviet Pacific coast.[222]

In November 1982, Yasuhiro Nakasone was elected prime minister and began almost immediately to call for increased Japanese defense efforts. As a former head of the Japanese Defense Agency, Nakasone had the reputation of being hawkish on security matters which went a long way toward endearing him to President Reagan.[223] Nakasone stressed the Soviet threat and agreed that Japan must forge a closer relationship with both the US and ROK.[224] In a visit to Washington in January 1983, Nakasone referred to the Japanese-US relationship as an "unshakeable alliance," a much stronger statement than that contained in the joint communiqué less than two years before which resulted in the dismissal of a foreign minister.[225]

As a result of the closer relationship between Reagan and Nakasone, the United States and Japan have moved toward greater military coordination and cooperation in Northeast Asia. Various administration officials have labeled Japan as "our most important defense relationship in Asia" and "one of our primary partners on the world stage."[226] The approximately 26,000 US combat forces assigned in Japan not only serve as a deterrent to an attack on Japan, but these forces and their support bases would be extremely valuable in responding to a contingency on the Korean peninsula.[227] The Seventh Fleet operating in the Sea of Japan presents a definite threat to the Soviet Pacific Fleet which is home-ported at Vladivostok and Petropavlovsk. Control of the Tsushima, Tsugaru, and Soya Straits leading out of the Sea of Japan would be critical in restricting Soviet naval forces in the region in a conflict.[228] Japanese naval forces have participated in joint naval exercises with the US designed to exploit Soviet vulnerability in the Sea of Japan.[229]

Clearly, the Reagan administration valued the security relationship with Japan, particularly when confronted with an increased Soviet challenge. The trade deficit which reached $36.8 billion in 1984 became an increasingly serious problem, but the president attempted to attenuate the protectionist sentiments in Congress by convincing Nakasone to open Japanese markets to more US exports. Their meeting in January 1985 was specifically designed for this purpose.[230] Japan eventually exceeded the 1 percent of GNP limit for defense in the late 1980s through the implementation of a five-year, $100 billion military spending plan that significantly expanded Japanese sea lane and air defense capabilities.[231] While there is always the possibility of adverse public reaction in Japan, it appears that the US-Japanese security relationship remains close, despite continuing tensions in the bilateral economic relationship.

KOREA

Concerning US-ROK security issues, President Reagan was determined to channel American policy in a different direction than his predecessor. President Chun Doo Hwan, who seized power shortly after the assassination of President Park in late 1979, was not by accident the first foreign leader received by President Reagan. In the joint communiqué issued after Chun's visit, Reagan affirmed that the United States "had no plans to withdraw US ground combat forces from the Korean peninsula."[232]

In November 1983, President Reagan visited Seoul and traveled to the Joint Security Area at

Panmunjom on the 38th parallel. He emphasized again that US ground forces would remain and declared Northeast Asia to be a "region of critical strategic significance."[233] President Chun returned to Washington in April 1985, and the American president described US-ROK security ties as "the linchpin of peace in Northeast Asia."[234] More important, at the conclusion of the annual Security Consultative Meeting in May 1985, the final communique stated that the security of the ROK was "pivotal" to peace and stability in the Northeast Asian region and "vital to the security of the United States," one of the strongest statements made concerning the importance of the ROK-US security relationship.[235] The Reagan administration was also far less vocal publicly than its predecessor about human rights issues in Korea. Administration spokesmen were quick to point out that there still was an active interest in human rights, but that the emphasis was on quiet diplomacy to effect change in the ROK rather than through more public means.

The ROK provides a good example of the administration's three-part strategy in Asia. US ground forces deployed along crucial invasion routes remained in place throughout the Reagan years. Even though reductions are likely in the 1990s, a total withdrawal of US military presence is unlikely. The bilateral relationship was strengthened through the exchanges of official visits and American support for Seoul's political and economic initiatives including the hosting of the 1988 Olympics. Finally, US security assistance to Korea was increased to enable the ROK to contribute even more to its own security in the future.[236] As with Japan, the US strengthened its security ties with Korea largely because, if conflict occurs in the Korean peninsula, it will be extremely difficult to keep it from expanding to the superpower level.

CHINA

The last two years of the Carter administration resulted in much improved US relations with the PRC as political, economic, cultural, scientific, and educational ties were established. However, cerain serious problems remained, including the Taiwan issue and also differences over what the proper security relationship between the two countries should be. Carter and his National Security Adviser Zbigniew Brzezinski favored a "protoalliance" against the Soviet Union while the PRC was less sanguine about such a formal relationship. Ronald Reagan, who had a long record of support for Taiwan, however, criticized Carter's China policy during the 1980 presidential campaign.[237] Sensitive to the expressed concerns in Beijing, the new president dispatched Gerald Ford early in his administration to inform the Chinese that the US would continue efforts to improve the relationship and expand it if possible.[238]

During a visit by Secretary of State Alexander Haig in June 1981, the US agreed to sell military equipment to the PRC on a "case by case" basis, including defensive hardware such as antitank weapons and antiaircraft missiles.[239] This was a significant change of policy from the Carter administration which had been reluctant to supply weapons of any sort. However, Taiwan continued to be a problem difficult to resolve. In late 1981, the Taiwanese sought to purchase new jet fighters known as the F-X, a term descriptive of both the F-5G and F-16 aircraft. The PRC violently objected to this sale, setting the stage for a policy debate within the administration. Finally, in July 1982, the president made the decision to proceed with coproduction of the F-5E version and not to supply the more sophisticated F-5G or F-16.[240] In August 1982, the US and PRC entered into an agreement whereby the US pledged not to send weapons to Taiwan which exceeded in quantity or quality those which had been supplied in the years since normalization. Further, the US promised to reduce gradually the sale of arms over time to a "final resolution" of the issue. In return, the Chinese stated their intent to resolve the Taiwan issue peacefully and established this intent as "fundamental policy" of the government.[241]

Despite this agreement, Taiwan remains a major problem in the relationship. The Taiwanese still have influence in the US Congress as recently evidenced by their opposition to the sale of sonar equipment, turbine engines, and other new technologies to the PRC's navy.[242] Other difficulties remain as well. President Reagan and many of his advisers valued the relationship with the PRC in large measure because of the difficulties this relationship presents to the Soviet Union. However, China is intent upon not entering into an alliance with the US not only because of the obvious adverse effects such action would have on Sino-Soviet relations, but also because of its desire to retain its position as a leader of the nonaligned movement. In many ways, Chinese-American security relations will continue to present challenges to American policymakers in the years to come.

SOUTHEAST ASIA

An additional important Reagan security policy initiative in Asia was the continued support for ASEAN in its efforts to force the Vietnamese to withdraw from Cambodia. The administration was content to offer political support to these efforts in the United Nations and elsewhere but only provided humanitarian support to the

Cambodian resistance forces opposing the Vietnamese militarily. However, ASEAN urged the US to begin providing military aid directly to the resistance.[243]

While the US was willing to provide military assistance to Thailand as a frontline state against Vietnamese expansion, the Reagan administration was reluctant to provide such assistance to Cambodian opposition groups.[244] This reluctance was based on fears concerning another US involvement in Indochina. Nonetheless, Congress took steps which could result in direct military assistance. The House Asian and Pacific Subcommittee of the Foreign Affairs Committee voted in March 1985 to provide $5 million to the noncommunist resistance groups.[245]

PHILIPPINES

As might be expected, the Reagan administration was even more vocal than its Ford and Carter predecessors in expressing support for the retention of the Philippine military bases as a means to thwart Soviet expansion in Southeast Asia. Several high-ranking administration officials made visits to the Philippines early in the first Reagan term and restated the importance of ASEAN in general and the Philippines in particular in meeting American security goals. Vice President George Bush represented the United States at the Marcos inauguration in June 1981 and, in a toast to the Philippine president, praised his "adherence to democratic principles and to the democratic processes." Bush indicated the US intended to remain an Asian power and would not leave the Philippines in isolation in the face of increased Soviet military power in the region.[246] Perhaps the strongest endorsement came from Secretary of Defense Weinberger during a visit in April 1982 when he toured both Clark Air Base and Subic Naval Base, highlighting the importance of these bases in meeting American security responsibilities in Asia and in the Indian Ocean area. He was unequivocal that the US was going to retain its military presence in the Philippines.[247]

The Marcos visit to the United States in September 1982 was also important as an indicator of improving US-Philippine relations. Marcos had wanted to make such a trip for several years, but for a variety of reasons, particularly President Carter's criticism of human rights violations in the Philippines, he was not invited. During his meeting with Reagan at the White House, the American leader complimented Marcos and discussed the impending negotiations to review the US retention of the military bases.[248] This review was completed in early June 1983, and the US agreed to pay the Philippines $900 million to be divided among FMS credits, economic support funds, and military assistance programs. This arrangement went into effect in October 1984 and ensured the US of continued use of the bases until 1989, two years before the original MBA is to expire.[249]

The Philippine military bases do contribute to the achievement of important military and political objectives in Asia. They allow the US to project its military forces north toward Korea and Japan, south toward the critical straits of Southeast Asia, and on into the Indian Ocean/Persian Gulf region. No other American facilities are so strategically located and have the capabilities to support the accomplishment of so many military missions from one general location. Similarly, the bases serve as a symbol of American resolve and commitment which reassures allies and friends in Asia, especially during a period when Soviet regional capabilities and initiatives are increasing. The political upheaval in the Philippines, culminating in the resignation of Marcos in February 1986, raised serious questions about the future of US military presence.

PHILIPPINE UPHEAVAL, ELECTION, AND AFTERMATH

The final chapter in the Marcos presidency began with the assassination of former Senator Benigno Aquino on his return to Manila from the US in August 1983.[250] The alleged military involvement in this murder, including the indictment of Armed Forces of the Philippines (AFP) Chief of Staff Fabian Ver, cast doubt on President Marcos and his regime, particularly because Ver was so close to Marcos. The president's declining health further exacerbated the political instability.[251] As in the past, the political opposition selected the military bases as tangible evidence of American support for the Marcos regime without which they believed he could not survive.[252] As a result, calls for the removal of the military bases increased, and the New People's Army (NPA), the military arm of the Philippine Communist party, gained in strength, presenting a formidable problem for the Marcos government in several parts of the country.

The Reagan adminstration increasingly encouraged the Marcos regime to initiate political and economic reforms, particularly within the military.[253] As political turmoil in the Philippines increased, the adminstration faced a dilemma. By pushing Marcos too hard or by reducing American assistance, it might bring the government down. On the other hand, to support Marcos in his resistance to reform could lead to a violent confrontation and the assumption of power by those individuals radically opposed to the retention of the bases.

In early November 1985, President Marcos announced that he would advance the presidential election scheduled for 1987 to 7 February 1986, believing that the opposition was divided and that he would easily win this election. As a reelected president with a popular mandate, Marcos thought he would be in a stronger position to deflect US pressure to make the economic, political, and military reforms which many in the US Congress in particular believed were essential.

During his 20 years in office, President Marcos had established himself as one of the world's most astute political leaders, particularly in understanding and manipulating the Philippine political system and the US-Philippine relationship. However, in his decision to conduct early elections, he made at least three miscalculations which led directly to his eventual resignation. First, while there were significant differences within the opposition camps, they were able to form a united ticket with Corazon C. Aquino, the wife of the assassinated Benigno Aquino, running for president and Salvador Laurel agreeing to run for vice president.[254] During the course of the campaign Aquino and Laurel proved to be formidable opponents who were able to strike responsive chords within the Philippine electorate.

Marcos's second miscalculation was closely associated with his first. He had simply become isolated from the Philippine people during the last few years of his regime and did not realize how unpopular he and his government had become. The actions of two particular institutions or factions within them were particularly important. The Catholic church, under the leadership of Cardinal Jaime Sin, issued a number of pastoral letters during the campaign which were critical of the Marcos regime and its policies. In addition to these criticisms, the Church urged the government to hold free and fair elections, and in the final stages of the campaign came out strongly in support of the Aquino-Laurel ticket.[255] The support of the Church was absolutely essential for the opposition candidates not only because approximately 85 percent of the population is Catholic, but also because the Church provided a source of communication to the people since Aquino and Laurel were denied access to government-controlled television, radio, and newspapers.

The other institution which played an important role in the election and the events thereafter was the Armed Forces of the Philippines (AFP). Two factions developed which represented differences over personal loyalties and military reforms. One faction associated itself with General Ver, the chief of staff, and the other with Lieutenant General Fidel Ramos, deputy chief of staff and head of the Philippine

Constabulary. The former faction was comprised of senior officers, many of them referred to as "overstaying generals" because Marcos had extended them beyond the mandatory retirement age. The latter faction included those officers who were members of the "Reform the AFP Movement" (RAM).[256] This group believed Marcos retained many of the "overstaying generals" because of their loyalty to the president rather than their military competence. RAM members also espoused reforms in the military to make it an effective combat force against the communist insurgents. The significance of this military factionalism was that one of Marcos's strongest pillars of support was weakening.

The third Marcos miscalculation concerned his belief that the United States would continue to support him because of US security interest in the Philippines. However, many American congressional leaders were convinced that Marcos was incapable or unwilling to initiate needed reforms, and favored reducing US military assistance.[257] President Reagan agreed to send an election observer team to the Philippines, headed by Senator Richard Lugar, chairman of the Senate Foreign Relations Committee. After observing the election and the difficulties involved with counting the votes by two different organizations, Senator Lugar reported to President Reagan that there had been massive vote fraud and intimidation, primarily by Marcos supporters, and he viewed the election as "fatally flawed."[258]

The Philippine Constitution designates the National Assembly as the final arbiter in a presidential election. The Marcos political party enjoyed a two-thirds majority in the legislature, so it came as no surprise that the Assembly declared Marcos the winner almost one week after the election occurred. The reaction within the Philippines and US was swift. Aquino vilified the National Assembly action and called upon her supporters to protest peacefully by holding rallies and boycotting those businesses which supported Marcos.[259] President Reagan eventually blamed Marcos for the fraud and violence associated with the election.[260] Shortly thereafter, the US Senate in a vote of 85 to 9 condemned the Marcos election tactics, and measures were introduced in both houses of Congress to suspend military assistance and redirect economic programs through the Catholic church.[261]

The demonstrations in Manila against Marcos increased in size and scope. However, by far the most significant event was the defection to the Aquino side by Minister of National Defense Juan Ponce Enrile and Lieutenant General Ramos on 23 February. The split within the military erupted, and units either declared their allegiance to Enrile and Ver or remained loyal to Marcos.[262] As time went on, it became clear that the tide of events was running against Marcos. Finally, on 25 February 1986, after more than 20 years in power, Ferdinand Marcos fled the Philippines and Corazon Aquino was sworn in as president.[263]

At the time of this writing, it is difficult to evaluate in any detail what effects this change in the Philippines will have on US security interests, particularly the retention of the major military bases. During the campaign, Aquino took the position that bases would remain until the Military Bases Agreement expires in 1991. After that, however, she indicated that the Philippines would keep its options open.[264] Following a sometimes acrimonious review of the Bases Agreement in 1988, in which the US agreed to increase military assistance to the Philippines, the US began to explore other options for its military presence at Clark AFB and Subic Bay. As the decade ended and President Aquino's own political position became threatened, the future of the bases remained unresolved. Nevertheless, the new Bush administration agreed to the limited use of military forces to support Mrs. Aquino in a coup attempt, thus demonstrating the importance of the US presence for her own position. At the same time, however, US military bases were seen as tangible evidence of earlier US support of Marcos, thus creating a residual anti-American sentiment in the Philippines. If President Aquino is able to initiate necessary economic, political, and military reforms and the United States supports her in these efforts, the bases may continue to serve US and Philippine security interests well into the future.

SUMMARY: THE REAGAN YEARS

In summarizing the Reagan administration's Asian security policy, the argument can be made that it continued Carter's shift back to the symmetrical style of containment reminiscent of the Kennedy-Johnson years. Opposition to Soviet expansion, a buildup of US military forces and equipment in the region, and strong political, economic, and military support for friends and allies are all symptomatic of this containment style. However, the Reagan adminstration remained cautious about such an extensive American involvement. The "Vietnam Trauma" persisted and restricted US initiatives. As such, Reagan's emphasis on providing economic and military assistance so allies can assume more responsibility for regional peace and security was more similar to the asymmetric style of containment. In conclusion, it seems fair to say that President Reagan steered a somewhat moderate course in Asia providing assistance where possible but remaining cognizant of the definite limits to what the US can accomplish.

CONCLUSION

US security policy in Asia during the post–World War II era has been and remains characterized by containment despite changes in Soviet policy. However, the implementation of this strategy against the Soviet Union and its allies has oscillated over the years between asymmetrical to symmetrical styles. Two important distinctions need to be made in understanding how the containment policy has evolved. First, some countries, notably China, which were originally to be contained, have now become at least partial partners in the containment of the Soviet Union. Without question, the threat of a Sino-Soviet alignment in Asia was a primary factor in the evolution of security policies from Truman through Johnson. Second, Asia has taken on increased importance to the United States during the 1970s and 1980s. As a region, it has surpassed Europe as the primary trading partner for the United States, and the political stability which characterizes most Asian countries stands in stark contrast with the Middle East, Latin America, and Africa. While problems remain, as evidenced in the Philippines, the political and economic growth and development apparent from Japan and Korea in the north to Malaysia in the south is impressive.

Further, if the US goal in Asia has been to contain Soviet influence, then American security policy has achieved a fair measure of success. Clearly the Soviets have increased their ability to project military power into all areas of Asia, but their political and economic initiatives have largely failed. Soviet access to Vietnamese bases contributes to this power projection capability, but Vietnam is also a drain on the Soviet economy, and the Soviets would do well to remember the strength of Vietnamese nationalism in relying on indefinite access to those facilities. In the final analysis, the dominoes did not fall in Asia as President Eisenhower had feared, largely due to US security policies. But the costs have been high as the Vietnam involvement attests. Nonetheless, Asia has achieved a measure of peace and prosperity which few would have predicted in the late 1940s and early 1950s. As American political leaders look to the future, they can feel somewhat reassured that US allies in Asia will be able, and hopefully willing, to share more of the costs and responsibilities for protecting and advancing the progress so far achieved.

NOTES

1. For an excellent historical treatment of the different containment strategies, see John Lewis Gaddis, *Strategies of Containment: A Critical Appraisal of Postwar American National Security Policy* (New York: Oxford University Press, 1982), and, for a shorter treatment, his "Containment: Its Past and Fu-

ture," *International Security* 5, no. 4 (1981): 74–102.

2. In addition to the Philippines, the other members of the Association of Southeast Asian Nations (ASEAN) are Indonesia, Malaysia, Thailand, Singapore, and Brunei. The Indochina countries are Vietnam, Laos, and Cambodia.

3. A copy of the Truman Doctrine can be found in Henry Steele Commager, ed., *Documents of American History*, 6th ed. (Englewood Cliffs, N.J.: Prentice-Hall, 1973), pp. 704–6.

4. George F. Kennan, *Memoirs, 1925–1950*, (Boston: Little, Brown, 1967), pp. 320–21.

5. Thomas H. Etzold and John Lewis Gaddis, eds., *Containment: Documents on American Policy and Strategy, 1945–1950* (New York: Columbia University Press, 1962), p. 225.

6. JCS-1769/1, "United States Assistance to Other Countries from the Standpoint of National Security," 27 Apr. 1947, contained in ibid., pp. 71–84.

7. Kennan also believed in 1947 that Japan and the Philippines were most important to the US; see Kennan, *Memoirs*, p. 381.

8. Etzold and Gaddis, *Containment Documents*, p. 79.

9. NSC-48/1, "The Position of the United States with Respect to Asia," 23 Dec. 1949, in ibid., pp. 252–69.

10. Ibid., esp. pp. 255, 263–64.

11. "Crisis in Asia—An Examination of U.S. Policy," *Department of State Bulletin*, 22, no. 551 (23 Jan. 1950): 111–18. Acheson's speech became controversial after the beginning of the Korean War because some of his critics charged that, by not including Korea within the US defensive perimeter, he gave the "green light" for the North Korean invasion. Acheson vehemently denied this charge. See Dean Acheson, *Present at the Creation* (New York: W. W. Norton, 1969), pp. 357–58.

12. Roosevelt's speech to the Philippines on 28 Dec. 1941, *Department of State Bulletin* 6, no. 132 (3 Jan. 1942): 5.

13. Joint Resolution No. 93, 29 June 1944, *U.S. Statutes at Large*, p. 265 (1944).

14. For an expression of this view, see Senator Millard Tydings's speech in June 1945 in *Congressional Record*, 79th Cong., 1st sess., p. 5697.

15. *New York Times*, 18 Nov. 1946, p. 1.

16. *Papers Relating to the Foreign Relations of the United States* [hereafter *FRUS*] *1946*, 8:939–40.

17. Esp. note 19, ibid., p. 935.

18. Roxas's views are reported in ibid., pp. 939–40.

19. *FRUS, 1947*, 6:1102–3.

20. "Agreement between the United States of America and the Republic of the Philippines Concerning Military Bases," 14 Mar. 1947, *Treaties and Other International Agreements Series* [hereafter *TIAS*] 1775 (1947–48). For a detailed history of the negotiations leading to this agreement, see William E. Berry, Jr., "American Military Bases in the Philippines, Base Negotiations, and Philippine-American Relations: Past, Present, and Future," Ph.D. diss., Cornell University, May 1981, pp. 110–62.

21. *New York Times*, 10 Feb. 1949, p. 9.

22. On 21 Mar. 1947 the US and Philippines signed an agreement entitled "Military Assistance to the Philippines." This agreement provided for the transfer of US military equipment and supplies to the Philippines. See *TIAS* 1662 (1947–48).

23. *FRUS, 1949* 7:597–99 and *New York Times*, 9 Aug. 1949, p. 1.

24. An extract of Johnson's letter is found in *FRUS, 1950* 7:1410–11.

25. Warner R. Schilling, "The Politics of National Defense: Fiscal 1950," in Warner R. Schilling, Paul Y. Hammond, and Glenn H. Snyder, eds., *Strategy, Politics, and Defense Budgets* (New York: Columbia University Press, 1962), pp. 153–55.

26. Ibid., p. 52.

27. "The Situation in the Philippines," *FRUS, 1950* 6:1461–63.

28. Ibid., pp. 1462–63.

29. For those interested in the debate over who was responsible for the "Loss of China" see Anthony Kubek, *How the Far East Was Lost: American Policy and the Creation of Communist China, 1941–1949* (Chicago: Henry Regnery, 1963), and John S. Service, *The American Papers: Some Problems in the History of U.S.-China Relations* (Berkeley and Los Angeles: University of California Press, 1963).

30. *The China White Paper: United States Relations with China*, Department of State Publication 3573, July 1949 (hereafter *China White Paper*), pp. 210–13.

31. William Whitney Stueck, Jr., *The Road to Confrontation: American Policy toward China and Korea, 1948–1950* (Chapel Hill: University of North Carolina Press, 1981), pp. 36 and 53.

32. Ibid., p. 65, and Donald F. Lach and Edmund S. Wehrle, *International Politics in East Asia since World War II* (New York: Praeger, 1975), p. 29.

33. Stueck, *Road to Confrontation*, p. 45.

34. *China White Paper*, pp. 773–74.

35. Lach and Wehrle, *International Politics*, p. 41 and Stueck, *Road to Confrontation*, p. 55.

36. Lach and Wehrle, *International Politics*, pp. 42–43.

37. James C. Thomson, Jr., Peter W. Stanley, and John Curtis Perry, *Sentimental Imperialists: The American Experience in East Asia* (New York: Harper and Row, 1981), p. 225.

38. NSC-22, "Possible Courses of Action for the U.S. with Respect to the Critical Situation in China," 26 July 1948, in Etzold and Gaddis, *Containment Documents*, pp. 236–38.

39. NCS-34, "United States Policy toward China," 13 Oct. 1948, in Etzold and Gaddis, *Containment Documents*, pp. 240–47.

40. For a good account of how emotional this issue was, see Thomson, Stanley, and Perry, *Sentimental Imperialists*, p. 217.

41. Acheson, *Present*, p. 358.

42. Lach and Wehrle, *International Politics*, p. 65.

43. Ibid., pp. 64–65.

44. Ibid., pp. 67–68.

45. Stueck, *Road to Confrontation*, p. 140.

46. Acheson, *Present*, p. 349.

47. A copy of the Cairo Communique is found in Commager, *Documents*, p. 20.

48. Stueck, *Road to Confrontation*, p. 20.

49. Lach and Wehrle, *International Politics*, pp. 73–74. For an argument that the division of Korea at the 38th parallel made no historical or cultural sense, see Thomson, Stanley, and Perry, *Sentimental Imperialists*, p. 237.

50. Gregory Henderson, *Korea: The Politics of*

the Vortex (Cambridge: Harvard University Press, 1968), p. 122.

51. For two good sources expressing this view, see the State, War, Navy Coordinating Committee Report 176/30 dated 4 Aug. 1947 in *FRUS, 1947* 6:738 and Lt. Gen. Wedemeyer's "Resume of United States Policy toward Korea," *FRUS, 1947* 6:803.

52. This information is contained in a memorandum written by Secretary of Defense Forrestal, 26 Sept. 1947, in *FRUS, 1947* 6:817–18.

53. NCS-8, "Report by the National Security Council with Respect to Korea," 2 Apr. 1948, in *FRUS, 1948* 6:1167–68.

54. Stueck, *Road to Confrontation*, pp. 97–98.

55. Ibid., p. 105.

56. NSC-8/2, "Position of the United States with Respect to Korea," 22 Mar. 1949, in *FRUS, 1949* 7:977 and Acheson, *Present*, p. 358. Acheson indicates a 500-man advisory unit was envisioned to complete the training and equipping of ROK military units.

57. Claude A. Buss, *The United States and the Republic of Korea: Background for Policy* (Stanford: Hoover Institution Press, 1982), p. 60.

58. Edwin O. Reischauer, *The Japanese* (Cambridge, Mass.: Belknap Press, 1977), p. 104, and Herbert Feis, *Contest over Japan* (New York: W. W. Norton, 1967), pp. 7–8.

59. A copy of this document is found in Feis, *Contest*, pp. 167–75.

60. The 1947 Constitution is contained in Hugh Borton, *Japan's Modern Century*, 2d ed. (New York: Ronald Press, 1970), pp. 569–88. For Article 9, see p. 572.

61. Bradley M. Richardson and Scott C. Flanagan, *Politics in Japan* (Boston: Little, Brown, 1984), p. 66.

62. Ibid., pp. 64–66, and Reischauer, *Japanese*, pp. 106–9.

63. Frederick S. Dunn, *Peace-Making and the Settlement with Japan* (Princeton: Princeton University Press, 1963), pp. 72–73, Richardson and Flanagan, *Politics*, p. 68, and Lach and Wehrle, *International Politics*, p. 140.

64. NSC-49, "Strategic Evaluation of United States Security Needs in Japan," 15 June 1949, in Etzold and Gaddis, *Containment Documents*, pp. 231–33.

65. NSC-49/1, "Department of State Comments on NSC-49," 30 Sept. 1949, in ibid., pp. 233–36. See also Kennan, *Memoirs*, p. 394, for another evaluation of the internal threat in Japan and an argument for an early peace treaty.

66. Dunn, *Peace-Making*, pp. 84–85 and 87–88, and Lach and Wehrle, *International Politics*, p. 147. For the Soviet view of the peace treaty issue, see Max Beloff, *Soviet Policy in the Far East, 1944–1951* (New York: Oxford University Press, 1953), pp. 145–54.

67. The full text of NSC-68, "United States Objectives and Programs for National Security," dated 14 Apr. 1950, is available in Etzold and Gaddis, *Containment Documents*, pp. 385–442. For accounts of this document, see Samuel P. Huntington, *The Common Defense* (New York: Columbia University Press, 1961), pp. 49–53, Franz Schurmann, *The Logic of World Power* (New York: Pantheon Books, 1974), pp. 155–56, and Gaddis, *Strategies*, pp. 89–109.

68. Huntington, *Common Defense*, p. 51.

69. Gaddis, *Strategies*, pp. 96–98.

70. Ibid., p. 100, and Etzold and Gaddis, *Containment Documents*, p. 384.

71. Gaddis. *Strategies*, p. 113. President Truman did not officially approve the recommendations of NSC-68 until 30 Sept. 1950.

72. Acheson, *Present*, p. 405.

73. Stueck, *Road to Confrontation*, pp. 180–81.

74. For Truman's 27 June 1950 statement, see *Public Papers of the Presidents of the United States: Harry S. Truman, 1950*, p. 492. For more detail see Acheson, *Present*, pp. 402–13. For the reaction in the Philippines to Truman's announcement, see the *New York Times*, 29 June 1950, p. 3.

75. Gaddis, *Strategies*, pp. 114–15.

76. Robert E. Osgood, *Alliances and American Foreign Policy* (Baltimore: Johns Hopkins University Press, 1968), p. 46.

77. While it is true that Truman was influenced by the "Munich analogy," he also feared that if the United Nations did not respond effectively, it could suffer the same fate as the League of Nations. See Merle Miller, *Plain Speaking: An Oral Biography of Harry Truman* (New York: Berkeley, 1974), p. 294. Domestic factors were important, too, since anticommunist sentiments were high in the US at this time as evidenced by the increased power wielded by Sen. Joseph R. McCarthy.

78. Dunn, *Peace-Making*, p. 134.

79. Mutual Defense Treaty Between the United States of America and the Republic of the Philippines, *TIAS* 2529. Security Treaty Between the United States of America, Australia, and New Zealand, *TIAS* 2429 (1952). Security Treaty Between the United States of America and Japan, *TIAS* 2491 (1952).

80. Treaty of Peace with the Allied Powers, *TIAS* 2490 (1952). The San Francisco Peace Conference was attended by fifty-two countries. Only the Soviet Union, Poland, and Czechoslovakia refused to sign the treaty.

81. Jerome H. Kahan, *Security in the Nuclear Age* (Washington, D.C.: Brookings Institution, 1975), p. 11.

82. Dulles's massive retaliation speech can be found in *Department of State Bulletin* 30 (Jan. 1954): 108. (Excerpts of Dulles's speech are included in Selected Readings II.)

83. Gaddis, *Strategies*, p. 148.

84. It is not clear whether Eisenhower or Dulles coined this phrase. In an April 1954 press conference, however, the president made the following comment: "You have a row of dominoes set up, you knock over the first one, and what will happen to the last one is the certainty that it will go over very quickly. So you could have a beginning of a disintegration that would have the most profound influences." Quoted in Paul M. Kattenburg, *The Vietnam Trauma in American Foreign Policy, 1945–75* (New Brunswick, N.J.: Transaction Books, 1980), p. 66, note 20.

85. Mutual Defense Treaty Between the United States and the Republic of Korea, *TIAS* 3097 (1954). Interestingly the Senate, during the ratification debate, added an understanding that the provisions of this treaty were applicable only in cases of an external attack against territory which the US recognized as lawfully under the administrative control of the ROK. Apparently some senators remained concerned that Syngman Rhee would initiate an attack on North Korea and then seek US assistance.

86. Mutual Defense Treaty Between the United

States of America and the Republic of China, *TIAS* 3178 (1955).

87. "Southeast Asia Treaty Organization: Response to the Communist Threat" in Marvin E. Gettleman, ed., *Vietnam: History, Documents, and Opinions* (Greenwich, Conn.: Fawcett, 1965), pp. 121–24. The document establishing SEATO is commonly referred to as the Manila Pact.

88. Guenter Lewy, *America in Vietnam* (New York: Oxford University Press, 1978), p. 11.

89. Kahan, *Security*, pp. 18–19. Both Dulles and Eisenhower later indicated they believed this threat was instrumental in concluding the Korean negotiations in the summer of 1953. See Dwight Eisenhower, *Mandate for Change* (Garden City, N.Y.: Doubleday, 1963), p. 180.

90. Chalmers M. Roberts, "The Day We Didn't Go to War" in Gettleman, *Vietnam*, pp. 124–33; Paul Y. Hammond, *Cold War and Détente* (New York: Harcourt Brace Jovanovich, 1975), p. 99; and Kahan, *Security*, p. 19.

91. Kahan, *Security*, pp. 20–21, and Hammond, *Cold War*, pp. 126–28.

92. Lewy, *America*, pp. 4–5.

93. The Geneva Agreements and the Final Declaration are included in Gettlemen, *Vietnam*, pp. 164–88. See also Lewy, *America*, pp. 7–10, and Kattenburg, *Vietnam Trauma*, pp. 47–48.

94. Kattenburg, *Vietnam Trauma*, pp. 58–60 and Lewy, *America*, pp. 13–14.

95. See Gaddis, *Strategies*, pp. 193–95 and Donald S. Zagoria, *The Sino-Soviet Conflict, 1956–1961* (Princeton: Princeton University Press, 1962), pp. 214–15.

96. Martin E. Weinstein, *Japan's Postwar Defense Policy, 1947–1968* (New York: Columbia University Press, 1971), p. 74.

97. Ibid., pp. 75–77.

98. Security Treaty Between the United States of America and Japan, 8 Sept. 1951, *TIAS* 2491 (1952).

99. Treaty of Mutual Cooperation and Security Between the United States and Japan, 19 Jan. 1960, *TIAS* 4509 (1960).

100. For a good account of the negotiations of the new treaty between 1958 and 1960, see Weinstein, *Japan's Policy*, pp. 91–103. Despite US concessions, this treaty was very controversial in Japan particularly for those who wanted no security relationship between Japan and the US. Demonstrations occurred in May and June 1960 to the extent that President Eisenhower was forced to cancel his visit to Japan. Prime Minister Kishi resigned as a result of these demonstrations.

101. Kahan, *Security*, pp. 16–17.

102. Gaddis, *Strategies*, pp. 203–4. For a good critique of massive retaliation from the military perspective, see Maxwell D. Taylor, *The Uncertain Trumpet* (Westport, Conn.: Greenwood Press, 1974).

103. Gaddis, *Strategies*, p. 204. The similarities here to NSC-68 are evident, partly because individuals such as Paul Nitze, Dean Acheson, and Dean Rusk who were associated with NSC-68 were also important advisers to President Kennedy.

104. Kahan, *Security*, p. 75 and Gaddis, *Strategies*, p. 216.

105. An interesting commentary on the writing of the inaugural address as well as a copy of it is found in Theodore C. Sorensen, *Kennedy* (New York: Harper and Row, 1965), pp. 240–48.

106. For a good account of Kennedy's appeal in the Third World as well as his goals toward these countries, see Roger Hilsman, *To Move a Nation* (Garden City, N.Y.: Doubleday, 1967), pp. 362–67.

107. Leslie H. Gelb with Richard K. Betts, *The Irony of Vietnam: The System Worked* (Washington, D.C.: Brookings Institution, 1979), pp. 70–71.

108. Ibid., p. 71.

109. For accounts of the situation in Laos, see Hilsman, *Move a Nation*, pp. 125–55 and Sorensen, *Kennedy*, pp. 639–48.

110. Hilsman, *Move a Nation*, p. 151, and Sorensen, *Kennedy*, pp. 643–48.

111. National Security Memorandum 52 in *The Pentagon Papers* (New York: Bantam Books, 1971) as published by the *New York Times*, pp. 126–27. See also Gelb, *Irony*, p. 72.

112. *Pentagon Papers*, pp. 127–30.

113. Ibid., pp. 141–43.

114. Ibid., pp. 150–53.

115. Lewy, *America*, p. 24.

116. This comparison is also made by Gelb, *Irony*, p. 84.

117. The Kennedy administration tacitly approved the coup against Diem. One of the clearest statements of this support is contained in a State Department message to Ambassador Henry Cabot Lodge in Saigon on 24 Aug. 1963. See *Pentagon Papers*, pp. 194–95.

118. Hilsman, *Move a Nation*, pp. 524–37, addresses this question directly and believes there was a good chance Kennedy would have changed course.

119. *Pentagon Papers*, pp. 127–30.

120. Lyndon B. Johnson, *The Vantage Point* (New York: Holt, Rinehart and Winston, 1971), p. 43.

121. Ibid., pp. 135–36, 151–52. Johnson was very concerned about the PRC particularly as the Chinese improved their relationship with Sukarno in Indonesia and Kim Il-Sung in North Korea.

122. *Pentagon Papers*, pp. 277–83.

123. Ibid., pp. 283–85.

124. Johnson, *Vantage Point*, pp. 66–67.

125. A copy of the Gulf of Tonkin Resolution is included as app. 1 in George Kahin and John Lewis, *The United States in Vietnam* (New York: Dial Press, 1969), pp. 477–78.

126. Kattenburg, *Vietnam Trauma*, pp. 122–23.

127. Herbert Y. Schandler, *Lyndon Johnson and Vietnam* (Princeton: Princeton University Press, 1977), pp. 11–15.

128. Gelb, *Irony*, p. 121.

129. Schandler, *Lyndon Johnson*, pp. 115–17. For the best source on the Tet offensive, see Don Oberdorfer, *Tet!* (Garden City, N.Y.: Doubleday, 1971).

130. See Schandler, *Lyndon Johnson*, pp. 266–319, for an excellent analysis of Johnson's decision not to run for reelection. See also Townsend Hoopes, *The Limits of Intervention* (New York: D. McKay, 1969).

131. Testimony of James M. Wilson, Jr., deputy chief of mission, US Embassy, Manila, at hearings before the Subcommittee on U.S. Security Agreements and Commitments Abroad of the Committee on Foreign Relations (Senate) 30 Sept.–30 Oct. 1969 (hereafter the Symington Hearings), "United States Security Commitments Abroad, the Republic of the Philippines," 91st Cong., 2d sess., p. 355.

132. Sungjoo Han, "South Korea's Participation in the Vietnam Conflict: An Analysis of the U.S.–Korean Alliance," *Orbis* (Winter 1978): 893.

133. Symington Hearings, p. 358.

134. Han, *South Korea*, p. 908.

135. *Far Eastern Economic Review* (hereafter *FEER*), 27 Oct. 1966, pp. 221–26.

136. Han, *South Korea*, pp. 901–2.

137. Ibid., p. 902.

138. For Humphrey's visit, see the Symington Hearings, p. 147.

139. John Lewis Gaddis, "The Rise, Fall, and Future of Détente," *Foreign Affairs* 62, no. 2 (1983–84): 364.

140. Kahan, *Security*, p. 147.

141. Cecil V. Crab, Jr., and Pat M. Holt, *Invitation to Struggle: Congress, the President, and Foreign Policy* (Washington, D.C.: Congressional Quarterly Press, 1980), pp. 125–26.

142. "Resolution Concerning the War Powers of Congress and the President," *Legislation on Foreign Relations with Explanatory Notes*, Committee on Foreign Relation, US Senate and Committee on International Relations, US House of Representatives, June 1975, pp. 973–77.

143. Stanley Hoffman, *Gulliver's Troubles, or the Setting of American Security Policy* (New York: McGraw-Hill, 1968), pp. 33–35.

144. Harold C. Hinton, *Three and a Half Powers: The New Balance in Asia* (Bloomington: Indiana University Press, 1975), p. 174.

145. For a copy of the Nixon Doctrine, see *U.S. Foreign Policy for the 1970s: A New Strategy for Peace* (The President's Report to the Congress, 18 Feb. 1970). Harold Hinton suggests that, in addition to Nixon's desire not to repeat the Vietnam experience, he was also influenced by Indonesia's ability to suppress a major coup attempt in 1965–66. Hinton argues that Nixon became convinced that other Asian states could deal with communist subversion without the assistance of American ground forces. See Hinton, *Three and a Half*, pp. 128–29.

146. Hinton, *Three and a Half*, p. 54.

147. Ibid., pp. 55–56.

148. Ibid., p. 56. For more on the Nixon Doctrine, see Ralph N. Clough, *East Asia and U.S. Security*, pp. 2–5.

149. Kahan, *Security*, pp. 147–48.

150. Henry Brandon, *The Retreat of American Power* (Garden City, N.Y.: Doubleday, 1973), pp. 213–15.

151. Robert E. Osgood, "The Diplomacy of Allied Relations: Europe and Japan" in Robert E. Osgood, ed., *Retreat from Empire* (Baltimore: Johns Hopkins University Press, 1973), p. 194.

152. For an excellent account of the negotiations which occurred during the first Nixon administration, see Tad Szulc, "Behind the Vietnam Cease-Fire Agreement," *Foreign Policy*, no. 15 (Summer 1974): 21–69.

153. For a detailed analysis of Vietnamization, see Lewy, *America*, pp. 162–222.

154. The Joint Communique, dated 22 Aug. 1969, is found in the Symington Hearings, pp. 1724–25.

155. Buss, *United States*, pp. 143–44.

156. Richard H. Solomon, "The China Factor in America's Foreign Relations" in Richard H. Solomon, ed., *The China Factor* (Englewood Cliffs, N.J.: Prentice Hall, 1981), p. 1.

157. Gaddis, *Strategies*, pp. 295–96.

158. Brandon, *Retreat*, pp. 186 and 194. For more detail on the PRC's decision to improve relations with the US see John W. Garver, *China's Decision for Rapprochement with the United States, 1968–1971* (Boulder, Colo.: Westview Press, 1982), William R. Heaton, *A United Front against Hegemonism* (Washington, D.C.: National Defense University Press, 1980), and Jonathan D. Pollack, *The Sino-Soviet Rivalry and Chinese Security Debate* (New York: Praeger, 1966).

159. A copy of the Shanghai Communique is located in Solomon, *China Factor*, pp. 296–300.

160. Lach and Wehrle, *International Politics*, p. 276.

161. Clough, *East Asia*, p. 106.

162. For more detail on the New Economic Policy, see Joan E. Spero, *The Politics of International Economic Relations*, 2d ed. (New York: St. Martin's Press, 1977), pp. 90–92.

163. See President Nixon's Jan. State of the World message, cited in Lach and Wehrle, *International Politics*, p. 277.

164. Buss, *United States*, p. 108.

165. Brandon, *Retreat*, p. 198.

166. For an excellent article which develops this thesis see Gaddis, "Rise, Fall, and Future," pp. 359–61.

167. Marcos's speech entitled "A Matter of Survival" is found in *Philippines Official Gazette* 71, no. 17 (1975): 2420–31.

168. The Joint Communique, dated 22 Nov. 1974, is found in *Department of State Bulletin* 71, no. 1852 (23 Dec. 1974): 877–78.

169. *Department of State Bulletin* 73, no. 1905 (29 Dec. 1975): 913–16.

170. Kissinger's speech entitled "America and Asia" is found in *Department of State Bulletin* 75, no. 1938 (16 Aug. 1976): 217–26.

171. The 1947 MBA was to remain in effect for 99 years; however, the Rusk-Romulo Agreement in 1966 reduced the duration of the agreement to 25 years. For more detail on the 1976 negotiations, see Berry, "American Military Bases," pp. 301–8.

172. *New York Times*, 4 December 1976, p. 1.

173. Ibid., 5 Dec. 1976, p. 1.

174. In Sept. 1972, President Marcos declared martial law in the Philippines. By 1976, there were rather widespread reports of human rights violations in the Philippines. See for example "Human Rights and Philippine Political Prisoners," statement by Reverend Paul Wilson before the Subcommittee on Foreign Assistance of the Committee on Foreign Relations, US Senate, 94th Cong., 1st sess. (Washington, D.C.: G.P.O., 1975).

175. Holbrooke's testimony is in *Department of State Bulletin* 77, no. 1971 (4 Apr. 1977): 322–26. For another early administration statement, see Secretary of State Cyrus Vance's address before the Asia Society in June 1977 contained in *Department of State Bulletin* 77, no. 1988 (1 Aug. 1977): 141–45.

176. For Carter's speech on human rights at Notre Dame University in early 1977, see Jimmy Carter, *Keeping Faith* (New York: Bantam Books, 1982), p. 141.

177. "U.S. Troop Withdrawal from the Republic of Korea," A Report to the Committee on Foreign Relations, U.S. Senate, 95th Cong., 2d sess., 9 Jan. 1978, by Senators Hubert H. Humphrey and John Glenn (hereafter the Humphrey and Glenn Report), p. 1.

178. *Public Papers of the Presidents of the United States, Jimmy Carter*, 1977, Book 1, p. 343.

179. Humphrey and Glenn Report, p. 20.

180. *Public Papers, Jimmy Carter,* Book 1, p. 1018.

181. "Force Planning and Budgetary Implications of U.S. Withdrawal from Korea," Background Paper, Congressional Budget Office (hereafter CBO Report), May 1978, p. 11, and Philip Habib's statement in *Department of State Bulletin* 67, no. 1985 (11 July 1977): 49.

182. CBO Report, pp. 19–29. For a good discussion, see Franklin B. Weinstein and Fuji Kamiya, eds. *The Security of Korea* (Boulder, Colo.: Westview Press, 1980), pp. 81–84. See also Frank Gibney, "The Ripple Effects in Korea," *Foreign Affairs* 56, no. 1 (1977): 167–68, on the human rights issue.

183. For domestic criticisms, see the testimony of Maj. Gen. John K. Singlaub and Prof. Donald S. Zagoria in "Review of the Policy Decision to Withdraw United States Ground Forces from Korea," Hearings before the Investigations Subcommittee, Committee on Armed Forces, U.S. House of Representatives, 95th Cong., 1st and 2d sess. (hereafter Troop Withdrawal Hearings). For Singlaub, see pp. 5, 14, 31, and 55 esp. and, for Zagoria, pp. 161–72.

184. Humphrey and Glenn Report, pp. 20–23.

185. Chae Jin Lee and Hideo Sato, eds., *U.S. Policy toward Japan and Korea* (New York: Praeger, 1982), p. 107, and Zagoria's testimony in Troop Withdrawal Hearing, p. 167.

186. "Impact of Intelligence Reassessment on Withdrawal of U.S. Troops from Korea," Hearings before the Investigations Subcommittee of the Committee on Armed Forces, House of Representatives, 96th Cong., 1st sess., 21 June and 17 July 1979, p. 6 (hereafter Intelligence Reassessment Hearing).

187. *Department of State Bulletin* 79, no. 2029 (Aug. 1979): 15.

188. *Department of State Bulletin* 79, no. 2030 (Sept. 1979): 37.

189. For the details of these negotiations, see Berry, "American Military Bases," pp. 319–58.

190. See Michael Leifer, "Conflict and Regional Order in Southeast Asia," *Adelphi Paper* 162 (Winter 1980): 16.

191. TIAS 9224 (1980), "Military Bases in the Philippines," 7 Jan. 1979, pp. 1–38.

192. *New York Times,* 24 Nov. 1977, p. 14.

193. David W. P. Elliott, *The Third Indochina Conflict* (Boulder, Colo.: Westview Press, 1981), pp. 106–9.

194. Carter, *Keeping Faith,* pp. 194–95.

195. *New York Times,* 7 Feb. 1979, p. 1, and Liefer, "Conflict," p. 28.

196. Carter, *Keeping Faith,* p. 190.

197. *New York Times,* 17 Dec. 1978, p. 1.

198. Carter, *Keeping Faith,* p. 195.

199. Ibid., pp. 194, 201.

200. *New York Times,* 16 Dec. 1978, p. 8.

201. For the PRC unilateral statement dated 15 Dec. 1978, see Solomon, *China Factor,* p. 304. Secretary of State Vance also indicated that no pledge was given by the PRC on the use of force issue. See *New York Times,* 18 Dec. 1978, p. 12.

202. The US unilateral statement dated 15 Dec. 1978 is in Solomon, *China Factor,* p. 303.

203. Carter, *Keeping Faith,* p. 206.

204. *New York Times,* 19 Dec. 1978, p. 1, and ibid., 20 Dec. 1978, p. 27.

205. A copy of the Taiwan Relations Act is in Solomon, *China Factor,* pp. 304–14.

206. John E. Endicott, "The Defense Policy of Japan" in Douglas J. Murray and Paul R. Viotti, eds., *The Defense Policies of Nations* (Baltimore: Johns Hopkins University Press, 1982), p. 454.

207. See Carter's laudatory remarks about Prime Minister Ohira in Carter, *Keeping Faith,* p. 113.

208. For Weinberger's speech in May 1981, see *Department of State Bulletin* 81, no. 2052 (July 1981): 46–48. Schultz's comments are in *Current Policy,* no. 459, US Department of State Bureau of Public Affairs, 5 Mar. 1983.

209. Robert E. Osgood, "The Revitalization of Containment," *Foreign Affairs: America and the World, 1981* 60, no. 3 (1981): 483–84.

210. For some good comparisons between Carter and Reagan foreign policy initiatives, see Coral Bell, "From Carter to Reagan," *Foreign Affairs: American and the World, 1984* 63, no. 3 (1984): 490–510.

211. Richard H. Solomon, "East Asia and the Great Power Coalitions," *Foreign Affairs: America and the World, 1981* 60, no. 3 (1981): 690.

212. US Foreign Policy Objectives Study, pp. 166–67. See also Henry Rowen, "American Security Interests in Northeast Asia," *Daedalus* (Fall 1980): 81–96.

213. Assistant Secretary of Defense for International Security Affairs Richard Armitage in an address to the US-Asia Institute, 19 June 1984. For other comments by Armitage on Soviet access to Vietnamese bases, see *FEER,* 6 Sept. 1984, pp. 40–46.

214. For a good statement of this strategy, see James A. Kelly, deputy assistant secretary of defense, in testimony before the Senate Subcommittee on East Asia and Pacific Affairs, 21 Mar. 1985.

215. *New York Times,* 29 Apr. 1981, p. 7.

216. Richard L. Sneider, *U.S.-Japanese Security Relations* (New York: East Asian Institute, Columbia University, 1982), pp. 90–91.

217. For a draft of the Joint Communique, see *New York Times,* 9 May 1981, p. 7, and for Ito's dismissal, *New York Times,* 17 May 1981, p. 1.

218. *New York Times,* 9 May 1981, p. 7.

219. For other domestic factors, see Robert F. Reed, *The U.S.-Japan Alliance: Sharing the Burden of Defense* (Washington, D.C.: National Defense University Press, 1983), pp. 15–20.

220. See the statement of then Philippine Foreign Minister Carlos Romulo on this subject in the *Chicago Tribune,* 30 Dec. 1982, p. 1.

221. Sneider, *Security Relations,* pp. 42–43, and Reed, *U.S.-Japan Alliance,* pp. 4–5.

222. *New York Times,* 1 Oct. 1982, p. 3.

223. Ibid., 28 Nov. 1982, p. 1.

224. Ibid., 7 Jan. 1983, p. 4.

225. Ibid., 21 Jan. 1983, p. 4.

226. James A. Kelly, deputy assistant secretary of defense, *Wireless File,* USIS-Embassy Seoul, 22 Mar. 1985, p. 8, and Richard Armitage, assistant secretary of defense for International Security Affairs, *Wireless File,* USIS-Embassy Seoul, 29 June 1984, p. 4.

227. Anthony H. Cordesman, "The Military Balance in Northeast Asia," *Armed Forces Journal,* Nov. 1983, p. 80.

228. For two good articles on the strategic importance of the Japan-US relationship, see the *Wall Street Journal,* 1 Dec. 1982, p. 26, and *Christian Science Monitor,* 3 May 1983, p. 3.

229. *FEER*, 11 Apr. 1985, pp. 36–37.

230. *International Herald Tribune*, 4 Jan. 1985, p. 1.

231. Ibid., 30 Apr. 1985, p. 1.

232. Joint Communique dated 2 Feb. 1981, *Department of State Bulletin* 81, no. 2048, p. 14. For additional coverage of the Chun visit and its implications, see *FEER*, 6–12 Feb. 1981, p. 18. Secretary of State Schultz reaffirmed the American commitment to the ROK at a news conference in Seoul on 6 Feb. 1983. See *Department of State Bulletin* 83, no. 2072 (March 1983): 59.

233. *New York Times*, 14 Nov. 1983, p. 1. For more on the Reagan visit and its significance, see ibid., 15 Nov. 1983, p. 2.

234. *International Herald Tribune*, 29 Apr. 1985, p. 4.

235. The Final Communique was included in *Wireless File*, USIS-Embassy, Seoul, 8 May 1985, pp. 4–7. See esp. paragraph 5.

236. In Secretary Weinberger's 1985 *Annual Report*, the administration pledged to raise FMS credits to ROK from $220 million to $228 million for the ROK Force Improvement Plan. This figure represents more than half of the proposed FMS credits for the East Asia and Pacific region.

237. Steven I. Levine, "China and the United States: The Limits of Interaction," in Samuel S. Kim, ed., *China and the World* (Boulder, Colo.: Westview Press, 1984), pp. 115–18.

238. *New York Times*, 28 Mar. 1981, p. 3.

239. Ibid., 17 June 1981, p. 1.

240. Ibid., 30 Oct. 1981, p. 5, and ibid., 17 July 1982, p. 1. See also A. Doak Barnett, *The FX Decision: Another Crucial Moment in U.S.-China-Taiwan Relations* (Washington, D.C.: Brookings Institution, 1981).

241. This agreement is found in the *New York Times*, 18 Aug. 1982, p. 4. For a critical appraisal of US-Chinese relations, see A. James Gregor and Maria Hsia Chang, *The Iron Triangle* (Stanford: Hoover Institution Press, 1984).

242. *International Herald Tribune*, 24 Jan. 1985, p. 3.

243. Ibid., 13 Feb. 1985, p. 6.

244. *New York Times*, 10 Apr. 1983, p. 6. Even though the Southeast Asian Treaty Organization no longer exists, the 1954 Manila Treaty which brought SEATO into existence remains. The United States and Thailand are allied through this treaty which provides justification for American military assistance programs in support of Thailand.

245. For two articles in support of this military assistance, *International Herald Tribune*, 16–17 Mar. 1985, p. 6, and ibid., 26 Mar. 1985, p. 8.

246. *New York Times*, 1 July 1981, p. 13. The Philippine Constitution was amended in 1981 providing for a presidential term of 6 years. Marcos was reelected in June 1981 and was scheduled to serve until 1987 under this amendment. There was no prohibition against his running for another term.

247. Ibid., 3 Apr. 1982, p. 3, and ibid., 5 Apr. 1982, p. 2.

248. For a series of articles during the Marcos visit see ibid., 16 Sept. 1982, p. 3; 17 Sept. 1982, p. 3; 21 Sept. 1982, p. 9; and 27 Sept. 1982, p. 8.

249. *FEER*, 16 June 1983, pp. 30–32.

250. On the Aquino assassination, see *FEER*, 1 Sept. 1982, pp. 10–15.

251. On the economic and political crises in the Philippines, see Charles W. Lindsey, "Economic Crisis in the Philippines" and Richard J. Kessler, "Politics Philippine Style, Circa 1984," in *Asian Survey* 24, no. 12 (1984): 1185–1208 and 1209–28, respectively.

252. For an example see the *International Herald Tribune*, 1–2 Dec. 1984, p. 5.

253. Ibid., 15 Jan. 1985, p. 5, and *FEER*, 31 Jan. 1985, pp. 30–31.

254. Mrs. Aquino's decision to run for president was made easier when General Fabian Ver and the other twenty-four defendants implicated in the Aquino assassination were acquitted by a Philippine court and Marcos reinstated Ver as Chief of Staff of the Armed Forces of the Philippines. See the *International Herald Tribune*, 3 Dec. 1985, p. 1, and *FEER*, 12 Dec. 1985, pp. 12–13.

255. *FEER*, 26 Dec. 1985, pp. 17–18; and 16 Jan. 1986, pp. 12–13; *International Herald Tribune*, 20 Jan. 1986, p. 1; and 30 Jan. 1986, p. 5.

256. *FEER*, 6 Feb. 1986, pp. 25–27, and *International Herald Tribune*, 17 Dec. 1985, p. 1.

257. *International Herald Tribune*, 10 Dec. 1985, p. 4.

258. Ibid., 14 Feb. 1986, p. 5.

259. Ibid., 13 Feb. 1986, p. 1.

260. For Reagan's initial statement, see ibid., 13 Feb. 1986, p. 1. For the latter, ibid., 17 Feb. 1986, p. 1.

261. Ibid., 20 Feb. 1986, p. 1.

262. Ibid., 24 Feb. 1986, p. 1.

263. Ibid., 26 Feb. 1986, p. 1.

264. Ibid., 4 Feb. 1986, p. 2.

CHAPTER 8

DEFENDING GLOBAL INTERESTS

E. DOUGLAS MENARCHIK

American policymakers and military planners focus primarily on the Soviet threat to the American homeland, Western Europe and Northeast Asia. The threat is real, and losing a high-technology, high-intensity conflict in any of these vital areas would indeed be catastrophic for America and the West. The United States has structured its security system to defend these vital interests by allocating significant resources and by orchestrating its political, military, and economic instruments of power to be prepared to wage high-technology, high-intensity conflict against adversaries who can credibly threaten these interests. High-technology, high-intensity warfare is potentially so destructive that no rational state is likely to risk such a form of conflict, as no conceivable gain seems worth the cost. Nuclear powers can inflict destruction on others, but they are also vulnerable to that destruction. Since the end of World War II, the international system has been spared the trauma of a major war involving the great powers. Yet, the international system is experiencing a growing frequency of political violence, especially low-intensity conflict. Of the various forms of political violence, terrorism and guerrilla warfare are the principal forms of conflict in the world today.[1]

It is risky for major powers to compete directly with each other. Rather, major powers compete in the less risky areas of what is known as the Third World. In addition to major power competition in the Third World, nation building there brings with it the frictions of modernization, providing fertile ground for low-intensity conflict. America is increasingly faced with "dirty little wars" and other forms of low-intensity conflict "in the shadows" with guerrillas and terrorists, on the periphery of the developed world where less-than-vital interests are threatened. Guerrillas and terrorists are not likely to be deterred by nuclear weapons, aircraft carriers, wings of fighter aircraft, or airborne divisions. This kind of high-technology military power is often irrelevant in low-intensity conflict which typically is waged for the "hearts and minds" of Third World peoples.

The US is a superpower with political, military, and economic interests in all geographic areas of the globe, and American interests are threatened by a variety of antagonists. American policymakers and military planners seek to defend a multiplicity of global interests but are constrained by limited resources, public opinion, domestic needs, and international pressures. US policymakers must make tough choices in allocating scarce resources: what is vital to defend, what is "less-than-vital," what strategy best defends US global interests? Matching national interests with limited resources requires that US leaders be selective. Frederick the Great observed two centuries ago, "He who attempts to defend everywhere, defends nothing."[2] The dilemma inherent in his statement is pertinent today as America wrestles with determining which global interests have priority and how best to defend them.

The US must be prepared to face two kinds of conflict. One is high-technology, high-intensity conflict—nuclear war and large-scale conventional war like World War II. The other kind of war consists of lesser battles on the periphery of the "heartland" of Eurasia, most often in the Third World. These "small wars" are conflicts waged by and against forces of lesser powers and include indigenous guerrilla-type movements, terrorism, and conflicts between the clients and proxies of superpowers. They tend to threaten less-than-vital interests and tend to be more limited in terms of both geographic scope and commitment of resources. Such conflicts are variously called "small wars," "low-level conflict," or "low-intensity warfare." There is no agreement on the definition of such terms and, as a result, it is not uncommon to find these terms used imprecisely and interchangeably. Low-intensity conflict, as used here, refers to a range of activities and operations on the lower end of the spectrum of conflict involving the use of force or security assistance on the part of the antagonist to influence an adversary to accept a certain condition.[3] Low-intensity conflict comprises a variety of forms of political violence and nonviolent activities. Each is unique in form and requires a unique response.

Clausewitz warned long ago that statesmen and military strategists must observe international politics "in the round." They must know what kind of war they are to fight and what kind of enemy they are to face.[4] Should the US prepare for nuclear Armageddon or the "dirty little wars" fought in the shadows, or both? How does America best defend global interests against an array of enemies and threats?

This chapter looks at the ability of the US to defend its interests and to project its global power outside its direct alliance commitments in Western Europe and Northeast Asia. It addresses the factors which account for the US focus on "high-technology, high-intensity" conflict, then addresses the nature of low-intensity conflict as a part of a broader spectrum of conflict. It examines US interests in the Third World and illustrates the challenges which low-intensity conflict poses to the US ability to defend its global interests. The chapter concludes by considering the use of force as an instrument of US power in dealing with those challenges.

AMERICAN DEFENSE PRIORITIES AND LOW-INTENSITY CONFLICT

The contemporary international system has unique characteristics. It is characterized by the presence of nuclear weapons and the threat of nuclear war on a regional or global scale; the rapid rise in the number of states and the number and types of non-state actors; an increasingly decentralized state system and increased interdependence among the actors in the state system; the increased vulnerability of states to external intrusions; and the rising importance of non-state actors. These characteristics have affected dramatically the conduct of international relations and reoriented international security planning.

The threat of nuclear holocaust has had a strange effect on the conduct of strategic international relations. The contemporary international system has demonstrated a marked degree of stability, largely because the fear of nuclear war or a general conventional war on the grand scale of World War II is an effective psychological deterrent to super- and major-power conflict. Conflicts involving two or more major nuclear powers could potentially be global in scope. But, not only is the number of such possibilities small, the magnitude of the associated costs and risks is so high that conflicts at this level are likely to be averted and supplanted by less direct, less provocative forms of competition.

The defense of the American homeland is the first strategic priority of the US security system. Because strategic nuclear war with the Soviet Union represents the only level of conflict that would *immediately* threaten the survival of the US, the major security concern of the US is deterrence of strategic nuclear conflict. The cost of failing to deter strategic nuclear war is so high that US security planners have rightly given

priority to the precise management of force to ensure strategic deterrence.[5]

Another strategic priority is to defend Western Europe, demonstrated by such commitments as the Truman Doctrine, the Marshall Plan, the Berlin airlift, and the North Atlantic Treaty. A major conflict in Europe involving the superpowers would involve the massive use of conventional or tactical nuclear weapons and has the potential to escalate to a strategic exchange between the US and USSR. A military conflict would be high intensity because both the Warsaw Pact and NATO maintain large conventional and nuclear forces ready for employment in such a contingency.

There is a paradox of US and Soviet interest and power in Europe where the stakes are high, but where military conflict has been nonexistent since World War II. Both superpowers have large forces stationed in Europe in close proximity, yet they have not engaged each other militarily. While the presence of formidable forces reflects strong interests and an implied intent to use force if necessary, modern weaponry and nuclear arsenals make the potentially catastrophic risks outweigh potential gains. The great powers still compete, however, and have looked to other, less risky forms of conflict and less vital interests over which to compete.

The US has commitments and interests in the Middle East, Africa, Asia, and Latin America. Major conventional conflict involving the superpowers could occur here, but neither the US nor the Soviet Union deploys forces in these regions on the scale of European deployment. It is in these regions, however, that the major powers have attempted to make marginal gains—on the periphery of their vital strategic areas. Much of major power competition resides in the Third World where indigenous problems also abound. There, poverty is endemic, political institutions are fragile, and dissatisfaction with the status quo is widespread.

The great powers' fear of high-intensity conflict, combined with inherent Third World instability, appears to have increased the probability of more frequent conflict in the Third World, at levels of much lower intensity, and consequently, lower superpower risk. While the state system has not experienced global conflict in 40 years, the frequency and pervasiveness of "peacetime" violence indicates that the system is not inherently stable. In addition to the wars in Korea and Vietnam, the US has called upon its armed forces in contingency operations more than 220 times since World War II. In 1983, for instance, 40 conflicts were being waged: 10 in the Middle East and Southwest Asia; 10 in Africa; 3 in Europe; and 7 in Latin America. No fewer than a fourth of the nations around the globe were caught up in some form of armed conflict. Forty-five of the world's 165 nation-countries were involved in hostilities that claimed as many as 5 million lives, conflicts which included 8 wars and more than 30 revolutionary and separatist insurgencies.[6] In 1981 alone, these wars cost an estimated $528 billion.[7] The prevailing condition of the contemporary international system is one of non-war and non-peace, and the US and USSR have been involved, in various ways, in many of these conflicts.

Internal conflicts in the Third World constitute the most pervasive and prevalent form of conflict due to the ease with which such conflicts can be started, the limited scope and resources required, their limited consequences, and the immense number of military options open to the antagonists. The potential for conflict remains strong across Asia, Africa, Latin America, and the Middle East. In addition, internal conflict is most likely to be below the conventional warfare threshold, at the unconventional or irregular warfare level. This level of conflict is by far the most likely to occur, for it is within this domain that the relatively weak can test, and attempt to gain concessions from, the strong. Insurgency and terrorism will remain the principal vehicles for growing revolutionary movements, especially those with the patience to pursue protracted struggle. The continued existence of unpopular regimes in politically unstable and economically underdeveloped societies will provide pretexts for sustaining such conflict.[8]

THE US EXPERIENCE WITH LOW-INTENSITY CONFLICT: THE LEGACY OF VIETNAM

In his 1961 inauguration speech, President Kennedy declared, "Let every nation know, whether it wishes well or ill, that we shall pay any price, bear any burden, meet any hardship, support any friend, oppose any foe, to assure the survival of liberty." He implied the US would be the policeman of the world, representative of an idealistic thread in American foreign policy. There is also a realist or pragmatic thread: "Simply because (the US) is a major Free World power does not mean we have a responsibility for everything that goes wrong in the world."[9] These are expressions of conflicting and competing views of US idealism and *Realpolitik*. Both approaches are relevant to the pursuit of US defense policy. This dichotomy manifests itself in ambiguous and changing attitudes on how best to conduct US security policy in the Third World, and how best to construct US capabilities to manage low-intensity conflict.

When the US first ventured into conflict at the lower end of the conflict spectrum, in counterinsurgency warfare in Southeast Asia, the US did so with the belief that Soviet support for wars of national liberation must be opposed by US counterinsurgency efforts. That experience has remained an important influence in shaping the world view of American military and civilian leaders. The "Vietnam syndrome" has had a profound effect on US political-military strategy. The result has been a very cautious approach to conflict, especially conflict on the "low frontiers."

In the case of Vietnam, conventional methods increasingly replaced counterinsurgency tactics. The war continued to escalate as the US attempted to deny a victory to the Viet Cong insurgents and their North Vietnamese benefactors. Large-force operations, fighter-bomber attacks against military targets in North Vietnam, and B-52 saturation bombing missions made the Vietnam conflict look like a localized conventional conflict. Commandos were replaced by regular forces; the tactics and strategy for a counterinsurgency situation lost out to the prevailing military paradigm in America—high-technology, high-intensity conflict. This was, and is, the "American way of war."

US involvement in Vietnam rekindled a military preoccupation with the conventional environment of European wars, manifested in an emphasis on hardware and tactics as well as in professional military education and training. Further, perception of military capability and of the imperatives of political-military policy appear to have become closely wedded to a preference for conventional conflict in which issues appear clearer and where military capability and policy seem to have a more understandable goal. The US popular disaffection with Vietnam indicated serious questioning of the premises about the use of limited war as an instrument of American policy, premises that originally motivated the proponents of limited-war strategy and that underlay the Kennedy administration's original confidence in America's ability to cope with local communist incursions of all kinds. Vietnam demonstrated that the use of force for any length of time requires public support. Without it, military intervention of any type may quickly lose its legitimacy.

The US experience in Vietnam graphically illustrated to many an inability to cope successfully with, and win, one form of low-intensity conflict, counterinsurgency war. More recent failures of the Iranian rescue operation, of the US peacekeeping mission in Lebanon, and in other forms of low-intensity conflict such as combating terrorism indicated that US preparations for an increasingly important sector of the conflict spectrum remained inadequate. To many observers, the US has, since Vietnam, displayed a decreasing capability and credibility to influence political-military matters outside the European theater. The US has strengthened its NATO commitment and has developed an increasingly effective nonnuclear capability for large-scale battles likely to occur on the central plains of Europe. Few would argue the need for increasing non-nuclear capabilities in Europe, because, without a strong political-military base in Europe, US interests in other areas would be seriously jeopardized. Nonetheless, the US has given less attention and resources to developing a capability to project power outside Europe, in areas with a potential for serious security problems for the US. Despite the reawakened interest in low-intensity conflict because of the Iranian crisis, the Soviet incursion into Afghanistan, threats in Central America, and the growth of terrorism, the US has barely begun to build a military posture to respond to threats in the Third World. One indication of this was the report of the Long Commission, which investigated the 1983 terrorist attack against the Marines in Beirut. The commission concluded that "the US . . . is inadequately prepared to deal with this threat. Much needs to be done, on an urgent basis, to prepare US military forces to defend against and counter terrorist warfare."[10]

In sum, the ability to deter is not exclusively based on nuclear or conventional war capability. Successful deterrence policy requires a credible nonnuclear capability as well. It does not necessarily follow that nonnuclear forces must engage in conflict, but forces structured for low-intensity conflict must be credible and capable of deterring threats to less-than-vital US interests. The social, political, and military upheavals of recent decades are likely to be relived for the remainder of the century. As the number and type of actors increase and compete for scarce resources, as the international system becomes more decentralized, as authority erodes, as more international actors get access to increasingly sophisticated weaponry, as communications and transportation systems expand, the frequency and forms of conflict are likely to increase. New modes of conflict, such as transnational and international terrorism, are taxing the creativity and resources of those responsible for international security, particularly as the distinctions between war and peace become blurred in this increasingly complex political-military environment. Subsequent sections in this chapter thus examine this evolving spectrum of conflict, focusing on the challenges of low-intensity conflict to American defense policy.

THE SPECTRUM OF CONFLICT

There are many forms of conflict. One way of analyzing these forms is to place them on a spectrum of conflict, or an array of levels of warfare along a continuum, according to the intensity of conflict. *Intensity of conflict* is, of course, a matter of perspective. For the individual combatant engaged in any active fighting, the fight for physical survival on any battlefield is by definition "intense." Intensity may also refer to the potential levels of destruction for any given conflict. From a policymaker's perspective, however, intensity refers more broadly to levels of combat expected, limits placed on force employment, geographic scope for the conflict, numbers of participants, or objectives. This is the perspective used here.[11]

The forms of conflict along the spectrum can thus be aggregated into high, middle, and low levels of intensity. *High-intensity conflict* includes nuclear war such as a strategic nuclear exchange between the US and the USSR or theater nuclear war in Europe, as well as unlimited general conventional war such as World War II. *Mid-intensity conflict* includes limited conventional war such as the Korean conflict or unilateral intervention such as the US intervention in Grenada. Overlapping between levels of conflict also occurs. Vietnam-type conflict which combines conventional and counterinsurgency warfare straddles both mid-intensity conflict and low-intensity conflict.

Low-intensity conflict includes many variants. *Guerrilla III-type conflict* is a form in which intervening combat units integrate with indigenous forces. The US involvement in Vietnam, the Soviet intervention into Afghanistan, and the Cubans' activities in Angola and Ethiopia are examples. *Guerrilla II-type conflict* is a form in which the special forces of the intervening power provide cadre for indigenous forces. The use of US Special Forces in Bolivia against Che Guevara's *foco* (armed nucleus) is an example. In *guerrilla I-type conflict*, security and advisory assistance is provided to the host country. US involvement in El Salvador is an example. All three of these forms of low-intensity conflict occurred at various stages and times during the US involvement in Vietnam.

Terrorism is a form of conflict which can be an adjunct to any other form of political violence. Terrorism can be a tactic used at any level of conflict, but it can also be a strategy unto itself. Conducted by individuals or very small groups, terrorism involves the purposive use of systematic, arbitrary, and amoral violence in order to achieve both short- and long-term political objectives. *Counterterrorism* is a form of conflict directed against individual terrorist organizations as well as state-sponsored terrorist groups. This can include conventional or special-operations forces of the military, intelligence operatives conducting operations "in the shadows," and police operations. Many counterterrorist operations are surgical or special-operations missions. Israel's bombing raids into

Figure 8.1
The Spectrum of Conflict

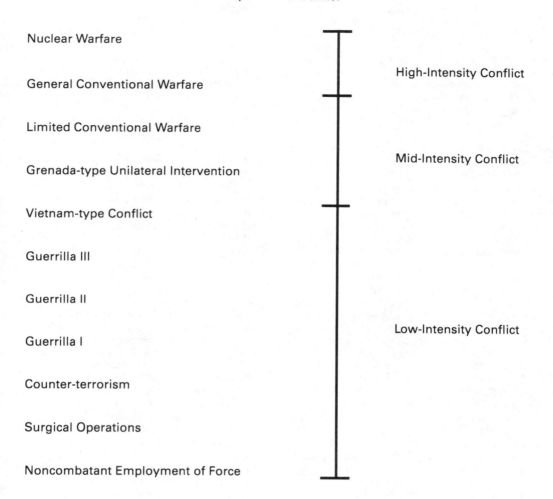

Nuclear Warfare

General Conventional Warfare — High-Intensity Conflict

Limited Conventional Warfare

Grenada-type Unilateral Intervention — Mid-Intensity Conflict

Vietnam-type Conflict

Guerrilla III

Guerrilla II

Guerrilla I — Low-Intensity Conflict

Counter-terrorism

Surgical Operations

Noncombatant Employment of Force

Lebanon against Palestinian encampments and commando raids against PLO headquarters and military facilities are examples of the conventional and unconventional use of military force in a counterterrorist mode. Israel's use of intelligence operatives to assassinate Palestinian terrorists during the 1970s is an example of a state's "war in the shadows," a covert intelligence war of "spy versus spy." The October 1977 rescue in Mogadishu, Somalia, by a West German counterterrorist police unit of passengers aboard a West German airliner hijacked by terrorists is an example of effectively using police forces in transnational counterterrorist operations.

Surgical operations are a form of conflict in which force is used in a narrowly defined context for a specific and limited objective. Examples of surgical operations include the Son Tay raid to rescue US prisoners of war in North Vietnam,

Israel's Entebbe raid in 1976 to rescue hostages held by terrorists, Israel's bombing attacks against the nuclear reactor in Iraq and the PLO headquarters in Tunisia, the US DESERT ONE rescue attempt in Iran to free hostages held in the US embassy in Teheran, and the US F-14 air interception of the Egyptian airliner to arrest Palestinian terrorists who earlier had murdered an American on board a hijacked Italian ocean liner.

When US vital interests are not directly engaged, the US may want to avoid employing its military forces, thereby avoiding the possibility of directly confronting Soviet or proxy forces and potentially escalating the level of conflict. It also avoids the logistical difficulty of moving and supporting large forces to distant lands. The US does not maintain a force structure and capabilities sufficient to defend all of its interests around the globe, but depends to a large extent upon friends and allies to deter local threats to common interests. *Noncombatant employment of force* occurs when a supporting nation provides security assistance to deter internal or external aggression against its allies or friends. This form includes show of force, peacekeeping, combat support, joint exercises, and security assistance to designated countries.

Security assistance is a vehicle to provide friends and allies the necessary equipment and training for self-defense. Such security assistance can take several forms: military sales, military grants, and military education and training.[12] Military sales is the purchase of US military equipment and training. Military grants provide US funds for equipment and services. Military training and education are used to bring allied military personnel to the US for specialized training. Peacekeeping operations is the smallest of the security assistance accounts and allows the US to participate in multilateral peacekeeping activities in some volatile areas, such as the Sinai, Cyprus, and Lebanon. By sharing costs and effort, many countries can achieve a level of mutual security unattainable through independent efforts. Security assistance provides for collective defense of common interests. It also is an economy-of-force measure which allows the US to concentrate its available forces in the areas of greatest threat. Between 1950 and 1984, the US negotiated $144 billion worth of foreign military sales agreements, of which $22.5 billion was allocated to East Asia and the Pacific region, $66.5 billion to the Near East and South Asia, $49.4 billion to Europe and Canada, $0.9 billion to Africa, and $2.7 billion to the Latin American republics.[13]

Actual involvement of US forces in a conflict would most likely first include noncombat missions, such as a show of force to demonstrate American concern. "Show of force" is a technique whereby the US can demonstrate its ability to project power without actually applying force. Naval or air forces are well suited for such employment because of their capability for rapid deployment (particularly in the case of air forces) and the flexibility of those forces in signaling US concern without commitment of forces on the ground.[14] The US has dispatched the Airborne Warning and Control System (AWACS) to the Middle East several times to deter Libyan attacks in the region. Such actions, known as "AWACS diplomacy," are analogous to the use of "gunboat diplomacy" during the nineteenth century. Joint military exercises in the Middle East, such as BRIGHT STAR, demonstrate an ability to project power and to wage war in that region. The US conducts similar exercises, called TEAM SPIRIT, in the Korean peninsula. Naval maneuvers and joint training exercises with Honduras likewise underscore the US commitment. Such exercises enhance the readiness of US and Honduran forces and serve as a credible assurance that the US has the ability and resolve to meet any challenge to the inter-American system.[15] Other possible low-level conflict missions that do not directly involve US combat forces include support for allied forces engaged in combat or peacekeeping missions. Examples of combat support include the strategic airlift of friendly forces to Shaba Province in Zaire in 1978.[16]

US DEFENSE POLICY AND THE SPECTRUM OF CONFLICT

The foregoing description of the spectrum of conflict focused on conflict intensity from the policymaker's perspective, in which categories were distinguished according to the modalities of force employment. This spectrum is useful in that it enables one to analyze the effectiveness with which a state can employ those different levels of force. In the case of the US, such an analysis suggests that both the credibility and capabilities of US force employment are the lowest in the mid-range of the low-intensity portion of the conflict spectrum, where the challenges to US security interests may be the most difficult. It also reveals some interesting conclusions about the paradoxical relationship between threats to national interest and the likelihood of conflict as one moves along that spectrum.

Theoretically, as the threat to a nation's vital interests increases, the more resources and resolve that nation will expend to protect its interests. Conversely, as the threat to a nation's interests decreases toward the lower end of the conflict spectrum and toward a nation's less-than-vital interests, the less intense will be its commitment in resources and resolve. Ironically, for the US, the greater the threat to US interests at the upper end of the conflict spectrum, the lower the likelihood of that form of conflict. Similarly, at the lower end of the conflict spectrum, where conflict is less intense and where a nation's interests tend to be less than vital, the higher the likelihood of that form of conflict.

The *credibility* of US force employment varies along the conflict spectrum. At the upper end of the conflict spectrum, US credibility is high. US nuclear and conventional forces are of high quality and of sufficient quantity to have been a major factor in deterring the Soviet and Warsaw Pact forces for the past forty years. Similarly, at the lower end of the spectrum, specifically in the noncombat employment of force and in surgical operations, US credibility is at least adequate. The US has provided high levels of security assistance to friends and allies. For example, the US-Israel "special" relationship, manifested in military and economic assistance,

Figure 8.2
The Threat: Probability Paradox along the Spectrum of Conflict

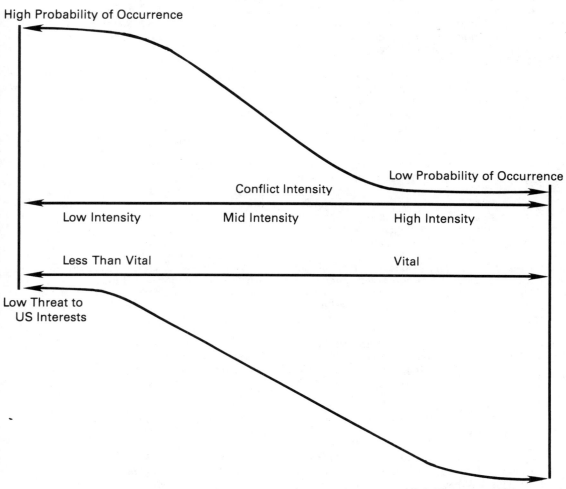

cost the US $5 billion in 1985. In military terms, the US has assured Israel a "qualitative edge" in armaments vis-à-vis potential Arab adversaries. Although US security assistance elsewhere is not as generous, US credibility is generally good.

US *capabilities* for force employment also vary along the spectrum of conflict. Again, US capabilities tend to be designed to wage war at the level where threats to US vital interests are highest. US capabilities are best in high-technology, high-intensity conflict, and the US maintains a credible deterrent capability both in its strategic nuclear posture and in its nuclear and conventional contributions to alliance defense in NATO.

The US likewise has the capability for effec-

tive surgical operations, although this policy option has been used infrequently. Such operations tend to be militarily and politically risky. The 1985 air interception of the Egyptian airliner transporting Palestinian terrorists who hijacked an Italian ocean liner and murdered an American citizen is illustrative of the risks inherent in surgical operations in combating terrorism. Although that operation was militarily successful, it had major political effects throughout the Mediterranean: the Italian government fell because of its handling of the crisis, and Egypt felt abused and itself "terrorized" by its American ally who had commandeered its aircraft.

Political considerations are much more intertwined with low-intensity conflict military operations in the contemporary international system. Political actions, and their relationship with military actions, are much more interdependent. Military defeat in the Third World can be translated into a political success, and conversely, military success can have adverse political consequences for the intervening major power. For example, Egypt's Gamal Nasser turned the military defeat of the 1956 Sinai war into a political success and became the preeminent Arab leader and a leading spokesman for the Third World. Nasser was able to say that the major powers and Israel "ganged up" on Egypt in a neo-imperialist drive for regional hegemony. Britain's military success along the Suez Canal brought no political gains, and the British government under Anthony Eden fell from power as a direct result of the 1956 Sinai war.

Even military success of "surgical" operations in combating terrorism may have temporary adverse political consequences in the Third World. Israel's operation against PLO headquarters in Tunisia in 1985 demonstrates the difficulty of assessing success or failure of "surgical" operations. Although Israeli aircraft delivered their ordnance on target with great precision, at least 15 Tunisians were killed in the attack, in addition to over 40 Palestinians who were specifically targeted. Many in the international community branded the attack as an act of state-sponsored terrorism, an act of aggression against Tunisia, and an incursion into Tunisian sovereign territory which resulted in the deaths of 15 innocent people. Thus, despite a precise military plan and almost flawless execution, Israel was still held responsible for killing innocent people.

In the vast middle range of low-intensity conflict, and in counterterrorism, however, US credibility and capability are relatively low. The US has an adequate capability to conduct short-duration surgical operations such as the 1986 raid on Libya and limited conventional wars in the fashion of Korea. The US is most capable in the high- and mid-intensity levels of conflict, is adequate at the noncombatant employment-of-force portion of the low-intensity conflict area, but is most limited in the counter-terrorism and revolutionary and guerrilla warfare portion of the conflict spectrum, where most future conflicts are likely to occur.

THE CHALLENGE OF LOW-INTENSITY CONFLICT

The low-intensity spectrum of conflict—ranging from Vietnam-type conflict through different forms of guerrilla or counterguerrilla warfare, counterterrorism, surgical operations, and non-combatant use of force—represents different ways in which the US might expect to employ force below the threshold of conventional conflict. It is possible as well to examine low-intensity conflict from a different perspective—that of the threats posed to the US which might require such unconventional employment of force. Revolutionary warfare, insurgency, and terrorism are distinct forms of low-intensity conflict. Each is distinct, not by virtue of differing intensity, but because each poses different threats to US interests and different challenges to the US ability to employ force to defend those interests.

Revolutionary warfare is the most virulent and potentially destructive form of low-intensity conflict. A revolution is an acute, prolonged crisis in one or more of the traditional class, status, and power systems of stratification within a political community. It involves a purposive, often elite-directed, attempt to abolish or to reconstruct one or more of these systems by means of an intensification of political power and recourse to violence.[17] This implies that an acute, prolonged crisis is neither spasmodic nor transient. A revolutionary situation explodes because of the aggregated discontent of the masses that has become incompatible with the existing order. A revolution leaves permanent scars and imposes fundamental changes on the order in which it erupts. The revolutionary crisis is distinguished from the lesser and more frequent crises manifested in coups, revolts, or secessions. In these lesser forms of political violence, the effects of violence are generally absorbed within the existing political order without substantive change. Revolutionaries, in contrast, attempt to accomplish a basic transformation of the political, economic, or social values and institutions in a state system.

Once discontent becomes widespread in a society and a cohesive group can organize and focus that discontent against the established order, revolutionary violence can become self-sustaining. Revolution feeds on discontent and gets its energy from a supporting ideology, whether that ideology be one of nationalism, utopian universalism, or religious fundamentalism. The purpose of revolutionary warfare is the overthrow of an established order and its substitution by another. To be successful, a revolution requires adequate strength on the part of the revolutionary group, the organized support of a considerable number of a nation's citizenry, and weakness or lack of commitment on the part of the existing government.

A true revolution must be "purposive." A revolution feeds on the energy of an ideology which indicts the status quo by illuminating what is wrong with it, posits a utopia for a better post-revolutionary order, and outlines a blueprint to follow to achieve the transition. Lesser forms of

Figure 8.3
US Defense Policy and the Spectrum of Conflict

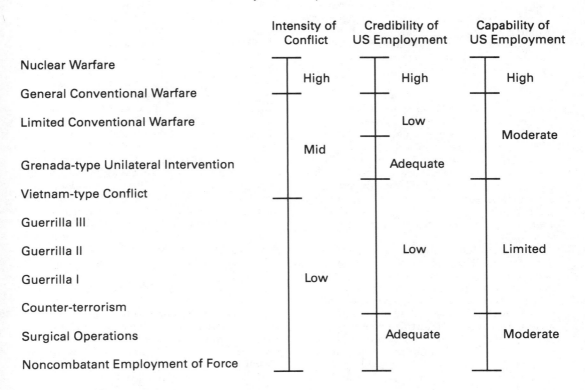

political violence do not possess this purposive quality. Revolution often needs elite direction. The "true believer," the fanatic who is willing to sacrifice all for a holy cause, gives the revolution direction and is not necessarily controlled by events.[18] The revolutionary leader brands the existing order as evil, articulates goals of the revolution, specifies the means for achieving them, and fashions the new order. The attempted restructuring or destruction of a class, status, or power system cannot get very far without effective leadership.

Insurgency is another form of low-intensity warfare involving an armed revolt against an established order. Insurgency is a struggle between a nonruling group and the ruling authorities, in which the former consciously employs political resources and instruments of violence to establish legitimacy for its political cause.[19] The objective of insurgency is the transfer of national authority. It entails a fragmentation of the society, in which citizens lose allegiance to the symbols of the existing order. An insurgency attempts to erode the legitimacy of the government in power. If the society is substantially polarized into pro-government and anti-government forces, it may become a civil war. If one element of society is trying to rid

itself of an external social structure imposed upon it, it may take the form of a colonial war.

Insurgency falls short of a full-scale revolution because the insurgents often do not possess sufficient strength or popular support to use more conventional tactics for toppling existing governments from power. While the weapons for insurgent actions may be the same as those causing revolution, lack of numbers and organization dictates different strategies and tactics. The immediate insurgent goal is not to win, but to wear down existing governmental forces over time, thus allowing insurgent forces additional time to gather strength. In this context, success is measured not in terms of victory, but in the insurgents' continued challenge to the established political order.

The causes for modern insurrections and revolution are many and varied. Decolonization helped destabilize the international order. Within a few years, virtually the entire African continent and Asia emerged as independent entities, changing the character of the United Nations and bringing to the international system a large number of sovereign states. What has made this a particularly important factor in international politics is the fact that many of these states were, and are, institutionally fragile and politically unstable. Developing nations provide fertile soil for numerous potentially violent confrontations. Weak or corrupt national governments, foreign influences, inequitable distribution of wealth, the presence of minority groups, external wars, competing ideologies, and the revolution of rising expectations can all contribute to those preconditions necessary for widespread social unrest.

The third and most recent form of low-intensity conflict is *international terrorism*.[20] Political terrorism is the calculated use or threat of violence employed in a culturally unacceptable form to achieve political goals through fear, intimidation, or coercion to manipulate the attitudes, opinions, emotions, or behavior of the observing mass of society. Political terrorism is not mindless, senseless, nor irrational violence, but a violent form of graffiti aimed at a global audience rather than at its immediate victims. It is "theater for effect."

Most civilized nations have identified, through law, modes of conduct that are criminal. Even in war, there are rules that proscribe the use of certain weapons and tactics. Terrorism, however, is defined by the nature of the act, not by the identity of the perpetrator or the nature of the cause. All terrorist acts are crimes. Yet, because of the political nature of the violence, "one man's freedom fighter has become another man's terrorist." There is no agreement on the lexicon of terrorism. It remains fundamentally a political issue.

The terrorist uses violence in an antisocial context. The apparently random, indiscriminate violence of the terrorist is neither sanctioned nor approved by the mass of society. Terrorist violence is beyond the acceptable bounds and limits of "normal" behavior. The cultural unacceptability of terrorist violence gives it a quality perhaps best described as "extra-normal" or "culturally deviant." The quality of the terrorists' violent acts is more important than the magnitude of the destruction. Target selection is more symbolic than pragmatic in nature. The quality of the violence must be such that a watching audience is drawn to it. Terrorists select targets which are random or indiscriminate only insofar as the immediate target provides a symbolic relationship with the terrorist organization's ultimate target, while other users of political violence employ force as a means of directly affecting the resolution of political problems. Because the terrorist is less concerned with winning the "hearts and minds" of the population or expanding his power base and forming a government, he is not restricted to those targets "acceptable" to the public. The terrorist incident must leave a lasting effect. This quality, in part, distinguishes the terrorist from other users of political violence.

Contemporary terrorist violence is a means to specific political goals. It is most often a tactic of the weak. Individual terrorist groups simply cannot generate enough force to seek the immediate resolution of perceived political problems. They use violence as an indirect and intermediate step toward achieving their ultimate goals. In hostage situations, for example, the terrorist designs the bargaining situation to send a message of apparent irrationality to instill fear and to intimidate the immediate victims. But while intimidation, coercion, or fear may be the immediate result of terrorist violence or threats, they are not the ultimate desired result. The final or ultimate goal of the terrorist is attaining a political objective. Since most terrorist groups are relatively weak and unable to generate massive violence, they can seldom hope to attain their goals by coercing an entire population. Terrorism is therefore primarily a psychological weapon, a form of unrestricted psychological warfare. The terrorist uses violence as a lever to manipulate the attitudes, emotions, and opinions of the mass of society hoping to elicit a desired change in behavior.

The terrorist exposes the apathetic bystander to the terrorist cause not only through acts of political violence, but also through media coverage. Consciously or unconsciously, the observer begins to form attitudes, opinions, and emotions regarding the individual terrorists, the group, and their espoused cause. This psychological manipulation of mass society makes terrorism potentially effective. If the terrorist can manipulate the masses of any society into accepting, to any degree, the terrorists as bona fide combatants contesting a legitimate grievance, the tactic has been effective. Once the masses have accepted the terrorist action, the terrorist has built the groundwork for recruitment, support, and an increased power base. The key to this mass effect is that elusive quality which makes terrorism newsworthy, reflected in the increasingly frequent description of terrorism as theater and terrorists as actors on the world stage. The media provides the lens through which society views the terrorists' acts.

Contemporary terrorism has aspects that differ markedly from traditional terrorism. Transnational terrorism is a relatively new form of terrorist violence which, because of technology and the nature of the international system, has created a type of conflict unique to the past decade. Transnational terrorism is the planned threat or use of extra-normal violence for long-range political purposes where the action is intended to influence the attitude and behavior of a target group wider than the immediate victims and with ramifications that transcend national boundaries. The terrorist has become a "techno-guerrilla" who exploits the technological advances in communications, transportation, and weaponry.

Transnational terrorists are particularly effective against open, industrialized societies. Such societies offer abundant targets. The transnational terrorist attempts to convert the strength of his opponent into a weakness by converting the outward power of industrialized society into a liability. The vulnerable nature of open, industrialized societies, with their highly complex and interdependent infrastructure, enables the transnational terrorist actor to magnify the effect of his actions far beyond his capacity to bring havoc to an agrarian or closed society. Israel, for example, ties down almost 40,000 people daily in activities directed at combating terrorism, representing a diversion of manpower from security tasks more directly associated with the military survival of the state. Consequently, the terrorist is inherently more cost-effective in maximizing his resources than national authorities who must respond to both the threat and the act of terrorism.

As N. C. Livingstone has pointed out, terrorism is distinct from other forms of warfare in several important respects.[21] First, traditional warfare is a form of "institutionalized violence." There is a certain legitimacy in conflict between states, recognized in the laws and traditions of international law. Terrorism, on the other hand, is generally regarded as illegitimate violence, both in terms of the nature of its actions and its targets—essentially a form of indiscriminate violence—and because it is violence directed against society and the state by non-state actors. As such, terrorist acts are generally regarded as criminal activity rather than a form of warfare. Terrorists are accountable to the laws of society—punishable for murder, arson, etc.— rather than to the laws of warfare. Nonetheless, terrorists have in recent years sought protection under laws governing treatment of prisoners of war, although Western states, including the US, have resisted such legitimization of terrorist activity.

Second, the essence of war is to preserve oneself and annihilate the enemy. Terrorism, by contrast, is essentially a political act designed not necessarily to destroy the enemy but to demoralize him. Rather than overthrowing the existing political system, the terrorist seeks to intimidate the state. Terrorists may not be capable of defeating the state, but they can highlight a political cause. Furthermore, the terrorists' willingness to sacrifice themselves to enhance the political visibility of a cause contradicts the traditions of warfare which emphasize preservation of force.

Because the terrorist seeks ultimately to communicate rather than to destroy, Livingstone continues, terrorism tends to be less destructive than other forms of warfare, although that violence tends also to be more spontaneous and indiscriminate. Terrorism is, however, a more "democratic" form of warfare. It is relatively easy for terrorists to muster the means to attack and to gain access to targets. Weaponry is available and affordable. Political and commercial targets in developed, industrialized societies are ubiquitous and vulnerable. And, since terrorists require neither the collapse of the state nor the sympathy of society to achieve success, "victory" is a relatively cheap commodity.

STRATEGIES FOR LOW-INTENSITY CONFLICT

Formulas for successful revolution come in many forms and have evolved over time and adapted to changing circumstances. Strategies for rebellion incorporate a variety of forms of political violence, each of which emphasizes particular weapons, tactics, and targets. Governments facing a legitimacy crisis may confront one or more political challenges or forms of violence, including propaganda-organizational activity, terrorism, guerrilla warfare, and limited conventional warfare.[22] Dealing with these challenges to national security and political stability requires an understanding of the ideological and operational foundation of the various forms of low-intensity conflict.

THE LENINIST APPROACH

Proponents of political violence generally claim a universally applicable ideological base from which operational strategies are derived. The most pervasive strategy is the Leninist approach, based on Marxism-Leninism. Marxism-Leninism called for an elite vanguard to lead the revolutionary proletarian (working) masses into revolution in industrialized societies. Lenin believed revolution had to be well coordinated by a central revolutionary authority standing above the masses. He believed that a small, tightly knit, disciplined, and highly organized conspiratorial group that had obtained support from major discontented social groups was the most effective way to achieve the goal of the movement. While basing mass action of the proletariat on the central thrust of revolutionary action, Lenin introduced an elite vanguard, imbued with the necessary understanding of Marxian ideology, to lead the revolution both politically and militarily. Lenin's greatest contribution to revolutionary practice was his conspiratorial organization. The vanguard organization was hierarchical and pyramidal in structure and authoritarian in nature due to Lenin's belief in "democratic centralism." Decisions were arrived at democratically *within* the party apparatus, but all elements of society were then expected to adhere to this centrally made decision. The target of the revolution was the capitalist-imperialist system and the "bourgeois" state in general. Lenin assumed that a government that was alienated from its population would capitulate when confronted by low-level terrorism, subversion of the military and police, and the final seizure of radio stations, government offices, and other state institutions.[23]

Lenin attempted to consolidate the position of socialism in Russia when the universal revolution failed to materialize in war-torn Europe following World War I. Lenin's disciple, Leon Trotsky, was totally dedicated to the concept of "permanent revolution" and believed that it was the duty of "socialist" Russia to encourage revolutions. Trotsky stressed internationalism, a linking of revolutionary movements worldwide, each of which had a common goal of establishing a socialist order, providing ideological justification for "wars of national liberation."

THE ANARCHIST APPROACH

The Russian anarchists, the "philosophers with bombs," were socialists, like the communists, but were fundamentally opposed to a "vanguard organization" or any form of authority from above. The Father of Russian nineteenth-century revolutionary anarchism was Mikhail Bakunin and his fiery disciple, Sergei Nechayev. Bakunin agreed with Marx in advocating a dictatorship of the workers over the exploiting classes, but held that this dictatorship had to be a "spontaneous dictatorship" of the entire working class and not leaders comprising an elite vanguard set in authority over the masses, such as Lenin's later elitist "vanguard of the proletariat." Bakunin gave modern-day revolutionaries the justification for individuals to attack the existing system of authority, ". . . the voluntary and considered agreement of individual effort toward the common aim. Hierarchical order and promotion do not exist, so that the commander of yesterday can become a subordinate tomorrow. No one rises above the other, or if one does rise, it is only to fall back again a moment later, like the waves of the sea returning to the salutary level of equality."[24] Consequently, the organizational structure of anarchist groups tended to be decentralized, horizontal, and loosely federated.

THE MAOIST APPROACH

Mao Tse-tung (Ze-dong) was the foremost exponent of Third World socialist revolution. Mao focused on the peasant masses, the vanguard, and the strategy of protracted people's war as the components of successful revolution. Mao's contributions went beyond his theory and ideology. The concept of a protracted people's war provided a practical blueprint for a suppressed people who lacked resources to overthrow an established government. Mao's protracted people's war consisted of three phases. Initially, the established government had the advantage because of its recognized authority, resources, and institutions. Those opposing the government, on the other hand, had few resources with which to begin a struggle. In phase I, which Mao called the "strategic defensive," the insurgent forces built an organization in a "base" area, a sanctuary in which a political army could be trained. In this organizational stage, cellular networks were created around which the guerrilla built political-propaganda groups to win popular support and trained guerrilla teams to engage in selective intimidation. Occasional guerrilla attacks designed to infiltrate enemy institutions, foment strikes and riots, and carry out sabotage and hit-and-run missions supplemented this consolidation effort.

In phase II, government and insurgent forces had reached a stalemate, and guerrilla warfare became the central thrust of the movement. Small bands operated in rural areas where terrain was rugged and government control was weak. The guerrilla's aim was to isolate the people from the government. Scattered attacks throughout the countryside were intended to force the government forces into adopting a static defensive posture that stressed the dispersal of forces in order to protect many potential targets. The insurgents controlled the countryside, while the government controlled the cities. The base area was expanded, and guerrilla attacks became more common. The guerrilla forces became more institutionalized. The guerrillas' political organization was also expanded in the regions they controlled. A "shadow government," or parallel hierarchy, was developed and became more visible to the people. The purpose of this parallel hierarchy was to demonstrate the legitimacy of the movement and that the rebels were capable of ruling. As phase II matured, occasional, more conventional pitched battles occurred with government forces as the guerrilla bands become more organized.

In phase III, the strategic offensive, the insurgent forces had built their forces into a conventional army and were prepared to confront directly the military forces of the government. In this phase, the insurgency evolved into a civil war in which guerrilla forces joined regular forces in conducting mobile conventional warfare and pitched battles. Maoist theory was effective both in China and in Southeast Asia.[25]

THE CUBAN APPROACH

The Latin American variant is an adaptation of Maoist-Marxist-Leninist theory to the conditions in Latin America. Regis Debray observed

Che Guevera's armed struggle in Cuba, and Debray's and Che's writings provide the major body of revolutionary thought for rural insurgency in Latin America. Che Guevera derived three revolutionary principles from the revolution in Cuba which he felt could be universally applied. First, popular forces could win against the regime's army. Second, a vanguard or *foco* (nucleus) could foment revolution. Finally, the revolution would take place in the rural areas. Debray believed that, in Latin American rural societies, an armed nucleus, or *foco*, was all that was necessary for the revolt. The military members of the armed *foco* who proved their mettle in military action would become the postrevolutionary leaders. Debray stated, "under certain conditions, the political and military are not separate, but form one organic whole, consisting of the people's army, whose nucleus is the guerrilla army. The vanguard party can exist in the form of the guerrilla *foco* itself. The guerrilla force is the party of embryo."[26] Those who defeated the government would be the new leaders of the postrevolutionary society. Political organization and party politics were trappings.

The *foco* theory maintained that the elite, or vanguard, of the revolution can inject itself into any society to spark the revolution, since contradictions already exist in any society. Like Mao, Che believed that violent acts were necessary to force the government into responding in a repressive manner to unmask its inherent "capitalistic or fascist nature." Debray and Che contended that, rather than wait for the emergence of an organization, it would be necessary to proceed from the armed guerrilla nucleus. The *foco* was seen as "the small motor" that sets "the big motor of the masses" in action. Military priorities had to take precedence over politics. The mere fact of taking up arms in situations where grievances existed would create suitable conditions for revolution. Insurgent leaders did not have to wait for the preconditions of insurgency to appear. A guerrilla *foco* could act as a catalyst to foment existing grievances into positive revolutionary action against the government. The theory of the *foco* was successful only in the revolution in Cuba. Che Guevera failed in fomenting revolution in Africa and died in his attempt to inject the armed nucleus into the rural countryside of Bolivia.[27] In 1987, however, a non-Marxist *foco*, known as the Mozambique National Resistance (RENAMO), was waging successful guerrilla warfare against the Marxist (FRELIMO) Government of Mozambique. RENAMO has all the organizational and military characteristics of the Che Guevera *foco*, but without the Marxist underpinnings. The contest in Mozambique will be a test case for the effectiveness of the *foco* to "roll back" communism.

URBAN GUERRILLA WARFARE AND TERRORISM

Abraham Guillen observed Guevera's notable lack of success in injecting revolution into Latin America. Cuba's Fidel Castro has said that the countryside was the proper place for revolution in Latin America and that the cities were the "graveyard of revolutionaries." Guillen discounted the strategic importance of the country and believed rural revolutions would fail in Latin America. Guillen observed that 50 percent of the people of Latin America lived primarily in the cities. The epicenter of Latin American political, economic, and cultural life was the metropolis, not the countryside. By projection, Guillen believed that the locus of future revolution in all modern, industrialized societies should be the city.

Guillen called for Latin American revolutions to move to the cities. Carlos Marighella, who was a Latin American revolutionary, applied Guillen's theories and espoused urban terrorism as a necessary adjunct to revolutionary warfare. Marighella proposed to "turn political crisis into armed conflict by performing violent actions that will force those in power to transform the political situation of the country into a military situation."[28] He believed the center of conflict during the initial phases of revolution should be in the cities, not the countryside, because the increased size and socioeconomic differentiation of urban centers would make them especially vulnerable to terrorism and sabotage. The concentration of people and buildings in the cities rendered government assets, resources, and firepower unusable. Marighella proposed using small-unit tactics, each cell being a separate entity capable of independent action. If the government did, indeed, attack the insurgents, the collateral damage would, in Marighella's estimation, alienate the masses who, from then on, "will revolt against the army and the police and thus blame them for this state of things."[29] The aim of such "armed propaganda" was to create havoc and insecurity, eventually producing a loss of confidence in the government. Marighella acknowledged that the function of the urban terrorists was to tie down government forces in the cities so as to permit the emergence and survival of rural guerrilla warfare, "which is destined to play the decisive role in the revolutionary war."[30] Marighella's major contribution was thus to bring terrorism and insurgency into the cities.

Although Marx had predicted that the proletarian revolution would occur in the advanced industrialized societies, communist revolution escaped the industrialized West. Herbert Marcuse, an American philosopher, provided the ideological justification for revolution in advanced societies. He believed that Marx's theories were essentially correct, but had been applied too early, to the wrong historical epoch. The proletarians had not been the true revolutionaries, but had been coopted or "bought off" by the economic benefits of the system. To Marcuse, the power elite used technology and science to produce welfare-state capitalism in which the basic needs of all workers were satisfied. The new historical class, intellectuals, students, and those crushed by the "system," were alienated and would become the new generation to wage revolution against the power elites and continue the revolution toward a libertarian socialist utopia.

This "New Left" ideology provided the rationale for student rioting, turmoil, and urban terrorism in industrialized societies. Out of the student riots in America, which were directed against the war in Vietnam, students, intellectuals, and politicized radicals carried out acts of violence against the American government. West German students, among others, followed their example. Out of the German student unrest, urban terrorist groups developed into decentralized organizational structures with an ability to inflict sporadic terrorist attacks against the government. German terrorist groups linked up with other European groups, as well as with Palestinian groups which provided training and financial support. Because of the international linkages, a new form of terrorism emerged, which was transnational in character.

Worldwide terrorism used as an instrument of policy against the US continues to pose a formidable challenge. More than 40 percent of all recorded international terrorist events have been directed against US interests—mostly against official US personnel and facilities overseas. During the past decade, a terrorist event was directed against US officials or installations abroad approximately every seventeen days. US interests in Europe and the Middle East have often been targeted by terrorists. These areas, as well as Latin America, are the settings of the greatest number of terrorist activities.[31]

Through the late 1980s, international terrorism continued to increase, resulting in over 4000 people killed and almost 8300 wounded since official statistics were first compiled in 1968. Additionally, the lethality of terrorist violence has increased markedly. The 1983 bombings of the US embassy and Marine Headquarters in Lebanon, and the bombings of the US embassy and airport in Kuwait, illustrated the vastly increased destructiveness of modern terrorism resulting from relatively simple means and techniques.

Along with the renewed activity of terrorists indigenous to countries in Western Europe, the threat is growing from Muslim traditional groups originating in the Middle East and supported by countries such as Iran, Libya, and

Syria. These groups have benefited from relaxed controls at Western European borders and along lines of communication within the region. Their relative freedom of movement allows them to smuggle explosives and related paraphernalia for use in terrorist activities. These groups possess a proven ability to plan, equip, organize, and execute their operations. The task of combating terrorism in Western Europe and elsewhere has become complicated and more serious.

Future terrorism will likely be more lethal and more frequent. More sophisticated weapons and tactics may be employed by terrorists. Terrorists could target "chokepoints" such as power grids, waterways, command and communication systems, mass transportation systems, and centralized computer facilities. International connections among terrorists may also increase. State support for wars of national liberation and international terrorist organizations will be of special concern. Support from the Soviet Union, North Korea, Cuba, and their allies, and the provisions of financial aid, weapons, and training from Syria, Iran, Libya, and the People's Democratic Republic of Yemen may continue. Terrorists may or may not be centrally controlled by their patrons. Regardless, the instability they create in the industrialized West and Third World nations threatens the security interests of the US and its allies.

THE SETTINGS OF LOW-INTENSITY CONFLICT

THE MIDDLE EAST

The Middle East is the concourse of the Eurasian and African land masses, and, as such, has witnessed the ebb and flow of major civilizations and population movements. It is a land rich in natural and human resources, and consequently, prone to perennial conflict and endemic unrest. The US and USSR have strategic interests in the region which often conflict with the regional interests of the indigenous states. The consequences of this strategic and regional competition have great importance for the superpowers, the regional powers, and global stability.

Official US contacts with the region began in the early nineteenth century, but major interest in the region developed only after World War II. The US has geostrategic interests in the region because the area is a land, sea, and air bridge. The Suez Canal, Bab el-Mandeb, and the Strait of Hormuz are "global chokepoints" because through these narrow waterways pass the region's most valuable commercial commodity—oil. About 60 percent of the world's proven oil reserves are located in the region, the great majority of it concentrated around the

Persian Gulf littoral states. The region accounts for about half of world oil production. The US consumes 30 percent of the world's oil production and imports approximately 40 percent of its daily oil needs. Of that import total, about 20 percent is Middle East oil. More significantly, US strategic allies, especially Europe and Japan, import between 50 and 60 percent of their oil needs from the region.[32] A serious interruption of the oil supply could mean economic collapse for those countries. Since the economic health and political stability of Western Europe and Japan are vital to US national security, it is very much in the US interest to insure that the West has adequate supplies of Middle East oil at reasonable prices. Free navigation of the region's international waterways and access to its resources are vital to the West.

Primary US interests in the region are the maintenance of US influence, oil, and Arab-Israeli peace. The US is committed to reducing regional tension which could escalate to major-power involvement. In addition, the US has made a "special relationship" with Israel. Nine US presidents have committed the US to Israel's well-being.[33] To attain these interests, the US has articulated specific objectives. US policy is designed to deter conflict with the USSR and to minimize regional conflict. If deterrence fails, the US is prepared to wage conflict for terms favorable to the US at any point on the conflict spectrum. The diplomatic corollary to this military policy is to deny Soviet expansion into, and control over, the area, as well as to facilitate positive relations among the indigenous states. The economic objective is to safeguard US and allied access to the region's oil.[34] US interests are best served by peace; hence, the overall US policy objective is to assure the security of allies and friendly states in the region and to promote political stability and economic development.

The Soviets face many of the same problems in the region which confront the US. The Middle East is the most complex region on earth because of its multiplicity of ethnic, religious, political, and economic cleavages. US and Soviet assets and liabilities in the region have been directly affected by the inherent instability of many of the indigenous regimes and the overarching regional conflicts, such as the Arab-Israeli dispute, the Iran-Iraq war and the rise of Islamic fundamentalism. In addition, the indigenous peoples and their leaders have historically resisted outside influence. Because of the region's historical development, Middle East political systems tend to be xenophobic, rejecting many of the values of Western civilization while embracing the West's technology. These regional characteristics affect superpower efforts to influence regional affairs.

Besides these political constraints, military constraints also abound. Any military contingency outside Europe would require the US to deploy a rapidly deployable combat force. Middle East contingencies are conceptually, tactically, and strategically different from a European contingency because of the distances involved between the US and the theater of operations; the lack of sustained US presence in the area; the inadequate infrastructure in terms of roads, rail, air, and communication modes; and the harshness of climate and terrain. For the US to be able to sustain military operations in the region as a whole, the US would need to develop a basing infrastructure, a rapidly deployable combat force to protect US interests in the region, and a functioning economic and security assistance program.[35]

The physical location of the region and its geostrategic environment pose a major liability for the US. The area is larger than the US, and the distances from the US are staggering. The Persian Gulf is 7000 miles from the US East Coast by air, 8000 miles by sea through the Suez Canal, and 12,000 miles around the Cape of Good Hope. Large amounts of airlift and sealift would be necessary for a contingency in the area. Besides the huge inter-theater distances involved, the logistics infrastructure in the region is inadequate to support a major US military involvement. For example, the paved roads in the area amount to only two-thirds of the mileage found in Florida. To address these monumental logistical problems, the US would require a large capital investment in developing a military-force structure for the area. The US is currently developing a capability to insert forces for contingency operations. Sustaining a presence in the area also presents a problem for the US. The US is building and improving the indigenous infrastructure and basing systems in the area and is negotiating with friendly regional powers for pre-positioning of war stocks and host nation support.[36]

Thus far, friendly Arab states have been reluctant to provide the US substantial access to the area. The US is currently confined to facilities in Bahrain, Diego Garcia in the Indian Ocean, and Mahe Island in the Seychelles. The US has agreements only to pre-position war stocks and to use regional facilities during crises and exercises. No nation in the region is a formal ally of the US, and the US must deal with differing strategic priorities, the lack of common command and operational concepts, long lines of communications, major geographical barriers, political instability, economic uncertainties, and cultural biases. An emerging US military presence is manifested in the BRIGHT STAR exercises in Egypt and the fact that over half of the $5.1 billion in US foreign military sales credits goes to the region.[37] Nonetheless, the US has more military liabilities than assets to achieve its objectives in the region.

Analysts generally predict that the political environment in the region will deteriorate in the near-term and remain volatile throughout the remainder of the century. Because of the ethnic, sectarian, religious, and cultural divisions in such countries as Syria, Iran, Lebanon, and Iraq, domestic political instability is a chronic condition. Soviet penetration of the region could increase as the Soviet bloc begins to depend on the region's natural resources. Islamic fundamentalism will exacerbate existing differences among the indigenous faiths.

Specific US political liabilities in the region account for the stymied negotiations with the Arabs for US access and regional presence. The overarching obstacle is the US "special relationship" with Israel and its collateral effects on the Arab-Israeli dispute. Arabs tend to see Israel, and not the USSR, as the cause of the majority of the problems in that zone of conflict. US financial and military support for Israel, a state viewed by Arabs as an "outpost for neo-colonialism, imperialism, and Zionist racism," place the US squarely in the Israeli camp in the view of many Arab leaders. The US, despite this "tilt" toward Israel, espouses an even-handed approach toward the issue. The US has attempted, with a modicum of success, to cultivate relations with moderate Arab states, such as Egypt, Saudi Arabia, and Jordan, while seeking to facilitate resolution of the most critical issue of the Arab-Israeli dispute, the Palestinian problem.

Conflict has been incessant in the region since World War II and abounds at many different levels of intensity. The Iran-Iraq war was a relatively high-intensity, high-technology conflict which threatens to spill over into other areas of the Persian Gulf. Threats to neutral shipping in the Gulf—particularly tankers carrying oil to the US, Europe, and Japan—stimulated the US in 1987 to expand substantially its long-time naval presence in the Persian Gulf region in defense of the principle of free navigation in a key international waterway. On the ground, hundreds of thousands of soldiers and civilians were killed and billions of dollars spent since hostilities started in 1980 as a border dispute. The war was typified by large-scale frontal attacks along stationary borders, armored and infantry thrusts in perceived gaps in enemy lines, artillery barrages, and bomber and missile attacks on cities and economic targets such as oil refineries and transshipment facilities. These tactics of attrition did not result in significant gains for either side. Iraqi use of chemical weapons in that conflict, however, set an important and unfortunate precedent for regional conflicts.

In many respects, the 1982 Israeli invasion of Lebanon may be more typical of future conflict in the region. The war in Lebanon has been a war without any definite pattern, with shifting actors and strategies. Part conventional war, part insurgency, and part terrorism, the war has involved regular armies, guerrillas, private militias, and terrorist gunmen operating independently, some sponsored by political or religious factions and some by other states. The Israelis have fought three enemies in Lebanon—the PLO, Syrians, and Lebanese fundamentalists and nationalists. Each adversary fought a different type of war, and each pursued a different strategy.[38] In 1982, the Israelis were overwhelmingly successful against Syrian conventional forces and pushed back PLO conventional and guerrilla units to Beirut within a few days. Israeli forces ringed the Lebanese capital, finally entered the city, and forced PLO forces to evacuate under the protective auspices of a multinational force comprised of West European and US forces. But once the static fronts disappeared, Lebanese fundamentalist and nationalist forces, supported by independent PLO conventional and guerrilla units and Syrian units, began attacking Israeli forces occupying southern Lebanon. Most attacks against the Israelis were by terrorists. By the time the Israelis finally withdrew from Lebanon, Israeli forces had sustained over 500 killed. Thousands of Palestinian, Lebanese and Syrians, moreover, have been killed in the Lebanese conflict since civil war began in 1975.

In Afghanistan, fundamentalist rebels controlled most of the country and conducted insurgent warfare against over 100,000 Soviet troops propping up a Moscow-installed Marxist government in Kabul. Soviet forces conducted large area "sweep" operations using conventional and unconventional forces, resulting in major civilian casualties. The Soviets, too, discovered in Afghanistan the difficulties associated with achieving political and military objectives in low-intensity conflict, even though Afghanistan is contiguous to the Soviet Union. Unable to achieve "victory," they increasingly shifted to small-unit tactical operations to sustain their military position, and much of their conventional armor and air defense units became largely irrelevant to that conflict. Military failure, plus a recognition of the political, economic, and social costs of the continuing Soviet military presence, prompted the USSR to begin withdrawing its forces in May 1988.

Other low-intensity conflicts abound in the region. Kurdish monitors have conducted hit-and-run raids against the Iraqi, Turkish, and Iranian governments, and scattered border clashes occasionally flare up with Marxist South Yemen. The Marxist Polisario is waging a war of national liberation against the Moroccan government. Both the Moroccan government and the Polisario claim control of the Western Sahara. The Moroccans have been somewhat successful in limiting the effects of Polisario military incursions by building a *cordon sanitaire*, an earthen beam extending the entire length of the disputed desert border. This novel idea of a desert "maginot line" has thus far effectively constrained the military efforts of the Polisario. Political reconciliation, however, remains elusive. This array of low-intensity conflicts characteristic of the Middle East will continue to challenge the ability of powers like the US to pursue their interests in the region.

AFRICA

Africa is likely to be an arena for low-intensity conflict for the remainder of the century. US strategic interests in the region are of relatively low priority compared to those in other regions. The area is neither a hub of world power or trade, nor a national battlefield for the great powers. The region is underdeveloped, and famine and hunger are endemic. No African nation constitutes a military threat to the US or to US allies. Only a few African nations can be said to possess even significant regional power. Nevertheless, the US has critical concerns in Africa on several levels.

Africa is important because of its strategic location. The huge African continent sits astride key shipping lanes, the route for Middle Eastern oil to Europe and America. Thus, its littoral states assume considerable geopolitical importance to Western oil-importing powers and, consequently, to their potential Soviet adversary. The African coastline is similarly important for communications with the Indian Ocean, where superpower rivalry continues to grow.

Africa is a natural cauldron for low-order power rivalries. Nearly a third of the world's sovereign nations are on the continent. Since the mid-1960s, African nations have played a large role in seeking a new international order through increasingly active participation in conferences of nonaligned nations; in the UN and its specialized agencies; in the North-South dialogues on aid, trade, and investment; and in numerous other fora. The US has an interest in African support in such functional areas as management of the international environment, the future of oceans, nuclear nonproliferation, energy, and population growth.

Africa's role in the world economy is small by US, European, or East Asian standards. Africa accounts for only 5 percent of world trade, and the US investment in Africa constitutes a very small proportion of America's national product. However, America's European allies have a significant stake in the African economies. West European interest in trade and investment in Africa is five to ten times that of the US. Second, all of the Western industrialized nations have become more dependent on African fuels and

nonfuel minerals. Nigeria, for example, has become a leading supplier of American oil imports. But the more crucial concern south of the Sahara is Western access to Africa's underdeveloped mineral deposits. In the noncommunist world, African countries yield over half the production of seven critical minerals—chromium, cobalt, industrial diamonds, germanium, manganese, platinum, and vanadium—and are a major source of several other commodities. Europe is more dependent on these minerals than the US, which has sizable strategic stockpiles and mineral wealth. However, the location of most of the mineral deposits in the troubled areas of southern Africa poses difficult policy constraints on all Western powers.[39]

Ironically, the very weaknesses that preclude African states from posing a security threat to the US contribute to situations conducive to regional strife and outside intervention. The general instability of African regimes rests upon a host of problems such as poverty, tribalism, secessionist and irredentist movements, and inadequate political institutions. While Africans are trying to find African solutions to their problems, the Soviets have exploited the instability inherent in the African modernization process. The US is particularly concerned about Soviet and Cuban willingness to intervene militarily in local conflicts such as Angola and Ethiopia-Somalia. Jonas Savimbi's insurgent forces are pressing the Marxist Angolan government supported by the Soviets and Cubans. Savimbi's guerrillas are active throughout Angola and have successfully brought Angola's economy to a halt, except for government revenues derived from offshore oil.

US interests in Africa are not only economic and geostrategic, but are social, moral, and humanitarian as well. Fundamental American norms of independence and patriotism are compatible with African anticolonialism, nationalism, and Pan-African solidarity. Other values, however, such as democracy, human rights, and free enterprise, encounter few parallels in the numerous single-party and military and socialist-oriented regimes of Africa. Thus, some US and African values are compatible, while others clash. The issue of majority rule in South Africa, however, brings these values into focus. Both the size of the American black population and the postwar emergence of the civil-rights struggle give the US a moral interest in Southern Africa. Strategic and economic interests accentuate the urgency US policymakers attach to finding solutions to the struggles in Namibia and Mozambique.

The Horn of Africa and North Africa have an array of ongoing political problems which makes that area a natural incubator for low-intensity conflict. In 1985, a political coup in Sudan ousted Jafaar Nimieri after fifteen years of autocratic rule. While Khartoum is unsettled and awaiting the focus of the new regime's political orientation, a virulent insurgency in Southern Sudan continues to threaten to divide the country into ethnic and religious enclaves. Marxist Ethiopia, although undergoing a catastrophic period of drought, famine, and hunger, is simultaneously facing an insurgency in its provinces of Tigray and Eritrea. Libya, while supporting terrorist activity worldwide, is undergoing its own internal pressures to remove the charismatic leadership of Qadhaffi. Meanwhile, Qadhaffi continues to foment turmoil in Chad and Tunisia. Chad, too, offers an interesting example of internal civil war, complicated by external intervention. Chadian government forces are fighting insurgents heavily supported by Libyan conventional forces. In late 1986–early 1987, Chadian President Habre stunned Libyan forces with a dazzling display of boldness, courage, and panache. His forces, despite being outnumbered and outgunned, took the offensive against superior Libyan forces. Habre demonstrated how a low-technology force, possessing high mobility, morale, and raw courage, could overwhelm poorly led, low-spirited military forces tied to heavy equipment.

LATIN AMERICA

A former military leader has said that "Latin America's geographical proximity to the US, its economic potential, and its existing commercial relationships with the US make it an area of prime strategic significance."[40] The US has significant economic, strategic, and political interests in the region. Latin America possesses important raw materials, such as oil, copper, tin, and bauxite. Although many of the raw materials of strategic significance have been stockpiled by the US or are available from other sources, Latin America's importance as a source cannot be easily dismissed. For example, the region continues as a low-cost producer of raw materials and is an important supplier of strategic items to some of America's allies who do not possess the same degree of self-sufficiency as the US. American private capital investment in Latin America was valued at approximately $22.2 billion in 1975. These investments traditionally have been in raw-materials extraction, but increasingly are concentrated in manufacturing, commerce, and banking. American private enterprise also seeks growth in the transfer of technology to Latin America. Despite the political problems of previous nationalizations of American foreign investment, the prospect that future investments will be forthcoming is an important factor in keeping Latin American governments more friendly to Washington than they otherwise might be. Strategically, insuring continued access to the Panama Canal for the

deployment of warships and the movement of strategic goods remains an important US interest.

Maintenance of southern flank security in the Western hemisphere remains a vital US interest. Resources employed in constructing warning systems facing an opponent based in the Western Hemisphere would not be available for the defense of allies in other regions. Consequently, the US remains adamant that the Soviet Union live up to the 1962 understanding that it not place "offensive" weapons in Cuba. The same rationale of defense opportunity costs also helps to explain US interest in preventing historic rivalries between Latin American countries from degenerating into armed conflict. War in Latin America would not only divert American energies away from the delicate European and Asian balances, but might also create situations in which Latin American countries might seek substantial military assistance from the Soviet Union.

For most of the twentieth century, the US has had a secure southern perimeter. Secure hemispheric borders facilitated the rise of the US as a world power and enabled the US to focus on strategic threats emanating from Europe, Asia, and the Middle East. But Central America forms an integral part of the larger Caribbean Basin, and this region is the US "strategic backyard." In 1982, the Caribbean Basin supplied the US with 1.8 billion barrels of petroleum per day—over 11 percent of total US oil consumption, and 35 percent of total gross US imports of oil. Moreover, the Caribbean Basin forms part of the US strategic rear area. More than 50 percent of all US seaborne trade passes through the Caribbean Basin, including half of all crude oil imports. In a European crisis, roughly half of all US supplies for NATO would pass through this region by sea.

The US is the oldest democracy in the Western Hemisphere. For two centuries, the US has sought to promote a set of political, economic, and humanitarian values in Latin American that derive from Western philosophical traditions. At times, the US has supported authoritarian regimes in the region, but it has generally wanted to bring about the evolution of democratic states as natural US allies. As such, the US has an obvious and legitimate national interest in the further evolution of free, progressive, and secure nations throughout Latin America. The issues of democracy, development, diplomacy, and defense are thus inextricably linked in Latin America. The area, however, has a long tradition of oligarchic regimes using repressive rule. Dissident elements in Latin America have sought an alternative, and many have turned to radical ideologies. It may be fruitless to oppose Marxist-Leninist insurgency in Latin America if the single alternative to left-wing totalitarianism is to suffer continued economic and political injustice under repressive, right-wing regimes. Similarly, economic justice, respect for human rights, and the establishment of genuine democracy are impossible if security requirements are ignored and if externally supported guerrilla movements are allowed to exploit existing instability with impunity. Indigenous revolutions, especially these aided by Soviet proxy forces with interests inimical to the US, threaten US hemispheric interests. Failure to defend legitimate US interests in Latin America would undermine the confidence of friends and allies elsewhere in the US commitment. Victory by guerrilla forces in Central America could produce totalitarian states on the Cuban model and extinguish prospects for democracy, pluralism, and human rights. The resulting social upheaval would produce a massive exodus of political refugees. The region is already the largest source of legal and illegal immigration to the US.

Several potential threats to US interests are developing in Latin America. Low-intensity conflict is spreading by means of guerrilla warfare, leftist and rightist terrorism, government repression, and border conflict. A hostile alignment continues between Cuba and Nicaragua. In addition, Soviet-Cuban capabilities for power projection continue to pose potential threats to US lines of communication in the Caribbean and mid-Atlantic Ocean. As the Soviet Union turns inward and deemphasizes support for communist adventurism in remote regions, however, the threats to US interests in Latin America are increasingly indigenous. A more familiar threat is the continued support by Cuba and Nicaragua for communist insurgency in countries such as El Salvadore. Growing in importance, however, is the problem of drug trafficking as a substantial source of income for both goverments (such as Panama's Noriega) and private groups (such as Colombia's Medellin cartel). For many Americans, drugs have become a vital national security, as well as an important domestic, policy issue. One can expect drug interdiction efforts—involving the close cooperation of the military services, the Coast Guard, and law enforcement officials both at home and abroad—to feature prominently in US military planning.

THE USABILITY OF MILITARY FORCE AS AN INSTRUMENT OF US POLICY

Two major hemorrhages in the twentieth century killed over 60 million people in inter-state warfare. This total war reflected an evolution in the history of warfare whereby whole societies became participants in "total war." Such a development was reminiscent of Clausewitz's century-old view that war in its "pure" form tended to "absolute violence," the aim of which

was "the destruction of the enemy's armed forces."[41] The world has not witnessed a repeat of global war, a fact often attributed to the advent of nuclear weapons. In enabling the mutual annihilation of states, nuclear weapons may have had the paradoxical effect of fulfilling Clausewitz's vision of "absolute violence" while at the same time seeming to deny any rational political purpose which would justify resort to such destructive capacity.

For the most part, contemporary international political relationships are characterized by bargaining, persuasion, or reward, rather than by violent means. Nonetheless, the use of force has not abated in the international system despite the nuclear "balance of terror." Military force remains a necessary instrument of state policy, both for national self-defense and for defending international interests. As such, policymakers and military strategists must answer several relevant questions: what are the more likely scenarios in which force might be required, what forms will that force assume, and to what extent is force a usable instrument of policy to achieve political ends?

Wars and other forms of armed conflict tend to reflect the age in which they are fought. Global wars reflected the industrial age. Today, terrorism and other forms of political violence reflect the postindustrial age in which international interdependence and technological advancements in communications, transportation, and weaponry have increased the capability for small groups of people to influence political events on a scale that is out of all proportion to their size and resources. The threat of high-intensity conflict remains. However, new modes of conflict, with new political actors, crowd the spectrum of conflict, blurring the distinctions between war and peace and the differentiation between combatant and noncombatant. Recent patterns of conflict indicate warfare will be less coherent and less decisive. Because of the constraints on the application of total military power, wars seldom end in conquest or capitulation. Instead, external intervention or exhaustion of the combatants often results in cease-fires, while underlying political disputes fester unresolved, erupting into periodic confrontations and crises.

States will pursue interests through violent means short of nuclear war. Although conventional war tends to be more limited in scope than the two world wars of this century, high-intensity conflict involving substantial commitments of conventional forces persists. While states possessing nuclear weapons have not waged high-intensity conflict with each other, they continue to compete with each other on the margins of their vital issues, on the periphery of their spheres of influence, and at lower levels of risk. Because of the increasing destructiveness and costs of modern warfare, the increasing indispensability of highly trained military personnel to wage modern warfare, the complexity of superpower relations, and the limitations posed by public opinion, states and non-state actors have relied increasingly on more indirect forms of conflict and the use of surrogates to achieve their political objectives.[42]

New actors and modes of political conflict increase the complexity of armed conflict to the point where the traditional Western approach to warfare may not be applicable. This approach tends to view war and peace, combatant and noncombatant as clearly distinct entities. These distinctions were important in defining the legitimate bounds of warfare as an action between sovereign states which possessed the inherent right of self-defense. That approach involved an agreement to respect the basic social order of any enemy state while attacking the enemy's armed forces which were seen as the center of power. Revolutionary warfare, insurgency, and terrorism, on the other hand, blur these distinctions. These modes of conflict aim to destroy the social fabric of a society but without employing armed forces in a traditional sense. These modes of conflict often render the traditional elements of military response—organized forces delivering superior firepower against opposing armed forces—ineffective if not irrelevant.

In defending itself and its allies and in protecting its global interests, the US must maintain the credible capabilities to use force along the entire spectrum of conflict. Substantial political and military constraints, however, hamper the effective use of these diverse capabilities. The US is most likely to find itself involved militarily in contingencies in the Third World, in the realm of low-intensity conflict, where the traditional forms of military power may be ineffective in achieving anything but the most immediate operational objectives. As a result of increased public wariness of armed conflict and unilateral military intervention, almost any application of American military power in circumstances that do not involve a direct attack on the US or its formal allies may be subject to intense domestic debate. Whereas the "American way of warfare" is well adapted psychologically and materially to the defense of core US values and interests—defense of the American homeland and strategically vital allies and preservation of fundamental American values and well-being—American military power may not always be clearly usable in other, more ambiguous, contexts.[43]

The usability and effectiveness of American military force varies, therefore, as one moves along the spectrum of military conflict. American military power comes from its ability to employ force effectively in a high-intensity nuclear or conventional conflict. The bulk of US military force, as well as US alliance commitments, is designed expressly to win high-intensity conflicts. The absence of such conflicts involving the US and its allies in Europe and Asia may be directly attributed to the deterrent effect of the US strategic nuclear arsenal and the forward deployment of US conventional and nuclear forces in Western Europe and the Pacific region. While the threat to vital US interests is high, due to the catastrophic consequences should deterrence fail, the likelihood of global nuclear or conventional conflict is slight. The prospect of assured destruction remains a powerful incentive to avoid conflict.

Although theoretically more likely, nuclear conflict confined to a particular region or theater of operations remains improbable for much the same reason. US nuclear weapons serve deterrent purposes, with the clear implication that an initially limited nuclear conflict could escalate to general nuclear war between the superpowers. The same applies to conventional war in a theater, such as Western Europe, which is vital to the US. There, the prospect of escalation to general nuclear war remains the ultimate deterrent, even as conventional forces in theater are designed to provide flexible response options.

In the future, similar threats to core US interests may occur by virtue of the proliferation of nuclear weapons, in which the potential for a regional conflict to escalate into general nuclear war may become greater. One report speculates, "If there is to be nuclear war, it will begin with one of the emerging nuclear powers, where command and control systems may not be as refined or the government as stable, as ours . . . begun by a country with a fairly small population using nuclear weapons probably against its neighbors."[44]

In the defense of core interests, the usability and effectiveness of US military power are in little doubt. The employment of US military power in this context is as an established deterrent, against threats that are fairly well understood and with forces that are viewed as legitimate for the purposes for which they are intended. In the realms of mid- and low-intensity conflict, however, the usability of US military force is less clearly established beyond the same broad principle of deterrence, and the effectiveness of that force is more contingent upon unforeseeable circumstances.

Limited conventional conflict is confined to a region and has circumscribed objectives. Because the threats to core values and interests tend to be lower, limited conventional conflict offers a less risky option for regional powers which might seek to challenge US interests in that region. While the US is capable of waging this type of warfare successfully, logistical problems could be monumental, depending on the location of the conflict. Political constraints are especially relevant in determining whether the US would be able to see a protracted limited conventional conflict through to a successful conclusion.

Unilateral military intervention may become more commonplace. The US is developing a rapidly deployable force for contingencies where less-than-vital but important US interests are involved. If the US is able to project power with a reasonable chance of obtaining its limited objectives—as in Grenada and Panama—the US may consider such military intervention when other alternatives are foreclosed. Unilateral interventions may be risky, however, especially if the issue is politically complex and the military operations difficult. The American public is likely to support such actions if they are successful, if the government can demonstrate a real need for such action, and if the operation can be efficiently carried out and combat terminated with minimal loss of life and collateral damage. As was the case with limited conventional conflict, the US ability to carry out such operations will depend heavily on the magnitude of political and logistical constraints, as well as the ability to tailor military forces, weapons, and doctrine to their assigned purpose. In the case of the US invasion of Panama in December 1989 to oust Manuel Noriega, the success of the operation combined with significant public concern over drug trafficking provided important domestic legitimacy for the application of the military force. To project such power successfully on a global scale, however, requires a basing structure and transport capability which presently are underdeveloped.

Conflicts such as Vietnam, involving both conventional and unconventional forms of warfare, pose special problems for the US. Such power projection is difficult, due to both logistical and operational problems as well as the requirement to sustain domestic political resolve in a more protracted conflict. While the US is more prepared to fight in the conventional environment which such conflicts present, the more an adversary de-escalates the conflict to lower levels of intensity, the more problematic the US ability to wage war successfully becomes. Escalating the conflict to higher levels may prove too costly, given the limited objectives which would likely be at stake.

The direct employment of US military power may be the least effective in guerrilla conflicts in which lesser powers seek to gain concessions from stronger states. Among the three levels of guerrilla conflict described earlier, the US seems more able to sustain its involvement the more indirect that involvement is. US forces can be effective in a purely advisory role (guerrilla I) in which military advisory teams work with the host government to enhance the proficiency of indigenous forces to conduct small-unit operations applicable to the combat environment. In the guerrilla II model, US forces would serve as cadre to bolster the effectiveness of indigenous units. In the guerrilla III level of involvement, however, the completed integration (or substitution) of US forces into the counterinsurgency effort brings with it higher political as well as military risks. The applicability of each of these three levels will depend on the level of US interests involved and the degree to which the US can rely on the political efficacy of the host government as well as the military effectiveness of indigenous forces.

Terrorism poses a growing challenge to US security interests and is demanding an increasing investment in counterterrorism measures as the US and other Western countries struggle to define appropriate response options. The dramatic rise in casualty rates for US citizens reflects the exposure and vulnerability of US noncombatants in a global war zone. International cooperation and unilateral passive defense measures offer only partial remedies. The US lacks a fully capable military or paramilitary instrument to target terrorist cadres due to intelligence gaps. The more likely response is resort to surgical operations in a counterterrorist mode, involving the discrete use of a highly trained military force to achieve specified and limited objectives.

The US ability to defend its global interests rests largely on the projection of US power short of employing military force in combat. Whether in the form of "AWACS diplomacy," the maneuvering of naval forces to signal special concern about events in remote regions, or the more benign and routine provision of security assistance, the US has been largely successful in demonstrating its ability to project a military presence throughout the globe in ways that enhance deterrence without provoking escalation of local conflicts. This noncombatant employment of force will appropriately remain a key element in US strategy, particularly given the constraints on the more direct employment of combatant forces in low-intensity conflict situations.

This overview of the US capability for defending global interests underlines three essential conclusions about the contemporary nature of military power. First, military force remains a usable instrument of US policy, if for no other reason than other states which threaten US interests view military force as usable and appropriate for achieving their ends. Second, while the most important threats to US interests will

remain in the high-intensity portion of the spectrum of conflict, where US credibility and capability are clearest, the most likely threats will
involve low-intensity conflict in the Third
World, where the full power of American military force may be less usable. Finally, the most
important use of force in the defense of US interests will remain, ironically, its nonuse, in
which the threat of use—either overt or tacit—
can enhance the effectiveness of other political
and diplomatic processes. This is the common
denominator of the two ends of the spectrum of
conflict where US military power is most demonstrably effective. As in nuclear deterrence,
the clear capacity to employ military power in
limited conflict situations when it is appropriate
to the defense of US interests is the essential
vehicle for not having to employ those forces in
combat in the end.

NOTES

1. See Bard O'Neill's *Insurgency in the Modern
World* (Boulder, Colo.: Westview Press, 1980), p. ix.
2. Quoted in Richard Nixon's article, "How to
Win: Putting the Big Stick Back," *New York Times*,
7 Oct. 1985, D-1. Nixon argues that the US tends
not to recognize vital interests in the Third World,
that US fixation on defense of the homeland and
Western Europe, at the expense, and in his opinion,
exclusion, of defending legitimate interests in the
Third World, undermines US status as a superpower.
3. See Sam C. Sarkesian, "American Policy and
Low-Intensity Conflict: An Overview," in Sam C.
Sarkesian and William L. Scully, eds., *US Policy and
Low-Intensity Conflict: Potentials for Military Struggles in the 1980s*, (New Brunswick, N.J.: Transaction
Books, 1981), p. 3.
4. "(T)he greatest and the most decisive act of
judgment which a statesman and commander perform
is that of recognizing correctly the kind of war in
which they are engaged; of not taking it
for . . . something . . . it cannot be . . . We must
proportion our effort to (the enemy's) power of resistance." Quoted in Frank N. Trager and William L.
Scully, "Low-Intensity Conflict: The US Response,"
in Sarkesian, *US Policy and Low-Intensity Conflict*,
p. 178.
5. See Howard D. Graves, "US Capabilities for
Military Intervention," in Sarkesian, *US Policy and
Low-Intensity Conflict*, p. 69.
6. See "Even in 'Peacetime,' 40 Wars Are Going
On," in *US News and World Report*, 11 July 1983,
pp. 44–45.
7. See General Wallace H. Nutting, "World in
Conflict," in *Low-Intensity Conflict* (Fort Leavenworth: US Army Command and General Staff College, 1983), p. 694.
8. Gregory D. Foster, "Conflict to the Year
2000," in *Air University Review*, Sept.–Oct. 1985,
pp. 27–28.
9. Nixon, "How to Win."
10. Quoted in Foster, "Conflict," p. 29.
11. See Sarkesian, *US Policy and Low-Intensity
Conflict*, pp. 5–6. Figure 1 is adapted from Sarke-

sian's conflict of spectrum in ibid., p. 6. Many other
schemes are also available.
12. See *US Military Posture: FY 1985*. The Organization of the Joint Chiefs of Staff, 1982, pp.
82–84.
13. See *Foreign Military Sales, Foreign Military
Construction Sales and Military Assistance Facts*,
published by Data Management Division, Comptroller, Defense Security Assistance Agency, Sept.
1984, pp. 2–7.
14. See Graves, "US Capabilities," p. 71.
15. US Information Agency, *Overview, US Policy
in Central America*, no date, p. 2.
16. See Graves, "US Capabilities," p. 71.
17. Mark N. Hagopian, *The Phenomenon of Revolution* (New York: Dodd, Mead, 1974), p. 1.
18. For a discussion of the types of leaders in a
revolutionary movement, see Eric Hoffer's classic
study, *The True Believer* (New York: Harper and
Row, 1951), pp. 130–34.
19. See O'Neill, *Insurgency*, p. 19.
20. The discussion of terrorism is based on Doug
Menarchik's "The Politics of the Israeli Rescue
Operation at Entebbe: Crisis Resolution Between
State and Terrorist Organizations," Ph.D. diss.,
The George Washington University, 1983, pp. 11–
46.
21. See N. C. Livingstone, "Fighting Terrorism
and 'Dirty Little Wars,' " in William A. Buckingham,
ed., *Defense Planning for the 1990s and the Changing
International Environment* (Washington, D.C.: National Defense University Press, 1984), pp. 168–70,
for a fuller discussion of these distinctions.
22. See O'Neill, *Insurgency*, p. 19.
23. Ibid., pp. 26–28.
24. See Menarchik, *Political Violence and Revolutionary Change*, USAF Academy, Colo., 1981,
mimeo.
25. See O'Neill, *Insurgency*, pp. 28–30.
26. See ibid., p. 32.
27. See ibid., pp. 32–33.
28. See Menarchik, *Political Violence*.
29. Ibid.
30. Ibid.
31. See *Patterns of International Terrorism: 1983*,
US Department of State, Sept. 1984.
32. R. K. Ramazani, *Beyond the Arab-Israeli Settlement: New Directions for US Policy in the Middle
East* (Cambridge: Institute for Foreign Policy Analysis, 1977), pp. 7–12.
33. Robert McFarlane, "What Are America's Principal Interests in the Middle East?" *Selected Statements*, News Clippings and Analysis Service, Aug.
1984, 84–2, p. 39.
34. US Department of Defense, *Annual Report to
Congress: 1985*, Office of the Secretary of Defense
(Washington, D.C.: G.P.O., 1984), pp. 210–14.
35. US Department of Defense, *United States Military Posture, FY 1985*, Organization of the Joint
Chiefs of Staff (Washington, D.C.: G.P.O., 1985), p.
210.
36. Thomas L. McNaugher, "Balancing Soviet
Power in the Persian Gulf," *Brookings Review* (Summer 1983): 20–24.
37. Department of Defense, *Annual Report FY
85*, p. 215.
38. See Brian Jenkins, *New Modes of Conflict*
(Santa Monica, Calif.: RAND Corporation, 1983),

pp. 7–8. Also Livingstone, "Fighting Terrorism,"
p. 167.

39. See George Osborn and William Taylor, Jr.,
"The Employment of Force: Political-Military Con-
siderations," in Sarkesian, *US Policy and Low-
Intensity Conflict*, pp. 34–36.

40. Quoted in ibid., p. 36.

41. Carl von Clausewitz, *On War*, trans. Col.
J. J. Graham (New York: Barnes and Noble, 1968),
1:44–45, 284.

42. Jenkins, *New Modes*, passim.

43. See Russell F. Weigley, *The American Way
of War: A History of United States Military Strategy and Policy* (New York: Macmillan, 1977), pp.
xiv–xv.

44. Foster, "Conflict," pp. 26–27.

BIBLIOGRAPHIC ESSAY

US COMMITMENTS AND STRATEGIES

FRANK L. ROSA

The literature on the evolution of American defense policy is rich and diverse. That literature addresses a broad spectrum of issues encompassing the strategic environment that emerged out of World War II and the American response to the perceived threats and dangers. This essay is restricted principally to books, and, as such, serves as a useful starting point for comprehending the scope of the policy as it evolved and in offering further source documents.

It is readily apparent in the defense literature that security policy is developed in neither an intellectual nor a political void. Security policy reflects fundamental debates about the nature of the threat and how properly to respond to that threat. In addition, there are varying opinions even regarding the nature of stability in the nuclear age, as well as the utility of military force as an instrument of political policy. Various bureaucratic and political actors often see the world differently and act in accordance with their world outlook. The result is policy that often contains apparently conflicting elements. This is the first concept of security studies that students of security studies should grasp.

THE NATURE OF THE THREAT

A nation's security policy is a reaction to a threat, real or perceived. In American diplomatic history, World War II and the immediate postwar period is a crucial era for understanding the formulation of American defense policy. Certain fundamental assumptions about the world and the appropriate role for the United States were accepted. From this world view emerged a policy of containment toward the Soviet Union, the second dominant power in the postwar period. Containment, as a policy direction, has remained a consistent feature of Soviet-American relations.

An understanding of the policymaking environment, the policy actors, and major policy issues in American diplomacy is relevant to the student of defense policy. An excellent overview is provided by several texts, including *American Foreign Policy: Pattern and Process*, 3d ed., by Charles Kegley, Jr., and Eugene R. Wittkopf

(New York: St. Martin's Press, 1987). Kegley and Wittkopf observe a pattern of continuity in American foreign policy behavior. The goals are constant; it is the methods employed to realize them that change.

A pattern of continuity is also observed in *American Foreign Policy: A Search for Security*, 3rd ed., by Henry T. Nash (Homewood, Ill.: Dorsey Press, 1985). Especially noteworthy is Nash's discussion of events that shaped the development of American perceptions and, consequently, American policy in the postwar period.

The advent of the nuclear age and the emergence of the United States as the dominant international actor occurred almost simultaneously. While continuity in American foreign policy has been noted, the evolving nuclear technology has had a major impact on American foreign policy. This thesis is noted in *American Foreign Policy and the Nuclear Dilemma*, by Gordon C. Schloming (Englewood Cliffs, N.J.: Prentice-Hall, 1987). Schloming is critical of American policy direction, noting that no administration has achieved a permanent improvement in this nation's relationship with the Soviet Union. For Schloming, the threat of nuclear war is the overriding consideration. Yet, the American policy of containment was conceived as a clear attempt to modify the behavior of the Soviet Union to avoid such an apocalyptic event.

When George F. Kennan developed and advanced the concept of containment in 1946–47, he envisioned the strengths and capabilities of the United States—moral, economic, political, and military—containing the advance of Soviet power. This policy, pursued consistently and coherently, could lead to internal changes within the Soviet Union. This was Kennan's argument, and it provided the intellectual framework for the Truman administration's response to Soviet actions throughout the world. Rather than prompting internal changes within the Soviet system, what emerged was a cold war between the two dominant powers.

An extensive literature has evolved over the years on the United States' application of "containment." On the 40th anniversary of the "Mr. X" article, the publishers of Foreign Affairs have revisited that policy (65, no. 4, [1987]). This includes Kennan's "The Sources of Soviet Conduct" and excerpts from Walter Lippmann's serial response, "The Cold War." Kennan has also provided an updated assessment, "Containment Then and Now." This retrospective look offers an excellent starting point for a study of the cold war. A similar, but expanded, historical review of the period is offered in an edited work by Terry L. Deibel and John Lewis Gaddis, Containment: Concept and Policy, 2 vols. (Washington, D.C.: National Defense University Press, 1986).

The cold war can be analyzed from two distinct perspectives: an "orthodox" view or a "revisionist" view. The proponents of the orthodox theory see the cold war as the inevitable result of the confrontation of two opposed political-economic systems within the power vacuum that emerged in Central Europe at the end of World War II. This view of the cold war is presented in The United States and the Origins of the Cold War, 1941–47, by John Lewis Gaddis (New York: Columbia University Press, 1972). This work presents both descriptive and analytical discussion of this period.

Walter Lafeber presents a detailed examination of this same period in America, Russia, and the Cold War, 1945–1984, 5th ed. (New York: Alfred A. Knopf, 1985). This work describes the events of the immediate postwar period and the implications for the last decades of the twentieth century. While accepting the orthodox view of the origins of the cold war, Lafeber stresses the changes that have taken place in the international environment. These changes—the growth of Soviet power and the destructiveness of war—dictate a change in policy to stress diplomacy and negotiation.

In contrast to the orthodox view, the revisionist view argues that the cold war was the result of either deliberate American policy choices or the inherent structural demands of a capitalist system. The first explanation is advanced by Barton J. Bernstein in Politics and Policies of the Truman Administration (Chicago: Quadrangle, 1970) and D. F. Fleming in The Cold War and Its Origins, 1917–1960, 2 vols. (Garden City, N.Y.: Doubleday, 1961). The second explanation—structural determinism—is central to Gabriel Kolko's argument in The Roots of American Foreign Policy (Boston: Beacon Press, 1969), and to the thesis advanced in William Appelman Williams' The Tragedy of American Diplomacy, 2d ed. (New York: Delta, 1972).

The orthodox/revisionist debate has raged for decades and is based upon different assessments of the motivations behind the policies of the Soviet Union and the United States. A partial explanation for the divergence is found in Shattered Peace, The Origins of the Cold War and the National Security State, by Daniel Yergin (Boston: Houghton Mifflin, 1979). Yergin presents the concept of the Riga and Yalta Axioms. Each is a framework for understanding US policymakers' views of the Soviets. With the Riga Axiom the Soviet Union is viewed as an implacable enemy; with the Yalta Axiom, policymakers are more likely to concede to the Soviet Union the status of a more traditional great power. With the first, conflict is the norm. The second offers the possibility of accommodation, cooperation, and convergence. This fundamental distinction, often drawn by other authors using different terms, clearly illustrates the impact on policy formulation of perceptions held by decisionmakers in the United States about the intentions and capabilities of the Soviet leadership.

POLICY CHOICES AND GROWING COMMITMENTS

The security policy choices of President Truman in dealing with the crisis situations that developed in Iran, Greece, Turkey, the Central European theater, the Korea peninsula, and Indochina set the pattern for American relations with the Soviet Union and America's global role in the postwar era. This pattern (and its associated security commitments) was based upon two widely accepted beliefs. First, the United States must take an active role in international affairs, rejecting any return to the prewar isolationist policy. Second, the greatest danger to world peace and security was communism. The United States, as part of its activist role, must limit the political influence and geographic expansion of the Soviet Union—communism's fountainhead. Clearly, the threat was recognized; the response required an expanded American peacetime commitment to the security of various regions of the world.

The history of this period is described in Adam Ulam's seminal work, The Rivals: America and Russia since World War II (New York: Penguin Books, 1971). Covering the period from 1945 to 1970, Ulam offers a balanced treatment of the foreign policies of the two superpowers for dealing with each other. An updated assessment, dealing with the rise and fall of détente throughout the 1970s and into the first term of the Reagan administration, is provided in Détente and Confrontation, Soviet-American Relations from Nixon to Reagan, by Raymond L. Garthoff (Washington, D.C.: Brookings Institution, 1985). Similarly, Seyom Brown's updated and comprehensive review of postwar US foreign policy, The Faces of Power: Constancy and Change in United States Foreign Policy from Truman to Reagan (New York: Columbia

University Press, 1983), grapples with the problems the US has had in wielding power in the international system, particularly after the breakdown of the bipartisan consensus on the goals and means of America's containment policy.

This breakdown of a policy consensus in the US, beginning in the late 1960s, stemmed largely from the legitimate concern that extensive international security commitments have led to tremendous strains on diminishing American resources with few perceived benefits or improvements in the world situation. In this context, John Lewis Gaddis's *Strategies of Containment—A Critical Appraisal of Postwar American National Security Policy* (New York: Oxford University Press, 1982) is an important contribution to the literature on evolving US national security policy. He offers a basic definition of strategy: "The process by which ends are related to means, intentions to capabilities, objectives to resources." Gaddis highlights the problems of balancing means and ends in the defense strategies selected by the postwar presidents. He analyzes successive US policies through a framework which differentiates between "symmetrical" and "asymmetrical" forms of containment which he contrasts between a universalistic and particularistic definition of the threat; between a perimeter and a strong-point defense; and between a presumption of unlimited and limited resources. The concern over resources and the limits of military power have led to a scale-back in the commitments that the United States has been willing to undertake—especially in the aftermath of the disastrous Vietnam experience.

While President Truman charted the path for America's world role, each subsequent administration contributed something to the further development and expansion of the containment policy. The extensive literature on the problems that faced the United States and the programs that were advanced to deal with them by postwar presidents provides a ready source of information for the ambitious student of defense policy. Such works offer a more complete picture of the hectic and dynamic nature of the decisionmaking environment. Additionally, the memoirs, reflections, and political observations of presidents, cabinet officers, and policy advisers are a rich source of material, though some of these works may be self-serving rather than illuminating.

For an understanding of the Truman era, the following works should be noted: Dean Acheson, *Present at the Creation: My Years at the State Department* (New York: Norton, 1969); James F. Burns, *Speaking Frankly* (New York: Harper and Row, 1947); and Harry S Truman, *Memoirs*, 2 vols. (Garden City, N.Y.: Doubleday, 1955–56). Assessments of the foreign policy initiatives of President Truman are provided in such works as Timothy Ireland, *Creating the Entangling Alliance: The Origins of the North Atlantic Treaty Organization* (Westport, Conn.: Greenwood Press, 1981); Charles L. Nee, *The Marshall Plan: The Launching of Pax Americana* (New York: Simon and Schuster, 1984): Avi Schlaim, *The United States and the Berlin Blockade, 1948–1949* (Berkeley and Los Angeles: University of California Press, 1983); and, William W. Stueck, Jr., *The Road to Confrontation: American Policy toward China and Korea, 1947–1950* (Chapel Hill: University of North Carolina Press, 1981).

The Eisenhower era saw a continuation and expansion of the Truman policies. This period is described by the president himself in two works: *Mandate for Change* (Garden City, N.Y.: Doubleday, 1963), and *Waging Peace* (Garden City, N.Y.: Doubleday, 1965). Traditionally, the intellectual driving force of this administration was considered to be Secretary of State John Foster Dulles, whose role is assessed in John R. Beal, *John Foster Dulles: A Statesman and His Times* (New York: Columbia University Press, 1972); and Townsend Hoopes, *The Devil and John Foster Dulles* (Boston: Little, Brown, 1973). A reassessment of the Eisenhower presidency, however, has taken place in the 1980s. Some political analysts increasingly depict Eisenhower in a more positive light, recognizing the quiet decisiveness of this president. The following works are representative of this trend: Stephen E. Ambrose, *Eisenhower: The President*, vol. 2 (New York: Simon and Schuster, 1984); Fred I. Greenstein, *The Hidden Hand Presidency: Eisenhower as Leader* (New York: Basic Books, 1982); and W. W. Rostow, *Europe after Stalin: Eisenhower's Three Decisions of March 11, 1953* (Austin: University of Texas Press, 1982). Further assessments of the international events and crises that occurred in the decade of the 1950s include Charles C. Alexander, *Holding the Line: The Eisenhower Era, 1952–1961* (Bloomington: Indiana University Press, 1975); Robert A. Divine, *Eisenhower and the Cold War* (New York: Oxford University Press, 1981); and Donald Neff, *Warriors at Suez: Eisenhower Takes America into the Middle East* (New York: Linden Press/Simon and Schuster, 1981).

President Kennedy opened his administration with an inaugural address that pledged "that we shall pay any price, bear any burden, meet any hardship, support any friend, oppose any foe to assure the survival and success of liberty." America's commitment to world security through the containment of communism had reached its zenith. But America also recognized the high costs of such a commitment—in terms of human costs, expended resources, lost credibility, and the strain on the political and social fabric of the nation. This nation's commitment to the Republic of South Vietnam became the catalyst for a reassessment of American foreign and military policy.

The goal of the Kennedy administration was to tailor foreign policy and military capabilities to deal with a broader range of issues and threats. The United States would have the ability to challenge Soviet or Soviet-sponsored aggression in low-intensity conflicts, as well as at the strategic level. Flexible response was the codeword. This was not a new concept. In 1959, Maxwell Taylor had called for a flexible response strategy in his book, *The Uncertain Trumpet* (New York: Harper, 1960). With Kennedy, there was now a presidential willingness to spend more on defense to match the requirements for this new strategy.

The strategic policy direction during this period is examined by Lawrence Martin in *Strategic Thought in the Nuclear Age* (Baltimore: Johns Hopkins University Press, 1979) and Jerome Kahan in *Security in the Nuclear Age* (Washington, D.C.: Brookings Institution, 1975). The American strategic buildup caused immediate concern in the Kremlin. Premier Khrushchev responded with a parallel construction program for the long term. Moscow's short-term response, however, created the most dangerous superpower confrontation in the nuclear age—the Cuban missile crisis. This incident has been analyzed extensively, but the preeminent works remain Graham Allison's *Essence of Decision: Explaining the Missile Crisis* (Boston: Little, Brown, 1971) and Robert F. Kennedy's *Thirteen Days: A Memoir of the Cuban Missile Crisis* (New York: W. W. Norton, 1968). In 1987, commemorating the 25th anniversary of the Cuban missile crisis, several excellent studies also appeared in various journals that took a retrospective look at US decisionmaking in that crisis. These include "The Cuban Missile Crisis Revisited" by James G. Blight, Joseph S. Nye, Jr., and David A. Welch in *Foreign Affairs* (66, no. 1, [1987]: 170–88) and "The Eleventh Hour of the Cuban Missile Crisis: An Introduction to the ExComm Transcripts" also by David Welch and James Blight in *International Security* (Winter 1987–88).

On the conventional level, the Kennedy administration increased the number of military advisers to the South Vietnamese government. The pattern of increased involvement leading to the actual American direction of the war effort was set into operation. The American experience in Vietnam has been extensively covered, with the notable works being Frances Fitzgerald's *Fire in the Lake: The Vietnamese and Americans in Vietnam* (Boston: Little, Brown, 1972); David Halberstam's *The Best and the Brightest* (New York: Fawcett Crest, 1972); Guenter Lewy's *America in Vietnam* (New York:

Oxford University Press, 1978); and Stanley Karnow's *Vietnam: A History* (New York: Viking Press, 1983).

The literature on the overall legacy of the Kennedy period is encompassed in several exceptional, though highly partisan, works. These include Roger Hilman *To Move a Nation* (Garden City, N.Y.: Doubleday, 1967); Arthur Schlesinger, Jr., *A Thousand Days: John F. Kennedy in the White House* (Boston: Houghton Mifflin, 1965); and Theodore C. Sorensen, *Kennedy* (New York: Harper and Row, 1965) and *The Kennedy Legacy* (New York: Mentor, 1969).

The presidency of Lyndon B. Johnson proved in many respects to be just as tragic as that of Kennedy; it fell as one of the casualties of the Vietnam war. Johnson entered office with the desire to complete the political and economic programs envisioned in Franklin Roosevelt's New Deal, seeking to create the Great Society as his enduring legacy. His strength was in domestic issues, but it was foreign policy that demanded Johnson's attention. When faced with the real possibility of an imminent South Vietnamese military defeat, he opted to take direct charge of the war effort. It was now America's war. The Johnson administration sought to keep the best possible light on the American efforts in Vietnam. This carefully managed effort was to collapse in the wake of the 1968 Tet offensive. Though a military disaster for the North Vietnamese, it was the United States that lost the most. The administration's credibility was called into question by increasing numbers of congressmen as well as the American public.

The trauma of the Vietnam conflict continues to be assessed and analyzed in the literature. The purpose is to examine and learn from the lessons of this extended conflict. One of the first retrospective examinations of the war was undertaken by prominent statesmen and military leaders of this period in *The Lessons of Vietnam*, edited by W. Scott Thompson and Donaldson D. Frizzell (New York: Crane, Russak, 1977). Another notable work in this same vein is Paul M. Kattenburg's *The Vietnam Trauma in American Foreign Policy, 1945–75* (New Brunswick, N.J.: Transaction Books, 1980). According to Kattenburg, the main consequence of the loss of the war was a reduction in America's international posture resulting from the loss of domestic consensus on foreign policy. The most current analysis is contained in *Assessing the Vietnam War*, edited by Lloyd Matthews and Dale Brown (McLean, Va.: Pergamon-Brassey's International Defense Publishers, 1987). The central theme of these three works is the realization of the limits of American power.

The literature on the Johnson presidency reflects the sadness of the lost dream—of the lost American innocence. This attitude is seen in *The Lost Crusade*, by Chester Cooper (New

York: Dodd, Mead, 1970); *The Tragedy of Lyndon Johnson,* by Eric F. Goldman (New York: Dell, 1968); and *Lyndon Johnson and the American Dream,* by Doris Kerns (New York: Signet, 1977). As with other former presidents, Lyndon Johnson gave his view in *The Vantage Point: Perspectives of the Presidency, 1963–1969* (New York: Holt, Rinehart and Winston, 1971).

Subsequent presidents were forced to reassess America's world role. A more forceful Congress and a more wary American public sought limits on American commitments. President Nixon attempted to reduce this nation's commitments without losing American power and influence. This particular theme, from a British perspective, is poignantly examined in Henry Brandon's *The Retreat from American Power* (London: Bodley Head, 1973). Containment of the Soviet Union was still accepted, but this was to be achieved through diplomatic initiatives rather than force of arms. The Nixon-Kissinger view was that the Soviet leadership could be co-opted into the existing international political-economic system. As noted by Roger P. Labrie in *SALT Handbook* (Washington, D.C.: American Enterprise Institute for Public Policy Research, 1979), President Nixon's goal was to ensure that by acquiring "a stake in this network of relationships with the west, the Soviet Union may become more conscious of what it would lose by a return to confrontation."

Détente was the new catch word in Soviet-American relations. While moving from one foreign policy success to another—opening relations with the People's Republic of China, the success of arms control negotiations with the USSR, and the finalization of a peace treaty ending the Vietnam war, President Nixon became the victim of a domestic political crisis—Watergate. The purpose of this essay is not to review events that led to Nixon's resignation. The focus, rather, is on the foreign and military policy initiatives of this presidency.

The two primary actors in this administration—President Nixon and Secretary of State Henry A. Kissinger—have offered their own memoirs of their experience: Kissinger's *The White House Years* (Boston: Little, Brown, 1979) and *Years of Upheaval* (Boston: Little, Brown, 1982), and Richard Nixon's *RN: The Memoirs of Richard Nixon* (New York: Grosset and Dunlap, 1978). More critical assessments are provided in such diverse works as Seymour M. Hersh's *The Price of Power* (New York: Summit Books, 1983); Roger Morris's *Uncertain Greatness: Henry Kissinger and American Foreign Policy* (New York: Harper and Row, 1977); Tad Szulc's *The Illusion of Peace: Foreign Policy in the Nixon-Kissinger Years* (New York: Viking, 1978); and Gary Will's *Nixon Agonistes* (Boston: Houghton Mifflin, 1970).

In *A Time to Heal* (New York: Harper and Row, 1979), Gerald Ford explained the necessary but controversial decision to pardon President Nixon while elaborating on his efforts to continue the policy focus of his predecessor. President Ford, however, was faced with a more hostile Congress, no longer content to play a secondary role in the formulation of foreign and defense policy.

The election of Jimmy Carter in 1976 set the stage for a return to the more traditional, moralistic policy consistent with what has been termed the "American character." This American character is explored by Arthur M. Schlesinger, Jr., in "Foreign Policy and the American Character," *Foreign Affairs* (60, no. 1 [1983]: 1–16) and by George F. Kennan in *American Diplomacy, 1900–1950* (New York: New American Library, 1951).

While continuing to pursue a policy of détente with the Soviet Union, the Carter administration showed a marked orientation toward North-South rather than East-West issues. By supporting and advancing the concerns of the Third World, Carter hoped to stabilize this area and increase American influence throughout the developing world. The goals were noble; the actors were sincere; and the policy initiatives were carefully developed. Nevertheless, the Carter presidency seemed unable to wield American power effectively. There was a crisis of confidence. While President Carter pursued a policy of détente with the Soviet leadership, the Kremlin appeared to be exploiting crisis situations in Africa, the Middle East, and the Far East to advance its own position and undermine that of the United States. The Soviet invasion of Afghanistan proved to be the turning point for relations between the two superpowers. The diplomatic strains were seen as the start of a new cold war. At this point, the administration acted more decisively, but events in Iran were to prove fatal to the president's reelection plans.

The four years of the Carter presidency are explored in significant works by several of the primary actors in the administration. Zbigniew K. Brzezinski's *Power and Principle: Memoirs of the National Security Advisor, 1977–1981* (New York: Farrar, Straus, and Giroux, 1983) offers an insightful look at some of the significant events, programs, and policies of the Carter years. Brzezinski provides a coherent framework for the analysis of this presidency. Cyrus Vance, in *Hard Choices: Critical Years in America's Foreign Policy* (New York: Simon and Schuster, 1983), focuses the reader's attention to the range of tough issues that the Carter administration was forced to tackle. These were the sad consequences of the entire pattern of past American foreign policies. The impact of the Iran hostage crisis on the president is the theme of Hamilton Jordan's *Crisis: The Last Year of the Carter Presidency* (New York: Putnam, 1982). The inner strength of Carter, the man, is captured in the former president's own book, *Keeping Faith: Memoirs of a President* (New York: Bantam Books, 1982).

It will be left to future political analysts and historians to assess with greater objectivity President Carter's overall impact on America's status in the world. The electorate in 1980 passed judgment on the policies of the 1970s: it was unwilling to accept a reduction in the status or power of the United States. Instead, it accepted the Reagan view that more must be done to reassert American influence in the world.

Reagan entered the White House committed to a program of rearming America. His administration seemed committed to action—even if the United States had to act unilaterally. Some analysts saw the administration overcommitting the United States. This is the view expressed in *Eagle Defiant: United States Foreign Policy in the 1980s*, edited by Kenneth Oye, Robert Lieber, and Donald Rothchild (Boston: Little, Brown, 1983). The assessments of the various compiled articles are often harsh and affected by ideological biases. This points out the problems of evaluations of a current president's policies. A far better analysis of the events that have occurred during President Reagan's tenure is contained in the collective annual reports, *America and the World*, published by *Foreign Affairs*.

CONCLUSION

Toward the end of the Reagan administration, the dominant question remained the proper extent to which the United States should be responsible for the defense of the Western nations and Western interests. While President Reagan sought to maintain free passage in the Persian Gulf, critics argued that the Allies should play a greater role. There are shared interests, indeed, but the modalities of sharing the responsibilities between a superpower and its allies

defy clear definition. This has resulted in a definite strain within the various alliance relationships that the United States had entered into in the post–World War II period. *Estrangement: America and the World*, edited by Sanford J. Ungar (New York: Oxford University Press, 1985) highlights the numerous problems that have faced the United States in its dealings with traditional allies and the emerging nations of the Third World. What is crucial to understand is that the interests and world view of the United States do not always coincide with those of America's allies. To some, it is the American world view itself that is the real problem. A critical assessment of America's policy direction is offered by Louis Rene Beres' *America Outside the World: The Collapse of U.S. Foreign Policy* (Lexington, Mass.: Lexington Books, 1987). Beres argues that "America exists outside the world, in defiance of its global obligations. It betrays both its traditions and potentialities in a frenzied search for power. As a result, it displays only impotence." The real culprit, according to Beres, is America's obsession with the "evil empire"—the Soviet Union. Policy must be built upon concrete realities and foresight, not caricatures and vague ideals. For Beres, much is demanded of American power.

The truth is that there are clear limits to American power, just as there are unique responsibilities that fall to the United States as a superpower. John G. Stoessinger's *Crusaders and Pragmatists* (New York: W. W. Norton, 1985, 2d ed.), for example, portrays the role of personalities in making American foreign policy. In a broader sense, Stoessinger's work reflects the enduring conflict between, on the one hand, the tendency to act unilaterally to defend American interests and, on the other hand, the reality of America's security interdependence with its allies who often condition the exercise of American power. This is a central theme to much of the literature discussed in this essay.

SELECTED READINGS II

THE SOURCES OF SOVIET CONDUCT

GEORGE KENNAN

The political personality of Soviet power as we know it today is the product of ideology and circumstances: ideology inherited by the present Soviet leaders from the movement in which they had their political origin, and circumstances of the power which they now have exercised for nearly three decades in Russia. There can be few tasks of psychological analysis more difficult than to try to trace the interaction of these two forces and the relative role of each in the determination of official Soviet conduct. Yet the attempt must be made if that conduct is to be understood and effectively countered.

It is difficult to summarize the set of ideological concepts with which the Soviet leaders came into power. Marxian ideology, in its Russian-Communist projection, has always been in process of subtle evolution. The materials on which it bases itself are extensive and complex. But the outstanding features of Communist thought as it existed in 1916 may perhaps be summarized as follows: (a) that the central factor in the life of man, the factor which determines the character of public life and the "physiognomy of society," is the system by which material goods are produced and exchanged; (b) that the capitalist system of production is a nefarious one which inevitably leads to the exploitation of the working class by the capital-owning class and is incapable of developing adequately the economic resources of society or of distributing fairly the material goods produced by human labor; (c) that capitalism contains the seeds of its own destruction and must, in view of the inability of the capital-owning class to adjust itself to economic change, result eventually and inescapably in a revolutionary transfer of power to the working class; and (d) that imperialism, the final phase of capitalism, leads directly to war and revolution.

The rest may be outlined in Lenin's own words: "Unevenness of economic and political development is the inflexible law of capitalism. It follows from this that the victory of Socialism may come originally in a few capitalist countries or even in a single capitalist country. The victorious proletariat of that country, having expropriated the capitalists and having organized Socialist production at home, would rise against the remaining capitalist world, drawing to itself in the process the oppressed classes of other countries."[1] It must be noted that there was no assumption that capitalism would perish without proletarian revolution. A final push was needed from a revolutionary proletariat movement in order to tip over the tottering structure. But it was regarded as inevitable that sooner or later that push be given.

For 50 years prior to the outbreak of the Revolution, this pattern of thought had exercised great fascination for the members of the Russian revolutionary movement. Frustrated, discontented, hopeless of finding self-expression—or too impatient to seek it—in the confining limits of the Tsarist political system, yet lacking wide popular support for their choice of bloody revolution as a means of social betterment, these revolutionists found in Marxist theory a highly convenient rationalization for their own instinctive desires. It afforded pseudoscientific justification for their impatience, for their categoric denial of all value in the Tsarist system, for their yearning for power and revenge and for their inclination to cut corners in the pursuit of it. It is therefore no wonder that they had come to believe implicitly in the truth and soundness of the Marxian-Leninist teachings, so congenial to their own impulses and emotions. Their sincerity need not be impugned. This is a phenomenon as old as human nature itself. It has never been more aptly described than by Edward Gibbon, who wrote in *The Decline and Fall of the Roman Empire*: "From enthusiasm to imposture the step is perilous and slippery; the demon of Socrates affords a memorable instance how a wise man may deceive himself, how a good man may deceive others, how the conscience may slumber in a mixed and middle state between self-illusion and voluntary fraud." And it was

Reprinted with minor revisions and by permission from Foreign Affairs 25, no. 4 (1947): 566–82. *Copyright © 1947 by the Council on Foreign Relations, Inc.*

with this set of conceptions that the members of the Bolshevik party entered into power.

Now it must be noted that through all the years of preparation for revolution, the attention of these men, as indeed of Marx himself, had been centered less on the future form which Socialism[2] would take than on the necessary overthrow of rival power which, in their view, had to precede the introduction of Socialism. Their views, therefore, on the positive program to be put into effect, once power was attained, were for the most part nebulous, visionary, and impractical. Beyond the nationalization of industry and the expropriation of large private capital holdings there was no agreed program. The treatment of the peasantry, which according to the Marxist formulation was not of the proletariat, had always been a vague spot in the pattern of Communist thought; and it remained an object of controversy and vacillation for the first ten years of Communist power.

The circumstances of the immediate postrevolution period—the existence in Russia of civil war and foreign intervention, together with the obvious fact that the Communists represented only a minority of the Russian people—made the establishment of dictatorial power a necessity. The experiment with "war Communism" and the abrupt attempt to eliminate private production and trade had unfortunate economic consequences and caused further bitterness against the new revolutionary regime. While the temporary relaxation of the effort to communize Russia, represented by the New Economic Policy, alleviated some of this economic distress and thereby served its purpose, it also made it evident that the "capitalistic sector of society" was still prepared to profit at once from any relaxation of governmental pressure, and would, if permitted to continue to exist, always constitute a powerful opposing element to the Soviet regime and a serious rival for influence in the country. Somewhat the same situation prevailed with respect to the individual peasant who, in his own small way, was also a private producer.

Lenin, had he lived, might have proved a great enough man to reconcile these conflicting forces to the ultimate benefit of Russian society, though this is questionable. But be that as it may, Stalin, and those whom he led in the struggle for succession to Lenin's position of leadership, were not the men to tolerate rival political forces in the sphere of power which they coveted. Their sense of insecurity was too great. Their particular brand of fanaticism, unmodified by any of the Anglo-Saxon traditions of compromise, was too fierce and too jealous to envisage any permanent sharing of power. From the Russian-Asiatic world out of which they had emerged they carried with them a skepticism as to the possibilities of permanent

and peaceful coexistence of rival forces. Easily persuaded of their own doctrinaire "rightness," they insisted on the submission or destruction of all competing power. Outside of the Communist party, Russian society was to have no rigidity. There were to be no forms of collective human activity or association which would not be dominated by the Party. No other force in Russian society was to be permitted to achieve vitality or integrity. Only the Party was to have structure. All else was to be an amorphous mass.

And within the Party the same principle was to apply. The mass of Party members might go through the motions of election, deliberation, decision, and action; but in these motions they were to be animated not by their own individual wills but by the awesome breath of the Party leadership and the overbrooding presence of "the word."

Let it be stressed again that subjectively these men probably did not seek absolutism for its own sake. They doubtless believed—and found it easy to believe—that they alone knew what was good for society and that they would accomplish that good once their power was secure and unchallengeable. But in seeking that security of their own rule they were prepared to recognize no restrictions, either of God or man, on the character of their methods. And until such time as that security might be achieved, they placed far down on their scale of operational priorities the comforts and happiness of the peoples entrusted to their care.

Now the outstanding circumstance concerning the Soviet regime is that down to the present day this process of political consolidation has never been completed and the men in the Kremlin have continued to be predominantly absorbed with the struggle to secure and make absolute the power which they seized in November 1917. They have endeavored to secure it primarily against forces at home, within Soviet society itself. But they have also endeavored to secure it against the outside world. For ideology, as we have seen, taught them that the outside world was hostile and that it was their duty eventually to overthrow the political forces beyond their borders. The powerful hands of Russian history and tradition reached up to sustain them in this feeling. Finally, their own aggressive intransigence with respect to the outside world began to find its own reaction; and they were soon forced, to use another Gibbonesque phrase, "to chastise the contumacy" which they themselves had provoked. It is an undeniable privilege of every man to prove himself right in the thesis that the world is his enemy; for if he reiterates it frequently enough and makes it the background of his conduct he is bound eventually to be right.

Now it lies in the nature of the mental world of the Soviet leaders, as well as in the character of their ideology, that no opposition to them can be officially recognized as having any merit or justification whatsoever. Such opposition can flow, in theory, only from the hostile and incorrigible forces of dying capitalism. As long as remnants of capitalism were officially recognized as existing in Russia, it was possible to place upon them, as an internal element, part of the blame for the maintenance of a dictatorial form of society. But as these remnants were liquidated, little by little, this justification fell away; and when it was indicated officially that they had been finally destroyed, it disappeared altogether. And this fact created one of the most basic of the compulsions which came to act upon the Soviet regime: since capitalism no longer existed in Russia and since it could not be admitted that there could be serious or widespread opposition to the Kremlin springing spontaneously from the liberated masses under its authority, it became necessary to justify the retention of the dictatorship by stressing the menace of capitalism abroad.

This began at an early date. In 1924 Stalin specifically defended the retention of the "organs of suppression," meaning, among others, the army and the secret police, on the ground that "as long as there is capitalist encirclement there will be danger of intervention with all the consequences that flow from that danger." In accordance with that theory, and from that time on, all internal opposition forces in Russia have consistently been portrayed as the agents of foreign forces of reaction antagonistic to Soviet power.

By the same token, tremendous emphasis has been placed on the original Communist thesis of a basic antagonism between the capitalist and Socialist worlds. It is clear, from many indications, that this emphasis is not founded in reality. The real facts concerning it have been confused by the existence abroad of genuine resentment provoked by Soviet philosophy and tactics and occasionally by the existence of great centers of military power, notably the Nazi regime in Germany and the Japanese government of the late 1930s, which did indeed have aggressive designs against the Soviet Union. But there is ample evidence that the stress laid in Moscow on the menace confronting Soviet society from the world outside its borders is founded not in the realities of foreign antagonism but in the necessity of explaining away the maintenance of dictatorial authority at home.

Now the maintenance of this pattern of Soviet power, namely, the pursuit of unlimited authority domestically, accompanied by the cultivation of the semi-myth of implacable foreign hostility, has gone far to shape the actual machinery of Soviet power as we know it today. Internal organs of administration which did not serve this purpose withered on the vine. Organs which did serve this purpose became vastly swollen. The security of Soviet power came to rest on the iron discipline of the Party, on the severity and ubiquity of the secret police, and on the uncompromising economic monopolism of the state. The "organs of suppression," in which the Soviet leaders had sought security from rival forces, became in large measure the masters of those whom they were designed to serve. Today the major part of the structure of Soviet power is committed to the perfection of the dictatorship and to the maintenance of the concept of Russia as in a state of siege, with the enemy cowering beyond the walls. And the millions of human beings who form that part of the structure of power must defend at all costs this concept of Russia's position, for without it they are themselves superfluous.

As things stand today, the rulers can no longer dream of parting with these organs of suppression. The quest for absolute power, pursued now for nearly three decades with a ruthlessness unparalleled (in scope at least) in modern times, has again produced internally, as it did externally, its own reaction. The excesses of the police apparatus have fanned the potential opposition to the regime into something far greater and more dangerous than it could have been before those excesses began.

But least of all can the rulers dispense with the fiction by which the maintenance of dictatorial power has been defended. For this fiction has been canonized in Soviet philosophy by the excesses already committed in its name; and it is now anchored in the Soviet structure of thought by bonds far greater than those of mere ideology.

II

So much for the historical background. What does it spell in terms of the political personality of Soviet power as we know it today?

Of the original ideology, nothing has been officially junked. Belief is maintained in the basic badness of capitalism, in the inevitability of its destruction, in the obligation of the proletariat to assist in that destruction and to take power into its own hands. But stress has come to be laid primarily on those concepts which relate most specifically to the Soviet regime itself: to its position as the sole truly Socialist regime in a dark and misguided world, and to the relationships of power within it.

The first of these concepts is that of the innate antagonism between capitalism and Socialism. We have seen how deeply that concept has become inbedded in foundations of Soviet power. It has profound implications for Russia's conduct as a member of international society. It means that there can never be on Moscow's side any sincere assumption of a community of aims between the Soviet Union and powers which are regarded as capitalist. It must invariably be assumed in Moscow that the aims of the capitalist world are antagonistic to the Soviet regime, and therefore to the interests of the peoples it controls. If the Soviet government occasionally sets its signature to documents which would indicate the contrary, this is to be regarded as a tactical maneuver permissible in dealing with the enemy (who is without honor) and should be taken in the spirit of *caveat emptor*. Basically, the antagonism remains. It is postulated. And from it flow many of the phenomena which we find disturbing in the Kremlin's conduct of foreign policy: the secretiveness, the lack of frankness, the duplicity, the wary suspiciousness, and the basic unfriendliness of purpose. These phenomena are there to stay, for the foreseeable future. There can be variations of degree and of emphasis. When there is something the Russians want from us, one or the other of these features of their policy may be thrust temporarily into the background; and when that happens there will always be Americans who will leap forward with gleeful announcements that "the Russians have changed," and some who will even try to take credit for having brought about such "changes." But we should not be misled by tactical maneuvers. These characteristics of Soviet policy, like the postulate from which they flow, are basic to the internal nature of Soviet power, and will be with us, whether in the foreground or the background, until the internal nature of Soviet power is changed.

This means that we are going to continue for a long time to find the Russians difficult to deal with. It does not mean that they should be considered as embarked upon a do-or-die program to overthrow our society by a given date. The theory of the inevitability of the eventual fall of capitalism has the fortunate connotation that there is no hurry about it. The forces of progress can take their time in preparing the final *coup de grace*. Meanwhile, what is vital is that the "Socialist fatherland"—that oasis of power which has been already won for Socialism in the person of the Soviet Union—should be cherished and defended by all good Communists at home and abroad, its fortunes promoted, its enemies badgered and confounded. The promotion of premature, "adventuristic" revolutionary projects abroad which might embarrass Soviet power in any way would be an inexcusable, even a counter-revolutionary act. The cause of Socialism is the support and promotion of Soviet power, as defined in Moscow.

This brings us to the second of the concepts important to contemporary Soviet outlook. That is the infallibility of the Kremlin. The Soviet concept of power, which permits no focal points of organization outside the Party itself, requires that the Party leadership remain in theory the sole repository of truth. For if truth were to be found elsewhere, there would be justification for its expression in organized activity. But it is precisely that which the Kremlin cannot and will not permit.

The leadership of the Communist party is therefore always right, and has been always right ever since in 1929 Stalin formalized his personal power by announcing that decisions of the Politburo were being taken unanimously.

On the principle of infallibility there rests the iron discipline of the Communist party. In fact, the two concepts are mutually self-supporting. Perfect discipline requires recognition of infallibility. Infallibility requires the observance of discipline. And the two together go far to determine the behaviorism of the entire Soviet apparatus of power. But their effect cannot be understood unless a third factor be taken into account: namely, the fact that the leadership is at liberty to put forward for tactical purposes any particular thesis which it finds useful to the cause at any particular moment and to require the faithful and unquestioning acceptance of that thesis by the members of the movement as a whole. This means that truth is not a constant but is actually created, for all intents and purposes, by the Soviet leaders themselves. It may vary from week to week, from month to month. It is nothing absolute and immutable—nothing which flows from objective reality. It is only the most recent manifestation of the wisdom of those in whom the ultimate wisdom is supposed to reside, because they represent the logic of history. The accumulative effect of these factors is to give to the whole subordinate apparatus of Soviet power an unshakeable stubbornness and steadfastness in its orientation. This orientation can be changed at will by the Kremlin but by no other power. Once a given party line has been laid down on a given issue of current policy, the whole Soviet governmental machine, including the mechanism of diplomacy, moves inexorably along the prescribed path, like a persistent toy automobile wound up and headed in a given direction, stopping only when it meets with some unanswerable force. The individuals who are the components of this machine are unamenable to argument or reason which comes to them from outside sources. Their whole training has taught them to mistrust and discount the glib persuasiveness of the outside world. Like the white dog before the phonograph, they hear only the "master's voice." And if they are to be called off from the purposes last dictated to them, it is the master who must call them off. Thus the foreign representative cannot hope that his words will make any impression on them. The most that he can hope is that they will be transmitted to those at the top, who are capable of changing the party line. But even those are not likely to be swayed by any normal logic in the words of the bourgeois representative. Since there can be no appeal to common purposes, there can be no appeal to common mental approaches. For this reason, facts speak louder than words to the ears of the Kremlin; and words carry the greatest weight when they have the ring of reflecting, or being backed up by, facts of unchallengeable validity.

But we have seen that the Kremlin is under no ideological compulsion to accomplish its purposes in a hurry. Like the Church, it is dealing in ideological concepts which are of long-term validity, and it can afford to be patient. It has no right to risk the existing achievements of the revolution for the sake of vain baubles of the future. The very teachings of Lenin himself require great caution and flexibility in the pursuit of Communist purposes. Again, these precepts are fortified by the lessons of Russian history: of centuries of obscure battles between nomadic forces over the stretches of a vast unfortified plan. Here caution, circumspection, flexibility and deception are the valuable qualities; and their value finds natural appreciation in the Russian or the oriental mind. Thus the Kremlin has no compunction about retreating in the face of superior force. And being under the compulsion of no timetable, it does not get panicky under the necessity for such retreat. Its political action is a fluid stream which moves constantly, wherever it is permitted to move, toward a given goal. Its main concern is to make sure that it has filled every nook and cranny available to it in the basin of world power. But if it finds unassailable barriers in its path, it accepts these philosophically and accommodates itself to them. The main thing is that there should always be pressure, unceasing constant pressure, toward the desired goal. There is no trace of any feeling in Soviet psychology that that goal must be reached at any given time.

These considerations make Soviet diplomacy at once easier and more difficult to deal with than the diplomacy of individual aggressive leaders like Napoleon and Hitler. On the one hand it is more sensitive to contrary force, more ready to yield on individual sectors of the diplomatic front when that force is felt to be too strong, and thus more rational in the logic and rhetoric of power. On the other hand it cannot be easily defeated or discouraged by a single victory on the part of its opponents. And the patient persistence by which it is animated means that it can be effectively countered not by sporadic acts which represent the momentary whims of democratic opinion but only by intelligent long-range policies on the part of Russia's adversaries—policies no less steady in their purpose, and no less variegated and resourceful in their application, than those of the Soviet Union itself.

In these circumstances it is clear that the main element of any United States policy toward the Soviet Union must be that of a long-term, patient but firm and vigilant containment of Russian expansive tendencies. It is important to note, however, that such a policy has nothing to do with outward histrionics: with threats or blustering or superfluous gestures of outward "toughness." While the Kremlin is basically flexible in its reaction to political realities, it is by no means unamenable to considerations of prestige. Like almost any other government, it can be placed by tactless and threatening gestures in a position where it cannot afford to yield even though this might be dictated by its sense of realism. The Russian leaders are keen judges of human psychology, and as such they are highly conscious that loss of temper and of self-control is never a source of strength in political affairs. They are quick to exploit such evidences of weakness. For these reasons, it is a *sine qua non* of successful dealing with Russia that the foreign government in question should remain at all times cool and collected and that its demands on Russian policy should be put forward in such a manner as to leave the way open for a compliance not too detrimental to Russian prestige.

III

In the light of the above, it will be clearly seen that the Soviet pressure against the free institutions of the western world is something that can be contained by the adroit and vigilant application of counterforce at a series of constantly shifting geographical political points, corresponding to the shifts and maneuvers of Soviet policy, but which cannot be charmed or talked out of existence. The Russians look forward to a duel of infinite duration, and they see that already they have scored great successes. It must be borne in mind that there was a time when the Communist party represented far more of a minority in the sphere of Russian national life than Soviet power today represents in the world community.

But if ideology convinces the rulers of Russia that truth is on their side and that they can therefore afford to wait, those of us on whom that ideology has no claim are free to examine objectively the validity of that premise. The Soviet thesis not only implies complete lack of control by the West over its economic destiny; it likewise assumes Russian unity, discipline, and patience over an infinite period. Let us bring this apocalyptic vision down to earth, and suppose that the Western world finds the strength and resourcefulness to contain Soviet power over a period of 10 to 15 years. What does that spell for Russia itself?

The Soviet leaders, taking advantage of the contributions of modern technique to the arts of despotism, have solved the question of obedience within the confines of their power. Few challenge their authority; and even those who do are unable to make that challenge valid as against the organs of suppression of the state.

The Kremlin has also proved able to accomplish its purpose of building up in Russia, regardless of the interests of the inhabitants, an industrial foundation of heavy metallurgy, which is, to be sure, not yet complete but which is nevertheless continuing to grow and is approaching those of the other major industrial countries. All of this, however, both the maintenance of internal political security and the building of heavy industry, has been carried out at a terrible cost in human life and in human hopes and energies. It has necessitated the use of forced labor on a scale unprecedented in modern times under conditions of peace. It has involved the neglect or abuse of other phases of Soviet economic life, particularly agriculture, consumers' goods production, housing, and transportation.

To all that, the war has added its tremendous toll of destruction, death, and human exhaustion. In consequence of this, we have in Russia today a population which is physically and spiritually tired. The mass of the people are disillusioned, skeptical, and no longer as accessible as they once were to the magical attraction which Soviet power still radiates to its followers abroad. The avidity with which people seized upon the slight respite accorded to the Church for tactical reasons during the war was eloquent testimony to the fact that their capacity for faith and devotion found little expression in the purposes of the regime.

In these circumstances, there are limits to the physical and nervous strength of people themselves. These limits are absolute ones, and are binding even for the cruelest dictatorship, because beyond them people cannot be driven. The forced labor camps and the other agencies of constraint provide temporary means of compelling people to work longer hours than their own volition or mere economic pressure would dictate; but if people survive them at all they become old before their time and must be considered as human casualties to the demands of dictatorship. In either case their best powers are no longer available to society and can no longer be enlisted in the service of the state.

Here only the younger generation can help. The younger generation, despite all vicissitudes and sufferings, is numerous and vigorous; and the Russians are a talented people. But it still remains to be seen what will be the effects on mature performance of the abnormal emotional strains of childhood which Soviet dictatorship created and which were enormously increased by the war. Such things as normal security and placidity of home environment have practically ceased to exist in the Soviet Union outside of the most remote farms and villages. And observers are not yet sure whether that is not going to leave its mark on the overall capacity of the generation now coming into maturity.

In addition to this, we have the fact that Soviet economic development, while it can list certain formidable achievements, has been precariously spotty and uneven. Russian Communists who speak of the "uneven development of capitalism" should blush at the contemplation of their own national economy. Here certain branches of economic life, such as the metallurgical and machine industries, have been pushed out of all proportion to other sectors of the economy. Here is a nation striving to become in a short period one of the great industrial nations of the world while it still has no highway network worthy of the name and only a relatively primitive network of railways. Much has been done to increase efficiency of labor and to teach primitive peasants something about the operation of machines. But maintenance is still a crying deficiency of all Soviet economy. Construction is hasty and poor in quality. Depreciation must be enormous. And in vast sectors of economic life it has not yet been possible to instill into labor anything like that general culture of production and technical self-respect which characterizes the skilled worker of the West.

It is difficult to see how these deficiencies can be corrected at an early date by a tired and dispirited population working largely under the shadow of fear and compulsion. And as long as they are not overcome, Russia will remain economically a vulnerable, and in a certain sense an impotent, nation, capable of exporting its enthusiasms and of radiating the strange charm of its primitive political vitality but unable to back up those articles of export by the real evidences of material power and prosperity.

Meanwhile, a great uncertainty hangs over the political life of the Soviet Union. That is the uncertainty involved in the transfer of power from one individual or group of individuals to others.

This is, of course, outstandingly the problem of the personal position of Stalin. We must remember that his succession to Lenin's pinnacle of preeminence in the Communist movement was the only such transfer of individual authority which the Soviet Union has experienced. That transfer took 12 years to consolidate. It cost the lives of millions of people and shook the state to its foundations. The attendant tremors were felt all through the international revolutionary movement, to the disadvantage of the Kremlin itself.

It is always possible that another transfer of preeminent power may take place quietly and inconspicuously, with no repercussions anywhere. But again, it is possible that the questions involved may unleash, to use some of Lenin's words, one of those "incredibly swift transitions" from "delicate deceit" to "wild violence" which characterize Russian history, and may shake Soviet power to its foundations.

But this is not only a question of Stalin himself. There has been, since 1938, a dangerous congealment of political life in the higher circles of Soviet power. The All-Union Congress of Soviets, in theory the supreme body of the Party, is supposed to meet not less often than once in three years. It will soon be eight full years since its last meeting. During this period membership in the Party has numerically doubled. Party mortality during the war was enormous; and today well over half of the Party members are persons who have entered since the last Party congress was held. Meanwhile, the same small group of men has carried on at the top through an amazing series of national vicissitudes. Surely there is some reason why the experiences of the war brought basic political changes to every one of the great governments of the West. Surely the causes of that phenomenon are basic enough to be present somewhere in the obscurity of Soviet political life, as well. And yet no recognition has been given to these causes in Russia.

It must be surmised from this that even within so highly disciplined an organization as the Communist party there must be a growing divergence in age, outlook and interest between the great mass of Party members, only so recently recruited into the movement, and the little self-perpetuating clique of men at the top, whom most of these Party members have never met, with whom they have never conversed, and with whom they can have no political intimacy.

Who can say whether, in these circumstances, the eventual rejuvenation of the higher spheres of authority (which can only be a matter of time) can take place smoothly and peacefully, or whether rivals in the quest for higher power will not eventually reach down into these politically immature and inexperienced masses in order to find support for their respective claims? If this were ever to happen, strange consequences could flow for the Communist party: for the membership at large has been exercised only in the practices of iron discipline and obedience and not in the arts of compromise and accommodation. And if disunity were ever to seize and paralyze the Party, the chaos and weakness of Russian society would be revealed in forms beyond description. For we have seen that Soviet power is only a crust concealing an amorphous mass of human beings among whom no independent organizational structure is tolerated. In Russia there is not even such a thing as local government. The present generation of Russians have never known spontaneity of collective action. If, consequently, anything were ever to occur to disrupt the unity and efficacy of the Party as a political instrument, Soviet Russia might be changed overnight from one of the strongest to one of the weakest and most pitiable of national societies.

Thus the future of Soviet power may not be by any means as secure as Russian capacity for self-delusion would make it appear to the men in the Kremlin. That they can keep power themselves, they have demonstrated. That they can quietly and easily turn it over to others remains to be proved. Meanwhile, the hardships of their rule and the vicissitudes of international life have taken a heavy toll of the strength and hopes of the great people on whom their power rests. It is curious to note that the ideological power of Soviet authority is strongest today in areas beyond the frontiers of Russia, beyond the reach of its police power. This phenomenon brings to mind a comparison used by Thomas Mann in his great novel *Buddenbrooks*. Observing that human institutions often show the greatest outward brilliance at a moment when inner decay is in reality farthest advanced, he compared the Buddenbrook family, in the days of its greatest glamour, to one of those stars whose light shines most brightly on this world when in reality it has long since ceased to exist. And who can say with assurance that the strong light still cast by the Kremlin on the dissatisfied peoples of the Western world is not the powerful afterglow of a constellation which is in actuality on the wane? This cannot be proved. And it cannot be disproved. But the possibility remains (and in the opinion of this writer it is a strong one) that Soviet power, like the capitalist world of its conception, bears within it the seeds of its own decay, and that the sprouting of these seeds is well advanced.

IV

It is clear that the United States cannot expect in the foreseeable future to enjoy political intimacy with the Soviet regime. It must continue to regard the Soviet Union as a rival, not a partner, in the political arena. It must continue to expect that Soviet policies will reflect no abstract love of peace and stability, no real faith in the possibility of a permanent happy coexistence of the Socialist and capitalist worlds, but rather a cautious, persistent pressure toward the disruption and weakening of all rival influence and rival power.

Balanced against this are the facts that Russia, as opposed to the Western world in general, is still by far the weaker party, that Soviet policy is highly flexible, and that Soviet society may well contain deficiencies which will eventually weaken its own total potential. This would of itself warrant the United States' entering with reasonable confidence upon a policy of firm containment, designed to confront the Russians with unalterable counterforce at every point where they show signs of encroaching upon the interests of a peaceful and stable world.

But in actuality the possibilities for American

policy are by no means limited to holding the line and hoping for the best. It is entirely possible for the United States to influence by its actions the internal developments, both within Russia and throughout the international Communist movement, by which Russian policy is largely determined. This is not only a question of the modest measure of informational activity which this government can conduct in the Soviet Union and elsewhere, although that, too, is important. It is rather a question of the degree to which the United States can create among the peoples of the world generally the impression of a country which knows what it wants, which is coping successfully with the problems of its internal life and with the responsibilities of a world power, and which has a spiritual vitality capable of holding its own among the major ideological currents of the time. To the extent that such an impression can be created and maintained, the aims of Russian Communism must appear sterile and quixotic, the hopes and enthusiasm of Moscow's supporters must wane, and added strain must be imposed on the Kremlin's foreign policies. For the palsied decrepitude of the capitalist world is the keystone of Communist philosophy. Even the failure of the United States to experience the early economic depression which the ravens of the Red Square have been predicting with such complacent confidence since hostilities ceased would have deep and important repercussions throughout the Communist world.

By the same token, exhibitions of indecision, disunity, and internal disintegration within this country have an exhilarating effect on the whole Communist movement. At each evidence of these tendencies, a thrill of hope and excitement goes through the Communist world; a new jauntiness can be noted in the Moscow tread; new groups of foreign supporters climb on to what they can only view as the band wagon of international politics; and Russian pressure increases all along the line in international affairs.

It would be an exaggeration to say that American behavior unassisted and alone could exer-cise a power of life and death over the Communist movement and bring about the early fall of Soviet power in Russia. But the United States has it in its power to increase enormously the strains under which Soviet policy must operate, to force upon the Kremlin a far greater degree of moderation and circumspection than it has had to observe in recent years, and in this way to promote tendencies which must eventually find their outlet in either the breakup or the gradual mellowing of Soviet power. For no mystical, messianic movement—and particularly not that of the Kremlin—can face frustration indefinitely without eventually adjusting itself in one way or another to the logic of that state of affairs.

Thus the decision will really fall in large measure in this country itself. The issue of Soviet-American relations is in essence a test of the overall worth of the United States as a nation among nations. To avoid destruction the United States need only measure up to its own best traditions and prove itself worthy of preservation as a great nation.

Surely, there was never a fairer test of national quality than this. In the light of these circumstances, the thoughtful observer of Russian-American relations will find no cause for complaint in the Kremlin's challenge to American society. The observer will rather experience a certain gratitude to a Providence which, by providing the American people with this implacable challenge, has made this entire security as a nation dependent on their pulling themselves together and accepting the responsibilities of moral and political leadership that history plainly intended them to bear.

NOTES

1. "Concerning the Slogans of the United States of Europe," Aug. 1915. Official Soviet edition of Lenin's works.
2. Here and elsewhere in this paper "Socialism" refers to Marxist or Leninist Communism, not to liberal Socialism of the Second International variety.

US NATIONAL SECURITY OBJECTIVES

NSC-68 (1950)

BACKGROUND ON THE PRESENT CRISIS

Within the past 35 years, the world has experienced two global wars of tremendous vio-lence. It has witnessed two revolutions—the Russian and the Chinese—of extreme scope and intensity. It has also seen the collapse of five empires—the Ottoman, the Austro-Hungarian,

---

Excerpts from NSC-68 (no longer classified) and an official response (no longer classified) from William F. Schaub are reprinted here with minor revisions.

German, Italian, and Japanese—and the drastic decline of two major imperial systems, the British and the French. During the span of one generation, the international distribution of power has been fundamentally altered. For several centuries it had proved impossible for any one nation to gain such preponderant strength that a coalition of other nations could not in time face it with greater strength. The international scene was marked by recurring periods of violence and war, but a system of sovereign and independent states was maintained, over which no state was able to achieve hegemony.

Two complex sets of factors have now basically altered this historical distribution of power. First, the defeat of Germany and Japan and the decline of the British and French Empires have interacted with the development of the United States and the Soviet Union in such a way that power has increasingly gravitated to these two centers. Second, the Soviet Union, unlike previous aspirants to hegemony, is animated by a new fanatic faith, antithetical to our own, and seeks to impose its absolute authority over the rest of the world. Conflict has, therefore, become endemic and is waged, on the part of the Soviet Union, by violent or nonviolent methods in accordance with the dictates of expediency. With the development of increasingly terrifying weapons of mass destruction, every individual faces the ever-present possibility of annihilation should the conflict enter the phase of total war.

On the one hand, the people of the world yearn for relief from the anxiety arising from the risk of atomic war. On the other hand, any substantial further extension of the area under the domination of the Kremlin would raise the possibility that no coalition adequate to confront the Kremlin with greater strength could be assembled. It is in this context that this Republic and its citizens in the ascendancy of their strength stand in their deepest peril . . .

THE UNDERLYING CONFLICT IN THE REALM OF IDEAS AND VALUES BETWEEN THE US PURPOSE AND THE KREMLIN DESIGN

Objectives

. . . In a shrinking world, which now faces the threat of atomic warfare, it is not an adequate objective merely to seek to check the Kremlin design, for the absence of order among nations is becoming less and less tolerable. This fact imposes on us, in our own interests, the responsibility of world leadership. It demands that we make the attempt, and accept the risks inherent in it, to bring about order and justice by means consistent with the principles of freedom and democracy. We should limit our requirement of the Soviet Union to its participation with other nations on the basis of equality and respect for the rights of others. Subject to this requirement, we must with our allies and the former subject peoples seek to create a world society based on the principle of consent. Its framework cannot be inflexible. It will consist of many national communities of great and varying abilities and resources, and hence of war potential. The seeds of conflict will inevitably exist or will come into being. To acknowledge this is only to acknowledge the impossibility of a final solution. Not to acknowledge it can be fatally dangerous in a world in which there are no final solutions.

All these objectives of a free society are equally valid and necessary in peace and war. But every consideration of devotion to our fundamental values and to our national security demands that we seek to achieve them by the strategy of the cold war. It is only by developing the moral and material strength of the free world that the Soviet regime will become convinced of the falsity of its assumptions and that the preconditions for workable agreements can be created. By practically demonstrating the integrity and vitality of our system the free world widens the area of possible agreement and thus can hope gradually to bring about a Soviet acknowledgment of realities which in sum will eventually constitute a frustration of the Soviet design. Short of this, however, it might be possible to create a situation which will induce the Soviet Union to accommodate itself, with or without the conscious abandonment of its design, to coexistence on tolerable terms with the non-Soviet world. Such a development would be a triumph for the idea of freedom and democracy. It must be an immediate objective of United States policy.

There is no reason, in the event of war, for us to alter our overall objectives. They do not include unconditional surrender, the subjugation of the Russian peoples or a Russia shorn of its economic potential. Such a course would irrevocably unite the Russian people behind the regime which enslaves them. Rather these objectives contemplate Soviet acceptance of the specific and limited conditions requisite to an international environment in which free institutions can flourish, and in which the Russian peoples will have a new chance to work out their own destiny. If we can make the Russian people our allies in the enterprise we will obviously have made our task easier and victory more certain . . .

Means

Practical and ideological considerations therefore both impel us to the conclusion that we have no choice but to demonstrate the superiority of the idea of freedom by its constructive application, and to attempt to change the world situation by means short of war in such a way as to frustrate the Kremlin design and hasten the decay of the Soviet system . . .

But if war comes, what is the role of force? Unless we so use it that the Russian people can perceive that our effort is directed against the regime and its power for aggression, and not against their own interests, we will unite the regime and the people in the kind of last-ditch fight in which no underlying problems are solved, new ones are created, and where our basic principles are obscured and compromised. If we do not in the application of force demonstrate the nature of our objectives we will, in fact, have compromised from the outset our fundamental purpose. In the words of the Federalist (No. 28), "The means to be employed must be proportioned to the extent of the mischief." The mischief may be a global war or it may be a Soviet campaign for limited objectives. In either case we should take no avoidable initiative which would cause it to become a war of annihilation, and if we have the forces to defeat a Soviet drive for limited objectives it may well be to our interest not to let it become a global war. Our aim in applying force must be to compel the acceptance of terms consistent with our objectives, and our capabilities for the application of force should, therefore, within the limits of what we can sustain over the long pull, be congruent to the range of tasks which we may encounter.

SOVIET INTENTIONS AND CAPABILITIES

Military

. . . We do not know accurately what the Soviet atomic capability is, but the Central Intelligence Agency intelligence estimates, concurred in by State, Army, Navy, Air Force, and Atomic Energy Commission, assign to the Soviet Union a production capability giving it a fission bomb stockpile within the following ranges:

By mid-1950	10–20
By mid-1951	25–45
By mid-1952	45–90
By mid-1953	70–135
By mid-1954	200

This estimate is admittedly based on incomplete coverage of Soviet activities and represents the production capabilities of known or deducible Soviet plants. If others exist, as is possible, this estimate could lead us into a feeling of superiority in our atomic stockpile that might be dangerously misleading, particularly with regard to the timing of a possible Soviet offensive. On the other hand, if the Soviet Union experiences operating difficulties, this estimate would be reduced. There is some evidence that the Soviet Union is acquiring certain materials essential to research on and development of thermonuclear weapons.

The Soviet Union now has aircraft able to deliver the atomic bomb. Our intelligence estimates assign to the Soviet Union an atomic bomber capability already in excess of that needed to deliver available bombs. We have at present no evaluated estimate regarding the Soviet accuracy of delivery on target. It is believed that the Soviets cannot deliver their bombs on target with a degree of accuracy comparable to ours, but a planning estimate might well place it at 40 to 60 percent of bombs sortied. For planning purposes, therefore, the date the Soviets possess an atomic stockpile of 200 bombs would be a critical date for the United States, for the delivery of 100 atomic bombs on targets in the United States would seriously damage this country . . .

US INTENTIONS AND CAPABILITIES— ACTUAL AND POTENTIAL

Political and Psychological

. . . As for the policy of "containment," it is one which seeks by all means short of war to (1) block further expansion of Soviet power, (2) expose the falsities of Soviet pretensions, (3) induce a retraction of the Kremlin's control and influence and (4) in general, so foster the seeds of destruction within the Soviet system that the Kremlin is brought at least to the point of modifying its behavior to conform to generally accepted international standards.

It was and continues to be cardinal in this policy that we possess superior overall power in ourselves or in dependable combination with other like-minded nations. One of the most important ingredients of power is military strength. In the concept of "containment," the maintainance of a strong military posture is deemed to be essential for two reasons: (1) as an ultimate guarantee of our national security and (2) as an indispensable backdrop to the conduct of the policy of "containment." Without superior aggregate military strength, in being and readily mobilizable, a policy of "containment"—which is in effect a policy of calculated and gradual coercion—is no more than a policy of bluff.

At the same time, it is essential to the successful conduct of a policy of "containment" that we always leave open the possibility of negotiation with the USSR. A diplomatic freeze—and we are in one now—tends to defeat the very purposes of "containment" because it raises tensions at the same time that it makes Soviet retractions and adjustments in the direction of moderated behavior more difficult. It also tends to inhibit our initiative and deprives us of opportunities for maintaining the moral ascendency in our struggle with the Soviet system.

In "containment" it is desirable to exert pressure in a fashion which will avoid so far as possible directly challenging Soviet prestige, to keep open the possibility for the USSR to retreat before pressure with a minimum loss of face and to secure political advantage from the failure of the Kremlin to yield or take advantage of the openings we leave it.

We have failed to implement adequately these two fundamental aspects of "containment." In the face of obviously mounting Soviet military strength ours has declined relatively. Partly as a byproduct of this, but also for other reasons, we now find ourselves at a diplomatic impasse with the Soviet Union, with the Kremlin growing bolder, with both of us holding on grimly to what we have and with ourselves facing different decisions . . .

ATOMIC ARMAMENTS

Stockpiling and Use of Atomic Weapons

1. From the foregoing analysis it appears that it would be to the long-term advantage of the United States if atomic weapons were to be effectively eliminated from peacetime armaments; the additional objectives which must be secured if there is to be a reasonable prospect of such effective elimination of atomic weapons are discussed in chapter 9. In the absence of such elimination and the securing of these objectives, it would appear that we have no alternative but to increase our atomic capability as rapidly as other considerations make appropriate. In either case, it appears to be imperative to increase as rapidly as possible our general air, ground and sea strength and that of our allies to a point where we are militarily not so heavily dependent on atomic weapons . . .

. . . 3. It has been suggested that we announce that we will not use atomic weapons except in retaliation against the prior use of such weapons by an aggressor. It has been argued that such a declaration would decrease the danger of an atomic attack against the United States and its allies.

In our present situation of relative unpreparedness in conventional weapons, such a declaration would be interpreted by the USSR as an admission of great weakness and by our allies as a clear indication that we intended to abandon them. Furthermore, it is doubtful whether such a declaration would be taken sufficiently seriously by the Kremlin to constitute an important factor in determining whether or not to attack the United States. It is to be anticipated that the Kremlin would weigh the facts of our capability far more heavily than a declaration of what we proposed to do with that capability.

Unless we are prepared to abandon our objectives, we cannot make such a declaration in good faith until we are confident that we will be in a position to attain our objectives without war, or, in the event of war, without recourse to the use of atomic weapons for strategic or tactical purposes.

International Control of Atomic Energy

. . . The above considerations make it clear that at least a major change in the relative power positions of the United States and the Soviet Union would have to take place before an effective system of international control could be negotiated. The Soviet Union would have had to have moved a substantial distance down the path of accommodation and compromise before such an arrangement would be conceivable. This conclusion is supported by the third report of the United Nations Atomic Energy Commission to the Security Council, 17 May 1948, in which it is stated that "the majority of the Commission has been unable to secure . . . their acceptance of the nature and extent of participation in the world community required of all nations in this field . . . As a result, the Commission has been forced to recognize that agreement on effective measures for the control of atomic energy is itself dependent on cooperation in broader fields of policy." . . .

POSSIBLE COURSES OF ACTION

The Role of Negotiation

. . . A sound negotiation position is, therefore, an essential element in the ideological conflict. For some time after a decision to build up strength, any offer of, or attempt at, negotiation of a general settlement along the lines of the Berkeley speech by the secretary of state could only be a tactic. Nevertheless, concurrently with a decision and a start on building up the strength of the free world, it may be desirable to pursue this tactic both to gain public support for the program and to minimize the immediate risks of war. It is urgently necessary for the United States to determine its negotiating position and to obtain agreement with its major allies on the purposes and terms of negotiation . . .

. . . This seems to indicate that for the time being the United States and other free countries would have to insist on concurrent agreement on the control of nonatomic forces and weapons and perhaps on the other elements of a general settlement, notably peace treaties with Germany, Austria, and Japan and the withdrawal of Soviet influence from the satellites. If, contrary to our expectations, the Soviet Union should accept agreements promising effective control of atomic energy and conventional armaments, without any other changes in Soviet policies, we would have to consider very carefully whether we could accept such agreements. It is unlikely that this problem will arise . . .

The Remaining Course of Action—A Rapid Buildup of Political, Economic, and Military Strength in the Free World

. . . Moreover, the United States and the other free countries do not now have the forces in being and readily available to defeat local Soviet moves with local action, but must accept reverses or make these local moves the occasion for war—for which we are not prepared. This situation makes for great uneasiness among our allies, particularly in Western Europe, for whom total war means, initially, Soviet occupation. Thus, unless our combined strength is rapidly increased, our allies will tend to become increasingly reluctant to support a firm foreign policy on our part and increasingly anxious to seek other solutions, even though they are aware that appeasement means defeat. An important advantage in adopting the fourth course of action lies in its psychological impact—the revival of confidence and hope in the future. It is recognized, of course, that any announcement of the recommended course of action could be exploited by the Soviet Union in its peace campaign and would have adverse psychological effects in certain parts of the free world until the necessary increase in strength had been achieved. Therefore, in any announcement of policy and in the character of the measures adopted, emphasis should be given to the essentially defensive character, and care should be taken to minimize, so far as possible, unfavorable domestic and foreign reactions.

Political and Economic Aspects. The immediate objectives—to the achievement of which such a buildup of strength is a necessary though not a sufficient condition—are a renewed initiative in the cold war and a situation to which the Kremlin would find it expedient to accommodate itself, first by relaxing tensions and pressures and then by gradual withdrawal . . .

. . . The threat to the free world involved in the development of the Soviet Union's atomic and other capabilities will rise steadily and rather rapidly. For the time being, the United States possesses a marked atomic superiority over the Soviet Union which, together with the potential capabilities of the United States and other free countries in other forces and weapons, inhibits aggressive Soviet action. This provides an opportunity for the United States, in cooperation with other free countries, to launch a buildup of strength which will support a firm policy directed to the frustration of the Kremlin design. The immediate goal of our efforts to build a successfully functioning political and economic system in the free world backed by adequate military strength is to postpone and avert the disastrous situation which, in light of the Soviet Union's probable fission bomb capability and possible thermonuclear bomb capability, might arise in 1954 on a continuation of our present programs. By acting promptly and vigorously in such a way that this date is, so to speak, pushed into the future, we would permit time for the process of accommodation, withdrawal and frustration to produce the necessary changes in the Soviet system. Time is short, however, and the risks of war attendant upon a decision to build up strength will steadily increase the longer we defer it.

COMMENTS OF THE BUREAU OF THE BUDGET ON NSC-68

TOP SECRET memorandum by the Deputy Chief of the Division of Estimates, Bureau of the Budget, William F. Schaub to the Executive Secretary of the National Security Council (Lay). [Washington] 8 May 1950.

WHAT, SPECIFICALLY, DOES THE PAPER MEAN?

Military

a. Do we anticipate that Russia will strike in 1954 and we should prepare to mobilize by that date?

If so, do we prepare the country and organize all of our resources to meet that contingency? This would require wartime controls in this country and be tantamount to notifying Russia that we intended to press war in the near future. Would this force Russia to retreat from the satellite countries and other areas of influence, or would it force them to take direct military action on them? Is this the kind of national policy that we want to present to the world? What relative emphasis do we place on the abilities of our allies and the rest of the free world?

b. Do we anticipate that Russia will be *sufficiently* capable of successfully attacking the US by 1954 to require us to have a program for complete preparation for defending the US and successfully striking back and delaying Russian advances to permit our mobilization and the maintainance of advance positions in Europe, Africa, the Near East, and other strategic areas?

If so, do we prepare the country to accept limited controls and increased taxation? What would be the effect on our relations with Russia?

Our allies? The rest of the world? To what extent do we rely on the abilities of our allies and the rest of the free world?

c. Do we estimate that Russia's strength is increasing to a point which is dangerous to our security and that the US should improve its own defense and attack capabilities and those of our allies in order to keep pace with Russia's increasing strength?

This would probably mean a rounding out and firming up of our military structure and could probably be done without domestic controls and would probably not create a much greater fiscal and economic problem than now exists.

d. Do we want to change the trend towards economy at the expense of national security programs and present a firmer and stronger military posture accompanied by a more intensive program for approaching our international problems?

General

1. What is the "sharp disparity between our actual military strength and our commitments"? What are our commitments?

2. Do we have a so-called "war plan" or "mobilization plan"? If not, what is being done to develop one? Is such planning being related to the potential strength being planned for our allies? How is such planning being related to peacetime forces and equipment, current procurement and training programs and war reserve materiel? Are industrial production facilities being related to planned requirements?

3. At what point do we intend to use military force to protect our "basic values"? What authority, short of a declaration of war, do we have for using force? Should our resources go to assist the preparation of our allies and other "fringe" countries as a first priority? Do we move ahead building forces in allied countries without regard to their ability to maintain them on a continuing basis, thus requiring our assistance indefinitely?

4. If our danger is from Soviet influence on vulnerable segments of society—generally large masses of subjugated, uneducated peoples—what is our program to reach these masses and prevent Soviet influence? How do you promise them and insure for them a chance for freedom and improvement?

Our policies in the past have armed our enemies. How do we insure against this in the future?

POLITICAL AND PSYCHOLOGICAL

A. NSC-68 emphasizes "the present polarization of power" to an extent which underemphasizes the fact that, while the two "poles" (US and USSR) are each possessed of great power, each is dangerous to the other only to the extent that it can attract and keep allies.

Would not an all-out program for civil defense and military defense of this country with all that it entails in stirring up public opinion and support tend to defeat our objectives with our allies?

This would appear to be an important weakness of NSC-68.

B. Throughout NSC-68 appear such statements as "The idea of freedom is the most contagious idea in history, more contagious than the idea of submission to authority"; "The greatest vulnerability of the Kremlin lies in the basic nature of its relations with the Soviet people"; "The Kremlin's relations with its satellites and their peoples are likewise a vulnerability."

These statements reach toward the core of the problem dealt with by NSC-68, yet reference to policies and programs in the ideological war or war for men's minds is subordinated to programs of material strength; in fact, the only program dealt with in any detail is the military program.

NSC-68 deals with this problem as being one involving "the free world" and "the slave world." While it is true that the USSR and its satellites constitute something properly called a slave world, it is not true that the US and its friends constitute a free world. Are the Indochinese free? Can the peoples of the Philippines be said to be free under the corrupt Quirino government? Moreover, what of the vast number of peoples who are in neither the US nor the USSR camp, and for whom we are contesting? By and large, by our standards, they are not free. The free world versus slave world treatment obscures one of the most difficult problems we face—the fact that many peoples are attracted to Communism because their governments are despotic or corrupt or both. And they are not going to become the friends of a major power simply because of that power's military strength. Rather, their friendship is to be had at the price of support of moves which will improve or, failing that, replace their present governments.

Finally, the point that is touched upon in NSC-68 and then lost sight of in preoccupation with the USSR itself, is that were it not for the recent successes and possible further successes of the Russian-controlled international Communist movement, we would have small reason to fear the imperialism of the USSR. To illustrate: The US is stronger militarily and economically in relation to the USSR than was the case just before World War II. We hardly gave Russia a second thought then. What makes for the difference today? A most important difference is that today many peoples are striving actively to better themselves economically and politically and have thus accepted or are in danger of accepting the leadership of the Communist movement.

Just what types of political and psychological actions have we proposed to meet this situation?

C. NSC-68 is based on the assumption that the military power of the USSR and its satellites is increasing in relation to that of the US and its allies. In view of the vast preponderance of US and allied assets in every respect except that of manpower that assumption needs more documentation than is contained in NSC-68. In particular no attention seems to have been given to the question of the possible drain which recent developments may have placed on Soviet military strength. Tightening of controls at home and in particular in the satellites would tie down military manpower and equipment. The furnishing of military technicians to China in any number would constitute an important drain on the USSR whose supply is relatively limited. Put another way, it is hard to accept a conclusion that the USSR is approaching a straight-out military superiority over us when, for example, (1) our Air Force is vastly superior qualitatively, is greatly superior numerically in the bombers, trained crews, and other facilities necessary for offensive warfare; (2) our supply of fission bombs is much greater than that of the USSR, as is our thermonuclear potential; (3) our navy is so much stronger than that of the USSR that they should not be mentioned in the same breath; (4) The economic health and military potential of our allies is, with our help, growing daily; and (5) while we have treaties of alliance with and are furnishing arms to countries bordering on the USSR, the USSR has none with countries within thousands of miles of us.

SUPPLEMENTARY BUDGET COMMENTS ON NSC-68

These comments are directed primarily at the nonmilitary aspects of the document. There is an inadequate description of objectives and means, and a failure to assess—or to make possible an assessment of—the implications of the proposed courses of action. It is not enough to say that objectives should be adopted and then their implementation spelled out, since the objectives are so general that they cannot be given meaningful content except in more substantive terms.

DISCUSSION OF "THE UNDERLYING CONFLICT"

This section of the paper lays an unsound basis for the document as a whole. The neat dichotomy between "freedom" and "slavery" is not a realistic description either of the situation today or of the alternatives as they appear to present themselves to large areas of the world. There are diverse types and degrees of freedom and slavery, and it is doubtful that the extent of hegemony of the United States or even the extent of national independence is considered the predominant measure by many peoples.

To classify as "free" all those peoples whose governments oppose Russia, or we seek to have oppose Russia, is a travesty on the word. Freedom as we know it is a highly developed concept, frequently of little meaning and less use in dealing with backward or disorganized peoples. The most potent weapon of the Russians outside of Eastern Europe has been and is revolt against social and economic as well as political inequities. To think of freedom in primarily political terms is itself grossly inadequate. But to imply—as this report seems to do, despite occasional references of a broader nature—that its most important meaning today is the simple ability to preserve national existence, is a highly dangerous matter. An upsurge of unadulterated nationalism might for the time being lessen or remove the military threat of Russia, but it would over time tend to accentuate the subtle undermining of our own system and guarantee the eventual loss of the cold war through the proliferation and subsidization of unstable little tyrants.

The gravest error of NSC-68 is that it vastly underplays the role of economic and social change as a factor in "the underlying conflict." Tyranny is not new or strange, even on the Russian scale and manner; nor is it unusual for tyranny to ride the crest of swelling social and economic pressures, as the Russians are successfully doing in many parts of the world. The test of survival for an established civilization is its ability, not only to defend itself in a military sense, but also to handle these pressures by removing or alleviating the causes—a most difficult task of adjustment since it frequently requires removal of ruling groups or injury to vested interests. One might generalize that the degree of underlying success in the cold war to date has been in direct ratio to the success in adjusting social and economic structures to the twentieth century wave of economic egalitarianism—even though the methods have frequently been inept and have violated our concepts of a desirable and efficient economic system.

These adjustments are not being made in any of the critical areas of the world today. We are being increasingly forced into associations which are exceedingly strange for a people of our heritage and ideals. It can be persuasively argued that there is no alternative course. If so, we should not be blind to the gaping weakness which is forced upon us, which will grow rather than decline as time passes, and of which above all others the Russians, with their talents for subversion, are able to take advantage. This is a major dilemma of American foreign policy, and deals with a subject much more difficult than making guns. In many countries today, for example, there is a simple test question: Is there no way to attain thoroughgoing land reform except through Communist revolution? It is highly doubtful that we are actually so handicapped in our choice of friends or limited in our influence on policies. At any rate, we will never make use of our opportunities as long as the issue is submerged, as it is in NSC-68. Indeed, we seem today to be exerting decisive influence in the wrong direction in some places, such as Western Germany.

The above comments do not detract from the seriousness of the military situation, nor necessarily weaken the case for increasing and reorienting our military strength and for assisting other countries to defend themselves. But unless we are prepared to undertake extensive military occupation, we cannot win the cold war by a predominant reliance on military force even if combined with large-scale dollar assistance. Nor is it sufficient to add preachments of the concepts of democracy in terms too sophisticated for understanding or too remote from the particular issues foremost in the minds of the peoples. Only as we develop methods for capitalizing on the emerging social pressures can we beat the Russians at their most dangerous game and safely take advantage of a rising tide of nationalism.

A revealing commentary on NSC-68 is that it does not basically clarify or utilize the Chinese experience in the discussion of issues and risks, nor does it point toward a course of action which can effectively deal with probable repetitions of that experience in the future. There is no follow-through on the social and economic schisms which today provide the basic ground swell for disorder and weakness, which make our task so difficult, and for which we have not developed guidelines and techniques adequate to cope with the vicious "ideological pretensions" and methods of the Communists. A revolutionary movement taking advantage, however cynically, of real elements of dissatisfaction cannot be stopped by the threat of force alone.

DISCUSSION OF ECONOMIC FACTORS

Lacking any indication of the magnitude of the proposed increase in security expenditures, it is impossible to assess the economic impact of this document and the economic risks which it might involve. There is no doubt that a larger share of resources could be devoted to security purposes, but such a course is not without its cost under any circumstances, and the extent of diversion is crucial to an analysis of consequences.

The comparison of the present situation with

that of the peak of World War II is misleading. Apart from statistical difficulties in computing GNP in wartime on a basis comparable to peacetime, the effort achieved in 1944 was possible only under wartime conditions, with widespread controls, heavy deterioration in many types of capital assets, and bulging inflationary pressure subject only to short-range restraints. Under a total war effort the US might, in time and barring internal destruction, exceed its World War II performance, but this effort would not be sustained for a long period and is hardly relevant to the task of a long drawn-out cold war.

Unless the risks of war are considered sufficiently grave to require moving now toward large-scale mobilization, determination of the size of our military posture should be heavily influenced by its sustainability over an indefinite period and by a balancing of the military risks to our society and to the prospects for economic growth. Expansion of military expenditures involves an economic cost, particularly if sustained for a substantial period, and it also involves a cost in terms of the psychology and orientation of our society. This is always true, and temporary factors such as unemployment should not be permitted to obscure the issue.

At the moment there are some 3.5 million unemployed and certain industries are operating below capacity. However, at present levels of activity there are signs of inflationary pressure, particularly in heavy industries and construction. It would be difficult to conclude categorically that under current conditions substantial further armament demands could be placed upon durable goods industries without requiring a diversion from present civilian purposes either through inflation or through taxes or direct controls. The result might be little or no net increase in total output depending upon the methods used. It is thus necessary to assess the impact of increased security expenditures on specific sectors of the economy as well as in terms of aggregates.

More importantly, over a period of time, it is neither necessary nor desirable to regard military expenditures per se as a method of maintaining high employment. Large and growing military expenditures not only would divert resources from the civilian purposes to which they should be put but also would have more subtle effects on our economic system. Higher taxes, if necessary, would have a proportional dampening effect on incentives and on the dynamic nature of the economy, without any offsetting productive impact from the expenditures. The rate of private investment might be slowed down unless special measures or controls were undertaken. There would be a continuing tendency to reduce public expenditures for developmental purposes which are highly desirable for the continual strengthening of our economy.

The document gives figures indicating a much higher investment rate in Russia than in the US at the present time. Aside from doubts as to the feasibility of constructing estimates for Russia which are comparable with US statistics, it is generally agreed that the present rate of investment in the US is itself still abnormally high for our economic system. It is true that much of it is for luxury or other purposes with a low security priority. If it is proposed to alter significantly this situation, the implications of attempting to redirect the flow of investment should be frankly faced. At some point, direct controls on a continuing basis may become necessary if inflationary pressure in some areas is not to be restrained by methods which create unemployment in others.

The implications of higher military expenditures are of course mainly a matter of degree. It cannot be said that at any point such expenditures are "too high." They must be sufficient to meet minimum requirements for the security of the nation. But security rests in economic as well as military strength, and due consideration should be given to the tendency for military expenditures to reduce the potential rate of economic growth, and at an advanced stage to require measures which may seriously impair the functioning of our system.

In the immediate situation and outlook, it seems probable that a moderate increase in security expenditures, partially or wholly offset by the prospective decline in ECA, can be undertaken without serious economic consequences. As the document points out, the potential growth in the economy can permit some increase while still permitting a rise in the civilian sector. However, this would not be without cost in preventing either an otherwise possible tax decrease or an increase in productive programs.

The document, however, is subject to criticism for inconsistency in proposing that higher security expenditures be counteracted by increased taxes and a curtailment of domestic programs. This seems hardly a program for stimulating economic growth. It is suggested as a general guideline that any security program which requires either a significant increase in the tax base or the curtailment of domestic programs which have an investment or developmental effect, should be considered as raising serious questions on the economic side.

No course of action is without risks, but the risks in the proposed course are not adequately considered. The type of military program seemingly implied on pp. 54–55 most certainly raises serious questions. This is even more true of the document as a whole which appears basically, despite general statements in other directions, to point down the road of principal reliance on military force which can only grow in its demands over time, as well as scarcely fail to lose the cold war.

WILLIAM F. SCHAUB

MASSIVE RETALIATION

JOHN FOSTER DULLES

It is now nearly a year since the Eisenhower administration took office. During that year I have often spoken of various parts of our foreign policies. Tonight I should like to present an overall view of those policies which relate to our security.

First of all, let us recognize that many of the preceding foreign policies were good. Aid to Greece and Turkey had checked the Communist drive to the Mediterranean. The European Recovery Program had helped the peoples of Western Europe pull out of the postwar morass. The Western powers were steadfast in Berlin and overcame the blockade with their airlift. As a loyal member of the United Nations, we had reacted with force to repel the Communist attack in Korea. When that effort exposed our military weakness, we rebuilt rapidly our military establishment. We also sought a quick buildup of armed strength in Western Europe. These were the acts of a nation which saw the danger of Soviet communism; which realized that its own safety was tied up with that of others; which was capable of responding boldly and promptly to emergencies. These are precious values to be acclaimed. Also, we can pay tribute to congressional bipartisanship which puts the nation above politics.

But we need to recall that what we did was in the main emergency action, imposed on us by our enemies.

Let me illustrate.

1. We did not send our army into Korea because we judged in advance that it was sound

Excerpted with minor revisions from address by Secretary of State John Foster Dulles to the Council on Foreign Relations, 12 Jan. 1954, Department of State Bulletin, 25 Jan. 1954, pp. 107–8.

military strategy to commit our army to fight land battles in Asia. Our decision had been to pull out of Korea. It was Soviet-inspired action that pulled us back.

2. We did not decide in advance that it was wise to grant billions annually as foreign economic aid. We adopted that policy in response to the Communist efforts to sabotage the free economies of Western Europe.

3. We did not build up our military establishment at a rate which involved huge budget deficits, a depreciating currency, and a feverish economy because this seemed, in advance, a good policy. Indeed, we decided otherwise until the Soviet military threat was clearly revealed.

We live in a world where emergencies are always possible, and our survival may depend upon our capacity to meet emergencies. Let us pray that we shall always have that capacity. But, having said that, it is necessary also to say that emergency measures—however good for the emergency—do not necessarily make good permanent policies. Emergency measures are costly; and they imply that the enemy has the initiative. They cannot be depended on to serve our long-term interests.

This "long time" factor is of critical importance.

The Soviet Communists are planning for what they call "an entire historical era," and we should do the same. They seek, through many types of maneuvers, gradually to divide and weaken the free nations by overextending them in efforts which, as Lenin put it, are "beyond their strength, so that they come to practical bankruptcy." Then, said Lenin, "our victory is assured." Then, said Stalin, will be "the moment for the decisive blow."

In the face of this strategy, measures cannot be judged adequate merely because they ward off an immediate danger: It is essential to do this, but it is also essential to do so without exhausting ourselves.

When the Eisenhower administration applied this test, we felt that some transformations were needed.

It is not sound military strategy permanently to commit US land forces to Asia to a degree that it leaves us no strategic reserves.

It is not sound economics or good foreign policy to support permanently other countries; for in the long run, that creates as much ill will as good will.

Also, it is not sound to become permanently committed to military expenditures so vast that they lead to "practical bankruptcy."

Change was imperative to assure the stamina needed for permanent security. But it was equally imperative that change should be accompanied by understanding of our true purposes. Sudden and spectacular change had to be avoided. Otherwise, there might have been a panic among our friends and miscalculated aggression by our enemies. We can, I believe, make a good report in these respects.

We need allies and collective security. Our purpose is to make these relations more effective, less costly. This can be done by placing more reliance on deterrent power and less dependence on local defensive power.

This is accepted practice so far as local communities are concerned. We keep locks on our doors, but we do not have an armed guard in every home. We rely principally on a community security system so well equipped to punish any who break in and steal that, in fact, would-be aggressors are generally deterred. That is the modern way of getting maximum protection at a bearable cost.

What the Eisenhower administration seeks is a similar international security system. We want, for ourselves and the other free nations, a maximum deterrent at a bearable cost.

Local defense will always be important. But there is no local defense which alone will contain the mighty landpower of the Communist world. Local defenses must be reinforced by the further deterrent of massive retaliatory power. A potential aggressor must know that it cannot always prescribe battle conditions that suit it. Otherwise, for example, a potential aggressor, who is glutted with manpower, might be tempted to attack in confidence that resistance would be confined to manpower. It might be tempted to attack in places where its superiority was decisive.

The way to deter aggression is for the free community to be willing and able to respond vigorously at places and with means of its own choosing.

So long as our basic policy concepts were unclear, our military leaders could not be selective in building our military power. If an enemy could pick its time and place and method of warfare—and if our policy was to remain the traditional one of meeting aggression by direction and local opposition—then we needed to be ready to fight in the Arctic and in the tropics; in Asia, the Near East, and in Europe; by sea, by land, and by air; with old weapons and with new weapons.

The total cost of our security efforts, at home and abroad was over $50 billion per annum, and involved, for 1953, a projected budgetary deficit of $9 billion; and $11 billion for 1954. This was on top of taxes comparable to wartime taxes; and the dollar was depreciating in effective value. Our allies were similarly weighed down. This could not be continued for long without grave budgetary, economic, and social consequences.

But before military planning could be

changed, the president and his advisers, as represented by the National Security Council, had to make some basic policy decisions. This has been done. The basic decision was to depend primarily upon a great capacity to retaliate, instantly, by means and at places of our choosing. Now the Department of Defense and the Joint Chiefs of Staff can shape our military establishment to fit what is *our* policy, instead of having to try to be ready to meet the enemy's many choices. That permits of a selection of military means instead of a multiplication of means. As a result, it is now possible to get, and share, more basic security at less cost.

DEFENSE ARRANGEMENTS OF THE NORTH ATLANTIC COMMUNITY

ROBERT S. MCNAMARA

What I want to talk to you about today are some of the concrete problems of maintaining a free community in the world today. I want to talk to you particularly about the problems of the community that binds together the United States and the countries of Western Europe . . .

THE NORTH ATLANTIC TREATY ORGANIZATION

NATO was born in 1949 out of the confrontation with the Soviet Union that ensued from the breakdown in relations between the former wartime allies. The Soviet Union had absorbed the states of Eastern Europe into its own political framework, most dramatically with the Czechoslovakian coup of 1948. It had been fomenting insurrection in Greece, menacing Turkey, and encouraging the Communist parties in Western Europe to seize power in the wake of postwar economic disorder. The sharpest threat to Europe came with the first Berlin crisis, when the Russians attempted to blockade the western sectors of the city. Our response was immediate and positive. President Truman ordered an airlift for the isolated population of West Berlin which, in time, denied the Soviets their prize. The Marshall Plan, then in full swing, was assisting the economic recovery of the Western European nations. The Truman Doctrine had brought our weight to bear in Greece and Turkey to prevent the erosion of their independence.

But Western statesmen concluded that it would be necessary to secure the strength and growth of the North Atlantic community with a more permanent arrangement for its defense. The effective defense of Western Europe could not really be accomplished without a commitment of the United States to that defense for the long term. We made this commitment without hesitation. Arthur Vandenberg, one of the chief architects of NATO, expressed the rationale of the organization in the Senate debate preceding passage of the treaty.

The North Atlantic Alliance is a unique alignment of governments. The provision for the common defense of its members has led to a remarkable degree of military collaboration and diplomatic consultation for a peacetime coalition. The growth of the alliance organization has accelerated in scope, size, and complexity. NATO has had its stresses and strains, but it has weathered them all.

Today NATO is involved in a number of controversies, which must be resolved by achieving a consensus within the organization in order to preserve its strength and unity. The question has arisen whether Senator Vandenberg's assertion is as true today as it was when he made it 13 years ago. Three arguments have raised this question most sharply:

It has been argued that the very success of Western European economic development reduces Europe's need to rely on the United States to share in its defenses.

It has been argued that the increasing vulnerability of the United States to nuclear attack makes us less willing as a partner in the defense of Europe and hence less effective in deterring such an attack.

It has been argued that nuclear capabilities are alone relevant in the face of the growing nuclear threat and that independent national nuclear forces are sufficient to protect the nations of Europe.

I believe that all of these arguments are mistaken. I think it is worthwhile to expose the US views on these issues as we have presented them to our allies. In our view, the effect of the

Excerpted with minor revisions from an address made at commencement exercises at the University of Michigan by Secretary of Defense Robert S. McNamara, Ann Arbor, Mich., 16 June 1962, Department of State Bulletin, 9 July 1962, pp. 64, 66–69.

new factors in the situation, both economic and military, has been to increase the interdependence of national security interests on both sides of the Atlantic and to enhance the need for the closest coordination of our efforts.

NUCLEAR STRATEGY

A central military issue facing NATO today is the role of nuclear strategy. Four facts seem to us to dominate consideration of that role. All of them point in the direction of increased integration to achieve our common defense. First, the Alliance has overall nuclear strength adequate to any challenge confronting it. Second, this strength not only minimizes the likelihood of major nuclear war but makes possible a strategy designed to preserve the fabric of our societies if war should occur. Third, damage to the civil societies of the Alliance resulting from nuclear warfare could be very grave. Fourth, improved nonnuclear forces, well within Alliance resources, could enhance deterrence of any aggressive moves short of direct, all-out attack on Western Europe.

Let us look at the situation today. First, given the current balance of nuclear power, which we confidently expect to maintain in the years ahead, a surprise nuclear attack is simply not a rational act for any enemy. Nor would it be rational for an enemy to take the initiative in the use of nuclear weapons as an outgrowth of a limited engagement in Europe or elsewhere. I think we are entitled to conclude that either of the actions has been made highly unlikely.

Second, and equally important, the mere fact that no nation could rationally take steps leading to a nuclear war does not guarantee that a nuclear war cannot take place. Not only do nations sometimes act in ways that are hard to explain on a rational basis, but even when acting in a "rational" way they sometimes, indeed disturbingly often, act on the basis of misunderstandings of the true facts of a situation. They misjudge the way others will react and the way others will interpret what they are doing.

We must hope—indeed I think we have good reason to hope—that all sides will understand this danger and will refrain from steps that even raise the possibility of such a mutually disastrous misunderstanding. We have taken unilateral steps to reduce the likelihood of such an occurrence. We look forward to the prospect that through arms control the actual use of these terrible weapons may be completely avoided. It is a problem not just for us in the West but for all nations that are involved in this struggle we call the cold war.

For our part we feel we and our NATO allies must frame our strategy with this terrible contingency, however remote, in mind. Simply ig-

noring the problem is not going to make it go away.

The United States has come to the conclusion that, to the extent feasible, basic military strategy in a possible general nuclear war should be approached in much the same way that more conventional military operations have been regarded in the past. That is to say, principal military objectives, in the event of a nuclear war stemming from a major attack on the Alliance, should be the destruction of the enemy's military forces, not of his civilian population.

The very strength and nature of the Alliance forces make it possible for us to retain, even in the face of a massive surprise attack, sufficient reserve striking power to destroy an enemy society if driven to it. In other words, we are giving a possible opponent the strongest imaginable incentive to restrain from striking our own cities.

The strength that makes these contributions to deterrence and to the hope of deterring attack upon civil societies even in wartime does not come cheap. We are confident that our current nuclear programs are adequate and will continue to be adequate for as far into the future as we can reasonably foresee. During the coming fiscal year the United States plans to spend close to $15 billion on its nuclear weapons to assure their adequacy. For what this money buys, there is no substitute.

In particular, relatively weak national nuclear forces with enemy cities as their targets are not likely to be sufficient to perform even the function of deterrence. If they are small, and perhaps vulnerable on the ground or in the air, or inaccurate, a major antagonist can take a variety of measures to counter them. Indeed, if a major antagonist came to believe there was a substantial likelihood of its being used independently, this force would be inviting a preemptive first strike against it. In the event of war, the use of such a force against the cities of a major nuclear power would be tantamount to suicide, whereas its employment against significant military targets would have a negligible effect on the outcome of the conflict. Meanwhile the creation of a single additional national nuclear force encourages the proliferation of nuclear power with all of its attendant dangers.

In short, then, limited nuclear capabilities, operating independently, are dangerous, expensive, prone to obsolescence, and lacking in credibility as a deterrent. Clearly the United States' nuclear contribution to the Alliance is neither obsolete nor dispensable.

IMPORTANCE OF CENTRAL CONTROL

At the same time, the general strategy I have summarized magnifies the importance of unity of planning, concentration of executive authority, and central direction. There must not be competing and conflicting strategies to meet the contingency of nuclear war. We are convinced that a general nuclear war target system is indivisible and if, despite all of our efforts, nuclear war should occur, our best hope lies in conducting a centrally controlled campaign against all of the enemy's vital nuclear capabilities, while retaining reserve forces, all centrally controlled.

We know that the same forces which are targeted on ourselves are also targeted on our allies. Our own strategic retaliatory forces are prepared to respond against these forces, wherever they are and whatever their targets. This mission is assigned not only in fulfillment of our treaty commitments but also because the character of nuclear war compels it. More specifically, the United States is as much concerned with that portion of Soviet nuclear striking power that can reach Western Europe as with that portion that also can reach the United States. In short, we have undertaken the nuclear defense of NATO on a global basis. This will continue to be our objective. In the execution of this mission, the weapons in the European theater are only one resource among many.

There is, for example, the Polaris force, which we have been substantially increasing and which, because of its specially invulnerable nature, is peculiarly well suited to serve as a strategic reserve force. We have already announced the commitment of five of these ships, fully operational, to the NATO Command.

This sort of commitment has a corollary for the Alliance as a whole. We want and need a greater degree of Alliance participation in formulating nuclear weapons policy to the greatest extent possible. We would all find it intolerable to contemplate having only a part of the strategic force launched in isolation from our main striking power.

We shall continue to maintain powerful nuclear forces for the Alliance as a whole. As the president has said, "Only through such strength can we be certain of deterring a nuclear strike, or an overwhelming ground attack, upon our forces and allies."

But let us be quite clear about what we are saying and what we would have to face if the deterrent should fall. This is the almost certain prospect that, despite our nuclear strength, all of us would suffer deeply in the event of major nuclear war.

We accept our share of this responsibility within the Alliance. And we believe that the combination of our nuclear strength and a strategy of controlled response gives us some hope of minimizing damage in the event that we have to fulfill our pledge. But I must point out that we do not regard this as a desirable prospect,

nor do we believe that the Alliance should depend solely on our nuclear power to deter actions not involving massive commitment of any hostile force. Surely an Alliance with the wealth, talent, and experience that we possess can find a better way than extreme reliance on nuclear weapons to meet our common threat. We do not believe that if the formula $E = MC^2$ had not been discovered, we should all be Communist slaves. On this question I can see no valid reason for a fundamental difference of view on the two sides of the Atlantic.

STRENGTHENING NATO'S NONNUCLEAR POWER

With the Alliance possessing the strength and the strategy I have described, it is most unlikely that any power will launch a nuclear attack on NATO. For the kinds of conflicts, both political and military, most likely to arise in the NATO area, our capabilities for response must not be limited to nuclear weapons alone. The Soviets have superiority in nonnuclear forces in Europe today. But that superiority is by no means overwhelming. Collectively, the Alliance has the potential for a successful defense against such forces. In manpower alone, NATO has more men under arms than the Soviet Union and its European satellites. We have already shown our willingness to contribute through our divisions now in place on European soil. In order to defend the populations of the NATO countries and to meet our treaty obligations, we have put in hand a series of measures to strengthen our nonnuclear power. We have added $10 million for this purpose to the previously planned level of expenditures for the fiscal years 1962 and 1963. To tide us over while new permanent strength was being created, we called up 158,000 reservists. We will be releasing them this summer, but only because in the meantime we have built up on an enduring basis more added strength than the call-up temporarily gave us. The number of US combat-ready divisions has been increased from 11 to 16. Stockpiled in Europe now are two full sets of equipment for two additional divisions: the men of these divisions can be rapidly moved to Europe by air.

We expect that our allies will also undertake to strengthen further their nonnuclear forces and to improve the quality and staying power of these forces. These achievements will complement our deterrent strength. With improvements in Alliance ground force strength and staying power, improved nonnuclear air capabilities, and better equipped and trained reserve forces, we can be assured that no deficiency exists in the NATO defense of this vital region and that no aggression, small or large, can succeed . . .

DETERRENCE, ASSURED DESTRUCTION, AND STRATEGIC OPTIONS

JAMES R. SCHLESINGER

I frankly doubt that our thinking about deterrence and its requirements has kept pace with the evolution of these threats. Much of what passes as current theory wears a somewhat dated air—with its origins in the strategic bombing campaigns of World War II and the nuclear weapons technology of an earlier era when warheads were bigger and dirtier, delivery systems considerably less accurate, and forces much more vulnerable to surprise attack.

The theory postulates that deterrence of a hostile act by another party results from a threat of retaliation. This retaliatory threat, explicit or implicit, must be of sufficient magnitude to make the goal of the hostile act appear unattainable, or excessively costly, or both. Moreover, in order to work, the retaliatory threat must be credible: that is, believable to the party being threatened. And it must be supported by visible, employable military capabilities.

The theory also recognizes that the effectiveness of a deterrent depends on a good deal more than peacetime declaratory statements about retaliation and the existence of a capability to do great damage. In addition, the deterrent must appear credible under conditions of crisis, stress, and even desperation or irrationality on the part of an opponent. And since, under a variety of conditions, the deterrent forces themselves could become the target of an attack, they must be capable of riding out such an attack in a second strike.

The principle that nuclear deterrence (or any form of deterrence, for that matter) must be based on a high-confidence capability for second-strike retaliation—even in the aftermath of a well-executed surprise attack—is now well established. A number of other issues remain outstanding, however. A massive, bolt-out-of-the-blue attack on our strategic forces may well be the worst possible case that could occur, and therefore extremely useful as part of the force-sizing process. But it may not be the only, or even the most likely, contingency against which we should design our deterrent. Furthermore, depending upon the contingency, there has been a long-standing debate about the appropriate set of targets for a second strike which, in turn, can have implications both for the types of war plans we adopt and the composition of our forces.

This is not the place to explore the full history and details of that long-standing strategic debate. However, there is one point to note about its results. Although several targeting options, including military only and military plus urban-industrial variations, have been a part of US strategic doctrine for quite some time, the concept that has dominated our rhetoric for most of the era since World War II has been massive retaliation against cities, or what is called assured destruction. As I hardly need emphasize, there is a certain terrifying elegance in the simplicity of the concept. For all that it postulates, in effect, is that deterrence will be adequately (indeed amply) served if, at all times, we possess the second-strike capability to destroy some percentage of the population and industry of a potential enemy. To be able to assure that destruction, even under the most unfavorable circumstances—so the argument goes—is to assure deterrence, since no possible gain could compensate an aggressor for this kind and magnitude of loss.

The concept of assured destruction has many attractive features from the standpont of sizing the strategic offensive forces. Because nuclear weapons produce such awesome effects, they are ideally suited to the destruction of large, soft targets such as cities. Furthermore, since cities contain such easily measurable contents as people and industry, it is possible to establish convenient quantitative criteria and levels of desired effectiveness with which to measure the potential performance of the strategic offensive forces. And once these specific objectives are set, it becomes a relatively straightforward matter—given an authoritative estimate about the nature and weight of the enemy's surprise attack—to work back to the forces required for second-strike assured destruction.

The basic simplicity of the assured destruction calculation does not mean that the force planner is at a loss for issues. On the contrary, important questions continue to arise about the assumptions from which the calculations proceed. Where, for the sake of deterrence, should we set the level of destruction that we want to assure? Is it enough to guarantee the ruin of several major cities and their contents, or should we—to assure deterrence—move much further and upward on the curve of destruction?

Excerpted with minor revisions from Secretary of Defense James R. Schlesinger, Department of Defense Annual Report, FY 1975, *pp. 32–38.*

Since our planning must necessarily focus on the forces we will have five or even ten years hence, what should we assume about the threat—that is, the nature and weight of the enemy attack that our forces must be prepared to absorb? How pessimistic should we be about the performance of these forces in surviving the attack, penetrating enemy defenses (if they exist), and destroying their designed targets? How conservative should we be in buying insurance against possible failures in performance?

Generally speaking, national policymakers for more than a decade have chosen to answer these questions in a conservative fashion. Against the USSR, for example, we tended in the 1960s to talk in terms of levels of assured destruction at between a fifth and a third of the population and between half and three-quarters of the industrial capacity. We did so for two reasons:

—Beyond these levels very rapidly diminishing increments of damage would be achieved for each additional dollar invested;
—It was thought that amounts of damage substantially below those levels might not suffice to deter irrational or desperate leaders.

We tended to look at a wide range of threats and possible attacks on our strategic forces, and we tried to make these forces effective even after their having been attacked by high but realistically constrained threats. That is to say, we did not assume unlimited budgets or an untrammeled technology on the part of prospective opponents, but we were prudent about what they might accomplish within reasonable budgetary and technological constraints. Our choice of assumptions about these factors was governed not by a desire to exaggerate our own requirements but by the judgment that, with so much at stake, we should not make national survival a hostage to optimistic estimates of our opponents' capabilities.

In order to ensure the necessary survival and retaliatory effectiveness of our strategic offense, we have maintained a *triad* of forces, each of which presents a different problem for an attacker, each of which causes a specialized and costly problem for his defense, and all of which together currently give us high confidence that the force as a whole can achieve the desired deterrent objective.

That, however, is only part of the explanation for the present force structure. We have arrived at the current size and mix of our strategic offensive forces not only because we want the ultimate threat of massive destruction to be really assured, but also because for more than a decade we have thought it advisable to test the force against the "higher-than-expected" threat. Given the built-in surplus of warheads

generated by this force-sizing calculation, we could allocate additional weapons to nonurban targets and thereby acquire a limited set of options, including the option to attack some hard targets.

President Nixon has strongly insisted on continuing this prudent policy of maintaining sufficiency. As a result, I can say with confidence that in 1974, even after a more brilliantly executed and devastating attack than we believe our potential adversaries could deliver, the United States would retain the capability to kill more than 30 percent of the Soviet population and destroy more than 75 percent of Soviet industry. At the same time we could hold in reserve a major capability against the PRC.

Such reassurances may bring solace to those who enjoy the simple but arcane calculations of assured destruction. But they are of no great comfort to policymakers who must face the actual decisions about the design and possible use of the strategic nuclear forces. Not only must those in power consider the morality of threatening such terrible retribution on the Soviet people for some ill-defined transgression by their leaders; in the most practical terms, they must also question the prudence and plausibility of such a response when the enemy is able, even after some sort of first strike, to maintain the capability of destroying our cities. The wisdom and credibility of relying simply on the preplanned strikes of assured destruction are even more in doubt when allies rather than the United States itself face the threat of a nuclear war.

THE NEED FOR OPTIONS

President Nixon underlined the drawbacks to sole reliance on assured destruction in 1970 when he asked:

Should a President, in the event of a nuclear attack, be left with a single option of ordering the mass destruction of enemy civilians, in the face of the certainty that it would be followed by the mass slaughter of Americans? Should the concept of assured destruction be narrowly defined, and should it be the only measure of our ability to deter the variety of threats we may face?

The questions are not new. They have arisen many times during the nuclear era, and a number of efforts have been made to answer them. We actually added several response options to our contingency plans in 1961 and undertook the retargeting necessary for them. However, they all involved large numbers of weapons. In addition, we publicly adopted to some degree the philosophies of counterforce and damage-limiting. Although differences existed between those two concepts as then formulated, particularly in their diverging assumptions about cities as likely targets of attack, both had a number of features in common.

—Each required the maintenance of a capability to destroy urban-industrial targets, but as a reserve to deter attacks on US and allied cities rather than as the main instrument of retaliation.

—Both recognized that contingencies other than a massive surprise attack on the United States might arise and should be deterred; both argued that the ability and willingness to attack military targets were prerequisites to deterrence.

—Each stressed that a major objective, in the event that deterrence should fail, would be to avoid to the extent possible causing collateral damage in the USSR, and to limit damage to the societies of the United States and its allies.

—Neither contained a clear-cut vision of how a nuclear war might end, or what role the strategic forces would play in their termination.

—Both were considered by critics to be open-ended in their requirement for forces, very threatening to the retaliatory capabilities of the USSR, and therefore dangerously stimulating to the arms race and the chances of preemptive war.

—The military tasks that each involved, whether offensive counterforce or defensive damage-limiting, became increasingly costly, complex, and difficult as Soviet strategic forces grew in size, diversity, and survivability.

Of the two concepts, damage-limiting was the more demanding and costly because it required both active and passive defenses as well as a counterforce capability to attack hard targets and other strategic delivery systems. Added to this was the assumption (at least for planning purposes) that an enemy would divide his initial attack between our cities and our retaliatory forces, or switch his fire to our cities at some later stage in the attack. Whatever the realism of that assumption, it placed an enormous burden on our active and passive defenses—and particularly on antiballistic missile (ABM) systems—for the limitation of damage.

With the ratification of the ABM treaty in 1972, and the limitation it imposes on both the United States and the Soviet Union to construct no more than two widely separated ABM sites (with no more than 100 interceptors at each), an essential building-block in the entire damage-limiting concept has now been removed. As I shall discuss later, the treaty has also brought into question the utility of large, dedicated antibomber defenses, since without a defense against missiles, it is clear that an active defense

against bombers has little value in protecting our cities. The salient point, however, is that the ABM treaty has effectively removed the concept of defensive damage limitation (at least as it was defined in the 1960s) from contention as a major strategic option.

Does all of this mean that we have no choice but to rely solely on the threat of destroying cities? Does it even matter if we do? What is wrong, in the final analysis, with staking everything on this massive deterrent and pressing ahead with a further limitation of these devastating arsenals?

No one who has thought much about these questions disagrees with the need, as a minimum, to maintain a conservatively designed reserve for the ultimate threat of large-scale destruction. Even more, if we could all be guaranteed that this threat would prove fully credible (to friend and foe alike) across the relevant range of contingencies—and that deterrence would never be severely tested or fail—we might also agree that nothing more in the way of options would ever be needed. The difficulty is that no such guarantee can be given. There are several reasons why any assurance on this score is impossible.

Since we ourselves find it difficult to believe that we would actually implement the threat of assured destruction in response to a limited attack on military targets that caused relatively few civilian casualties, there can be no certainty that, in a crisis, prospective opponents would be deterred from testing our resolve. Allied concern about the credibility of this particular threat has been evident for more than a decade. In any event, the actuality of such a response would be utter folly except where our own or allied cities were attacked.

Today, such a massive retaliation against cities, in response to anything less than an all-out attack on the US and its cities, appears less and less credible. Yet as pointed out above, deterrence can fail in many ways. What we need is a series of measured responses to aggression which bear some relation to the provocation, have prospects of terminating hostilities before general nuclear war breaks out, and leave some possibility for restoring deterrence. It has been this problem of not having sufficient options between massive response and doing nothing, as the Soviets built up their strategic forces, that has prompted the president's concerns and those of our allies.

Threats against allied forces, to the extent that they could be deterred by the prospect of nuclear retaliation, demand both more limited responses than destroying cities and advanced planning tailored to such lesser responses. Nuclear threats to our strategic forces, whether limited or large-scale, might well call for an option to respond in kind against the attacker's military forces. In other words, to be credible, and hence effective over the range of possible contingencies, deterrence must rest on many options and on a spectrum of capabilities (within the constraints of SALT) to support these options. Certainly such complex matters as response options cannot be left hanging until a crisis. They must be thought through beforehand. Moreover, appropriate sensors to assist in determining the nature of the attack, and adequately responsive command-control arrangements, must also be available. And a venturesome opponent must know that we have all of these capabilities.

Flexibility of response is also essential because, despite our best efforts, we cannot guarantee that deterrence will never fail; nor can we forecast the situations that would cause it to fail. Accidents and unauthorized acts could occur, especially if nuclear proliferation should in-

crease. Conventional conflicts could escalate into nuclear exchanges; indeed, some observers believe that this is precisely what would happen should a major war break out in Europe. Ill-informed or cornered and desperate leaders might challenge us to a nuclear test of wills. We cannot even totally preclude the massive surprise attack on our forces which we use to test the design of our second-strike forces, although I regard the probability of such an attack as close to zero under existing conditions. To the extent that we have selective response options—smaller and more precisely focused than in the past—we should be able to deter such challenges. But if deterrence fails, we may be able to bring all but the largest nuclear conflicts to a rapid conclusion before cities are struck. Damage may thus be limited and further escalation avoided.

THE COUNTERVAILING STRATEGY

HAROLD BROWN

A significant achievement in 1980 was the codification of our evolving strategic doctrine, in the form of Presidential Directive No. 59 (PD-59). In my report last year, I discussed the objectives and the principal elements of this countervailing strategy, and in August 1980, after PD-59 had been signed by President Carter, I elaborated it in some detail in a major policy address. Because of its importance, however, the countervailing strategy warrants special attention in this report as well.

Two basic points should underlie any discussion of the countervailing strategy. *The first is that, because it is a strategy of deterrence, the countervailing strategy is designed with the Soviets in mind.* Not only must we have the forces, doctrine, and will to retaliate if attacked, we must convince the Soviets, *in advance*, that we do. Because it is designed to deter the Soviets, our strategic doctrine must take account of what we know about Soviet perspectives on these issues, for, by definition, deterrence requires shaping Soviet assessments about the risks of war—assessments they will make using their models, not ours. We must confront these views and take them into account in our planning. We may, and we do, think our models are more accurate, but theirs are the reality deterrence drives us to consider.

Several Soviet perspectives are relevant to the formulation of our deterrent strategy. First, Soviet military doctrine appears to contemplate the possibility of a relatively prolonged nuclear war. Second, there is evidence that they regard military forces as the obvious first targets in the nuclear exchange, not general industrial and economic capacity. Third, the Soviet leadership clearly places a high value on preservation of the regime and on the survival and continued effectiveness of the instruments of state power and control—a value at least as high as that they place on any losses to the general population, short of those involved in a general nuclear war. Fourth, in some contexts, certain elements of Soviet leadership seem to consider Soviet victory in a nuclear war to be at least a theoretical possibility.

All this does not mean that the Soviets are unaware of the destruction a nuclear war would bring to the Soviet Union; in fact, they are explicit on that point. Nor does this mean that we cannot deter, for clearly we can and we do.

The second basic point is that, because the world is constantly changing, our strategy evolves slowly, almost continually, over time to adapt to changes in US technology and military capabilities, as well as Soviet technology, military capabilities, and strategic doctrine. A stra-

Excerpted with minor revisions from Secretary of Defense Harold Brown, Department of Defense Annual Report, FY 1982, *pp. 38–43.*

tegic doctrine that served well when the United States had only a few dozen nuclear weapons and the Soviets none would hardly serve as well unchanged in a world in which we have about 9000 strategic warheads and they have about 7000. As the strategic balance has shifted from overwhelming US superiority to essential equivalence, and as ICBM accuracies have steadily improved to the point that hard-target kill probabilities are quite high, our doctrine must adapt itself to these new realities.

This does not mean that the objective of our doctrine changes; on the contrary, deterrence remains, as it always has been, our basic goal. Our countervailing strategy today is a natural evolution of the conceptual foundations built over a generation by men like Robert McNamara and James Schlesinger.

The United States has never—at least since nuclear weapons were available in significant numbers—had a strategic doctrine based simply and solely on reflexive, massive attacks on Soviet cities and populations. Previous administrations, going back almost twenty years, recognized the inadequacy as a deterrent of a targeting doctrine that would give us too narrow a range of options. Although for programming purposes, strategic forces were sometimes measured in terms of ability to strike a set of industrial targets, we have always planned both more selectively (for options limiting urban-industrial damage) and more comprehensively (for a wide range of civilian and military targets). The unquestioned Soviet attainment of strategic parity has put the final nail in the coffin of what we long knew was dead—the notion that we could adequately deter the Soviets solely by threatening massive retaliation against their cities.

The administration's systematic contributions to the evolution of strategic doctrine began in the summer of 1977, when President Carter ordered a comprehensive review of US strategic policy to ensure its continued viability and deterrent effect in an era of strategic nuclear parity. Over the next 18 months, civilian and military experts conducted an extensive review, covering a wide range of issues, including US and Soviet capabilities, vulnerabilities, and doctrine. As soon as the report was ready, implementation began. The broad set of principles this review yielded constitutes the essence of the countervailing strategy. I outlined these in my FY 1981 Defense Report and reviewed them at the NATO Nuclear Planning Group meeting in Norway in June 1980. Three years after he ordered the initial review, President Carter signed the implementing directive—PD-59—formally codifying the countervailing strategy and giving guidance for the continuing evolution of US planning, targeting, and systems acquisition. In September 1980, Secretary of State Muskie and I testified on the countervailing strategy and PD-59 before the Senate Foreign

Relations Committee. Again, in November of 1980, I engaged in extensive and intensive discussions of the countervailing strategy with our NATO allies, this time at the fall Nuclear Planning Group meeting.

Our countervailing strategy—designed to provide effective deterrence—tells the world that no potential adversary of the United States could ever conclude that the fruits of his aggression would be worth his own costs. This is true whatever the level of conflict contemplated. To the Soviet Union, our strategy makes clear that no course of aggression by them that led to use of nuclear weapons, on any scale of attack and at any stage of conflict, could lead to victory, however they may define victory. Besides our power to devastate the full target system of the USSR, the United States would have the option for more selective, lesser retaliatory attacks that would exact a prohibitively high price from the things the Soviet leadership prizes most—political and military control, nuclear and conventional military force, and the economic base needed to sustain a war.

Thus, the countervailing strategy is designed to be fully consistent with NATO's strategy of flexible response by providing options for appropriate response to aggression at whatever level it might occur. The essence of the countervailing strategy is to convince the Soviets that they will be successfully opposed at any level of aggression they choose, and that no plausible outcome at any level of conflict could represent "success" for them by any reasonable definition of success.

Five basic elements of our force employment policy serve to achieve the objectives of the countervailing strategy.

FLEXIBILITY

Our planning must provide a continuum of options, ranging from use of small numbers of strategic and/or theater nuclear weapons aimed at narrowly defined targets, to employment of large portions of our nuclear forces against a broad spectrum of targets. In addition to preplanned targeting options, we are developing an ability to design other employment plans—in particular, smaller scale plans—on short notice in response to changing circumstances.

In theory, such flexibility also enhances the possibility of being able to control escalation of what begins as a limited nuclear exchange. I want to emphasize once again two points I have made repeatedly and publicly. First, I remain highly skeptical that escalation of a limited nuclear exchange can be controlled, or that it can be stopped short of an all-out, massive exchange. Second, even given that belief, I am convinced that we must do everything we can to make such escalation control possible, that

opting out of this effort and consciously resign-
ing ourselves to the inevitability of such escalation is a serious abdication of the awesome
responsibilities nuclear weapons, and the un-
believable damage their uncontrolled use would
create, thrust upon us. Having said that, let me
proceed to the second element, which is escalation control.

ESCALATION CONTROL

Plans for the controlled use of nuclear weapons,
along with other appropriate military and political actions, should enable us to provide leverage for a negotiated termination of the fighting.
At an early stage in the conflict, we must convince the enemy that further escalation will not
result in achievement of its objectives, that it
will not mean "success," but rather additional
costs. To do this, we must leave the enemy with
sufficient highly valued military, economic, and
political resources still surviving but still clearly
at risk, so that it has a strong incentive to seek
an end to the conflict.

SURVIVABILITY AND ENDURANCE

The key to escalation control is the survivability
and endurance of our nuclear forces and the
supporting communications, command and control, and intelligence (C^3I) capabilities. The supporting C^3I is critical to effective deterrence,
and we have begun to pay considerably more
attention to these issues than in the past. We
must ensure that the United States is not placed
in a "use or lose" situation, one that might lead
to unwarranted escalation of the conflict. That
is a central reason why, while the Soviets cannot
ignore our *capability* to launch our retaliatory
forces before an attack reaches its targets, we
cannot afford to rely on "launch on warning" as
the long-term solution to ICBM vulnerability.
That is why the new MX missile should be deployed in a survivable basing mode, not in
highly vulnerable fixed silos, and that is why we
spend considerable sums of money to ensure
the continued survivability of our ballistic missile submarine fleet. Survivability and endurance are essential prerequisites to an ability to
adapt the employment of nuclear forces to the
entire range of potentially rapidly changing and
perhaps unanticipated situations and to tailor
them for the appropriate responses in those situations. And, without adequate survivability
and endurance, it would be impossible for us
to keep substantial forces in reserve.

TARGETING OBJECTIVES

In order to meet our requirements for flexibility
and escalation control, we must have the ability
to destroy elements of four general categories
of Soviet targets.

Strategic Nuclear Forces

The Soviet Union should entertain no illusion
that by attacking our strategic nuclear forces, it
could significantly reduce the damage it would
suffer. Nonetheless, the state of the strategic
balance after an initial exchange—measured
both in absolute terms and in relation to the
balance prior to the exchange—could be an important factor in the decision by one side to
initiate a nuclear exchange. Thus, it is important—for the sake of deterrence—to be able to
deny to the potential aggressor a fundamental
and favorable shift in the strategic balance as a
result of a nuclear exchange.

Other Military Forces

"Counterforce" covers much more than central
strategic systems. We have for many years
planned options to destroy the full range of Soviet (and, as appropriate, non-Soviet Warsaw
Pact) military power, conventional as well as
nuclear. Because the Soviets may define victory
in part in terms of the overall postwar military
balance, we will give special attention, in implementing the countervailing strategy, to more
effective and more flexible targeting of the full
range of military capabilities, so as to strengthen
deterrence.

Leadership and Control

We must, and we do, include options to target
organs of Soviet political and military leadership
and control. As I indicated earlier, the regime
constituted by these centers is valued highly by
the Soviet leadership. A clear US ability to destroy them poses a marked challenge to the essence of the Soviet system and thus contributes
to deterrence. At the same time, of course, we
recognize the role that a surviving supreme
command could and would play in the termination of hostilities, and can envisage many
scenarios in which destruction of them would
be inadvisable and contrary to our own best
interests. Perhaps the obvious is worth emphasizing: possession of a capability is not tantamount to exercising it.

Industrial and Economic Base

The countervailing strategy by no means implies
that we do not—or no longer—recognize the
ultimate deterrent effect of being able to
threaten the full Soviet target structure, including the industrial and economic base. These targets are highly valued by the Soviets, and we
must ensure that the potential loss of them is
an ever-present factor in the Soviet calculus regarding nuclear war. Let me also emphasize that
while, as a matter of policy, we do not target
civilian population per se, heavy civilian fatalities and other casualties would inevitably occur
in attacking the Soviet industrial and economic
base, which is collocated with the Soviet urban

population. I should add that Soviet civilian casualties would also be large in more focused attacks (not unlike the US civilian casualty estimates cited earlier for Soviet attacks on our ICBM silos); indeed, they could be described as limited only in the sense that they would be significantly less than those resulting from an all-out attack.

RESERVE FORCES

Our planning must provide for the designation and employment of adequate, survivable, and enduring reserve forces and the supporting C³I systems both during and after a protracted conflict. At a minimum, we will preserve such a dedicated force of strategic weapon systems.

Because there has been considerable misunderstanding and misinterpretation of the countervailing strategy and of PD-59, it is worth restating what the countervailing strategy is *not*.

—It is *not* a new strategic doctrine; it is *not* a radical departure from US strategic policy over the past decade or so. It *is* a refinement, a recodification of previous statements of our strategic policy. It *is* the same essential strategic doctrine, restated more clearly and related more directly to current and prospective conditions and capabilities—US and Soviet.

—It does *not* assume, or assert, that we can "win" a limited nuclear war, nor does it pretend or intend to enable us to do so. It *does* seek to convince the Soviets that they could not win such a war, and thus to deter them from starting one.

—It does *not* even assume, or assert, that a nuclear war could remain limited. I have made clear my view that such a prospect is highly unlikely. It *does*, however, prepare us to respond to a limited Soviet nuclear attack in ways other than automatic, immediate, massive retaliation.

—It does *not* assume that a nuclear war will in fact be protracted over many weeks or even months. It *does*, however, take into account evidence of Soviet thinking along those lines, in order to convince them that such a course, whatever its probability, could not lead to Soviet victory.

—It does *not* call for substituting primarily military for primarily civilian targets. It *does* recognize the importance of military and civilian targets. It does provide for increasing the number and variety of options available to the president, covering the full range of military and civilian targets, so that he can respond appropriately and effectively to any kind of an attack, at any level.

—It is *not* inconsistent with future progress in arms control. In fact, it *does* emphasize many features—survivability, crisis stability, deterrence—that are among the core objectives of arms control. It does *not* require larger strategic arsenals; it *does* demand more flexibility and better control over strategic nuclear forces, whatever their size.

—Lastly, it is *not* a first-strike strategy. Nothing in the policy contemplates that nuclear war can be a deliberate instrument for achieving our national security goals, because it cannot be. The premise, the objective, the core of our strategic doctrine remains unchanged—deterrence. The countervailing strategy, by specifying what we would do in response to any level of Soviet attack, serves to deter any such attack in the first place.

THE STRATEGIC DEFENSE INITIATIVE

RONALD REAGAN

My fellow Americans, thank you for sharing your time with me tonight.

The subject I want to discuss with you, peace and national security, is both timely and important. Timely, because I've reached a decision which offers a new hope for our children in the twenty-first century, a decision I'll tell you about in a few minutes. And important because there's a very big decision that you must make for yourselves. This subject involves the most basic duty that any president and any people share, the duty to protect and strengthen the peace . . .

Tonight, I want to explain to you what this defense debate is all about and why I'm convinced that the budget now before the Congress is necessary, responsible, and deserving of your support. And I want to offer hope for the future . . .

Excerpted with minor revisions from address to the nation by President Ronald Reagan, 23 Mar. 1983, Weekly Compilation of Presidential Documents, *28 Mar. 1983, pp. 442–45, 447–48.*

Since the dawn of the atomic age, we've sought to reduce the risk of war by maintaining a strong deterrent and by seeking genuine arms control. "Deterrence" means simply this: making sure any adversary who thinks about attacking the United States or our allies, or our vital interests, concludes that the risks to him outweigh any potential gains. Once he understands that, he won't attack. We maintain peace through our strength; weakness only invites aggression.

This strategy of deterrence has not changed. It still works. But what it takes to maintain deterrence has changed. It took one kind of military force to deter an attack when we had far more nuclear weapons than any other power; it takes another kind now that the Soviets, for example, have enough accurate and powerful nuclear weapons to destroy virtually all of our missiles on the ground. Now, this is not to say that the Soviet Union is planning to make war on us. Nor do I believe that a war is inevitable—quite the contrary. But what must be recognized is that our security is based on being prepared to meet all threats.

There was a time when we depended on coastal forts and artillery batteries, because, with the weaponry of that day, any attack would have had to come by sea. Well, this is a different world, and our defenses must be based on recognition and awareness of the weaponry possessed by other nations in the nuclear age.

We can't afford to believe that we will never be threatened. There have been two world wars in my lifetime. We didn't start them and, indeed, did everything we could to avoid being drawn into them. But we were ill-prepared for both. Had we been better prepared, peace might have been preserved.

For 20 years the Soviet Union has been accumulating enormous military might. They didn't stop when their forces exceeded all requirements of a legitimate defensive capability. And they haven't stopped now. During the past decade and a half, the Soviets have built up a massive arsenal of new strategic nuclear weapons—weapons that can strike directly at the United States . . .

There was a time when we were able to offset superior Soviet numbers with higher quality, but today they are building weapons as sophisticated and modern as our own.

As the Soviets have increased their military power, they're emboldened to extend that power. They're spreading their military influence in ways that can directly challenge our vital interests and those of our allies . . .

Some people may still ask: Would the Soviets ever use their formidable military power? Well, again, can we afford to believe they won't? There is Afghanistan. And in Poland, the Soviets denied the will of the people and in so doing demonstrated to the world how their military power could also be used to intimidate.

The final fact is that the Soviet Union is acquiring what can only be considered an offensive military force. They have continued to build far more intercontinental ballistic missiles than they could possibly need simply to deter an attack. Their conventional forces are trained and equipped not so much to defend against an attack as they are to permit sudden, surprise offensives of their own . . .

Now, thus far tonight I've shared with you my thoughts on the problems of national security we must face together. My predecessors in the Oval Office have appeared before you on other occasions to describe the threat posed by Soviet power and have proposed steps to address that threat. But since the advent of nuclear weapons, those steps have been increasingly directed toward deterrence of aggression through the promise of retaliation.

This approach to stability through offensive threat has worked. We and our allies have succeeded in preventing nuclear war for more than three decades. In recent months, however, my advisers, including in particular the Joint Chiefs of Staff, have underscored the necessity to break out of a future that relies solely on offensive retaliation for our security.

Over the course of these discussions, I've become more and more deeply convinced that the human spirit must be capable of rising above dealing with other nations and human beings by threatening their existence. Feeling this way, I believe we must thoroughly examine every opportunity for reducing tensions and for introducing greater stability into the strategic calculus on both sides.

One of the most important contributions we can make is, of course, to lower the level of all arms, and particularly nuclear arms. We're engaged right now in several negotiations with the Soviet Union to bring about a mutual reduction of weapons. I will report to you in a week from tomorrow my thoughts on that score. But let me just say, I'm totally committed to this course.

If the Soviet Union will join us in our effort to achieve major arms reductions, we will have succeeded in stabilizing the nuclear balance. Nevertheless, it will still be necessary to rely on the specter of retaliation, on mutual threat. And that's a sad commentary on the human condition. Wouldn't it be better to save lives than to avenge them? Are we not capable of demonstrating our peaceful intentions by applying all-out abilities and ingenuity to achieving a truly lasting stability? I think we are. Indeed, we must. After careful consultation with my advisers, including the Joint Chiefs of Staff, I believe there is a way. Let me share with you a vision of the future which offers hope. It is that we embark on a program to counter the awesome Soviet missile threat with measures that are defensive. Let us turn to the very strengths in technology that spawned our great industrial base and that have given us the quality of life we enjoy today.

What if free people could live secure in the knowledge that their security did not rest upon the threat of instant US retaliation to deter a Soviet attack, that we could intercept and destroy strategic ballistic missiles before they reached our own soil or that of our allies?

I know this is a formidable, technical task, one that may not be accomplished before the end of this century. Yet, current technology has attained a level of sophistication where it's reasonable for us to begin this effort. It will take years, probably decades of effort on many fronts. There will be failures and setbacks, just as there will be successes and breakthroughs. And as we proceed, we must remain constant in preserving the nuclear deterrent and maintaining a solid capability for flexibile response. But isn't it worth every investment necessary to free the world from the threat of nuclear war? We know it is.

In the meantime, we will continue to pursue real reductions in nuclear arms, negotiating from a position of strength that can be assured only by modernizing our strategic forces. At the same time, we must take steps to reduce the risk of a conventional military conflict escalating to nuclear war by improving our nonnuclear capabilities.

America does possess—now—the technologies to attain very significant improvements in the effectiveness of our conventional, nonnuclear forces. Proceeding boldly with these new technologies, we can significantly reduce any incentive that the Soviet Union may have to threaten attack against the United States or its allies.

As we pursue our goal of defensive technologies, we recognize that our allies rely upon our strategic offensive power to deter attacks against them. Their vital interests and ours are inextricably linked. Their safety and ours are one. And no change in technology can or will alter that reality. We must and shall continue to honor our commitments.

I clearly recognize that defensive systems have limitations and raise certain problems and ambiguities. If paired with offensive systems, they can be viewed as fostering an aggressive policy, and no one wants that. But with these considerations firmly in mind, I call upon the scientific community in our country, those who gave us nuclear weapons, to turn their great talents now to the cause of mankind and world peace, to give us the means of rendering these nuclear weapons impotent and obsolete.

Tonight, consistent with our obligations of the ABM treaty and recognizing the need for closer consultation with our allies, I'm taking an important first step. I am directing a comprehensive and intensive effort to define a long-term research and development program to begin to achieve our ultimate goal of eliminating the threat posed by strategic nuclear missiles. This could pave the way for arms control measures to eliminate the weapons themselves. We seek neither military superiority nor political advantage. Our only purpose—one all people share—is to search for ways to reduce the danger of nuclear war.

My fellow Americans, tonight we're launching an effort which holds the promise of changing the course of human history. There will be risks, and results take time. But I believe we can do it. As we cross this threshold, I ask for your prayers and your support.

VIETNAM AND THE AMERICAN THEORY OF LIMITED WAR

STEPHEN PETER ROSEN

The idea of limited conventional war fought by American troops outside Europe is no longer unthinkable, but there has been no recent analysis of the whole idea of limited war. As a result, the concepts that we use when we think about this problem are, by and large, the concepts we inherited from a small group of scholars and policy analysts who did their most important writing in the middle 1950s. There is, of course, no automatic need for new strategic concepts every decade. Still, in the generation that has gone by since Robert Osgood and Thomas Schelling analyzed this subject, much has been learned while the American strategy of limited war has remained the same. As Osgood wrote in 1979, "the strategy transcended the Vietnam War and not only survived it but continued to expand in application and acceptance."[1] Much of what their strategy stated was and is true, but much is not, and much about the nature of

©1982 by the President and Fellows of Harvard College and of the Massachusetts Institute of Technology. Reprinted with minor revisions and by permission from International Security 7, no. 2 (1982): 83–113.

war was simply ignored. A reconsideration of limited war strategy in light of what we can learn from historical experience leads to a new strategy to supplement the old. Limited wars are not only political wars, as the original theorists wrote, but *strange* wars. The general problem of limited war is not only the *diplomatic* one of how to signal our resolve to our enemy, but the *military* one of how to adapt, quickly and successfully, to the peculiar and unfamiliar *battlefield* conditions in which our armed forces are fighting. Diplomatic success will depend on military success since resolve cannot survive repeated failure on the battlefield. Finally, the factors that determine whether this adaptation is successful or not are largely *moral* factors: the presence or absence of political courage at the central levels of command that enables men to make clear decisions about the missions and resources allocated to the theater commander, and to delegate responsibility to the local commanders. Military courage is required of the officers, to earn and keep the confidence of their men, as soldiers die without winning in the early stages of the war, and to adopt operational changes that necessarily stake the lives of the soldiers on untried and possibly incorrect tactics. Intellectual ability is obviously necessary but in some ways secondary; solutions to military problems have often been recognized but not implemented because men, with very good reason, are afraid of what would happen if they are wrong. In war the easiest thing is difficult, not because soldiers are stupid, but because they are human and do not regard human life as a resource to be expended as needed.

EARLY THEORISTS OF LIMITED WAR

Between 1957 and 1960, two books were published by professors that set the terms of discussion about limited war. Robert Osgood's book *Limited War* argued that politics is primary. What is special about limited war is that resources and goals are constrained by policy, not capabilities. The object of the war is political, to be obtained by negotiation and compromise, and not military, involving the physical destruction of the enemy. Therefore, the special problem of limited war is "more broadly, the problem of combining military power with diplomacy and with the economic and psychological instruments of power."[2] While it is true that limited wars deal with smaller problems than those found in total wars, in both kinds of wars the objects have been *political*. Both world wars had explicitly political objectives and were the extension of politics just as much as in any smaller war. Nor was resource allocation determined by the physical capacity of the nation. The Allied commanders in Italy and Burma were painfully aware that they were fighting "limited wars" in order to permit the operations in other more important theaters. Those disputes over "limited" resources were analogous to the disputes over how to limit the Korean War so as to allow the simultaneous building of the US position in Europe,[3] and to those within the American Joint Chiefs of Staff as the Vietnam war drained their military power in the United States and Europe.[4]

Osgood focused on the primacy of politics and ignored the other elements of limited war. In particular, he slighted the peculiar political problems inherent in limited war. The second theme of his book was that military problems had no proper place in a theory of limited war. This was because limited war was, essentially, a diplomatic instrument, a tool for bargaining with the enemy. Earlier students of limited war, whom we will consider later, assumed that limited war was a form of a war, a variety of combat. But for Osgood, war was "an upper extremity of a whole scale of international conflict of ascending conflict and scope . . . [N]o *definition can determine precisely at what point on the scale conflict becomes 'war.'* "[5] War, as such, did not deserve study, Osgood thought, because war was like peace, only more so. This is contrary to the traditional and commonsense view that war is governed by a set of special rules. Even if it does not set its own goals, as Clausewitz wrote, "it has its own grammar."[6] This attitude led Osgood to an important policy conclusion. If limited war is to be a diplomatic tool, it must be centrally directed by the political leadership. The special needs of the military should not affect the conduct of the war. The military should be "the controllable and predictable instruments of national policy . . . [I]t would be a dangerous error to apply to the whole complex problem of harmonizing military policy with national policy . . . the far simpler imperatives of the battlefield."[7] If war is just another form of coercive diplomacy, then it should be run by the political leadership in Washington, not by the generals in the field. Indeed, in 1979 Osgood wrote that the president of the United States "must be provided with a reliable communications, command, and control system that would enable him to tailor force to serve political purposes under varied conditions of combat."[8] The insistence on centralized control of a limited war derives from his definition of limited war. An appreciation of the problems of the battlefield might have modified this insistence on centralization.

The third theme in Osgood's work was the unimportance of domestic politics. When Osgood says that politics is primary, he meant international politics, *Realpolitik*. After much discussion, Osgood concluded that even though the American people will be hostile, because of their national traditions and ideology, to the kind of strategy he proposes, that strategy must still be adopted. This attitude resulted in some very odd conclusions. For example, he wrote that, "if we anticipate a 'war of attrition,' that would be precisely the kind of war in which our superior production and economic base would give us the greatest advantage." He quotes with approval Henry Kissinger's 1955 assertion that "a war of attrition is the one war China could not win."[9] After a war of attrition in Vietnam produced a wave of popular revulsion we can easily see the error of this statement. This mistake was apparent even in the 1950s to the Eisenhower administration, which was determined to end the war in Korea quickly and to avoid an involvement in Indochina. In both cases, protracted war promised political suicide.

Thomas Schelling began from a point of view much less historical and political than Osgood's, but he arrived at much the same conclusions. He too argued that the study of limited war in no way depended on any actual knowledge of war. In his book *The Strategy of Conflict*, Schelling explained that "the theory is not concerned with the efficient *application* of violence or anything of the sort; it is not essentially a theory of aggression or of resistance of war. *Threats* of war, yes, or threats of anything else."[10] The strategy of conflict is about bargaining, about conditioning someone else's behavior to one's own. It is, therefore, about communication of a certain kind. From these simple assumptions proceed all of Schelling's concepts of deterrence, the communication of resolve, war-limitation by tacit communication, the rationality of looking irrational, and so forth, that are by now so familiar to us. But in the end, it does not look that much different from Osgood's strategy. Neither "limited war" nor "the strategy of conflict" is about war, but about diplomacy and bargaining. The conference table and not the battlefield is the center of the action. Looking back, Osgood summed up the theoretical consensus by saying, "The theory of limited war came to be seen as part of a general 'strategy of conflict' in which adversaries would bargain with each other through the mechanism of graduated military responses . . . in order to achieve a negotiated settlement."[11]

Political scientists have kept this consensus alive and well for a generation. When Kenneth Waltz recently discussed a strategy for limited war in the Persian Gulf area, the same focus on communications, the same rejection of military considerations, was obvious. Though the United States does face the threat of war with the Soviet Union for Iran, "the problem is not to develop a strategy that will help enable us to fight such a war. Instead, the problem is to develop a strategy that will help us to avoid having to do so."[12] Military forces are not for fighting, but for signaling. "If, in a crisis, we were to put our troops

in the oil fields, it would make the depth of our interest, the extent of our determination, and the strength of our will manifest."[13] Minor military problems, such as the fact that the Rapid Deployment Force could not presently keep itself supplied with water in the desert, are apparently irrelevant.[14] Robert Jervis, in an article about limited *nuclear* war, pauses to make the familiar point about the real purpose of *conventional* forces fighting in Africa or Asia: "using large armies . . . [is] less important for influencing the course of the battle than for showing the other side that . . . things will get out of hand."[15]

Other political scientists have begun to notice that all is not well with the theory of limited war. Samuel Huntington suggests that one lesson of the Vietnam war is that limited wars must be acceptable to domestic opinion, and this means they must be short. He is supported by Stanley Hoffmann, who asks, naturally enough, what should policymakers do if strategic communication fails?[16] How can they proceed beyond the old strategy?

VIETNAM AS A TEST CASE

Wars are complex, and the Vietnam war was no exception. Still, it is useful to ask whether the central tenets of the old theory of limited war tended to be confirmed or falsified by the information supplied by that war. In the conduct of the war, was there an emphasis on strategic communication, a consequent focus on tight central control, and a relative disregard for the military and domestic political problems of waging war? If there was, what consequences can legitimately be attributed to that emphasis? It is also necessary to look at the causes of US military successes and failures that had nothing to do with the decisions made in Washington. To sum up the answers to these two sets of questions it is fair to say that, first, the greater the costs and risks of a military measure, actual or contemplated, the greater the tendency for the men at the higher levels of government to talk and act as if they were guided by the academic theory of limited war. This approach to the problem seemed to minimize risk and offer victory without combat. This is true for civilians but, to a surprising extent, also true of military men, particularly General Maxwell Taylor. This tendency was reinforced by mutual distrust between civilians and the military. Together, these factors produced inattention and irresolute behavior that hampered the formation and implementation of an effective military strategy. Second, there were local military mistakes committed by the American forces in South Vietnam that were partially the result of the lack of a clear national strategy but which were also the result of the desire of General William Westmoreland, commander of American ground forces in South Vietnam, to avoid dramatic failure.

Examination of the *Pentagon Papers* can leave no doubt that in 1961 and 1962 American leaders, from the president on down, did *not* think in terms of limited war theory. They focused instead on the military problem of how to beat a guerrilla enemy in a counterinsurgency war. They realized that such a war could only be won by a combination of military action and political and administrative reform, but they nonetheless thought in terms of a war-winning strategy, not in terms of deference, signaling, limitations, or bargaining. In August 1962, McGeorge Bundy, special assistant to the president for national security affairs, circulated a National Security Action Memorandum that instructed the relevant departments to draw up plans of action consistent with the doctrine of counterinsurgency.[17] A bit later that year Michael V. Forrestal, the White House aide for Far Eastern affairs, in a memo to the president dwelled at some length on the specific problems of the battlefield. In light of the subsequent tendency of the army to emphasize "search and destroy" missions and of the government tendency to rely on military signals and diplomacy, this memo is startling both for its attention to military detail and for the quality of the military analysis provided by the soldiers advising the Vietnamese. The advisors said that in the operations by the South Vietnamese army "the proportion of 'clear and hold' operations . . . is too low in proportion to the 'hit and withdraw' operations designed to destroy regular Viet Cong units." Both kinds of action were necessary, but there was too much emphasis on the latter. This is exactly the same criticism that was to be leveled at the United States army five or six years later. The American advisers said the South Vietnamese went in too much for large unit, "elaborate, set piece operations" that chewed up the countryside, but which the communists could easily evade. The Vietnamese tended not to patrol at night, and spent too little time on extended patrols. Air power was being misused and was possibly causing unnecessary civilian damage.[18] Many of the mistakes that the American army would later make were visible on a small scale to Forrestal and to the president.

Nor was this memo an aberration. Roger Hilsman, director of the State Department Bureau of Intelligence and Research, had intensively studied the phenomenon of counterguerrilla warfare by the United States in the Philippines at the turn of the century, and by the French and British,[19] and his memo to Secretary of State Dean Rusk in December 1962 evaluating the performance of the South Vietnamese showed the same attention to military problems, as well as to those of political and administrative reform, and none to signaling the North. Specifically, he warned that the South Vietnamese army would not have much success unless it "appreciably modifies military tactics (particularly those relating to large unit actions and tactical use of airpower and artillery),"[20] and concentrated more on "clear and hold" operations followed by civil and political reforms.[21]

But when an increase in the stakes was considered, the question the leaders in Washington asked themselves was not "how will this affect the resolution of combat" but "what signal are we sending the enemy?" This concern with the question of signaling applied, at least in part, to the American military. In May 1961, Deputy Secretary of Defense Roswell Gilpatrick asked the Joint Chiefs of Staff if they would recommend a plan to send two 1600-man American combat units to South Vietnam. The Chiefs said yes, they would, because it would deter the North Vietnamese and the Chinese, but also because it would release ARVN (Army of the Republic of Vietnam) troops for patrolling.[22]

When America had 685 soldiers in South Vietnam, 3200 more men represented a small increase in the stakes. The Chiefs wanted to communicate resolve and commitment to their enemies, neutrals, and allies, but also considered the plain military advantages.

As the stakes grew larger, so did the perceived importance of an action as a signal. In October of that year, General Maxwell Taylor recommended to the president that 6000–8000 combat troops be sent to South Vietnam. The purpose was to show America's commitment. "[T]here can be no action so convincing of US seriousness of purpose and hence so reassuring to the people and Government of SVN and to our other friends and allies in SEA." The fact that an increment of force this small would have little practical military effect was unimportant since a lot of troops were not "necessary to produce the desired effect on national morale in SVN and on international opinion."[23] Defense Secretary Robert McNamara, moreover, rejected the proposal, largely because it was likely to be an inadequate signal and unlikely to "tip the scales decisively . . . [I]t will not convince the other side (whether the shots are called from Moscow, Beijing, or Hanoi) that we mean business." This could be done "only if we accompany the initial force introduction by a clear commitment to the full objective accompanied by a warning through some channel to Hanoi that continued support of the Viet Cong will lead to punitive retaliation against North Vietnam."[24]

Here was the idea of a mix of military action and diplomacy in order to communicate with the enemy in a way that was favored by the theorists. There is no doubt that others watch American actions and draw conclusions from them, but this was one of the first suggestions that leaders should decide what military actions should be taken on the basis of their value as a signal.

This tendency became unmistakable early in 1964. The proximate cause was the disastrous military conditions within South Vietnam. Following the decline and fall of the Diem regime, almost all counterinsurgency efforts, including the strategic hamlet program, came to a halt or went into reverse. Simultaneously, the Viet Cong stepped up their side of the war and, by November, regiment-size Viet Cong units were conducting conventional attacks.[25] Nothing the United States could do in the short term, other than the immediate dispatch of large numbers of ground troops, could help the actual military situation in the South. McNamara had estimated that it would take 208,000 American troops to make a difference and to deal with the inevitable overt introduction of North Vietnamese troops that would follow in response. Any more than 208,000 would interfere with US plans for the defense of Europe.[26] In 1965, Lyndon Johnson was not prepared to lose South Vietnam, but neither was he prepared to send 208,000 men. But he could send signals and avoid making a decision. It was a cheap, low-risk approach.

This is harsh judgment, but it is justified by the record in the *Pentagon Papers*. Various measures to coerce North Vietnam were considered. Covert actions against the North were continued despite the judgment of the review committee chaired by General Krulak, former Pentagon counterinsurgency expert and then commanding general, Fleet Marine Force, Pacific, that the communist leaders were tough and would not respond to this "coercive diplomacy" unless "the *damage* visited upon them is of great magnitude."[27] Walt Rostow, chairman of the State Department Policy Planning Council, disagreed. By November of 1964, Rostow was complaining to McNamara that "too much thought is being given to the actual damage we do in the North, not enough to the signal we wish to send." He recommended that any use of force against the North "should be as limited and insanguinary as possible."[28] The State Department had recommended to the president in February that twelve F-100 fighters be sent to Thailand, not for their military utility (they would have no significant effect on infiltration through Laos, said the US local embassy in Laos), but "with a view toward . . . potential deterrence and signaling impacts on communist activities in Laos."[29] It was generally acknowledged that the kinds of attacks against the North that were being contemplated would be militarily ineffective. The National Intelligence Estimate explicitly stated, in May 1964, that the combination of bombing and negotiation under

consideration "would not seriously affect communist *capabilities* to continue that insurrection," but it would affect Hanoi's *will* to some extent, and it would also signal America's intention to limit the extent of the war.[30] The Joint Chiefs of Staff were aware of and unhappy with the coercive diplomacy. "We should not waste critical time and more resources in another protracted series of messages, but rather we should take positive, prompt, and meaningful military action."[31]

If the civilians were wrong, the military was not necessarily right. The full-scale bombing effort recommended by the military would *not* have ended the war on terms favorable to the United States, and McNamara was justified when he asked the Joint Chiefs if their recommended 94 target plan would end North Vietnamese support to the Viet Cong. If not, what would?[32] It is, however, legitimate to attribute to this limited war attitude three practical consequences. Concentrating on the dispatch of signals diverted attention from a search for military measures that could have been successful. It led decisionmakers into actions they knew the American people would not like and might reject. And it was consciously used as a way of avoiding or deferring risky decisions. These conclusions are best supported by one of the most important decisions of this period, the decision to begin the ROLLING THUNDER bombing campaign against North Vietnam. A working group chaired by William P. Bundy, assistant secretary of state of East Asian and Pacific affairs, convened in November 1964 to review future American policy for Southeast Asia. It came up with three options. Option A was more of the same: United States aid and advisers plus tit-for-tat strikes against the North if American soldiers were attacked in the South. Option B quickly acquired the nickname "the full squeeze": enemy bridges, lines of communication, and industry were to be bombed and enemy harbors mined. It was to be "a systematic program of progressively heavier pressures against the North Vietnamese, to be continued until current objectives were met. Negotiations were to be resisted" in order to prevent communist peace offensives from halting American action. Option C was the "progressive squeeze and talk," which was a program to increase US pressure against the North gradually, coupled with a stated willingness to stop the pressure and negotiate.[33] This was the policy which Lyndon Johnson's conduct of the war—the slow expansion of the target list plus the 1967 San Antonio formula for unconditional negotiations[34] plus the repeated bombing pauses—most closely resembled. It was a limited war strategy of signaling by military means US commitment and then proceeding diplomatically. Assistant Secretary of Defense for International Security

Affairs John McNaughton, the drafter of the options, unconsciously highlighted this resemblance, "To change DRV [North Vietnam] behavior [change can be tacit] US should 'negotiate' by an optimum combination of words and deeds." At the same time, "It is important that USSR and China understand the limited nature of our deeds—i.e., not for a colony or base and not to destroy NVN."[35]

What were the consequences of this attitude? First, it caused the military problem of how to win the war on the ground in the South to be neglected. The anonymous author of the section of the *Pentagon Papers* that discusses the beginning of the American combat troop commitment comments on the absence of any documents discussing the proper role or rationale for American *ground* forces in South Vietnam until they were actually sent in March. "In other words, it appears that the key decision-makers in Washington are not focusing on the importance of deployment. The attention getter as the [2/7/65] Bundy memo [to LBJ] indicates was the impending air war against North Vietnam."[36] The author is not quite correct. There was some discussion of the role of ground troops during this period. They were considered as part of a limited war-bargaining strategy. Within Bundy's working group, the *Pentagon Papers* notes, "It was the recognized lack of strong bargaining points that led the working group to consider the introduction of ground forces into the northern provinces of South Vietnam." Troop deployment signaled commitment, and, as the representative from the State Department Policy Planning Staff pointed out, Hanoi's price for negotiations was likely to be an end to the bombing. In that event, troops on the ground would be "a valuable bargaining piece."[37] William Bundy wrote in January 1965 that he liked the idea of sending troops to the northern provinces of South Vietnam. "It would have a real stiffening effect in Saigon, and a strong signal effect to Hanoi."[38]

Because troops were there primarily to signal, it did not matter so much what their combat qualities were. The first troops not to be sent as advisers were the Marines who guarded the air base at Da Nang. They were chosen because, as Marines, they had the ability to keep themselves supplied "over the beach" in an area in which the logistics network was not yet developed. They also had some heavy equipment to help defend themselves and the base. At the last moment, John McNaughton tried to halt their dispatch. The *Pentagon Papers* infer that he thought they were *too* heavy. They would signal the North that Americans were coming with heavy offensive units that were there to stay. Instead the 173d Airborne Brigade should be sent. It was light, and would signal the willingness to move them out if necessary.[39]

McNaughton ultimately was unsuccessful in this instance but his way of thinking did prevail, Westmoreland claims, in another case. In 1965, the United States observed the construction of the first surface-to-air missile sites in North Vietnam, and the military sought permission to attack them before they were completed, to save American casualties. "McNaughton ridiculed the idea. 'You don't think the North Vietnamese are going to use them!' he scoffed to General Moore. 'Putting them in is just a political ploy by the Russians to appease Hanoi.'

"It was all a matter of signals said the clever civilian theorists in Washington. We won't bomb the SAM sites, which signals the North Vietnames not to use them."[40]

There is a good deal of bitterness in the story, and Westmoreland seems to be reporting it secondhand. McNaughton's earlier action is, however, well established though his motives are not. There is also the matter of the one effort made to figure out what military role the American troops would be *able* to perform, *before* large numbers of them were sent to fight. Late in March 1965 Maxwell Taylor, then ambassador to South Vietnam, cabled Washington with a natural request. What was to be the strategy for the use of troops that were coming? They could be used either offensively or defensively to establish enclaves, they could conduct "clear and hold" operations, or they could be used as a general reserve to backstop the South Vietnamese. Taylor preferred a combination of the first and the last, but more than anything he wanted some kind of decision. He proposed that the first Marine units be used for 60 days in an experiment to see whether conventional American ground troops could successfully adapt to the requirements of a counterinsurgency war, *before* more troops were sent to try and fight that war.[41] His questions were not answered, and at the urging of Lyndon Johnson the United States, less than a month after Taylor sent his cable,[42] authorized the deployment of 82,000 troops with more to follow.

It is not correct to say that the Washington leadership was insensitive to public opinion. There is the famous example of Johnson explicitly approving a change in mission for the Marines sent to Vietnam "to permit their more active use" and, in the same document, calling on officials to avoid any publicity and to "minimize any appearance of sudden changes in policy."[43] The men who formulated the three bombing options, however, were aware that what they were suggesting was not going to receive the blessing of the American people. William Bundy wrote a report, based on drafts by McNaughton, summing up the problems with option C. "This course of action is inherently likely to stretch out and to be subject to major pressures both within the United States and internationally. As we saw in Korea, an 'in-between' course of action will always arouse a school of thought that believes things should be tackled quickly and conclusively. On the other side, the continuation of military action and a reasonably firm posture will arouse sharp criticism in other political quarters."[44] Looking back on this aspect of the war, Dean Rusk said "We never made any effort to create a war psychology in the United States during the Vietnam affair. We didn't have military parades through cities . . . We tried to do in cold blood perhaps what can only be done in hot blood, when sacrifices of this order are involved. At least that's a problem that people have to think of if any such thing, God forbid, should happen again."[45]

The national leaders wanted to keep public opinion quiet in order to keep control over the war and to avoid escalation that might lead to a large conventioinal or nuclear war. The need to avoid nuclear war was and is unquestionable. It quickly produced, however, a tendency to choose plans that were controllable over plans that would be militarily successful. Option C was preferred because, as McNaughton put it, it was "designed to give the US the option at any time to proceed or not, to escalate or not, and to quicken the pace or not."[46] The desire to keep the limited war limited also increased the pressure to centralize control of the war in the hands of the president. Johnson proudly told Doris Kearns that, "by keeping a lid on all the designated targets, *I* knew *I* could keep control of the war in my own hands. If China reacted to our slow escalation by threatening to retaliate, we'd have plenty of time to ease off the bombing."[47] The ultimate result was that, by 1968, General Westmoreland needed special authorization to use antipersonnel rounds in the artillery pieces defending Khe Sanh. Johnson and his senior advisers would insistently interrogate Westmoreland about the details of his defense plans. What would he do if there was bad weather around Khe Sanh? Is his long-range artillery effective?[48] The desire for central control became excessive as the war dragged on.

At the same time, the Washington leadership failed to do the one thing that the central leadership must do. It did not define a clear military mission for the military, and it did not establish a clear limit to the resources to be allocated for that mission. Because the Johnson administration could not bring itself to make two big decisions, it intruded itself into the making of innumerable little decisions. This hindered the military from carrying out the mission that it had, of necessity, defined for itself. The limited war attitude, with its emphasis on signals, central direction, and war limitation by means of flexible policies, contributed to, if it did not cause, this situation.

This combination of high-level indecision and

micromanagement first arose when combat troops were sent to Westmoreland. Taylor had tried unsuccessfully to get a strategy defined, but now the practical question was—how many troops to Vietnam? In the case of the Korean war, the decision was made in a clear-cut military manner. MacArthur was initially given all the troops in the Pacific Command plus just about all the reserves in the continental United States, Puerto Rico, and the Canal Zone. The troops in Europe were not touched, and, in December 1950, as mobilization made more troops available, the Joint Chiefs of Staff made a difficult decision: no more reinforcements for Korea. The United States had the responsibility to defend Europe in a general war. If Korea became part of a bigger war, the defense of Europe would have to come first. Korea would be defended with the existing commitment, or it would be evacuated.[49]

No such decision as to where the Vietnam war lay on the US list of national security priorities was ever made. McNamara tried in 1961, when he said 208,000 was the maximum that could be spared, as we have seen. But that level was passed by 1967. Instead of making that decision, the buck was passed downward. Johnson did not want a big war, but neither did he wish to be accused of losing the war by denying his field commander what he needed. So Westmoreland was to be given whatever he asked for, short of force levels that would require mobilizing the reserves. In July 1965, McNaughton instructed the staff of the Joint Chiefs that the president was willing to keep adding ground troops "as required and as our capabilities permit."[50] Over 400,000 men were authorized in the spring of 1966, but one can search the *Pentagon Papers* in vain for a rationale justifying that level. The author of that section himself notes, "The question of where the numbers . . . came from provokes much speculation."[51] The figure came from nowhere because the administration abdicated its responsibility to set priorities. Johnson never formulated a clear policy, either global or regional, so he could never say what was or was not needed in Vietnam. All he could do was send a memo to McNamara in June 1966 stating "As you know, we have been moving our men to Vietnam on a schedule determined by General Westmoreland's requirements."[52] How Westmoreland's requirements should fit in with the leaders' other requirements was never resolved. Because McNamara had no policy line, all *he* could do was send a memo to the generals in August and demand detailed accounting of what they were up to. He would send everything Westmoreland required. "Nevertheless I desire and expect a detailed, line-by-line analysis of these requirements to determine that each is truly essential to the carrying out of our war plan."

This was nonsense. The administration could either send everything Westmoreland asked for, or better, it could send what was required by the war plan consistent with its other objectives; but this was just harassment. The best McNamara could do to set guidelines for the generals was to warn them that sending too many troops to Vietnam would "weaken our ability to win by . . . raising doubts concerning the soundness of our planning."[53] This was a far cry from the detailed military discussions of counterinsurgency of 1961, which set concrete goals and courses of actions against which the military could be measured but within which the military would be left alone to deal with local conditions as they saw fit. This failure of strategic thinking grew worse until the Tet crisis brought matters to a head, and Clark Clifford to the Pentagon, where he exclaimed in despair, "I couldn't get hold of a plan to win the war . . . [W]hen I attempted to find how long it would take to achieve our goal, there was no answer. When I asked how many more men it would take . . . no one could be certain."[54]

THE ABSENCE OF STRATEGY

What had happened to American military strategy? Its absence became painfully obvious in 1968, but actually it was missing from 1964 on. The answer, in part, is that America had adopted a limited war signaling strategy. The military measures adequate for that strategy were not adequate for successfully prosecuting the war in South Vietnam. As the Bundy working group admitted, the measures it proposed "would almost certainly not destroy DRV capability to continue supporting the insurrection . . . should Hanoi so wish."[55] Hanoi *did* so wish, the signaling strategy collapsed, and the leaders were left without a policy. The importance of a clear policy and a clear allocation of resources is made apparent by looking at what happened when, pressed to the wall, the Johnson administration did set a troop ceiling. By late 1967, 525,000 men had gradually come to be authorized for Vietnam. Many had been taken away from units defending Europe. No more would be available without calling up the reserves, and this Johnson would not do. Westmoreland did not like this situation, but finally he knew where he stood. He could no longer hope to get the number of troops he wanted to wage his kind of war, but he could and did formulate a strategy to make the best use of the resources he had. He began a serious program of Vietnamization. "It was the only strategy that I could come up with that was viable if there was no change in policy . . . It was my strategy, and I portrayed it as such. The administration was totally noncommittal on it. They kind of nodded their heads and did not disagree."[56] But

this was better than nothing, and it began the program that, by 1972, would allow the South Vietnamese to defeat an armored invasion with the help of American logistics and air power, but without American combat troops.

Why was the warfighting, counterinsurgency attitude of 1961 replaced by the bargaining, limited war attitude by 1964? While it was natural for Washington leaders to try and increase their political control over the military as American commitment grew, it would have been equally natural for them also to increase their concern with problems of military strategy, as opposed to signaling, once US forces were actually involved in combat. This did not happen. Why?

The nature of the people handling the war in Washington changed. As the war grew bigger, it drew in more senior people, as more high-level officials became involved in the direction of the war. These men simply had had little experience in the direction of a war, and had not studied military problems. What *did* they know? Walt Rostow described himself and his colleagues who came to work in Washington for John Kennedy: "Some had been trained in modern economic theory . . . In the 1950s they had focused their minds sharply on problems of nuclear deterrence and arms control; on the need for highly mobile conventional forces in a nuclear world; on how to organize the Pentagon and military budget to produce a rational force structure. And when they took posts of responsibility, they felt comfortable with this array of problems, even in such acute forms as the Berlin and Cuba missile crises of 1961–1962.

"But they found themselves caught up in a problem for which they were ill-prepared—guerrilla warfare."[57] In short, these men knew limited war theory and defense economics, but not military strategy. Rostow argues that, "instead of constructing an alternative, systematic analysis of the cause of the battle, they tended to do something more limited but wholly legitimate: that is, to debate critically the views (or believed views) of the military."[58] Men found limited war theory a quick and easy way to become fluent participants in a crucial debate. Other officials had had direct experience with the skillful *nonuse* of force in the missile crisis. From the documents, it seems that some of them carried this way of thinking into an area where it was less helpful, into the realm of the actual conduct of war.

If the civilian leadership did not have such knowledge, where could they get it? They could go to school, and at the instigation of President Kennedy, the Foreign Service Institute set up a course for senior- and middle-level officials to teach them about counterinsurgency warfare. Henry Cabot Lodge delayed his departure to South Vietnam as ambassador so that he could take the course. Hundreds of other officials eventually joined him. Lectures were given by Walt Rostow, Edward Lansdale, and MIT professor Lucian Pye. It was a failure. The course lasted six weeks and dealt with the whole range of counterinsurgency problems in underdeveloped countries all over the world. This simply could not be done, or done well. Douglas Blaufarb, a CIA official who attended the first six-week session, described the character of the course: "It was highly generalized and often left the officers at a loss as to how to translate the generalities into policies, and, even more difficult, into practical actions."[59] It could not have been otherwise.

If the civilian leadership could not acquire the necessary expertise in a hurry, to whom could they turn for help? The obvious answer is "to the military." This was not done because the military was not trusted. Civil-military relations in the United States were, and are, on the surface, satisfactory. Civilian control is a universally accepted principle. Below the surface, relations were bad. In the back of everyone's mind in the 1960s was the memory of General Douglas MacArthur's insubordination in Korea. Lyndon Johnson, never a man to leave something in the back of his mind, told Westmoreland flat out in February 1966: "General, I have a lot riding on you . . . I hope you don't pull a MacArthur on me."[60] Johnson and Westmoreland got along reasonably well, but the suspicion was there. In private, Johnson was vivid: "And the generals. Oh, they'd love the war, too. It's hard to be a military hero without a war . . . That's why I am so suspicious of the military. They're always so narrow in their appraisal of everything."[61]

This general suspicion had been increased for the men who worked for John Kennedy by the experience of the Cuban missile crisis. According to his brother Robert, John Kennedy was "distressed" with his military advisers. "They seemed always to assume that if the Russians and Cubans would not respond, or [even] if they did, that a war was in our national interest." This remark was probably prompted by the recommendation that Air Force General Curtis Lemay had made to bomb the missiles in Cuba *after* the Soviets had begun to withdraw them. "[T]his experience pointed out for all of us the importance of civilian direction and control."[62] A somewhat less tactful remark by John Kennedy has been recorded: "The first advice I'm going to give my successor is to watch the generals and to avoid the feeling that just because they are military men their opinion on military matters is worth a damn."[63] There was the famous encounter between McNamara and the Chief of Naval Operations during the blockade of Cuba that ended with McNamara shouting that the object of the operation was not to shoot Russians but to communicate a political message: "I don't give a damn what John Paul Jones would have done."[64]

Relations had not improved with time. McNamara's obsession with getting control of the defense budget made things worse. In Lyndon Johnson's words, "Why, no military men could spend a dime without McNamara's approval. He fought and bled for the principle that the Joint Chiefs of Staff could not get a mandate without a specific request. Otherwise, we would be giving them money based on pie-in-the-sky figures."[65] All in all, the civilians were not men who would turn easily to the generals and say, "Teach us about strategy."

It must be said that generals were and are often wrong. Their advice in the Vietnam war was often bad. The military, however, was fighting the war and had the data and personal experience that were crucial to the formulation of good strategy. Bad relations meant that the civilians and the soldiers were less likely to work together to develop good strategy. Instead the civilians were inclined to turn to limited war theory. It enabled them to make strategy of a sort without help from the generals. It gave them power over the generals, which is what they wanted.

This state of affairs came to a head with the 1968 Tet offensive. To sum up the state of affairs before the communists launched their attack, we can say that: 1) there was no generally agreed upon comprehensible military strategy for winning the war, and no clear definition of the amount of resources to be devoted to the war; 2) there was a limited war theory of signaling, but it had been a complete failure; and 3) as a result of the limited war attitude and other causes, decisionmaking had become centralized in Washington. This combination of factors had brought the civilian leadership in Washington close to collapse and the Tet offensive pushed them over the edge. The kind of "war" they understood had not produced results. They had no theory to help them understand the macro-course of the war. They had masses of detailed data, but no way to understand the micro-course of the war. They were in Washington and not in the field so they had no way to see through the statistics to the reality the numbers were supposed to represent. And they did not trust the judgment of their military men in the field. An enemy "spectacular," no matter how catastrophic for the enemy, was likely to be seen in the worst possible light. Henry McPherson, a speechwriter for Johnson, was disturbed by Tet and talked to Rostow:

Well, I must say that I mistrusted what [Rostow] said, although I don't say with any confidence that I was right to mistrust him, because . . . I had the feeling that the country had just about had it, that they would simply not take any more . . . I suppose, from a social scientist point of view, it is particularly interesting that people like me . . . could be so affected by the media as everyone else was, while downstairs, within fifty yards of my desk, was that enormous panoply of intelligence-gathering devices—tickers, radios, messages coming in from the field. I assume the reason this is so . . . [was] I was fed up with the optimism that seemed to flow without stopping from Saigon.[66]

William Bundy remembers a memo about the effect of Tet on pacification: "It was a poignant memo which said in effect, 'They've had it.' That memo reflected my view for a period."[67] Daniel Ellsberg, then on loan to the Defense Department from the RAND Corporation, was more precise.

In a 28 February 1968 memo, he wrote, "I think that the war is over; [our] aims are lost . . . The Tet offensive and what is shortly to come do not mark a 'setback' to pacification; it is the death of pacification . . . I am forced to predict not only that the 'blue' areas will contract in the next few months and the 'red' zone expand, but that the new red on the maps will *never go back.*"[68]

This was proven to be wrong. The optimism flowing from Saigon was, for once, justified. The data on pacification that was available to the men at the center looked bad. There was a sevenpoint drop in the number of secure or relatively secure hamlets in South Vietnam after Tet. The men in the field were frantically reporting that this was true, but that they were also rapidly wiping out the Viet Cong forces that had made the countryside insecure. The VC had come into the open to attack the cities, and they were being killed. Pacification figures temporarily looked bad because Allied forces were being drawn away from the countryside to kill the communists in the cities. Pacification would be in great shape in a while. They were absolutely correct—by 1970 over 90 percent of the hamlets were at least relatively secure. These figures were verified by students of the peasantry hostile to US policy such as Samuel Popkin.[69] But in Washington, men were neither ready to believe their men in the field, nor able to understand the war themselves.

REDEFINING THE THEORY

There is another way to think about limited war. It emphasizes the construction of clear military objectives and military limits by the government, as well as the political objectives and limits that were the concern of the old theory. It emphasizes the need for decentralized, rather than centralized, control of the war, to the extent possible, once fighting actually begins.

War is uncertain, and so there is no certain road to success. There are, however, better and worse ways to begin. Stanley Hoffmann has noted an important fact: "When one is talking about limited war . . . each one is *sui generis.*

Forces designed to fight a major technological war against the main opponent are fairly fungible; forces supposed to fight low intensity wars are not. A force that would have been perfectly equipped to fight in Vietnam is not usable as such in the Persian Gulf . . . Those who talk about the primary role of military force have never really faced that problem."[70]

Small wars are *sui generis,* but so are big wars. The difference is we generally have only one or two big enemies. A nation usually has ample time to study its big enemy, and to review its past wars with it. Countries are familiar with the terrain because they have fought there before, and indeed, often live on it. They will usually design and build their armed forces to beat this one, specific enemy. And even then, there can be surprises, as the French found out in 1940. If big wars are hard to understand in advance, how much more difficult is the task of fighting little wars for those countries who must do so? Little enemies are legion. They are located all around the world, and have different climates, terrains, and armies. The special military problems of small wars, Hoffmann to the contrary notwithstanding, have long been studied by military thinkers. G. F. R. Henderson, the foremost British military historian of the nineteenth century, listed the maxims that guided military strategy for wars in Europe, but then acknowledged their irrelevance to wars in India.[71]

Another British officer, Charles Callwell, wrote a book at the beginning of the century entitled *Small Wars,* in which he addresses exactly our problem. "In great campaigns, the opponent's system is understood . . . it is only when some great reformer of the art of war springs up that it is otherwise. But each small war presents new features . . . Small wars break out unexpectedly and in unexpected places . . . The nature of the enemy . . . can be only very imperfectly gauged."[72] Closer to our own time, the commander of US Marine Corps forces in South Vietnam captured the essence of the problem in the title of his book *Strange War, Strange Strategy.*[73]

There are two conceivable ways of dealing with the problem of small, strange wars. The first is to foresee all possible contingencies and tailor separate forces for each of these wars. This is near to impossible. The United States army is currently tying itself into knots over the problem of how to structure itself for two different kinds of war, one against the Warsaw Pact and one in the Persian Gulf area. Everything policymakers know about war, everything they know about large organizations, argues against having, in effect, two separate forces within one organization. Now multiply this problem by the number of different conceivable little wars. To start, there are at least three distinct types of terrain merely in the Persian Gulf region: dry desert in the interior of Saudi Arabia and Iran, heavily forested mountains in Oman and along the Caspian Sea, and humid, malarial swamp along the Iran-Iraq border. Three different kinds of enemy could be encountered: Soviet armor and forces, local armored forces, and light infantry. War in each terrain and against each enemy has its own set of requirements. If you have only one enemy, you can afford to tailor your forces. To the extent that strategists can prethink operations in Southwest Asia, they should do so, but they cannot create little armies for every little war.

The other conceivable approach to the problem is to increase the speed with which America's one army can adapt to local circumstances. Specialized advance planning can backfire. The British army in India was trained for desert warfare in the 1930s as a result of its experience in World War I, but was then suddenly called upon to fight in the jungle to defend Burma against the Japanese.

What do armies need to deal with this problem of unfamiliar combat? The first and most basic necessity is confidence in themselves and confidence in their leaders. This is because, in these difficult circumstances, there will be some, perhaps many, initial defeats. This will crack some units. Some men will not obey officers, and some officers will fear to undertake offensive operations or to run risks. If this goes too far, it will be fatal. A professional army with long-service men is more likely to have this confidence simply because everyone will know and trust everybody else. The bad ones can be weeded out, and trust and cohesion established. The British army in Burma was such an army. It was repeatedly mauled by the Japanese because it had not learned to conduct operations in the jungle. It was roadbound, and so was outmaneuvered. It had to retreat from Burma into India through hundreds of miles of jungle and mountains. When they emerged, they passed in review before their commander, General William Slim. "All of them, British, Indian, Gurkha, were gaunt and as ragged as scarecrows. Yet, as they trudged behind their surviving officers in groups pitifully small, they still carried their arms and kept their ranks, they were still recognizable as fighting units. They might look like scarecrows, but they looked like soldiers, too."[74] They gave a cheer for Slim. They were, in short, ready to learn from their experiences and to follow their commanders.

Courage is necessary to keep the army going, but it is necessary for another reason. This point requires explanation. By 1964, the war in Vietnam was not a guerrilla and antiguerrilla war only. The communists could and did assemble from time to time large units up to the size of a division. William Westmoreland was perfectly correct when he said that you could practice counterinsurgency with all the success in the world, but unless you also dealt with the big enemy units, they could bust up the pacification program at will. At the end of 1964 a Viet Cong division, in what, as Westmoreland put it, was "probably the most portentous ARVN defeat of 1964,"[75] overran a hamlet and destroyed two elite South Vietnamese battalions. For this reason, Westmoreland consistently rejected the advice of many (and the actual example of the US Marine Corps) to use a large portion of the army in small unit patrols, though this was the most effective means of locating and killing enemy forces.[76] Instead he conducted large-unit (i.e., battalion-sized and up) patrols to locate and fight the enemy. Large units could better defend themselves against large enemy units. They could not, however, bring the communists to battle against their will because the large units could be detected and evaded. A good strategist in Washington would have given Westmoreland the mission: "Deny the enemy the ability to operate large units in the South. Then we can get on with the anti-guerrilla war." Large-unit actions, the infamous "search and destroy" missions, could not and did not do this. In 1965 and 1966, the communists would often stand and fight against US units when their base areas were invaded. After some very costly defeats, the communists gave up this strategy and evaded large US units. By late 1966, the *Pentagon Papers* reports, "the VC/NVA avoided initiating actions which might result in large and unacceptable casualties from the firepower of Allied Forces. During the year, the enemy became increasingly cautious in the face of increased Allied strength . . . VC tactics were designed to conserve main strength for the most opportune targets."[77]

The US Army, in its *Vietnam Studies* series, has written that the enemy "normally defended by evading . . . The enemy's combat forces were lightly equipped so that they could move more freely and quickly."[78] Big enemy units successfully evaded American forces by breaking up into small units out in the wilderness.[79] But even when the enemy came out of hiding and attacked en masse as they did in February, May, and August 1968, their casualty rate did not exceed their ability to replace their men— 291,000 were lost; 298,000 were brought in or locally recruited.[80] High casualties, even when replaced, meant the loss of experienced cadres and battle-tested soldiers. But they could protract the war indefinitely.

This was not an unsoluble problem for us. The key was intelligence. Police work and the usual tools of counterinsurgency warfare worked against guerrillas, but the big enemy units often operated away from the population. The traditional tools of conventional military intelligence

—interrogation of prisoners, use of captured documents, aerial reconnaissance, communications intercepts—worked, but were also deficient. The enemy could move fast. Both police work and conventional military intelligence did yield results, but they were *slow*. A village might learn of the presence of an enemy unit, but by the time he went to town and reported it, and this information had trickled up and down the chain of command, three days, on average, had gone by.[81] The enemy unit had long gone. The same was true for other techniques. American large-unit operations aggravated this problem. Large units moved more slowly on the ground. If they moved by helicopter, they required a long period of advance planning to assemble the necessary aircraft, plan the resupplies, coordinate the artillery fire, and so forth.[82] If the South Vietnamese units were involved, it was almost certain that warning would be given to the enemy long in advance. Planning took weeks. General Julian Ewell, who commanded an American division in the Mekong Delta, found that he could stop the enemy from escaping when his reaction time was reduced from 60 to 10 *minutes*.[83] The big American units could only cover a certain amount of ground. They would kill the enemy soldiers they found, but, as Westmoreland admits, the "enemy often escaped."[84] The worst problem with big-unit operations was that military leaders could not have many of them. They could not check out all intelligence leads or pursue every contact. There were only about 100 American maneuver battalions in Vietnam at the peak of US involvement. Base defense, rest, and replenishment cut down on the number available for patrol. As a result, American troops were concentrating on fighting the enemy's big units, which was all right, but they were losing, which was not. Even so, large-unit sweeps were the rule through 1968.[85]

The answer is obvious in retrospect, but it was apparent to some at the time as well. If the enemy escaped by breaking into small units, American troops would break down into small units to keep after him. If he assembled a superior force, they would use their superior firepower and mobility to reinforce their patrol, and do it quickly, before the enemy had time to disperse and escape one more time. It was not always the right tactic, but it kept the pressure on the enemy. General Julian Ewell applied this tactic with an air mobile force against enemy guerrilla forces in the Mekong Delta in 1969.[86] Marines applied it against North Vietnamese units in the northern provinces of South Vietnam until 1967.[87]

Why was this method not employed more extensively? Why did Westmoreland not change his tactics when they did not produce results? If the small unit was not rapidly and adequately assisted after it made contact, it would be wiped out. Let us backtrack to the beginning of Westmoreland's tour of duty.

When Westmoreland first arrived in South Vietnam, he *did* advocate these tactics. Then something happened. Late in 1964, "the ARVN incurred a serious defeat *for which I bear a measure of responsibility. At my urging,* ARVN leaders broke down their forces into small units, parceling them out to district chiefs to provide protection throughout the province and to patrol extensively in hope of inhibiting VC movement. The tactics worked fine for awhile, but in November 1964 two mainforce VC regiments came out of the hills and opened a general offensive.

"One by one the big VC units defeated the small ARVN and militia units. Lacking an adequate reserve, ARVN leaders were powerless to strike back."[88] Westmoreland never again advocated small-unit operations on a large scale. It did not matter that by 1967 America *did* have the ability to reinforce patrols. Westmoreland himself gives examples of how air power or air reinforcement saved ARVN and American forces in the same places and circumstances where French units, without air support, had been wiped out.[89] *A strategy of small-unit patrols ran too many risks.* American generals, Westmoreland included, are aggressive and proud of it. "Nobody ever won a war sitting on his ass" is the remark that sums up the attitude of the American army. But Westmoreland was interested in avoiding disaster. He is proud of this too. In Vietnam, he writes, no sizeable American unit "ever incurred what could fairly be called a setback. That is a remarkable record." He later repeats himself. "I could take comfort in the fact that in the Highlands [scene of the search and destroy missions] . . . the American fighting men and his commander had performed without the setbacks that have sometimes marked first performances in other wars." He repeats himself again—Americans had none of the catastrophes experienced by the French in Vietnam.[90] We cannot help but remember the old saying "one cannot learn unless one makes mistakes," and one cannot win wars by avoiding risks.

Westmoreland's predicament was a painful one, and it points out why learning in war is different from other kinds of organization learning. Mistakes in war mean the wasteful death of men who trusted one to make the right decision. No one would want US officers ever to forget that. Small units *were* occasionally badly hurt in the Delta.[91] This does make operational military innovation difficult. Westmoreland was not the first officer to hesitate before trying a new tactic that would be disastrous if incorrect. One of the longest-lived failures of military adaptation occurred in the battle of the Atlantic

in World War I. The German submarine force
had interfered with British shipping, but had
not created an intolerable situation until the
start of the unrestricted U-boat campaign in
February 1917. Merchant ship losses quickly
mounted to the point where shipping capacity
was predicted to be only 60 to 70 percent of
what would be needed in the period April to
August 1917.[92] The problem was in one way
analogous to that facing Westmoreland. Enemy
submarines could see destroyers before the destroyers could see them. If one area was pa-
trolled, the U-boats would operate elsewhere.
There were not nearly enough destroyers to
blanket the zones in which the U-boats could
operate. The answer was the convoy. If you put
the destroyers with the merchant ships, they
would be able to respond to the appearance of
a submarine in good time. Many people at the
time, up to and including Prime Minister David
Lloyd-George, saw the merits of the convoy,
and still it was resisted by the Royal Navy. The
reason is no secret. *The risk was too great until
all else had failed, and national defeat was the
only visible alternative.* Admiral John Jellicoe
was quite frank after the war: "Until unrestricted submarine warfare was instituted, the
losses in the Mercantile Marine from submarine
attack were not sufficiently heavy to cause the
Admiralty *to take upon themselves the very
grave responsibility of attempting to introduce
the Convoy System,* because of its many disadvantages combined with *the fear* that an insufficiently protected convoy, if seriously
attacked by submarines, might involve *such
heavy losses as to be a real calamity.*[93] The Admiralty finally did adopt the convoy system before there was an actual confrontation with
Lloyd-George, but they waited until they calculated they could lose three ships out of every
convoy to submarines and still be no worse off
than they already were.[94] They waited, in other
words, until the *relative cost of making a mistake* by adopting the new tactic was low. These
men were anything but cowards, but the unique
demands of this kind of innovation required a
kind of courage they did not have.

The need for this courage is the reason why
learning in war is so difficult. When the need
for this courage is removed, the task is easier.
An outstanding success story in the history of
military learning and adaptation was in the "war
of the beams," the war between the Germans,
who built the radio navigation aids that guided
German night bombers, and the Britons who
tried to thwart them. R. V. Jones gives a splendid account of his personal victories in this battle. What his account makes clear is that his
countermeasures were rapidly invented *and implemented,* because *nobody would be worse off
if he were wrong.* His work did not require the
diversion of large amounts of resources. If his
interference with the German system failed, the
bombers were coming anyway, and it did no
harm to try. In the one case where matters
would have been worse if he were wrong, things
were quite different. He was an advocate of the
use of chaff to help British bombers penetrate
German radar. He saw no reason to "be squeam-
ish" as he put it, that the Germans might learn
about chaff from the British and turn around
and use it to increase their bombing of England.
If Jones were wrong, hundreds *more* people
would die. The final decision was made by
Churchill in consultation with Leigh-Mallory,
head of Fighter Command who, in Churchill's
words, "would have to 'carry the can' " if British
defenses broke down because of German chaff.
In the decisive meeting, Churchill turned to
Leigh-Mallory, who "very decently gave the
opinion that even though his defenses might be
neutralized he was now convinced that the advantage lay with saving the casualties in Bomber
Command, and *that he would take the responsibility.*"[95] Leigh-Mallory, not the scientists,
should get the credit for having the courage to
take risks.

Technical innovation is easier than *tactical*
innovation because new equipment can be tried
out before it gets to the battlefield. The German
tank commander Heinz Guderian put his finger
on the problem. He was discussing the merits
of tanks with another German commander who
finally cast some doubt on this technical marvel.
"All technicians are liars," he warned. Guderian
replied, "I admit they do tell lies, but their lies
are generally found out after a year or two when
their technical ideas can't be put into concrete
shape. Tacticians tell lies, too, but in their case
the lies only become evident after the next war
has been lost."[96]

The process of military adaptation and innovation requires courage in one other way. In a
strange war, the new data is first encountered
by the men in the field, but the process of adaptation can proceed from the top down or the
bottom up. Information can be transferred up
the line to the central command where it is
evaluated, and where new solutions are formulated. Then the new orders are sent back
down the line, where they are finally implemented. This process is sometimes necessary.
It may be necessary to put together pieces of
the puzzle coming from widely separated men
in the field before it is clear what to do. If the
men in the field have neither the competence
nor a self-interest in making the necessary
changes, central direction is in order. But it is
slow.

If we are brave enough to trust our local commanders and if we have given them a well-
defined mission, we can delegate responsibility
to them. This speeds up enormously the process
of innovation. In both the Philippines in 1901

and in the Mekong Delta in 1968, the decisive tactical innovations were developed by low-level commanders, by captains, and were only then picked up by the higher-ranking officers.[97] For the last 100 years, the German army has been good at fighting precisely because it selects officers who can make decisions under pressure, and then trains them to take the initiative within the framework of their missions. This principle and the success it brings have long been apparent to observers.[98] But before this can be done political leaders must trust them, and not have to worry about whether they will "pull a Douglas MacArthur on you."

CONCLUSIONS

The implications for policy are simple. Limited war is strange war, and policymakers will have to adapt to new circumstances. They will be better able to do so if the civilian leadership has the courage to make clear decisions as to resources and missions. The military should *not* be given a free hand, but it must be allowed the freedom to solve the military problem within the limits set for it. The military will be able to begin solving the problem only after it receives meaningful instructions and parameters. The military itself should be staffed at the highest levels with men who have demonstrated the ability to command and adapt to difficult circumstances in combat and who are respected for that ability within the army. These measures cannot be taken until the civilian leadership learns enough about military problems to set meaningful missions for the military. It is not enough to say America's goal is "a free South Vietnam" or "the free flow of oil from the Persian Gulf." Plans must be sufficiently precise that a military commander knows what to do and sufficiently well defined for him to go back to the president and say "I can't do it with these forces." The president can then make an informed decision. Neither he nor his generals can make useful choices if the mission is to "deter the Russians" or "defend the oil fields."

Civil-military relations must be improved. The civilian leadership in the Pentagon for the most part does not trust the military to wage war properly, and the military has vivid and painful memories of the Vietnam war. It hates the sound of the term "limited war." It will tend to recommend *against* any war in which it is not given a free hand. The education of the civilians and the cultivation of mutual trust will be helped by intensive peacetime exercises that involve both civilians and soldiers, by war games involving civilians, and by the revision of the theory of limited war. The military must respond by placing men in command who have

demonstrated the ability to command and innovate under fire.

In domestic politics, no one today would dare to expect the bureaucracy to be the neutral executor of, for example, a guaranteed annual income. Americans have had too much experience, and they have paid attention to that experience. They know better. They ought to know better than to expect their military to be the neutral executor of diplomatic policy. The military, particularly when engaged in combat, has its own special needs and ideals. Leaders must know what these are if they wish to make effective military policy. Trust and courage are what is needed to win strange wars. The old theory of limited war rejected the traditional wisdom about war, one maxim of which held that, in war, the moral virtues are at least as important, and probably more important, than the intellectual virtues. It is past time that we recalled this obvious truth.

NOTES

1. Robert Osgood, *Limited War Revisited* (Boulder, Colo.: Westview Press, 1979), p. 10.
2. Robert Osgood, *Limited War* (Chicago: University of Chicago Press, 1957), p. 7.
3. Dean Acheson, *Present at the Creation* (New York: Norton, 1969), p. 514; Samuel Huntington, *The Common Defense* (New York: Columbia University Press, 1961), p. 55.
4. Herbert Y. Schandler, *The Unmaking of a President: Lyndon Johnson and Vietnam* (Princeton: Princeton University Press, 1977), pp. 94–101.
5. Osgood, *Limited War*, p. 20, emphasis added.
6. Carl von Clausewitz, *On War*, ed. and trans. Michael Howard and Peter Paret (Princeton: Princeton University Press, 1976), p. 605.
7. Osgood, *Limited War*, p. 14.
8. Osgood, *Limited War Revisited*, p. 11.
9. Osgood, *Limited War*, pp. 271–72, citing Kissinger's article, "Military Policy and Defense of the 'Grey Areas,' " *Foreign Affairs* 33, no. 3 (1955): 416–28.
10. Thomas Schelling, *The Strategy of Conflict* (Cambridge: Harvard University Press, 1963), p. 15.
11. Osgood, *Limited War Revisited*, p. 11.
12. Kenneth Waltz, "A Strategy for the Rapid Deployment Force," *International Security* 5, no. 4 (1981): 57.
13. Ibid., pp. 64, 67.
14. Gen. Edward Meyer, *The Posture of the Army and the Department of the Army Budget Estimates, FY 1982* (Washington, D.C.: Department of the Army, 1981), p. 17.
15. Robert Jervis, "Why Nuclear Superiority Doesn't Matter," *Political Science Quarterly* 94, no. 4 (1979–80): 618, 619.
16. Stanley Hoffmann et al., "Vietnam Reappraised," *International Security*, 6, no. 1 (1981): 8, 10.
17. *The Pentagon Papers*, vol. 2 (Boston: Gravel ed., n.d.), p. 689.
18. Ibid., pp. 719–23.
19. Douglas Blaufarb, *The Counterinsurgency*

Era: US Doctrine and Performance 1950 to Present (New York: Free Press, 1977), p. 60.

20. *Pentagon Papers*, vol. 2, p. 691.

21. Ibid., pp. 701–2.

22. Ibid., p. 49.

23. Ibid., pp. 90–91.

24. Ibid., p. 108.

25. William C. Westmoreland, *A Soldier Reports* (New York: Dell, 1980), p. 126.

26. *Pentagon Papers*, vol. 2, p. 108.

27. Ibid., vol 3, pp. 152–53, emphasis in the original.

28. Ibid., pp. 632–33.

29. Ibid., pp. 157, 515.

30. Ibid., p. 169.

31. Ibid., p. l73.

32. Ibid., p. 555.

33. Ibid., pp. 221–24.

34. On 29 Sept. 1967, Johnson made a speech in San Antonio, Texas, in which he stated that the United States would cease all bombing of North Vietnam when it was convinced that doing so would lead promptly to "productive discussions." Johnson also said he expected North Vietnam not to take advantage of the bombing halt. See Lyndon Baines Johnson, *The Vantage Point* (New York: Holt, Rinehart, Winston, 1971), p. 267.

35. *Pentagon Papers*, vol. 3, pp. 580–82.

36. Ibid., p. 431.

37. Ibid., p. 226.

38. Ibid., p. 606.

39. Westmoreland, *Soldier Reports*, p. 158; *Pentagon Papers*, vol. 3, pp. 421–24.

40. Westmoreland, *Soldier Reports*, p. 154.

41. *Pentagon Papers*, vol. 3, pp. 453, 455.

42. Ibid., p. 457.

43. Ibid., p. 703, citing NSAM-328, 4/6/65.

44. Ibid., p. 617.

45. Michael Charlton and Anthony Moncrieff, *Many Reasons Why: The American Involvement in Vietnam* (New York: Hill and Wang, 1978), p. 115.

46. *Pentagon Papers*, vol. 3, p. 224.

47. Doris Kearns, *Lyndon Johnson and the American Dream* (New York: Harper and Row, 1976), p. 264.

48. Schandler, *Unmaking of a President*, pp. 90, 88.

49. Matthew B. Ridgway, *The Korean War* (New York: Doubleday, 1967), p. 91.

50. *Pentagon Papers*, vol. 4, p. 291.

51. Ibid., p. 320.

52. Ibid., p. 323.

53. Ibid., p. 326.

54. Schandler, *Unmaking of a President*, p. 162.

55. *Pentagon Papers*, vol. 3, p. 653.

56. Schandler, *Unmaking of a President*, p. 62.

57. Walt Whitman Rostow, *The Diffusion of Power* (New York: Macmillan, 1972), p. 494.

58. Ibid., p. 495.

59. Blaufarb, *Counterinsurgency*, pp. 72–73.

60. Westmoreland, *Soldier Reports*, p. 208.

61. Kearns, *Lyndon Johnson*, p. 262.

62. Robert Kennedy, *Thirteen Days* (New York: New American Library, 1969), p. 119.

63. John Keegan, "The Human Face of Deterrence," *International Security* 6, no. 1 (1981): 147.

64. Graham T. Allison, *Essence of Decision: Explaining the Cuban Missile Crisis* (Boston: Little, Brown, 1971), p. 131.

65. Kearns, *Lyndon Johnson*, p. 298.

66. Schandler, *Unmaking of a President*, pp. 81–82.

67. Ibid., p. 83.

68. Quoted in Rostow, *Diffusion of Power*, pp. 519–20.

69. Blaufarb, *Counterinsurgency*, pp. 270–71.

70. Hoffmann et al., "Vietnam Reappraised," p. 10.

71. G. F. R. Henderson in *The Science of War*, ed. Neill Malcom (London, 1930), p. 102.

72. C. E. Callwell, *Small Wars: Their Principles and Practice*, 3d ed. (1906; Totowa, N.J.: Rowman and Littlefield, 1976, rpt.), pp. 33, 43.

73. Lewis Walt, *Strange War, Strange Strategy* (New York: Funk and Wagnalls, 1970).

74. William Slim, *Defeat Into Victory* (London: Cassell, 1956), pp. 109–10.

75. Westmoreland, *Soldier Reports*, p. 132.

76. Ibid. pp. 214–15.

77. *Pentagon Papers*, vol. 4, p. 321.

78. Lt. Gen. John H. Hay, Jr., *Vietnam Studies: Tactical and Material Innovations* (Washington, D.C.: Department of the Army, 1974), p. 5.

79. Francis J. West, *Small Unit Action in Vietnam, Summer 1966* (Washington, D.C.: Department of the Army, 1967), p. 15.

80. "National Security Study Memorandum 1" (hereafter NSSM 1) 21 Feb. 1969, inserted in the 12 May 1972 *Congressional Record* by Rep. Ronald Dellum, p. 16751.

81. Lt. Gen. Julian Ewell, *Vietnam Studies: Sharpening the Combat Edge: The Use of Analysis to Reinforce Military Judgment* (Washington, D.C.: Department of the Army, 1974), pp. 96–97, 103.

82. Hay, *Innovations*, pp. 29–30.

83. Ewell, *Analysis*, p. 94.

84. Westmoreland, *Soldier Reports*, p. 197.

85. Ewell, *Analysis*, p. 78; NSSM 1, p. 16754.

86. Ewell, *Analysis*, pp. 76–78.

87. West, *Small Unit Action*, p. 15; Walt, *Strange War*, p. 48.

88. Westmoreland, *Soldier Reports*, p. 126, emphasis added.

89. Ibid., pp. 141, 175.

90. Ibid., pp. 202, 204, 205.

91. Ewell, *Analysis*, p. 92.

92. Arthur Jacob Marder, *From the Dreadnought to Scapa Flow: The Royal Navy in the Fisher Era, 1904–1919*, vol. 4, *1917: Year of Crisis* (London: Oxford University Press, 1969), p. 105.

93. Ibid., p. 147, emphasis added. Also, p. 162.

94. Ibid., p. 163.

95. R. V. Jones, *The Wizard War* (New York: Coward, McCann and Geoghegan, 1978), p. 297, emphasis added.

96. Heinz Guderian, *Panzer Leader* (New York: Dutton, 1952), p. 32.

97. For the Philippine campaign, see John M. Gates, *Schoolbook and Krags: The United States Army in the Philippines, 1898–1902* (Westport, Conn.: Greenwood Press, 1973), pp. 199–200. In the South Vietnam Mekong Delta area, see Ewell, *Analysis*, p. 78.

98. See Henderson, *Science of War*, "Military Criticism and Modern Tactics" and "The Training of Infantry for the Attack"; BDM Corporation, *Generals Balck and von Mellenthin on Tactics* (McLean, Va.: BDM Corporation, 1980), W-81-077-TR for OSD Net

Assessment; Martin van Creveld, *Fighting Power* (Bethesda, Md.: C and L Associates, 1980); W. von Los- sow, "Mission Type Tactics Versus Order Type Tactics," *Military Review* 57, no. 6 (1977): 87–91.

US MILITARY FORCES AS A POLITICAL INSTRUMENT SINCE WORLD WAR II

BARRY M. BLECHMAN and STEPHEN S. KAPLAN

On 11 November 1944, the Turkish ambassador to the United States, Mehmet Munir Ertegün, died in Washington; not a very important event at a time when Allied forces were sweeping across France and Eastern Europe toward Germany, and Berlin and Tokyo were approaching *Götterdämmerung*. Sixteen months later, however, the ambassador's remains were the focus of world attention as the curtain went up on a classic act in the use of armed forces as a political instrument. On 6 March 1946, the US Department of State announced that the late Ambassador Ertegün's remains would be sent home to Turkey aboard the *USS Missouri*, visibly the most powerful warship in the US Navy and the ship on board which General Douglas MacArthur had recently accepted Japan's surrender.

Between the ambassador's death and this announcement, not only had World War II ended, the cold war—as yet untitled—had begun. In addition to conflicts between the United States and the Soviet Union over Poland, Germany, Iran, and other areas, the Soviet Union had demanded from the Turkish government the concession of two of its provinces in the east and, in the west, a base in the area of the Dardanelles.

On March 22, the *Missouri* began a slow journey from New York harbor to Turkey. At Gibraltar the British governor had a wreath placed on board. Accompanied by the destroyer *Power*, the great battleship was met on April 3 in the eastern Mediterranean by the light cruiser *Providence*. Finally, on the morning of April 5, the *Missouri* and her escorts anchored in the harbor at Istanbul.[1]

The meaning of this event was missed by no one; Washington had not so subtly reminded the Soviet Union and others that the United States was a great military power and that it could project this power abroad, even to shores far distant. Whether the visit of the *Missouri*, or subsequent US actions, deterred the Soviet Union from implementing any further planned or potential hostile acts toward Turkey will probably never be known. What is clear is that no forceful Soviet actions followed the visit.

Moreover, as a symbol of American support for Turkey vis-á-vis the Soviet Union, the visit of the *Missouri* was well received and deeply appreciated by the government of Turkey, the Turkish press, and presumably by the Turkish citizenry at large. The American ambassador stated that to the Turks the visit indicated that "the United States has now decided that its own interests in this area require it to oppose any effort by the USSR to destroy Turk[ey's] independence and integrity."[2]

In this incident, as in hundreds of others since 1945, US military forces were used without significant violence to underscore verbal and diplomatic expressions of American foreign policy. Recently we concluded a study concerned with some of these uses of the armed forces: those in which the various branches of the US military were used in a discrete way for specific political objectives in a particular situation.[3] Historically, of course, the United States has not been the only nation to use its armed forces for political objectives; all the great powers have engaged in such activity, and in the last three decades the Soviet Union has been a frequent practitioner of the political use of the armed forces.

A principal objective of our study was to evaluate the effectiveness of the US military as a political instrument, in the short term and over a longer period, by analyzing the consequences of such factors as: the size, type, and activity of military units involved in the incident; the nature of the situation at which they were directed; the character of US objectives; the international and domestic context in which the incident occurred; and the extent and type of diplomatic activity that accompanied the use of the armed forces. We concluded, generally, that discrete, demonstrative uses of the military were often effective political instruments in the short term, but that effectiveness declined when situations were reexamined after longer periods of time had elapsed. We also found that each of the variables mentioned above had important consequences for the effectiveness of these uses of the military.

Reprinted with minor revisions and with permission from Political Science Quarterly *94, no. 2 (1979): 193–209.*

METHODOLOGY

Armed forces serve foreign policy functions in many ways—by their existence and character alone, by their location abroad, by their routine exercises and visits, by their provision of military assistance and other forms of support. The United States has used military units often and in a wide variety of ways since World War II. Most of these uses have had a political dimension; that is, they were liable to influence the perceptions and behavior of political leaders in foreign countries to some degree.

For the purposes of the study described in this article, we define a political use of the armed forces as a physical action taken by one or more components of the uniformed military services as part of a deliberate attempt by the national authorities to influence, or to be prepared to influence, specific behavior of individuals in another nation without engaging in a continuing contest of violence.

Using this definition, 215 incidents were identified in which the United States used its armed forces for political objectives between 1 January 1946, and 31 December 1975—an arbitrary cutoff date. We are confident that the list of incidents adequately represents all those instances in which US armed forces were used in a way that would fit the terms of the definition.[4]

The use of military power as a tool of diplomacy can risk the security and well being of the United States and importantly affect US and other international relationships both immediately and for a long time afterward. Although on some occasions the risks of intervention may be small, in many instances it is difficult to determine all possible dangers and the likelihood of their being realized. Military action in still other situations may clearly entail great risk—particularly if the USSR is an actor and is committed to a different result, or if another opponent is prepared to use violence as a last resort. The discrete political use of military power may also lead to unwanted dependency on the United States and the hostility, not only of antagonists, but of other nations in the affected region. US relations with the USSR, China, and uninvolved US allies may be made more difficult. In light of the Vietnam war experience, the effects of US military activities abroad on the political culture of the United States and the fabric of American politics should also be considered. Before reaching a judgment that military intervention is necessary for the preservation of important US interests and accepting the immediate and long-term risks that may be apparent, policymakers would be wise to consider also the utility of past political-military operations.

To evaluate this past effectiveness, two approaches, aggregate analysis and case study, which present macro- and micro-views of the same phenomena, were adopted. The value of the first is that it permits broad generalizations that might be applicable in the future when the armed forces are used for political objectives. Individual case studies can confirm or disprove these generalizations, allow the inference of propositions related to the peculiarities and complexities of specific situations, and provide a sense of the psychological climate and individual concerns that condition the choices of policymakers. In short, the two approaches are complementary; each has advantages and disadvantages, but together they afford greater understanding than either can provide separately.

For the aggregate analysis, a sample of 33 incidents, 15 percent of the total number of cases examined, was selected for systematic and rigorous analysis of outcomes. For each of the incidents in the sample, the available literature, documents, and newspaper accounts were investigated so as to determine:

—US objectives vis-á-vis each participant, and whether those objectives were satisfied within six months and retained over three years following the use of US armed forces;

—The size, type, and activity of US armed forces involved in the incident;

—The character of the targets in relation to US objectives;

—Other activities (for example, diplomatic) undertaken in support of US objectives along with the use of the armed forces;

—Certain US domestic conditions.

In the analysis, the degree of satisfaction of US objectives was related to each of these factors with the aim of highlighting the crucial variables determining whether a political use of the armed forces was likely to be successful or not. Five specialists made detailed assessments of the specific mechanisms through which military operations affected the perceptions and decisions of foreign policymakers in ten case studies.[5]

What have been the results of discrete US political-military operations? Are the prospects for success in such ventures good enough that policymakers should consider the armed forces an important option in these situations? Or should they view use of the military with great caution, since this type of activity often fails to meet its objectives and sometimes backfires? More to the point, under what circumstances are discrete uses of the armed forces for political objectives more likely to succeed, and when are they more likely to fail?

THE UTILITY OF MILITARY FORCE

The evidence supports the hypothesis that discrete uses of the armed forces are often an effective way of achieving short-term foreign policy objectives. The aggregate analyses show clearly that the outcomes of United States political-military activities were most favorable from the perspective of US decisionmakers—at least in the short term.

In a very large proportion of the incidents, however, this "success rate" eroded sharply over time. Thus, it would seem that, to the degree that they did influence events, discrete uses of military forces for political objectives served mainly to delay, rather than resolve, unwanted developments abroad. Though there is some value in "buying time"—that is, keeping a situation open and flexible enough to prevent an adverse fait accompli—it should be recognized that these military operations cannot substitute for more fundamental policies and actions that can form the basis either for sound and successful alliances or for stable adversary relations, such as diplomacy, close economic and cultural relations, or an affinity of mutual interests and perceptions. What political-military operations perhaps can do is provide a respite, a means of postponing adverse developments long enough to formulate and implement new policies that may be sustained over the longer term. Or, if that is not possible, the political use of armed forces may serve to lessen the consequences of detrimental events. However, some of the case studies suggest that even these delaying and minimizing accomplishments do not always obtain.

In some cases the discrete use of force clearly has been ineffectual. For example, US support for Pakistan during the 1971 war with India was a relatively empty gesture because the target actors recognized that under virtually no circumstances would the United States become militarily involved in the war. Consequently, the deployment of the *Enterprise* task force to the Indian Ocean had almost no effect on the decisions of either the immediate actors, India and Pakistan, or the actors indirectly involved, China and the Soviet Union.

In other cases, a discrete political application of the armed forces seems to have been associated with the creation of a situation that remained tolerable for a period of months or, in some cases, years. The intervention in Laos in 1962 was such an incident; the landing of US Marines in Thailand coincided with the negotiation of a settlement that kept the peace in Laos for several years. US actions to oust the Trujillo family from the Dominican Republic following Rafael Trujillo's assassination, and subsequent actions to support the new govern-

ment, were of a similar character. These actions precipitated a more acceptable political situation in the Dominican Republic that persisted for several years, even though it ultimately foundered in 1965.

Was it worthwhile to use armed forces to obtain positive outcomes that could be sustained only temporarily, whether the duration was of several months or several years? We believe the answer is yes, insofar as an opportunity was gained for diplomacy.

When a positive outcome did not endure, it was usually attributable not to an absence of follow-up diplomatic effort, but to the fact that the internal situation within a state, or a specific interstate relationship, was strong and durable and thus not subject to penetration by either single or periodic US military actions or by the use of other US policy instruments. Realizing that this was the situation in Vietnam in 1964–65, American policymakers chose war as the solution, which contrasts with US policy in the late 1940s in China when Communist forces there triumphed and forced the Kuomintang government of Chiang Kai-shek to flee to Taiwan; or in the early 1960s over Cuba when Fidel Castro identified his regime with communism and the Soviet Union; or in the late 1960s against North Korea after the *Pueblo* was seized and a US Navy EC-121 aircraft was shot down.

Finally, in some cases with very special circumstances, discrete political uses of the armed forces contributed to the establishment of new international relationships, such that US interests were protected for decades. For example, following the 1946 visit to Turkey by the battleship *Missouri* and further displays of US military support for Ankara, Soviet pressures on Turkey declined; they have not been renewed in a serious way since. Displays of American military support for Italy prior to the 1948 elections seem to have contributed, along with such other instruments of policy as economic aid and covert support for democratic political parties, to the defeat of the Italian Communist party. The dominance of the Christian Democratic party, which resulted from that election, persisted for more than two decades. Political uses of the armed forces during the Berlin crises of 1958–59 and 1961 helped create stable conditions in Central Europe that continue to exist.

THE EFFECTS OF THE MILITARY DEMONSTRATION

What is the mechanism at work here? How do the armed forces play their special role? The process begins when a given framework of relations among several countries, or a domestic political configuration abroad, is disrupted by something unexpected or at least unwelcome: a domestic upheaval, a new departure in a major power's foreign policy, or perhaps an unexpected armed clash between the military units of hostile states. Regardless of cause, this development often creates uncertainties and a distinct psychological unease among interested parties; at other times it leads directly to an unraveling of the fabric of relations that previously had been established and maintained by existing policies. Under such circumstances (and mindful of the fact that even when favorable outcomes do occur they are likely to persist only over the short term), a discrete demonstration of US military capability can have a stabilizing and otherwise beneficial effect, perhaps persuading the target that the wise course of action is to alter the undesirable policy.

The effect of the military demonstration will depend to a large extent on whether the target finds the threat credible. A prime example was the arrival of US military forces off the coast of the Dominican Republic in November 1961, an event that changed the perceptions of the Trujillo family and their lieutenants as to the United States' willingness to act. Coupled as it was with a clear ultimatum, the action seems to have exerted a powerful influence.

Such a military demonstration may be particularly effective when the target state is not fully committed to the course from which the United States hopes to dissuade it. We will never know for sure, but US actions in support of Yugoslavia following President Tito's break with Stalin may have lessened Stalin's inclination to take more aggressive measures against Yugoslavia. Moscow may also have been somewhat deterred from taking action against Romania following the intervention in Czechoslovakia in 1968. We will return to these distinctions. First, however, it is helpful to examine the mechanisms at work more closely.

In some of the cases studied, the insertion (even symbolically) of US military forces may have provided leverage to US decisionmakers where previously there had not been any. A US military presence or operation may furnish an incentive to a foreign leader to consider the wishes of US policymakers. After US Marines landed in Thailand in 1962, and the threat of a US intervention in Laos became credible, for example, the United States may have gained a decided edge at the negotiating table. In such a case the prime result is to lessen the potential US loss. Foreign decisionmakers may act to avoid those extreme choices that they fear would precipitate a violent US response. Unless the United States is willing to extend its new military presence into a permanent operation, however, any such demonstrative action is likely to be successful, to the degree it is successful at all, only for a limited time.

The intervention in Lebanon in 1958 provides a pointed example of this phenomenon. Although President Eisenhower authorized the landing of US Marines and other forces in Lebanon in July 1958, the president—or at least Secretary of State Dulles—recognized that an American-imposed solution in Lebanon would be unacceptable to most of the Lebanese actors and therefore short-lived. Thus the United States adopted a twofold approach. On the one hand, a massive demonstration of American military power was staged, involving the landing of thousands of American troops. On the other hand, Deputy Under Secretary of State Robert Murphy was dispatched to Lebanon, where, recognizing the political realities of the situation, he negotiated a settlement that was probably more favorable to the actors who opposed US policy than to the presumed American client, President Camille Chamoun. In other words, to the degree that it did influence the situation, the US military demonstration seems to have bought sufficient time to reach new political arrangements that more realistically reflected the distribution of power among competing ethnic and political groups in Lebanon. Without the more realistic political solution negotiated by Murphy, the 1958 US intervention in Lebanon would probably have been associated with a much less favorable outcome. At the same time, without the leverage provided by the US military presence in Lebanon, with its implied threat of greater violence, Murphy might well have failed in his attempts to negotiate a political solution.

Military demonstrations also can ease domestic political pressures on the president from groups demanding more forceful action. In less serious incidents, these pressures—which can originate from ethnic groups, the Congress, friends and political associates of the decisionmakers, and executive branch uniformed and civilian officials, among others—are directed at lower-ranking foreign policy managers. Insofar as they can help to stabilize a situation, thereby postponing perceived adverse consequences of unexpected changes in international relations or the internal politics of foreign nations, discrete uses of the armed forces may diminish calls for more decisive action. The postponement provides the time needed to gather support within the bureaucracy and in the Congress and the public for the fundamental changes in policy required to accommodate developments. In the absence of the time bought by the military demonstration, these fundamental changes in policy may be more difficult to bring about, and the president might fear that his constituencies, both at home and abroad, will see him as bowing to foreign pressures.

This phenomenon was also demonstrated pointedly by the intervention in Lebanon in 1958. During the spring of 1958, President Eisenhower resisted several requests from President Chamoun for American military assistance. An unexpected event—the coup in Iraq in July—made it impossible to avoid the request any longer. Eisenhower feared that further inaction, after the seeming wholesale defeat of American clients in the Middle East, would have negative effects both on the perceptions of decisionmakers in foreign nations and on domestic opinion in this nation. American support for the realistic solution to the Lebanese problem negotiated by Murphy might have been difficult to muster without the symbolism of American strength suggested by the military intervention. In the absence of the intervention, President Eisenhower may have feared the effects of such an apparent concession on opinion both at home and abroad.

Similarly, following American intervention in the Dominican Republic in 1965, Ambassador Ellsworth Bunker negotiated an agreement in which it was all but explicit that the president elected the following year would be either Joaquin Balaguer or Juan Bosch. Balaguer was the last president during the Trujillo era, and a person the United States took pains to keep out of the Dominican Republic in later years; Bosch was strongly disliked by many American officials and suspected of tendencies that might allow an eventual takeover by the extreme left. The emplacement of US forces in the Dominican Republic thus not only made an election possible, it also made it possible for the United States to accept results of that election which it previously could not have tolerated.

None of this is meant to imply that the gains seemingly associated with discrete uses of the armed forces for political objectives have been fraudulent. The amelioration of pressures for extreme actions is an important benefit, as is the provision of time necessary to build a consensus for a new US policy; so too is the provision of leverage to negotiators.

CORRELATES OF SUCCESS

Four groups of factors have been associated with the relative success or failure of political uses of the US armed forces: the type of objective; involvement by the Soviet Union; the context of the incident; and the nature and activity of the US military forces involved. (Unless noted otherwise, these conclusions refer mainly to short-term outcomes.)

US OBJECTIVES

The nature of US objectives may be an important determinant of whether a political use of force is successful. Favorable outcomes occurred most often when the objective of US policymakers was to maintain the authority of a

specific regime abroad. Such was the case, for example, when naval movements and other activities were undertaken in support of King Hussein during the 1970 civil war in Jordan. Indeed, the aggregate analyses suggest that maintenance of regime authority was the one type of objective associated with the persistence of a favorable outcome over the longer term. The armed forces were least often associated with favorable outcomes when the objectives concerned the provision of support by actors to third parties—for example, the many incidents in which US military activities were undertaken in order to persuade the Soviet Union to cease supporting hostile political initiatives by its allies or clients.

Between these two extremes there were a reasonable number of favorable outcomes when discrete political uses of the armed forces were undertaken to offset the use of force by another actor. Over the longer term, however, frequencies of favorable outcomes were low when discrete political uses of the military were aimed at either the use of force by another actor or another actor's support of a third nation's use of force. Illustrative are the futile attempts during the late 1950s and early 1960s to convince the Pathet Lao and Viet Cong to terminate their insurgencies in Laos and South Vietnam.

Perhaps more significant, the mode in which the armed forces are used as a political instrument may also be an important determinant of success. It is evident from the aggregate analyses that discrete uses of the armed forces for political purposes were more often associated with favorable outcomes when the US objective was to reinforce, rather than modify, the behavior of a target state. This stands to reason. Nikita Khrushchev, no doubt, found it much easier not to follow through on the various threats he made concerning Europe and the Middle East than he did to withdraw Soviet missiles from Cuba; just as deterring the outbreak of violence is usually an easier task than bringing violence to a satisfactory conclusion.

Human behavior is difficult to change. Individuals tend to be more aware of the risks of change than they are of the dangers of continuing their prevailing course. After all, established policies are known entities; even when the risks in continuing established policies are evident, the dangers of change will often appear more threatening. More to the point, no one—least of all the head of a nation—can afford to be told publicly what he should be doing. Thus, national leaders will resist demands for policy modifications most strenuously when such demands are made publicly, which is usually unavoidable when military power is used.

In short, whether a discrete application of military power was made in order to coerce a hostile target state to change its behavior or to encourage a friendly target state to change its behavior, the outcomes were similar; most often they were unfavorable from the US perspective. On the other hand, when US policymakers used the armed forces to coerce a hostile target state to continue to do something (for example, stay at peace), or to encourage a friendly state to remain on the same course, military demonstrations were relatively more often associated with favorable outcomes.

Consider, for example, the starkly different outcomes of the Berlin crisis of 1958–59 and 1961 and the several incidents in Southeast Asia in the late 1950s and early 1960s. In the former situation the United States sought essentially to assure allies and to deter certain threatened actions by the USSR and East Germany. Both types of objectives required only that the targets not change their behavior, and both were achieved. Not so favorable, especially over the longer term, were the outcomes of the Southeast Asian incidents, in which the United States sought to compel various Communist actors to stop using force and to induce the government of South Vietnam to behave differently (that is, more assertively). Here both types of objectives required a change of behavior on the part of target states, and only rarely were these outcomes achieved.

To some extent this conclusion may reflect Tolstoy's view that the only decisions that are carried out are those corresponding to what would have happened if the decisions had not been made. In many of the incidents, although US decisionmakers may have thought—or feared—that a target state was prepared or intending to change its behavior, and thus used the armed forces to reinforce existing behavior, the target state actually may have had not such intention. A good example is the US military activity that followed the Soviet occupation of Czechoslovakia, consisting primarily of increases in the readiness for war of US forces in Europe. One objective of that action was to deter a Soviet invasion of Romania. No invasion occurred; hence the military demonstration appears to have been effective. The question, however, is whether Soviet leaders ever seriously contemplated such an invasion. The same may be said about the Soviet Union's not taking violent action against President Tito and Yugoslavia almost two decades earlier. This is not necessarily to discount the importance of the US military activity. US policymakers had other important objectives in mind as well. Still, the proportion of favorable "reinforcement" outcomes that are accounted for by the "unreality" of the feared target state behavior may be high. Unfortunately, that proportion is impossible to determine empirically.

Political uses of the armed forces were often associated with favorable outcomes when US objectives were at least loosely consistent with prior US policies. The purpose of discrete political uses of force must fit within a fundamental framework of expectations held by decisionmakers both in this country and abroad if the military activity is to be associated with a favorable outcome. With regard to the incidents in the sample, although prior diplomacy was closely associated with positive outcomes, diplomacy during the course of the incidents themselves was not.

When a treaty exists, or when policymakers have taken pains to make clear that the United States perceives itself to have a commitment, antagonists are less likely to probe, or to probe only within narrower limits than would have been the case otherwise, and to leave themselves a way open for retreat. In other words, antagonists may be less likely to try to present the United States with a fait accompli that could be reversed only through a major use of the armed forces. Prior diplomacy may have reinforced the antagonists' continued performance of desired behavior, and thus lessened the significance of a breakdown in relations or the severity of a crisis. Rather than have to cope with hostile actions, policymakers in these cases might have only had to respond to hostile rhetoric; alternatively, the degree of desired behavior modification of foreign decisionmakers might have been less.

Hence, although China initiated the 1958 Offshore Islands crisis by shelling Quemoy and Matsu, and Khrushchev threatened Berlin on several occasions, both Peking and Moscow carefully controlled their behavior during these incidents because they perceived a prior US commitment (or at least feared the strength of announced US commitments). In the absence of a prior US commitment, as in Korea in 1950, we might surmise that not only would China and the USSR have gone further in their initial actions, but also that sudden US diplomatic action, even if supported by a discrete political use of force, might have had a lesser effect. Skilled diplomacy during incidents—for example, by Robert Murphy in Lebanon, Averell Harriman in Laos, and Ellsworth Bunker in the Dominican Republic—has typically borne fruit only after ambiguous US military commitments have been clarified by the movement of major military units. Thus, the United States should not count on skilled diplomacy as being effective in controlling crises in the absence of prior commitments and reinforcing uses of the armed forces; it should however, aim to avoid such difficult tests in the first place by being quite clear as to what its commitments are.

Similarly, the aggregate analyses suggest that prior US military engagement in conflicts in a region was often associated with favorable outcomes when subsequent demonstrative uses of military force took place. The fact that the United States previously had been willing to engage in violence in the region may have made the threats or assurances implied by the subsequent military activity more credible. Much less often associated with favorable outcomes were previous demonstrative uses of force in the region; the willingness to engage in violence seems to have been the key. Previous discrete political uses of force were associated with favorable outcomes more often, however, when the US objective was to assure that a target state would continue to do something. Good examples are provided by US naval demonstrations in the Mediterranean. US military forces have not fought in that region since 1945, yet as concerns the assurance of Israel (and less frequently, Jordan), these displays of naval power were often associated with favorable outcomes. US objectives in situations were far less often attained, however, when the political use of force was meant to modify the behavior of Israel's (or Jordan's) enemies.

In short, prior demonstrative uses of force or even prior military engagements did not seem to be sufficient to compensate for the previously noted difficulty of modifying a target state's behavior. Indeed, very little seemed to compensate for the difficulty of modifying behavior; more than any other factor this was the basic determinant of when a discrete political use of force would or would not be associated with the attainment of foreign policy objectives. It overshadowed the diplomacy that accompanied the military activity, the nature of the situation, and the timing, size, composition, and activity of the military units themselves.

SOVIET ACTIVITY

A second group of factors that seemed to be associated with favorable outcomes in the aggregate analyses included the character of United States-Soviet relations and the specific role played by the Soviet Union in an incident.

One conclusion that runs counter to prevailing views concerns the possible effects of the United States-Soviet strategic nuclear balance on the relative fortunes of the superpowers. We did not find that the United States was less often successful as the Soviet Union closed the US lead in strategic nuclear weapons that had been maintained for the first 20 or so years following World War II. Whether discrete political uses of US armed forces were associated with positive outcomes seems to have been independent of relative United States-Soviet aggregate strategic capabilities. Of course, the United States may have engaged less often in these incidents

since the late 1960s precisely because it understood that its chances of success were smaller; as the USSR closed the nuclear gap, the United States may have chosen to participate in incidents more selectively, choosing only those cases in which its chances of success were greatest.

Since this study was not designed to test the effects of the strategic balance, the findings in this regard are clearly tentative. Still, both the aggregate analyses and the case studies provide little support for the notion that decisions during crises are strongly influenced by aggregate strategic capabilities. Studies of Lebanon and Jordan, for example, indicate that, to the extent that evaluations of the military balance played any role, decisionmakers in the United States and Soviet Union, as well as those in the nations directly involved, were more concerned with the local balance of conventional power. More to the point, most local actors in these incidents seem to have had only a rudimentary and impressionistic sense of relative military capabilities in general.

Soviet political and/or military involvement in the incident itself, on the other hand, was clearly associated with the frequency of outcomes favorable to the United States, which were less often favorable when the Soviet Union was involved, particularly when the Soviet Union threatened to employ, or actually employed, its own armed forces in the incident. The seemingly pernicious effect of Soviet involvement was tempered at times when broader United States-Soviet relations had been improving, and outcomes were more often favorable when overall United States-Soviet relations were characterized by greater cooperation. As in the previous finding, this conclusion is stronger when just those incidents in which the Soviet Union participated were considered, and stronger still when just those incidents were considered in which Soviet military forces were involved.

NATURE OF THE SITUATION

Outcomes were favorable more frequently when discrete political uses of force were directed at intranational, as contrasted to international, situations. We are not confident about this finding, however, because two other factors that are also closely associated with favorable outcomes are highly correlated with intranational situations: lesser amounts of force tend to be used in intranational situations; and the US objective in these situations is more often reinforcement, not modification, of behavior.

In the international situations, a positive outcome was most likely if the United States was involved from the very onset of a conflict.

This finding complements the previous finding concerning the need for the US objective and the specific use of force to be consistent with the prior framework of relations between the United States and the nations involved in the incident. In those situations in which the United States was involved initially, such as the Berlin crises of 1958–59 and 1961, US statements of aims and objectives were more likely to be considered seriously. Similarly, US threats or promises implied by the use of armed forces were more likely to be perceived as credible. In international situations where the United States intervened in affairs that did not concern it directly (or at least not initially), however, there seems to have been some question in the minds of the other actors whether US threats or promises were credible. Consider the perceptions of Hanoi in the late 1950s and early 1960s and those of the Gandhi government in India during the crisis and then war with Pakistan in 1971, for example. The US military demonstrations then were not taken too seriously at first. Why, North Vietnamese and Indian leaders might have asked, would the United States become involved militarily? Such questions were less likely to be raised in those situations in which US interests were directly and obviously threatened.

Boding ill for the future, the proportion of incidents involving hostility between states appears to be increasing, while there is a decline in situations of an intrastate nature. Less and less is the United States being called upon by governments for support against internal dissidents; rather, the trend is toward being asked by one state for support against another. Insofar as such a shift is discernible, the risks of involvement to the United States, especially in a situation of violence, can only increase. The shift away from an intrastate focus means that the instruments brought to bear by regional actors will often be more powerful, both diplomatically and militarily. Allies usually act more overtly when they are supporting a state rather than a subnational group; hence, the facing-off of states is more likely to occasion the facing-off of alliances, whether formal or otherwise. Most important, the likelihood of superpower confrontation is increased.

States, unlike subnational groups, also have air forces, navies, and heavily armed ground forces. Thus the level of violence that can be threatened in a crisis or manifested in a conflict is much greater. For a threat by US policymakers to be credible in these circumstances, large and technologically sophisticated forces must be available. Should these forces be committed in a conflict, they might have to be used in strength and be prepared to take significant casualties. The danger that US action of this sort might stimulate a Soviet military response is obvious.

SIZE, ACTIVITY, AND TYPE OF US MILITARY FORCES

The firmer the commitment implied by the military operation, the more often the outcome of the situation was favorable to the United States. Forces actually emplaced on foreign soil were more frequently associated with positive outcomes than were deployments of naval forces, which can be withdrawn almost as easily as they can be moved toward the disturbed area. The movement of land-based forces, on the other hand, involves both real economic costs and a certain psychological commitment that are difficult to reverse, at least in the short term. This is an interesting finding, not so much because of its novelty—after all, it only confirms the common perception—but because its implications run counter to common US practice. The Navy has been the preeminent military force in discrete political operations. Naval forces participated in more than 80 percent of the incidents; and reliance on the Navy was the case regardless of region, time period, type of situation, and whether or not the Soviet Union participated in the incident.

Naval forces can be used more subtly to support foreign policy initiatives—to underscore threats, warnings, promises, or commitments—than can land-based units, and they can do so without inalterably tying the president's hands. But it is precisely this last fact that probably diminishes the effectiveness of naval forces in a political role. Foreign decisionmakers also recognize that warships can be withdrawn as easily as they can enter a region of tension and, hence, that the commitment they imply is not so firm as that implied by land-based units.

Positive outcomes were particularly frequent when land-based combat aircraft were involved in an incident. This would suggest, particularly in view of the much greater mobility of modern land-based tactical air units, that the air force might be used more frequently in political-military operations than has been the case in the past.

Such a shift in US practice would not be without its costs, because the use of land-based forces is perceived by foreign decisionmakers as greater evidence of commitment. If the US objective is not so certain, if all that is desired is to take an action that signifies interest and concern but leaves room for maneuver—such as US naval deployments during the Cyprus crises of 1964, 1967, and 1974—then the use of land-based forces would not be advisable. Moreover, in situations in which a military move is intended as a bluff or to screen a political defeat—as seems the case in the Indo-Pakistani war in 1971—the use of land-based air forces would not be advisable. In all these types of situations, naval forces would provide greater flexibility to decisionmakers and thus would be more appropriate even if their probability of succeeding may be less.

There are other ways to enhance the effectiveness of the armed forces. Outcomes were more often favorable when the units involved actually did something, instead of merely emphasizing their potential capability to intervene—for example, by reducing the time delay between a decision to intervene and the actual operation by moving toward the scene of the incident or by increasing their state of alert. The inolvement of the military unit in a specific operation, such as mine-laying, or mine-clearing, or patrolling—and certainly when the actual exercise of firepower was involved—seems to have indicated a more serious intent on the United States' part. The movement of the force toward the region of concern by itself could be an ambiguous signal; it might not be clear to foreign decisionmakers what the United States had in mind, or the movement might pass unobserved. A more specific action, one that gave a clearer signal, thus was more often associated with favorable outcomes.

Positive outcomes were also more likely when the forces involved included strategic nuclear forces. Foreign decisionmakers seem to have perceived the use of strategic nuclear forces—whether or not it was accompanied by a specific threat to use nuclear weapons—as an important signal that the United States perceived the situation in a most serious way. Thus, the employment of nuclear-associated forces, such as Strategic Air Command aircraft or Sixth Fleet carriers when they were central to US plans for nuclear war, served the same purpose as the involvement of military units in a specific activity or the use of ground forces as compared to naval forces: they bolstered US credibility.

The risks of such a policy should be evident. There is no guarantee that any military demonstration will be successful. When nuclear weapons are involved, and the demonstration is not successful, the result could be disastrous; US policymakers may be faced with the choice of admitting the emptiness of the nuclear threat, and thus undermining the credibility of fundamental US commitments, or actually employing nuclear weapons.

Moreover, it may be that positive outcomes have more often occurred when nuclear forces were involved simply because these weapons have been used infrequently. The more US decisionmakers turn to nuclear forces, even demonstratively, to ensure the credibility of signals in incidents such as we have described, the more quickly the special message now associated with nuclear weapons might erode. Eventually the movement of nuclear forces would not receive much more attention and would not convey any more credibility than movements of conventional forces.

A LAST WORD

The discrete use of the armed forces for political objectives should not be an option that decisionmakers turn to frequently or quickly to secure political objectives abroad; it should be used only in very special circumstances. We have found that over the longer term such uses of the armed forces were not often associated with positive outcomes. Decisionmakers thus should not expect them to serve as substitutes for broader and more fundamental policies tailored to the realities of politics abroad, and incorporating diplomacy and the many other potential instruments available to US foreign policy.

Moreover, there are dangers in using the armed forces as a discrete political instrument. Symbolic low-level uses of force may be disregarded by antagonists or friends in a situation in which US policymakers have not seriously contemplated the need for, or the consequences of, using larger forces in a more manifest way. Foreign decisionmakers may not perceive important US interests to be involved; the initial US military action may be seen as symbolic of US interest but not of a commitment; foreign decisionmakers may calculate that they will be able to cope successfully with the forces that they expect the United States to bring to bear; or, when an actor feels its very existence is at stake, the calculus may not matter at all. In all these situations there is a risk that lesser military actions may lead to pressures for greater US involvement.

Case studies bear this out: The Castro regime did not yield at the Bay of Pigs; Hanoi and the Viet Cong were not swayed in the early 1960s; and India dismembered Pakistan in 1971. In each case, either the United States suffered humiliation or the fear of exposure embarrassed decisionmakers into escalation and war.

Still, in particular circumstances, discrete political uses of the armed forces often were associated—at least in the short term—with the securing of US objectives or the stabilization of adverse situations while more fundamental policies could be formulated. Thus, at times, and although decisionmakers should view this option with some caution, the discrete use of the armed forces for political objectives seems to have been a useful step in shoring up situations enough to avoid dramatic setbacks, to mitigate domestic and international pressures for more forceful and perhaps counterproductive actions, and to gain time for sounder policies to be formulated and implemented.

To reach this conclusion about the apparent effectiveness of the armed forces as a political instrument is not to reach any judgment about the wisdom of using the armed forces for these purposes. That is a more difficult question, which can only be answered in the context of the specific choices—and the various costs and benefits associated with each choice—facing decisionmakers at that time.

NOTES

The research for this article was supported by the Advanced Research Projects Agency of the Department of Defense and was reviewed by the Office of Naval Research under Contract N00014-75-C-0140. The views expressed are the authors' alone and should not be interpreted as representing the official policies, either expressed or implied, of the Advanced Research Projects Agency, the Department of the Navy, or the US government; nor should they be ascribed to the officers, trustees, or other staff members of the Brookings Institution.

1. Log of the USS *Missouri*, Washington National Records Center, Suitland, Md.

2. US Department of State, *Papers Relating to the Foreign Relations of the United States, 1946*, vol. 7, *The Near East and Africa* (Washington, D.C.: G.P.O., 1969), p. 822.

3. Barry M. Blechman and Stephen S. Kaplan, *Force without War: US Armed Forces as a Political Instrument* (Washington, D.C.: Brookings Institution, 1978).

4. Analysts undertaking a similiar study on a classified basis have indicated that there is a correlation of 0.89 between our incident list and a list of incidents which, under the terms of the definition employed in this study, their data would indicate took place. See Robert B. Mahoney, Jr., "A Comparison of the Brookings and CNA International Incidents Projects," Center for Naval Analyses, Professional Paper 174 (Washington, D.C., Feb., 1977), processed.

5. David Hall, William Quandt, Jerome Slater, Robert Slusser, and Phillip Windsor.

PART III

ARMS CONTROL AND DEFENSE POLICY

CHAPTER 9

ARMS CONTROL AND DEFENSE POLICY
An Overview

SCHUYLER FOERSTER

At a summit meeting in Washington in December 1987—their third summit in less than two years—President Ronald Reagan and Soviet Communist Party General Secretary Mikhail Gorbachev signed the treaty on the elimination of their intermediate-range and shorter-range missiles. This INF Treaty culminated a process that began eight years earlier, in December 1979, when the United States and its NATO allies commited themselves to their "dual track" decision. In response to Soviet deployments of SS-20 intermediate range ballistic missiles (IRBMs), NATO had decided that the US should pursue arms control negotiations with the Soviet Union in hopes of creating a stable balance in what became known as intermediate nuclear forces (INF), and that, beginning in 1983, the US would deploy 108 Pershing II and 464 ground-launched cruise missiles (GLCMs) in Western Europe unless an arms control agreement rendered that deployment unnecessary.

In the intervening years, NATO governments demonstrated their political determination to fulfill plans to deploy the missiles while calling, in 1981, for the elimination of all land-based long-range INF missiles (with ranges over 1000 kilometers)—the US Pershing II and GLCMs plus Soviet SS-20s and older SS-4s and SS-5s. Ultimately, the Soviets agreed to a global ban on all land-based INF missiles, including shorter-range INF missiles with ranges between 500 and 1000 kilometers, of which the US had none deployed. In addition, the INF Treaty included two protocols that are unprecedented in postwar arms control agreements. The first protocol defines procedures for eliminating the missiles over a three-year period—unprecedented because, unlike previous arms control agreements, this treaty actually mandates the dismantlement and destruction of an entire class of missile systems which were deployed. The second protocol defines procedures to allow each side to inspect bases, missile support facilities, and certain production facilities to monitor compliance with the treaty—providing an unprecedented verification regime based on on-site and short-notice inspection rights.[1]

In the Joint US-Soviet Summit Statement, issued on 10 December, the president and Gorbachev characterized the treaty as "historic," noting that "this mutual accomplishment makes a vital contribution to greater stability."[2] In addition, they agreed on certain parameters for a treaty to reduce and limit strategic offensive arms—including a ceiling of 6000 warheads and 1600 strategic offensive delivery systems plus subceilings on ballistic missile warheads (4900) and 1540 warheads on 154 "heavy" missiles—which would build on the "innovative" verification provisions incorporated in the INF Treaty.

In their review of other arms control issues, the two leaders also welcomed the beginning of negotiations on improved verification measures to facilitate agreements to limit nuclear testing; reaffirmed their continued commitment to the nonproliferation of nuclear weapons; welcomed the agreement, signed in 1987, to establish nuclear risk reduction centers in their capitals; expressed their commitment to negotiate a "verifiable, comprehensive and effective international convention on the prohibition and destruction of chemical weapons"; and discussed the importance of reducing conventional forces in Europe and enhancing stability through further agreement on confidence and security building measures (CSBMs) in Europe.

Viewed in comparison with the early years of the Reagan administration, in which the administration was often criticized for being an "opponent" to "arms control," the combination of agreement on INF, the prospect of agreement in Strategic Arms Reduction Talks (START), and apparent movement in a variety of other arms control fora is nothing less than dramatic. As evidenced by the growing debate that surrounded ratification of the INF Treaty and on prospects for START agreement—a debate colored as well by a presidential electoral campaign—it is also clear that there was no consensus within the US on the merits of US arms control policy. Presented with an unprecedented Soviet willingness to make dramatic progress in reducing nuclear, conventional, and chemical weapons—and faced with tremendous domestic pressures to cut defense spending— the Bush administration accelerated and expanded that arms control agenda. Yet the debate over the merits of arms control continued.

As the next two chapters suggest, this debate over the very utility of arms control and the role arms control should play in enhancing national security is not new. At the heart of this controversy is a fundamental debate over the nature of security in the nuclear age, the relationship between nuclear and conventional weapons in US defense policy, and, indeed, the character of the relationship between the US and its principal strategic and political adversary, the Soviet Union. In focusing on the evolution of US strategy as a world power, in meeting its alliance commitments, and in defending its global interests, previous chapters have not addressed arms control directly. Nonetheless, those discussions have not been able to avoid the issue entirely since arms control has been a persistent element in US policy. This and the next two chapters deal directly with the theoretical and practical issues posed by arms control in both the strategic and theater domains.

STRATEGIC ARMS CONTROL

Strategic arms control, as it has been practiced over the past two decades, has been an exclusively bilateral process involving the two superpowers in the regulation of what have become known as "central strategic systems." These systems include land-based intercontinental ballistic missiles (ICBMs), submarine-launched ballistic missiles (SLBMs), and intercontinental bombers—essentially the US "strategic triad" and corresponding Soviet systems. In addition, strategic arms control has also encompassed strategic defense—the means of defending one's territory against missile attack. Both aspects of strategic arms control have produced agreements, but only the 1972 Treaty on the Limitation of Anti-Ballistic Missile (ABM) Systems remains in force as a ratified treaty. The SALT I Interim Agreement on Strategic Offensive Missiles, signed in 1972, expired in 1977. Its 1979 successor, SALT II, was never ratified by the US Senate.

Current negotiating efforts—the Strategic Arms Reduction Talks (START) and the Defense and Space Talks (DST) in Geneva—represent in many respects an entirely new arms control agenda for the US. As the next chapter elaborates, the "SALT era" reflected the premise that, if both sides refrained from strategic defense and codified the condition of "parity" in the strategic offensive relationship, then a stable nuclear balance could be established that avoided the perils and costs of an arms race. Such a prescription for strategic stability—derived from notions of mutual assured destruc-

tion (MAD) in the 1960s—ultimately proved to be unattainable in SALT, largely because the Soviets did not share the US view that such a prescription was indeed desirable. By the 1980s, it was clear that codifying the existing strategic relationship in offensive arms was potentially destabilizing since it would institutionalize Soviet superiority in heavy ICBMs with multiple, independently targetable, reentry vehicles (MIRVs) that could threaten the US landbased retaliatory capability. The revised US strategic arms control policy thus sought to reduce dramatically the size of each side's offensive arsenal and, in the process, to restructure the Soviet arsenal to reduce the threat posed by its substantial heavy ICBM force. At the same time, the US—having embarked on a major initiative to explore the possibilities of strategic ballistic missile defense—sought to preserve that option against Soviet challenges in the Defense and Space Talks.

Chapter 10 thus traces the development of US arms control theory as an essential foundation for understanding the US experience with strategic arms control and, with it, the nature of the contemporary debate. Its focus is confined to the bilateral nuclear relationship between the US and the Soviet Union—clearly the central element of US arms control policy. It begins with the premise that the US has, since 1945, followed two simultaneous tracks in its security policy—"arms buildup" and "arms control"—and that these two tracks reflect complementary as well as contradictory features of the US search for security.

THEATER ARMS CONTROL

Chapter 11 focuses more specifically on the role of arms control in European security. Here the scope of effort is much broader, spanning theater nuclear systems, the full array of conventional forces deployed between the Atlantic and the Ural Mountains, and a range of military activities including maneuvers, exercises, and force deployments. European arms control efforts bring into play not only the superpowers but also their allies, both collectively and individually, and, in some fora, the neutral and nonaligned states that share that same geographical space.

It is in Europe that the postwar confrontation between East and West has been most evident. It is also in Europe that the pursuit of arms control agreements is so heavily laden with competing images of what "security" means and of how arms control might contribute not only to the security of the status quo but also to the resolution of the broader political conflicts manifested by Europe's continued division. Thus, nuclear arms control is inherently related to the conventional security relationship, and negotiations on conventional forces are bound up in broader discussions of confidence and security building measures.

These inherent complexities in theater arms control account, in large measure, for the fact that there is little in the way of concrete results to show for all that intense diplomatic effort. The 1975 Helsinki Final Act and subsequent review conferences in the Conference on Security and Cooperation in Europe (CSCE) and the 1986 Stockholm Agreement from the Conference on Disarmament in Europe (CDE) have set important precedents by injecting some transparency into the East-West military relationship, but they have had no effect on the structure of opposing forces. Attempts to negotiate a balance in conventional forces in the Mutual and Balanced Force Reductions (MBFR) negotiations virtually stagnated during their fourteen-year tenure. The INF Treaty, for all of its historic success and its important precedents on verification, will in the long run arguably have a greater effect on the political climate than on the strategic equation. Even the tremendous potential of the Conventional Armed Forces in Europe (CFE) negotiations for reducing NATO and Warsaw Pact forces has been overshadowed and almost overcome by independent political changes and economic pressures.

These same complexities also account for the great anxieties that surround the future of theater arms control efforts. The political dynamics in the post-INF environment touch on core security questions, such as theater nuclear force modernization and the nature of the US extended deterrent guarantee. At the same time, those dynamics have given additional impetus to the pursuit of conventional arms control negotiations, the outcome of which could be revolutionary. Nonetheless, the implications of these negotiations remain more political than military, involving questions of whether the very structure of East-West relations can—or should—be altered.

THEORETICAL UNDERPINNINGS

These chapters, each in its own way, testify to the fact that arms control involves more than just an attempt to regulate the "military balance" between strategic adversaries. Whether applied to the superpower relationship or to the European theater, arms control involves fundamental questions about the nature of the postwar political order and how security is to be maintained in that order.

There are two discernible trends in US arms control thinking that apply both to the strategic and theater domains. One is a trend away from nominal limitations on existing force structures and towards insistence on radical reductions in

offensive weaponry. Consequently, US objectives involve not just a freeze in existing deployments but a significant restructuring of Soviet nuclear and conventional force deployments to preclude preemptive offensive actions.

Similarly, there is a growing tendency to think in terms of a jointly managed transition to a potentially more stable security environment based on defense. Certainly this is the case with notions of sharing the fruits of the US Strategic Defense Initiative (SDI) with the Soviet Union.[3] A related element is the more far-reaching proposition of not only reducing but ultimately eliminating all ballistic missiles which, as the Reagan administration reiterated, are "the most destabilizing and dangerous system of all."[4] Comparable thinking is reflected in the conventional arms control arena in attempts to favor ostensibly "defensive" weaponry while restricting elements of offensive combined arms formations. Complementing such reduction and redeployment proposals would be a comprehensive system of "stability enhancement" measures that would create greater transparency in force deployments to increase warning time.

Such thinking is nothing less than revolutionary in its implications for US security policy. To be sure, it has not been without controversy, not least because it involves assumptions about whether the Soviet Union's own approach to security and arms control is undergoing a metamorphosis. In setting the stage for the next chapters, this chapter discusses the nature of arms control and its relationship to national security, examines the various purposes that arms control has been viewed to serve, and highlights the difficulties that have limited the effectiveness of arms control in achieving the objectives set out for it. Finally, this chapter reviews some of the policy choices that have dominated the development of arms control policy.

ARMS CONTROL AND NATIONAL SECURITY

In the opening chapter to this volume, we distinguished between *realist* and *idealist* traditions that have competed with each other in the development of US national security policy. The clash between these two traditions was perhaps most marked in the aftermath of World War I, when states sought to pursue both utopian notions of *disarmament* as well as more limited aspirations in the form of *arms control*. Attempts like the 1928 Kellogg-Briand Pact to outlaw war proved futile. In a system of sovereign states—in which definitions of security and self-defense differ, and fundamental political values are often at odds—there can be no enforceable edict against resort to force. Popular notions in the interwar years that wars were caused by the possession of arms were equally sterile. The Washington and London naval conferences of the 1920s, on the other hand, were more limited arms control efforts to regulate the size and nature of great powers' navies. It was soon clear, however, that such mechanistic and partial solutions could easily be circumvented and could not offer a panacea to security problems.

In the nuclear age, the United States has tended to blend the two traditions of realism and idealism, fashioning its arms control policies as an element of national security policy that has, at the same time, both the narrower purpose of maintaining a balance of military power and the broader purpose of managing the political relationship with the other superpower. The US has consistently argued that the essential prerequisite of successful arms control is the deterrent strength of the United States. From Truman to Reagan, "negotiation from strength" has been a central principle of US arms control policy.[5] Because arms control is fundamentally a contractual process between independent (sovereign) actors, each party to the contract prefers to approach the process with a sense of *strength* with which to negotiate, lest one be compelled to make unacceptable concessions. On the other hand, the nuclear age and the destructiveness of modern warfare have cast a shadow of *vulnerability* over the relations of the superpowers and their allies. This has provided an impetus to arms control, even as strength has been a precondition for entering into negotiations. Fundamentally, arms control is an essentially political process that, to be successful, must be consistent with the nature of the international political system. To the extent that arms control has, both in theory and practice, become distinct from the more idealistic notions of "disarmament," it has sought to accommodate its inherent contradictions.

The distinction between arms control and disarmament is an important one. As Thomas Schelling and Morton Halperin pointed out over 25 years ago, arms control is "concerned less with reducing national *capabilities* for destruction in the event of war than in reducing the *incentives* that may lead to war or that may cause war to be more destructive in the event it occurs."[6] Similarly, Hedley Bull noted at the same time that disarmament—the reduction or abolition of arms—need not be controlled, while arms control involves the necessary element of restraint in arms policies.[7] Such restraint may apply to the character of weapons, to their deployment, or to their employment. It need not involve a reduction in the level of armaments; indeed, an increase in certain types of armaments may not necessarily be incompatible with arms control as long as that increase is within a framework of restraint and contributes to the stability and predictability of the security relationship.

The US approach to arms control has tended to focus on enhancing the stability of the relationship between strategic adversaries rather than achieving "disarmament," although, in the last decade, reduction in levels of armaments has replaced the establishment of notional ceilings as the preferred way to enhance stability through arms control. There are, to be sure, other possible benefits of arms control, such as reducing the effects of war, building mutual confidence between adversaries, and lowering the costs of defense. Yet, of these three possible benefits, the first is, fortunately, an untested proposition, the second is a question of political perceptions, and the third has not yet occurred. Rather, the central utility of arms control in the nuclear age lies in the potential for reducing the chances of war by minimizing the likelihood of miscalculation, misperception, and anxiety in a crisis and by reducing the incentives for starting a war. To the extent that the US and the USSR—as the principal participants in arms control—share a common interest in avoiding war and especially in avoiding nuclear confrontation, arms control can potentially offer benefits to each in a non-zero-sum game. Thus, arms control includes both "realist" and "idealist" notions as it attempts to deal with a fundamentally competitive strategic relationship.

DEFINING THE PURPOSES OF ARMS CONTROL

The logic of arms control—focusing on precepts such as strategic stability and the potential for mutual gain between adversaries at the negotiating table—must, of course, be compatible with the real world antagonisms that exist among its participants. As noted in earlier chapters, and elaborated further in the next, a consistent element of the debate over the role of nuclear weapons in the postwar world is whether nuclear weapons provide a uniquely powerful instrument to serve political ends or whether—as an "absolute weapon"—nuclear weapons are essentially apolitical, rendering traditional thinking on strategy irrelevant. In the context of arms control, this debate has often taken the form of a question: whether limiting nuclear weapons could be pursued regardless of the broader superpower political antagonism or whether nuclear arms control had to be viewed in the context of that competitive relationship, with *linkages* to other security and political issues.

After Soviet rejection of the US "Baruch Plan" in 1945 to vest control of atomic energy in the UN, the growing polarization between the superpowers in the cold war seemed to preclude any prospect for success in arms control. On the eve of the Korean War, the Truman administration had overruled those like George Kennan who had argued that the US should content itself with an imperfect system of control and verification to facilitate agreement.[8] As expressed in NSC-68, Truman's review of US national security policy in 1950, regulation of atomic weapons had become part of a broader fabric of political disputes in the cold war rather than an apolitical issue subject to isolated agreement:

For the time being, the US and other free countries would have to insist on *concurrent agreement* on the control of *nonatomic* forces and weapons *and* perhaps on other elements of a *general settlement*, notably *peace treaties* with Germany, Austria, and Japan, and the *withdrawal* of Soviet influence from the satellites. *If*, contrary to our expectations, the Soviet Union should accept agreements promising effective control of atomic energy *and* conventional armaments, without any other changes in Soviet policies, we would have to *consider very carefully* whether we could accept such agreements. It is unlikely that this problem will arise.[9]

After 1953, Stalin's death and the subsequent interregnum in the Soviet leadership suggested a "thaw" in East-West relations. The West had weathered the tumultuous storms associated with the consolidation of NATO and German rearmament within the Alliance. There was, therefore, an apparent basis of strength from which the West could negotiate. There remained, as well, a sense of vulnerability in the West both to the destructive power of thermonuclear weapons and to the domestic political pressures for reducing East-West tensions. In many respects, the series of East-West conferences in 1954–55 that culminated in the "Spirit of Geneva" reflected a form of "dual-track" decisionmaking. Final agreement among NATO allies on German rearmament, for example, required visible attempts to seek an alternative solution to European security through negotiation with the Soviet Union.

It was in this context that the US, UK, and French foreign ministers met in 1953 in Washington and agreed to invite the Soviets to a series of conferences to discuss outstanding issues. Each government's invitation to the Soviet leadership noted a desire "to dispose now of those problems which are capable of solution."[10] Ultimately, the Four Powers agreed to the Austrian State Treaty in May 1955, following which the four heads of state and government met in Geneva in July 1955. The Geneva summit produced no substantive agreement on the principal issues on their agenda, such as German reunification. At the summit, however, President Eisenhower presented to the Soviets a surprise bilateral arms control offer, the "Open Skies" proposal.

In rejecting "Open Skies" just as they had rejected the "Baruch Plan," the Soviets declined the invitation to engage in a process based on the premise that arms control could provide increased security to both sides even though the overarching political relationship remained antagonistic. As the next chapter describes more fully, the Open Skies proposal reflected an attempt to address, on the margins, the character of the strategic relationship, in the hope that incremental improvement in each side's confidence in its own security could provide the foundation for future improvement in the broader political relationship.

Prior to the 1960s, arms control efforts had tended to focus on two levels. At the nuclear level, the dominant rhetoric was of "disarmament," but the principal negotiating issue was ways to limit nuclear testing, since the ability to test nuclear weapons was increasingly viewed as a "chokepoint" for nuclear proliferation. If one could ban testing, states with nuclear weapons would not be able to modernize their arsenals with confidence (thereby stemming "vertical proliferation"), and states without nuclear weapons would be restrained from joining the nuclear club (thereby discouraging "horizontal proliferation"). At the conventional level, several proposals emanated from both governmental and private commentators, designed to achieve a "disengagement" of US and Soviet forces in Central Europe.[11] As stressed in chapter 11, such proposals were fundamentally concerned with the structure of the *political* order in Europe, a problem even more formidable than attempting to regulate the military relationship between two opposing alliances.

The deep antagonisms of the cold war proved largely impervious to these early attempts to improve the East-West security relationship through arms control. The late 1950s and early 1960s were years of successive crises in East-West relations, constant reminders that the relations were plagued with fundamental political conflict and not merely "misunderstandings" that could be worked out on a technical level over a negotiating table. UN disarmament conferences seemed to have deteriorated into either insubstantial platitudes or opportunities for diplomatic finger-pointing. Four Power conferences concerned with constructing a European political order that did not entail East-West confrontation foundered on the intractable "German question." Discussions on ways to ban nuclear testing continued, but they tended to be one-sided in the face of Soviet intransigence. The combination of public anxiety prompted by the Soviet launch of *Sputnik* in 1957, plus a wide-ranging defense debate in the US that spilled over into partisan politics in congressional and presidential electioneering, however, provided an impetus for reshaping US thinking on arms control.

The increased focus on "arms control" in US strategic thinking coincided with the major reevaluation of defense policy in the late 1950s. Not surprisingly, therefore, arms control became an integral element of an evolving strategic theory that drew an increasingly clear distinction between *arms control* and *disarmament*. That theory focused exclusively on the nuclear relationship, on the assumption that, despite the enduring political conflict between the superpowers, each shared a common interest in avoiding nuclear war even though each relied on nuclear weapons as an important element of its power base. Not until the prospect of a SALT agreement a decade later focused attention on the conventional force imbalance in Europe did arms control thinking begin to address the problems in regulating that domain.

Beginning in the 1960s, therefore, arms control efforts tended to focus on three principal facets of the nuclear relationship: (1) limiting conflict and preventing miscalculation in a crisis; (2) limiting superpower strategic arsenals; and (3) curbing the proliferation of nuclear weapons. All had as their centerpiece the notion of *stability*, which was quickly becoming the dominant criterion for measuring the utility of an arms control scheme. Of particular significance, arms control and arms modernization became viewed as complementary approaches to the same end—not disarmament, but an increase in the stability of the nuclear relationship in which mutual restraint, codified in treaty form, reduced both the incentives for war and the incentives for arms racing. The relative emphasis on each of these two approaches—arms modernization and arms control—often varied over time, and the precise relationship between the two was not always clear.

MANAGING CRISES AND CONFLICT

The first area of emphasis in arms control thinking had less to do with the control of armaments but was more concerned with what became known as *crisis management* and *conflict management*. The dawn of the missile age coincided with a period of intense crises in East-West relations. Whatever Khrushchev's motives— some argue a bloated sense of Soviet superiority; others argue bluff and bluster to hide Soviet weakness—the launch of *Sputnik* was followed by successive Soviet challenges in Berlin and elsewhere, political posturing with numerous and massive thermonuclear tests in the atmosphere (up to 50 and 100 megatons), and generally erratic and bellicose Soviet behavior. The political climate was hardly conducive to

negotiations aimed at limiting arms, but it heightened the incentive in the West to search for ways to control the effects of conflict and to reduce the likelihood that crisis would lead to war.

In terms of crisis management, arms control theory embraced the notion, popularized within the scientific community in particular, that conflict represented, at least in part, misperception and miscalculation between adversaries. One solution, therefore, was to increase the opportunities for direct communication in a crisis. In this respect, *arms control theory represented an extension of deterrence theory* in that the essential act in deterrence is communication of intent to an adversary. Even if fundamental conflict could not be resolved by direct communication, at least crises could be restrained. Such notions eventually led to the 1963 US-Soviet "Hot Line" agreement, signed in the wake of the Cuban missile crisis and later updated in 1971 and 1984. They later appeared as well in proposals for "confidence and security building measures" (CSBMs) in Europe and the 1987 US-Soviet agreement on "nuclear risk reduction centers."

Similarly, arms control could serve conflict management purposes by limiting the scope of conflicts that might develop. In this respect, *arms control also represented an extension of US limited war theory* as the US sought to reduce its reliance on threats of massive retaliation and to enhance the role of military force as a usable instrument of political policy. Thus, in assessing the Eisenhower administration's security policy in the wake of *Sputnik*, the Second Rockefeller Report in 1958 argued:

We must face the fact that a meaningful reduction of armaments must be *preceded* by a reduction in tensions and a settlement of the outstanding issues that have divided the world since World War II. *At the same time, concrete proposals to limit such wars as might be forced upon us* should be introduced into negotiations on reductions in force.[12]

Even if armaments themselves could not be limited, it seemed possible to reduce the likelihood that conflict would escalate automatically to general nuclear war.

The linkage of arms control to theories of deterrence and limited war broadened the scope of arms control theory to encompass a variety of *unilateral* measures which could be taken in the framework of "stability." Defense Secretary McNamara's suggestion in 1962 that the Soviets join the US in refraining from targeting cities, for example, represents a form of unilateral arms control.[13] Similarly, efforts to refine mechanisms for the command and control of nuclear weapons and increasing attention to survivable basing modes for nuclear missiles were potential areas of both Soviet-American cooperation and unilateral action to reduce incentives for preemptive or accidental war. Ultimately, even the increase of certain weapons systems—"stabilizing" by virtue of their invulnerability—was viewed as conducive to arms control as the notion of mutual assured destruction began to dominate strategic thinking in the 1960s.[14]

LIMITING SUPERPOWER ARSENALS

The linkage of arms control theory to broader strategic thinking on nuclear force requirements likewise dominated the second area of emphasis for arms control: proposals for bilateral agreements to limit superpower deployments of ballistic missiles. The central theme was that the absence of a defense against offensive missiles offered a framework for stability in the nuclear relationship. As early as 1956, the secretary of the Air Force had noted:

Neither side can hope by a mere margin of superiority in airplanes or other means of delivery of atomic weapons to escape the catastrophe of such a war. Beyond a certain point, the prospect is *not* the result of *relative* strength of the two opposed forces. It is the *absolute* power in the hands of each, and the substantial *invulnerability* of the powers to interdiction.[15]

As this notion developed into later theories of mutual assured destruction, it implied that there was a "minimum" level of weapons necessary to pose a credible deterrent. It also implied that, beyond a certain level, additional weaponry was superfluous. Provided that each superpower possessed a minimum level of invulnerable weapons capable of holding each other's vulnerable societies at risk, it was argued, more weapons offered no additional increment of strategic advantage or even security. Given the existence of some hypothetical optimum level of armaments, arms control was not only possible but, in its own right, enabled a greater degree of strategic stability.

In a world in which only the US and the USSR possessed strategic nuclear arsenals, it might make sense to construct a hypothetical balance that, in effect, institutionalized a nuclear stalemate based on strategic parity and mutual vulnerability. Reality, however, did not conform to a theory that restricted itself to the nuclear relationship between two superpowers. Each superpower had allies and depended on more than just the nuclear dimension of military power to defend its interests. In the West particularly, the search for a deterrent formula had to encompass allied security interests and incorporate the means to wage war on a limited scale.

There were some who recognized more readily than others the inherent limitations that political reality imposed. An arms control seminar in 1960, for example, addressed the possibility of a hypothetical US-Soviet agreement to limit each side to 200 ICBMs, representing a "minimum" deterrent. In considering the security implications of such an agreement, Morton Halperin acknowledged that it would cause serious problems within the NATO Alliance:

The European members of NATO have become increasingly aware of the possibility that the American strategic deterrent will not prevent a limited attack on Europe . . . The United States might have to follow up this agreement [limiting each to 200 missiles] by stationing larger forces on the European continent and substantially improving NATO's capacity to fight limited nuclear and conventional wars. The question would have to be faced as to *whether this activity would destroy the Russian-American confidence necessary to maintain the arms control agreement.*[16]

Halperin's argument highlighted a dilemma that has persisted throughout the SALT era and the Reagan administration's attempt to negotiate major reductions in both intermediate and strategic nuclear forces. Bilateral arms limitations aimed at stabilizing a balance in superpower nuclear arms require that the US take steps—to include even the buildup of arms outside the scope of the agreement—to reassure allies of the continuing effectiveness and reliability of the US extended deterrent.

RESTRAINING NUCLEAR PROLIFERATION

The 1960 *Summer Study on Arms Control* also highlighted the fact that attempts to construct a stable superpower balance based on a "minimum deterrent" ignored the inherent dangers of nuclear proliferation.[17] Thus, the third area of emphasis in arms control thinking—*efforts to restrain the proliferation of nuclear weapons*—reflected a view that the nuclear dimension of arms control could not be confined to the superpowers. Britain had tested its first thermonuclear device in May 1957. France's first thermonuclear test followed in April 1960, and President de Gaulle gave every indication that he would not subject French nuclear programs to international control. Soviet nuclear assistance to the People's Republic of China abruptly ended as the Sino-Soviet relationship ruptured, but China continued to pursue its independent program and tested its own device in October 1964.

The prognosis that the "nuclear club" would expand to encompass multiple actors, each with its own political agenda, suggested that future nuclear relationships would be more complex, unpredictable, unmanageable, and dangerous. This suggested as well that the US and the Soviet Union might share a common interest in restricting membership in the nuclear club. Walt Rostow, consultant to the Eisenhower administration and later a key foreign policy adviser to presidents Kennedy and Johnson, wrote in 1958 that the Soviets might wish "*to exploit the transient primacy* of the Soviet Union and the United States *to create a system of armaments control* so solid and so secure that it would guarantee a world of reasonable and orderly politics by the time the new nations come into maturity."[18] Rostow's underlying premise was that, at least in the realm of nuclear weapons, the Soviet Union had become a status quo rather than a revisionist power, content with preserving its international status rather than trying to alter the international political order. Such a superpower nuclear "condominium" was politically appealing because it preserved the prerogatives of the superpowers and avoided the prospect of smaller powers having "a bargaining position disproportionate to their industrial capacity and military potential."[19]

The desire to restrain nuclear proliferation complemented efforts to define a basis for superpower arms limitation since the latter would only be complicated by an increasing number of nuclear actors. Hence, an implicit sequencing of the arms control agenda developed in the early 1960s. A US-Soviet agreement setting ceilings on offensive missiles could potentially stabilize bilateral nuclear relationships, but only after or in conjunction with an agreement on nuclear nonproliferation. Such a nonproliferation agreement would require two elements. First, the nuclear powers—the "haves"—would have to agree not to transfer nuclear capabilities to the "have-nots." Second, nonnuclear weapons states had to agree to remain in the have-not category. The essential prelude to a nonproliferation agreement, however, seemed to be a test ban agreement, in which the development of nuclear technology itself would be restricted.

While the logic of this sequence seemed clear enough, political realities continued to plague its execution. Where this logic prescribed agreements—such as a comprehensive test ban involving effective compliance verification, or adherence to a nonproliferation agreement by those regional powers most likely to resist such restraints—fundamental political conflicts precluded progress. Those agreements that did ultimately transpire—such as the 1963 Partial Test Ban Treaty and the 1968 Nuclear Non-Proliferation Treaty—only partly fulfilled the requirements of the sequence. Despite the theory, arms control remained constrained by superpower competition, alliance anxieties, and international political realities.

ARMS CONTROL AGREEMENTS: THE RECORD

Perhaps the most remarkable characteristic of the US experience with arms control is that so much activity over the past quarter century has apparently had so little effect on national armaments programs. In many respects, this reflects the fact that the dominant approach to arms control has been, at least until recently, to focus more on the regulation of arms competition than on the actual reduction of arms. It also reflects the underlying political realities that characterize the broader East-West relationship. The superpowers may share a common interest in controlling the more dangerous aspects of their arms competition, but those common interests remain conditioned by their broader political competition, and both sides often differ in definitions of their security interests. It is not surprising, therefore, that arms control has arguably been most successful in areas where it is needed least and least successful in areas where it is needed most.

The first postwar arms control agreement actually occurred in 1959, when the US, the UK, and the USSR, as the only nuclear powers, agreed not to "militarize" the Antarctic continent. Like its "successor" agreements—the 1967 treaty prohibiting the deployment of weapons of mass destruction in outer space, the 1968 treaty creating a Latin American Nuclear Weapons Free Zone, and the 1971 treaty banning deployment of nuclear weapons in the seabed—the effect was largely cosmetic. None of these treaties had any discernible impact on East-West relations. The signatories were merely agreeing to refrain from activities in which they had no intention of engaging in the first place. They involved no behavioral change. Compliance was not an issue; verification was a relatively simple proposition, requiring no intrusive inspection mechanism. More to the point, national security interests were not particularly sensitive to compliance.

In much the same vein, the 20 June 1963 signing of the Hot Line agreement between the US and the USSR—like its subsequent modernization agreements—reflected "positive-sum" arms control. Neither side gave up anything, and neither risked anything. The agreement clearly reflected the fact that there was at least some overlap in superpower security interests—the avoidance of war by miscalculation or accident—although the Hot Line offered no restraint on an adversary intent on waging war.

Nonetheless, these early agreements contributed to what some viewed as a climate conducive to further progress in arms control, particularly in light of the fact that the Hot Line agreement followed almost effortlessly from the brinkmanship that characterized the Cuban missile crisis in October 1962. The virtually simultaneous termination of the Berlin and Cuban crises seemed to indicate that Khrushchev had suddenly acknowledged the perils of the nuclear age. In December 1962, Khrushchev warned Peking that "this 'paper tiger' [the US] has atomic teeth . . . and it must not be treated lightly."[20] Moreover, in admonishing the Chinese that "atomic bombs do not recognize class distinctions," Khrushchev seemed to many to have accepted a central thesis of Western arms control thinking—that nuclear weapons were essentially apolitical.

Regardless of Khrushchev's actual views— and whatever relevance they might have had in the long run for Soviet thinking—the Kennedy administration seized on this seemingly new Soviet attitude by stressing the "positive sum" nature of arms control:

It is wrong . . . that any US gain in security necessarily implies a concomitant Soviet loss . . . Indeed, it may be that the one distinguishing characteristic of all "arms control" measures . . . is that of a design to achieve *mutual* improvement of security.[21]

President Kennedy's later "Strategy of Peace" speech at American University in June 1963 clearly reflected the prevailing view that a "practical, attainable peace" could be sought if "total peace" were unattainable.[22] In short, arms control could proceed to the extent that it was *possible*. Other East-West conflicts that were not so amenable to resolution would have to be dealt with in more traditional ways, but they should not be allowed to interfere with the need to make progress on nuclear issues. There was no "linkage" to other issues; nuclear arms control would proceed on its own merit, in the recognition, as Kennedy noted, that "no government or social system is so evil that its people must be considered as lacking in virtue."

In light of this rhetoric in the wake of profound and dangerous East-West crises the year before, it is perhaps surprising that the subsequent achievements in arms control were not more dramatic than they were. The Partial Test Ban Treaty (PTBT), initialed in Moscow on 25 July 1963 by the three nuclear powers, is commonly heralded as a cornerstone of détente and arms control, but it is significant more for the narrowness of its application. After almost a decade of negotiations on a comprehensive test ban—which defied agreement on provisions that could verify compliance—the PTBT banned nuclear testing only in the atmosphere and in outer space. It thus avoided the intractable issues of compliance verification and accepted more modest objectives. Those verification issues have continued to plague subsequent efforts to limit nuclear testing. The Threshold Test Ban Treaty and Peaceful Nuclear Explosions Treaty—signed in the 1970s to limit underground testing to 150 kilotons—remain unratified by the US Senate because of doubts about verification.

The limited nature of the PTBT represented as well a cautious approach to arms control on the part of policymakers. In his Senate ratification testimony, Secretary of State Dean Rusk bluntly argued, "This treaty does not rest on the element of trust. The Soviet Union does not trust the United States. We do not trust the Soviet Union. But that is not the point."[23] The "point," presumably, was that the PTBT was only a first step in an attempt to engage the Soviet Union in a broader process of arms control in which both superpowers would pursue a "rational" definition of mutual interests in a domain that could be kept isolated from the political conflicts that dominated the rest of their competitive relationship. In the meantime, Defense Secretary McNamara assured conservative critics during Senate ratification hearings that the "principal military effect" of the PTBT was the "prolongation of our technical superiority."[24]

Underlying the PTBT and the negotiations on the Nuclear Non-Proliferation Treaty (NPT) that followed was an assumption that the Soviet Union would cooperate in codifying a nuclear status quo. This assumption was perhaps most valid with respect to the NPT. The Soviets were particularly anxious that the FRG might join the nuclear club through the back door, either through Franco-German nuclear cooperation, joint US-FRG control arrangements, or a NATO control-sharing scheme such as the Multilateral Force (MLF). US and Soviet drafts of those articles of the treaty concerning transfer of nuclear weapons or their control did not begin to converge until after NATO resolved its internal debates over nuclear sharing.

The internal US policy debate over the NPT reflected as well an interesting clash between those in the arms control community who favored an NPT that preserved the nuclear status quo and those in the foreign policy establishment who wanted to preserve the MLF as a vehicle for stimulating greater nuclear cooperation and self-reliance among the NATO allies. As noted in chapter 6, the MLF had been proposed as a medium for greater NATO integration, as an alternative to independent nuclear deterrents (especially that of the French), and as a way of providing the FRG with a sense of participation in Alliance nuclear matters. Even though President Johnson had, by 1964, effectively killed any prospect of developing an MLF, US draft NPT texts in 1965 continued to prohibit transfer of nuclear control only to "states," leaving open the option that a nuclear power might share control with another state or group of states.[25] The issue within NATO was whether the NPT would preclude the possibility of a "European" nuclear force, a revolutionary development which the Soviets preferred not to entertain, especially if it involved a West German share in nuclear control.

The issue for the FRG, on the other hand,

was not so much an insistence on a "hardware" solution to Alliance nuclear sharing, but a broader political agenda that could not accept such a status quo in the European political order. Bonn had, in conjunction with its accession to the NATO treaty, already unilaterally renounced to its allies the manufacture, possession, or control of nuclear weapons. But the pivotal role of the FRG in the NPT discussions offered an opportunity for Bonn to advance its own desire for German reunification. As the FRG foreign minister noted in July 1965:

If [German security] will be satisfied through the form of a Multilateral Atlantic Deterrent Force or a similar solution, Germany will renounce *vis-à-vis her allies* the acquisition of her own nuclear weapons. Should the Soviet Union be prepared . . . to agree to essential and irrevocable steps toward German reunification, the security question would change. The accession of a *unified Germany* to a *worldwide agreement* would be possible.[26]

Bonn's demand ultimately disappeared amid the domestic and allied pressures for a new *Ostpolitik*, or "Eastern Policy." Bonn's accession to the NPT in 1969 was the first step in that *Ostpolitik* and a clear precondition to Soviet cooperation in normalizing political relationships in Central Europe.

The final NPT text, opened for signature on 1 July 1968, not only prohibited transfer of nuclear control to states but also ruled out any multilateral co-ownership scheme. It did not, however, prejudice US "programs of cooperation" with NATO allies (including the FRG) whereby the US controlled warheads for delivery systems owned by allies. Neither did the treaty, in the unchallenged unilateral interpretation, "bar succession by a new federated European state to the nuclear status of one of its members." Although the FRG did not sign the NPT until November 1969—having deferred decision after the Soviet invasion of Czechoslovakia—the US had consulted closely with Bonn and accommodated a variety of technical objections to the final drafting sessions with the Soviets. Finally, reflecting both West German pressures and the original vision for sequencing arms control endeavors, the NPT committed the nuclear weapons states to negotiate "in good faith" on the reduction of nuclear weapons.

The NPT marked the end of multilateral *nuclear* arms control. The long process associated with reaching an agreement in which the superpowers shared substantial common interests had highlighted the difficulties of multilateral arms control. Even if the principal powers involved agreed that nuclear issues could be addressed in isolation from their mutual antagonisms, the intertwining of US and allied security interests imposed a burden on the negotiating process. (The Soviet Union, which enjoys control over its Eastern European allies, presumably does not have the same problem.) Subsequent arms control endeavors relating to both strategic and theater nuclear systems have been restricted to US and Soviet weapons. Even there, the US has discovered that consultation with allies is a necessary process that can be as arduous as negotiation with the Soviets.

By the late 1960s, the prospects for success in the previously intractable domain of superpower limitations on strategic arsenals seemed real. In part, this was due to resolution of the central issues surrounding the NPT. More fundamentally, this reflected the fact that the US and the Soviet Union had reached "parity" in strategic arms, defined by President Nixon in early 1970 as "the ability [of each superpower] to inflict unacceptable damage on the other, no matter who strikes first."[27] Establishment and modernization of the strategic "triad" in the 1960s, essentially completed by 1967, theoretically enabled the pursuit of strategic arms control, the objective of which—like the PTB and the NPT—was to codify the status quo and institutionalize a stable balance of superpower nuclear armaments. Agreement in SALT I was to lay the foundation for subsequent negotiations not only in strategic intercontinental systems but also on theater nuclear systems and on conventional force reductions in Europe.

Beginning in the late 1960s—ushering in a period characterized as détente—the US and the Soviet Union negotiated and, in 1972, signed the SALT I agreements limiting both strategic offensive missiles and antiballistic missile systems. As the next chapter describes, these agreements were to provide the basis for subsequent negotiations on strategic intercontinental systems. The signing of the SALT II Treaty in 1979, however, occurred in the midst of a vociferous debate in the US on the merits of strategic arms control, a debate that ultimately precluded ratification of the treaty. This debate reflected disillusionment and frustration over the failure of détente to effect any change in Soviet international political behavior in general and the failure of SALT to reverse the growth of a Soviet nuclear posture that contradicted the theoretical foundations of SALT and posed an increasing threat to the stability of the nuclear relationship. Although the outlines of a prospective START treaty clearly reflect a fundamentally different approach in attempting to redress that strategic relationship, the debate about the merits of *any* arms control approach to the problem continues.

That same debate is also evident in assessments of the INF Treaty, with the added dimension that the INF Treaty—although a bilateral agreement between the superpowers—involves the framework of *European* security most of all. As such, the debate in the

US over the merits of an INF agreement insofar as it affects the US-Soviet relationship occurred alongside a debate in Europe over the impact such an agreement will have on the US commitment to Europe.

Just as SALT led to a parallel effort to negotiate conventional force reductions in Europe, so too has the prospect of reductions in both theater and strategic nuclear systems highlighted the relationship between nuclear and conventional forces in the continuing East-West security relationship. To the extent that the US and the Soviet Union actively pursue "the goal of the reduction of nuclear weapons and, ultimately, their reduction"—as noted in the December 1987 Joint Summit Statement—arms control will ultimately demand a reassessment of over four decades of thinking on the role of nuclear weapons as a deterrent and whether conventional deterrence could ever substitute for nuclear deterrence, particularly in Europe.

In that respect, recent efforts to move beyond the stagnant Mutual and Balanced Force Reductions (MBFR) talks to negotiate a framework for stability in the *conventional* force relationship in Europe are significant. The 1986 Stockholm Agreement offers a useful precedent for "challenge inspections" of military maneuvers in an area stretching from the Atlantic Ocean to the Ural Mountains in the Soviet Union. The Conventional Armed Forces in Europe (CFE) negotiations, along with negotiations on confidence- and security-building measures (CSBMs) in Europe, will likely expand that "openness" in conjunction with deep force reductions. As political change in Europe proceeds independently, the *process* of arms control takes on greater political significance just as its concrete results are militarily significant. As chapter 11 points out, the implications of such agreements involve the very nature of the European political order.

THE LIMITS OF ARMS CONTROL: POLICY CHOICES

Arms competition is partly a manifestation of the underlying antagonisms among sovereign states that persist in a world that has not yet been able to maintain peace and security without relying, at least in part, on national capabilities for self-defense. That fact makes disarmament a utopian vision. Arms control—as distinct from disarmament—nonetheless has a role to play as a complement to arms development in the pursuit of security. Probably the most that arms control can do, however, is to regulate that arms competition in a way that enhances the stability of the security relationship, tempers the incentives for arms racing by injecting a greater degree of confidence and predictability in the way adversaries view each other, and reduces the incentives for war, especially preemptive war.

The pursuit of even these ambitious objectives for arms control is plagued by the fact that significant choices have to be made in defining the parameters of any particular agreement. Two obvious issues stand out. The first is *what to control*; the second, *how to control it*. The first focuses attention on the object of arms control—those weapons and defense-related activities to be captured in an agreement and, just as significant, what is left outside the agreement. The second involves the modalities of control—whether, for example, to freeze weapons or reduce them and, if the latter, by how much. These two issues are, of course, interrelated. Nonetheless, arms control involves choices that are inherently limiting. They can also determine whether a particular arms control agreement contributes to or detracts from security.

DECIDING WHAT TO CONTROL

Arms control negotiations can focus on a variety of weapons categories and defense activities. At the nuclear level, for example, one can focus on different types of weapons according to range (strategic, theater, battlefield, etc.) or basing mode (air-, sea-, or land-based). Negotiations on conventional forces can focus on manpower levels or attempt to single out certain kinds of military equipment (such as main battle tanks or certain types of naval vessels). These more common efforts tend to address weapons delivery systems. One can also focus on what is delivered by such systems, such as chemical or biological munitions, just as earlier treaties sought to prohibit "dum-dum" bullets or the use of gas.

At the heart of such choices is the assumption that, despite the continued necessity of military defenses, certain kinds of weapons systems are less desirable than others. They may be less desirable because of the threats they pose—the preemptive capacity of hard-target-kill capable ballistic missiles or the similar offensive capability of massive forward-deployed combined-arms ground formations, for example—or because the munitions themselves are abhorrent to the ethical norms of warfare.

Defining weapons as more or less desirable, however, is hardly a precise business. For example, arms control efforts have long grappled unsuccessfully with distinctions between offensive and defensive weapons, on the premise that, if one could eliminate the means of aggression while retaining the means of self-defense, one could construct a stable regime in which no one felt threatened. The problem, however, is that there is no clear distinction between the two, except in terms of the user's intentions. Weapons that are ostensibly offensive can also provide for an effective defense. At the nuclear level in particular, deterrence remains based on threats of offensive retaliatory strikes. Similarly, weapons that are ostensibly defensive can also support—or be viewed as intended to support—offensive aggression.

The dominant criterion for desirability, therefore, has been the notion of stability rather than a clear distinction between offense and defense. As the subsequent chapters testify, however, effectively defining the requirements for stability has been a frustrating process, not only because the requirements change over time and are affected by technological developments but also because the antagonists facing each other across the negotiating table do not necessarily define stability in compatible terms.

In addition, the significance of certain weapons can rarely be viewed in isolation. Rather, they need to be assessed in relation to other kinds of weapons systems and in terms of the impact that combination of weapons systems has on the military relationship. Regulation of strategic nuclear arsenals, for example, may be useful in creating a stable nuclear relationship between the superpowers, but it also has serious implications for European security. Popular notions of eliminating all nuclear weapons—or even strategic nuclear systems, as Gorbachev proposed in 1986 in Reykjavik—must be assessed in view of the fact that nuclear weapons do not exist in isolation. Their elimination may be detrimental to Western security if it were to create opportunities for aggression that had not otherwise existed since the Second World War.

In the same way, barring nuclear weapons from certain regions may have broader security implications. The 1968 treaty prohibiting nuclear weapons in Latin America, like the earlier Antarctica Treaty, sought to exclude nuclear weapons from areas where they did not already exist and concerned areas that were largely peripheral to the East-West military relationship. Proposals for a Balkan or Nordic nuclear weapons–free zone, on the other hand, would involve changes in the deployment (including maritime deployment) of nuclear weapons that affect not only the states concerned but also their respective alliance partners.

It also matters whose forces are being controlled. Negotiations to limit nuclear arsenals—SALT, START, INF—have, to date, been confined to the superpowers. In part, this is a process of self-selection, since none of the other nuclear powers have indicated a willingness to negotiate on their limited national nuclear arsenals. Conventional arms control, on the other hand, has been a multilateral endeavor. One of the impediments to the Mutual and Balanced Force Reductions (MBFR) negotiations was that the area under consideration was confined to the FRG and the Benelux countries in NATO and the GDR, Poland, and Czechoslovakia in the Warsaw Pact. France refused to participate. As a result, any reductions of US and Soviet forces in the region would essentially require US forces to return to North America while Soviet forces need only return to the Western USSR, thus giving the Soviets a greater opportunity to reintroduce forces in times of crisis and to do so over land, with much smaller distances, and with more secure lines of communication.

Since March 1989, negotiations on conventional force reductions from the "Atlantic to the Urals"—involving all twenty-three members of NATO and the Warsaw Pact—have provided opportunities for the West to deal with the geographic asymmetries between NATO and the Warsaw Pact. In addition, by emphasizing military equipment such as tanks, artillery, and armored combat vehicles, the CFE talks have focused on critical elements of an offensive attack capability rather than simply on manpower levels. Establishing ceilings on "stationed" forces in Europe—particularly US and Soviet forces outside their own territory—raises as well the political issue of what role each superpower will continue to play in Europe.

Arms control, by its nature, has tended to focus on those military capabilities most amenable to regulation. Agreements such as SALT were possible in part because the units of account, delivery vehicles, could be seen and counted. US-Soviet agreement on the global elimination of land-based INF missiles reflected, in part, the relative ease in dealing with US and Soviet systems without the complexities of negotiating on shorter-ranged systems. Dealing with shorter-ranged systems (below 500 kilometers) would not only pose a more difficult verification problem but would also open up the negotiations to "third-country" systems. The global nature of the INF ban reflected as well the relative ease in verifying total elimination of a weapons system rather than a partial ban with provision for production and modernization.

Similarly, the Partial Test Ban Treaty had to be content with prohibitions on nuclear testing in mediums that were visible. Continuing efforts to negotiate a Comprehensive Test Ban or even to finalize the Threshold Test Ban Treaty on underground nuclear testing have been plagued by uncertainties about verification. Efforts to ban chemical weapons have also been stymied by the immense difficulties of verifying such a ban.

Efforts to regulate, if not ban, the development of certain weapons technologies have also been complicated by the fact that such technologies often cannot readily be divorced from legitimate peaceful pursuits. Some of the most contentious issues in the Nuclear Non-Proliferation Treaty, for example, concerned the mechanisms by which nonnuclear states could still reap the benefits of peaceful application of nuclear energy. Similarly, chemical and biological research pervades the scientific establishments of all powers, precluding any state from having total confidence that weapons programs might not proceed clandestinely. Against that reality, research and development of chemical defense or biological antidotes, while prudent, provide further opportunities to mask weapons programs. By the same token, nuclear weapons testing can serve useful purposes by enhancing the reliability and safety of existing stockpiles.

In sum, the pursuit of arms control is, in the first instance, plagued by problems of defining what military capabilities should be controlled. That decision is complicated by the fact that there are no problem-free candidates. Moreover, those areas most amenable to potential regulation do not exist in isolation, and regulation of such military capabilities needs to be approached with an understanding of their broader implications.

DECIDING HOW TO CONTROL

Having decided what kinds of weapons systems one wants to control, a corollary issue is how they should be controlled—whether there should be a ceiling that encompasses existing or planned deployments or whether weapons should be reduced and, if so, by how much. SALT was an attempt to freeze the deployment of strategic offensive systems, while the subsequent START negotiations sought to reduce substantially (by roughly 50 percent) those arsenals. Agreement to reduce to zero all US and Soviet land-based INF missiles represents, of course, an ultimate form of reduction. Totally eliminating a category of weapons, however, presupposes that no such weapons is better than a balance at some level above zero.

The choice between freeze and reductions reflects in part a judgment about the nature of the existing security relationship before an agreement. President Johnson's 1964 proposal to "freeze" strategic missile deployments was certainly advantageous to the US and was, not surprisingly, rejected by the Soviet Union, whose strategic arsenal at the time was nothing close to that of the US in either quantity or quality. Superpower agreement in the SALT I treaty on offensive weapons, on the other hand, did reflect the belief that the two superpowers were at parity, or "essential equivalence." Hence, a "balance" was seen to exist which merely needed to be codified by a regime that preserved that balance. Advocates of SALT II similarly argued that a ratification of the existing strategic relationship was in the US interest.

Critics of the SALT process, on the other hand, argued that Soviet strategic weapons developments had destabilized the strategic relationship, that no "balance" existed, and that freezing that destabilizing imbalance merely perpetuated a disadvantageous US security process. If strategic arms control were to be effective, therefore, it would at a minimum have to enable a restoration of a stable balance. That requires reductions, substantial reductions, of those aspects of the Soviet arsenal which are most destabilizing. Those reductions should also be asymmetrical. In addition to reducing the size of the arsenals, subsequent US START proposals also sought to reduce the quantitative disparity in certain categories of weapons systems (particularly land-based MIRVed ICBM warheads) and, in the process, restructure the Soviet strategic arsenal away from a heavy reliance on first-strike capable systems.

Similar issues have dominated efforts to negotiate conventional arms control agreements so as to achieve a more stable balance in Europe. The long-standing Warsaw Pact superiority over NATO in conventional forces suggests that asymmetrical reductions rather than notional freezes at existing levels would be necessary to achieve this end. Negotiations on MBFR, however, never got past the point of defining the data base necessary to ascertain, for purposes of a treaty, the existing force levels from which one would reduce—an essential point if one is talking in terms of percentage reductions in force levels. Whether focusing on manpower or equipment such as tanks and artillery, "ceilings" rather than percentage reductions would seem to be a more manageable approach. In either case, the West's aim in seeking asymmetrical conventional force reductions would be, as in START, to restructure Soviet forces so as to remove their most destabilizing characteristics.

The question of reductions, however, begs the question of the scale of reductions. This, in turn, involves a judgment not about the balance before an agreement but, more importantly, about the balance after an agreement. In the nuclear domain, the call for a 50 percent reduction in strategic warheads and associated delivery vehicles reflects a desire, on the one hand, to reduce opportunities for a preemptive first strike and, on the other hand, to retain the deterrent capability which those weapons afford. Proposals to eliminate all strategic ballistic missiles, or all land-based INF missiles, reflect a judgment that nuclear deterrence is still effective without those particular systems.

In the conventional domain, the issue of scale of reductions assumes greater complexity, owing to the increased importance of geographic realities and the array of weapons systems involved. In the European theater, for instance, the requirement for NATO to maintain a cohesive forward defense and to sustain that defense over time while awaiting transatlantic reinforcements suggests a minimum force level below which reductions would be deleterious. By the same token, the Warsaw Pact's capacity for mobilization and forward deployment in crisis and in the early stages of conflict, along with its substantial numerical superiority, suggests the need for sizeable reductions—or at least major redeployments back to military districts in the Soviet Union—in order to impair significantly its offensive capability.

Quite apart from achieving such large and asymmetrical Eastern reductions, even defining the desired scale is a difficult undertaking. In 1970, for example, a West German study group in the early phases of preparing for MBFR argued against the prevailing concept of a 10 to 30 percent reduction on both sides as destabilizing: the Soviets would retain an offensive capability while NATO's defensive posture would be undermined. The group instead recommended a 75 percent reduction on both sides on the grounds that, although NATO's defensive capability would be largely depleted, the Warsaw Pact would at least be deprived of its offensive capability.[28] Similarly, recent studies have suggested that, unless reductions of Soviet/Warsaw Pact forces were both substantial and significantly asymmetrical (on the order of at least 5:1), an agreement to reduce conventional force levels in Europe might decrease rather than increase NATO's capacity to mount a cohesive defense.[29]

The obvious difficulty of achieving asymmetrical reductions, in either nuclear or conventional forces, is that, unless both sides view it in their interests to establish a mutually agreed balance at lower levels, the side seeking greater reductions must have sufficient leverage at the bargaining table to entice the other to make the requisite concessions. This underscores one of the fundamental limitations of arms control. Arms control is arguably most possible when a balance is perceived to exist, thus theoretically facilitating a mutually agreeable freeze or reduction to lower levels to maintain the balance. Arms control is less likely to succeed in creating a balance when a balance does not already exist. Yet perceived imbalances are more likely to be the norm in an adversarial security relationship. As a result, efforts to achieve results in arms control are, not surprisingly, often accompanied by complementary efforts to remove perceived imbalances by arms enhancements.

CONCLUSION

Arms control is essentially a contractual process between or among states to regulate aspects of their security relationship. As such, states—presumably out of some intersection of mutual interests—oblige themselves to restrain armaments development and related defense programs. Almost by redefinition, however, the parties to an arms control agreement have an adversarial security relationship. While the form of that security competition may be altered by arms control, it is unlikely that the nature of the adversarial relationship will be transformed by arms control. To the extent that arms competition is a tangible symptom of underlying political conflict, arms control is inherently limited in what it can accomplish. As experience suggests, arms control can, at best, be only a complementary path to security as one element in a nation's defense policy. Evidence does not suggest that arms control can substitute for a viable defense posture in the pursuit of national security.

In the nuclear age, notions of absolute disarmament are manifestly no more achievable or even desirable than in previous eras. Arms competition remains a symptom, not a cause, of fundamental political antagonisms and is likely to endure for as long as nations continue to define their interests in conflict with each other and to be compelled to defend their vision of security against political challenge. The mere possession of military power remains a usable instrument of political influence, even if that use is passive. This is no less true—and is perhaps truer—with respect to strategic nuclear weapons. The need both to deter the active use of military force and to deny political intimidation that can potentially derive from the possession of military power remains a central characteristic of the superpower relationship.

This is not to say that arms control has no role to play in helping to manage that competitive relationship and to embue that relationship with a degree of stability. It does suggest, however, that arms control cannot be relied upon to change these enduring realities. Indeed, arms control has become a more or less permanent feature in US defense policy. For the most part, the contemporary debate is about the proper role of arms control in defense policy and, in particular, about the criteria for defining "good" arms control and distinguishing it from arms control pursuits that are potentially deleterious in their effect on national security. That debate will persist.

In addition, the practical difficulty with arms control is that one has to make more deliberate choices in determining its boundaries. Intellectually, it is certainly easier to assert that *all* weapons should be controlled or limited or even eliminated altogether and done so in a comprehensive package. In practice, however, one has to make choices regarding both what to control and how to control it. These choices are complicated because the US and the Soviet Union both approach the security dilemmas of the nuclear age with different geostrategic outlooks, different doctrinal perspectives on the utility of nuclear weapons and of military force in general, and, accordingly, different force structures. Even in the relatively clear-cut domain of strategic arms control—with its narrow boundaries that confine the participants to the superpowers and that clearly circumscribe definable categories of weapons systems—the process has not been able to shed its essentially political nature.

Obviously, an arms control agreement has to be negotiable in that it constitutes the negotiated outcome of a bargaining process. In many respects, the choices one makes in defining the parameters of a negotiation reflect compromises in determining what is negotiable. They also reflect a recognition that an arms control agreement should be verifiable. Verification is necessary because arms control remains a contractual process between adversaries, in which it is neither expected nor prudent for adversaries to base their security on trust, and because there is no alternative to self-help in enforcing international agreements.

Although verification is often viewed as a largely technical process by which parties to an agreement monitor compliance, it is, significantly, also a political process. Arguably, the more far-reaching an arms control agreement is in affecting the security relationship, the more important and the more difficult verification becomes. Verification is very rarely, if ever, a perfectly precise endeavor. Ambiguity—even if only because of the limitations of language in crafting a treaty—is inevitable. Indeed, as the next chapter suggests, notwithstanding Soviet violations of arms control agreements, the *real* strategic problems for the US stem as much from Soviet activities that have been *in compliance* with existing agreements. Moreover, while one can monitor certain defense activities, the determination of whether such activities are or are not in compliance with treaty provisions remains a subjective political decision. If noncompliance is suspected, the question then is what to do about it. In the absence (as in civil law) of any external enforcement mechanism, states are left to decide whether they wish to continue in that contractual relationship or to decide, instead, that they are likewise freed of their own treaty obligations. As arms control increasingly encompasses forces and activities less amenable to precise monitoring, a legalistic approach to verification will have to give way to a more subjective political judgment in the face of ambiguity. This further suggests that reliance on arms control as the sole or principal means to achieve security objectives is imprudent.

In any assessment of arms control, there is a temptation to bemoan its limitations and to im-

pose on it the burden of resolving fundamental security dilemmas. Yet arms control can no more provide an "absolute solution" than past attempts to resolve security dilemmas through the development of an "absolute weapon." Weapons technologies are a dynamic, and the histories of both warfare and arms control are littered with efforts that have been rendered obsolete or inadequate by the evolution of technology. Whether in the domain of offensive nuclear weapons, of strategic defenses, or of conventional forces, arms control negotiations deal with weapons technologies that are not static and which are difficult to accommodate in a legal document. Ultimately, while cast in the form of legal instruments, arms control remains a political undertaking in which the interpretation and enforcement continue as political processes long after the ink has dried.

In politics there are few, if any, blacks and whites but, rather, varying shades of grey. The limitations of arms control discussed in this and the next two chapters are in many ways inherent to the broader political realities of the contemporary international system. At the same time, however, there is ample evidence to suggest that arms control, if pursued with realistic expectations, can potentially contribute to the stability of a security relationship. But it will likely do so only when it is coupled with the necessary programs of arms enhancements and defense modernization that enable a stable deterrent to underpin that security relationship. Ideally, both of these "tracks" will be consistent within a strategic framework that embodies a clear vision of how security is to be preserved. The key here, as in much of politics, is to find the proper balance and to avoid either "track" from being held hostage to the other. Finding that balance is a continuous process, subject to the dynamic pressures that characterize not only the adversarial relationship, but also one's relationship with allies and within one's domestic political environment.

NOTES

1. In addition to the text of the INF Treaty, Protocols, and detailed Memorandum of Understanding on Data, see the Fact Sheet issued by the Arms Control and Disarmament Agency (ACDA), "The INF Treaty: What's in It?" USIS Wireless File, 8 Dec. 1987.

2. Joint US-Soviet Summit Statement, USIS Wireless File, 10 Dec. 1987.

3. For example, President Reagan's speech to the nation after his meeting with Soviet General Secretary Gorbachev at Reykjavik, USIS Wireless File, 13 Oct. 1986; and his later speech to the Town Hall of California and the Chautauqua Conference in New York, USIS Wireless File, 6 Aug. 1987.

4. For an authoritative discussion, see Richard Perle, "Reykjavik as a Watershed in U.S.-Soviet Arms

Control," *International Security* 12, no. 1 (1987): 175–78.

5. This is a central theme of virtually every administration statement on the requirements for the successful pursuit of arms control agreements. See Seyom Brown, *The Faces of Power: Constancy and Change in United States Foreign Policy from Truman to Reagan* (New York: Columbia University Press, 1983).

6. Thomas C. Schelling and Morton H. Halperin, *Strategy and Arms Control*, reprinted. (New York: Pergamon Press, 1985), p. 3.

7. Hedley Bull, *The Control of the Arms Race: Disarmament and Arms Control in the Missile Age*, 2d ed. (New York: Praeger, 1965), pp. vii–viii.

8. See Kennan's memorandum to Secretary of State Acheson, 20 Jan. 1950 and Nitze's critique, in Department of State, *Papers Relating to the Foreign Relations of the United States, 1950 [hereafter FRUS]*, vol. I, "National Security Affairs; Foreign Economic Policy" (Washington, D.C.: G.P.O., 1977), pp. 13–17 and 22–24. Also George F. Kennan, *Memoirs: 1925–1950* (Boston: Little, Brown, 1967), pp. 471–76.

9. *FRUS, 1950* 1:271. Emphasis added.

10. Identical notes from the Governments of the US, UK, and France, sent to the Government of the USSR, 15 July 1953, in Royal Institute for International Affairs (RIIA), *Documents, 1953*, pp. 77–78.

11. See the discussion in chapter 6 on "disengagement," plus Michael Howard's review of various proposals by the same title, published for Chatham House by Pelican Books, 1958.

12. Cited in Walt W. Rostow, *The United States in the World Arena: An Essay in Recent History* (New York: Harper and Row, 1960), pp. 370ff. Emphasis added.

13. Bull, *Control of the Arms Race*, pp. 77–91.

14. For a summary of the role of "stability" in evolving arms control policy in the 1960s and early 1970s, see Jerome H. Kahan, "Arms Interaction and Arms Control," in his *Security in the Nuclear Age: Developing U.S. Strategic Arms Policy* (Washington, D.C.: Brookings Institution, 1975), pp. 263–327.

15. Quoted in Samuel P. Huntington, *The Common Defense* (New York: Columbia University Press, 1961), p. 101. Emphasis added.

16. Morton Halperin, "Implications for Limited War," in American Academy of Arts and Sciences (AAAS), *Summer Study on Arms Control, 1960: Collected Papers* (Boston: AAAS, 1961), pp. 158ff. Emphasis added.

17. See the discussions by Arthur Barber, Jay Orear, Dr. Bernard T. Feld, and Donald Brennan, in AAAS, *Summer Study on Arms Control*, pp. 160–61, 183, and 359.

18. Rostow, *United States in the World Arena*, pp. 429–30. Emphasis added.

19. Ibid., p. 412.

20. See Khrushchev's 12 Dec. 1962 speech to the USSR Supreme Soviet, in RIIA, *Documents, 1962*, p. 257.

21. Assistant Secretary of Defense for Arms Control, James T. McNaughton, in an address before the International Arms Control Association, Ann Arbor, Michigan, 19 Dec. 1962, generally regarded as a response to Khrushchev's speech cited note 20, a week before. See *New York Times*, 20 Dec. 1962.

22. Kennedy's speech, 10 June 1963, at American University, in RIIA, *Documents, 1963*, pp. 14–19.

23. US Congress. Senate Committee on Foreign Relations, Committee on Armed Services, and Senate Members, Joint Committee on Atomic Energy, *Hearings: Nuclear Test Ban Treaty*. 88th Cong., 1st sess., 1963, p. 5.

24. Ibid., p. 105.

25. *Baltimore Sun*, 27 July 1965; *New York Times*, 18 Aug. 1965.

26. Interview by FRG Foreign Minister Gerhard Schroeder in *Duesseldorfer Nachrichten*, 9 July 1965, reprinted in Heinrich Siegler, ed., *Dokumentation zur Abruestung und Sicherheit*, vol. 3, 1964–1965 (Bonn: Seiger, 1966), pp. 246–47. Emphasis added.

27. *United States Foreign Policy for the 1970s: A New Strategy for Peace* (The White House, 18 Feb. 1970), p. 2.

28. Martin Mueller, "Konzeption und Akteur: Die Entwicklung der MBFR-Politik der Bundesrepublik zwischen 1968 und 1971," in Helga Haftendorn et al, eds., *Verwaltete Aussenpolitik: Sicherheits—und entspannungspolitische Entscheidungsprozesse in Bonn* (Cologne: Wissenschaft und Politik, 1978), pp. 167–90.

29. For example, James A. Thomson and Nanette C. Gantz, "Conventional Arms Control Revisited: Objectives in the Next Phase" (RAND Note N-2697-AF, Santa Monica, Calif., Dec. 1987). Also Philip A. Karber, "Conventional Arms Control Options, or Why Nunn is Better than None," in Uwe Nerlich and James A. Thomson, eds., *Conventional Arms Control and the Security of Europe* (Boulder, Colo.: Westview Press, 1988), pp. 158–65. For a broader discussion of CFE, see Schuyler Foerster et al., *Defining Stability: Conventional Arms Control in a Changing Europe* (Boulder, Colo.: Westview Press, 1989).

CHAPTER 10

STRATEGIC ARMS CONTROL
Theory and Practice

R. JOSEPH DeSUTTER

The notion of a "dual-track" security policy is generally associated with NATO's 1979 decision to deploy intermediate-range nuclear forces (INF), while pursuing negotiations that would limit or render unnecessary those deployments. In one cause, that was a unique decision, in that actual deployments were held "hostage" to possible outcomes in arms control negotiations.

In a broader sense, however, such dual-track security planning was not new in 1979, but has characterized US policy since the beginning of the nuclear age in 1945. According to the theory behind this approach, diplomacy and negotiations accompany the modernization and deployment of weapons. The purpose of both tracks is to enhance national security. In a 1951 radio address to the nation, President Truman explained that:

The buildup of the defenses of the free world is one way to security and peace. As things now stand it is the only way open to us. But there is another way to security and peace—a way we would much prefer to take. We would prefer to see that the nations cut down their forces on a balanced basis that would be fair to all . . . [T]here is nothing inconsistent about these two things. Both have the same aim—the aim of security and peace. If we can't get security and peace one way, we must get it the other way.[1]

For Truman, "security and peace," or world peace on terms compatible with American political values, would be the principal goal of "good" arms control. Thus, arms control was "another way" (indeed the *preferred* way) of achieving the same security goals otherwise sought by the "buildup" track.

At least at the rhetorical level, this has remained the purpose of arms control throughout the nuclear age for the United States—to enhance national security without all of the economic, social, political, and environmental burdens that necessarily accompany an arms buildup. In 1955, President Eisenhower repeated Truman's logic in his famous "Open Skies" address to the Geneva Conference of Heads of Government.[2] In his 1961 inaugural address, President Kennedy employed much the same reasoning by arguing that we should neither "negotiate out of fear" nor "fear to negotiate."[3] It was in this spirit that Kennedy sponsored the 1961 Arms Control and Disarmament Act,[4] which created the Arms Control and Disarmament Agency (ACDA). ACDA's purpose, according to this legislation, was to formulate US arms control and disarmament policy "in a manner that will promote the national security." Kennedy's commitment to arms control was nevertheless accompanied by an arms buildup that produced the basic structure of today's strategic nuclear triad.

Similarly, Richard Nixon defended his SALT I agreements in 1972 on grounds that they were "in the interest of the security of the United States."[5] Cyrus Vance, President Carter's secretary of state, argued in 1979 that "the SALT II Treaty will greatly assist us in maintaining a stable balance of nuclear forces."[6] Likewise, President Reagan's secretary of state, George Shultz, frequently reminded audiences that "the prerequisite of successful arms control—and world peace—is the deterrent strength of the United States."[7]

There is a definite "Jekyll and Hyde" character to this dual approach to security, an ambivalence that is deeply rooted in the American personality. As noted in the previous chapter, the arms buildup approach responds to the call for "realism" in managing a reduction of power in an anarchic international system. The arms control approach, on the other hand, portrays our more "idealistic" side—the optimistic belief that law and reason can resolve differences among states. The United States has historically been so secure that we have come to regard the freedom to live in peace as a natural condition. Yet we have also felt it necessary several times this century to cross the very oceans that secured our borders in order to defend our interests. There are powerful conflicting forces in our national psyche, therefore, both to heed the demands of maintaining a balance of power and to avoid the burdens of participating in its management. The result, obviously, is that we have tried to do both.

Following both tracks more or less simultaneously, of course, obscures some of their contradictory premises. The justification for an arms buildup derives from a realistic belief that states compete for political influence through military power. By this logic, the nuclear arms race is a contemporary version of traditional balance of power competition. The justification for arms control, by comparison, derives from the more idealistic hope that a community of interests among nations can reduce if not eliminate altogether this competition. One approach focuses on competition, the other on tolerance and accommodation. Theoretically, however, both seek the same end—what Truman called "security and peace"—and both tracks have been ongoing elements of US security policy for over 40 years.

The following sections trace the evolution of US arms control theory from 1945 through the end of the Reagan presidency and examine the practice of strategic arms control in the context of that theoretical development. While the US has indeed sought to pursue both the arms buildup and arms control tracks more or less simultaneously, each chronological period has been marked by differing orientations toward the utility of one track or the other. Over time, however, the US approach to arms control has changed, reflecting its imperfect nature as a means of enhancing security in an imperfect world.

The first decade following World War II witnessed the development of what can be characterized as an "absolute" theory of arms control. In short, the US sought "absolute" solutions to the security dilemma—nuclear disarmament—as the essential prerequisite to regulating armaments. That theory, manifested in the Baruch Plan, failed in the face of political incompatibility between the US and USSR.

Beginning in 1955, the US embarked on a new approach, illustrated by Eisenhower's "Open Skies" proposal, by which arms control would aim toward more modest and "achievable" objectives. Such "partial" arms control efforts were guided by the criterion that they would be pursued only to the extent to which they could "assure results." At the same time, a new element began to enter arms control theory: as "disarmament" gave way to "arms control," such partial regulatory measures were expected to engender greater mutual confidence in the political relationship, thereby facilitating further progress.

This evolving theory did produce results in the 1960s, but the resulting treaties—such as the Limited Test Ban and the Nuclear Non-Proliferation Treaties—had little if any substantive effect on the security relationship. Nevertheless, their success prompted the formulation of even more far-reaching expectations for arms control, not only to regulate Soviet and American strategic arsenals but also to codify principles for their underlying political relation-

ship. This laid the foundation for the SALT era.

The SALT process was marked by efforts to regulate the buildup of strategic weapons to agreed levels, a far cry from the original absolutist notions of disarmament. That theoretical formula persisted through the 1970s, despite results that challenged the very strategic stability that the agreements were supposed to achieve. This chapter argues that arms control became, in the minds of many US policymakers, an end in itself—or at least a means to nonstrategic ends—rather than a means to greater security. Despite the early efforts by the Reagan administration to restore a balance between the arms buildup and arms control tracks of US security policy, the notion that arms control is a policy objective in its own right has endured.

As discussed in the previous chapter, there are real limits to arms control, despite a tendency to burden arms control with expectations it cannot meet. The dilemmas associated with the choice between arms buildup and arms control are significant, as are the dilemmas between "absolutist" and "partial" arms control choices. At the heart of these choices is a conflict between "realism" and "idealism" in US national security policy. An understanding of the contemporary debate about the utility of arms control, and its ability to temper the inherent political-military competition between the US and USSR, requires that we understand its theoretical premise and the problems that have plagued the practice of arms control since 1945.

1945–1955: THE RISE AND FALL OF DISARMAMENT THEORY

"The fact that we can release atomic energy ushers in a new era in man's understanding of nature's forces." So proclaimed Harry Truman after announcing, on 6 August 1945, that "[t]he force from which the sun draws its power has been loosed against those who brought war to the Far East."[8] Almost immediately, the bomb began to affect thinking on strategy. Massive strategic bombing of enemy population centers had precedent in the cases of London, Coventry, Dresden, Hamburg, Berlin, and Tokyo during World War II, but the new weapon was perceived as more than a difference in degree. The great efficiency with which destructiveness could now be packaged, the speed with which it could be delivered at great range, and the vast areas over which it could wreak havoc were virtually unprecedented in the history of warfare. Ironically, even though the atom bomb's use against two Japanese cities had produced nearly instant military victory, it also fostered the contrary impression of a new weapon capable of only terror, blackmail, and vengeance—everything *except* military victory. Douhet's theory of strategic bombing, it was said, had been carried to its logical extreme, and warfare had reached an apocalyptic stage where death and destruction would be self-evidently disproportionate to any political purpose.

In a particularly influential collection of articles edited in 1946 by Bernard Brodie, under the noteworthy title *The Absolute Weapon*,[9] a group of civilian strategic theorists thus argued that war no longer represented the rational policy choice it had been throughout history. Military superiority had no meaning by this logic because, as one author put it, "when dealing with the absolute weapon, arguments based on relative advantage have lost their point."[10] As discussed in chapter 3, modern understandings of "deterrence" are often traced to Brodie's own conclusion: "Thus far the chief purpose of our military establishment has been to win wars. From now on its chief purpose must be to avert them. It can have almost no other purpose."[11]

The relationship between politics and military strategy could not be severed without having a similarly dramatic effect on diplomacy. Efforts to control arms competition had generally failed, most recently in the 1920s and 1930s, but the nuclear factor and the alleged end of strategy had introduced a whole new set of variables. If relative strategic advantage had lost meaning, then rules of conflict resolution would seemingly become negotiable among nuclear powers, and a peaceful status quo would finally become possible. The principal task remaining, according to the emergent theory, was to negotiate verifiable agreements. Of particular significance, such a contract among nuclear powers was to be based on the collective self-interests of its participants rather than on moralism or surrender. Realism seemed to have joined idealism in eliminating the motivation for traditional competition.

ABSOLUTISM AND REALISM: A THEORY BEGINS

As discussed in chapter 1, the seventeenth-century philosopher Thomas Hobbes had used similar logic to explain why individuals entered society. According to Hobbes's classical realist assumptions, competitive self-interest is human nature, but constant fear causes a collective recalculation of the costs and benefits of individuals' natural sovereignty. Hobbes concluded that a reciprocal contractual agreement offered the needed alternative to this insecurity of nature—if it was collectively enforceable. Enforcement of the social contract, however, could be "guaranteed" only if the transfer of individual rights to a superordinate authority was absolute.

Early arms control theorists incorporated many of these realist assumptions and absolutist solutions. The international political system, it was agreed, is anarchic by nature. Sovereign states must, like Hobbes's natural man, provide for their own security without the help of a superior authority. No higher power was capable of adjudicating disputes that arise from states' competitive assertions of self-interest.

The "absolute weapon," by this analogy, introduced a new dimension of *societal* insecurity, with no prospect of defense on the horizon. This, according to arms control theory, would provide the motivation for states to create a "nuclear social contract." As with Hobbes's model, the contract would evolve because each state viewed it as beneficial in cost-benefit terms to relinquish national power in exchange for a new form of security in the nuclear age. In other words, national sovereignty would become a negotiable quantity.

As with Hobbes's model, the solution could be what game theorists call a "positive-sum game." Instead of one state losing sovereignty to another state—with the number of winners equaling the number of losers—the possibility existed for all participants to win. Furthermore, if governments were willing to negotiate that portion of their sovereignty that was associated with nuclear weaponry, then the newly created United Nations might represent a natural repository for that relinquished authority. The United Nations, however, would have to possess sufficient authority to enforce reciprocal international compliance with the new rules of nuclear disarmament.

As with the Hobbes model, therefore, it was clear to the 1940s theorists that the ultimate *solution* to insecurity arising from the "absolute weapon" would have to be absolute as well. This was true for several reasons unique to the postwar period. The first of these—an American advantage—was the outright US monopoly on nuclear weaponry between 1945 and 1949. This factor endowed the US government with political choices ranging from military coercion to diplomatic magnanimity. This US advantage, however, was offset by an equally absolute Soviet advantage—a tradition of societal closure and political secrecy that was deeply anchored in Russian and Soviet history. Because of the unprecedented advantages that would accrue to the successful violator of a total nuclear disarmament regime, verification provisions would therefore have to be virtually absolute as well.

A third characteristic of the era, and clearly the most formidable of all, was the overarching political antagonism that existed between the US and the USSR. American arms control theorists assumed correctly that the two principal victors of World War II would be the most relevant players in any postwar power distribution scheme with or without arms control. But the "other party" being dealt with was not some theoretical "country B," and the Stalin regime had already made clear its intention to dominate Eastern Europe—despite Soviet commitments to the contrary under the Yalta Agreements. The Soviets, in fact, bluntly defined their own political values as "antithetical" to those held throughout the West. And yet even so stark a divergence as this at the formal ideological level did not *necessarily* preordain hostility in the conduct of day-to-day international politics. Even if it did, one could argue that this simply made nuclear disarmament all the more urgent.

Thus, the emergent theory of verifiable nuclear arms control did not regard the problem of political incompatibility as prohibitive. In the context of Brodie's argument, since the absolute nature of the weapon had disrupted the traditional relationship between politics and nuclear strategy, the diplomacy of nuclear disarmament could proceed even if political competition *did* complicate relations in other areas. We would later label this argument the linkage problem, which would resurface again and again as a topic of debate in nuclear arms control.

For the 1945 arms control theorists, however, the need for serious political compromise as a precondition for nuclear disarmament was categorically rejected. Nuclear arms control, like nuclear strategy, was to be above politics, or apolitical. The powerful incentives for Soviet noncompliance, while troublesome, ought to be suppressed by the political pointlessness of nuclear strategy in the first place, and would be subject to pervasive verification provisions in any case. In short, the Soviets were expected to set aside any Marxist-Leninist principles that might stand in the way of compliance with international security contracts. They would do so for the same (realistic) reasons that brought them to the negotiating table in the first place— an overriding self-interest that happened to coincide with that of the US. Even if it were only the security of the Communist party and the preservation of their revolution that compelled them to do so, the Soviets would see the same self-evident value in nuclear disarmament as the US saw in its own quest for "national" security, and no change in the essence of Soviet politics would be necessary to bring this about.

The spirit that moved the US toward policies of accommodation regarding the interests of other nations was also seen by the theorists as the product of political realism. Many American political interests, it was argued, were served by the postwar status quo, and anything that would institutionalize its peaceful perpetuation would serve US self-interest as well. All of the lessons of history, particularly those associated with "peace" settlements after World War I, counseled victors to treat former adversaries with dignity and restraint in the formulation of postwar policy. Instead of disgracing and suppressing Germany and Japan, the US had there-

fore welcomed their new governments into the community of nations and contributed hefty sums from its own treasury toward their reconstruction. In the same spirit, the US and Britain had acknowledged Soviet interests in Eastern Europe as legitimate security concerns at the Yalta conference, and had treated their battered ally as a coequal power in structuring the new United Nations. If major conflicts were to be avoided for the remainder of the twentieth century—conflicts that could well involve nuclear weaponry again—then the security interests of potential contestants would have to be understood and accommodated, and new international mechanisms like the UN would have to institutionalize peaceful means of resolving future disputes.

While these lessons of history could guide the US in the more or less traditional problems of international relations, no such past experiences were available in the area of nuclear arms control. Nor could students of diplomacy point to comprehensively successful analogies in the area of conventional arms control when advising the practitioners of policy how to proceed. Since the Manhattan Project itself had been conducted in such great secrecy, the community of informed analysts was a small one, and, as a result, early arms control theory was largely a product of the atomic scientists themselves. This was regarded as appropriate for several reasons. First, if nuclear disarmament was apolitical and distinct from strategy, then it made sense for the technical experts to play a dominant role in controlling, just as they had in creating, the nuclear weapon. Secondly, if revolutionary diplomatic solutions were called for by the emergence of an "absolute weapon," then it also made sense to tap minds that were unburdened by the assumptions behind more traditional diplomacy. As chief Manhattan scientist J. Robert Oppenheimer put it in June 1946:

it may be permitted that men who have no qualifications in statecraft concern themselves with the control of atomic energy. For I think the control of atomic energy is important, in part, because it enables us to get away from patterns of diplomacy which are, in some respects at least, unsatisfactory as a model for the relations between nations.[12]

Oppenheimer began by rejecting two of the traditional "old models" of arms control that he regarded as unsatisfactory. The first of these, which Oppenheimer dismissed as inadequate, was what he called the "regulatory" method— piecemeal agreements sanctioned by periodic inspection. Since the US and the USSR could not trust one another in the extant political environment, he argued, such agreements would require extremely rigid protocols for inspection. These would never be adequate to assure compliance. The result would be what he called a cops-and-robbers scheme. He rejected such an arrangement because:

it seemed to us the robbers always have the advantage and the cops are always dumb cops . . . There is very much more than one way of going from raw material to the bomb that we know of, perhaps four or five that work today, and we are quite sure that new ones will be discovered. I'm afraid the cops would never know about the new ones, only the robbers.[13]

Oppenheimer's reference to the cops-and-robbers analogy became a recurring theme over the ensuing decades.[14] In the first place, there is a logical problem with inspecting another nation's homeland to *confirm* compliance or non-compliance. One might find compliance at various times and places of inspection, explained Oppenheimer, but no number of compliance discoveries could ever be enough to prove the absence of noncompliance, and no number of inspections would ever reveal non-compliance by a closed society intent upon violating a treaty clandestinely. This was all the more true under arrangements of partial rather than total controls, particularly given the originating premise of distrust. In short, the US, as an open society of "finders," could never totally unburden itself of anxiety over the prospect of noncompliance by a closed Soviet society of "hiders," especially given a powerful Soviet incentive to "hide" things. Secondly, and more in line with the realism underlying the argument, such a cops-and-robbers or hiders-and-finders scheme had been employed with disastrous results during German rearmament under the Versailles Treaty after World War I. As Richard Barnet has described that experience:

First, inspectors did not, both because of political obstacles and lack of personal initiative, exercise most of the inspection rights accorded under the treaty. Second, the inspection process did not deter wide-spread evasion. Third, within a short time after systematic evasion of the arms restrictions commenced, inspection was not needed for detection; independent of inspection, governments obtained sufficient information about the violations to enable responses and countermeasures. Failure to act was unrelated to the failure of the inspection system to uncover violations.[15]

Having rejected the regulatory approach to arms control, Oppenheimer turned next to the so-called retaliatory method of atomic weapons control—the "collective security" concept whereby "aggressors" are identified by an international consensus and punished by the combined powers of the aggrieved nations. Once again, recent experience had demonstrated the difficulty of ever achieving such a consensus, because this was precisely the mechanism designed by the League of Nations after World War I, and it had hardly inhibited aggression

by Germany, Italy, and Japan. The League's last official act, in fact, was its expulsion (not punishment) of the Soviet Union for invading Finland.

In short, realism in a theory of verifiable nuclear arms control required a degree of absolutism that neither the regulatory nor retaliatory methods could deliver. Yet as a scientist, Oppenheimer also realized that the limitless possibilities for peaceful use of nuclear energy ruled out outright prohibitions on the technology, and that peaceful nuclear research would always complicate controls on its military-related development. Thus, "controlled development" and shared learning were necessary preconditions to any meaningful compliance controls on nuclear disarmament. Moreover, since knowledge regarding the technology would be necessary for the discovery and enforcement of compliance, a charter for research and development of the new energy source should reside under the unchallenged purview of the UN, which would thereafter supervise all nuclear energy–related activities anywhere on earth. This total transfer of individual nations' decision-making authority in the nuclear area, to a superior enforcement entity, completed the Hobbesian "leviathan" logic by hypothesizing an "enforcer." The idea of great powers of enforcement in the hands of a UN agency also completes our theoretical journey leading to the early arms control proposals. From absolutist premises about the nuclear fear dilemma, to absolute disarmament with absolute means of verification, the early theorists concluded that, without absolute enforcement provisions, the requisite "contract" would be an irrational act.

ABSOLUTISM AND REALISM IN PRACTICE

The arms control and disarmament proposals of the late 1940s followed theory closely. The first of these, the Baruch Plan, named for Ambassador Bernard Baruch, who proposed it formally to the UN General Assembly in June 1946, was a case in point. The proposal formally called for the complete transfer by the US of all atomic weapons, atomic power facilities, and atomic know-how to international control. Accompanying this act of unilateral nuclear disarmament would be the creation of an International Atomic Development Authority "to which should be entrusted *all* phases of the development and use of atomic energy," including the ownership of *all* atomic energy activities potentially dangerous to world security; licensing authority over all other atomic activities; and research and development responsibilities to discover peaceful applications of nuclear energy and to supplement legal authority with the power inherent in leadership in knowledge.[16]

Since the US had already developed, tested,

and operationally employed atomic weapon technology, American facilities were to be opened to the proposed Authority's inspectors, and all raw materials were to be yielded to international control. Since inspectors would always be burdened with confusion in distinguishing atomic energy for peaceful use from that with purely military intent, both types of research would be outlawed unless licensed and operated under UN auspices. There would be no veto power by which UN Security Council members might protect violators. Qualified representatives of the Authority were to be granted "adequate ingress and egress" by all nations as necessary to assure compliance. Perhaps most significant of all, the international agency would actually be empowered to administer what the proposal called "condign punishments" in the event of serious violation—presumably to include nuclear explosions in the violator's homeland—so that compliance could be not simply verified but also enforced.

According to theory, the risk and uncertainty of disarmament—incumbent upon the West as a result of Soviet secrecy—were to be offset by these inspection and enforcement provisions. Similarly, the risk and uncertainty incumbent upon the USSR as a result of the American monopoly in atomic technology were to be offset by the open sharing of that know-how and the international ownership of existing weapons.

Contrary to theory, however, the Soviet response, delivered at the next meeting of the Atomic Energy Commission, was a counterproposal amounting to outright rejection. The Soviet version would have required American nuclear disarmament *prior to* either the establishment of inspection schemes or the creation of the UN Authority. Enforcement thereafter was to be a private, domestic, rather than international, responsibility; peaceful atomic research was to proceed unhindered and uninspected within individual nations; and Security Council veto power would remain intact for the five permanent members. As Britain's UN Representative Michael Wright summarized:

What in fact was the Soviet reaction to the Baruch Plan? With remarkable candour they rejected it summarily and made it clear they did not want a ban on nuclear weapons except on terms that would have given no semblance of satisfactory verification arrangements.[17]

The only aspect of the Baruch Plan found to be acceptable to the USSR, then, was that the US should internationalize its nuclear know-how as well as its arsenal.

In effect, the theory from which the Baruch Plan was derived had been tested in the real world by its proposal at the UN, and a number of its assumptions had been found to be flawed.

Principal among these was the belief that fear of war in the nuclear age would activate the same sense of urgency toward arms control among other nations as it had among the group of American thinkers who had devised the theory. In fact, whatever fear there may have been among Soviet decisionmakers, the relief from anxiety offered by the American proposal was clearly not worth the political compromises it entailed. From a Marxist-Leninist standpoint, the transfer of nuclear decisionmaking authority from the Soviet government to a veto-proof United Nations agency would have contradicted the very belief system from which the Communist party drew its legitimacy to govern. To a system that defines itself as the vanguard of a global proletariat, such an abdication of responsibility was too high a price to pay even when weighed against the potential consequences of war. The perceived benefits of nuclear disarmament were clearly not as mutual in practice as in theory.

THEORY REEVALUATED

In reaction to Soviet failure to behave in accordance with the American theory, Bertrand Russell argued in 1946 for a much more assertive approach—one that would raise the stakes for the Soviets by bringing the fear factor closer to home.[18] Normally recalled for his more powerful political positions, Russell argued that the US should exercise the advantages of its nuclear strength while its monopoly remained intact. Time was now of the essence, according to Russell. On the one hand, "if war comes it will begin with a surprise attack in the style of Pearl Harbor," followed by the "fierce revenge" of the attacked nation. On the other hand, he observed, "if there is *no war* in the near future, there will have been time for *Russia* to manufacture atomic bombs,"[19] with problems of secrecy and mutual suspicion likely to encumber all subsequent approaches to disarmament. This paradox led Russell to the conclusion that, "if utter and complete disaster are to be avoided, there must never again be a great war unless it occurs in the next few years." But, Soviet rejection of the Baruch proposal left this theorist with just one logical alternative—what he called "supremacy in one nation":

America, at this moment . . . could compel the rest of the world to disarm, and establish a worldwide monopoly of American armed forces. But in a few years the opportunity will be gone . . . American victory would no doubt lead to world government under the hegemony of the United States—a result which, for my part, I should welcome with enthusiasm.[20]

If Bernard Baruch were correct—that a strategy of nuclear warfare had no meaning in support of traditional diplomacy—the message was clearly lost on Bertrand Russell. In fact, Russell was among the first to apply the traditional pre–World War II model of diplomacy to the post-war security dilemma. Having revised his own prewar pacifism, he now addressed the problem quite squarely:

A policy most likely to lead to peace is not one of unadulterated pacifism. A complete pacifist might say: "Peace with Russia can always be preserved by yielding to every Russian demand." This is the policy of appeasement, pursued with disastrous results by the French and British governments in the years before the war that is now ended. I myself supported this policy on pacifist grounds, but I now hold that I was mistaken. Such a policy encourages continually greater demands on the part of the power to be appeased . . . It is not by giving the appearance of cowardice or unworthy submission that the peace of the world can be secured.[21]

Russell thus rejected the appeasement model as the proper route to arms control, because, much like the Soviets in rejecting the Baruch Plan, he did not consider "peace" worthy of the political price. Russell went on to advocate an Anglo-American ultimatum to Russia from its then current position of nuclear strength. In contrast to Brodie's "no other purpose" for nuclear power argument, Russell argued that, in the absence of Russian acquiescence to Western demands, it would be necessary to bring greater pressure "even to the extent of risking war, for in that case it is pretty certain that Russia would agree."

While it is difficult to imagine a democracy like Britain or the United States supporting such assertive means of international disarmament—either then or now—Russell had correctly pinpointed the core of the problem. It was political. Like the USSR, Britain and France would soon develop their own nuclear arsenals. Nonetheless, nuclear disarmament would have been easily accomplished had Britain, France, and America—whose shared value systems would have rendered armament *or* disarmament unnecessary—been the only members of the nuclear club. The problem was, as Russell put it, that the Soviets defined their own value system as incompatible with that of the West. It was this political incompatibility that caused arms competition, not vice versa. As Russell put it, Russia was "a dictatorship in which public opinion has no free means of expression," and where "persuasion cannot be effected by arguments of principle." In short, as this conundrum plays itself out, arms control on purely diplomatic grounds would be easiest where needed least—among powers that share common values—and impossible without powerful coercion where it is needed most—among powers that hold competing value orientations. If "arguments of principle" held sway, then the unlikelihood of war would radically diminish the urgency of dis-

covering new mechanisms for the settlement of disputes. It was this central paradox that confronted arms control theorists.

It was true, perhaps, that an absolute monopoly of nuclear power in the hands of the US made absolute hegemony a plausible political choice for the American government; but such a choice also called for absolute threats, which are all but ruled out by pluralistic methods of decisionmaking. Ironically, such a route to "peace" may have been altogether plausible had the tables been reversed—if fortune had placed such a choice in the hands of an absolute dictatorship—because policies of absolutism require absolutism in the machinery of decisionmaking.

Instead of forcing acceptance of its own conditions for peace in a world unburdened by fear of nuclear war, the Western democracies had long since chosen to include Soviet interests in the pluralistic deliberations that would reshape the postwar world. This meant that bargaining and negotiating (i.e., politics) would be necessary after all. This proved difficult, however, since the security of the international status quo, which the West defined as the necessary condition of peace, was by Soviet definition intolerable to the purpose of politics. Thus, the Soviets viewed as patently unacceptable the prospect of foreign inspectors trespassing the sovereign territory for which Russia had just paid so dear a price. As a result, the inspection issue became the presumed "sticking point" of negotiations to which theory would have to accommodate itself. More fundamentally, of course, arms control could not possibly bear the weight of the opposing real-world, core, political values that governed the bilateral relationship, because its underlying theory was based upon a premise of compatible security interests that existed purely in the abstract.

As a result of this basic political antagonism, both sides' commitments to disarmament could only be expressed rhetorically. The Soviets openly advocated disarmament as an "ultimate goal" throughout these years, yet they also sponsored the invasion of South Korea, stimulated a series of crises in Berlin, fostered unrest in the Third World, and repulsed one democratizing effort after another throughout Eastern Europe—as prescribed by their political orientations. Similarly, the US helped form the NATO Alliance and proclaimed containment goals—both of which would grow increasingly reliant on nuclear weapons—even while also serving as the *demandeur* of nuclear disarmament. Like the Soviet Union, the US, acting on behalf of its own ideals of "peace and freedom," would have felt irresponsible in abandoning its political goals. Neither side could have found negotiating room with regard to nuclear weapons so long as their political goals depended so thoroughly on the military prowess and political influence these weapons provided.

Thus, the US opted against assertive measures such as those advocated by Bertrand Russell. Instead, President Truman endorsed disarmament as advanced by the theorists, but only so long as provisions for inspection were (his words) "foolproof," by which he meant "adequate to give immediate warning of any threatened violation."[22] The Soviets, however, while also approving the ideals of disarmament in principle, defined inspection of any kind as espionage and as a quest for targets against which the West wished to employ its nuclear weapons. Disarmament was an important goal according to the Soviet view, but inspection was another issue altogether, which could become negotiable only *after* the West disarmed itself. The intransigence of the situation, symptomatic of the real-world incompatibility of two conflicting sets of political values, thus came to be expressed rhetorically as an inspection problem in the context of arms control theory.

As the initial decade of the nuclear age progressed, it therefore became increasingly clear that many of the assumptions underpinning early arms control theory lacked empirical validity. As expected, though sooner than most expected, the Soviets conducted their first successful atomic explosion in 1949 and became increasingly adventurous in their foreign policy. Since American inspection requirements had come to be regarded as the principal explanation for the failure of theory, demands for relaxing these once demanding standards of verification became inevitable. By 1953, Oppenheimer himself returned to the very measures he had earlier rejected as inadequate, calling now for "a very broad and robust regulation of armaments."[23] Despite a still "troublesome margin of uncertainty with regard to accounting," some combination of "defense and regulation" (i.e., dual-track policy) became, in his view, the only alternative to an arms race that would lead (inevitably according to the theory) to nuclear war. Oppenheimer's reorientation from absolutism to partialism, or regulation, exemplified the spirit of accommodation beginning to dominate the atomic scientists' community at the time.

Indeed, many scientists would simply mirror-image their sense of urgency about disarmament onto Soviet thinking and in this manner sustain the premise of apolitical mutual interest that was so essential to positive-sum arms control theory. Eugene Rabinowitch, a professor of botany and biophysics, for example, asserted in 1959, much as Oppenheimer had argued in 1946, that a "radical break with the traditional ways of all nations" was required, and that "only scientists can make this revision in values readily."[24] Rabinowitch was particularly encouraged by the participation of Soviet scientists in the Pugwash international conferences among scientists, which persuaded him that "they, too, are now concerned that man's mastery of nuclear forces has put an end to acceptance of war as a . . . rational means of settling international disputes . . . The situation is inevitably much less clear for those without scientific background, and this includes the national political leadership of all nations."[25] The Soviets were merely afraid that inspections would reveal the locations of their secret installations, explained Rabinowitch—taking the Soviets at face value— but to suppress these anxieties "would be a small price [for the Soviets] to pay for an agreement." Not to be outdone, Bertrand Russell expressed his own new conviction that even surrender by the West to tyrannical foreign domination would be preferable to the nuclear war to which the arms race would lead us.[26]

It is ironic that scientists in particular would appeal so blatantly to fear as the basis of universal interest in disarmament, instead of substantiating their political arguments with empirical observation; but traditionalism in politics was precisely what they were seeking to overhaul, and they were not without great influence on American foreign policy. Their belief that the political gap between the US and the USSR could be bridged by "rational" appeal to principle led to a demand for results rather than mere proposals, and for real rather than rhetorical compromises. As the opening decade of the nuclear age moved to a close, modifications in arms control theory began to stimulate modifications in the substance of American foreign policy. The belief that arms competition was the cause of political competition, rather than vice versa, persisted.

1955–1967: THEORY REVISED—FROM "DISARMAMENT" TO "ARMS CONTROL"

The precise beginnings and ends of the stages of theoretical development are impossible to pinpoint in any area of inquiry, and major changes in the evolution of theories can normally only be delineated years later with the benefit of hindsight. As we have seen, subtle changes in arms control theory were underway well before 1955, but several important developments, both technical and political, mark 1955 as something of a watershed in the chronology of the theory's evolution. As a recent US Arms Control and Disarmament Agency publication puts it:

By 1955 both the United States and the Soviet Union had come to acknowledge that it would be extremely difficult to verify any accounting for the nuclear weapons and materials already produced, and hence that *the complete elimination of such weapons was not a practicable goal* for the foreseeable future. Accordingly, *attention began to focus on the possibilities of partial disarmament* and measures that would facilitate it—such as a ban on nuclear testing—and on steps to avert the danger of a surprise attack.[27]

The year 1955 is therefore a useful point of departure for the discussion of altered thinking because by then the festering inconsistencies between the real and the hypothesized nature of arms control had crystalized sufficiently to necessitate the rejection or revision of theory. Specifically, the accumulation of nuclear arsenals by both sides, and the failure of initial guidelines to stem competition, marked the end of the era of absolutism. Nonetheless, the core assumption—that both Soviet and American political interests could still be served by arms control—endured.

THE "OPEN SKIES" PROPOSAL

President Eisenhower's 1955 explanation of this transition was quite clear: "It is our impression that many past proposals are more sweeping than can be insured by effective inspection." This was true, according to the president, because the US had been unable to discover "any scientific or other inspection effort which would make certain the *elimination* of nuclear weapons."[28] The solution, which he proposed to a gathering of heads of government (including Soviet President Bulganin), was twofold. First, technical rather than physical means of inspection:

[Let us] give each other a complete blueprint of our military establishments, from beginning to end, from one end of our countries to the other . . . Next, to provide within our countries facilities for aerial photography to the other country—we provide you the facilities within our country, ample facilities for aerial reconnaissance, where you can make all the pictures you choose, and take them to your own country to study, you provide exactly the same facilities for us and we too make these examinations . . . thus lessening the danger and relaxing tension.[29]

Second, since technical means of inspection would fall far short of the foolproof measures needed for *absolute* disarmament, future proposals would be limited "to the extent that the system *will* provide assured results."[30] The practice of arms control, following theory, would therefore emphasize partial disarmament with less than physical (i.e., technical) means of verification so as to "relax tension" and build confidence.

The modifications in Eisenhower's thinking were important. By correlating "stages" of disarmament directly with available means of verification ("to the extent the system will provide assured results"), the president had accepted a sort of linear relationship between American ability to confirm Soviet compliance on the one

hand, and a climate conducive to American negotiating flexibility on the other. Logically, since it was the USSR that was sensitive about its force posturing and political intentions, this would have the effect of granting them considerable voice in determining the *substance* of negotiations: the two sides could negotiate about whatever the Soviets felt comfortable "exposing." For Eisenhower, aerial photography was intended as "only a beginning," an opening gambit from which a process of confidence building might get underway. In one sense the revised theory essentially contradicted itself: if agreed arms limitations so clearly served mutual security interests, then why were physical *or* technical means of inspection so essential? In another sense, the theory was altogether logical: since the mutual interest premise had yet to materialize in practice, partial, low-risk, verifiable measures would be attempted; if mutual interest really was more than an abstraction, its confirmation would "build confidence" before any real security risks were undertaken.

And yet even this degree of caution assigned an asymmetrical burden of risk to the US. The Soviets knew very well that the international status quo was acceptable to the US, just as the US knew it was unacceptable to the USSR. Cheating on arms control agreements would not only have been very difficult in the wide-open American political system, it would also have been pointless so long as the Soviets took the live-and-let-live approach to international politics that the theorists expected. But, in order for the Soviet leadership to take such an approach, they would have had to abandon Marxism-Leninism's central tenets—the very basis of their legitimacy—which did not appear likely. The problem, therefore, was not just distrust. Rather, it stemmed from a fundamentally antagonistic political relationship in which the Soviets viewed the Western political order as exploitive and unjust and its eventual overthrow as their responsibility.

Partial disarmament sanctioned by partial inspection, therefore, was more than a change in *degree* from earlier approaches to arms control. If the solution could be less than absolute, then the problem must be less than absolute as well. Policies based on absolutism had required, as a minimum, that the two antagonists pursue their political goals by nonmilitary—or at least nonnuclear—means. A policy based on partial solutions represented a change in *kind*, one that legitimized the accumulation of weaponry—so long as it was of the unregulated/uninspected variety.

The incentive for entering into a restraint contract under theories of absolutism presumably had been to achieve relief from the overwhelming anxiety associated with political competition between nuclear powers. Partial solutions, however, offered no such relief. Compliance was to be virtually assured under the more realistic enforcement provisions of earlier theories, but would be much less certain under partial inspection provisions. This would be particularly burdensome for the US, where an open political and social system made noncompliance easy to discover, and where the Constitution granted inviolable legal status to a ratified treaty. Most significant, therefore, was the question of how such agreements would genuinely serve the stated purpose of arms control— to enhance US security. The Soviets would have a dominant voice in specifying the means, times, and objects of inspection (looking conciliatory in the process), leaving competition unregulated in the areas they called intrusive. Above all, the threat of nuclear warfare would remain intact.

In effect, Eisenhower had associated US arms control policy with a new theory for which partial inspection would not only serve verification purposes but would also serve broader political purposes of relaxing tensions and building confidence between the antagonists. In the nuance of diplomacy, three major changes had been introduced. First, disarmament had given way to arms control in both vocabulary and eventually in US policy. Second, "foolproof inspection" to confirm the absence of violations had begun moving toward "verification of compliance" to confirm the presence of legal behaviors (even if it could not confirm the absence of illegal behavior). Third, the goal of absolute security by international contract had begun yielding to vague notions of confidence building and stability as ultimate goals of what was now understood as a "process." Not until 1963 did the new thinking begin producing actual agreements, but the Soviets were immediately eager to solemnize its intellectual content. As Premier Bulganin promptly reported to the Supreme Soviet: "As the United States President justly points out, every disarmament plan boils down to the quality of control and inspection. The question is indeed very serious and we should find a solution to it which would be mutually acceptable."[31]

The Soviets then vehemently rejected Eisenhower's aerial reconnaissance proposal as, again, legalized espionage and "inspection without disarmament." In 1956, the Soviets claimed (falsely) to have surpassed the US in nuclear weapons,[32] and in 1957, after developing the world's first intercontinental ballistic missile (ICBM), began proposing a series of "first steps" toward uninspected nuclear test bans. Although these "first steps" were unacceptable to the US, Eisenhower responded in 1958 with a counterproposal recommending a conference of technical experts under the UN—a group of specialists whose purpose would be to certify

adequate means of test ban verification and whose conclusions the US would commit itself in advance to accept. Soviet rejections of both the 1955 Open Skies proposal and this 1958 "Conference of Experts" proposal were instructive regarding the new theory. These partial measures, which came about because previous proposals required inspections that were too intrusive, were now repudiated by the Soviets not simply as intrusive, but more significantly because they would consign essentially political questions to technical arbitration.

The idea of arms control as a purely political process—a position the Soviets have held then and now with remarkable consistency—was an enormously burdensome one in the context of Western arms control theory. What it suggested—that arms control was just another arena in which the two sides' political and security interests compete—was difficult to reconcile with Western theory's mutual interest (positive-sum) axiom. The Western proposition that neutral technical processes could arbitrate compliance-related disagreements that arose during the implementation of treaties, and the Soviet response, brought these conflicting perspectives into clearer view. For the Soviets, acceptance of a given balance of power in deployed weapons would mean acceptance of the political conditions those weapons were deployed to achieve. Acceptance of "disarmament" in the earlier period would have meant absolute acceptance of an international status quo by this view—and was therefore unacceptable to the Soviets. Transition to partial measures in the new era permitted the Soviets to select the political conditions of which they approved, and to dignify by treaty the associated weapons balance. Neutral technical arbitration devices, however, flew in the face of this logic by removing the powers of selectivity and control from the Soviets themselves and placing it in the hands of a more objective, value-free process.

To the Western theorists, of course, such an objective enforcement arrangement seemed altogether reasonable because the theory defined nuclear weapons in nonstrategic terms, and controls on nuclear weapons competition were seen as mutually beneficial—completely apart from the political conditions the resultant power relationship would codify. So long as the political status quo was to their liking, and so long as the terms of negotiation and enforcement were absolute, it had also been easy for the *practitioners* of Western security policy to align themselves with this theory. For Harry Truman during the 1950 Korean War buildup, the goals of disarmament and US security had been identical: "security and peace." World peace, on terms compatible with American political values, was the only acceptable standard for "good" arms control. The standard was nonnegotiable, and the means of achieving it were absolute.

Now that partial measures of control would govern some, but not all, of the two sides' nuclear arsenals, the issue was no longer so simple. The new approach called not just for peace and security, but also for the building of confidence in the process itself. Now arms control was not simply "another way"—or a "means"—to security, but also an end in itself. Furthermore, in place of the unequivocal ends and means associated with absolutism, partialism introduced new uncertainties at each juncture: which weapons should be built and which should be constrained to achieve peace and security? How much verification was "enough" to provide "assured results"? And how much compliance was "enough" to build confidence? Quite apart from the fact that Soviets would now participate in answering these questions, their resolution would add several increasingly problematical variables to the domestic equation as well. Like any political issue, these questions would portend an element of partisan debate that had been alien to the era of absolutism. All that was needed to complete the transition to partialism was a goal for "arms control" short of "disarmament."

A NEW STANDARD EMERGES

The solution was soon forthcoming as arms control theory began building on the notion of stability. A stable nuclear arms relationship, it was said, would reduce motivations toward conflict, ease tensions, and create the test-bed from which confidence in arms control might mature into confidence in disarmament. Now that it had been disproved that both sides feared war sufficiently to motivate the absolute international security regime once envisioned, the more modest assumption came of age that both sides shared a common interest in a stable balance of arms that would at least reduce the *risks* of war. Stability, it was said, would be the logical goal of both tracks of security policy—both arms building and arms control. But arms control in particular, if properly oriented toward stability, would presumably diminish the incentives for arms building and arms racing.

The essence of the stability theory, which arose during the initial deployment of ICBMs by both sides, was that the risk of surprise attacks, preemptive attacks, and wars by accident was actually reduced by the presence of vulnerable societies and survivable retaliatory forces on both sides. An outpouring of academic texts and articles in the 1960–63 time frame thus argued that the goal of arms control, as one author put it, was to "reduce the hazards of present armament policies by a factor greater than the amount of risk introduced by the con-

trol measures themselves."[33] Another author of the era observed that American "force as a whole has to be secure enough to remove the possibility that the Russians can reduce [US] force to a level where they could avoid effective retaliation."[34] The notions of deterrence, mutual deterrence, and assured destruction flowed logically from this line of thought.[35] While they had moved far away from their predecessors of 1945–55 in many respects, the new thinkers still held to the fundamental belief that there were significant areas of mutual security interest between the US and the USSR. For the earlier theorists these mutual interests had been essentially absolute, leading to disarmament. For the new theorists, the mutual interests involved survivable nuclear retaliatory forces that, once achieved, would facilitate arms control. In either case there could be security negotiations despite political differences, and negotiations could still proceed above politics.

Because they agreed with their predecessors, going back to Bernard Brodie, about the nonstrategic nature of nuclear weapons, these theorists saw "realism" in their premise about arms control's apolitical character. Again, it was said to be mutual self-interest that would energize negotiations, not agreements for their own sake. In many ways, however, the latter theory embraced an idealism that the former theory had avoided. For the absolutists of 1945–55, the assumption of negotiations' apolitical character had been an hypothesis to be tested in the real world. For their successors, who had seen that theory disproved, it was largely an article of faith—albeit one that was essential to a rational theory of arms control—that the process of negotiating security was above or apart from politics. Thus, widely accepted explanations of the era sought to substantiate a more or less mechanical view of an arms race. Chief among these was the "action-reaction" model advanced by the Quaker physicist Lewis F. Richardson.[36] Using differential equations to simulate the two governments' security decisions, Richardson sought to discover the "reaction coefficient" under varying conditions. If side A increases defense spending, for example, how does side B respond? Does B react slowly? Moderately? Or at a pace still faster than A's? The latter, argued Richardson, would cause A to react even faster, setting in motion a cycle of response and counterresponse by both sides that would lead axiomatically to war. Despite herculean efforts by analysts from a variety of disciplines, no study has yet substantiated Richardson's hypothesis. Yet it endures, and continues to influence arms control theory to the present.

Equipped with Richardson's action-reaction model of a purely mechanistic arms race, and with stability seen as a mutual Soviet-American security goal, Western arms control theorists completed their transition from disarmament to arms control. Significantly, these apolitical theories were almost exclusively Western in origin, and their accompanying projection of motives onto the Soviets was a speculative exercise that existed *only* in theory. Defense intellectuals of the ensuing years, however, took these mechanistic explanations very seriously, raising their associated mathematics to an art form. Any understanding of the agreements actually reached over the following decade requires a prior understanding of how this thinking came into being and how widespread its acceptance became. In the following sections, we will see how the application of these ideas fared in the real world.

PARTIAL MEASURES AND ASSURED RESULTS

We have seen that from 1945 to 1955 real-world outcomes failed to conform to theoretical expectations. When this happens, as shown for the period from 1955 to 1963, the theory that guides policy must be modified or else rejected altogether. This is merely another way of saying we learn from our mistakes. The relationship between theory's abstract expectations, and diplomacy's observable outcomes, however, evolves continually: one is always calling for adjustments in the other. It is difficult to specify under these conditions precisely when a theory has been sufficiently tested. As in the natural sciences, the relationship between the observer and the observed influences students and practitioners alike. Differing preconceptions can cause them to see differing phenomena in the same "factual" observations, or to disagree about how to categorize them. In the case of arms control theory, as with the social sciences in general, there is no laboratory in which other variables can be held constant while a hypothesized relationship is subjected to isolated scrutiny. Instead, arms control theory's testing environment involves negotiations for national security. Evidence and criteria for evaluating a theory's validity often lie in the eye of the beholder under these conditions. Is success differentiated from failure in relation to whether or not agreements are reached? Are more agreements axiomatically better than fewer? Does failure to achieve agreements establish the invalidity of the theory? These ultimately unanswerable questions had to be faced as an evolving theory of arms control guided policymakers into the early 1960s.

Early Realism Produces Treaties

Beginning in the late 1950s, negotiations no longer focused solely on absolute notions of dis-

armament. Instead of seeking to change the reality of a nuclear armed world—in which bipolar ideological tension was at least as relevant as fear of war—theorists and practitioners began to focus on more modest and "achievable" goals. This meant partial rather than total disarmament—that is, "arms control"—or, as Eisenhower had put it in 1955, "to the extent that the system *will* provide assured results." Given the hostility of their cold war political relationship, Soviet-American security agreements that provided "assured results" would have to be modest indeed. Yet, once this hard reality came to be accepted, a series of mutually acceptable armament-related agreements actually became possible.

Initially, both theorists and policymakers embraced the belief that agreements could not significantly *change* a political-military status quo. There were, however, a number of features characterizing the status quo that were either mutually beneficial (positive-sum) in their own right, or potentially capable of producing mutual benefits in the future. By striving to codify these existing realities rather than pursuing a nonexistent motivation to change them, arms control theory for a time found a formula that could provide "assured results." Agreements emanating from this approach began with the Limited Test Ban Treaty of 1963 and ended when the SALT process began to follow a much different set of expectations. In particular, three categories of agreements, discussed below, followed the logic of "partial measures and assured results."

The first category, of which the Nuclear Non-Proliferation Treaty (NPT) of 1968 and the Hot Line agreement of 1963 (updated in 1971 and 1984) are good examples, involved actual Soviet and American security interests in the here and now. Both sides agreed, according to the 1968 NPT, for example, not to transfer nuclear weapons, nuclear explosive devices, or weapons-related nuclear know-how to other states. In agreeing to restrain themselves this way, the US and USSR reduced the likelihood of nuclear threats from third parties and did so at neither side's expense. As the theory suggested, this was mutually beneficial and, for that reason, unlikely to create serious compliance problems.

It is true, of course, that the USSR could clandestinely transfer nuclear weapons to an ally while the US complied scrupulously, but the US had no interest in expanding membership in the nuclear club whether the Soviets complied or not. Nor would the Soviets if the tables were reversed. In other words, noncompliance by either side would have been counterproductive, and thus enormously unlikely. Besides, who is to say whether it was the US or the USSR that would be penalized more severely by a Soviet ally like Poland, Czechoslovakia, or Hungary becoming a nuclear power? Compliance was therefore self-enforced. Like the Hot Line agreements, the NPT was a realistic application of Eisenhower's maxim—limiting goals to agreements whose beneficial results could be assured despite imperfect means of verification. Not surprisingly, the question of compliance has never yet been seriously raised with regard to this category of agreements.

A second category of agreements reached in the 1960s also involved mutually beneficial Soviet and American restraint provisions regarding nuclear weaponry, but, unlike the first category, category two agreements involved no immediately significant security issues. The best example, the Limited Test Ban Treaty (LTBT) of 1963, prohibited nuclear testing in the atmosphere and in outer space. After almost a decade of negotiations toward a Comprehensive Test Ban (CTB), the LTBT became a benchmark of arms control's shift toward partial and achievable rather than absolute goals. Since no mutually agreeable verification provisions could assure compliance under a CTB, negotiations had shifted toward more modest objectives. Because it merely restricted nuclear testing to underground areas, the LTBT, unlike a CTB, involved no substantive security risk. As a result, it required less than perfect verification to "assure results" and "build confidence." It may have advanced bilateral, indeed multilateral, interests in environmental protection, but it imposed no substantive restrictions on nuclear weapons' development regardless of the compliance record. It was, therefore, judged to be beneficial politically and tolerable militarily despite its verification imperfections. Although some have made a good case that the LTBT has been asymmetrically burdensome to the US from a direct security standpoint,[37] experience has generally borne out the validity of the 1963 political judgments upon which it was based.

The LTBT was the only agreement of this period that required outright behavior changes by the nuclear signatories because atmospheric testing had been the order of the day until this agreement took effect. Additionally, while the US has complied quite scrupulously with its intent, the Soviets have vented nuclear debris beyond their boundaries, in violation of the LTBT, on several occasions.[38] Furthermore, the LTBT did not advance strategic or crisis stability or even create an environment conducive to further testing restrictions, but these were not among the partial goals sought by the treaty's negotiators. Thus, most observers regard the treaty as mildly harmful at worst from an overall US security standpoint *despite* its demonstrably uneven compliance record. If anything, Soviet noncompliance has been more perplexing precisely because the military benefits are so minimal—all of which testifies to the vision of the policymakers who insisted that the treaty's goals must be limited.

Following the same logic, the third category of nuclear arms control agreements during the 1960s involved mutual interests like the first two categories, and were generally irrelevant from a security standpoint like the second. But category three agreements portended some possibility of affecting the strategic relationship sometime in the *future*. To the extent that their near-term security relevance was nil, these agreements could fulfill the minimalist "assured results" criteria specified by Eisenhower even without cooperative means of verification. But, like the other two categories, this was true of category three agreements because there would be no discernible security risk and thus no political price associated with unilateral compliance. These treaties, which actually began in 1959 with the agreement not to "militarize" the Antarctic continent, went on into the 1960s to include the 1967 Outer Space Treaty, the 1968 Latin American Nuclear Weapons Free Zone Treaty, and the 1971 Seabed Arms Control Treaty. Like the other treaties we have examined, these required restraint from activities in which neither side intended to engage in the first place. Thus, like the other two categories, compliance was either self-fulfilling or irrelevant to national security.

Agreements throughout the 1960s therefore followed the modest but realistic conditions set down by President Eisenhower in 1955. There were no serious compliance concerns with them—until later with the LTBT—because they were designed to accommodate noncompliance. As we have seen, the US fulfilled the mandate for "assured results" not with high standards of verification but with a low tolerance for security risk in *anticipation* of enforcement difficulties. Arms control was simply not looked upon as a mechanism that could accommodate a hostile political relationship. Indeed, there was a fairly widespread consensus in the US that the best treaties could do was to institutionalize segments of political life that both governments considered acceptable. There were precious few of these where the USSR was involved.

This was the same political environment, after all, in which American intelligence agencies had surfaced a partly mythical missile gap in reaction to Soviet disinformation about their emergent ICBM strength. It was the same environment that had produced a series of direct challenges to the independence of Berlin by the Soviets, the deployment of Soviet ballistic missiles in Cuba, the beginnings of US involvement in Vietnam, the Soviets' suppression of various uprisings in Eastern Europe, and the initial buildup of strategic nuclear force levels on both sides. To say that international politics constrained the reach of arms control would clearly understate the case.

THE BUILDUP TRACK CONTINUES

While the agreements reached during the 1960s generated no major compliance controversies, they also failed to control competitive armament patterns. As the Soviets expanded their long-standing conventional weapons superiority in the European region and supplemented it with hundreds of theater-based nuclear armed missiles, the US countered with increased intercontinental-range strength reflecting a continuing emphasis on nuclear weapons for deterrence. The US had made this choice primarily on the basis of economic motivations since nuclear weapons are inexpensive by comparison with comparable conventional strength, and yet capable, it was argued, of deterring aggression of either kind. Thus, the US had approximately 4000 nuclear weapons deliverable at intercontinental range in 1960, which grew to perhaps 4500 by 1967.[39] The Soviets, by comparison, expanded their own intercontinental-range nuclear arsenal from roughly 200 to perhaps 1000 weapons over the same period. The US, having relied on long-range bombers throughout the 1950s for its doctrine of massive retaliation, spent the early 1960s modernizing B-47s with B-52s, Minuteman I ICBMs, and Polaris submarines. In all, the US inventory of long-range delivery systems rose from approximately 1500 such vehicles in 1960 to some 2300 more modern systems by 1967. The Soviets, who continued to increase both conventional and nuclear forces at the regional level, invested heavily in strategic air defenses, but also enlarged their own strategic offensive arsenal from 166 delivery vehicles in 1960 to over 800 by 1967.

Thus, by 1967, several influential developments had become clear. On the one hand, it had been demonstrated that when Soviet and American interests do converge, as in the case of our three categories of partial measures, bilateral and multilateral agreements could codify those interests without serious risk to national security. Having limited themselves to the low-risk goals for which arms control could provide "assured results," arms control advocates could correctly point to the fulfillment of these agreements' minimal promise. Far from relaxing tensions or building confidence as the Eisenhower thesis had portended, however, the superpower conflict had intensified in breadth and depth during this same period. Throughout the decade, the US and the USSR found their interests clashing not simply in Europe, where postwar political agreements had more or less preordained competition, but also in the Middle East, the Caribbean, and in Southeast Asia. Although the US still held a substantial advantage over the Soviet Union in most quantitative and qualitative measures of strategic nuclear strength by 1967, the Soviets were now in the early phases of an arms buildup that was only beginning to gain momentum; and for a combination of political and economic reasons, the US strategic modernization program was about to level off.

It was at this juncture in postwar history that the urge to go beyond the partial measures/assured results formula, and to consider previously unacceptable security risks for the sake of arms control, became increasingly attractive to US policymakers. In the first place, the unarguable asymmetry in strategic nuclear forces favoring the US could, in theory, offset the asymmetric risk associated with imperfectly verifiable agreements. Given the vastly uneven challenge faced by the open and democratic American system over the cloistered, highly centralized Soviet system, the margin of error provided by US superiority would be a seemingly essential ingredient to the acceptance of compliance uncertainties. But beyond these realities of the strategic balance, theorists and practitioners of arms control alike were buoyed by the tangible evidence now at hand that carefully structured agreements could indeed sanction a mutually agreeable status quo.

Although these diplomatic exercises had failed to foster the political confidence previously deemed essential for more far-reaching negotiations, there were additional factors now motivating politicians to move in that direction anyway. Bilateral arms competition was about to take a new turn with the pending development of the ballistic missile defense (BMD) and multiple independently targetable reentry vehicles (MIRVs) by both sides. Military spending was becoming increasingly unpopular in the West as opposition to the Vietnam war became more and more visceral. If indeed an agreement could be fashioned that would freeze the strategic balance where it stood, these forces could be silenced, without Soviet noncompliance on the margins upsetting the status quo in any dramatic way. In short, a myriad of reinforcing pressures had begun to converge by 1967, and their collective influence disrupted the once solid American political consensus and drove policy toward much more ambitious arms control ventures. With all of these pressures coinciding to make a strategic nuclear weapons freeze a seemingly self-evident US interest, all that was needed was for the Soviets to agree that such an arrangement would serve their security interests as well.

ARMS CONTROL TURNS AWAY FROM REALISM

The solution to this dilemma involved a two-part process. First, the notion of stability achieved a status in arms control theory (once reserved for disarmament itself) as the natural goal of a world in fear of nuclear war. This thesis depended, of course, on a Soviet appreciation for both the danger of war and the benefits of achieving "stability" through arms control. Secondly, the medium through which this lesson could most easily be communicated to the Soviets was said to be readily available in the form of the negotiating table itself. Not only did these beliefs guide the US in the negotiation of SALT I, but after the agreements were ratified in 1972, John Newhouse, the principal chronicler of their negotiating history, cited these two generally accepted beliefs as the basis of the new harmony of strategic interests:

Soviet leaders, like America's, hope to head off another major offensive weapons cycle. They know that to succeed they must inhibit ballistic missile defense, an insight acquired from the Americans. Baldly, this means that defending people is the most troublesome of all strategic options, for stability demands that each of the two societies stand wholly exposed to the destructive power of the other. Acceptance of this severe and novel doctrine illustrates the *growing sophistication* of Soviet thinking and some willingness to break with fixed attitudes, including the *old Russian habit* of equating security with territorial defense. And it points up the American interest in *raising the Russian learning curve*—in creating a dialogue that will encourage, however gradually, a *convergence of American and Russian thinking about stable deterrence.*[40]

This belief that the US could "raise the Russian learning curve" through arms control reflected a broader assumption that "a convergence of American and Russian thinking about stable deterrence" had occurred at last. It was this presumed convergence that created the natural and long-sought environment for *real* rather than cosmetic arms control. Any evidence of the validity of such thinking would be quickly interpreted as "growing sophistication" on the Soviets' part. Evidence to the contrary, routinely dismissed as a case of "the old Russian habit of equating security with territorial defense," was abundant. Soviet Premier Alexei Kosygin, for example, was quite "unsophisticated" on the subject in February 1967: "Maybe an antimissile system is more expensive than an offensive system, but it is designed not to kill people but to preserve human lives."[41]

The fact that territorial defense had been a high-priority Soviet objective for a generation, and a central component of strategy throughout human history, was essentially ignored. Instead, President Johnson and Secretary of Defense Robert S. McNamara began the process of "educating" the Soviets at their June 1967 summit meeting in Glassboro, New Jersey. There, as Secretary of State Dean Rusk has described the scene, the Americans "tackled" Premier Kosygin "in a go for broke fashion."[42] When Kosygin responded with disbelief over the prospect of outlawing territorial defense, the Americans chose to interpret his reluctance as evidence that he simply lacked the necessary authority to negotiate.

Convinced that he understood the Soviets' strategic self-interests better than the Soviets themselves did, McNamara proceeded to restructure American strategic force planning around his own vision of stability, convergence, and arms control. His effort to raise their learning curve commenced shortly after the Glassboro meeting. As he explained in a remarkable address in 1967 in San Francisco,[43] the US currently enjoyed strategic nuclear superiority over the USSR in terms of gross megatonnage, numbers of missile launchers, and (by "three or four to one") in the number of warheads "capable of being reliably delivered with accuracy and effectiveness on the appropriate target sets." This overall condition of US strategic superiority meant several things to McNamara. First, he explained, the US motivation in achieving this advantage had resulted from an overreaction to the Soviet force buildup as estimated through 1961. Second, given the limited contribution of nuclear weapons to stability, both the US *and* the USSR had acquired nuclear arsenals greatly in excess of their needs. McNamara's explanation for this development—a classic case of "mirror imaging"—provided a foundation for the redirected US approach to arms control:

These arsenals have reached the point of excess in each case for precisely the same reason: We each have reacted to the other's buildup with very conservative calculations. We have, that is, each built a greater arsenal than either of us needed for a second-strike capability, simply because we each wanted to be able to cope with the "worst plausible case."

McNamara's formula—known as "mutual assured destruction" (or MAD), whereby the USSR pursues only a secure retaliatory deterrent force like the US, and for "precisely the same reason"—also guided the US in SALT I negotiations.

Each side, for identical reasons, had acquired a deterrent in excess of its needs, according to the MAD theory. Thus, both would benefit from "a properly safeguarded agreement, first to limit, then to reduce, both offensive and defensive strategic nuclear forces." The timing was right for such an agreement, according to McNamara, because the Soviets had also become lethargic about strategic force expansion. They had decided, in fact, that they had "lost the quantitative race . . . and are not seeking to engage us in that contest."[44] Like the US, by this logic, the Soviets had come to realize that the pursuit of first-strike capability was pointless:

It would not be sensible for either side to launch a maximum effort to achieve a first-strike capability. It would not be sensible because, the intelligence gathering capability of each side being what it is and the realities of lead time from technological breakthrough to operational readiness being what they are, neither of us would be able to acquire a first strike capability in secret.[45]

Thus, by 1967, a remarkable coincidence of events—a presumably natural convergence of national security interests—had emerged at long last according to this theory. At precisely the time of maximum US strategic superiority, at the very moment in history when the war in Vietnam (against a Soviet ally) was becoming most unpopular, when a costly US conventional arms buildup was clearly needed, when public approval of defense spending was declining, when domestic social spending was placing its greatest demand in history on the US budget, and when congressional support for the nascent ABM was becoming questionable, the *Soviet Union* had allegedly joined the US in looking for a way out of the arms race. Bilateral competition, which had stifled meaningful arms control for a generation, could now be explained as simply a combination of misunderstanding and worst-case planning as if both sides suffered from the same limited access to one another's force acquisition plans. The arms race, which McNamara understood in terms of Richardson's apolitical action-reaction hypothesis rather than the traditional political hostility model that for so long had constrained strategic diplomacy, could finally be tamed by arms control. This mutual yearning for stability, it was argued, would lead the two sides into a new era, and arms control would be the mechanism both for communicating with and for educating one another.

In effect, the theorists and policymakers, by and large two separate groups in the past, had become one in the McNamara Pentagon. The theoretical vocabularies surrounding assured destruction, stability, action-reaction, and strategic sufficiency began appearing in McNamara's annual "Reports to Congress" as the MAD hypothesis grew in stature to justify accommodation in arms control policy and restraint in force posturing. In the past, theorists had argued merely that nuclear weapons were incompatible with traditional strategy. Now, the belief that outright vulnerability was strategically beneficial became a focal point of policy.

In short, the mechanical, apolitical worldview had reached the central corridors of dual-track defense planning. Accordingly, defense planning models began rejecting ballistic missile defense, air defense, and civil defense, not just as destabilizing and incompatible with arms control, but as strategically and economically counterproductive in their own right.

By 1969, when Richard Nixon replaced Johnson in the White House, and Henry Kissinger became his principal architect of strategic policy, the language, grammar, and belief system associated with MAD had taken hold throughout the US Government. Arms control had become an outright objective of policy rather than just the test-bed for confidence building it had previously been. As the strategic balance began moving from clear US superiority toward parity, the Nixon administration began defining vague notions of parity as its natural goal. As US bomber dismantlements continued and ballistic missile deployments were capped, the US arsenal of long-range nuclear delivery vehicles actually declined below 2200 from 1967 to 1969, while the Soviet increase continued apace from around 800 to some 1500 during the same two-year period. By 1969, the two sides had almost exactly the same number of ICBMs, but Soviet ICBM throw-weight, roughly equal to that of the US in 1967, had doubled. The US retained its lead in SLBMs and ballistic-missile submarines in 1969, while the Soviets intensified their buildup in those areas. Although the Soviet air defense network continued to expand, the US retained its lead of around 600 to some 200 Soviet strategic bombers. By 1969, the two sides approached numerical and technological parity and had reached the ideal conditions for arms control by all the logic of MAD, action-reaction, and stability theory. For the next three years, the US and USSR formally negotiated SALT I.

SALT I: THE APPLICATION OF MAD

From the standpoint of convergence theory, SALT I was a stunning diplomatic breakthrough. Composed of the ABM Treaty and the Interim Offensive Agreement, its provisions institutionalized MAD perhaps as thoroughly as could be expected from a negotiated instrument. The ABM Treaty outlawed territorial defense against ballistic missiles, and the Interim Offensive Agreement promised, first, a freeze, and, eventually, major reductions in strategic offensive forces. To whatever extent "action" was what had generated "reaction," SALT I had taken a giant step toward "inaction" and a disruption in the "cycle" of the "arms race." Most importantly, SALT I was the centerpiece of a much larger redefinition of the bilateral relationship. As its principal architect, Henry Kissinger, described it, "détente" was stability writ large—a convergence not just in strategic weapons thinking but in the entire political relationship between the US and USSR:

By the end of the 1960s . . . it was clear that the international structure formed in the immediate postwar period was in fundamental flux, and that a new international system was emerging. America's historic opportunity was to help shape a new set of international relationships—more pluralistic, less dominated by military power, less susceptible to confrontation, more open to genuine cooperation among the free and diverse elements of the globe . . . Finally, a breakthrough was made in 1971 on several fronts—in the Berlin settlement, in the SALT talks, in other arms control negotiations—that generated the process of détente. It consists of these elements: an elaboration of principles; political discussions to solve outstanding issues and to reach cooperative agreements; economic relations; and arms control negotiations, particularly those concerning strategic arms.[46]

The "elaboration of principles" to which Kissinger referred took the form of a statement of "Basic Principles of US-Soviet Relations" that accompanied SALT I.[47] In this document, which formally chartered détente, the two governments spelled out the mutual objectives of their strategic relationship—stability—and emphasized that "efforts to obtain unilateral advantage at the expense of the other, directly or indirectly, are inconsistent with these objectives." Aside from a broad pattern of economic, cultural, and social ties, this document stipulated the convergence of mutual interests that arms control theory had postulated as essential to the success of peacetime security negotiations. In short, it proceeded from the assumption that "enough" political confidence had now been built, and that substantive security risks could now expand the reach of arms negotiations.

The SALT "breakthrough" to which Kissinger referred took place in 1971 when the Soviets agreed to discuss offensive arms limits if an agreement to constrain ABM deployment were also included.[48] This in itself represented a major turnabout from the position Premier Kosygin had taken four years earlier in Glassboro. Although many observers interpreted this reversal as evidence of converging world views—centered on MAD and stability—self-serving Soviet motives were at least as plausible. US ABM technology, vastly superior to anything the Soviets could develop in the near-term, represented a potential counter to the growing threat of a now huge Soviet ICBM force. Once the US agreed to stifle strategic defense—not to protect its population centers or its missile fields—strategic strength would be measured by offensive capability alone, in which the Soviets held numerical superiority. This ratified the Soviets' long sought achievement of coequal superpower status.

Throughout the three years of negotiations, while the balance of offensive forces continued to shift in Soviet favor, the US sought to finalize an agreement that would freeze the offensive status quo. By 1972, when the agreement was signed, the Soviets had some 1618 ICBM launchers operational or under construction to an American force which had been frozen at 1054 since 1967. The agreement also codified the shifting sea-based strategic force balance at 44 submarines (SSBNs) for the US to 62 for the Soviets, and 710 US SLBM launchers to the Soviets' 956. Neither bombers nor bomber defenses were included, nor were there limits on MIRVs, which the US had already begun to deploy. The resulting license to fractionate missile payloads, however, only offered a short-term US advantage until the Soviets began deploying their own MIRVs in 1974. Subsequently, their vastly superior missile throw-weights began to pay strategic dividends. The roughly 3 to 2 Soviet advantages in ICBM launchers, SSBNs, and SLBM launchers were tolerable, according to Kissinger, for a variety of reasons. Primarily, there was the explicit Soviet agreement promptly to negotiate follow-on offensive arms reductions. There were also the Soviets' lack of warm-water seaports, and the geographic benefits associated with US nuclear weapons based in Europe. Above all, however, the agreed balance was a mutually acceptable, low-risk undertaking, according to its proponents, because it represented "stability."

Both sides, of course, also agreed to cap ABM deployments, first at two sites each and later (1974) at one each, to permit unconstrained ABM research and testing at approved sites, and to eschew the testing of air defenses in an (undefined) "ABM mode." ABM launchers, interceptors, and tracking radars were carefully defined in terms of 1972 technology, with new technologies subject to further discussion as they arose. All of these ABM restrictions took the form of an international treaty, ratified by the US Senate in 1972 and subject to bilateral reviews at five-year intervals. Overall, the framework of SALT I on paper followed the basic rules of MAD. But the picture was far different in 1972 from what it had been in 1967, when the talks began in Glassboro, and when the strategic balance appeared to offset US strategic risks in such an agreement, because the US had lost its clear superiority in most categories of strategic-range offensive weaponry during the three years of negotiations.

In general, SALT and its intellectual underpinnings found little opposition throughout the US political system. Among skeptics, however, two principal shortcomings stood out. First, the offensive agreement accepted numerically uneven force levels and codified several Soviet advantages such as sole possession of "heavy"

ICBMs, the right to modernize and MIRV them, and the similar right to enlarge their "light" ICBMs with only the most ambiguous and unenforceable limitations. While the strategic significance of Soviet numerical superiority was considerable in itself, the psychological importance of uneven force levels was particularly startling. It is true that proponents claimed only to have codified the existing status quo; it was also true that, given MAD motivations and "worst-case" planning on both sides, even an uneven status quo would be acceptable if it meant an end to the hypothetical action-reaction cycle. Still, to accept such an obvious force imbalance carried politically relevant image factors that seemed to reward the Soviets for a buildup that continued unabated throughout negotiations.

A second factor buried within the first, however, posed a more serious challenge to the convergence theory's overall validity. The 1972 strategic status quo had represented merely a snapshot in time in a highly dynamic process of competition between the US and USSR. Given the ebb and flow of such interactions, perhaps one side had to have more of this or that at any one time than the other. But the principal Soviet advantages lay in the only items whose further growth was frozen by the offensive agreement—numbers of ICBM and SLBM launchers and SSBNs. Furthermore, they had achieved that numerical superiority by building rapidly throughout the years of negotiation and *after* the US had quit deploying new launchers. What remained unexplained was the extent to which US security objectives had changed during these negotiations, because the agreements failed to assure survivable US retaliatory forces—initially described as the purpose of arms control. [49]

In effect, the US agreed in SALT I not only to permit potentially superior Soviet means of attacking US assets, but also not to deploy the only existing weapons that might plausibly defend its retaliatory forces. So concerned was the US government about the implications of this outcome that the American delegation was directed formally to say so in the agreements themselves. Thus, the ABM Treaty included a US Unilateral Statement[50] lamenting the failure of SALT to achieve more complete offensive limits. Looking to SALT I's hypothetical follow-on agreement to correct this failure, the US made the following unilateral assertion as to what such an agreement must entail and when it must be achieved:

If an agreement providing for more complete strategic offensive arms limitations were not achieved within five years, US supreme interests could be jeopardized. Should that occur, it would constitute a basis for withdrawal from the ABM Treaty . . . The US Executive will inform the Congress, in connection with Congressional consideration of the ABM Treaty and the Interim Agreement of this statement of the US position.

Recognizing that the strategic balance in 1967, or even 1969, may have rendered marginal shifts in the Soviets' favor relatively unthreatening for the US, this oft-forgotten statement resulted from serious concern about US vulnerability now that such a margin for error could no longer be cited. Regarding the US verification problem, SALT I limited only the large, fixed-site weapon components that could be seen and counted by "National Technical Means" (NTM)—a euphemism for overhead reconnaissance. This meant that virtually no qualitative factors, such as ICBM modernization, ABM research and development, MIRV upgrades, or missile throw-weight could be limited by the agreements—all of which made sense from the theoretical standpoint of controlling those items for which "assured results" could be achieved. Thus, SALT's offensive agreement did not constrain actual weapons, reentry vehicles, or even missiles, but only the large silos they required for launch in 1972. Similarly, ballistic missile defense (BMD) or ABM modernization could not be "counted," but 1972-vintage large ABM launchers and radars could. Significantly, given congressional reluctance in the US and technologically inferior progress in the USSR, neither side was said to be eager to compete in the ABM realm for the next several years in the first place. If true, this, more than the use of overhead reconnaissance for verification, would be the most plausible guarantor of compliance.

In light of the MAD and détente world views that had guided the US into the SALT years, however, architects of the American arms control policy had come to expect far more than that for which results could be assured. There was a "spirit" of mutual obligation implied by the statement of Basic Principles of Relations, whereby both sides' retaliatory power was to remain secure. There was also the promise of follow-on offensive reductions and a cooperative approach to international problem solving on the horizon, according to this view, as the two sides would come to terms with one another's superpower status. It was on the basis of these beliefs—that SALT I represented more than just a rough ceiling on large launchers, extant ABM hardware, and SSBNs under construction—that American officials defended SALT I before Congress and the American public. [51] Based on his confidence in the Soviets' respect for Unilateral Statements appended to the agreements by the US, for example, Henry Kissinger argued in 1972 that the Soviets would not develop or deploy the larger ICBMs that so thoroughly redefined their forces beginning two years later.

It was Kissinger's view in 1972, just as it had been Robert McNamara's hope in 1967, that:

SALT became one means by which we and the Soviet Union could enhance stability by setting mutual constraints on our respective forces and by gradually reaching an understanding of the *doctrinal considerations* that underlie the deployment of nuclear weapons. Through SALT the two sides can reduce the suspicions and fears which fuel strategic competition.[52]

SALT I ENCOUNTERS THE REAL WORLD

The proposition that US and Soviet security interests had converged, thus lowering the security risk associated with arms control, became difficult to substantiate on grounds of Soviet behavior. Although proponents claimed that SALT I had prohibited "significant" increases in the dimensions of the largest "light" Soviet ICBM silo (the SS-11), for example, the Soviets increased that missile's silo volume by over 50 percent and replaced it with the SS-19 just two years after SALT I's ratification. The newer missile nearly quadrupled its predecessor's single warhead throw-weight, enabling up to six independently retargetable warheads, and markedly improved on its predecessor's accuracy. Similarly, the Soviets replaced their SS-9s—already the largest missile in either side's inventory—with SS-18s, increasing throw-weight by another 30 percent with up to ten reentry vehicles and impressive new accuracies. As these two new ICBMs entered the Soviet inventory beginning in 1974, and brought the "ride-out" survivability of American retaliatory forces into serious doubt, Soviet Communist Party Leader Leonid Brezhnev proclaimed boldly to the Twenty-Fifth Party Congress:[53] "We see détente as the way to create more favorable conditions for peaceful Socialist and Communist construction." Détente, Brezhnev explained, referred to *interstate* relations; it "does not in the slightest abolish, and cannot abolish the laws of the class struggle." In fact, so far as the class struggle itself was concerned, it was "gaining in intensity" according to Brezhnev, and "the scale of the revolutionary-democratic anti-imperialist movement is steadily growing," a process that would continue, he predicted, because "life itself has refuted the inventions about 'freezing the status quo.'"

While rhetorical contradictions like these continued throughout the 1970s in the aftermath of SALT I, the strategic imbalance continued to grow. By 1979, even while he was *defending* the new SALT II agreement, Secretary of Defense Harold Brown finally acknowledged the dismal consequences beginning to emerge from a Soviet strategic force buildup now underway for over a decade.[54] He

described to the Senate a "hypothetical ability of the Soviets to destroy even 90 percent or more of our ICBM warheads." Although these conditions had developed throughout the years governed by SALT, it was Secretary Brown's contention in 1979 that SALT II would ease this burden for the US. Unexplained, of course, was why the Soviets would now negotiate a treaty that would reverse the advantages they had gone to such trouble and expense to achieve.

The idea that SALT I's shortcomings could be, indeed *must* be, resolved by SALT II had been literally written into the ABM Treaty in the form of a unilateral US statement noted earlier. This alleged capacity of arms control for self-correction, in fact, increasingly became a primary defense against growing empirical evidence that the convergence, or MAD, theory might simply be invalid. Even smaller agreements during the 1970s exceeded the boundaries of "assured results" by investing ambitious confidence in the presumed similarity of the US and Soviet security motives. Unlike the cautious realism manifested by the Nuclear Non-Proliferation Treaty, the Hot Line agreement, and the Limited Test Ban Treaty, for example, President Nixon described the 1972 Biological Weapons Convention (BWC) as "the first international agreement since World War II to provide for the actual elimination of an entire class of weapons from the arsenals of nations."[55] Similarly, the 1976 Threshold Test Ban Treaty (TTBT), which limited nuclear-armed signatories to a ceiling of 150 kilotons in their underground tests, remained unratified because it could not be verified. Although the US has complied scrupulously with these agreements, a large body of evidence suggests that the Soviets have violated both of them.[56]

Meanwhile, the post–SALT I environment failed to produce the follow-on offensive reductions promised to Congress within five years. As the Soviet strategic arms buildup began producing outright *instability* by the very terms that had guided SALT's negotiation, a debate emerged as to whether a differently structured agreement could correct its evident deficiencies or whether, indeed, the entire theory on which SALT had been based was even valid. In the post-Watergate White House, President Ford moved to downplay strategic arms control. After agreeing to negotiate toward equal ceilings on delivery vehicles and MIRVed systems at a 1974 summit meeting in Vladivostok with Brezhnev, Ford took no further steps toward this SALT formula. The growing debate over the purposes of arms control continued throughout the Carter administration's negotiation of SALT II and culminated in 1979 with the intense ratification debate when SALT II was submitted to the Senate.

CONVERGENCE THEORY: THE DEBATE REVISITED

AN OPPOSITION COMMUNITY COMES OF AGE

The opposition community that emerged during SALT II's ratification debate was not new in 1979. Its origins might be associated with Herman Kahn or Alblert Wohlstetter in the 1950s, or with Fred Iklé, Paul Nitze, and Senator Henry "Scoop" Jackson in the 1960s. This community's participation in the SALT debates first crystalized with William Van Cleave's 1972 testimony against SALT I. Van Cleave's testimony marked the opening of this round in arms control theory's evolution because it anticipated many of the problems that were to characterize SALT I's implementation. By the time of Ronald Reagan's presidency, which brought into public office many representatives of the community that had opposed the SALT process, the "SALT era" had, in effect, been politically rejected. Subsequently, Reagan reinvigorated the weapons modernization aspect of dual-track security policy and established actual reductions in destabilizing offensive weapons as the primary criterion of useful arms control agreements.

Van Cleave's 1972 critique of SALT I had challenged the central contributions of the agreements to stability itself:

It should be made very clear that the agreements do not solve or even ease our strategic force problems. They do not arrest the expected development of the threat or competition in strategic arms. They do, unfortunately, accept higher numerical levels of the threat than we ever before contemplated and do restrict at the same time US ability to cope with the threat. Their tendency, therefore, is toward less rather than more stability.[57]

Van Cleave, a member of the US delegation that negotiated SALT I, tracked US proposals from early ones calling for equal levels of offensive forces to the higher, unequal balances eventually agreed upon. "One way of putting this," he testified, "is that in two-and-a-half years of SALT [negotiations] the US has managed to trade away the Safeguard [ABM], and most of the important options to assure retaliatory force survivability, for a doubling of the threat." He then shifted gears and attacked the core beliefs surrounding stability and convergence theory. Rejecting these assumptions, Van Cleave suggested that actual Soviet goals called for a "future capability for a substantially disarming first strike with a fraction of their total force, enabling an overwhelming assured destruction capability to be held in reserve." He concluded his testimony, in response to questions from Senator Jackson, by arguing that there would be no difference between the Soviet offensive forces over the ensuing five years with or without the agreement, and predicted a Soviet foreign policy that would be "much more adventuresome and willing to take risks greater than anything we have had in the past."

Others began to employ similar logic during the ensuing decade. In 1974, for example, Secretary of Defense Schlesinger publicly highlighted the Soviets' continuing development of destabilizing counterforce capability and, in 1975, called for a "wider set of much more selective targeting options" in the US strategic arsenal.[58] These were needed, he argued, "to shore up deterrence across the entire spectrum of risk," and also because:

In recent years, the USSR has been pursuing a vigorous strategic R&D program. This we had expected. But its breadth, depth, and momentum as now revealed come as something of a surprise to us.[59]

In fact, explained the Secretary, if the Soviets were to deploy the three heavier missiles (the SS-17, SS-18, and SS-19) then under development, "Soviet throw-weight in their ICBM force will increase from 6–7 million pounds to an impressive 10–12 million pounds," which, combined with increased accuracy and MIRVs, "could give the Soviets on the order of 7000 one- to two-megaton warheads in their ICBM force alone."[60] As Schlesinger concluded:

I frankly doubt that our thinking about deterrence and its requirements has kept pace with the evolution of these threats. Much of what passes as current theory wears a somewhat dated air—with its origins in the strategic bombing campaigns of World War II and the nuclear weapons technology of an earlier era, when warheads were bigger and dirtier, delivery systems considerably less accurate, and forces much more vulnerable to surprise attack.[61]

What Schlesinger was criticizing, of course, was the McNamara era belief that minimal assured destruction capability would be sufficient for deterrence. McNamara's anticipation that the Soviets would accept "parity" and "stability" had clearly been proved incorrect.

In the same vein, historian Richard Pipes and others began arguing in the mid-1970s that the Soviets sought nothing less than a capability to "fight and win" a strategic nuclear war.[62] Edward Luttwak laid the blame for US failure to appreciate such Soviet thinking on the doorstep of arms control theory itself as it had developed during the 1960s:

Above all, the United States has misused arms control in the attempt to dampen the strategic competition in itself, as if the growth of strategic arsenals were the cause of Soviet-American rivalry rather than merely one of its symptoms, and incidentally a much less dangerous symptom than the growth of nonnuclear forces, whose warlike use is much more likely.[63]

In short, argued Luttwak, arms control theorists had chosen to ignore a pattern of Soviet behavior that stood in direct contrast with their core expectations, and had lost all contact with realism.

One poignant manifestation of this tendency was the reluctance of SALT advocates to come to terms with a growing debate over whether or not the Soviets were complying with SALT I. Instead, theory was defended by its appeal to fine distinctions. The Soviets, for example, were said to be bound only by the strict letter of the agreement, but—ignoring the high expectations surrounding détente—not its spirit. This argument in particular struck many critics as disingenuous. The spirit of détente had been the central basis for the entire claim surrounding converged national security interests. If confidence building had not yet progressed even this far, why were we involved in such a substantive treaty in the first place?

Additionally, Unilateral Statements (upon which congressional approval of SALT I had rested) were now said to have no meaning in international law. Thus, many alleged violations could be dismissed as legally irrelevant by appealing to undefined or ambiguous treaty terminology. As the debate over Soviet noncompliance with SALT I intensified through the 1970s, it became clear that verification was no less political than negotiations had been, that "ambiguities" were the inevitable product of conflicting national security goals, and that, as a result, treaty negotiations merely carried forward (in the form of compliance discord) into the implementation phase of the agreements. Like arms control theory itself, verification had long since come to be regarded by American theorists as an essentially technical undertaking. Now, as conflicting treaty interpretations became the rule rather than the exception, the strictly political character of both processes became impossible to deny.

Other central assumptions of the theory that had guided SALT I's negotiators came under attack during the years between SALT I and SALT II as well. One of these was the Richardson action-reaction theorem that had been central to the McNamara belief system of the late 1960s. Another was the fundamental teachings of Bernard Brodie, whereby a theory of strategic warfighting was inapplicable from a practical standpoint to the nuclear era. For those who challenged the tenets of contemporary arms control theory, the experience of SALT I served as a "reality check" between hypothesized expectations and observed outcomes.

If the Richardson-McNamara explanation for interactive patterns of nuclear armament had been correct, for example, then frequent or at least periodic overestimations of the Soviet threat would be the rule. McNamara, after all, had argued that the US had assured destruction capability in excess of its deterrent requirements because of its overreaction to a perceived missile gap in the early 1960s, and that the Soviets had overarmed themselves for precisely the same reasons. Arms control, it was argued, could disrupt this recurring, mutually reinforcing series of misperceptions by instilling confidence through verifiable limitations in arms deployments. Albert Wohlstetter, in a series of articles beginning in 1974, however, demonstrated that US intelligence—even in its worst-case forecasts—had not once overestimated a Soviet buildup since 1962.[64] Using declassified National Intelligence Estimates (NIEs) and comparing the annual forecasts of various agencies within the intelligence community, Wohlstetter showed that forecasts had actually been well below actual Soviet ICBM deployment levels since 1962. Long-term US projections had not only *underestimated* the Soviet ICBM buildup, but had done so by increasing amounts from one year to the next as the intelligence community (guided by theory, according to Wohlstetter) failed to learn from its mistakes.[65]

Wohlstetter also demonstrated that, while US force planning responded to underestimates of the Soviet threat rather than to overestimates, Soviet ICBM deployment continued with a momentum of its own. He thereby rebutted both the action-reaction theory and McNamara's explanation for why the Soviets build ICBMs. It was a product of "wishful myopia" and mirror imaging that had led the US to wrong predictions about Soviet actions, according to Wohlstetter, a tendency which "made inaction on our part seem reasonable." Wohlstetter concluded from all of this simply that "dogma dies hard."

In a similar attack on the standing theory of arms control in 1976, Paul Nitze challenged the MAD-related proposition that an American assured destruction capability was all that was needed to deter nuclear war.[66] Before further agreements grew from this theory, suggested Nitze, the following assumptions, defined as irrelevant by arms control theory, must be reexamined:

Whether the Soviet side comes to have more or bigger offensive warheads; the degree to which they improve their weapons technologically; the extent of the asymmetrically better Soviet defenses, both active and passive; or whether one side or the other strikes first, provided only that we maintain strategic offensive forces for retaliation approximately as numerous and powerful as those the US now has and has programmed for the future.[67]

Nitze suggested that "no more serious question faces this country than whether these propositions are true or false." He went on to demonstrate that, after a Soviet first strike and a US response, the USSR would hold substantial advantages in all indices of strategic warfighting capability except static numbers of ballistic missile warheads, forecasting that even this sole remaining US advantage would be gone in a few years. Furthermore, argued Nitze, neither SALT I nor the projected SALT II agreements would have any discernible effect in arresting this trend towards Soviet superiority.

Nitze then asked whether, given an American capacity to attack Soviet population and industry with surviving forces, any of this made any difference. In answering this question affirmatively, Nitze looked at Soviet civil defense manuals, which estimated that casualties could be held at "3 or 4 percent of population after a Soviet first strike." This may or may not represent "acceptable losses" to Soviet leaders, argued Nitze, but it demonstrated that the common assumption, whereby the US possesses vast population overkill, was without foundation. Still, did any of these asymmetries really matter short of an all-out global war? Yes, argued Nitze again, because US strategic nuclear preponderance had long been relied upon to offset Soviet conventional superiority, to deter its offensive employment, and to enable the US to use the seas to project supporting power despite the Soviets' very real sea-denial capabilities. An imbalance in the Soviets' favor in the strategic nuclear relationship, he concluded, reverses these factors. Thus, the Soviets' motivation in deploying the type of force they had deployed since SALT I was clear to Nitze. A potentially disarming first-strike capability, combined with a reserve force of high megatonnage on unused delivery systems, had the effect of "deterring our deterrent." In other words: "[T]hey wish to be able, after a counterforce attack, to maintain sufficient reserve megatonnage to hold US population and industry hostage in a wholly asymmetrical relationship."[68]

Nitze's conclusion stood in stark contrast with the still influential Bernard Brodie maxim (which had been central to standing arms control theory) according to which a strategy of victory could serve no reasonable political purpose in nuclear war planning. It suggested that, whether or not such a victory was physically possible, Soviet military planners had acquired and deployed forces as if they *thought* it was. This conclusion further suggested that Soviet force planning could not simply be explained by action-reaction and worst-possible-case premises. Moreover, instead of converging, Soviet arms control motives diverged widely from the favored American notions of stability and mutual assured destruction.

The principal lessons on arms control when the Carter administration assumed office in 1977 were that the Soviets had failed to behave in accordance with the theory that had guided the US in SALT I, and that follow-on offensive reductions promised within five years had failed

to materialize. The Soviets had continued their offensive force buildup, with particular emphasis on heavy land-based missiles—precisely the destabilizing weaponry singled out for reduction by the logic of MAD. Additionally, while the case for Soviet violations of SALT I was tedious and legalistic, no one could argue that compliance was as thorough as the MAD theorists and treaty architects had anticipated. In fact, stability itself—the central criterion of success for SALT I as spelled out by McNamara and Kissinger—had eroded more from an American standpoint during the years after SALT I than at any previous time period of the nuclear era. Arms control theory's most basic assumption—that mutual security interests unique to the nuclear age could be served by agreements—had been brought into serious question by SALT I's real-world outcome. In the minds of many, these assumptions had been disproved altogether. At a minimum, the time had come for arms control to reflect in practice the new understanding of the strategic relationship between the US and the USSR.

REVISED THINKING FACES AN EARLY SETBACK

The first and most crucial arms control decision facing President Carter when he took office in 1977 was whether to heed the new thinking calling for verifiable reductions in destabilizing forces or stay within the SALT framework of mutually agreed strategic force increases as capped by the 1974 Vladivostok ceilings (2400 delivery systems; 1320 MIRVs). Carter promptly surrounded himself with articulate advocates of both schools of thought. The "reductions" school was represented by National Security Adviser Zbigniew Brzezinski; the "SALT/Vladivostok process" school was spearheaded by Secretary of State Cyrus Vance. Although the Vance faction eventually dominated this debate, Brzezinski's view proved influential during the early days of the administration.

Evidence that the president had taken note of widespread disappointment with SALT I surfaced when the administration's first major arms control proposal followed the logic of the "reductions" school. The March 1977 proposal, as Vance explained before delivering it in Moscow, consisted of a "comprehensive" version and a fall-back or "deferral" version.[69] In his predeparture press briefing, however, Vance essentially anticipated its rejection by the Soviets:

I've never said I was very optimistic, but the President has said that we have high hopes that we would be able to reach agreement with the Soviets. I don't want to characterize ourselves as very optimistic. We will be hopeful and thus hope to achieve the objective which I mentioned . . .[70]

Armed with this degree of enthusiasm, Vance carried a proposal to Moscow which, if accepted, would have oriented arms negotiations toward stabilizing reductions rather than toward the still higher ceilings associated with the Vladivostok formula. Recognizing that a fourth-generation Soviet ICBM modernization program was now nearing completion, the March 1977 proposal offered to permit these systems and simply ban newer ones from that point forward. The effect would have been to cancel the MX missile, the Small ICBM and the Trident SLBM programs, in exchange for a ban on *fifth*-generation Soviet systems. The plan would have limited cruise missiles, and reduced the Vladivostok launcher and MIRV ceilings (from 2400 to 1800–2000 SNDVs; from 1320 to 1100–1200 MIRVs) while counting ALCM-carrying aircraft as MIRVs. In exchange for these modernization constraints, the proposal envisioned a reduction from 308 to 150 heavy ICBMs (of which the Soviets held a monopoly) and a ceiling of 550 (the existing US count) on MIRVed ICBMs.

The Soviets, for whom throw-weight, MIRVed ICBMs, and overall preemptive targeting advantages served important strategic purposes, rejected the proposal out of hand. They also rejected Vance's so-called deferral option that would have merely "set aside" the contentious Backfire and cruise missile issues while a simpler framework was established around the Vladivostok agreement. In rejecting without counterproposing, Soviet Foreign Minister Gromyko characterized it as one-sided and detrimental to Soviet security precisely *because* the US was attempting to depart from the SALT-Vladivostok formula. Similarly, the deferral option would have reopened competition in cruise missiles after it had already been resolved, according to Gromyko, by the Vladivostok accords.[71] As Roger Labrie later characterized the situation:

The agreement reached in Vladivostok, together with the principles agreed to in subsequent discussions with the Ford Administration, were the only basis for negotiations, according to Gromyko, and any attempt by the United States to alter that basis could lead to the reopening of the issue of American forward-based systems in Europe that had been resolved in Vladivostok. Gromyko finished his press conference with a denunciation of President Carter's outspoken stance on human rights violations in the Soviet Union.[72]

Soviet rejection of the March 1977 proposal marked another watershed in the development of nuclear era arms control thinking. A decade of experience with SALT I had persuaded many, in and out of the Carter administration, that arms control required a different theoretical base if it were to contribute to genuine security. By seeking reductions in the most destabilizing weapons, the March 1977 proposal represented an effort to move in that direction. Soviet in-

sistence that the process could continue only on the basis of the SALT/Vladivostok formula put the Carter administration at a crossroads. Carter had to decide whether to continue down the new path he had initiated in 1977 or return to the arms control framework that had led to the current situation. This policy dilemma reflected the intellectual debate that had persisted throughout the 1970s. The Soviets had not simply rejected a *proposal* in March 1977, but an entire negotiating formula—one that held the promise of enhancing stability if all parties complied. Carter's subsequent decision to revert to the SALT/Vladivostok formula guided US policy until he was defeated by Ronald Reagan four years later.

POLICY RETURNS TO MAD GUIDELINES

The SALT II Treaty, signed by the US and the USSR at a Moscow summit meeting in 1979, reflected the Vladivostok formula. It set higher ceilings of 2250 strategic nuclear delivery vehicles (SNDVs) for both sides, and equal sub-ceilings on MIRVed delivery vehicles (1320), MIRVed ballistic missiles (1200), and MIRVed ICBMs (820). Each side was permitted one "new type" ICBM, with "counting rules" establishing an upper limit on the number of warheads carried by each system. Unlike SALT I, SALT II constrained long-range bombers (but still not air defenses) and indirectly imposed strict numerical limits on deployed air-launched cruise missiles. Treaty advocates claimed to have followed congressional guidance by establishing "equal limits" on the US and the USSR, but they had failed to reverse the reductions called for by SALT I. More significantly, SALT II failed to retard the growing Soviet threat to US retaliatory forces which had emerged in the aftermath of SALT I. SALT II's opponents focused on the treaty's shortcomings in the terms of verification, its failure to include the new Soviet Backfire bomber as a strategic vehicle, its failure to reduce or offset the unilateral Soviet advantage in heavy ICBMs, and its numerous conceptual ambiguities, such as the definition of the "new types" of ICBMs constrained by the treaty.

These criticisms, which surfaced recurrently throughout Senate ratification hearings, proved sufficiently damaging to draw 17 opposing votes (8 supporting) from within the Foreign Relations and Armed Services Committees in a Senate dominated by Carter's own Democratic party.[73] Realizing that it would take only twice that number of dissenting votes to defeat ratification on the floor of the Senate, Carter used the occasion of the Soviets' New Year's Eve invasion of Afghanistan to ask the Senate majority leader on 3 January 1980 to delay consideration of SALT II while he and the Congress assessed Soviet intentions. Ratification hearings never re-opened.[74]

Ironically, withdrawal of the treaty from Senate consideration was a severe rebuff to the anti-SALT school on arms control. Throughout the decade, attacks against SALT had reflected opposition more to the underlying theory of converged security interests than to the short-comings of any particular agreement. Opposition to the SALT II Treaty has been directed more against standing arms control doctrine, its appeal to stability in MAD terms, its assumption that action-reaction explained arms competition, and its pursuit of arms control as an end rather than a means. Thus, it was these more far-reaching intellectual themes that the treaty's opponents wanted politically defeated, not just the treaty itself. It could still be argued that stability was a worthy goal; but there was considerable disagreement as to what that term actually meant, and few could argue that arms control had retarded its erosion. Now, having put the larger intellectual debate on hold, however, Carter could publicly declare his intention to abide by SALT II's provisions as long as the Soviets did likewise.[75]

THE NEW REALIST SCHOOL TAKES OVER

Having barred SALT II's legal ratification, the new realist school set about the business of defeating the SALT process in a political sense. The 1980 presidential election provided the mechanism. In rejecting the "fundamentally flawed" SALT II Treaty, the Republican platform outlined its own approach to arms control—a return to dual-track planning:

First, before arms negotiations may be undertaken, the security of the United States must be assured by the funding and deployment of strong military forces sufficient to deter conflict at any level or to prevail in battle should aggression occur;

second, negotiations must be conducted on the basis of strict reciprocity of benefits—unilateral restraint by the US has failed to bring reductions by the Soviet Union; and

third, arms control negotiations, once entered, represent an important political and military undertaking that cannot be divorced from the broader political and military behavior of the parties.[76]

The document went on to critique Carter administration policies as "diametrically opposed to these principles," and as guilty of "unilateral disarmament" which "removed any incentives for the Soviets to negotiate for what they could obviously achieve by waiting." The platform concluded its treatment of arms control by deploring "the attempts of the Carter Administration to cover up Soviet noncompliance with arms control agreements."

As the Reagan presidency got underway in 1981, it thus did so after a thorough public airing of the central themes associated with the arms control theory debate. The new president had said that a resolute new emphasis on the arms buildup route was not only a legitimate means of achieving objectives promised by the diplomatic track, but also was necessary both to security and arms control. He had clearly rejected the action-reaction axiom or unilateral restraint as an unacceptable guideline in the effort to bring reductions by the Soviet Union. He had articulated an approach to arms control that "cannot be divorced from broader political and military behavior" by the Soviets, and he had served notice of a compliance policy that would openly identify what he saw as a pattern of Soviet treaty violations.

Despite the apparently powerful political backing for a new approach to arms control, however, the Reagan "solution" was slow in developing. Some of this confusion could be attributed to intense political conflicts with the Soviet Union over developments in Poland, the Middle East, and Central America. Some of it has been explained by rapid turnover in key administration personnel—including the secretary of state, the national security adviser, the director of the Arms Control and Disarmament Agency, and the chief architect of tense negotiations in Europe as the 1983 deadline for intermediate-range nuclear force (INF) deployments drew near. Some of it might also be attributed to the similarly abrupt changes of leadership in the USSR, where four different Party chairmen held office between 1981 and 1985. While each of these factors would seriously complicate US and Soviet relations, several additional problems were the result of difficult, divisive policy choices within the administration itself. Three issue areas in particular invoked initial defense/arms control positions staked out by Ronald Reagan during the presidential election campaign.

The first of these involved the question of ICBM modernization. The president had boldly forecast a "window" during the late 1980s when ICBM vulnerability would represent serious security problems for the US if left uncorrected. He had said that the Soviet ICBM buildup had advanced both quantitatively and qualitatively in the aftermath of SALT I, and that SALT II would do nothing to retard this situation. But Reagan had also criticized his predecessor's plan to disperse some 200 movable, MIRVed MX missiles among 4600 hardened shelters in Utah and Nevada, and Congress was unreceptive to all of his alternative basing schemes. Two years into his first term, the president therefore chartered a Blue-Ribbon "Commission on Strategic Forces" to develop an "integrated solution" to the problem.[77] This Scowcroft Commission—

named for its chairman, Ford-era National Security Adviser Brent Scowcroft—was widely credited with saving the MX in Congress, but the political price for the administration was high.

The commission recommended limited acquisition and deployment of the MX missile, but for reasons the administration would not have emphasized. One was "to provide an inducement for . . . an arms control environment"; another was to provide the Soviets with "an incentive to move toward small ICBMs." This "bargaining-chip" rationale for the MX ran counter to much of what the president had been arguing over the years. The commission's position, unlike the administration's, downplayed the MX's contribution to strategic stability and elevated arms control to a position of primacy in any decision to deploy the new missile. Furthermore, regarding the question of where to deploy the new missile, the commission worked from the premise that the US "probably cannot return to systems as invulnerable as were Minuteman and Polaris in their day," a position that called into question whether there even *was* a window of vulnerability, never mind how it should be closed.

As an unusually broad political consensus began to develop around these positions, it became clear that the Scowcroft Commission's integrated solution had resurrected the MX missile and had shown the administration a way out of its political conundrum. Nonetheless, the fifty missiles ultimately approved by Congress were far fewer than Carter's 200-missile program that Reagan had criticized so vehemently. Furthermore, the proposal to base the missiles in existing silos directly challenged the administration's claim that there was an emerging US ICBM vulnerability crisis. Moreover, the success of this solution was to be contingent upon the new missile's indirect contributions to arms control. A president who had criticized arms control as a major cause of US ICBM vulnerability found himself pressed to solve those problems with more arms control—and to use as leverage a missile-basing scheme he had called unacceptable.

The MX decision process coincided with a second major issue—what to do with the unratified SALT II agreement—which further complicated the new administration's arms control policy. The Carter administration had already withdrawn the SALT II Treaty from Senate consideration. Reagan, who had called the treaty "fatally flawed," proclaimed after taking office that he had no intention to seek ratification of SALT II. In 1982, however, on the eve of his own Strategic Arms Reduction Talks (START), the president decided that he too would take no steps to undercut the SALT Agreements as long as the Soviets did likewise.[78]

This policy of "no undercut" reflected the reality that it would take years to deploy weapons systems that had not been in the acquisition pipeline for the previous six years. In 1982, the Soviets, who had negotiated SALT II so as to accommodate their own defense plans, pledged a similar obligation not to undercut SALT II.[79]

The mutual commitment of no undercut was political rather than legal in nature. It meant that neither side could engage in activities that would "defeat the object and purpose of a signed but unratified treaty."[80] Since the treaty permitted only one new type of ICBM, for example, a second new missile—which, once tested, could not be "untested"—would permanently violate the treaty and thus "defeat its object and purpose." But failure to dismantle systems in order to comply with numerical ceilings was another matter. Since these could be dismantled later, the no-undercut principle did not apply. Thus, precedent meant that both sides would either freeze their SNDV deployments at the numbers they held when the treaty was signed or (if higher) build up to the treaty's 2250 ceiling. Ironically, Reagan's no-undercut policy meant that the Soviets could retain hundreds more SNDVs than the US without violating their political commitment.[81] This pledge of mutual restraint would not impact major US force deployment decisions for several years, but it nonetheless appeared to accept a SALT process that Reagan and his supporters had repudiated without equivocation since 1979.

Along with the administration's hesitation over the SALT II withdrawal and the difficult choices it faced regarding the MX, a third issue involved what to do about Soviet noncompliance. In response to formal requests from Congress and in keeping with his campaign commitments, President Reagan began submitting annual reports to Congress documenting his concerns about Soviet noncompliance with several arms control agreements. The first of these reports, which were unprecedented in the postwar era, drew the following conclusions:

January 1984:
The United States Government has determined that the Soviet Union is violating the Geneva Protocol on Chemical Weapons, the Biological Weapons Convention, the Helsinki Final Act, and two provisions of SALT II: telemetry encryption and a rule concerning ICBM modernization. In addition we have determined that the Soviet Union has almost certainly violated the ABM Treaty, probably violated the SALT II limit on new types, probably violated the SS-16 deployment prohibition of SALT II, and is likely to have violated the nuclear testing yield limit of the Threshold Test Ban Treaty.[82]

Subsequent reports confirmed these findings and cited additional violations of the LTBT, SALT II, and ABM treaties.[83]

Having taken the historic step of identifying

Soviet noncompliance, however, Reagan faced the similarly unprecedented challenge of deciding what to do about it. Weapons development programs can take as long as 15 years from the time they are conceptualized until they are fielded, and SALT II had presumably taken account of all US programs in an advanced stage of that process. Furthermore, the president himself had been complying scrupulously with SALT I and SALT II as promised by his no-undercut commitment. The US could, of course, simply increase quantities of systems being deployed, thereby violating SALT's numerical ceilings. But the funds to do so would either have to be appropriated anew, which Congress was reluctant to do, or else taken from other components of the defense budget, which the military services were reluctant to do. Furthermore, the Soviets, who had already added more than 3400 strategic missile warheads to their inventory during the first Reagan term,[84] had "hot" production lines that would permit them to build SALT-limited systems much faster than the US. As newer Soviet strategic weapon systems began being fielded in large numbers, however, the question of continued US restraint ultimately called for concrete rather than just rhetorical policy choices by the Reagan administration.

A NEW COMPLIANCE POLICY TAKES HOLD

When the June 1985 deployment of the seventh Ohio-class Trident submarine, with 24 MIRVed SLBMs on board, threatened to broach SALT II's ceiling of 1200 MIRVed ballistic missiles, the president had to decide whether to abandon SALT's limits or to continue his policy of compliance by dismantling compensatory numbers of older MIRVed missiles. In a statement to Congress, the president concluded that "the Soviet Union was not exercising the equal restraint on which the US interim restraint policy had been conditioned," and that he could not accept a "double standard" of US compliance coupled with Soviet noncompliance.[85] Still, with the administration's gearing up for a November summit meeting with the Soviet leadership, the president decided that it was in the US interest to "go the extra mile" with the Soviets. Accordingly, he decided to deactivate and dismantle an older Poseidon submarine as the seventh Trident put to sea in August 1985. By the spring of the following year, however, he was faced with the same dilemma as the eighth Trident was scheduled for deployment, and as he looked ahead to late 1986 when the 131st ALCM-carrying US bomber would challenge SALT II's (1320) ceiling on MIRVed systems.[86] It was then that he advanced the policy spelled out in his election platform six years earlier:

The US cannot continue to support unilaterally a SALT structure that Soviet noncompliance has so grievously undermined and that the Soviets appear unwilling to repair. Therefore, in the future, the United States will base decisions regarding its strategic forces on the nature and magnitude of the threat posed by the Soviet Union, rather than on standards contained in the expired SALT agreements unilaterally observed by the United States.[87]

This decision drew public outrage from the Soviets, anxieties from many in Congress and the media, criticism from allies, and scorn from those in the arms control community who remained committed to SALT as the foundation of arms control. But the president's position was consistent with his earlier arguments and was significant more for the logic of the policy than for its immediate impact on US force structure. The US defense program, according to this logic, should be tied to the threat itself rather than to treaties that would have expired if ratified and that had been violated in any case. The chorus of opposition, however, reflected the political price tag for Reagan's years of hesitation. Thus, he added in his 27 May 1986 Statement, "I do not anticipate any appreciable numerical growth in US strategic offensive forces" as a result of this policy. And even this move would be structured to facilitate arms control:

Assuming no significant change in the threat we face, as we implement the strategic modernization program the United States will not deploy more strategic nuclear delivery vehicles than does the Soviet Union . . . In sum, we will continue to exercise the utmost restraint, while protecting strategic deterrence, in order to help foster the necessary atmosphere for significant reductions in the strategic arsenals of both sides.[88]

After threatening to do something about Soviet noncompliance since his own party platform in 1980, Reagan had finally implemented his long-standing belief that "a country simply cannot be serious about effective arms control unless it is equally serious about compliance." As Secretary of State Shultz put it when asked by the press how the US could cope with a much greater Soviet weapon construction capacity to break out of the SALT II regime, "The Soviets have already broken out of SALT II; that's the point."[89]

With the MX, SALT II, and Soviet noncompliance problems essentially driving US policy on strategic arms control, direct negotiations with the Soviets in Geneva were virtually mandated by the surrounding political environment. These began in June 1982. Believing as he did that SALT II's basic flaw was that it permitted substantial increases in both sides' strategic forces, the president declared that his approach would be called START—to emphasize strategic arms *reductions*—rather than SALT. By the same logic, he specified early on that the "unit of account" for these talks would be warheads rather than launchers, with emphasis on those warheads based in the most destabilizing deployment modes. The objective was substantial reductions in ballistic missile warheads, because of their short flight times, rather than on bomber warheads, and specifically on land-based, rather than sea-based missile warheads because of their superior counterforce or war-fighting capabilities. Since ICBMs were most capable of threatening preemptive attack, reduction of ICBM warheads made the most sense in terms of strategic stability, but this objective also aimed the administration's arms control policies squarely at those forces in which the Soviets had made their greatest investments over the years. In 1982, the US had some 6949 ballistic missile warheads, about 300 fewer than when SALT II was signed. The Soviets, by comparison, had increased their 1979 arsenal by almost 2000 ballistic missile warheads and by 1982 had a total of 7727. In ICBM warheads, however, the Soviets had 5862 to a US total of 2149.[90] This disparity in ICBM warheads, coupled with a Reagan political commitment to correct it, generated the first line of debate at the negotiating table.

The president first spelled out his new approach to strategic arms control in a May 1982 commencement address to his alma mater, Eureka College, in Peoria, Illinois:

At the first phase, or end of the first phase of START, I expect ballistic missile warheads, the most serious threat we face, to be reduced to equal levels, equal ceilings, at least a third below the current levels. To enhance this stability, I would ask that no more than half those warheads be land-based . . . In a second phase, we'll seek to achieve an equal ceiling on other elements of our strategic nuclear forces including limits on ballistic missile throw-weight at less than current American levels.[91]

A month later, when the START talks began in Geneva, the president's formula acquired specific numbers. His one-third reductions translated into a proposal for both sides to cut their total missile warhead arsenals to 5000, with no more than half (2500) of these on ICBMs, and to reduce total ballistic missiles to 850 each. Phase two of the proposal involved reductions in "overall destructive power," which meant missile throw-weight, to an equal level below the existing US total.[92]

The Soviets criticized the proposal on familiar grounds—with which many American commentators agreed—focusing on the one-sided demand for the Soviets to reduce their ICBM warheads to 2500, while the US would actually be able to add some 350 warheads. The administration responded that *total* land- and sea-based warhead reductions were roughly equal, and that Soviet ICBMs would become vulnerable themselves when new American systems,

such as MX, came on line. Congressional opposition to MX, however, did little to enhance this argument, and a massive movement by administration critics advocating a freeze at existing force levels, to be *followed* by reductions, further deflected pressure from the president's START formula. The argument that MIRVed ICBMs deserved special attention in arms control because of their destabilizing characteristics was muted in the public debate over the urgency of a freeze.

Soviet counterproposals also called for reductions, but in delivery vehicles, as in SALT, rather than in warheads, and in proportions that would preserve Soviet first-strike targeting advantages. With the two sides unable to agree on a unit-of-account, and with the Soviet leadership crisis beginning, negotiations remained at a stalemate. Finally, in November 1983, when US INF missile deployments began in Europe, the Soviets terminated both the INF and START talks in protest.[93]

FULL CIRCLE: STRATEGIC DEFENSE REENTERS THE ARMS CONTROL EQUATION

The recess in START lasted 15 months, during which administration policies on MX, SALT II, and Soviet noncompliance continued to fester. On 23 March 1983, however, the president had added another issue to the strategic debate— one that shifted it directly back to its pre–SALT I terms. In a nationally televised speech the president proposed his Strategic Defense Initiative (SDI), asking, "Are we not capable of demonstrating our peaceful intentions by applying our abilities and our ingenuity to achieving a truly lasting stability?"[94] He then answered his own rhetorical question: "I think we are— indeed we must." The president's avenue to that lasting stability is said to have surprised many of his closest advisers:

It is that we embark on a program to counter the awesome Soviet missile threat with measures that are defensive. Let us turn to the very strengths in technology that spawned our great industrial base . . . I know this is a formidable task, one that may not be accomplished before the end of the century. Yet current technology has attained a level of sophistication where it is reasonable for us to begin this effort.

The president launched SDI as a long-range research effort to discover means of reversing militarily the destabilizing trends in Soviet ICBM strength that arms control had thus far failed to curb. The association of SDI with research, restrictions on its testing, and compliance with the ABM Treaty, much like Reagan's decisions on MX, SALT II, and Soviet violations, preserved the long-standing US emphasis on arms control. And yet the ultimate possibility of SDI's deployment—if research were to produce the answers it sought—underpinned the administration's effort to assemble modest political support for its funding. The proposal drew strong popular approval, but also strong and vocal opposition from foreign and domestic devotees of MAD. SDI was controversial because it challenged such central tenets of stability theory as societal vulnerability, and reoriented the strategic arms debate back onto the relationship between offense and defense.[95]

As with the SALT I breakthrough of 1971, it was this reorientation by the Reagan administration back toward ballistic missile defense that helped draw the Soviets back to the negotiating table two years after the president's announcement. When Secretary of State Shultz met with Soviet Foreign Minister Gromyko to reestablish negotiations in January 1985, their very divergent views of arms control were embodied in a joint statement that framed the debate for the next several years. Instead of just two separate negotiating tables—one for European-based nuclear arms (INF) and one for strategic range weapons (START)—a third forum, labeled the Defense and Space Talks (DST), became the focus of much discussion in the other two. The Shultz-Gromyko meeting had resolved that talks would involve a "complex of questions concerning space and nuclear arms both strategic and intermediate range with all the questions considered and resolved in their interrelationship."[96]

The carefully crafted ambiguity of this statement guided negotiators to debate over the ensuing years about the same issues said to have been resolved by SALT I thirteen years earlier. By insisting once again that a ban on defenses must precede any agreement on offenses, the Soviets aimed their efforts squarely at SDI—i.e., squarely at the advanced technology on which the US would rely to offset Soviet offensive superiorities. By countering that reductions in destabilizing offensive weapons must proceed toward equal levels with or without further defense-related agreements, the US sought to impose the framework already established in SALT I, and to diminish the strategic advantages deriving from a Soviet ICBM force already capable of delivering over 50 percent more warheads than when SALT II was signed.

At the START table, both sides proposed actual weapons reductions. The Soviets' offensive reduction proposals sought to protect their advantages in land-based warheads, on the precondition that the Defense and Space Talks establish a ban on what they called "space strike arms."[97] US proposals sought to protect SDI and to engage the Soviets in a discussion as to how a "cooperative transition to a more defense reliant regime" might be effected.[98] In February 1988, for example, the US proposed a reduction in ballistic missile warheads to 4900 each, with

a subceiling of 3300 on ICBMs. This required a reduction of about 2500 warheads in the Soviet ICBM force—to over a thousand fewer than they had when they signed SALT II—and would permit about a thousand *more* than the US had in its ICBM force when the proposal was made. The US proposal also called for a large reduction in Soviet heavy ICBMs and in the Soviet throw-weight advantage and would ban mobile ICBMs.[99]

The Reagan administration's emphasis on reductions, aside from whether they were negotiable or fair, represented several important modifications long developing in arms control. First, his insistence that agreements must clearly serve US national security interests, in the face of considerable political pressures over the years, purported to reorient policy away from agreements for the sake of a process or for the sake of agreements themselves. Instead, such proposals, according to the administration, were designed to reduce the threat to US retaliatory force survivability. Second, in structuring his proposals around ICBM warhead reductions in particular, the president took account of the warfighting strategy behind the force planning *and* arms control policies of the Soviets. The dual-track Soviet approach—both parts of which served the same destabilizing objective—would have to be answered with greater emphasis on warfighting in US dual-track planning if deterrence were to be served by either.[100] The president's SDI and strategic modernization programs on the force-planning side thus dovetailed with his emphasis on ICBM reductions in pursuit of that objective. Through the SALT years, arms control had proceeded with a momentum of its own toward a negotiated management of a buildup. The Reagan arms control policy, having rejected the SALT framework as destabilizing in itself, sought agreements designed to enhance security if successful and weapons programs designed to serve that same end if negotiations failed.

Striving to align his strategic modernization track with his arms control track, President Reagan said that the goal of arms control should be the same as the goal he initially assigned to his Strategic Defense Initiative: to eliminate the political and military utility of ballistic missiles. At his October 1986 summit meeting with Secretary General Gorbachev in Reykjavik, Iceland, he took this position to its extreme by proposing a 50 percent reduction in strategic nuclear weapons in five years, followed by the *complete elimination* of both sides' ballistic missiles within ten years. While advancing schemes like this one at the negotiating table, the president sustained his commitment to SDI as a means of devaluing the preemptive potential of Soviet ICBMs. By this logic, according to the president, SDI would force the outcome that SALT had promised, but had failed to deliver. If deployed in conjunction with a drawdown in ballistic missiles, however, SDI would provide an "insurance policy to hedge against cheating or other contingencies."[101] Thus, stability would be enhanced even if, as expected, the Soviets were to deploy similar systems. Finally, echoing McNamara's argument that arms control provides the mechanism for educating the Soviets, Reagan argued that the Defense and Space Talks could represent an ideal forum in which the two sides could discuss how to share the benefits of strategic defenses while managing a joint transition to a deterrent relationship based on defense.[102]

Of course, the fulfillment of such a sweeping strategic vision is hardly a foregone conclusion. The Soviets continued to hold progress in START hostage to a version of the ABM Treaty that would cripple the SDI program. In response to the president's proposals at Reykjavik, for example, the Soviets proposed the elimination of not just ballistic missiles, but all nuclear weapons which, as Gorbachev argued, would negate any requirement for strategic defenses. Given the extent to which the US and its allies rely on nuclear technology to offset the Soviet conventional threat, such an agreement would be profoundly unacceptable for the West. The president declined this offer on grounds that strategic defense was the linchpin of security for the US, with or without offensive reductions. In short, he rejected, as he said he would, the proposition that arms control could serve as a singular vehicle to alter unstable conditions in the strategic balance.

CONCLUSION

The US entered the postwar era with an international power base that was unprecedented in human history. A virtually unrivaled military superiority derived not just from victory on two fronts, but from a monopoly on nuclear strength. An economic base, bolstered by its wartime surge, was uniquely intact after global war had ravaged all other participants. Above all, US political strength was characterized by a cohesive bipartisan commitment to containment for domestic security and to industrial reconstruction abroad.

This American strength was a blessing and a curse in the realm of national security policy. Choices available to the American government ranged from world domination to benevolent magnanimity in the international arena. To grasp the possibilities of what this implied, one need only contemplate how Joseph Stalin might have employed such power had the Soviet Union emerged from the war in this position. The US neither sought nor welcomed this position or preeminence, and was innocently un-

prepared for it. For most Americans, peace was synonymous with the mere absence of war, and was the normal state of affairs. Once the four-year disruption of normalcy was over, the continued production of armed strength was hardly attractive from a political standpoint. Thus, even if its postwar power position did permit the US to compel its own terms for international peace, its pluralistic and legalistic political traditions essentially ruled out such an approach. International political conditions, however, featured no less animosity after the war than before it. The Soviet Union, also militarily victorious, had its own vision of how the postwar international order should be shaped and employed its military power to pursue that vision.

Faced with a clear and unrelenting new adversary, and with demonstrably powerful military means at its disposal, the US hesitated for the next several decades while the Soviets first matched then surpassed the US in most categories of military strength. Throughout these years, the US opted for a two-track approach to its national security. In over four decades, however, the US has never fully reconciled the contradictory elements of that approach. For the first decade of the postwar period during which the Soviets developed and deployed their own nuclear arsenal, the US sought an elusive form of absolute disarmament designed to put the nuclear genie back in its bottle and to bring the Soviet Union into a peaceful, pluralistic, political relationship. Failing to achieve this on terms that would in any way unburden itself or its allies from the threat of war, the US turned to a partial approach to disarmament that came to be known as arms control. To date, this approach has failed either to check the growth of the Soviet nuclear threat or to build confidence that could facilitate resolution of future security challenges.

Dual-track policy for the US has been a series of fits and starts much like the "boom-bust" cycle that economists use to describe patterns of Western productivity. We have seen this pattern recur over and over. In the 1940s, Robert Oppenheimer rejected regulatory arms control as inadequate, then embraced it in the 1950s as the only alternative to a nuclear holocaust; meanwhile Bertrand Russell advocated first a nuclear ultimatum, then unilateral Western disarmament. In the 1960s, Robert McNamara first demanded a resolute US buildup emphasizing realistic, flexible, strategic-targeting options, then advocated a minimally armed US strategic arsenal, focused on Soviet population centers and regulated by arms control, as the natural determinant of strategic sufficiency. In the early 1970s, Henry Kissinger and Richard Nixon moved from a robust US deterrent structure heavily reliant on advanced ABM technologies to a MAD policy, institutionalized by SALT I, in which the Soviets held offensive nuclear superiority, and in which territorial defense against ballistic missiles was outlawed. In the late 1970s, Jimmy Carter first advanced a forward-thinking proposal for significant reductions in destabilizing force levels, then agreed to a managed buildup to predeterminend ceilings of these very same weapons in the form of SALT II.

In much the same way, the administration of Ronald Reagan went through a similar shift in emphasis—from arms buildup to arms control. Criticized in the 1984 presidential campaign for having pursued a strategic force modernization program without ever having met with a Soviet leader, Reagan met with Gorbachev with unprecedented frequency in his second term as defense spending declined in real terms. By the closing years of his administration, the president seemed guided by a set of principles much different from those behind his 1980 platform. Domestic budgetary pressures and the difficulties described earlier with his strategic modernization program only partially explain this shift. Beyond these very relevant political realities, Mr. Reagan's own personal judgment of Soviet motivations appeared to have changed as well.

This change could be seen in several key areas. Early reports to Congress on Soviet noncompliance, for example, insisted that "compliance with past arms control agreements is an essential prerequisite for future arms control agreements."[103] By the end of 1987, however, when agreements were approaching fruition in several areas, such language began to disappear from these reports. Instead, the president began saying that he could sum up his position with the Russian proverb *Dovorney no provorney*. "It means trust but verify," according to the president,[104] implying that shortcomings in verification had been the principal cause of arms control's failure all along.

More generally, the president suggested that the nature of the bilateral political struggle itself was different from what it had been when he assumed office. Thus, in 1983, while cautioning against the ground swell of popular support behind a nuclear freeze, the president had described Soviet motives in unmistakable terms to a gathering of American religious leaders.

I urge you to speak out against those who would place the United States in a position of military and moral inferiority . . . [I]n your discussions of nuclear freeze proposals, I urge you to beware the temptation of pride—the temptation to blithely declare yourselves above it all and label both sides equally at fault, to ignore the facts of history and the aggressive impulses of an evil empire, to simply call the arms race a giant misunderstanding and thereby remove yourself from the struggle between right and wrong and good and evil.[105]

But in 1988, when asked if he still viewed Soviet purposes in these terms, the president conceded "I no longer feel that way about Soviet leaders."[106] He then justified this turnaround by appealing to what the current Soviet leader had *not* said:

The Soviet Union has, back through the years, made it plain, and certainly leader after leader has declared his pledge that they would observe the Marxian concept of expansionism: that the future lay in a one world Communist State . . . [W]e now have a leader . . . that has never made that claim, but is willing to say that he's prepared to live with other philosophies in other countries.[107]

Indeed, there have been encouraging signs that Gorbachev's *glasnost* and *perestroika* policies may portend more than just rhetorical change in both Soviet domestic and foreign policy. Nonetheless, American political leaders are hardly immune to wishful thinking in their approach to arms control. We have seen how others changed their entire worldviews in anticipation of a Soviet transformation to a more benign security policy. McNamara, for example, altered the central tenets of US security policy on his belief that the arms race was just a mechanistic and cyclic habit resulting from misunderstood intentions and overestimated threat assessments. In turn, Kissinger enshrined his pursuit of arms control as an "historic opportunity" to create "a new set of international relationships—more pluralistic, less dominated by military power, less susceptible to confrontation, [and] more open to the free and diverse elements of the globe." Such visions have produced rhetorical bridges across a political chasm, but have had no discernible effect on the growth of Soviet military power.

While we might hope that Mr. Gorbachev will endure and redirect Soviet foreign policy away from its Leninist premises, our own national security must be oriented by realities, not hopes. President Reagan's reorientation appeared to be another iteration in the recurring pattern of alternating emphasis between modernizing the deterrent force and negotiating its regulation. Einstein was wrong when he said that "the unleashed power of the atom has changed everything save our modes of thinking."[108] As shown in this chapter, our "modes of thinking" tend to change every five or six years. This "boom-bust" pattern may be endemic to pluralistic democracies, but it may also be the central flaw of their security policies.

Rather than using the same national security strategies to guide both tracks of policy simultaneously, governments have tended to choose one or the other based on which wing of which political party is in power, or which agency of the Federal Government is bureaucratically ascendent. As a result, the shifting winds of national security planning tend to be driven by the exigencies of domestic politics at least as much as by the realities of the threat from abroad. Thus, presidents have looked to arms control for relief from budgetary pressures, for enhanced reelection prospects, or for historical legacy; for justification of new or improved weapons systems, or simply for something to sign during ceremonial portions of an upcoming summit. In the face of such pressures, the real purposes of arms control—to preserve what Truman called "peace and security" in a politically competitive world, to reduce the likelihood of war, or the threat of war, and to reduce the destructiveness that would occur if deterrence were to fail—are all too easily forgotten.

NOTES

1. US Department of State *Bulletin* 25 (19 Nov. 1951): 799–803.

2. "Statement by President Eisenhower at the Geneva Conference of Heads of Government: Aerial Inspection and Exchange of Military Blueprints," 21 July 1955, in Trevor N. Dupuy and Gay M. Hammerman, eds., *A Documentary History of Arms Control and Disarmament* (New York: R. R. Bowker, 1973), p. 379.

3. John F. Kennedy, "Inaugural Address," in *The Annals of America*, vol. 18, 1961–68 (Chicago: Encyclopædia Britannica), pp. 6–7.

4. Public Law 87–294. "The Arms Control and Disarmament Act," *Congressional Record*, 26 Sep. 1961 (75 Stat), p. 631.

5. Remarks by President Nixon at Congressional Briefing: Strategic Arms Limitation Agreements, 15 June 1972. *Weekly Compilation of Presidential Documents*, 19 June 1972, p. 1043.

6. US Congress, Senate, Committee on Foreign Relations, 96th Cong., 1st sess., Statement by Hon. Cyrus R. Vance, Secretary of State, 9 July 1979, p. 89.

7. Current Policy No. 676, "Arms Control: Objectives and Prospects," US Department of State, Bureau of Public Affairs, Washington, D.C., 28 Mar. 1985, p. 1.

8. Harry S. Truman, "Announcement of the Dropping of an Atomic Bomb on Hiroshima," *Annals of America*, vol. 16, 1940–49 (Chicago: Encyclopædia Britannica), pp. 334–36.

9. Bernard Brodie, ed., *The Absolute Weapon* (New York: Harcourt Brace, 1946).

10. William T. R. Fox, in *Absolute Weapon*, p. 181.

11. Brodie, in *Absolute Weapons*, p. 76.

12. J. Robert Oppenheimer, "International Control of Atomic Energy," *Bulletin of the Atomic Scientists* (June 1946). Like many other significant articles published in this magazine between 1945 and 1963, this article is conveniently reprinted in Morton Grodzins and Eugene Rabinowitch, eds., *The Atomic Age* (New York: Basic Books, 1964), pp. 53–63.

13. Ibid., p. 56.

14. See, for example, any of the works of Amrom Katz over the past quarter century. Katz's work on the compliance control problem is probably the most insightful on the subject over the past decades. But as early as the mid-1950s he wrote: "we don't need [an inspection] system which works well against a careless, uninformed, unimaginative opponent, but one that works well against an opponent who is smart,

careful, and imaginative." Amrom H. Katz, "Hiders and Finders," reprinted in *Bulletin of the Atomic Scientists* 17 (Dec., 1961): 423–24.

15. Richard J. Barnet, "Inspection: Shadow and Substance," in Richard J. Barnet and Richard A. Falk, eds., *Security in Disarmament* (Princeton: Princeton University Press, 1965), p. 20.

16. The Baruch Plan: Statement By United States Representative Baruch to the United Nations Atomic Energy Commission, 14 June 1946, in Dupuy and Hammerman, eds., *Documentary History*, p. 302. Emphasis added.

17. Michael Wright, *Disarm and Verify* (New York: Praeger, 1964), p. 19.

18. Bertrand Russell, "The Prevention of War," *Bulletin of the Atomic Scientists* (Oct. 1946). Reprinted in Grodzins and Rabinowitch, *Atomic Age*, pp. 100–106.

19. Ibid., p. 100. Emphasis added.

20. Ibid., p. 102.

21. Ibid.

22. Truman's address to the UN General Assembly, Oct. 1950, in Bernard G. Bechhoefer, *Postwar Negotiations for Arms Control* (Washington, D.C.: Brookings Institution, 1961), p. 153.

23. Oppenheimer, "Atomic Weapons and American Policy," *Bulletin of the Atomic Scientists* (July 1953). Reprinted in Grodzins and Rabinowitch, *Atomic Age*, pp. 188–96.

24. Eugene Rabinowitch, "Status Quo with a Quid Pro Quo," *Bulletin of the Atomic Scientists* (Sept. 1959). Reprinted in Grodzins and Rabinowitch, *Atomic Age*, pp. 197–205.

25. Ibid., p. 198.

26. As cited by Rabinowitch, ibid.

27. *Verification, The Critical Element of Arms Control*, Publ. no. 85 (US Arms Control and Disarmament Agency, 1976), pp. 11–12. Emphasis added.

28. Dwight D. Eisenhower, "Aerial Inspection and Exchange of Military Blueprints," Statement at Geneva Conference of Heads of Government, 21 July 1955, in Dupuy and Hammerman, eds., *A Documentary History*, pp. 379–81. Emphasis added.

29. Ibid., p. 378.

30. Ibid. Emphasis added.

31. "Geneva: Bulganin Report, Aug 4, 1955," in Bernard G. Bechhoefer, *Postwar Negotiations*.

32. Henry W. Forbes, *The Strategy of Disarmament* (Washington, D.C.: Public Affairs Press, 1962), p. 96.

33. Donald G. Brennan, "Setting and Goals of Arms Control," in Brennan, ed., *Arms Control, Disarmament, and National Security* (New York: George Braziller, 1961), p. 37.

34. Arthur T. Hadley, *The Nation's Security and Arms Control* (New York: Viking, 1961), p. 113.

35. Others include but are not limited to Thomas Schelling and Morton Halperin, *Strategy and Arms Control* (New York: Twentieth Century Fund, 1961); Hedley Bull, *The Control of the Arms Race* (London: Weidenfeld and Nicolson, 1961); and Ernest W. Lefever, ed., *Arms and Arms Control* (New York: Praeger, 1962). For useful summaries of the "stability" literature of the early 1960s see Jerome H. Kahan, "Arms Interaction and Arms Control," in John E. Endicott and Roy W. Stafford, eds., *American Defense Policy*, 4th ed. (Baltimore: Johns Hopkins University Press, 1965), pp. 102–19. See also Kahan, *Security in the Nuclear Age* (Washington, D.C.: Brookings Institution, 1975).

36. See Lewis F. Richardson, *Arms and Insecurity* (Pittsburgh: Boxwood, 1960) and *Statistics and Deadly Quarrels* (Chicago: Quadrangle Books, 1960).

37. See for example William R. Van Cleave, "The Arms Control Record: Successes and Failures" in Richard F. Staar, ed., *Arms Control: Myth vs. Reality* (Stanford: Hoover Institution Press, 1984), pp. 1–23. Van Cleave points out that the LTBT not only failed to fulfill its supporters' long-term promises, but limited knowledge about Soviet weapons design and other critical nuclear effects such as electromagnetic pulse. In this sense, the LTBT has been costly in terms of knowledge that would contribute to force survivability with no commensurate contributions to national security.

38. Soviet violations of the LTBT have been a particularly interesting aspect of the noncompliance debate because they provide no discernible military benefit and because the technology of underground containment is so advanced. The US has privately offered to share this technology with the Soviets, but the offer has been refused. See "Adherence to and Compliance with Agreements," prepared by the US Arms Control and Disarmament Agency, 30 Jan. 1986, p. 8.

39. Numerical approximations used throughout this chapter for US and Soviet weapons strengths come from a variety of unclassified sources. These include: John M. Collins, *US-Soviet Military Balance, Concepts, and Capabilities, 1960–1980* (Washington, D.C.: Library of Congress, 1980); *The Military Balance* (London: The International Institute for Strategic Studies, annual); and *A Review of Trends in US and USSR Nuclear Force Levels and Related Measures* (Alexandria: Santa Fe Corporation, 30 June 1983).

40. John Newhouse, *Cold Dawn: The Story of SALT* (New York: Holt, Rinehart, and Winston, 1973), pp. 3–4. Emphasis added.

41. As quoted by *Izvestiya*, 11 Feb. 1967.

42. Newhouse, *Cold Dawn*, p. 94.

43. Robert S. McNamara, "The Dynamics of Nuclear Strategy," *US Department of State Bulletin*, 9 Oct. 1967, pp. 443–51.

44. *US News and World Report*, 12 Apr. 1965.

45. McNamara, "Dynamics of Nuclear Strategy."

46. "Kissinger Statement on Détente and SALT, 19 September 1974," in Roger P. Labrie, ed., *SALT Handbook* (Washington, D.C.: American Enterprise Institute, 1979), pp. 270–71. Emphasis added.

47. "Basic Principles of Relations between the United States of America and the Union of Soviet Socialist Republics," *Department of State Bulletin*, 26 June 1972, pp. 898–99.

48. See his elaboration of this argument in Henry Kissinger, *White House Years* (Boston: Little, Brown, 1979). Esp. chap. 20, "US-Soviet Relations: Breakthrough on Two Fronts," pp. 788–841.

49. For an insightful description of this process, see US Congress, Senate, Hearing before the Committee on Armed Services, "Statement by Dr. William R. Van Cleave," in *Military Implications of the Treaty on the Limitations of Anti-Ballistic Missile Systems and the Interior Agreement on the Limitation of Strategic Offensive Arms*, 25 July 1972, p. 570.

50. Unilateral Statement A, "Withdrawal from the ABM Treaty," to the Treaty Between the United States of America and the Union of Soviet Socialist Republics on the Limitation of Anti-Ballistic Missile

Systems, in *Arms Control and Disarmament Agreements, Texts and Histories of Negotiations* (Washington, D.C.: US Arms Control and Disarmament Agency, 1982), p. 146.

51. Evidence that Nixon administration officials believed SALT I represented a whole new era in US-Soviet relations is myriad. Examples can be found by the most cursory reviews of the ratification testimonies of Gerard Smith and Melvin Laird. But perhaps the most classical underpinning for this point can be found by reviewing any of Kissinger's briefings to the media or to the congressional leadership after the signing summit in May 1972. These are nicely collected and easily reviewed by Roger P. Labrie in *SALT Handbook*, pp. 32–49, 169–75, and 176–83.

52. Henry A. Kissinger, "Détente with the Soviet Union: The Reality of Competition and the Imperative of Cooperation," *Department of State Bulletin* (14 Oct. 1974), p. 513.

53. Leonid I. Brezhnev, "Report to the Twenty-Fifth Congress of the Communist Party of the Soviet Union, 24 Feb. 1976," in Robert J. Pranger, ed., *Détente and Defense* (Washington, D.C.: American Enterprise Institute, 1977), pp. 178–89.

54. Harold Brown, US Department of Defense, *Annual Report, FY 1981*, 29 Jan. 1980, p. 6.

55. *Arms Control and Disarmament Agreements*, p. 122.

56. *Soviet Noncompliance*, US Arms Control and Disarmament Agency, 1 Feb. 1986.

57. "Statement by Dr. William R. Van Cleave" (see note 53), p. 572.

58. James Schlesinger, *1975 DoD Report to Congress*, p. 4.

59. Ibid., p. 5.

60. Ibid.

61. Ibid., p. 32.

62. Richard Pipes, "Why the Soviet Union Thinks It Could Fight and Win a Nuclear War," in Douglas Murray and Paul Viotti, eds., *The Defense Policies of Nations* (Baltimore: Johns Hopkins University Press, 1982), pp. 134–46.

63. Edward N. Luttwak, "Why Arms Control Has Failed," *Commentary* (Jan. 1978): 27.

64. See Wohlstetter's articles in *Foreign Policy*, no. 15 (Summer 1974); no. 16, (Fall 1974); and no. 20 (1974). Quotations used here are drawn from Albert Wohlstetter, "Racing Forward or Ambling Back?" in *Defending America* (New York: Basic Books, 1977), pp. 110–68.

65. Ibid., p. 126.

66. Paul H. Nitze, "Nuclear Strategy: Détente and American Survival," in *Defending America* (see note 71), pp. 97–109. The same article appeared as "Deterring Our Deterrent," in *Foreign Policy* (Winter 1976–77).

67. Ibid., p. 97.

68. Ibid., pp. 107–8.

69. Press briefing of Secretary of State Cyrus R. Vance, 26 Mar. 1977. Department of State Release no. 130, 28 Mar. 1977.

70. Ibid.

71. Foreign Minister Andrei Gromyko, Press Conference of 31 Mar. 1977. News from the USSR, Soviet Embassy Information Dept., Washington, D.C. Cited by Labrie, *The SALT Handbook*, p. 435.

72. Labrie, *SALT Handbook*, p. 386.

73. The votes were 10–0 against SALT II in the Senate Armed Services Committee, and 8–7 in favor

of a renegotiated version of the Treaty in the Foreign Relations Committee.

74. John M. Collins, *US-Soviet Military Balance, 1980–1985* (Washington, D.C.: Permagon Press, 1985), p. 9.

75. Ibid.

76. "Republican Platform: Family, Neighborhood, Work, Peace, Freedom." Proposed by the Committee on Resolutions to the Republican National Convention, 14 July 1980, Detroit, Michigan, p. 62.

77. Report to the President by the President's Commission on Strategic Forces, Brent Scowcroft, chairman, 21 Mar. 1982, pp. 224–26.

78. Strobe Talbott, *Deadly Gambits* (New York: Vintage Books, 1982), pp. 224–26.

79. *Soviet Noncompliance*, p. 7.

80. These words come from Article 18 of the Vienna Convention on the Law of Treaties. See for elaboration William R. Harris, "Breaches of Arms Control Obligations and Their Implications," in Richard F. Staar, ed., *Arms Control*, pp. 134–53.

81. *Soviet Noncompliance*, p. 8.

82. US Congress, Senate, *Congressional Record*, Proceedings and Debates of the 98th Congress, 2d Sess., vol. 30, no. 8, 1 Feb. 1984, reprint.

83. See "Soviet Noncompliance with Arms Control Agreements," Department of State Special Report no. 136, Dec. 1985, and subsequent reports in March 1986 and December 1987. See also US Congress, Senate, Committee on Armed Services, Testimony of the Honorable Richard Perle, Assistant Secretary of Defense, International Security Policy, 7 May 1985, transcript.

84. Collins, *US-Soviet Military Balance*, p. 183.

85. The White House, Office of the Press Secretary, Fact Sheet, "US Interim Restraint Policy: Responding to Soviet Arms Control Violations," 27 May 1986, transcript.

86. Ibid.

87. Ibid., p. 4.

88. Ibid., p. 1.

89. "Meet the Press," NBC Television, 1 June 1986.

90. Collins, *US-Soviet Balance*, p. 183.

91. "Arms Control and the Future of East-West Relations," 9 May 1982, Eureka College, Peoria, Illinois, *Weekly Compilation of Presidential Documents*, 10 May 1982.

92. Richard Burt, "Evolution of the US START Approach," *NATO Review* 30 (Sept. 1982). Reprinted in US Department of State, Current Policy no. 436.

93. "START in a Historical Perspective," Address by Edward L. Rowny, 10 April 1984, US Department of State, Current Policy no. 563, April 1984, p. 1.

94. Ronald Reagan, Address to the Nation, 23 Mar. 1983.

95. For more on the origins and directions of SDI, and a dispassionate analysis of its surrounding debate, see Keith B. Payne, *Strategic Defense: "Star Wars" in Perspective* (Lanham: Hamilton Press, 1985).

96. US-USSR Joint Statment, 8 Jan. 1985, transcript.

97. Negotiations on Nuclear and Space Arms, Address by Ambassador Paul H. Nitze, 13 Mar. 1986, US Department of State, Current Policy no. 807.

98. Ibid.

99. Ibid.

100. The best descriptive analysis of "warfighting" is Colin S. Gray, "Nuclear Strategy: A Case for a Theory of Victory," *International Security* (Summer 1979): 54–87.

101. See Secretary of State George Shultz's remarks at the *Chicago Sun-Times* Forum, USIS Wireless File, 17 Nov. 1986.

102. For example, President Reagan's speech to the nation after his meeting with Soviet General Secretary Gorbachev at Reykjavik, USIS Wireless File, 13 Oct. 1986, and his later speech to the Town Hall of California and the Chautauqua Conference in New York, USIS Wireless File, 26 Aug. 1987.

103. *Soviet Noncompliance*, Mar. 1986.

104. President Ronald Reagan, "Interview with Television Network Anchormen," 3 Dec. 1987, in *Weekly Compilation of Presidential Documents*, vol. 23, no. 48, p. 1426.

105. President Ronald Reagan, "Remarks at the Annual Convention of the National Association of Evangelicals in Orlando, Florida," 8 Mar. 1983, in *Public Papers of the President of the United States, 1983*, Book I (Washington, D.C.: Superintendent of Documents, G.P.O.), pp. 363–64.

106. William Safire, "The Fawning After," *New York Times*, 13 Dec. 1987, E-25.

107. Reagan, "Interview with Television Network Anchormen."

108. The Einstein quote is frequently cited, in and out of context; see, for example, Jonathan Schell, *The Fate of the Earth* (New York: Avon Books, 1982), p. 188.

CHAPTER 11

ARMS CONTROL AND SECURITY IN EUROPE

PAUL R. VIOTTI

This chapter addresses arms control and its relation to security in Europe. The first section contains a brief overview of the different arms control negotiations dealing with European security questions. The importance to most Europeans of continuing arms control efforts, even when they show very slow or little progress, is underscored by the deep concern that Europe never again suffer the devastation of any war, conventional or nuclear. Perhaps not surprisingly, given the experience of World War II, this feeling is particularly strong among Germans whose territory is in the very center of Europe and thus likely to be a principal battlefield in any future war.

The next three sections deal with alternative images of European security. The first is the familiar East-West image in which the Soviet Union and its Warsaw Pact allies in the East confront the United States and its NATO allies in the West. This image is one of a balance of power between East and West. A second image, less familiar to American readers, portrays the balance of power somewhat differently. The focus in this second image is on the threat to European security that had historically emanated from the center of the continent, particularly Germany. In this geopolitical analysis of the balance of power, the center had become too strong relative to states on the European periphery, and two world wars were the result. From this point of view and apart from East-West considerations, the balance of power in Europe today is also one in which Germany is divided so as not to become too strong and thus not become a threat to the countries on its borders. Moreover, alliance with West or East Germany is a further hedge against a resurgent German threat to those states on the European periphery that suffered invasion or occupation by Germany in past wars.

A third image, the subject of the fourth section, departs from the idea that security comes only from the balance of power, whether cast in terms of an East-West balance or one in which the center of Europe is kept from threatening the periphery. Instead, the third image offers a vision of an increasingly interdependent Europe in which national boundaries are frequently crossed through trade, investment, travel, other forms of commerce, mass communications media, and countless other ways. This is a cosmopolitan and pluralist view in which efforts are made to build trust and confidence among Europeans of all countries. Although some see it as utopian, others see this cosmopolitanism as displacing ultimately the threat and counterthreat behaviors that stem from more traditional considerations of balance of power.

These three sometimes competing, sometimes overlapping views are underlying considerations that influence the way arms control and other security discussions are conducted. Equipped with only an East-West view—which tends to be the dominant image, particularly among Americans—one is not likely to understand the full significance of some of the issues being negotiated.

The final section of the chapter takes a look at the future of arms control in Europe, underlining the point that arms control negotiations are, in effect, discussions about the various alternative foundations on which a secure European political order might be based.

ARMS CONTROL NEGOTIATIONS AND THE QUEST FOR SECURITY IN EUROPE

To an even greater extent than Americans, European leaders and their populations have a continuing and deep concern with limiting armaments and managing, if not eliminating, conflicts. The utter devastation wrought by two world wars in this century produced in Europeans across the political spectrum a common and deep commitment to finding ways to maintain the peace. There are, of course, great differences in view among Europeans, as there are among Americans, as to the best means to reduce the likelihood of war. Anxiety about and, in some, an outright fear of war is the pervasive emotion more intensely felt by Europeans than among populations further removed from the center of East-West conflict. Surveys confirm this underlying Angst, or fear of war, that is by no means confined to pacifist and neutralist elements or any other "specific segments of the population."[1]

In particular, many Germans see themselves as a single nation divided into eastern and western states on whose territory a future war would be fought. Speeches by West German government leaders frequently use the expression "two German states in one German nation." Be that as it may, the types of weapons used in any future war (nuclear, chemical, or conventional) would make differences only in the degree of devastation wrought. No point on this spectrum of destruction is acceptable to Europeans, least of all to the Germans who would likely be at the center of hostilities.

Although peace is the commonly desired end, there is no unanimity on appropriate means to that end. Avoiding war or maintaining peace through deterrence and military strength is the traditional mode: *si vis pacem, para bellum* (if you want peace, then prepare for war). A complementary track is peace through conflict management, negotiating arms reductions or limitations agreements, and relying on such confidence- and security-building measures (CSBMs) as advance notifications and observation of military exercises. Many Europeans expect the United States to provide reassurance that the lid is being kept on conflicts in addition to providing the forces and commitment for deterrence.[2]

In recent years arms control negotiations related to military forces in Europe have been held in such cities as Geneva, Helsinki, Vienna, and Stockholm. Strategic Arms Limitations Talks (SALT) in Helsinki and Vienna (SALT I) and Geneva (SALT II) in the late 1960s and 1970s and Strategic Arms Reductions Talks (START) in Geneva in the 1980s have focused primarily on bombers and land- and sea-based missiles of intercontinental range deployed or under development by the United States and the Soviet Union. Although these so-called central strategic forces are not deployed in Europe, they are part of the overall military balance between East and West. For NATO, the US strategic arsenal is tied or coupled through *extended deterrence* to the security of Western Europe. Talks on the use of space date from the 1960s in the multilateral, Geneva-based Eighteen Nation Disarmament Conference that has become a 40-nation Conference on Disarmament (CD), loosely associated with the United Nations and based in Geneva. Given the development of antisatellite weapons and other new technologies with space applications and with implications for both superpower and European security, bilateral negotiations between the United States and the Soviet Union on space began anew in the mid-1980s. Meanwhile, the CD continues to address such issues as nuclear weapons testing and chemical, biological, and radiological warfare, which pose a threat to European security.

Negotiations were also conducted in the 1980s in Geneva on intermediate-range nuclear (INF) missiles in Europe and, between 1973 and 1989, in Vienna on conventional force reductions in the European central front (the MBFR or mutual and balanced force reductions talks).[3] In the 1970s, congressional pressures in the United States to reduce US troop strength in Europe (a view associated at the time with Senate Majority Leader Mike Mansfield) influenced the Nixon administration to pursue MBFR negotiations. If these pressures could not be turned off, at least any reduction of conventional forces in Europe would be accomplished as part of a negotiated agreement aimed at maintenance of an East-West balance.

The 1975 Helsinki Final Act and subsequent review conferences in Belgrade (1977–78), Madrid (1980–83), and Vienna (1986–1989) are all part of the Conference on Security and Cooperation in Europe (CSCE) process. An outgrowth of the CSCE process was a mandate for convening in Stockholm in 1984 a Conference on Disarmament in Europe (CDE) that reported to the subsequent CSCE review conference in Vienna. The CDE included 35 countries covering all of Europe, thus going well beyond the MBFR's focus on central Europe. The CSCE gave the CDE its charter to construct confidence- and security-building measures among the parties, attempting to establish the requisite trust for future substantial arms reductions. New negotiations on Conventional Forces in Europe (CFE), following upon progress made in the CDE, have replaced MBFR. Conducted within the framework of the CSCE process, these negotiations began in March 1989 and include the 23 NATO and Warsaw Pact members, but not the 12 neutral or nonaligned states. The latter continue to negotiate on confidence-building measures and other issues within a separate 35-country CSCE forum.

INF, MBFR, CSCE, and CDE have had the most direct impact on European security, although the CD, periodic UN Special Sessions on Disarmament, and bilateral negotiations between the superpowers also affect European security interests. Before dealing with the specifics of any of these negotiations, however, it is essential to understand the cognitive contexts—the diverse images of European security—that have affected these discussions. Recognition of these alternative images is usually neglected in studies on arms control in Europe. These studies usually focus on the more specific issues being negotiated (such as force levels and weapon types) and how to respond to the other side's offers and gambits. Moreover, the issues are presented almost exclusively in East-West terms.

This chapter is about arms control, but in a more fundamental sense it is about security in

Table 11.1

Arms Control Negotiations in Europe

Abbreviation	Name	Site	Scope/Function/Results
1. CD	Committee on Disarmament	Geneva	Loosely affiliated with United Nations. 40 country participants.
2. MBFR	Mutual and Balanced Force Reductions	Vienna (1973–1989)	Territorial scope limited to European central front. MBFR has been superseded by talks on conventional armed forces in Europe (CFE) within the CSCE process (see below).
3. CSCE	Conference on Security and Cooperation in Europe	Begun in Helsinki; review conferences in Belgrade, Madrid, Vienna	Territorial scope extends from the Atlantic to the Urals. 35-country process. Provides forum in Vienna for negotiations on confidence- and security-building measures begun in March 1989.
4. CDE	Conference on Confidence- and Security-Building Measures and Disarmament in Europe	Stockholm	Conference concluded its work in 1986 and reported to CSCE. Superseded (see 3 above and 5 below).
5. CFE	Conventional Armed Forces in Europe	Vienna	Begun in March 1989 as part of CSCE process. 23 country participants (members of NATO and Warsaw Pact alliances). Purpose of removing imbalances and achieving reductions in conventional forces.
6. DST	Defense and Space Talks	Geneva	Begun in March 1985 in conjunction with START and INF negotiations to address possible limitations on strategic defense programs.
7. START	Strategic Arms Reduction Talks	Geneva	Successor to earlier Strategic Arms Limitation Talks (SALT I and SALT II).
8. INF	Intermediate-Range Nuclear Forces	Geneva	Treaty concluded in 1987 to eliminate all US and Soviet INF missiles on a global basis.

Europe. A central thesis to be argued here is that there are three alternative images of security in Europe that overlay each other and, indeed, are sometimes held simultaneously by the same statesmen. Political realism, with its focus on states, alliances, and the balance of power, presents European security in terms either of "East-West" or "center-periphery" imagery—the latter reflecting a concern among non-German states that the center of Europe not become too strong. A third image is that security is served by increasing interdependence among European states and societies—more trade, investment, travel, cultural exchange, and other contacts. This is a pluralist view favorably disposed toward growing interdependence and interconnectedness. If barriers between East and West cannot be broken down easily, then they can at least be crossed more often by increasing transnational interactions and exchange.

Although one can focus endlessly on the intricate details of the various arms control negotiations, it is not really possible to understand the issues under discussion (much less the diverse national positions on these questions) without an appreciation of the different interpretations of security in Europe held by European decisionmakers. Force reductions and confidence-building measures are only partly what the MBFR and CDE negotiations in Vienna and Stockholm were about. The fundamental issue in these and the CFE talks is the political order in Europe—a conversation that began among the victorious Soviet, British, American, and French allies during and immediately after World War II. The number of parties to the discussions has increased significantly in recent years with the addition of allies on both sides, the two Germanys, and even nonaligned states as formal participants in the CSCE process. Nonetheless, the underlying questions of how Europe is to be ordered politically and how international security is to be maintained have remained essentially the same. To understand the process, one must delve beneath the formal agenda to uncover the competing and sometimes overlapping images of international relations and security in Europe.

Table 11.2

CDE Negotiations in Stockholm

· *Formal Conference Title:* "Conference on Confidence- and Security-Building Measures and Disarmament in Europe" (CDE)
· *Participants:* 35 Western, Eastern, and neutral/nonaligned states
· *Conference Scope:* All of Europe ("Atlantic to the Urals")

Confidence-Building Measures (CBMs) in Helsinki Final Act (1975)
· Notification 21 days or more in advance of major military maneuvers exceeding 25,000 troops on the territory in Europe of any participating state as well as the adjoining sea and air space.
· Voluntary prior notification of other military maneuvers.
· Voluntary exchange of observers of military maneuvers.

Western Proposals for Expanding CSBMs
1. Military information exchange on armed forces units.
2. Annual notification of planned military exercises.
3. Forty-five-day advance notifications of out-of-garrison land activities of division-level units and larger.
4. Mutual observation of exercises.
5. Inspection of activities states suspect were not notified but should have been.
6. Improvement of communications means and procedures between the two sides.

Eastern Proposals for CSBMs
1. Nonuse of force treaty.
2. Nuclear no-first-use pledge.
3. Ban chemical weapons.
4. Expand CSBMs.

Concluding Agreements (1986)
· Six weeks advance, detailed notice of all land exercises and troop concentrations involving at least 13,000 troops or 300 tanks (numbers of troops, tanks, and artillery as well as location and purpose of exercise must be announced); however, readiness exercises may be conducted without notice, but foreign observers must be invited if they involve more than 13,000 troops for longer than 72 hours.
· Observers must be invited to maneuvers with 17,000 or more troops; sea and air exercises are not included.
· Up to three no-notice inspections a year of military activity any signatory suspects should have been notified; all such inspections can be conducted on land or in the air with the inspected country providing the aircraft.
· Exercises involving at least 75,000 troops must be announced two years in advance; those with 40,000–75,000 troops are subject to one year notice.

AN EAST-WEST IMAGE: THE WARSAW PACT VERSUS NATO

The dominant view, particularly among American policymakers, is that security in Europe is based on a balance of military forces between East and West. Thus, the US and other Western governments have resisted European peace movements that oppose nuclear weapons and other military deployments to the extent that these efforts could undermine the East-West military balance. Both East and West fear the encroachment by the other that might occur in the absence of countervailing power. Peace is to be maintained by posing an obstacle to hegemony by the other side. Both alliances combine the resources of member states, effectively maintaining a balance of power. The West, following George Kennan's classic prescription, aims to contain Soviet expansionism.[4]

Delayed entry into two world wars is evidence of America's traditional desire to stay out of entangling European alliances.[5] Nevertheless, two world wars and recognition that technology had reduced the security once provided by ocean frontiers led to a major revision in American national security and foreign policy. Security in Europe had become central to the security of the United States, hence the very large American commitment of resources and manpower to the NATO Alliance. Isolation from Europe and the rest of the world is no longer a serious option. Maintaining the security of the West, given the Yalta and Potsdam agreements and the postwar division across Europe between East and West, has been the dominant theme in American thinking since World War II.

Today, two alliances face each other—the North Atlantic and Warsaw Treaty Organizations—in a stalemate called peace. In the East-West image, this balance of power between East and West maintains the status quo in Europe. The "extended deterrence" of an American security umbrella over the West persists in the face of Soviet preeminence in the East. Indeed, the foreign and domestic policies of European states are directly influenced by the orientations and climate of relations between the United States and the Soviet Union.

According to this East-West image, force reductions and confidence- and security-building measures are justified only to the extent that a balance between East and West can be maintained. The stability of that balance is the principal arms control criterion, although East and West may view the requirements of stability in different ways. Confidence- and security-building measures are useful only if they reduce East-West tensions and if they are not the vehicle either for deception or penetration of one side by the other. In negotiations on reducing defense expenditures and manpower, each side is extremely careful to maintain its position in the East-West balance lest it be exposed to the threat of hegemony by the other.

THE EAST-WEST BALANCE

Measuring this balance in military terms or as an overall "correlation of forces"—the preferred Soviet term that includes military, social, economic, and other factors—is difficult. Using the concept of NATO and Warsaw Pact "triads"— the three-part analytical division of military forces into (1) central strategic nuclear, (2) European-based or "theater" nuclear, and (3) conventional or nonnuclear categories—is one way to assess the military balance between NATO and the Warsaw Pact. Table 11.3 summarizes measures of these forces organized according to this three-part division. To be sure, merely counting forces or units—so-called static measures—is not a sufficient indicator of military capabilities. At best it is a first estimate. A "dynamic" analysis would consider likely performance in combat, based on geography, logistics, warning time, speed of mobilization, weather, technology, training, readiness, reliability, quality and flexibility of command, and the ability to sustain combat operations over time. Nevertheless, the static balance or numbers of forces or units on each side is typically the focus of arms control negotiations.

STRATEGIC NUCLEAR FORCES

The first component of the NATO triad is the strategic nuclear leg composed of another "triad"—the US strategic triad of submarine-launched ballistic missiles (SLBMs), land-based bombers, and land-based intercontinental ballistic missiles (ICBMs). The UK has its own submarine-based nuclear deterrent. France has not been integrated militarily with the NATO Alliance since 1966, but it maintains its own triad of land-based bombers, intermediate-range ballistic missiles (IRBMs), and SLBMs—forces based on French soil or operating from French ports. French military forces are not part of the Alliance's integrated structure, but French nuclear forces constitute a substantial contribution to NATO's deterrent posture. Although these British and French nuclear systems could be counted as part of NATO's central strategic

forces, they are shown in table 3 as part of the second component of the NATO triad—European-based or so-called theater nuclear forces (TNF).

Indeed, in bilateral talks between the United States and the Soviet Union on intermediate-range nuclear forces, the Soviets initially argued for inclusion of British and French systems, while the United States and its allies resisted such pressures. The United States consistently refused to negotiate on the nuclear weapons systems belonging to third countries because such systems are not under US control.[6] That same argument was applied to other theater nuclear systems (such as the FRG's Pershing I missiles) whose launchers are under allied control even though the warheads remain under US control. During the INF negotiations in 1987, the United States let the West Germans decide whether they would comply with the Soviet demand that they would relinquish their Pershings. Indeed, the West Germans decided that they would not disband their Pershing I units prematurely and that they would do so only after the United States and the Soviet Union actually had eliminated their own intermediate-range nuclear missile forces.

THEATER NUCLEAR FORCES (TNF)

The European theater nuclear forces (TNF) shown in table 11.3 are composed of all British and French nuclear systems as well as American and other allied artillery, short- and intermediate-range missiles, and carrier- and land-based, air-delivered systems. Many are "dual capable," possessing either nuclear or conventional ordnance. Nuclear weapons are to be found with local or allied ground and tactical air forces in France, the United Kingdom, the Federal Republic of Germany, Holland, Belgium, Italy, Greece, and Turkey; however, the nuclear warheads themselves are under exclusive American, British, or French control. Similarly, the Soviet Union retains control of all nuclear weapons deployed with Warsaw Pact units.[7]

There are several categories of theater nuclear forces whose distinctions have begun to figure prominently in nuclear arms control negotiations. TNF refers to *all* nuclear forces in the European theater of whatever range, whereas INF refers only to those of intermediate range. Intermediate-range nuclear forces (INF) encompass systems with ranges between 500 and 5500 kilometers (300 and 3300) miles), beyond which is the category of intercontinental strategic forces. Within INF, however, are the two categories of long-range INF (LRINF)— above 1000 kilometers (600 miles)—and short-range INF (SRINF)—between 500 and 1000 kilometers (300 and 600 miles). The distinction between LRINF and SRINF was significant

Table 11.3
NATO and Warsaw Pact (WTO) Triads

Central Strategic Forces
(British and French nuclear forces included below)

	Launchers		Warheads	
	US	USSR	US	USSR
ICBM	1,000	1,418	2,450	6,657
SLBM	608	942	6,208	3,806
Bombers	360	195	5,872	1,940
Totals	1,967	2,588	14,530	12,403

European-based/Theatre Nuclear Forces (TNF)
(excludes Soviet ICBMs counted above)

	NATO	WTO
Land-based, intermediate-range (INF) missiles:		
(500–5500 km range)[a]	262	1,122
Land-based, short-range missiles:		
(less than 500 km range)[b]	125	849
Artillery pieces (nuclear-capable)	5,209	5,517
Land-based aircraft (nuclear-capable)	2,190	2,299

European-based/Theater Conventional (Nonnuclear) Forces

	Central Europe		Total in Europe	
	NATO	WTO	NATO	WTO
Active duty ground forces (thousands)	793	975	2,243	2,317
Reserve ground forces (thousands)	969	920	4,136	3,908
Peacetime divisions	29⅔	50⅓	94⅔	114
Total wartime divisions	40⅔	68⅓	135⅔	220⅔
Main battle tanks	13,100	19,800	21,900	58,500
Artillery, multiple rocket launchers	6,100	14,000	18,100	49,600
Land-based bombers and attack aircraft	1,010	1,140	3,210	2,510
Land-based interceptors/fighter aircraft	340	1,230	1,200	4,240
Sea-based bombers and attack aircraft			650	360
Combatant ships: European and Atlantic waters				
Carriers			13	2
Cruisers, destroyers, and frigates			354	209
Submarines			185	191

[a]Deployments prior to implementation of INF Treaty: see table 11.4 for list by INF missile type and warheads.
[b]See table 11.4 for list by missile type and warheads.
Sources: International Institute for Strategic Studies (London).

in INF Treaty negotiations. For NATO, the LRINF category encompassed systems (such as Pershing II or ground-launched cruise missiles) that, if launched from Western Europe, could strike targets in the Soviet Union. On the other hand, NATO's SRINF systems (such as Pershing I) could not reach the Soviet Union, although they held targets in Eastern Europe at risk. With respect to Warsaw Pact forces, the USSR's SS-20 and SS-4 were categorized as LRINF missiles while the SS-22 and SS-23 missiles were SRINF.

Below INF—systems with ranges below 500 kilometers (300 miles)—are short-range nuclear forces (SNF), often called "battlefield" or "tactical" nuclear weapons. For the most part these systems have ranges measured in tens of kilometers (such as nuclear-capable artillery). The SNF category also includes short-range ballistic

missiles (SRBMs) such as the Soviet SS-21 or American Lance missiles capable of reaching targets beyond the forward battle area. The abbreviations and categories of weapons systems can be confusing, but the distinctions have both strategic and political significance in the context of current and future arms control efforts. Table 11.4 provides an overview of NATO and Warsaw Pact land-based missile deployments within these categories (LRINF, SRINF, SRBMs) prior to reductions under the INF Treaty.

INF TREATY NEGOTIATIONS: A REVIEW

In 1979 NATO called for a direct response to its lack of long-range INF missiles, a commitment underscored by new Soviet deployments of SS-20 long-range INF mobile missiles (added to the existing stocks of older SS-4 and SS-5

Table 11.4

Surface-to-Surface Nuclear Missile Deployments in Europe (Atlantic to Urals)

Category	WARSAW PACT		NATO	
	System	Launchers/ Warheads	System	Launchers/ Warheads
Long-range INF (LRINF)[a] (1,000–5,500 kilometer range)	SS-20	270/810	Pershing II	108/108
	SS-4	112/112	Cruise (GLCM) SSBS-S3 (France)	64/256[b] 18/18
Short-range INF (SRINF)[a] (500–1,000 kilometer range)	SS-12/22 Scud/SS-23	90/90[c] 863/863[e]	Pershing I-A	72/72[d]
Short-range nuclear missiles (Less than 500 kilometer range)	Frog/SS-21[f] Sepal/SS-C-1b	1158/1158 40/40	Lance Pluton (France)	123/123 32/32

[a] To be removed pursuant to INF Treaty
[b] Total deployment by end of 1980s was to have been 464 launchers/1856 warheads.
[c] SS-22 as replacement for SS-12
[d] West German Pershing I-A to be withdrawn after US and USSR have withdrawn all INF missiles.
[e] SS-23 as replacement for "Scuds."
[f] SS-21 as replacement for "Frogs."
Source: International Institute for Strategic Studies and Royal United Services Institute (London).

LRINF missiles). In December 1983, in the absence of progress in US-Soviet arms control negotiations, deployment began under NATO auspices of American ground-launched cruise missile (GLCM) units in the United Kingdom, Italy, West Germany, Belgium, and Holland and Pershing II ballistic missiles in West Germany. According to the deployment schedule, by the end of the 1980s NATO was to have 108 Pershing IIs in West Germany and 464 GLCMs distributed among the five deployment countries. As mentioned earlier, the United States and the Soviet Union agreed in 1987 to eliminate *all* (both short- and long-range) land-based INF missiles. As part of the arrangement, West Germany also agreed to relinquish its 72 Pershing I units, whose nuclear warheads have been under American control.

Because both the Pershing II and GLCMs could strike targets in the western USSR, some Europeans have noted the coupling effects these US-controlled forces have had.[8] In the event of Warsaw Pact aggression leading to a US nuclear response with European-based INF missiles, it would be unlikely that the Soviets would differentiate between a nuclear attack against the Soviet Union launched by American units in Europe and one emanating from the United States or from US forces at sea. Thus, a nuclear response launched by American units in Europe would have the effect of committing the full force of US central strategic units. Given this coupling of American central strategic and theater nuclear forces to the defense of Europe, the Warsaw Pact would be deterred from ever undertaking aggression against the West in the first place. From this point of view, theater nuclear force modernization was important not because these weapons were ever intended for actual use, but rather because of their increased deterrent effect against the Soviet Union and its Warsaw Pact allies.

Some Europeans have argued that removal of INF missiles weakens deterrence because it undermines this coupling effect. The American response to this argument is that the US guarantee is still assured by the continuing presence in Europe of US ground forces, land-based aircraft capable of striking targets in the USSR with either nuclear or conventional weapons (dual-capable aircraft), as well as the continuing threat posed by sea-based systems.

Because the INF missile issue was the dominant European arms control concern throughout the 1980s, a brief review of the INF deployment–arms control, "two-track" approach may be useful. Recognizing the imbalance caused by Soviet SS-20 deployments, NATO members agreed in 1979 on a so-called two-track approach—preparing to deploy long-range INF missiles of their own while at the same time having the superpowers conduct arms control negotiations on the issue. Were NATO not to modernize its theater nuclear forces, proponents argued, it would concede an advantage to the Warsaw Pact in two of the three triad components, theater nuclear and conventional, thus weakening the credibility of the NATO deterrent and potentially undermining NATO's defensive posture. On the other hand, meaningful progress in these talks would either eliminate the need for deployments altogether or at least allow for deployments at lower levels.

US-Soviet INF negotiations that began in Geneva in 1981 quickly reached an impasse. The United States offered a "zero" option (often referred to at the time as a "zero-zero" option—no US or NATO INF deployments would be made if the USSR were to remove and dismantle its SS-20 and older SS-4 and SS-5 missiles targeted against Western Europe. The Soviets indicated they would agree to reduce the number of their missiles to 162, the number of British and French ICBMs and SLBMs, thus precluding any US INF deployments. The United States did not accept this linkage to allied forces outside of American control. Even if they had, the 162 number would have left the Soviets with a considerable warhead advantage since each SS-20 carried three independently targetable reentry vehicles, while in the early 1980s many of the 162 British and French missiles carried only single warheads. (British and French modernization plans going well into the 1990s do call for a substantial increase in the number of warheads per launcher, mainly on ballistic missiles carried by submarines.)

In a celebrated "walk in the woods" just outside Geneva in July 1982, the US and Soviet ambassadors tried to find a middle ground between the extremes of the zero option and 162 missiles.[9] In the tentative formula they worked out, each side would be limited to 75 launchers; the Soviets were to reduce to this number while the United States was to be allowed to build up to it. Because each GLCM launcher carried four single-warhead missiles while each SS-20 missile carried three warheads, this would have left the United States with a 300:225 warhead advantage. On the other hand, the United States was to be allowed to field only its slower-moving, air-breathing GLCMs, not its Pershing II, a ballistic missile with a short flight time to targets in the USSR and with sufficient accuracy to give it capability against underground and other "hardened" targets (a so-called hard target kill capability). Because of these time and accuracy characteristics, the Pershing II was considered to be more of a threat to the Soviets than the GLCM.

As it turned out, neither ambassador's home government accepted this compromise as a basis for agreement. Notwithstanding a warhead advantage in the US favor, American officials objected to allowing the Soviet SS-20 ballistic missile deployment, while denying the US any LRINF ballistic missiles. There was also objection on the American side (echoed by US allies in the Far East) to allowing the Soviet Union to maintain some SS-20s east of the Urals in the Asian part of the USSR. As for the Soviets, the idea of reducing forces already deployed while allowing the West to build up its missile force from zero and give it an INF warhead advantage proved unpalatable. Moreover, there was undoubtedly some hope in Moscow that domestic political pressures and differences among the NATO allies might lead to an unraveling of their

commitment to deploy LRINF missiles. In any event, negotiations were suspended in December 1983 when the United States began deploying GLCMs in the UK and Italy and Pershing IIs in West Germany. Given the lack of progress in arms control talks, a decision by each country to go ahead with planned INF deployments had become a political litmus test of NATO's solidarity. The deploying countries met the test, and US-Soviet negotiations on INF resumed in 1985. Along with the zero option and the proposal to reduce Soviet INF missiles to the number of British and French missiles, the "walk in the woods" formula became yet another yardstick for comparing and measuring proposals raised by either side in subsequent negotiations.

The most dramatic of these was the Soviet proposal in 1987 for elimination of all US and Soviet land-based INF missiles. The term zero-zero, originally referring to zero LRINF missiles on both sides, had assumed a new, expanded meaning—zero LRINF and zero SRINF missiles for both sides. Indeed, since Soviet leader Mikhail Gorbachev's accession to power in 1985, negotiations on INF witnessed a series of Soviet concessions, bringing the Soviet position more into line with Western preferences.

In January 1986, for example, Gorbachev conceded that French and British systems need not be reduced. At the same time, however, he argued that their deployments should be frozen and that planned modernization of these systems should be abandoned. Although the agreement would apply to US and Soviet LRINF missiles in Europe, the Soviets were to retain their SS-20s in Asia. When Gorbachev and President Reagan met at Reykjavik, Iceland, in October 1986, both moved toward a formula that would have eliminated all US and Soviet LRINF missiles from Europe, allowing the two countries to retain 100 LRINF missiles on home territories, but outside of Europe. British and French systems would be unaffected and, for its part, the Soviet Union would deploy its remaining LRINF arsenal east of the Urals, out of range of targets in Western Europe. Although this represented a greater degree of agreement between the parties than had previously been achieved, Gorbachev held agreement on INF hostage to US acceptance of constraints on the Reagan administration's Strategic Defense Initiative (SDI), concessions President Reagan was not about to make. In time, however, Gorbachev dropped the linkage between INF and SDI, declaring in July 1987 that the USSR would accept a "global zero" on both LRINF and SRINF missiles.

The prospect of a zero LRINF agreement had stimulated ambivalent reactions, particularly among some of the NATO allies. Given the controversy over the INF deployment in the early

1980s, considerable political capital had been invested in securing approval for the deployments in West Germany, Britain, Italy, Belgium, and Holland that began in December 1983. Despite the fact that NATO had endorsed the zero option for LRINF in 1981, many pointed to the increased Soviet deployments of SRINF missiles in Eastern Europe. Indeed, the SRINF missiles covered many of the same targets as SS-20s, missiles that would be withdrawn under a zero LRINF agreement. The United States, by contrast, had no remaining SRINF missiles in Europe (having replaced in 1983 and 1984 its Pershing I SRINF missiles with the LRINF Pershing II). The FRG's Pershing I missiles, equipped with US-controlled nuclear warheads, were aging. Moreover, in agreeing on zero LRINF, NATO was giving up a significant capability to strike Soviet territory from Western Europe. As a result of these considerations, the US and its NATO allies initially insisted that a zero LRINF accord also incorporate constraints on SRINF.[10]

During Secretary of State Shultz's trip to Moscow in March 1987, Gorbachev offered to expand the scope of an INF agreement to encompass both LRINF and SRINF land-based missiles (the so-called double zero). He also called for further negotiations on "operational-tactical nuclear systems of shorter ranges." NATO accepted the INF double zero in June, but insisted that further negotiations on nuclear systems in Europe be deferred pending the elimination of imbalances in conventional and chemical weapons in which the Soviets have had a substantial numerical superiority.[11]

A double-zero INF regime offers substantial advantages to NATO. On a political level, Soviet agreement provides evidence of the success of NATO's earlier two-track approach in which NATO's collective resolve in fulfilling its LRINF deployment commitments had influenced the Soviets to agree to reduce INF systems to zero. On a military level, the Soviet Union is dismantling substantially more missiles—not only its LRINF force of over 100 remaining SS-4s and the 441 mobile SS-20s with three warheads each, but also several hundred SRINF missiles. Aside from the West German Pershing I, NATO is relinquishing only 572 Pershing II and ground-launched cruise missiles, some of which would not have been deployed and fully operational until late in the decade.[12]

The US did rebuff Soviet attempts to force inclusion of the FRG's 72 Pershing I missiles into a bilateral US-Soviet INF agreement, but the FRG—faced with substantial domestic pressure—offered to facilitate conclusion of an INF agreement by dismantling unilaterally its Pershing I missiles after the US and USSR have implemented the agreement and eliminated their INF missiles.[13] Once the issue of German

Pershing I missiles was resolved, the remaining problem facing negotiators was an agreement on verification measures. To the surprise of many, the two sides established a regime allowing for on-site inspections not only of missile destruction but also of production facilities in both countries. Some inspections were to be scheduled, but others could be "challenge" or short-notice inspections.

The debate within NATO over "double zero" reflects an even deeper concern—how the future arms control agenda should proceed. Anticipating Soviet pressure for "triple zero"—encompassing short-range ballistic missiles (SRBMs), if not all short-range nuclear forces (SNF) below 500 kilometers (300 miles)—NATO members revealed their worry that Gorbachev's penchant for surprise proposals would generate political pressures for the denuclearization of Europe. Zero-LRINF alone had already stimulated exaggerated speculation that US nuclear power was being decoupled from Western Europe.[14] Some argued that successive "zeros" in nuclear arms control might strip NATO of the nuclear response capability that was fundamental to its two-decades-old strategy of flexible response. Many feared that, if the Soviets succeeded in removing NATO's nuclear capability, the Soviet superiority in conventional forces would create unprecedented opportunities for coercion and intimidation. Thus, as dramatic nuclear reductions began to look like a real possibility, achieving a balance in conventional forces assumed greater urgency.

THE CONVENTIONAL EAST-WEST BALANCE

The third component of military forces in Europe consists of conventional (nonnuclear) ground, air, and naval forces. NATO's ground forces are concentrated along the "central front" adjacent to the bulk of Warsaw Pact units and, should war occur, opposite the likely attack routes from East Germany and Czechoslovakia. Air forces committed to NATO are deployed throughout Western Europe. In addition, France maintains air and ground forces deployed at home and in West Germany opposite the French province of Lorraine and across the Rhine from the French province of Alsace. NATO naval forces are to be found primarily in the Mediterranean and the North Atlantic, important sea lines of communication that would be used to send forces and supplies to Europe from the United States and Canada.

The Soviet presence in the Mediterranean has increased substantially in recent years, but the presence of the US Sixth Fleet and other allied naval units leaves NATO in a relatively strong position in naval and naval air forces. Both alliances concentrate their ground and air forces in the central region and are less well equipped in ground and ground-based air forces in either the southern or northern regions. The Soviet Union and its allies have the geographic and logistical advantage of both shorter and contiguous land lines of communication. Sending reinforcements and supplies to the front is far easier for the USSR than for the United States, since US cargo aircraft and ships must cross the Atlantic Ocean. The American wartime commitment to field ten divisions in ten days through massive air and sea lift, drawing on supplies in storage or pre-positioned on European soil, is a major undertaking, even in peacetime. In wartime, ships crossing the Atlantic would be vulnerable to submarine attack, and cargo and troop transport aircraft would likely face hostile fire in many landing areas. Moreover, moving troops and supplies over land would pose major problems for both sides. Although shallow rear areas are a major NATO problem, the "unsinkable aircraft carrier" offered by UK territory and the strategic depth offered by France and Spain, assuming they joined their NATO allies in the fray, are compensating factors.

If one compares the NATO triad to a similar Warsaw Pact triad, one discovers marked asymmetries in types, numbers, quality, and readiness of forces. Nevertheless, most observers would acknowledge rough parity or essential equivalence at the strategic nuclear level, Soviet numerical superiority in theater nuclear forces, and a substantial Warsaw Pact superiority in terms of the mass of ground and air forces that the USSR and its allies could bring to bear, particularly on the central front. Whatever arguments one makes about the quality of Western equipment and training, most would concede that the preponderance of nonnuclear forces in the East translates into a decided Warsaw Pact advantage in these conventional forces. There is also considerable allied concern over NATO's inferior offensive and defensive chemical warfare capabilities compared to those of the Warsaw Pact.

The overall balance of military capabilities, if not equivalent in all respects, does not necessarily translate, however, into the kind of advantage either side would need to start a war against the other. Maintaining this overall balance, while reducing those asymmetries that are potentially destabilizing, is seen by many as an essential, minimum objective for any arms control negotiations.

MBFR, CFE, AND THE EAST-WEST CONVENTIONAL FORCE BALANCE

Dispute in the early 1970s in Vienna over whether conventional force reductions should be equal or balanced illustrates this point. Given its preponderance of mass or numbers in most categories of conventional forces, the East favored mutual force reductions (MFR) equal in numbers whereas the West held out for mutual and *balanced* force reductions (MBFR). Equal reductions from the Western view would only increase the existing conventional force imbalance.

NATO's position is that its theater-nuclear forces are necessary precisely because of this imbalance. Indeed, the deterrent threat of a NATO "first-use" nuclear response is designed to offset the inability of the Western alliance to defend for very long against a full Warsaw Pact invasion with conventional forces alone.

If the ultimate aim were to reduce the current NATO reliance on nuclear weapons, the West has two broad alternatives. One is to rearm so as to match Eastern deployments of conventional forces or at least to provide a credible conventional defense—an expensive and politically difficult alternative that could lead to a new East-West arms race and, some argue, destabilize existing deterrence relations between East and West.

Another approach is to achieve a reduction in the conventional force threat from the East through negotiations. The East would have to reduce more since its starting position numerically is substantially greater. To achieve an overall balance in conventional forces at lower levels, the Soviet Union and other Warsaw Pact countries would have to make greater reductions than the West, not just in absolute numbers, but also *proportionately* greater. After lengthy negotiations in MBFR talks, East and West did agree on the goal of "balanced" reductions to a common ceiling of 900,000 ground and air force personnel with a sublimit of 700,000 ground forces.[15]

Even with this agreed goal, however, negotiations were stalemated over the so-called data problem, with both sides disagreeing on the numbers of forces presently stationed in the central front, the obvious starting point for negotiating reductions. Western estimates put the Warsaw Pact well ahead with just under 1.2 million air and ground troops in the central region compared to 990,000 for NATO. Within this total, the Warsaw Pact was also given an edge of 960,000 ground force personnel to less than 800,000 for NATO. The Warsaw Pact would only admit to 800,000 troops, substantially less than Western estimates. Indeed, the Warsaw Pact countries claimed that the West had under-counted NATO military forces, a problem compounded by the extensive Western use of civilians for military support purposes, functions still performed largely by military personnel in the Warsaw Pact countries.[16]

Seemingly endless discussions over several years failed to resolve this data problem, leading negotiators to put it aside for the time being.

Table 11.5

The MBFR Negotiations in Vienna (1973–1989)

Area of reductions: Territory of Belgium, Netherlands, Luxembourg, Federal Republic of Germany, German Democratic Republic, Poland, and Czechoslovakia (but *not* Hungary).

Direct participants: 11 states with forces in central Europe.
West: US, UK, FRG, Canada, Belgium, the Netherlands, and Luxembourg but *not* France.
East: USSR, Poland, GDR, Czechoslovakia.

Special participants: 8 states with no forces in central Europe.
West: Denmark, Greece, Italy, Norway, Turkey.
East: Bulgaria, Hungary, Romania.

The "Data Problem"

Forces	Figure source	Total	Ground	Air
Warsaw Pact	Warsaw Pact	980,000	800,000	180,000
Warsaw Pact	NATO	1,190,000	960,000	230,000
NATO	NATO	990,000	790,000	200,000

MBFR "Associated Measures": Western Proposals

1. Advance notification of out-of-garrison activity by one or more division-size formations (except for alert activities that would be announced at onset).
2. Each side would allow representatives from the other to observe these out-of-garrison activities.
3. Advance notification by direct participants outside of the reduction area of major military movements of ground forces into the reduction area.
4. Quota of annual inspections (by each side) of the reduction area by air, ground, or both.
5. Observers would be stationed at permanent exit and entry points to monitor military movements into and out of the reduction area.
6. Periodic information exchanges on the reduction area concerning personnel strength, organization of forces, and forces to be withdrawn.
7. No interference by either side with national, technical means of verification.

One alternative to counting personnel is counting equipment such as tanks or aircraft, but this idea is fraught with its own difficulties. Some argued that, instead of counting troops or equipment, counting units—divisions, regiments, and brigades—might resolve the data problem, even though the numbers of soldiers assigned to these units also differ substantially. Later sessions focused on negotiating an initial reduction of relatively small proportions just to get the process started. Thus, in a relatively modest move, the US proposed to withdraw 5000 troops if the Soviets would withdraw 11,500. "Associated measures" to an MBFR agreement included establishment of observation posts and notification procedures for normal rotations (withdrawals and replacements). Verifying withdrawals and monitoring the subsequent freeze on personnel numbers through such measures would be a necessary part of any reductions regime, particularly if reductions were to be greater than the modest numbers proposed in this initial step.

Despite the fact that MBFR was a protracted and largely fruitless negotiating effort, conventional arms control has assumed increasing importance to the West. Europeans—and Germans in particular—have long argued that advances in nuclear arms control increased the significance of the imbalance between East and West in conventional forces. In 1986, the NATO allies agreed to begin new negotiations with members of the Warsaw Pact on conventional forces—now referred to as the "Conventional Armed Forces in Europe" (CFE) talks—not merely with regard to central Europe (the scope of MBFR), but with a broader "Atlantic to the Urals" scope. In addition, the NATO allies agreed to negotiate on improved CSBMs within the context of the Stockholm CDE agreement and the CSCE process as a whole.[17] Both negotiations began in earnest in March 1989, with both East and West pushing for early agreement on military, as well as politically, significant reductions.

The process of substantial reductions in the nuclear arsenals of both superpowers has given new impetus to Western thinking on conventional armaments. Rather than focusing only on manpower levels, CFE negotiations have aimed at reducing the capacity of conventional forces to launch offensive operations. From the West's perspective, the objective is to reduce the Warsaw Pact's capability to mount a potentially decisive surprise attack against NATO.[18] The target of such negotiations is not Soviet manpower as such, but rather the elements of Soviet military power (armor, artillery, armored combat vehicles, etc.) that are forward-deployed in Eastern Europe and that are essential to combined arms, offensive operations. The issue is thus more than just reducing force levels. It is also a question of relocating critical types of military forces so that any offensive against NATO would have to be preceded by substantial mobilization and force integration efforts—activities that would provide increased warning time to NATO.

To the surprise of many, the Warsaw Pact has accepted the need to redress military imbalances through CFE and has indicated its willingness to take large and asymmetrical force reductions to a common and equal ceiling below current NATO force levels. Opening proposals in CFE by both alliances revealed differences in data bases and equipment counting rules, but also striking similarity in each alliance's approach to regulating ground forces. Nonetheless, Soviet desires to incorporate nuclear-capable weapons systems—particularly strike aircraft—clash with desires within NATO to preserve its nuclear response capability, especially in the wake of INF reductions. While NATO seeks to reduce Soviet forces forward-deployed in Eastern Europe, the United States and its allies seek to maintain US military presence in Western Europe. An agreement that emasculated both the Soviet military presence in Eastern Europe and the US military presence in Western Europe would raise fundamental political issues about the nature of the future European security order. As such, the arms control process involves not only an East-West image of security, but also alternative images as well.

AN ALTERNATIVE IMAGE OF SECURITY IN EUROPE: CENTER VERSUS PERIPHERY

Through a somewhat different set of lenses, peace in Europe is maintained not only by keeping the Soviet Union at bay, but also by keeping Germany constrained. The passage of time may have lessened this concern, particularly among younger generations. Nevertheless, it is a view with deep historical and geopolitical roots which reflects concern about maintaining a balance between center and periphery on the European continent. In the past, this view has been expressed frequently and openly by Soviet leaders much to the dismay of the West Germans, but concern about the German center is by no means confined to the Soviets. It is a worry held even by some of West Germany's allies, although diplomatic concerns for favorable allied relations usually preclude such open admissions. Indeed, fear of Germany was a principal motive for formation under the 1948 Brussels Treaty of the Western Union—later renamed the Western European Union (WEU)—by Belgium, France, Luxembourg, the Netherlands, and the United Kingdom. It was a motive for

keeping Germany divided and thus weaker than it would otherwise be. Even though East Germany became a full member of the Warsaw Pact and West Germany was admitted to both NATO and the WEU in 1955 at the formal end of the postwar occupation, European fear of an overly strong Germany has remained. Quite apart from common concerns with the Soviet and Warsaw Pact threat further to the East, one reason for Dutch, Belgian, and French membership in NATO—countries on the German periphery— is the security derived from alliance with their former adversary. The sense or degree of security derived by East European members of the Warsaw Pact from their alliance with East Germany is difficult to gauge. Although the number of divisions is to be reduced in the 1990s, some interpret Soviet stationing of 19 divisions in East Germany not just for allied defense against a NATO threat from the West, but also as a demonstration of Moscow's continuing effort to keep both East and West Germany in check.

HISTORICAL LEGACY

Keeping the geopolitical center of Europe at bay as the key to maintaining peace is hardly a new idea. Although it has earlier roots, the nineteenth- and twentieth-century historical experience has particular relevance to understanding this image of European security. Because this center-periphery image is less familiar these days than the East-West view that dominates most American thinking on European security questions, a brief review of its historical legacy seems in order.

Following the Napoleonic wars and the defeat of France, statesmen met in 1815 in Vienna to reestablish a balance conducive to maintaining peace and avoiding another conflagration. In a stunning feat of diplomacy, France's position in Europe was restored with territorial borders even slightly larger than they had been in 1789 when the revolution began and the old order was overthrown.

Such favorable treatment of France was by no means due to charitable feelings by her victorious conquerors. British and Russian diplomats understood that too strong a center in Europe (Prussia, the German states, and the Austrian empire) was not in their interest. The security interest of Britain and Russia, powers on the periphery of the European continent, would be served by restoring the territorial integrity of France as a counterweight to Austria and Prussia—powers in the European center. The concern was for maintaining a balance of power, not in East-West terms, but rather one between center and periphery. Indeed, another general war was avoided for 99 years, although numerous smaller wars were fought.

Failure to maintain the peace and the breakdown that occurred in 1914 have been explained by many in these geopolitical terms: the European center had become too strong. Unification of the German states had followed the defeat of France in the Franco-Prussian War of 1870. The industrial revolution transformed Germany's economy, making it possible for her to build a strong army—a threat to France and Russia—and a navy that challenged the British fleet. The subsequent alliance of the European center—from Germany in the north across Austria and Hungary to Turkey in the southeast— countered by an alliance of states on the European periphery—Britain, France, and Russia—provided the battle lines of World War I, with the United States finally joining the Allies against the Central Powers in 1917.[19] In 1918, the new Bolshevik government made its peace with the center in the Treaty of Brest-Litovsk, an accord that enabled Germany to concentrate its remaining efforts against states on the Western periphery.

In the Versailles negotiations the victorious states on the German-Austrian periphery sought to weaken the center militarily and economically to prevent the German-speaking countries from ever posing a new threat to the peace. The peripheral powers, now joined by the United States, agreed to construct a new balance in which the center was to be kept at bay. If too strong a center had upset the balance of power, then a weak center would be maintained. The Austrian and Ottoman empires were broken up and, although a unified Germany was allowed to exist, the country was demilitarized with the expectation of nullifying it as a threat to the security of Europe.

Those who constructed the post–World War I order did not anticipate that a rearmed Germany would unite with Austria and Italy to pose yet another challenge to the periphery.[20] Germany was kept depressed throughout the 1920s when much of the rest of the world was experiencing a period of relative prosperity. The worldwide financial collapse in 1929 only exacerbated Germany's problem. Depression on top of depression created conditions conducive to extreme measures and, in Germany, politics took a sharp turn to the right.

Under National Socialism, Germany rebuilt its economy, fostered the social cohesiveness of German-speaking peoples to include the Austrians, and, contrary to provisions of the postwar settlements, rearmed. Once again, the European center had become very strong and, once again, general war was the outcome. The Soviets signed a nonaggression pact with the Germans in 1939 (allowing the latter to concentrate efforts against states to the West), but even this accommodation failed. The same alliance of the European periphery against the center emerged in 1939 soon after the outbreak of war. The United States again delayed entry into the war, but ultimately threw its weight into an alliance with the Soviet, British, and Free French governments.

After the defeat of the German Reich, the American, Soviet, British, and French armies established separate zones of occupation in Germany and Austria. Austria was reunified in 1955 by agreement among the occupying allies that the country be neutral and virtually disarmed, thus posing no threat to any of the states on its borders or beyond. No such solution worked for Germany, given both the deep distrust of Germany by countries on its periphery and Germany's pivotal position in the emerging cold war. Demilitarization of Germany had not worked following World War I, and the danger that a militarized Germany could once again pose to European security was uppermost in the minds of those statesmen tasked with ordering postwar Europe. The Soviets had lost 20 million people in World War II, and Tsarist Russia had lost more than 9 million before the Bolsheviks came to power in 1917—losses the Soviets blamed on German military aggression. Losses by the French, British, and other populations on the German periphery, as well as wartime occupation of many of these states by Germany, made them equally distrustful of any settlement that allowed for any possibility of the European center rising again. Rather than rely on a united, demilitarized Germany, an approach that had clearly failed after World War I, Germany was initially to be divided among the four wartime allies, now in occupation of the defeated Reich. For many, the imperative in 1945 was division, not unity, in the center of Europe as the key to maintaining security on the continent. In contrast to the East-West image evoked by the American George Kennan's later containment article, the British scholar, Sir Halford Mackinder, wrote of what amounted to a center-periphery balance of power.[21]

The emerging East-West conflict combined with this image of maintaining a weak center by dividing Germany between East and West. Thus, the American, British, and French zones of occupation in the west combined in 1949 to form the Federal Republic of Germany with admission to NATO in 1955. Similarly, the Soviet zone of occupation in the east became the German Democratic Republic. Only Berlin, capital of the defeated German Reich, remained under Four Power occupation, a status it retains to this day. Germany remained divided as a hedge against the resurgence of a strong center that might once again threaten the European periphery. Yet, occupation was replaced by opposing East-West alliances, and foreign military forces in both Germanys have remained in place.[22]

Maintaining a center-periphery balance was a long-established conceptual foundation for the European security order. The new postwar focus on East-West conflict did not displace center-periphery concerns, but was superimposed upon them. Both images existed simultaneously, even though the East-West image tended to dominate policymaking.

Center versus periphery is a fundamentally different view of the balance of power than that offered by the East-West image. According to center-periphery imagery, security in Europe is maintained by dividing Germany and neutralizing a virtually disarmed Austria completely separate from both East and West Germany. Even the German advocate of reunification who is sensitive to the concerns of states on Germany's periphery realizes that German unification will only be possible if the outside powers, particularly the Soviet Union, can be assured that the center will not rise again. A unified Germany in the Austrian model—virtually demilitarized and neutralized, posing no threat to the periphery, East or West—is one alternative to a divided Germany. The failure of this approach following World War I and the heavy costs paid by the periphery states for this failure account for their reluctance to try this approach once again. Adding East-West conflict to these center-periphery concerns makes German reunification even more difficult to achieve. Thus, attempts in the late 1950s to negotiate a disengagement of allied forces in Germany to fulfill allied promises of reunifying Germany and to defuse the cold war in Europe were unsuccessful. For many, resolving East-West conflict in Europe and achieving reunification of Germany are integrally related to finding another solution to the center-periphery issue.[23]

ARMS LIMITATIONS AS A MEANS OF CONSTRAINING THE CENTER

Beyond force reductions, confidence-building measures, and related questions, the MBFR, CDE, and CFE talks really have been addressing the political order in Europe (whether present arrangements and spheres of influence are to be maintained or revised), and this is of significance to negotiators in such matters as how large a West German army (Bundeswehr) will be allowed in any reductions program. If keeping a check on the European center (and West Germany in particular) were not such a Soviet concern, the size of the Bundeswehr would not have been emphasized so much by Soviet negotiators in MBFR talks. Whenever the issue arose, the Soviets insisted that the US not draw down its forces in West Germany without similar reductions in the size of the Bundeswehr. This obsession cannot be explained purely in East-West terms. Instead it reflects center-

periphery concerns or, more specifically, the periphery keeping the center at bay.

The Soviets, and initially the French, voiced strenuous opposition to German rearmament in the 1950s. Washington had proposed rearmament, justifying its position largely in East-West terms—the necessity for a West German defense contribution to NATO. The armed forces that ultimately emerged were limited to just under half a million in size with no long-range deployment capability outside the central front. Moreover, West Germany cannot acquire an independent nuclear capability. All weapons in nuclear-equipped units remain under US control until released in wartime by American authorities.

The WEU was formed in 1948 as a fifty-year alliance by the UK, France, Belgium, the Netherlands, and Luxembourg to counter any future threat of German militarism. The focus changed, of course, when the Federal Republic of Germany was admitted in 1955 as a full-fledged member. Although the new focus was on fostering West European defense cooperation in an East-West context, the WEU remained the vehicle by which specific limits on the size and nature of West German military forces and weapons production capability were maintained. In 1984, when the WEU terminated certain constraints on West German conventional arms development and production, the USSR protested, reminding the West of the original intent of its WEU. WEU prohibitions against German development or acquistion of nuclear weapons, however, did remain in place.

In more recent years, the French in particular have worried about the prospect of the West Germans being too accommodating toward the East. Reminiscent of the Brest-Litovsk Treaty that ended German participation in World War I against the Russians and the Hitler-Stalin Nonaggression Pact of 1939, alignment or other accommodation by the center with the East merely accentuates the threat to states on the Western part of the continent. Thus, collaboration by the French and Germans in the WEU and formation of a Franco-German brigade are, at least in part, an attempt by France to ensure that West Germany remains closely tied to the West, avoiding any drift toward the East that might tend to occur in periods of détente between East and West.

The West Germans acknowledge their constrained position in the European security equation. Recognizing external sensitivities by states on their borders as well as domestic political pressures, the West Germans have been disposed to accept constraints on their foreign and defense policies. Assured they would remain under the American nuclear umbrella, they subscribed to the nuclear nonproliferation treaty. Americans control the nuclear weapons assigned for use by German units. Not only does West Germany have no nuclear weapons under its control, thus posing no direct nuclear threat to the USSR and other states, but also nuclear-capable Pershing ballistic missiles *in German units* have not been allowed to have sufficient range to reach targets in the Soviet Union. (Only those land-based INF missiles that have been assigned to US units—the Pershing II and cruise missiles—or land-based missiles belonging to the French have had sufficient range to strike targets in the western USSR.)

Not only do the West Germans not seek their own nuclear weapons capability, but they also refrain from deploying forces in contingencies beyond their borders. Moreover, signing the 1975 Helsinki accords (which, among other things, legitimized borders in Europe as they have come to exist since World War II) was interpreted as an acceptance, at least for the time being, of a divided Germany, as well as being a renunciation of German claims to territories now within the borders of Poland and the USSR. As the Berlin Wall went down, however, German reunification emerged as an active political issue, demonstrating that the underlying structure of central Europe is not yet resolved.

At the same time, the West Germans have been unwilling to accept what they consider to be excessive security risks as Alliance members. Because most of NATO's nuclear weapons are stationed on West German soil, the FRG has insisted that other allies share the nuclear risks. Thus, the five-country, multiyear deployment of LRINF missiles that began in December 1983 in the UK, Italy, and the FRG (with plans for later deployments in Belgium and Holland) represented not only the military desire to enhance survivability of these weapons through dispersal, but also the FRG's desire that it not be the only nonnuclear NATO ally with missiles on its soil capable of hitting the Soviet Union. The zero-zero INF agreement between the US and the USSR also underscored West German sensitivity about bearing the bulk of allied nuclear risks. In particular, Germans have advocated negotiations on short-range nuclear weapons (the so-called third zero), complaining that eliminating both LRINF and SRINF missiles leaves only those nuclear weapons systems that can be launched from, and targeted on, German soil.[24] This notion that Germany bears a singular nuclear risk is a manifestation of the overlapping images of European security that involve not only East-West strategic relations, but also relations within NATO itself—between the FRG and the Alliance partners on its periphery. One cannot begin to comprehend the complexity of these security relations without some understanding of both East-West and center-periphery images of security in Europe.

A PLURALIST VIEW OF EUROPEAN SECURITY

Apart from the East-West and center-periphery views is a third view that sees the division of Europe across Germany as an unfortunate state of affairs best addressed by increasing commercial, social, and cultural contacts and exchanges. If walls exist, then it is best to cross them as often as possible, fostering transnational ties. Plans for even greater economic liberalization by the end of 1992 are consistent with this image. Reducing tensions by making East and West and their societies more interdependent is seen as making war less likely. Just as increasing commercial ties and other contacts within the European Communities (EC) in the West have led to historically unprecedented cooperation among former adversaries, so Europe as a whole, from this perspective, could benefit if such relations were allowed to develop across the entire continent.

Not surprisingly, this view is particularly popular among many Germans. As previously mentioned, official West German government statements repeatedly refer to two German states in one German nation, an indicator of a commitment, contained in the West German Basic Law or Constitution, to eventual reunification. Apart from far-off dreams, making it possible for individuals to transit the political barriers that divide East and West has been a concern of every West German government, especially since former Chancellor Willy Brandt inaugurated his *Ostpolitik* or Eastern policy in the late 1960s.

Integral to this thinking is the importance of arms control, particularly to the West Germans who see their territory as the principal battlefield in any future European war. The devastation of World War II caused by conventional, nonnuclear weapons gives the Germans little hope that they could survive any future war, be it conventional or nuclear. Profound anxiety about the likelihood of war crosses the political spectrum and is the basis for a national consensus supportive of what they see as any genuine arms control effort—whether between the superpowers in strategic arms talks or multilaterally in CFE and the earlier MBFR talks in Vienna, in the several CSCE meetings since the mid-1970s, or the CDE in Stockholm. Indeed, the CDE's first task was to devise confidence-building measures to reduce anxiety about the threat of war, building a foundation of trust that would make meaningful arms reductions possible in later stages. These confidence-building measures include rules or routines of mutual observation, notification of military maneuvers, and constraint on size and scope of military exercises. Consistent with efforts to increase commercial and cultural ties, these confidence-building measures are seen as contributing directly to security in Europe, reducing the likelihood of war. They provide a vehicle both to break down East-West divisions and to alleviate center-periphery concerns.

The Helsinki Final Act of 1975 was the culmination of two years of discussion in the CSCE on three broad areas or "baskets"—security, economic cooperation, and human rights. In return for economic and other concessions from the West, the East made certain human rights pledges. At the same time, these concessions facilitated increased commercial and cultural contacts and ties between East and West, consistent with the view that increasing interdependence or interconnectedness is conducive to peace. Over time, it is hoped, fostering common interests between the societies of East and West builds a basis for greater European unity by overcoming the ideological, historical, and national barriers to social integration.

The idea of confidence- and security-building measures (CSBMs or simply CBMs) has intellectual roots in the behavioral science and decisionmaking literature. Misperception and miscalculation, particularly in crises, are seen as significant causes of war. One remedy, then, is to improve communications. At the central strategic level, the superpowers established a hot line or emergency direct communications network in 1963 that has been improved technologically over the years. Agreements have also been made to confer so as to avoid war due to accident or miscalculation. Most recently, the superpowers have agreed to establish nuclear risk reduction centers for the routine exchange of data required by various arms control and related agreements.

Beyond these arrangements, the Helsinki Final Act mandated several measures in the security basket to include notification of exercises involving more than 25 thousand troops and provision for observation of these maneuvers. The logic was similar in some respects to the hot line and other bilateral arrangements between the superpowers—that such measures ideally reduce the uncertainty that could lead one side to misperceive even the benign acts of an adversary as offensive or aggressive.

After three years of rancor in dealing with allegations against the East on human rights violations with denials and countercharges by the East against the West, the Madrid CSCE review conference ended in 1983 with a mandate for convening a Conference on Security and Confidence-Building Measures and Disarmament in Europe (CDE)—originally a French idea posed as an alternative to MBFR with its East-West, Warsaw Pact-NATO orientation. Particularly contentious, however, was debate prior to agreement on the CDE's geographic scope. Careful diplomatic wording acknowl-edged that all of Europe (from the Atlantic to the Urals) would be subject to any measures agreed to in the CDE, but the CDE area would exclude the sea and air approaches to the continent, particularly the north Atlantic and the Mediterranean.

In coming years, multilateral negotiations on European security will likely continue to be conducted within the CSCE framework. As noted earlier, the full 35-member CSCE forum will continue work on constructing, expanding, and maintaining the regime of confidence- and security-building measures. The 23 members of the two alliances are holding separate talks on conventional forces, a task performed previously within the MBFR negotiations. The geographic scope of the CFE expands from the central front focus on MBFR to include all of Europe, from the Atlantic to the Urals. Given French policy against integrating its military forces with NATO's formal command structure, France had opposed the NATO versus Warsaw Pact format of MBFR and refused to participate. By contrast, the French are full participants in the CSCE process and in the CFE talks within that framework. Even though only members of the two competing alliances are voting participants in these conventional armed forces reductions talks, results of these efforts will be reported to the CSCE as a whole. At the same time, confidence- and security-building measures and approaches to enhancing stability are being discussed in Vienna in a separate 35-member CSCE-sponsored forum. As political change in Europe accelerates, these negotiating fora take on added importance as vehicles for managing their participants' security interdependence and for avoiding miscalculation.

ARMS CONTROL AND THE FUTURE OF SECURITY IN EUROPE

Given diverse images of European security, the absence of a consensus on what a future European order should or can be is not surprising. The range of alternatives is wide.

Perpetuation of the status quo with division of the European center between East and West is one possibility. In the Atlanticist view, the United States will remain strongly committed to NATO as a counterweight to Soviet influence in Eastern Europe. A modified version of the Atlanticist view, however, sees greater West European assertiveness within the NATO Alliance, particularly in a period of reduced East-West tensions. Proponents of this alternative point to greater military collaboration in recent years between France and West Germany. There has been movement beyond joint exercises to joint military planning and even to the formation of Franco-German units. Greater French military coordination with the West

Germans not only strengthens its position against a Warsaw Pact threat to France, but also gives France an additional assurance against the increasingly unlikely prospect of any resurgence of Germany as a threat. Some criticize this greater European independence of the United States as a more dilute form of Atlanticism; however, others contend that greater European participation constitutes Atlanticism on a firmer foundation.

There is no assurance, of course, that greater West European assertiveness in NATO would in any way be matched by East European countries. Even under the more open and flexible leadership of Gorbachev that has allowed states in the East greater freedom to determine the nature of their domestic political economies, the Soviets set very real limits to the independence of the Warsaw Pact allies. Apart from any predisposition to be hegemonic, Soviet reluctance to loosen the reins too far can also be explained by that country's continuing security concerns both with the West and with the Germanic center of Europe. So long as the Soviets retain hegemony in the East and thus pose a direct threat to the West, most of the NATO allies will want to keep the United States and its nuclear and conventional forces coupled to the defense of Europe as the most credible means for deterring war and maintaining the peace.

A second option is an all-West European defense against the East that would largely exclude the United States—an alternative to Atlanticism that has been entertained, but as opponents are quick to point out, national differences would make any such collaboration difficult at best. Assuming that West Germany would continue to defer to the Soviet Union and other neighbors on the periphery of Europe and thus decline to acquire nuclear weapons, the absence of an American nuclear pledge would make the Germans effectively dependent on a French or British nuclear guarantee. However assured the West Germans may feel about the American nuclear guarantee they now have, they are not likely to be sanguine about relying exclusively on the French and British. Moreover, an all-European defense exclusive of the United States would suggest a division of labor requiring West Germany, in exchange for a Franco-British nuclear contribution, to provide the bulk of the ground forces, the most expensive component both financially and politically. An alternative formulation of a European nuclear force in which the FRG would share nuclear control is equally unlikely, given French and British desires to retain exclusive control over their nuclear arsenals.[25]

Third, a nuclear-free and chemical-free Europe divided across the center between East and West is yet another possibility. Some have proposed establishing nuclear-free zones in the Nordic, Baltic, and Balkan areas, and eventually removing nuclear forces from Germany. If one assumes that deployment of nuclear weapons is NATO's response to an inferior conventional force posture in the European balance, then negotiations that reduce conventional forces on both sides to equivalent levels would in principle make possible the reduction, or even elimination, of nuclear weapons from Europe.

Because the West relies on nuclear weapons as a substitute for the mass of conventional forces needed to match the East, Western defense establishments have been unreceptive to Soviet calls for a nuclear no-first-use pledge. If overwhelmed by a conventional force attack from the East, the West would be forced either to capitulate or to escalate to first use of nuclear weapons. In the view of most Western officials, of course, a credible threat of nuclear escalation is precisely the right prescription for deterring any use of force by the East, conventional or nuclear. Should war ever occur in central Europe, the use of nuclear weapons against the USSR would deny the Soviets sanctuary they might otherwise enjoy, a fact that contributes to the stability of deterrence. Indeed, even if a stable conventional force balance between East and West is established, Western reliance on nuclear weapons seems likely to continue.

Reduced birth rates and a resulting sharp decline in both East and West of available European military manpower is a factor encouraging both sides to seek reductions in conventional forces. Even if forces were reduced to equivalent levels, many see a nuclear-based deterrence as more stable than a conventionally based one. Indeed, such advocates of nuclear weapons point to World Wars I and II as clear examples of the failure of conventionally based deterrence. Moreover, not everyone would agree that conventional forces alone provide a deterrent comparable in credibility or stability to one with nuclear weapons. Conventional or nonnuclear deterrence may be neither as credible nor as stable as deterrence based on nuclear weapons.

Creating a chemical-weapons-free Europe is another challenge faced by arms control negotiators. Like nuclear and conventional weapons, chemical weapons (CW) are an integral part of the Soviet and Warsaw Pact arsenal. Most observers concede that the East has a substantial advantage in this type of warfare. Although the West also has chemical weapons in the field, such weapons have not been central to national or allied force employment doctrines. Indeed, the only reason for Western CW deployments is to deter their use by the East. However repugnant these weapons may be to the West, establishing a chemical-free Europe would require that the Soviets relinquish their current lead and accept verification methods that would be very intrusive. Negotiations on chemical, biological, and radiological weaponry are part of the agenda at the CD in Geneva. Progress has been slow, but the USSR has indicated somewhat greater willingness to accept on-site verification measures.

Fourth, and increasingly popular among several West European left-of-center political parties, is an alternative approach to defense policy while staying within the NATO Alliance. Examining the defense policies of such present-day neutral or nonaligned states as Austria, Finland, Sweden, and Switzerland, some advocate reliance on "nonprovocative defense" (fielding only those weapons systems with little or no offensive capability) and a territorial defense with rapid and mass mobilization of reserves. This is also a reaction by the left against proposals within NATO for not confining attacks to frontline forces, but for attacking secondary echelons or "follow-on" Warsaw Pact forces. A Western threat to take any war to the East and deployment of forces capable of doing so is seen, particularly on the left, as unduly provocative.

A fifth alternative is a Europe without competing hegemonies or spheres of influence of the superpowers—an end to division between East and West. Neutralization of a reunited Germany, given geopolitical realities in the present world, would likely result in somewhat greater German accommodation of, or orientation toward, the East. Attainment of this outcome would, in part, also represent the triumph of pluralist notions of increasing interdependence within Europe. A unified but militarily weak Germany (after the model of present-day Austria) is part of this vision. A neutralized and militarily weak Germany would reduce worries by non-German states about too strong a geopolitical center in Europe. Pluralism would displace the balance of power whether understood in East-West or center-periphery terms. Security in Europe would rely on a cosmopolitan principle—a harmony of common interest. This image, of course, contrasts sharply with the present order in which considerations of power and balance of power still dominate the thinking of statesmen.

Of the alternative security orders discussed above, perpetuation of Atlanticism, albeit with increasing commerce in central Europe and a greater degree of autonomy by states in both alliances, seems most likely, if for no other reason than the political obstacles confronting any of the other alternatives. Arms control has the potential for charting the direction that change in Europe might take, but the right political conditions are needed to propel that change. The political pressures necessary to change the legacy of World War II will likely continue to develop slowly. Nonetheless, the arms control agenda in Europe is a dynamic one. Eliminating an entire class of weapons (land-based INF missiles), expanding negotiations on conventional forces to the territory between the Atlantic and the Urals, and establishing regimes with an unprecedented degree of inspection, notification, and observation rights are all noteworthy developments.

NOTES

The views expressed here are the author's and do not necessarily reflect those of any agency of the US government. The author would like to acknowledge the suggestions made by colleagues and friends who reviewed earlier drafts of this chapter. Comments made by Mark Kauppi, Schuyler Foerster, James Busey, Jeffrey Larsen, Joseph Rallo, Jay Lorenzen, Kenneth Rogers, Larry Madsen, and Edward Wright were particularly helpful. My focus on alternative cognitive images as an important dimension for understanding policy choice reflects earlier work with Ernst Haas. Moreover, the contrasting pluralist and balance-of-power or realist images can be traced to work with both Haas and Kenneth Waltz respectively.

1. Bruce Russett and Donald Deluca, "Theater Nuclear Forces: Public Opinion in Western Europe," *Political Science Quarterly* 98, no. 2 (1983): 180. Cf. Wallace J. Thies, *The Atlantic Alliance, Nuclear Weapons and European Attitudes: Reexamining the Conventional Wisdom.* Policy Papers in International Affairs, no. 19. (Berkeley: University of California Institute of International Studies, 1983).

2. See Michael Howard, "Reassurance and Deterrence: Western Defense in the 1980s," *Foreign Affairs* (1982–83): 118.

3. The European "central front" refers primarily to the border between East and West Germany and between West Germany and Czechoslovakia.

4. See George Kennan, "Sources of Soviet Conduct," *Foreign Affairs* 25, no. 4 (1947): 566–82.

5. This was the advice in President Washington's farewell address. It remained a key element in American foreign policy until World War I. In the interwar period US isolationism was a return to avoidance of European balance-of-power politics.

6. One of the best collections of the politics associated with INF is Richard K. Betts, ed., *Cruise Missiles: Technology, Strategy, Politics* (Washington, D.C.: Brookings Institution, 1981).

7. The best single source for the development of these arrangements in the Federal Republic of Germany is Catherine McArdle Kelleher, *Germany and the Politics of Nuclear Weapons* (New York: Columbia University Press, 1975). For the UK, see Lawrence Freedman, *Britain and Nuclear Weapons* (London: Macmillan and the Royal Institute of International Affairs, 1980). For extensive and detailed discussion of politics and nuclear weapons relations in NATO, see Leon V. Sigal, *Nuclear Forces in Europe* (Washington, D.C.: Brookings Institution, 1984); David N. Schwartz, *NATO's Nuclear Dilemmas* (Washington, D.C.: Brookings Institution, 1983); and John D. Steinbruner and Leon V. Sigal, eds., *Alliance Security: NATO and the No-First-Use Question* (Washington, D.C.: Brookings Institution, 1983). Cf. Andrew J. Pierre, ed., *Nuclear Weapons in Europe* (New York: Council on Foreign Relations, 1984).

8. This is the view expressed by former West German Chancellor Helmut Schmidt in a 1977 speech to the International Institute for Strategic Studies in London. The text is in *Survival*, Jan.–Feb. 1978, pp. 2–10. A later statement of his views is in Helmut Schmidt, *A Grand Strategy for the West* (New Haven: Yale University Press, 1985). On defense issues, esp. pp. 10–13, 19–21, 33–43, 52–55, 139–45, and 152–55.

9. Jed C. Snyder, "European Security, East-West Policy, and the INF Debate," *Orbis* 27, no. 4 (1984): 960–62.

10. See "NATO Endorses Reagan in Arms Talks," *Baltimore Sun*, 23 Oct. 1986. See also the Gleneagles Nuclear Planning Group Communique, NATO Press Information Service, 22 Oct. 1986.

11. See the June 1987 North Atlantic Council Communique, NATO Press Information Service.

12. See ACDA Director Kenneth Adelman's speech to Brandeis University, 28 Apr. 1987. Notwithstanding apparent advantages for NATO as a whole, some Germans expressed concern that, apart from nuclear-capable aircraft having the range to strike targets in the USSR and elsewhere in Eastern Europe, elimination of INF missiles leaves short-range nuclear weapons that can be used only on West or East German soil. This, of course, evokes the recurring German nightmare—the prospect of a nuclear war being confined to German territory and leaving the superpowers and other countries relatively unscathed.

13. FRG Chancellor Helmut Kohl's press conference statement in Bonn, 26 Aug. 1987.

14. See, for example, "Nervous NATO Allies Seek US Assurances on Superpower Arms Deal," *Christian Science Monitor*, 23 Oct. 1986. See also "General Sees Missile Plans as a Mistake," *New York Times*, 24 June 1987.

15 On MBFR, see Richard F. Staar, "The MBFR and Its Prospects," *Orbis* (Winter 1984): 999–1009 and Jonathan Dean, "MBFR: From Apathy to Accord," *International Security* 7, no. 4 (1983): 116–39.

16. See Dean, "MBFR," pp. 124–25.

17. See the Brussels Declaration from the December 1986 North Atlantic Council Ministerial Meeting, NATO Press Information Service.

18. Senator Sam Nunn's speech in Brussels, 3 Apr. 1987 in *NATO Review* 3 (June 1987): 1–8, is a representative argument.

19. Germany allied with the Austro-Hungarian Empire and later with the Ottoman Empire as well. This "central" alliance was countered by the allies on the "periphery" of Europe—Britain, France, and Russia—who would later be joined by the United States toward the end of World War I.

20. Then a young British economist, John Maynard Keynes opposed the draconian measures taken against Germany in the postwar settlements. He remarked that no effort was spared to impoverish the Germans. Though not demanding reparations payments directly from Germany, the United States required repayment of Allied war debts, a cost the British and French imposed on the Germans. France eventually occupied German territory to extract mineral resources in settlement of the German reparations debt.

21. See Sir Halford J. Mackinder, "The Round World and the Winning of the Peace," *Foreign Affairs* 21, no. 4 (1943): 595–605.

22. As such, some argue that their continuing presence serves a secondary, if not often publicly stated, occupation function for those worried about any potential German threat, given the historical legacy of two world wars in the first half of the twentieth century. Such matters are, of course, an extremely delicate aspect of diplomatic relations among the West European allies, although their delicacy is well understood by the parties.

23. See Michael Howard, ed., *Disengagement in Europe* (London: Penguin Books, 1958).

24. See, for example, "Bonn, Irked by US, Shifts on NATO Plan," *International Herald Tribune*, 13 July 1987.

25. Hedley Bull, "European Self-Reliance," *Foreign Affairs* 61, no. 4 (1983): 874–92.

BIBLIOGRAPHIC ESSAY

ARMS CONTROL

CHARLES E. COSTANZO and JAY L. LORENZEN

The cycle of war and peace runs as a common thread throughout human history. Breaking this cycle has been the passion of thoughtful and anxious people. Two diverse avenues stand out. One avenue acknowledges the wisdom of the old adage, "If you want peace, prepare for war." The other avenue follows the simple and alluring logic that war is not possible without weapons. The former argues for armaments, the latter sees disarmament and arms control as more appropriate paths to peace.

Technological advances in weaponry during the last hundred years have given disarmament and arms control a new imperative. This concern has led to a great debate about the advantages and disadvantages of such destructive forces. Naturally, this debate has elicited a vast literature addressing the various arguments for and against particular approaches to arms control. We have attempted to sift out some of the more relevant publications and representative works for inclusion in this essay. The reader must understand that this essay cannot exhaust the available arms control literature. Nevertheless, the essay will hopefully serve as a starting point for understanding the scope and general content of that literature.

ARMS CONTROL BIBLIOGRAPHIES

In view of the volume of arms control literature written, various bibliographies have been compiled to facilitate research. Richard Dean Burns's *Arms Control and Disarmament: A Bibliography* (Santa Barbara, Calif.: ABC-Clio, 1977) may be the most comprehensive bibliography on arms control and disarmament theory and practice. In addition to an exhaustive coverage of the nuclear age, Burns has referenced many works that predate the nuclear era. Almost 9000 books and articles are arranged topically, and each topic is introduced with a short preface. *The SALT Era: A Selected Bibliography* (Los Angeles, Calif.: Center for the Study of Armament and Disarmament, 1979) by Richard D. Burns and Susan Hoffman Hutson specifically addresses the background to the SALT

negotiations. Sources covering superpower, Chinese, and European views of the SALT process; nonproliferation; the strategic balance; civil defense; and the consequences of nuclear war are included in this concise bibliography. Together, these bibliographies serve as excellent, comprehensive source surveys of the arms control literature, particularly as it pertains to strategic nuclear arms control through the signing of SALT II.

ARMS CONTROL HANDBOOKS

The Arms Control and Disarmament Agency (ACDA) periodically publishes a volume of treaty texts and their negotiating backgrounds entitled *Arms Control and Disarmament Agreements: Texts and Histories of Negotiations*. It incorporates arms control agreements beginning with the Geneva Protocol of 1925. The *SALT Handbook: Key Documents and Issues: 1972–1979* (Washington, D.C.: American Enterprise Institute, 1979), edited by Roger P. Labrie, includes the texts of the agreements, the history of the SALT negotiations, and the evolution of the major issues in strategic nuclear arms policy since the signing of the first SALT agreements. Also included are congressional testimony and statements of policymakers, annotated bibliographies, and a glossary of technical terms. This volume is an invaluable aid to individuals researching this period.

Jozef Goldblat's *Arms Control Agreements* (New York: Praeger, 1983) is also a comprehensive and useful handbook. The author provides a well-informed historical overview, analysis, and critique of arms control agreements and negotiations from the Hague Declaration of 1899 to the SALT accords. Besides including the texts of more than 50 arms control agreements, Goldblat appraises the extent to which these agreements have influenced the arms race and the likelihood of war.

Since 1968, the Stockholm International Peace Research Institute (SIPRI) has annually published the *World Armaments and Disarmament: SIPRI Yearbook*. These yearbooks review annual military activity, conflicts, military expenditures, arms trade, weapons development, and the production and deployment of nuclear weapons. In addition, major sections are dedicated to arms control and disarmament initiatives. The yearbooks devote space to various background studies in which eminent scholars address particular issues in the arms control field.

JOURNALS AND NEWSLETTERS

Numerous journals and newsletters address arms control issues. *Arms Control Today*, published monthly by the Arms Control Association, is dedicated to public appreciation of the need for arms limitation and reduced international tension. Published in newsletter format, it provides information on national security issues for the nonexpert. Perhaps most useful to the researcher is an annotated bibliography of recent literature.

Although they are not dedicated exclusively to arms control, many academic journals routinely contain articles on the subject. Such journals include *International Security*, published quarterly by the Center for Science and International Affairs at Harvard University; *Foreign Policy*, published quarterly by the Carnegie Endowment for International Peace; *Foreign Affairs*, published five times a year by the Council on Foreign Relations; *Orbis*, published quarterly by the Foreign Policy Research Institute; and *Strategic Review*, published quarterly by the United States Strategic Institute. Obviously this list of purely American publications is not exhaustive. These journals are, however, some of the more pertinent publications reflecting much of the debate among the finest scholars in the field. Many other journals, both American and foreign, also contain articles in the area of arms control.

Scientific American has compiled its articles dealing with nuclear arms, arms control, and strategy in three volumes. *Arms Control* (San Francisco: W. H. Freeman, 1973) focuses on the history and technology of nuclear weaponry, efforts to limit nuclear weapons, and the peaceful uses of nuclear technology. *Progress in Arms Control?* (San Francisco: W. H. Freeman, 1979) traces the development of US and Soviet nuclear policies from the development of the hydrogen bomb to the SALT negotiations. The problems of a changing technology, nuclear proliferation, and nuclear strategy are among the issues discussed in this collection. A third collection, *Arms Control and the Arms Race* (New York: W. H. Freeman, 1985), contains articles on the evolution of Soviet and American nuclear policy through SALT I and II, a broad range of current strategic issues, and European security.

GENERAL WORKS ON ARMS CONTROL

Although it encompasses many nuclear issues, *Living with Nuclear Weapons* (New York: Bantam Books, 1983) by the Harvard Nuclear Study Group provides a realistic and objective look at the problem of nuclear arms control. This distinguished group of scholars has accomplished its purpose by providing the public with a highly readable discussion of the nuclear debate and arms control.

The major tenets of American arms control theory had their provenance in the late 1950s and early 1960s. Inspired by the post-*Sputnik* anxiety in the US, a fervor of intellectual activity in this period produced an outpouring of literature on the subject of arms control. There are several important works from this period. Bernard B. Bechhoeffer's *Postwar Negotiations for Arms Control* (Washington, D.C.: Brookings Institution, 1961), Donald G. Brennan's *Arms Control, Disarmament, and National Security* (New York: George Braziller, 1961), Hedley Bull's *The Control of the Arms Race* (New York: Praeger, 1961), David H. Frisch's *Arms Reduction: Program and Issues* (New York: Twentieth Century Fund, 1961), Arthur T. Hadley's *The Nation's Safety and Arms Control* (New York: Viking, 1961), Louis Henkin's *Arms Control Issues for the Public* (New York: Prentice Hall, 1961), and Thomas C. Schelling and Morton H. Halperin's *Strategy and Arms Control* (New York: Twentieth Century Fund, 1961).

Hedley Bull's work, *The Control of the Arms Race*, has become a classic in the field. More so than other authors, Bull was successful at fitting his conceptualizations into the political and historical context of the modern world. He analyzed, for example, the prenuclear attempts at arms control such as the Washington Naval Conference of the early twenties and the League of Nations' efforts of the thirties. Then, in light of the blossoming arms race, Bull examined both disarmament and arms control as measures by which the "race" can be controlled. Placing each in the context of strategies, weapons, and the political tensions of the international system, he concluded that arms control is achievable, but disarmament is not. Also, Bull's argument that international security is the chief objective of arms control provided the basis against which arms control initiatives have been, and are being, judged. Because of its theoretical standard-setting role, this classic work deserves thoughtful reading. *The Control of the Arms Race*, as well as the other works of the period, possessed additional merit in that one can apply a quarter century of hindsight to test the validity of the major theoretical tenets proposed during that period.

Arms, Defense Policy, and Arms Control

(New York: W. W. Norton, 1976), edited by Franklin A. Long and George W. Rathjens, analyzes the 15 years of change in the arms control field since those initial works. Twelve excellent essays address the interconnections between arms control and international political relations, crisis stabilization, and weapons procurement. These essays are notable in that they analyze the intricate and complex context of international political relations and the domestic political and military-industry pressures influencing the arms race. For example, Marshall Shulman's essay, "Arms Control in an International Context," demonstrates that technical approaches alone cannot solve the problems inherent in arms control. Changes in the international political environment also bear on arms control initiatives. Similarly, the other essays provide an in-depth analysis of the issues underlying the arms race and arms control.

A general principle of arms control, seen by almost all scholars and policymakers since the late 1950s, is the principle of strategic stability. According to Jerome H. Kahan in his book, *Security in the Nuclear Age: Developing U.S. Strategic Arms Policy* (Washington, D.C.: Brookings Institution, 1975), strategic stability is the imperative of the nuclear age and the principal objective of arms control. Providing a comprehensive summary of the complex elements of US and Soviet nuclear policy from the Eisenhower administration to the end of the Nixon presidency, Kahan furnishes the reader with a historical perspective of the central issues influencing America's strategic posture through 1975. In a particularly illuminating chapter, "Arms Interaction and Arms Control," Kahan addresses the rationale for arms decisions, risks of the arms race, purposes of arms control, verification issues, and negotiating strategies. Noting an "action-reaction" relationship between US and Soviet strategic forces, Kahan argues the necessity of negotiated arms control to break the cycle.

Albert Wohlstetter's classic two-part article, "Is There a Strategic Arms Race?" and "Rivals, But No Race" in *Foreign Policy* (Summer and Fall 1974) warns that too much emphasis has been placed on the "action-reaction" cycle in the arms race debate. Theories on the strategic weapons race, many of which Wohlstetter believes are based on myths, have failed to clarify strategic force changes and have not aided "thoughtful national choice or agreement with adversaries." Accordingly, these myths have only fueled public debate. Graham Allison's "What Fuels the Arms Race?" in *Contrasting Approaches to Strategic Arms Control* (Lexington, Mass.: Lexington Books, 1974) argues that the arms race is fueled by forces within each country. "Internal, bureaucratic impulses" may be more significant in understanding the arms race than the "action-reaction" premise offered by many arms control theorists.

The theoretical tenets of arms races and arms control initiatives are inextricably linked, and the researcher must confront both areas. Placing defense and arms control policies in the context of a broad understanding of the nature of the arms race is a contribution made by Colin Gray's *The Soviet-American Arms Race* (Westmead, England: Gower, 1981). In this book, Gray attempts to assemble diverse elements of the explanatory theorems and argues for an improved arms race theory. To Gray, the arms race is an "inescapable product" imposed by the structure of international hierarchy, where the US and Soviet Union are political and military rivals. Therefore, Gray maintains that arms control should serve the needs of defense and not vice versa. Arms control is but one of the criteria for a "good strategic posture." No doubt many would disagree with his analysis; however, his book is a profitable introduction to the arms race and arms control in a world of interstate political conflict.

Expanding our understanding of arms race theory, Ian Bellany, in "An Analogy for Arms Control," in *International Security* (Winter 1981–82), claims that arms control "is an arrangement between superpowers to avoid the costs of military competition." Arms control is analogized as a group of producers of near identical commodities agreeing to an "orderly marketing" arrangement, realizing that none seems likely to dominate the market, and that "competition is, therefore, a waste of their resources." This insightful piece elucidates a number of features characteristic of the SALT negotiations. Also, arms control arrangements, like "orderly marketing" arrangements, will inevitably be unstable; long periods of arms agreements may abruptly degenerate into a renewed arms race. A provocative and stimulating article, it is must reading for those analyzing the rationale behind arms control initiatives.

The preceding discussion illustrates the diverse nature of arms control. While the works cited above afford students a broad framework for understanding the arms control debate, the following sections focus on selected readings pertaining to specific arms control processes.

A SELECTIVE REVIEW OF ARMS CONTROL LITERATURE

Bilateral and multilateral efforts to control the instruments of armed conflict have a long history and have taken many forms. However, in the post–World War II era these efforts have become less intermittent undertakings and more urgent enterprises. Technological advances and the quantitative growth of arsenals, particularly those of the superpowers, have transformed traditional notions about national security and created a compelling incentive for arms control. More armaments no longer guarantee more security. In fact, the acquisition of more or improved weapons may have the opposite effect. Expanded arsenals are often viewed suspiciously and can act as the source of increased tension between contending nations.

Arms control initiatives over the past four decades have addressed various classes of armaments and several geographic regions. The remainder of this essay, however, will be devoted to only three categories of arms control: strategic nuclear, theater nuclear, and conventional forces. This focus was adopted not to diminish the importance of other arms control endeavors, but to highlight those fora that have been at the fountainhead of East-West negotiations and intra-alliance debates. Few other arms control deliberations encompass the capacity to mold future military strategies or generate wide official and public debate.

SALT AND START: NARRATIVE WORKS

Although the SALT I agreements were signed almost 16 years ago by President Nixon and General Secretary Brezhnev, they remain of continuing importance for two principal reasons. First, the bold provisions of SALT I heralded the beginning of a process aimed at limiting the strategic offensive and defensive forces of the US and the Soviet Union. The superpowers were not only signaling their intentions to engage in a long-term process to limit strategic forces, but in a real sense their willingness to relinquish a degree of sovereignty to bilateral arms restraint. Secondly, the accords also marked the beginning of intense, and frequently acrimonious, public debate in the West concerning strategic arms control. While the SALT I Interim Agreement expired in 1977 (the Anti-Ballistic Missile Treaty is of indefinite duration), the debate about the purposes, merits, and weaknesses of strategic arms control continued. This debate attended the SALT II negotiations and in no small measure contributed to the eventual withdrawal of the treaty from Senate ratification. The same issues that accompanied the SALT agreements from the negotiations in Vienna to the hearings on Capitol Hill will no doubt visit the Strategic Arms Reduction Talks (START). Only by studying the lessons of the SALT process can we hope to avoid the pitfalls and disappointments of the last decade.

Much has been written during the intervening years about SALT issues and the political dynamics of the negotiations; however, several works are notable. Thomas W. Wolfe's *The SALT Experience* (Cambridge, Mass.: Ballinger, 1979) is one of the most informative and comprehensive volumes available. Noted for its clarity, Wolfe's book is perhaps the best starting point for acquiring a thorough understanding of SALT. Beyond Wolfe's book, other accounts of the SALT process are useful. *Doubletalk: The Story of the First Strategic Arms Limitation Talks* (New York: Doubleday, 1980) by Gerard Smith offers an illuminating discussion of SALT I. Appointed as the chief US delegate to SALT I by President Nixon, Smith presents a sympathetic assessment of the accord's achievements and makes a forceful argument in behalf of its contributions to American security and superpower stability. Complementing Smith's book, John Newhouse's *Cold Dawn: The Story of SALT I* (New York: Holt, Rinehart, and Winston, 1973) is an interesting and insightful addition to the SALT I literature. Newhouse adeptly unravels the background preceding the talks, the weapons issues, and the complexities of domestic and diplomatic bargaining. In a similar vein, Strobe Talbott's *Endgame: The Inside Story of SALT II* (New York: Harper and Row, 1979) offers a superb account of these negotiations. Written while the SALT II talks were active, the book is based on interviews with officials from the Nixon, Ford, and Carter administrations; congressmen; and West European and Soviet diplomats. Talbott's later book, *Deadly Gambits* (New York: A. A. Knopf, 1984), elaborates further on the political dimensions of arms control. Talbott depicts candidly the bureaucratic maneuvering and ideological conflicts within the Reagan administration during START, as well as the Intermediate-range Nuclear Force (INF) talks.

SALT AND BEYOND

As the criticisms of SALT II accumulated in the late 1970s and the prospects for its ratification declined, advocates of the accord were championing its accomplishments. These proponents, determined to ensure a future for strategic arms control, sought to rehabilitate SALT and portray a world endangered by unrestrained arms competition.

Walter Slocombe's "A SALT Debate: Hard but Fair Bargaining" in *Strategic Review* (Fall 1979) refutes the argument that SALT facilitated Soviet strategic goals at the cost of American security. Aside from discounting many technical criticisms aimed at SALT, and particularly SALT II, Slocombe asserts that several lessons emerged from the arms control dialogue with the Soviets. Slocombe urges, for instance, the importance of precise treaty language to preclude definitional disputes such as the one that surrounded the Soviet-American debate concerning whether the SS-19 ICBM should be classified as a "heavy" or "light" missile. Since neither party seemed wholly satisfied with the final resolution of the issue, it is imperative that the language of future accords not leave unresolved sources of recrimination. While Slocombe's favorable assessment of SALT is rooted in more technical grounds, Michael Nacht's "In the Absence of SALT" in *International Security* (Winter 1978–79) endorses the process for its contributions to Soviet-American stability. Nacht concedes that there could be positive alternatives following a cessation of SALT, depending upon how the process was terminated. However, an arms competition could also ensue that would stimulate mutual tension and increase the possibility of crisis instability.

Insofar as the superpowers pursue strategic arms control as one method to maintain stability, Jan M. Lodal's article "SALT II and American Security" in *Foreign Affairs* (Winter 1978–79) presents a positive judgment of the treaty. Lodal argues that SALT II would have constrained Soviet offensive forces and prevented certain areas of strategic force competition perceived as destabilizing by the US. The treaty, in Lodal's estimation, would have permitted US strategic force modernization while eliminating "worst-case" estimates. Accordingly, SALT II provided a genuine opportunity for the US to further its security interests. John M. Lee, in "An Opening 'Window' for Arms Control," *Foreign Affairs* (Fall 1979), expands on this proposition and suggests that when certain variables exist, both parties to negotiations will seek agreement. These variables include the technical status of progress, weapons balances, verification capacities, and budgetary and political forces. When some of these variables are neutral and others pressing, the so-called window of bilateral opportunity prevails and each side perceives benefits from reaching an agreement. Although the interaction of these variables during the SALT II negotiations seemingly militated against the treaty, Lee's analysis is applicable to other arms control initiatives, particularly START. Lee's article is instructive because it relates the importance of technical and political variables to the success or failure of arms negotiations and presents "lessons" of future applicability.

While many were lauding SALT's achievements and campaigning for its continuation, others were offering critical appraisals of the process and condemning specific failures in each treaty. In *SALT II: How Not to Negotiate with the Russians* (Miami: Advanced International Studies Institute, 1979) Foy D. Kohler maintains that the US has consistently misinterpreted Soviet intentions. Utilizing a facade of sincerity, the Soviets have adroitly marshalled American and West European sentiment, exhorted those hopeful for strategic arms control, and exploited emotionalism while seeking unilateral advantage. Kohler asserts that this tactic

manifested American strategic inferiority that could not be amended by SALT II. Consistent with this theme, Colin Gray, in "SALT: Time to Quit," *Strategic Review* (Fall 1976), implores the US to eschew agreements that masquerade as arms control and to assume the initiative in rectifying the imbalances in its strategic arsenal.

The adverse effects of SALT on US strategic forces are examined by Paul H. Nitze, James E. Dougherty, and Francis X. Kane in *The Fateful Ends and Shades of SALT* (New York: Crane, Russak, 1979). This penetrating inquiry is noteworthy for Nitze's measurement of SALT II against its implications for the Soviet-American strategic balance. Nitze traces American strategic forces since 1972 and concludes that the provisions of SALT II threaten the US's ability to maintain stability and essential equivalence. In a final section of the book, Kane urges that an essential concomitant to SALT II be a prudent US strategic weapons development and deployment program. An equally notable discussion of the impact of SALT II on US force modernization and strategic stability is available in *SALT II and U.S.-Soviet Strategic Forces* (Cambridge, Mass.: Institute for Foreign Policy Analysis, 1979) by Jacquelyn K. Davis, Patrick J. Friel, and Robert L. Pfaltzgraff, Jr.

Although many criticized SALT for not constraining the Soviet strategic buildup or strengthening strategic stability, others assailed it because of inadequate verification and compliance standards. A caustic assessment of SALT's inadequacies in these areas is Jake Garn's "The SALT II Verification Myth," in *Strategic Review* (Summer 1979). Garn argues that fear of detection does not deter Soviet noncompliance with arms control agreements, and he cites several instances of alleged violations to substantiate his claims. Moreover, according to Garn, Soviet gains achieved through violations were legitimized in SALT II by American acquiescence to Soviet actions. A more balanced treatment is available in James A. Schear's "Arms Control Treaty Compliance: Buildup to a Breakdown" in *International Security* (Fall 1985). Schear probes the SALT I and II compliance issues and after ascertaining that none are intractable, offers solutions to restore public confidence in the arms control process.

Several books are also useful in providing clarity to the study of verification and compliance. William C. Potter's *Verification and SALT: The Challenge of Strategic Deception* (Boulder, Colo.: Westview Press, 1980) is perhaps the most extensive collection of SALT-related verification essays obtainable. Potter's latest effort, *Verification and Arms Control* (Lexington, Mass.: D. C. Heath, 1985), is a collection of essays addressing several areas of arms control. It is especially beneficial for its discussion of the Reagan administration's allegations of Soviet noncompliance and Soviet responses to these charges. William F. Rowell's *Arms Control Verification* (Cambridge, Mass.: Ballinger, 1986) is an excellent verification and compliance primer for inquiries into the role of political actors in the arms control process, policy options, and the impact of technology. The technology of verification is the subject of *Arms Control Verification: The Technologies That Make It Possible* (Washington, D.C.: Pergamon-Brassey's, 1986) edited by Kosta Tsipis, David W. Hafemeister, and Penny Janeway. The volume succeeds in rebutting arguments that verification is the principal obstacle to arms control; however, the essays are often highly technical and require that the reader possess more than a superficial understanding of the physical sciences.

In addition to the more polarized views presented above, there are a number of works that dissect both the merits and weaknesses of the SALT process and treaties. Raymond L. Garthoff's "SALT I: An Evaluation" in *World Politics* (Oct. 1978) identifies several of the agreement's military and political shortcomings, but also enumerates its achievements. Garthoff posits that SALT I contributed to the Soviet and American dialogue and facilitated more openness between the two countries. The central point of his conclusion seems to be that the results of a strategic arms control accord should not be measured against only technical criteria or excessively ambitious goals. Similarly, Richard Burt, in "The Scope and Limits of SALT," *Foreign Affairs* (July 1978), concludes that SALT should not be judged a failure. Burt asserts that expectations for SALT were too ambitious, and when the treaty emerged as an accommodation to reality, it was widely regarded as an accommodation to the Soviet Union.

Also critical of the inability of some to adapt to existing realities is Aaron L. Friedberg's "What SALT Can (and Cannot) Do" in *Foreign Policy* (Winter 1978–79). Specifically, he faults those who adhere indefatigably to technically sophisticated but politically naive arms control proposals. That is, SALT should not have been designed to end the arms race, educate the Soviets, or achieve grand schemes. SALT's limited, albeit important, role should have been to enhance strategic stability and thereby reduce the possibility of nuclear conflict. A fine comment on strategic stability is available in Richard Burt's "The Future of Arms Control: A Glass Half Empty" in *Foreign Policy* (Fall 1979). Burt contends that the SALT negotiations could not obviate sources of instability since there was no consensual definition of strategic stability. As an alternative to using arms control as the sole solution, Burt urges the US reconsider its approach and formulate a defense policy to eliminate sources of military instability. In this sense arms control must be employed as only one aspect of a sound defense policy. Burt's piece is a rejoinder to Leslie H. Gelb's "The Future of Arms Control: A Glass Half Full" in the same issue of *Foreign Policy*. Aware that arms control has been a disappointment, Gelb nonetheless argues for its continuation, but with a narrower scope designed to seek methods to maintain stability as well as reduce arms. This approach could achieve uniform progress and convince the skeptics of the overall value of arms control.

After assuming office, the Reagan administration began exploring arms control alternatives and in June 1982 entered into START with the Soviet Union. This departure from the SALT process represented a formidable challenge of the new administration. No longer would the US attempt simply to "limit" offensive arsenals. The new initiative called for phased "reductions" of the most destabilizing elements of Soviet and American nuclear forces. Recalling the political miscalculations of the Carter administration's handling of SALT II, Alan Platt, in "STARTing on SALT III" in *Washington Quarterly* (Spring 1982), urged the structuring of a domestic political campaign to engender support before signing any strategic arms control accord in order to preclude the potentially disastrous outcome of another unratified treaty.

The initial US START proposal sought deep reductions in "heavy" Soviet land-based missiles. The Soviets rejected this formula, contending that it would undermine the mainstay of their military posture. Under mounting pressure from congressional critics, the Reagan administration proposed the so-called double build-down formula at START in October 1983. This scheme would require each side to destroy two or more older nuclear weapons for each new weapon added to its arsenal. Again, the Soviets rejected the proposal, arguing that it was an unbalanced arrangement and weighted against them. Alton Frye's in-depth analysis in "Strategic Build-Down: A Context for Restraint" in *Foreign Affairs* (Winter 1983–84) evaluates the merits of build-down. Frye concludes that such regulation of strategic modernization would indirectly control menacing qualitative improvements and could contribute to strategic stability. An innovative approach to reduce threatening force developments is offered by Glenn A. Kent (with Randall J. DeValk and Edward L. Warner III) in *A New Approach to Arms Control* (Santa Monica, Calif.: RAND Corporation, 1984). Kent et al. assert the principal objective of arms control must be the reduced likelihood of nuclear war. Toward this end, the superpowers must reject first-strike capabilities and reduce counterforce potential. If the US and USSR can agree on these goals, both must constrain missile throw-weight in order to reduce offensive destructive capacity to a level that enhances strategic stability. Three approaches to achieve this end are assessed and a scheme allowing a series of trade-offs and force mix between ICBMs and strategic bombers is chosen as the optimal apportionment of constraint between the US and Soviet Union.

In December 1983, following completion of the fifth round of talks, the Soviets declined to set a resumption date for START, a move precipitated by the arrival in Western Europe of ground-launched cruise missiles and Pershing II intermediate-range missiles. Stanley Kober, in "Swapping with the Empire," *Foreign Policy* (Spring 1984), identifies two lessons of the strategic arms control enterprise that are as instructive now as they were when the article was written. First, the intricacy of previous negotiations precluded arms control from keeping pace with quantitative and qualitative weapons advancements. Secondly, the US effort to seemingly dictate the structure of strategic arsenals was resisted by the Soviets. Kober advises avoidance of these pitfalls and recognition that each side's unique requirements should determine the features of strategic inventories.

Following a prolonged hiatus, the START negotiations resumed in 1985 immediately after Mikhail Gorbachev became the Soviet Communist Party leader. Some have argued the Soviets returned to the talks because they realized the decision to suspend START was counterproductive. Others have suggested that renewed Soviet interest in START was in part precipitated by President Reagan's Strategic Defense Initiative (SDI) and the challenges it poses to Soviet strategic planning. As prospects for arms control agreements increased, the debate over the relationship between SDI and arms control intensified. One of the most pessimistic appraisals of the SDI's consequences for arms control is "The President's Choice: Star Wars or Arms Control," *Foreign Affairs* (Winter 1984–85), by McGeorge Bundy, George F. Kennan, Robert S. McNamara, and Gerard Smith. The authors contend that if the program continues along its present course, it can only increase Soviet suspicions, intensify the superpower arms race, and destroy the ABM Treaty. The alternative is to "choose the pursuit of agreement over the demands of Star Wars." Adopting a divergent position on the SDI's implications for arms control, Keith B. Payne and Colin S. Gray in "Nuclear Policy and the Defensive Transition," *Foreign Affairs* (Spring 1984), argue the merits of the program. Payne and Gray assert that the transition to a defense-oriented strategy creates an incentive for a Soviet commitment to arms control because of the diminished utility of an offensive missile competition. This debate continued in the literature throughout the 1980s.

The issues in the debate involve disparate interpretations of ABM Treaty restrictions, disagreement over ABM Treaty nonwithdrawal pledges, and the linkage between a START accord and future SDI deployments. Although, in 1989, the Soviets dropped their demand for SDI restrictions as a prerequisite for a START agreement, the issue will continue to be a central feature of the superpower relationship.

THEATER NUCLEAR ARMS CONTROL

When theater nuclear forces (TNF) were originally deployed in Western Europe, they were intended to serve two principal purposes: to deter Soviet aggression by offsetting a perceived imbalance in conventional forces and to function as a tangible and direct link between NATO conventional forces and US strategic nuclear forces. This "extended" deterrence was designed to communicate an explicit message to the Soviet Union: aggression against Western Europe could escalate to the nuclear level, ultimately resulting in attacks against the USSR. So long as the US possessed strategic superiority this threat seemed credible. However, with the advent of nuclear parity many Americans and West Europeans questioned this strategy. Additionally, West Europeans have increasingly pointed to the devastating effects of using nuclear weapons on the continent if deterrence fails for unforeseen reasons.

A growing awareness of these issues caused many on both sides of the Atlantic to urge the limitation of TNF. A thorough discussion of arms control issues and TNF strategy is available in *Challenges for U.S. National Security: Nuclear Strategy Issues of the 1980s* (New York: Carnegie Endowment for International Peace, 1982). Also, Robert Metzger and Paul Doty's "Arms Control Enters the Gray Area" in *International Security* (Winter 1978–79) offers a detailed conceptualization of the TNF problem. Occasionally preoccupied with quantitative limits, this trenchant analysis does reflect the need to limit nuclear forces not constrained by other arms control agreements.

TNF and arms control questions were propelled to the forefront of the Western Alliance when, in December 1979, NATO adopted a dual strategy of arms control and intermediate-range missile modernization. The so-called dual-track decision involved arms control measures to achieve a more stable balance of theater nuclear weapons and a deployment of 108 Pershing II missiles and 464 ground-launched cruise missiles. The decision to adopt a parallel strategy of arms control and arms deployment was not without problems. West European governments that opted to permit deployment on their territories were subjected to domestic turmoil and Soviet criticism. A particularly informative study of the dual-track controversy is H. J. Neuman's *Nuclear Forces in Europe: A Handbook for the Debate* (London: The International Institute for Strategic Studies, 1982). The monograph provides a succinct discussion of the dual-track decision and its arms control implications. It also contains an extensive glossary of terms and summaries of key points in the long debate of European strategy and nuclear weapons.

More analytical studies of the dual-track strategy elaborate on the background and themes presented by Neuman and are most helpful for a thorough understanding of the subject. David N. Schwartz's *NATO's Nuclear Dilemma* (Washington, D.C.: Brookings Institution, 1983) explains how NATO arrived at the 1979 decision and identifies the problems facing the Atlantic Alliance as it endeavored to fulfill the strategy. *Nuclear Weapons in Europe: Modernization and Limitation* (Lexington, Mass.: D. C. Heath, 1983), edited by Marsha McGraw Olive and Jeffrey D. Porro, is useful for comprehending the paradoxical nature of NATO nuclear doctrine. The product of a conference sponsored by the Arms Control Association in 1981, this collection of essays is invaluable for untangling the complex technical and political demands confronting NATO as it attempted to forge nuclear policy in the 1980s. The literature also benefits from Leon V. Sigal's *Nuclear Forces in Europe: Enduring Dilemmas, Present Prospects* (Washington, D.C.: Brookings Institution, 1984). Sigal examines the dilemmas created by NATO's nuclear doctrines and West European ambivalence about the role of nuclear weapons. Further, he identifies the paradoxes and inconsistencies in the logic of NATO's nuclear doctrine, and he assesses the contradictions in the military rationale for Pershing II and ground-launched cruise missiles. Sigal also looks at the West European political parties and movements that influenced the dual-track decision.

While some leaders and scholars were considering the contours and content of the arms control accord that would coexist with missile modernization, others were proposing alternative regimes to constrain theater nuclear forces. In "Nuclear Weapons and the Atlantic Alliance," *Foreign Affairs* (Spring 1982), McGeorge Bundy, George F. Kennan, Robert S. McNamara, and Gerard Smith argued forcefully for NATO to renounce the first use of nuclear weapons and adopt a formal declaration of no first use. In this way, the authors believe NATO could reduce its reliance on nuclear weapons. A reply to this proposal was provided by Karl Kaiser, George Leber, Alois Mertes, and Franz-Josef Schulze in "Nuclear Weapons and the Preservation of Peace: A German Response to No First Use," *Foreign Affairs* (Summer 1982).

These prominent West Germans object to the declaration of no first use and provide a persuasive argument that it would weaken deterrence in Central Europe. An excellent work that articulates the major aspects of the first-use debate is *Alliance Security: NATO and the No-First-Use Question* (Washington, D.C.: Brookings Institution, 1983) edited by John D. Steinbruner and Leon V. Sigal. This compilation of essays, written by American and West European scholars, does not resolve the issues of NATO's threat of nuclear first use. Indeed, as the editors state in the book's introduction, the essay should promote discussion of this regime.

As disputes raged over the dual-track deployment decision, the Intermediate-range Nuclear Forces (INF) negotiations between the US and Soviet Union began in November 1981. At this meeting the American delegation tabled the now famous zero-zero proposal. This proposal entailed American and West European willingness to cancel deployment of the Pershing II and ground-launched cruise missiles in exchange for the dismantling of Soviet SS-20, SS-4, and SS-5 missiles. The goal of the proposal was the elimination of an entire class of nuclear weaponry. However, the moment was not propitious for such a sweeping plan, and the Soviets rejected the proposal.

Skeptics of Reagan administration arms control policy believed the zero-zero proposal was doomed since the Soviets would not barter one of their most modern systems (the SS-20) for as yet undeployed NATO weapons. Strobe Talbott suggests in "Buildup and Breakdown" (*Foreign Affairs: America and the World: 1983*) that this expectation was a ploy by some in the administration to portray the Soviets as the obstacle to theater arms control. Further, Talbott presents an intriguing story of the bureaucratic maneuvering and dissension within the Reagan administration, culminating in the president's refusal to accept the so-called walk-in-the-woods agreement formulated by the American and Soviet negotiators, Paul Nitze and Yuri Kvitsinsky. The adamant position of the administration concerning deployment of the Pershing II and cruise missiles, so long as the Soviets retained SS-20s, had an immediate impact. Talbott believes the stance depicted President Reagan as insincere about theater arms control. Moreover, steadfast insistence to deploy the controversial Pershing II as an implicit test of West German allegiance to NATO insulted former Chancellor Helmut Schmidt. The need for responsiveness to West European, particularly West German, interests is the theme of William E. Griffith's, "Bonn and Washington From Deterioration to Crisis," in *Orbis* (Spring 1982). Griffith urges the US to consider Bonn's attitude toward nuclear weapons, stop talking about limited nuclear conflict, and pursue tangible negotiations with the Soviets.

During the following months the American and Soviet delegations evaluated alternative INF proposals; however, scant progress occurred toward concluding an agreement. The Soviets walked out of the INF talks in November 1983, fulfilling a threat if NATO began deployment of Pershing II and cruise missiles. The US and Soviet Union resumed confidential talks in March 1985 aimed at controlling space and nuclear arms, including longer-range theater missiles.

Subsequent INF negotiations involved a series of Soviet and American proposals and counterproposals designed to reduce or possibly eliminate intermediate-range missiles. It was during the latter stage of these negotiations that the zero-zero proposal was resurrected and the resolve of the parties to reach agreement was subjected to its most demanding test. By late 1987, both sides achieved resolution on several vexing issues, including the exclusion of British and French nuclear forces from any INF accord, the disposition of West German Pershing Ia nuclear missiles, and stringent verification provisions that will involve on-site inspections. The double-zero INF arms control accord was signed by President Reagan and General Secretary Gorbachev in Washington on 8 December 1987. The treaty mandates the destruction of all land-based American and Soviet intermediate-range missiles in the range of 300 to 3400 miles (500 to 5500 kilometers) within three years after the treaty goes into force. The treaty also includes extensive and unprecedented verification measures that can be applied to future arms control fora.

Although the INF agreement is a landmark achievement, the final chapter of theater nuclear arms control has yet to be written. One question is whether, without the intermediate-range missiles deployed in West Europe, the US could be less prone to employ nuclear responses to Soviet aggression. While there are other measures (shorter-range nuclear weapons, for example) that can reinforce trans-Atlantic defense commitments, these are not without inherent problems. The INF accord does not constrain several thousand shorter range nuclear weapons (missiles, bombs, and artillery shells) which await later negotiations by the US and Soviet Union and perhaps the French and British. Neither does the treaty constrain air- or sea-launched weapons. Finally, the removal of NATO's intermediate-range nuclear missiles highlighted the question of how to offset Warsaw Pact conventional force superiority. Barring significantly increased NATO military expenditures, which seem remote for economic and political reasons, an obvious alternative was conventional arms control.

CONVENTIONAL ARMS CONTROL: THE EUROPEAN CONTEXT

Since the end of World War II, various plans for the demilitarization of Central Europe have emerged; however, disparate East-West aims and intra-alliance interests have inhibited the attainment of wide or lasting results. The most prolonged arms control negotiations between NATO and the Warsaw Pact are the Mutual and Balanced Force Reduction (MBFR) talks. These talks, begun in October 1973 and ended in February 1989, sought to reduce the conventional forces arrayed in Central Europe to significantly lower levels. However, as the history of European conventional arms control suggests, different East-West postures and conceptual outlooks have often infused negotiations with intricacies that make any discussion of force reductions a complex and tedious undertaking.

One of the earliest and most insightful examinations of the postwar European military quandary is Michael Howard's *Disengagement in Europe* (Baltimore: Penguin Books, 1958). Professor Howard offers no specific proposals for arms reduction or limitation, but this book is a necessary starting point for understanding the basis of conventional arms control in Europe. One can then turn to works that address the MBFR forum specifically. William B. Prendergast's *Mutual and Balanced Force Reductions: Issues and Prospects* (Washington, D.C.: American Enterprise Institute, 1978) provides a background to the Vienna negotiations and summarizes the broad issues of MBFR, assessing not only the quantitative aspects but also the negotiating process from American, Soviet, and European perspectives. Also, *The Negotiations on Mutual and Balanced Force Reductions: The Search for Arms Control in Central Europe* (New York: Pergamon Press, 1980) by John G. Keliher contains an extensive discussion of the data base issue, residual forces, and "associated measures" (verification and confidence-building measures), as well as an analysis of MBFR proposals advanced between 1973 and 1979. An equally thoughtful review of MBFR approaches proposed over the years is Lothar Ruehl's *MBFR: Lessons and Problems*, Adelphi Paper 176 (London: IISS, Summer 1982). *Force Reductions in Europe: Starting Over* (Washington, D.C.: Institute for Foreign Policy Analysis, 1980) by Jeffrey Record is particularly useful for its critical appraisal of MBFR. Record examines the Warsaw Pact military buildup during the MBFR period and concludes that this arms control regime has done little to promote European stability.

While analyses of the negotiating forum warrant study to grasp and apply legacies applicable to the future of conventional arms control in Europe, sorting out intra- and inter-alliance dynamics is of inestimable importance in the search for an acceptable reduction formula. In an early attempt to decipher this puzzle, John Yochelson, in "MBFR: The Search for an American Approach," *Orbis* (Spring 1973), pointed out that there are no obvious solutions in MBFR, only subtle tradeoffs between gains and risks. An MBFR approach that yields to domestic pressures could adversely affect NATO and Warsaw Pact military and political relationships. Conversely, a purely pragmatic MBFR policy could precipitate the erosion of intra-NATO and domestic support. A detailed examination of the political context of European arms control, with relevance to MBFR, is provided by Theodor H. Winkler in *Arms Control and the Politics of European Security*, Adelphi Paper 177 (London: IISS, Autumn 1982). Winkler suggests that the complexity of competing interests, both within and between alliances, may preclude bold, broad initiatives. Similarly, Christoph Bertram's *Mutual Force Reductions in Europe: The Political Aspects*, Adelphi Paper 84 (London: IISS, January 1972), measures the political significance of MBFR on East-West relations and its likely impact on NATO and Western Europe.

The absence of any substantive progress in MBFR generated pessimism about the ability of the participants to achieve mutual force reductions. Perhaps, as Jonathan Dean suggests in "MBFR: From Apathy to Accord," *International Security* (Spring 1983), short-sighted concern for nuclear arms control distracted the US and the USSR from the risk of conventional war in Europe. After progress on INF seemed apparent, both NATO and the Warsaw Pact turned their attention to conventional arms control in a different forum than MBFR.

In late 1986, 35 nations in the Conference on Confidence and Security Building Measures and Disarmament in Europe (CDE) agreed in Stockholm on measures for observation and inspection of military maneuvers. In Feburary 1987, NATO and Warsaw Pact countries gathered in Vienna to discuss a mandate for negotiating reductions in conventional forces from the Atlantic Ocean to the Ural Mountains. The Soviets had signaled their willingness to accept Western demands for stringent verification measures, force strength information exchanges, and asymmetrical force reductions. In "The Future of Conventional Arms Control in Europe," in *Survival* (Jan./Feb. 1987), Richard Darilek discusses the CDE Stockholm agreement and its implications for future arms control negotiations. Similarly, John Borawski looks at the prospects for future agreement in "Toward Conventional Stability in Europe," in *The Washington Quarterly* (Fall 1987).

One of the objectives of the West in conventional arms control has been to achieve substantial reductions in Soviet offensive capability without undermining the cohesiveness of NATO defense. The RAND Corporation has proposed a useful framework for assessing various arms control proposals in James A. Thomson and Nanette C. Gantz, "Conventional Arms Control Revisited: Objectives in the New Phase" (Santa Monica, Calif.: RAND Note N-2697-AF, Dec. 1987). The authors caution that the Soviets would have to withdraw forces on a ratio of at least 5 to 1 for the result to be beneficial to the stability of the Central European conventional balance. This and other insightful analyses of the problems of conventional arms control are included in Uwe Nerlich and James A. Thomson's edited volume, *Conventional Arms Control and the Security of Europe* (Boulder, Colo.: Westview Press, 1988).

In March 1989, the members of NATO and the Warsaw Pact began negotiations on Conventional Armed Forces in Europe (CFE). Initial proposals by both alliances envisioned dramatic reductions in ground force equipment, tactical aircraft, and manpower. In December 1988, Secretary General Gorbachev had already fueled speculation about major changes in the European military balance by announcing unilateral reductions of Soviet forces from Eastern Europe that far exceeded anything NATO had ever sought in MFBR. Coupled with important changes in the political and economic dynamics in both Eastern and Western Europe, CFE negotiations could serve as a vehicle for revolutionizing the postwar East-West security relationship in Europe. An important issue, then, is how one understands stability—the goal of CFE negotiations—and how one achieves that through arms control. This is the focus of *Defining Stability: Conventional Arms Control in a Changing Europe* (Boulder, Colo.: Westview Press, 1989), written by Schuyler Foerster, William A. Barry III, William R. Clontz, and Harold F. Lynch, Jr., all of whom are military officers.

CONCLUSION

The intricacies and complexities of arms control are the subject of a rich and extensive literature. The literature surveyed here suggests themes and lessons across a broad spectrum of arms control fora and provides a foundation for further investigation and research. It is important to keep in mind that arms control exists not only in a security context but also in a broader political contest. To help place arms control into perspective, Joseph Kruzel's "From Rush-Bagot to START: The Lessons of Arms Control," *Orbis*

(Spring, 1986), develops several propositions about the preconditions of arms control and suggests why some negotiations end in agreement and others result in failure. In addition, *Superpower Arms Control: Setting the Record Straight* (Cambridge, Mass.: Ballinger Press, 1987), edited by Richard Haass and Albert Carnesale, surveys a variety of US-Soviet arms control efforts, examining the conditions for success and the effects that arms control has had on the superpower military relationship.

Perhaps the most striking aspect of the arms control literature, however, is how enduring many of the initial principles are that were developed decades ago by the pioneers in the field. Works by Thomas Schelling, Morton Halperin and Hedley Bull are included in the Selected Readings that follow. They merit reexamination as arms control becomes such a central feature of states' efforts to reduce the dangers of war.

SELECTED READINGS III

STRATEGY AND ARMS CONTROL
THOMAS C. SCHELLING and MORTON H. HALPERIN

This study is an attempt to identify the meaning of arms control in the era of modern weapons, and its role in the pursuit of national and international security. It is not an advertisement for arms control; it is as concerned with problems and difficulties, qualifications and limitations, as it is with opportunities and promises. It is an effort to fit arms control into our foreign and military policy, and to demonstrate how naturally it fits rather than how novel it is.

This is, however, a sympathetic exploration of arms control. We believe that arms control is a promising, but still only dimly perceived, enlargement of the scope of our military strategy. It rests essentially on the recognition that our military relation with potential enemies is not one of pure conflict and opposition, but involved strong elements of mutual interest in the avoidance of a war that neither side wants, in minimizing the costs and risks of the arms competition, and in curtailing the scope and violence of war in the event it occurs.

Particularly in the modern era, the purpose of military force is not simply to win wars. It is the responsibility of military force to deter aggression, while avoiding the kind of threat that may provoke desperate, preventive, or irrational military action on the part of other countries. It is the responsibility of military policies and postures to avoid the false alarms and misunderstandings that might lead to a war that both sides would deplore.

In short, while a nation's military force opposes the military force of potentially hostile nations, it also must collaborate, implicitly if not explicitly, in avoiding the kinds of crises in which withdrawal is intolerable for both sides, in avoiding false alarms and mistaken intentions, and in providing—along with its deterrent threat of resistance or retaliation in the event of unacceptable challenges—reassurance that restraint on the part of potential enemies will be matched by restraint on our own. It is the responsibility of military policy to recognize that, just as our own military establishment is largely a response to the military force that confronts us, foreign military establishments are to some extent a response to our own, and there can be a mutual interest in inducing and reciprocating arms restraint.

We use the term "arms control" rather than "disarmament." Our intention is simply to broaden the term. We mean to include all the forms of military cooperation between potential enemies in the interest of reducing the likelihood of war, its scope and violence if it occurs, and the political and economic costs of being prepared for it. The essential feature of arms control is the recognition of the common interest, of the possibility of reciprocation and cooperation even between potential enemies with respect to their military establishments. Whether the most promising areas of arms control involve reductions in certain kinds of military force, increases in certain kinds of military force, qualitative changes in weaponry, different modes of deployment, or arrangements superimposed on existing military systems, we prefer to treat as an open question.

If both sides can profit from improved military communications, from more expensive military forces that are less prone to accident, from expensive redeployments that minimize the danger of misinterpretation and false alarm, arms control may cost more, not less. It may by some criteria seem to involve more armament, not less. If we succeed in reducing the danger of certain kinds of war, and reciprocally deny ourselves certain apparent military advantages (of the kind that cancel out for the most part if both sides take advantage of them), and if in so doing we increase our military requirements for other dangers of warfare, the matter must be judged on its merits and not simply according to whether the sizes of armies go up or down. If it appears that the danger of accidental war can be reduced by improved intelligence about each other's military doctrines and modes of deployment, or by the provision of superior communication between governments in the event

Excerpted with minor revisions and with permission from Thomas C. Schelling and Morton H. Halperin, Strategy and Arms Control, © *1961, The Twentieth Century Fund, New York.*

of military crisis. These may have value independently of whether military forces increase, decrease, or are unaffected.

This approach is not in opposition to "disarmament" in the more literal sense, involving the straightforward notion of simple reductions in military force, military manpower, military budgets, aggregate explosive power, and so forth. It is intended rather to include such disarmament in a broader concept. We do not, however, share the notion, implicit in many pleas for disarmament, that a reduction in the level of military forces is necessarily desirable if only it is "inspectable" and that it necessarily makes war less likely. The reader will find that most of the present study is concerned less with reducing national *capabilities* for destruction in the event of war than in reducing the *incentives* that may lead to war or that may cause war to be the more destructive in the event it occurs. We are particularly concerned with those incentives that arise from the character of modern weapons and the expectations they create.

An important premise underlying the point of view of this study is that a main determinant of the likelihood of war is the nature of present military technology and present military expectations. We and the Soviets are to some extent trapped by our military technology. Weapon developments of the last fifteen years, especially of the last seven or eight, have themselves been responsible for some of the most alarming aspects of the present strategic situation. They have enhanced the advantage, in the event war should come, of being the one to start it, or of responding instantly and vigorously to evidence that war may have started. They have inhumanly compressed the time available to make the most terrible decisions. They have almost eliminated the expectation that a general war either could be or should be limited in scope or brought to a close by any process other than the sheer exhaustion of weapons on both sides. They have greatly reduced the confidence of either side that it can predict the weapons its enemy has or will have in the future. In these and other ways the evolution of military technology has exacerbated whatever propensities towards war are inherent in the political conflict between us and our potential enemies. And the greatly increased destructive power of weapons, while it may make both sides more cautious, may make the failure to control these propensities extremely costly.

THE STRATEGIC BALANCE

What kinds of arms arrangements would, and what kinds would not, increase the security of the participants and the rest of the world against the dangers of war and aggression? Much current discussion suggests that there are but two

principal desiderata. The first is that the *general level of armaments*, somehow measured, be reduced; the second is that the *ratio* of strengths of the two blocs (or relative strengths of the main participants) not become too disaligned, that an appropriate "balance" be maintained in the reduction. Both considerations are implied in such phrases as "balanced reduction," "phased reduction," or "proportionate reduction" of armaments. It is of course recognized that a distinction has to be made between strategic and tactical forces, nuclear and conventional forces, missiles and manpower, and so forth; but even in this connection, the problem seems to be widely viewed as one of maintaining an appropriate balance between adversaries, and something like proportionate reductions across the board is often referred to as a crude but satisfactory way around this complication.

The earlier discussion should have made clear that the strategic evaluation of an arms-control proposal, particularly of a comprehensive proposal, involves a complex of considerations that are not easily summarized in the simple notion of a "balanced reduction" of military forces. While it seems almost certain that any comprehensive and explicit arms accommodation between the two main power blocs would, as a practical matter, entail a reduction in military forces and that the relative strengths of the two blocs will be a major consideration, it is not at all certain—in fact, it is unlikely—that these two considerations should be dominant. There are many others: the vulnerability of strategic weapons to attack; the susceptibility of weapon systems to accident or false alarm; the reliability of command and communication arrangements; the susceptibility of weapon systems to sudden technological obsolescence; the confidence with which each side can estimate the capabilities of the opponent's weapons; the reaction time that weapon systems allow to decisionmakers in a crisis; the susceptibility of weapon systems to control and restraint in the event of war; the suitability of weapon systems for blackmail, intimidation, wars of nerves, and general mischief; and the effects of different weapons systems on the internal relations within alliances. These are important considerations and are not closely enough correlated with the general *level* of weapons on both sides or the simple arithmetical *ratio* of strength between the power blocs to permit "balanced reduction of forces" to be an adequate description of the strategic objective of arms control.

STABILITY

The concept of "stability" is often adduced as a third consideration. It is useful, though still incomplete. A "balance of deterrence"—a situation in which the incentives on both sides to initiate war are outweighed by the disincentives—is described as "stable" when it is reasonably secure against shocks, alarms and perturbations. That is, it is "stable" when political events, internal or external to the countries involved, technological change, accidents, false alarms, misunderstandings, crises, limited wars, or changes in the intelligence available to both sides, are unlikely to disturb the incentives sufficiently to make mutual deterrence fail. This concept of "stability" is a fuzzy one, partly because there is no clear-cut agreement among those who use the term on precisely what the trends and events are that might upset the balance and what the relative likelihoods are. The concept is also inadequate to cover all the considerations left out of the two usual ones (*level* and *ratio*) mentioned above; it could not, for example, be extended to cover the scale of unintended damage in the event of war, without its becoming just a synonym for everything important and desirable in the field of arms control. It does, nevertheless, cover a set of important considerations ranging from the vulnerability of retaliatory weapons themselves to the security of command and control arrangements, dependence on fallible warning systems, the urgency of decisions based on incomplete information, and the potential for misunderstanding that may be involved in the mode of deployment of weapons.

Qualitative and Quantitative Limitations

To some extent a distinction can be made between the quantitative and the qualitative aspects of arms control. In earlier eras there was interest in the limitation of offensive weapons in contrast to defensive weapons, of weapons that were unnecessarily cruel, or of weapons that involved civil damage all out of proportion to their military accomplishments. The present era is one in which important distinctions can be made; but the terms "offensive" and "defensive" are misleading, and the concept of "retaliation" is largely punitive rather than military. The distinction between a "first-strike" and a "second-strike" military capability, however, is a crude but useful distinction.

The distinction is crude because almost any weapon capable of firing back in retaliation is worth something in a first strike, or can be adapted to the purpose. And almost any weapon capable of launching an attack aimed at disarming the enemy would have some prospect of delivering retaliatory damage in the event it were attacked first. One can kill a rabbit with an elephant gun, and can probably kill an elephant with .22 rifles if enough of them are skillfully used; but there is a distinction between a rabbit gun and an elephant gun, and there is a distinction, though unfortunately not as striking, between the modern strategic weapons particularly suitable for a disarming attack on the enemy, and those that are particularly suitable for threatening retaliation if attacked first. On the whole, it costs money to make a retaliatory system reasonably secure against attack; for a given outlay, one gets less of a first-strike capability as a by-product of procuring a second-strike capability than if one ignores the second-strike qualities and invests the available military resources in a maximum first-strike capability. This distinction refers, it must be emphasized, to the entire weapon system, not just to the missiles or the aircraft: to the base configuration, the hardening or the mobility, the communications, the warning system, and everything else that goes to make up the strategic force. It furthermore is a distinction that is not absolute, but relative to the enemy's own forces: a missile-carrying submarine may be a useful first-strike weapon if the enemy has fixed land-based missiles near his shores, but not if he, too, has his missiles under water.

The distinction should not be pressed too far. A "second-strike" capability is not necessarily just a punitive retaliatory capability; it might include provision for attacking such enemy weapons as had not yet been launched. It might also include active and passive defense of the homeland, of the kind that would be involved (and would be more effective) in a first-strike force. Similarly, a first-strike force would not necessarily be one that depended solely or even mainly on surprise; certainly, a first-strike force that is used to threaten a major war in a hope of achieving objectives without a war must be one that, if war comes, enjoys some security against a preemptive blow. Furthermore, to the extent that a general or strategic war could be limited in its scope or intent instead of being an indiscriminately "all-out" affair, the distinction between first strike and second strike is blurred. These qualifications are important, but do not wholly detract from the usefulness of the distinction.

Offensive and Defensive Weapons

It is important to note that the old distinctions between offensive and defensive weapons are quite inapplicable in the present era, and are more nearly applicable in reverse. Weapons that are particularly effective against enemy weapons, and capable of launching a "disarming" attack, are precisely the weapons needed for the initiation of war. Weapons that are potent against populations, urban complexes, and economic assets have essentially a punitive rather than military quality. They are capable of retaliating, and of threatening retaliation, but are incapable of disarming the enemy and thus give their possessor little incentive or none at all to launch an attack. In that sense they may be reassuring to the other side.

In this connection, there are some logical similarities between weapons designed to destroy enemy forces before they are launched and active defenses designed to destroy them as they approach one's own country. To launch a major attack with some assurance against insupportable retaliation, one needs reasonable prospects either of destroying enemy weapons before they are committed, or of destroying them in flight, or of evading their effects by mobility, sheltering, or evacuation. In that sense, *defensive* measures may be at least as characteristic of a first-strike strategic force as of a purely retaliatory force. They do, of course, differ from the so-called counterforce weapons (those that seek to destroy enemy weapons before they are committed) in that defensive weapons themselves have usually tended to be the type that can only respond to an enemy initiative and not take initiative themselves.

It is worth observing in this connection that even passive defenses of the population, like fallout shelters and evacuation procedures, food stockpiles or organizational arrangements for the aftermath of war, are as natural a component of a first-strike force as they are a supplement to a purely retaliatory force. While this observation cannot do justice to the complex and important question of civil defense, it does help to illustrate that the traditional distinction among weapon systems, the traditional qualitative arms-control categories, are misleading in the present era of deterrence.

The missile-carrying submarine is an important illustration of the complexities involved. In some discussions this weapon system has been viewed as a deplorable extension of the arms race into a new medium. The fact that an enemy submarine can be close to one's own borders unobserved is especially disturbing; and the difficulties of antisubmarine warfare are viewed with alarm. But there is a growing recognition that the Polaris submarine may embody many of the qualities that we and our potential enemies would be seeking through arms control to embody in our strategic-weapon systems. If in fact the submarine proves to be reasonably invulnerable, capable of deliberate response, able to maintain communications, and not susceptible to accidents of a provocative sort nor prone to create false alarm, it may prove to be an ideally "retaliatory" and "deterrent" weapon, particularly if possessed by both sides (or if the side that does not possess them enjoys some equivalent invulnerability). It is probably also true, roughly speaking, that the submarine achieves these qualities at some expense: either side may be able to afford this kind of weapon system better if the other's weapons are not so

threatening as to stimulate a strong desire for a preemptive capability.

ASYMMETRIES IN THE LIMITATIONS

The Polaris system also reminds us that a comprehensive arms-control program may involve important asymmetries, or striking differences between the weapon systems allowed to the two power blocs. There is no reason to believe that we and the Soviets would be equally interested in this particular weapon system. For reasons that range from geography to military tradition, from intelligence and secrecy to technology, from the nature of our alliances to the character of our space program, the Western alliance and the Soviet bloc may develop very different interests in, or attitudes towards, the submarine as a strategic weapon. An implication of this is that any comprehensive arms-control program might have to recognize that the basic strategic weapons on both sides would be quite dissimilar. This in turn would make it difficult to arrive at any simple, durable, and reasonable comparison between the weapons on both sides; a submarine with sixteen missiles of a certain accuracy and reliability, carrying a certain nuclear warhead, on station a certain fraction of time, with a certain reaction time depending on its communications and its distance from an ideal launching point, admits of no simple obvious comparison with, say, a base complex of hard or soft, large, land-based missiles, or a railroad train with a certain number of smaller missiles spending a certain fraction of its time in a state of readiness, with a certain vulnerability to detection and destruction. This is emphasized not because it makes arms control more difficult (which it probably does, but not strikingly so) but because it reminds us that certain assymmetries under arms control are inevitable.

DIVERSIFICATION OF WEAPONS

While the qualities of particular weapon systems deserve careful attention, it is important not to push too far the search for the "ideal" weapon, under arms control as under the uncontrolled arms race. An important reason is that diversification itself may be strategically important, and even more so in a partially disarmed world. If we want strategic military forces on both sides that lend "stability" to the balance of deterrence—weapons that are unlikely to be substantially destroyed in an enemy attack, weapons that are unlikely to become suddenly impotent because of a technological revolution—there are powerful reasons for believing that a diversified retaliatory system is better than one built around a single weapon or a single operating concept. To restate this in a way that makes its arms-control implications clear: If the participants in a comprehensive arms-control arrangement want security, simultaneous with a reduction in the level of armaments, a diversified mix of retaliatory weapons may permit a lower level of *total* armaments than would be acceptable if each has to rely on a single weapon system.

The reasons why this is so are many. One is that the different weapon systems may pose sufficiently different kinds of targets as to complicate the coordination problem. Synchronizing a no-warning strike may be harder when the targets are in different places, with different characteristics, different environments, etc. Second, by requiring a diversity of attack, they may increase the ways that the defender can get intelligence. Third, one who exploits this possibility of diversity may design a weapon system, particularly with respect to its size and scope, in order to optimize the cost advantage involved—to maximize the enemy's costs in being able to deal with all the defender's weapon systems simultaneously. Fourth, by moving into new systems and environments one may force the attacker to prepare attacking weapons subject to serious uncertainty about the target they will have to be used on. Fifth, it increases the amount and kinds of intelligence the attacker needs. And sixth—a point that deserves to be emphasized—it must reduce greatly the confidence of any political decision-maker in the estimates that experts give about the outcome of a war.

It is important to stress that the diversified weapon system might well be more expensive than one that concentrates on the single "best" system. There are overhead costs in the development, production, and operation of strategic weapons; greater variety would entail smaller-scale exploitation of each weapon system. This may be a price worth paying, and is emphasized here as a further reminder that, while fewer weapons may be cheaper than more weapons, "better" weapons may not be cheaper than "worse" ones.

STABILITY AND SIZE OF FORCES

The foregoing discussion has treated the qualitative issues separately from the quantitative. It has suggested that the "stability" and other important characteristics of weapons are separate from the question of numbers or size. There is a connection, however, between the qualitative and the quantitative characteristics. The stability of the weapon systems on both sides may itself be a function of the numbers of weapons and the numbers of targets they present to each other. Whether a strategic force is, and appears to be, essentially designed to carry out a threat of retaliation, and hence suitable for deterrence, or instead is designed for maximum

success in a sudden attack on the enemy, to some extent depends on the very size of the force itself.

This is obvious when one side tries, or each tries, to achieve such numerical dominance that it could virtually preclude retaliation by an overwhelming attack. But the point to be emphasized here is that even if some rough equality is achieved and maintained in the balance of forces between the two sides, through arms control or otherwise, the *level* of forces on both sides may be an important determinant of the stability of that balance.

The reason is that the outcome of a general war, as perceived by either side, is not simply a function of the ratio of forces. That ratio may determine, very crudely speaking, what *fraction* of the victim's forces are capable of retaliating if one side launches a sudden attack on the other; but that fraction would be large or small in absolute size, and would imply larger or smaller retaliatory destruction, depending on the absolute level at the outset.

To illustrate the point, consider a situation in which one side has, say, half again as many weapons as the other. Suppose that—taking into account how the strategic forces are clustered in their bases and what kinds of targets they offer to each other, and taking into account the accuracy of the weapons, target location, and intelligence—every weapon fired in a well-coordinated surprise attack has about a 75 percent chance of destroying an enemy weapon. Two weapons fired at the same target would have a fifteen-sixteenths chance of destroying the target. Then 150 weapons fired at 100 weapons would be expected to destroy about 84 weapons on the other side, leaving about 16 with which the victim could attempt retaliation in the hostile and disorganized environment in which he then found himself. Fifteen hundred weapons fired at 1000 could, crudely speaking, then be expected to destroy a similar proportion, say, about 840, leaving about 160 with which the victim could attempt retaliation. The ratio of weapons in both cases is the same; the retaliatory damage to be anticipated by the attacker, however, may be strikingly different, since the attacker faces a residual force of 16 or 160 depending on the initial scale of military forces on both sides.

This crude illustration is hardly a model of modern warfare, and ignores especially the interdependencies among weapons that rely on the same communications, warning, and so forth. Nevertheless it illustrates a valid and important point: that the *level* of forces on both sides, not simply the *ratio*, is an important determinant of the prospective scale of retaliation, and of the potency of deterrence.

This is hardly an argument for encouraging a maximum of weapons on both sides. (The point elaborated here has dealt only with the *deterrent* qualities of the strategic forces; there are also the questions of whether larger forces are conducive to keeping limited a war once it starts, and whether larger forces might be more susceptible to "accidents" of some sort. Neither of these questions admits of a simple answer.) It is an argument, however, that must be weighed in the balance with other considerations in deciding how low a level of forces on both sides might be contemplated in an arms-control arrangement that wished to preserve and stabilize the deterrent balance.

It is essentially this consideration and the fact that very small forces are more vulnerable to a clandestine attacking force that lead many who concern themselves with arms control to think of a goal well short of the complete elimination of strategic weapons. It is not simply that a reduction to modest levels is a less ambitious goal. It is that the situation may become safer in the event of war, and more stable with respect to the likelihood of war, if forces are substantially reduced, but that *beyond a certain point* further reduction may increase both the fears and the temptations that aggravate the likelihood of war.

This is not said to settle the question of whether total disarmament with respect to strategic weapons is the wrong goal. The point is rather that, beyond a certain point in the reduction of retaliatory weapons on both sides, one must recognize that the balance of deterrence is being dismantled in exchange for the advantages of total disarmament. One has to consider whether the exchange is a profitable one, and to recognize that the rationale for proceeding towards zero must be different from the original rationale for reducing to moderate levels. One may also have to recognize that, no matter how nicely the balance *between* the opposing forces may be kept—no matter how well some *ratio* of forces is preserved in the process of arms reduction—the greater security sought at some extremely low level of forces may require passing through a region of greater insecurity, a region where forces are too small for mutual deterrence and too large for the hoped-for benefits of a totally (or nearly) disarmed world to compensate for the disappearance of mutual deterrence.

STABILITY VERSUS DISARMAMENT

The concept of "stabilized deterrence" has recently come into vogue as a characterization of a particular school of arms control. Those who wish to reduce armaments and those who wish to stabilize deterrence are contrasted. Up to a point the contrast is useful, and identifies a difference in emphasis. But the two objectives are not alternatives. Whatever the level of arma-

ments, more stability is better than less. Even with zero armaments, there is still a problem of deterring rearmament; and the stability of this deterrence may depend on much the same considerations that stabilized nuclear deterrence depends on in an armed world—namely, the elimination of the advantage in going first, in starting rearmament as in starting a war; increasing the tolerance of the system to errors in judgment or mistaken intentions; minimizing the haste with which decisions to initiate rearmament or to initiate war must be taken; structuring the incentives so that, whatever the capabilities, the mutually destructive or risky action is not initiated.

"Stabilized deterrence" is also sometimes equated with stabilizing the balance of terror. But the greater the stability, the less terror there is. The fear is a function of the instability, of the lack of confidence in the incentives that exist, of anxiety about the need to react with violence to every alarm and threat. Efforts to stabilize deterrence are efforts to tranquilize anxieties and decisions, to strengthen the incentives towards deliberate rather than hasty action, to minimize the alarms and mistakes. To make the initiation of war manifestly profitless, and to provide each side sufficient control over its own decisions to eliminate the inadvertent initiation of war, may substantially deflate the fearsomeness of the balance.

One abstains from crossing the street in the face of onrushing traffic not out of "fear" but simply because one knows better; he takes for granted that the consequences would be disastrous, but he knows that he can control his own actions. Successfully stabilized deterrence, to the extent that it could be achieved, might better be described as an effort to replace the balance of fear with a "balance of prudence."

The search for stability has also been criticized, and can justly be criticized, for expecting too much. We live in a world of technological as well as political uncertainties; the stabilized deterrence of tomorrow may become less stable the day after that. This is a splendid caution about what to expect, and a reminder that to stabilize deterrence is not equivalent to guaranteeing freedom from war. This is no objection to pursuing stabilization vigorously and persistently; we can never be guaranteed against household accidents, but that does not make a safety program futile.

There is an incompatibility, of course, between the notion that war is prevented by some kind of deterrence, and that war is prevented by the eradication of military establishments, military tradition, and military thinking. Stabilizing deterrence is not therefore compatible with all motives for disarmament. On the other hand, some proposals for stabilizing deterrence do involve substantial reduction of force levels;

and stabilized deterrence as a concept is not committed in advance to any particular level of destructive power.

The relation between stabilized deterrence and reduced force levels can be indirect as well as direct. One may hope directly to reduce force levels in the interest of minimizing the possible violence in the case of war. One may propose reducing force levels in the interest of stabilizing deterrence. Or one may propose vigorous measures to help stabilize deterrence, in the expectation that force levels will naturally and unilaterally be reduced as a result. Certainly a powerful motivation toward the indefinite buildup of destructive power is a preoccupation with the need to preempt the enemy's attack; another is the fear that one's own forces might be so destroyed in the surprise attack that deterrence itself requires an enormous margin for safety. If forces on both sides become fairly invulnerable to preemptive attack, the incentive toward larger and larger numbers might be greatly deflated. Reductions in force levels might be reciprocated, in something of a downward spiral. The more the arms race can be diverted into those kinds of qualitative improvements in military force that reduce the danger of war, the less the incentive may be to maintain exorbitant levels of destructive power. Furthermore, the likelihood that all existing explosive potential would be used exhaustively and indiscriminately in the event of war becomes less and less, the more each side designs its military forces for deliberate rather than for hasty actions, the better the control that it maintains over them, and the less obvious advantage there is in attempts to saturate the enemy's country in the hope of nullifying his retaliatory forces.

STABILITY AND TOTAL DISARMAMENT

It is often said or implied that many problems that remain with us under arms control, or that arms control itself may usher in, are due solely to the incompleteness of the disarmament. In a wholly disarmed world, it is asserted, especially one subject to an organized international disciplinary force, the blackmailing threats would be hollow, limited war would by physically unavailable, and even the power to reverse the trend of history and rearm would have been put beyond the reach of national decision.

The argument is quite unpersuasive. In the absence of some effective policing force, primitive war is still possible, rearmament is possible, and primitive wars that last long enough may convert themselves by rapid mobilization into very modern warfare. Nor is primitive warfare necessarily a very attractive alternative to the more modern type, unless the sides are so well balanced and the supply lines so critical

that the tactical defense automatically assumes dominance over the offense.

The "police force" itself poses some genuine problems. Will it rely on "containment" and limited war; or will it rely on the threat of retaliation to keep nations not only confined within their own borders but deprived of weapons and their productive facilities? How will it cope with "creeping rearmament" or with national rearmament that is motivated by the fear that others already are, or soon will be, rearming? Will nations be coerced by other nations' threats of rearmament; and will the international force's deterrent threat (to intervene forcibly, or to retaliate from a distance) against a rearming nation be considered adequate protection? Would it permit defensive alliances among states that are too small to resist alone even technologically primitive aggression by large powers; if so, what rules of "balance of power" will it permit or encourage in respect of military alliances? If a decisive and effective monopoly of military force is brought into existence, one that can act quickly and decisively and against the will of its opponents and the vacillation of its well-wishers, can it be prevented from acting as either a benevolent or a tyrannical despot? And how does it distinguish between external warfare and civil war?

These are not secondary problems to be solved after total disarmament is agreed on, nor are they problems whose solution is held up at the present time only because we lack the will and the political environment to make disarmament possible. These questions are as fundamental as the political and strategic military problems that characterize the present world. They may be easier to manage in an environment of "total disarmament" (including a world with a unified monopoly of military force) than they are in the present world. But they will not be absent.

In this respect, "total disarmament" may not differ as much in character from "arms control" as is sometimes implied. In any of these contingencies, conflict of interest will occur, potential force will always be at hand, and the military technology of the present era and new technology still to come will not have been erased from the records and men's memories.

CONCLUSIONS

This has been an effort to take arms control seriously. Like any serious business, arms control is complicated and uncertain; it involves problems and risks. Like any substantially new business, it is commonly the object of oversimplification. Like any business involving the security of the nation and of the world, it gives rise to excessive hopes and naive formulations, or naive criticism and uncritical rejection. Most seriously of all, there is a tendency to think of "arms control" as a well-defined subject, one whose nature we perceive and understand, and one that stands in rather clear antithesis to the world that obtains, a world alleged to be without arms control.

We have attempted to show that arms control is a rich and variegated subject whose forms and whose impact on security policy and world affairs have been only dimly perceived. It can be as formal as a multilateral treaty or as informal as a shared recognition that certain forms of self-control will be reciprocated. It may involve some "cops and robbers" activities like cheating and detection, but may also involve many of the continuing regulatory and negotiatory processes that we associate with bureaucracy and diplomacy. It may involve the straightforward elimination of armaments or rather subtle changes in the character of armaments—even improvements in certain kinds of arms—or may involve communications, traffic rules, or other arrangements superimposed on existing military establishments. It may be as "political" as the demilitarization of a disputed country, or as little political as an understanding about noninterference in each other's military communications. And, like many reciprocal restraints that we take for granted, arms limitations may exist without our being actively aware of them.

What we have tried to emphasize more than anything else is that arms control, if properly conceived, is not necessarily hostile to, or incompatible with, or an alternative to, a military policy properly conceived. The view we have taken is that arms control is essentially a means of supplementing unilateral military strategy by some kind of collaboration with the countries that are potential enemies. The aims of arms control and the aims of a national military strategy should be substantially the same. Before one considers this an excessively narrow construction of arms control, he should consider whether it cannot just as well be viewed as a very broad statement of what the aims of military strategy should be.

Surely arms control has no monopoly of interest in the avoidance of accidental war; anyone concerned with military policy must be concerned to minimize the danger of accident, false alarm, unauthorized action, or misunderstanding, that might lead to war. Arms control has no monopoly of interest in reducing the destructiveness of war if war occurs; military policy, too, should be concerned with the survival and welfare of the nation. Arms control need have no monopoly of virtue, in being less concerned with the nation and more concerned with humanity; a responsible military policy should not, and certainly would not, value at zero the lives and welfare of other populations, even enemy populations.

Military strategy is no longer concerned with simply the conduct of a war that has already started towards some termination that is taken for granted. Especially since World War II, military strategy has been as concerned with *influencing* potential enemies as with defeating them in combat. The concept of "deterrence" is itself a recognition that certain outcomes are worse for both ourselves and our potential enemies than other outcomes, and that a persuasive threat of military action coupled with a promise to withhold such action if the other country complies may be more significant than the military action itself. The concept of limited war is a recognition that we and our potential enemies may have a common interest, even after war starts, in limiting our objectives and checking war. Thus, military policy itself recognizes that we have a common interest with our potential enemies in avoiding a mutually destructive war, and a common interest in limiting war even if it occurs.

But sophistication comes slowly. Military collaboration with potential enemies is not a concept that comes naturally. Tradition is against it.

What we call "arms control" is really an effort to take a long overdue step towards recognizing the role of military force in the modern world. The military and diplomatic worlds have been kept unnaturally apart for so long that their separation came to seem natural. Arms control is a recognition that nearly all serious diplomacy involves sanctions, coercion, and assurances involving some kind of power or force, and that a main function of military force is to influence the behavior of other countries, not simply to spend itself in their destruction.

It is the conservatism of military policy that has caused "arms control" to appear as an alternative, even antithetical, field of action. Perhaps arms control will eventually be viewed as a step in the assimilation of military policy in the overall national strategy—as a recognition that military postures, being to a large extent a response to the military forces that oppose them, can be subject to mutual accommodation. Adjustments in military postures and doctrines that induce reciprocal adjustments by a potential opponent can be of mutual benefit if they reduce the danger of a war that neither side wants, or contain its violence, or otherwise serve the security of the nation.

This is what we mean by arms control.

CONTROL OF THE ARMS RACE

HEDLEY BULL

THE OBJECTIVES OF ARMS CONTROL

It is commonly assumed that the only important questions that arise in connection with disarmament or arms control concern how it may be brought about. But the question must first be asked, what is it for? Unless there can be some clear conception of what it is that disarmament or arms control is intended to promote, and to what extent and in what ways it is able to do so, no disciplined discussion of this subject can begin.

The demand for disarmament is not always accompanied by any close definition of the objectives it might promote, or by any recognition of conflict among these objectives, in cases where this exists. The greater part of what has been said about disarmament is the product of public movements of agitation and protest against governments for their failure to disarm; or the product of official pronouncements designed to allay such protest, or to exonerate the government at home and direct the protest towards governments abroad. It is not to be expected of protest movements or of governments that, in their pronouncements about disarmament, they will set more store on precision of expression and the recognition of difficulty and complexity than on the political effects these pronouncements will have. Nor, having regard for the essential character of protest movements about disarmament and of government disarmament policies, and for what they are intended to achieve, is it any reproach to them that they do not.

There may be distinguished, however, three main grounds upon which disarmament or arms control has been held to be desirable. The first ground is that disarmament contributes to international security: that armaments, or particular kinds or levels of armaments, or armaments races, are a cause of war, which disarmament or arms control will remove ("stop the drift to war"). The second ground is that it releases economic resources: that armaments, or armaments races, are economically ruinous or profligate, and that disarmament or arms control would make possible the diversion of resources now squandered in armaments into other and worthier channels ("reduce the burden of armament"). The third ground is the moral one: that war, or certain kinds of war, or preparing for and threatening war, is morally wrong . . .

Before the First World War disarmament doctrine placed its chief emphasis on the economic objective. But since then it has had as its core the notion that armaments and, in particular, armaments races are a cause of war; and that to abolish or reduce and limit armament, and to halt or otherwise control armaments races, is to contribute to international security.

The extent to which armaments and armaments races provoke international tension and war, and the extent to which arms control can provide guarantees of international security, are often exaggerated. The contribution of arms control to international security may be a modest one, and we must first reflect upon the source of its limitations.

The military factor is an important one in international politics, but it is not the only one. The view that international security is a matter of disarmament or arms control belongs to a group of opposed but closely related views, whose common theme is *peace through the manipulation of force*. According to one of these views, the maintenance of peace is a matter of securing a preponderance of force in the hands of a central authority, a world government or universal state. According to another, it is a matter of the pooling of force among right-minded nations, in a system of collective security. According to another, it is a matter of securing an equilibrium of force or balance of power throughout international society. According to yet another, security depends on neutralizing the advantages of resorting to force, in a system of mutual deterrence or general terror. Finally, there is the view that it depends on abolishing or reducing force, through arms control.

All of these doctrines can be pursued a certain distance, and there are political and strategic circumstances in which each of them carries conviction. But the claims of each become absurd when they are pressed to the point of considering military in abstraction from political considerations. What happens in international politics is a matter of will as well as of weapons; or, in the language of the recent American strategic debate, of intentions as well as of capability. Armed power and the will to employ it is each itself and not the other thing. Each affects and limits the other, but neither can be regarded as the mere expression of the other.

This is as true of the doctrines of disarmament and arms control as of the others. Whatever contribution such phenomena as disarmament or arms control may have made in the past, or may come to make in the future, to international peace and order, it is a contribution whose bounds can be no greater than those set by the limited importance of the military factor in international relations. We may illustrate this limitation by considering two of the characteristic assertions of disarmament doctrine: that arms races lead to war, and that disarmament or arms control halts arms races.

Arms races are intense competitions between opposed powers or groups of powers, each trying to achieve an advantage in military power by increasing the quantity or improving the quality of its armaments or armed forces.[1] Arms races are not peculiar to the present time or the present century, but are a familiar form of international relationship. Arms races which are qualitative rather than quantitative, which proceed more by the improvement of weapons or forces than by the increase of them, have grown more important with the progress of technology: and what is peculiar to the present grand Soviet-Western arms race is the extent to which its qualitative predominates over its quantitative aspect. However, there have not always been arms races, and they are in no way inherent in international relations, nor even in situations of international political conflict. Where the political tensions between two powers are not acute; where each power can gain an advantage over the other without increasing or improving its armaments, but simply by recruiting allies or depriving its opponent of them; where the economic or demographic resources to increase armaments, or the technological resources to improve them, do not exist, we do not find arms races. The prominent place which the Soviet-Western arms race occupies in international relations at the present time arises from the circumstance that the opportunities available to each side for increasing its relative military power lie very much more in the exploitation and mobilization of its own military resources, than in the attempt to influence the direction in which the relatively meager military resources of outside powers are thrown, by concluding favorable alliances or frustrating unfavorable ones. It arises from the circumstance that the balance of military power can at present be affected very much more by armaments policy than by diplomacy.

It is seldom that anything so crude is asserted as that all wars are caused by arms races, but the converse is often stated, that all arms races cause wars: that in the past they have always resulted in war; or even that they lead inevitably to war.

It is true that, within states, armaments tend to create or to shape the will to use them, as well as to give effect to it; and that, between states, arms races tend to sustain or to exacerbate conflicts of policy, as well as to express them. Within each state, the military establishment, called into being by the policy of competitive armament, develops its own momentum: it creates interests and diffuses an ideology favorable to the continuation of the arms race, and generates pressures which will tend to resist any policy of calling it off. In this respect, that they display a will to survive, the armed forces, the armaments industries, the military branches of science and technology and of government, the settled habits of mind and those who think about strategy and defense, are like any great institution involving vast, impersonal organizations and the ambitions and livelihood of masses of men. Apart from these internal pressures of armaments establishments upon policy, it is possible to see in the pressures which each state exerts upon the other, in the action and reaction that constitute an arms race, a spiraling process in which the moving force is not, or not only, the political will or intention of governments, but their armaments and military capability. One nation's military security can be another's insecurity. One nation's military capability of launching an attack can be interpreted by its opponent as an intention to launch it, or as likely to create such an intention, whether that intention exists or not. Even where neither side has hostile intentions, nor very firmly believes the opponent to have them, military preparations may continue, since they must take into account a range of contingencies, which includes the worst cases. Insofar as the competitors in an arms race are responsive to each other's military capabilities rather than to each other's intentions, insofar as they are led, by estimates of each other's capabilities, to make false estimates of each other's intentions, the arms race has a tendency to exacerbate a political conflict, or to preserve it where other circumstances are making towards its alleviation.

But the idea that arms races obey a logic of their own and can only result in war, is false; and perhaps also dangerous. It is false because it conceives the arms race as an autonomous process in which the military factor alone operates. The chief source of this error in recent years has been the belief in the importance of the various armaments competitions among the European powers, and especially the Anglo-German naval race, in contributing to the outbreak of the First World War. Even in this case, the importance of this factor is a matter of controversy. In the case of the origins of the Second World War, the autonomous arms race cannot be regarded as having been important. On the contrary, we should say that the military factor which was most important in bringing it about was the failure of Britain, France, and the Soviet Union to engage in the arms race with sufficient vigor, their insufficient response to the rearmament of Germany. In general, arms races arise as the result of political conflicts, are kept alive by them, and subside with them. We have only to reflect that there is not and has never been such a thing as a general or universal arms race, a war of all against all in military preparation. The context in which arms races occur is that of a conflict between particular powers or groups of powers, and of military, political, economic, and technological circumstances in which an armaments competition is an appropriate form for this conflict to take.

Arms races have not always led to war, but have sometimes come to an end, like the Anglo-French naval races of the last century, and the Anglo-American naval race in this one, when, for one reason or another, the parties lost the will to pursue them. There is no more reason to believe that arms races must end in war than to believe that severe international rivalries, which are not accompanied by arms races, must do so. It is of course the case that wars are made possible by the existence of armaments. But the existence of armaments, and of sovereign powers commanding them and willing to use them, is a feature of international society, whether arms races are in progress or not. What is sometimes meant by those who assert that arms races cause wars is the quite different proposition that armaments cause wars. While armaments are among the conditions which enable wars to take place, they do not in themselves produce war, or provide in themselves a means of distinguishing the conditions of war from the conditions of peace. All international experience has been accompanied by the existence of armaments, the experience of peace as much as the experience of war. To show why, in a context in which armaments are endemic, wars sometimes occur and sometimes do not; to show why, in this context, arms races sometimes arise, and either persist, subside, or end in war, it is necessary to look beyond armaments themselves to the political factors which the doctrine of the autonomous arms race leaves out of account.

These observations have not been made with a view to inducing complacency about the present arms race, or to minimizing its risks. But the idea of an inevitable connection between arms races, or armaments, and the eruption of war, as well as being false, contains a fatalism with risks and dangers of its own. It is natural for the advocate of extreme and drastic remedies, whether in the direction of disarmament, world government, or elsewhere, to strengthen his case by asserting that if the world moves on without these remedies, disaster is certain to follow. It cannot be said that this is an optimistic view, nor one which leaves us with much hope.

For however desirable it may be that the world should be radically different from what it is, or that it should be made so by a sudden act of will, we must contemplate the possibility that changes of this kind will not occur. In particular, however desirable it may be to erase international armaments competition from the world, and however great the priority any feasible means of doing so must have over lesser measures, we must have regard for the possibility that it will continue in some form. Given the pressures that operate against drastic disarmament, and against that reversal of the main political tendencies of the world with which it is so closely connected, not to do so would be blindness indeed. We should not therefore lightly deny ourselves the possibility of asking whether, if it does continue, there are steps that can be taken which might reduce its dangers. It is the concern of this inquiry to present views which are true, not views which are hopeful. If it were true that arms races are bound to lead to war, this is something which we should not fail to recognize. But as it is false, we may draw comfort from the reflection that even in the absence of sudden and drastic changes in the military and political structure of international society, there may be policies that can be shaped, and international arrangements arrived at, that will reduce the dangers of a world still armed and divided.

The limitations of the view that arms races lead to war apply also to the view that the halting or reversing of arms races is a matter of arms control or disarmament. Just as the history of armaments and arms races is only one aspect of the history of international politics, so the history of disarmament and arms control can be only one aspect of the history of armament. We cannot expect that a system of arms control will be brought into operation, nor that, if it is, it will persist, unless certain political conditions are fulfilled. It is, I believe, quite erroneous to suggest that disarmament cannot begin until political disputes have been removed, or that disarmament is something which follows automatically once they have been. On the contrary, it is only in the presence of political disputes and tensions serious enough to generate arms competitions that arms control has any relevance. Arms control is significant only among states that are politically opposed and divided, and the existence of political division and tension need not be an obstacle to it. On the other hand, the political conditions may allow of a system of arms control, or they may not. Unless the powers concerned want a system of arms control; unless there is a measure of political détente among them sufficient to allow of such a system; unless they are prepared to accept the military situation among them which the arms control system legitimizes and preserves, and can agree and remain agreed about what this situation will be, there can be little place for arms control.

This is indicated by the recent history of arms control agreements. The Rush-Bagot Treaty of 1817, under which the British Empire and the United States reduced and limited their naval armaments on the Great Lakes, made an important contribution to the development of confidence between these powers. The Washington Naval Treaty of 1922, under which the British Empire, the United States, Japan, France, and Italy declared a ten-year "naval holiday" in the construction of new capital ships and aircraft carriers, limited each others' permitted total tonnage in these classes of vessel, and limited the size and armament of vessels in these classes and in the cruiser class, had a notable effect both on the course of armaments competition and on the course of international politics generally. So did the London Naval Treaty of 1930, under which the British Empire, the United States, and Japan extended the "naval holiday" till 1936, and limited the maximum displacement of vessels in the destroyer and submarine classes, and the total tonnages permitted in the destroyer, submarine, and cruiser classes. The treaty systems which preserved and made legitimate these levels and kinds of armaments or armaments ratios confirmed and strengthened the wider areas of agreement and mutual accommodation among these powers, on which peaceful and stable relations among them rested. But they also expressed, and were part of, these wider areas of agreement. Moreover, it was only in favorable political conditions that, once established, they continued to flourish. The Rush-Bagot Treaty, though it had to weather some political storms, survived in the temperate climate of Anglo-American, and Canadian-American, relations. On the other hand, when Japan determined to overthrow the political and territorial status quo in the Pacific, the Washington and London naval systems collapsed. The dependence of arms control systems for their survival on the continuance of those wider areas of political agreement of which they are a part is best illustrated by the most common form of arms control system, that contained in the military clauses of treaties of peace. It is characteristic of such systems that, like the disarmament clauses binding Germany in the Treaty of Versailles, they are part of wider political situations which the treaty makes legitimate and seeks to preserve, but are dictated by the victor and are inherently temporary.[2]

Arms control systems therefore do not guarantee their own perpetuity. Theories of arms control often try to circumvent this difficulty by building into their projected systems some feature which purports to guarantee the permanence of the system, come what may politically.

Thus the abolition of armaments is sometimes held to afford a means of making war physically impossible, whether nations have the will to wage it or not. Systems of inspection, supervision, and sanctions are sometimes conceived of as means whereby arms control can be kept in operation, whatever the intention of the inspectors and the inspected. There are certainly means, formal and informal, of curbing or restraining the will to challenge or upset the military situation preserved in the system: if there were not, there would be no such thing as arms control. But arms control does not provide a technique of insulating a military situation from the future will of states to change it: it cannot bind, or settle in advance, the future course of politics. There are no technical means of excluding the political factor.

If armaments and arms races are a threat to international security, and disarmament or arms control a means of removing it, there are limits beyond which neither proposition can be pressed. However, these are limits which circumscribe not only the doctrine of disarmament, but also those military policies and doctrines with which it most frequently comes into collision. That armaments are causes as well as effects, and shape political motives and intentions as well as express them, is as much an assumption of the policy of "peace through strength" as it is of the policy of "peace through the reduction of armaments." It is no less a part of the policy of "deterrence" or the idea of "peace through terror," than it is a part of the ideas of disarmament or arms control. The limitations of disarmament doctrine to which I have drawn attention are not peculiar to this doctrine, but are the limitations of the whole field of strategic studies.

Moreover, within these limits, there is a vast area which it is the purpose of this study to explore. If armaments do not of themselves produce war, certain kinds of levels of deployments of armament may be more likely to give rise to the decision to go to war than others. If arms races do not necessarily lead to war, there are directions they can take which undermine security against it.

This much has been said not in order to discredit the doctrine of disarmament, but to suggest a sense of sobriety in place of the intoxicating hopes to which it sometimes gives rise. What disarmament and arms control may have to contribute to international security is considered at length later.

THE CONDITIONS OF ARMS CONTROL

However desirable it may be, arms control can occur only if circumstances are such that governments both want it and can agree on its terms. In this sense, that it is brought about and maintained by the policies of sovereign governments, the conditions of arms control are political. Though this is an elementary point, it is one which is not sufficiently taken into account in public discussion of disarmament or arms control, which is inclined to view the latter as waiting upon the evolution of a method or the discovery of a technique. The assumption being that the powers want disarmament and subordinate their policies to the pursuit of an agreement about it, "the problem of disarmament" is seen to consist of procedural questions concerning what preparations should be made for the negotiations, what powers should be represented at them and by whom, whether they should consider arms control as a whole or piecemeal, before security or after it, and so on. Or else it is seen to consist of technical questions concerning the characteristics of weapons and armed forces, the design of systems of inspection and supervision, and the elaboration of administrative machinery. Problems of this kind exist, and systems of arms control require solutions to them. But unless the political conditions for arms control are present, the question of what method or procedure is appropriate in arms control negotiations, and the question of how the technical problems involved in arms control can be solved, are of minor importance, and attempts to solve them in abstraction from political circumstances are of no significance. The view that international negotiations about arms control are concerned with the search for solutions to these problems, or that their failure to issue in agreements arises from the difficulty of finding these solutions, is, on the whole, mistaken. For the protracted public conversation of the powers about arms control should be viewed not as a cooperative attempt to solve a problem, but as a theme in their political relations.

Since the First World War, the ritual pursuit of disarmament has been part of the foreign policy of every important state. A nation's disarmament policy is, however, a subordinate part of its foreign policy, and expresses the general character of that policy. It is not necessarily shaped primarily, nor even at all, with a view to bringing about disarmament. At the World Disarmament Conference which began in 1932, for example, French disarmament policy was directed towards security against Germany, and German disarmament policy was directed towards equality with France. There was common ground between them as the policies of each included the alternative of some form of disarmament: an agreement, to this extent, was negotiable. But what determined the stance of each power in the conference chamber was not the range of ideas constituted by "the problem of disarmament," but the range of pressures and demands constituting its national policy. The most persistent objective of any nation's disarmament policy is that of demonstrating to opinion at home and abroad that efforts are being made towards disarmament, and that the reason why no agreement is arrived at lies in the policies of other nations, not in its own. The more radical and grandiose a disarmament proposal, the more it will satisfy this objective. Proposals which are radical and grandiose, moreover, are advanced in the knowledge that they will be rejected, and are an indication that the policy of the power advancing them is not directed towards disarmament.

But it would be quite mistaken to regard arms control negotiations as an elaborate charade enacted by the governments of the world for the benefit of the peoples of the world. Though they want other things more, most governments in most negotiations do want an agreement, if it can be had on their own terms. Usually, each government is divided as to how seriously an agreement should be sought: as the United States government has been regularly and publicly divided over the desirability of a ban on nuclear tests.

However, the question of establishing a system of arms control does not arise unless the powers concerned regard the pursuit of arms control, as distinct from the conduct of political warfare about arms control, as among the objectives of their policy. It cannot by any means be assumed that they do (nor should it be suggested without regard for other considerations, that they always should); and in fact many of them do not, except in the sense that they want other states to disarm. The conduct of political warfare about arms control, the cultivation of a favorable public image in relation to disarmament, is a constant preoccupation of all governments concerned in arms control negotiations or in making public pronouncements about them. The serious pursuit of arms control agreements, on the other hand, may be part of a nation's foreign policy, or it may not. Whether or not it is will depend upon the particular system of arms control, the general character of the nation's foreign policy, and the circumstances of the moment. We can say that powers which are bent at all costs on achieving or maintaining military supremacy over their opponents cannot contemplate the restraints of a comprehensive system of arms control. Powers that are weak, ambitious, and frustrated, as Germany and the Soviet Union were in the years of the League disarmament negotiations, and as many Asian, African, and Latin American nations are now, are adamant in their demands that the Great Powers should disarm, and would welcome a system of arms control which brought such disarmament about. But they would be hostile to the reduction of their own armaments or to the imposition of controls on the local arms races to which they are parties. The judgment that the serious pursuit of a particular arms control system does, or does not, form part of a nation's policy can be made only imprecisely and temporarily: for governments, as we know from public debates in Western countries on these matters, are often divided about the pursuit of a system of arms control, and the support for any such system within a government may ebb and flow.

If, however, the powers want a system of arms control, and present proposals which they believe might be accepted, arms control negotiations move out of the sphere of political warfare and into that of diplomacy. The obstacles to agreement in arms control negotiations are the same as in other kinds of negotiations, that in them each party seeks to promote its own advantage. It contemplates its own advantage, however, with a sensitivity and jealous regard that are peculiar even in diplomacy. For the subject matter of arms control negotiations is the balance of military power. Treaties of arms control do not abolish military power, but stabilize a military situation. The central issue of arms control negotiations is: what military situation? No power is prepared to contemplate a treaty unless the situation that results from it is one in which its own military interest is firmly secured. Two facts stand persistently in the way of agreement: the inherent uncertainty as to what constitutes an equal balance between opponents, and the determination inherent in all military policy to err on the safe side. The imprecise and constantly changing balance that exists in the real world, and in the calculations and anxieties of statesmen, must be translated into the precision and fixity of a treaty. In all serious arms control negotiations, the proposals of any power express a closely reasoned (though not necessarily correct) estimate of its military interests. Thus in the League period, the negotiators were unable to agree about what levels of military manpower a disarmament treaty should preserve, and about what categories of armament should be abolished or restricted by a scheme of qualitative disarmament, each power wishing to restrict those weapons most useful to its opponents. Britain and the United States, as the great naval powers, wished to abolish the submarine but retain battleships and aircraft carriers; the lesser naval powers wished to retain the submarine. France, as the great land power, wished to retain tanks and heavy guns, the lesser land powers favored restriction of them, and so on. In the years of United States monopoly or superiority in nuclear weapons, there was a long debate concerning whether nuclear weapons should be abolished before the establishment of controls, as the Soviet Union insisted, or after, as the United States insisted. At the present time the United States, decisively inferior in military intelligence, gives priority to the idea of inspection; the Soviet Union, suffering a geographical disadvantage from her encirclement by the Western system of alliances, gives priority to the abolition of foreign military bases; France, with the makings of a stock of nuclear explosives, but without an advanced means of delivering them, gives priority to the abolition of military missiles and long-range bombing aircraft.

The examples that have been mentioned do not do justice to the great subtlety and complexity of the relationship between disarmament proposals and estimates of military advantage. That relationship can be best studied in the debates of the League of Nations Disarmament Conference, which, unlike any serious negotiations of the United Nations period, was attended by nearly all the nations then in existence. All kinds of governments were forced to think carefully about disarmament, and their proposals reflect their calculations and anxieties about the most elusive and imponderable ingredients in war potential, from communications, geography, climate, industry, and wealth, to military commitments, social structure, political ideology, and national character. The proceedings of the League of Nations Disarmament Conference represent the most searching analysis that has ever been made of the nature of military power.

These reflections should not be the occasion of, nor do they express, cynicism about arms control negotiations, or despair about the possibility of agreements emerging from them. If governments did not consider disarmament proposals with a view to how they would affect their military security, they would be failing in their duty. Nor does their pursuit of military advantage mean that in principle agreement cannot be reached. All that follows from this interpretation is that an agreement, if it is reached, represents not the discovery of the solution to a problem, but *the striking of a bargain*. As in other kinds of bargaining agreements may emerge when proposals are worked out that advance the interests of both without injuring the interests of either.

NOTES

1. For a pioneering study of arms races see the essay by Samuel P. Huntington: "Arms Races: Prerequisites and Results," *Public Policy*, Yearbook of the Graduate School of Public Administration (Harvard University, 1958).

2. The political conditions of arms control are further explored in chapter 3, "The Conditions of Arms Control."

STRATEGIC MODERNIZATION AND ARMS CONTROL

THE SCOWCROFT COMMISSION

I. DETERRENCE AND ARMS CONTROL

The responsibility given to this commission is to review the purpose, character, size, and composition of the strategic forces of the United States. The members of the commission fully understand not only the purposes for which this nation maintains its deterrent, but also the devastating nature of nuclear warfare, should deterrence fail. The commission believes that effective arms control is an essential element in diminishing the risk of nuclear war—while preserving our liberties and those of like-minded nations. At the same time the commission is persuaded that as we consider the threat of mass destruction we must consider simultaneously the threat of aggressive totalitarianism. Both are central to the political dilemmas of our age. For the United States and its allies the essential dual task of statecraft is, and must be, to avoid the first and contain the second.

It is only by addressing these two issues together that we can begin to understand how to preserve both liberty and peace. Although the United States and the Soviet Union hold fundamentally incompatible views of history, of the nature of society, and of the individual's place in it, the existence of nuclear weapons imbues that rivalry with peril unprecedented in human history. The temptation is sometimes great to simplify—or oversimplify—the difficult problems that result, either by blinking at the devastating nature of modern full-scale war or by refusing to acknowledge the emptiness of life under modern totalitarianism. But it is naive, false, and dangerous to assume that either of these, today, can be ignored and the other dealt with in isolation. We cannot cope with the efforts of the Soviet Union to extend its power without giving thought to the way nuclear weapons have sharply raised the stakes and changed the nature of warfare. Nor can we struggle against nuclear war or the arms race in some abstract sense without keeping before us the Soviet Union's drive to expand its power, which is what makes those struggles so difficult.

We should face both problems directly.

Our words, policies, and actions should all make clear the American conviction that nuclear war, involving few or many nuclear weapons, would be tragedy of unparalleled scope for humanity. It is wrong to pretend or suggest otherwise. Neither the American people, our allies, nor the Soviets should doubt our abhorrence of nuclear war in any form.

By the same token, however, our task as a nation cannot be understood from a position of moral neutrality toward the differences between liberty and totalitarianism. These differences proceed from conflicting views regarding the rights of individuals and the nature of society. Only if Americans believe that it is worth a sustained effort over the years to preserve liberty in the face of challenge by a system that is the antithesis of liberal values can our task be seen as a just and worthy one in spite of its dangers.

We do have many strengths in such an effort. Over the long run, the strengths lent by liberty itself are our greatest ones—our abilities to adapt peacefully to political change, to improve social justice, to innovate with technology, to produce what our people need to live and prosper. What we have most to fear is that confusion and internal divisions—sometimes byproducts of the vigorous play of our free politics—will lead us to lose purpose, hope, and resolve.

We have good reason to maintain all three. Neither time nor history is on the side of large, centralized, autocratic systems that seek to achieve and maintain control over all aspects of the lives of many diverse peoples. We should, with calm persistence, limit the expansion of today's version of this sort of totalitarian state, the Soviet Union. We should persuade its leaders that they cannot successfully divert attention from internal problems by resorting to international blackmail, expansion, and militarism—rationalized by alleged threats posed by us or our allies. We should also be ready to encourage the Soviets to begin to settle differences between us, through equitable arms control agreements and other measures. But moral neutrality and indifference or acquiescence in the face of Soviet efforts to expand their military and political power do not hasten such settlements—they delay them, make them less likely, and ultimately increase the risk of war.

Deterrence is central to the calm persistence we must demonstrate in order to reduce these risks. American strategic forces exist to deter attack on the United States or its allies—and the coercion that would be possible if the public or decisionmakers believed that the Soviets might be able to launch a successful attack. Such a policy of deterrence, like the security policy

Reprinted with minor revisions from the Report of the President's Commission on Strategic Forces, Brent Scowcroft, chairman, 21 Mar. 1982.

of the West itself, is essentially defensive in nature. The strategic forces that are necessary in order to support such a policy by their very existence help to convince the Soviet Union's leaders: that the West has the military strength and political will to resist aggression; and that, if they should ever choose to attack, they should have no doubt that we can and would respond until we have so damaged the power of the Soviet state that they will unmistakably be far worse off than if they had never begun.

There can be no doubt that the very scope of the possible tragedy of modern nuclear war, and the increased destruction made possible even by modern nonnuclear technology, have changed the nature of war itself. This is not only because massive conventional war with modern weapons could be horrendously destructive— some fifty million people died in "conventional" World War II before the advent of nuclear weapons—but also because *conventional* war between the world's major power blocs is the most likely way for *nuclear* war to develop. The problem of deterring the threat of nuclear war, in short, cannot be isolated from the overall power balance between East and West. Simply put, it is war that must concern us, not nuclear war alone. Thus we must maintain a balance between our nuclear and conventional forces, and we must demonstrate to the Soviets our cohesion and our will. And we must understand that weakness in any one of these areas puts a dangerous burden on the others as well as on overall deterrence.

Deterrence is not, and cannot be, bluff. In order for deterrence to be effective we must not merely have weapons, we must be perceived to be able, and prepared, if necessary, to use them effectively against the key elements of Soviet power. Deterrence is not an abstract notion amenable to simple quantification. Still less is it a mirror image of what would deter ourselves. Deterrence is the set of beliefs in the minds of the Soviet leaders, given their own values and attitudes, about our capabilities and our will. It requires us to determine, as best we can, what would deter them from considering aggression, even in a crisis—not to determine what would deter us.

Our military forces must be able to deter war even if the Soviets are unwilling to participate with us in equitable and reasonable arms control agreements. but various types of agreements can, when the Soviets prove willing, accomplish critical objectives. Arms control can: reduce the risk of war; help limit the spread of nuclear weapons; remove or reduce the risk of misunderstanding of particular events or accidents; seal off wasteful, dangerous, or unhelpful lines of technical development before either side gets too committed to them; help channel modernization into stabilizing rather than destabilizing paths; reduce misunderstanding about the purpose of weapons developments and thus reduce the need to over-insure against worst-case projections; and help make arsenals less destructive and costly. To achieve part or all of these positive and useful goals, we must keep in mind the importance of compliance and adequate verification—difficult problems in light of the nature of the Soviet state—and the consequent importance of patience in order to reach fair and reasonable agreements.

This is a vital and challenging agenda. In some of these areas of arms control our interests coincide closely with those of the Soviets. In others, their efforts to undermine the effectiveness of our deterrent and to use negotiations to split us from our allies will make negotiations difficult.

But whether the Soviets prove willing or not, stability should be the primary objective both of the modernization of our strategic forces and of our arms control proposals. Our arms control proposals and our strategic arms programs should thus be integrated and be mutually reinforcing. They should work together to permit us, and encourage the Soviets, to move in directions that reduce or eliminate the advantage of aggression and also reduce the risk of war by accident or miscalculation. As we try to enhance stability in this sense, the commission believes that other objectives should be subordinated to the overall goal of permitting the United States to move—over time—toward more stable strategic deployments, and giving the Soviets the strong incentive to do the same. Consequently it believes, for the reasons set forth below, that it is important to move toward reducing the value and importance of individual strategic targets.

II. SOVIET OBJECTIVES AND PROGRAMS

Effective deterrence and effective arms control have both been made significantly more difficult by Soviet conduct and Soviet weapons programs in recent years. The overall military balance, including the nuclear balance, provides the backdrop for Soviet decisions about the manner in which they will try to advance their interests. This is central to our understanding of how to deter war, how to frustrate Soviet efforts at blackmail, and how to deal with the Soviets' day-to-day conduct of international affairs. The Soviets have shown by word and deed that they regard military power, including nuclear weapons, as a useful tool in the projection of their national influence. In the Soviet strategic view, nuclear weapons are closely related to, and are integrated with, their other military and political instruments as a means of advancing their interests. The Soviets have concentrated enormous effort on the development and modernization of nuclear weapons, obviously seeking to achieve what they regard as important advantages in certain areas of nuclear weaponry.

Historically, the Soviets have not been noted for taking large risks. But one need not take the view that their leaders are eager to launch a nuclear war in order to understand the political advantages that a massive nuclear weapons buildup can hold for a nation seeking to expand its power and influence, or to comprehend the dangers that such a motivation and such a buildup hold for the rest of the world.

Although there is legitimate debate about the exact scope of Soviet military spending in recent years, it is nonetheless clear that the Soviet leaders have embarked upon a determined, steady increase in nuclear (and conventional) weapons programs over the last two decades— a buildup well in excess of any military requirement for defense.

For example, as a result of this determined investment the Soviet ICBM force has grown to nearly 1400 launchers carrying over 5000 warheads, with a throw-weight about four times that of the US ICBM force. The US ICBM force has 1047 launchers and about 2150 warheads. More than half of the Soviet ICBMs—the SS-17, SS-18, and SS-19 missiles—have been deployed since the last US ICBM was deployed. These new Soviet ICBMs are equipped with multiple, independently targetable reentry vehicles (MIRVs). Over 600 of these recently deployed missiles, the SS-18s and SS-19s, have payloads as large or larger than the MX and have excellent accuracy. Many Soviet launchers can be reloaded. The Soviets are now pushing forward with tests of two even newer ICBMs.

While Soviet operational missile performance in wartime may be somewhat less accurate than performance on the test range, the Soviets nevertheless now probably possess the necessary combination of ICBM numbers, reliability, accuracy, and warhead yield to destroy almost all of the 1047 US ICBM silos, using only a portion of their own ICBM force. The US ICBM force now deployed cannot inflict similar damage, even using the entire force. Only the 550 MIRVed Minuteman III missiles in the US ICBM force have relatively good accuracy, but the combination of accuracy and yield of their 3 warheads is inadequate to put at serious risk more than a small share of the many hardened targets in the Soviet Union. Most Soviet hardened targets—of which ICBM silos are only a portion—could withstand attacks by our other strategic missiles.

The Soviet ballistic missile submarine force currently consists of 62 modern submarines; these are armed with 950 missiles, with a total of almost 2000 nuclear warheads. The US has fewer such submarines (34) and missiles (568), but more warheads (about 5000), in its submarine force. Our submarines, moreover, are quieter than those of the Soviets. Recent Soviet ballistic missile submarine–building programs have been vigorous: four times that of the US rate. While the US has a substantial present advantage in the overall capability of its ballistic missile submarine force, this gap is narrowing. The US also has a present advantage in anti-submarine warfare and submarine quietness, but the Soviets appear to be giving high priority to these areas.

Soviet heavy strategic bombers (not including the Backfire) now number about 150, around half equipped with air-to-surface missiles. This force is considerably less capable than the total active US bomber force, which numbers about 270 B52 G and H bombers and about 60 FB-111 bombers. The US bomber force has just begun to be equipped with long-range cruise missiles. Both US and Soviet bombers have carried short-range missiles for many years. A new Soviet intercontinental bomber (the Blackjack) is now being flight-tested. It is similar in appearance to, but larger than, the US B-1B now in production. The Blackjack will probably begin to enter service during the mid-to-late 1980s.

Soviet strategic defenses are extensive, consisting of a dense nationwide air defense network and a limited ballistic missile defense at Moscow. Both are undergoing modernization. Their vigorous research and development programs on ballistic missile defense provide a potential, however, for a rapid expansion of Soviet ABM defenses; should they choose to withdraw from or violate the ABM treaty. Such a potential is enhanced by the continued deployment of modern and capable Soviet air defense missile systems. At least one new Soviet defense system is designed to have capability against short-range ballistic missiles; it could perhaps be upgraded for use against the reentry vehicles of some submarine-launched missiles and even ICBMs. Proliferation of such Soviet air defense missile systems thus creates a need for us to have enough throw-weight to carry sufficient numbers of warheads, and penetration aids such as decoys, in order to be assured of maintaining a deterrent. The US has dismantled its ABM system and has minimal continental air defenses.

These Soviet programs do not, in and of themselves, indicate plans to initiate nuclear attacks. But they do confirm the value that Soviet leaders place on military programs across the board, both to provide an essential backdrop for their political purposes and—should circumstances dictate—to give them the capability to fight effectively. They also understand that the success of their efforts depends upon the outside world's perception. If comparative military trends were to point toward their becoming superior to the West in each of a number of military areas, they might consider themselves able to raise the risks in a crisis in a manner that could not be matched.

In a world in which the balance of strategic nuclear forces could be isolated and kept distinctly set apart from all other calculations about relations between nations and the credibility of conventional military power, a nuclear imbalance would have little importance unless it were so massive as to tempt an aggressor to launch nuclear war. But the world in which we must live with the Soviets is, sadly, one in which their own assessments of these trends, and hence their calculations of overall advantage, influence heavily the vigor with which they exercise their power.

III. PREVENTING SOVIET EXPLOITATION OF THEIR MILITARY PROGRAMS

In our effort to make a strategy of deterrence and arms control effective in preventing the Soviets from political or military use of their strategic forces, we must keep several points in mind.

The Soviets must continue to believe what has been NATO's doctrine for three decades: that if we or our allies should be attacked—by massive conventional means or otherwise—the United States has the will and the means to defend with the full range of American power. This by no means excludes the need to make improvements in our conventional forces in order to have increased confidence in our ability to defend effectively at the conventional level in many more situations, and thus to raise the nuclear threshold. Certainly mutual arms control agreements to reduce both sides' reliance on nuclear weapons should be pursued. But effective deterrence requires that early in any Soviet consideration of attack, or threat of attack, with conventional forces or chemical or biological weapons, Soviet leaders must understand that they risk an American nuclear response.

Similarly, effective deterrence requires that the Soviets be convinced that they could not credibly threaten us or our allies with a limited use of nuclear weapons against military targets, in one country or many. Such a course of action by them would be even more likely to result in full-scale nuclear war than would a massive conventional attack. But we cannot discount the possibility that the Soviets would implicitly or explicitly threaten such a step in some future crisis if they believed that we were unprepared or unwilling to respond. Indeed lack of preparation or resolve on our part would make such blackmail distinctly more probable.

In order to deter such Soviet threats we must be able to put at risk those types of Soviet targets—including hardened ones such as military command bunkers and facilities, missile silos, nuclear weapons and other storage, and the rest—which the Soviet leaders have given every indication by their actions they value most, and which constitute their tools of control and power. We cannot afford the delusion that Soviet leaders—human though they are and cautious though we hope they will be—are going to be deterred by exactly the same concerns that would dissuade us. Effective deterrence of the Soviet leaders requires them to be convinced in their own minds that there could be no case in which they could benefit by initiating war.

Effective deterrence of any Soviet temptation to threaten or launch a massive conventional or a limited nuclear war thus requires us to have a comparable ability to destroy Soviet military targets, hardened and otherwise. If there were ever a case to be made that the Soviets would unilaterally stop their strategic deployments at a level short of the ability seriously to threaten our forces, that argument vanished with the deployment of the SS-18 and SS-19 ICBMs. A one-sided strategic condition in which the Soviet Union could effectively destroy the whole range of strategic targets in the United States, but we could not effectively destroy a similar range of targets in the Soviet Union, would be extremely unstable over the long run. Such a situation could tempt the Soviets, in a crisis, to feel they could successfully threaten or even undertake conventional or limited nuclear aggression in the hope that the United States would lack a fully effective response. A one-sided condition of this sort would clearly not serve the cause of peace.

In order, then, to pursue successfully a policy of deterrence and verifiable, stabilizing arms control we must have a strong and militarily effective nuclear deterrent. Consequently our strategic forces must be modernized, as necessary, to enhance to an adequate degree their overall survivability and to enable them to engage effectively the targets that Soviet leaders most value.

Also, as described below, we should seek to use arms control agreements to reduce instabilities and to channel both sides' strategic modernization toward stabilizing developments, deployments, and reductions. Regardless of what we are able to accomplish with arms control agreements, however, two aspects of deterrence are crucial. The problems of maintaining an effective deterrent and of reaching stabilizing and verifiable arms control agreements cannot be addressed coherently without keeping in mind the nature of Soviet expansionism. Second, the deterrent effect of our strategic forces is not something separate and apart from the ability of those forces to be used against the tools by which the Soviet leaders maintain their power. Deterrence, on the contrary, requires military effectiveness.

IV. US STRATEGIC FORCES AND TRENDS

A. STRATEGIC FORCES AS A WHOLE

The development of the components of our strategic forces—the multiplicity of intercontinental ballistic missiles (ICBMs), submarine-launched ballistic missiles (SLBMs), and bombers—was in part the result of an historical evolution. This triad of forces, however, serves several important purposes.

First, the existence of several strategic forces requires the Soviets to solve a number of different problems in their efforts to plan how they might try to overcome them. Our objective, after all, is to make their planning of any such attack as difficult as we can. If it were possible for the Soviets to concentrate their research and development efforts on putting only one or two components of US strategic forces at risk—e.g., by an intensive effort at antisubmarine warfare to attempt to threaten our ballistic missile submarines—both their incentive to do so and their potential gains would be sharply increased. Thus the existence of several components of our strategic forces permits each to function as a hedge against possible Soviet successes in endangering any of the others. For example, at earlier times uncertainties about the vulnerability of our bomber force were alleviated by our confidence in the survivability of our ICBMs. And although the survivability of our ICBMs is today a matter of concern (especially when that problem is viewed in isolation), it would be far more serious if we did not have a force of ballistic missile submarines at sea and a bomber force. By the same token, over the long run it would be unwise to rely so heavily on submarines as our only ballistic missile force that a Soviet breakthrough in antisubmarine warfare could not be offset by other strategic systems.

Second, the different components of our strategic forces would force the Soviets, if they were to contemplate an all-out attack, to make choices which would lead them to reduce significantly their effectiveness against one component in order to attack another. For example, if Soviet war planners should decide to attack our bomber and submarine bases and our ICBM silos with simultaneous detonations—by delaying missile launches from close-in submarines so that such missiles would *arrive* at our bomber bases at the same time the Soviet ICBM warheads (with their longer time of flight) would arrive at our ICBM silos—then a very high proportion of our alert bombers would have escaped before their bases were struck. This is because we would have been able to, and would have, ordered our bombers to take off from their bases within moments after the launch of the first Soviet ICBMs. If the Soviets, on the other hand, chose rather to *launch* their ICBM and SLBM attacks at the same moment (hoping to destroy a higher proportion of our bombers with SLBMs having a short time of flight), there would be a period of over a quarter of an hour after nuclear detonations had occurred on US bomber bases but before our ICBMs had been struck. In such a case the Soviets should have no confidence that we would refrain from launching our ICBMs during that interval after we had been hit. It is important to appreciate that this would not be a "launch-on-warning," or even a "launch under attack," but rather a launch *after* attack—after massive nuclear detonations had already occurred on US soil.

Thus our bombers and ICBMs are more survivable together against Soviet attack than either would be alone. This illustrates that the different components of our strategic forces should be assessed collectively and not in isolation. It also suggests that whereas it is highly desirable that a component of the strategic forces be survivable when it is viewed separately, it makes a major contribution to deterrence even if its survivability depends in substantial measure on the existence of one of the other components of the force.

The third purpose served by having multiple components in our strategic forces is that each component has unique properties not present in the others. Nuclear submarines have the advantage of being able to stay submerged and hidden for months at a time, and thus the missiles they carry may reasonably be held in reserve rather than being used early in the event of attack. Bombers may be launched from their bases on warning without irretrievably committing them to an attack; also, their weapons, though they arrive in hours, not minutes, have excellent accuracy against a range of possible targets. ICBMs have advantages in command and control, in the ability to be retargeted readily, and in accuracy. This means that ICBMs are especially effective in deterring Soviet threats of massive conventional or limited nuclear attacks, because they could most credibly respond promptly and controllably against specific military targets and thereby promptly disrupt an attack on us or our allies.

B. TECHNOLOGICAL TRENDS FOR STRATEGIC FORCES

1. Accuracy

The accuracy of strategic weapons in the foreseeable future will continue to increase. There are lower limits, perhaps a few hundred feet, to the accuracy of strategic weapons that do not rely on some kind of terminal guidance. For weapons using terminal guidance, accuracy should be even better. Accuracy is most advanced today in the ICBM forces, but in the 1990s SLBMs should have sufficient accuracy seriously to threaten hardened targets. Nevertheless, ICBM accuracy should remain somewhat better than that for submarine-launched missiles.

These accuracy developments and the ability of an attacker to use more than one warhead to attack each fixed target on the other side increasingly put at risk targets of high value such as fixed launchers for MIRVed ICBMs. Although such fixed targets may retain some survivability for a number of years—because of problems of operational accuracies, planning uncertainties (as discussed at Section V.E. below), and the previously described need to coordinate ICBM and SLBM attacks—their survivability will nevertheless continue to decline over time. Thus reasonable survivability of fixed targets, such as ICBM silos, may not outlive this century, even when one considers them together with the rest of our strategic forces. In time, even nonnuclear weapons with excellent accuracy may be able to attack effectively some fixed targets previously thought to be vulnerable only to nuclear weapons.

2. Superhardening

New concepts and developments in hardening are quite promising. They could lead to the capability to harden such targets as ICBM silos far in excess of what was thought possible only a short time ago. Eventually the survival of even the hardest such targets would be doubtful in light of the accuracy improvements described above. Nonetheless increased hardness would raise the weapons requirements and the risk of attack for some years. Hardening will also be able to postpone vulnerability to, and therefore the probability of, attack by submarine-launched ballistic missiles.

3. Mobility

New techniques in guidance, miniaturization of electronic components, hardening against nuclear effects, and solid fuels will continue to make mobile strategic systems more feasible. Strategically useful hardening of land-based mobile launchers appears more feasible than in the past.

4. Antisubmarine Warfare

The problem of conducting open-ocean search for submarines is likely to continue to be sufficiently difficult that ballistic missile submarine forces will have a high degree of survivability for a long time. Nevertheless, the prospect of concentrating all of the submarine-launched missiles at sea in a few very large submarines raises some concern. Communication links with submarines, while likely to improve, will still offer problems not present for land-based systems.

5. Ballistic Missile Defense

Substantial progress has been made in the last decade in the development of both endoatmospheric and exoatmospheric ABM defenses. However, applications of current technology offer no real promise of being able to defend the United States against massive nuclear attack in this century. An easier task is to provide ABM defense for fixed hardened targets, such as ICBM silos. However even this will be a difficult feat if an attacker can use a large number of warheads against each defended target. The effectiveness of such a defense could be enhanced by some types of bunching and close spacing of the defended targets, in order to reduce the number of ABM systems required. It could also be enhanced by having multiple shelters for each missile and preferentially defending only the shelter containing the missile while facing an attacker with the need to attack all shelters.

Improvements in Soviet air defense systems—to give them some capability against some submarine-launched ballistic missile warheads, and even against some warheads fired by ICBMs—are likely to continue as such air defenses are made capable of dealing with modern aircraft, cruise missiles, and shorter-range ballistic missiles. The 1972 ABM treaty, however, has provisions prohibiting the testing of air defense systems as ABMs.

V. STRATEGIC MODERNIZATION PROGRAMS

Although there is room for improvement and adjustments in the several strategic programs discussed below, the commission noted that these programs are—in the main—proceeding reasonably well. Therefore this report concentrates on the current issues presented by the ICBM force (Section E below) and its relation to arms control (Section VI). The current and recommended programs, taken as a package, should give us high confidence in maintaining an effective deterrent in the years to come.

A. COMMAND, CONTROL, AND COMMUNICATIONS

Our first defense priority should be to ensure that there is continuing, constitutionally legitimate, and full control of our strategic forces under conditions of stress or actual attack. No attacker should be able to have any reasonable confidence that he could destroy the link between the president and our strategic forces.

The commission urges that this program continue to have the highest priority and urges the investigation of ways in which the planned improvements could be augmented by low-cost back-up systems.

B. Sea-based Missile Programs

1. Deployment

The commission supports the continuation of the Trident submarine construction program. It also supports the continued development and the deployment of the Trident II (D-5) missile as rapidly as its objectives of range, accuracy, and reliability can be attained. The Trident submarine's significantly reduced noise level and the D-5 missile's greater full-payload range will add importantly to the already high degree of survivability of the ballistic missile submarine force. Given the increased importance of that force, both programs are essential. The D-5 missile's greater accuracy will also enable it to be used to put some portion of Soviet hard targets at risk, a task for which the current Trident I (C-4) missile is not sufficiently accurate. The commission also stresses the importance of the command, control, and communication improvements of particular relevance to the submarine force—namely the ELF communication system, the ECX aircraft, and the MILSTAR satellite.

The commission does not recommend the development and deployment of a system for the launch of ballistic missiles from surface ships. Such a system appears to have no net advantage over submarine basing and would have vulnerabilities that submarines do not possess.

For the reasons stated in section IV.A., above, the commission recommends strongly against adopting a strategic force posture relying solely on submarines and bombers to the exclusion of ICBM modernization; it recognizes, however, the increasing importance of the ballistic missile submarine force.

2. Research

The commission notes that—although it believes that the ballistic missile submarine force will have a high degree of survivability for a long time—a submarine force ultimately consisting solely of a relatively few large submarines at sea, each carrying on the order of 200 warheads, presents a small number of valuable targets to the Soviets. Vigorous pursuit of the longstanding program to avoid technological surprise by the Soviets in antisubmarine warfare is thus of vital importance.

Consistent with the long-term program recommended for the ICBM force, below, to reduce the value of individual targets, the commission recommends that research begin now on smaller ballistic missile–carrying submarines, each carrying fewer missiles than the

Trident, as a potential follow-on to the Trident submarine force. The objective of such research should be to design a submarine and missile system that would, as much as possible, reduce the value of each platform and also present radically different problems to a Soviet attacker than does the Trident submarine force. This work should proceed in such a way that a decision to construct and deploy such a submarine force could be rapidly implemented should Soviet progress in antisubmarine warfare so dictate.

C. Bomber and Air-Launched Cruise Missile Programs

Our bomber and air-launched cruise missile force is of vital importance to the maintenance of an effective deterrent. As long as its ability to survive and penetrate Soviet defenses can be maintained, it provides unique advantages of its own as a strategic system. It also provides mutual support to the survivability of the ICBM force, as discussed in Section IV.A., above. Furthermore the commission bases its other recommendations on the assumption that a strong bomber and cruise missile program is continued. The commission is unanimous in these views although it recognizes that there are legitimate differences about the best and least expensive way to provide for the necessary modernization of the bomber and cruise missile force. Since these modernization decisions, although not wholly independent of other strategic force decisions, may reasonably be considered within their own framework, the commission—having concentrated its efforts on the ballistic missile forces and related issues—has no changes to recommend in these bomber and cruise missile programs.

D. Ballistic Missile Defense

Vigorous research and development on ABM technologies—including, in particular, ways to sharpen the effectiveness of treaty-limited ABM systems with new types of nuclear systems and also ways to use nonnuclear systems—are imperative to avoid technological surprise from the Soviets. Such a vigorous program on our part also decreases any Soviet incentive—based on an attempt to achieve unilateral advantage—to abrogate the ABM treaty. At this time, however, the commission believes that no ABM technologies appear to combine practicality, survivability, low cost, and technical effectiveness sufficiently to justify proceeding beyond the stage of technology development.

Of particular importance, however, is the ability to counter any improvement in Soviet ABM capability by being able to maintain the effectiveness of our offensive systems. The possibility of either a sudden breakthrough in ABM technology, a rapid Soviet breakout from the ABM treaty by a quick further deployment of their current ABM systems, or the deployment of air defense systems also having some capability against strategic ballistic missiles all point to the need for us to be able to penetrate some level of ABM defense. This dictates continued attention to having sufficient throw-weight for adequate numbers of warheads and of decoys and other penetration aids.

E. ICBM PROGRAMS

The problem that led to the establishment of this commission is the same one that has been at the heart of much of the controversy concerning strategic forces and arms control for over a decade—the future of our ICBM force. As described above (Section IV.A.) our ICBM force has three main strategic purposes: (1) serving as a hedge against possible vulnerabilities in our submarine force; (2) introducing complexity and uncertainty into any plan of Soviet attack, because of the different types of attacks that would have to be launched against our ICBMs and our bombers; and (3) helping to deter Soviet threats of massive conventional or limited nuclear attacks by the ability to respond promptly and controllably against hardened military targets.

ICBM modernization is also particularly important now in order to encourage the Soviets to reach stabilizing arms control agreements and to redress perceived US disadvantages in strategic capability.

The commission believes that, because of changing technology, arms control negotiations, and our own domestic political process, this issue—the future of our ICBM force—has come to be miscast in recent years.

To many the problem has become: "How can a force consisting of relatively large, accurate land-based ICBMs be deployed quickly and be made survivable, even when it is viewed in isolation from the rest of our strategic forces, in the face of increasingly accurate threatened attacks by large numbers of warheads—and how can this be done under arms control agreements that limit or reduce launcher numbers?" It is this complex problem that many, inside and outside the government, have sought to solve for a variety of reasons. These reasons fall into five main groups.

First, in order to serve one of the necessary purposes of a strategic force—namely to hedge against possible failure by the others, such as would be caused by a Soviet breakthrough in antisubmarine warfare—many have felt that any new ICBM deployment should be almost totally survivable even when viewed in isolation from our bomber force and the rest of our strategic forces. The threat now posed by accurate Soviet ICBMs to the Minuteman force, viewed in isolation, has also led many to argue that this particular survivability problem has to be solved quickly.

Second, the overall perception of strategic imbalance caused by the Soviets' ability to destroy hardened land-based targets—with more than 600 newly deployed SS-18 and SS-19 ICBMs—while the United States is clearly not able to do so with its existing ballistic missile force, has been reasonably regarded as destabilizing and as a weakness in the overall fabric of deterrence. In particular, since the ICBM force helps to deter massive conventional or limited nuclear attack against us or our allies, this has led many to believe that the serious imbalance between US and Soviet capabilities should be rectified quickly in the overall interest of the alliance.

Third, arms control agreements—in part to be verifiable without resort to the sorts of cooperative measures such as on-site inspection typically opposed by the Soviets—have concentrated to a significant degree on limiting or reducing strategic missile launchers rather than warheads. This is in some measure because launchers are more easily counted by satellite reconnaissance than are other ICBM characteristics and because launcher numbers provide relatively unambiguous terms for a treaty. Launcher or missile limits have the indirect effect, however, of encouraging both sides to build large ICBMs with many warheads.

Fourth, if one sets aside survivability, basing, and other cost considerations and looks solely at the cost of the missiles themselves, it is cheaper to deploy a given number of warheads in a few relatively large missiles than to deploy the same number of warheads on a larger number of smaller missiles. Fewer expensive guidance systems need to be purchased, for example.

Fifth, for almost two decades our Minuteman ICBM force had virtually all of the positive characteristics desirable for any strategic system. It was survivable, even when an attack on it was viewed in isolation, because Soviet accuracies were not good enough to threaten silos. Command and control were comparatively easy. ICBMs were more accurate than submarine-based missiles and could reach their targets faster than bombers. And, when compared to either submarine-based missiles or bombers, silo-based ICBMs, once purchased, had strikingly low annual operating costs. This history has led many to continue to seek to replicate those two decades of Minuteman history, and in so doing to try not only to meet these objectives, but to do so with a single way of basing a single type of ICBM that would have all of these desirable characteristics.

These five sets of considerations, different ones of them of greater importance to different decisionmakers at different times, have led us as a nation in recent years to try to re-create all of the desirable characteristics that Minuteman possessed during the sixties and much of the seventies. We have tried to do so by deploying a few relatively large missiles as quickly as possible, in a single basing mode, on land, under arms control agreements limiting or reducing launcher numbers, in the face of a threat of attack by increasingly accurate and numerous warheads—and to do so in a manner that seeks to preserve ICBM survivability for the long term, even when the ICBM force is viewed in isolation. But by trying to solve all ICBM tasks with a single weapon and a single basing mode in the face of the trends in technology, we have made the problem of modernizing the ICBM force so complex as to be virtually insoluble.

In arriving at its recommendations regarding ICBM programs, the commission was mindful of the following criteria. For the near term, it would concentrate on possible deployments and basing modes that appeared to have straightforward and achievable technical and military value. For the long term, compatibility of ICBM programs with the need for flexibility and innovation in responding to possible Soviet actions would be of great importance. Economic cost would be considered carefully. The commission would not insist on seeking a single solution to all the problems—near-term and long-term—with which the ICBM force must cope. Finally, and of great importance, our ICBM programs should support pursuit of a stable regime of arms control agreements.

The commission has concluded that the preferred approach for modernizing our ICBM force seems to have three components: initiating engineering design of a single-warhead small ICBM, to reduce target value and permit flexibility in basing for better long-term survivability; seeking arms control agreements designed to enhance strategic stability; and deploying MX missiles in existing silos now to satisfy the immediate needs of our ICBM force and to aid that transition.

A more stable structure of ICBM deployments would exist if both sides moved toward more survivable methods of basing than is possible when there is primary dependence on large launchers and missiles. Thus from the point of view of enhancing such stability, the commission believes that there is considerable merit in moving toward an ICBM force structure in which potential targets are of comparatively low value—missiles containing only one warhead. A single-warhead ICBM, suitably based, inherently denies an attacker the opportunity to destroy more than one warhead with one attacking warhead. The need to have basing flexibility, and particularly the need to keep open the option for different types of mobile

basing, also suggests a missile of small size. If force survivability can be additionally increased by arms control agreements which lead both sides toward more survivable modes of basing than is possible with large launchers and missiles, the increase in stability would be further enhanced.

In the meantime, however, deployment of MX is essential in order to remove the Soviet advantage in ICBM capability and to help deter the threat of conventional or limited nuclear attacks on the Alliance. Such deployment is also necessary to encourage the Soviets to move toward the more stable regime of deployments and arms control outlined above.

The commission stresses that these two aspects of ICBM modernization and this approach toward arms control are integrally related. They point toward the same objective—permitting the United States and encouraging the Soviets to move toward more stable ICBM deployments over time in a way that is consistent with arms control agreements having the objective of reducing the risk of war. The commission is unanimous that no one part of the proposed program can accomplish this alone.

1. ICBM Long-term Survivability: Toward the Small, Single-Warhead ICBM

The commission believes that a single-warhead missile weighing about 15 tons (rather than the nearly 100 tons of MX) may offer greater flexibility in the long-run effort to obtain an ICBM force that is highly survivable, even when viewed in isolation, and that can consequently serve as a hedge against potential threats to the submarine force.

The commission thus recommends beginning engineering design of such an ICBM, leading to the initiation of full-scale development in 1987 and an initial operating capability in the early 1990s. The design of such a missile, hardened against nuclear effects, can be achieved with current technology. It should have sufficient accuracy and yield to put Soviet hardened military targets at risk. During that period an approach toward arms control, consistent with such deployments, should also seek to encourage the Soviets to move toward a more stable ICBM force structure at levels which would obviate the need to deploy very large numbers of such missiles. The development is feasible.

Decisions about such a small missile and its basing will be influenced by several potential developments: the evolution of Soviet strategic programs, the path of arms control negotiations and agreements, general trends in technology, the cost of the program, operations considerations, and the results of our own research on specific basing modes. Although the small missile program should be pursued vigorously, the way these uncertainties are resolved will inevitably influence the size and nature of the program. We should keep in mind, however, that having several different modes of deployment may serve our objective of stability. The objective for the United States should be to have an overall program that will so confound, complicate, and frustrate the efforts of Soviet strategic war planners that, even in moments of stress, they could not believe that they could attack our ICBM forces effectively.

Different ICBM deployment modes by the United States would require different types of planned Soviet attacks. Deployment in hardened silos would require the Soviets to plan to use warheads that are large, accurate, or both. Moreover, for those silos or shelters holding a missile with only one warhead, each would present a far less attractive target than would be the case for a silo containing a large missile with many MIRVs. Mobile deployments of US missiles would require the Soviets to try to barrage large areas using a number of warheads for each of our warheads at risk, to develop very sophisticated intelligence systems, or both. In this context, deployment of a small single-warhead ICBM in hardened mobile launchers is of particular interest because it could permit deployment in peacetime in limited areas such as military reservations. Land-mobile deployments without hard launchers could be threatened by a relatively small attack—in the absence of an appropriate arms control agreement—unless our own missiles were distributed widely across the country in peacetime. The key advantages of a small single-warhead missile are that it would reduce the value of each strategic target and that it is also compatible with either fixed or mobile deployments, or with combinations of the two.

As discussed below (Section VI), deployment of such small missiles would be compatible with arms control agreements reducing the number of warheads, in which case only a small number of such missiles would probably need to be deployed. If the Soviets proved unwilling to reach such agreements, however, the United States could deploy whatever number of small missiles were required—in whatever mix of basing modes—to maintain an adequate overall deterrent.

2. Immediate ICBM Modernization: Limited Deployment of the MX Missile

a. The MX in Minuteman Silos.   There are important needs on several grounds for ICBM modernization that cannot be met by the small, single-warhead ICBM.

First, arms control negotiations—in particular the Soviets' willingness to enter agreements that will enhance stability—are heavily influenced by ongoing programs. The ABM Treaty of 1972, for example, came about only because the United States maintained an ongoing ABM program and indeed made a decision to make a limited deployment. It is illusory to believe that we could obtain a satisfactory agreement with the Soviets limiting ICBM deployments if we unilaterally terminated the only new US ICBM program that could lead to deployment in this decade. Such a termination would effectively communicate to the Soviets that we were unable to neutralize their advantage in multiple-warhead ICBMs. Abandoning the MX at this time in search of a substitute would jeopardize, not enhance, the likelihood of reaching a stabilizing and equitable agreement. It would also undermine the incentives to the Soviets to change the nature of their own ICBM force and thus the environment most conducive to the deployment of a small missile.

Second, effective deterrence is in no small measure a question of the Soviets' perception of our national will and cohesion. Cancelling the MX, when it is ready for flight testing, when over $5 billion have already been spent on it, and when its importance has been stressed by the last four presidents, does not communicate to the Soviets that we have the will essential to effective deterrence. Quite the opposite.

Third, the serious imbalance between the Soviets' massive ability to destroy hardened land-based military targets with their ballistic missile force and our lack of such a capability must be redressed promptly. Our ability to assure our allies that we have the capability and will to stand with them, with whatever forces are necessary, if the Alliance is threatened by massive conventional, chemical or biological, or limited nuclear attack is in question as long as this imbalance exists. Even before the Soviet leaders, in a grave crisis, considered using the first tank regiment or the first SS-20 missile against NATO, they must be required to face what war would mean to them. In order to augment what we would hope would be an inherent sense of conservatism and caution on their part, we must have a credible capability for controlled, prompt, limited attack on hard targets ourselves. This capability casts a shadow over the calculus of Soviet risk taking at any level of confrontation with the West. Consequently, in the interest of the Alliance as a whole, we cannot safely permit a situation to continue wherein the Soviets have the capability promptly to destroy a range of hardened military targets and we do not.

Fourth, our current ICBM force is aging significantly. The Titan II force is being retired for this reason and extensive Minuteman rehabilitation programs are planned to keep those missiles operational.

The existence of a production program for an ICBM of approximately 100 tons[1] is important for two additional reasons. As Soviet ABM modernization and modern surface-to-air missile development and deployment proceed—even within the limitations of the ABM treaty—it is important to be able to match any possible Soviet breakout from that treaty with strategic forces that have the throw-weight to carry sufficient numbers of decoys and other penetration aids; these may be necessary in order to penetrate the Soviet defenses which such a breakout could provide before other compensating steps could be taken. Having in production a missile that could effectively counter such a Soviet step should help deter them from taking it. Moreover, in view of our coming role reliance on space shuttle orbiters, it would be prudent to have in production a booster, such as MX, that is of sufficient size to place in orbit at least some of our most strategically important satellites.

These objectives can all be accomplished, at reasonable cost, by deploying MX missiles in current Minuteman silos.

In the judgment of the commission, the vulnerability of such silos in the near term, viewed in isolation, is not a sufficiently dominant part of the overall problem of ICBM modernization to warrant other immediate steps being taken such as closely spacing new silos or ABM defense of those silos. This is because of the mutual survivability shared by the ICBM force and the bomber force in view of the different types of attacks that would need to be launched at each, as explained above (Section IV.A.). In any circumstances other than that of a particular kind of massive surprise attack[2] on the United States by the Soviet Union, Soviet planners would have to account for the possibility that MX missiles in Minuteman silos would be available for use, and thus they would help deter such attacks. To deter such surprise attacks we can reasonably rely both on our other strategic forces and on the range of operational uncertainties that the Soviets would have to consider in planning such aggression—as long as we have underway a program for long-term ICBM survivability such as that for the small, single-warhead ICBM to hedge against long-term vulnerability for the rest of our forces.

None of the short-term needs for ICBM force modernization set forth above would be met by deploying any missile other than the MX.

The commission examined the concept of a common missile to serve the function of both the Trident II (D-5) missile, now under development for the Trident submarine, and of MX. At this point such a common missile would essentially be a modified Trident II. But deployment of that missile as an ICBM would not only lag several years behind the MX, its payload at the full ICBM would be reduced. Since a larger number of Trident II missiles would need to be deployed in order to have the same number of warheads as the MX force, there would be no cost savings.

The commission also assessed the possibility of improving the guidance on the Minuteman ICBM to the level of accuracy being developed for the MX. Such a step, however, would take some two to three years longer than production of the MX and would not redress the perceived imbalance between US and Soviet capabilities. The wisdom of placing new guidance systems on the front ends of aging 1960s-era missiles is highly questionable. Moreover, shifting to such a program at this point would not provide the increased throw-weight needed to hedge either against Soviet ABM improvements or against the need to launch satellites in an emergency. Most importantly, a Minuteman modification program would not provide the incentive to the Soviets to negotiate that would be provided by production of the MX.

A program of deploying on the order of 100 MX missiles in existing Minuteman silos would, on the other hand, accomplish the objectives set forth in this section, and it would do so without threatening stability. The throw-weight and megatonnage carried by the 100 MX missiles are about the same as that of the 54 large Titan missiles now being retired plus that of the 100 Minuteman III missiles that the MXs would replace. Such a deployment would thus represent a replacement and modernization of part of our ICBM force. It would provide a means of controlled limited attack on hardened targets but not a sufficient number of warheads to be able to attack all hardened Soviet ICBMs, much less all of the many command posts and other hardened military targets in the Soviet Union. Thus it would not match the overall capability of the recent Soviet deployment of over 600 modern ICBMs of MX size or larger. But a large deployment of several hundred MX missiles should be unnecessary for the limited but very important purposes set forth above. Should the Soviets refuse to engage in stabilizing arms control and engage instead in major new deployments, reconsideration of this and other conclusions would be necessary.

b. Other Possible MX Basing Modes. The commission assessed several basing modes for the MX missile as a way of solving the problem of long-term ICBM survivability.

Deploying the MX missile in Multiple Protective Shelters (MPS) meets the need of long-term survivability reasonably well. It would have a similar advantage to the deployment of small, single-warhead missiles in silos or shelters—namely it would force an attacker to plan to deal with a multiplicity of targets. It would not, however, have the advantages of the missile's being able to move, in the event of an attack, outside its basing complex—a capability that is potentially available in some types of small missile deployments. The basing complex required for MPS necessarily affects a land area sufficiently large that local political opposition to it has been significant. There is also a possibility that, over the long run, even if the SALT II Agreement were ratified, a Soviet abrogation or refusal to renew the limits on ICBM launchers or on the number of warheads per missile contained therein could create difficulties for MPS basing. It could lead to the need either to add shelters (and not clearly at a lower cost than the Soviets' cost of adding warheads) or the need to defend the MPS basing complex with an ABM deployment in excess of that permitted under the ABM treaty.

Another alternative MX deployment that has some attractiveness for long-run survivability is closely spaced basing (CSB). Such a deployment—e.g., 100 missiles in 100 new closely spaced silos—would sharply reduce the land area required by the MPS system and could cause significant difficulties for some types of planned Soviet attacks by forcing the attacker to take account of the circumstances under which one of his attacking warheads would destroy others ("fratricide"). This basing scheme would require newly developed techniques for hardening silos in order to avoid the possibility that one attacking warhead could destroy more than one silo. It would also, by its close spacing, make several potential types of ABM defense of the ICBM deployment more feasible. Some of the ABM defenses, countering some potential types of Soviet attacks, could be deployed within the numerical limits of the 1972 ABM treaty, but other more generally effective ones could not. The effectiveness of a CSB deployment in preserving the survivability of the ICBM force over the long run would depend significantly upon advances in hardening silos; the effectiveness of this is yet to be demonstrated, and the cost is as yet uncertain. It also would depend upon fratricide effects that are not fully understood.

These uncertainties would not be eliminated by adding multiple hardened shelters for each missile to a CSB deployment to permit deceptive basing—a combination of MPS and CSB. Beginning a hardened shelter deployment immediately would be a concurrent program, involving a commitment to construction before new hardening techniques are fully understood or developed. In addition, although a greater number of shelters could improve survivability, constructing a number of very hard shelters would be expensive. Each shelter would be considerably more costly than the shelters in the original MPS system. Since more shelters would be needed than in the original CSB proposal, the total program would also be more costly than CSB basing.

Other basing modes for a large missile involve longer delays than those for MPS or CSB deployment (or a hybrid of the two). Thus, the improvement in survivability that might be offered by, for example, basing MX in continuous patrol aircraft or in deep underground deployments—given the time it would now require to design and develop these basing modes—would not permit deployment in this decade. Moreover, the large size of the MX missile could complicate these and other longer-term deployments.

c. Research and Development Work on ICBM Basing. The commission recognizes that a series of phased decisions involving both the executive branch and the Congress will be necessary in order to determine the future shape of our ICBM force. Not all decisions can or should be made in 1983. The commission believes, however, that it is important to pursue the following research and development programs now in order to allow the US government to make intelligent future decisions about ICBM basing.

The commission believes that the work done to date (much of it in connection with designing CSB) is impressive on the technology for dramatic improvements in hardening ICBM silos or shelters. It thus recommends that vigorous research should proceed on new techniques for hardening silos and shelters generally. A specific program to resolve the uncertainties regarding hardness should be undertaken under the leadership of the Defense Nuclear Agency, and with the cooperation of the Air Force and of those Department of Energy laboratories with expertise in the relevant technology. In the event that such hardening proves sufficiently effective and affordable it may later prove useful for some or all of the silos containing MX to be hardened appropriately. In any case, such hardening techniques could prove useful for small missile deployments in the 1990s. Research on the circumstances in which there could be mutual destruction of one attacking warhead by another (fratricide) should be continued.

Vigorous investigation should proceed on different types of land-based vehicles and launchers, including hardened vehicles, for mobile deployment of small ICBMs. Depending on the hardening level achievable for such mobile launchers, it may be possible, for example, to obtain adequate survivability by deploying small ICBMs on military facilities in vehicles alone or in vehicles in simple shelters, with the added advantages of wider mobility if there is warning of an attack. This would avoid the need to disperse the missiles beyond such areas in normal peacetime conditions. For the longer run, other types of mobile basing should also be explored.

The above ICBM programs should contribute to stability and point toward—and be compatible with—a responsible set of arms control principles that can be sustained over the years during negotiations and new agreements.

F. SUMMARY OF MODERNIZATION RECOMMENDATIONS

1. Strategic Forces Other Than ICBMs

a. As first priority, vigorous programs should continue to improve the ability of the president to command, control, and communicate with the strategic forces under conditions of severe stress or actual attack.

b. The Trident submarine construction program and the Trident II (D-5) ballistic missile development program should continue with high priority; the work recommended on small submarines to avoid technological surprise in antisubmarine warfare should begin now.

c. No changes are recommended in the bomber and air-launched cruise missile programs.

d. Vigorous research and technology development on ABM should be pursued. The development of decoys and other penetration aids for our ballistic missiles is also recommended.

2. ICBM Programs

a. Engineering design should be initiated, now, of a single-warhead ICBM weighing about fifteen tons; this program should lead to the initiation of full-scale development in 1987 and an initial operating capability in the early 1990s. Deploying such a missile in more than one mode would serve stability. Hardened silos or shelters and hardened mobile launchers should be investigated now.

b. One hundred MX missiles should be deployed promptly in existing Minuteman silos as a replacement for those 100 Minutemen and the Titan II ICBMs now being decommissioned and as a modernization of the force.

c. A specific program to resolve the uncertainties regarding silo or shelter hardness should be undertaken, leading to later decisions about hardening MX in silos and deploying a small single-warhead ICBM in hardened silos or shelters. Vigorous investigation should proceed on different types of land-based vehicles and launchers, including particularly hardened vehicles.

d. Costs. The long-term costs of major programs are necessarily subject to uncertainty. Moreover, the standard of comparison is not clear in this case because, in order to compare costs, one should assess programs of equal effectiveness. Effectiveness of various types of ICBM deployments, especially with regard to long-term survivability, is precisely the issue which is most in controversy. For comparative purposes, the commission has considered only evolutionary expansions of CSB basing of MX under which there would be some effort, in light of possible Soviet reactions, to preserve long-term ICBM force survivability.

The commission compared the costs of the program that it recommends to the current program and to other possible strategic programs over the years of the Department of Defense's Five Year Defense Program (FY 1984–88).

The comparison, displayed in table A1, shows that the recommended program is about $1 billion per year less than the CSB program for each of the next four years, and that the total net savings during the five-year period is about $3 billion.

There will be significant costs incurred by the commission's recommended program beyond these five years, but the exact amount would depend heavily on: the type of basing modes chosen for the small, single-warhead ICBM; the number of US ICBMs deployed; whether the silos in which MX missiles are deployed are hardened, and to what degree; the evolution of the Soviet threat; and the terms of any arms control agreement at that time. For those programs which the commission considered as reasonable alternatives, such as the further evolution of ABM defenses or multiple shelters for CSB of MX, there would similarly be significant additional costs beyond the five-year period; the magnitude of these would be affected by similar uncertainties.

VI. ARMS CONTROL

It is a legitimate, ambitious, and realistic objective of arms control agreements to channel the modernization of strategic forces, over the long term, in more stable directions than would be the case without such agreements. Such stability supports deterrence by making aggression less likely and by reducing the risk of war by accident or miscalculation. The strategic modernization program recommended herein and the arms control considerations contained in this report are consistent with an important aspect of such stability. In light of the developments in technology set forth in Section IV.B., above, they seek to enhance survivability by moving both sides, in the long term, toward strategic deployments in which individual targets are of lower value. The recommended strategic program thus proposes an evolution for the US ICBM force in which a given number of ballistic missile warheads would, over time, be spread

Table A1

Costs in Billions of Fiscal Year 1982 Dollars[a]

	1984	1985	1986	1987	1988	Total 5-Year
Alternatives						
100 MX in CSB (current program)	5.6	6.0	4.9	4.1	2.3	22.9
100 MX in CSB/MPS (300 shelters)	5.6	6.0	4.9	4.1	2.3	22.9[b]
100 MX in CSB/MPS, with treaty-limited ABM defense, for initial operating capability in 1993	5.9	6.5	5.7	5.6	4.2	27.9[b]
Commission recommendations						
100 MX in Minuteman Silos	3.9	4.1	3.0	2.2	1.4	14.6
Development of small, single-warhead ICBM and basing R&D, for initial operating capability in 1993	0.5	0.5	0.5	1.0	2.8	5.3[b]
Total	4.4	4.6	3.5	3.2	4.2	19.9

[a] Constant FY 1982 dollars are used in this comparison, since these were the units used in December 1982 to present CSB costs to the Congress. Using either constant dollars of a later fiscal year or "then-year" dollars would show higher numbers for all alternatives. Figures were provided by the Department of Defense. FY 1983 costs are not included
[b] All involve significant costs beyond the five-year period.

over a larger number of launchers than would otherwise be the case.

This evolution is important for long-term strategic stability, but it is not without its costs. Spreading a given number of ICBM warheads, whatever the number, over greater numbers of ICBM launchers would normally mean added operating costs, for example. But in the judgment of the commission, permitting our forces to evolve in this direction and encouraging the Soviets to do likewise is worth such costs. Moreover, if such programs can lead to mutually agreed lower levels of warhead deployments in time, then ultimately the net cost may be less.

Such an evolution marks a sound principle to guide our own long-term strategic force modernization and arms control proposals, but it is neither necessary nor wise to move precipitously in that direction. In part this is because time is required to develop such new systems properly, in part it is because continued efforts on our current strategic programs are needed to encourage the Soviets to move in a stabilizing direction. Absent such encouragement there is no realistic hope that the Soviets will join such an evolution and forego the current advantages they have in the ability to attack hard targets and to barrage large areas with their preponderance in throw-weight.

Over the long run, stability would be fostered by a dual approach toward arms control and ICBM deployments which moves toward encouraging small, single-warhead ICBMs. This requires that arms control limitations and reductions be couched, not in terms of launchers, but in terms of equal levels of warheads of roughly equivalent yield. Such an approach could permit relatively simple agreements, using appropriate counting rules, that exert pressure to reduce the overall number and destructive power of nuclear weapons and at the same time give each side an incentive to move toward more stable and less vulnerable deployments.

Arms control agreements of this sort—simple and flexible enough to permit stabilizing development and modernization programs, while imposing quantitative limits and reductions—can make an important contribution to the stability of the strategic balance. An agreement that permitted modernization of forces and also provided an incentive to reduce while modernizing, in ways that would enhance stability, would be highly desirable. It would have the considerable benefit of capping both sides' strategic forces at levels that would be considerably lower than they would otherwise reach over time. It would also recognize, realistically, that each side will naturally desire to configure its own strategic forces. Simple aggregate limits of this sort are likely to be more practical, stabilizing, and lasting than elaborate, detailed limitations on force structure and modernization whose ultimate consequences cannot be confidently anticipated.

Encouraging stability by giving incentives to move toward less vulnerable deployments is more important than reducing quickly the absolute number of warheads deployed. Reductions in warhead numbers, while desirable for long-term reasons of limiting the cost of strategic systems, should not be undertaken at the expense of influencing the characteristics of strategic deployments. For example, warhead reductions, while desirable, should not be proposed or undertaken at a rate that leads us to limit the number of launching platforms to such low levels that their survivability is made more questionable.

For a variety of historical, technical, and verification reasons, both the SALT II unratified treaty and the current START proposal contain proposals to limit or reduce the number of ICBM launchers or missiles. Unfortunately this has helped produce the tendency to identify arms control with launcher or missile limits, and to lead some to identify successful arms control with low or reduced launcher or missile limits. This has, in turn, led to an incentive to build launchers and missiles as large as possible and to put as many as possible into each missile. Such an incentive has been augmented by the cost savings involved in putting a given number of warheads on a few large missiles rather than on a number of smaller ones. Although reasonable efforts have been made to constrain warheads through arms control (e.g., by the payload-fractionation limits in the negotiated SALT II treaty), these types of limits have still not produced an incentive mutually to move away from large land-based missiles. They will not do so as long as launcher or missile limitations are seen, in and of themselves, as primary arms control objectives.

We will have for some time strategic forces in which the numbers of launchers on one side are outnumbered many times over by the number of warheads on the other. Under such circumstances, it is not stabilizing to use arms control to require mutual reductions in the number of launching platforms (e.g., submarines or ICBM launchers) or missiles. Such a requirement further increases the ratio of warheads to targets. It does not promote deterrence and reduce the risk of war for the Soviets to have many more times the number of accurate warheads capable of destroying hard targets than the US has ICBM launchers.

In time we should try to promote an evolution toward forces in which—with an equal number of warheads—each side is encouraged to see to the survivability of its own forces in a way that does not threaten the other. But if the Soviet Union chooses to retain a large force of large missiles, each with many warheads, the US must be free to match this by the sort of deployment it chooses. Any arms control agreement equating SS-18s and small single-warhead ICBMs because each is one missile or because each is on one launcher would be destabilizing in the extreme.

The approach toward arms control suggested by the commission, moreover, is compatible with the basic objectives and direction of several other current arms control proposals.

For example, the negotiated SALT II treaty indirectly limited warheads by its limits on launchers and on the fractionation of payloads. It also barred deployments of new large ICBMs or the construction of additional fixed launchers. And it pointed toward further reductions in a follow-on SALT III agreement. These broad purposes of SALT II are wholly compatible with the arms control approach suggested here.

However, it should be noted that, as a method of restricting ICBM modernization, the negotiated SALT II Treaty, which would have expired in 1985, would have prohibited testing of more than one new ICBM. The two-part ICBM modernization program suggested by the commission would not violate that negotiated agreement because testing of a small, single-warhead ICBM could not begin before this expiration date. Of more long-term importance, however, the approach toward arms control and force modernization suggested here is fundamentally compatible with the sort of stability that SALT II sought to achieve. SALT II specifically contemplated the negotiation of extension agreements with improved terms, and there is no reason to doubt that future extension agreements would have allowed the testing and deployment of a second new ICBM missile with the stabilizing potential of a small, single-warhead ICBM. Moreover, the Soviets have tested two new ICBMs since October 1982.

The current administration's START proposal is centered on warhead limitations and reductions, with some attention to throw-weight limitations. These are consistent with the commission's recommended program. It also contains a proposed limit on launchers that the commission believes should be reassessed since it is not compatible with a desirable evolution toward small, single-warhead ICBMs.

Some current arms control proposals in Congress concentrate on warhead limitations in which reductions are forced in warhead numbers as a price of modernization; others seek explicitly to encourage movement toward small, single-warhead ICBMs on both sides. These general directions are also consistent with the approach suggested in this report.

The commission urges the continuation of vigorous pursuit of arms control; it is beyond the scope of this report, however, for the commission to recommend specific arms control proposals, the size of numerical limits, or the pace and scope of reductions. Of course any arms control proposal must be carefully designed with a view to compliance and verification—often particularly difficult questions in agreements with the Soviets. Some proposals may require innovation in verification techniques.

Finally, the commission is particularly mindful of the importance of achieving a greater degree of national consensus with respect to our strategic deployments and arms control. For the last decade, each successive administration has made proposals for arms control of strategic offensive systems that have become embroiled in political controversy between the executive branch and Congress and between political parties. None has produced a ratified treaty covering such systems or a politically sustainable strategic modernization program for the US ICBM force. Such a performance, as a nation, has produced neither agreement among ourselves, restraint by the Soviets, nor lasting mutual limitations on strategic offensive weapons.

The commission realizes that its recommendations will probably not fully satisfy any one of the many contending groups and individuals, inside and outside government, that have staked out claims to particular approaches to strategic modernization or arms control—much less all of them. In the interest of producing a national consensus on these two large issues, however, the commission has developed an approach that is different in kind from what has gone before.

The commission believes that all of the difficult issues discussed in this report—including the devastating nature of the modern war and the totalitarian and expansive character of the Soviet system—must be considered fairly in trying to reach a national consensus about a broad approach to strategic force modernization and arms control that can set a general direction for a number of years. Clearly there will be, and should be, many different views about specific elements in that approach. But the commission unanimously believes that such a new consensus—requiring a spirit of compromise by all of us—is essential if we are to move toward greater stability and toward reducing the risk of war. If we can begin to see ourselves, in dealing with these issues, not as political partisans or as crusaders for one specific solution to a part of this complex set of problems, but rather as citizens of a great nation with the humbling obligation to persevere in the long-run task of preserving both peace and liberty for the world, a common perspective may finally be found.

NOTES

1. MX weighs 195,000 pounds. Thus it is a "light ICBM" under the terminology of SALT II, approximately the same size as the 330 newly deployed Soviet SS-19 ICBMs. The MX is well under half the dimensions of the much larger 308 newly deployed SS-18s; the latter are designated as "modern heavy ICBMs" under SALT II.

2. An attack in which thousands of warheads were targeted at our ICBM fields, but there were no early detonations on our bomber bases from attacks by Soviet submarines.

ARMS CONTROL AFTER REYKJAVIK

GEORGE P. SHULTZ

I have come here to the University of Chicago to talk about nuclear weapons, arms control, and our national security. These issues have been given special timeliness by the president's recent meeting with Soviet General Secretary Gorbachev in Reykjavik. In years to come, we may look back at their discussions as a turning point in our strategy for deterring war and preserving peace. It has opened up new possibilities for the way in which we view nuclear weapons and their role in ensuring our security.

We now face a series of questions of fundamental importance for the future: How can we maintain peace through deterrence in the midst of a destabilizing growth of offensive nuclear weapons? How can we negotiate a more stable strategic balance at substantially lower levels of offensive forces? How can we use new defensive technology to contribute to that stability? How can the West best seek to reduce its reliance on offensive nuclear weapons without running new risks of instability arising from conventional imbalance? . . .

I'm not here tonight to announce the end of the nuclear era, but I will suggest that we may be on the verge of important changes in our approach to the role of nuclear weapons in our defense. New technologies are compelling us to think in new ways about how to ensure our security and protect our freedoms. Reykjavik served as a catalyst in this process. The president has led us to think seriously about both the possible benefits—and the costs—of a safer strategic environment involving progressively less reliance on nuclear weapons. Much will now depend on whether we are far-sighted enough to proceed towards such a goal in a realistic way that enhances our security and that of our allies.

It may be that we have arrived at a true turning point. The nuclear age cannot be undone or abolished; it is a permanent reality. But we can glimpse now, for the first time, a world freed from the incessant and pervasive fear of nuclear devastation. The threat of nuclear conflict can never be wholly banished, but it can be vastly diminished by careful but drastic reductions in the offensive nuclear arsenals each side possesses. It is just such reductions—not limitations in expansion, but reductions—that is the vision President Reagan is working to make a reality.

Excerpted with minor revisions from a speech at the University of Chicago, 17 November 1987.

Such reductions would add far greater stability to the US-Soviet nuclear relationship. Their achievement should make other diplomatic solutions obtainable, and perhaps lessen the distrust and suspicion that have stimulated the felt need for such weapons. Many problems will accompany drastic reductions: problems of deployment, conventional balances, verification, multiple warheads, and chemical weapons. The task ahead is great but worth the greatest of efforts.

This will not be a task for Americans alone. We must engage the collective effort of all of the Western democracies. And as we do, we must also be prepared to explore cooperative approaches with the Soviet Union, when such cooperation is feasible and in our interests.

THE EVOLUTION OF OUR THINKING ABOUT NUCLEAR WEAPONS

Let me start by reviewing how our thinking has evolved about the role of nuclear weapons in our national security.

In the years immediately after Fermi's first chain reaction, our approach was relatively simple. The atomic bomb was created in the midst of a truly desperate struggle to preserve civilization against fascist aggression in Europe and Asia. There was a compelling rationale for its development and use.

But since 1945—and particularly since America lost its monopoly of such weapons a few years later—we have had to adapt our thinking to less clear-cut circumstances. We have been faced with the challenges and the ambiguities of a protracted global competition with the Soviet Union. Nuclear weapons have shaped, and at times restrained, that competition; but they have not enabled either side to achieve a decisive advantage.

Because of their awesome destructiveness, nuclear weapons have kept in check a direct US-Soviet clash. With the advent in the late 1950s of intercontinental-range ballistic missiles—a delivery system for large numbers of nuclear weapons at great speed and with increasing accuracy—both the United States and the Soviet Union came to possess the ability to mount a devastating attack on each other within minutes.

The disastrous implications of such massive attacks led us to realize, in the words of President Kennedy, that "total war makes no sense." And as President Reagan has reiterated many times: "A nuclear war cannot be won and must never be fought"—words that the president and General Secretary Gorbachev agreed on in their joint statement at Geneva a year ago.

Thus, it came to be accepted in the West that a major role of nuclear weapons was to deter their use by others, as well as to deter major conventional attacks, by the threat of their use in response to aggression. Over the years, we sought through a variety of means and rationales—beginning with "massive retaliation" in the 1950s up through "flexible response" and "selective nuclear options" in the 1970s—to maintain a credible strategy for that retaliatory threat.

At the same time, we also accepted a certain inevitability about our own nation's vulnerability to nuclear-armed ballistic missiles. When nuclear weapons were delivered by manned bombers, we maintained air defenses. But as the ballistic missile emerged as the basic nuclear delivery system, we virtually abandoned the effort to build defenses. After a spirited debate over anti-ballistic missile systems in the late 1960s, we concluded that—on the basis of technologies now twenty years old—such defenses would not be effective. So our security from nuclear attack came to rest on the threat of retaliation and a state of mutual vulnerability.

In the West, many assumed that the Soviets would logically see things this way as well. It was thought that once both sides believed that a state of mutual vulnerability had been achieved, there would be shared restraint on the further growth of our respective nuclear arsenals.

The ABM Treaty of 1972 reflected that assumption. It was seen by some as elevating mutual vulnerability from technical fact to the status of international law. That treaty established strict limitations on the deployment of defenses against ballistic missiles. Its companion Interim Agreement on Strategic Offensive Arms was far more modest. SALT I was conceived of as an intermediate step towards more substantial future limits on offensive nuclear forces. It established only a cap on the further growth in the numbers of ballistic missile launchers then operational and under construction. The most important measures of the two sides' nuclear arsenals—numbers of actual warheads and missile throw-weight—were not restricted.

But controlling the number of launchers without limiting warheads actually encouraged deployment of multiple warheads—called MIRVs—on a single launcher. This eventually led to an erosion of strategic stability as the Soviets—by proliferating MIRVs—became able to threaten all of our intercontinental ballistic missiles with only a fraction of their own. Such an imbalance makes a decision to strike first seem all the more profitable.

During this postwar period, we and our allies hoped that American nuclear weapons would serve as a comparatively cheap offset to Soviet conventional military strength. The Soviet Union, through its geographic position and its massive mobilized conventional forces, has powerful advantages it can bring to bear against

Western Europe, the Mideast and East Asia—assets useful for political intimidation as well as for potential military aggression. The West's success or failure in countering these Soviet advantages has been, and will continue to be, one of the keys to stability in our postwar world.

Our effort to deter a major Soviet conventional attack through the existence of opposing nuclear forces has been successful over the past four decades. It gave the industrialized democracies devastated by the Second World War the necessary "breathing space" to recover and thrive. But there has also been recurring debate over the credibility of this strategy, as well as controversy about the hardware required for its implementation.

Over time, we and our allies came to agree that deterrence required a flexible strategy combining both conventional and nuclear forces. This combined strategy has been successful in avoiding war in Europe. But our reliance for so long on nuclear weapons has led some to forget that these arms are not an inexpensive substitute—mostly paid for by the United States—for fully facing up to the challenges of conventional defense and deterrence.

SOURCES OF STRATEGIC INSTABILITY

The United States and our allies will have to continue to rely upon nuclear weapons for deterrence far, far into the future. That fact, in turn, requires that we maintain credible and effective nuclear deterrent forces.

But a defense strategy that rests on the threat of escalation to a strategic nuclear conflict is, at best, an unwelcome solution to ensuring our national security. Nuclear weapons, when applied to the problem of preventing either a nuclear or conventional attack, present us with a major dilemma. They may appear a bargain—but a dangerous one. They make the outbreak of a Soviet-American war most unlikely; but they also ensure that should deterrence fail, the resulting conflict would be vastly more destructive, not just for our two countries, but for mankind as a whole.

Moreover, we cannot assume that the stability of the present nuclear balance will continue indefinitely. It can deteriorate and has. We have come to realize that our adversary does not share all of our assumptions about strategic stability. Soviet military doctrine stresses warfighting and survival in a nuclear environment, the importance of numerical superiority, the contribution of active defense, and the advantages of preemption.

Over the past fifteen years, the growth of Soviet strategic forces has continued unabated—and far beyond any reasonable assessment of what might be required for rough equivalency with US forces. As a result, the Soviet Union has acquired a capability to put at risk the fixed land-based missiles of the US strategic triad, as well as portions of our bomber and in-port submarine force and command and control systems, with only a fraction of their force, leaving many warheads to deter any retaliation.

To date, arms control agreements along traditional lines—such as SALT I and II—have failed to halt these destabilizing trends. They have not brought about significant reductions in offensive forces, particularly those systems that are the most threatening to stability. By the most important measure of destructive capability—ballistic missile warheads—Soviet strategic forces have grown by a factor of four since the SALT I Interim Agreement was signed. This problem has been exacerbated by a Soviet practice of stretching their implementation of such agreements to the edge of violation—and sometimes, beyond. The evidence of Soviet actions contrary to SALT II, the ABM Treaty and various other arms control agreements is clear and unmistakable.

At the same time, technology has not stood still. Research and technological innovation of the past decade now raise questions about whether the primacy of strategic offense over defense will continue indefinitely. For their part, the Soviets have never neglected strategic defenses. They developed and deployed them even when offensive systems seemed to have overwhelming advantages over any defense. As permitted by the ABM Treaty of 1972, the Soviets constructed around Moscow the world's only operational system of ballistic missile defense. Soviet military planners apparently find that the modest benefits of this system justify its considerable cost, even though it would provide only a marginal level of protection against our overall strategic force. It could clearly be a base for the future expansion of their defenses.

For well over a decade—long before the president announced three years ago the American Strategic Defense Initiative—the Soviet Union has been actively investigating much more advanced defense technologies, including directed energy systems. If the United States were to abandon this field of advanced defensive research to the Soviet Union, the results ten years hence could be disastrous for the West.

THE PRESIDENT'S APPROACH: SEEKING GREATER STABILITY

President Reagan believes we can do better. He believes we can reverse the ever-increasing numbers and potency of nuclear weapons that are eroding stability. He believes we can and should find ways to keep the peace without basing our security so heavily on the threat of nuclear escalation. To those ends, he has set in motion a series of policies which have already brought major results.

First, this administration has taken much needed steps to reverse dangerous trends in the military balance by strengthening our conventional and nuclear deterrent forces. We have gone forward with their necessary modernization.

Second, we have sought ambitious arms control measures—agreements that seriously contribute to the goal of stabilizing reductions in offensive forces. In 1981, the president proposed the global elimination of all Soviet and American longer-range INF nuclear missiles. Not a freeze or token reductions, as many urged at the time, but the complete elimination of this class of weapons.

The following year, at Eureka College, the president proposed major reductions in strategic offensive forces, calling for cuts by one-third to a level of 5000 ballistic missile warheads on each side. Again, this was a major departure from previous negotiating approaches—both in the importance of the weapons to be reduced and in the magnitude of their reduction. Critics claimed he was unrealistic, that it showed he was not really interested in arms control. But the president's call for dramatic reductions in nuclear warheads on the most destabilizing delivery systems has been at the core of our negotiating efforts. The Soviets have finally begun to respond to the president's approach, and are now making similar proposals.

Finally, the president also set out to explore whether it would be possible to develop an effective defense against ballistic missiles, the central element of current strategic offensive arsenals. To find that answer, he initiated in 1983 the SDI program, a broad-based research effort to explore the defensive implications of new technologies. It is a program that is consistent with our obligations under the ABM Treaty. He set as a basic goal the protection of the United States and our allies against the ballistic missile threat.

Since then, we have been seeking both to negotiate deep reductions in the numbers of those missiles, as well as to develop the knowledge necessary to construct a strategic defense against them. It is the president's particular innovation to seek to use these parallel efforts in a reinforcing way—to reduce the threat while exploring the potential for defense.

REYKJAVIK: A POTENTIAL WATERSHED IN NUCLEAR ARMS CONTROL

All of these efforts will take time to develop, but we are already seeing their first fruits. Some became apparent at Reykjavik. Previously, the prospect of 30, let alone 50, percent reductions in Soviet and American offensive nuclear arsenals was considered an overly ambitious goal.

At Reykjavik, the president and General Secretary Gorbachev reached the basis for an agreement on a first step of 50 percent reductions in Soviet and American strategic nuclear offensive forces over a five-year period. We agreed upon some numbers and counting rules—that is, how different types of weapons would count against the reduced ceilings.

For INF nuclear missiles, we reached the basis for agreement on even more drastic reductions, down from a current Soviet total of over 1400 warheads to only 100 on longer-range INF missiles worldwide on each side. This would represent a reduction of more than 90 percent of the Soviet SS-20 nuclear warheads now targeted on our allies and friends in Europe and Asia. There would also have to be a ceiling on shorter-range INF missiles, the right for us to match the Soviets in this category, and follow-on negotiations aimed at the reduction in numbers of these weapons.

Right there is the basis for an arms control agreement that does not just limit the future growth of Soviet and American nuclear arsenals, but which actually makes deep and early cuts in existing force levels. These cuts would reduce the numbers of heavy, accurate multiple-warhead missiles that are the most threatening and the most destabilizing. These ideas discussed at Reykjavik flowed directly from the president's longstanding proposals. They are a direct result of his vision of major offensive reductions as a necessary step to greater stability.

At Reykjavik, the president and the general secretary went on to discuss possible further steps towards enhanced stability. The president proposed to eliminate all ballistic missiles over the subsequent five years. Mr. Gorbachev proposed to eliminate all strategic offensive forces. They talked about these and other ideas, including the eventual elimination of all nuclear weapons. The very scope of their discussion was significant. The president and the general secretary set a new arms control agenda at Reykjavik, one that will shape our discussion with the Soviets about matters of nuclear security for years to come.

Make no mistake about it. Tough, and probably drawn-out, negotiations will still be required if we are to nail down any formal agreement on offensive force reductions. For example, the Soviets are now linking agreement on anything with agreement on everything. But the fact that we now have such reductions clearly on the table has only been made possible by:

—our steps to restore America's military strength;

—our firm and patient negotiating efforts over the past five years;

—the sustained support of our allies; and not the least,

—our active investigation into strategic defenses.

The prospect of effective defenses, and our determined force modernization program, have given the Soviet Union an important incentive to agree to cut back and eventually eliminate ballistic missiles. Within the SDI program, we judge defenses to be desirable only if they are survivable and cost-effective at the margin. Defenses that meet these criteria—those which cannot be easily destroyed or overwhelmed—are precisely the sort which would lead Soviet military planners to consider reducing, rather than continuing to expand, their offensive missile force.

But only a dynamic and ongoing research program can play this role. And for their part, the Soviets are making every effort to cripple our program. Thus, there were major differences over strategic defenses at Reykjavik. The president responded to Soviet concerns by proposing that, for ten years, both sides would not exercise their existing right of withdrawal from the ABM Treaty and would confine their strategic defense program to research, development, and testing activities permitted by the ABM Treaty. This commitment would be in the context of reductions of strategic offensive forces by 50 percent in the first five years and elimination of the remaining ballistic missiles in the second five years, and with the understanding that at the end of this ten-year period, either side would have the right to deploy advanced defenses, unless agreed otherwise.

But at Reykjavik, the Soviet Union wanted to change existing ABM Treaty provisions to restrict research in a way that would cripple the American SDI program. This we cannot accept.

Even after the elimination of all ballistic missiles, we will need insurance policies to hedge against cheating or other contingencies. We don't know now what form this will take. An agreed-upon retention of a small nuclear ballistic missile force could be part of that insurance. What we do know is that the president's program for defenses against ballistic missiles can be a key part of our insurance. A vigorous research program will give the United States and our allies the options we will need to approach a world with far fewer nuclear weapons—a world with a safer and more stable strategic balance, one no longer dependent upon the threat of mutual annihilation . . .

THE CHALLENGES OF A LESS NUCLEAR WORLD

The longer-term implications of the Reykjavik discussion may prove even more challenging for us. Thus far in the nuclear age, we have become accustomed to thinking of nuclear weapons in terms of "more bang for the buck"—and of the high price for any possible substitute for these arms. But to my mind, that sort of bookkeeping approach risks obscuring our larger interests. We should begin by determining what is of value to us, and then what costs we are prepared to pay to attain those ends.

The value of steps leading to a less nuclear world is clear—potentially enhanced stability and less chance of a nuclear catastrophe. Together with our allies, we could enjoy a safer, more secure strategic environment.

But we would not seek to reduce nuclear weapons only to increase the risks of conventional war, or more likely, of political intimidation through the threat of conventional attack. Therefore, a central task will be to establish a stable conventional balance as a necessary corollary for any less nuclear world.

How would a less nuclear world, one in which ballistic missiles have been eliminated, work? What would it mean? It would not mean the end of nuclear deterrence for the West. With a large inventory of aircraft and cruise missiles, the United States and NATO would retain a powerful nuclear capability. In a sense, we would return to the situation of the 1950s, when strategic bombers served as our primary nuclear deterrent force. But there would be an important difference in the 1990s and beyond. Our aircraft would now be supplemented by a host of new and sophisticated technologies as well as cruise missiles launched from the air and sea. It would be a much more diverse and capable force than in previous decades.

In such circumstances, both the United States and the Soviet Union would lose the capacity provided by ballistic missiles to deliver large numbers of nuclear weapons on each others' homelands in less than thirty minutes time. But Western strategy is, in fact, defensive in nature, built upon the pledge that we will only use our weapons, nuclear and conventional, in self-defense. Therefore, the loss of this quick-kill capability—so suited to preemptive attack—will ease fears of a disarming first strike.

For our friends and allies in Europe and Asia, the elimination of Soviet ballistic missiles and its many SS-20s, but also the shorter-range missiles for which we currently have no deployed equivalent—would remove a significant nuclear threat. But it would also have nonnuclear military benefits as well. Today, the Soviet Union has ballistic missiles with conventional and chemical warheads targeted on NATO airfields, ports, and bases. The elimination of ballistic missiles would thus be a significant plus for NATO in several respects.

The nuclear forces remaining—aircraft and cruise missiles—would be far less useful for first-strike attacks, but would be more appropriate for retaliation. They would be more flexible in use than ballistic missiles. The slower-flying aircraft can be recalled after launch. They can be retargeted in flight. They can be reused for several missions. We currently have a major advantage in the relative sophistication of our aircraft and cruise missiles; the Soviets have greater numbers and are striving hard to catch up in quality. They have given far more attention to defense, where we have a lot of catching up to do. But our remaining nuclear forces would be capable of fulfilling the requirements of the Western alliance's deterrent strategy.

THE WEST'S ADVANTAGES IN A LESS NUCLEAR WORLD

The prospect of a less nuclear world has caused concern in both Europe and America. Some fear that it would place the West at a grave disadvantage. I don't think so.

In any competition ultimately depending upon economic and political dynamism and innovation, the United States, Japan, and Western Europe have tremendous inherent advantages. Our three-to-one superiority in GMP over the Warsaw Pact, our far greater population, and the Western lead in modern technologies—these are only partial measure of our advantages. The West's true strength lies in the fact that we are not an ideological or military bloc like the Warsaw Pact—we are an alliance of free nations, able to draw upon the best of the diverse and creative energies of our peoples.

But dramatic reductions in nuclear weapons and the establishment of stronger conventional defenses will require a united Alliance effort. In light of the president's discussions in Reykjavik, we must join with our allies in a more systematic consideration of how to deal with a less nuclear world. To my mind, that sort of process of joint inquiry is healthy for the Alliance, particularly since we remain firmly agreed on the basics—the Alliance's fundamental principle of shared risks and shared burdens on behalf of the common defense.

All of these steps—deep reductions of nuclear weapons, a strong research program in strategic defense, improvements in conventional defenses, and negotiations with the Soviet Union and Warsaw Pact—will have to be closely synchronized. This will require a carefully coordinated political strategy on the part of the Alliance to deal with these interrelated aspects of the larger problem of stability and Western security. We will begin a preliminary discussion of just such an approach during my next meeting with my NATO counterparts in Brussels at the December session of the North Atlantic Council.

CONCLUSION

This is a full and complex agenda for all of us to consider. Is it ambitious? Yes. Unrealistic? No. I think that, on the basis of the progress made at Reykjavik, substantial reductions in Soviet and American nuclear forces are possible, and they can be achieved in a phased and stabilizing way.

But we need to think hard about how to proceed. We are taking on a difficult task as we seek to create the conditions in which we can assure the freedom and security of our country and our allies without the constant threat of nuclear catastrophe.

And, of course, our work to achieve greater strategic stability at progressively lower levels of nuclear arms is only part of our larger effort to build a more realistic and constructive relationship with the Soviet Union. We cannot pursue arms control in isolation from other sources of tension. We will continue to seek a resolution of the more fundamental sources of political distrust between our nations, especially those in the areas of human rights and regional conflicts.

Progress—whether in science or foreign affairs—often has to do with the reinterpretation of fundamental ideas. That's no easy task. It requires challenging conventional wisdom. And often we find that gaining new benefits requires paying new costs.

PART IV

DEFENSE POLICY AND FORCE PLANNING

CHAPTER 12

THE DEFENSE POLICY PROCESS AND THE ROLE OF THE MILITARY OFFICER
An Overview

EDWARD N. WRIGHT

Defense policy *is* public policy. Consuming more than $300 billion annually, or approximately one quarter of the federal budget, defense programs comprise one of the most significant and complex challenges confronting the president and the executive branch, the Congress, and the American people. As discussed in the introductory chapter, the domestic political system, responsible for formulating and implementing defense policy, is no less complex than the international environment from which the security dilemma arises. Within the broader range of foreign policy, the central element of national security and defense policy is the creation of forces and their use. Therefore, Part 4 focuses on the domestic political system, the decisionmakers, or actors, and the process through which decisions are made about military forces and their use.

THE DECISIONMAKERS

Chapter 13 focuses on the executive branch and its primary actors in the definition of US policy interests, and the competition among those actors in developing coordinated policy initiatives in the foreign and national security policy area. The central actor in the process is the president. How he chooses to receive advice and make decisions is important in determining the organizational structure and operational style of his White House staff, as well as how he will use the National Security Council and its staff. Likewise, both of these characteristics significantly influence the interrelationships among the principal participants in defense policy decisions: the secretaries of state and defense and the assistant to the president for national security affairs. Accordingly, this chapter examines organizational influences on defense decisionmaking including the National Security Council and its staff as agents of the president in coordinating policy and brokering the competing interests of executive departments and other actors in the policy process. Focusing on the organization of the State Department, the

traditional competition between the assistant to the president for national security affairs and the secretary of state is addressed as a consequence, in part, of organizational factors.

Finally, chapter 13 provides a detailed examination of the Defense Department—the evolution of defense organization, the nature of policymaking within the Department, and the consequences of the Goldwater-Nichols Defense Reorganization Act of 1986. Three major components of reform are evaluated: strengthening the roles of the chairman of the Joint Chiefs of Staff and the commanders-in-chief of the unified commands, restructuring the Office of the Secretary of Defense, and reform of acquisition program management. The impact and effectiveness of the reforms mandated by the Congress during President Reagan's second term will, in large measure, be determined by the managerial approach taken by Secretary of Defense Richard Cheney and his leadership team.

Chapter 14 examines the importance of the Congress. The roles of the Congress in defense decisionmaking are the same important ones it performs in other areas of public policy: lawmaking to include declaring war; powers of the purse; oversight of the executive; executive counsel, debate, and deliberation; and facilitating consensus. The president and secretary of defense may be the most visible contributors to defense decisions, however, the Congress makes fundamentally important contributions in the performance of its traditional roles and constitutional authority. Chapter 14 is intended for the student of defense policy who is not particularly familiar with the structure and organization of the US Congress. Therefore, it first focuses on the traditional and constitutional congressional roles and the legislative process—which are essentially the same for all areas of public policy. The chapter then examines more explicitly the application of the norms of the Congress to issues of defense policy and analyzes why different kinds of defense issues get handled differently. Structural policies that involve weapons systems procurement, the distribution of military forces, and basing decisions are essentially distributive policies and the subject of intense congressional interest and lobbying. On the other hand, strategic policies and programs directed at long-term strategy, and relationships with foreign allies and potential adversaries, as well as crisis policies, are dominated by the president, executive branch agencies, and presidential advisers. As is the case with domestic policy, congressional power is most evident in the budget process, which begins when the president submits his defense budget in January for the following fiscal year.

The examination of the more visible actors in the defense decisionmaking process provided by these two chapters is supplemented in the Selected Readings by a more detailed treatment of the National Security Council and the Office of Management and Budget. The fundamental importance of understanding the proper, lawful role of the NSC is underscored by the revelations and controversies surrounding the Iran-contra affair. Two specific articles address different aspects of the NSC. Philip Odeen's "Role of the National Security Council in Coordinating and Integrating US Defense and Foreign Policy" examines the NSC's policy roles of creating policy frameworks that define basic policies and priorities, forcing decisions on major issues, managing the decision process, and ensuring that decisions are implemented. Odeen concludes that in the absence of forward-looking planning capabilities in the Departments of State and Defense, it appears that the most important function that should be ascribed to the NSC is long-range planning. Excerpts from the *Report of the President's Special Review Board*, or the Tower Commission Report, are included to emphasize the commission's assessment of "what went wrong" in the unfortunate and politically debilitating consequences of the linkage of arms sales to Iran and aid to the Nicaraguan contras. One of the principal investigators and authors of the commission's report was previously assistant to the president for national security affairs under President Ford and was selected for that same position by President George Bush. Interestingly, Brent Scowcroft was also one of the editors of the first edition of *American Defense Policy* in 1965.

POLICY PROCESSES

In addition to the structural and organizational roles of decisionmakers, the process whereby defense decisions are made is equally important. Chapter 15, "Force Planning: Bridge Between Doctrine and Forces," is new to the subjects addressed in this text during the more than 25 years it has been in publication. The output of the defense policy process consists of

military forces in the form of weapon systems, materiel, facilities, and manpower as well as strategies and plans for their use. Chapter 15 sets out to introduce students of defense policy to the argument "that a rational basis for the design of US military forces does exist" and in exploring that thesis exposes the underlying tension between experience and rationality, short-term interests and long-term requirements, and rational actor and bureaucratic politics models.[1] The conclusion reached in Chapter 15 is that in the absence of rational force planning and its analytical approach, "the larger process of formulating defense policy fragments into separate, uncoordinated, and irrational programs"[2] which are the "irrational outcome of a bureaucratic political process."[3]

In contrast with the rational actor decisionmaking model, many argue that "the bureaucratic politics model is perhaps the best intellectual construct available for understanding national security policymaking."[4] From this perspective, the policy process is described as "unfolding in a governmental structure more akin to a confederacy than a hierarchy."[5] In such a construct

Decisionmakers are viewed as actors or players in a game of politics, promoting bureaucratic interests in competition for various stakes and prizes. Bureaucratic positions on policy issues are determined by bureaucratic interests (or, where one stands depends on where one sits). Policy outcomes, more often than not, reflect a synthesis, or compromise among different positions.[6]

Graham Allison's three-model decisionmaking paradigm, that includes rational actor, organizational process, and bureaucratic politics, is a classic in the literature of decisionmaking and is included in the Selected Readings. "It is worth remembering that Allison's prescription for improving explanation and prediction of national security problems was to use the models together, not to each other's exclusion."[7]

PLURALISM AND POLICY PROCESS

One of the reasons it is difficult to sustain a "rational" planning or policy process is the pluralistic and fragmented nature of the American political system. In Chapter 1, pluralism, or the competition in the policy process among multiple elites or groups holding particularistic interests, was described as the strongest characteristic of the American political system. David Kozak's "American Pluralism and Defense Policy" in the Selected Readings describes five models of American politics. These models illuminate political structures and developments that have strongly affected the defense policy process. *Madisonian pluralism* shows the effects of decisions by the founders to constrain the power of the central government by fragmenting authority and creating institutional conflict. *Extraconstitutional developments* account for the rise of political parties and the increasing importance of public opinion and mass preferences that have important consequences for the defense decisionmaker. The development of *interest group subgovernments*, comprised of representatives of the bureaucracy, Congress and its staff, and representatives of interest groups, has provided subgovernment specialization and policy dominance by subgovernments and inhibited political leadership by those elected to government. *Destabilizing transformations* is the fourth contributor to increased pluralism. As a consequence of Vietnam and Watergate in the 1960s and 1970s, major changes occurred in the American political system, strengthening the forces of fragmentation while weakening the forces of concentration. The centralizing powers of the presidency were weakened by a resurgent Congress, while the rise of single-issue politics undercut the traditional roles of political parties, adding more actors to the policy process. For a time the Reagan presidency restored some of the *centripetal powers of the presidency*, which resulted in early successes in reversing the influence of subgovernments through Cabinet Councils, consolidation and contraction in social program funding and federal divestiture, and significant increases in defense spending. However, as Kozak concludes,

The government of 1787 was not intended to be neat and orderly. It did not place a premium on tidiness, harmony or efficiency. It was designed to be chaotic and disorderly for the purposes of advancing self government, limited government and conflict resolution. Although such a system at times is very frustrating, its values are precious. They are American, they are pluralistic (as opposed to dictatorial), and they are worth defending.

In addressing students of defense policy, military and civilian, Kozak argues that military officers need to increase their understanding of the American political system in order to operate in the policy process more effectively. This raises an important question, what is the proper role of the military officer within the political system and in the policy arena?

THE ROLE OF THE MILITARY OFFICER

Recently, the frequent debate over the proper role of military officers in formulating defense and foreign policy reached new heights. Actions of military members of the National Security Council staff in the Iran-contra affair, especially the diversion of funds to contra forces in Nicaragua despite Congressional prohibitions, re-

newed public concern with the relationship between the military and politics. Public concern focused on

the propriety and legality of actions by the staff of the National Security Council and other officers of the government. As first a Presidential Special Review Board (The "Tower Commission") and then the Congress conducted investigations into what has come to be known as the "Iran-contra Affair," the mass media tended to dwell upon the question of whether or not there was presidential involvement in the affair. Yet, within governmental circles, another area of concern with longer range implications began to emerge, relating to the prominent role played by military officers in the unfolding Iran-contra drama.[8]

The educational process for professional officers was called into question, and the House Armed Services Committee announced it would conduct an investigation "into the quality of US service academy education."[9] One committee member, Representative Joseph Brennan (Dem., Me.), observed that "the military has a responsibility to conduct itself in a manner consistent with laws passed by the United States Congress, even the ones they don't like."[10] Brennan was outraged that the actions of senior military officers demonstrated a "disregard for the Constitution."[11] One of the tragedies of this event is that the very military officers accused by Representative Brennan and others "defended their conduct as being fully justified in the interest of national security by their oaths of office to 'support and defend' the very Constitution they were accused of disregarding."[12]

The two Selected Readings, John Garrison's "Political Dimension of Military Professionalism," and Zeb B. Bradford, Jr., and James R. Murphy's "New Look at the Military Profession," provide a time-tested basis for exploring the differing perspectives on the proper role for military officers in our political system.

As is the case with nearly all academic discussions of this issue, both articles are influenced by Samuel Huntington's classic work, *The Soldier and the State*.[13] Huntington's benchmark study established three criteria that distinguished the military officer as a professional. These criteria—expertise, responsibility, and corporateness—combined with Huntington's definition of the officer's vocation as the management of violence defined the military profession for decades. All other functions, such as leading troops, debating strategy, and motivating subordinates, were viewed as necessary actions in support of the larger professional expertise. As Huntington describes it, the officer's responsibility is to the State. Indeed the officer corps is monopolized by the State whereas other professions are merely regulated by it.

Bradford and Murphy would argue that Huntington's narrow definition of the professional officer's expertise is responsible for the "traditional," or "apolitical," view of the military officer's role in policymaking. This apolitical model, and the contrasting "fusionist" perspective, are examined by Garrison, who concludes that if professional officers are not involved in the formulation of national security policy, then the policy they are to implement "runs the risk of being less appropriate and less effective than it should, or could, be." But if officers are involved, what is to prevent situations like the Iran-contra case where both accusers and defenders cite the Constitution or their constitutional oaths as justification for their beliefs and actions? What is to prevent the military officer from becoming a policy entrepreneur? What safeguards are there to dampen policy zealots and to condition a "can do" attitude and "have done" results?

In their critique of Huntington's original description of the military profession, Bradford and Murphy provide powerful prescriptions for the issues raised above. At the center of their argument, and underlying disagreement with Huntington, is that the military profession is "more than a uniformed structure incorporating a functional expertise." It is true that the military officer is first an officer, and a pilot, or infantryman, or personnel manager second. The characteristic which most strongly establishes the common core of the profession is *responsibility*. Bradford and Murphy argue that

Responsibility is more than a means of insuring that the military exercise its expertise in the service of the state. Far more essential to military professionalism is an internalized *sense* of responsibility, of allegiance to duly constituted authority.

The professional officer's expertise is not the "management of violence," but *officership* and a profound understanding of the nature of "unconditional service to the lawful authority of the State." In the case of the Iran-contra affair, the loss of this professional character produced military officers who became a *political* force of their own. Their goal was to achieve national security objectives which they had defined. Their failure was both personal and institutional. The renewed debate as to the proper role of the military officer is essential in the education of the officer corps, the civilian leaders they serve directly, and the society they serve collectively.

NOTES

1. Robert P. Haffa, Jr., *Rational Methods, Prudent Choices: Planning U.S. Forces* (Washington, D.C.: National Defense University Press, 1988), pp. 9–12. Chapter 15 is excerpted from this larger work that provides excellent case studies on strategic nuclear forces, general purpose forces, and rapidly deployable forces.
2. Ibid., p. 9.
3. Ibid., p. xiii.
4. David C. Kozak and James M. Keagle, *Bureaucratic Politics and National Security: Theory and Practice* (Boulder, Colo.: Lynne Rienner, 1988), p. 1.
5. Ibid.
6. Ibid.
7. Kozak and Keagle, *Bureaucratic Politics*, p. 54.
8. S. Nelson Drew, "Bearing 'True Faith and Allegiance' to the Constitution: The Imperative for Integrity in National Security Policymaking." Paper presented at the Joint Annual Convention of the British International Studies Association and the International Studies Association, London, 1 Apr. 1989, p. 1.
9. As cited in Drew, "Bearing 'True Faith,' " p. 2.
10. Ibid.
11. Ibid.
12. Ibid.
13. Samuel P. Huntington, *The Soldier and the State: The Theory and Politics of Civil-Military Relations* (Cambridge: Harvard University Press, 1957).

CHAPTER 13

THE STRUCTURE OF AMERICAN DEFENSE POLICYMAKING

EDWARD N. WRIGHT

Reagan administration policies concerning strategic deterrence, the use of force as an instrument of policy, and the relative priorities of defense and social spending resulted in one of the most important debates about defense organization since the 1950s. The Iran-contra affair and a National Security Council system seemingly out of control, executive conflict with the Congress over continued support for the contras and policy in Nicaragua, failed policy initiatives in Panama, and continued debate about war powers surrounding US actions in the Persian Gulf, are but a few examples of the complexity of defense policymaking and both the scope and limits of presidential authority.

This chapter addresses the principal executive branch actors in the American defense policy process and concludes with a detailed examination of the organization of the Department of Defense and the prospects of reforms mandated by the Congress late in the Reagan presidency, the utility of which will not be known, perhaps, for several years.

DEFENSE POLICYMAKERS

Domestic, economic, and foreign and defense policies are characterized by distinctive patterns of who dominates the policymaking process. In comparison with domestic policy, foreign and defense policy are characterized by the primacy of executive authority. In no other area of public policy is the president's constitutional authority as strong as it is in the superintendence of foreign and defense policy. As the commander-in-chief and chief diplomat, the president encounters significantly less formidable competition for control of the policy process. Presidents, and their advisers, respond to crises and develop long-range strategies with little competition from Congress and control the priorities contained in the defense budget. The Congress shares in the constitutional authority for defense policy and the budget. Through the authorization and appropriations process, Congress debates and modifies the president's budget request. However, Congress's strongest influence is exercised over specific defense programs

or weapons systems and results in marginal or incremental changes in the overall budget. By its nature, the Congress is ill-equipped to affect significantly the president's overall strategy and objectives.

The president is the premier symbol of civilian control of the military. As commander-in-chief, he exercises the supreme national command authority over military forces, yet, the Constitution similarly grants the authority to declare war to the Congress. As a consequence, a constant state of tension exists between the exercise of presidential authority and prerogative and congressional constitutional and oversight powers.

THE PRESIDENT AND THE EXECUTIVE OFFICE

The president's most important responsibility is to ensure the nation's survival by providing for the common defense. The objective is achieved, in large measure, by his political leadership and coordination of the actions of the executive branch.

Central to the president's abilities as a leader and the way policy is developed is the president's conception of the presidency and his approach to decisionmaking. As described by Alexander George,

the first and foremost task that a new President faces is to learn to define his own role in the policymaking system; only then can he structure and manage the roles and relationships within the policymaking system of his secretary of state, the special assistant for national security affairs, the secretary of defense, and other cabinet and agency heads with responsibilities for the formulation and implementation of policy.[1]

The structure of advisory systems responsible for developing options for presidential decisionmaking and how the White House and executive branch departments and agencies interact in formulating policy must, of necessity, be compatible with the president's management style.

Most important in deciding how advisory systems and coordination among the various policy advisors and executives will operate is the president's own cognitive style. As Alexander George points out, the president's approach will be based on his previous executive experience and how he has come to evaluate and use information for the purpose of making decisions. His previous experience with staff, and how he relates to that staff, likewise affects how he structures the advisory systems of the White House.[2]

Equally important in how the president approaches decisionmaking is his sense of self-confidence and his conception of the presidency. If the president understands and is comfortable with issues involved in foreign and defense policy, he is more likely to project self-confidence. This self-confidence then easily extends to his political subordinates and results in their greater latitude of action, and the president places more confidence in their actions and advice.[3] A third, and especially important, characteristic is the degree to which the president permits or encourages policy debate or political conflict.

The personal attitude toward conflict that a president brings into office is likely to determine his orientation to the phenomena of "Cabinet politics" and "bureaucratic politics" within his administration, as well as the larger, often interlinked game of politics surrounding the Executive Branch. Individuals with a pronounced distaste for "dirty politics" and for being exposed to face-to-face disagreements among advisors are likely to favor policy making systems that attempt to curb these phenomena or at least shield them from direct exposure. They also are likely to prefer staff and advisory systems in which teamwork or formal analytical procedures are emphasized in lieu of partisan advocacy and debate.[4]

These three characteristics—cognitive style, sense of self-confidence and role, and the degree to which the president permits or encourages debate of policy options—all "combine to determine how a new president will structure the policymaking system around him and how he will define his own role and that of others in it."[5]

How the president chooses to use his innermost circle of policy advisers and staffs—the assistant to the president for national security affairs and the NSC staff, the Office of Management and Budget, the science adviser and the Office of Science and Technology Policy—and their relationship with the executive departments and agencies and cabinet officers are important dimensions of presidential politics and power.

The National Security Council

The principal White House staff charged with responsibility for coordinating national security policy is the assistant to the president for national security affairs and the National Security Council staff. As noted above, the role and in-

terrelationship between the assistant for national security affairs and the secretaries of defense and state are important in the formulation and execution of defense policy. The national security adviser's client is the president— it is the president he and the NSC staff serve. The NSC staff's principal purpose is to work to achieve lateral, or horizontal, coordination across the particularistic, or vertical, interests of executive branch departments and agencies in the formulation of a consistent and coherent foreign and defense policy.

The involvement of the NSC staff, as well as the assistant to the president for national security affairs, in selling arms to Iran and using the profits of that sale to assist the contras in Nicaragua severely damaged the Reagan administration's credibility in national security affairs and resulted in irreparable political damage to the president it served. Importantly, the Iran-contra affair precipitated a national debate and reexamination of the purpose of the NSC. A presidential commission was followed by an 11-month congressional inquiry to assess how the NSC could have become so operationally involved in the conduct of foreign policy, apparently without the knowledge or approval of the president. Another debate addressed in the previous chapter was the proper role of the military officer in the policy process.[6]

In response to the experiences of World Wars I and II, the Congress enacted the National Security Act of 1947 "to provide for the establishment of integrated policies and procedures for the departments, agencies, and functions of the government relating to national security."[7] In addition to creating the Department of Defense, the three military services, and the Central Intelligence Agency, the National Security Act created the National Security Council

to advise the president with respect to the integration of domestic, foreign, and military policies relating to the national security so as to enable the military services and the other departments and agencies of the government to cooperate more effectively in matters involving the national security.

Statutory members of the council are the president, the vice president and the secretaries of state and defense. The director of the Central Intelligence Agency and the chairman of the Joint Chiefs of Staff are statutory advisers to the NSC. How the council is structured and operates is at the discretion of the president, and various presidents have used the NSC in different ways. On occasion, other officials, to include members of the White House staff and agency heads with responsibilities for peripheral aspects of national security policy, also participate, at the direction of the president. Practicing a politics of inclusion, President Reagan included as fully participating members the secretary of the treasury, the attorney general, and the White House chief of staff. In addition, a number of subordinate staff also attended NSC meetings. In the second term, President Reagan began using a narrower group of advisers when the most sensitive policy issues were to be considered. This body, the National Security Planning Group, was comprised of the statutory NSC members and advisers, the secretary of treasury, attorney general, White House chief of staff, and the assistant to the president for national security affairs. Though larger than the statutory NSC, this group represented a smaller group than was routinely included in full NSC meetings earlier in the administration.

The assistant to the president for national security affairs directs the NSC staff and has risen in prominence to be the manager of the national security advisory system. The assistant to the president and the NSC staff serve the interest of the president in managing and coordinating foreign and national security policy. The process designed to develop and shape the competing views of policy issues which eventually reach the NSC and the president for decision, typically consists of several tiers of functional and regional interagency groups and decisionmakers.

Figure 13.1 depicts the structure of the Reagan NSC which was put in place following the Iran-contra affair. The members of the Policy Review Group and Senior Interagency Groups are normally subcabinet political appointees. These officials represent the political leadership of their departments and agencies and are expected to be responsive to the president's priorities and political leadership. Senior Interagency Groups are supported by subordinate interagency working groups whose task is the initial development of policy options and recommendations for the president. These groups are normally populated by department and agency representatives at the program-level of the bureaucracy. Where coordinated policy recommendations cannot be achieved due to conflicting interests of the participating departments and agencies, resolution is sought by referral to the next higher level. The recommendations eventually agreed to in the interagency process are then forwarded to the NSC and the president. The most contentious issues may eventually require the president to make the final decision.

As the Tower Commission report emphasized, the purpose of the assistant to the president for national security affairs and the NSC staff is to act as an advisory and coordinating body. Although each president since its creation has used the NSC staff in a way that reflects his personal working style, "it has developed an important role within the executive branch of coordinating policy review, preparing issues for

presidential decision, and monitoring implementation."[8] One of the conclusions of the commission's report was that in the case of the arms transfers to Iran the actions of the NSC staff to support the contras resulted in a breakdown in the "coordination process" for which the staff was created. Principal criticisms were that the policy initiatives undertaken "were not adequately vetted below the cabinet level" which is the purpose of the successive reviews by interagency groups depicted in figure 13.1. Secondly, the report concluded that "the NSC staff assumed direct operational control," violating its traditional role as a coordinator and adviser and impinging on "the traditional jurisdictions of the Departments of State, Defense and CIA." The activities of the NSC staff and the failed policies involved in the Iran-contra affair provide interesting "case studies in the perils of policy pursued outside the constraints of orderly process." Excerpts from the Tower Commission Report and the reading by Philip Odeen are useful starting points to understand better the history of the NSC, its organization, and the enormous tragedy of the Iran-contra affair.

THE DEPARTMENT OF STATE

The State Department is the senior executive department in the US government. The secretary of state is described as "the principal foreign policy adviser to the president"[9] and is the ranking member of the cabinet. Accordingly, a review of the literature on foreign policymaking illustrates that "most scholars and ex–State officials who have addressed the subject use as their point of departure the belief that foreign policymaking should be centered at State."[10] Yet, paradoxically, the locus of relative power has gravitated over time to the White House. It can, however, be argued that the Reagan administration restored significant importance to the State Department and its career staff when compared to the periods of very visible national security advisers such as Henry Kissinger and Zbigniew Brzezinski. State's improved stature under Reagan can be attributed to the succession of six different national security advisers during the Reagan presidency as well as the discredited standing of the NSC staff resulting from the Iran-contra affair and the disarray which followed.[11]

President George Bush demonstrated his intention to be an activist president in foreign affairs and moved quickly to establish the authority of the assistant to the president for national security affairs and the NSC staff in a White House–centered process for foreign and national security policy. The president's close personal relationship with Secretary of State James Baker may be the key to insuring that the traditional tension between the secretary of

Figure 13.1
The Reagan National Security Council System

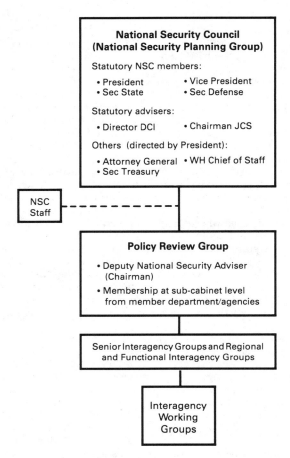

**National Security Council
(National Security Planning Group)**

Statutory NSC members:

- President   • Vice President
- Sec State   • Sec Defense

Statutory advisers:

- Director DCI   • Chairman JCS

Others (directed by President):

- Attorney General   • WH Chief of Staff
- Sec Treasury

NSC
Staff

Policy Review Group

- Deputy National Security Adviser
  (Chairman)
- Membership at sub-cabinet level
  from member department/agencies

Senior Interagency Groups and Regional
and Functional Interagency Groups

Interagency
Working
Groups

Source: National Security Decision Directive 266.

state and the national security adviser and their staffs is productive rather than destructive. The issue of competition for policy dominance, raised by Leslie Gelb in his "Why Not the State Department,"[12] persists. Gelb and others argue that whether the secretary of state or the national security adviser will have the strongest influence on the substance of foreign policy will eventually depend on the one with whom the president comes to feel most comfortable. In this regard, the national security adviser has an advantage because of proximity and also because his sole client is the president. The secretary of state has to be responsive to the needs of his department as well as to his president. And, the more he represents the department's interests, where they conflict with other policy players, the more the president is likely to turn to his in-house expertise and the assistant for national security affairs.

In suggesting that policymaking ought to be

centered in the White House, some have argued that State is better organized for carrying out foreign policy than it is for formulating it. The State Department's organization reflects its unparalleled responsibilities for implementing foreign and national security policy around the world and for serving its foreign constituency. Figure 13.2 shows the State Department is organized around bureaus that, by their nature, compete for resources and representation in the policymaking process. The top-level decision-making matrix consists of the secretary, his personal staff, the deputy secretary, four under-secretaries and the counselor. At the next level two potentially competing power centers are the five regional bureaus, each headed by an assistant secretary, and no less than a dozen functional bureaus and offices. The interests of the functional bureaus cut across those of the regional bureaus and frequently conflict with the immediate objectives of the regional bureaus.

As an example, refugee quotas and funds for the refugee program are inadequate to meet the collective needs of all the regional bureaus, which results in significant competition and conflict with regard to both policy and the more pragmatic allocation of resources among the many claimants. An example more germane to defense policy is the existence of conflict between the Bureau for Politico-Military Affairs and the European Bureau, which has its own politico-military affairs section known as EUR/RPM, with respect to NATO and European arms control issues.

This redundant overlapping structure of bureaus belies another characteristic of the organization. The department is organized to oversee the day-to-day conduct of foreign policy with other nations. The function of the regional assistant secretaries is to advise the secretary on the formulation of policy concerning the countries in their area of responsibility and to supervise the operation of US diplomatic establishments in the countries in their region. Collectively, there are more than 150 missions and embassies, more than 73 consulates general and 29 consulates throughout the world, as well as US representatives in nearly 50 different international organizations.

Three other agencies closely involved in foreign affairs are loosely aligned under the secretary of state for general direction. With regard to national security policy, the Arms Control and Disarmament Agency (ACDA) is the most notable. ACDA was created by the Congress in 1961 in an effort to centralize the primary responsibility for arms control and disarmament. However, the agency's importance depends directly on what role the president and his principal advisers allow. Under the authorizing legislation, ACDA "conducts studies and provides advice relating to arms control and disarmament policy formulation."[13] In theory, the director of ACDA reports directly to the president. However, the agency's organizational distance from the White House and administrative relationship in the State Department insures that agency views will not be far removed from those of the Secretary of State. The record of the Reagan administration is that when arms control was an important issue, ACDA was an important, but second-tier player.

The other two agencies with significant foreign policy responsibilities aligned under the secretary of state are the United States Information Agency (USIA) and the International Development Cooperation Agency (IDCA). USIA is responsible for cultural and informational activities directed at overseas audiences while IDCA and its principal operational unit, the Agency for International Development (AID), provide economic assistance to developing countries.

Unfortunately, the State Department is not organized to accomplish effectively the important function of planning and formulating US foreign and defense policy. Instead it is organized to represent US interests abroad and the interests of other countries to the US government. Leslie Gelb, a former assistant secretary of state for politico-military affairs, offers a useful example of the difference between the kind of advice offered the president by State and the NSC staff which results from this organizational and operational principle.

The European bureau is at the very heart of the State department in prestige and influence, and the Europeanists often find reasons for the United States not to take issue with their clients on economic matters, security questions, or East-West relations. The National Security Council staff often argues that the department is "babying" the Europeans.[14]

In addition, as Gelb points out, the State Department is at a practical disadvantage in the competition for policy influence:

Precisely because the secretary and the department are engaged in and have primary responsibility for the conduct of foreign policy, i.e., the day-to-day business of diplomacy and congressional appearances, as a practical matter there is little time to make policy. It seems inconceivable that such day-to-day tasks should take precedence over policy-making, but they must be done; there is no choice.[15]

The Department of State plays an overarching role in the national security policy arena and competes with the NSC and the Department of Defense for influence over policy. The Department of Defense is plagued by problems similar to those at State. The Defense Department, too, is better organized to execute policy and to react to crises than to plan national security policy.

THE DEPARTMENT OF DEFENSE

The *United States Government Manual* describes the Department of Defense as the "successor agency to the National Military Establishment created by the National Security Act of 1947."[16] Since that time, defense officials have grown fond of noting that there are "only two modes of life in the Pentagon: preparation for the next reorganization and recovery from the last one."[17] Prior to the Goldwater-Nichols Defense Reorganization Act of 1986, four major reorganizations of the Defense Department had taken place, with the most recent one occurring in 1958. The legacy of all reorganizations has been to provide greater centralization of authority in the Office of the Secretary of Defense to enable individual secretaries to manage the defense program more effectively and to play a more prominent role in the development and articulation of defense policy. The most recent reforms also sought to strengthen the chairman

Figure 13.2
Department of State

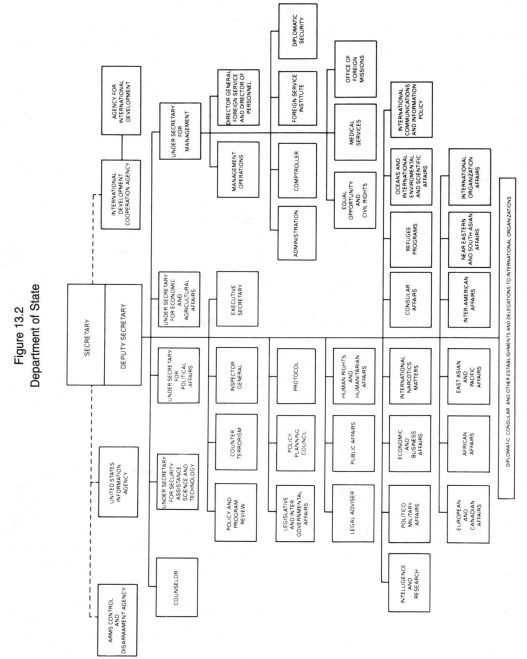

Source: *The United States Government Manual, 1987–88.*

of the Joint Chiefs of Staff and the role of the Joint Chiefs in the formulation of defense policy, especially with regard to the budget and joint operations. The remainder of this chapter examines the evolution of defense organization and the nature of policymaking within the Department of Defense. Patterns of decisionmaking and power in the bureaucracy are observed in organizational structure, and changes to that structure are intended to change the nature of the process and the relative advantages and disadvantages of the players. To understand the growing call for reform in the mid-1980s, it is important to review what preceded it.

Evolution of Defense Organization

Following the First World War, calls for a separate Air Force initiated a national debate on the organization of the War and Navy Departments and on the unification of the armed forces. A separate Air Force was vehemently opposed by the Navy and traditional elements of the Army. Although 60 unification bills were introduced in the Congress, only a more autonomous Army Air Corps and two independent bodies—an Aeronautical Board and a joint Army-Navy Munitions Board—were approved. By the late 1920s, "everyone was exhausted by the controversy, and further unification was delayed for two decades."[18]

The combined and joint operations of World War II again focused attention on the issue of unification and joint planning. President Truman proposed a unified Department of National Defense headed by a cabinet-level secretary. An assistant secretary would head each of the three coordinate branches—Land, Navy, and Air Forces. Each of the "services" would be headed by a military commander, who along with a chief of staff of the Department of National Defense, would advise the secretary of national defense and the president.[19]

The Congress opposed the Truman plan, "fearing a loss of power by the Congressional units although it had itself 'unified' naval and military affairs committees in the House and Senate in the Legislative Reorganization Act of 1946."[20] The compromise legislation instead created the National Military Establishment, headed by the secretary of defense and three separately organized and administered executive Departments of the Army, Navy, and Air Force. Further, "all powers and duties not relating to such departments not specifically conferred upon the secretary of defense" were vested in the separate services.[21]

The act also established the Joint Chiefs of Staff supported by a Joint Staff, responsible for preparing plans for the "strategic direction of the military forces." The Joint Chiefs of Staff were further designated "the principal military advisers to the president and secretary of defense."[22]

Far different from President Truman's concept of a unified defense establishment, the National Security Act of 1947 "created not a unified department, or even a federation, but a confederation of three military departments presided over by a secretary of defense with carefully enumerated powers."[23] These deficiencies were evident even as the first secretary of defense, James Forrestal, took office.

The first reordering of the defense confederacy was accomplished by the National Security Act Amendments of 1949.[24] Secretary Forrestal, an ardent opponent of President Truman's unification plan, recommended in his first annual report that "the statutory authority of the secretary of defense should be materially strengthened."[25] The original responsibilities of the secretary of defense were too general in nature, Forrestal noted; revisions "making it clear that the secretary of defense has responsibility for exercising 'direction, authority, and control' over the departments and agencies of the National Military Establishment" were necessary.[26]

The authority of the secretary of defense was increased dramatically by the 1949 amendments. The reforms emphasized the importance of the secretary as the principal assistant to the president on defense matters by directing that the service secretaries administer their services "under the direction, authority, and control of the secretary of defense."[27] The services were redesignated as military departments, and lost their status as executive departments. The Office of the Secretary of Defense was expanded to include a deputy and four assistant secretaries, one of whom would serve as comptroller and administer a uniform budget. The position of the chairman of the Joint Chiefs of Staff was also established; although the presiding officer for the JCS, the chairman, was not given a vote in their deliberations. In addition, the size of the Joint Staff was doubled.

The size and responsibility of the Office of the Secretary of Defense were expanded further in the 1953 defense reorganization. The number of assistant secretaries was increased to nine, and OSD absorbed the functions of the Munitions Board and Research and Development Board.[28] The service secretaries were designated "operating managers" subordinate to the secretary of defense in the administration of their respective departments.[29]

The structural evolution required to solidify the secretary's control over the department was accomplished by the Defense Reorganization Act of 1958. The position of director of defense research and engineering was created to "direct and control" research and development activities requiring centralized management. The secretary of defense's authority was further strengthened by removing the military departments from the operational chain of command.[30]

The reforms of 1949, 1953, and 1958 incrementally gave the secretary of defense the authority and staff resources necessary to control the department. Robert S. McNamara, the secretary of defense from 1961 until 1968, solidified that control; more importantly, he established expectations for management responsibility for all who followed him. With the creation of a systems analysis unit, Secretary McNamara "strove with determination if not success for one-man rule of the Pentagon."[31] Even if leadership styles of individual secretaries vary and manifest a more decentralized decisionmaking style, the structure of the organization, norms of bureaucracy, nature of the resources involved, and importance of national security objectives now preclude any substantial increase in the authority of the military services, their secretary, or their staffs.[32]

The result of continued reform through 1958 was ever-increasing centralization of authority in the Office of the Secretary and is reflected in figure 13.3. Changes after 1958 were designed primarily to deal with specific managerial issues, such as the creation of an undersecretary for policy and an under secretary for research and engineering. The latter was subsequently redesignated as the undersecretary of defense for acquisition.

The organization of the Department of Defense centers around two basic functions: the development and training of forces and the integration of forces to conduct joint operations. The military services are responsible for training and equipping their respective forces to support the requirements of the operational, combatant commands—the unified and specified commands. In fulfilling their responsibilities, the military services receive direction from the secretary of defense through the service secretaries to the chiefs of staff of their respective services. In this endeavor, the secretary of defense is responsible for managing manpower and equipment requirements and the budget necessary to sustain US warfighting capabilities and deterrent strength in the face of challenges to US power.

The employment of forces in combat or joint military operations requires that forces be integrated with clear and centralized command of operations. This responsibility rests with the commanders of the unified and specified commands, the Joint Chiefs of Staff, and the secretary of defense. Unified commands are assigned by geographic region to a senior commander-in-chief (CINC) whose job it is to develop plans for the integration of the forces made available to him by the component commands of the military services. The chain of command in wartime or for joint operations is from the president to the secretary of defense to the unified and specified commanders through the Joint Chiefs of Staff. The chiefs of staff of the military services are collectively responsible, as members of the Joint Chiefs, for overseeing the development of coordinated plans for joint operations. However, they have no individual, or collective, authority for command in joint operations or wartime.

The Call for Reform

In October 1985, the Senate Armed Services Committee issued a report, *Defense Organization: The Need for Change*, calling for significant change in the structure of defense organization. The report also gave new meaning to the long-running, but low-intensity debate about defense organization sparked by retiring Chairman of the Joint Chiefs General David C. Jones in 1982.

One set of concerns addressed in the Senate report was how the Department of Defense and JCS are organized to make decisions, to advise the president on national security matters, and to implement decisions and exercise command. A second set of concerns centered on management of the approximately $100 billion annual defense procurement program.

In his final year as JCS chairman, General Jones openly discussed what he believed were the great failings of the Joint Chiefs of Staff: the "enforced diffusion of military authority."[33] Jones's criticism was taken seriously enough to result in congressional hearings in 1983 and the introduction of multiple pieces of legislation intent on modifying the JCS structure. More importantly, perhaps, these developments initiated a public debate on defense organization.

Two studies of defense organization were initiated in mid-1983. The first, the Defense Organization Project, was a private endeavor by Georgetown University's Center for Strategic and International Studies. Involving more than 70 people with long experience in defense affairs, they described their efforts as based

on the premise that the national defense debate had maintained a myopic focus on the overall level of defense expenditures and a few major weapons systems, neglecting the more fundamental issues of how to give coherent direction to the complex military operations and large-scale scientific and industrial processes that have come to characterize the national defense effort.[34]

Working groups were created dealing with military command structure, defense planning and resource allocation, the congressional defense budget process, and weapons acquisition. Important here is the breadth of the effort to examine more than just the JCS.

Figure 13.3
Department of Defense

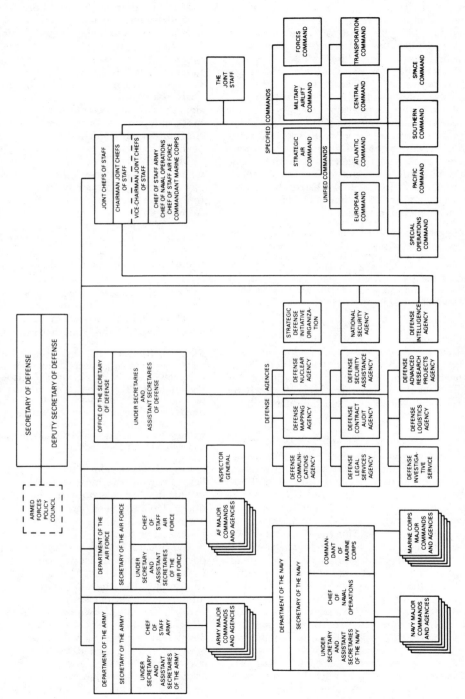

Source: *The United States Government Manual, 1987–88.*

The second study, also begun in 1983, was initiated by Senators John Tower and Henry Jackson, the chairman and ranking minority member of the Senate Armed Services Committee. That study resulted in the report released in 1985 with the aggressive support of Senators Goldwater and Nunn who had since replaced Senators Tower and Jackson. The Senate report called for sweeping changes in defense organization. All elements of the defense establishment were affected, including the JCS, the Office of the Secretary of Defense, the military services, and Congress.[35]

Subsequently, in response to growing criticism of procurement and weapons acquisitions policies, President Reagan created the President's Blue-Ribbon Commission on Defense Management in July, 1985. Chaired by former Deputy Secretary of Defense David Packard, its charter included review of

the budget process, the procurement system, legislative oversight, and the organizational and operational arrangements, both formal and informal, among the Office of the Secretary of Defense, the Organization of the Joint Chiefs of Staff, the Unified and Specified Command system, the Military Departments, and the Congress.[36]

Reforming the Joint Chiefs of Staff

Criticism of the Joint Chiefs of Staff, as an institution, has essentially developed along three lines: (1) the quality of its advice and staff work, (2) the individual behavior and loyalties of the members of the "corporate" body of the JCS, and (3) its ability to plan and conduct military operations.

Quality of Advice. Prior to Goldwater-Nichols, the collective Joint Chiefs of Staff were designated as the principal advisers to the president, the secretary of defense, and the National Security Council. Accordingly, the JCS were expected to subordinate "individual service interests . . . [and] provide broader cross-service perspective . . . for the effective direction and management of the defense establishment."[37] Although the advice of individual service chiefs of staff has been frequently praised, civilian leaders have been "highly critical of corporate JCS advice" as lacking independent, cross-service perspective.[38]

The Joint Staff too suffered from an inability to shed their service-specific biases, functioning more as a confederation of service representatives than as an integrated joint staff. "Staff procedures are designed to protect the interests of each of the four services rather than [to] seek the best cross-service solutions."[39] Consequently, the result is that "each service has a veto over every joint recommendation, forcing joint advice toward the level of common assent."[40]

To improve the quality of advice to the civilian leadership, numerous studies suggested, and Goldwater-Nichols directed, that the chairman of the Joint Chiefs would provide his own, personal advice to the president, secretary of defense, and National Security Council, replacing the corporate JCS in that role. Consequently, the chairman is expected to provide independent advice without the necessity of unanimity which was rigidly adhered to previously. Goldwater-Nichols also created a four-star deputy, or vice chairman, and a joint staff more independent of the service chiefs—measures intended to enhance the independent role of the chairman.

Although it is too soon to tell the degree to which military advice to the civilian leadership has improved, initial indications are that the first chairman to be provided this opportunity, Admiral Crowe, did not hesitate to voice his independence from the corporate Joint Chiefs.[41] President Bush's appointment of Gen. Colin Powell to succeed Admiral Crowe is sure to strengthen further the chairman's independence.

Individual Behavior versus Corporate Decisionmaking. One of the most controversial aspects of defense organization and the Joint Chiefs of Staff is the "dual hatting" issue.

Each member of the JCS, except the Chairman [and now vice chairman], faces an inherent conflict between his joint role and his responsibility to represent the interests of his service. As the senior military planning and advisory body, the JCS is charged with providing military advice that transcends individual service concerns. At the same time, each chief is the military leader of his service and its primary spokesman to civilian policymakers.[42]

Proponents of reform conclude that "it is difficult for even the most well-intentioned chief to abandon service positions in JCS deliberations." Doing so necessarily costs him "the support and loyalty of his service, thus destroying his effectiveness."[43]

The Senate Armed Services Committee Report recommended disestablishing the JCS and replacing it with a Joint Military Advisory Council composed of the chairman and a four-star military officer from each service on his last tour of duty. The council would develop joint plans and analyses to pose as alternatives to proposals developed by the individual services. Opponents of reform argued that because of their operational responsibilities, the service chiefs are in the best position to make collective joint decisions. Further, they argued that creating a joint staff above the service chiefs raises the specter of a "General Staff" and threatens civilian control of the military.[44] This recommendation was not included in the final version of Goldwater-Nichols.[45]

Planning and Conducting Military Operations.
The corporate nature of the JCS and the fact
that the chairman is the only member "unconstrained by current service responsibilities" im-
pairs the Joint Chiefs' ability to address "issues
that involve competing service interests and
prerogatives."[46] Bureaucratic imperatives to
protect one's service are very strong because
these issues "include the most important on the
JCS agenda: strategic planning, the distribution
of military roles and missions among the four
services, and the command arrangements for
combat forces."[47]

Strategic planning is the process of linking
ends (national objectives) and means (forces and
weapons).[48] Accordingly, the JCS's Joint Strategic Planning Document (JSPD), which recommends the forces necessary to attain
established national objectives, is potentially an
important strategic planning document. However, prior to Goldwater-Nichols, the document
had minimal utility because its recommendations were not constrained by the defense
budget. As a result the JSPD was of little or no
use to the secretary of defense in the allocation
of defense resources since it made no recommendations as to priorities and did not take positions "on the tradeoffs necessary to construct
a force structure within the bounds of available
financial resources."[49] The Goldwater-Nichols
Reorganization Act directed that the chairman
of the JCS submit a fiscally constrained JSPD
to the president and secretary of defense identifying the tradeoffs required by the limits on
defense resources. This independent, constrained JSPD supplements the unconstrained
assessments of US defense requirements that
are already proposed by the JCS.

Just as the JCS, as a group, is incapable of
establishing priorities among competing requirements, its ability to assign and coordinate
service responsibilities in combined-arms operations is inhibited by the norm of corporate
decisionmaking. Conducting joint military operations is the responsibility of the commanders-in-chief (CINCs) of the unified and specified
commands. As noted earlier, command of the
CINCs is exercised by the president, and
the secretary of defense on the authority of the
president, through the JCS. The difficulty,
however, has been that while the CINCs are
accountable to the corporate JCS for their com-
bined-forces missions, they must do so with
component forces furnished by the military
services. Under normal operations, these forces
are accountable to their own service chiefs. In
the area of "such vital matters as logistical
support, training, and maintenance, the component commands report directly to their respective military departments, leaving the CINCs
with only limited and indirect influence."[50] As
a result, the CINCs have had little voice in

developing the capabilities their commands
possess.

The Goldwater-Nichols Reorganization Act
did expand and strengthen the authority of the
CINCs over their service component commands, especially in the areas of training, logistics, and planning. As recommended in the
Packard Commission Report, the CINCs "have
greater latitude and full authority to organize
assigned forces as they deem necessary to accomplish their missions."[51] Similarly, CINCs
now exercise far greater authority in the selection of key personnel and subordinate, component commanders, which previously was
almost solely the prerogative of the services.
One of the most important aspects of reorganization is the increasingly important role the
CINCs have played in the programming phase
of the budget process. According to the JCS,
"This evolving role has successfully moved
CINC warfighting requirements to the forefront
of resource allocation deliberations."[52]

Reforming the Office of the Secretary of Defense

Although the three principal studies preceding
the Goldwater-Nichols Reorganization Act went
well beyond the Joint Chiefs of Staff and recommended restructuring the Office of the Secretary of Defense as well, it is here that the least
amount of change resulted. The Defense Organization Project and Senate Armed Services
Committee concluded that (1) OSD's internal
structure failed to conform to the department's
strategic purposes, impeding its ability to provide clear, consistent policy direction; (2) there
was an inadequate institutional voice for such
operational concerns as readiness and sustainability; and (3) that the OSD staff was too large
and too involved in the details of program management.[53]

The greatest deficiency identified was that
OSD is organized along functional lines, such
as manpower, health affairs, etc., rather than in
conformance with the department's output—
capabilities to perform military missions. "As a
result," the study groups concluded, "OSD is
ill-equipped to translate mission-oriented planning and programming guidance into force requirements and weapon programs."[54] An
example is that of the defense manpower program. OSD contains an office responsible for
the overall defense manpower program. That
office, however, cannot address service manpower requirements in coordination with logistics, procurement, construction, and training
programs intended to meet the needs of a particular theater of operations.[55]

Both the Senate Armed Services Committee
and Defense Organization Project identified
three fundamental missions which the Office of

the Secretary ought to influence. These are (1) deterrence of an attack on the US and its allies, (2) defense of the US and Western Europe should deterrence fail, and (3) projection of US forces to defend vital global interests. Both reports noted, however, the "absence of a senior official with exclusive responsibility for any of these three strategic missions."[56] Both argued that this concern with functional aspects of defense programs, rather than the outputs generated, has contributed to the OSD's over-involvement in the management of service programs.

Recommended reforms included expansion of the undersecretary for policy's responsibilities "for program integration on a mission basis."[57] Three new assistant secretaries for nuclear deterrence, NATO and North Atlantic Defense, and regional defense, were recommended. A third undersecretary, for readiness and sustainability, to coordinate manpower, logistics, and installations functions was also proposed. None of these proposals were included in the final Goldwater-Nichols bill. However, the position of assistant secretary for low-intensity conflict was created by separate legislation. The most serious deficiencies recognized in the Office of the Secretary of Defense dealt with the acquisition process and were the principal focus of the Packard Commission.

Reform of Acquisition Program Management. A principal concern behind President Reagan's appointment of a Blue Ribbon Commission on Defense Management was growing criticism of the weapon acquisition process. The Packard Commission, Defense Organization Project, and Senate Armed Services Committee reports all addressed deficiencies in the acquisition process. The most direct and strongest recommendations, however, came from the Packard Commission:

The nation's defense programs lose far more to inefficient procedures than to fraud and dishonesty. The truly costly problems are those of overcomplicated organization and rigid procedure, not avarice or connivance.

Chances for meaningful improvement will come not from more regulation but only with major institutional change.[58]

Among the reforms resulting from the Goldwater-Nichols Act, was creation of the position of undersecretary of defense for acquisition, whose occupant would "set overall policy for procurement and research and development, supervise the performance of the entire acquisition system, and establish policy for administrative oversight and auditing of defense contractors."[59] The service secretaries were essentially cut out of the acquisition process by the creation of a senior acquisition official in each of the services who would report directly to the undersecretary of defense for acquisition. Other proposals, such as directly involving the new vice chairman of the Joint Chiefs of Staff in the acquisition programs and significant changes to streamline and simplify the very cumbersome, and therefore costly, procurement process, went unattended.

One evaluation of the success of the Goldwater-Nichols reorganization is that the roles of the Joint Chiefs and CINCs were appropriately strengthened and that further centralization of the Office of the Secretary of Defense was achieved at some cost to the service secretaries and their staffs. However, in the weapons acquisition area considerable work remains. The entire defense procurement system was impugned when the Justice Department revealed it was investigating massive charges of fraud and illegal activities in the acquisition process involving senior Pentagon officials and defense contractors. Throughout 1989 Congress again clamored for more much needed reform, claiming that Goldwater-Nichols had been too little, too late when it came to weapons procurement programs.

David Packard reiterated that the Congress and the Defense Department had been urged repeatedly to correct what he described as five fundamental deficiencies:

—Setting requirements for the most sophisticated systems attainable often regardless of cost.

—Underestimating schedules and costs of major programs, distorting the decisionmaking process for the allocation of the national budget.

—Changes in programs and contract requirements caused by changes in military user preferences, leading to annual or more frequent changes in program funding levels by Congress and the Department.

—Lack of incentives for contractors and government personnel to reduce program costs.

—Failure to develop sufficient numbers of military and civilian personnel with training and expertise in business management and in overseeing the development and production of enormous, highly technical industrial programs.[60]

Packard concluded that even after all the studies that were conducted in the mid-1980s, and the extensive hearings and debate in the Congress, "there is still no rational system whereby the executive branch and Congress can reach coherent and enduring agreement on national military strategy, the forces to carry it out and the funding that should be provided—in light of the overall economy and competing claims on national resources."[61] This persistent weakness in executive-congressional relations "contributes substantially to the instability and uncertainty that plague our defense program" and "increases the costs of procuring military equipment."[62]

In response to these criticisms and calls for further reforms to include an independent procurement agency, Secretary of Defense Carlucci cautioned that, "Our challenge is to make certain that whatever reforms we make are truly reforms and actually improve the system—instead of saddling it with a new and different set of problems."[63] Carlucci proposed his own five-point program of reforms directed at the Congress as an equal partner in the lackluster performance of defense procurement.

—Congress should combine the separate authorizations and appropriations committees.

—Congress should reduce the number of committees and subcommittees that now exercise overlapping authority and oversight roles in the defense budgeting process.

—Congress should revise its procedures to make it impossible for individual members to force the president to purchase an item not included in his defense request to do so by burying an amendment within the overall defense budget package.

—Congress should follow the Pentagon's lead in shifting to a biennial budget for defense programs.

—Congress should adopt an additional reform that will further stabilize the procurement process, funding more defense programs on a multiyear basis.[64]

The Legacy and Prospects of Reform

The greatest challenges facing President Bush and his team of national security decisionmakers are clear. First and foremost, they must determine the proper allocation of the nation's resources among the competing objectives of national security and social programs in an era of enormous deficits, while taking into account the changing security equation in Europe and with the Soviet Union—employing, perhaps, a cautious optimism. Within the defense budget the president must decide how to accommodate the rising costs already locked in for weapons in the pipeline. If the defense budget remains flat, or grows only with the rate of inflation, pressure on other accounts, such as personnel and maintenance, poses the prospect of reducing essential readiness.

Clearly, there are calls for significant reductions; however, prudence would dictate that such reductions only follow a strategic analysis and underpinning. Equally important, the defense procurement system must be brought under control—addressing both the procurement process internal to the Defense Department and defense contractors, and the role of the Congress as well. Much of what happens will depend on President Bush's leadership and how firmly Secretary of Defense Cheney takes control of the process. It will be for subsequent editions of this text to assess how well they measured up to the job.

NOTES

1. Alexander L. George, *Presidential Decisionmaking in Foreign Policy: The Effective Use of Information and Advice* (Boulder, Colo.: Westview Press, 1980), p. 146.

2. Alexander George provides a useful and detailed examination of personality characteristics and their influence on various models of management styles and of presidential decisionmaking. These include formal, competitive, and collegial approaches to White House management. In chapter 11 George provides a theory of multiple advocacy, arguing that multiple advocacy will improve the quality of information provided to the president and thereby improve the options available for decisionmaking.

3. George, *Presidential Decisionmaking*, p. 148.

4. Ibid.

5. Ibid.

6. *Report of the President's Special Review Board*, 26 February 1987.

7. The National Security Act of 1947 is Public Law 253, 80th Cong., 26 July 1947; 61 Stat. 495.

8. *Report of the President's Special Review Board*, p. II–4.

9. *United States Government Manual, 1987–88*, p. 423.

10. Richard Brown, "Toward Coherence in Foreign Policy: Greater Presidential Control of the Foreign Policymaking Machinery," in R. Gordon Hoxie, ed., *The Presidency and National Security Policy* (New York: Center for the Study of the Presidency, 1984), p. 326.

11. See James T. Hackett and Robert M. Soofer, "Understanding the State Department," *Heritage Foundation Backgrounder No. 605* (Washington: Heritage Foundation, 1987), for an account of the State Department's improved standing in the foreign policymaking process and especially that of the career foreign service.

12. *Washington Quarterly* (Autumn 1980): 25–40.

13. *United States Government Manual, 1987–88*, p. 690.

14. Leslie Gelb, "Why Not the State Department," *Washington Quarterly* (Autumn 1980): 46.

15. Ibid.

16. *United States Government Manual, 1987–88*, p. 174.

17. William E. Depuy, "Unification: How Much More?" *Army* (Apr. 1961): 30.

18. Charles J. Hitch, "Evolution of the Department of Defense," in Richard G. Head and Ervin J. Rokke, eds., *American Defense Policy*, 3d ed. (Baltimore: Johns Hopkins University Press, 1973), p. 346.

19. *Public Papers of the Presidents, Harry S. Truman, 1945* (Washington, D.C.: G.P.O., 1961), p. 546.

20. Harry Ransom, "Department of Defense:

Unity or Confederation?" in Mark E. Smith and Claude J. Johns, Jr., eds., *American Defense Policy*, 2d ed. (Baltimore: Johns Hopkins University Press, 1965), p. 363. Also see Ransom, *Can America Survive Cold War?* (New York: Doubleday, 1968).

21. 61 Stat. 495.

22. Ibid.

23. Hitch, "Evolution," p. 547.

24. Public Law 216, 81st Cong., 10 Aug. 1949, 63 Stat. 578.

25. *First Report of the Secretary of Defense, 1948* (Washington, D.C.: G.P.O., 1948), p. 3.

26. Ibid.

27. 63 Stat. 578.

28. Reorganization Plan no. 6 of 1953, 67 Stat. 638.

29. *Public Papers of the Presidents, Dwight D. Eisenhower, 1958* (Washington, D.C.: G.P.O., 1959), pp. 225–38.

30. Department of Defense Reorganization Act of 1958, Public Law 85–599, 85th Cong., 6 Aug. 1958, 72 Stat. 514.

31. James M. Roherty, "The Office of the Secretary of Defense: The Laird and McNamara Styles," in John E. Endicott and Roy W. Stafford, Jr., eds., *American Defense Policy*, 4th ed. (Baltimore: Johns Hopkins University Press, 1977), p. 289. Also see Roherty, *New Civil Military Relations* (New York: Transaction, 1974).

32. Testimony to this fact is observed in the Senate Armed Services Committee Report's suggestion that the Service Secretaries and Chiefs of Staff organizations be combined. See Senate Armed Services Committee Report, *Defense Organization: The Need for Change*, 1985.

33. David C. Jones, "Why the Joint Chiefs of Staff Must Change," *Armed Forces Journal International* (March 1982): 62.

34. Barry M. Blechman and William J. Lynn, eds., *Toward a More Effective Defense: Report of the Defense Organization Project* (Cambridge, Mass.: Ballinger, 1985), p. xi.

35. *Defense Organization*, p. iii.

36. The President's Blue Ribbon Commission on Defense Management, *An Interim Report to the President* (Washington, D.C.: G.P.O., 1986), p. 1.

37. Blechman and Lynn, *Toward a More Effective Defense*, pp. 5–6.

38. William J. Lynn and Barry R. Posen, "Reorganizing the Joint Chiefs of Staff," *International Security*, (Winter 1985–86): 76. This is well documented in the Senate Armed Services Committee Report, pp. 158–60.

39. Lynn and Posen, "Reorganizing the JCS," p. 77.

40. Ibid.

41. Discussion with Senate Armed Services professional staff, November 1988.

42. Ibid., p. 76.

43. Ibid.

44. See MacKubin Thomas Owens, "The Hollow Promise of JCS Reform," *International Security*, (Winter 1985–86): 98–111; Robert J. Murray, "JCS Reform: A Defense of the Current System," *Naval War College Review* (Sept.–Oct. 1985): 20–27; "Is JCS Reorganization Really Needed?" *Sea Power* (Dec. 1985): 21–29; and V.H. Krulak, "Defense Reorganization—or Tinkering?" *Strategic Review* (Winter 1986): 5–7.

45. Senate Committee on Armed Services Report.

46. Lynn and Posen, "Reorganizing the JCS," p. 77. It is not yet clear what the norms of behavior will be for the vice chief who does not have an official position in a service, as do the chiefs of staff. However, if the vice chief aspires to be the chairman, or to return to a service, then this could potentially cause less than objective positions on issues pending before the JCS.

47. Ibid.

48. See Michael Howard, "The Forgotten Dimensions of Strategy," *Foreign Affairs* (Summer 1979): 975–86.

49. Blechman and Lynn, *Toward a More Effective Defense*, p. 10. This criticism is found in the Senate Armed Services and Packard commission reports as well.

50. Ibid.

51. *Organization of the Joint Chiefs of Staff, Military Posture Statement, FY 1988* (Washington, D.C.: G.P.O., 1988) p. 87.

52. Ibid., p. 88.

53. Blechman and Lynn, *Toward a More Effective Defense*, p. 14.

54. Ibid.

55. Ibid.

56. Ibid.

57. Ibid., p. 18.

58. *Interim Report to the President*, p. 15. A discussion of issues related to the acquisition process is found on pages 529–52 in the Senate Armed Services Committee Report.

59. *Interim Report to the President*, p. 16.

60. David Packard, "Cure Procurement Ills Carefully," *Electronic Engineering Times*, 24 Oct. 1988, pp. M27–M28.

61. Ibid.

62. Ibid.

63. Frank Carlucci, "Managing Defense Procurement," *Electronic Engineering Times*, 24 Oct. 1988, pp. M13–M14.

64. Ibid.

CHAPTER 14

THE DEFENSE POLICY PROCESS IN CONGRESS
Roles, Players and Setting, Trends, and Evaluations

DAVID C. KOZAK

Events of the past two decades illustrate well the need for the defense establishment to reflect on the roles and operations of the US Congress. During these decades Congress has influenced national security decisions in ways unmatched by other legislatures. It is not an exaggeration to state that an understanding of Congress's role in national security is vital for a full appreciation of the defense policy process in America.

To date, political scientists have failed to systematically examine the "defense policy" process in Congress. Although there have been sophisticated published studies of congressional policymaking in other areas—budgeting,[1] appropriations,[2] agriculture,[3] foreign policy,[4] commerce,[5] finance,[6] environment,[7] education and labor,[8] public works,[9] banking,[10] and government operations,[11]—with few exceptions,[12]—there are no comparable analyses of how Congress processes defense issues. Most of the existing studies that review Congress and defense decisionmaking focus on very limited topics, such as appropriations,[13] constituency-oriented behavior,[14] voting coalitions,[15] or distributive impacts of committee membership.[16] This article aspires to (1) bring together the diverse threads of our understanding of Congress and defense policy, and (2) examine Congress's defense behavior in light of the many changes that have transpired in the past two decades.

ROLES OF CONGRESS CONCERNING DEFENSE POLICY (OR, WHY LEARN ABOUT CONGRESS WHEN STUDYING DEFENSE POLICY?)

Congressional roles in defense decisionmaking are the same important ones it performs in other areas of public policy. Specifically, Congress performs the following six roles in the development of defense policy: (1) making the laws and legislating (up to, and including declaring war), (2) caretaking of the purse, (3) overseeing the executive, (4) offering executive counsel, (5) providing debate and deliberation and (6) facilitating legitimation. Although the president and secretary of defense may be the most visible contributors to defense policy, the Congress, through the performance of these roles, makes significant contributions to defense policy that must be understood if one is to grasp the total defense policy process.

1. The *lawmaking* function of Congress cannot be overrated in terms of its impact on defense policy. The statement in Article I, Section 1 of the Constitution that "All legislative powers herein granted shall be vested in a Congress of the United States" confers enormous power on the legislative body. In practice, this provision means that no defense program can be undertaken without the approval of Congress. Throughout American history, Congress in exercising this power has shaped significantly the basic infrastructure of American defense: manning and personnel, weapons systems, organization, force structure, and doctrine. For example, the landmark National Security Act of 1947 established the basic structure of defense forces. Section 412(b) of the Military Construction Act of 1960—which requires annual congressional authorization of naval vessels, missiles, and ships—gives Congress enormous power over weapons acquisitions.[17] Although Congress frequently defers to the advice and input of the executive branch concerning defense matters, legislative authority insures that Congress will be at least a partner in the initiation and formulation of defense policy.

The warmaking power of Congress, although eclipsed somewhat by two major "undeclared" US engagements since World War II, is nonetheless a major source of constitutional authority that militates in favor of presidential consultations with Congress in times of hostility.

2. The *"power of the purse"* places the Congress in a key position with regard to defense policy, for the Constitution requires that "No money shall be drawn from the treasury, but in consequence of appropriations made by law" (Article I, Section 7, Para. 7), and "no appropriation of money to that use . . . (to raise and

support armies) . . . shall be for a longer term than two years" (Article I, Section 8, Para. 11).

3. The *oversight* role of Congress has given it the opportunity to make an impact on defense policy. As Holbert Carroll has stated, a lesson from history "is that the way the Constitution distributes power assures tension in foreign-military matters."[18] The Constitution intrudes the Congress into defense matters. Pursuant to its statutory and purse roles, Congress investigates (or oversees) the executive. The 1946 Legislative Reorganization Act, in fact, directed that each standing committee exercise "continuous watchfulness" over the implementation of law by the administrative agencies under their jurisdiction. The famous Truman investigations of corruption in defense contracting during World War II (led by then Senator Harry Truman from Missouri) and the Army–Senator Joseph McCarthy hearings are historic examples from the 1940s and 1950s of congressional investigations of defense business. More contemporary examples are the investigations by Senator Henry Jackson into the national security decision apparatus, and the inquiries of various Senate committees into intelligence activities. Oversight may occur formally (e.g., hearings by a congressional panel) or in a more latent,[19] episodic manner (e.g., information gathering actions and inquiries of individual members). Regardless of the mode, such investigations can have a big impact on the country's defense policies. Previously, Congress also exercised oversight with the "legislative veto." With this latter technique, one or both houses of Congress (depending on the enabling legislation) sought to block executive decisions. The 1981 presidential conflict with Congress on the sale of Airborne Warning Aircraft Systems (AWACS) to the Saudi Arabian government is an example of congressional attempts to influence and oversee through the legislative veto. However, in the summer of 1983, in the "Chadha" case, the Supreme Court declared such vetoes an unconstitutional abrogation of executive power by the Congress.

4. The Constitution, by requiring the president to seek congressional *"advice and consent"* for executive appointments and ratification of treaties, further assures Congress a role in the defense-policy process. The Senate, in exercising advice and consent authority, will inevitably raise questions concerning the qualifications and views of presidential nominees to major security jobs.

5. An important contribution of Congress to the American political system is its role as a *forum for debate* and deliberation. Congress, as a representative assembly, holds high debate concerning national priorities. The defense policy analyst must study Congress, for in Congress major issues of national defense are debated and discussed by representatives of divergent positions and interests. There are congressmen who are strongly supportive of defense programs and others who are less inclined. Debates among them serve to inform and to frame and focus conflict. By following Congress, one gets a feel for the issues and contestants of national political dialogue concerning national defense.

6. Finally, perhaps the most important contribution made by Congress is that of *consensus building*.[20] All policies—including defense matters—must be legitimated there. It is where support is developed and where deals are worked out between hawks and doves.

In sum, Congress plays an extremely important and viable role in defense decisionmaking. In the words of Roger Hilsman,

the peculiar and somewhat elusive power of Congress lies . . . (in) limit setting—the power of deterrence and the threat of deterrence and the threat of retaliation. On some specific detail of a foreign policy, the President may frequently ignore the Congress with complete impunity, but on the overall, fundamental issues that persist over time, he must have their cooperation and acquiescence, even if he is not required to have their formal and legal consent. Although he may ignore individual congressmen, even powerful committee chairmen, collectively he must bring them along in any fundamental policy. Here again, he is the 'President-in-sneakers' trying to induce the congressmen and senators to 'climb aboard.'[21]

Huntington, in *The Common Defense*, detects a similar role. He writes "The Administration can never be sure of its policy so long as potential sources of opposition exist outside the executive branch."[22] The existence of such processes more than warrant a study of Congress in a survey of defense policy actors.

FEATURES AND CHARACTERISTICS: ACTORS AND THE SETTING (OR, WHO MAKES DEFENSE POLICY AND HOW DO THEY DO IT?)

To understand how Congress processes defense issues, one must be aware of certain "initial realities"[23] concerning Congress as an organization. There are 18 significant structural aspects of the Congress which must be known for an understanding of how Congress does its business. They reveal the distinctive organizational properties of the Congress which make certain kinds of congressional responses to policy issues—including defense issues—inevitable.

CONGRESS IS NONHIERARCHICAL

The US Congress, unlike other organizations such as military units, businesses, or universities is nonhierarchical. Although certain congressional leaders are more powerful than others, they are "first among equals." The political party organizations over which leaders preside are notoriously decentralized and undisciplined. Leaders lack control over entry to or exit from the Congress. In other words, congressional leaders have no say over who serves in Congress. Almost any brand of self-starting Democrat or Republican can get elected. Therefore, although congressional leadership has "fragments of power"[24] with which to work, it cannot automatically deliver the majorities needed for making policy.

CONGRESS IS DRIVEN BY AN "ELECTORAL CONNECTION"

One of the peculiarities of legislative life is "the electoral connection."[25] Members of Congress are elected to office and must stand for reelection. According to David Mayhew, in his widely cited *Electoral Connection*, members are driven by a desire to get reelected. For most, it is their overarching goal. The electoral imperative colors all they do, leading them, as Mayhew argues, constantly to engage in reelection activities such as advertising, credit claiming, and position taking.[26] Congressman Les Aspin has written that the reelection imperative and the localism it engenders are especially relevant to defense decisionmaking. In his words, "Congressmen vote the way they do primarily because of their constituents, and this is particularly true when it comes to votes pertaining to defense."[27] Aspin argues that, for most members, a paramount concern is "defense related jobs." He concludes that, "Congress is essentially a political institution and responds to political stimuli."[28]

CONGRESS IS A BICAMERAL LEGISLATURE

An important feature of the US Congress that is frequently overlooked in general commentary is that Congress is comprised of two very different legislative bodies: the House and the Senate, and both must act before a bill becomes a law. It has been said, without tongue in cheek, that the House and Senate have only four things in common: (1) their members are elected, (2) they both make the laws, (3) they share the same building, and (4) both have fragmented and decentralized structures.

Certainly, the differences between House and Senate outnumber their commonalities. They differ in membership, constituency, time perspective, structure, process, functions, and as organizations (see table 14.1). The House and Senate each do business in a distinctive way, and it is simplistic to speak of "the Congress." As Charles O. Jones has written, the Senate is an institution of functional self-indulgence or permissive egocentrism.[29] In the words of Nelson Polsby,

Where the House of Representatives is a large, impersonal, and highly specialized machine for processing bills and overseeing the executive branch, the Senate is, in a way a theater where dramas—comedies and tragedies, soap operas and horse operas—are staged to enhance the careers of its members and to influence public policy by means of debate and public investigation. The essence of the Senate is that it is a great forum, an echo chamber, a publicity machine. Thus 'passing bills' which is central to the life of the House, is peripheral to the Senate.[30]

Frequently, there is very meaningful bicameralism in defense policy as House and Senate, perhaps responding to differences in constituencies, stake out different positions on issues. For example, in the 96th Congress, even when the Democrats controlled both houses, the Senate and House were frequently at loggerheads on questions concerning the B-1, binary munitions, the role of reserve destroyers, and general funding levels. The relative defense orientation of each house largely reflects the composition of each committee and the full membership. Each must be dealt with differently. Of course, bicameral differences have been accentuated since 1980 with split party control of House and Senate, with Senate Republicans more likely to support the Reagan administration's defense buildup while House Democrats were more likely to oppose.

STANDING COMMITTEES ARE THE WORKHORSES OF CONGRESS

Most of the work of Congress is accomplished in standing committees that Woodrow Wilson once referred to in his classic *Congressional Government* as "Little Legislatures." Committees decide which of the many proposed bills will be seriously considered. It is in those panels that proposed legislation is studied and scrutinized. For bills under consideration, committees hold hearings and call various witnesses to give testimony. Before a bill goes to the floor, its exact wording is hammered out in a "markup" session of the committee, and the committee issues a report on the legislation to its parent house.

Although many committees consider matters relevant to the military, the House and Senate armed services committees are most influential in defense policymaking. Both committees have jurisdiction over general defense matters and the national military establishment. As Patterson, Davidson, and Ripley note concerning the committees,

annually, they authorize the Pentagon's spending for research, development, and procurement of all weapon systems, construction of military facilities and levels of civilian and uniformed personnel. They also

Table 14.1

The House and Senate Compared: A Perspective from Literature

House	Item	Senate
Larger (435) Less prestigious More incumbency	Membership	Smaller (100) More prestigious Less incumbency
More homogenous Narrower	Constituency	More heterogeneous Broader
Shorter (elected every 2 years) Noncontinuous body (2-year cycle)	Time perspective	Longer (elected every 6 years) Continuing body (staggered terms of office)
More hierarchical Less equitable Speaker is power broker Fusion in speaker's office of presiding powers and head of majority party Committees more important Members serve on fewer committees Less staff reliance	Structure	Less hierarchical More equitable Majority leader is power broker Separation of presiding officer and majority party head Committees less important Members serve on more committees More staff per member
Formal, rigid Less comity Rules Committee structures debate Most floor work done in "committee of the whole" Five calendars Limited debate (5 min. rule) Nongermane amendments (riders) not allowed Acts more quickly Electronic voting TV coverage	Process	Informal, flexible, (much business done through unanimous consent) More comity Rules Committee does not structure debate No counterpart to "committee of the whole" Two calendars Unlimited debate until cloture invoked or unless a unanimous consent limitation Nongermane amendments (riders) allowed Acts more slowly No electronic voting No TV coverage
Originates revenue measures Expertise, specialization	Functions	Executive counsel (appointments and treaties), debate, deliberation, discussion of national priorities
Representative assembly Less visibility/media coverage	Characteristics	Forum of functional self-indulgence; Breeding grounds for presidential candidates More visibility/media coverage

authorize the disposal of strategic and critical materials from the national stockpile.[31]

The House Armed Services Committee is comprised of seven subcommittees: Investigations, Military Personnel and Compensation; Military Installations and Facilities; Readiness; Procurement and Military Nuclear Systems; Research and Development; Seapower; and Strategic and Critical Materials.

The Senate committee breaks down into six subcommittees: Manpower and Personnel; Military Construction; Preparedness; Seapower and Force Projection; Strategic and Theater Nuclear Forces; and Tactical Warfare.

Both committees have been described as prestigious, nonpartisan, formal, and constituency-serving. They are considered prestigious in that in the Senate, the Armed Services Committee is considered a "major" one while the House committee is classified as "semi-exclusive." An assignment to the Armed Services Committee in either house is considered a good one, actually courted by many members. However, unlike the situation one finds on the more prestigious committees of the House, members of House Armed Services Committee also serve on at least one other committee.

The nonpartisanship of the two committees—

especially the House Armed Services Committee—is a conspicuous trait. The staff of the House Armed Services Committee is nonpartisan (i.e., the majority and minority have a common staff, in contrast to the conventional committee practice where each side has its own). One member emphasized the nonpartisanship of the House Armed Services Committee, "There are philosophical divisions to be sure but you don't find the party-bickering, dissent, and divisions that you run into on so many other committees."

The formality of the House and Senate committees was remarked by another member, "Because you deal with national defense matters, there is an aura of dignity and decorum that never seems to get punctured. You just don't have the occasional levity that you have elsewhere."

The constituency-orientation of the armed services committee has frequently been noted. Patterson, Davidson, and Ripley have concluded that

Armed services committee members are less interested in questions of global strategy than they are about . . . force levels, military installations, and the distribution of defense contracts. Thus, military policy is in many respects an extension of constituency politics.[32]

Arnold has concluded that "the armed services committee has become more and more dominated by congressmen representing military districts."[33] The reason for this, as Patterson et al. are quick to emphasize, is that "in the last several decades, the two committees have tended to attract members favorable toward military spending and House-district installations."[34] Congressman Aspin concurs. He writes that the way a congressman gets reelected

is through constituent service—and to do this well,—the Congressman should first gain membership on a committee relevant to his constituents' basic economic needs. A Congressman with defense bases, installations, or contractors in his district will obviously want to, and usually does, get himself appointed to a defense committee.[35]

One member of the House Armed Services Committee drew an analogy between the committee and its counterpart in agriculture. "Everyone seems to be driven by a concern for 'What can I get for my district?' Everybody is out for as much as they can get for back home." Another senior member was quick to discount that interpretation, "Sure we are motivated by reasons of constituency but there are a lot of strong defense-minded legislators on that committee who would be on the committee and vote the same way regardless of what they had in their districts."[36]

There is no doubt that the armed services committees are major actors in defense decisionmaking. As one staffer emphasized, "If an armed services committee is for it, it's got a chance. If a panel is opposed, the fight's over. They can make a policy go or make it stop."

CONGRESSIONAL DECISIONMAKING IS LABYRINTHINE, INVOLVING MULTIPLE, SUCCESSIVE STAGES

One of the major features of the legislative process is its abundance of decision points. For a bill to become a law, it must successfully navigate numerous obstacles and hurdles. A flow chart of the legislative process (table 14.2) clearly shows the numerous opportunities for delay, deadlock, and defeat of a proposed bill. Once a bill is introduced in both houses, it must be referred to committee, favorably reported by the relevant subcommittee and full armed services committee, scheduled on the floor, debated and voted on the floor, and, if necessary, sent to conference to iron out differences between the two houses before being forwarded for presidential approval. So complicated is the congressional process that Woodrow Wilson was moved to write, "Once begins the dance of legislation, and you must struggle through its mazes as best you can to the breathless end—if any there be."[37]

CONGRESSIONAL DECISIONMAKING IS A TWO-TRACK PROCESS, REQUIRING BOTH AUTHORIZING AND APPROPRIATING ACTIONS

For any governmental program to materialize, Congress must act twice (i.e., at least two separate legislative processes must be completed in each house). There must be an authorization (entitling the government to undertake certain actions) and an appropriation (an allocation of public money funding such an action). This "two-track" process has two implications for defense decisionmaking in Congress. First, the Defense Appropriations Subcommittees of both the House and the Senate Appropriations Committees are important actors in the congressional decision process. Both committees interact extensively with representatives of the president's Office of Management and Budget (OMB) and with budget officers of the armed services. As keepers of the purse strings, these panels are important centers of power. As several have noted, there is patterned role-playing in the appropriations process. The subcommittees act as budget cutters or conservers with the Senate subcommittee serving as a court of last appeal for appropriation proponents.[38] Second, the appropriation requirement provides opponents of a proposed defense program yet another opportunity to have their way. There are

Table 14.2
Stages of the Legislative Process

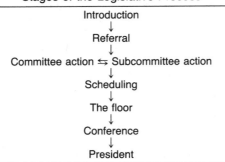

Introduction
↓
Referral
↓
Committee action ⇆ Subcommittee action
↓
Scheduling
↓
The floor
↓
Conference
↓
President

numerous examples of authorized expenditures being scuttled during the appropriations process or of new authorizations being attached to a bill on the Senate side as a rider to an appropriations bill.

THE MULTIPLICITY OF CONGRESSIONAL LEADERS

A criticism frequently leveled against the Congress is that "there are no leaders anymore—certainly not of the vintage of Rayburn or Johnson." Although it may be true that men of their legislative and bargaining skills have not materialized in recent years (although some argue that even Rayburn and Johnson would have difficulty these days), more accurately there are far too many leaders and not enough followers in the contemporary Congress. Each party in each house has a leadership structure (see table 14.3). Further, committee and subcommittee chairmen also serve as formal leaders when bills from their panels are considered.

The multiplicity of leadership has two important consequences for decisionmaking. First, because the cooperation of so many leaders is necessary for bringing legislation to the floor, proponents of legislation must concert among multiple, frequently nonaligned centers of power within each party. Second, because the leadership in both parties in both houses is so heterogeneous, different leaders will take different positions, making the process of coalition building extremely difficult.

MEMBERS FREQUENTLY ASSOCIATE WITH AND TAKE THEIR BEARINGS FROM INFORMAL GROUPS

Due to several of the above-mentioned factors—the nonhierarchical organization, the electoral connection, the multiplicity of leaders and weak political parties—a number of informal ideological groups and cliques have devel-

oped among members of Congress. Many of the groups have their own leadership, staff, and information processes. The best-known examples in the House are various state delegations that meet concerning state interests, the Democratic Study Group (a group of liberal Democrats), the Wednesday Club (a group of liberal Republicans), and the Black Caucus. There are also regional groupings. Recently, in the House and Senate, members have organized according to classes (i.e., the year when members first came to Congress). All of these groups provide multiple sources of information and cue-taking that further compound the process of majority building. Two such groups noteworthy in defense matters are the "Members of Congress for Peace Through Law" and the Military Reform Group, initiated by Senators Gary Hart (Dem., Colo.) and Sam Nunn (Dem., Ga.) and Congressmen Dickinson (Rep., Ala.) and Whitehurst (Rep., Va.).

CONGRESSIONAL POLITICS ARE STRUCTURED BY VARIOUS INFORMAL NORMS

As with all human organizations, behavior within the Congress is affected by certain unwritten rules. The informal "dos and don'ts" or what Donald Matthews calls "folkways"[39] entail incentives for compliance and stricture for noncompliance. The norms of Congress most frequently identified are: specialization, attention to committee duties, apprenticeship, civility, reciprocity or logrolling, and constituency-oriented voting. The significance of these norms, as many analysts have emphasized, is that they reinforce the congressional committee system by fostering the twin notions that (1) members on a committee should work hard specializing and (2) members not on a particular committee should defer to those who are. Thus, in defense policymaking, a system of "contained specialization"[40] works to keep many defense issues within the confines of the armed services committees.

BURGEONING CONGRESSIONAL STAFFS HOLD IMPORTANT SWAY OVER PUBLIC POLICY

Contrary to conventional wisdom, the fastest growing bureaucracy in Washington is the staff of Congress, not the Executive Office of the President. As of 1981, more than 30,000 staffers were employed by the House and Senate.[41] This figure includes the personal staffs of members in both Washington and district offices, committee staffers, and employees of various congressional support agencies such as the General Accounting Office and Congressional Re-

Table 14.3
Multiple Formal Leadership Positions in Congress

HOUSE Speaker		SENATE President Pro Tempore	
Democrats	*Republicans*	*Democrats*	*Republicans*
Floor Leader	Floor Leader	Floor Leader	Floor Leader
Whip	Whip	Assistant Floor Leader (Whip)	Assistant Floor Leader
Chief Deputy Whip	Conference Chairman	Policy Committee	Conference Chairman
Caucus Chairman	Research Committee Chairman		Policy Committee Chairman
Caucus Secretary	Policy Committee Chairman		Chairman, Committee on Committees
State Delegation Deans			
Zone Whips			

Note: Democrats in the Senate tend to concentrate their leadership while Republicans are more dispersed.

search Service, plus clerical and housekeeping personnel.

There is no doubt that congressional staff exercise influence over congressional policymaking. Obviously, members use staffers to gather information and, on occasion, members will defer to the opinion and judgment of staffers. John Allsbrook has documented the important role of committee staffs in weapons systems acquisition.[42] The influence wielded by staffers has led many to conclude that they are "unelected representatives."[43] Others have expressed alarm that huge staffs, attempting to justify their positions, tend to overwhelm members with information and requests for delegated responsibility, thus increasing rather than decreasing the member's workload. However, the prevailing view seems to be that staffers do not constitute a threat to the power of legislators. To the contrary, many view staffers as extensions of a member's power, working within parameters of acceptability carefully defined by their employers. Because staffers are constantly anticipating the boss's reaction to their actions, it is the members, not the staffers, who wag the tail.

Congress Is Lobbied by a Nonmonolithic Executive Branch

When visualizing lobbying activity in Congress, one automatically thinks of representatives of interest groups and of the president, and certainly many different interests lay claim to defense policy. However, as all seasoned veterans of Capitol Hill can attest, Congress also is lobbied intensely by representatives of the various public agencies. All agencies, including defense, have a legislative liaison staff.[44] Ripley and Franklin describe the Department of Defense liaison network as follows:

Each service has its own very large liaison staff that works directly with relevant committees, subcom-

mittees, and individual Senators and Representatives . . . By contrast, the central DoD liaison operation in the Office of the Secretary had . . . hardly enough to compete with the services if they were pushing policy views and making alliances not in direct accord with policies favored by the Secretary of Defense.[45]

Individual Congressmen on and off the Armed Services and Defense Committees Can Influence Defense Legislation

Congress is inherently a place of unequals. Some members will always have more influence than others. Some members will be powerful in one policy area or in one issue domain while others will be influential in other areas.

Disproportionate influence is readily observable in committee decisionmaking. As Holbert Carroll has argued, the membership of each committee can be divided into formal and efficient parts. The formal part is comprised of all members of the committee. The efficient part of a committee

consists of a core of members, usually only a handful of men representing both political parties. These men actively participate in the hearings, propose the amendments that are accepted, and shape the legislation. They write parts of the committee's report, or at least take the time to slant it to their satisfaction. The efficient element then takes the bill to the floor and fights for it. Their knowledge of the subjects within the committee's jurisdiction may be more specialized than that of witnesses from the executive branch who appear before them.[46]

With the armed services committees, it is obvious that certain members—such as Senator Sam Nunn (Dem., Ga.)—are more associated with, dedicated to, interested in, and informed on defense matters than other members of the committee who might be more interested in their other committee assignments.

Certain members (especially senators) not on the relevant congressional committees can have an impact by serving as catalyst, pointman, and lobbyist (yes, members also are intensely lobbied by each other) on behalf of a particular cause. For example, the late Senator Hubert Humphrey led a crusade against Defense Department medical research involving beagles. Several House members have repeatedly pressed for the reactivation of certain mothballed ships. As Hilsman has written, "the individual Congressman can often dominate the headlines."[47] This is especially true if the member has acknowledged expertise or a publicized association within the Congress on a particular topic or if he or she is considered a major power broker.

MOST MEMBERS MAKE UP THEIR MINDS WHEN VOTING ON THE FLOOR ON THE BASIS OF THEIR OWN IDEOLOGY (POLICY POSITIONS)

The single best predictor of how House members and Senators vote is their personal ideology. Although some votes may be determined by constituency or power struggles among leaders, a member's ideology, or policy predisposition, better than any other factor predicts (1) the voting behavior of a member and (2) the sources of input and information on which they rely. Members must cast 1500 or so recorded roll-call votes each session of Congress. To cope, they develop shortcuts and routines that simplify the process. For most, voting becomes an effort to achieve a coherent record in each policy domain. In other words, members try to consistently vote for or against certain programs and policies. When they are unclear what the implications of a particular vote are, members turn for advice to a trusted colleague (cue-giver), usually a friend on a parent committee, with whom there is ideological compatibility.[48] This process of cue-taking, it should be emphasized, is used primarily on amendments and esoteric votes and is essentially an extension of ideological voting.

For defense decisionmaking, the implication of this initial reality is clear. Most members already have made up their minds before coming to Congress (usually during campaigns) about major defense issues. Although there is some play in every congressional vote, the basic inclinations of members significantly affect the final outcome. Defense matters are usually supported by legislators who are conservative and from the West or South. Opposition is frequently encountered from more liberal legislators from the Northeast.[49] For there to be a major change in the "standing vote," one of two things must occur: either there must be a change in the international system or the per-

ceptions of it or there must be a significant turnover of congressional seats, with outgoing members replaced by new members with significantly different views.[50]

APPROPRIATION DECISIONS ARE INCREMENTAL

The annual defense appropriations bill contains more than 5000 separate items. In his many important votes on the budgetary process, Aaron Wildavsky has eloquently emphasized that appropriations are processed according to a pattern of "budgetary incrementalism." In *The Politics of the Budgetary Process* he writes,

Budgeting . . . is incremental, not comprehensive. The beginning of wisdom about an agency budget is that it is almost never actively reviewed as a whole every year, in the sense of reconsidering the value of all existing programs as compared to all possible alternatives. Instead, it is based on last year's budget with special attention given to a narrow range of increases or decreases. Thus, the men who make the budget are concerned with relatively small increments to an existing base.[51]

For defense appropriations, this means that this year's appropriation is the best guide to spending in the future. Routine requests that constitute only gradual increases in past funding levels normally will be approved in a perfunctory way. Intense scrutiny will occur at the "margins" that involve significant increases or decreases in past funding levels. It should be emphasized that this "mixed scanning"[52] approach constitutes a sense of rationality in the very complex business of public and defense budgeting. With time and information scarce, congressional budgeters judiciously use their time to address major changes in policy. Also, as Arnold Kanter has stressed in his study of budgets between 1960 and 1970, although the defense budget as an aggregate reflects incrementalism, an analysis that disaggregates the budget reveals meaningful changes in priorities.[53]

PROPONENTS OF LEGISLATION MUST WORK WITHIN A "PROCEDURAL" MAJORITY, BUILDING COALITIONS

The name of the game in congressional policymaking is coalition building. Legislation is "the art of the possible" as Lyndon Johnson was fond of saying during his tenure as Senate majority leader.

To successfully build coalitions, proponents of legislation must work within "procedural majorities." In the words of Charles O. Jones, "Procedural majorities are those necessary to organize the House for business and maintain that organization."[54] Party leaders serve as caretakers of procedural majorities. Again, to quote from Jones, this requires that "House leaders must take care not to lose touch with any sizeable segment of their procedural majorities."[55]

The complexities of legislative policymaking stem from the requirements to build a different procedural majority within different decision units at various decision stages for each of the many policy issues in two houses for both authorizations and appropriations.

Most Policy Innovations and Changes Passed by Congress Evolve through an Incubation Process

Major changes in policy do not occur out of the blue. They evolve and develop through a rather lengthy process of incubation. As Nelson Polsby emphasizes, "Many of our important policy innovations take years from initiation to enactment." Polsby argues that major policy innovations require "incubation—a process in which men of Congress often play significant roles."[56] In the defense realm, most suggestions for alternative and improved defense systems have been kicking around Washington for some time. Hence, Congress is likely to pass programs "whose time has come" rather than radical, "spanking new" programs. For example, the notion of modernizing nuclear delivery systems has been an idea thrashed about for some time among senators, House members, Armed Services Committee staffers, think tank consultants (the RAND Corporation and Brookings Institution), defense-minded lobbyists, and Defense Department analysts. Like other policy proposals, defense programs come to the forefront only when a favorable policy climate emerges.

There Are a Plethora of Defense Issues upon Which the US Congress Must Act

Congress grapples with a prodigious number of defense issues, such as military personnel, operations and maintenance, pay, procurement, and research and development. Blackman attempted to list various kinds of programmatic and fiscal decisions made by the Senate Armed Services Committee. His categorization illustrates the extreme diversity of issues. According to Blackman, the different kinds of programmatic defense decisions faced by the committee include: DoD action, foreign policy, force modernization, strategic programs, tactical programs, and development. Different kinds of fiscal issues pertain to questions of cost effectiveness and program management.[57] In any given session of Congress, defense decisionmakers must decide on thousands upon thousands of defense requests ranging from routine reauthorizations and appropriations for ongoing programs to questions of major new programs such as the MX weapon system, B-1 bomber, and cruise missiles.

Different Kinds of Defense Issues Get Handled Differently in Congress

Congressional decisionmakers, when faced with a multitude of policy questions, devise categorizations or issue contexts with which to classify all the diverse issues. In Frederick Cleaveland's words, issue contexts affect "the way members of Congress perceive a policy proposal that comes before them, how they consciously or unconsciously classify it for study, and what group of policies they believe it related to."[58] To Cleaveland, "Such issue contexts strongly influence legislative outcomes because their structure helps determine the approach for analysis . . . as well as the advice and expertise that enjoys privileged access."[59]

Several efforts have been made to identify the different issue contexts with which members of Congress categorize defense issues. The most useful appears to be one developed by Huntington[60] and further elaborated on by Ripley and Franklin.[61] Their typology differentiates between structural, strategic, and crisis decisions:

Structural policies and programs aim primarily at procuring, deploying, and organizing military personnel and material . . . Examples . . . include specific decisions for individual weapon systems; (and) the placement, expansion, and closing of military bases . . .[62]

Strategic policies and programs are designed to assert and implement the basic military and foreign policy stance of the U.S. . . . Examples . . . include decisions about the mix of military forces (for example, the ratio of ground-based missiles to submarine-based missiles) . . . and the level and location of U.S. troop overseas . . .[63]

Crisis policies are short-run responses to immediate problems that are perceived to be serious, that have burst on the policymakers with little or no warning, and that demand immediate action . . . Examples . . . include the U.S. response to the Japanese attack on Pearl Harbor . . . the Soviet Union's placement of missiles in Cuba in 1962 . . . and the Cambodian seizure of a U.S. merchant ship in 1975.[64]

Their distinction shows that congressional policymaking is not undifferentiated. These authors emphasize that different kinds of defense issues get processed differently. In the words of Ripley and Franklin, "The basic notion behind our categorizations is that each type of policy generates and therefore is surrounded by its own different set of political relationships."[65] Their summary of the different kinds of policy relationships associated with each policy arena (see table 14.4) emphasizes the differences in

Table 14.4

Political Relationships for Policymaking

Policy Type	Primary Actors	Relationship Among Actors	Stability of Relationship	Visibility of Decision
Structural	Congressional sub-committees and committees; executive bureaus; small interest groups	Logrolling (everyone gains)	Stable	Low
Strategic	Executive agencies; president	Bargaining; compromise	Unstable	Low until publicized then low to high
Crisis	President and advisers	Cooperation	Unstable	Low until publicized then generally high

Source: Randall B. Ripley and Grace A. Franklin, *Congress, the Bureaucracy, and Public Policy*, 3d ed. (Homewood, Ill.: Dorsey Press, 1980), pp. 24–25.

how each type of policy is handled. Structural policies are hammered out in a normal manner at a low level in a subgovernment comprised of congressional committees, subdivisions of the Defense Department, and contractors. Strategic issues are decided at a somewhat higher level, while crisis decisions are acted on at the highest reaches of government.

It seems that the reason for this dynamic process is that structural issues are low-grade and not of interest to a broad segment of society. Thus, they are relegated for resolution to low levels in the system. Strategic issues, somewhat controversial in nature and involving "high politics," are of concern to a broad public, and handled at a higher level in the system. Crises, on the other hand, require immediate action and are highly visible and very salient. As such, they require upper-level attention.

Interviews with members and staffers corroborate this notion of multiple decision arenas and variable congressional decisionmaking. As one Senate committee staff member emphasized,

There are issues and there are *issues.* Normally, subcommittees make policy with the full committee and the full membership of the Senate ratifying decisions made in subcommittee. Other issues, however, such as the introduction of new weapon systems, strength of key units, and nerve gas, raise a different specter. On these there is likely to be a Senate floor fight and disagreement between the two houses.

Another staffer noted that "Most of the 3000 line items in the budget are noncontroversial. They sail through." Or, on the other hand, as a ranking member of the House Armed Services Committee emphasized, "Nuclear carriers, personnel matters, chemical warfare, and consideration of the proper role of Congress in defense issues cause the damnedest fuss." A Republican on the same committee contended that "Year

after year we fight over new systems, shipbuilding, and level of funding. The rest goes through routinely and without much controversy."

In conclusion, these 18 "initial realities" significantly affect both the style and outcomes of the congressional decision process.[66] Table 14.5 diagrams the causal influence various conditions of congressional organization and structure have on policy content. It is argued that the unique structural configuration of Congress as an organization has led to the development of certain facilitating norms. For example, the norm of specialization reinforces and makes functional the committee system. With regard to the norm of logrolling, or reciprocity, Barbara Hinckley has written that, "A norm of reciprocity seems crucial to such a collegial, representative assembly, divided as it is by the claims of individual constituencies and compartmentalized by arenas of specialization."[67]

The combination of structural aspects and norms results in an approach to decisionmaking that features compromise and negotiation. Because of all the strategic choke points in the legislative process at which a bill can be blocked, proponents must attempt to appease and compromise with intense, would-be opponents. Of course, this causes certain inevitable incremental, watered-down responses. As Charles Lindblom has argued, in a system featuring fragmented authority, decisions will be made with "successive limited comparisons" and "muddling through" as opposed to a totally rational "from the roots up" approach.[68] Or, to borrow from March and Simon, decisions are made in an effort to "satisfice"—that is, to come up with a solution that satisfies minimum criteria, appeases outspoken interests and provides a satisfactory—rather than an optimum—alternative.[69]

Table 14.4
Continued

Influence Patterns				
President, Centralized Bureaucracy	Bureaus	Congress as a Whole	Congressional Subcommittees	Private Sector
Low	High	Low (supports subcommittees)	High	High (subsidized groups and corporations)
High	Low	High (often responsive to executives)	Low	Moderate (interest groups, corporations)
High	Low	Low	Low	Low

For defense policy, the implications are clear. Congress will rarely make sweeping changes in defense programs in a comprehensive sense. Congress is concerned less with developing an ideal or optimum defense policy than with developing a defense program politically feasible within the Congress. Exceptions to incremental and segmental decisionmaking occur only during crisis, emergency, major turnover in membership, prolonged presidential honeymoon, or "textbook" presidential leadership and acumen.

TRENDS AND CHANGES IN CONGRESS DURING THE SEVENTIES AND EIGHTIES

The contemporary Congress is an institution in change. During the late 1970s and into the 1980s, many changes transpired on Capitol Hill. The significance of these recent trends cannot be overstated, for as Lawrence C. Dodd and Bruce I. Oppenheimer have written in the preface to their authoritative *Congress Reconsidered*, "What we have witnessed, in fact, was the culmination of a substantial period of congressional reform and a consequent creation of a new structure of congressional power."[70]

For defense policymaking, it appears that five major changes have occurred, altering the way Congress conducts its internal business and how it relates to forces external to it.

THE BREAKDOWN OF DEFENSE SUBGOVERNMENTS AND THE CONTAINED SPECIALIZATION NORM

In the 1950s and 1960s, Congress made defense policy in what H. Bradford Westerfield dubbed "a closed system." The essence of the closed system was that, in Westerfield's words, "the policy position which was emerging dominant within the administration before the issue was thrown open to 'outside politics' would still pre-vail after all the turmoil had died down, as the position of the U.S. government."[71] Other additional characteristics of the "closed system period" were:

—bipartisan defense decisionmaking;

—consensual, nonconflictual armed services committees populated by conservative, promilitary members;

—deferrence to the expertise of the armed service committee by those not on the committee.

The breakdown of the closed system is perhaps the most significant recent trend affecting defense decisionmaking in Congress. The breakdown has occurred simultaneously at three different levels.[72]

First, there were challenges within the Congress to the expertise of the committees. Bills brought to the floor by the armed services committees were no longer automatically ratified by the general membership of the parent house. Within the Congress, approximately 170 senators and representatives formed an informal group called Members of Congress for Peace Through Law (MCPL). For many members, this group served as a counterforce and alternative information source to the armed services committee. During the late 1960s and early 1970s, as Edward J. Laurance points out, floor debate and amendments increased significantly.[73] One member of the Senate Armed Services Committee described the breakdown in committee hegemony.

In the old days, the only trouble we got was on the total size of the budget and questions of cost overruns, necessity, and efficiency. Then, after Vietnam, every new weapon system—ABM, B-1, MX, Cruise Missile, neutron warhead—was subject to a floor fight as various members not on the committee raised questions concerning the morality of war and social costs.

A second level of military subgovernment breakdown is the decline in consensus among

Table 14.5
An Organization Theory of the US Congress

Structural Characteristics →	→ Norms →	→ Decisionmaking Style →	→ Inevitable Outcomes
Lack of hierarchy	Specialization	Piecemeal problem	Succession of limited
Fragmentation of	Deference to expertise	solving	comparisons
authority/multiple	Logrolling	Contained specializa-	Incrementalism
centers of power	Coalition-building	tion	Muddling through
Bicameralism	Civility	Compromise	Satisficing
Committee system	Serving and voting the	Accommodation	Lukewarm, watered-
Labyrinthine process	constituency	Bargaining	down outputs
Two-track system			Avoidance of bold,
Multiple leaders			radical, comprehen-
Informal groups			sive policy
Staff			
Electoral connection			

members serving on the armed services committees. Although the committees continue to be among the least partisan, they are less integrated than they used to be. As Laurance has demonstrated, the length of hearings and debates and the number of minority reports and dissenting votes have all increased. Some have concluded that an increase in conflict on the committee is the result of committee diversification. The membership of both committees appears to better reflect the membership of the parent houses than it used to.[74] As one senior Democrat on the House committee stated, "Up to a few years ago, most of the members on the committees were mainly military advocates. Now we have a lot of very skeptical people on the committee, and it becomes a tough fight to get things passed and funded through the committee." The recent deposing of House Armed Services Committee Chairman Melvin Price and the replacement of him as chairman by Congressman Les Aspin further strained the consensus of that committee.

The third level of subgovernment decline can be seen in the increased influence of nonmilitary groups. It appears that defense decisionmaking in Congress has undergone what Cobb and Elder refer to as "issue expansion."[75] Defense issues are now debated in an arena comprised of interests broader than the "affected publics" or normal military constituency of reserve and retiree organizations and contractors. Issues previously defined as technical, complex, and noncontroversial have become visible and politically salient. Now, as Ornstein and Elder have argued in their case study of the 1977 decision not to produce the B-1 bomber, there is "grassroots" lobbying on defense matters.[76] Also, Laurance documents an increase in nonmilitary testimony during armed services hearings.[77] Certainly the debate concerning the MX missile augurs a new defense policy process. As one staffer emphasized, "Discussions concern-

ing the MX missile system have invoked political implications concerning the environment, local economies, jobs, and intergovernmental relations in addition to national strategy. These other considerations have been relevant in the past but not to the extent to which they are now."

The reasons for the breakdown of the defense subgovernment and the expansion of defense issue controversiality stem from changes in the national political environment in the late 1960s and the early 1970s. George C. Gibson has attributed these shifts directly to American involvement in Vietnam, concern over domestic needs, excessive cost overruns, the My Lai incident, and opposition to the draft.[78] Concern over these issues contributed to factionalism in the Congress, demands for a "more balanced" committee, and conflict on the floor concerning defense issues.

CONGRESSIONAL ASSERTIONS CONCERNING PERCEIVED PRESIDENTIAL DOMINANCE IN THE DEFENSE AND FOREIGN POLICY ARENA

Aaron Wildavsky in his "Two Presidencies Theory" differentiates policymaking in the foreign/defense and domestic realms, characterizing the former as an arena of presidential control and decision latitude due to the absence of intense, politically influential interest groups.[79]

The experience since the 1960s, however, points to the erosion of presidential dominance if indeed it ever existed.[80] Congress has become more of a competitor with the president for authority to set foreign and defense priorities. As a legacy of the Vietnam war and Watergate, Congress reared up on its hind legs, asserting that it expects to be neither a silent partner nor a rubber stamp. Harvey Zeidenstein has catalogued recent, post-Watergate assertions of congressional prerogatives in defense policy. They are indeed impressive. They include: (1) the Case Act that prohibits the use of executive agreements for making international agreements which might lead to the employment of US military forces, requiring instead that such commitments be submitted to the Senate ratification process in the same manner as a treaty; (2) the establishment of a system for delaying and terminating national emergencies under 470 different wartime and emergency laws; (3) more aggressive oversight of intelligence operations; (4) the establishment of a procedure whereby Congress can veto US foreign arms sales; and (5) the War Powers Act that attempts "to insure that the collective judgment of both the Congress and the President will apply."[81] This last assertion is most significant. By its terms, the president is required to "consult" with Congress before he introduces armed forces into hostilities. The president can deploy troops no longer than 60 days (with an additional 30 for withdrawal) without specific authorization from Congress. Congress, with a concurrent resolution, can direct the termination of conflict and withdrawal anytime within the first 60 days of conflict[82] (see Appendix I).

The War Powers Act in many ways epitomizes this era of congressional assertiveness and the perennial yet cyclical nature of congressional-presidential rivalry in national security affairs. Passed in an override of President Nixon's veto in 1973, it has been and continues to be a topic of heated debate and controversy. Presidents Ford, Carter and Reagan, in crises involving the Mayaguez rescue, Iranian hostage rescue attempt, combat missions for US Marines in Lebanon, and the US invasion of Grenada, complied with various provisions of the act. However, they, their assistants, and many pundits publicly doubted the wisdom and constitutionality of the processes established pursuant to the act. Contemporary opinion ranges from condemnation of the act as an unwise constraint on the powers of the commander-in-chief to criticism that the act goes "too far" in empowering presidential use of military force and thus needs to be "tightened up" even further by more constraining amendments. Other commentary focuses on the problems of constitutionality raised by the 1983 *Chadha* decision of the US Supreme Court invalidating the legislative veto as a congressional policy tool. Some argue that this decision calls into question Congress's ability to discontinue hostility prior to a 60-day period with a concurrent resolution.

In addition to procedural alterations, Congress has asserted itself concerning the substance of military hardware and arms shipments. Carroll writes about the significance of the ABM votes: "In 1969, only by a tie vote, 50-50, had the Senate failed to delete entire authority for deployment of the ABM. At no time in history had the Congress so vigorously challenged a basic strategic decision."[83] Congressional challenges to US aid to Turkey, Angola, and El Salvador also illustrate strong assertions. As Ripley and Franklin have written concerning the 1974 Turkish embargo:

Despite the fervent protests of the President, Secretary of State, and other high-ranking executive branch officials, Congress made the strategic policy decision that a cutoff to Turkey (presumably coupled with U.S. mediation of the Greek-Turkey dispute over Cyprus) would be a better means of forcing a solution to the military situation on Cyprus than quiet diplomatic moves alone.[84]

A NEW CONGRESSIONAL BUDGET PROCESS

Before 1974, Congress's processing of the budget could be explained best in terms of extreme fragmentation. There was (1) no single coordinating body concerting authorizations and appropriations; (2) little concern with a total budget figure; (3) significant trust-fund spending outside the normal appropriations channel; (4) excessive dependence on the executive branch, specifically OMB, for policy analysis; and (5) a general untidy, commonly violated series of deadlines. In response to a perceived erosion of congressional power during the "imperial" Johnson and Nixon presidencies and, more specifically, the abundant use of impoundments (refusal to spend congressionally appropriated monies) by the Nixon administration, Congress was driven to overhaul and modernize its budgetary process. As was the case with defense and foreign policy assertions, the impetus for the new congressional budget process was a desire on the part of Congress to strengthen itself and to regain power perceived to be lost to the president.

The 1974 Congressional Budget and Anti-Impoundment Act engendered four major changes in the way Congress handled the budget. First, budget committees have been established in both the House and Senate to provide a comprehensive view of proposed government spending. Presumably, appropriation subcommittees were to handle "micro" spending within their jurisdictions while the budget committees would focus on "macro" budgetary considerations. Second, a Congressional Budget Office (CBO) was constituted to provide Congress with its own independent source of policy evaluation and budgetary information, thus lessening Congress's dependence on the executive. Third, a new fiscal year and congressional budget timetable was set into effect (see table 14.6). The major change over the older process is that with the budget resolutions reported out by the budget committees, a congressional ceiling on total expenditures and targets within

Table 14.6

Congressional Budget Timetable

Deadline	Action to Be Completed
November 10	Current services budget received
January 18[a]	President's budget received
March 15	Advice and data from all congressional committees to budget committees
April 1	CBO reports to budget committees
April 15	Budget committees report out first budget resolution
May 15	Congressional committees report new authorizing legislation
May 15	Congress completes action on first budget resolution
Labor Day +7[b]	Congress completes action on all spending bills
September 15	Congress completes action on second budget resolution
September 25	Congress completes action on reconciliation bill
October 1	Fiscal year begins

[a] Or 15 days after Congress convenes.
[b] Seven days after Labor Day.
Source: US Congress, Senate, Committee on the Budget, *Congressional Budget Reform*, 93d Cong., 2d sess., 4 March 1975, p. 70.

each major policy sector (urban aid, agriculture, education, human services, defense, foreign aid, etc.) is established. The reconciliation process permits the budget committees to enforce these limits on the standing committees (HASC, SASC) with substantive authority. In theory, the first budget resolution "guides" the actions of the authorization and appropriations committees as they act on the next year's budget. The second resolution has "the force" of law, establishing "firm" ceilings and targets. In practice, in most years, Congress has passed only a first resolution. Fourth, the act placed firm limits on presidential impoundment authority. At the heart of impoundment control is the act's distinction between deferrals and recisions. Deferrals pertain to a president's refusal to immediately spend money within a 60-day period. Recisions refer to presidential non-spending beyond 60 days. Deferrals are permitted providing one house of Congress does not "veto" with a negative vote. Recisions *are not* permitted unless Congress specifically rescinds its earlier appropriation.

The budget process, as Robert Lockwood has argued, was an attempt to counter fragmentation in congressional budgetmaking. In his words, "The new procedures were cast with an eye toward efficiently processing and formulating a timely budget."[85] It's a bit early to ascertain the full impact of the new congressional budget, although the many missed deadlines, failures to get a second budget resolution, and uses of continuing resolutions for funding must be listed as disappointments. For defense policymaking, however, it seems safe to conclude with some certainty that at least the following appear to be true: Congressional budget committees and the CBO have become important actors in defense policy with the budget committees emerging as "super committees"; proposed defense expenditures have been and will continue to be weighed against proposed spending in other functional areas (urban aid, health care, transportation) especially with the Gramm-Rudman-Hollings deficit reduction package of 1985; and, because, of the anti-impoundment provisions, Congress is more involved in budget implementation, as it makes numerous required recision rules.

MISCELLANEOUS CHANGES WITHIN THE CONGRESSIONAL ENVIRONMENT ALTERING THE POWER STRUCTURE WITHIN CONGRESS

In addition to the previously mentioned major trends and changes concerning defense policy in Congress, sundry changes have occurred within Congress that portend changes in the way in which Congress does its business.

Despite the major turnover in Congress as a result of the 1980 elections, there is more careerism in Congress than previously was the case. Members have a longer tenure, and incumbency has become an important campaign resource, especially in the House during off years (election years when there is not a presidential race). This means that certain congressmen with longevity will be very powerful and will have expertise within their area of interest.[86]

Congress has become a more open institution. Committee markup and conference committee proceedings are now open to the public. Congress has become more accountable, providing for roll-call votes in the committee of the whole (the forum within which the House does most of its amending business). This was not possible prior to the Legislative Authorization Act of 1970.

The obstructionalistic potential of congressional bogeymen—filibustering senators, a tyrannical rules committee, and autocratic committee chairmen—has been reduced. It is now easier to invoke cloture to curtail debate and delaying tactics in the Senate. The Speaker of the House (when a Democrat) now controls the selection of Democratic members to the rules committee. Committee chairmen no longer are chosen automatically on the basis of seniority. Their initial appointment and subsequent reappointment in each Congress must be approved by party forums, thus making committee leaders more accountable to their party in their house.[87] In fact, two of the past chairmen of the House Armed Services Committee—Chairmen Herbert and Price— have been replaced in revolts by House Democrats, many of whom perceived them to be out of step with their party.

There are four party organizations in the Congress: Democrats and Republicans in both the House and Senate. Organizations of each have become more active, assertive, and broader based.[88]

Authority within the Congress has been splintered further among subcommittees, especially in the House with its 1973 "subcommittee Bill of Rights" which makes subcommittees the real locus of work and power. Some have concluded that because of the strengthened subcommittees, Congress has become a highly "individualized" legislative system.[89]

The 1980 election aside, there has been a decline in the "coattail effect" of presidential candidates on congressional elections.[90] Candidates for Congress run on their own, frequently aloof from their candidates for the presidency. As a result, congressmen do not feel exceptionally loyal or beholden to the president. Instead, as *Time* magazine emphasized in a 1978 article, "the contemporary Congress has become bold, balky, and independent."[91] Also, in some recent Congresses, the incidence of party loyalty has declined while the proclivity toward cross-party coalition-building has increased.

Single-issue interest groups have proliferated. In addition to an increase in numbers, it is generally acknowledged that their influence has increased. These groups normally focus on only one issue (e.g., gun control, abortion, environmentalism, or women's rights) to which they are intensely devoted. Combined with the subcommittee bill of rights, their impact is to make the basis of policymaking more narrow, impeding a comprehensive conception of public interest, national priorities, and policy spillovers. The involvement of one of these groups— the antiabortion "Right-to-Life" lobbyists—in defense decisionmaking was cited by a staff member of the Senate Appropriations Defense Subcommittee: "Every time we do the defense appropriations, you can bet we will hear from the anti-abortion crowd. They are constantly at work to prohibit abortions in military hospitals."

Congressional support agencies such as the Office of Technological Assessment, Congressional Research Service, and the General Accounting Office (together with the Congressional Budget office) as well as "think tank" consultants (RAND, and the Brookings and Hoover Institutions) are having a bigger impact on congressional deliberations. For example, Ralph Sanders in *The Politics of National Defense* highlights the role of outside consultants in the 1969 ABM decision.[92]

Last, the workload of members of both houses of Congress has grown immensely in recent years. As Hilsman notes,

Any discussion of the role of Congress in foreign affairs must begin with an acknowledgement of the load of work carried by most congressmen, and their busyness. The individual congressman carries a formidable burden.[93]

Congressmen must split their time among committee assignments, floor responsibilities, and office and constituency work. These burdens are increasing significantly.

Because the dust has yet to settle, it is difficult to pinpoint what the exact consequences of these changes have been for defense policy. Contradictory tendencies are at work. Most of the trends—such as careerism, declining party loyalty and cohesion, and increased subcommittee power—point to a more fragmented Congress. Others—such as the reduction of obstructionalism and party reinvigoration—perhaps indicate a concentration of authority and a counter to dispersion. William J. Keefe has emphasized that, for the most part, these changes have exacerbated sharply the dispersed structure of Congress. As he writes, "The changes made in the name of democratization have contributed significantly to the further devolution of power in Congress."[94] But, as Samuel Patterson notes, despite the many significant changes that have occurred, Congress still retains its "peculiar" character. In Patterson's words,

If Henry Clay were alive today, and he were to serve again in the House and Senate to which he was chosen so many times in the nineteenth century, he would find much that was familiar. He would certainly recognize where he was, and perhaps after some initial shock he surely would be reasonably comfortable in the modern congressional envelope.[95]

The Reagan Administration and the Exercise of Presidential Leadership

Perhaps the most recent significant development has been the Reagan administration's successful exercise of presidential leadership on defense issues before Congress. Although some—including members of the president's own party—were critical of Reagan defense priorities, increased defense spending and the social costs of rearming America, there was almost universal agreement about the Reagan mastery of Congress. The Reagan administration followed our failed presidencies: those of Johnson, Nixon, Ford, and Carter. Toward the end of their administrations, each of these presidents suffered major defeats in Congress on national security votes and issues. This trend was reversed in the Reagan administration as the president scored victory after victory on defense and national security issues that had assumed critical importance and involved the president's prestige. Among the major successes were votes on budget priorities and relative defense spending (more for defense and less for social programs), sale of AWACS planes to Saudi Arabia, MX basing and deployment, and aid to contra rebels fighting in Nicaragua. The significance of these victories for the defense policy process cannot be overstated. They prove that the fragmentation and dispersion of power within the Congress can be overridden by determined and adroit exercise of presidential power, supported by a string of legislative victories, a perceived public consensus and skillful congressional leaders. Things can be coordinated; the system can be made to work.

To conclude this discussion of recent developments and trends, changes in how the Congress handles defense policy issues are the result of factors both internal and external to it. Gibson discusses in depth the process of congressional change, identifying the major variables affecting the defense policy process (see table 14.7). It should be emphasized that changes in both structure and personnel will lead not only to changes in the congressional process but also, more importantly, to changes in the substance of congressional output. Finally, remember that Congress is most likely to adapt, reform, and change, not in response to decline in public esteem, but in response to perceived encroachment by the president or internal leadership excesses.

EVALUATION OF CONGRESSIONAL ACTION AND ASSERTIVENESS

Debate on the adequacy of congressional involvement in the defense policy process abounds. It is as old as the republic itself, for throughout *The Federalist Papers*—the oratory in support of the proposed new constitution of 1787—numerous references are made to the presumed strengths and weaknesses in foreign and defense affairs of Congress and the presidency.

Today, evaluations of the congressional role in defense policymaking are of two types: the

Table 14.7
Factors Affecting Congressional Attitudes on Defense

House Internal Parts	Senate Internal Parts	External Environment
Leadership	Leadership	Events
Committees	Committees	International
Membership	Membership	Domestic
		Executive
		President
		OSD
		Services
		Public opinion
		Interest groups/ mass media

Source: George C. Gibson, "Congressional Attitudes Toward Defense," in *American Defense Policy,* 3d ed., ed. Richard F. Head and Ervin J. Rokke (Baltimore: Johns Hopkins University Press, 1973), p. 359.

impact of recent congressional assertions and general assessments of congressional defense roles. The purpose here is not to make an evaluation or to take sides in the debate, but to inventory the various positions, pro and con, that have been offered. A survey of them reveals the strengths and weaknesses of congressional involvement in national security affairs.

THE IMPACT OF RECENT CONGRESSIONAL ASSERTIONS

There are two schools of thought concerning the impact on defense policy of the new congressional procedures and attitudes.

One view in this healthy disagreement is that these changes have not been all that significant. Illustrative of this position is Lawrence J. Korb's conclusion that:

Congress has attempted to play a more active role in the defense decisionmaking process . . . (However) although it is difficult to measure the exact influence of the legislature on policy, and while one must always be cognizant of the law of anticipated reaction, it becomes fair to conclude that, even in the post-Vietnam period, the Congress has had only a marginal impact on policy and force structure.[96]

Along the same lines, some have pointed to the failure of presidents to "consult" with the Congress during recent deployments of US troops as indicative of the inconsequence of the War Powers Act. The act requires presidential consultation with Congress when deployments are made. However, President Ford in the Mayaguez affair, President Carter in the ill-fated attempt to rescue American hostages in Iran, and President Reagan during the Grenada invasion, merely notified the Congress after efforts were underway, giving little opportunity for congressional input.

The opposing view—that substantial changes have ensued as a result of congressional assertions—is put forth by Richard Haass. Haass sees a constant struggle throughout US constitutional history between the Congress and the presidency for control of national security policy. He concludes that:

Although the reforms of recent years have strengthened Congress at the expense of the presidency, they have not settled in any permanent way the continuing struggle over the distribution of foreign and defense powers in the American government. For the foreseeable future however, this struggle will be waged by branches more equal than at any time since the end of World War Two.[97]

In the same vein, one staffer commented: "More defense policies and force structure are being formulated on the Hill than ever before. This is because of the requirements of the new budgetary process."

ASSESSMENTS OF CONGRESSIONAL ROLES IN DEFENSE POLICY

Negative Case

Many of the commentaries on congressional involvement are quite negative. They stress the weaknesses of congressional institutions for processing important national security questions. Negative critiques of Congress can be grouped as follows: general criticisms, specific defense-oriented critiques, criticisms by hawks, and criticisms by doves.

General indictments by prominent students of Congress illustrate weaknesses in the congressional system. Sundquist identifies four such weaknesses: parochialism, irresponsibility (or lack of collective responsibility), sluggishness, and amateurism.[98] David R. Mayhew emphasizes the reelection imperatives and sense of shared fates that pervade Congress and lead to advertising, credit claiming, case work, and posturing among the membership. According to Mayhew, this system results in delay, particularism, serving of the organized, and symbolism. Others have criticized the system for minority rule, the lack of policy coherence, and the lack of persistence. In this view, the presidency is much better prepared to develop foreign commitments and coherent policy objectives.

Specific criticisms of Congress's defense role have focused on problems of security, multitudinous actors, the lack of internal coordination, segmentalism, inadequate staff and analytical capability, excessive localism, and porousness.

Most discussions of the strengths of the presidency vis-à-vis the Congress in defense policy emphasize congressional security problems. Critics underscore leaked information concerning the Pentagon Papers, US involvement in the overthrow of the Chilean regime of Salvadore Allende, and CIA operations. Given the large number of actors and their staff, it becomes very difficult to maintain security on Capitol Hill. One critic emphasized, "It's like a sieve up here. You tell one person, you tell twenty."

A persistent criticism of Congress is that the congressional process involves too many actors. Carroll alludes to this when he states: "Fragmentation and diffusion erode its (the Congress's) powers."[99] Because of the numerous decision points and multiple actors, there is constant danger that Congress will immobilize itself. Or, as one old-time staffer emphasized, "It is amazing that the system works at all. There are just too many actors, and the number of players have increased significantly in recent years." With the consent of so many required, there is the possibility that an intensely opposed, strategically located legislator may thwart the majority will. As the same staffer stated, "While one can say no, many must say yes."

It has been argued that the same multiplicity that harbors potential for obstructionalism also precludes coordination. "Congress speaks with too many voices on security matters" was a criticism offered by a House member. He added,

the different parts of Congress rarely talk to each other. As a result, you frequently find the different power centers with different positions, taking different public stands. One hand appears to be trying to undo what the other is trying to accomplish. It is plain chaotic. We are our own worst enemies.

To many, the current system is akin to rugged "individualism." Individualism in defense policy is most readily seen in the numerous riders to defense appropriations bills by members motivated by personal interest, ideology, or constituency.[100]

The criticism is made that Congress is excessively segmented when processing defense policy. The argument is that the subcommittee arrangement leads to a piecemeal focus on defense matters culminating in policy particularism. Huntington's criticism reflects this argument. He concludes that "the most prominent congressional role is that of prodder or goad of the administration on behalf of specific programs or activities."[101] This piecemeal and individualistic handling of defense matters is thought by many to detract from policy coherence.

Congress is faulted for inadequate staff and lack of analytical ability when processing defense policy. One staffer lamented, "There is only a handful of us in the Congress compared with the thousands of analysts that work for the

Pentagon. We have neither the time nor the resources." Moreover, Bearg and Deagle have argued pursuasively that Congress lacks systematic procedures for congressional examination of defense issues in a multiyear budget.[102]

"Excessive localism" is perhaps what Congress is most criticized for in its making of defense policy. Dexter harshly concludes that "military policy is not considered,"[103] referring to the House Armed Services Committee as "primarily a real estate committee"[104] whose members try to get as much boodle for their districts as they can. One staffer concurred, "Although some guys are motivated by a desire for good defense policy, most have a parochial view whereby they serve local interests back home." In other words, the US scheme of congressional elections yields a Congress that is very "representative" and "responsible" but not responsive in a collective manner.

Some have concluded that, because of its fragmentation of authority and the local orientation of many congressmen, Congress is a porous institution, constantly penetrated and affected by special interest groups in the making of defense policy. The lobbying efforts of Greek nationals in the United States for the Turkish arms embargo during the Cyprus crisis is cited as an illustration of how a determined, intense interest can affect the outcome of the defense policy process in Congress.[105]

Some of the criticisms leveled at Congress come from a decidedly hawkish (or prodefense) perspective. Among them are the arguments that (1) Congress plays only a negative (naysaying) role in defense policy, (2) recent congressional assertions hamstring the president in his dealings with foreign nations, (3) the new budget process distorts and "forces" defense policy, and (4) the congressional process often does not input the "true" feelings of ranking military men.

Congressional negativism is emphasized by Hilsman. He states that the

command of both information and expertise gives the executive the intellectual initiative in making foreign policy. The Congress as a whole can criticize; it can add to, amend, or block an action by the Executive. But Congress can succeed only occasionally in forcing Executive attention to the need for a change in policy, and rarely can it successfully develop and secure approval from the public for a policy of its own.[106]

Some have argued that recent congressional assertions have constrained, to an undesirable degree, the president's ability to act during international crises. Dodd and Schott feel that these assertions raise a fundamental question, "Can Congress oversee the executive without reducing it, and government in general, to impotence?"[107] Some observers would answer no to that question with regard to the War Powers

Act. Neuchterlein, in *National Interest and Presidential Leadership*, takes this approach. He concludes that recent congressional actions undercut the president and the nation. To him, "the real issue in American foreign policy . . . remains the authority of the President to decide U.S. national interests in consultation with Congress without being hamstrung by the legislative branch."[108]

The new budget process is indicted by many as detrimental to sound defense policymaking. One staff member of the House was most critical. He stated: "Now, policy must be made to fit within budgetary constraints. We become more interested in meeting budgetary ceilings than in achieving the best defense posture."

A final criticism from a hawkish perspective is that Congress frequently does not get candid advice from military leaders. A provision of the National Defense Act of 1947, "authorizes the members of the Joint Chiefs of Staff to express their views on national security matters directly to the president and Congress."[109] Despite this, as a House Republican pointed out, the problem is that "it is difficult to differentiate between the opinion of the administration and career military officers, between official and personal positions."

Criticisms from a dovish (antimilitary) perspective fault (1) Pentagon hegemony, (2) contractor influence and regional competition in weapons acquisitions, and (3) "domination" of congressional defense policymaking by hawks.

The dominance of the Pentagon in the congressional decision process, especially on complex and technical matters, is highlighted by Dexter. He refers to the congressional defense policy process as "the tyranny of information and ideas,"[110] with little intense congressional scrutiny of routine operations. Dexter argues that the military specialist monopolizes the presentation of the defense alternatives that the Congress considers.[111] Moreover, "few in Congress want to challenge the experts in a highly specialized field."[112] Thus, with regard to the Pentagon, many members take the approach "who are we to say no?"[113]

Several liberal members of Congress have criticized the inordinate influence of contractor lobbying in congressional defense policymaking, raising the specter of a self-serving, military-industrial complex. One liberal member of the House Armed Services Committee developed this argument:

Because of the dynamics of politics, the PAC's (Political Action Committees) of various defense contracting interests have more sway over defense policy than anyone else. Once members get campaign money from them, they lose their free vote. It becomes hard to say no. Because of the budget ceiling, money spent to placate defense contractors must be squeezed out of other accounts—such as maintenance, spare parts, pay and benefits—adversely affecting defense preparedness.

Another equally liberal member was quick to point out that direct lobbying by contractors sets off regional competition, leading to a symbiotic relationship between members and contractors. In this member's words,

The policy approach used these days is similar to what you find in the competition for federal grant money to localities. Congressmen from different areas, such as New York and California, compete for contracts regardless of their need or impact on a total defense picture. They get political pluses back home for doing this. The name of the game has become jobs, facilities and payrolls. Under this system, good defense gets short-changed.

Both members acknowledged that the military is frequently, but not always, an innocent bystander in the process. The military is appreciative of lobbying efforts by contractors, but aware of the motivations that underlie contractors' efforts and the potential adversities to defense and national strategy that might result.

Alleged hawkish domination is a commonly heard criticism of the congressional defense role. It is alleged that committees of both House and Senate, despite recent efforts at diversification, still are more promilitary than their respective Houses. It is alleged that the military gets a free ride, that the Pentagon is not sufficiently surveilled by an objective, critical Congress. A member's staffer expressed it this way: "Those committees are ultra pro-defense. Unusual, contrary, unorthodox, imaginative points of view rarely are considered seriously. Social tradeoffs are not considered."

Positive Case

Although arguments supporting congressional involvement do not seem as abundant as criticisms, their adherents are no less vocal. Those who view congressional involvement in defense policy favorably, stress Congress's role as (1) a source of multiple input and (2) an independent counter to professional mind-sets.

Yarmolinsky draws an overall favorable conclusion about the defense policy process in Congress. He states that:

in fact the legislative process . . . is a good deal more disciplined than it is generally given credit for. Compared, for example, to the Rivers and Harbors Bill, the defense authorization and appropriations are handled as models of decorum. The Rivers and Harbors Bill is annually pushed and pummelled through the Congress until it emerges in a shape scarcely recognizable as what it was when first introduced by the Administration. Defense legislation, on the other hand, generally emerged virtually unscathed, with few, if any, deletions, to be sure, but with few, if any, additions as well. An eager defense contractor, pressing for the adoption of a new weapons system,

can almost never persuade the Congress (much less the White House) to put an add-on in the budget no matter how many powerful allies he can muster, if he has failed to persuade the authorities within the Pentagon itself.[114]

An evaluation by Bearg and Deagle is likewise favorable. They conclude that "congressional oversight of both broad foreign policy issues and details of weapon systems acquisition and financial management is reasonably effective and continually improving."[115]

One positive view contends that congressional involvement in security matters enhances the prospects for a good policy because an active Congress is assurance that (1) a proposed policy will be deliberated, (2) many points of view will be "brought to bear" and considered, and (3) support for the proposal will be mobilized. As Francis O. Wilcox observes, "Strengthening the congressional role is likely to result in a more viable foreign policy, if for no other reason than it will command broader public support, based on a broader national consensus."[116] Yarmolinsky points to a multiple advocacy system in Congress concerning defense matters that enhances policy. In his words, "the built-in difference among the military service and the official spokesmen . . . (in Congress) . . . create some of the best opportunities for movement within the bureaucracy."[117]

Another positive assessment stresses that Congress, as an additional input, provides an opportunity for correcting and overcoming shortcomings and mistakes in defense policy that might develop during executive planning. One staff member felt that congressional assertions "stimulate good management and decisionmaking in the executive branch." He concluded that "the congressional stage affords another opportunity to make wrong, right." A Democratic member of the armed services committee who is a subcommittee chairman argued that "those of us who are defense-minded have an opportunity to be assertive and to insure that defense needs, not bureaucratic or political ones, are met. These are good checks and balances." One Republican representative stressed that Congress can bring

a broad, fresh view to things. The expertise, information, hearings, and understanding of the armed services committee are better than what you find on other committees. I really think we have something to contribute. We have the ability to see beyond the tunnel-vision and blinders that eventually arise in all bureaucracy.

Congressional activism in defense policy is applauded as a relatively independent, quality response, unencumbered by the political considerations that so strongly influence domestic policymaking. Although, as noted above, the realms of security and domestic policy are converging, for some there is enough of a distinction to warrant a positive comparison. This is emphasized by Carroll who states that

In reacting to international matters, the member enjoys greater freedom in expressing himself than is deemed prudent in dealing with much domestic business. Few constituents are likely to mount worrisome challenges about how the congressman reacts to events in India, Africa, Europe, or Japan, or about his attitude on the price of gold, foreign aid, or the United Nations.[118]

Hilsman concurs by saying that

Congress has a greater freedom in the field of foreign policy than might ordinarily be supposed. For one thing there seem to be relatively few pressure groups at work on foreign policy, quite unlike the solid array that plead increasingly for their points of view on domestic matters.[119]

He concludes that "by and large, then, Congress seems somewhat freer from organized pressure when it comes to decide on an issue of foreign policy than it is on domestic policy."[120]

Finally, further congressional activism is encouraged by those who desire an "alternative national defense budget" and "a sound look." This is the preference underlying Congressman Aspin's call for "procedural requirements" that would permit more legislative leadership.[121]

CONCLUSIONS

Congress is a peculiar institution. It is chaotic, non-hierarchical, fragmented, and lacking in neatness and order. It is "a political institution and responds primarily to political stimuli."[122] It is a "forge of democracy," a representative forum of high debate and deliberation. The distinctive contributions of Congress are defined by Charles O. Jones, a dean of congressional scholars.

Congress is not now and never has been well designed to create its own agenda and act on it in a coordinated way to produce a unified domestic and/or foreign policy program. It is particularly well structured to react to many publics (including other governmental institutions) and, in reacting, to criticize, refine, promote alternative proposals, bargain, and compromise.[123]

It is unrealistic, unfair, and perhaps unwise to expect more of the Congress. The performance of these functions more than outweighs the inefficiencies, limitations, foibles, and irrationalities of Congress.

NOTES

The author would like to thank Congressman Jim Lloyd (D-Calif.) for sponsoring this research, for providing his office as a base for operations, and for his many insights as a member of the House Armed Services Committee. Gratitude is also owed to seven other members and ten staffers for their off-the-record comments, liberally used throughout this analysis.

This article is written for academic purposes and does not necessarily reflect the views of the Department of Defense, the United States Air Force, or the National Defense University.

1. Aaron Wildavsky, *The Politics of the Budgetary Process*, 3d ed. (Boston: Little, Brown, 1979).

2. Richard F. Fenno, Jr., "The House Appropriations Committee as a Political System: The Problem of Integration," *American Political Science Review* 56 (1962): 310–24, and *The Power of the Purse* (Boston: Little, Brown, 1966).

3. Charles O. Jones, "Representation in Congress: The Case of the House Agriculture Committee," *American Political Science Review* 55 (1961): 358–67 and Norman J. Ornstein and David W. Rohde, "Shifting Forces, Changing Rules and Political Outcomes: The Impact of Congressional Change on Four House Committees," in *New Perspectives on the House of Representatives*, 3d ed., ed. Robert L. Peabody and Nelson W. Polsby (Chicago: Rand McNally, 1977), pp. 186–96.

4. Holbert N. Carroll, *The House of Representatives and Foreign Affairs*, rev. ed. (Boston: Little, Brown, 1966), and Ornstein and Rohde, "Shifting Forces."

5. David E. Price, "Policy-making in Congressional Committees: The Impact of Environment Factors," *American Political Science Review* 72 (1978): 548–74, and *Who Makes the Laws?* (Cambridge: General Learning Press, 1972) and Ornstein and Rohde, "Shifting Forces."

6. John F. Manley, "The House Committee on Ways and Means: Conflict Management in a Congressional Committee," *American Political Science Review* 59 (1965): 927–39, and *The Politics of Finance* (Boston: Little, Brown, 1970).

7. Charles O. Jones, *Clean Air: The Policies and Politics of Pollution Control* (Pittsburgh: University of Pittsburgh Press, 1975) and "Speculative Augmentation in Federal Air Pollution Policymaking," *Journal of Politics* 36 (1974): 438–64.

8. Richard F. Fenno, Jr., *Congressmen in Committees* (Boston: Little, Brown, 1973).

9. James T. Murphy, "Political Parties and the Pork Barrel: Party Conflict and Cooperation in House Public Works Committee Decisionmaking," *American Political Science Review* 68 (1974): 169–85.

10. John F. Bibby and Roger H. Davidson, *On Capitol Hill*, 2d ed. (Hinsdale, Ill.: Dryden, 1972), pp. 183–206.

11. Ornstein and Rohde, "Shifting Forces."

12. Holbert N. Carroll, "The Congress and National Security Policy" in *The Congress and America's Future*, 2d ed., ed. David B. Truman (Englewood Cliffs, N.J.: Prentice-Hall, 1973).

13. Arnold Kanter, "Congress and the Defense Budget: 1960–1970," *American Political Science Review* 66 (1972): 129–43.

14. Lewis Anthony Dexter, "Congressmen and the Making of Military Policy" in *New Perspectives*, pp. 305–24.

15. See esp. Bruce A. Ray, "The Responsiveness of the U.S. Congressional Armed Services Committees to Their Parent Bodies," *Legislative Studies Quarterly* 5 (1980): 501–15.

16. See esp. Lawrence Ritt, "Committee Position, Seniority, and the Distribution of Government Expenditures," *Public Policy* 24 (1976): 437–62.

17. For a discussion of the origin and impact of

the 412 process see Raymond H. Dawson, "Innovation and Intervention in Defense Policy" in *New Perspectives*, pp. 273–303.

18. Carroll, "Congress," p. 181.

19. For an explication of the concept of "latent" oversight and how it is different from "manifest" oversight, see Morris S. Ogul, *Congress Oversees the Bureaucracy* (Pittsburgh: University of Pittsburgh Press, 1976), chap. 6.

20. For a broad discussion of Congress's consensus-building role in national security policy see Carroll, "Congress," p. 186.

21. Roger Hilsman, *The Politics of Policymaking in Defense and Foreign Affairs* (New York: Harper and Row, 1971), p. 83.

22. Samuel P. Huntington, *The Common Defense* (New York: Columbia University Press, 1961), p. 145.

23. This term was coined by Charles O. Jones, *An Introduction to the Study of Public Policy*, 2d ed. (North Scituate, Mass.: Duxbury, 1977), p. 8.

24. This term is used by David Truman, *The Congressional Party* (New York: Wiley, 1959), pp. 104–5.

25. David R. Mayhew, *Congress: The Electoral Connection* (New Haven: Yale University Press, 1974).

26. Mayhew, *Congress*, pp. 49–61.

27. Hon. Les Aspin, "The Defense Budget and Foreign Policy, the Role of Congress," in *American Defense Policy*, 4th ed., ed. John E. Endicott and Roy W. Stafford, Jr. (Baltimore: Johns Hopkins University Press, 1977), p. 321.

28. Ibid., p. 333.

29. Charles O. Jones, "Will Reform Change Congress?" in *Congress Reconsidered*, ed. Lawrence C. Dodd and Bruce I. Oppenheimer (New York: Praeger, 1977), p. 249.

30. Nelson W. Polsby, "Policy Analysis and Congress," *Public Policy* 17 (1969): 61.

31. Samuel C. Patterson, Robert H. Davidson, and Randall B. Ripley, *A More Perfect Union: Introduction to American Government*, 1st ed. (Homewood, Ill.: Dorsey, 1979), p. 742.

32. Ibid.

33. R. Douglas Arnold, *Congress and the Bureaucracy* (New Haven: Yale University Press, 1979), p. 128.

34. Patterson et al., *A More Perfect Union*, p. 742.

35. Aspin, "The Defense Budget," p. 325.

36. Arnold, in fact, concludes that "The Armed Services Committee tends to attract two types of congressmen: those interested in defense policy and those interested in district interests." Arnold, *Congress and the Bureaucracy*, p. 125.

37. Woodrow Wilson, quoted from *Congressional Government*, in Eric Redman's *The Dance of Legislation* (New York: Simon, 1973), p. 9.

38. For a discussion of patterned role playing see: Aaron Wildavsky, *Budgeting: A Comparative Theory of Budgetary Processes* (Boston: Little, Brown, 1975), pp. 3–9. For a discussion of the Senate's role in appeals see Fenno, *Power of the Purse*.

39. Donald Matthews, *U.S. Senators and Their World* (New York: Vintage Books, 1960), pp. 92–117.

40. The term was coined by Ira Sharkansky, *The Politics of Taxing and Spending* (Indianapolis: Bobbs-Merrill, 1969), p. 38.

41. Harrison W. Fox, Jr., and Susan Webb Hammond, *Congressional Staffs* (New York: Free Press, 1977), p. 168.

42. John W. Allsbrook, "The Role of Congressional Staffs in Weapon System Acquisition," *Defense Systems Management Review* 1, no. 2 (1977): 34–71.

43. For this point of view see a symposium on "Our Unelected Representatives," M. J. Malbin, "Congressional Committee Staff: Who's in Charge Here?" and M. A. Scully, "Reflections of a Senate Aid," *Public Interest* (1977): 16–48.

44. For a detailed discussion of agency legislative liaison see Abraham Holtzman, *Legislative Liaison: Executive Leadership in Congress* (Chicago: Rand, McNally, 1970).

45. Randall B. Ripley and Grace A. Franklin, *Congress, the Bureaucracy and Public Policy*, rev. ed. (Homewood, Ill.: Dorsey, 1980), p. 190.

46. Holbert N. Carroll, *The House of Representatives and Foreign Affairs*, rev. ed. (Boston: Little, Brown, 1966), p. 28.

47. Hilsman, *Politics of Policymaking*, p. 74.

48. For an elaboration of this argument see Aage R. Clausen, *How Congressmen Decide: A Policy Focus* (New York: St. Martin's Press, 1973); John W. Kingdon, *Congressmen's Voting Divisions*, 2d ed. (New York: Harper and Row, 1981); Donald R. Matthews and James A. Stimson, *Yeas and Nays* (New York: Wiley-Interscience, 1975); and David C. Kozak, "Decisionmaking on Roll Call Votes in the House of Representatives," *Congress and Presidency* 9 (1982): 51–78.

49. These patterns were documented by George C. Gibson, "Congressional Voting on Defense Policy: An Examination of Voting Dimensions, Determinants and Change," Ph.D. diss. Ohio State University, 1975.

50. See Marvin G. Weinbaum and Dennis R. Judd, "In Search of a Mandated Congress," *Midwest Journal of Political Science* 14 (1970): 276–302.

51. Aaron Wildavsky, *The Politics of the Budgetary Process*, 2d ed. (Boston: Little, Brown, 1974), p. 216.

52. This concept is borrowed from Anatai Etzioni, "Mixed Scanning: A 'Third' Approach to Decisionmaking," *Public Administration Review* 27 (1967): 385–92.

53. Arnold Kanter, "Congress and the Defense Budget: 1960–1970," *American Political Science Review* 66 (1972): 129–43.

54. Charles O. Jones, "Joseph G. Cannon and Howard W. Smith: An Essay on the Limits of Leadership in the House of Representatives," *The Journal of Politics* 30 (Aug. 1968): 617.

55. Ibid.

56. Polsby, "Policy Analysis." For additional commentary on policy incubation see James L. Sundquist, *Politics and Policy* (Washington, D.C.: Brookings Institution, 1968).

57. L. Blackman, "An Application of Content Analysis to the Budgetary Behavior of the Senate Armed Services Committee," M.A. Thesis, Naval Postgraduate School, 1980.

58. Frederick N. Cleaveland, "Legislating for Urban Areas: An Overview," in *Congress and Urban Problems*, ed. F.N. Cleaveland (Washington, D.C.: Brookings Institution, 1969), pp. 356–57.

59. Ibid.

60. Huntington, *Common Defense*, pp. 3–7.

61. Ripley and Franklin, *Congress, the Bureaucracy*, pp. 26–28.

62. Ibid., pp. 26–27.

63. Ibid., p. 27.

64. Ibid., pp. 27–28.

65. Ibid., p. 20.

66. The discussion in this summary relies heavily on rich conversation with Professor Nelson W. Polsby, University of California at Berkeley. It is hoped that this section does justice to the thoughts and insights he provided.

67. Barbara Hinckley, *Stability and Change in Congress* (New York: Harper and Row, 1971), p. 68.

68. Charles E. Lindblom, "The Science of Muddling Through," *Public Administration Review* 19 (1959): 79–88.

69. James G. March and Herbert A. Simon, *Organizations* (New York: Wiley, 1958), pp. 140–41.

70. Dodd and Oppenheimer, eds., *Congress Reconsidered* pp. v–vi.

71. H. Bradford Westerfield, "Congress and Closed Politics in National Security Affairs," *Orbis* (1966): 738.

72. For an excellent summary of the characteristics of the open system see Edward J. Laurance, "The Changing Role of Congress in Defense Policymaking," *Journal of Conflict Resolution* 20 (1976): 213–55.

73. Ibid., p. 245.

74. George C. Gibson, "Congressional Attitudes Toward Defense," in *American Defense Policy*, 3d ed., ed. Richard G. Head and Ervin J. Rokke (Baltimore: Johns Hopkins University Press, 1973), pp. 358–69.

75. Rodger W. Cobb and Charles D. Elder, *Participation in American Politics: The Dynamics of Agenda Building* (Boston: Allyn and Bacon, 1972), chap. 7.

76. Norman J. Ornstein and Shirley Elder, *Interest Groups, Lobbying and Policymaking* (Washington, D.C.: Congressional Quarterly Press, 1978), chap. 7.

77. Laurance, "The Changing Role," pp. 235, 245.

78. George C. Gibson, "Congressional Attitudes," pp. 359–69.

79. Aaron Wildavsky, "The Two Presidencies," *Transaction* 4 (1966).

80. See Donald A. Peppers, "The Two Presidencies: Eight Years Later," in *Perspectives on the Presidency*, ed. Aaron Wildavsky (Boston: Little, Brown, 1975), pp. 462–71.

81. Harvey G. Zeidenstein, "The Reassertion of Congressional Power: New Curbs on the President," *Political Science Quarterly* 93 (1978): 393–409.

82. This discussion of the provisions of the War Powers Act strongly relies on Zeidenstein, ibid.

83. Carroll, "Congress," p. 196.

84. Ripley and Franklin, *Congress, the Bureaucracy*, p. 201.

85. Robert Lockwood, "Congressional Budget Reform," unpublished paper, the National War College, 1980, p. 11.

86. See Albert D. Cover and David R. Mayhew, "Congressional Dynamics and the Decline of Competitive Congressional Elections," *Congress Reconsidered*, pp. 54–72; David R. Mayhew, "Congressional Elections: The Case of the Vanishing Marginals," *Polity* 6 (Spring 1974): 295–317; Walter Dean Burnham, "Insulation and Responsiveness in Congressional Elections," *Political Science Quarterly*

90 (1975): 411–35; John A. Forejohn, "On the Decline in Competition in Congressional Elections," *American Political Science Review* 71 (Mar. 1977): 1666–76; Albert D. Cover, "One Good Term Deserves Another: The Advantage of Incumbency in Congressional Elections," *American Journal of Political Science* 21 (Aug. 1977): 523–41; Warren Lee Kostrowski, "Party and Incumbency in Postwar Senate Elections," *American Political Science Review* 67 (Dec. 1973): 1213–34.

87. See Norman J. Ornstein, Robert L. Peabody, and David W. Rohde, "The Changing Senate: From the 1950s to the 1970s," and Lawrence C. Dodd and Bruce I. Oppenheimer, "The House in Transition," both in *Congress Reconsidered*, pp. 3–53.

88. Ibid.

89. Ibid.

90. See Milton C. Cummings, *Congressmen and the Electorate* (New York: Macmillan, 1966); Charles M. Tidmarch and Douglas Carpenter, "Congressmen and the Electorate, 1968 and 1972," *Journal of Politics* 40 (1978): 479–87.

91. "A Bold and Balky Congress," *Time*, 23 Jan. 1978, pp. 8–10.

92. Ralph Sanders, *The Politics of Defense Analysis* (New York: Duellan, 1973), p. 238.

93. Hilsman, *Politics of Policymaking*, p. 67.

94. William J. Keefe, *Congress and the American People* (Englewood Cliffs, N.J.: Prentice-Hall, 1980), p. 160.

95. Samuel C. Patterson, "The Semi-Sovereign Congress," in *The New American Political System*, ed. Anthony King (Washington, D.C.: American Enterprise Institute, 1979), p. 132.

96. Lawrence J. Korb, "Defense Policy of the United States," in *Defense Policies of Nations*, ed. Douglas J. Murray and Paul Viotti (Baltimore: Johns Hopkins University Press, 1981), p. 69.

97. Richard Haass, "Congressional Power: Implications for American Security Policy," *Adelphi Paper* 153 (London: International Institute for Strategic Studies, 1979), p. 34.

98. James L. Sundquist, "Congress and the President: Enemies or Partners," in *Congress Reconsidered*, p. 229.

99. Carroll, "Congress," p. 200.

100. The concept of congressional individualism is developed by Randall B. Ripley, *Power in the Senate* (New York: St. Martin's Press, 1969), p. 53.

101. Huntington, *Common Defense*, p. 145.

102. Nancy J. Bearg and Edwin A. Deagle, Jr., "Congress and the Defense Budget," *American Defense Policy*, 4th ed., pp. 335–53.

103. Dexter, "Congress and Military Policy," p. 306.

104. Ibid., p. 310.

105. Ripley and Franklin, *Congress, the Bureaucracy*, p. 2.

106. Hilsman, *Politics of Policymaking*, p. 79.

107. Lawrence C. Dodd and Richard L. Schott, *Congress and the Administrative State* (New York: Wiley, 1979), p. 10.

108. Donald E. Neuchterlein, *National Interests and Presidential Leadership: The Setting of Priorities* (Boulder, Colo.: Westview, 1978), p. 228.

109. Adam Yarmolinsky, *Paradoxes of Power: The Military Establishment in the Eighties* (Bloomington: Indiana University Press), 1983.

110. Dexter, "Congress and Military Policy," p. 312.

111. Ibid., p. 314.
112. Ibid., p. 313.
113. Ibid.
114. Yarmolinsky, *Paradoxes*, pp. 290–91.
115. Bearg and Deagle, "Congress and the Defense Budget," p. 349.
116. Francis D. Wilcox, *Congress, the Executive and Foreign Policy*, Council on Foreign Relations Policy Book (New York: Harper and Row, 1971), p. 144.

117. Yarmolinsky, *Paradoxes*, p. 292.
118. Carroll, "Congress," p. 184.
119. Hilsman, *Politics of Policymaking*, p. 69.
120. Ibid., p. 71.
121. Aspin, "The Defense Budget," pp. 324, 327.
122. Ibid., p. 333.
123. Jones, "Will Reforms Change Congress," p. 250.

APPENDIX: THE WAR POWERS RESOLUTION

SHORT TITLE

SECTION 1. This joint resolution may be cited as the "War Powers Resolution."

PURPOSE AND POLICY

SEC. 2. (a) It is the purpose of this joint resolution to fulfill the intent of the framers of the Constitution of the United States and insure that the collective judgment of both the Congress and the President will apply to the introduction of United States Armed Forces into hostilities, or into situations where imminent involvement in hostilities is clearly indicated by the circumstances, and to the continued use of such forces in hostilities or in such situations.

(b) Under article I, section 8, of the Constitution, it is specifically provided that the Congress shall have the power to make all laws necessary and proper for carrying into execution, not only its own powers but also all other powers vested by the Constitution in the Government of the United States, or in any department or officer thereof.

(c) The constitutional powers of the President as Commander-in-Chief to introduce United States Armed Forces into hostilities, or into situations where imminent involvement in hostilities is clearly indicated by the circumstances, are exercised only pursuant to (1) a declaration of war, (2) specific statutory authorization, or (3) a national emergency created by attack upon the United States, its territories or possessions, or its armed forces.

CONSULTATION

SEC. 3. The President in every possible instance shall consult with Congress before introducing United States Armed Forces into hostilities or into situations where imminent involvement in hostilities is clearly indicated by the circumstances, and after every such introduction shall consult regularly with the Congress until United States Armed Forces are no longer engaged in hostilities or have been removed from such situations.

REPORTING

SEC. 4. (a) In the absence of a declaration of war, in any case in which United States Armed Forces are introduced—

(1) into hostilities or into situations where imminent involvement in hostilities is clearly indicated by the circumstances;

(2) into the territory, airspace or waters of a foreign nation, while equipped for combat, except for deployments which relate solely to supply, replacement, repair, or training of such forces; or

(3) in numbers which substantially enlarge United States Armed Forces equipped for combat already located in a foreign nation; the President shall submit within 48 hours to the Speaker of the House of Representatives and to the President pro tempore of the Senate a report, in writing, setting forth—

(A) the circumstancs necessitating the introduction of United States Armed Forces;

(B) the constitutional and legislative authority under which such introduction took place; and

(C) the estimated scope and duration of the hostilities or involvement.

(b) The President shall provide such other information as the Congress may request in the fulfillment of its constitutional responsibilities with respect to committing the Nation to war and to the use of United States Armed Forces abroad.

(c) Whenever United States Armed Forces are introduced into hostilities or into any situation described in subsection (a) of this section, the President shall, so long as such armed forces continue to be engaged in such hostilities or

---

The full text of the War Powers Resolution, Public Law 93–148, is presented here.

situation, report to the Congress periodically on the status of such hostilities or situation as well as on the scope and duration of such hostilities or situation, but in no event shall he report to the Congress less often than once every six months.

CONGRESSIONAL ACTION

SEC. 5. (a) Each report submitted pursuant to section 4(a)(1) shall be transmitted to the Speaker of the House of Representatives and to the President pro tempore of the Senate on the same calendar day. Each report so transmitted shall be referred to the Committee on Foreign Affairs of the House of Representatives and to the Committee on Foreign Relations of the Senate for appropriate action. If, when the report is transmitted, the Congress has adjourned sine die or has adjourned for any period in excess of three calendar days, the Speaker of the House of Representatives and the President pro tempore of the Senate, if they deem it advisable (or if petitioned by at least 30 percent of the membership of their respective Houses, shall jointly request the President to convene Congress in order that it may consider the report and take appropriate action pursuant to this section.

(b) Within sixty calendar days after a report is submitted or is required to be submitted pursuant to section 4(a)(1), whichever is earlier, the President shall terminate any use of United States Armed Forces with respect to which such report was submitted (or required to be submitted), unless the Congress (1) has declared war or has enacted a specific authorization for such use of United States Armed Forces, (2) has extended by law such sixty-day period, or (3) is physically unable to meet as a result of an armed attack upon the United States. Such sixty-day period shall be extended for not more than an additional thirty days if the President determines and certifies to the Congress in writing that unavoidable military necessity respecting the safety of United States Armed Forces requires the continued use of such armed forces in the course of bringing about a prompt removal of such forces.

(c) Notwithstanding subsection (b), at any time that United States Armed Forces are engaged in hostilities outside the territory of the United States, its possessions and territories without a declaration of war or specific statutory authorization, such forces shall be removed by the President if the Congress so directs by concurrent resolution.

CONGRESSIONAL PRIORITY PROCEDURES FOR JOINT RESOLUTION OR BILL

SEC. 6. (a) Any joint resolution or bill introduced pursuant to section 5(b) at least thirty calendar days before the expiration of the sixty-day period specified in such section shall be referred to the Committee on Foreign Affairs of the House of Representatives or the Committee on Foreign Relations of the Senate, as the case may be, and such committee shall report one such joint resolution or bill, together with its recommendations, not later than twenty-four calendar days before the expiration of the sixty-day period specified in such section, unless such House shall otherwise determine by the yeas and nays.

(b) Any joint resolution or bill so reported shall become the pending business of the House in question (in the case of the Senate the time for debate shall be equally divided between the proponents and the opponents), and shall be voted on within three calendar days thereafter, unless such House shall otherwise determine by yeas and nays.

(c) Such a joint resolution or bill passed by one House shall be referred to the committee of the other House named in subsection (a) and shall be reported out not later than fourteen calendar days before the expiration of the sixty-day period specified in section 5(b). The joint resolution or bill so reported shall become the pending business of the House in question and shall be voted on within three calendar days after it has been reported, unless such House shall otherwise determine by yeas and nays.

(d) In the case of any disagreement between the two Houses of Congress with respect to a joint resolution or bill passed by both Houses, conferees shall be promptly appointed and the committee of conference shall make and file a report with respect to such resolution or bill not later than four calendar days before the expiration of the sixty-day period specified in section 5(b). In the event the conferees are unable to agree within 48 hours, they shall report back to their respective Houses in disagreement. Notwithstanding any rule in either House concerning the printing of conference reports in the Record or concerning any delay in the consideration of such reports, such report shall be acted on by both Houses not later than the expiration of such sixty-day period.

CONGRESSSIONAL PRIORITY PROCEDURES FOR CONCURRENT RESOLUTION

SEC. 7. (a) Any concurrent resolution introduced pursuant to section 5(c) shall be referred to the Committee on Foreign Affairs of the House of Representatives or the Committee on Foreign Relations of the Senate, as the case may be, and one such concurrent resolution shall be reported out by such committee together with its recommendations within fifteen calendar days, unless such House shall otherwise determine by the yeas and nays.

(b) Any concurrent resolution so reported shall become the pending business of the House in question (in the case of the Senate the time for debate shall be equally divided between the proponents and the opponents) and shall be voted on within three calendar days thereafter, unless such House shall otherwise determine by yeas and nays.

(c) Such a concurrent resolution passed by one House shall be referred to the committee of the other House named in subsection (a) and shall be reported out by such committee together with its recommendations within fifteen calendar days and shall thereupon become the pending business of such House and shall be voted upon within three calendar days, unless such House shall otherwise determine by yeas and nays.

(d) In the case of any disagreement between the two Houses of Congress with respect to a concurrent resolution passed by both Houses, conferees shall be promptly appointed and the committee of conference shall make and file a report with respect to such concurrent resolution within six calendar days after the legislation is referred to the committee of conference. Notwithstanding any rule in either House concerning the printing of conference reports in the Record or concerning any delay in the consideration of such reports, such report shall be acted on by both Houses not later than six calendar days after the conference report is filed. In the event the conferees are unable to agree within 48 hours, they shall report back to their respective Houses in disagreement.

INTERPRETATION OF JOINT RESOLUTION

SEC. 8. (a) Authority to introduce United States Armed Forces into hostilities or into situations wherein involvement in hostilities is clearly indicated by the circumstances shall not be inferred—

(1) from any provision of law (whether or not in effect before the date of the enactment of this joint resolution), including any provision contained in any appropriation Act, unless such provision specifically authorizes the introduction of United States Armed Forces into hostilities or into such situations and states that it is intended to constitute specific statutory authorization within the meaning of this joint resolution; or

(2) from any treaty heretofore or hereafter ratified unless such treaty is implemented by legislation specifically authorizing the introduction of United States Armed Forces into hostilities or into such situations and stating that it is intended to constitute specific statutory authorization within the meaning of this joint resolution.

(b) Nothing in this joint resolution shall be construed to require any further specific statutory authorization to permit members of United States Armed Forces to participate jointly with members of the armed forces of one or more foreign countries in the headquarters operations of high-level military commands which were established prior to the date of enactment of this joint resolution and pursuant to the United Nations Charter or any treaty ratified by the United States prior to such date.

(c) For purposes of this joint resolution, the term "introduction of United States Armed Forces" includes the assignment of members of such armed forces to command, coordinate, participate in the movement of, or accompany the regular or irregular military forces of any foreign country or government when such military forces are engaged, or their exists an imminent threat that such forces will become engaged, in hostilities.

(d) Nothing in this joint resolution—

(1) is intended to alter the constitutional authority of the Congress or of the President, or the provisions of existing treaties; or

(2) shall be construed as granting any authority to the President with respect to the introduction of United States Armed Forces into hostilities or into situations wherein involvement in hostilities is clearly indicated by the circumstances which authority he would not have had in the absence of this joint resolution.

SEPARABILITY CLAUSE

SEC. 9. If any provision of this joint resolution or the application thereof to any person or circumstance is held invalid, the remainder of the joint resolution and the application of such provision to any other person or circumstance shall not be affected thereby.

EFFECTIVE DATE

SEC. 10. This joint resolution shall take effect on the date of its enactment.

CHAPTER 15

FORCE PLANNING
Bridge between Doctrine and Forces

ROBERT P. HAFFA, JR.

Since the end of 1979, with its dual crises of US hostages in Iran and Soviet troops in Afghanistan, American defense policy has once again become the object of public concern, academic interest, and government effort. The reluctance to use the military as an instrument of US foreign policy lasted a decade. Effects of the American withdrawal from an unsuccessful military engagement in Vietnam included a retrenchment in Southeast Asia, a reduction in the defense budget, and a rollback in US military capabilities.

But great powers cannot so easily forswear their political responsibilities and military commitments. Thus, when US global interests became threatened in the late 1970s, the return to a military option came sooner than many expected. The president who entered office dedicated to reducing force deployments and defense spending left office emphasizing anew the military instrument and declaring an American willingness to use force in regions far from the US mainland.

In 1981, that proposed defense buildup gained both momentum and money. Though a stagnant industry in the 1970s, defense became the growth industry of the 1980s (see figure 15.1). Unsurprisingly, the direction federal funds flow tends to attract attention in America. Some line up for their fair share; others question the course being set. The debate surrounding American defense policy in the 1980s featured several contradictory pairings: quality versus quantity, attrition versus maneuver, strategists versus managers. The debates have been complex, the results inconclusive.

Throughout the early 1980s, the military buildup continued. Events in Southwest Asia contributed to a permissiveness in public opinion allowing increased defense spending for greater military capability. However, by the mid-1980s, growing budget deficits gave pause to the planning of forces and the procurement of weapon systems. These criticisms of US defense policy based on the budget deficit are not only often misplaced, they also fail to consider the fundamentals that underlie the planning of our military forces. This primer seeks to foster understanding of force planning basics by focusing on the following:

The baseline force: What are current force levels? How were they reached?

The adequacy of the force: How capable are current US forces of meeting anticipated contingencies? How can we test those capabilities?

The future of the force: If deficiencies are demonstrated, on what basis should the United States plan its forces to remedy those deficiencies?

An answer to the first question calls for a description and explanation of the basis for the planning behind existing US military forces. In a time when force planning is too often thought of in relation to some percent of the gross national product, this methodology is frequently overlooked. Explaining this methodology highlights the rational methods and prudent choices used to construct the baseline force. The central thesis of this study answers the second question. A rational framework for planning military forces based on tests of their adequacy—threat assessment, campaign analysis, and quantitative modeling—exists and has been used effectively. A realistic conclusion addresses the third question: methods used successfully in the past should not be carelessly cast aside as the United States embarks on a major military improvement program—but reemphasized as balance is restored to the nation's budget.

Although the fundamentals of force planning are admittedly incomplete, they are all too often forgotten in analyses that stress bureaucratic outcomes of defense decisionmaking and international perceptions of force capability. Today, far too many defense debates are cast in terms of dollars rather than in terms of objectives, missions, and forces. My purpose here is to argue that a more satisfactory method of understanding the baseline force and evaluating programs to improve it is to estimate the extent

Excerpted/edited from Robert Haffa, Jr., Rational Methods, Prudent Choices: Planning US Forces *(Washington, D.C.: NDU Press, 1988), with permission.*

Figure 15.1
Defense Budget Authority and Outlays, FY 1950–1992 (in billions of 1988 dollars)

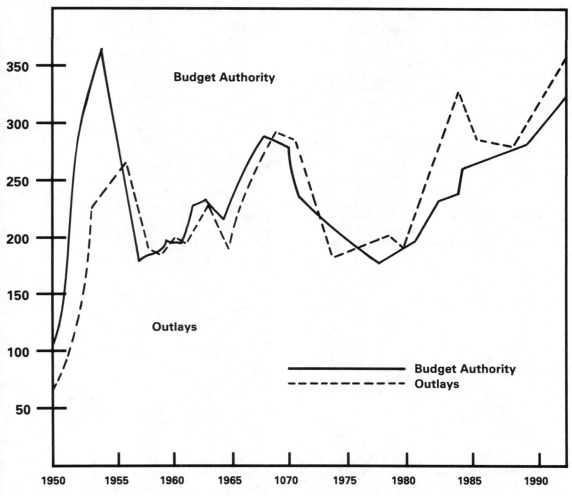

Source: Unpublished material furnished by the Office of Management and Budget. The numbers for 1987–92 reflect the president's
fiscal 1988 budget request.

to which existing and planned forces can meet national objectives and commitments. Planned increases or reductions in those forces must be related to their capability to meet those goals.

WHAT FORCE PLANNING IS

Force planning is subsumed under defense policy, which in turn acts in support of US national security policy and foreign policy. One of the best ways to distinguish between force planning and other elements of defense policy is to differentiate among policy levels. This approach is not new. Writing in *Foreign Affairs* in 1956, Paul Nitze distinguished between declaratory policy—statements of political objectives with intended psychological effects—and action or employment policy—concrete military objectives and plans employing current forces in support of those objectives.[1] Nitze also saw the requirement to match the two levels closely, lest declaratory policy appear hollow or employment capability inadequate. But that fit has never been perfect.

Nitze's concept has since been refined. Donald Snow and others have inserted a policy level between declaration and employment: force development and deployment.[2] Force planning is the development of forces flowing from the requirements of declaratory policy or the shortfalls in employment policy. Force development planning should, therefore, unite a declared strategy and the means to implement it. Snow also noted some important differences between the levels. Declaratory policy is the strategy of the elected political leaders. Therefore, it is the most political level, subjected to cosmetic if not comprehensive change during election cycles. Declaratory policy is also the level at which most academic debate is focused; it is the most public decision level. Employment or action policy, on the other hand, lies in the domain of the military. It demonstrates remarkable continuity, even in the face of political change. Often it is highly compartmentalized and protected from public view; strategic nuclear employment policy is the most secret decision level.

What of force planning? A split exists in this middle kingdom of decision levels. Those with a microperspective on force planning tend to concentrate on weapons system acquisition. Case studies abound in documentation of the difficulties of weapons system development. The majority of these analyses ultimately explain the acquisition process as a nonrational, political-military-budgetary compromise. This, of late, has been the most maligned decision level.

But the macroperspective on planning US military forces may be the most ignored decision level. The concern at this level is not with what individual weapons systems are procured but what military forces are required to meet specific contingencies. The units of analysis are not single M-1 tanks or F-111 aircraft but collective forces of strategic missiles, army divisions, or carrier battle groups. Judgments are required not only on the size and structure of the force, but also on the mix of force modernization, readiness, and sustainability. Force planning must be related to declaratory and employment policy in a rational way. This study assesses the importance of such an approach to force planning.

Given these three policy levels—declaratory, force development, employment—what would a rational force planning process look like? While there is an obvious difficulty in attempting to link these levels, an ideal process can be designed. The declaratory policy should come first: incorporating objectives formulated by political leaders enjoying popular support. The employment policy should follow: utilizing existing forces to accomplish the declared strategy. Force planning is third: developing forces in support of declared policy and designing forces to overcome shortfalls in contingency war plans. Finally a budget emerges: within given constraints, supporting the planned and programmed force.

Even dilettantes of defense policy who do little more than glance at newspaper headlines will recognize that the current force planning process does not work this way. There is not a close link between declaratory and employment policy: there is, instead, a great gap. To a certain extent, this gap will always exist. The danger is that, in a strategy-force mismatch, the choices too often become too stark: reduce US commitments abroad or expand defense budgets to meet all comers. When defense policy is reduced to such simplistic terms, force planning becomes programming. Programming ends with a budgetary battle in which arbitrary across-the-board cuts are made to fit fiscal guidance fashioned by committee. This emphasis on defense spending cannot be characterized as a rational way to plan US military forces.

There is another way. It has been applied and applied successfully. The baseline force in both strategic nuclear and general purpose forces derives from rational methods of contingency planning. When tested with these methods and models, much of the baseline force appears adequate. Planners may reach prudent choices by employing these rational methods when they consider improvements to documented deficiencies in the force structure. Programmers may also wish to return to these methods as they seek more modest defense growth in the late 1980s and 1990s.

WHAT FORCE PLANNING IS NOT

If the rational chain of the force planning process is continued, the result should be a budget.

Since 1961, the Defense Department has constructed its budget through the Planning, Programming, and Budgeting System, referred to as the PPBS. Force planning is not in the PPBS, and this study does not attempt to explain that system.[3] Indeed, Washington wags have suggested that the first "P" in PPBS is silent. They may be close to the truth. In theory, knowledge of the PPBS helps to understand the force planning process and its need for rational input.

The planning cycle is actually the longest of the series of events in the PPBS. Joint and service staff force planning starts as much as a full year before the initial development of the OSD-directed Defense Guidance. Thus there is plenty of time for planning and analysis. But a major discontinuity occurs within this process. The service and joint force planners submit a "planning force"—levels required to meet declared policy with reasonable assurance of success. Defense Guidance responds with fiscal constraints to be used by the services in preparing their programs.

Consequently, Defense Guidance ends the planning process and turns the attention of would-be defense planners to programs and budgets. The concept of moving from a minimum-risk force to a constrained force can be valuable if accompanied by a rational process to consider and make explicit the political cost of increasing risk as the dollar cost declines. But there is no meaningful coordinated joint force-planning process that makes tradeoffs within these areas. Instead, we usually find a bureaucratic battle of service-oriented programs and across-the-board budgetary cuts, revising, as Lawrence Korb writes, the would-be rational policy of moving from declared goals to forces, to funding.[4]

The actual process is too familiar. The administration decides on the size of the total federal budget and allocates some of that to defense. Further distributions apportion money required to execute the program among the services (called Total Obligational Authority, or TOA). Programming takes place based on allocation. Planning is left behind. Such an emphasis on funding focuses on "how much" and disregards the important questions of "what for, how many, and how well." Thus, once the planning cycle has been put to rest, the real business of defense takes over. The services dominate the programming cycle with their separate roles, missions, and agendas. Here decisions are made, not only on weapons programs, but on what kind and on how many forces can be acquired based on service desires and resource limits.

In the budgetary phase, the president and the Office of Management and Budget review the programs passed on by the secretary of defense. Ultimately, the Congress gets several shots in what has by now become merely a fiscal exercise. First a budget resolution establishes a ceiling on expenditures. Congress then embarks on a lengthy and multicommitteed exercise to authorize and appropriate the funds. It micromanages the remnants of the planning process.

By the time Congress has matched its target with its final budget, the planning force, or some rational version of it, has been forgotten. Although the final budget may vary considerably (higher or lower for separate line items, depending on the political and economic climate) from the president's budget, the alterations often demonstrate little consideration of those forces planned against assessments of enemy threats. Rather, these changes in expenditures are often based on constituent interests and short-term political perceptions, and often evince a tendency to reduce the "fast money" of pay, allowances, and operations and maintenance costs rather than to cut "big ticket" items. Thus programming, primarily, and budgeting, ultimately, rule the PPBS. Rational force planning, a realistic matching of national commitments and forces based on threat assessment and resource constraints, is unachievable in a system dominated by parochial interests. Fiscally constrained strategy does not have to be developed; it is already a reality.

Parenthetically, force planning is not simply systems analysis, although analysis is an important part of rational force planning. Systems analysis–bashing is considered good sport these days, principally by agencies and individuals feeling threatened by force and budgetary analyses emanating from the Office of the Secretary of Defense.[5] These laments frequently culminate in the accusation that systems analysts are closet strategists who, given any encouragement at all, will eagerly fill the void between declaratory and employment policy. Analysts have, at times, played that role, but probably more because other actors abdicate that function than because of their own ambitions. Most analysts would agree that they should not make decisions and should not set goals or assign priorities. But analysis can assist the force-planning decision process.

Analysts pursuing rational force planning as a link between declared policy objectives and their implementation will also admit to shortcomings. At times, economic graphs have been allowed to sweep over less quantitative political judgments and military expertise. Analysis has been used in pursuit of political goals as well. But analysts also will argue that their influence is less than they desire or their opponents fear.[6]

From this parenthetical discussion, a larger question emerges: what could substitute for analysis in the systematic planning of military forces? Some argue for military experience, yet force planners assert that decisions must be a judgmental blend of analysis and experience, with civilian control imperative. Others argue that analysis omits important, unquantifiable variables. Yet there are indices, imperfect but important, that can weigh the training, morale, and maneuver of opposing forces over varying terrain. The most difficult issue to reconcile is that of perception. Notions of strategic superiority or conventional parity have great political appeal but little meaning to the force planner. Static measures of comparison and fuzzy ideas of how forces are perceived by allies and adversaries are poor guides to rational force planning, and likely guarantors of unlimited budgetary claims.

PLANNING US MILITARY FORCES

Rational force planning is an analytical process designed to link declaratory and employment policy. To that end it assesses the military balance in possible contingencies, measures force capabilities in relation to requirements, and, after cross-program evaluation, establishes broad priorities for allocating resources. The task is big, but planning is essential in determining whether the US defense effort coherently supports US goals. Without planning, the larger process of formulating defense policy fragments into separate, uncoordinated, and irrational programs.

This chapter introduces the concept and process of force planning. The process is complex. Describing and explaining that process, and attendant arguments for rational relationships among decision levels, are made more difficult by the number of foreign policy commitments the United States has undertaken and by the variety of military forces that the defense of so many agreements requires. We must, therefore, examine the planning of US military forces under broad headings.[7] I have selected three: strategic nuclear forces, general-purpose forces, and rapidly deployable forces.

Although this chapter seeks to make a case for the application of analysis to force planning in a rational process, it also seeks a middle ground between contending theses and antitheses. One of the dialectics attending the defense debate is the dichotomy between experience and rationality or, as defense analyst Edward Luttwak has coined it, between "military effectiveness" and "civil efficiency." The argument against rational analytic methods asserts that, paradoxically, they do not reflect reality since there are simply too many variables and too many unknowns in the fog and friction of war. The middle ground is that both experience and reason are required. Churchill's dictum that "it is sometimes necessary to take the enemy into consideration" applies. But enemy considerations are difficult to assess. The more problematic decision is how much attention to give both analysis and experience.

Other sets of opposing forces act as *leitmotifs*

in this force planning debate. In *The Politics of the Budgetary Process*, Aaron Wildavsky called the budget the "life blood of the government." In *The Air Force Plans for Peace*, Perry Smith termed doctrine "the life blood of the service." Much of both types of blood gets spilled in annual bureaucratic combat; there has even been some practice bleeding. The flow can be stanched and transfusions obviated if service budget allocations become rational outputs of the system rather than priority-setting inputs.

A third issue is one of management style within the Defense Department, often framed in arguments for centralization or decentralization. There have been definite swings in management style—more centralized under Secretaries McNamara and Brown, more decentralized under Secretaries Laird and Weinberger. Many observers contend that the defense secretary's reliance on his staff of systems analysts to aid defense decisionmaking has removed force-planning power from the services, the commanders-in-chief (CINCs) and the Joint Chiefs of Staff. But the services have their analysts too, and there are advantages and disadvantages to each style. Some students of the issue have suggested that the management of defense is likely to depend on the political and military climate; others have posited that the structure of decisionmaking makes little difference.[8] Such contentions aside, only a strong secretary of defense can make the force-planning process work well. He ultimately makes the hard choices on force development, regardless of the managerial style used to generate options.

A fourth theme that runs through this force-planning study is the conflict between the realities of the short term versus the implications of the long term. Lord Keynes warned us that the long term ceases to compel attention when the out-years begin to exceed our life expectancy. A political corollary to Keynes's dictum contends that the long term fails to compel when it exceeds the watch of those in power. But force planners, owing to the long lead times of weapons acquisition, must reject such corollaries. The tyranny of annual cycles in formulating budgets, programs, and even strategies pressures planners to perform in the present. Force planners are also susceptible to an MBA mindset that demands a quick profit, a rapid promotion, and an immediate transfer. Planning for the year 2000 with five-year defense plans, three-year assignments and one-year budgets is unwise, if not irrational. As Lawrence Korb has noted, a future-oriented perspective becomes impossible to maintain when policy is shaped by short-term interests.[9] The Packard Commission suggested in 1986 that the United States needs to pay more attention to long-range planning and reward long-range planners for their efforts.

In sum, these themes may offer little optimism to those seeking to add rationality to a force-planning process that at times seems out of control. But if optimism is not the message in this medium, education is. An explanation of the existence, worth, and influence of a rational process in force planning can add relevance and coherence to future debates over the size, structure, and purpose of US military forces.

Although incomplete and imperfect, a rational basis for the planning of US military forces exists. An analytical foundation has been laid to size and structure the US strategic, general-purpose, and rapidly deployable forces. Over the last 25 years, despite some alteration in strategic concepts, these forces have remained relatively constant in size if not in capability. In the 1980s the United States embarked on a major effort to shore up these forces which had been designed and procured more than two decades ago. The question to be addressed here is, "Are the rational procedures that helped guide force planning in the past operative and influential today?" The answer is important, for the decisions of force planners now will affect the ability of US forces to deter and defend in the next century.

Recent emphasis on American defense policy has generated an enormous amount of literature about US military strategy, tactics, and weapons acquisition practices. The concentration here will remain on the requirements for rational force planning. In a time of relatively permissive public support for robust defense budgets, force-sizing exercises appear less compelling. In a time of reduced defense spending, a basis for rational and prudent choice is required. This chapter examines the contemporary thrust of force planning in three main categories: strategic, conventional, and limited contingency operations. In each case the conclusion offers recommendations about how the planning process can be strengthened to ensure the future effectiveness of US military forces.

STRATEGIC NUCLEAR FORCES

The policy statements, force development plans, and weapons employment decisions made in the 1960s held through most of the 1970s. Midway through the period the United States made three important choices ensuring that its strategic force posture would continue to resemble closely the triad of forces designed by McNamara.[10] The first decision was made against a comprehensive antiballistic missile (ABM) system. An agreement with the USSR in 1972 to limit the deployment of such systems appeared to enshrine the declared concept of assured destruction as one of mutual interest.

Secondly, the United States elected *not* to build new strategic forces to match the Soviet buildup in intercontinental missiles, well under way in the early 1970s. Instead, from its leading position, the United States chose to accept parity or sufficiency, to limit the total number of strategic launchers on both sides through arms control agreements, and to ensure the deterrent value of its strategic forces through the modernization of each leg of the triad. Finally, and simultaneously, the decision was made not to restrict the deployment of MIRVed systems. Thus, although the size of the US strategic force appeared to remain the same, the number of warheads that could be delivered by that force—and the number of targets covered—was significantly increased.

These decisions were not made irrationally. Rather, exchange simulations acted as the guide to and rationale for force planning. As Warner Schilling has argued, considering alert rates, weapons system survivability, and weapons yield, the 1980 US second-strike force was approximately double the retaliatory force that might have been delivered in 1964[11] The basic criteria guiding US strategic forces, that they must be capable of assured destruction and damage limitation, while appearing stable in a crisis, remained.[12] In an era of emerging strategic force parity such a policy appeared wise.

A more problematic issue attending US agreement to superpower strategic parity is how that balance has been perceived. Often the focus of the contemporary strategic debate is on the differences between the asymmetrical forces of the United States and USSR and where each "leads" in the "race." Such a comparison, accompanied by the now-familiar rising-and-falling red-and-blue bar charts, is easily absorbed by the attentive public. An acknowledgment that these static measures have not changed the deterrent balance, accompanied by a discussion of war outcomes, has been absent. Yet an emphasis on planning as opposed to perception could help restore rationality to the debate. Strategic force planners well understand that rational and fiscally realistic choices cannot be based on perception. Congress and the American public, as evidenced by the favorable votes cast on the Peacekeeper (MX) in March 1985 to strengthen the US bargaining position at the Geneva arms control talks, apparently do not.

The Peacekeeper stands as a major strategic force initiative. Proposed and accepted by both the Carter and Reagan administrations, this MIRVed ICBM reveals force-planning assumptions of strategic planners over the last 10 years. When the MX was first proposed in the mid-1970s as a modernization and replacement of 10-year-old ICBMs, its contributions emanated from its combination of throw-weight and accuracy, equating to a prompt hard-target kill capability.

Such qualifications raised again the old strategic argument about which comes first, the strategic doctrine or the means to implement it. While the MX was faithful to employment doctrines of counterforce targeting, deployed in a fixed mode it appeared to violate declaratory guidance relying on crisis stability. Defense Secretary Brown judged that a mobile MX could unite declaratory and employment policies. And, as land-based ICBMs became increasingly vulnerable, both sides would have to face the considerable costs of making them survivable. But the deployment of Peacekeeper in fixed silos is troublesome. From a force-planning perspective, such a deployment can be seen as a victory of perceived "equality" of strategic forces over crisis stability and rational planning. Where the original multiple protective shelter (MPS) scheme for basing the MX posed a significant targeting problem for the Soviets, placing the 10-warhead Peacekeeper in a fixed Minuteman silo provides a more tempting target for a Soviet first strike.

While it cannot be shown conclusively that a mobile MX deployment will greatly enhance deterrence—approximately 50 percent of US ICBM capability will remain in fixed silos, while the enormous retaliatory potential of the SLBM and bomber forces are unaltered—the rejection of mobile basing appeared to lessen the importance of crisis stability as a criterion for the planning of strategic forces. If the decision to base Peacekeeper in Minuteman silos holds, and mobile or rail-garrison basing is eschewed, it may be said that the political concept of perceived equality has triumphed over analytical measures of deterrent value.[13]

But if a *perception* of what is required for strategic parity becomes a guide to the acquisition of US strategic forces, then the analytical basis of force planning is lost. Although they may have preceded, rather than followed, the declared countervailing targeting policy of Presidential Directive 59, the strategic forces planned in the late 1970s appeared to match employment and declared doctrine while maintaining crisis stability. These forces were composed of the mobile MX, the Trident SLBM, air-launched cruise missiles attached to an aging but adequate B-52 bomber force, and a follow-on advanced technology bomber (B-2).

These substantial force improvements have been supplemented to a considerable degree by the architects of strategic planning in the 1980s. With the Peacekeeper planned to be deployed in Minuteman silos, the Scowcroft Commission called for the development of a new, smaller, mobile ICBM. Besides the follow-on "Stealth" bomber, the B-1 was resuscitated. In addition to the B-52 ALCM program, sea-launched cruise missiles supplement the ongoing Trident program and an advanced, "Stealth" air-launched cruise missile was proposed. Yet as extensive as these programs are, the perception of continued growth in Soviet nuclear arsenals

and capabilities resulted in a call for the research, development, and deployment of strategic defensive systems.

The purpose here is not to question the worth—or the costs—of these new strategic offensive and defensive programs; it is to inquire into the basis for such planning. Are the goals that were claimed by the force-planning process of the 1960s and 70s—first, assured destruction and damage limitation, then strategic parity and crisis stability—now being replaced? Should they be? Are the aims of strategic defense and strategic superiority—deliberated on and put aside by the force planners in the 60s and 70s—now being reclaimed by the planners of the 1980s? Are those goals based on perception rather than analysis? How would such forces be limited?

Answering these questions requires a return to the basic policies of strategic declaration, force development, and weapons employment. Decision levels must be more closely linked if a rational force posture is to be planned and programmed. While it can be argued that the forces were planned in the 1960s on a rational (assured-destruction) basis, a considerable gap between declared and employment policy has continually characterized the planning process.

An opportunity to reconnect the three levels of strategic force planning was lost in the 1970s. There are a number of reasons for this failure.[14] Dollar constraints in the post-Vietnam defense budget played a role. Arms control agreements were seen as being of much greater value in limiting strategic forces than has been the case. The attention given to political perceptions of the balance rather than to analyses of outcomes had done more harm than good. The influence of these factors on current strategic forces has been considerable. The general tendency was to replace the strategic force on a one-for-one basis rather than to adopt a more comprehensive approach.[15]

In addition to these externalities, it has also proved difficult to draw together these decision levels in an organizational sense. Declaratory policy, enunciated by the president and the secretary of defense, must retain the flexibility that high-level policymakers demand. And while US strategic nuclear policy at this level can be described as both consistent and evolutionary since the 1960s, presidential elections periodically interrupt the process with the review and renaming of declared doctrine. In the 80s, we have returned, once again, to a declared policy of flexible response.

Force-development policy has often been separated from declaratory policy. History suggests that, regardless of the strategy declared, the armed services are apt to argue for more forces to support it. Thus under the relatively finite strategy of massive retaliation, President

Eisenhower received a report that the Strategic Air Command was "manipulating calculations on the probability of target damage so as to provide a powerful argument for massive increases in SAC forces."[16] One of the reasons McNamara allegedly backed away from declared counterforce policy was the force requirement it generated. In 1979 the commander-in-chief of SAC informed Secretary Brown that US strategic forces could not implement the strategy embodied in PD-59 without significant increases. The services and CINCs stepped forward smartly with their "wish lists" in the early 80s. Often, the ultimate arbiters of these policy debates proved to be program analysts. But decisions on force structure based on arbitrary budget goals do little to help relate forces to strategy and are hardly a rational way to design or acquire forces.

If politics dominates the first level and economics the second, it is not until we reach the third level of employment planning that the military comes into its own. That the military should rule here seems proper, as the decisions on targets, tactics, and campaign strategies require the expertise of the professional. There is also support for the position that employment policy should be the preeminent element of nuclear policy. Rational policymaking implies that decisions on how US forces would be used should precede decisions on force acquisition and deployment.[17] But since the time Secretary McNamara sketched a baseline strategic force based on a redundant delivery capability rather than on employment doctrine, this has not been the case.

Moreover, employment planning is probably the least understood and the most autonomous of the three policy levels. The blame must be shared. Until Richard Nixon received a SIOP briefing in 1969, leading him to provide better presidential guidance on strategic force employment, planners had been required to infer employment policy from administration declaratory policy. But force capability far exceeded the requirements of a countervalue second strike. Defense Secretary Schlesinger ultimately provided explicit guidance in the form of requirements for "Limited Nuclear Options," and Jimmy Carter was probably the first president to consider seriously his role as commander-in-chief of a nuclear warfighting force.[18] On the other hand, nuclear warfighting documents, as employment plans, have been closely held by the strategic planners.

A rational basis for the planning of US strategic forces therefore calls for greater cohesion in US declaratory, development, and employment planning. President Carter's PD-59 suggested that US declaratory policy was then approaching, if not directing, the targeting plans that have included limited nuclear options and

attempted escalation control for more than a decade. As Desmond Ball has pointed out, better connectivity between declaratory and employment policy demands improvement in the existing machinery. Employment planners must pay increasing attention to policy pronouncements such as the Nuclear Weapons Employment Plan, while decisionmakers must appreciate the uncertainties that dominate the domain of the strategic targeters.[19]

A better fit between public policy and targeting doctrine can reveal the difficulties of planning strategic forces, but it cannot determine how much force is enough. Force programs in the Carter and Reagan administrations reflected differing judgments and perceptions about how the United States can best plan its forces to assure the credibility of deterrence while including a warfighting capability. Shrinking defense budgets will demand that those planning methods be revisited.

A number of approaches can help rationalize the strategic force planning process. The first step—reconnecting policy formulation and employment planning—is already under way and should be continued. The second step should be an examination of the systems themselves. Strategic weapons systems should be planned and procured according to their deterrent value in contributing to the overall US strategic posture, not with the principal concern being how such an acquisition will appear to our allies or our adversaries. History provides good evidence of the limited worth of nuclear weapons, while analytical and hypothetical force exchanges provide us with an empirical base on which to make a reasoned judgment on the size and type of these forces.

A third principle is to emphasize planning rather than budgeting. Just as the planning of strategic forces must be divorced from perceptions, so also must it be separated from the programming and budgetary cycles of the process that will tend to dominate it if left unchecked. The defense budget should be an output of the process, not an input to important strategic decisions. Just as throwing money at the problem does not guarantee that the force generated will be desirable or suitable, the paring of the defense budget by some fixed or compromised percentage in order to reach a fiscal rather than a security goal undercuts the rationally planned force. This is not to imply that the planning of strategic forces should be immune from program/budget cycles, but rather to insist that choices between a minimum-risk force and existing capabilities are made explicit in terms of reduced expenditures and increased risk.

A fourth principle to be emphasized is the need for long-range planning. When one considers that both the MX and the B-1 were proposed in the early 1970s, it becomes apparent that major force acquisition efforts *are* long-range projects. The adaptability of the Minuteman, Polaris, and B-52 forces over two decades should be kept in mind as the United States plans its forces in the 1980s. These new forces, too, will likely be required to last longer and to be deployed later than originally planned.

These four themes—coherent policy relationships, a reliance on empirical rather than data-free analysis, emphasis on planning as opposed to budgeting, and a need to look to the long term—can serve as guideposts in the planning of a strategic force that genuinely contributes to greater US security.[20]

GENERAL-PURPOSE FORCES

This planning process is similar to the three-tiered decision levels that act in the planning of strategic nuclear forces. In the conventional case, declaratory policy is composed of the defense commitments the US has made to its allies and the strategic concept that defines these interests for planning purposes. Force-development planning results from an assessment of the forces needed to support those agreements; employment planning designs war plans to execute the strategy.

The actors involved in this planning process also parallel the three nuclear policy levels. Declaratory policy continues to lie at the highest levels of the government, although here the legal and moral continuity of US commitments allows the president and his advisers less opportunity for reappraisals of US interests. The military again rules at the third level where wartime tactics are planned. While the middle ground of force planning has often been occupied by the systems analysts in OSD, their influence has not proved as great here as in the nuclear realm. Conventional force exchanges are more difficult to quantify than atomic ones, and the menu of options is considerably more complex. Moreover, with the general-purpose forces absorbing almost 80 percent of the defense budget and capturing the organizational essence of the armed services, the domestic stakes in this game are considerably higher. Little wonder that case studies in weapons-system acquisition frequently turn out to be after-the-fact accounts of bureaucratic political maneuvering.

This diversity found within the conventional force planning process is not unique to the 1980s. The methodology, however, seems to have changed. Traditionally the United States matched conventional objectives and resources through a series of analytical force-sizing exercises such as those outlined earlier. By allowing the testing of planned conventional forces against anticipated threats in given contingencies, these quantitative methods helped plan-

ners judge the appropriate size of the forces to deter and defend. As the United States began a major, general force-planning initiative in the early 1980s, some of these time-tested practices were set aside. This section inquires into the wisdom of that approach.

At first glance it seems that contemporary conventional planners are not deviating significantly from the course set by their predecessors. Efforts continue to enhance the sustainability of ground and air forces deployed forward in Europe and Asia, both through the buildup of war-readiness materiel and with the allocation of naval forces to defend the sea lanes of reinforcement. But if the practice is consistent with the past, the theory tends to diverge. In the fashioning of a new strategic concept, defense planners in the early 1980s called for the preparation of US general-purpose forces to fight a "global war" against Soviet or proxy aggression on several fronts simultaneously.[21] According to this formulation, the "mechanistic assumptions" calling for forces to be planned to fight "two-and-one-half" or "one-and-one-half" wars neglected both risks and opportunities. The risks were that US forces planned for known contingencies would be unprepared to meet unforeseen ones. The opportunities were that these forces properly equipped could engage in counteroffensive operations in areas not of the adversary's choosing—the so-called horizontal escalation concept.[22]

Although some of this rhetoric has been softened in subsequent and sequential secretary of defense *Annual Reports to Congress*, this less-structured approach to general-purpose force planning has had an effect. One organizational result has been the decentralization of the planning process. In the last few years the employment planners have been able to generate force requirements for the contingencies they conceive of as likely and important. With contingency guidance from above given less emphasis, the planning profile of the OSD systems analysts was lowered, and the planning responsibilities were passed to the armed service staffs.

If the military services have been granted a larger role in the planning process, it is useful to consider the effects of employment planning on force development. A rational model would consider an employment input valuable in deriving a prudent and practical force. But the force resulting from the process does not seem to have those properties. Seeking planning guidance from the bottom up acts to turn the traditional strategy-force mismatch on its head. Now the problem may not be insufficient forces to meet the strategy, but too many strategies. Without a meaningful process that makes enforceable joint judgments on force priorities and contingency planning, the services may well pursue separate and unrelated doctrines that enhance their organizational interests and support their desires for larger forces. Thus, the Navy is concerned primarily with attacking enemy naval forces, the Air Force with defeating the enemy in the air, the Army with the land battle and the occupation of enemy territory, and the Marine Corps with whatever is left—beachheads and islands. Such a simplistic description of service interests and implied interservice rivalries can be patently unfair if taken to the extreme. Yet it also can contain an element of truth, particularly if service strategy is driven more by competition for a share of the market than by some higher calling of grand strategy.[23]

A former member of the Joint Chiefs observed that interservice rivalry declines as defense budgets rise. But times of relatively unconstrained defense dollars have been measured in moments. It ultimately falls to the JCS to prepare contingency plans for unified action and make the hard choices between a planning force that would meet the threat with acceptable risk and the programmed force which must make prudent choices with limited resources. These choices the JCS has been reluctant to make, as documented by a series of former defense secretaries and retired chiefs of service.

Many reasons exist for this situation—and as many proposed solutions to the problem. A 1985 work chastised the JCS for being unable to rise above their individual service interests to provide cross-service counsel on effective force planning. The result, the study concluded, is "diluted advice wedded to the status quo, reactive rather than innovative."[24]

If the JCS remains unwilling or unable to make the hard choices and the systems analysts in DOD have abandoned the force-planning arena, then conventional force planning in the 1980s has been delegated to the services. But contingency planning has lost its primacy as the methodological mainstay of force planning. What are the services using in its stead? How rational are the force-planning practices in the 1980s?

Looking first to the ground forces, we might expect that the major contingency in Europe would continue to drive US Army force planning. However, the majority of force planning in the Army appears to be devoted to the creation of the "Light Division"—a unit oriented to other-than-NATO contingencies. The construction of the light division makes the traditional index of Army combat strength, the division, less meaningful. For although the number of divisions in the Army recently rose from 16 to 18, the number of personnel in that service remained capped by Congress. Thus the new light divisions are composed of about 10,000 men rather than the 18,000 or so that make up a standard division.[25]

It is too early to judge the contributions of the light division to the existing force. According to former Army Chief of Staff General Wickham, the light divisions were formed to meet a need for highly trained, rapidly deployable light forces.[26] But the manner in which these forces will be used is less clear. Certainly a smaller division, with less heavy and outsized equipment, will be easier to move. Where it will move to, and how, is another question. The light division is touted as a multi-purpose force capable of fighting in a range of contingencies: from low to high intensity, from mountains to urban areas, from quick raids to antiarmor tactics, from Europe to Southwest Asia. Perhaps this force can be used in all of these events, but past US attempts at creating such a flexible organization suggest otherwise, and unified commanders have expressed their doubts. It has proved more profitable to plan forces for specific contingencies and to train, equip and exercise them accordingly. More will be said about this approach in the following section.

The major concentration of Air Force planning has been placed on the NATO-oriented strategies known variously as "Airland Battle," "Follow-On-Forces Attack," and "Deep Attack."[27] These plans to attack and disrupt enemy reinforcements in rear echelons place a premium on the roles of counterair and interdiction rather than on close air support. The force-planning implications are for long-range, high-speed aircraft armed with precision guided munitions as opposed to close air support systems that might make up for lightened army firepower in the target-rich environment at the forward line of troops. Both missions are important. If an effort is not made at some planning level to integrate strategy and forces, the separate services will tend to plan forces for different wars, even in the same theater.

To counter this inclination, a 1984 Memorandum of Agreement between the chiefs of the Army and the Air Force framed a series of initiatives in joint force development, including the study of certain roles and missions and cross-service participation in program design. The memorandum was labeled historic—perhaps in recognition that little had been accomplished in the arena of Army–Air Force agreement since 1965.[28] But while that has been too long between drinks from the fountain of joint doctrine, and while the hard issues in the document have yet to be reconciled, the sheer existence of such an effort gives encouragement to those who plead that rational general-purpose force planning must become a joint undertaking.

Unfortunately, talk of unified doctrine and joint initiatives often is drowned out when it comes to force planning. The Navy, long an opponent of contingency-based force planning, has taken the ball of decentralized management and attempted to run with it to the score of 15 carrier battle groups and 600 ships. There is no national policy requiring a naval presence in certain oceans or dictating the number of carriers to be deployed in peacetime. If the services are allowed to plan their own campaigns and contingencies, peacetime deployments have a way of becoming justification for larger warfighting forces.[29]

Is the right number for carrier battle groups 12, 15, or 24? The answer may depend on the credibility of rational force-exchange models, the political power of the secretary of the navy, and the dollar amount that Congress decides it can afford. But a strategic issue larger than the budgetary one lies just beneath the bow wave of force modernization and acquisition. Former Under Secretary of Defense for Policy Robert Komer has argued that early budgetary commitment to the 600-ship Navy may result in fiscal shortfalls for other defense needs. Decentralization of force planning may not cause four or more incomplete strategies, but it certainly has occasioned two different approaches. Komer has identified these as the "Maritime Strategy" —a unilateral, sea-based policy that emphasizes naval forces and attacks—and "Coalition Defense"—an air-land battle linked with allies to deter and defend in Europe and Southwest Asia. The strategies need not be mutually exclusive. The concern is that they can become so if the rush to a naval buildup robs NATO-oriented air and ground forces and rapidly deployable forces of their share of the resources. Backing into a maritime strategy, Komer warns, may be the consequence of rejecting coherent contingency planning and allowing piecemeal planning by the services.[30]

Given the above appraisal of contemporary force-planning trends, the perceived need for change may depend on the colors of one's old school tie. While the Army is lightening its divisions to deploy to out-of-area contingencies, the Air Force is seeking high-tech systems to attack the enemy supply lines in East Europe. Meanwhile the Navy is preparing to sail in harm's way against enemy coastal defenses, far from the central fray. Surely some of these capabilities are desirable; they certainly stand a chance of confusing the adversary. But the nation cannot afford a force-planning process that favors individual services or irrational methods. The solution to the strategy-force mismatch, particularly under balanced-budget legislation, is not separate strategies and more forces.

Like some other remedies, the medicine prescribed to restore rationally based planning to US general-purpose forces may be as difficult to administer as to swallow. One also wonders at which end to begin the treatment. Starting at the top, it is clear that a stronger link between

policy commitments and employment planning is called for. Disregarding DoD management style, there is a continuing need to make difficult choices on the number of contingencies for which conventional forces should be prepared. Disregarding service parochialism, no substitute is available for the analysis of plausible campaigns in areas of vital interest to help develop capable and affordable forces.

As in the strategic case, there is a need for greater dialogue between the policymakers and the war planners. A 1980 attempt to subject JCS contingency plans to review by the under secretary of defense for policy might have educated the civilian leadership on resource shortfalls, while updating the combatant commanders, or CINCs, on declaratory policy.[31] That opportunity was lost with decentralization; recent reorganization legislation has attempted to regain it. But when the secretary of defense loses influence over the planning process, a fracturing and a fractioning of planning power within the Defense Department results. This fragmenting brings with it the ascendancy of service-oriented programs and budgets.

There is, of course, a legitimate role for the services and the JCS in the force-planning process. That these agencies' employment plans should be influencing force planning is proper. That the influence appears to take the form of championing service agendas rather than supporting unified war plans or joint guidance is at best inefficient. Strengthening the role of the CINCs in the planning and programming process is a positive trend that must continue. Also heartening is that the JCS and unified commanders are placing greater emphasis on analytical capabilities and long-range planning.

Such incremental steps are likely to prove inadequate in reducing the overwhelming influence of the services and incapable of adding to the joint responsibility for force planning from a national perspective. One answer to this dilemma is, of course, JCS reform, and the recent passage of the Goldwater-Nichols DoD Reorganization Act offers some hope of a strengthened joint input to force planning. But rationality cannot be legislated and, although this new emphasis on joint action is welcome— as are a number of DoD initiatives such as a two-year defense budget—the results of such reforms lie in the long term.[32]

In the near term we must return to the erstwhile conventional force planners—the systems analysts residing in OSD. Their task is to make analysis more relevant to force planning. Richard Kugler has suggested that analysts place a greater effort on the achievement of military goals, construct more dynamic force balance assessments to aid cross-program evaluation, and involve themselves deeply in military operations and strategy.[33] Such an effort would help complete the loop between general-purpose force policy, employment, and planning.

Certainly the force-planning process would be strengthened by more effective linkage between policy and war plans, by a more unified military voice, and by better analysis and long-range planning. But the serious business of general-purpose force planning requires high-level decisions on appropriate forces to deter and defend in regions vital to US interests. In the end, there appears to be no substitute for a strong secretary of defense to restore order and rationality to the conventional force planning process.

RAPIDLY DEPLOYABLE FORCES

Recent events in the Persian Gulf, including an apparently inadvertent Iraqi attack on a US warship, the "reflagging" of Kuwaiti tankers and the accidental shoot-down of an Iranian Airbus by a US vessel in the gulf, have again focused attention on rapidly deployable forces. The planning of general-purpose forces to meet a limited contingency resulting from increased hostilities or commitments in Southwest Asia or elsewhere can profit from lessons learned in the past. From an examination of those unsuccessful experiences in planning for the "half war," the requirements for the planning of a coherent limited-contingency force appear to be:

STRATEGY

If the region is judged worthy by committing US troops in its defense, planning must be devoted to designing a force capable of defending these interests. A focus on a particular region facilitates force planning by allowing forces to be sized against a specific threat, to be trained in appropriate tactics, and to be exercised within the region. Forces planned under strategies that failed to discriminate among service interests and instead called for versatile organizations were denied these advantages. Thus a strategic concept that seeks a limited-contingency capability should not be interpreted as demanding a global half-war deployment capability. While the United States must face the prospect of meeting multiple contingencies simultaneously, it must also decide which of those contingencies are most important to US interests and plan accordingly.

ORGANIZATION

It follows from the strategy that a coherent limited-contingency force must include multiservice forces under a unified command. Inclusion of each component in the limited-contingency force does not require that all units from each service be employed in every situation. The contingency force, however, must have the capability of operating flexibly in the region and sequentially employing its own forces under centralized command and control.

SUPPORT

A limited-contingency force must possess organic or dedicated air- and sealift, a program of appropriate pre-positioning, a means to gain access to needed facilities, and a power-projection capability. If the contingency is judged worthy of defending, forces assigned to the command must be guaranteed adequate strategic mobility systems, even during a major reinforcement of another region.

If these are the half-war force-planning lessons, how are they being applied in the 1980s? The strategic concept governing the planning of general-purpose forces was just discussed. It appears to contain both advantages and disadvantages for the planning of rapidly deployable forces. In the first case, the United States has explicitly recognized the goal of developing a capability for deploying forces to several contingencies simultaneously. Thus, the limited contingency seems to have been rescued from its secondary status as a lesser included case. This revision in strategic thinking was based to a significant extent on a reappraisal of the threat. If the Soviets were capable of military intervention in regions outside the more commonly conceived regions of US-USSR conflict, then the United States was forced to respond. Southwest Asia has been central to this concern after events in Iran and Afghanistan.

But realizing that limited contingencies can escalate into high-intensity US-USSR conflict is only the first strategic step. Forces now need to be planned and allocated in support of those contingencies. Here the new strategic concept seems to short-change limited-contingency operations. Claims about a global threat, the possibility of a prolonged conventional war, and the deterrent attractions of horizontal escalation do little to aid a coherent or rational force-planning process. The concentration of general-purpose force planning on Europe and the Pacific in the 1960s and 1970s led to ineffective rapid-deployment forces. And although a military buildup across the board also enhances the capabilities of rapidly deployable forces, a rational plan for which forces will fight where and against whom is also required.

Turning to the organization of rapidly deployable forces in the 1980s, it might have been expected, given the strategic concept just described, that a rapid-deployment force would be modeled in the image of Strike Command—

a go-anywhere, do-anything force. Even though it is not clear that this concept has been completely abandoned, it does not apply to the establishment in January 1983 of Central Command. USCENTCOM is composed of all four services with a single commander. The creation of the Central Command required a change in the Unified Command Plan and the assignment to USCENTCOM of an area of responsibility previously shared by the US European Command and the Pacific Command, as well as some areas previously unassigned. One principal advantage of having a single command in the region is that the countries affected are asked to deal with only one organization on most security issues.

The allocation of the Southwest Asia region to a single unified command appears as a rational geopolitical choice, but there was a bureaucratic rationale as well. The Rapid-Deployment Joint Task Force (RDJTF) had been plagued by conflicting and controversial command arrangements. The separate services were reluctant to surrender units earmarked for other areas of responsibility to a contingency task force. As a unified command, USCENTCOM's span of control is more definitive than that of the joint task forces, although control of its forces could be characterized as somewhat looser than that possessed by the RDJTF.[34]

Although the RDJTF commander was given day-to-day operational control over certain Army and Air Force units, the USCENTCOM commander has forces available for planning purposes only.[35] However, USCENTCOM has access to a reservoir of forces that could be assigned depending on the nature of the contingency. Although Central Command's control over its to-be-assigned force is not substantially different from other unified commands depending on CONUS-based reinforcements, USCENTCOM is unique in that its headquarters and its component commands are not located within the area of responsibility. Also, as in other unified commands, CINC-CENTCOM has to live with the fact that the daily operations and training allegiances of his component commands lie with their separate services. But the presence of these components clearly enhances USCENTCOM's capability for mission accomplishment and encourages interservice cooperation to a greater extent than existed under the old RDJTF.

It is somewhat surprising that force planners favoring the maneuverability of a maritime, as opposed to a coalition, strategy would move in the direction of a unified command with an assigned area of responsibility. More than an organizational or strategic rationale, it was probably the perception of the threat in the region that led to the multiservice composition of the force: the scenario envisioned to guide force

Table 15.1

Combat Forces Earmarked for the US Central Command

Service and Type	Number
Army	
Airborne division	1
Airmobile-air assault division[a]	1
Mechanized infantry division	1
Infantry division	1
Marine Corps	
Marine amphibious force[b]	1⅓
Air Force	
Tactical fighter wing[c]	7
Strategic bomber squadron (B-52)	2
Navy	
Carrier battle group	3
Surface action group	1
Maritime air patrol squadron[d]	5

[a]Although these forces are described as being initially available to USCENTCOM, that would be true only if no other major contingency had previously occurred.
[b]A Marine amphibious force typically consists of a reinforced Marine division, a support group, and a Marine air wing (containing roughly twice as many tactical aircraft as an Air Force fighter wing as well as a helicopter unit).
[c]Typically 72 fighter and attack aircraft.
[d]Usually consists of 9 P-3 long-range antisubmarine warfare aircraft.
Source: Department of Defense Annual Report, FY 1987, p. 272.

planning was increasingly Soviet-centered. The primary mission of USCENTCOM is to deter Soviet aggression and to protect US interests in Southwest Asia.[36] However, this is not the most likely case, but the worst. The Central Command is also engaged in planning for more likely lesser contingencies. Focused planning on a key region, undertaken by a well-staffed joint headquarters, may be the most significant contribution of CENTCOM to US force planning for a limited contingency. Most revealing of CENTCOM'S role will be its continuing control over US forces deployed to Southwest Asia to enhance US presence and protect freedom of navigation in the Persian Gulf. Keeping these forces under the operational control of CENTCOM would signify a significant departure from past practice. The more likely event of such forces remaining under their peacetime CINC/service control would suggest that little joint organization reform of contingency forces has occurred since the early days of STRIKE Command.[37]

In any case, the problem of force availability remains. Forces that could be assigned to USCENTCOM might also be assigned elsewhere and are therefore trained and exercised in other contingencies. Barring the creation of new and separate forces for each unified command—an unlikely event in the face of current resource constraints—steps to free forces for deployment to Southwest Asia will be scenario-dependent. The creation of Central Command therefore moves away from the unattainable organizational goal that characterized US planning for limited contingencies in the past—a central reserve intended to respond to any global contingency. But in assigning the same units to multiple tasks, the issues of force versatility and concurrent availability have not been resolved.

In addition, two recent organizational reforms may hold some promise for the deployment and employment of limited-contingency forces. Goldwater-Nichols legislation removed the prohibition against the establishment of a unified Transportation Command, and responded to the President's Blue Ribbon Commission on Defense Management which had urged the establishment of a "single unified command to integrate global air, land and sea transport."[38] The new command will absorb the deployment-planning functions of the Joint Deployment Agency, which will be phased out over a two-year period. Like the operational commands that attempted to guide limited-contingency deployments before it, the JDA never had the power to carry out its responsibilities and ultimately surrendered to the service transportation commands. This new organization, along with increased attention to deployment goals, the attention of a new CINC and the attendant budgetary participation now granted to CINCs, may bring increased coherence to the deployment of US military forces to all contingencies.

The United States Special Operations Command is the second new unified command resulting from reorganization legislation, taking the place of another outdated attempt at organizing for rapid deployment, the Readiness Command. USSOCOM's active forces include the SOF units of the 23rd Air Force, the Army's 1st Special Operations Command and the JFK Special Warfare Center, and the Navy's Special Warfare Center. Established by Congress over some service opposition, the Special Operations Command was designed to guarantee the budgetary support to special operations the Congress demanded and the services granted grudgingly. But there is danger in confusing the mission of special forces with counterterrorism, low-intensity conflict, and lesser contingencies. Special forces are structured based on the warfighting objectives of the theater commanders and are not planned to be employed primarily in very low intensity, counterterrorist operations. The new SOCOM must not be perceived as a STRIKE Command.

Finally it should be admitted that this focus on USCENTCOM and Southwest Asia should not be interpreted as the final resting place of US rapidly deployable forces. The unified command approach appears to be the best for planning the use of US general-purpose forces in regions of vital interest. But rapid-deployment forces—in small letters—still exist in the more traditional guises of the Army's Airborne and Air Assault Forces and the Marine Corps. Other units pledged to the Central Command could also be deployed to regions far from Southwest Asia. Thus, while USCENTCOM probably never will be directed to deploy its forces to contingencies outside its area of responsibility, the forces themselves are vulnerable to out-of-area assignments. In this regard, the Central Command can be seen as a case study for the organization and management of limited-contingency forces, but not as a panacea for rapid-deployment requirements.

Despite this overall favorable rating given the organizational side of planning for limited contingencies, particularly when compared with the less-capable RDJTF and JDA, the formation of organizations is far easier to accomplish than the rapid deployment of general-purpose forces. A rapid-deployment capability remains fundamental to a contingency force, particularly if the model of the contingency in the 1980s is Southwest Asia—where US forces most likely will not be predeployed. In the case of rapidly deployable forces, advances in strategic concept and organization have far outpaced the modes and monies available for strategic lift. Those planning rapid-deployment forces in the 1980s have had to be content with the classic forms of projecting power overseas: sealift, airlift, prepositioning, and access to foreign facilities.[39]

SEALIFT

Force planning for sealift was given focus by the congressionally mandated Mobility Study in 1981. On the basis of a two-war scenario, analysts concluded that one million tons of sealift capacity would meet the immediate needs of a European and Southwest Asian contingency, and would be able to sustain forces in both theaters for more than 30 days. Yet government-controlled shipping appears to be able to reach barely more than half of that goal by the early 1990s, and the prospects for commercial sealift making up the shortfall are dim.[40]

This is not to say that the issue has been avoided. The Navy has elevated strategic sealift to one of its three principal functions, now sharing top billing—if not budgetary clout—with sea control and power projection. A bureaucratic push was also provided in 1984 with the establishment of a Strategic Sealift Division in Naval Headquarters. This policy initiative has resulted in significant gains: a Ready Reserve Force of 83 ships, 13 maritime pre-positioning ships, and the purchase and conversion of eight SL-7 fast sealift ships. But the sealift force is still regarded as "marginally inadequate" by those in the know, and improvements are still required to meet the demanding goals for lift if a major conflict and a limited contingency occur simultaneously.[41] Much will depend on the continued success of pre-positioning programs in Europe and Southwest Asia, the future of the US Merchant Marine, the continued interest and largesse of Congress, and the influence of the Unified Transportation Command.

AIRLIFT

As it laid down a marker for sealift, so the Congressionally Mandated Mobility Study also established a goal for airlift—66 million ton-miles per day. And as in the case of sealift, the airlift shortfalls that remain belie the significant progress made throughout the 1980s toward that goal. For although the goal remains elusive—it will not be reached even with a full acquisition of the C-17—such aggregate measures of capability do not provide the force planner with the specifics of the contingency to be provided for and do not identify the tradeoffs between intra- and inter-theater airlift, or the demands for an airlifter capable of operating in austere low-intensity environments.

Meanwhile, procurement of C-5Bs and KC-10s, along with modifications to the C-141 and C-5A fleets, have boosted airlift capacity considerably. The planned advantages of the C-17—its great maneuverability on the ground when compared with the C-5, its direct delivery capability and its accessibility to smaller, shorter runways—make it the ideal airlifter for out-of-NATO contingencies. But its hefty price tag makes the C-17 a likely target for delay and program stretch-out in a defense budget seeking substantial near-term savings.[42]

PRE-POSITIONING

In Europe the equation is simple—unit equipment that is not pre-positioned forward will add to the above-mentioned air- and sealift requirements. In Southwest Asia the problem is more complex owing to the limited facilities available for the pre-positioning of US equipment. Nevertheless, some pre-positioning of Army and Air Force equipment is occurring, and on-going negotiations may yield a still better situation.

In the meantime, a significant amount of pre-positioned equipment continues to be maintained afloat, with 12 near-term prepositioning ships still serving the Army, Navy, and Air Force at Diego Garcia. The new maritime pre-positioning ships are there also—as well as in the Atlantic and Pacific—in support of Marine Expeditionary Brigades. The phrase in *DoD's Annual Report to Congress* that the "MPS brigades can be rapidly deployed to any trouble spot worldwide" suggests that the spirit of a "go anywhere, do anything" force once supported by the FDL still lives.[43]

ACCESS TO FOREIGN FACILITIES

Negotiations with key nations, particularly those in or near the Southwest Asia region, and arrangements to pre-position materiel, conduct training exercises, and use regional facilities in the event of a crisis have continued. The emphasis here is not on constructing new bases or raising the visibility of the United States in the region, but on improving host-nation support and infrastructure. A most notable change of attitudes has been seen within the Gulf Cooperation Council as a result of US air and naval forces escorting shipping in the Gulf. Careful management of this delicate relationship could allow not only increased access to the region but also a forward headquarters ashore for CENT-COM.[44]

This brief review of planning for limited contingencies has served only to reemphasize lessons learned in previous decades. The strategic concept that must guide limited-contingency planning remains faithful to the larger sphere of general-purpose force planning. A strategic concept too narrowly drawn can invite aggression, but too grand a concept will be forced to rely on the shadow, rather than the substance, of power. For limited-contingency forces, too, hard choices must be made about the importance and the affordability of specific capabilities. When the limited contingency focuses on a specific region and a particular adversary, forces can be designed for the threat, tailored to the area, and exercised appropriately. The command structure that has proved over time to be the most effective in leading multiservice forces is the unified command.

But whatever the strategy and organization, the support system is likely to determine the winner on the battlefield. Along with the need for greater and dedicated strategic mobility are also requirements for political, logistical, and military support from the states in the region. Therefore, the formulation of a multilateral strategy that weaves together programs of host-nation support, foreign military sales, and security assistance should accompany US unilateral force improvements and plans.

Ultimately, of course, it will not be an abstract strategic concept, a single organization, or improved mobility systems that alter, direct, or transport US rapid-deployment forces in the 1990s, but the willingness of the American government and people to support the force structure required to meet US security interests. To earn that support, force planning for a limited contingency must remain sound and rational.

Until the end of this century at least, the planning of US general-purpose forces to meet a limited contingency will play an important role in supporting commitments abroad. The strategic, organizational, and logistic experiences of the past have generally informed us how not to go about it. But steps taken in the 1980s show more promise. The challenge is to continue the effort of constructing coherent limited-contingency forces that match rational deployment strategies, organization, and support.

Reemphasis on Rational Force Planning

In summing up the requirements for planning rapidly deployable forces, the above paragraph is relevant to the conclusion of this study. Ultimately, force planning will be constrained by the willingness of the Congress and the American public to support the force structure requested by the Department of Defense and the executive branch. But the halcyon days of the early 1980s have passed, and the public mood has altered perceptibly. And while the US treasury check made out to force planning is not likely to be cancelled, it is likely to be delayed in the future owing to insufficient funds and lengthy questioning of credit references. "The same factors that led Congress to reduce the defense budget for fiscal years 1986 and 1987 are all present again this year;" writes Bruce MacLaury, "it is difficult to forecast how long this cycle will last."[45]

Such a fiscal environment argues all the more for a return to the rational force-planning approach outlined in this study. *Return*, however, is probably not the appropriate word. This work was not meant to imply that planners and analysts are not working on the problems suggested here. They are, armed with tools far more sophisticated than those outlined in this primer. It is a *reemphasis* on rational force planning that is needed.

To stop and listen for rustlings that may suggest a reemphasis of rational force-planning methods seems prudent at this point. There are many issues that demand their application. In the strategic arena the plans have been made, but the modernization of the force requires several years of continuing effort. Exchange calculations demonstrate that the greatest shortfall remains in the US capability to respond promptly against hardened Soviet ICBM silos, launch control facilities, and command and control elements. Despite its checkered history, the Peacekeeper ICBM still appears to be the lowest-cost, most effective approach to filling that shortfall. The application of rational methods should assist in the hard choices that will have to be made on Peacekeeper basing and

among competing strategic systems constrained by arms reduction agreements.

Achieving stable deterrence at the strategic level does not relieve force planners of the need for a strong conventional deterrent at the theater level. Now, particularly as we see progress toward reducing intermediate- and shorter-range nuclear forces, adequate conventional forces become even more important in deterring Soviet adventurism. Considerable progress has been made; weighty decisions remain. How should ground-force responsibilities be allocated between Central Europe and Southwest Asia? What is the correct fit between the airland battle and maritime strategies—should we be stretching our procurement of tanks to allow an earlier buy of aircraft carriers? Rational force-planning methodologies can illuminate the consequences of such choices.

In the realm of the half war, organizational efforts continue to outpace operational concepts and logistic capabilities. Targets set by the Congressionally Mandated Mobility Study—initially fiscally constrained—may have to be revisited in even less permissive budgetary climes, and tradeoffs among airlift, sealift, and pre-positioning must be continually reevaluated. Meanwhile, misguided proposals advocating the withdrawal of US troops from Europe to be absorbed into a new Rapid Deployment Force, or fashioning the new Special Operations Command into the latest version of STRIKE command can be enlightened by analysis and force-planning history.

Rational force planning has been used to establish the baseline force and to test the adequacy of the existing force. The real question is: Can these methods be reapplied and reemphasized to plan future US military forces? Our security and solvency may depend on the answer.

NOTES

1. Paul H. Nitze, "Arms, Strategy and Policy," *Foreign Affairs* (Jan. 1956): 187–98.
2. Donald M. Snow, "Levels of Strategy and American Strategic Nuclear Policy," *Air University Review*, no. 1 (1983): 63–73.
3. The intellectual roots of the PPBS are developed by Alain Enthoven and K. Wayne Smith in *How Much Is Enough?* (New York: Harper and Row, 1971). Arguments favoring and opposing the system are collected in *An Analysis and Evaluation of Public Expenditures: The PPB System* (Washington, D.C.: G.P.O., 1969). For an inside-the-building look at the Air Force process see *A Primer, the Planning, Programming and Budgetary System* (Washington, D.C.: DCS/Programs and Resources, Department of the Air Force, Jan. 1987).
4. Lawrence J. Korb, "Planning, Programming and Budgeting System" (mimeographed document, 1983).
5. Two recent examples are Eliot A. Cohen,

"Guessing Game: A Reappraisal of Systems Analysis," in Samuel P. Huntington, ed., *The Strategic Imperative* (Cambridge: Ballinger, 1982), pp. 163–92, and Stephen Rosen, "Systems Analysis and the Quest for Rational Defense," *The Public Interest*, no. 76 (1984): 3–17.

6. For an argument that the "McNamara Revolution" failed to achieve drastic improvements in DoD organizational control, see Arnold Kanter, *Defense Politics* (Chicago: University of Chicago Press, 1975).

7. The planning of tactical or theater nuclear forces will not be addressed here. While a rational basis exists for the planning and deployment of some of these systems—such as the 108 Pershing 2 and 464 ground-launched cruise missiles once planned for Europe—the major purpose of these systems has been political: the linking of NATO's conventional defense with US strategic nuclear forces. The recently signed INF treaty tends to bear this out. Also, weapons systems able to carry out these limited nuclear missions were usually "dual capable," and were therefore originally planned as general-purpose forces. See Enthoven and Smith, *How Much Is Enough?* pp. 125–32, and James A. Thompson, "Nuclear Weapons in Europe: Planning for NATO's Nuclear Deterrent in the 1980s and 1990s," *Survival* 25, no. 3 (1983): 98–109.

8. See Ralph Sanders, *The Politics of Defense Analysis* (New York: Dunellen, 1973), esp. chapter 3, and Laurence E. Lynn and Richard I. Smith, "Can the Secretary of Defense Make a Difference?" *International Security* 7 (Summer 1982): 45–69.

9. Lawrence J. Korb, "On Making the System Work," in Philip S. Kronenberg, ed., *Planning US Security* (Washington, D.C.: National Defense University, 1981), pp. 139–46. Bob Kennedy has pointed out to me that short-term interests are also important. The problem is that in choosing options to resolve short-term issues, the system fails to provide a long-term guide to evaluate those options. Only with a long-term framework can the future value of present decisions be assessed.

10. See Warner R. Schilling, "US Strategic Nuclear Concepts in the 1970s," *International Security* 6 (Fall 1981): 48–79.

11. Ibid.

12. The criterion for assured destruction also remained the same. The percentages of population and industrial damage equating to assured destruction were essentially the same in 1979 as they were in 1965. See Defense Secretary Brown's FY 1979 Posture Statement, pp. 49–55.

13. Another interpretation is that MX acquisition has been driven by the need for increased hard-target kill capability, and that it is "war-fighting" that has triumphed over "assured destruction" concepts of deterrence rather than "perceived equality."

14. For expansion on some of these points see Kevin N. Lewis, "US Strategic Force Planning: Restoring the links between Strategy and Capabilities," (Santa Monica, Calif.: RAND P-6742, 1982).

15. Ibid. Lewis notes that the original B-1 buy of 240 equated to the B-52 G and H models in the strategic offensive role. Two hundred MX with 10 RVs approximated the Minuteman force. Twenty-five Trident submarines, with higher on-station rates, came close to the number of alert tubes in the Polaris/Poseidon fleet.

16. George Kistiakowsky's report is quoted from Desmond Ball's, "Targeting for Strategic Deterrence," *Adelphi Paper* 185 (London: International Institute for Strategic Studies), p. 41.

17. Ibid. p. 37.

18. Thomas Powers, "Choosing a Strategy for World War III," *Atlantic Monthly*, Nov. 1982, p. 95. But see Scott D. Sagan, "SIOP-62: The Nuclear War Plan Briefing to President Kennedy," *International Security* 12, no. 1 (1987): 22–51.

19. Ball, "Targeting for Strategic Deterrence," p. 41.

20. For one image of what such a strategic force might look like, see Joshua Epstein, *The 1988 Defense Budget* (Washington, D.C.: Brookings Institution, 1987), pp. 15–32.

21. See Caspar W. Weinberger, *Annual Report FY 1983*, pp. 1–15.

22. For early formulations of the horizontal escalation concept see Fred Iklé, "The Reagan Defense Program: A Focus on the Strategic Imperative," *Strategic Review*, Spring (1982): 11–34, and also his "Strategic Principles of the Reagan Administration," *Strategic Review* (Fall 1983): 13–18.

23. Certainly such a thesis is supported by the staff study emanating from the Senate Armed Services Committee led by defense stalwarts Senator Nunn and former Senator Goldwater. See *Armed Forces Journal*, Oct. 1985 (extra issue), and the study itself, "Defense Organization: The Need for Change." The study formed the basis for the subsequent Goldwater-Nichols legislation.

24. The 1985 study, "Toward a More Effective Defense," was conducted by Georgetown University's Center for Strategic and International Studies. See Michael Weisskopf, "Defense Buying Systems and Command Faulted," *Washington Post*, 26 Feb. 1985, p. A7. For opposing views on JCS reform see John G. Kester, "The Role of the Joint Chiefs of Staff" in Reichart and Sturm, eds., *American Defense Policy*, 5th ed. (Baltimore: Johns Hopkins University Press, 1982), pp. 527–45, and Victor H. Krulak, *Organization for National Security* (Washington, D.C.: US Strategic Institute, 1983).

25. By 1985, the Army was planning for five light divisions. Early plans for significant manpower increases were dropped in 1984 and have not been returned to. See "Army Reported Ready to Seek a New Division," *Philadelphia Inquirer*, 24 Dec. 1983, p. 8, and Walter Andrews, "Defense Approves Army's Plan to Form 5 Light Infantry Divisions," *Washington Times*, 23 Aug. 1985, p. 5. An attendant problem is the dependence on civilian crews, particularly in the Navy, as force structure expands without comparable manpower growth. See Fred Hiatt, "Civilian Officers' Strike Immobilized Navy Ship," *Washington Post*, 7 Apr. 1985, p. A5.

26. General John A. Wickham, Jr., "White Paper on the Light Division," 16 April 1984.

27. See "Deep Attack in Defense of Central Europe: Implications for Strategy and Doctrine" in *Essays on Strategy* (Washington, D.C.: National Defense University Press, 1984), pp. 29–75.

28. "Army and Air Force Sign Agreement" OSD News Release, 22 May 1984.

29. See William W. Kaufmann, *The 1986 Defense Budget* (Washington, D.C.: Brookings Institution, 1985), p. 35.

30. Robert Komer, "Maritime Strategy vs. Coa-

lition Defense," *Foreign Affairs* 60 (Summer 1982): 1124–44. For an argument that there lies a middle ground between the two strategies see Keith A. Dunn and William D. Staudenmaier, "Strategy for Survival," *Foreign Policy* no. 52 (1983): 22–41.

31. Robert Komer, "The Neglect of Strategy," *Air Force* (Mar. 1984): 51–59.

32. A glimpse of the political magnitude of the task of DoD reorganization was revealed by its architect in Archie D. Barrett, *Reappraising Defense Organization* (Washington, D.C.: National Defense University, 1983), pp. 279–84.

33. Richard L. Kugler, "Whither Defense Analysis: Toward a New Gestalt," May 1984 (unpublished).

34. See Thomas L. McNaugher, "Balancing Soviet Power in the Persian Gulf," *The Brookings Review* (Summer 1983): 20–24.

35. Caspar W. Weinberger, *Annual Report FY 1983*, p. 11–103.

36. Caspar W. Weinberger, *Annual Report FY 1984*, p. 194.

37. See Bernard E. Trainor, "Concern Reported in U.S. Military on Gulf Command Structure," *New York Times*, 15 Aug. 1987, p. 3.

38. James A. Russell, "Deployment: Will TRANSCOM Make a Difference?" *Military Logistics Forum* 3, no. 9 (1987): 39.

39. See P. M. Dadant, *Improving U.S. Capability to Deploy Ground Forces to Southwest Asia in the 1990s* (Santa Monica, Calif.: RAND Corporation, 1983).

40. Scott C. Truver, "Sealift Manning: Critical Period, Critical Choices," *Armed Forces Journal International* 124 (July 1987): 34.

41. Interview with Vice Admiral Walter T. Piotti, Commander, Military Sealift Command, *Armed Forces Journal International* 124 (July 1987): 48–52.

42. See Jeffrey Record, *U.S. Strategic Airlift: Requirements and Capabilities*, National Security Paper no. 2 (Washington, D.C.: Institute for Foreign Policy Analysis, 1985). Preliminary results of the JCS Revised Intertheater Mobility Study (RIMS) suggest that the airlift requirements reflected in the 1981 Congressionally Mandated Mobility Study will not diminish.

43. Caspar W. Weinberger, *Annual Report to Congress FY 1988*, p. 230.

44. See Patrick E. Tyler, "Kuwait May Offer Support Facilities," *Washington Post*, 21 July 1987, p. A1, and John H. Cushman, Jr., "Weinberger Urges Bases in Gulf Area for U.S. Air Patrol," *New York Times*, 25 May 1987, p. 1.

45. MacLaury, in his introduction to Joshua Epstein's *The 1988 Defense Budget* (Washington, D.C.: Brookings Institution, 1987), p. vii.

BIBLIOGRAPHIC ESSAY

THE CHANGING FACE OF NATIONAL SECURITY POLICYMAKING

MARILYN H. GARCIA and HENRY O. JOHNSON III

The National Security Act of 1947 is the foundation of American defense policy formulation as it exists today. The postwar shift of world power to the US signaled the end of isolationism as a coherent approach to security. It became necessary to create a rational method of dealing with defense policy formulation, but the Act of 1947 was only the beginning. Although its inception marked a radical departure from the system that existed before, the process established was not invulnerable to change and reform, and neither is the system that exists today. The object of this bibliographic essay is to identify material available on all facets of defense policy formulation keeping in mind that it is, and always has been, an evolutionary process. Material concerning national security and foreign policy, as well as that aimed more specifically at defense policy, is addressed.

This essay is divided into three main sections. The first deals with the US defense policy process as a whole. The involved literature describes the process, in both its current and past forms, and covers different writers' prescriptions for better policy creation. Case studies of famous decisions and what led to them are included in this section.

The second part covers the material that focuses on the different actors in the policy process. This part is divided to cover the president, the Executive Office and its agencies, and the Congress. The section on Congress deals with the role and mechanics of the legislative body, the budgetary process, executive-Congressional relations, and the War Powers Resolution.

The final part of the essay examines the role and influence of nonofficial actors, such as interest groups and political parties, in the context of the domestic environment. It also introduces a case study on the MX/MPS-basing plan in Nevada and Utah, an instance when citizens were able to impact a major defense program.

A defense policy researcher will be assailed with a broad spectrum of material on how change has affected the policy process, and what further reform might be taken. Because the process is constantly changing, finding current, accurate material becomes an added difficulty. The intent of this essay is to provide a useful guide through the labyrinth of that literature and shorten the researcher's path to useful material.

THE DEFENSE POLICY PROCESS: AN OVERVIEW

Developing theories on defense policy formulation requires a thorough understanding of the process. A good starting point is the literature that examines the process as a whole. In *National Security Policy: The Decision Making Process* (Hamden, Conn.: The Shoe String Press, Archon Book series, 1984), Robert Pfaltzgraff, Jr., and Uri Ra'anan have compiled a group of articles which go a long way toward explaining both current and historical decision-making associated with national security policy. It begins with a look at historical patterns of decisionmaking, both in the US and abroad. The book then shifts to an examination of the various actors in the process: Congress, the executive, the media, the military-industrial complex, and other interest groups and concludes with recommendations for future national security policymakers.

Another introductory book, *The Politics of Decision Making in Defense and Foreign Affairs* (New York: Harper and Row, 1971), by Roger Hilsman, is an extremely comprehensive, explanatory discussion of the policymaking process, its actors, inputs, and outputs. It includes Hilsman's personal experiences from his service as an under secretary of state in the Kennedy administration, and insights from other prominent participants in the policymaking process as well.

In a revised edition, Amos A. Jordan, William J. Taylor, Jr., and Lawrence Korb examine the theory, mechanics, and future of US security policy. Their book is entitled *American National Security: Policy and Process* (Baltimore: Johns Hopkins University Press, 1989). Equally valuable is Charles W. Kegley, Jr., and Eugene R. Wittkopf's collection of essays found in *Perspectives on American Foreign Policy* (New York: St. Martin's Press, 1983). This book explores the important issues of post–World War II American foreign policy. The articles are informative and address the various sources that have influenced the direction of American foreign policy.

The bureaucratic politics model provides the architecture and approach taken by David C. Kozak and James M. Keagle (eds.) in *Bureaucratic Politics and National Security: Theory and Practice* (Boulder, Colo./London: Lynne Rienner, 1988). This collection of essays supports the editors' contention that "the bureaucratic policies model is perhaps the best intellectual construct available for understanding national security policymaking."

An excellent historical account of American foreign policy since the beginning of the cold war is John Spanier's *American Foreign Policy Since World War II*, 10th ed. (New York: Holt, Rinehart, and Winston, 1985). He examines the containment of Soviet power as the centerpiece of postwar US foreign policy. His account is a rich blend of facts and analysis. The latest edition includes an assessment of the first Reagan administration's handling of foreign affairs.

Barry H. Steiner's article, "Policy Organization in American Security Affairs: An Assessment" (*Public Administration Review* 37, July/Aug. [1977]), is a comprehensive examination of the organization of American national security policy. It also offers an evaluation of the system's ability to meet national security objectives. He establishes a framework, based on five uses of policy organization, with which to judge effectiveness, and then examines the evolution of security policy organization since World War II. His conclusion, based on this framework, is that despite a vast amount of change in the system, there is still room for improvement. In addition to its analytical utility, the Steiner article offers an organized, informative history of the changes in the policy organization since World War II.

A different perspective on the organization of American defense policy processes is provided in *The Management of Defense* (Baltimore: Johns Hopkins University Press, 1964). In this work, John C. Ries provides an analysis of the origin and development of centralization as a concept of unification.

A study that is useful not only as a historical examination of US doctrine but also because of its excellent analysis of decisionmaking is "American Atomic Strategy and the Hydrogen Bomb Decision," by David Allen Rosenberg (*Journal of American History* 66, no. 1 [1979]). Rosenberg traces the evolution of American nuclear doctrine from the end of World War II until the early 1950s, concentrating on the factors that led to the decision to make the hydrogen bomb.

A comprehensive review and discussion of the impact of nuclear weapons on the development of American national security policy can be found in Richard Smoke's *National Security and the Nuclear Dilemma* (New York: Random House, 1984). Smoke argues that the US is less secure today than it was several decades ago. He focuses on the evolution of the nuclear threat facing the US and how America responds to it.

More ideas on organizational problems can be found in "Organizational and Political Dimensions of the Strategic Posture: The Problems of Reform" (*Daedalus* 104, no. 3 [1975]) by John Steinbruner and Barry Carter. The authors use the acquisition process for several different weapons (including the Trident submarine and the F-111) to point out the absence of effective organization in the development of defense policy. Due to political and organizational problems, weapons are not always bought to suit real defense needs.

Robert F. Coulomb uses a similar approach to reach the same conclusion in a case study on the F-111 project. He describes the politics involved in the acquisition process and why he believes the program was a failure. In "The Importance of the Beginning: Defense Doctrine of the F-111 Fighter Bomber" (*Public Policy* 23, no. 1 [1975]), Coulomb concludes that the program was doomed from very early on by impractical decisions concerning desired aircraft capabilities. He argues that decisions were made to satisfy all involved parties, ignoring realistic engineering capabilities. The F-111 case shows vividly the weaknesses of the decision-making processes of the early 1960s, many of which still exist. Politics and rivalries entangled the process until mission was no longer a consideration. As a result, the projected Air Force–Navy "fighter of the seventies" was never bought by the Navy, and used much differently and less extensively than the Air Force had originally planned.

OFFICIAL ACTORS

The Constitution and subsequent legislation identify and prescribe the official actors in government decisionmaking. The Constitution itself, however, does not give specific guidance on the national security powers of the different branches of the federal government. Some insight into the original intent of the Framers can be gained by examination of *The Federalist Papers*. Critical insights into defense policymaking can be found in number 23 (on the importance of a strong national government), numbers 37, 47, 48, 49, and 51 (on the separation of powers as a key to stability and energy in government), number 64 (on Senatorial treaty power), numbers 74 and 75 (concerning the president's treaty-making powers and role as the nation's commander-in-chief), and in the conclusion, number 85. On that philosophical base, one can proceed to an examination of the evolution of the struggle between the separate branches and the rise of specialized agents in the formulation of policy.

James A. Nathan and James K. Oliver help bridge the gap between the Founding Fathers' perspective and the modern process of formulating and administering foreign policy. In *Foreign Policy Making and the American Political System* (Boston: Little, Brown, 1983), the authors define and illuminate the limitations of American foreign and national security policy-making. They examine barriers to the formulation and implementation of foreign policy under three broad categories. The first deals with the limits and obstacles imposed by the constitutional and institutional framework of the American political system. The second focus is on bureaucratic power and politics, and the final section of the book explores the relationship between democracy and the policy process.

A vast amount of literature explores the relationships between, and among, key actors in the American political system and in the conduct of American foreign policy. These actors are generally found in the executive or legislative branches of US government, although the Court can also be a powerful figure in the policy arena. This was demonstrated by its historic decision in *US v. Curtis-Wright Export Corporation* (1936).

THE PRESIDENCY

The presidency (comprised of the president, the Executive Office of the President, and those top-level political appointees in the executive branch who act for the president) plays a significant role in the foreign policy process. There are many good texts on the role of the presidency in foreign affairs. Classics include Edward S. Corwin's *The President: Office and Powers 1987–1948* (New York: New York University Press, 1948) and Clinton Rossiter's *The American Presidency* (New York: Harcourt, Brace, 1956). Both books provide an excellent foundation for understanding the powers, duties, and responsibilities of the chief executive. Thomas Cronin gives an excellent analysis of the constraints on presidential leadership in *The State of the Presidency* (Boston: Little, Brown, 1980). In another book, *Rethinking the Presidency* (Boston: Little, Brown, 1982), Cronin assembles a fine collection of articles by various politicians, political scientists, and journalists, in reappraising the presidency. A more recent book examining the foreign policy decisionmaking process in the White House is Robert E.

DiClerico's *The American President* (Englewood Cliffs, N.J.: Prentice-Hall, 1983). DiClerico, in this revised edition, includes an expanded treatment on the role of the presidency in foreign affairs including such factors as personality, decisionmaking style, and leadership.

In *Presidential Power* (New York: John Wiley, 1980), Richard E. Neustadt describes the president as one man wearing many hats and performing many roles, each role affecting the others. Neustadt illuminates what we know as the "rational-presidential" leadership style, without limiting the discussions to the president's role in defense. One of his conclusions is that the president has limited power to actually command, but rather a power to persuade. This 1980 edition of *Presidential Power* updates the 1960 edition of the same name.

An important work by Colin Campbell, S.J., *Managing the Presidency: Carter, Reagan and the Search for Executive Harmony* (Pittsburgh, University of Pittsburgh Press, 1987), examines the exercise of presidential power and the structure of administrative and advisory options in the White House during the Carter and Reagan presidencies. His analysis of central-agency bureaucrats is especially useful in understanding presidential policymaking.

Sam Sarkesian focuses more directly on presidential power in national security policy in his *Presidential Leadership and National Security: Style, Institution and Politics* (Boulder, Colo.: Westview Press, 1984). By compiling each section from works of other authors, Sarkesian is able to introduce a wide variety of viewpoints into his study of presidential decisionmaking in security policy. Different factors such as ideology, domestic political forces, external forces, and the president's own staff are examined in order to synthesize a successful model for the making of security policy. Presidential style is found to be one of the keys to strong, sound security policy. The impact of that style on the various factors involved is what leads to a consistent, well-thought-out policy.

R. Gordon Hoxie examines the role of the presidency in shaping national strategy in his book *Command Decision and the Presidency* (New York: Reader's Digest Press, 1977). He studies national security policy and organization with special focus on the Truman and Eisenhower years. Hoxie advocates a strategy of flexibility, deployability, and readiness. He argues for a strong president able to act at all times and to meet any threat.

In *Presidents, Bureaucrats, and Foreign Policy* (Princeton: Princeton University Press, 1972), I. M. Destler analyzes the presidential bureaucracy and recommends changes necessary to overcome what he concludes is our government's apparently irrational behavior in foreign policy. Destler examines the influence of organization on policy, in theory and in practice, in the Kennedy, Johnson, and Nixon administrations and proposes a method of government organization for coherent foreign policy. He concludes that the key to strategy is strong presidential leadership oriented toward national objectives and a broad perspective on foreign policy. His view is that leadership flows from the president, or it doesn't flow at all, and that presidential confidence is the strongest source of influence for high officials.

John Allen Williams examines constraints on presidential actions. In "Defense Policy: The Carter-Reagan Record" (*Washington Quarterly* 6, no. 4 [1983]), Williams argues that due to the realities of the system, domestic politics, and international events, the actual national security policy of Reagan does not differ as widely from that of Carter as their two ideologies and rhetorics would suggest.

In *Presidential Decisionmaking in Foreign Policy: The Effective Use of Information and Advice* (Boulder, Colo.: Westview Press, 1980), Alexander L. George provides a classic examination of presidential decisionmaking. He examines various ways in which the judgments of foreign policy decisionmakers are distorted by the misuse or nonuse of information and policy analysis. Methods of organizing and managing the policymaking process to minimize such distortions is discussed. George analyzes the different presidential styles and personality characteristics of FDR, Truman, Eisenhower, JFK, and Nixon. The book focuses on improving the quality of information, analysis, and advice available to a president and leading presidential advisers for making important foreign policy decisions.

A broad spectrum of literature on the role of the Executive Office places less importance on presidential style and more on the organization of the Executive Office and its agencies. A pioneering study of the Executive Office is *The President and the Management of National Security* (New York: Praeger, 1969), a product of the Institute for Defense Analysis, edited by Keith C. Clark and Laurence J. Legere. It thoroughly reviews decisionmaking in national security affairs in the executive branch and examines the roles of the National Security Council, the Department of Defense, the Department of State, the White House staff, and the Bureau of the Budget. This work focuses on structure and interagency relationships through a historical look at the Eisenhower, Kennedy and Johnson administrations.

The *Report to the Secretary of Defense on the National Military Command Structure* (Washington, D.C.: G.P.O., July 1978), the results of a study conducted by Richard C. Steadman at the request of President Carter, deals extensively with the National Military Command Structure (NMCS). The report first summarizes the organization of the NMCS. It explains the Unified Command Plan, defining all of the defense commands and assessing their individual strengths and weaknesses. Based on a study of ten crises beginning with the 1967 Middle East War, it makes a thorough report of US crisis-meeting capabilities. The roles of the Joint Chiefs of Staff and unified and specified commanders are examined, as are the policy, planning, and advisory functions of the NMCS. Steadman's recommendations argue for a stronger role for the Joint Chiefs and the unified and specified commanders, a theme repeated in numerous later studies.

The Philip Odeen report, *National Security Policy Integration* (Washington, D.C.: G.P.O., Sept. 1979), also based on a study requested by President Carter, examined the actors and agencies involved in the national security policy formulation, decisionmaking, and implementation of the Carter administration. Many of the problems that Odeen highlights were unique to the Carter administration because of the President's personal management style, something that every president must face. One problem was a cabinet strong in its advisory role, but weaker in its implementation of the resulting policies. This was not as serious a problem at the beginning of Carter's term, when the advisory role was especially important, but later it caused the administration to be labeled as inconsistent and lacking coherence in policy and action. The Department of Defense was shown to be, if not the cause, at least involved in many of the problems of coordination between agencies (the State Department and ACDA in particular). There are several instances of after-the-fact notification of policies and changes by the DoD which had implications in other agencies (again, most often State and ACDA) that were not consulted. Also, Odeen states that there was not enough DoD involvement by the civilians working there, and almost none by the president, the State Department, or the NSC. This resulted in an apparent lack of coherency in foreign (and sometimes domestic) policy. Odeen finishes the report with a number of recommendations for improvement in the areas of the roles and functions of the President's staff as well as interagency coordination.

Archie D. Barrett analyzes the Defense Organization Study of 1977–1980 in his book *Reappraising Defense Organization* (Washington, D.C.: National Defense University Press, 1983). Barrett examines the organizational problems of the Department of Defense and various reorganizational configurations. He concludes that limited reorganization is warranted. Richard J. Daleski reaches the same conclusion in a monograph entitled *Defense Management in the*

1980s: The Role of the Service Secretaries (National Security Affairs Monograph Series 80–8, National Defense University Press, 1980). Daleski argues that the effects of reorganization on efficiency and effectiveness are usually exaggerated. He makes the case for enhancing the role of the service secretaries in policy formulation.

A more recent effort to examine proposals for improving the organization and management of the defense establishment is the Final Report for the Center for Strategic and International Studies (CSIS) Defense Organization Project, Georgetown University, Washington, D.C. Phillip A. Odeen was chairman and led a group of nearly 70 individuals studying and debating the issues of defense organization. The final report, "Toward a More Effective Defense," was published in February 1985. The group made specific recommendations in three areas: defense planning and military advice, resource allocation and congressional oversight, and program execution.

An excellent collection of essays on defense reform in the United States is *Reorganizing America's Defense* (Washington, D.C.: Pergamon-Brassey's International Defense Publishers, 1985), edited by Robert J. Art, Vincent Davis, and Samuel P. Huntington. Another highly acclaimed work on defense reform is the Goldwater and Nunn report entitled *Defense Organization: The Need for Change*. The staff of the Committee on Armed Services prepared a comprehensive study of the organization and decisionmaking procedures of the Department of Defense. Although the vast majority of the study's 91 specific recommendatons for reforming DoD were not adopted, its analysis formed the basis for the Goldwater-Nichols Reorganization Act of 1986.

In *The President and National Security Policy* (New York: Center for the Study of the Presidency, 1984), editor R. Gordon Hoxie has gathered contributions from several key authors, including Brent Scowcroft, George Shultz, Caspar Weinberger, and John Vessey, on national security policy organization and the executive. The first two parts of the book examine issues and conditions in the world today, with a historical look at some major presidential involvements since the end of the 19th century. The third part, entitled "Organization, Reform and Strategy," addresses the role of the commander-in-chief, the NSC, national emergency powers, foreign policy coherence, national security policy considerations, domestic politics, and the War Powers Act. The articles tend to favor a strong presidential role in policy formulation. This book is especially relevant to the Reagan administration, and since many of its contributors are members of that administration, it gives an inside look at national security issues.

Lawrence Korb assesses budgetary decisions from the executive's side in "The Budget Process in the Department of Defense, 1947–77: The Strengths and Weaknesses of Three Systems" (*Public Administration Review* 37, no. 7 [1977]). He divides recent DoD budgetary history into three periods and evaluates the strengths and weaknesses of each. Korb feels that considerable improvement has been made since 1947 but that there is still room for a great deal more. The budget process will always be a difficult one, he concludes, due to its political nature and the fact that planning is not always made relevant to budgetary decisionmaking.

Arnold Kanter's *Defense Politics* (Chicago: University of Chicago Press, 1979) gives another perspective on the budget process. Kanter suggests that the process which generates the annual defense budget involves much of the interaction between the military services and the incumbent presidential administration. He attempts to discover how, and how successfully, incumbent administrations have sought convergence between their policies and the behavior of the military. The defense policies of the Eisenhower and Kennedy-Johnson administrations are evaluated from a budgetary perspective in their efforts to achieve their national security policy objectives.

A major actor in the executive branch is the secretary of defense. Laurence E. Lynn and Richard I. Smith explore the limitations of the secretary's power in "Can the Secretary of Defense Make a Difference?" (*International Security* 7, no. 1 [1982]). Beginning with the Kennedy administration, much influence over security policy and defense budgeting was shifted from the secretaries of state and treasury to the secretary of defense. Smith examines in detail the very different styles of two secretaries, McNamara and Laird, but concludes that the factor that determined how much influence each was able to exert was the amount of time each personally invested in a project.

Harold Brown, a former secretary of defense, has addressed the most pressing political and military elements of US national security policy. In his book, *Thinking about National Security* (Boulder, Colo.: Westview Press, 1983), Brown develops a framework to approach the problems of formulating and implementing national security policy. He views nuclear weapons and the growing economic interdependence of the world as key factors that have altered US relations with its allies, the Soviet Union and the Third World. Brown argues for an increased emphasis on the maintenance of alliances and alignments for the protection of US national security interests. He advocates an integrated approach of political, economic, and military strategies, to include arms control, as the basis for a sound national security policy.

Much has been written about the limitations of the Joint Chiefs of Staff and reform of that organization. Literature dealing with the decisionmaking process as a whole, such as the Steadman report, is a good source of information on the JCS. A view of the JCS as one author would like to see it is found in "An Armed Forces Staff," by William G. Hanne (*Parameters* 12, no. 3 [1982]). Hanne writes on the current role of the JCS, and why it was created with little real power. He sees its shortcomings as a result of this lack of real authority. He categorizes and explains the many reform proposals made since 1947, and then makes his own. An excellent understanding of the place of the Joint Chiefs of Staff within the American political system and their role in the policymaking process is found in Lawrence J. Korb's *The Joint Chiefs of Staff* (Bloomington: Indiana University Press, 1976).

A review of proposals to reform the Joint Chiefs of Staff is found in *Reorganization Proposals for the Joint Chiefs of Staff—1985*. (Hearings before the Investigations Subcommittee of the Committee on Armed Services, House of Representatives, 99th Cong., 1st sess., June 13, 19, and 26, 1985.)

Useful and informative insights on the making and implementation of American foreign and national security policy can be obtained from the numerous accounts written by key officials of the various administrations since World War II. Dean Acheson, a key official in the Truman administration, traces the development of the Truman Doctrine and the Marshall Plan and the American shift from isolation to collective security in his memoir, *Present at the Creation* (New York: W. W. Norton, 1969).

In the late '50s, Maxwell D. Taylor led the charge against the Eisenhower administration's strategy of massive retaliation. Taylor in *The Uncertain Trumpet* (New York: Harper and Bros., 1959) calls for a reappraisal of US military strategy and advocates a policy of flexible response.

The architect of American foreign policy in the 1970s was Henry A. Kissinger. In *The White House Years* (Boston: Little, Brown, 1979) and *Years of Upheaval* (Boston: Little, Brown, 1982), Kissinger chronicles his days as national security assistant and secretary of state in the Nixon White House. Although memoirs have a tendency to be self-serving, they still can be useful sources of scholarly information.

THE CONGRESS

Due to the Constitutional separation of powers, the role of Congress in the defense process is not easily discussed without reference to the executive. Textbooks are a good source of information on the details of congressional procedures and politics. The researcher should be aware that significant procedural changes were made in the mid-70s.

This section on Congress is divided into three parts. The first two parts, which introduce literature available on congressional rules and procedures and the budget, are not specific to national defense policy. They are intended to familiarize the researcher with the role and mechanics of Congress as it performs all of its functions. The third part covers the role and problems of Congress in the national security policy process. This last part deals extensively with executive-Congressional relations in that area, ending with the War Powers Act.

Rules and Procedures

Two books that thoroughly cover congressional rules and procedures are *Congressional Procedures and the Policy Process* (Washington, D.C.: Congressional Quarterly Press, 1978) by Walter J. Oleszek and *The Politics of Congress* (Boston: Allyn and Bacon, 1978) by David Vogler. Neither book is specific to national security policy, but each gives a good view of public policy formulation in Congress. Oleszek's book explains the path legislation travels and details the strategic way in which rules and procedures are used to affect legislation. Vogler's book looks at how congressional politics produce policy, and discusses committees, political parties, budgeting, rules and norms, pressure groups, and the president.

An in-depth discussion of congressional committees is found in Richard Fenno's *Congressmen in Committees* (Boston: Little, Brown, 1973). Although published in 1973 it is still a good work on the role and importance of committees, decisionmaking, and members' goals. Fenno researched six committees, including Foreign Affairs, to find out how committees differed and why. He found major differences in areas such as the environmental constraints faced by each committee and the goals of its members.

Randy Huwa examines two big influences on congressional decisions: political action committees and lobbyists. In his article, "The Special Interest State" (*Parameters* 12, no. 3 [1982]), Huwa concludes that there is too much concentration on single issues which leads to incremental, and often ineffective, decisionmaking.

The Budget

Congress has a major effect on defense through budget allocation. One work that sheds light on the budgetary process and bureaucratic politics is Aaron Wildasky's *The Politics of the Budgetary Process*, 4th ed. (Boston: Little, Brown, 1984). Wildasky gives a good, brief overview of the budget process in Congress as it existed in

1984. He discusses recent and proposed reforms, including the balanced budget amendment, and explains the politics of reform. Although parts of the book are slightly outdated (because they were held over from an earlier edition), it is still useful in understanding bureaucratic politics.

Allen Schick's book, *Reconciliation and the Congressional Budget Process* (Washington, D.C.: American Enterprise Institute for Public Policy Research, 1981), is short, yet thorough. It analyzes the impact of budgetary reconciliation, the process by which Congress is able to have budget changes and cuts meet targets set by the first budget resolution and the Budget Committee. Harvey C. Mansfield's *Congress Against the President* (New York: Academy of Political Science, 1975) contains an article by Schick called "Battle of the Budget" which explains the budget reforms and the strained executive-congressional relations of that time.

Security Policy

The role of Congress in national security policy has changed dramatically in the last 20 years. Many writers consider this a resurgence of congressional influence in the policy process. This has strained the relationship between the executive and the legislature. Most of the available literature on security policy at least touches on this resurgence and executive-congressional relations.

One such book is *A Responsible Congress* (New York: McGraw-Hill, 1975) in which author Alton Frye traces the emergence of Congress as a power in the formulation of security policy in the early '70s, culminating with the War Powers Resolution in 1973. He discusses what he considers to be the successful congressional inputs to security policy as well as the failures. Frye sees Congress as an essential player in the national security policy process complementing the president potentially to form better national security policy. Another positive discussion of the role of Congress in conducting foreign policy can be found in "Congress of Foreign Policy: The Nixon Years," by Edward A. Kolodziej in *Congress Against the President*, edited by Harvey C. Mansfield (New York: Academy of Political Science, 1975).

A less hopeful book is *Congress, The Presidency, and American Foreign Policy* (New York: Pergamon Press, 1981), edited by John Spanier and Joseph Nogee. The introduction, in which Spanier details congressional weaknesses, is an argument favoring a strong presidential role in policy creation. The remainder of the book is a collection of case studies that blame executive-congressional relations for blunders in many post-Vietnam national security issues. For these authors, foreign policy requires coherence, flexibility, speed, and a strong center that can only be supplied by the president.

"Congressional Power: Implications for American Security Policy" (*Adelphi Paper* 153 [London: International Institute for Strategic Studies, 1979]), by Richard Haass, addresses the Constitutional division of powers and the evolution of the legislature's input to defense policy. Haass sums up the procedural changes of the '70s and recent congressional moves to limit executive power and expand its own in the security policy process. He examines the impact on security policy, specifically in the areas of defense, foreign assistance, war powers, nuclear proliferation, military transfers, and intelligence and concludes that there will always be a healthy tension between the executive and Congress over foreign and defense policy, with each capable of being the dominant influence.

Charles W. Whalen suggests that reforms introduced to help Congress better review and modify foreign policy have actually led to an excess of bad decisions. In his book, *The House and Foreign Policy* (Chapel Hill: University of North Carolina Press, 1982), Whalen discusses congressional resurgence and reform, relations with the president, and war powers. He argues that Congress relies on amendments to exercise its foreign policy prerogative and that congressional voting is generally not informed.

A useful article on congressional involvement since the National Security Act is Edward J. Laurance's "The Changing Role of Congress in Defense Policy-Making" (*Journal of Conflict Resolution* 20, no. 2 [1976]). Laurance asserts that such significant changes have occurred that two very different systems have existed. He describes the "old" system of 1947–67, painting a picture of an executive-dominated system where Congress was neither well informed nor organized enough to be an effective participant. At that time Congress tended to support decisions made by the executive branch. The change began in 1968 with defense cutbacks initiated by the president and a public that demanded action from every congressman on controversial issues. Laurance examines the changes in four areas (inputs, outputs, the conversion process, and feedback), and summarizes what he believes are the causes for the change.

Some writers believe that the increase in congressional participation in foreign policy is overstated. One is Frank B. Feigert, author of "Congressional Response to Presidential Actions in Foreign Policy and Defense Policy: Truman to Ford," in *Interaction: Foreign Policy and Public Policy*, edited by Donald C. Piper and Ronald J. Telchak (Washington, D.C.: American Enterprise Institute for Public Policy Research, 1983). Feigert argues that Congress rarely challenges presidential leadership in foreign policy when a presidential veto is likely. He considers this inadequate for forming policy and makes suggestions for improvement.

Lee Aspin's "Defense Budget and Foreign Policy: The Role of Congress" (*Daedalus* 104, no. 3, [1975]) is a good discussion of constraints on Congress. Aspin's premise is that even with the powerful tool of the budget, Congress is reluctant to be assertive in foreign policy creation. The congressman lacks the time and information to develop a sense of expertise. The size of the Executive Office and the technical sophistication of modern weapons make it hard for Congress to be a leader in foreign policy. Aspin also discusses the impact of constituencies, interest groups, and committees on congressional decisionmaking.

Along the same vein, authors Leslie H. Gelb and Anthony Lake think Congress is not as assertive as it should be. In their article, "Congress: Politics and Bad Policy" (*Foreign Policy* no. 20, 1975), they lament what they see as a return to the "old ways," that is, a legislature approving everything the executive does. This is evidenced by congressional approval during the Mayaguez affair and other emergency executive actions. By doing this, Congress runs the risk of losing its part in the foreign policy process, erasing all the "gains" made during the Vietnam conflict. The authors feel the legislature should set national priorities.

One of the most controversial congressional acts in recent history is the War Powers Resolution, passed on the heels of the Vietnam conflict. Almost any literature that deals with executive-congressional relations will at least mention this legislation. Executive-oriented literature such as the Hoxie book mentioned earlier include discussions of presidential war powers. An examination of the act from the view of its creators can be found in a publication by the Committee on Foreign Affairs called, simply enough, *The War Powers Resolution* (Washington, D.C.: G.P.O., 1982). This publication covers the legislative history of the act and its impact, using original documents, Congressional daily records and interviews. A historical account of the battle between the president and Congress over the War Powers Resolution is in *The President's War Powers: From the Federalists to Reagan* (New York: Academy of Political Science, 1984). Demetrios Caraley has edited this volume of reprinted *Political Science Quarterly* articles addressing the many times in history when there have been disagreements over the president's constitutional authority as commander-in-chief. Another favorable article is Stephen L. Carter's "The Constitutionality of the War Powers Resolution" (*Virginia Law Review* 70, no. 1 [1984]). Law journals are a good source of opinion and legal analysis, not only for the War Powers Act but for other questions on separation of powers as well.

NONOFFICIAL ACTORS

Not all influences on security policy can be assigned to the Office of the President or Congress. Many persons and conditions are involved, including interest groups, political parties, and public opinion, that can change or constrain final outputs.

One such constraint, according to Albert Wohlstetter, is the chaos caused by misperception, a usual participant in any human interaction. In "Optimal Ways to Confuse Ourselves" (*Foreign Policy*, no. 20 [1975]), Wohlstetter uses the strategic arms race and its deceptive language to point out some sources of the confusion that makes policy planning difficult.

Robert Mandell looks at one way of dealing with misperceptions policymakers have—political gaming. His article, "Political Gaming and Foreign Policy Making" (*World Politics* 29, no. 4 [1977]) evaluates gaming as an antidote to distortion for decisionmakers.

Interaction: Foreign Policy and Public Policy (Washington, D.C.: American Enterprise Institute for Public Policy Research, 1983) by Ronald J. Telchack and Donald C. Piper contains two useful sections on outside influences on policymaking. "Foreign-Domestic Policy Linkages: Constraints and Challenges to the Foreign Policy Process" by Telchack focuses on interest group activity. The author sees interest groups as a growing power and warns that the strength of individual groups could become more important than international events or administrations in determining policy. "Public Policy and Foreign Policy: A Pragmatic View" by Norman Grabner deals with public opinion and the important role that government exercises in shaping public opinion.

Barry B. Hughes's *Domestic Content of American Foreign Policy* (San Francisco: W. H. Freeman, 1978) addresses the domestic environment created by public opinion, interest groups, and political parties, and its impact on foreign policy. Hughes finds little influence exerted by the public attitude and attributes this to the minimal control of the electorate. The impact of political parties is mixed: they have less influence on the president and the general public than on Congress and high elected officials. Interest groups have real influence, but it is generally overstated.

One case study showing the effect that states and citizens can have on defense policy is "The Use of NEPA in Defense Policy Politics: Public and State Involvement" (*The Social Science Journal* 21, no. 3 [1984]). In it, Lauren Holland examines the influence Nevada and Utah residents had on plans for MX/MPS deployment. Although Holland is not asserting that the citizens' actions were solely responsible for defeating deployment, he examines extensively the manner in which they were able to infiltrate the policy process. Conditions were right in this case, and the NEPA was available to provide a direct, formal opportunity for participation.

Samuel P. Huntington in his book *The Soldier and the State* (Cambridge: Harvard University Press, 1957) presents a theory of civilian-military relations, a totally different approach to the study of national security policymaking.

CONCLUSION

This bibliographic essay is by no means an exhaustive examination of the sources available to the student of national security policymaking. The effort has been directed at identifying useful and informative literature that will facilitate further research into the defense policy process. Much of the literature on national security policymaking focuses on the presidency, Congress, and bureaucracy. More empirical analysis and research is needed on how institutions and staff have actually functioned both within and across different administrations. There is a tendency to focus on the individual actors at the expense of the policy process as a series of inputs, political decisions, and outputs that are implemented and continually monitored.

FORCE PLANNING
GREGORY J. RATTRAY

Force planning encompasses a vast range of widely discussed subjects. His work falling somewhere between declaratory policy and employment doctrine, the force planner must consider factors ranging from alliance politics to the combat effectiveness of the individual rifleman. Force planners operate in a political environment of limited defense budgets and overseas commitments which constrain their possible choices. The very nature of our pluralistic political system limits the rationality of the force planning process. At the other end of the spectrum, the force planner must concern himself with the minutia of weapons systems, both ours and our potential adversaries'. Given the complexities of this field, any bibliographical list can only scratch the surface of and open doors to material relevant to force planning. This essay focuses on works conducted according to rational planning models, the methodologies involved, and the data needed to conduct such analyses.

GENERAL WORKS

Serious efforts at analytical force planning really started in the early 1960s under Secretary of Defense Robert McNamara, after the earlier attempt of Paul Nitze in the guise of NSC-68 failed. The first major work to directly embrace the analytical approach to force planning, Alain Enthoven and K. Wayne Smith's *How Much Is Enough? Shaping the Defense Program, 1961–1969* (New York: Harper and Row, 1971), examines these early efforts. Other works which deal with the political environment of these early force planning efforts are Sanuel P. Huntington's *The Common Defense: Strategic Programs in National Security* (New York: Columbia University Press, 1961); Warner Schilling, Paul Hammond and Glen Snyder's *Strategy, Politics, and Defense Budgets* (New York: Columbia University Press, 1962), and William W. Kaufmann's *The McNamara Era* (Princeton: Princeton University Press, 1962). Stephen Rosen provides a historical overview of how rational planning has fit into our approach to national security in his article "Systems Analysis and the Quest for Rational Defense" (*Public Interest*, no. 76 [1984]).

Henry C. Bartlett's article, "Approaches to Force Planning" (*Naval War College Review*, Spring [1983], pp 37–48), outlines a number of different approaches to rational force planning. Bartlett demonstrates how focusing on different aspects of the overall problem such as the threat, scenario, or mission can lead to very different conclusions about the proper structure for US forces. Desmond Ball has taken an updated look at how declaratory policy and national strategy affect force planning in *Politics and Force Levels* (Berkeley and Los Angeles: University of California Press, 1980). A RAND study by Glenn A. Kent, "Concepts of Operations: A More Coherent Framework for Defense Planning" (RAND N-2026-AF, Aug. 1983), points out the dichotomy that often exists between the goals of force structure planners and those responsible for employment doctrine. Two other recent works have leveled cogent criticisms at the US force planning process: Edward L. Luttwak's *The Pentagon and the Art of War* (New York: Simon and Schuster, 1984) and John M. Collins' *U.S. Defense Planning: A Critique* (Boulder, Colo.: Westview Press, 1982). Both works focus on how problems with the military bureaucracy, especially the officer corps, limit the effective planning and employment of US armed forces.

Under the auspices of the Brookings Institution, William Kaufmann has continued to examine the US force planning process. In *Planning Conventional Forces 1950–1980* (Washington, D.C.: Brookings Institution, 1982), he reviews how the US has planned and deployed forces to meet its declared commitments and objectives. Kaufmann projects his view of US force planning requirements into the future with *Defense in the 1980's* (Washington, D.C.: Brookings Institution, 1981). He also produces a yearly review of the defense budget evaluated in force planning terms. Other general Brookings Studies in Defense Policy are Barry M. Blechman's *The Soviet Military Buildup and U.S. Military Spending* (Washington, D.C.: Brookings Instituion, 1977) and Richard K. Betts's *Surprise Attack: Lessons for Defense Planning* (Washington, D.C.: Brookings Institution, 1982).

The US government also publishes much background material related to force planning The secretary of defense publishes a comprehensive *Annual Report to the Congress* (Washington, D.C.: Department of Defense) in support of his defense program. Similarly, the Joint Chiefs of Staff publish annually the *United States Military Posture* (Washington, D.C.: G.P.O.) outlining US military requirements and strengths. The Congressional Budget Office (CBO) prepared a study entitled *Resources for Defense: A Review of Key Issues for Fiscal Years 1982–1986* (Washington, D.C.: CBO, 1981) outlining the fiscal considerations for force planning for the first half of this decade.

METHODOLOGY

Available literature can only provide a general introduction to the analytic approach to force planning. These analytic models have not been fully developed nor their intricate details explained. For those who wish to delve more deeply, several important sources can enhance their understanding of the assumptions and applicability of force planning models.

Strategic force planners base their estimates on exchange models requiring knowledge of targets and capabilities of nuclear weapons systems. Much of this material is classified, although the physical effects of nuclear detonations have been well documented. The DoD publication edited by Samuel Glasstone and Phillip J. Dolon, *The Effects of Nuclear Weapons* (Washington, D.C.: G.P.O., 1977), is the handbook in this area. A good quantitative analysis of the targeting problem can be found in Donald M. Snow's *The Nuclear Future* (Tuscaloosa, Ala.: University of Alabama Press, 1983). The issues of kill probabilities, fratricide, and other complex calculations important to strategic force planning are discussed by Lynn Davis and Warner Schilling in "All You Ever Wanted to Know About MIRV and ICBM Calculations but Were Not Afraid to Ask," *Journal of Conflict Resolution* 17, no. 2 [1973]) and by John D. Steinbrunner and Thomas M. Garwin in "Strategic Vulnerability: The Balance Between Prudence and Paranoia" (*International Security*, 1, no. 1 [1976] 138–181).

Planners of general-purpose forces have access to much more unclassified literature. Lanchester equations provide the basis for most general-purpose force planning, and James C. Taylor fully explains the development and relevance of these equations in his two volumes of *Lanchester Models of Warfare* (Monterey, Calif.: Naval Postgraduate School, 1983). The US Army Concept Analysis Agency publishes *Weapon Effectiveness Index/Weighted Unit Values* (WIE/WUV) (Washington, D.C.: G.P.O., 1974) which provides the force planner with a comprehensive guide to the effectiveness of ground force units and weapons systems, and how these values are derived. William Kaufmann provides an excellent step-by-step guide to applying general-purpose force planning models to the Central Front in Europe in the appendix of John D. Steinbrunner and Leon V. Sigal's *Alliance Security: NATO and the No First Use Question* (Washington, D.C.: Brookings Institution, 1983). Similarly, Robert L. Fisher lays out the quantitative factors involved in evaluating ground force contingencies in "Defending the Central Front: The Balance of Forces" (Adelphi Paper 127, London: IISS, Autumn, 1976).

Joshua Epstein develops a more detailed force model in *Measuring Military Power: The Soviet Air Threat to Europe* (Princeton: Princeton University Press, 1984) which provides an excellent example of how to measure military power in terms of dynamic performance rather than static comparison of numbers of weapons systems on each side. In *The Calculus of Modern War: Dynamic Analysis without Lanchester Equations*, Epstein develops his model further, contending that Lanchester fails to capture warfare's basic dynamics. He presents a set of highly mathematical equations to be used in a dynamic analysis of opposing forces.

Many have criticized the quantitative approach to force planning. The Comptroller General's Report to Congress, *Models, Data and War: A Critique of the Foundation for Defense*

Analysis, contains good explanations of the models used by US force planners as well as a critique of those models. Col. Trevor N. Dupuy has been a consistent critic of the quantitative approach to modeling combat in his two works, *The Evolution of Weapons and Warfare* (Indianapolis: Bobbs-Merrill, 1980) and *Numbers, Predictions and War: Using History to Evaluate Combat Factors and Predict the Outcomes of Battles* (Indianapolis: Bobbs-Merrill, 1979). The RAND Corporation furnishes another critical look in J. A. Storchfitsch's "Models, Data and War: A Critique of the Study of Conventional Forces" (Santa Monica, Calif.: R-1526-PR, March 1976).

Yu. V. Chuyev and Yu. B. Mikhaylov's *Forecasting in Military Affairs: A Soviet View* (Washington, D.C.: G.P.O., 1980) presents an interesting account of the Soviet perspective on force planning.

MEASURING CAPABILITIES

The force planner needs data to use the analytic models. These data include information not only on the weapons systems and unit effectiveness of all four services of the US but also those of our likely enemies. The Army's WEI/WUV approach evaluates weapons systems' overall effectiveness in terms of the power of the individual rifleman.

Several general sources provide accurate information about the military strength and capabilities of the world's nations. The International Institute for Strategic Studies publishes *The Military Balance* (London: IISS) yearly, a widely used guide to the world's military forces. Another good survey work is Trevor N. Dupuy, Grace P. Hayes, and John A. C. Andrew's *Almanac of World Military Power* (San Rafael, Calif.: Presidio Press, 1980). John Keegan's *World Armies* (New York: Facts on File, 1979) provides useful information on the organization, history, commitments, and deployments of each nation's ground forces. Jane's Publishing Company each year puts out very comprehensive, detailed listings of the combat weapons systems of all nations, including *Jane's Fighting Ships* and *All the World's Aircraft*.

Planners can easily gain access to the necessary information on US forces. The Department of Defense, Congressional Budget Office and other government agencies publish many documents presenting the numbers and capabilities of our armed forces in great detail. The most important of these documents have been mentioned earlier. For unclassified information on nuclear weapons, Thomas B. Cochran, William M. Arkin and Milton M. Hoenig's *Nuclear Weapons Databook, Volume 1: U.S. Nuclear Forces and Capabilities* (Cambridge, Mass.: Ballinger, 1984) yields useful information.

The issues of the size and composition of the armed forces of primary concern to the US, those of the Soviet Union and its allies are crucial to US force planners attempting to evaluate the sufficiency of and future requirements for US forces. US forces are often compared to those of the Soviet Union. For good comparisons of the US and Soviet forces, one can turn to John M. Collins' *U.S.-Soviet Military Balance, 1980–1985* (New York: Pergamon-Brassey's International Defense Publishers, 1985) and *NATO and the Warsaw Pact Force Comparisons* (Brussels: NATO Information Service, 1982). Much of the information produced by the Central Intelligence Agency (CIA) and the Defense Intelligence Agency (DIA) is classified. The Department of Defense, however, has over the past few years published a yearly evaluation of Soviet forces called *Soviet Military Power* (Washington, D.C.: DoD) based on DIA estimates. Force planners do have a large amount of information on Soviet forces available to them, and independent analyses of the Soviet armed forces and the US-Soviet military balance should be encouraged.

A number of people have produced studies on the Soviet military. Harriet F. and William F. Scott's *The Armed Forces of the USSR* provides an excellent overview of the development and organization of the Soviet military effort. Other important works detailing the size and composition of Soviet and Warsaw Pact forces include: John Erikson's *Soviet Military Power* (London: Royal United Services Institute for Defense Studies, 1971); Friedrich Wiener and William J. Lewis's *The Warsaw Pact Armies* (Vienna: Carl Ueberreuter, 1977); David C. Isby's *Weapons and Tactics of the Soviet Army* (London: Jane's Publishing Company, 1981), and *Soviet Aerospace Handbook* (Washington, D.C.: HQ USAF, 1978). The Brookings Studies in Defense Policy also have included a series of works on the Soviets composed of Jeffery Record's *Sizing up the Soviet Army* (Washington, D.C.: Brookings Institution, 1975), Robert P. Berman's *Soviet Air Power in Transition* (Washington, D.C.: Brookings Institution, 1978), and, Barry M. Blechman's *The Changing Soviet Navy* (Washington, D.C.: Brookings Instituioin, 1973). An often revealing criticism of recent studies by DoD and others of the Soviet military capabilities can be found in Andrew Cockburn's *The Threat: Inside the Soviet Military Machine* (New York: Random House, 1983).

PLANNING STRATEGIC FORCES

While much has been written about nuclear weapons strategies, systems, employment, and morality, the available unclassified literature concerning the planning of these forces is limited.

The administration of President Reagan and

Secretary of Defense Weinberger has emphasized both arms control and strategic nuclear strength. This emphasis has spurred major programs to enhance all three legs of the US strategic triad as well as a strategic defense initiative. The US needs to take a rational, analytical approach to the planning of these new forces. David A. Rosenberg provides a historical review of strategic force planning in "The Origins of Overkill" (*International Security*, Spring [1983]: 3–71, 140–62). Studies of the current US strategic force planning situation include: *Challenges for U.S. National Security, Defense Spending and the Economy: The Strategic Balance and Strategic Arms Control* (Washington, D.C.: Carnegie Endowment for International Peace, 1981); Desmond Ball, "The Future of the Strategic Balance" (Strategic and Defense Studies Center, Australian National University, Nov. 1980), Jeffery Richelson, "PD-59, NSDD-13 and the Reagan Modernization Program" *Journal of Strategic Studies* 6, no. 2 [1983]: 125–46), and Kevin N. Lewis, "U.S. Strategic Force Planning: Restoring Links Between Strategy and Capabilities" (Santa Monica, Calif.: RAND P-6742, Jan. 1982). Lewis's work in particular stresses the lack of coordination between force structure planning and employment doctrine. He believes that the US must take a coherent long-range approach to force planning based on employment themes. The Brookings Institution has produced two relevant works along these lines, Alton H. Quanbeck and Archie L. Wood's *Modernizing the Strategic Bomber Force* (Washington, D.C.: Brookings Institution, 1976) and Robert P. Berman and John C. Baker's *Soviet Strategic Forces: Requirements and Responses* (Washington, D.C.: Brookings Institution, 1982). In a recently published speech, Paul H. Nitze looks to the future of these forces in the light of the Strategic Defense Initiative in "U.S. Strategic Force Structure: The Challenges Ahead," US Department of State, Current Policy no. 794 (Washington, D.C.: Bureau of Public Affairs, Feb. 1986).

Force planners in general have ignored the subject of military space systems. Two exceptions, offering tentative efforts in this crucial new area, are worth noting: Neil E. Lamping and Richard P. MacLead's "Space—A National Security Dilemma—Key Years of Decision," Unpublished Report (Washington, D.C.: National Defense University, July 1979) and Robert B. Giffen's *U.S. Space System Survivability: Strategic Alternatives for the 1990's* (Washington, D.C.: National Defense University, 1982).

PLANNING GENERAL-PURPOSE FORCES

Much more analytical work has been done on the planning of these forces. The Congressional Budget Office publishes an official guide in the report "Planning General Purpose Forces: Overview (Washington, D.C.: CBO, Jan. 1976). Other works which take a comprehensive view of general-purpose forces include: *Challenges for U.S. National Security: Defense Spending and the Economy: Assessing the Balance Defense Spending and Conventional Forces* , two reports by the Carnegie Endowment for International Peace (Washington, D.C., 1981); John Glenn, Barry E. Carter and Robert W. Komer's *Rethinking Defense and Conventional Forces* (Washington, D.C.: Center for National Policy, 1983), and Seymour J. Deitchman's *New Technology and Military Power: General Purpose Military Forces for the 1980's and Beyond* (Boulder, Colo.: Westview Press, 1979).

Turning to the individual services, President Reagan and Secretary of the Navy John Lehman have emphasized the Navy and its 600-ship program. Force planners have difficulty analyzing the Navy in terms of specific contingencies. As a result, most analytical studies of our Navy's requirements and capabilities consider the entire service. The CBO published a guide in "Planning General Purpose Forces: The Navy" (Washington, D.C.: 1976). Paul Nitze and others have published an excellent study entitled *Securing the Seas: The Soviet Naval Challenge and Western Alliance Options* (Boulder, Colo.: Westview Press, 1980). The Marine Corps was the subject of the Brookings study by Martin Binkin and Jeffrey Record, *Where Does the Marine Corps Go From Here?* (Washington, D.C.: Brookings Institution, 1976).

The Congressional Budget Office has done a number of reports dealing with the Navy: "U.S. Naval Force Alternatives," Staff Working Paper (Washington, D.C.: CBO, Mar. 1976); "Shaping the General Purpose Navy of the 1980's: Issues for FY 1981–1986" (Washington, D.C.: CBO, Jan. 1981); "U.S. Maritime Strategy and Naval Shipbuilding Program" (Washington, D.C.: CBO, Aug. 1976); "Naval Surface Combatants for the 1990's: Possibilities and Prospects" (Washington, D.C.: CBO, Apr. 1981); "Costs of Expanding and Modernizing the Navy's Carrier Based Forces" (Washington, D.C.: CBO, May 1982), and, "U.S. Naval Forces: The Peacetime Presence Mission" (Washington, D.C.: CBO, 1978). The CBO has also published a series of studies on the 600-ship initiative, specifically: "Building a 600-ship Navy: Costs, Timing and Alternative Approaches" (Washington, D.C.: CBO, Mar. 1982); "Future Budget Requirements for the 600-ship Navy" (Washington, D.C.: CBO, Sept. 1985); and, "Manpower for a 600-ship Navy: Costs and Policy Alternatives" (Washington, D.C.: CBO, Aug. 1983).

Several general planning studies have focused on the tactical air forces. The CBO works in this area include "Planning General Purpose Forces: The Tactical Air Forces" (Washington, D.C.:

CBO, July 1977) and "U.S. Tactical Air Forces: Overview and Alternative Forces, 1976–1981" (Washington, D.C.: CBO, Apr. 1976). William D. White has produced another study, *U.S. Tactical Air Power* (Washington, D.C.: Brookings Institution, 1974). The recent article by Benjamin S. Lambeth, "Pitfalls in Force Planning: Structuring America's Air Arm" in *International Security* (Fall [1985]: 84–120), does an excellent job of detailing the factors besides sheer numbers which determine the effectiveness of tactical air forces and should properly be considered in planning them.

Force planners usually consider the Army's capabilities in terms of geographical contingencies such as Central Europe and the Persian Gulf (which will be covered in the next section of the essay). The CBO, however, provides general guides in "Planning General Purpose Forces: Army Procurement Issues" (Washington, D.C.: CBO, 1976) and in another study by Patrick Hillier and Norma Slatkin, *U.S. Ground Forces: Design and Cost Alternatives for NATO and Non-NATO Contingencies* (Washington, D.C.: CBO, 1980). Further, the Army's light division is the subject of General John A. Wickam's "White Paper on the Light Division" (Washington, D.C.: Department of the Army, 16 April 1984).

The Brookings Institution also provides two valuable force planning studies on the subjects of US reserve forces and manpower, Martin Binkin's *U.S. Reserve Forces: The Problem of the Weekend Warrior* (Washington, D.C.: Brookings Institution, 1974) and Martin Binkin and Irene Kyriakapoulos, *Youth or Experience: Manning the Modern Military* (Washington, D.C.: Brookings Institution, 1979).

PLANNING FOR CONTINGENCIES

The US has responsibilities and commitments all over the world. Thus, in evaluating a specific contingency, the force planner must first identify the force plausibly available to both sides. Using the appropriate methodology, the force planner then evaluates our existing forces and assesses the possibilities and costs of improvement. The list of possible contingencies is long. The Army and tactical air forces must be ready to fight campaigns in areas including Central Europe, the Persian Gulf, Korea, and Central America/Caribbean. The Navy's responsibilities are just as wide-ranging, including protection of sea lines of communication in the North Atlantic, Pacific, and Indian Oceans and possible power projection missions against a set of targets as diverse as Murmansk, Thrace, Vladivostok, and Cam Ranh Bay. The International Institute of Strategic Studies' yearly *The Strategic Survey* (London: IISS) and James E. Dunnigon and Austin Bay's *A Quick and Dirty Guide to War*

(New York: William Morrow, 1985) provide planners with an evaluation of current and emerging contingencies.

EUROPE

Most force planning work evaluates the balance of forces between NATO and the Warsaw Pact forces in Central Europe, because of the importance the US places on this contingency. The International Institute for Strategic Studies has published a number of *Adelphi Papers* on the subject: Robert Lucas Fisher's "Defending the Central Front: Balance of Forces" (London, IISS, 1976); Trevor Cliffe's "Military Technology and the European Balance" (London: IISS, 1972); and two of a series on the Alliance and Europe, Part 2, Kenneth Hunt's *Defence with Fewer Men* (London: IISS, 1973) and Part 4, Steven L. Canby's *Military Doctrine and Technology* (London: IISS, 1975). Canby also produced two studies for the RAND Corporation entitled *NATO Military Policy: Obtaining Conventional Capability vs. The Warsaw Pact*, R-1088-ARPA (Santa Monica, Calif.: RAND Corporation, 1973) and *Short (and Long) War Responses: Restructuring Border Defense and Reserve Mobilization for Armored Warfare* (Santa Monica, Calif.: RAND Corporation, 1978).

A general work by Richard D. Lawrence and Jeffery Record, *U.S. Force Structure in NATO: An Alternative* (Washington, D.C.: Brookings Institution, 1974) provides a fine example of the force planning approach applied to Central Europe. Barry M. Posen's article, "Measuring the European Conventional Balance: Coping With the Complexity in Threat Assessment" (*International Security*, Winter [1984–85]: 47–88) also does an excellent job of laying out the background factors such as mobilization rates and geography which would affect a campaign in Europe. The CBO has produced two studies of its own: James Blaker and Andrew Hamilton, "Assessing the NATO/Warsaw Pact Military Balance" (Washington, D.C.: CBO, Dec. 1977) and "U.S. Air and Ground Conventional Forces for NATO: Firepower Issues" (Washington, D.C.: CBO, Mar. 1978).

Little open work has been done on force planning for tactical/theater nuclear forces. However, tactical nuclear forces are most often treated on a contingency basis. The Brookings Institution has done two studies of nuclear weapons in Europe, Jeffery Record's *U.S. Nuclear Weapons in Europe: Issues and Alternatives* (Washington, D.C.: Brookings Institution, 1974) and *Alliance Security and the No First Use Option*, edited by John Steinbrunner and Leon V. Sigal (Washington, D.C.: Brookings Institution, 1983).

A great number of books have been written

about how the Warsaw Pact and NATO might wage a conflict in Europe. Sir John Hackett's book, *The Third World War: A Future History* (New York: Macmillan, 1978), provides an excellent discussion of factors that might influence the campaign in Central Europe. William P. Mako's *U.S. Ground Forces and the Defense of Central Europe* (Washington, D.C.: Brookings Institution, 1983) considers a ground campaign in the analytical force planning framework.

US reinforcement of its force deployments will also play a critical part role in any conflict in Europe. Sherwood S. Cordier's *Air and Sea Lanes of the North Atlantic* (Boulder, Colo.: Westview Press, 1981) studies the safety of NATO's lines of communication. The CBO has reviewed the reinforcement issue in depth through a number of reports: "Army Ground Combat Mobilization for the 1980's: Potential Costs and Effects for NATO" (Washington, D.C.: CBO, Nov. 1982); Patrick Hillier's *Strengthening NATO: POMCUS and Other Approaches* (Washington, D.C.: CBO, 1979); and Norma Slatkin, *Costs of Prepositioning Additional Army Divisions in Europe* (Washington, D.C.: CBO, 1980).

RAPID DEPLOYMENT FORCES AND OTHER CONTINGENCIES

The invasion of Afghanistan by the Soviet Union in December 1979 vigorously renewed US interest in our ability to deploy forces and project power in areas outside Europe, particularly in the Persian Gulf. Colonel Robert P. Haffa's work, *The Half War: Planning U.S. Rapid Deployment Forces to Meet a Limited Contingency 1960–1983* (Boulder, Colo.: Westview Press, 1984), provides an excellent overview of how, historically, the US has planned to meet such contingencies. Two other general works projecting US needs for future rapid deployment forces are Sherwood S. Cordier's *U.S. Military Power and Rapid Deployment: Requirements for the 1980's* (Boulder, Colo.: Westview Press, 1983) and John J. Hamre's *U.S. Airlift Forces: Enhancement Alternatives for NATO and Non-NATO Contingencies* (Washington, D.C.: CBO 1979). The RAND Corporation produced an early study on the lift requirements for such contingencies by Richard B. Raney, Jr., "Mobility—Airlift, Sealift and Prepositioning" (Santa Monica, Calif.: RAND Corporation, 1966).

Much has been written about US involvement and strength in the Persian Gulf in the post–Afghanistan invasion era. Most of this work focuses on the political, security, and geographic factors at work in the region. Albert Wohlstetter's "Meeting the Threat in the Persian Gulf" (*Survey*, Spring [1980]: 128–88), and Thomas L. McNaughter's "Balancing Soviet Power in the Persian Gulf" *Brookings Review*, Summer [1981]: 20–34) deal with the planning of US forces to handle the Persian Gulf contingency. In addition, a number of important congressional hearings have dealt with the US role and capabilities in the Persian Gulf and can provide some very useful information.

The force planning literature on other important contingencies, such as Korea, is much weaker. Two Brookings studies deal with our commitments in Northeast Asia: Stuart E. Johnson and Joseph A. Yager's *The Military Equation in Northeast Asia* (Washington, D.C.: Brookings Institution, 1979) and Ralph N. Clough's *Deterrence and Defense in Korea: The Role of U.S. Forces* (Washington, D.C.: Brookings Institution, 1976). Very little information can be found concerning US force planning for contingencies in Latin America or Southeast Asia.

RECURRING SOURCE MATERIALS

Almost any journal that deals with issues of national and international security will contain articles relevant to the force planning process. Some of these deal with force planning more indirectly, concentrating on the broader issues of US foreign policy and international affairs in general. But, these journals are worth consulting regularly because the specifics of force planning are often better understood in this wider context. They include *Foreign Affairs*, published five times a year by the Council on Foreign Relations; *International Security*, published quarterly by the joint Harvard/MIT Center for Science and International Affairs; *The Journal of Conflict Resolution*, published quarterly by Sage Publications; *Orbis*, published quarterly by the Foreign Policy Research Institute; and *Strategic Review* published quarterly by the United States Strategic Institute.

A number of journals deal more specifically with the doctrine and weapons systems of the forces which planners must evaluate. Each of the service schools of the Army, Navy, and Air Force publishes a journal on defense policy issues: *Military Review* (monthly), *Naval War College Review* (quarterly) and *Air University Review* (bimonthly). In addition, each of the services puts out a monthly magazine dealing with service-related issues: *Air Force Magazine*; *Army; Marine Corps Gazette*; and *Proceedings*. Other journals providing important information on weapons systems and force posture are: *Aviation Week and Space Technology*, published weekly by McGraw-Hill; the *RUSI Journal*, published quarterly by the Royal United Services Institute for Defense Studies in London; *Jane's Defense Weekly, International Defense Review*, published nine times a year by Interavia in Geneva.

A few scholarly institutions do a large volume of force planning analysis on a continuing basis, and their recent publications should be checked thoroughly. These include: the Brookings Institution in Washington, D.C., which also publishes the monthly *Brookings Review*; the International Institute for Strategic Studies in London which publishes a bimonthly journal, *Survival*, in addition to the previously cited Adelphi Papers, *Strategic Survey* and *Military Balance*; and the RAND Corporation in Santa Monica, Calif., which produces numerous reports and studies for the US government and others.

Finally, the congress provides a great wealth of information. Some of the most important sources are the yearly appropriations and authorizations for the Department of Defense as well as the Hearings of the Senate and House of Representatives Armed Services and Intelligence Committees. Statements by leading congressional figures such as Senators Sam Nunn and Barry Goldwater, or Representative Les Aspin also provide useful information. The Congressional Budget Office, as already mentioned, produces numerous studies on force planning, and the Office of Technology Assessment does useful studies of weapons systems and emerging technologies.

Where Do We Go from Here?

Much work has been done on methodologies to evaluate our conventional capabilities and how effective our large forces committed to Europe would be in a conflict. While the Persian Gulf and the Middle East region in general have received greatly increased attention since 1979, other areas ouside Europe such as Latin American and East Asia remain largely ignored.

Similarly, while evaluations of the capabilities and future needs of our offensive strategic forces are strong, other areas have been largely missed. Little quantitative work is available concerning the capabilities and utility of our tactical nuclear systems, particularly outside of Europe. Force planners have yet to address the whole arena of the military uses of space. Newly developed models and assessments of our strategic space systems would prove invaluable given the increasing emphasis placed on this new frontier for military activity.

This essay does not claim to be exhaustive, but it does attempt to present a cross-section of readings dealing with the major, relevant aspects of force planning. Most of the souces identified above have bibliographical references which can be pursued in further depth. The field of force planning is ripe for fresh work.

SELECTED READINGS IV

CONCEPTUAL MODELS AND THE CUBAN MISSILE CRISIS
GRAHAM T. ALLISON

The Cuban missile crisis is a seminal event. For thirteen days of October 1962, there was a higher probability that more human lives would end suddenly than ever before in history. Had the worst occurred, the death of 100 million Americans, over 100 million Russians, and millions of Europeans as well would make previous natural calamities and inhumanities appear insignificant. Given the probability of disaster— which President Kennedy estimated as "between 1 out of 3 and even"—our escape seems awesome.[1] This event symbolizes a central, if only partially thinkable, fact about our existence. That such consequences could follow from the choices and actions of national governments obliges students of government as well as participants in governance to think hard about these problems.

Improved understanding of this crisis depends in part on more information and more probing analyses of available evidence. To contribute to these efforts is part of the purpose of this study. But here the missile crisis serves primarily as grist for a more general investigation. This study proceeds from the premise that marked improvement in our understanding of such events depends critically on more self-consciousness about what observers bring to the analysis. What each analyst sees and judges to be important is a function not only of the evidence about what happened but also of the "conceptual lenses" through which he looks at the evidence. The principal purpose of this essay is to explore some of the fundamental assumptions and categories employed by analysts in thinking about problems of governmental behavior, especially in foreign and military affairs.

The general argument can be summarized in three propositions:

(1.) Analysts think about problems of foreign and military policy in terms of largely implicit conceptual models that have significant consequences for the content of their thought.[2]

Though the present product of foreign policy analysis is neither systematic nor powerful, if one carefully examines explanations produced by analysts, a number of fundamental similarities emerge. Explanations produced by particular analysts display quite regular, predictable features. This predictability suggests a substructure. These regularities reflect an analyst's assumptions about the character of puzzles, the categories in which problems should be considered, the types of evidence that are relevant, and the determinants of occurrences. The first proposition is that clusters of such related assumptions constitute basic frames of reference or conceptual models in terms of which analysts both ask and answer the questions: What happened? Why did the event happen? What will happen?[3] Such assumptions are central to the activities of explanation and prediction, for in attempting to explain a particular event, the analyst cannot simply describe the full state of the world leading up to that event. The logic of explanation requires that he single out the occurence.[4] Moreover, as the logic of prediction underscores, the analyst must summarize the various determinants as they bear on the event in question. Conceptual models both fix the mesh of the nets that the analyst drags through the material in order to explain a particular action or decision and direct casting that net in select ponds, at certain depths, in order to catch the desired fish.

(2.) Most analysts explain (and predict) the behavior of national governments in terms of various forms of one basic conceptual model, here entitled the Rational Policy Model.[5]

In terms of this conceptual model, analysts attempt to understand happenings as the more or less purposive acts of unified national governments. For these analysts, the point of an explanation is to show how the nation or government could have chosen the action in question, given the strategic problem that it faced.

Reprinted with minor revisions and with permission from The American Political Science Review *no. 3, (1969):* 689–718.

For example, in confronting the problem posed by the Soviet installation of missiles in Cuba, rational policy model analysts attempt to show how this was a reasonable act from the point of view of the Soviet Union, given Soviet strategic objectives.

(3.) Two "alternative" conceptual models, here labeled an Organizational Process model (model II) and a Bureaucratic Politics model (model III) provide a base for improved explanation and prediction.

Although the standard frame of reference has proved useful for many purposes, there is powerful evidence that it must be supplemented, if not supplanted, by frames of reference which focus upon the large organizations and political actors involved in the policy process. Model I's implication that important events have important causes, i.e., that monoliths perform large actions for big reasons, must be balanced by an appreciation of the facts (a) that monoliths are black boxes covering various gears and levers in a highly differentiated decisionmaking structure, and (b) that large acts are the consequences of innumerable and often conflicting smaller actions by individuals at various levels of bureaucratic organizations in the service of a variety of only partially compatible conceptions of national goals, organizational goals, and political objectives. Recent developments in the field of organization theory provide the foundation for the second model. According to this organizational process model, what model I categorizes as "acts" and "choices" are instead *outputs* of large organizations functioning according to certain regular patterns of behavior. Faced with the problem of Soviet missiles in Cuba, a model II analyst identifies the relevant organizations and displays the patterns of organizational behavior from which this action emerged. The third model focuses on the internal politics of a government. Happenings in foreign affairs are understood, according to the bureaucratic politics model, neither as choices nor as outputs. Instead, what happens is categorized as *outcomes* of various overlapping bargaining games among players arranged hierarchically in the national government. In confronting the problem posed by Soviet missiles in Cuba, a model III analyst displays the perceptions, motivations, positions, power, and maneuvers of principal players from which the outcome emerged.[6]

A central metaphor illuminates differences among these models. Foreign policy has often been compared to moves, sequences of moves, and games of chess. If you were limited to observations on a screen upon which moves in the chess game were projected without information as to how the pieces came to be moved, you would assume—as model I does—that an individual chess player was moving the pieces with reference to plans and maneuvers toward the goal of winning the game. But a pattern of moves can be imagined that would lead the serious observer, after watching several games, to consider the hypothesis that the chess player was not a single individual but rather a loose alliance of semi-independent organizations, each of which moved its set of pieces according to standard operating procedures. For example, movement of separate sets of pieces might proceed in turn, each according to a routine, the king's rook, bishop, and their pawns repeatedly attacking the opponent according to a fixed plan. Furthermore, it is conceivable that the pattern of play would suggest to an observer that a number of distinct players, with distinct objectives but shared power over the pieces, were determining the moves as the resultant of collegial bargaining. For example, the black rook's move might contribute to the loss of a black knight with no comparable gain for the black team, but with the black rook becoming the principal guardian of the "palace" on that side of the board.

The space available does not permit full development and support of such a general argument.[7] Rather, the sections that follow simply sketch each conceptual model, articulate it as an analytic paradigm, and apply it to produce an explanation. But each model is applied to the same event: the US blockade of Cuba during the missile crisis. These "alternative explanations" of the same happening illustrate differences among the models—*at work.*[8] A crisis decision, by a small group of men in the context of ultimate threat, this is a case of the rational policy model *par excellence.* The dimensions and factors that models II and III uncover in this case are therefore particularly suggestive. The concluding section of this paper suggests how the three models may be related and how they can be extended to generate predictions.

MODEL I: RATIONAL POLICY

RATIONAL POLICY MODEL ILLUSTRATED

Where is the pinch of the puzzle raised by the *New York Times* over Soviet deployment of an antiballistic missile system?[9] The question, as the *Times* states it, concerns the Soviet Union's objective in allocating such large sums of money for this weapon system while at the same time seeming to pursue a policy of increasing détente. In former President Johnson's words, "the paradox is that this [Soviet deployment of an antiballistic missile system] should be happening at a time when there is abundant evidence that our mutual antagonism is beginning to ease."[10] This question troubles people primarily because Soviet antiballistic missile deployment, and evidence of Soviet actions towards détente, when juxtaposed in our implicit model, produce a question. With reference to what objective could the Soviet government have rationally chosen the simultaneous pursuit of these two courses of action? This question arises only when the analyst attempts to structure events as purposive choices of consistent actors.

How do analysts attempt to explain the Soviet emplacement of missiles in Cuba? The most widely cited explanation of this occurrence has been produced by two RAND Sovietologists, Arnold Horelick and Myron Rush.[11] They conclude that "the introduction of strategic missiles into Cuba was motivated chiefly by the Soviet leaders' desire to overcome . . . the existing large margin of US strategic superiority."[12] How do they reach this conclusion? In Sherlock Holmes style, they seize several salient characteristics of this action and use these features as criteria against which to test alternative hypotheses about Soviet objectives. For example, the size of the Soviet deployment, and the simultaneous emplacement of more expensive, more visible intermediate-range missiles as well as medium-range missiles, it is argued, exclude an explanation of the action in terms of Cuban defense—since the objective could have been secured with a much smaller number of medium-range missiles alone. Their explanation presents an argument for one objective that permits interpretation of the details of Soviet behavior as a value-maximizing choice.

How do analysts account for the coming of the First World War? According to Hans Morgenthau, "the first World War had its origin exclusively in the fear of a disturbance of the European balance of power.[13] In the period preceding World War I, the Triple Alliance precariously balanced the Triple Entente. If either power combination could gain a decisive advantage in the Balkans, it would achieve a decisive advantage in the balance of power. "It was this fear," Morgenthau asserts, "that motivated Austria in July 1914 to settle its accounts with Serbia once and for all, and that induced Germany to support Austria unconditionally. It was the same fear that brought Russia to the support of Serbia, and France to the support of Russia."[14] How is Morgenthau able to resolve this problem so confidently? By imposing on the data a "rational outline."[15] The value of this method, according to Morgenthau, is that "it provides for rational discipline in action and creates astounding continuity in foreign policy which makes American, British, or Russian foreign policy appear as an intelligent, rational continuum . . . regardless of the different motives, preferences, and intellectual and moral qualities of successive statesmen."[16]

Stanley Hoffmann's essay, "Restraints and Choices in American Foreign Policy," concentrates, characteristically, on "deep forces": the international system, ideology, and national character—which constitute restraints, limits, and blinders.[17] Only secondarily does he consider decisions. But when explaining particular occurrences, though emphasizing relevant constraints, he focuses on the choices of nations. American behavior in Southeast Asia is explained as a reasonable choice of "downgrading this particular alliance (SEATO) in favor of direct U.S. involvement," given the constraint: "one is bound by one's commitments; one is committed by one's mistakes."[18] More frequently, Hoffmann uncovers confusion or contradiction in the nation's choice. For example, US policy towards underdeveloped countries is explained as "schizophrenic."[19] The method employed by Hoffmann in producing these explanations as rational (or irrational) decisions, he terms "imaginative reconstruction."[20]

Deterrence is the cardinal problem of the contemporary strategic literature. Thomas Schelling's *Strategy of Conflict* formulates a number of propositions focused upon the dynamics of deterrence in the nuclear age. One of the major propositions concerns the stability of the balance of terror: in a situation of mutual deterrence, the probability of nuclear war is reduced not by the "balance" (the sheer equality of the situation) but rather by the *stability* of the balance, i.e., the fact that neither opponent in striking first can destroy the other's ability to strike back.[21] How does Schelling support this proposition? Confidence in the contention stems not from an inductive canvass of a large number of previous cases, but rather from two calculations. In a situation of "balance" but vulnerability, there are values for which a rational opponent could choose to strike first, e.g., to destroy enemy capabilities to retaliate. In a "stable balance" where no matter who strikes first, each has an assured capability to retaliate with unacceptable damage, no rational agent could choose such a course of action (since that choice is effectively equivalent to choosing mutual homicide). Whereas most contemporary strategic thinking is driven *implicitly* by the motor upon which this calculation depends, Schelling explicitly recognizes that strategic theory does assume a model. The foundation of a theory of strategy is, he asserts: "the assumption of rational behavior—not just of intelligent behavior, but of behavior motivated by conscious calculation of advantages, calculation that in turn is based on an explicit and internally consistent value system."[22]

What is striking about these examples from the literature of foreign policy and international relations are the similarities among analysts of various styles when they are called upon to produce explanations. Each assumes that what must be explained is an action, i.e., the realization of some purpose or intention. Each assumes that the actor is the national government. Each assumes that the action is chosen as a calculated response to a strategic problem. For each, explanation consists of showing what goal the government was pursuing in committing the act and how this action was a reasonable choice, given the nation's objectives. This set of assumptions characterizes the rational policy model. The assertion that model I is the standard frame of reference implies no denial of highly visible differences among the interests of Sovietologists, diplomatic historians, international relations theorists, and strategists. Indeed, in most respects, differences among the work of Hans Morgenthau, Stanley Hoffmann, and Thomas Schelling could not be more pointed. Appreciation of the extent to which each relies predominantly on model I, however, reveals basic similarities among Morgenthau's method of "rational reenactment," Hoffman's "imaginative reconstruction," and Schelling's "vicarious problem solving"; family resemblances among Morgenthau's "rational statesman," Hoffmann's "roulette player," and Schelling's "game theorist."[23]

Most contemporary analysts (as well as laymen) proceed predominantly—albeit most often implicitly—in terms of this model when attempting to explain happenings in foreign affairs. Indeed, that occurrences in foreign affairs are the *acts* of *nations* seems so fundamental to thinking about such problems that this underlying model has rarely been recognized: to explain an occurrence in foreign policy simply means to show how the government could have rationally chosen that action.[24] These brief examples illustrate five uses of the model. To prove that most analysts think largely in terms of the rational policy model is not possible. In this limited space it is not even possible to illustrate the range of employment of the framework. Rather, my purpose is to convey to the reader a grasp of the model and a challenge: let the readers examine the literature with which they are most familiar and make a judgment.

The general characterization can be sharpened by articulating the rational policy model as an "analytic paradigm" in the technical sense developed by Robert K. Merton for sociological analyses.[25] Systematic statement of basic assumptions, concepts, and propositions employed by model I analysts highlights the distinctive thrust of this style of analysis. To articulate a largely implicit framework is of necessity to caricature. But caricature can be instructive.

RATIONAL POLICY PARADIGM

Basic Unit of Analysis: Policy as National Choice

Happenings in foreign affairs are conceived as actions chosen by the nation or national government.[26] Governments select the action that will maximize strategic goals and objectives. These "solutions" to strategic problems are the fundamental categories in terms of which the analyst perceives what is to be explained.

Organizing Concepts

National Actor. The nation or government, conceived as a rational, unitary decision-maker, is the agent. This actor has one set of specified goals (the equivalent of a consistent utility function), one set of perceived options, and a single estimate of the consequences that follow from each alternative.

The Problem. Action is chosen in response to the strategic problem which the nation faces. Threats and opportunities arising in the "international strategic market place" move the nation to act.

Static Selection. The sum of activity of representatives of the government relevant to a problem constitutes what the nation has chosen as its "solution." Thus the action is conceived as a steady-state choice among alternative outcomes (rather than, for example, a large number of partial choices in a dynamic stream).

Action as Rational Choice. The components include:

1. *Goals and Objectives.* National security and national interests are the principal categories in which strategic goals are conceived. Nations seek security and a range of further objectives. (Analysts rarely translate strategic goals and objectives into an explicit utility function; nevertheless, analysts do focus on major goals and objectives and trade off side effects in an intuitive fashion.)

2. *Options.* Various courses of action relevant to a strategic problem provide the spectrum of options.

3. *Consequences.* Enactment of each alternative course of action will produce a series of consequences. The relevant consequences constitute benefits and costs in terms of strategic goals and objectives.

4. *Choice.* Rational choice is value-maximizing. The rational agent selects the alternative whose consequences rank highest in terms of his goals and objectives.

Dominant Inference Pattern

This paradigm leads analysts to rely on the following pattern of inference: if a nation performed a particular action, that nation must have had ends towards which the action constituted an optimal means. The rational policy model's explanatory power stems from this inference pattern. Puzzlement is relieved by revealing the purposive pattern within which the occurrence can be located as a value-maximizing means.

General Propositions

The disgrace of political science is the infrequency with which propositions of any generality are formulated and tested. "Paradigmatic analysis" argues for explicitness about the terms in which analysis proceeds, and seriousness about the logic of explanation. Simply to illustrate the kind of propositions on which analysts who employ this model rely, the formulation includes several.

The basic assumption of value-maximizing behavior produces propositions central to most explanations. The general principle can be formulated as follows: the likelihood of any particular action results from a combination of the nation's (1) relevant values and objectives, (2) perceived alternative courses of action, (3) estimates of various sets of consequences (which will follow from each alternative), and (4) net valuation of each set of consequences. This yields two propositions.

(1) An increase in the cost of an alternative, i.e., a reduction in the value of the set of consequences which will follow from that action, or a reduction in the probability of attaining fixed consequences, reduces the likelihood of that alternative being chosen.

(2) A decrease in the costs of an alternative, i.e., an increase in the value of the set of consequences which will follow from that alternative, or an increase in the probability of attaining fixed consequences, increases the likelihood of that action being chosen.[27]

Specific Propositons

Deterrence. The likelihood of any particular attack results from the factors specified in the general proposition. Combined with factual assertions, this general proposition yields the propositions of the subtheory of deterrence.

(1) A stable nuclear balance reduces the likelihood of nuclear attack. This proposition is derived from the general proposition plus the asserted fact that a second-strike capability affects the potential attacker's calculations by increasing the likelihood and the costs of one particular set of consequences which might follow from attack—namely, retaliation.

(2) A stable nuclear balance increases the probability of limited war. This proposition is derived from the general proposition plus the asserted fact that though increasing the costs of a nuclear exchange, a stable nuclear balance nevertheless produces a more significant reduction in the probabililty that such consequences would be chosen in response to a limited war. Thus this set of consequences weighs less heavily in the calculus.

Soviet Force Posture. The Soviet Union chooses its force posture (i.e., its weapons and their deployment) as a value-maximizing means of implementing Soviet strategic objectives and military doctrine. A proposition of this sort underlies Secretary of Defense Laird's inference from the fact of 200 SS-9s (large intercontinental missiles) to the assertion that, "the Soviets are going for a first-strike capability, and there's no question about it."[28]

VARIANTS OF THE RATIONAL POLICY MODEL

This paradigm exhibits the characteristics of the most refined version of the rational model. The modern literature of strategy employs a model of this sort. Problems and pressures in the "international strategic marketplace" yield probabilities of occurrence. The international actor, which could be any national actor, is simply a value-maximizing mechanism for getting from the strategic problem to the logical solution. But the explanations and predictions produced by most analysts of foreign affairs depend primarily on variants of this "pure" model. The point of each is the same: to place the action within a value-maximizing framework, given certain constraints. Nevertheless, it may be helpful to identify several variants, each of which might be exhibited similarly as a paradigm. The first focuses upon the national actor and his choice in a particular situation, leading analysts to further constrain the goals, alternatives, and consequences considered. Thus, (1) national propensities or personality traits reflected in an "operational code," (2) concern with certain objectives, or (3) special principles of action, narrow the "goals" or "alternatives" or "consequences" of the paradigm. For example, the Soviet deployment of ABMs is sometimes explained by reference to the Soviets' "defense-mindedness." Or a particular Soviet action is explained as an instance of a special rule of action in the Bolshevik operational code.[29] A second, related, cluster of variants focuses on the individual leader or leadership group as the actor whose preference function is maximized and whose personal (or group) characteristics are allowed to modify the alternatives, consequences, and rules of choice. Explanations of the US involvement in Vietnam as a natural consequence of the Kennedy-Johnson administration's axioms of foreign policy rely on this variant. A third, more complex variant of the basic model recognizes the existence of several actors within a government, for example, hawks and doves or military and civilians, but attempts to explain (or predict) an occurrence by reference to the objectives of the victorious actor. Thus, for example, some revisionist histories of the cold war recognize the forces of light and the forces of darkness within the US government, but explain American actions as a result of goals and perceptions of the victorious forces of darkness.

Each of these forms of the basic paradigm constitutes a formalization of what analysts typically rely upon implicitly. In the transition from implicit conceptual model to explicit paradigm much of the richness of the best employments of this model has been lost. But the purpose in raising loose, implicit conceptual models to an explicit level is to reveal the basic logic of analysts' activity. Perhaps some of the remaining artificiality that surrounds the statement of the paradigm can be erased by noting a number of the standard additions and modifications employed by analysts who proceed *predominantly* within the rational policy model. First, in the course of a document, analysts shift from one variant of the basic model to another, occasionally appropriating in an ad hoc fashion aspects of a situation which are logically incompatible with the basic model. Second, in the course of explaining a number of occurrences, analysts sometimes pause over a particular event about which they have a great deal of information and unfold it in such detail that an impression of randomness is created. Third, having employed other assumptions and categories in deriving an explanation or prediction, analysts will present their product in a neat, convincing rational policy model package. (This accommodation is a favorite of members of the intelligence community whose association with the details of a process is considerable, but who feel that by putting an occurrence in a larger rational framework, it will be more comprehensible to their audience.) Fourth, in attempting to offer an explanation—particularly in cases where a prediction derived from the basic model has failed—the notion of a "mistake" is invoked. Thus, the failure in the prediction of a "missile gap" is written off as a Soviet mistake in not taking advantage of their opportunity. Both these and other modifications permit model I analysts considerably more variety than the paradigm might suggest. But such accommodations are essentially appendages to the basic logic of these analyses.

The US Blockade of Cuba: A First Cut[30]

The US response to the Soviet Union's emplacement of missiles in Cuba must be understood in strategic terms as simple value-maximizing escalation. American nuclear superiority could be counted on to paralyze Soviet nuclear power; Soviet transgression of the nuclear threshold in response to an American use of lower levels of violence would be wildly irrational since it would mean virtual destruction of the Soviet Communist system and Russian nation. American local superiority was overwhelming: it could be initiated at a low level while threatening with high credibility an ascending sequence of steps short of the nuclear threshold. All that was required was for the United States to bring to bear its strategic and local superiority in such a way that American determination to see the missiles removed would be demonstrated, while at the same time allowing Moscow time and room to retreat without humiliation. The naval blockade—euphemistically named a "quarantine" in order to circumvent the niceties of international law—did just that.

The US government's selection of the blockade followed this logic. Apprised of the presence of Soviet missiles in Cuba, the president assembled an Executive Committee (ExCom) of the National Security Council and directed them to "set aside all other tasks to make a prompt and intense survey of the dangers and all possible courses of action."[31] This group functioned as "fifteen individuals on our own, representing the President and not different departments."[32] As one of the participants recalls, "The remarkable aspect of those meetings was a sense of complete equality."[33] Most of the time during the week that followed was spent canvassing all the possible tracks and weighing the arguments for and against each. Six major categories of action were considered.

1. Do nothing. US vulnerability to Soviet missiles was no new thing. Since the US already lived under the gun of missiles based in Russia, a Soviet capability to strike from Cuba too made little real difference. The real danger stemmed from the possibility of US overreaction. The US should announce the Soviet action in a calm, casual manner thereby deflating whatever political capital Khrushchev hoped to make of the missiles.

This argument fails on two counts. First, it grossly underestimates the military importance of the Soviet move. Not only would the Soviet Union's missile capability be doubled and the US early warning system outflanked. The Soviet Union would have an opportunity to reverse the strategic balance by further installations, and indeed, in the longer run, to invest in cheaper, shorter-range rather than more expensive longer-range missiles. Second, the political importance of this move was undeniable. The Soviet Union's act challenged the American president's most solemn warning. If the US failed to respond, no American commitment would be credible.

2. Diplomatic pressures. Several forms were considered: an appeal to the UN or OAS for an inspection team, a secret approach to Khrushchev, and a direct approach to Khrushchev, perhaps at a summit meeting. The United States would demand that the missiles be removed, but the final settlement might include neutralization of Cuba, US withdrawal from the Guantanamo base, and withdrawal of US Jupiter missiles from Turkey or Italy.

Each form of the diplomatic approach had its own drawbacks. To arraign the Soviet Union before the UN Security Council held little promise since the Russians could veto any proposed action. While the diplomats argued, the missiles would become operational. To send a secret emissary to Khrushchev demanding that the missiles be withdrawn would be to pose untenable alternatives. On the one hand, this would invite Khrushchev to seize the diplomatic initiative, perhaps committing himself to strategic retaliation in response to an attack on Cuba. On the other hand, this would tender an ultimatum that no great power could accept. To confront Khrushchev at a summit would guarantee demands for US concessions, and the analogy between US missiles in Turkey and Russian missiles in Cuba could not be erased.

But why not trade US Jupiters in Turkey and Italy, which the president had previously ordered withdrawn, for the missiles in Cuba? The US had chosen to withdraw these missiles in order to replace them with superior, less vulnerable Mediterranean Polaris submarines. But the middle of the crisis was no time for concessions. The offer of such a deal might suggest to the Soviets that the West would yield and thus tempt them to demand more. It would certainly confirm European suspicions about American willingness to sacrifice European interests when the chips were down. Finally, the basic issue should be kept clear. As the president stated in reply to Bertrand Russell, "I think your attention might well be directed to the burglars rather than to those who have caught the burglars."[34]

3. A secret approach to Castro. The crisis provided an opportunity to separate Cuba and Soviet Communism by offering Castro the alternatives, "split or fall." But Soviet troops transported, constructed, guarded, and controlled the missiles. Their removal would thus depend on a Soviet decision.

4. Invasion. The United States could take this occasion not only to remove the missiles but also to rid itself of Castro. A Navy exercise had long been scheduled in which Marines, ferried from Florida in naval vessels, would liberate the imaginary island of Vieques.[35] Why not simply shift the point of disembarkment? (The Pentagon's foresight in planning this operation would be an appropriate antidote to the CIA's Bay of Pigs!)

Preparations were made for an invasion, but as a last resort. American troops would be forced to confront 20,000 Soviets in the first cold war case of direct contact between the troops of the superpowers. Such brinksmanship courted nuclear disaster, practically guaranteeing an equivalent Soviet move against Berlin.

5. Surgical air strike. The missile sites should be removed by a clean, swift conventional attack. This was the effective counteraction which the attempted deception deserved. A surgical strike would remove the missiles and thus eliminate both the danger that the missiles might become operational and the fear that the Soviets would discover the American discovery and act first.

The initial attractiveness of this alternative was dulled by several difficulties. First, could the strike really be "surgical"? The Air Force could not guarantee destruction of all the missiles.[36] Some might be fired during the attack; some might not have been identified. In order to assure destruction of Soviet and Cuban means of retaliating, what was required was not a surgical but rather a massive attack—of at least 500 sorties. Second, a surprise air attack would of course kill Russians at the missile sites. Pressures on the Soviet Union to retaliate would be so strong that an attack on Berlin or Turkey was highly probable. Third, the key problem with this program was that of advance warning. Could the president of the United States, with his memory of Pearl Harbor and his vision of future US responsibility, order a "Pearl Harbor in reverse"? For 175 years, unannounced Sunday morning attacks had been an anathema to our tradition.[37]

6. Blockade. Indirect military action in the form of a blockade became more attractive as the ExCom dissected the other alternatives. An embargo on military shipments to Cuba enforced by a naval blockade was not without flaws, however. Could the US blockade Cuba without inviting Soviet reprisal in Berlin? The likely solution to joint blockades would be the lifting of both blockades, restoring the new status quo, and allowing the Soviets additional time to complete the missiles. Second, the possible consequences of the blockade resembled the drawbacks which disqualified the air strike. If Soviet ships did not stop, the United States

would be forced to fire the first shot, inviting retaliation. Third, a blockade would deny the traditional freedom of the seas demanded by several of our close allies and might be held illegal, in violation of the UN Charter and international law, unless the United States could obtain a two-thirds vote in the OAS. Finally, how could a blockade be related to the problem, namely, some 75 missiles on the island of Cuba, approaching operational readiness daily? A blockade offered the Soviets a spectrum of delaying tactics with which to buy time to complete the missile installations. Was a fait accompli not required?

In spite of these enormous difficulties the blockade had comparative advantages: (1) It was a middle course between inaction and attack, aggressive enough to communicate firmness of intention, but nevertheless not so precipitous as a strike. (2) It placed on Khrushchev the burden of choice concerning the next step. He could avoid a direct military clash by keeping his ships away. His was the last clear chance. (3) No possible military confrontation could be more acceptable to the US than a naval engagement in the Caribbean. (4) This move permitted the US, by flexing its conventional muscle, to exploit the threat of subsequent nonnuclear steps in each of which the US would have significant superiority.

Particular arguments about advantages and disadvantages were powerful. The explanation of the American choice of the blockade lies in a more general principle, however. As President Kennedy stated in drawing the moral of the crisis:

Above all, while defending our own vital interests, nuclear powers must avert those confrontations which bring an adversary to a choice of either a humiliating retreat or a nuclear war. To adopt that kind of course in the nuclear age would be evidence only of the bankruptcy of our policy—of a collective death wish for the world.[38]

The blockade was the United States' only real option.

MODEL II: ORGANIZATIONAL PROCESS

For some purposes, governmental behavior can be usefully summarized as action chosen by a unitary, rational decisionmaker: centrally controlled, completely informed, and value maximizing. But this simplification must not be allowed to conceal the fact that a "government" consists of a conglomerate of semifeudal, loosely allied organizations, each with a substantial life of its own. Government leaders do sit formally, and to some extent in fact, on top of this conglomerate. But governments perceive problems through organizational sensors. Governments

define alternatives and estimate consequences as organizations process information. Governments act as these organizations enact routines. Government behavior can therefore be understood according to a second conceptual model, less as deliberate choices of leaders and more as *outputs* of large organizations functioning according to standard patterns of behavior.

To be responsive to a broad spectrum of problems, governments consist of large organizations among which primary responsibility for particular areas is divided. Each organization attends to a special set of problems and acts in quasi-independence on these problems. But few important problems fall exclusively within the domain of a single organization. Thus government behavior relevant to any important problem reflects the independent output of several organizations, partially coordinated by government leaders. Government leaders can substantially disturb, but not substantially control, the behavior of these organizations.

To perform complex routines, the behavior of large numbers of individuals must be coordinated. Coordination requires standard operating procedures: rules according to which things are done. Assured capability for reliable performance of action that depends upon the behavior of hundreds of persons requires established "programs." Indeed, if the eleven members of a football team are to perform adequately on any particular down, each player must not "do what he thinks needs to be done" or "do what the quarterback tells him to do." Rather, each player must perform the maneuvers specified by a previously established play which the quarterback has simply called in this situation.

At any given time, a government consists of *existing* organizations, each with a *fixed* set of standard operating procedures and programs. The behavior of these organizations—and consequently of the government—relevant to an issue in any particular instance is, therefore, determined primarily by routines established in these organizations prior to that instance. But organizations do change. Learning occurs gradually, over time. Dramatic organizational change occurs in response to major crises. Both learning and change are influenced by existing organizational capabilities.

Borrowed from studies of organizations, these loosely formulated propositions amount simply to *tendencies*. Each must be hedged by modifiers like "other thngs being equal" and "under certain conditions." In particular instances, tendencies hold—more or less. In specific situations, the relevant question is: more or less? But this is as it should be. For, on the one hand, "organizations" are no more homogeneous a class than "solids." When scientists tried to generalize about "solids," they achieved similar results. Solids tend to expand when heated, but some do and some don't. More adequate categorization of the various elements now lumped under the rubric "organizations" is thus required. On the other hand, the behavior of particular organizations seems considerably more complex than the behavior of solids. Additional information about a particular organization is required for further specification of the tendency statements. In spite of these two caveats, the characterization of government action as organizational output differs distinctly from model I. Attempts to understand problems of foreign affairs in terms of this frame of reference should produce quite different explanations.[39]

ORGANIZATIONAL PROCESS PARADIGM[40]

Basic Unit of Analysis: Policy as Organizational Output

The happenings of international politics are, in three critical senses, outputs of organizational processes. First, the actual occurrences are organizatinal outputs. For example, Chinese entry into the Korean War—that is, the fact that Chinese soldiers were firing at UN soldiers south of the Yalu in 1950—is an organizational action: the action of men who are soldiers in platoons which are in companies, which in turn are in armies, responding as privates to lieutenants who are responsible to captains and so on to the commander, moving into Korea, advancing against enemy troops, and firing according to fixed routines of the Chinese Army. Government leaders' decisions trigger organizational routines. Government leaders can trim the edges of this output and exercise some choice in combining outputs. But the mass of behavior is determined by previously established procedures. Second, existing organizational routines for employing present physical capabilities constitute the effective options open to government leaders confronted with any problem. Only the existence of men, equipped and trained as armies and capable of being transported to North Korea, made entry into the Korean War a live option for the Chinese leaders. The fact that fixed programs (equipment, men, and routines which exist at the particular time) exhaust the range of buttons that leaders can push is not always perceived by these leaders. But in every case it is critical for an understanding of what is actually done. Third, organizational outputs structure the situation within the narrow constraints of which leaders must contribute their "decision" concerning an issue. Outputs raise the problem, provide the information, and make the initial moves that color the face of the issue that is turned to the leaders. As Theodore Sorensen has observed: "Presidents rarely, if ever, make decisions—particularly in foreign affairs—in the

sense of writing their conclusions on a clean slate . . . The basic decisions, which confine their choices, have all too often been previously made."[41] If one understands the structure of the situation and the face of the issue—which are determined by the organizational outputs—the formal choice of the leaders is frequently anticlimactic.

Organizing Concepts

Organizational Actors. The actor is not a monolithic "nation" or "government" but rather a constellation of loosely allied organizations on top of which government leaders sit. This constellation acts only as component organizations perform routines.[42]

Factored Problems and Fractionated Power. Surveillance of the multiple facets of foreign affairs requires that problems be cut up and parceled out to various organizations. To avoid paralysis, primary power must accompany primary responsibility. But if organizations are permitted to do anything, a large part of what they do will be determined within the organization. Thus each organization perceives problems, processes information, and performs a range of actions in quasi-independence (within broad guidelines of national policy). Factored problems and fractionated power are two edges of the same sword. Factoring permits more specialized attention to particular facets of problems than would be possible if government leaders tried to cope with these problems by themselves. But this additional attention must be paid for in the coin of discretion for *what* an organization attends to, and *how* organizational responses are programmed.

Parochial Priorities, Perceptions, and Issues. Primary responsibility for a narrow set of problems encourages organizational parochialism. These tendencies are enhanced by a number of additional factors: (1) selective information available to the organization, (2) recruitment of personnel into the organization, (3) tenure of individuals in the organization, (4) small group pressures within the organization, and (5) distribution of rewards by the organization. Clients (e.g., interest groups), government allies (e.g., Congressional committees), and extranational counterparts (e.g., the British Ministry of Defense for the Department of Defense, ISA, or the British Foreign Office for the Department of State, EUR) galvanize this parochialism. Thus organizations develop relatively stable propensities concerning operational priorities, perceptions, and issues.

Action as Organizational Output. The preeminent feature of organizational activity is its programmed character: the extent to which behavior in any particular case is an enactment of preestablished routines. In producing outputs, the activity of each organization is characterized by:

1. *Goals: Constraints Defining Acceptable Performance.* The operational goals of an organization are seldom revealed by formal mandates. Rather, each organization's operational goals emerge as a set of constraints defining acceptable performance. Central among these constraints is organizational health, defined usually in terms of bodies assigned and dollars appropriated. The set of constraints emerges from a mix of expectations and demands of other organizations in the government, statutory authority, demands from citizens and special interest groups, and bargaining within the organization. These constraints represent a quasi-resolution of conflict—the constraints are relatively stable, so there is some resolution. But conflict among alternative goals is always latent; hence, it is a quasi-resolution. Typically, the constraints are formulated as imperatives to avoid roughly specified discomforts and disasters.[43]

2. *Sequential Attention to Goals.* The existence of conflict among operational constraints is resolved by the device of sequential attention. As a problem arises, the subunits of the organization most concerned with that problem deal with it in terms of the constraints they take to be most important. When the next problem arises, another cluster of subunits deals with it, focusing on a different set of constraints.

3. *Standard Operating Procedures.* Organizations perform their "higher" functions, such as attending to problem areas, monitoring information, and preparing relevant responses for likely contingencies, by doing "lower" tasks, for example, preparing budgets, producing reports, and developing hardware. Reliable performance of these tasks requires standard operating procedures (hereafter SOPs). Since procedures are "standard" they do not change quickly or easily. Without these standard procedures, it would not be possible to perform certain concerted tasks. But because of standard procedures, organizational behavior in particular instances often appears unduly formalized, sluggish, or inappropriate.

4. *Programs and Repertoires.* Organizations must be capable of performing actions in which the behavior of large numbers of individuals is carefully coordinated. Assured performance requires clusters of rehearsed SOPs for producing specific actions, e.g., fighting enemy units or answering an embassy's cable. Each cluster comprises a "program" (in the terms both of drama and computers) which the organization has available for dealing with a situation. The list of programs relevant to a type of activity, e.g., fighting, constitutes an organizational repertoire. The number of programs in a repertoire is always quite limited. When properly triggered, organizations execute programs; programs cannot be substantially changed in a particular situation. The more complex the action and the greater the number of individuals involved, the more important are programs and repertoires as determinants of organizational behavior.

5. *Uncertainty Avoidance.* Organizations do not attempt to estimate the probability distribution of future occurrences. Rather, organizations avoid uncertainty. By arranging a *negotiated environment*, organizations regularize the reactions of other actors with whom they have to deal. The primary environment, relations with other organizations that comprise the government, is stabilized by such arrangements as agreed budgetary splits, accepted areas of responsibility, and established conventional practices. The secondary environment, relations with the international world, is stabilized between allies by the establishment of contracts (alliances) and "club relations" (US State and UK Foreign Office of US Treasury and UK Treasury). Between enemies, contracts and accepted conventional practices perform a similar function, for example, the rules of the "precarious status quo" which President Kennedy referred to in the missile crisis. Where the international environment cannot be negotiated, organizations deal with remaining uncertainties by establishing a set of *standard scenarios* that constitute the contingencies for which they prepare. For example, the standard scenario for Tactical Air Command of the US Air Force involves combat with enemy aircraft. Planes are designed and pilots trained to meet this problem. That these preparations are less relevant to more probable contingencies, e.g., provision of close-in ground support in limited wars like Vietnam, has had little impact on the scenario.

6. *Problem-directed Search.* Where situations cannot be construed as standard, organizations engage in search. The style of search and the solution are largely determined by existing routines. Organizational search for alternative courses of action is problem-oriented: it focuses on the atypical discomfort that must be avoided. It is simple-minded: the neighborhood of the symptom is searched first; then, the neighborhood of the current alternative. Patterns of search reveal biases which in turn reflect such factors as specialized training or experience and patterns of communication.

7. *Organizational Learning and Change.* The parameters of organizational behavior mostly persist. In response to nonstandard problems, organizations search and routines evolve, assimilating new situations. Thus learning and change follow in large part from existing procedures.

But marked changes in organizations do sometimes occur. Conditions in which dramatic changes are more likely include: (1) Periods of budgetary feast. Typically, organizations devour budgetary feasts by purchasing additional items on the existing shopping list. Nevertheless, if committed to change, leaders who control the budget can use extra funds to effect changes. (2) Periods of prolonged budgetary famine. Though a single year's famine typically results in few changes in organizational structure but a loss of effectiveness in performing some programs, prolonged famine forces major retrenchment. (3) Dramatic performance failures. Dramatic change occurs (mostly) in response to major disasters. Confronted with an undeniable failure of procedures and repertoires, authorities outside the organization demand change, existing personnel are less resistant to change, and critical members of the organization are replaced by individuals committed to change.

Central Coordination and Control. Action requires decentralization of responsibility and power. But problems lap over the jurisdictions of several organizations. Thus the necessity for decentralization runs headlong into the requirement for coordination. (Advocates of one horn or the other of this dilemma—responsive action entails decentralized power versus coordinated action requires central control—account for a considerable part of the persistent demand for government reorganization.) Both the necessity for coordination and the centrality of foreign policy to national welfare guarantee the involvement of government leaders in the procedures of the organizations among which problems are divided and power shared. Each organization's propensities and routines can be disturbed by government leaders' intervention. Central direction and persistent control of organizational activity, however, are not possible. The relation among organizations, and between organizations and the government leaders depends critically on a number of structural variables including: (1) the nature of the job, (2) the measures and information available to government leaders, (3) the system of rewards and punishments for organizational members, and (4) the procedures by which human and material resources get committed. For example, to the extent that rewards and punishments for the members of an organization are distributed by higher authorities, these authorities can exercise some control by specifying criteria in terms of which organizational output is to be evaluated. These criteria become constraints within which organizational activity proceeds. But constraint is a crude instrument of control.

Intervention by government leaders does sometimes change the activity of an organization in an intended direction. But instances are fewer than might be expected. As Franklin Roosevelt, the master manipulator of government organizations, remarked:

The Treasury is so large and far-flung and ingrained in its practices that I find it is almost impossible to get the action and results I want . . . But the Treasury is not to be compared with the State Department. You should go through the experience of trying to get any changes in the thinking, policy, and action of the career diplomats and then you'd know what a real problem was. But the Treasury and the State Department put together are nothing compared with the Na-a-vy . . . To change anything in the Na-a-vy is like punching a feather bed. You punch it with your right and you punch it with your left until you are finally exhausted, and then you find the damn bed just as it was before you started punching.[44]

John Kennedy's experience seems to have been similar: "The State Department," he asserted, "is a bowl full of jelly."[45] And lest the McNamara revolution in the Defense Department seem too striking a counterexample, the Navy's recent rejection of McNamara's major intervention in Naval weapons procurement, the F-111B, should be studied as an antidote.

Decisions of Government Leaders. Organizational persistence does not exclude shifts in governmental behavior. For government leaders sit atop the conglomerate of organizations. Many important issues of governmental action require that these leaders decide what organizations will play out which programs where. Thus stability in the parochialisms and SOPs of individual organizations is consistent with some important shifts in the behavior of governments. The range of these shifts is defined by existing organizational programs.

Dominant Inference Pattern

If a nation performs an action of this type today, its organizational components must yesterday have been performing (or have had established routines for performing) an action only marginally different from this action. At any specific point in time, a government consists of an established conglomerate of organizations, each with existing goals, programs, and repertoires. The characteristics of a government's action in any instance follow from those established routines, and from the choice of government leaders—on the basis of information and estimates provided by existing routines—among existing programs. The best explanation of an organization's behavior at t is $t - 1$; the prediction of $t + 1$ is t. Model II's explanatory power is achieved by uncovering the organizational routines and repertoires that produced the outputs that comprise the puzzling occurrence.

General Propositions

A number of general propositions have been stated above. In order to illustrate clearly the type of proposition employed by model II analysts, this section formulates several more precisely.

Organizational Action.
Activity according to SOPs and programs does not constitute far-sighted, flexible adaptation to "the issue" (as it is conceived by the analyst). Detail and nuance of actions by organizations are determined predominantly by organizational routines, not government leaders' directions.

SOPs constitute routines for dealing with *standard* situations. Routines allow large numbers of ordinary individuals to deal with numerous instances, day after day, without considerable thought, by responding to basic stimuli. But this regularized capability for adequate performance is purchased at the price of standardization. If the SOPs are appropriate, average performance, i.e., performance averaged over the range of cases, is better than it would be if each instance were approached individually (given fixed talent, timing, and resource constraints). But specific instances, particularly critical instances that typically do not have "standard" characteristics, are often handled sluggishly or inappropriately.

A program, i.e., a complex action chosen from a short list of programs in a repertoire, is rarely tailored to the specific situation in which it is executed. Rather, the program is (at best) the most appropriate of the programs in a previously developed repertoire.

Since repertoires are developed by parochial organizations for standard scenarios defined by that organization, programs available for dealing with a particular situation are often ill-suited.

Limited Flexibility and Incremental Change.
Major lines of organizational action are straight, i.e., behavior at one time is marginally different from that behavior at $t - 1$. Simpleminded predictions work best: Behavior at $t + 1$ will be marginally different from behavior at the present time.

Organizational budgets change incrementally—both with respect to totals and with respect to intraorganizational splits. Though organizations could divide the money available each year by carving up the pie anew (in the light of changes in objectives or environment), in practice, organizations take last year's budget as a base and adjust incrementally. Predictions that require large budgetary shifts in a single year between organizations or between units within an organization should be hedged. Once undertaken, an organizational investment is not dropped at the point where "objective" costs outweigh benefits. Organizational stakes in adopted projects carry them quite beyond the loss point.

Administrative Feasibility.
Adequate explanation, analysis, and prediction must include administrative feasibility as a major dimension. A considerable gap separates what leaders choose (or might rationally have chosen) and what organizations implement.

Organizations are blunt instruments. Projects that require several organizations to act with high degrees of precision and coordination are not likely to succeed.

Projects that demand that existing organizational units depart from their accustomed functions and perform previously unprogrammed tasks are rarely accomplished in their designed form.

Government leaders can expect that each organization will do its "part" in terms of what the organization knows how to do.

Government leaders can expect incomplete and distorted information from each organization concerning its part of the problem.

Where an assigned piece of a problem is contrary to the existing goals of an organization, resistance to implementation of that piece will be encountered.

Specific Propositions
Deterrence. The probability of nuclear attack is less sensitive to balance and imbalance, or stability and instability (as these concepts are employed by model I strategists) than it is to a number of organizational factors. Except for the special case in which the Soviet Union acquires a credible capability to destroy the US with a disarming blow, US superiority or inferiority affects the probability of a nuclear attack less than do a number of organizational factors.

First, if a nuclear attack occurs, it will result from organizational activity: the firing of rockets by members of a missile group. The enemy's *control system*, i.e., physical mechanisms and standard procedures which determine who can launch rockets when, is critical. Second, the enemy's programs for bringing his strategic forces to *alert status* determine probabilities of accidental firing and momentum. At the outbreak of World War I, if the Russian tsar had understood the organizational processes which his order of full mobilization triggered, he would have realized that he had chosen war. Third, organizational repertoires fix the range of effective choice open to enemy leaders. The menu available to Tsar Nicholas in 1914 has two entrees: full mobilization and no mobilization. Partial mobilization was not an organizational option. Fourth, since organizational routines set the chessboard, the training and deployment of troops and nuclear weapons is crucial. Given that the outbreak of hostilities in Berlin is more probable than most scenarios for nuclear war, facts about deployment, training, and tactical nuclear equipment of Soviet troops stationed in

East Germany—which will influence the face of the issue seen by Soviet leaders at the outbreak of hostilities and the manner in which choice is implemented—are as critical as the question of "balance."

Soviet Force Posture. Soviet Force posture, i.e., the fact that certain weapons rather than others are procured and deployed, is determined by organizational factors such as the goals and procedures of existing military services and the goals and processes of research and design labs, within budgetary constraints that emerge from the government leader's choices. The frailty of the Soviet Air Force within the Soviet military establishment seems to have been a crucial element in the Soviet failure to acquire a large bomber force in the 1950s (thereby faulting American intelligence predictions of a "bomber gap"). The fact that missiles were controlled until 1960 in the Soviet Union by the Soviet Ground Forces, whose goals and procedures reflected no interest in an intercontinental mission, was not irrelevant to the slow Soviet buildup of ICBMs (thereby faulting US intelligence predictions of a "missile gap"). These organizational factors (Soviet Ground Forces' control of missiles and that service's fixation with European scenarios) make the Soviet deployment of so many MRBMs that European targets could be destroyed three times over, more understandable. Recent weapon developments, e.g., the testing of a Fractional Orbital Bombardment System (FOBS) and multiple warheads for the SS-9, very likely reflect the activity and interests of a cluster of Soviet research and development organizations, rather than a decision by Soviet leaders to acquire a first-strike weapon system. Careful attention to the organizational components of the Soviet military establishment (Strategic Rocket Forces, Navy, Air Force, Ground Forces, and National Air Defense), the missions and weapons systems to which each component is wedded (an independent weapon system assists survival as an independent service), and existing budgetary splits (which probably are relatively stable in the Soviet Union as they tend to be everywhere) offer potential improvements in medium- and longer-term predictions.

THE US BLOCKADE OF CUBA: SECOND CUT

Organizational Intelligence. At 7:00 P.M. on 22 October 1962, President Kennedy disclosed the American discovery of the presence of Soviet strategic missiles in Cuba, declared a "strict quarantine on all offensive military equipment under shipment to Cuba," and demanded that "Chairman Khrushchev halt and eliminate this clandestine, reckless, and provocative threat to world peace."[46] This decision was reached at the pinnacle of the US government after a critical week of deliberation. What initiated that precious week were photographs of Soviet missile sites in Cuba taken on October 14. These pictures might not have been taken until a week later. In that case, the president speculated, "I don't think probably we would have chosen as prudently as we finally did."[47] US leaders might have received this information three weeks earlier—if a U-2 had flown over San Cristobal in the last week of September.[48] What determined the context in which American leaders came to choose the blockade was the discovery of missiles on October 14.

There has been considerable debate over alleged American "intelligence failures" in the Cuban missile crisis.[49] But what both critics and defenders have neglected is the fact that the discovery took place on October 14, rather than three weeks earlier or a week later, as a consequence of the established routines and procedures of the organizations which constitute the US intelligence community. These organizations were neither more nor less successful than they had been the previous month or were to be in the months to follow.[50]

The notorious "September estimate," approved by the United States Intelligence Board (USIB) on September 19, concluded that the Soviet Union would not introduce offensive missiles into Cuba.[51] No U-2 flight was directed over the western end of Cuba (after September 5) before October 4.[52] No U-2 flew over the western end of Cuba until the flight that discovered the Soviet missiles on October 14.[53] Can these "failures" be accounted for in organizational terms?

On September 19 when USIB met to consider the question of Cuba, the "system" contained the following information: (1) shipping intelligence had noted the arrival in Cuba of two large-hatch Soviet lumber ships, which were riding high in the water; (2) refugee reports of countless sightings of missiles, but also a report that Castro's private pilot, after a night of drinking in Havana, had boasted: "We will fight to the death and perhaps we can win because we have everything, including atomic weapons"; (3) a sighting by a CIA agent of the rear profile of a strategic missile; (4) U-2 photos produced by flights of August 29, September 5 and 17 showing the construction of a number of SAM sites and other defensive missiles.[54] Not all of this information was on the desk of the estimators, however. Shipping intelligence experts noted the fact that large-hatch ships were riding high in the water and spelled out the inference: the ships must be carrying "space consuming" cargo.[55] These facts were carefully included in the catalogue of intelligence concerning shipping. For experts sensitive to the Soviets' shortage of ships, however, these facts carried no special signal. The refugee report of Castro's private pilot's remark had been received at Opa Locka, Florida, along with vast reams of inaccurate reports generated by the refugee community. This report and a thousand others had to be checked and compared before being sent to Washington. The two weeks required for initial processing could have been shortened by a large increase in resources, but the yield of this source was already quite marginal. The CIA agent's sighting of the rear profile of a strategic missile had occurred on September 12; transmission time from agent sighting to arrival in Washington typically took 9 to 12 days. Shortening this transmission time would impose severe cost in terms of danger to subagents, agents and communication networks.

On the information available, the intelligence chiefs who predicted that the Soviet Union would not introduce offensive missiles into Cuba made a reasonable and defensible judgment.[56] Moreover, in the light of the fact that these organizations were gathering intelligence not only about Cuba but about potential occurrences in all parts of the world, the informational base available to the estimators involved nothing out of the ordinary. Nor, from an organizational perspective, is there anything startling about the gradual accumulation of evidence that led to the formulation of the hypothesis that the Soviets were installing missiles in Cuba and the decision on October 4 to direct a special flight over western Cuba.

The 10-day delay between that decision and the flight is another organizational story.[57] At the October 4 meeting, the Defense Department took the opportunity to raise an issue important to its concerns. Given the increased danger that a U-2 would be downed, it would be better if the pilot were an officer in uniform rather than a CIA agent. Thus the Air Force should assume responsibility for U-2 flights over Cuba. To the contrary, the CIA argued that this was an intelligence operation and thus within the CIA's jurisdiction. Moreover, CIA U-2s had been modified in certain ways which gave them advantages over Air Force U-2s in averting Soviet SAMs. Five days passed while the State Department pressed for less risky alternatives such as drones and the Air Force (in Department of Defense guise) and CIA engaged in territorial disputes. On October 9 a flight plan over San Cristobal was approved by COMOR, but to the CIA's dismay, Air Force pilots rather than CIA agents would take charge of the mission. At this point details become sketchy, but several members of the intelligence community have speculated that an Air Force pilot in an Air Force U-2 attempted a high altitude overflight on October 9 that "flamed out", i.e., lost

power, and thus had to descend in order to restart its engine. A second round between Air Force and CIA followed, as a result of which Air Force pilots were trained to fly CIA U-2s. A successful overflight took place on October 14.

This 10-day delay constitutes some form of "failure." In the face of well-founded suspicions concerning offensive Soviet missiles in Cuba that posed a critical threat to the United States' most vital interest, squabbling between organizations whose job it is to produce this information seems entirely inappropriate. But for each of these organizations, the question involved the issue: "*Whose* job was it to be?" Moreover, the issue was not simply, which organization would control U-2 flights over Cuba, but rather the broader issue of ownership of U-2 intelligence activities—a very long standing territorial dispute. Thus though this delay was in one sense a "failure," it was also a nearly inevitable consequence of two facts: many jobs do not fall neatly into precisely defined organizational jurisdictions; and vigorous organizations are imperialistic.

Organizational Options. Deliberations of leaders in ExCom meetings produced broad outlines of alternatives. Details of these alternatives and blueprints for their implementation had to be specified by the organizations that would perform these tasks. These organizational outputs answered the question: What, specifically, *could* be done?

Discussion in the ExCom quickly narrowed the live options to two: an air strike and a blockade. The choice of the blockade instead of the air strike turned on two points: (1) the argument from morality and tradition that the United States could not perpetrate a "Pearl Harbor in reverse"; (2) the belief that a "surgical" air strike was impossible.[58] Whether the United States *might* strike first was a question not of capability but of morality. Whether the United States *could* perform the surgical strike was a factual question concerning capabilities. The majority of the members of the ExCom, including the president, initially preferred the air strike.[59] What effectively foreclosed this option, however, was the fact that the air strike they wanted could not be chosen with high confidence of success.[60] After having tentatively chosen the course of prudence—given that the surgical air strike was not an option—Kennedy reconsidered. On Sunday morning, October 21, he called the Air Force experts to a special meeting in his living quarters where he probed once more for the option of a "*surgical*" air strike.[61] General Walter C. Sweeny, commander of Tactical Air Forces, asserted again that the Air force could guarantee no higher than 90 percent effectiveness in a surgical air strike.[62] That "fact" was false.

The air strike alternative provides a classic case of military estimates. One of the alternatives outlined by the ExCom was named "air strike." Specification of the details of this alternative was delegated to the Air Force. Starting from an existing plan for massive US military action against Cuba (prepared for contingencies like a response to a Soviet Berlin grab), Air Force estimators produced an attack to guarantee success.[63] This plan called for extensive bombardment of all missile sites, storage depots, airports, and, in deference to the Navy, the artillery batteries opposite the naval base at Guantanamo.[64] Members of the ExCom repeatedly expressed bewilderment at military estimates of the number of sorties required, likely casualties, and collateral damage. But the "surgical" air strike that the political leaders had in mind was never carefully examined during the first week of the crisis. Rather, this option was simply excluded on the grounds that since the Soviet MRBMs in Cuba were classified "mobile" in US manuals, extensive bombing was required. During the second week of the crisis, careful examination revealed that the missiles were mobile, in the sense that small houses are mobile: that is, they could be moved and reassembled in 6 days. After the missiles were reclassified "movable" and detailed plans for surgical air strikes specified, this action was added to the list of live options for the end of the second week.

Organizational Implementation. Ex-Com members separated several types of blockade: offensive weapons only, all armaments, and all strategic goods including POL (petroleum, oil, and lubricants). But the *"details"* of the operation were left to the Navy. Before the president announced the blockade on Monday evening, the first stage of the Navy's blueprint was in motion, and a problem loomed on the horizon.[65] The Navy had a detailed plan for the blockade. The president had several less precise but equally determined notions concerning what should be done, when, and how. For the Navy the issue was one of effective implementation of the Navy's blockade—without the meddling and interference of political leaders. For the president, the problem was to pace and manage events in such a way that the Soviet leaders would have time to see, think, and blink.

A careful reading of available sources uncovers an instructive incident. On Tuesday the British ambassador, Ormsby-Gore, after having attended a briefing on the details of the blockade, suggested to the president that the plan for intercepting Soviet ships far out of reach of Cuban jets did not facilitate Khrushchev's hard decision.[66] Why not make the interception much closer to Cuba and thus give the Russian leader more time? According to the public account and the recollection of a number of individuals involved, Kennedy "agreed immediately, called McNamara, and over emotional Navy protest, issued the appropriate instructions."[67] As Sorensen records, "in a sharp clash with the Navy, he made certain his will prevailed."[68] The Navy's plan for the blockade was thus changed by drawing the blockade much closer to Cuba.

A serious organizational orientation makes one suspicious of this account. More careful examination of the available evidence confirms these suspicions, though alternative accounts must be somewhat speculative. According to the public chronology, a quarantine drawn close to Cuba became effective on Wednesday morning, the first Soviet ship was contacted on Thursday morning, and the first boarding of a ship occurred on Friday. According to the statement by the Department of Defense, boarding of the *Marcula* by a party from the *John R. Pierce* "took place at 7:50 A.M., E.D.T., 180 miles northeast of Nassau."[69] The *Marcula* had been trailed since about 10:30 the previous evening.[70] Simple calculations suggest that the *Pierce* must have been stationed along the Navy's original arc which extended 500 miles out to sea from Cape Magsi, Cuba's easternmost tip.[71] The blockade line was *not* moved as the president ordered, and the accounts report.

What happened is not entirely clear. One can be certain, however, that Soviet ships passed through the line along which American destroyers had posted themselves before the official "first contact" with the Soviet ship. On October 26 a Soviet tanker arrived in Havana and was honored by a dockside rally for "running the blockade." Photographs of this vessel show the name *Vinnitsa* on the side of the vessel in Cyrillic letters.[72] But according to the official US position, the first tanker to pass through the blockade was the *Bucharest*, which was hailed by the Navy on the morning of October 25. Again simple mathematical calculation excludes the possibility that the *Bucharest* and the *Vinnitsa* were the same ship. It seems probable that the Navy's resistance to the president's order that the blockade be drawn in closer to Cuba forced him to allow one or several Soviet ships to pass through the blockade after it was officially operative.[73]

This attempt to leash the Navy's blockade had a price. On Wednesday morning, October 24, what the president had been awaiting occurred. The 18 dry cargo ships heading towards the quarantine stopped dead in the water. This was the occasion of Dean Rusk's remark, "We are eyeball to eyeball and I think the other fellow just blinked."[74] But the Navy had another interpretation. The ships had simply stopped to pick up Soviet submarine escorts. The president became quite concerned lest the Navy—already riled because of presidential meddling in its affairs—blunder into an incident. Sensing the president's fears, McNamara became suspicious of the Navy's procedures and routines for making the first interception. Calling on the chief of Naval Operations in the Navy's inner sanctum, the Navy Flag Plot, McNamara put his questions harshly.[75] Who would make the first interception? Were Russian-speaking officers on board? How would submarines be dealt with? At one point McNamara asked Anderson what he would do if a Soviet ship's captain refused to answer questions about his cargo. Picking up the Manual of Navy Regulations the Navy man waved it in McNamara's face and shouted, "It's all in there." To which McNamara replied, "I don't give a damn what John Paul Jones would have done; I want to know what you are going to do, now."[76] The encounter ended on Anderson's remark: "Now, Mr. Secretary, if you and your Deputy will go back to your office the Navy will run the blockade."[77]

MODEL III: BUREAUCRATIC POLITICS

The leaders who sit on top of organizations are not a monolithic group. Rather, each is, in his own right, a player in a central, competitive game. The name of the game is bureaucratic politics: bargaining along regularized channels among players positioned hierarchically within the government. Government behavior can thus be understood according to a third conceptual model not as organizational outputs, but as outcomes of bargaining games. In contrast with model I, the bureaucratic politics model sees no unitary actor but rather many actors as players, who focus not on a single strategic issue but on many diverse intranational problems as well, in terms of no consistent set of strategic objectives but rather according to various conceptions of national, organizational, and personal goals, making government decisions not by rational choice but by the pulling and hauling that is politics.

The apparatus of each national government constitutes a complex arena for the intranational game. Political leaders at the top of this apparatus plus the men who occupy positions on top of the critical organizations form the circle of central players. Ascendancy to this circle assures some independent standing. The necessary decentralization of decisions required for action on the broad range of foreign policy problems guarantees that each player has considerable discretion. Thus power is shared.

The nature of problems of foreign policy permits fundamental disagreement among reasonable men concerning what ought to be done.

Analyses yield conflicting recommendations. Separate responsibilities laid on the shoulders of individual personalities encourage differences in perceptions and priorities. But the issues are of first-order importance. What the nation does really matters. A wrong choice could mean irreparable damage. Thus responsible men are obliged to fight for what they are convinced is right.

Men share power. Men differ concerning what must be done. The differences matter. This milieu necessitates that policy be resolved by politics. What the nation does is sometimes the result of the triumph of one group over others. More often, however, different groups pulling in different directions yield a result distinct from what anyone intended. What moves the chess pieces is not simply the reasons which support a course of action, nor the routines of organizations which enact an alternative, but the power and skill of proponents and opponents of the action in question.

This characterization captures the thrust of the bureaucratic politics orientation. If problems of foreign policy arose as discrete issues, and decisions were determined one game at a time, this account would suffice. But most "issues," e.g., Vietnam or the proliferation of nuclear weapons, emerge piecemeal, over time, one lump in one context, a second in another. Hundreds of issues compete for players' attention every day. Each player is forced to fix upon his issues for that day, fight them on their own terms, and rush on to the next. Thus the character of emerging issues and the pace at which the game is played converge to yield government "decisions" and "actions" as collages. Choices by one player, outcomes of minor games, outcomes of central games, and "foul-ups"—these pieces, when stuck to the same canvas, constitute government behavior relevant to an issue.

The concept of national security policy as political outcome contradicts both public imagery and academic orthodoxy. Issues vital to national security, it is said, are too important to be settled by political games. They must be "above" politics. To accuse someone of "playing politics with national security" is a most serious charge. What public conviction demands, the academic penchant for intellectual elegance reinforces. Internal politics is messy; moreover, according to prevailing doctrine, politicking lacks intellectual content. As such, it constitutes gossip for journalists rather than a subject for serious investigation. Occasional memoirs, anecdotes in historical accounts, and several detailed case studies to the contrary, most of the literature of foreign policy avoids bureaucratic politics. The gap between academic literature and the experience of participants in government is nowhere wider than at this point.

BUREAUCRATIC POLITICS PARADIGM[78]

Basic Unit of Analysis: Policy as Political Outcome

The decisions and actions of governments are essentially intranational political outcomes: outcomes in the sense that what happens is not chosen as a solution to a problem but rather results from compromise, coalition, competition, and confusion among government officials who see different faces of an issue; political in the sense that the activity from which the outcomes emerge is best characterized as bargaining. Following Wittgenstein's use of the concept of a "game," national behavior in international affairs can be conceived as outcomes of intricate and subtle, simultaneous, overlapping games among players located in positions, the hierarchical arrangement of which constitutes the government.[79] These games proceed neither at random nor at leisure. Regular channels structure the game. Deadlines force issues to the attention of busy players. The moves in the chess game are thus to be explained in terms of the bargaining among players with separate and unequal power over particular pieces and with separable objectives in distinguishable subgames.

Organizing Concepts

Players in Positions. The actor is neither a unitary nation, nor a conglomerate of organizations, but rather a number of individual players. Groups of these players constitute the agent for particular government decisions and actions. Players are men in jobs.

Individuals become players in the national security policy game by occupying a critical position in an adminstration. For example, in the US government the players include "Chiefs": the president, secretaries of state, defense, and treasury, director of the CIA, Joint Chiefs of Staff, and, since 1961, the special assistant for national security affairs;[80] "Staffer": the immediate staff of each Chief; "Indians": the political appointees and permanent government officials within each of the departments and agencies; and "Ad Hoc Players": actors in the wider government game (especially "Congressional Influentials"), members of the press, spokesmen for important interest groups (especially the "bipartisan foreign policy establishment" in and out of Congress), and surrogates for each of these groups. Other members of the Congress, press, interest groups, and public form concentric circles around the central arena—circles which demarcate the permissive limits within which the game is played.

Positions define what players both may and must do. The advantages and handicaps with which each player can enter and play in various games stems from his position. So does a cluster of obligations for the performance of certain tasks. The two sides of this coin are illustrated by the position of the modern secretary of state. First, in form and usually in fact, he is the primary repository of political judgment on the political-military issues that are the stuff of contemporary foreign policy; consequently, he is a senior personal adviser to the president. Second, he is the colleague of the president's other senior advisers on the problems of foreign policy, the secretaries of defense and treasury, and the special assistant for national security affairs. Third, he is the ranking US diplomat for serious negotiation. Fourth, he serves as an administration voice to Congress, the country, and the world. Finally, he is "Mr. State Department" or "Mr. Foreign Office," "leader of officials, spokesman for their causes, guardian of their interests, judge of their disputes, superintendent of their work, master of their careers."[81] But he is not first one, and then the other. All of these obligations are his simultaneously. His performance in one affects his credit and power in the others. The perspective stemming from the daily work which he must oversee—the cable traffic by which his department maintains relations with other foreign offices—conflicts with the president's requirement that he serve as a generalist and coordinator of contrasting perspectives. The necessity that he be close to the president restricts the extent to which, and the force with which, he can front for his department. When he defers to the secretary of defense rather than fighting for his department's position—as he often must—he strains the loyalty of his officialdom. The secretary's resolution of these conflicts depends not only upon the position, but also upon the player who occupies the position.

For players are also people. Men's metabolisms differ. The core of the bureaucratic politics mix is personality. How each man manages to stand the heat in his kitchen, each player's basic operating style, and the complementarity or contradiction among personalities and styles in the inner circles are irreducible pieces of the policy blend. Moreover, each person comes to his position with baggage in tow, including sensitivities to certain issues, commitments to various programs, and personal standing and debts with groups in society.

Parochial Priorities, Perceptions and Issues. Answers to the questions: "What is the issue?" and "What must be done?" are colored by the position from which the questions are considered. For the factors which encourage organizational parochialism also influence the players who occupy positions on top of (or within) these organizations. To motivate members of his organization, a player must be sensitive to the organization's orientation. The games into which the player can enter and the advantages with which he plays enhance these pressures. Thus propensities of perception stemming from position permit reliable prediction about a player's stances in many cases. But these propensities are filtered through the baggage which players bring to positions. Sensitivity to both the pressures and the baggage is thus required for many predictions.

Interests, Stakes, and Power. Games are played to determine outcomes. But outcomes advance and impede each player's conceptions of the national interest, specific programs to which he is committed, the welfare of his friends, and his personal interests. These overlapping interests constitute the stakes for which games are played. Each player's ability to play successfully depends upon his power. Power, i.e., effective influence on policy outcomes, is an elusive blend of at least three elements: bargaining advantages (drawn from formal authority and obligations, institutional backing, constituents, expertise, and status), skill and will in using bargaining advantages, and other players' perceptions of the first two ingredients. Power wisely invested yields an enhanced reputation for effectiveness. Unsuccessful investment depletes both the stock of capital and the reputation. Thus each player must pick the issues on which he can play with a reasonable probability of success. But no player's power is sufficient to guarantee satisfactory outcomes. Each player's needs and fears run to many other players. What ensues is the most intricate and subtle of games known to man.

The Problem and the Problems. "Solutions" to strategic problems are not derived by detached analysts focusing coolly on *the* problem. Instead, deadlines and events raise issues in games, and demand decisions of busy players in contexts that influence the face the issue wears. The problems for the players are both narrower and broader than *the* strategic problem. For each player focuses not on the total strategic problem but rather on the decision that must be made now. But each decision has critical consequences not only for the strategic problem but for each player's organizational, reputational, and personal stakes. Thus the gap between the problems the player was solving and the problem upon which the analyst focuses is often very wide.

Action-Channels. Bargaining games do not proceed randomly. Action-channels, i.e., regularized ways of producing action concerning types of issues, structure the game by preselecting the major players, determining their points of entrance into the game, and distributing particular advantages and disadvantages for each game. Most critically, channels determine "who's got the action," that is, which department's Indians actually do whatever is chosen. Weapon procurement decisions are made within the annual budgeting process; embassies' demands for action cables are answered according to routines of consultation and clearance from State to Defense and White House; requests for instructions from military groups (concerning assistance all the time, concerning operations during war) are composed by the military in consultation with the Office of the Secretary of Defense, State, and White House; crisis responses are debated among White House, State, Defense, CIA, and Ad Hoc players; major political speeches, especially by the President but also by other Chiefs, are cleared through established channels.

Action as Politics. Government decisions are made, and government actions emerge neither as the calculated choice of a unified group, nor as a formal summary of leaders' preferences. Rather the context of shared power but separate judgments concerning important choices, determines that politics is the mechanism of choice. Note the *environment* in which the game is played: inordinate uncertainty about what must be done, the necessity that something be done, and crucial consequences of whatever is done. These features force responsible men to become active players. The *pace of the game*—hundreds of issues, numerous games, and multiple channels—compels players to fight to "get others' attention," to make them "see the facts," to assure that they "take the time to think seriously about the broader issue." The *structure of the game*—power shared by individuals with separate responsibilities—validates each player's feeling that "others don't see my problem," and "others must be persuaded to look at the issue from a less parochial perspective." The *rules of the game*—he who hesitates loses his chance to play at that point, and he who is uncertain about his recommendation is overpowered by others who are sure—pressures players to come down on one side of a 51–49 issue and play. The *rewards of the game*—effectiveness, i.e., impact on outcomes, as the immediate measure of performance—encourages hard play. Thus, most players come to fight to "make the government do what is right." The strategies and tactics employed are quite similar to those formalized by theorists of international relations.

Streams of Outcomes. Important government decisions or actions emerge as collages composed of individual acts, outcomes of minor and major games, and foul-ups. Outcomes which could never have been chosen by an actor and would never have emerged from bargaining in a single game over the issue are fabricated piece by piece. Understanding of the outcome requires that it be disaggregated.

Dominant Inference Pattern

If a nation performed an action, that action was the *outcome* of bargaining among individuals and groups within the government. That outcome included *results* achieved by groups committed to a decision or action, *resultants* which emerged from bargaining among groups with quite different positions and *foul-ups*. Model III's explanatory power is achieved by revealing the pulling and hauling of various players, with different perceptions and priorities, focusing on separate problems, which yielded the outcomes that constitute the action in question.

General Propositions

Action and Intention. Action does not presuppose intention. The sum of behavior of representatives of a government relevant to an issue was rarely intended by any individual or group. Rather separate individuals with different intentions contributed pieces which compose an outcome distinct from what anyone would have chosen.

Where you stand depends on where you sit.[82] Horizontally, the diverse demands upon each player shape his priorities, perceptions, and issues. For large classes of issues, e.g., budgets and procurement decisions, the stance of a particular player can be predicted with high reliability from information concerning his seat. In the notorious B-36 controversy, no one was surprised by Admiral Radford's testimony that "the B-36 under any theory of war, is a bad gamble with national security," as opposed to Air Force Secretary Symington's claim that "a B-36 with a A-bomb can destroy distant objectives which might require ground armies years to take."[83]

Chiefs and Indians. The aphorism "where you stand depends on where you sit" has vertical as well as horizontal application. Vertically, the demands upon the president, Chiefs, Staffers, and Indians are quite distinct.

The foreign policy issues with which the president can deal are limited primarily by his crowded schedule: the necessity of dealing first with what comes next. His problem is to probe the special face worn by issues that come to his attention, to preserve his leeway until time has clarified the uncertainties, and to assess the relevant risks.

Foreign policy Chiefs deal most often with the hottest issue *de jour*, though they can get the attention of the president and other members of the government for other issues which they judge important. What they cannot guarantee is that "the President will pay the price" or that "the others will get on board." They must build a coalition of the relevant powers that be. They must "give the President confidence" in the right course of action.

Most problems are framed, alternatives specified, and proposals pushed, however, by Indians. Indians fight with Indians of other departments; for example, struggles between International Security Affairs of the Department of Defense and Political-Military of the State Department are a microcosm of the action at higher levels. But the Indian's major problem is how to get the *attention* of Chiefs, how to get an issue decided, how to get the government "to do what is right."

In policymaking then, the issue looking *down* is options: how to preserve my leeway until time clarifies uncertainties. The issue looking *sideways* is commitment: how to get others committed to my coalition. The issue looking *upwards* is confidence: how to give the boss confidence in doing what must be done. To paraphrase one of Neustadt's assertions which can be applied down the length of the ladder, the essence of a responsible official's task is to induce others to see that what needs to be done is what their own appraisal of their own responsibilities requires them to do in their own interests.

Specific Propositions

Deterrence. The probability of nuclear attack depends primarily on the probability of attack emerging as an outcome of the bureaucratic politics of the attacking government. First, which players can decide to launch an attack? Whether the effective power over action is controlled by an individual, a minor game, or the central game is critical. Second, though model I's confidence in nuclear deterrence stems from an assertion that, in the end, governments will not commit suicide, model III recalls historical precedents. Admiral Yamamoto, who designed the Japanese attack on Pearl Harbor, estimated accurately: "In the first six months to a year of war against the US and England I will run wild, and I will show you an uninterrupted succession of victories; I must also tell you that, should the war be prolonged for two or three years, I have no confidence in our ultimate victory."[84] But Japan attacked. Thus, three questions might be considered. One: could any member of the government solve his problem by attack? What patterns of bargaining could yield attack as an outcome? The major difference between a stable balance of terror and a questionable balance may simply be that in the first case most members of the government appreciate fully the consequences of attack and are thus on guard against the emergence of this outcome. Two: what stream of outcomes might lead to an attack? At what point in that stream is the potential attacker's politics? If members of the US government had been sensitive to the stream of decisions from which the Japanese attack on Pearl Harbor emerged, they would have been aware of a considerable probability of that attack. Three: how might miscalculation and confusion generate foul-ups that yield attack as an outcome? For example, in a crisis or after the beginning of conventional war, what happens to the information available to, and the effective power of, members of the central game.

THE US BLOCKADE OF CUBA: A THIRD CUT

The Politics of Discovery. A series of overlapping bargaining games determined both the *date* of the discovery of the Soviet missiles and the *impact* of this discovery on the administration. An explanation of the politics of the discovery is consequently a considerable piece of the explanation of the US blockade.

Cuba was the Kennedy administration's "political Achilles' heel."[85] The months preceding the crisis were also months before the Congressional elections, and the Republican Senatorial and Congressional Campaign Committee had announced that Cuba would be "the dominant issue of the 1962 campaign."[86] What the administration billed as a "more positive and indirect approach of isolating Castro from developing, democratic Latin America," Senators Keating, Goldwater, Capehart, Thurmond, and others attacked as a "do-nothing" policy.[87] In statements on the floor of the House and Senate, campaign speeches across the country, and interviews and articles carried by national news media, Cuba—particularly the Soviet program of increased arms aid—served as a stick for stirring the domestic political scene.[88]

These attacks drew blood. Prudence demanded a vigorous reaction. The president decided to meet the issue head-on. The administration mounted a forceful campaign of denial designed to discredit critics' claims. The president himself manned the front line of this offensive, though almost all administration officials participated. In his news conference on August 19, President Kennedy attacked as "irresponsible" calls for an invasion of Cuba, stressing rather "the totality of our obligations" and promising to "watch what happens in Cuba with the closest attention."[89] On September 4, he issued a strong statement denying any provocative Soviet action in Cuba.[90] On September 13 he lashed out at "loose talk" calling for an invasion of Cuba.[91] The day before the flight of the U-2 which discovered the missiles, he campaigned in Capehart's Indiana against those

"self-appointed generals and admirals who want to send someone else's sons to war."[92]

On Sunday, October 14, just as a U-2 was taking the first pictures of Soviet missiles, McGeorge Bundy was asserting:

I *know* that there is no present evidence, and I think that there is no present likelihood that the Cuban government and the Soviet government would, in combination, attempt to install a major offensive capability.[93]

In this campaign to puncture the critics' charges, the administration discovered that the public needed positive slogans. Thus, Kennedy fell into a tenuous semantic distinction between "offensive" and "defensive" weapons. This distinction originated in his September 4 statement that there was no evidence of "offensive ground to ground missiles" and warned "were it to be otherwise, the gravest issues would arise."[94] His September 13 statement turned on this distinction between "defensive" and "offensive" weapons and announced a firm commitment to action if the Soviet Union attempted to introduce the latter into Cuba.[95] Congressional committees elicited from administration officials testimony which read this distinction and the president's commitment into the *Congressional Record.*[96]

What the president least wanted to hear, the CIA was most hesitant to say plainly. On August 22 John McCone met privately with the president and voiced suspicions that the Soviets were preparing to introduce offensive missiles into Cuba.[97] Kennedy heard this as what it was: the suspicion of a hawk. McCone left Washington for a month's honeymoon on the Riviera. Fretting at Cap Ferrat, he bombarded his deputy, General Marshall Carter, with telegrams, but Carter, knowing that McCone had informed the president of his suspicions and received a cold reception, was reluctant to distribute these telegrams outside the CIA.[98] On September 9 a U-2 "on loan" to the Chinese Nationalists was downed over mainland China.[99] The Committee on Overhead Reconnaissance (COMOR) convened on September 10 with a sense of urgency.[100] Loss of another U-2 might incite world opinion to demand cancellation of U-2 flights. The president's campaign against those who asserted that the Soviets were acting provocatively in Cuba had begun. To risk downing a U-2 over Cuba was to risk chopping off the limb on which the president was sitting. That meeting decided to shy away from the western end of Cuba (where SAMs were becoming operational) and modify the flight pattern of the U-2s in order to reduce the probability that a U-2 would be lost.[101] USIB's unanimous approval of the September estimate reflects similar sensitivities. On September 13 the president had asserted that there were no Soviet offensive missiles in Cuba and committed his administration to act if offensive missiles were discovered. Before congressional committees, administration officials were denying that there was any evidence whatever of offensive missiles in Cuba. The implications of a National Intelligence estimate which concluded that the Soviets were introducing offensive missiles into Cuba were not lost on the men who constituted America's highest intelligence assembly.

The October 4 COMOR decision to direct a flight over the western end of Cuba in effect "overturned" the September estimate, but without officially raising that issue. The decision represented McCone's victory for which he had lobbied with the president before the September 10 decision, in telegrams before the September 10 estimate, and in person after his return to Washington. Though the politics of the intelligence community are closely guarded, several pieces of the story can be told.[102] By September 27, Colonel Wright and others in DIA believed that the Soviet Union was placing missiles in the San Cristobal area.[103] This area was marked suspicious by the CIA on September 29 and certified top priority on October 3. By October 4 McCone had the evidence required to raise the issue officially. The members of COMOR heard McCone's argument, but were reluctant to make the hard decision he demanded. The significant probability that a U-2 would be downed made overflight of western Cuba a matter of real concern.[104]

The Politics of Issues. The U-2 photographs presented incontrovertible evidence of Soviet offensive missiles in Cuba. This revelation fell upon politicized players in a complex context. As one high official recalled, Khrushchev had caught us "with our pants down." What each of the central participants saw, and what each did to cover both his own and the administration's nakedness, created the spectrum of issues and answers.

At approximately 9:00 A.M., Tuesday morning, October 16, McGeorge Bundy went to the president's living quarters with the message: "Mr. President, there is now hard photographic evidence that the Russians have offensive missiles in Cuba."[105] Much has been made of Kennedy's "expression of surprise,"[106] but "surprise" fails to capture the character of his initial reaction. Rather, it was one of startled anger, most adequately conveyed by the exclamation: "He can't do that to *me!*"[107] In terms of the president's attention and priorities at that moment, Khrushchev had chosen the most unhelpful act of all. Kennedy had staked his full presidential authority on the assertion that the Soviets would not place offensive weapons in Cuba. Moreover, Khrushchev had assured the president through the most direct and personal channels that he was aware of the president's

domestic political problem and that nothing would be done to exacerbate this problem. The chairman had *lied* to the president. Kennedy's initial reaction entailed action. The missiles must be removed.[108] The alternatives of "doing nothing" or "taking a diplomatic approach" could not have been less relevant to *his* problem.

These two tracks—doing nothing and taking a diplomatic approach—were the solutions advocated by two of his principal advisers. For Secretary of Defense McNamara, the missiles raised the specter of nuclear war. He first framed the issue as a straightforward strategic problem. To understand the issue, one had to grasp two obvious but difficult points. First, the missiles represented an inevitable occurrence: narrowing of the missile gap. It simply happened sooner rather than later. Second, the United States could accept this occurrence since its consequences were minor: "seven-to-one missile 'superiority,' one-to-one missile 'equality,' one-to-seven missile 'inferiority'—the three postures are identical." McNamara's statement of this argument at the first meeting of the ExCom was summed up in the phrase, "a missile is a missile."[109] "It makes no great difference," he maintained, "whether you are killed by a missile from the Soviet Union or Cuba."[110] The implication was clear. The United States should not initiate a crisis with the Soviet Union, risking a significant probability of nuclear war over an occurrence which had such small strategic implications.

The perceptions of McGeorge Bundy, the president's assistant for national security affairs, are the most difficult of all to reconstruct. There is no question that he initially argued for a diplomatic track.[111] But was Bundy laboring under his acknowledged burden of responsibility in Cuba I? Or was he playing the role of devil's advocate in order to make the president probe his own initial reaction and consider other options?

The president's brother, Robert Kennedy, saw most clearly the political wall against which Khrushchev had backed the president. But he, like McNamara, saw the prospect of nuclear doom. Was Khrushchev going to force the president to an insane act? At the first meeting of the ExCom, he scribbled a note, "Now I know how Tojo felt when he was planning Pearl Harbor."[112] From the outset he searched for an alternative that would prevent the air strike.

The initial reaction of Theodore Sorensen, the president's special counsel and "alter ego," fell somewhere between that of the president and his brother. Like the president, Sorensen felt the poignancy of betrayal. If the president had been the architect of the policy which the missiles punctured, Sorensen was the draftsman. Khrushchev's deceitful move demanded a strong countermove. But like Robert Kennedy, Sorensen feared lest the shock and disgrace lead to disaster.

To the Joint Chiefs of Staff the issue was clear. *Now* was the time to do the job for which they had prepared contingency plans. Cuba I had been badly done; Cuba II would not be. The missiles provided the *occasion* to deal with the issue: cleansing the Western hemisphere of Castro's communism. As the president recalled on the day the crisis ended, "An invasion would have been a mistake—a wrong use of our power. But the military are mad. They wanted to do this. It's lucky for us that we have McNamara over there."[113]

McCone's perceptions flowed from his confirmed prediction. As the Cassandra of the incident, he argued forcefully that the Soviets had installed the missiles in a daring political probe which the United States must meet with force. The time for an air strike was now.[114]

The Politics of Choice. The process by which the blockade emerged is a story of the most subtle and intricate probing, pulling, and hauling; leading, guiding, and spurring. Reconstruction of this process can only be tentative. Initially the president and most of his advisers wanted the clean, surgical air strike. On the first day of the crisis, when informing Stevenson of the missiles, the president mentioned only two alternatives: "I suppose the alternatives are to go in by air and wipe them out, or to take other steps to render them inoperable."[115] At the end of the week a sizeable minority still favored an air strike. As Robert Kennedy recalled: "The fourteen people involved were very significant . . . If six of them had been President of the U.S., I think that the world might have been blown up."[116] What prevented the air strike was a fortuitous coincidence of a number of factors— the absence of any one of which might have permitted that option to prevail.

First, McNamara's vision of holocaust set him firmly against the air strike. His initial attempt to frame the issue in strategic terms struck Kennedy as particularly inappropriate. Once McNamara realized that the name of the game was a strong response, however, he and his deputy Gilpatric chose the blockade as a fallback. When the secretary of defense—whose department had the action, whose reputation in the cabinet was unequaled, in whom the president demonstrated full confidence—marshalled the arguments for the blockade and refused to be moved, the blockade became a formidable alternative.

Second, Robert Kennedy—the president's closest confidant—was unwilling to see his brother become a "Tojo." His arguments against the air strike on moral grounds struck a chord in the president. Moreover, once his brother had stated these arguments so forcefully, the

president could not have chosen his initially preferred course without, in effect, agreeing to become what RFK had condemned.

The president learned of the missiles on Tuesday morning. On Wednesday morning, in order to mask our discovery from the Russians, the president flew to Connecticut to keep a campaign commitment, leaving RFK as the unofficial chairman of the group. By the time the president returned on Wednesday evening, a critical third piece had been added to the picture. McNamara had presented his argument for the blockade. Robert Kennedy and Sorensen had joined McNamara. A powerful coalition of the advisers in whom the President had the greatest confidence, and with whom his style was most compatible, had emerged.

Fourth, the coalition that had formed behind the president's initial preference gave him reason to pause. *Who* supported the air strike— the Chiefs, McCone, Rusk, Nitze, and Acheson—as much as *how* they supported it, counted. Fifth, a piece of inaccurate information, which no one probed, permitted the blockade advocates to fuel (potential) uncertainties in the president's mind. When the president returned to Washington Wednesday evening, RFK and Sorensen met him at the airport. Sorensen gave the president a four-page memorandum outlining the areas of agreement and disagreement. The strongest argument was that the air strike simply could not be surgical.[117] After a day of prodding and questioning, the Air Force had asserted that it could not guarantee the success of a surgical air strike limited to the missiles alone.

Thursday evening, the president convened the ExCom at the White House. He declared his tentative choice of the blockade and directed that preparations be made to put it into effect by Monday morning.[118] Though he raised a question about the possibility of a surgical air strike subsequently, he seems to have accepted the experts' opinion that this was no live option.[119] (Acceptance of this estimate suggests that he may have learned the lesson of the Bay of Pigs—"Never rely on experts"—less well than he supposed.)[120] But this information was incorrect. That no one probed this estimate during the first week of the crisis poses an interesting question for further investigation.

A coalition, including the president, thus emerged from the president's initial decision that something had to be done; McNamara, Robert Kennedy, and Sorensen's resistance to the air strike; incompatibility between the president and the air strike advocates; and an inaccurate piece of information.[121]

CONCLUSION

This essay has obviously bitten off more than it has chewed. For further developments and synthesis of these arguments the reader is referred to the larger study.[122] In spite of the limits of space, however, it would be inappropriate to stop without spelling out several implications of the argument and addressing the question of relations among the models and extensions of them to activity beyond explanation.

At a minimum, the intended implications of the argument presented here are four. First, formulation of alternative frames of reference and demonstration that different analysts, relying predominantly on different models, produce quite different explanations should encourage the analyst's self-consciousness about the nets he employs. The effect of these "spectacles" in sensitizing him to particular aspects of what is going on—framing the puzzle in one way rather than another, encouraging him to examine the problem in terms of certain categories rather than others, directing him to particular kinds of evidence, and relieving puzzlement by one procedure rather than another—must be recognized and explored.

Second, the argument implies a position on the problem of "the state of the art." While accepting the commonplace characterization of the present condition of foreign policy analysis—personalistic, noncumulative, and sometimes insightful—this essay rejects both the counsel of despair's justification of this condition as a consequence of the character of the enterprise, and the "new frontiersmen's" demand for a priori theorizing on the frontiers and ad hoc appropriation of "new techniques."[123] What is required as a first step is noncasual examination of the present product: inspection of existing explanations, articulation of the conceptual models employed in producing them, formulation of the propositions relied upon, specification of the logic of the various intellectual enterprises, and reflection on the questions being asked. Though it is difficult to overemphasize the need for more systematic processing of more data, these preliminary matters of formulating questions with clarity and sensitivity to categories and assumptions so that fruitful acquisition of large quantities of data is possible are still a major hurdle in considering most important problems.

Third, the preliminary, partial paradigms presented here provide a basis for serious reexamination of many problems of foreign and military policy. Model II and model III cuts at problems typically treated in model I terms can permit significant improvements in explanation and prediction.[124] Full model II and III analyses require large amounts of information. But even in cases where the information base is severely limited, improvements are possible. Consider the problem of predicting Soviet strategic forces. In the mid-1950s, model I style calculations led to predictions that the Soviets would rapidly deploy large numbers of long-range bombers. From a model II perspective, both the frailty of the Air Force within the Soviet military establishment and the budgetary implications of such a buildup, would have led analysts to hedge this prediction. Moreover, model II would have pointed to a sure, visible indicator of such a buildup: noisy struggles among the Services over major budgetary shifts. In the late 1950s and early 1960s, model I calculations led to the prediction of immediate, massive Soviet deployment of ICBMs. Again, a model II cut would have reduced this number because, in the earlier period, strategic rockets were controlled by the Soviet Ground Forces rather than an independent service, and in the later period, this would have necessitated massive shifts in budgetary splits. Today, model I considerations lead many analysts both to recommend that an agreement not to deploy ABMs be a major American objective in upcoming strategic negotiations with the USSR, and to predict success. From a model II vantage point, the existence of an on-going Soviet ABM program, the strength of the organization (National Air Defense) that controls ABMs, and the fact that an agreement to stop ABM deployment would force the virtual dismantling of this organization, make a viable agreement of this sort much less likely. A model III cut suggests that (a) there must be significant differences among perceptions and priorities of Soviet leaders over strategic negotiations, (b) any agreement will affect some players' power bases, and (c) agreements that do not require extensive cuts in the sources of some major players' power will prove easier to negotiate and more viable.

Fourth, the present formulation of paradigms is simply an initial step. As such it leaves a long list of critical questions unanswered. Given any action, an imaginative analyst should always be able to construct some rationale for the government's choice. By imposing, and relaxing, constraints on the parameters of rational choice (as in variants of model I) analysts can construct a large number of accounts of any act as a rational choice. But does a statement of reasons why a rational actor would choose an action constitute an explanation of the *occurrence* of that action? How can model I analysis be forced to make more systematic contributions to the question of the determinants of occurrences? Model II's explanation of t in terms of $t - 1$ is explanation. The world is contiguous. But governments sometimes make sharp departures. Can an organizational process model be modified to suggest where change is likely? Attention to organizational change should afford greater understanding of why particular programs and SOPs are maintained by identifiable types of organizations and also how a manager can improve organizational performance. Model III tells a fascinating "story." But its complexity is enormous, the information requirements are often overwhelming, and many of the details of the bargaining may be superfluous. How can such a model be made parsimonious? The three models are obviously not exclusive alternatives. Indeed, the paradigms highlight the partial emphasis of the framework—what each emphasizes and what it leaves out. Each concentrates on one class of variables, in effect, relegating other important factors to a *ceteris paribus* clause. Model I concentrates on "market factors": pressures and incentives created by the "international strategic marketplace." Models II and III focus on the internal mechanism of the government that chooses in this environment. But can these relations be more fully specified? Adequate synthesis would require a typology of decisions and actions, some of which are more amenable to treatment in terms of one model and some to another. Government behavior is but one cluster of factors relevant to occurrences in foreign affairs. Most students of foreign policy adopt this focus (at least when explaining and predicting). Nevertheless, the dimensions of the chess board, the character of the pieces, and the rules of the game—factors considered by international systems theorists—constitute the context in which the pieces are moved. Can the major variables in the full function of determinants of foreign policy outcomes be identified?

Both the outline of a partial, ad hoc working synthesis of the models, and a sketch of their uses in activities other than explanation can be suggested by generating predictions in terms of each. Strategic surrender is an important problem of international relations and diplomatic history. War termination is a new, developing area of the strategic literature. Both of these interests lead scholars to address a central question: *Why do nations surrender when?* Whether implicit in explanations or more explicit in analysis, diplomatic historians and strategists rely upon propositions which can be turned forward to produce predictions. Thus at the risk of being timely—and in error—the present situation (August, 1968) offers an interesting test case: Why will North Vietnam surrender when?[125]

In a nutshell, analysis according to model I asserts: nations quit when costs outweigh the benefits. North Vietnam will surrender when it realizes "that continued fighting can only generate additional costs without hope of compensating gains, this expectation being largely the consequence of the previous application of force by the dominant side."[126] US actions can increase or decrease Hanoi's strategic costs. Bombing North Vietnam increases the pain and thus increases the probability of surrender. This proposition and prediction are not without meaning. That—"other things being equal"— nations are more likely to surrender when the strategic cost-benefit balance is negative, is true. Nations rarely surrender when they are winning. The proposition specifies a range

within which nations surrender. But over this broad range, the relevant question is: why do nations surrender?

Models II and III focus upon the government machine through which this fact about the international strategic marketplace must be filtered to produce a surrender. These analysts are considerably less sanguine about the possibility of surrender *at the point* that the cost-benefit calculus turns negative. Never in history (i.e., in none of the five cases I have examined) have nations surrendered at that point. Surrender occurs sometime thereafter. *When* depends on process of organizations and politics of players within these governments—as they are affected by the opposing government. Moreover, the effects of the victorious power's action upon the surrendering nation cannot be adequately summarized as increasing or decreasing strategic costs. Imposing additional costs by bombing a nation may increase the probability of surrender. But it also may reduce it. An appreciation of the impact of the acts of one nation upon another thus requires some understanding of the machine which is being influenced. For more precise prediction, models II and III require considerably more information about the organizations and politics of North Vietnam than is publicly available. On the basis of the limited public information, however, these models can be suggestive.

Model II examines two subproblems. First, to have lost is not sufficient. The government must know that the strategic cost-benefit calculus is negative. But neither the categories, nor the indicators, of strategic costs and benefits are clear. And the sources of information about both are organizations whose parochial priorities and perceptions do not facilitate accurate information or estimation. Military evaluation of military performance, military estimates of factors like "enemy morale," and military predictions concerning when "the tide will turn" or "the corner will have been turned" are typically distorted. In cases of highly decentralized guerrilla operations, like Vietnam, these problems are exacerbated. Thus strategic costs will be underestimated. Only highly *visible* costs can have direct impact on leaders without being filtered through organizational channels. Second, since organizations define the details of options and execute actions, surrender (and negotiation) is likely to entail considerable bungling in the early stages. No organization can define options or prepare programs for this treasonous act. Thus, early overtures will be uncoordinated with the acts of other organizations, e.g., the fighting forces, creating contradictory "signals" to the victor.

Model III suggests that surrender will not come at the point that strategic costs outweigh benefits, but that it will not wait until the leadership group concludes that the war is lost.

Rather the problem is better understood in terms of four additional propositions. First, strong advocates of the war effort, whose careers are closely identified with the war, rarely come to the conclusion that costs outweigh benefits. Second, quite often from the outset of a war, a number of members of the government (particularly those whose responsibilities sensitize them to problems other than war, e.g., economic planners or intelligence experts) are convinced that the war effort is futile. Third, surrender is likely to come as the result of a political shift that enhances the effective power of the latter group (and adds swing members to it). Fourth, the course of the war, particularly actions of the victor, can influence the advantages and disadvantages of players in the loser's government. Thus, North Vietnam will surrender not when its leaders have a change of heart, but when Hanoi has a change of leaders (or a change of effective power within the central circle). How US bombing (or pause), threats, promises, or action in the South affect the game in Hanoi is subtle but nonetheless crucial.

That these three models could be applied to the surrender of governments other than North Vietnam should be obvious. But that exercise is left for the reader.

NOTES

1. Theodore Sorensen, *Kennedy* (New York: Harper and Row, 1965), p. 705.

2. In attempting to understand problems of foreign affairs, analysts engage in a number of related, but logically separable enterprises: (a) description, (b) explanation, (c) prediction, (d) evaluation, and (e) recommendation. This essay focuses primarily on explanation (and by implication, prediction).

3. In arguing that explanations proceed in terms of implicit conceptual models, this essay makes no claim that foreign policy analysts have developed any satisfactory, empirically tested theory. In this essay, the use of the term "model" without qualifiers should be read "conceptual scheme."

4. For the purpose of this argument we shall accept Carl G. Hempel's characterization of the logic of explanation: an explanation "answers the question, 'Why did the explanadum-phenomenon occur?' by showing that the phenomenon resulted from particular circumstances, specified in $C_1, C_2, \ldots C_k$, in accordance with laws $L_1, L_2, \ldots L_r$. By pointing this out, the argument shows that, given the particular circumstances and the laws in question, the occurrence of the phenomenon was to be *expected*; and it is in this sense that the explanation enables us to understand why the phenomenon occurred." *Aspects of Scientific Explanation* (New York: Harcourt, Brace and World, 1961), p. 337. While various patterns of explanation can be distinguished, *viz.*, Ernest Nagel, *The Structure of Science: Problems in the Logic of Scientific Explanation* (New York: Harcourt, Brace and World, 1961), satisfactory scientific explanations exhibit this basic logic. Consequently prediction is the converse of explanation.

5. Earlier drafts of this argument have aroused heated arguments concerning proper names for these models. To choose names from ordinary language is to court confusion, as well as familiarity. Perhaps it is best to think of these models as I, II, and III.

6. In strict terms, the "outcomes" which these three models attempt to explain are essentially actions of national governments (i.e., the sum of activities of all individuals employed by a government relevant to an issue). These models focus not on a state of affairs (i.e., a full description of the world) but upon national decision and implementation. This distinction is stated clearly by Harold and Margaret Sprout, "Environmental Factors on the Study of International Politics," in James Rosenau, ed., *International Politics and Foreign Policy* (Glencoe, Ill.: Free Press, 1961), p. 116. This restriction excludes explanations offered principally in terms of international systems theories. Nevertheless, this restriction is not severe, since few interesting explanations of occurrences in foreign policy have been produced at that level of analysis. According to David Singer, "The nation state—our primary actor in international relations . . . is clearly the traditional focus among Western students and is the one which dominates all of the texts employed in English-speaking colleges and universities." David Singer, "The Level-of-Analysis Problem in International Relations," Klaus Knorr and Sidney Verba, eds., *The International System* (Princeton: Princeton University Press, 1961). Similarly, Richard Brody's review of contemporary trends in the study of international relations finds that "scholars have come increasingly to focus on acts of nations. That is, they all focus on the behavior of nations in some respect. Having an interest in accounting for the behavior of nations in common, the prospects for a common frame of reference are enhanced."

7. For further development and support of these arguments see the author's larger study, *The Essence of Decision: Explaining the Cuban Missile Crisis* (Boston: Little, Brown, 1971). In its abbreviated form, the argument must, at some points, appear overly stark. The limits of space have forced the omission of many reservations and refinements.

8. Each of the three "case snapshots" displays the work of a conceptual model as it is applied to explain the US blockade of Cuba. But these three cuts are primarily exercises in hypothesis generation rather than hypothesis testing. Especially when separated from the larger study, these accounts may be misleading. The sources for these accounts include the full public record plus a large number of interviews with participants in the crisis.

9. *New York Times*, 18 Feb. 1967.

10. Ibid.

11. Arnold Horelick and Myron Rush, *Strategic Power and Soviet Foreign Policy* (Chicago: University of Chicago Press, 1965). Based on A. Horelick, "The Cuban Missile Crisis: An Analysis of Soviet Calculations and Behavior," *World Politics* 16 (Apr. 1964).

12. Horelick and Rush, *Strategic Power*, p. 154.

13. Hans Morgenthau, *Politics among Nations*, 3d ed. (New York: Knopf, 1960), p. 191.

14. Ibid., p. 192.

15. Ibid., p. 5.

16. Ibid., pp. 5–6.

17. Stanley Hoffmann, *Daedalus* 91 (Fall 1962); reprinted in Stanley Hoffmann, *The State of War* (New York: Praeger, 1965).

18. Ibid., p. 171.

19. Ibid., p. 189.

20. Following Robert MacIver; see Stanley Hoff-

mann, *Contemporary Theory in International Relations* (Englewood Cliffs, N.J.: Prentice-Hall, 1960), pp. 178–79.

21. Thomas Schelling, *The Strategy of Conflict* (New York: Harvard University Press, 1960), p. 232. This proposition was formulated earlier by A. Wohlstetter, "The Delicate Balance of Terror," *Foreign Affairs* 37 (Jan. 1959).

22. Schelling, *Strategy of Conflict*, p. 4

23. See Morgenthau, *Politics among Nations*, p. 5; Hoffmann, "Roulette in the Cellar," in *State of War*; Schelling, *Strategy of Conflict*.

24. The larger study examines several exceptions to this generalization. Sidney Verba's excellent essay "Assumptions of Rationality and Non-Rationality in Models of the International System" is less an exception than it is an approach to a somewhat different problem. Verba focuses upon models of rationality and irrationality of *individual* statesmen: in Knorr and Verba, *International System*.

25. Robert K. Merton, *Social Theory and Social Structures*, rev. and enl. ed. (New York: Free Press, 1957), pp. 12–16. Considerably weaker than a satisfactory theoretical model, paradigms nevertheless represent a short step in that direction from looser, implicit conceptual models. Neither the concepts nor the relations among the variables are sufficiently specified to yield propositions deductively. "Paradigmatic Analysis" nevertheless has considerable promise for clarifying and codifying styles of analysis in political science. Each of the paradigms stated here can be represented rigorously in mathematical terms. For example, model I lends itself to mathematical formulation along the lines of Herbert Simon's "Behavioral Theory of Rationality," *Models of Man* (New York: Wiley, 1957). But this does not solve the most difficult problem of "measurement and estimation."

26. Though a variant of this model could easily be stochastic, this paradigm is stated in nonprobabilistic terms. In contemporary strategy, a stochastic version of this model is sometimes used for predictions; but it is almost impossible to find an explanation of an occurrence in foreign affairs that is consistently probabilistic.

Analogies between model I and the concept of explanation developed by R. G. Collingwood, William Dray, and other "revisionists" among philosophers concerned with the critical philosophy of history are not accidental. For a summary of the "revisionist position" see Maurice Mandelbaum, "Historical Explanation: The Problem of Covering Laws," *History and Theory* 1 (1960).

27. This model is an analogue of the theory of the rational entrepreneur which has been developed extensively in economic theories of the firm and the consumer. These two propositions specify the "substitution effect." Refinement of this model and specification of additional general propositions by translating from the economic theory is straightforward.

28. *New York Times*, 22 Mar. 1969.

29. See Nathan Leites, *A Study of Bolshevism* (Glencoe, Ill.: Free Press, 1953).

30. As stated in the introduction, this "case snapshot" presents, without editorial commentary, a model I analyst's explanation of the US blockade. The purpose is to illustrate a strong, characteristic rational policy model account. This account is (roughly) consistent with prevailing explanations of these events.

31. Sorensen, *Kennedy*, p. 675.

32. Ibid., p. 679.

33. Ibid.

34. Elie Abel, *The Missile Crisis* (New York: J. B. Lippincott, 1966), p. 144.

35. Ibid., p. 102.

36. Sorensen, *Kennedy*, p. 684.

37. Ibid., p. 685. Though this was the formulation of the argument, the facts are not strictly accurate. Our tradition against surprise attack was rather younger than 175 years. For example, President Theodore Roosevelt applauded Japan's attack on Russia in 1904.

38. *New York Times*, 11 June 1963.

39. The influence of organizational studies upon the present literature of foreign affairs is minimal. Specialists in international politics are not students of organization theory. Organization theory has only recently begun to study organizations as decisionmakers and has not yet produced behavioral studies of national security organizations from a decision-making perspective. It seems unlikely, however, that these gaps will remain unfilled much longer. Considerable progress has been made in the study of the business firm as an organization. Scholars have begun applying these insights to government organizations, and interest in an organizational perspective is spreading among institutions and individuals concerned with actual government operations. The "decisionmaking" approach represented by Richard Snyder, R. Bruck, and B. Sapin, *Foreign Policy Decision-Making* (Glencoe, Ill.: Free Press, 1962), incorporates a number of insights from organization theory.

40. The formulation of this paradigm is indebted both to the orientation and insights of Herbert Simon and to the behavioral model of the firm stated by Richard Cyert and James March, *A Behavioral Theory of the Firm* (Englewood Cliffs, N.J.: Prentice-Hall, 1963). Here, however, one is forced to grapple with the less routine, less quantified functions of the less differentiated elements in government organizations.

41. Theodore Sorensen, "You Get to Walk to Work," *New York Times Magazine*, 19 Mar. 1967.

42. Organizations are not monolithic. The proper level of disaggregation depends upon the objectives of a piece of analysis. This paradigm is formulated with reference to the major organizations that constitute the US government. Generalization to the major components of each department and agency should be relatively straightforward.

43. The stability of these constraints is dependent on such factors as rules for promotion and reward, budgeting and accounting procedures, and mundane operating procedures.

44. Marriner Eccles, *Beckoning Frontiers* (New York: A. A. Knopf, 1951), p. 336.

45. Arthur Schlesinger, *A Thousand Days* (Boston: Houghton-Mifflin, 1965), p. 406.

46. US Department of State, *Bulletin* 47, pp. 715–20.

47. Schlesinger, *Thousand Days*, p. 803.

48. Sorensen, *Kennedy*, p. 675.

49. See US Congress, Senate, Committee on Armed Services, Preparedness Investigation Subcommittee, *Interim Report on Cuban Military Buildup*, 88th Cong., 1st sess., 1963, p. 2; Hanson Baldwin, "Growing Risks of Bureaucratic Intelligence," *The*

Reporter 29 (15 Aug. 1963): 48–50; Roberta Wohl-stetter, "Cuba and Pearl Harbor," *Foreign Affairs* 43 (July 1965): 706.

50. US Congress, House of Representatives, Committee on Appropriation, Subcommittee on Department of Defense Appropriations, *Hearings*, 88th Cong., 1st sess., 1963, 25ff.

51. R. Hilsman, *To Move a Nation* (New York: Doubleday, 1967), pp. 172–73.

52. Department of Defense Appropriations, *Hearings*, p. 67.

53. Ibid., pp. 66–67.

54. For (1) Hilsman, *Move a Nation*, p. 186; (2) Abel, *Missile Crisis*, p. 24; (3) Department of Defense Appropriations, *Hearings*, pp. 1–30.

55. The facts here are not entirely clear. This assertion is based on information from (1) "Department of Defense Briefing by the Honorable R. S. McNamara, Secretary of Defense, State Department Auditorium, 5:00 p.m., February 6, 1963." A verbatim transcript of a presentation actually made by General Carroll's assistant, John Hughes; and (2) Hilsman's statement, *Move a Nation*, p. 186. But see R. Wohlstetter's interpretation, "Cuba and Pearl Harbor," p. 700.

56. See Hilsman, *Move a Nation*, pp. 172–74.

57. Abel, *Missile Crisis*, pp. 26ff; Weintal and Bartlett, *Facing the Brink* (New York: Scribner's, 1967), pp. 62ff.; *Cuban Military Build-up*; J. Daniel and J. Hubbell, *Strike in the West* (New York: Scribner's, 1963), pp. 15ff.

58. Schlesinger, *Thousand Days*, p. 804.

59. Sorensen, *Kennedy*, p. 684.

60. Ibid., pp. 684ff.

61. Ibid., pp. 694–97.

62. Ibid., p. 697; Abel, *Missile Crisis*, pp. 100–101.

63. Sorensen, *Kennedy*, p. 669.

64. Hilsman, *Move a Nation*, p. 204.

65. See Abel, *Missile Crisis*, pp. 97ff.

66. Schlesinger, *Thousand Days*, p. 818.

67. Ibid.

68. Sorensen, *Kennedy*, p. 710.

69. *New York Times*, 27 Oct. 1962.

70. Abel, *Missile Crisis*, p. 171.

71. For the location of the orginal arc see ibid., p. 141.

72. *Facts on File*, vol. 22, 1962, p. 376, published by Facts on File, Inc., New York, yearly.

73. This hypothesis would account for the mystery surrounding Kennedy's explosion at the leak of the stopping of the *Bucharest*. See Hilsman, *Move a Nation*, p. 45.

74. Abel, *Missile Crisis*, p. 153.

75. See ibid., pp. 154ff.

76. Ibid., p. 156.

77. Ibid.

78. This paradigm relies upon the small group of analysts who have begun to fill the gap. My primary source is the model implicit in the work of Richard E. Neustadt, though his concentration on presidential action has been generalized to a concern with policy as the outcome of political bargaining among a number of independent players, the president amounting to no more than a "superpower" among many lesser but considerable powers. As Warner Schilling argues, the substantive problems are of such inordinate difficulty that uncertainties and differences with regard to goals, alternatives, and consequences are inevita-ble. This necessitates what Roger Hilsman describes as the process of conflict and consensus building. The techniques employed in this process often resemble those used in legislative assemblies, though Samuel Huntington's characterization of the process as "legislative" overemphasizes the equality of participants as opposed to the hierarchy which structures the game. Moreover, whereas for Huntington, foreign policy (in contrast to military policy) is set by the executive, this paradigm maintains that the activities which he describes as legislative are characteristic of the process by which foreign policy is made.

79. The theatrical metaphor of stage, roles, and actors is more common than this metaphor of games, positions, and players. Nevertheless, the rigidity connotated by the concept of "role" both in the theatrical sense of actors reciting fixed lines and in the sociological sense of fixed responses to specified social situations makes the concept of games, positions, and players more useful for this analysis of active participants in the determination of national policy. Objections to the terminology on the grounds that "game" connotes nonserious play overlook the concept's application to most serious problems both in Wittgenstein's philosophy and in contemporary game theory. Game theory typically treats more precisely structured games, but Wittgenstein's examination of the "language game" wherein men use words to communicate is quite analogous to this analysis of the less specified game of bureaucratic politics. See Ludwig Wittgenstein, *Philosophical Investigations*, 3d. ed. (New York: Macmillan, 1968), and Thomas Schelling, "What Is Game Theory?" in James Charlesworth, *Contemporary Political Analysis* (New York: Free Press, 1967).

80. Inclusion of the president's special assistant for national security affairs in the tier of "Chiefs" rather than among the "Staffers" involves a debatable choice. In fact he is both super-staffer and near-chief. His position has no statutory authority. He is especially dependent upon good relations with the president and the secretaries of defense and state. Nevertheless, he stands astride a genuine action-channel. The decision to include this position among the Chiefs reflects my judgment that the Bundy function is becoming institutionalized.

81. Richard E. Neustadt, Testimony, United States Senate, Committee on Government Operations, Subcommittee on National Security Staffing, *Administration of National Security*, 26 Mar. 1963, pp. 82–83.

82. This aphorism was stated first, I think, by Don K. Price.

83. Paul Y. Hammond, "Super Carriers and B-36 Bombers," in Harold Stein, ed., *American Civil-Military Decisions* (Birmingham: University of Alabama Press, 1963).

84. Roberta Wohlstetter, *Pearl Harbor* (Stanford: Stanford University Press, 1962), p. 350.

85. Sorensen, *Kennedy*, p. 670.

86. Ibid.

87. Ibid., pp. 670ff.

88. *New York Times*, Aug., Sept. 1962.

89. Ibid., 20 Aug. 1962.

90. Ibid., 5 Sept. 1962.

91. Ibid., 14 Sept. 1962.

92. Ibid., 14 Oct. 1962.

93. Cited by Abel, *Missile Crisis*, p. 13.

94. *New York Times*, 5 Sept. 1962.

95. Ibid., 14 Sept. 1962.

96. Senate Foreign Relations Committee; Senate Armed Services Committee; House Committee on Appropriation; House Select Committee on Export Control.

97. Abel, *Missile Crisis*, pp. 17–18. According to McCone, he told Kennedy, "The only construction I can put on the material going into Cuba is that the Russians are preparing to introduce offensive missiles." See also Weintal and Bartlett, *Facing the Brink*, pp. 60–61.

98. Abel, *Missile Crisis*, p. 23.

99. *New York Times*, 10 Sept. 1962.

100. See Abel, *Missile Crisis*, pp. 25–26; and Hilsman, *Move a Nation*, p. 174.

101. Department of Defense Appropriation, *Hearings*, p. 69.

102. A basic, but somewhat contradictory, account of parts of this story emerges in ibid., pp. 1–70.

103. Ibid., p. 71.

104. The details of the 10 days between the Oct. 4 decision and the Oct. 14 flight must be held in abeyance.

105. Abel, *Missile Crisis*, p. 44.

106. Ibid., pp. 44ff.

107. See Richard Neustadt, "Afterword," *Presidential Power* (New York: John Wiley, 1964).

108. Sorensen, *Kennedy*, 676; Schlesinger, *Thousand Days*, p. 801.

109. Hilsman, *Move a Nation*, p. 195.

110. Ibid.

111. Weintal and Bartlett, *Facing the Brink*, p. 67; Abel, *Missile Crisis*, p. 53.

112. Schlesinger, *Thousand Days*, p. 803.

113. Ibid., p. 831.

114. Abel, *Missile Crisis*, p. 186.

115. Ibid., p. 49.

116. Interview, quoted by Ronald Steel, *New York Review of Books*, 13 Mar. 1969, p. 22.

117. Sorensen, *Kennedy*, p. 686.

118. Ibid., p. 691.

119. Ibid., pp. 691–92.

120. Schlesinger, *Thousand Days*, p. 296.

121. Space will not permit an account of the path from this coalition to the formal government decision on Saturday and action on Monday.

122. Graham T. Allison, *Essence of Decision* (Boston: Little, Brown, 1971).

123. Thus my position is quite distinct from both poles in the recent "great debate" about international relations. While many "traditionalists" of the sort Kaplan attacks adopt the first posture and many "scientists" of the sort attacked by Bull adopt the second, this third posture is relatively neutral with respect to whatever is in substantive dispute. See Hedley Bull, "International Theory: The Case for a Classical Approach," *World Politics* 18 (Apr. 1966); and Morton Kaplan, "The New Great Debate: Traditionalism vs. Science in International Relations," *World Politics* 19 (Oct. 1966).

124. A number of problems are now being examined in these terms both in the Bureaucracy Study Group on Bureaucracy and Policy of the Institute of Politics at Harvard University and at the RAND Corporation.

125. In response to several readers' recommendations, what follows is reproduced verbatim from the paper delivered at the Sept. 1968 Association meetings (RAND P-3919). The discussion is heavily indebted to Ernest R. May.

126. Richard Snyder, *Deterrence and Defense* (Princeton: Princeton University Press, 1961), p. 11. For a more general presentation of this position see Paul Kecskemeti, *Strategic Surrender* (New York: Stanford University Press, 1964).

THE ROLE OF THE NATIONAL SECURITY COUNCIL IN COORDINATING AND INTEGRATING US DEFENSE AND FOREIGN POLICY

PHILIP A. ODEEN

The role of the National Security Council (NSC) staff in the national security process is so well accepted today that major influence exerted by the assistant to the president for national security affairs and his staff is rarely questioned. Since McGeorge Bundy, the NSC adviser has been seen as equal and often more than equal in importance to the secretary of state and he usually overshadows the role of statutory officials such as the directors of the Arms Control and Disarmament Agency (ACDA) and the Central Intelligence Agency (CIA). Despite this, the functions of the president's national security assistant and his staff have only been given limited systematic attention.[1]

This chapter examines the functions of the NSC adviser and his staff. Two principal roles are assessed. First, the NSC's role as the president's personal staff. Second, its institutional role as the executive branch staff with the power and position to coordinate and manage security-related issues that cut across the interests and responsibilities of two or more government departments or agencies. Particular attention is given to the NSC's role in managing the crises that administrations inevitably face, a critical

Reprinted with minor revisions and with permission from Public Policy and Political Institutions: Defense and Foreign Policy, *pp. 19–41. Copyright © 1985 by JAI Press, Inc.*

issue that deserves greater emphasis than it has thus far received.

BACKGROUND

There is no ideal system for managing national security policy. What is best at any particular point in time depends on the president's interests and management style; the experience, standing, and personalities of his key cabinet officers; and, to a degree, the nature of the issues they face. The experience, personality, and drive of the national security assistant are also factors of consequence, as was so well demonstrated during Henry Kissinger's tenure as President Nixon's key adviser.

Based on such considerations, each president will tailor the system to reflect his unique needs and style. Some presidents prefer a very structured system, others a more informal approach. But presidents should be cautious; too much informality poses risks both to the success of the administration and to the security interests of the nation. There are times when the president will be involved, in some depth, in major security decisions. To support this involvement, a degree of structure is essential in any president's national security system, regardless of his philosophy. Without some structure he is unlikely to get the facts, options, and advice he needs, and the decision process may be haphazard and shallow.

Experience suggests that every newly elected president, as he plans his administration, must consider the following in designing a national security system:

- Some questions must be decided by the president. Issues such as a basing mode for the MX missile or arms control policy critically affect the president and his administration and, therefore, are simply too important for cabinet officers alone to decide.
- Issues that reach the president for decision are inevitably complex. (The simple ones are decided at a lower level.) Both organizational structures and deliberative processes must ensure that a full range of options are developed and that the president addresses them early enough to have a meaningful choice. In addition, an informed presidential decision requires rigorous analysis of the options.
- On many issues, one department will have primary responsibility, but others will still have legitimate roles and interests. For example, decisions about theater nuclear weapons in Europe may be Department of Defense (DoD) business, but they have major implications for the Department of State and ACDA. If the structures and processes fail to integrate these agencies' views, poorer decisions may result. Moreover, unless they know

that their views have been fairly considered the level of discontent and discord within the administration will increase. The administration will then appear uncoordinated and at odds with itself, a problem that afflicted both the Reagan and Carter administrations.
- Policy execution as well as policymaking must receive attention. The NSC staff must ensure that the government performs in consonance with the president's decisions. White House staffs often feel that their only role is to "make policy," that execution is someone else's problem. The first requirement for sound execution is the clear communication of decisions and follow-up responsibilities. Also important is the expectation by the agencies that high-level review of the implementation will occur. In addition, the system should include the regular involvement of the bureaucracy in policymaking as a way of securing commitment to its successful execution. These conditions will increase the probability of effective policy execution and reduce the appearance of dissension and poor discipline in the affected agencies.

Given these considerations, some reasonable degree of structure in the NSC system is essential. Informal lunches and ad hoc meetings have their place. Issues can often be addressed more frankly and fully in such a setting. Unfortunately, leaks are so prevalent that frankness is constrained at formal sessions with their written papers, minutes, and "debriefs" to the staff in the aftermath. Thus, informal sessions are important. But they cannot replace a carefully structured system with clear roles and responsibilities for the key departmental and White House players. Such a structure is particularly important in the NSC's institutional role, a topic that will be discussed later.

THE NSC AND THE PRESIDENT

The most difficult role to define is the personal role that the NSC adviser and staff should play in supporting the president's management of defense and foreign policy issues. No firm set of rules can be prescribed since they will depend heavily on the president's own desires and style. Even the casual observer can visualize the sharply differing needs of Presidents Johnson, Nixon, Carter, and Reagan. Yet some criteria, let us call them "universal" guidelines, can be set. There are at least four areas where NSC support to the president is necessary.

- The president must be kept informed on key issues that he may be forced to address, be they potential crises or major matters that will gradually bubble to the top of the executive branch. If he appears uninformed, this will negatively affect public and media perceptions. Moreover, he may have a strong view on an issue that should be understood early by the bureaucracy, not when it eventually comes to him for decision.

- He must be brought into the issues in a timely fashion so that he, in fact, can make decisions. In many cases, if the president is not brought into the decision process in time, he is presented with a situation where he has little room for maneuver, or, if he does act, he risks a major controversy if he makes a choice other than that proposed by the bureaucracy. President Carter's decision to cancel the "neutron bomb" is illustrative.[2]
- The president must appear to be in charge; and, hopefully, it is more than appearance. The NSC staff should ensure that he is well prepared for press encounters and foreign visits. Major decisions should be announced by the White House, or at least an important White House role should be apparent. To the degree possible, divisive differences between agencies should be minimized or at least not made prominent as was the case in both the Carter administration (Cyrus Vance versus Zbigniew Brzezinski) and the first year of the Reagan administration (Alexander Haig versus Caspar Weinberger and the White House staff).
- Full support for the president during crises or conflicts is perhaps the most critical aspect of the NSC's responsibilities. When a crisis occurs, the president will be forced to play a central role and he will require a steady flow of information, analysis, and advice. As will be discussed in more detail later, this support should not be ad hoc or dependent on actions that take place after the crisis develops. Advance planning is essential.

These functions are essential and must be carried out regardless of the president's style. But style will play a major role in how the staff functions in support of the president. President Carter, for example, demanded detailed information and analysis and was not deterred by long reports. On the other hand, President Nixon expected three-page decision papers. Carter preferred to have things in writing while Reagan preferred verbal briefings. President Johnson searched out and dealt with staff at all levels in the White House. By contrast, Nixon seldom met with anyone on the staff except for Kissinger and, later, Kissinger's deputy, Alexander Haig. While these style differences significantly affected the attractiveness of the NSC as a place to work, they did not change, fundamentally, what the NSC staff needed to do to support the president.

THE NSC STAFF'S INSTITUTIONAL ROLE

While the NSC staff's role as the president's key White House body for foreign affairs is very visible, its institutional role is far more important. Thus, regardless of the personal demands made on the staff by the president or the manner in which he decides to organize the NSC staff and system, there are several institutional functions that must be carried out. The NSC staff must devote time and attention to five such responsibilities: setting out a policy framework, forcing decisions on major issues, managing the decision process, ensuring that decisions are implemented, and planning for crises.

SETTING OUT A POLICY FRAMEWORK

The first and most obvious task is to define the basic security policies and priorities for the administration. Such a policy framework is essential to guide the programs and actions of Defense, State, ACDA, and the other agencies involved in the nation's security and foreign policy affairs. This framework is normally the product of a series of studies to set the basic parameters for our security policy, such as our policy goals and priorities for Europe, the Middle East, and Japan; the principal objectives and missions of our military forces; and the administration's approach to major arms control issues. Interagency studies often serve as the basis for these policy decisions and the initial months of an administration are frequently periods of intense policy debate and decision. After this opening period, the outlines of the administration's policies are usually well defined and its priorities determined.

The early days of the Nixon and Carter administrations witnessed a blizzard of policy studies. The Nixon White House launched 85 studies in its first year, called National Security Study Memoranda (NSSMs). Carter undertook 32 in 1977, called Policy Review Memoranda (PRMs). By contrast, little such activity occurred in Reagan's first year. This partly reflected a greater degree of decentralization—Secretary Caspar Weinberger was given an unusual degree of latitude in setting defense policy and priorities. It also reflected the virtually exclusive focus of the president and his staff on budget and tax issues. In view of the subsequent attacks on the Reagan administration's arms control policy and the erosion of support for its defense program and budget, this was probably a major error.

FORCING DECISIONS ON MAJOR ISSUES

Helping the president cope with the heavy flow of issues coming to him for decision is also a basic function of the NSC staff. Yet, it has an equally important responsibility to identify the major issues in advance that will require the administration's attention, and then to see that they are adequately addressed. This is frequently more difficult than it would appear. The departments are often reluctant to expose certain issues to presidential scrutiny. Decisions on major weapons systems are a good example. DoD is seldom anxious to have "help" from the NSC, let alone State or ACDA, in deciding whether or not to produce a particular missile or aircraft. Another factor concerns the political costs that often are involved in addressing tough issues. Sensible management considerations may suggest that a problem be raised and decided. But politicians often feel otherwise. Base closures and the military retirement system are good examples of such issues. A final factor is the demanding pace of routine business, and this consideration is often the most significant. Vital issues and concerns are often neither immediate nor pressing; they are, therefore, not seen as critical. But the opposite is often true— the seemingly critical issues are frequently not very important to the nation's security interests, though they may seem so at the time. Thus, the challenge is to focus on those genuinely significant questions.

Issues that have received inadequate attention in recent years, where the NSC should have played a major role in forcing decisions, include the following:

- The future size and roles of the US Navy remain largely unresolved, or at least not widely agreed to. The problem is widely recognized and several efforts have been made to bring it into focus. But the tough questions have not been addressed adequately even within the DoD. Admittedly, it is probably the most difficult and contentious issue facing security planners, but it deserves greater attention and higher priority than it has received. Somewhat surprisingly, this issue has not been addressed by the Reagan administration—or even within DoD despite the heavy budget allocation and plans for a 600-ship Navy. The shots have been called by the Navy and to a large degree by the Secretary of the Navy John Lehman himself.
- There are force structure issues that affect our ability to execute foreign policy, such as the adequacy of our strategic airlift and sealift capabilities. Reagan's defense establishment gave this area greater priority than did Carter's, but it received little attention outside the Pentagon. Questions of this type deserve greater visibility either within the NSC system or between the White House staff and DoD so the president's priorities

will be considered when program choices are made.
- Some potentially critical aspects of arms control policy received little systematic focus during the early years of the Reagan administration. In the long term, however, they may have exceedingly important implications. Examples are the rapid development of military capabilities in space and the growing risks of nuclear proliferation.
- Similarly, the Reagan NSC gave little systematic focus to other issues that have potentially great significance. Perhaps the best example concerns the upgrading of military capabilities by our NATO allies, a central American policy thrust throughout the 1970s. Despite the heavy investment in U.S. military capability, ongoing efforts to strengthen allied forces were essentially ignored by the NSC staff and system. Even the Office of the Secretary of Defense (OSD) seemed to lose its enthusiasm for these programs.

MANAGING THE DECISION PROCESS

In defining policies or addressing specific issues, the NSC staff must manage the interagency process. The staff cannot guarantee that sound decisions will be made, but they can make the decision process more orderly and increase the flow of useful information, thereby increasing the likelihood of sensible decisions. Managing the process requires determining the appropriate forum for issue consideration—interagency, bilateral (between two departments or bureaucratic entities), or a single department; ensuring that all realistic options are considered, not just those proposed by the bureaucracy; pressing for sound analysis and exposing it to sharp criticism; and presenting the resulting options and analyses clearly and concisely to the decisionmakers.

Process management was a major priority for the NSC staff under Kissinger, especially in the first Nixon term. It received less attention under Brzezinski. And the Reagan NSC staff seemed to give this role little, if any, importance. Managing the decision process is a time-consuming and demanding, but important, task. Unless it is done well, poor decisions are likely to be made and excessive time required of the president and the cabinet. Inadequate process management may be a price President Carter paid for asking the NSC staff to give priority to policy advocacy and personal staff support. The Reagan administration suffered because of the lack of emphasis given to the NSC staff's institutional role and the failure of Richard Allen and, then, Judge William Clark to develop a strong and experienced staff.

Ensuring That Decisions Are Implemented

The fourth institutional task of the NSC staff is to ensure that the president's policies and decisions are carried out. First, this involves clearly communicating the decisions, and their rationale, to the rest of government. This may seem obvious, but it is a common fault. Fear of leaks is sometimes a factor, but in most cases it is neglect or the failure to recognize that implementation is at least as important as making policy and program decisions. Once the decision is communicated, oversight to see that the decision is executed in a timely and faithful fashion is necessary.

In the business world it is generally recognized that efficient execution is more critical than brilliant decisions. Any reasonable decision or choice that is well executed is a real plus for a company. But if poorly carried out, the best choice is of marginal value. Consequently, 80 percent of the attention of top management goes to operations and 20 percent to planning and policy. In government the percentages are reversed.

Implementation is the area where the Carter White House and staff were most constantly faulted. Similar criticism seems warranted of the Reagan NSC staff. A good illustration of the weakness of the Carter NSC staff in this area was its failure to obtain prompt action on the formation of a rapid deployment force, a priority goal of President Carter since the early days of his administration. Confusion and controversy over the execution of the president's policies on human rights, foreign arms sales, and nuclear test ban negotiations also presented problems. Other implementation shortcomings included such routine matters as failure of the Carter NSC staff to follow up meetings with a listing of the major conclusions and agreed actions, the further work to be done, assigned responsibilities, and due dates. Furthermore, overreliance on informal processes to make decisions (for instance, presidential breakfasts or Cyrus Vance-Harold Brown-Zbigniew Brzezinski lunches) made it all the more difficult for policy decisions to be systematically translated into action.

A further impediment to implementation in the Carter White House was the heavy emphasis the NSC staff gave to its personal support to the president and to policy formulation functions. This may have been appropriate in the early days of the administration. However, once the major decisions and policies were set, the White House staff should have shifted its priority to execution and follow-up. The president eventually recognized the weakness in domestic policy implementation. He appointed a chief of staff and took other steps to strengthen his domestic staff. The same priority to implementation should have been given to national security policy.

Planning for Crises

A final area where the NSC staff has an important role to play is in crisis planning. Inevitably, every president is faced with a significant crisis, usually involving the use, or threatened use, of military force. Almost as inevitably, the government is inadequately prepared. Later analyses invariably indicate that the crisis was either unanticipated or had not been planned for adequately. The result was a hasty, ad hoc reaction that was seen as ill advised. The departments must do the detailed preparatory work for crises, but the NSC staff also has a role. It should make sure that the planning is undertaken, provide guidance on the types of crises to be anticipated, outline the critical assumptions, and review the results.

Crisis planning is a demanding and often unrewarding effort. It is difficult to anticipate the locations or circumstances of specific future crises, let alone the sequence of events. Therefore, potential participants often underestimate the value of planning in the mistaken belief that forecasting is futile. The payoff in many cases is not in the plan itself but in the planning process. This process fosters questioning of assumptions, sharpened perceptions of US interests and options, and better understanding of other agencies' personnel and resources.

Within DoD, top civilians have little involvement in planning for the actual use of military force, especially when compared to their extensive involvement in policies, programs, and budgets. Military contingency planning is largely restricted to a limited group of military officers; civilians have little or no role. Outside of DoD, the NSC staff, State Department, and other nondefense agencies, planning for potential crises gets little attention, and planning is rarely done on an interagency basis. The ensuing lack of complementary policy options (political, economic, as well as military) or even sound military options, has at times slowed our response to a crisis or led to actions that, in retrospect, were seriously flawed.

The reasons for insufficient high-level attention to crisis planning go beyond the difficulty of the process and the normal preoccupations of the leadership with day-to-day business. Crisis planning is a sensitive process. Leaks can cause serious political problems or even trigger a crisis that the planning was intended to avoid or contain in the first place. The challenge is to improve and broaden planning by providing a greater and sustained leavening of political-economic-foreign policy considerations and options, without compromising the security of the plans or unduly complicating the problems of the military commands.

There are two separate but closely related aspects of crisis planning: providing national and foreign policy guidance for the military's planning effort, and adding political and economic options to military ones through interagency planning.

Planning for the use of military force is the province of the Joint Chiefs of Staff (JCS) and the Unified and Specified Commands. Given the long tradition of military responsibility for this activity, plus the sensitivity of the plans, the JCS and the commanders-in-chiefs (CINCs) vigorously defend their exclusive role. Though understandable, such exclusivity has serious drawbacks.

This military planning process has been sharply criticized. The Ignatius, Steadman, and Brehm (Nifty Nugget) reports all recommended broader civilian participation in military planning for setting the parameters and reviewing results.[3] These recommendations were not based on the view that the military planners had failed, but rather that the results would be improved through limited, but systematic, inclusion of political-economic-foreign policy advice. The shortcomings of the present system include the following:

- Planning may not cover a likely crisis situation unless it involves countries that face overt military threats or that are clearly allied with the United States.
- The military alternatives presented are too few. Political realities may dictate adding more limited military options, using allied forces, or considering what may be seen as less than optimal actions from a military point of view.
- Scenarios and assumptions for crises may not be accurately defined, the degree or nature of support expected from other nations may be unrealistic, or political constraints (overflight rights, use of bases, etc.) may not be fully taken into account.

In response to these criticisms, the Defense Department took steps to integrate foreign policy considerations into the military planning process, as well as to provide a limited review by responsible senior officials. Former Secretary of Defense Harold Brown directed the under secretary of defense for policy to initiate such a process which was pushed with some success by Robert Komer. This effort faded, however, during the Reagan administration. There appeared to be continued interest among some key civilian officers in OSD, but apparently they lacked Secretary Caspar Weinberger's strong support. The concerns of the various studies were also shared by Chairman of the JCS David C. Jones. He established an internal Crisis Planning and Assessment Group which was to consider political and foreign policy influences and tap the expertise of the Washington foreign policy and intelligence community. This was clearly a positive, though limited step. Selective, but serious and continuous participation by senior civilian officials is required.

Interagency planning is the second aspect of crisis planning requiring greater attention. Little crisis planning is done beyond DoD's military planning. Few efforts have been made to develop nonmilitary options or to force the White House, State, Treasury, and other agencies to engage in systematic contingency planning. In many geographic areas, such as in the less developed countries, US political and economic measures may often prove more effective than military actions. Even though political and economic approaches entail a lower level US commitment, and at times this may be congruent with the preferences of senior decisionmakers, they receive only limited attention.

Interagency planning has occasionally been undertaken, usually following a crisis for which we were ill prepared. All such efforts were short lived:

- In the mid-1960s, after the Dominican Republic intervention, a State/Defense/JCS planning group was established to develop a range of responses to crises. It functioned only briefly.
- After the EC-121 incident with North Korea in 1969, the NSC's Washington Special Action Group (WSAG) selected several potential crises for joint planning. A high-level JCS/Defense/State planning group was formed. At the same time, Defense Secretary Melvin Laird directed his staff to assess selected JCS contingency plans because of his dissatisfaction with the options available for meeting future crises. Again, it was effective for a period, but, in a year or so, attention turned to other matters.
- In 1973, the NSC Contingency Planning Working Group (CPWG), under the chairmanship of the secretary of state and including members of the JCS and the Central Intelligence Agency (CIA), was established. It, too, was short lived.
- Greater attention was also given to such matters by the Carter administration in the aftermath of the Iranian hostage situation. An ad hoc interagency group examined possible trouble spots, reviewed US objectives and policies, sorted out options, and took steps to ensure better preparation. This was a much needed initiative, but it quickly waned for lack of continuing high-level support.

Sustaining high-quality interagency planning for potential crises is difficult. Judging from past experiences, the pattern seems predictable: problem identification, a brief period of high-level attention, loss of interest, and eventual loss of momentum. To break this cycle, a new, more structured approach is needed.

A particular challenge in planning for future crises is to involve "domestic" agencies, where appropriate. Treasury should often be involved, given the importance of economic measures. But it is usually reluctant to join in such efforts. Its hesitancy is due partly to doubts that the government should even consider economic actions (e.g., freezing foreign assets, trade cutoffs, or embargoes) that are seen by internationally oriented financial officials as unthinkable. Yet, as the Iranian experience demonstrated, the United States may be forced to consider them.

Over the long term, an interagency planning effort would probably be more sustainable if primary responsibility was given to the State Department and if the process was managed by a State-chaired senior interagency committee. The actual planning would be done by regional interdepartmental groups led by the appropriate assistant secretary of state. These groups should have sufficient authority to consider the full range of governmental responsibilities and activities as well as to develop a spectrum of potential courses of action. Membership would include the NSC, Defense, JCS, CIA, Treasury, and other agencies (such as the Department of Commerce and Energy), as appropriate. The senior committee would provide these groups with a limited number of critical situations to consider. It would also periodically, at least annually, review the planning of each group to ensure its consistency with the views and policies of the principles and to make certain that the regional interdepartmental groups give this planning process a high priority.

FINAL CONSIDERATIONS

In performing these five institutional functions—setting a policy framework, forcing decisions, managing the process, monitoring implementation, and crisis planning—it is sometimes necessary for the White House to be involved in the details of departmental responsibilities. Meaningful, effective decisions cannot be made by senior policymakers if they only deal at the broad policy level. At times, they must "get their hands dirty." But deep involvement by the president's staff in some areas, at certain times, need not dictate such involvement across-the-board. The challenge in organizing a national security policy management system is to be selective in determining where centralization is essential and where decentralization makes sense. In making these choices, the pres-

ident must deal with the inevitable tension between his departmental leadership and his immediate staff. The departments will always seek some distance from the White House and plead for freedom of action; the White House staff will press for tight control and frequent, detailed involvement. How these tensions are resolved determines to a great extent the character of an administration's national security policy process.

THE NSC AND THE OFFICE OF MANAGEMENT AND BUDGET

While the NSC's role in defense and foreign policy issues gets primary attention, another key White House staff agency, the Office of Management and Budget (OMB), also plays a consequential role. Indeed, close OMB-NSC cooperation would give the president and his staff enormous clout over the entire national security bureaucracy. With the NSC providing the policy framework and program direction and the OMB having control over dollar resources, even DoD would be forced to be responsive to White House direction and priorities.

Such close cooperation, however, has never occurred on a sustained basis. In fact, the two units, located in close proximity to one another in the Old Executive Office Building, seldom speak, let alone coordinate their actions. The reasons are complex—in part jealousy of their own prerogatives, in part the focus on their own particular responsibilities. Another factor is the quite different people attracted to each organization. The NSC tends to recruit staff with broad foreign policy interests or detailed knowledge of a specific geographical region or functional area. OMB attracts financial or program analysts who are more comfortable with numbers and program data than with strategic concepts and complex, but "soft" issues of international relations. When the two units do work cooperatively, it is usually on a narrow issue—a request for supplemental military or economic aid, or the rare major DoD budget issue that goes to the president for decision. More commonly, they are on opposite sides of issues since agencies use the NSC as a court of last resort when they lose a budget issue to OMB. If a responsive chord is struck, the NSC may prevail on the OMB director to reconsider or, on occasion, ask the president to overrule his budget office.

To perform effectively, the NSC should be involved in broad budget policy, overall funding level decisions, and broad budgetary allocations by program. Several agency budgets, in addition to DoD's, are worthy of NSC concern. They include the Agency for International Development (AID), and the international financial institutions—International Monetary Fund

(IMF), World Bank, etc.—the intelligence community, and the nuclear weapons program of the Energy Department.

OMB participates actively in the Defense Department's budget review, working closely with OSD. Other elements of the White House staff only have a limited involvement. The NSC and science adviser's staffs attend the OMB Spring Planning Review as well as meetings where key presidential budget decisions are made. But with the exception of a few specialized areas, such as intelligence,[4] they play a passive or, at best, opportunistic role, on occasion seizing an issue that strikes their fancy. Other agencies with clear responsibility for security matters, for example, State and ACDA, are not formal participants in any phase of the defense budget process. Indeed, they are excluded from these issues at the behest of DoD as well as the NSC.

Despite the relative smoothness of the current process and the general satisfaction of DoD and OMB officials, a number of participants and observers question its adequacy.[5] Two basic questions have been raised. Does the White House staff ensure that the president is provided with a national security framework that is adequate for making the most critical defense program/budget decisions? What role should State and ACDA play in seeing that foreign policy and arms control implications of defense program/budget issues are adequately included in the process?

Regarding the first question, OMB plays the central role in managing discussions with the president that lead to his annual decisions on multiyear fiscal guidance and annual budget ceilings for defense, while the NSC evaluates the resources needed to carry out specific policies articulated in its studies and in crisis/contingency planning. Defense Department officials and some outside observers question the adequacy of OMB's input to the overall budget-level decisions, believing that its analysis concentrates excessively on how much we can afford and insufficiently on how much we need. Deciding "how much is enough" is a difficult, largely subjective determination. But the president seldom has a coordinated or systematic method for making this judgment. A regular, structured, integrated process is clearly in order.[6]

In some ways, the decisions early in the Reagan administration regarding the size and shape of its defense program may be the classic example of "how not to do it." Given the subsequent budget deficits, OMB apparently failed to address the issue of how much we could afford. On the other hand, given DoD's approach to using the funds, essentially "more of everything," there was little evidence that the NSC developed a policy framework, or any program priorities.

Concern has also been raised over the lack of presidential involvement in major decisions related to the broad allocation of DoD resources as well as selected weapons and force structure decisions, such as those involving troop deployments to NATO, Navy support of security policy, and tradeoffs between numbers and high-technology weapons. Unless the NSC and OMB undertake a close working relationship, these types of issues are unlikely to be addressed systematically outside the Department of Defense. It must be recognized that Executive Office staffs are seldom well-equipped to address such questions, and DoD will resist fiercely any White House involvement. Thus, the NSC and OMB must pick their issues carefully and selectively. They are in a weak position to dispute DoD on technical issues, although on an important matter—the B-1 bomber or MX missile—they can bring outside expertise to bear, such as using the president's science adviser's staff. The issues they can best address are broad, policy-oriented ones like major force structure questions such as the size and mix of weapons in our strategic forces, overseas deployments of military forces, and priorities for the allocation of resources (e.g., the emphasis on funds for NATO as opposed to the Persian Gulf, or tradeoffs between mobility and forward troop deployments). But the key is to be selective. If they are not, they will be unable to get into the issues in sufficient depth, and Defense will overwhelm them with details and analysis.

The second set of issues, related to the role of State and ACDA, can best be addressed in the context of the overall relationship of the NSC staff to DoD, which is discussed below.

THE NSC AND THE DEPARTMENT OF DEFENSE

The national security adviser and his staff have played a growing role in security affairs since 1961. Yet the NSC adviser has focused principally on foreign affairs rather than defense issues. State's role has diminished with the rise of the NSC, but not Defense's. (However, the Congress is much more involved in the details of DoD's activities than was the case prior to about 1969–1970. This greater congressional involvement has sharply constrained the freedom of action of OSD and the secretary of defense.)

This is not to say that DoD has been independent of White House oversight. Annual budget ceilings have clearly been presidential decisions, although during the Reagan administration Secretary Weinberger apparently had the primary influence on that decision. And the major issues (the B-1 bomber, neutron bomb, MX missile, etc.) usually go to the president for final decision. At times, particularly during Henry Kissinger's tenure as assistant for national security affairs, the NSC also played a substantial role in setting broad defense policy, although he admitted later that his influence in defense policy was far more limited than it was in the foreign policy arena.[7] But White House involvement in defense issues is usually limited and somewhat sporadic, a sharp contrast to the broad, intrusive, and continuous White House role in foreign affairs. There are at least three reasons for this difference.

First, the most compelling matters within State's purview are foreign policy issues: our relationships with El Salvador, China, and Iran; the Soviet gas pipeline to Western Europe; and our approach to the Arab-Israeli problem. These are issues that congressmen, academicians, and politicians feel comfortable in addressing (although this confidence may be misplaced). Moreover, the president's political philosophy may be particularly relevant on foreign policy issues (e.g., US policy toward the Soviet Union, Cuba, or Taiwan). On the other hand, defense issues (e.g., European force requirements, the Maverick missile, or military personnel compensation) seem far more technical and complex. They are normally left to the secretary of defense.

Second, participation in foreign affairs seems "presidential" and often generates positive publicity and recognition for the president. Economic summits and meetings with Soviet leaders are great television fare and a frequent White House response to falling public opinion polls or bad economic news. The public expects the president to call the shots on arms control negotiations, our position vis-à-vis the Israelis regarding settlement of the Palestinian question, and trade with the Soviets. He is not expected to be involved in M-1 tank production decisions, small versus large carriers, or the C-17/C-5 aircraft choice. Participating in these issues is unlikely to gain him many votes, only problems and grief.

Finally, for reasons not unrelated to the above considerations, national security advisers are almost always foreign policy, not defense, experts. McGeorge Bundy, Henry Kissinger, Zbigniew Brzezinski, and Richard Allen all made their marks as international affairs specialists. Even Judge Clark received his apprenticeship at State. Anyone with defense expertise is quickly dispatched to the Pentagon or, on occasion, to ACDA. Liberals often hope a defense expert at ACDA will "keep Defense honest." Conservatives, on the other hand, may hope to "keep ACDA honest."

As a result, there is little precedent and often little incentive for the NSC to address intricate, substantive military issues. Indeed, the NSC is seldom even involved in broader defense questions. To a degree this is proper. The NSC's defense role should be far more modest than its role in foreign affairs. DoD issues are usually much more technical, are often more programmatic than "policy-oriented," and the number of issues and their scope are broader. Thus, the NSC's involvement must be very selective. Nonetheless, present practices need strengthening for two reasons.

First, DoD needs a well-formulated policy framework. A coherent administration strategy and policy are essential if the defense program is to be focused, reasonably efficient, or, in the absence of very generous funding, adequate to handle our most vital security requirements. As the Reagan administration discovered to its dismay, the range of potential requirements is so great that without some reasonable policy focus and a set of priorities, even a string of large, real increases in resources seems inadequate to improve significantly our security posture.

Second, the government's fiscal crisis and ever-mounting deficits will be with us for years, if not decades. The long-term defense program must be undertaken with careful thought and analysis of its long-term affordability. The pace of budget growth and the start-up of new, costly programs must be carefully planned in light of realistic economic forecasts, the government's revenue outlook, and the resulting fiscal policy impact. This is clearly the kind of issue in which the NSC could and should be immersed. Indeed, it should probably lead in ensuring that the affordability of the defense program is addressed by the president and his senior advisers. This question is too often ignored. Consequently, expansions may be beyond the ability of the government to sustain over the long term. And when the "crunch" comes, security needs are ignored and little thought is given to the areas where the cuts are made—at least with the perspective of the president's security policy objectives. Over the long run our security does not benefit from a rollercoaster series of budget increases and cuts. A sustained level, even a flat or slow growth, may generate more capability over a period of years than larger sums spent erratically.

Given the need for greater NSC staff attention to defense, what type of issues should it address? In simple terms, the NSC should focus on those issues on which the president and his immediate advisers can play an effective role and where their involvement will improve defense management and policymaking. An active, aggressive NSC staff role need not be intrusive to the degree that it becomes a form of "micromanagement." The NSC should restrict itself to selected issues deserving of presidential attention. This is easy to say, but hard to define precisely, let alone put into practice. Nonetheless, it is possible to be more precise about an appropriate NSC role in defense management.

The goal of the NSC staff should be to help the president identify those defense issues that are of major importance to the nation and address them systematically. Let us recall President Eisenhower's purported dictum: What is important is seldom critical and what is critical is seldom important. It is easy for the NSC staff to be swept along with the flood of urgent policy papers, cables, and memos. They have seemingly tight deadlines and interested parties from the agencies clamoring for the NSC staff's attention and support. But while they cannot ignore this paper flow, they must rise above it and help the president identify those issues that are truly important to national security and deserving of his management attention. Lists of such critical issues were developed by President Nixon's staff during the transition period, and a similar agenda was developed by President Carter's staff. Once the issues are identified, the staff must focus on the decision process, bringing to bear the knowledge and talents of interested agencies. This is the most important role for the NSC assistant and his staff to play, helping the president address those issues that are truly presidential, rather than dissipating its energies editing the flood of State Department cables sent to the White House for clearance.

Within this broad framework let me be more specific. To define the proper role for the NSC on defense matters, I suggest using five criteria:

• The NSC should participate in issues so important that the president must make a decision or risk major public/congressional criticism (for example, the B-1 bomber decision and selective service registration during the Carter administration and the MX basing issue during the Reagan administration).
• The NSC, together with OMB, should be involved in setting overall resource levels for Defense for the upcoming budget year as well as the five-year programming period. They should also review the broad allocation of funds by mission and purpose (modernization, readiness, etc.). Clearly, this was not done systematically by the Reagan administration; equally clearly, this was a serious error by the NSC and OMB.
• NSC involvement is required on issues such as arms control and overseas troop deployments that impact other agencies significantly and where other agency views should be given appropriate consideration. This interagency coordination is unlikely to occur without White House intervention.
• The NSC must follow up on major program and policy decisions to see that the president's choices are carried out. This frequently fails to happen without careful follow-up. In some cases this is due to active efforts to sabotage presidential policies. But, usually, it is merely due to inertia and the failure of departmental leaders to see that the appropriate implementation occurred.
• The president has a central role to play in crises and actual conflicts, and the NSC must support him. This includes ensuring that sound military, diplomatic, and economic planning is done beforehand as well as advising him during a crisis.

The NSC adviser and his staff have an important role to play in defense management in the areas outlined above. In some cases the process can be bilateral—between the White House and the Pentagon. In others, a broader, interagency process is needed. Regardless, this responsibility can be exercised effectively without being unduly intrusive or greatly complicating the secretary of defense's already formidable management task. There are four keys to an effective interagency role in defense management.

• Address the issues early, before DoD's positions are hardened and it is necessary to override a secretary of defense decision or reverse a well-entrenched military service position.
• Keep the analyses at a level appropriate to a senior interagency group, that is, address the broad allocation of resources by program, not the funding for a particular weapon system.
• Define the NSC committee's role as advisory to the secretary of defense. It should not have a decisional role that challenges his authority and forces DoD to become defensive. If a serious difference with Defense arises, the appropriate agency head can ask the president to intervene. But this should be done only in exceptional circumstances. In most cases, the secretary's decision should be final, assuming he gives reasonable consideration to the input of other agencies.
• Keep the process at a senior level—assistant or under secretary level participants. If it slips below this level it often degenerates into "micromanagement," or the process becomes so intrusive that DoD shifts the focus to bureaucratic gamesmanship, rather than a serious addressal of the issues.

STAFFING AND STRUCTURAL IMPLICATIONS

The key to an effective NSC is people—the assistant to the president and staff—as well as the role given this group by the president. Each assistant, appropriately so, takes a somewhat different approach to selecting his staff. Yet some guidelines can be proposed. While the staff should be comfortable with the president's policy, it is a mistake to rely entirely on political supporters for key positions, regardless of their expertise. A healthy mix of able, experienced individuals from the key agencies (DoD, State, CIA, ACDA, etc.) is essential. They know the issues, have the contacts necessary to get needed information, and the knowledge of how to make things happen in the agencies. Outsiders, be they academics or politicians, lack this ability. Yet outsiders have a role—to inject new ideas into the process and to ensure that the president's policies and wishes are given appropriate focus. They should be included on the staff.

While people are crucial, structure is also important. A regularized, adequately staffed, interagency committee structure is essential to developing effective national security policies and programs. The committees can take many forms and be chaired by either senior agency or White House personnel. There is no one right way; it depends on the style of the president and his key subordinates. But there are some rules that should be observed.

• The interagency committees should not be too ad hoc. Several committees probably make sense, each with a clear area of responsibility (intelligence, arms control, crisis planning/management, defense, etc.). This approach was taken by Nixon, Ford, and Carter. While meetings often will be event-driven, each committee ought to have a policy and analytic agenda of major issues that are deserving of interagency review and possible presidential direction. The key is to select important issues worthy of careful high-level attention, and not to merely react to headlines and crises.
• The committees must be carefully staffed. A working group or interagency secretariat should be used to develop option papers and oversee the policy analysis effort. It is not wise to delegate the support responsibility to a single agency because that agency's views and approach will then dominate. As a result, the broader, interagency perspective will be lost. If ACDA prepares all the arms control policy papers or Defense all papers on major defense programs, the full spectrum of options will not surface, and the analysis will be unbalanced.
• The NSC staff must have authority to manage the analytic support for the interagency committees, or at least be able to see that a full range of options is considered and that solid, impartial analysis is done. The NSC staff is best positioned to ensure that all agency views are fully incorporated into the analysis, and not merely noted.

As noted earlier, there is no simple "right way" to organize the NSC system or for the NSC staff to carry out its functions. But when a president chooses his security assistant and lays out his system for managing defense, intelligence,

and foreign affairs, he should recall a few basic rules.

• The president's key staff capabilities lie on OMB and the NSC. These units should cooperate to see that the president's decisions match his policy goals.
• He must resist the tendency to get swept up with the rush of routine foreign policy matters, state visits, overseas trips, and day-to-day management of foreign affairs. Some time must be allotted to determining whether defense resource levels are adequate and whether these resources are allocated properly. In addition, the president must ensure that the arms control process focuses on important new issues, such as weapons in space, and not just on the more traditional areas.
• Finally, adequate contingency planning and preparations must be made for the crises that seem inevitable in every administration. Such contingency planning must encompass economic and diplomatic responses as well as military actions. Poor prior preparation usually results in poor execution when the crisis occurs.

One last observation is in order. For the forseeable future, economic issues will probably be as important to our security interests as military or diplomatic matters. This poses difficult new coordination and integration challenges to the White House. It is impractical for the NSC to be the predominant player in the economic arena. But it must work closely with whatever group the president creates to orchestrate trade, currency, and debt issues. Unless this integration is done effectively, the economic basis so necessary for a sound security policy will be lacking.

NOTES

1. Much of the literature regarding the NSC adviser revolves around personality differences (Haig versus Allen, Brzezinski versus Vance, etc.) or the role of key staff on specific issues such as Vietnam, Iran, or the Middle East. However, there are several more systematic studies, including: Philip A. Odeen, "Organizing for National Security," *International Security* 5 (Summer 1980): 111–29; I. M. Destler, *Presidents, Bureaucrats, and Foreign Policy* (Princeton: Princeton University Press, 1974); I. M. Destler, "National Security Advice to U.S. Presidents," *World Politics* 29 (Jan. 1977): 143–76; I. M. Destler, "A Job That Doesn't Work," *Foreign Policy* 38 (Spring 1980): 80–88; I. M. Destler, "National Security Management: What Presidents Have Wrought," *Political Science Quarterly* 95 Wiinter 1980–81): 573–88; Robert Hunter, *Presidential Control of Foreign Policy: Management or Mishap?* (New York: Praeger, 1982); US Senate, Committee on Foreign Relations, *Hearing: The National Security Adviser: Role and Accountability*, 96th Cong., 2d sess., 1980; Alexander George,

Presidential Decisionmaking in Foreign Policy (Boulder, Colo.: Westview Press, 1980); Peter Szanton, "Two Jobs, Not One," *Foreign Policy* 38 (Spring 1980): 89–91; William P. Bundy, "The National Security Process," *International Security* 7 (Winter 1982–83): 94–109.

2. Production of the enhanced radiation weapon (neutron bomb) had been a controversial issue since the early 1970s. Each time production was proposed by the Army, it was deferred. The reasons were complex. In part, the requirement had not been clearly established, and its cost was a problem since the US already had ample stocks of tactical weapons. Concern over starting a new arms race was another factor. In 1978, senior Defense Department managers were convinced the weapon was justified, in large part as an answer to the steady buildup of Soviet armored forces in Europe. Despite some initial reluctance, State agreed that it should proceed and, after clearing it with the NSC staff, began consultations with our European allies since the weapon was for use in Europe and would be deployed there when produced. There was little enthusiasm among European governments, but the plan was accepted. West German Chancellor Helmut Schmidt was reluctant, but he eventually agreed after much pressure was exerted. Once the allies were on board, the proposal was forwarded to President Carter for routine approval. (All nuclear weapons production plans must be approved by the president.) Unfortunately, no one had bothered to raise the issue with him prior to this time, and he disapproved. Despite several appeals, he made it clear that he had strong views on the topic and would not reconsider. The matter quickly reached the press and the administration looked foolish. Chancellor Schmidt was embarrassed and was sharply criticized by his own party. He was furious, and relations between the two leaders never really recovered from the episode.

3. Department of Defense, Paul Ignatius, *De-partmental Headquarters Study: A Report to the Secretary of Defense*, 1 June 1978; Department of Defense, Richard Steadman, *Report to the Secretary of Defense on the National Military Command Structure*, July 1978; Department of Defense, William Brehm, *Evaluation of Nifty Nugget Exercise*, Nov. 1978 (the Brehm report is classified, but passages are cited in *National Security Policy Integration*, President's Reorganization Project, Sept. 1979, p. 53).

4. In most administrations the intelligence budget, at least concerning the CIA and the National Security Agency, receives considerable attention from the NSC staff, usually working closely with OMB. In part, this reflects the extra sensitivity such programs have generated in recent years. Another factor is that the CIA director, who has the charter to manage the overall intelligence budget in his capacity as director of central intelligence, in fact needs help from the White House if he is to have an appreciable influence over DoD intelligence programs.

5. See, for instance: *National Security Policy Integration*; Odeen, "Organizing for National Security," Lawrence J. Korb and Keith D. Hahn, eds., *National Security Policy Organization in Perspective* (Washington, D.C.: American Enterprise Institute, 1981); Duncan L. Clarke, "Integrating Arms Control, Defense, and Foreign Policy in the Executive Branch of the U.S. Government," in Hans Guenter Brauch and Duncan L. Clarke, eds., *Decisionmaking for Arms Limitation: Assessments and Prospects* (Cambridge, Mass.: Ballinger, 1983), pp. 3–36.

6. See Odeen, "Organizing for National Security."

7. Henry A. Kissinger, *White House Years* (Boston: Little, Brown, 1979), p. 396. Kissinger created the Defense Program Review Committee (DPRC) in an effort to get the White House involved in major defense programs and budget decisions. Despite a sustained effort over several years, the DPRC had little impact.

COORDINATING NATIONAL SECURITY POLICY
The Role of the Office of Management and Budget

PETER L. SZANTON

The argument of this chapter can be stated in three propositions. First, measured against the standard of national needs, the role of the Office of Management and Budget (OMB) in the coordination of national security policy has been narrow in conception and small in result. Second, OMB's ambitions in national security decisionmaking—especially in helping assure the rationality of defense budgets—ought to be substantially greater. Third, with presidential support—but only with such support—such a change is achievable.

This chapter details these propositions. It first characterizes OMB performance in the coordination of national security policy and notes the persistent shortcomings in Department of Defense (DoD) performance that have been left undisturbed thereby; it then attempts to specify the major tasks not now being adequately performed; finally, it suggests how enlarged responsibilities, once assigned to OMB, might be discharged.

Reprinted with minor revisions and with permission from Public Policy and Political Institutions: Defense and Foreign Policy, *pp. 43–59. Copyright © 1985 by JAI Press, Inc.*

OMB'S PAST PATTERNS OF PERFORMANCE

OMB's role in the formulation and coordination of national security policy over the past several decades has not been smoothly constant; it has shifted with changes in the relative political and personal strengths of the relevant players and with the degree of presidential concern about the shape and (especially) the size of the defense budget. Since at least 1961, however, what has been striking about OMB's performance is not variability but constancy.

OMB's principal leverage is over budgets. Substantially more than 90 percent of what might be broadly defined as the US national security budget is composed of the budget of DoD.[1] Accordingly, OMB's main role in national security affairs is played out in establishing an administration's defense budget. It typically includes transmitting the president's budgetary target to the secretary of defense some seven or eight months before the administration's annual budgetary presentation to Congress; monitoring a limited set of resource-related issues as they are fought out within DoD in the months that follow; and reviewing DoD's requests, and the rationale for them, in preparing the full budget for presidential approval.

In outline, the process is similar to that employed by OMB with respect to the budgets of all other agencies of the executive branch. Indeed, the process appears more rigorous and better informed in the case of DoD because for many months each year members of OMB's national security division are physically present in the Pentagon, monitoring the development of DoD's program plans and budgetary requests before their formal submission to OMB. The results of the process, however, have been marked by major shortcomings.

No Appetite for the Biggest Issue

The quintessential OMB role is to assist the president to establish an overall budgetary level and to allocate that total among the various departments and agencies of government. Congress, of course, can greatly change both the budget total and its composition, and the budget any president submits to Congress has already been influenced by predictions of congressional reaction. But within those limits, OMB's views normally dominate the great majority of the decisions—all of the smaller ones, most of the larger ones—that are implicit in the president's budget. That influence arises from a combination of factors. One is presidential perspective; OMB's highest institutional loyalties attach to the presidency and to the programmatic goals of the administration in office, not to objectives

or programs of its own. Another is disinterestedness; OMB is not itself a contender for budget share. A third is expertise; OMB is the largest presidential staff by far, and its knowledge of the operations of the departments and agencies of the executive branch is unmatched outside of the departments themselves. The result is that OMB's comparative assessments of the relative worth of the contending budgetary claims of executive agencies are normally regarded by presidents as informed, balanced, and, with some politically imposed exceptions, authoritative.

But among all such comparative judgments, the most consequential is that which weighs domestic social claims against the claims of national security on the nation's resources. It should follow that assessing the relative strength of those claims—difficult as such a judgment is both substantially and politically—would be one of OMB's main responsibilities. But it is not and never has been. OMB takes an informed, skeptical, and independent look at the extent and priority of the national "needs" that major domestic programs address. It examines closely the effectiveness with which those programs operate. And it balances competing domestic programs and needs against each other. But no such serious examination is normally made either of military "requirements" or of the means chosen by DoD to meet them. Still less, then, is the result of such an examination used to compare domestic and national security claims against each other.

One result is the intellectually barren terms in which defense budgets are set. "This frantic game of bidding of percentage points is a deeply flawed analysis of needs. In fact, it is no analysis at all. We must begin to solve our military problems by asking the right questions—questions about the ends we want to achieve and the means required to achieve them. Beware of the practitioner of statecraft who designs his policies on a calculator."[2] So said Senate Budget Committee Chairman Edmund Muskie in 1980 about the defense budget of a president of his own party. His comments were hardly peculiar; responsible members of Congress and others had been voicing similar complaints for many years. It is clear both in theory and practice that such questions will rarely be asked—and, if asked, cannot be pressed home—within DoD. If raised at all within an administration, they must be asked and pursued by OMB. But OMB does not seriously address them.

Little Appetite for the Next Biggest Issues

Almost equally important are independent analyses of the tradeoffs among alternative national security investments. The DoD budget is not the only price we pay for national security. Some or all of the budgets of the Agency for International Development (AID), the United States Information Agency (USIA), and Central Intelligence Agency (CIA), the Department of State, and the Department of Energy are investments toward the same end. What pattern of investment makes most sense and how should it evolve over time? The answers to these questions must be judgmental and imprecise, constrained by history and by politics, the result of no science. But they are nonetheless worth arriving at thoughtfully. If 1 percent of DoD's current $270 billion budget were reallocated to economic assistance to Third World countries, for example, our AID budget would thereby increase by one-third. Even so modest a shift from defense to economic assistance would, of course, be hard to achieve politically. Yet over time it is achievable. OMB does not seriously raise or explore such alternatives.[3]

IMMERSION IN DETAIL

As suggested above, OMB has shown little initiative in probing the underlying issues of defense budgetmaking. It does not challenge assertions of military "need," or the relation of proposed force structures to such "needs." It has not independently sought to establish, for example, whether the "window of vulnerability" is real and, if so, whether the MX missile, in any deployment mode, is an appropriate solution. On the contrary, it has placed much of its national security staff physically inside the Pentagon during the Defense Department's budget preparation and immersed it in the details of that process. This is a practice that has brought early notice and detailed understanding of the nuts and bolts of defense budgetmaking, but at a price: loss of the outsider's freshness of perspective, implicit waiver of the opportunity to address later in the budgetary process issues not raised during DoD's deliberations, isolation from the rest of OMB and, all the more so, loss of contact with the remainder of the Executive Office of the President.

ISOLATION FROM POTENTIAL ALLIES

Defense programs affect the employment of roughly one out of every ten workers in the United States. The Department of Defense contracts with suppliers and operates installations in virtually every congressional district. It disposes of roughly a quarter of the federal budget. It performs an essential national function, and commands the elemental loyalties of a large portion of the electorate. Inevitably, then, defense programs generate lives and momentums of their own. Bases no longer needed by the Army may be politically uncloseable; aircraft not requested by the Air Force may be procured nonetheless at congressional insistence; equipment choices may be driven by what is technologically possible rather than militarily most useful. Meanwhile, the hard questions of which military capabilities meet national (rather than service) requirements, and which do not, get little attention; there are few officials, in uniform or out, who have the incentive, the authority, and the information to address them. So insuring that DoD budgets reflect national needs rather than contractor or service interests, or inertia, is a truly formidable task. It is a far larger undertaking than has or can be performed by any secretary of defense, and it is similarly beyond the capacities of OMB operating in isolation.

If OMB sought to pose effective challenges to questionable DoD assertions of "threat," or calculations of "need," or choice of weapons, it could gain important leverage from the perspectives, expertise, and influence of other agencies. Depending on the issue, the National Security Council Staff, the CIA, the president's science adviser, the Department of State, or even expert and prestigious persons outside the government might be enlisted as allies. In the Carter and Reagan administrations, OMB's process of budget review has in fact become a less intramural proceeding, with representatives of other presidential staff agencies invited to attend. But OMB has only, in very special situations, sought to build a consensus among presidential staff, or among relevant departments and agencies, about questionable DoD programs or budgets.

IGNORANCE OF THE END GAME

A final OMB weakness is that in focusing on the future—on the budgets to be presented to the president and the Congress next year and in the years after that—OMB devotes virtually no attention to monitoring how the budget for the current year is actually being spent. The combination of the great size of the defense budget and its relatively small number of separate categories or budget "lines" leaves DoD considerable flexibility as to what actually to purchase. OMB fails even to observe, let alone control, Defense's exercise of that broad budgetary discretion—an example of the general truth that most White House agencies, transfixed by concern for "policy," have little time for "implementation." Implementation is merely what actually happens.

The result of these factors, taken together, is profound and important: OMB exerts far less influence over national security issues generally and over the DoD budget in particular than over any other major function of the United States government.

PROBLEMS IN DEFENSE

The shortcomings in defense planning, budgeting, procurement, and operations that this placid performance of OMB leaves essentially unchallenged are widely recognized.[4] DoD's performance during the past two decades has been characterized by some six major defects.

NONCOMPLEMENTARITY AMONG THE SERVICES

In theory, the defense establishment receives from its political superiors specific guidance as to the kind of wars or other contingencies for which it should prepare. In fact, however, presidents have good bureaucratic and political reasons for avoiding such specificity. The result is that while the secretary of defense annually provides a loose framework for planning in his Defense Guidance (Consolidated Guidance in the Carter administration), each service remains effectively free to establish its own missions independently. In doing so, each seeks the most important and autonomous roles feasible. It is only a slight exaggeration to say that the Marines tend to prepare for something that resembles a short form of World War II in the Pacific, the Air Force anticipates a massive nuclear World War III, the Navy is designed for World War III followed by World War II, and the Army for a short form of World War II followed by World War III.

The Marines are a specialized, largely self-sufficient and, in any event, small service. Their choice of mission is probably appropriate, they are equipped to fulfill it, and high consistency between their roles and those of the other services is not required. But the other three services are intended to be complementary. Despite the establishment of a unified Department of Defense almost four decades ago, largely to achieve exactly that goal, it remains unachieved.

DISTORTED EXPENDITURES

Within each military service the strong tendency is to allocate budgets so as to maintain the largest possible future force structure. For reasons of pride, opportunities for promotion, and the long-run "good of the service," budgets tend to be used to procure the largest number of ships, or air wings, or divisions. And the tendency in all services is to equip these units with the most capable and sophisticated weapons that US manufacturers can devise.[5] The result, reinforced by congressional tendencies to impose modest current-year cost reductions—reductions most readily achieved through restrictions on training, exercises, maintenance, and stockage—is a systematic and persistent overstress

in the budget on procurement, and correspondingly inadequate support for operations and maintenance, training, and consumable supplies. The skeleton of a large-scale and very potent military is maintained, but the readiness of that skeleton to sustain heavy fighting at any point in the near future is normally poor.

MISESTIMATION OF COSTS AND CAPABILITIES

Spectacular cost overruns on individual procurement items often draw public notice but, at least in degree, they are exceptional. The more important defect is that the DoD planning and budgeting process systematically ignores four decades of evidence about the tendencies of all new weapon systems to cost, depending on their degree of technological novelty and the extent of specification changes made by the services during their design and production, two to ten times as much as was originally anticipated.[6]

And some of the weapons whose costs balloon between development and procurement provide limited additional capability. The $2 million Bradley infantry fighting vehicle, for example, boasts better armor than the $80,000 M-113 armored personnel carrier it replaces, and can fire antitank missiles. But it is too large to be airlifted by the standard military transport, and carries only six soldiers to the M-113's eleven. Similarly, the Navy's F/A-18 Hornet fighter bomber costing, at some $30 million each, more than three times the original estimates, will have a combat range approximately one-half of the A-7 aircraft it is replacing.

POOR OPERATIONAL PERFORMANCE

The limited performance of individual weapon systems is less important than the poor operational performance of the defense establishment as a whole. The following was written in 1976:

For fifteen years when the U.S. has resorted to military force, it has failed to achieve its objectives. The army's costly and unavailing strategy in Vietnam, the futile bombing of North Vietnam, the air drop to free U.S. prisoners from a deserted camp in the North, the invasion of Cambodia to capture a non-existent Vietcong headquarters, and most recently the Mayaguez incident in which 38 Americans were lost to rescue 39 whose release, it turned out, was already in progress—these are not isolated incidents. Sadly, they constitute the bulk of the recent record.[7]

Subsequent experience only confirms the story. The attempt to rescue American hostages in Iran was planned not on an emergency basis but over a long period of time and at the highest levels, free to draw upon the full resources of the US armed forces. It not only failed; it collapsed without having encountered any hostile force. The Marine "peacekeeping" force in Lebanon proved unable to protect itself from the most obvious and most dangerous threat it faced. Our sole recent military successes have occurred against the Libyan air force and the Grenadan army.

OVERESTIMATION OF THE THREAT

Like all healthy government organizations, DoD seeks to enlarge its budget. Like most military organizations, DoD is tempted to overstate the strength of forces it may have to fight. Overstatements can occur in the small or in the large. The Soviet MIG-25 fighter was consistently characterized by the US Air Force as an awesome aircraft capable of flying at Mach 3.2 or more and having combat range of 2000 miles. Such capabilities would threaten the global balance of air power and fully justify an all-out effort to develop even so expensive an aircraft as the American F-15. But when the first MIG-25 fell into US hands, it became clear that while it could achieve Mach 3.2 for very brief periods, that speed burnt out its engines. Soviet pilots were forbidden to fly at greater than Mach 2.5, and even at that speed combat range turned out to be less than 200 miles.[8] The same exaggeration of Soviet capacities characterizes assertions about Soviet submarines (numerous but noisy, prone to breakdown, and relatively easy to track and destroy), Soviet army divisions (numerous but typically understrength, poorly equipped, and far from combat-ready) and non-Soviet Eastern bloc capabilities generally (impressive in scale, much less so in equipment and motivation).

The same tendency to amplify the threat appears in discussions not of capabilities but of intentions. The asserted need for the MX missile, for example, was predicated on the existence of a so-called "window of vulnerability." Vulnerability was asserted to exist because in the late 1980s Soviet military planners and political leadership might calculate that a war was worth beginning (or worth threatening) because Soviet ICBMs might then be capable of destroying, before they could be launched, some 90 percent of American land-based intercontinental missiles, together with some smaller fractions of US airborne and submarine-launched weapons. What that calculation ignored, among many other factors, was:

- No Soviet ICBMs have ever been tested either in the trajectories all would have to fly in order to target US missile sites, or in the large simultaneous launches and near-simultaneous impacts that such an attack would require.
- Soviet leaders have even better reason than most executives to know that no complex and unprecedented operation is likely to work as planned the first time it is tried.
- Even if Soviet war planners had high confidence that such an attack would work exactly as planned, some 200 remaining US ICBMs, a substantial fraction of the 2750 US airborne nuclear weapons, something more than half of the 5000 warheads aboard US nuclear submarines, and a small proportion of the 6000 "tactical" nuclear weapons deployed by the US in Europe would remain available for counterattack. This is a force unquestionably capable of reducing the Soviet Union, in the course of a day, to a state of devastation and probable anarchy, while leaving it surrounded by China, Japan, Afghanistan, Turkey and Poland, all with historic scores to settle.

It does not follow that MX is a bad idea; it follows only that it is a very expensive idea whose rationale is clearly questionable. Yet the traditional relationship of OMB and DoD precludes the question, despite the expense.

WHY IT HAS WORKED THAT WAY

If the foregoing analysis is correct, DoD's performance over the past two decades has been marked by large and persistent failures, many of which would seem natural objects for sustained OMB attention. Yet OMB has shown an instinct for the capillaries rather than the jugular. Why should this be so?

The principal cause of OMB prudence is surely the recognition that OMB might lose a fight for higher stakes. Any determined cabinet member can appeal an OMB decision to the president. OMB's influence over the departments arises solely from the departments' perception that, in the event of such an appeal, the president would likely side with OMB. Accordingly, there is nothing more important to OMB than maintaining this perception. The institution is therefore, and necessarily, hesitant to engage in many fights it has a substantial chance of losing. And fights with Defense are low percentage affairs. This is true not only in administrations like that of President Reagan, in which a defense buildup was given high priority; it is true in all administrations. To all presidents, the DoD—unlike such Departments as Labor, Commerce, Agriculture, Housing and Urban Development, which he sees as advancing the claims of some particular constituency—performs an elemental and unarguable duty of the national government. And, unlike other departments, Defense provides the president with capabilities—ranging from the Marine Band and Air Force One through the military units he may deploy against domestic violence or foreign threat. As to all other departments, moreover, the president is simply the chief executive. To the military services he is commander-in-chief. Secretaries of defense, thus, tend to accord presidents a more nearly absolute form of loyalty than do secretaries of labor or agriculture, and presidents tend to reciprocate. The relationships between presidents and secretaries of defense over the last two decades have been, by and large, closer personally and professionally than relationships with most other department heads. James Schlesinger poses an exception to the rule; Robert McNamara, Donald Rumsfeld, Harold Brown and Caspar Weinberger exemplify it. Those relationships, coupled with the political sensitivity of the defense budget, make OMB properly wary of tangling with a secretary of defense.

A second constraint for OMB arises from the fact that defense issues appear to be highly technical and OMB is outside the profession that understands the technicalities. "When they say 'military judgment' the curtain just comes down," as a former OMB defense analyst has remarked.[9] That circumstance has been magnified by the fact that few OMB directors or deputy directors have come to their post with a substantial national security background. Domestic program experience in federal or state government, or a business background is far more typical.[10]

Another contributing factor is the continued domination of key decision processes within the Department of Defense by the uniformed services. In theory, the services are sharply limited in function. Their business is to organize, train, and equip forces, and nothing more. In theory, the operational control of forces, with all the influence over strategy, deployment, and readiness that should follow from that authority, is lodged in the joint and unified commanders, officers whose jurisdictions span service lines. In theory, responsibility for establishing military "requirements" and for determining service roles and missions is a function of the Joint Chiefs of Staff, whose jurisdiction also crosses services lines. And, in theory, ultimate responsibility for all defense decisionmaking, under the president, resides with the secretary of defense, an official superimposed on the previously autonomous services precisely to bring a unifying national perspective to defense decisionmaking.

In fact, however, the unified commanders have little grip on Pentagon resource-allocation decisions; the Joint Chiefs of Staff operate by implicit consensus, a process that preserves for the uniformed heads of each service the capacity to block most decisions strongly opposed by that service; and the Office of the Secretary of Defense lacks the means to impose a national perspective upon more than a few of the numerous and complex decision processes at work in the

services.[11] And the services, seeking to protect and expand their historic missions, remain the object of loyalty to those within them; it is the services that train, socialize, assign, and promote, or decline to promote. With respect to the behavior of everyone in uniform, therefore, it is the services, not joint staffs, not the secretary of defense, and still less OMB, which hold the cards. As Enthoven and Smith recounted:

In its budget request for fiscal years 1961–62, the Navy budgeted for only three Polaris submarines in each year. One of the first things that President Kennedy and Secretary McNamara did when they came into office was to speed up the Polaris Program and to authorize the building of 10 Polaris submarines in each of these fiscal years. Nobody, to our knowledge, has since questioned the necessity or the wisdom of that action. But at the time, senior Navy officers, when confronted with arguments for increasing the Polaris Program based on urgent national need, replied: "Polaris is a National Program not a Navy Program." By this was meant: the Polaris mission was not a traditional Navy mission and therefore should not be financed out of the Navy's share of the defense budget.[12]

That incident of more than 20 years ago remains typical of service performance today.[13]

A final contribution to this sorry situation is made by the Department of State. In theory, State's perspectives might be much like those of OMB: though far less interested in budget ceilings, State might share and reinforce an OMB concern for integrating the various elements of US national security policy, including effective military strength, into a coherent whole. With varying degrees of seriousness, several presidents have explicitly assigned State such a responsibility.[14] In fact, however, State has never convincingly attempted to assert such a role. Occasionally dressed as an umpire, State nonetheless performs as a fan. State is partially dependent upon Defense for communications and logistics support, it is poorly equipped to second-guess service judgments about military requirements and typically unwilling to challenge DoD assertions about the nature or extent of the Soviet military threat. Where an issue is cast as a question of more or less for defense, therefore, State has historically abstained or supported DoD requests for more.

THE TASKS THAT NEED PERFORMING

The shortcomings of defense cry out for treatment—preferably for treatment by the secretary of defense. The secretary is the president's man on the scene; he has a large and traditionally able staff; and insofar as any official does, he has the authority, responsibility, and the incentive to remedy these faults. But they are hardly new. The critique presented here is striking only for its familiarity; virtually identical observations have been commonly and convincingly made by respected observers for at least three decades. And over that period a series of forceful men have come and gone as secretaries of defense. The problems remain, not discernibly changed in nature or in degree. That is a strong argument for outside intervention—within limits.

OMB can appropriately question defense performance, as it can the performance of any department, under any of three conditions. The first is that the issue involves the overall size of a department's budget, with consequent implications for the size of the president's total budget, for the budget shares of other claimants, and for fiscal policy generally. The second condition arises where a program, perhaps unexceptionable when considered alone, operates at cross-purposes with related programs of other agencies, or falls outside the framework of presidential policy. The third, less commonly agreed, is simply that a program absorbing substantial funds is ineffectively managed.

None of these conditions creates a license to intervene without limit. But there should be no doubt in principle that when a department threatens the balance and size of the president's budget, or contributes to contradictions or incoherence across an administration as a whole, or simply manages a significant program poorly, then OMB, the only presidential agency capable of reviewing agency performance, serves the president (and the country) best by challenging that performance.

Correspondingly, of course, OMB must observe constraints. With respect to the Department of Defense, it should particularly avoid making longer or more cumbersome the already very lengthy and elaborate DoD planning process; it should avoid temptations to establish its own national security policies; and, as noted above, it had better refrain from frivolously second-guessing the secretary on any very large number of questions.

Consistent with those responsibilities, yet observing those constraints, OMB might most usefully undertake some four essential and currently underperformed tasks. One arises from OMB's primary responsibility for budget size, the other three from its secondary responsibility for budget composition. None are at all novel in conception; in each case, long histories of failure justify forceful OMB attention.

POLICY AND BUDGET INTEGRATION

OMB should seek to insure the consistency of proposed defense budgets and programs with both the general national security objectives established outside the Department of Defense,

and with the budget and programs of the related national security agencies: State, AID, CIA, Energy, USIA, ACDA. Defense expenditures are not ends in themselves. Neither are military capabilities. Both have meaning only as instruments, complementary to others, of considered national policy. It is not OMB's job to set security objectives. It should be OMB's job to see that, in each administration, such objectives are explicitly established, and that the major programs of DoD are measured against them.

The first half of that task is generally not difficult. Incoming administrations often initiate comprehensive and more or less formal reviews of security policy, and major changes in the international environment sometimes precipitate such reviews in mid-administration.[15] The second half is another matter. The fact is that there now exists little basis in theory or fact for sizing and shaping military forces to match specified missions. One reason for this is that the intellectual task of relating forces to capabilities and missions is formidable. Another is that it has served many interests to leave matters so foggy. Presidents could thus pretend that a constricted defense budget would provide great capabilities, and service chiefs could argue that generous budgets were still inadequate to meet a Soviet challenge. But some crude and partial rules of thumb for force sizing and force evaluation do exist, and far better ones can, over time, be developed. Given the enormous investment the United States must continue indefinitely to make in its military, and given OMB's responsibility and that of the secretary of defense to gauge the adequacy of that investment, there are few conceptual challenges more important to public decisionmaking.

INTERNAL CONSISTENCY OF DoD BUDGET AND PROGRAM

Secondly, OMB should assess the consistency, within DoD, of expenditures on force levels with those on readiness, of services with services, and of budget implementations with budget requests. Conceptually, these are far simpler tasks, but hardly less important. In no other way short of the test of war can the country be assured that defense expenditures are providing balanced, ready, and usable capabilities.

PRICING REALISM

Few phenomena have so powerfully distorted defense budgets as the unwillingness of the services to estimate procurement costs realistically. Predictable but unpredicted cost escalations produce even greater than programmed imbalances between force size and readiness, with the effect of keeping the services perpetually large but anemic. Skilled in cost analysis, fully familiar with the history of DoD cost escalation, and largely free of the incentives to underestimate costs that influence both defense contractors and the services, OMB is the preeminently appropriate agency to enforce on DoD the cost-projection realism that DoD is manifestly capable of enforcing on itself.

THE SYSTEMIC SOURCES OF POOR PERFORMANCE

Poor performance is the most difficult problem, and its solution would yield the highest payoff. Virtually every DoD function is now performed less well than the country might reasonably expect. Establishing military requirements, procuring equipment, maintaining readiness, planning and executing actual operations—all have shown grave and persistent deficiencies. The reasons for this poor record are not inadequate skill or dedication among military officers; by and large the professional military are among the most capable and devoted servants the country has. The reasons lie mainly in the conditions of service autonomy and interservice rivalry that reward parochialism, promote indifference to costs, and trade present strength for future force size. It is these conditions, consequently, that must sooner or later be changed if performance is to be substantially improved. The alternatives—both of them unsatisfactory—are to acquiesce in current levels of DoD performance, or to rely on OMB or the Congress or the press to impose specific ad hoc fixes from outside.

The proposition is as true now as when Admiral Alfred Thayer Mahan first asserted it that no military service can reform itself from within. Nor can DoD. The requirement must come from outside. And of all the agencies in the executive branch capable of engendering such outside pressure, OMB, if backed by a concerned president, is the only likely candidate. Should an administration give such reform genuine priority, it would assign to OMB responsibility for initiating and overseeing the process, not for administering it. OMB is well-positioned to challenge and disapprove departmental positions; it is poorly positioned to impose its own. OSD, the services, and various blue-ribbon task forces would have to take the direct responsibility for diagnosis and prescription. They would find available to them a rich history of organizational analysis of DoD and the services, and a long menu of systemic reforms, many directed at exactly the pathologies discussed above, and most proposed by military officers.[16] That history, to be sure, is cautionary. The long record of only glacial progress suggests the political difficulty of doing better. But it also testifies to the high importance of doing better.

GETTING THERE FROM HERE

If OMB were seriously to undertake such tasks, what resources would it bring to the job and how might it begin?

Aside from the relationship of the OMB director to the president, there are three major sources of OMB leverage. First and most important is OMB's role in the budget process. Over the last two decades OMB has lost leverage in domestic decisionmaking. Massive entitlement programs and the increasing use of off-budget financing now place many domestic outlays effectively beyond OMB's reach. But national security resources are still subject to annual budgetary decisionmaking; there are no military entitlements other than pensions and no substantial national security program is financed outside the budget. The formal budgetary decision process that OMB bestrides thus still offers real choices to be made.

Secondly, there is the persistent national unease about the wisdom of defense budgets and about national security decisionmaking in general. Wide current support for real growth in defense spending is accompanied by widespread public and congressional concern that DoD is grossly wasteful, and that defense decisionmaking takes place in the absence of any coherent framework of national security policy. A constituency for reform might therefore be assembled.

Finally, as suggested above, OMB may be able to reinforce its critique of DoD positions by strengthening its normally quite distant relationships with the NSC staff, the president's science adviser and, on some issues, with the Department of State, the CIA or others. Indeed, there is no reason why OMB might not commission blue-ribbon task forces composed in part of knowledgeable private citizens, including retired senior military officers, to examine selected defense issues in depth. OMB would thereby gain not merely greater expertise, but the additional legitimacy conferred by prestigious civilian and senior military views.

How might an OMB director intent on playing such a role go about it? The first requirement—to be met at the beginning of an administration, or the beginning of an OMB director's term—would be to get some assurance of presidential support. Presidents do not issue blank checks. Some initial presidential approval would provide OMB with the needed license to choose issues, prepare arguments and enlist allies. The absence of presidential support in principle would be a sure sign that no such posture was worth attempting.

A second step would simply be to establish good personal relations, at both leadership and staff levels, between OMB and the NSC staff. An OMB director intent on playing the role sketched here would take pains to sound out the national security assistant's interest in major defense questions and his willingness to let NSC staff explore them jointly with OMB's National Security Division. To build closer relations between the two staffs, persons might be rotated between them. Offices of the two staffs might be more closely colocated, and penalties might be imposed by the directors of each staff in the event of failures to coordinate. Quite apart from facilitating alliance-building on particular issues, such practices would broaden perspectives in both staffs, a desirable evolution on many grounds. To a lesser extent, similar arrangements might be attempted with other presidential staff.

Thirdly, the issues on which OMB intended to mount a serious challenge to DoD should be specified well in advance. Secretaries of defense complain often and sometimes rightly that OMB enters late into defense decisionmaking, seeking to upset, without adequate notice, complex and interdependent decisions already made. There is no need to remain vulnerable to such charges, especially since OMB will itself need many months for data gathering and analysis before its conclusions will be authoritative enough to affect presidential or congressional decisionmaking. Such examinations, therefore, should be begun long before the decisions they are intended to affect are scheduled. The secretary of defense should be promptly informed of them, and his staff should normally be invited to participate.

"Football is not a contact sport," someone has remarked, "dancing is a contact sport. Football is a collision sport." The defense budget game is also a collision sport, and many players outweigh OMB. The tasks here urged on OMB, therefore, are heroic ones, difficult even with presidential support and impossible without it. But two other truths are larger: the tasks badly need doing, and only OMB can do them.

NOTES

1. See Richard Stubbing, *The Defense Program*, (Washington, D.C.: The Urban Institute, 1986), pp. 29–30. Stubbing calculates the defense share of US defense and international affairs expenditures as averaging some 92 percent during the Carter administration and 96 percent under the Reagan administration.

2. *New York Times*, 11 Jan. 1981, p. 1F.

3. See Stubbing, *The Defense Program*, pp. 50–55.

4. For a recent discussion, with citations to prior studies, see Arch Barrett, *Reappraising Defense Organization* (Washington, D.C.: National Defense University Press, 1982); for a classic early study, see Alain Enthoven and Wayne L. Smith, *How Much Is Enough?* (New York: Harper and Row, 1971).

5. For recent popular discussions of this tendency see James Fallows, *National Defense* (New York: Random House, 1980); and the cover story, "Winds of Reform," *Time*, 7 Mar. 1983, pp. 12–30.

6. See generally, Franklin Spinney, in US Congress, Senate, Committee on Armed Services, *Hearings: The Plans/Reality Mismatch*, 98th Cong., 1st sess., 1983.

7. Graham Allison and Peter Szanton, *Remaking Foreign Policy* (New York: Basic Books, 1976), p. 170.

8. John Barron, *MIG Pilot* (New York: Reader's Digest Press, 1980), pp. 176–77; Fallows, *National Defense*, pp. 70–71.

9. Richard Stubbing, in conversation with author.

10. Oddly, however, the obvious cure may not work. The deputy director of OMB in the latter part of the Carter administration was a highly regarded administrator with extensive prior experience with defense issues. The view of some close observers of his performance was that he was at least as protective of DoD prerogatives as he was assertive of OMB's.

11. Barrett, *Reappraising Defense Organization*; and John N. Collins, *U.S. Defense Planning* (Boulder, Colo.: Westview Press, 1982).

12. Enthoven and Smith, *How Much Is Enough?* pp. 16–17.

13. See, for example, Gary Brewer, "Some Missing Pieces of the C3 I Puzzle," Unpublished paper, June 1983, pp. 8–9 (mimeo).

14. See I. M. Destler, ch. 5 in *Public Policy and Political Institutions: Defense and Foreign Policy*, JAI, 1985.

15. The so-called NSSM 3 Exercise of 1969 is a good example of the first; NSC-68 of 1951 is the classic example of the second.

16. The history of proposals to restructure the JCS so as to reduce the capacity of the services to block decisions with which any service disagrees, for example, goes back at least to the Rockefeller report of 1953, extends forward to the current proposals of former JCS Chairman David Jones, includes the work of a half dozen blue-ribbon commissions, and contains detailed suggestions to strengthen the authority of the chairman, to "re-doublehat" the service chiefs, and to create a single defense general staff. See John Kester, "The Future of the Joint Chiefs of Staff," *AEI Foreign Policy and Defense Review* (Feb. 1980): 2–23. Conspicuous inadequacies in costing, testing and development practices, and command structure have also generated rich histories of debate and proposal for reform. See, for example, David C. Jones, "What's Wrong with Our Defense Establishment?" *New York Times Magazine*, 7 Nov. 1982.

FROM THE REPORT OF THE PRESIDENT'S SPECIAL REVIEW BOARD

JOHN TOWER, EDMUND MUSKIE, AND BRENT SCOWCROFT

PART I: INTRODUCTION

In November 1986, it was disclosed that the United States had, in August 1985, and subsequently, participated in secret dealings with Iran involving the sale of military equipment. There appeared to be a linkage between these dealings and efforts to obtain the release of US citizens held hostage in Lebanon by terrorists believed to be closely associated with the Iranian regime. After the initial story broke, the attorney general announced that proceeds from the arms transfers may have been diverted to assist US-backed rebel forces in Nicaragua, known as contras. This possibility enlarged the controversy and added questions not only of policy and propriety but also violations of law.

These disclosures became the focus of substantial public attention. The secret arms transfers appeared to run directly counter to declared US policies. The United States had announced a policy of neutrality in the six-year-old Iran/Iraq war and had proclaimed an embargo on arms sales to Iran. It had worked actively to isolate Iran and other regimes known to give aid and comfort to terrorists. It had declared that it would not pay ransom to hostage-takers.

Public concern was not limited to the issues of policy, however. Questions arose as to the propriety of certain actions taken by the National Security Council staff and the manner in which the decision to transfer arms to Iran had been made. Congress was never informed. A variety of intermediaries, both private and governmental, some with motives open to question, had central roles. The NSC staff rather than the CIA seemed to be running the operation. The president appeared to be unaware of key elements of the operation. The controversy threatened a crisis of confidence in the manner in which national security decisions are made and the role played by the NSC staff.

It was this latter set of concerns that prompted the president to establish this Special Review Board on 1 December 1986. The president directed the board to examine the proper role of the National Security Council staff in national security operations, including the arms transfers to Iran. The president made clear that he wanted "all the facts to come out."

The board was not, however, called upon to assess individual culpability or be the final arbiter of the facts. These tasks have been properly left to others. Indeed, the short deadline set by the president for completion of the board's work and its limited resources precluded a separate and thorough field investigation. Instead, the board has examined the events surrounding the transfer of arms to Iran as a principal case study in evaluating the operation of the National Security Council in general and the role of the NSC staff in particular.

The president gave the board a broad charter. It was directed to conduct "a comprehensive study of the future role and procedures of the National Security Council (NSC) staff in the development, coordination, oversight, and conduct of foreign and national security policy."[1]

It has been forty years since the enactment of the National Security Act of 1947 and the creation of the National Security Council. Since that time the NSC staff has grown in importance, and the assistant to the president for national security affairs has emerged as a key player in national security decisionmaking. This is the first Presidential Commission to have as its sole responsibility a comprehensive review of how these institutions have performed. We believe that, quite aside from the circumstances which brought about the board's creation, such a review was overdue.

The board divided its work into three major inquiries: the circumstances surrounding the Iran/contra matter, other case studies that might reveal strengths and weaknesses in the operation of the National Security Council system under stress, and the manner in which that system has served eight different presidents since its inception in 1947.

At Appendix B is a narrative of the information obtained from documents and interviews regarding the arms sales to Iran. The narrative is necessarily incomplete. As of the date of this report, some key witnesses had refused to testify before any forum. Important documents located in other countries had yet to be released, and important witnesses in other countries were not available. But the appended narrative tells much of the story. Although more information will undoubtedly come to light, the record thus far developed provides a sufficient basis for evaluating the process by which these events came about.

During the board's work, it received evidence

Excerpts with minor editing from the report which is in the public domain.

concerning the role of the NSC staff in support of the contras during the period that such support was either barred or restricted by Congress. The board had neither the time nor the resources to make a systematic inquiry into this area. Notwithstanding, substantial evidence came before the board. A narrative of that evidence is contained at Appendix C.

The board found that the issues raised by the Iran/contra matter are in most instances not new. Every administration has faced similar issues, although arising in different factual contexts. The board examined in some detail the performance of the National Security Council system in 12 different crises dating back to the Truman administration.[2] Former government officials participating in many of these crises were interviewed. This learning provided a broad historical perspective to the issues before the board.

Those who expect from us a radical prescription for wholesale change may be disappointed. Not all major problems—and Iran/contra has been a major one—can be solved simply by rearranging organizational blocks or passing new laws.

In addition, it is important to emphasize that the president is responsible for the national security policy of the United States. In the development and execution of that policy, the president is the decisionmaker. He is not obliged to consult with or seek approval from anyone in the executive branch. The structure and procedures of the National Security Council system should be designed to give the president every assistance in discharging these heavy responsibilities. It is not possible to make a system immune from error without paralyzing its capacity to act.

At its senior levels, the National Security Council is primarily the interaction of people. We have examined with care its operation in the Iran/contra matter and have set out in considerable detail mistakes of omission, commission, judgment, and perspective. We believe that this record and analysis can warn future presidents, members of the National Security Council, and national security advisers of the potential pitfalls they face even when they are operating with what they consider the best of motives. We would hope that this record would be carefully read and its lessons fully absorbed by all aspirants to senior positions in the National Security Council system.

This report will serve another purpose. In preparing it, we contacted every living past president, three former vice presidents, and every living secretary of state, secretary of defense, national security adviser, most directors of central intelligence, and several chairmen of the Joint Chiefs of Staff to solicit their views. We sought to learn how well, in their experience, the system had operated or, in the case of past presidents, how well it served them. We asked all former participants how they would change the system to make it more useful to the president.[3]

Our review validates the current National Security Council system. That system has been utilized by different presidents in very different ways, in accordance with their individual work habits and philosophical predilections. On occasion over the years it has functioned with real brilliance; at other times serious mistakes have been made. The problems we examined in the case of Iran/contra caused us deep concern. But their solution does not lie in revamping the National Security Council system.

That system is properly the president's creature. It must be left flexible to be molded by the president into the form most useful to him. Otherwise it will become either an obstacle to the president, and a source of frustration; or an institutional irrelevance, as the president fashions informal structures more to his liking.

Having said that, there are certain functions which need to be performed in some way for any president. What we have tried to do is to distill from the wisdom of those who have participated in the National Security Council system over the past 40 years the essence of these functions and the manner in which that system can be operated so as to minimize the likelihood of major error without destroying the creative impulses of the president.

PART II: ORGANIZING FOR NATIONAL SECURITY

Ours is a government of checks and balances, of shared power and responsibility. The Constitution places the president and the Congress in dynamic tension. They both cooperate and compete in the making of national policy.

National security is no exception. The Constitution gives both the president and the Congress an important role. The Congress is critical in formulating national policies and in marshalling the resources to carry them out. But those resources—the nation's military personnel, its diplomats, its intelligence capability—are lodged in the executive branch. As chief executive and commander-in-chief, and with broad authority in the area of foreign affairs, it is the president who is empowered to act for the nation and protect its interests.

A. THE NATIONAL SECURITY COUNCIL

The present organization of the executive branch for national security matters was established by the National Security Act of 1947. That act created the National Security Council. As now constituted, its statutory members are the president, vice president, secretary of state, and secretary of defense. The president is the head of the National Security Council.

Presidents have from time to time invited the heads of other departments or agencies to attend National Security Council meetings or to participate as de facto members. These have included the director of cental intelligence (the "DCI") and the chairman of the Joint Chiefs of Staff (the "CJCS"). The president (or, in his absence, his designee) presides.

The National Security Council deals with the most vital issues in the nation's national security policy. It is this body that discusses recent developments in arms control and the Strategic Defense Initiative; that discussed whether or not to bomb the Cambodian mainland after the *Mayaguez* was captured; that debated the timetable for the US withdrawal from Vietnam; and that considered the risky and daring attempt to rescue US hostages in Iran in 1980. The National Security Council deals with issues that are difficult, complex, and often secret. Decisions are often required in hours rather than weeks. Advice must be given under great stress and with imperfect information.

The National Security Council is not a decisionmaking body. Although its other members hold official positions in the government, when meeting as the National Security Council they sit as advisers to the president. This is clear from the language of the 1947 Act:

The function of the Council shall be to advise the President with respect to the integration of domestic, foreign, and military policies relating to the national security so as to enable the military services and the other departments and agencies of the Government to cooperate more effectively in matters involving the national security.

The National Security Council has from its inception been a highly personal instrument. Every president has turned for advice to those individuals and institutions whose judgment he has valued and trusted. For some presidents, such as President Eisenhower, the National Security Council served as a primary forum for obtaining advice on national security matters. Other presidents, such as President Kennedy, relied on more informal groupings of advisers, often including some but not all of the council members.

One official summarized the way the system has been adjusted by different presidents:

The NSC is going to be pretty well what a President wants it to be and what he determines it should be. Kennedy—and these are some exaggerations and generalities of course—with an anti-organizational bias, disestablished all [the Eisenhower created] committees and put a tight group in the White House totally attuned to his philosophic approach . . . Johnson didn't change that very much, except certain difficulties began to develop in the informality which

was [otherwise] characterized by speed, unity of purpose, precision . . . So it had great efficiency and responsiveness. The difficulties began to develop in . . . the informality of the thing.

The Nixon administration saw a return to the use of the National Security Council as a principal forum for national security advice. This pattern was continued by President Ford and President Carter, and in large measure by President Reagan.

Regardless of the frequency of its use, the NSC has remained a strictly advisory body. Each president has kept the burden of decision for himself, in accordance with his constitutional responsibilities.

B. The Assistant to the President for National Security Affairs

Although closely associated with the National Security Council in the public mind, the assistant to the president for national security affairs is not one of its members. Indeed, no mention of this position is made in the National Security Act of 1947.

The position was created by President Eisenhower in 1953. Although its precise title has varied, the position has come to be known (somewhat misleadingly) as the national security adviser.

Under President Eisenhower, the holder of this position served as the principal executive officer of the council, setting the agenda, briefing the president on council matters, and supervising the staff. He was not a policy advocate.

It was not until President Kennedy, with McGeorge Bundy in the role, that the position took on its current form. Bundy emerged as an important personal adviser to the president on national security affairs. This introduced an element of direct competition into Bundy's relationship with the members of the National Security Council. Although President Johnson changed the title of the position to simply "special assistant," in the hands of Walt Rostow it continued to play an important role.

President Nixon relied heavily on his national security adviser, maintaining and even enhancing its prominence. In that position, Henry Kissinger became a key spokesman for the president's national security policies both to the US press and to foreign governments. President Nixon used him to negotiate on behalf of the United States with Vietnam, China, the Soviet Union, and other countries. The roles of spokesman and negotiator had traditionally been the province of the secretary of state, not of the national security adviser. The emerging tension between the two positions was only resolved when Kissinger assumed them both.

Under President Ford, Lt. Gen. Brent Scowcroft became national security adviser, with

Henry Kissinger remaining as secretary of state. The national security adviser exercised major responsibility for coordinating for the president the advice of his NSC principals and overseeing the process of policy development and implementation within the executive branch.

President Carter returned in large part to the early Kissinger model, with a resulting increase in tensions with the secretary of state. President Carter wanted to take the lead in matters of foreign policy, and used his national security adviser as a source of information, ideas, and new initiatives.

The role of the national security adviser, like the role of the NSC itself, has in large measure been a function of the operating style of the president. Notwithstanding, the national security adviser has come to perform, to a greater or lesser extent, certain functions which appear essential to the effective discharge of the President's responsibilities in national security affairs.

- He is an "honest broker" for the NSC process. He assures that issues are clearly presented to the president; that all reasonable options, together with an analysis of their disadvantages and risks, are brought to his attention; and that the views of the president's other principal advisers are accurately conveyed.
- He provides advice from the president's vantage point, unalloyed by institutional responsibilities and biases. Unlike the secretaries of state or defense, who have substantial organizations for which they are responsible, the president is the national security adviser's only constituency.
- He monitors the actions taken by the executive departments in implementing the president's national security policies. He asks the question whether these actions are consistent with presidential decisions and whether, over time, the underlying policies continue to serve US interests.
- He has a special role in crisis management. This has resulted from the need for prompt and coordinated action under presidential control, often with secrecy being essential.
- He reaches out for new ideas and initiatives that will give substance to broad presidential objectives for national security.
- He keeps the president informed about international developments and developments in the Congress and the executive branch that affect the president's policies and priorities.

But the national security adviser remains the creature of the president. The position will be largely what he wants it to be. This presents any president with a series of dilemmas.

- The president must surround himself with people he trusts and to whom he can speak in confidence. To this end, the national security adviser, unlike the secretaries of state and defense, is not subject to confirmation by the Senate and does not testify before Congress. But the more the president relies on the national security adviser for advice, especially to the exclusion of his cabinet officials, the greater will be the unease with this arrangement.
- As the "honest broker" of the NSC process, the national security adviser must ensure that the different and often conflicting views of the NSC principals are presented fairly to the president. But as an independent adviser to the president, he must provide his own judgment. To the extent that the national security adviser becomes a strong advocate for a particular point of view, his role as "honest broker" may be compromised, and the president's access to the unedited views of the NSC principals may be impaired.
- The secretaries of state and defense, and the director of central intelligence, head agencies of government that have specific statutory responsibilities and are subject to congressional oversight for the implementation of US national security policy. To the extent that the national security adviser assumes operational responsibilities, whether by negotiating with foreign governments or becoming heavily involved in military or intelligence operations, the legitimacy of that role and his authority to perform it may be challenged.
- The more the national security adviser becomes an "operator" in implementing policy, the less will he be able objectively to review that implementation—and whether the underlying policy continues to serve the interests of the president and the nation.
- The secretary of state has traditionally been the president's spokesman on matters of national security and foreign affairs. To the extent that the national security adviser speaks publicly on these matters or meets with representatives of foreign governments, the result may be confusion as to what is the president's policy.

C. The NSC Staff

At the time it established the National Security Council, Congress authorized a staff headed by an executive secretary appointed by the president. Initially quite small, the NSC staff expanded substantially under President Eisenhower.

During the Eisenhower administration, the NSC staff assumed two important functions: coordinating the executive departments in the development of national policy (through the NSC

Planning Board) and overseeing the implementation of that policy (through the Operations Coordination Board). A systematic effort was made to coordinate policy development and its implementation by the various agencies through an elaborate set of committees. The system worked fairly well in bringing together for the president the views of the other NSC principals. But it has been criticized as biased toward reaching consensus among these principals rather than developing options for presidential decision. By the end of his second term, President Eisenhower himself had reached the conclusion that a highly competent individual and a small staff could perform the needed functions in a better way. Such a change was made by President Kennedy.

Under President Kennedy, a number of the functions of the NSC staff were eliminated, and its size was sharply reduced. The Planning and Operations Coordinating Boards were abolished. Policy development and policy implementation were assigned to individual cabinet officers, responsible directly to the president. By late 1962 the staff was only 12 professionals, serving largely as an independent source of ideas and information to the president. The system was lean and responsive, but frequently suffered from a lack of coordination. The Johnson administration followed much the same pattern.

The Nixon administration returned to a model more like Eisenhower's but with something of the informality of the Kennedy/Johnson staffs. The Eisenhower system had emphasized coordination; the Kennedy-Johnson system tilted to innovation and the generation of new ideas. The Nixon system emphasized both. The objective was not interdepartmental consensus but the generation of policy options for presidential decision, and then ensuring that those decisions were carried out. The staff grew to 50 professionals in 1970 and became a major factor in the national security decisionmaking process. This approach was largely continued under President Ford.

The NSC staff retained an important role under President Carter. While continuing to have responsibility for coordinating policy among the various executive agencies, President Carter particularly looked to the NSC staff as a personal source of independent advice. President Carter felt the need to have a group loyal only to him from which to launch his own initiatives and to move a vast and lethargic government. During his time in office, President Carter reduced the size of the professional staff to 35, feeling that a smaller group could do the job and would have a closer relationship to him.

What emerges from this history is an NSC staff used by each president in a way that reflected his individual preferences and working style. Over time, it has developed an important role within the executive branch of coordinating policy review, preparing issues for presidential decision, and monitoring implementation. But it has remained the president's creature, molded as he sees fit, to serve as his personal staff for national security affairs. For this reason, it has generally operated out of the public view and has not been subject to direct oversight by the Congress.

D. THE INTERAGENCY COMMITTEE SYSTEM

The National Security Council has frequently been supported by committees made up of representatives of the relevant national security departments and agencies. These committees analyze issues prior to consideration by the Council. There are generally several levels of committees. At the top level, officials from each agency (at the deputy secretary or under secretary level) meet to provide a senior level policy review. These senior-level committees are in turn supported by more junior interagency groups (usually at the assistant secretary level). These in turn may oversee staff-level working groups that prepare detailed analyses of important issues.

Administrations have differed in the extent to which they have used these interagency committees. President Kennedy placed little stock in them. The Nixon and Carter administrations, by contrast, made much use of them.

THE REAGAN MODEL

President Reagan entered office with a strong commitment to cabinet government. His principal advisers on national security affairs were to be the secretaries of state and defense, and to a lesser extent the director of central intelligence. The position of the national security adviser was initially downgraded in both status and access to the president. Over the next six years, five different people held that position.

The administration's first national security adviser, Richard Allen, reported to the president through the senior White House staff. Consequently, the NSC staff assumed a reduced role. Mr. Allen believed that the secretary of state had primacy in the field of foreign policy. He viewed the job of the national security adviser as that of a policy coordinator.

President Reagan initially declared that the National Security Council would be the principal forum for consideration of national security issues. To support the work of the council, President Reagan established an interagency committee system headed by three Senior Interagency Groups (or "SIGs"), one each for foreign policy, defense policy, and intelligence. They were chaired by the secretary of state, the secretary of defense, and the director of Central Intelligence, respectively.

Over time, the administration's original conception of the role of the national security advisor changed. William Clark, who succeeded Richard Allen in 1982, was a long-time associate of the president and dealt directly with him. Robert McFarlane, who replaced Judge Clark in 1983, although personally less close to the president, continued to have direct access to him. The same was true for VADM John Poindexter, who was appointed to the position in December 1985.

President Reagan appointed several additional members to his National Security Council and allowed staff attendance at meetings. The resultant size of the meetings led the president to turn increasingly to a smaller group (called the National Security Planning Group or "NSPG"). Attendance at its meetings was more restricted but included the statutory principals of the NSC. The NSPG was supported by the SIGs, and new SIGs were occasionally created to deal with particular issues. These were frequently chaired by the national security adviser. But generally the SIGs and many of their subsidiary groups (called Interagency Groups or "IGs") fell into disuse.

As a supplement to the normal NSC process, the Reagan administration adopted comprehensive procedures for covert actions. These are contained in a classifed document, NSDD-159, establishing the process for deciding, implementing, monitoring and reviewing covert activities.

THE PROBLEM OF COVERT OPERATIONS

Covert activities place a great strain on the process of decision in a free society. Disclosure of even the existence of the operation could threaten its effectiveness and risk embarrassment to the government. As a result, there is strong pressure to withhold information, to limit knowledge of the operation to a minimum number of people.

These pressures come into play with great force when covert activities are undertaken in an effort to obtain the release of US citizens held hostage abroad. Because of the legitimate human concern all presidents have felt over the fate of such hostages, our national pride as a powerful country with a tradition of protecting its citizens abroad, and the great attention paid by the news media to hostage situations, the pressures on any president to take action to free hostages are enormous. Frequently to be effective, this action must necessarily be covert. Disclosure would directly threaten the lives of the hostages as well as those willing to contemplate their release.

Since covert arms sales to Iran played such a central role in the creation of this board, it has focused its attention in large measure on the role of the NSC staff where covert activity is involved. This is not to denigrate, however, the importance of other decisions taken by the government. In those areas as well the National Security Council and its staff play a critical role. But in many respects the best test of a system is its performance under stress. The conditions of greatest stress are often found in the crucible of covert activities.

PART IV: WHAT WAS WRONG

The arms transfers to Iran and the activities of the NSC staff in support of the contras are case studies in the perils of policy pursued outside the constraints of orderly process.

The Iran initiative ran directly counter to the administration's own policies on terrorism, the Iran/Iraq war, and military support to Iran. This inconsistency was never resolved, nor were the consequences of this inconsistency fully considered and provided for. The result taken as a whole was a US policy that worked against itself.

The board believes that failure to deal adequately with these contradictions resulted in large part from the flaws in the manner in which decisions were made. Established procedures for making national security decisions were ignored. Reviews of the initiative by all the NSC principals were too infrequent. The initiatives were not adequately vetted below the cabinet level. Intelligence resources were underutilized. Applicable legal constraints were not adequately addressed. The whole matter was handled too informally, without adequate written records of what had been considered, discussed, and decided.

This pattern persisted in the implementation of the Iran initiative. The NSC staff assumed direct operational control. The initiative fell within the traditional jurisdictions of the Department of State, Defense, and CIA. Yet these agencies were largely ignored. Great reliance was placed on a network of private operators and intermediaries. How the initiative was to be carried out never received adequate attention for the NSC principals or a tough working-level review. No periodic evaluation of the progress of the initiative was ever conducted. The result was an unprofessional and, in substantial part, unsatisfactory operation.

In all of this process, Congress was never notified.

As noted in Part III, the record of the role of the NSC staff in support of the contras is much less complete. Nonetheless, what is known suggests that many of the same problems plagued that effort as well.

The first section of this Part IV discusses the flaws in the process by which conflicting policies were considered, decisions were made, and the initiatives were implemented.

The second section discusses the responsibilities of the NSC principals and other key national security officials for the manner in which these initiatives were handled.

The third section discusses the special problem posed by the role of the Israelis.

The fourth section of this Part IV outlines the board's conclusions about the management of the initial public presentation of the facts of the Iran initiative.

A. A FLAWED PROCESS

1. Contradictory Policies Were Pursued.—The arms sales to Iran and the NSC support for the contras demonstrate the risks involved when highly controversial initiatives are pursued covertly.

Arms Transfers to Iran.—The initiative to Iran was a covert operation directly at odds with important and well-publicized policies of the executive branch. But the initiative itself embodied a fundamental contradiction. Two objectives were apparent from the outset: a strategic opening to Iran, and release of the US citizens held hostage in Lebanon. The sale of arms to Iran appeared to provide a means to achieve both these objectives. It also played into the hands of those who had other interests—some of them personal financial gain—in engaging the United States in an arms deal with Iran.

In fact, the sale of arms was not equally appropriate for achieving both these objectives. Arms were what Iran wanted. If all the United States sought was to free the hostages, then an arms-for-hostages deal could achieve the immediate objectives of both sides. But if the US objective was a broader strategic relationship, then the sale of arms should have been contingent upon first putting into place the elements of that relationship. An arms-for-hostages deal in this context could become counterproductive to achieving this broader strategic objective. In addition, release of the hostages would require exerting influence with Hizballah, which could involve the most radical elements of the Iranian regime. The kind of strategic opening sought by the United States, however, involved what were regarded as more moderate elements.

The US officials involved in the initiative appeared to have held three distinct views. For some, the principal motivation seemed consistently a strategic opening to Iran. For others, the strategic opening became a rationale for using arms sales to obtain the release of the hostages. For still others, the initiative appeared clearly as an arms-for-hostages deal from first to last.

Whatever the intent, almost from the beginning the initiative became in fact a series of arms-for-hostages deals. The shipment of arms in November 1985, was directly tied to a hostage release. Indeed, the August/September transfer may have been nothing more than an arms-for-hostages trade. By 14 July 1985, a specific proposal for the sale of 100 TOWs to Iran in exchange for Iranian efforts to secure the release of all the hostages had been transmitted to the White House and discussed with the president. What actually occurred, at least so far as the September shipment was concerned, involved a direct link of arms and a hostage.

The initiative continued to be described in terms of its broader strategic relationship. But those elements never really materialized. While a high-level meeting among senior US and Iranian officials continued to be a subject of discussion, it never occurred. Although Mr. MacFarlane went to Tehran in May of 1986, the promised high-level Iranians never appeared. In discussions among US officials, the focus seemed to be on the prospect for obtaining release of the hostages, not on a strategic relationship. Even if one accepts the explanation that arms and hostages represented only "bona fides" of seriousness of purpose for each side, that had clearly been established, one way or another, by the September exchange.

It is true that, strictly speaking, arms were not exchanged for the hostages. The arms were sold for cash; and to Iran, rather than the terrorists holding the hostages. Iran clearly wanted to buy the arms, however, and time and time again US willingness to sell was directly conditioned upon the release of hostages. Although Iran might claim that it did not itself hold the hostages, the whole arrangement was premised on Iran's ability to secure their release.

While the United States was seeking the release of the hostages in this way, it was vigorously pursuing policies that were dramatically opposed to such efforts. The Reagan administration in particular had come into office declaring a firm stand against terrorism, which it continued to maintain. In December of 1985, the administration completed a major study under the chairmanship of the vice president. It resulted in a vigorous reaffirmation of US opposition to terrorism in all its forms and a vow of total war on terrorism whatever its source. The administration continued to pressure US allies not to sell arms to Iran and not to make concessions to terrorists.

No serious effort was made to reconcile the inconsistency between these policies and the Iran initiative. No effort was made systematically to address the consequences of this inconsistency—the effect on US policy when, as it inevitably would, the Iran initiative became known.

The board believes that a strategic opening to Iran may have been in the national interest but that the United States never should have been a party to the arms transfers. As arms-for-hostages trades, they could not help but create an incentive for further hostage-taking. As a violation of the US arms embargo, they could only remove inhibitions on other nations from selling arms to Iran. This threatened to upset the military balance between Iran and Iraq, with consequent jeopardy to the Gulf States and the interests of the West in that region. The arms-for-hostages trade rewarded a regime that clearly supported terrorism and hostage-taking. They increased the risk that the United States would be perceived, especially in the Arab world, as a creature of Israel. They suggested to other US allies and friends in the region that the United States had shifted its policy in favor of Iran. They raised questions as to whether US policy statements could be relied upon.

As the arms-for-hostages proposal first came to the United States, it clearly was tempting. The sale of just 100 TOWs was to produce the release of all seven Americans held in Lebanon. Even had the offer been genuine, it would have been unsound. But it was not genuine. The 100 TOWs did not produce seven hostages. Very quickly the price went up, and the arrangements became protracted. A pattern of successive bargained exchanges of arms and hostages was quickly established. While release of all the hostages continued to be promised, in fact the hostages came out singly if at all. This sad history is powerful evidence of why the United States should never have become involved in the arms transfers.

NCS Staff Support for the Contras.—The activities of the NSC staff in support of the contras sought to achieve an important objective of the administration's foreign policy. The president had publicly and emphatically declared his support for the Nicaragua resistance. That brought his policy in direct conflict with that of the Congress, at least during the period that direct or indirect support of military operations in Nicaragua was barred.

Although the evidence before the board is limited, no serious effort appears to have been made to come to grips with the risks to the president of direct NSC support for the contras in the face of those congressional restrictions. Even if it could be argued that these restrictions did not technically apply to the NSC staff, these activities presented great political risk to the president. The appearance of the president's personal staff doing what Congress had forbade other agencies to do could, once disclosed, only touch off a firestorm in the Congress and threaten the administration's whole policy on the contras.

2. The Decisionmaking Process Was Flawed.— Because the arms sales to Iran and the NSC support for the contras occurred in settings of such controversy, one would expect that the decisions to undertake these activities would have been made only after intense and thorough consideration. In fact, a far different picture emerges.

Arms Transfers to Iran.—The Iran initiative was handled almost casually and through informal channels, always apparently with an expectation that the process would end with the next arms-for-hostages exchange. It was subjected neither to the general procedures for interagency consideration and review of policy issues nor the more restrictive procedures set out in NSDD 159 for handling covert operations. This had a number of consequences.

(i) The Opportunity for a Full Hearing before the President Was Inadequate.—In the last half of 1985, the Israelis made three separate proposals to the United States with respect to the Iran initiative (two in July and one in August). In addition, Israel made three separate deliveries of arms to Iran, one each in August, September, and November. Yet prior to 7 December 1985, there was at most one meeting of the NSC principals, a meeting which several participants recall taking place on August 6. There is no dispute that full meetings of the principals did occur on 7 December 1985, and on 7 January 1986. But the proposal to shift to direct US arms sales to Iran appears not to have been discussed until later. It was considered by the president at a meeting on January 17 which only the vice president, Mr. Regan, Mr. Fortier, and VADM Poindexter attended. Thereafter, the only senior-level review the Iran initiative received was during one or another of the president's daily national security briefings. These were routinely attended only by the president, the vice president, Mr. Regan, and VADM Poindexter. There was no subsequent collective consideration of the Iran initiative by the NSC principals before it became public 11 months later.

This was not sufficient for a matter as important and consequential as the Iran initiative. Two or three cabinet-level reviews in a period of 17 months was not enough. The meeting on December 7 came late in the day, after the pattern of arms-for-hostages exchanges had become well established. The January 7 meeting had earmarks of a meeting held after a decision had already been made. Indeed, a draft Covert Action Finding authorizing the initiative had been signed by the president, though perhaps inadvertently, the previous day.

At each significant step in the Iran initiative, deliberations among the NSC principals in the presence of the president should have been virtually automatic. This was not and should not have been a formal requirement, something prescribed by statute. Rather, it should have been something the NSC principals desired as a means of ensuring an optimal environment for presidential judgment. The meetings should have been preceded by consideration by the NSC principals of staff papers prepared according to the procedures applicable to covert actions. These should have reviewed the history of the initiative, analyzed the issues then presented, developed a range of realistic options, presented the odds of success and the costs of failure, and addressed questions of implementation, and execution. Had this been done, the objectives of the Iran initiative might have been clarified and alternatives to the sale of arms might have been identified.

(ii) The Initiative Was Never Subjected to a Rigorous Review below the Cabinet Level.—Because of the obsession with secrecy, interagency consideration of the initiative was limited to the cabinet level. With the exception of the NSC staff and, after January 17, 1986, a handful of CIA officials, the rest of the executive departments and agencies were largely excluded.

As a consequence, the initiative was never vetted at the staff level. This deprived those responsible for the initiative of considerable expertise—on the situation in Iran; on the difficulties of dealing with terrorists; on the mechanics of conducting a diplomatic opening. It also kept the plan from receiving a tough, critical review.

Moreover, the initiative did not receive a policy review below cabinet level. Careful consideration at the deputy/under secretary level might have exposed the confusion in US objectives and clarified the risks of using arms as an instrument of policy in this instance.

The vetting process would also have ensured better use of US intelligence. As it was, the intelligence input into the decision process was clearly inadequate. First, no independent evaluation of the Israeli proposals offered in July and August appears to have been sought or offered to US intelligence agencies. The Israelis represented that they for some time had had contacts with elements in Iran. The prospects for an opening to Iran depended heavily on these contacts, yet no systematic assessment appears to have been made by US intelligence agencies of the reliability and motivations of these contacts, and the identity and objectives of the elements in Iran that the opening was supposed to reach. Neither was any systematic assessment made of the motivation of the Israelis.

Second, neither Mr. Ghorbanifar nor the second channel seem to have been subjected to a systematic intelligence vetting before they were engaged as intermediaries. Mr. Ghorbanifar had been known to the CIA for some time, and the agency had substantial doubts as to his reliability and truthfulness. Yet the agency did not volunteer that information or inquire about the identity of the intermediary if his name was unknown. Conversely, no early request for a name check was made of the CIA, and it was not until 11 January 1986, that the agency gave Mr. Ghorbanifar a new polygraph, which he failed. Notwithstanding this situation, with the signing of the January 17 Finding, the United States took control of the initiative and became even more directly involved with Mr. Ghorbanifar. The issues raised by the polygraph results do not appear to have been systematically addressed. In similar fashion, no prior intelligence checks appear to have been made on the second channel.

Third, although the president recalled being assured that the arms sales to Iran would not alter the military balance with Iran, the board could find no evidence that the president was ever briefed on this subject. The question of the impact of any intelligence shared with the Iranians does not appear to have been brought to the president's attention.

A thorough vetting would have included consideration of the legal implications of the initiative. There appeared little effort to face squarely the legal restrictions and notification requirements applicable to the operation. At several points, other agencies raised questions about violations of law or regulations. These concerns were dismissed without, it appears, investigating them with the benefit of legal counsel.

Finally, insufficient attention was given to the implications of implementation. The implementation of the initiative raised a number of issues: should the NSC staff rather than the CIA have had operational control; what were the implications of Israeli involvement; how reliable were the Iranian and various other private intermediaries; what were the implications for the use of Mr. Secord's private network of operatives; what were the implications for the military balance in the region; was operational security adequate. Nowhere do these issues appear to have been sufficiently addressed.

The concern for preserving the secrecy of the initiative provided an excuse for abandoning sound process. Yet the initiative was known to a variety of persons with diverse interests and ambitions—Israelis, Iranians, various arms dealers and business intermediaries, and LtCol North's network of private operatives. While concern for secrecy would have justified limiting the circle of persons knowledgeable about the initiative, in this case it was drawn too tightly. As a consequence, important advice and counsel were lost.

In January of 1985, the president had adopted procedures for striking the proper balance between secrecy and the need for consultation on sensitive programs. These covered the institution, implementation, and review of covert operations. In the case of the Iran initiative, these procedures were almost totally ignored.

The only staff work the president apparently reviewed in connection with the Iran initiative was prepared by NSC staff members, under the direction of the national security adviser. These were, of course, the principal proponents of the initiative. A portion of this staff work was reviewed by the board. It was frequently striking in its failure to present the record of past efforts—particularly past failures. Alternative ways of achieving US objectives—other than yet another arms-for-hostages deal—were not discussed. Frequently it neither adequately presented the risks involved in pursuing the initiative nor the full force of the dissenting views of other NSC principals. On balance, it did not serve the president well.

(iii) The Process Was Too Informal.—The whole decision process was too informal. Even when meetings among NSC principals did occur, often there was no prior notice of the agenda. No formal written minutes seem to have been kept. Decisions subsequently taken by the president were not formally recorded. An exception was the January 17 Finding, but even this was apparently not circulated or shown to key US officials.

The effect of this informality was that the initiative lacked a formal institutional record. This precluded the participants from undertaking the more informed analysis and reflection that is afforded by a written record, as opposed to mere recollection. It made it difficult to determine where the initiative stood, and to learn lessons from the record that could guide future action. This lack of an institutional record permitted specific proposals for arms-for-hostages exchanges to be presented in a vacuum, without reference to the results of past proposals. Had a searching and thorough review of the Iran initiative been undertaken at any stage in the process, it would have been extremely difficult to conduct. The board can attest first hand to the problem of conducting a review in the absence of such records. Indeed, the exposition in the wake of public revelation suffered the most.

NSC Staff Support for the Contras.—It is not clear how LtCol North first became involved in activities in direct support of the contras during the period of the congressional ban. The board did not have before it much evidence on this point. In the evidence that the board did have, there is no suggestion at any point of any dis-

cussion of LtCol North's activities with the president in any forum. There also does not appear to have been any interagency review of LtCol North's activities at any level.

The latter point is not surprising given the congressional restrictions under which the other relevant agencies were operating. But the NSC staff apparently did not compensate for the lack of any interagency review with its own internal vetting of these activities. LtCol North apparently worked largely in isolation, keeping first Mr. McFarlane and then VADM Poindexter informed.

The lack of adequate vetting is particularly evident on the question of the legality of LtCol North's activities. The board did not make a judgment on the legal issues raised by his activities in support of the contras. Nevertheless, some things can be said.

If these activities were illegal, obviously they should not have been conducted. If there was any doubt on the matter, systematic legal advice should have been obtained. The political cost to the president of illegal action by the NSC staff was particularly high, both because the NSC staff is the personal staff of the president and because of the history of serious conflict with the Congress over the issue of contra support. For these reasons, the president should have been kept apprised of any review of the legality of LtCol North's activities.

Legal advice was apparently obtained from the president's Intelligence Oversight Board. Without passing on the quality of that advice, it is an odd source. It would be one thing for the Intelligence Oversight Board to review the legal advice provided by some other agency. It is another for the Intelligence Oversight Board to be originating legal advice of its own. That is a function more appropriate for the NSC staff's own counsel.[4]

3. Implementation Was Unprofessional.—The manner in which the Iran initiative was implemented and LtCol North undertook to support the contras are very similar. This is in large part because the same cast of characters was involved. In both cases the operations were unprofessional, although the board has much less evidence with respect to LtCol North's contra activities.

Arms Tranfers to Iran.—With the signing of the January 17 Finding, the Iran initiative became a US operation run by the NSC staff. LtCol North made most of the significant operational decisions. He conducted the operation through Mr. Secord and his associates, a network of private individuals already involved in the contra resupply operation. To this was added a handful of selected individuals from the CIA.

But the CIA support was limited. Two CIA officials, though often at meetings, had a relatively limited role. One served as the point man for LtCol North in providing logistics and financial arrangements. The other (Mr. Allen) served as a contact between LtCol North and the intelligence community. By contrast, George Cave actually played a significant and expanding role. However, Clair George, deputy director for operations at CIA, told the Board: "George was paid by me and on the paper was working for me. But I think in the heat of the battle, . . . George was working for Oliver North."

Because so few people from the departments and agencies were told of the initiative, LtCol North cut himself off from resources and expertise from within the government. He relied instead on a number of private intermediaries, businessmen, and the financial brokers, private operators, and Iranians hostile to the United States. Some of these were individuals with questionable credentials and potentially large personal financial interests in the transactions. This made the transactions unnecessarily complicated and invited kick-backs and payoffs. This arrangement also dramatically increased the risks that the initiative would leak. Yet no provision was made for such an eventuality. Further, the use of Mr. Secord's private network in the Iran initiative linked those operators with the resupply of the contras, threatening exposure of both operations if either became public.

The result was a very unprofessional operation.

Mr. Secord undertook in November, 1985, to arrange landing clearance for the Israeli flight bringing the HAWK missiles into a third-country staging area. The arrangements fell apart. A CIA field officer attributed this failure to the amateurish way in which Mr. Secord and his associates approached officials in the government from which landing clearance was needed. If Mr. Ghorbanifar is to be believed, the mission of Mr. McFarlane to Tehran was undertaken without any advance work, and with distinctly different expectations on the part of the two sides. This could have contributed to its failure.

But there were much more serious errors. Without adequate study and consideration, intelligence was passed to the Iranians of potentially major significance to the Iran/Iraq war. At the meeting with the second channel on 5–7 October 1986, LtCol North misrepresented his access to the president. He told Mr. Ghorbanifar stories of conversations with the president which were wholly fanciful. He suggested without authority a shift in US policy adverse to Iraq in general and Saddam Hussein in particular. Finally, in the nine-point agenda discussed on October 26–28, he committed the United

States, without authorization, to a position contrary to well-established US policy on the prisoners held by Kuwait.

The conduct of the negotiators with Mr. Ghorbanifar and the second channel was handled in a way that revealed obvious inexperience. The discussions were too casual for dealings with intermediaries to a regime so hostile to US interests. The US hand was repeatedly tipped and unskillfully played. The arrangements failed to guarantee that the US obtained its hostages in exchange for the arms. Repeatedly, LtCol North permitted arms to be delivered without the release of a single captive.

The implementation of the initiative was never subjected to a rigorous review. LtCol North appears to have kept VADM Poindexter fully informed of his activities. In addition, VADM Poindexter, LtCol North, and the CIA officials involved apparently apprised Director Casey of many of the operational details. But LtCol North and his operation functioned largely outside the orbit of the US government. Their activities were not subject to critical reviews of any kind.

After the initial hostage release in September 1985, it was over 10 months before another hostage was released. This despite recurring promises of the release of all the hostages and four intervening arms shipments. Beginning with the November shipment, the United States increasingly took over the operation of the initiative. In January 1986, it decided to transfer arms directly to Iran.

Any of these developments could have served as a useful occasion for a systematic reconsideration of the initiative. Indeed, at least one of the schemes contained a provision for reconsideration if the initial assumptions proved to be invalid. They did, but the reconsideration never took place. It was the responsibility of the national security adviser and the responsible officers on the NSC staff to call for such a review. But they were too involved in the initiative both as advocates and as implementors. This made it less likely that they would initiate the kind of review and reconsideration that should have been undertaken.

NSC Staff Support for the Contras.—As already noted, the NSC activities in support of the contras and its role in the Iran initiative were of a piece. In the former, there was an added element of LtCol North's intervention in the customs investigation of the crash of the SAT aircraft. Here, too, selected CIA officials reported directly to LtCol North. The limited evidence before the board suggested that the activities in support of the contras involved unprofessionalism much like that in the Iran operation.

iv. Congress Was Never Notified.—Congress was not apprised either of the Iran initiative or of the NSC staff's activities in support of the contras.

In the case of Iran, because release of the hostages was expected within a short time after the delivery of equipment, and because public disclosure could have destroyed the operation and perhaps endangered the hostages, it could be argued that it was justifiable to defer notification of Congress prior to the first shipment of arms to Iran. The plan apparently was to inform Congress immediately after the hostages were safely in US hands. But after the first delivery failed to release all the hostages, and as one hostage release plan was replaced by another, Congress certainly should have been informed. This could have been done during a period when no specific hostage release plan was in execution. Consultation with Congress could have been useful to the president, for it might have given him some sense of how the public would react to the initiative. It also might have influenced his decision to continue to pursue it.

v. Legal Issues.—In addition to conflicting with several fundamental US policies, selling arms to Iran raised far-reaching legal questions. How it dealt with these is important to an evaluation of the Iran initiative.

Arms Transfers to Iran.—It was not part of the board's mandate to consider issues of law as they may pertain to individuals or detailed aspects of the Iran initiative. Instead, the board focused on the legal basis for the arms transfers to Iran and how issues of law were addressed in the NSC process.

The Arms Export Control Act, the principal US statute governing arms sales abroad, makes it unlawful to export arms without a license. Exports of arms by US government agencies, however, do not require a license if they are otherwise authorized by law. Criminal penalties—fines and imprisonment—are provided for willful violations.

The initial arms transfers in the Iran initiative involved the sale and shipment by Israel of US-origin missiles. The usual way for such international retransfer of arms to be authorized under US law is pursuant to the Arms Export Control Act. This act requires that the president consent to any transfer by another country of arms exported under the act and imposes three conditions before such presidential consent may be given.

(a) the United States would itself transfer the arms in question to the recipient country;

(b) a commitment in writing has been ob-

tained from the recipient country against unauthorized retransfer of significant arms, such as missiles; and

(c) a prior written certification regarding the retransfer is submitted to the Congress if the defense equipment, such as missiles, has an acquisition cost of 14 million dollars or more. 22 U.S.C. 2753 (a), (d).

In addition, the act generally imposes restrictions on which countries are eligible to receive US arms and on the purposes for which arms may be sold.[5]

The other possible avenue whereby government arms transfers to Iran may be authorized by law would be in connection with intelligence operations conducted under the National Security Act. This act requires that the director of Central Intelligence and the heads of other intelligence agencies keep the two congressional intelligence committees "fully and currently informed" of all intelligence activities under their responsibility. 50 U.S.C. 413. Where prior notice of significant intelligence activities is not given, the intelligence committees are to be informed "in a timely fashion." In addition, the so called Hughes-Ryan Amendment to the Foreign Assistance Act requires that "significant anticipated intelligence activities" may not be conducted by the CIA unless and until the president finds that "each such operation is important to the national security of the United States." 22 U.S.C. 2422.

When the Israelis began transferring arms to Iran in August, 1985, they were not acting on their own. US officials had knowledge about the essential elements of the proposed shipments. The United States shared some common purpose in the transfers and received a benefit from them—the release of a hostage. Most importantly, Mr. McFarlane communicated prior US approval to the Israelis for the shipments, including an undertaking for replenishment. But for this US approval, the transaction may not have gone forward. In short, the United States was an essential participant in the arms transfers to Iran that occurred in 1985.

Whether this US involvement in the arms transfer by the Israelis was lawful depends fundamentally upon whether the president approved the transactions before they occurred. In the absence of presidential approval, there does not appear to be any authority in this case for the United States to engage in the transfer of arms or consent to the transfer by another country. The arms transfers to Iran in 1985 and hence the Iran initiative itself would have proceeded contrary to US law.

The attorney general reached a similar judgment with respect to the activities of the CIA in facilitating the November, 1985 shipment by the Israelis of HAWK missiles. In a letter to the board,[6] the attorney general concluded that with respect to the CIA assistance, "a finding under the Hughes-Ryan Amendment would be required."[7]

The board was unable to reach a conclusive judgment about whether the 1985 shipments of arms to Iran were approved in advance by the president. On balance the board believes that it is plausible to conclude that he did approve them in advance.

Yet even if the president in some sense consented to or approved the transactions, a serious question of law remains. It is not clear that the form of the approval was sufficient for purposes of either the Arms Export Control Act or the Hughes-Ryan Amendment. The consent did not meet the conditions in the absence of a prior written commitment from the Iranians regarding unauthorized retransfer.

Under the National Security Act, it is not clear that mere oral approval by the president would qualify as a presidential finding that the initiative was vital to the national security interests of the United States. The approval was never reduced to writing. It appears to have been conveyed to only one person. The president himself has no memory of it. And there is contradictory evidence from the president's advisers about how the president responded when he learned of the arms shipments which the approval was to support. In addition, the requirement for congressional notification was ignored. In these circumstances, even if the president approved of the transactions, it is difficult to conclude that his actions constituted adequate legal authority.

The legal requirements pertaining to the sale of arms to Iran are complex; the availability of legal authority, including that which may flow from the president's constitutional powers, is difficult to delineate. Definitive legal conclusions will also depend upon a variety of specific factual determinations that the board has not attempted to resolve—for example, the specific content of any consent provided by the president, the authority under which the missiles were originally transferred to Israel, the knowledge and intentions of individuals, and the like. Nevertheless, it was sufficient for the board's purposes to conclude that the legal underpinning of the Iran initiative during 1985 was at best highly questionable.

The Presidential Finding of January 17, 1986, formally approved the Iran initiative as a covert intelligence operation under the National Security Act. This ended the uncertainty about the legal status of the initiative and provided legal authority for the United States to transfer arms directly to Iran.

The National Security Act also requires notification of Congress of covert intelligence activities. If not done in advance, notification must be "in a timely fashion." The Presidential Finding of January 17 directed that congressional notification be withheld, and this decision appears to have never been reconsidered. While there was surely justification to suspend congressional notification in advance of a particular transaction relating to a hostage release, the law would seem to require disclosure where, as in the Iran case, a pattern of relative inactivity occurs over an extended period. To do otherwise prevents the Congress from fulfilling its proper oversight responsibilities.

Throughout the Iran initiative, significant questions of law do not appear to have been adequately addressed. In the face of a sweeping statutory prohibition and explicit requirements relating to presidential consent to arms transfers by third countries, there appears to have been at the outset in 1985 little attention, let alone systematic analysis, devoted to how presidential actions would comply with US law. The board has found no evidence that an evaluation was ever done during the life of the operation to determine whether it continued to comply with the terms of the January 17 Presidential Finding. Similarly, when a new prohibition was added to the Arms Export Control Act in August of 1986 to prohibit exports to countries on the terrorism list (a list which contained Iran), no evaluation was made to determine whether this law affected authority to transfer arms to Iran in connection with intelligence operations under the National Security Act. This lack of legal vigilance markedly increased the chances that the initiative would proceed contrary to law.

NSC Staff Support for the Contras.—The NSC staff activities in support of the contras were marked by the same uncertainty as to legal authority and insensitivity to legal issues as were present in the Iran initiative. The ambiguity of the law governing activities in support of the contras presented a greater challenge than even the considerable complexity of laws governing arms transfers. Intense congressional scrutiny with respect to the NSC staff activities relating to the contras added to the potential costs of actions that pushed the limits of the law.

In this context, the NSC staff should have been particularly cautious, avoiding operational activity in this area and seeking legal counsel. The board saw no signs of such restraint.

B. Failure of Responsibility

The NSC system will not work unless the president makes it work. After all, this system was created to serve the president of the United States in ways of his choosing. By his actions, by his leadership, the president therefore determines the quality of its performance.

By his own account, as evidenced in his diary notes, and as conveyed to the board by his principal advisers, President Reagan was deeply committed to securing the release of the hostages. It was this intense compassion for the hostages that appeared to motivate his steadfast support of the Iran initiative, even in the face of opposition from his secretaries of state and defense.

In his obvious commitment, the president appears to have proceeded with a concept of the initiative that was not accurately reflected in the reality of the operation. The president did not seem to be aware of the way in which the operation was implemented and the full consequences of US participation.

The president's expressed concern for the safety of both the hostages and the Iranians who could have been at risk may have been conveyed in a manner so as to inhibit the full functioning of the system.

The president's management style is to put the principal responsibility for policy review and implementation on the shoulders of his advisers. Nevertheless, with such a complex, high-risk operation and so much at stake, the president should have ensured that the NSC system did not fail him. He did not force his policy to undergo the most critical review of which the NSC participants and the process were capable. At no time did he insist upon accountability and performance review. Had the president chosen to drive the NSC system, the outcome could well have been different. As it was, the most powerful features of the NSC system—providing comprehensive analysis, alternatives and follow-up—were not utilized.

The board found a strong consensus among NSC participants that the president's priority in the Iran initiative was the release of US hostages. But setting priorities is not enough when it comes to sensitive and risky initiatives that directly affect US national security. He must ensure that the content and tactics of an initiative match his priorities and objectives. He must insist upon accountability. For it is the president who must take responsibility for the NSC system and deal with the consequences.

Beyond the president, the other NSC principals and the national security adviser must share in the responsibility of the NSC system.

President Reagan's personal management style places an especially heavy responsibility on his key advisers. Knowing his style, they should have been particularly mindful of the need for special attention to the manner in which this arms sale initiative developed and proceeded. On this score, neither the national security adviser nor the other NSC principals deserve high marks.

It is their obligation as members and advisers to the council to ensure that the president is adequately served. The principal subordinates to the president must not be deterred from urging the president not to proceed on a highly questionable course of action even in the face of his strong conviction to the contrary.

In the case of the Iran initiative, the NSC process did not fail, it simply was largely ignored. The national security adviser and the NSC principals all had a duty to raise this issue and insist that orderly process be imposed. None of them did so.

All had the opportunity. While the national security adviser had the responsibility to see that an orderly process was observed, his failure to do so does not excuse the other NSC principals. It does not appear that any of the NSC principals called for more frequent consideration of the Iran initiative by the NSC principals in the presence of the president. None of the principals called for a serious vetting of the initiative by even a restricted group of disinterested individuals. The intelligence questions do not appear to have been raised, and legal considerations, while raised, were not pressed. No one seemed to have complained about the informality of the process. No one called for a thorough reexamination once the initiative did not meet expectations or the manner of execution changed. While one or another of the NSC principals suspected that something was amiss, none vigorously pursued the issue.

Mr. Regan also shares in this responsibility. More than almost any chief of staff of recent memory, he asserted personal control over the White House staff and sought to extend this control to the national security adviser. He was personally active in national security affairs and attended almost all of the relevant meetings regarding the Iran initiative. He, as much as anyone, should have insisted that an orderly process be observed. In addition, he especially should have ensured that plans were made for handling any public disclosure of the initiative. He must bear primary responsibility for the chaos that descended upon the White House when such disclosure did occur.

Mr. McFarlane appeared caught between a president who supported the initiative and the cabinet officers who strongly opposed it. While he made efforts to keep these cabinet officers informed, the board heard complaints from some that he was not always successful. VADM Poindexter on several occasions apparently sought to exclude NSC principals other than the president from knowledge of the initiative. Indeed, on one or more occasions Secretary Shultz may have been actively misled by VADM Poindexter.

VADM Poindexter also failed grievously on the matter of contra diversion. Evidence indicates that VADM Poindexter knew that a diversion occurred, yet he did not take the steps that were required given the gravity of that prospect. He apparently failed to appreciate or ignored the serious legal and political risks presented. His clear obligation was either to investigate the matter or take it to the president— or both. He did neither. Director Casey shared a similar responsibility. Evidence suggests that he received information about the possible diversion of funds to the contras almost a month before the story broke. He, too, did not move promptly to raise the matter with the president. Yet his responsibility to do so was clear.

The NSC principals other than the president may be somewhat excused by the insufficient attention on the part of the national security adviser to the need to keep all the principals fully informed. Given the importance of the issue and the sharp policy divergences involved, however, Secretary Shultz and Secretary Weinberger in particular distanced themselves from the march of events. Secretary Shultz specifically requested to be informed only as necessary to perform his job. Secretary Weinberger had access through intelligence to details about the operation. Their obligation was to give the president their full support and continued advice with respect to the program or, if they could not in conscience do that, to so inform the president. Instead, they simply distanced themselves from the program. They protected the record as to their own positions on this issue. They were not energetic in attempting to protect the president from the consequences of his personal commitment to freeing the hostages.

Director Casey appears to have been informed in considerable detail about the specifics of the Iranian operation. He appears to have acquiesced in and to have encouraged North's exercise of direct operational control over the operation. Because of the NSC staff's proximity to and close identification with the president, this increased the risks to the president if the initiative became public or the operation failed.

There is no evidence, however, that Director Casey explained this risk to the president or made clear to the president that LtCol North, rather than the CIA, was running the operation. The president does not recall ever being informed of this fact. Indeed, Director Casey should have gone further and pressed for operational responsibility to be transferred to the CIA.

Director Casey should have taken the lead in vetting the assumptions presented by the Israelis on which the program was based and in pressing for an early examination of the reliance upon Mr. Ghorbanifar and the second channel as intermediaries. He should also have assumed responsibility for checking out the other intermediaries involved in the operation. Finally, because congressional restrictions on covert actions are both largely directed at and familiar

to the CIA, Director Casey should have taken the lead in keeping the question of congressional notification active.

Finally, Director Casey, and to a lesser extent, Secretary Weinberger, should have taken it upon themselves to assess the effect of the transfer of arms and intelligence to Iran on the Iran/Iraq military balance, and to transmit that information to the president.

C. THE ROLE OF THE ISRAELIS

Conversations with emissaries from the government of Israel took place prior to the commencement of the initiative. It remains unclear whether the initial proposal to open the Ghorbanifar channel was an Israeli initiative, was brought on by the avarice of arms dealers, or came as a result of an American request for assistance. There is no doubt, however, that it was Israel that pressed Mr. Ghorbanifar on the United States. US officials accepted Israeli assurances that they had had for some time an extensive dialogue that involved high-level Iranians, as well as their assurances of Mr. Ghorbanifar's bona fides. Thereafter, at critical points in the initiative, when doubts were expressed by critical US participants, an Israeli emissary would arrive with encouragement, often a specific proposal, and pressure to stay with the Ghorbanifar channel.

From the record available to the board, it is not possible to determine the role of key US participants in prompting these Israeli interventions. There were active and ongoing consultations between LtCol North and officials at the Israeli government, specifically David Kimche and Amiram Nir. In addition, Mr. Schwimmer, Mr. Nimrodi, and Mr. Ledeen, also in frequent contact with LtCol North, had close ties with the government of Israel. It may be that the Israeli interventions were actively solicited by particular US officials. With the benefit of the views of the Israeli officials involved, it is hard to know the facts.

It is clear, however, that Israel had its own interests, some in direct conflict with those of the United States, in having the United States pursue the initiative. For this reason, it had an incentive to keep the initiative alive. It sought to do this by interventions with the NSC staff, the national security adviser, and the president. Although it may have received suggestions from LtCol North, Mr. Ledeen, and others, it responded affirmatively to these suggestions by reason of its own interests.

Even if the government of Israel actively worked to begin the initiative and to keep it going, the US government is responsible for its own decisions. Key participants in US deliberations made the point that Israel's objectives and interests in this initiative were different from, and in some respects in conflict with, those of the United States. Although Israel dealt with those portions of the US government that it deemed were sympathetic to the initiative, there is nothing improper per se about this fact. US decisionmakers made their own decisions and must bear responsibility for the consequences.

D. Aftermath—The Efforts to Tell the Story

From the first hint in late October, 1986 that the McFarlane trip would soon become public, information on the Iran initiative and contra activity cascaded into the press. The veiled hints of secret activities, random and indiscriminate disclosures of information from a variety of sources, both knowledgeable and otherwise, and conflicting statements by high-level officials presented a confusing picture to the American public. The board recognized that conflicts among contemporaneous documents and statements raised concern about the management of the public presentation of facts on the Iran initiative. Though the board reviewed some evidence[8] on events after the exposure, our ability to comment on these events remains limited.

The board found evidence that immediately following the public disclosure, the president wanted to avoid providing too much specificity or detail out of concern for the hostages still held in Lebanon and those Iranians who had supported the initiative. In doing so, he did not, we believe, intend to mislead the American public or cover up unlawful conduct. By at least November 20, the president took steps to ensure that all the facts would come out. From the president's request to Mr. Meese to look into the history of the initiative, to his appointment of this board, to his request for an independent counsel, to his willingness to discuss this matter fully and to review his personal notes with us, the board is convinced that the president does indeed want the full story to be told.

Those who prepared the president's supporting documentation did not appear, at least initially, to share in the president's ultimate wishes. Mr. McFarlane described for the board the process used by the NSC staff to create a chronology that obscured essential facts. Mr. McFarlane contributed to the creation of this chronology which did not, he said, present "a full and completely accurate account" of the events and left ambiguous the president's role. This was, according to Mr. McFarlane, done to distance the president from the timing and nature of the president's authorization. He told the board that he wrote a memorandum on November 18, which tried to, in his own words, "gild the President's motives." This version was incorporated into the chronology. Mr. Mc-

Farlane told the board that he knew the account was "misleading, at least, and wrong, at worst." Mr. McFarlane told the board that he did provide the attorney general an accurate account of the president's role.

The board found considerable reason to question the actions of LtCol North in the aftermath of the disclosure. The board has no evidence to either confirm or refute that LtCol North destroyed documents on the initiative in an effort to conceal facts from threatened investigations. The board found indications that LtCol North was involved in an effort, over time, to conceal or withhold important information. The files of LtCol North contained much of the historical documentation that the board used to construct its narrative. Moreover, LtCol North was the primary US government official involved in the details of the operation. The chronology he produced has many inaccuracies. These "histories" were to be the basis of the "full" story of the Iran initiative. These inaccuracies lend some evidence to the proposition that LtCol North, either on his own or at the behest of others, actively sought to conceal important information.

Out of concern for the protection of classified material, Director Casey and VADM Poindexter were to brief only the congressional intelligence committees on the "full" story; the DCI before the committees and VADM Poindexter in private sessions with the chairmen and vice-chairmen. The DCI and VADM Poindexter undertook to do this on 21 November 1986. It appears from the copy of the DCI's testimony and notes of VADM Poindexter's meetings, that they did not fully relate the nature of events as they had occurred. The result is an understandable perception that they were not forthcoming.

The board is also concerned about various notes that appear to be missing. VADM Poindexter was the official note taker in some key meetings, yet no notes for the meetings can be found: The reason for the lack of such notes remains unknown to the board. If they were written, they may contain very important information. We have no way of knowing if they exist.

PART V: RECOMMENDATIONS

Not only . . . is the Federal power over external affairs in origin and essential character different from that over internal affairs, but participation in the exercise of the power is significantly limited. In this vast external realm, with its important, complicated, delicate and manifold problems, the President alone has the power to speak or listen as a representative of the nation." *United States* v. *Curtiss-Wright Export Corp.*, 299 U.S. 304, 319 (1936).

Whereas the ultimate power to formulate domestic policy resides in the Congress, the primary responsibility for the formulation and implementation of national security policy falls on the president.

It is the president who is the usual source of innovation and responsiveness in this field. The departments and agencies—the Defense Department, State Department, and CIA bureaucracies—tend to resist policy change. Each has its own perspective based on long experience. The challenge for the president is to bring his perspective to bear on these bureaucracies for they are his instruments for executing national security policy, and he must work through them. His task is to provide them leadership and direction.

The National Security Act of 1947 and the system that has grown up under it affords the president special tools for carrying out this important role. These tools are the National Security Council, the national security adviser, and the NSC staff. These are the means through which the creative impulses of the president are brought to bear on the permanent government. The National Security Act, and custom and practice, rightly give the president wide latitude in fashioning exactly how these means are used.

There is no magic formula which can be applied to the NSC structure and process to produce an optimal system. Because the system is the vehicle through which the president formulates and implements his national security policy, it must adapt to each individual president's style and management philosophy. This means that NSC structures and processes must be flexible, not rigid. Overprescription would, as discussed in Part II, either destroy the system or render it ineffective.

Nevertheless, this does not mean there can be no guidelines or recommendations that might improve the operation of the system, whatever the particular style of the incumbent president. We have reviewed the operation of the system over the past 40 years, through good times and bad. We have listened carefully to the views of all the living former presidents as well as those of most of the participants in their own national security systems. With the strong caveat that flexibility and adaptability must be at the core, it is our judgment that the national security system seems to have worked best when it has in general operated along the lines set forth below.

Organizing for National Security. Because of the wide latitude in the National Security Act, the president bears a special responsibility for the effective performance of the NSC system. A president must at the outset provide guidelines to the members of the National Security Council, his national security adviser, and the National Security Council staff. These guidelines, to be effective, must include how they will relate to one another, what procedures will be followed, what the president expects of them. If his advisers are not performing as he likes, only the president can intervene.

The National Security Council principals other than the president participate on the council in a unique capacity.[9] Although holding a seat by virtue of their official positions in the administration, when they sit as members of the council they sit not as cabinet secretaries or department heads but as advisers to the president. They are there not simply to advance or defend the particular positions of the departments or agencies they head but to give their best advice to the president. Their job—and their challenge—is to see the issue from this perspective, not from the narrower interests of their respective bureaucracies.

The National Security Council is only advisory. It is the president alone who decides. When the NSC principals receive those decisions, they do so as heads of the appropriate departments or agencies. They are then responsible to see that the president's decisions are carried out by those organizations accurately and effectively.

This is an important point. The policy innovation and creativity of the president encounters a natural resistance from the executing departments. While this resistance is a source of frustration to every president, it is inherent in the design of the government. It is up to the politically appointed agency heads to ensure that the president's goals, designs, and policies are brought to bear on this permanent structure. Circumventing the departments, perhaps by using the national security adviser or the NSC Staff to execute policy, robs the president of the experience and capacity resident in the departments. The president must act largely through them, but the agency heads must ensure that they execute the president's policies in an expeditious and effective manner. It is not just the obligation of the national security adviser to see that the national security process is used. All of the NSC principals—and particularly the president—have that obligation.

This tension between the president and the executive departments is worked out through the national security process described in the opening sections of this report. It is through this process that the nation obtains both the best of the creativity of the president and the learning and expertise of the national security departments and agencies.

The process is extremely important to the president. His decisions will benefit from the advice and perspectives of all the concerned departments and agencies. History offers numerous examples of this truth. President Kennedy, for example, did not have adequate consultation before entering upon the Bay of Pigs invasion, one of his greatest failures. He remedied this in time for the Cuban missile crisis, one of his greatest successes. Process will not always produce brilliant ideas, but history suggests it can at least help prevent bad ideas from becoming presidential policy.

The National Security Adviser. It is the national security adviser who is primarily responsible for managing this process on a daily basis. The job requires skill, sensitivity, and integrity. It is his responsibility to ensure that matters submitted for consideration by the council cover the full range of issues on which review is required; that those issues are fully analyzed; that a full range of options is considered; that the prospects and risks of each are examined; that all relevant intelligence and other information is available to the principals; that legal considerations are addressed; that difficulties in implementation are confronted. Usually, this can best be accomplished through interagency participation in the analysis of the issue and a preparatory policy review at the deputy or under secretary level.

The national security adviser assumes these responsibilities not only with respect to the president but with respect to all the NSC principals. He must keep them informed of the president's thinking and decisions. They should have adequate notice and an agenda for all meetings. Decision papers should, if at all possible, be provided in advance.

The national security adviser must also ensure that adequate records are kept of NSC consultations and presidential decisions. This is essential to avoid confusion among presidential advisers and departmental staffs about what was actually decided and what is wanted. Those records are also essential for conducting a periodic review of a policy or initiative, and to learn from the past.

It is the responsibility of the national security adviser to monitor policy implementation and to ensure that policies are executed in conformity with the intent of the president's decision. Monitoring includes initiating periodic reassessments of a policy or operation, especially when changed circumstances suggest that the policy or operation no longer serves US interests.

But the national security adviser does not simply manage the national security process. He is himself an important source of advice on national security matters to the president. He is not the president's only source of advice, but he is perhaps the one most able to see things from the president's perspective. He is unburdened by departmental responsibilities. The president is his only master. His advice is confidential. He is not subject to Senate confirmation and traditionally does not formally appear before congressional committees.

To serve the president well, the national security adviser should present his own views, but he must at the same time represent the views of others fully and faithfully to the president. The system will not work well if the national security adviser does not have the trust of the NSC principals. He, therefore, must not use his proximity to the president to manipulate the process so as to produce his own position. He should not interpose himself between the president and the NSC principals. He should not seek to exclude the NSC principals from the decision process. Performing both these roles well is an essential, if not easy, task.

In order for the national security adviser to serve the president adequately, he must have direct access to the president. Unless he knows first hand the views of the president and is known to reflect them in his management of the NSC system, he will be ineffective. He should not report to the president through some other official. While the chief of staff or others can usefully interject domestic political considerations into national security deliberations, they should do so as additional advisers to the president.

Ideally, the national security adviser should not have a high public profile. He should not try to compete with the secretary of state or the secretary of defense as the articulator of public policy. They, along with the president, should be the spokesmen for the policies of the administration. While a "passion for anonymity" is perhaps too strong a term, the national security adviser should generally operate off-stage.

The NSC principals of course must have direct access to the president, with whatever frequency the president feels is appropriate. But these individual meetings should not be used by the principal to seek decisions or otherwise circumvent the system in the absence of the other principals. In the same way, the national security adviser should not use his scheduled intelligence or other daily briefings of the president as an opportunity to seek presidential decision on significant issues.

If the system is to operate well, the national security adviser must promote cooperation rather than competition among himself and the other NSC principals. But the president is ultimately responsible for the operation of this system. If rancorous infighting develops among his principal national security functionaries, only he can deal with them. Public dispute over external policy by senior officials undermines the process of decisionmaking and narrows his options. It is the president's responsibility to ensure that it does not take place.

Finally, the national security adviser should focus on advice and management, not implementation and execution. Implementation is the responsibility and the strength of the departments and agencies. The national security adviser and the NSC staff generally do not have the depth of resources for the conduct of operations. In addition, when they take on implementation responsibilities, they risk compromising their objectivity. They can no longer act as impartial overseers of the implementation, ensuring that presidential guidance is followed, that policies are kept under review, and that the results are serving the president's policy and the national interest.

The NSC Staff. The NSC staff should be small, highly competent, and experienced in the making of public policy. Staff members should be drawn both from within and from outside government. Those from within government should come from the several departments and agencies concerned with national security matters. No particular department or agency should have a predominate role. A proper balance must be maintained between people from within and outside the government. Staff members should generally rotate with a stay of more than four years viewed as the exception.

A large number of staff action officers organized along essentially horizontal lines enhances the possibilities for poorly supervised and monitored activities by individual staff members. Such a system is made to order for energetic self-starters to take unauthorized initiatives. Clear vertical lines of control and authority, responsibility and accountability, are essential to good management.

One problem affecting the NSC staff is lack of institutional memory. This results from the understandable desire of a president to replace the staff in order to be sure it is responsive to him. Departments provide continuity that can help the council, but the council as an institution also needs some means to assure adequate records and memory. This was identified to the board as a problem by many witnesses.

We recognize the problem and have identified a range of possibilities that a president might consider on this subject. One would be to create a small permanent executive secretariat. Another would be to have one person, the executive secretary, as a permanent position. Finally, a pattern of limited tenure and overlapping rotation could be used. Any of these would help reduce the problem of loss of institutional memory; none would be practical unless each succeeding president subscribed to it.

The guidelines for the role of the national security adviser also apply generally to the NSC staff. They should protect the process and thereby the president. Departments and agencies should not be excluded from participation in that process. The staff should not be implementors or operators and staff should keep a low profile with the press.

PRINCIPAL RECOMMENDATION

The model we have outlined above for the National Security Council system constitutes our first and most important recommendation. It includes guidelines that address virtually all of the deficiencies in procedure and practice that the board encountered in the Iran/contra affair as well as in other case studies of this and previous administrations.

We believe this model can enhance the performance of a president and his administration in the area of national security. It responds directly to President Reagan's mandate to describe the NSC system as it ought to be.

The board recommends that the proposed model be used by presidents in their management of the national security system.

SPECIFIC RECOMMENDATIONS

In addition to its principal recommendation regarding the organization and functioning of the NSC system and roles to be played by the participants, the board has a number of specific recommendations.

1. The National Security Act of 1947. The flaws of procedure and failure of responsibility revealed by our study do not suggest any inadequacies in the provisions of the National Security Act of 1947 that deal with the structure and operation of the NSC system. Forty years of experience under that act demonstrate to the Board that it remains a fundamentally sound framework for national security decision-making. It strikes a balance between formal structure and flexibility adequate to permit each president to tailor the system to fit his needs.

As a general matter, the NSC staff should not engage in the implementation of policy or the conduct of operations. This compromises their oversight role and usurps the responsibilities of the departments and agencies. But the inflexibility of a legislative restriction should be avoided. Terms such as "operation" and "implementation" are difficult to define, and a legislative proscription might preclude some future president from making a very constructive use of the NSC staff.

Predisposition on sizing of the staff should be toward fewer rather than more. But a legislative restriction cannot foresee the requirements of future presidents. Size is best left to the discretion of the president, with the admonition that the role of the NSC staff is to review, not to duplicate or replace, the work of the departments and agencies.

We recommend that no substantive change be made in the provisions of the National Se-

curity Act dealing with the structure and operation of the NSC system.

2. Senate Confirmation of the National Security Adviser. It has been suggested that the job of the national security adviser has become so important that its holder should be screened by the process of confirmation, and that once confirmed he should return frequently for questioning by the Congress. It is argued that this would improve the accountability of the national security adviser.

We hold a different view. The national security adviser does, and should continue, to serve only one master, and that is the president. Further, confirmation is inconsistent with the role the national security adviser should play. He should not decide, only advise. He should not engage in policy implementation or operations. He should serve the president, with no collateral and potentially diverting loyalties.

Confirmation would tend to institutionalize the natural tension that exists between the secretary of state and the national security adviser. Questions would increasingly arise about who really speaks for the president in national security matters. Foreign governments could be confused or would be encouraged to engage in "forum shopping."

Only one of the former government officials interviewed favored Senate confirmation of the national security adviser. While consultation with Congress received wide support, confirmation and formal questioning were opposed. Several suggested that if the national security adviser were to become a position subject to confirmation, it could induce the president to turn to other internal staff or to people outside government to play that role.

We urge the Congress not to require Senate confirmation of the national security adviser.

3. The Interagency Process. It is the national security adviser who has the greatest interest in making the national security process work, for it is this process by which the president obtains the information, background, and analysis he requires to make decisions and build support for his program. Most presidents have set up interagency committees at both a staff and policy level to surface issues, develop options, and clarify choices. There has typically been a struggle for the chairmanship of these groups between the national security adviser and the NSC staff on the one hand, and the cabinet secretaries and department officials on the other.

Our review of the operation of the present system and that of other administrations where committee chairmen came from the departments has led us to the conclusion that the system generally operates better when the committees are chaired by the individual with the greatest stake in making the NSC system work.

We recommend that the national security adviser chair the senior-level committees of the NSC system.

4. Covert Actions. Policy formulation and implementation are usually managed by a team of experts led by policymaking generalists. Covert action requirements are no different, but there is a need to limit, sometimes severely, the number of individuals involved. The lives of many people may be at stake, as was the case in the attempt to rescue the hostages in Tehran. Premature disclosure might kill the idea in embryo, as could have been the case in the opening of relations with China. In such cases, there is a tendency to limit those involved to a small number of top officials. This practice tends to limit severely the expertise brought to bear on the problem and should be used very sparingly indeed.

The obsession with secrecy and preoccupation with leaks threaten to paralyze the government in its handling of covert operations. Unfortunately, the concern is not misplaced. The selective leak has become a principal means of waging bureaucratic warfare. Opponents of an operation kill it with a leak; supporters seek to build support through the same means.

We have witnessed over the past years a significant deterioration in the integrity of process. Rather than a means to obtain results more satisfactory than the position of any of the individual departments, it has frequently become something to be manipulated to reach a specific outcome. The leak becomes a primary instrument in that process.

This practice is destructive of orderly governance. It can only be reversed if the most senior officials take the lead. If senior decisionmakers set a clear example and demand compliance, subordinates are more likely to conform.

Most recent administrations have had carefully drawn procedures for the consideration of covert activities. The Reagan administration established such procedures in January, 1985, then promptly ignored them in their consideration of the Iran initiative.

We recommend that each administration formulate precise procedures for restricted consideration of covert action and that, once formulated, those procedures be strictly adhered to.

5. The Role of the CIA. Some aspects of the Iran arms sales raised broader questions in the minds of members of the board regarding the role of CIA. The first deals with intelligence. The NSC staff was actively involved in the preparation of the 20 May 1985, update to the Special National Intelligence Estimate on Iran. It is a matter for concern if this involvement and the strong views of NSC staff members were allowed to influence the intelligence judgments contained in the update. It is also of concern that the update contained the hint that the United States should change its existing policy and encourage its allies to provide arms to Iran. It is critical that the line between intelligence and advocacy of a particular policy be preserved if intelligence is to retain its integrity and perform its proper function. In this instance, the CIA came close enough to the line to warrant concern.

We emphasize to both the intelligence community and policymakers the importance of maintaining the integrity and objectivity of the intelligence process.

6. *Legal Counsel.* From time to time issues with important legal ramifications will come before the National Security Council. The attorney general is currently a member of the council by invitation and should be in a position to provide legal advice to the council and the president. It is important that the attorney general and his department be available to interagency deliberations.

The Justice Department, however, should not replace the role of counsel in the other departments. As the principal counsel on foreign affairs, the legal adviser to the secretary of state should also be available to all the NSC participants.

Of all the NSC participants, it is the assistant for national security affairs who seems to have had the least access to expert counsel familiar with his activities.

The board recommends that the position of legal adviser to the NSC be enhanced in stature and in its role within the NSC staff.

7. *Secrecy and Congress.* There is a natural tension between the desire for secrecy and the need to consult Congress on covert operations. Presidents seem to become increasingly concerned about leaks of classified information as their administrations progress. They blame Congress disproportionately. Various cabinet officials from prior administrations indicated to the board that they believe Congress bears no more blame than the executive branch.

However, the number of members and staff involved in reviewing covert activities is large; it provides cause for concern and a convenient excuse for presidents to avoid congressional consultation.

We recommend that Congress consider replacing the existing Intelligence Committees of the respective houses with a new joint committee with a restricted staff to oversee the intelligence community, patterned after the Joint Committee on Atomic Energy that existed until the mid-1970s.

8. *Privatizing National Security Policy.* Careful and limited use of people outside the US government may be very helpful in some unique cases. But this practice raises substantial questions. It can create conflict of interest problems. Private or foreign sources may have different policy interests or personal motives and may exploit their association with a US government effort. Such involvement gives private and foreign sources potentially powerful leverage in the form of demands for return favors or even blackmail.

The US has enormous resources invested in agencies and departments in order to conduct the government's business. In all but a very few cases, these can perform the functions needed. If not, then inquiry is required to find out why.

We recommend against having implementation and policy oversight dominated by intermediaries. We do not recommend barring limited use of private individuals to assist in United States diplomatic initiatives or in covert activities. We caution against use of such people except in very limited ways and under close observation and supervision.

Epilogue

If but one of the major policy mistakes we examined had been avoided, the nation's history would bear one less scar, one less embarrassment, one less opportunity for opponents to reverse the principles this nation seeks to preserve and advance in the world.

As a collection, these recommendations are offered to those who will find themselves in situations similar to the one we reviewed: under stress, with high stakes, given little time, using incomplete information, and troubled by premature disclosures. In such a state, modest improvements may yield surprising gains. This is our hope.

NOTES

1. See Appendix A, Executive Order No. 12575.
2. A list of those case studies is contained in Appendix F.
3. A list of the witnesses interviewed by the board is contained in Appendix F (to the full report).
4. The issue of legal advice to the NSC staff is treated in more detail in Part V of this report.
5. It may be possible to authorize transfers by another country under the Arms Export Control Act without obtaining the president's consent. As a practical matter, however, the legal requirements may not differ significantly. For example, section 614(2)

permits the president to waive the requirements of the Act. But this waiver authority may not be exercised unless it is determined that the international arms sales are "vital to the national security interests of the United States." Moreover, before granting a waiver, the president must consult with and provide written justification to the foreign affairs and appropriations committees of the Congress 22 U.S.C. 2374(3).

6. A copy of the letter is set forth in Appendix H.

7. Apparently no determination was made at the time as to the legality of these activities even though serious concerns about legality were expressed by the deputy director of CIA, a presidential finding was sought by CIA officials before any further CIA activities in support of the Iran initiative were undertaken, and the CIA counsel, Mr. Stanley Sporkin, advised that as a matter of prudence any new finding should seek to ratify the prior CIA activities.

8. See Appendix D.

9. As discussed in more detail in Part II, the statutory members of the National Security Council are the president, vice president, secretary of state, and secretary of defense. By the phrase "National Security Council principals" or "NSC principals," the Board generally means those four statutory members plus the director of central intelligence and the chairman of the Joint Chiefs of Staff.

AMERICAN PLURALISM AND DEFENSE POLICY

DAVID C. KOZAK

This chapter is about the role of US domestic political factors in the making of American defense policy. It is divided into two parts: (1) models for understanding American politics and (2) defense policy as public policy. The first part looks at concepts and approaches that help us understand how the American system works. The second looks at defense policy as the consequence of policy produced within a uniquely structured system.

The preeminent strategist Carl von Clausewitz writing in 1832 justified well a survey of domestic political factors in the study of the defense policy process:

A military commander-in-chief need not be a learned historian nor a political commentator, but he must be familiar with the higher affairs of state and its innate policies; he must know current issues, questions under consideration, the leading personalities, and be able to form sound judgments. He need not be an acute observer of mankind or a subtle analyzer of human character; but he must know the character, the habits of thought and action, the special virtues and defects of the men whom he is to command.[1]

There are three reasons why students of defense policy should study US domestic politics. First, as Clausewitz has illustrated in other writings, strategy is an extension of the political objectives of the state. Therefore, an American professional officer must appreciate the institutions and processes of US government through which political objectives are formulated. The more one knows about the broader political environment, the better one can appreciate the factors that have shaped current US strategy. Second, the defense policy manager is a player in the policy process. The more one knows about the game, the more effectively one will play. The American policy process is a unique one. Its democratic nature brings to bear on defense policy a host of factors—interest groups, media, public opinion, political party platforms, and congressional assertions—not usually influential in other types of political systems. Knowledge of these factors and a sophisticated anticipation of future trends and developments affecting them is indispensable for the effective strategist and defense planner. Third, since this text is used extensively in preparing future officers at the USAF Academy, all should recognize that American military leaders swear an oath of allegiance to the institutions, values, and spirit of the American constitutional system. An in-depth study of this system can serve to deepen one's sense of appreciation and commitment and is essential to a proper understanding of the value manifest in the ultimate sacrifice of life in the service of country. It should be emphasized that a study of domestic politics is undertaken not for the purposes of system manipulation that would violate the American tradition of an apolitical military, but for the purposes of broadening and enlightening a professional officer corps with regard to the higher affairs of state.

FIVE MODELS FOR UNDERSTANDING AMERICAN POLITICS

To appreciate the contemporary role of US domestic factors in defense policy, the defense policy analyst needs to understand five models of US government and politics. These models should be viewed as compatible and cumulative, not mutually exclusive. Each sheds light on an

---

Reprinted with minor revisions and by permission of the author.

important aspect of American government. The five models are: Madisonian pluralism, extra-constitutional developments, interest group subgovernments, recent destabilizing transformations, and presidential power (and the Reagan revolution). Taken together, they tell us how the system works and how it has evolved.

MADISONIAN PLURALISM

One way to understand the workings of any institution or organization is to examine what was intended by those who established it. Although organizations evolve, such an approach is most revealing, especially when applied to US government.

The intentions of the framers of US government can be described as Madisonian pluralism—"Madisonian" from James Madison, the principal architect of the American system, and "pluralism" meaning "multicentered" or of multiple parts.

To best understand the concept one must return to the spirit of the Constitutional Convention in Philadelphia in the summer of 1787 and the road that led to it. In the interlude between the end of the Revolutionary War and the Federal Constitution of 1789, America was governed under the Articles of Confederation with a weak central government, a unit veto by individual states, and no chief executive in any real sense.

The Articles reflected well the political philosophy of the American revolutionaries. Those who fought the war of separation from Great Britain had a great fear of and aversion to strong executive power, having suffered at the hands of arbitrary colonial governors and the Crown itself. To the revolutionaries, government should restrain the potential for tyranny.

Experience with the Articles showed the need for a stronger government. Problems abounded: debt, regulation of interstate commerce, revenue raising, and internal improvements. It became obvious that the Articles were ill-suited to the task of nation building. After initial meetings in Alexandria and Annapolis, the notables of the day decided to call a convention in Philadelphia for the purposes of "revising and updating" the Articles.

Most delegates realized the task that faced them: to strengthen the Articles of Confederation to provide a stronger central government without posing a threat to the rights of individuals. Individual liberty, safeguards against capricious authority, property rights, domestic tranquility, conflict resolution, and peaceful political change were the political values at the Constitutional Convention. But there was strong conflict among the delegates on how to promote these values and strike a delicate compromise between authority and liberty.

To resolve the disputes, the delegates turned to Madison and the federalist form of government he proposed. The federalist model features fragmented authority. Madison's explication of the federalist approach—found in both his journal of Convention deliberations and in *The Federalist Papers* (written with Jay and Hamilton, under pseudonyms, in defense of the proposed government)—constitutes the most authoritative rationale for the Constitution of 1787. To paraphrase, especially from *Federalist 10,*[2] freedom is constantly threatened by despotism and dictatorship. Men are not angels. They turn to dictatorship to safeguard themselves from the unrest that is engendered by tyranny of faction (to include majority faction). To promote freedom, stability, and conflict resolution, governmental authority should be fragmented and splintered, parceled out to separate, multiple centers of power through such means as a written constitution, the division of power between the national and state governments, the separation of power between the legislative, executive, and judicial branches, checks and balances, and bicameral (two-house) national legislature. In other words, as a bulwark against tyranny, power is dispersed to different institutions, each responding to different interests, each constituted differently, and each intruded into the business of the other. Multiple levels of power were created so that no one force could come to absolute control. Ambition will check ambition, power will countervail power.

To Madison and the Federalists, the purpose of government was to serve freedom and to resolve conflict in this heterogeneous nation. Government was not established to provide efficiency and effectiveness. Government was established to frustrate, not facilitate, majority rule. With the Madison model, government was deliberately engineered to be inefficient to preserve freedom. If Madison could be brought back today and held accountable for his form of government with its mounting deficits and other plaguing problems, undoubtedly he would ask one question in his defense—"Are you free?" One might be tempted to call Madison to task for stalemate and deadlock in government, for policy inefficiency, for unending problems, for sluggishness, incompetence, and obstructionalism. Again, he would ask "Are you free?" "How free is the rest of the world?" "Do you have troops imposing domestic order?" "Save for the tragedy of the Civil War, how many times have we faced a major, violent conflagration?" "How does this record compare to the rest of the world?" For Madison, inefficiency is an inevitable irritant, the price men pay for a free society.

The legacy of Madisonian pluralism is a system of government that is predicated on the desirability of institutional conflict as a way to preserve freedom and prevent pernicious factionalism. The structure is one of "separate institutions sharing power"[3] that provides, in national security affairs, what the late, eminent constitutional scholar Edward S. Corwin referred to as " 'an invitation to struggle' between the President and Congress for control over the foreign policy process of the United States."[4] In the words of Alexis de Tocqueville, the perceptive French visitor to US soil, these institutional arrangements ensure a "ceaseless agitation."[5]

EXTRACONSTITUTIONAL DEVELOPMENT: THE DEMOCRATIZATION OF A REPUBLIC

A constitution is but a blueprint of how a system was intended to operate—a basic yet general framework for policymaking. Constitutional systems evolve in an ever-changing political environment. Operation and practice, custom and tradition, and formal amendments fill in the details and fine print.

The Constitution of 1787 described a republic (indirect representation) rather than a democracy (direct representation). The institutions it established, such as the electoral college and an upper house legislature (Senate) selected by state legislatures, were predicated on a mistrust of mass politics and a preference for rule by political elites.

"The democratization of a republic" best describes American extraconstitutional development. After an initial period of nonpartisanship in the "era of good feeling," and beginning in the Jefferson and Jackson presidencies, the republican government of the federalists was made more democratic with the advent of political parties, widening enfranchisement, and popular election of presidential electors. Reforms of the populist era—specifically the seventeenth and nineteenth amendments which established, respectively, the direct election of senators and the enfranchisement of women— enlarged the scope and role of citizen involvement in government.

The political party system that has emerged in the almost two hundred years of constitutional development is one very different from what one finds in parliamentary democracies. The two major American political parties are much more decentralized and less ideological and less disciplined than their parliamentary counterparts. They are primarily concerned with getting candidates elected. They are not "responsible" for forming a government. They are temporary, loose, and supple coalitional confederations, manned by skeletal cadres and devoid of a strong mass base.

This democratized republic has two major implications for the defense policy manager. The first is that public opinion and mass preferences will have an impact, albeit an indirect one, on defense policy. American parties serve as brokers between people and policy. Although policy will never perfectly reflect mass sentiment, such sentiment will have an impact on policy decisions in government either by ruling out certain options or by militating others. And, given the reelection imperative of most politicians, policy will converge with a defined and intense public opinion. It should be emphasized, however, that politics is not central to life's activities for most Americans. There is not a prevalent ideological mass politics. The average voting American gets worked up about politics only when "lunch pail" and "pocket book" issues are pressing.

A second implication is best illuminated by Professor Samuel Huntington, a scholar of government equally noted for strategic thinking. Huntington argues that the American system has a congenital tension between ideas and institutions. While the political underpinnings of our system are very "republican," American institutions have evolved as quite "democratic" (i.e., favoring mass input to and influence over policy). This ideas-institutions incompatibility inevitably leads to a schizophrenic tension and disharmony in the American system.[6]

INTEREST GROUP SUBGOVERNMENTS

Due to both America's emergence as a world power and the advent of the welfare state, there has been an enormous growth in the size and scope of US government. The model of policymaking that seems best to capture the actual workings of the megasystem has been referred to as interest group liberalism, subsystem or subgovernment pluralism, issues networks, iron triangles of power, or "whirlpools" of policymaking.[7] Although the terminology varies and there has been some debate over who is a player, the concept is the same: policy, when business is usual in government, is generally thrashed out in a narrow arena comprised of proximate policymakers indigenous to any given issue area or policy sector. There is consensus that the proximate policymakers include members of Congress and their staffs from the relevant congressional committees and subcommittees, personnel from departments and agencies having administrative responsibility, representatives of affected groups and interests, representatives of the appropriate component of the Executive Office of the President (EOP) such as the Office of Management and Budget (OMB) or the National Security Council (NSC), media specialists, and analysts from think tanks. It is in the interplay of these actors—in the give and take and push and shove—that policy is forged on a day-to-day basis.

For defense policy, there are three important lessons. First, policy is usually handled within a defense subgovernment. On a routine basis, the most prevalent players include:

—members and staffs from four congressional committees: House Armed Services Committee (HASC), Senate Armed Services Committee (SASC), House Appropriations Committee Defense Subcommittee (HAC), and Senate Appropriations Committee Defense Subcommittee (SAC);

—assistant and under secretaries and bureau chiefs in the national security bureaucracy;

—lobbyists for defense contractors and service-related groups such as Retired Officers' Associations and the Navy League;

—staffs of the relevant national security component of the EOP;

—media reporters who specialize in defense issues;

—defense consultants from "beltway bandit" think tanks such as the RAND Corporation.

Only under extraordinary circumstances do the president, congressional party leaders, and the secretaries of state and defense get involved as players. Second, as is the case for each major policy sector, there is "contained specialization"[8] within the defense subgovernment. Issues there are rarely resolved with reference to the considerations of other subgovernments. Third, and most important, on most issues in the American system, interest groups are the major intermediary through which people approach government. Parties rarely have a role in day-to-day policy deliberations.

RECENT DESTABILIZING TRANSFORMATIONS: POLITICAL AND GOVERNMENTAL CHANGE IN THE 1960s AND 1970s

American politics is like a moving picture constantly changing and evolving. Underlying the almost two hundred years of political development in America is a perennial struggle between forces of dispersion (congressional dominance, committee government, interest groups, states rights) and forces of concentration and unity (national federalism, presidential power, and political parties). In certain periods, forces promoting fragmentation are dominant; in other periods forces of concentration are dominant.

In the 1960s and 1970s, in response to the dual crises of Vietnam and Watergate, a number of major changes occurred in the American political system which strengthened the forces of fragmentation and weakened the forces of concentration. So monumental were these changes that, in many ways, the political system of the 1980s was worlds apart from that of 1960—perhaps as different as 1960 was from politics in 1900. To grasp the magnitude of recent political change it is imperative that one employ a dynamic model stressing these recent transformations.

The most significant recent development is the decline of political parties, discussed in books such as David Broder's *The Party's Over*[9] and Byron Shafer's *Quiet Revolution*[10] and articles and commentary that stress party deterioration, decomposition, and disaggregation.[11]

As noted above, political parties in America are characteristically weak. In the course of American history, parties developed due to strong leadership and machine organizations, only to be undercut by the turn-of-the-century good government reform movement and the improved sophistication of education and communications. But although Richard Neustadt has written that "what the constitution separates our political parties do not combine"[12] and Ralph Huitt has stressed that a shortcoming of American party institutions has been their inability to promote party unity by closing the gap between ideology and constituency,[13] the "web of party" has provided cohesion in the American system of fragmented government. Parties have served as rallying and dividing points in Congress, as bridges between the president and Congress, as ordering and structuring premises for voters, as candidate recruiters and political educators, and as potential unifiers. Thus, their decline has served to exacerbate the disunities in American government.

The weakening has occurred in all three components of party: party-in-the-electorate, party-as-an-organization, and party-in-government. Ticket-splitting, switching party allegiance from one election to another, and independent status are more prevalent than before. Due to the proliferation of primaries and other reforms, party activists no longer enjoy the influence they once had in the selection and slating of nominees and the conduct of campaigns, producing what James Sundquist calls "haphazard presidential selection."[14] Party cohesion is down in Congress, and party leaders don't have the clout they once did. Party is not a bridge across the separation of powers. (In fact, much of a president's most intense congressional opposition comes from within his party.) Attending the decline of party has been a host of other noteworthy trends and changes. Some are concomitants of party decline, reflecting changing environmental conditions. Others are consequences and manifestations of party decline. Others, while consequences, are also causes of the general weakening of partisan institutions. Among the most important are the following:

—the rise of single-issue interest groups;

—the growing influence of the media in candidate recruitment and opinion formation;

—the development of a campaign industry of professional pollsters, image-makers, fund-raisers, and strategists;

—public financing of presidential primaries, conventions, and elections;

—a fickle, volatile electorate, characterized by Britisher Anthony King as comprised of "co-alitions made of sand,"[15] and promoting Theo-dore White to write of *America in Search of Itself*;[16]

—the increased role of ideology among the American electorate;

—the emergence of personality or cult politics involving single figures;

—growing political cynicism, alienation, pessi-mism, and mistrust.

Change has occurred not only in the general political environment but also within govern-mental institutions. Most change has exacer-bated the fragmentation of authority by weakening forces of centralization. Major recent trends affecting Congress include:

—increased assertiveness in national security affairs and foreign policy, symbolized by the War Powers Act, congressionally mandated cessation of US aid to Angola, the Turkish arms embargo, and investigations of the Cen-tral Intelligence Agency;

—the breakdown of "bipartisan" foreign policy;

—the enlargement of the defense subgovern-ment to include other than just pure "pro-defense" groups and interests;[17]

—a more aggressive congressional oversight of weapon systems procurement;

—a new breed of "independent" members of House and Senate, less inclined toward a congressional career but benefiting from higher incumbency retention;

—a growth of congressional staff and support agencies;

—an increase in workload and time in session, producing a much busier body;

—a miscellany of structural reforms, including an increase in sub-committee prerogatives, decline in the power of committee chairmen, an enhanced role for a proliferated party lead-ership, and more openness and accountabil-ity;

—a declining image of Congress;

—a blurring of House and Senate differences.[18]

For the presidency and presidential insitu-tions, the major trends have been:

—destabilization of the presidency with five presidents in the twenty-year period from 1960 to 1980, none of whom served two full terms;

—growth in the size of the White House staff and presidential support agencies (NSC, OMB, etc.) that comprise the Executive Of-fice of the President (EOP);

—constraints on the presidency in response to the perceived "imperial" presidency excesses of the Johnson and Nixon administrations;

—increased use of presidential commissions (such as the Social Security Commission, the Kissinger Commission on Latin America, and the Scowcroft Commission on MX), reflecting extrainstitutional attempts to forge consensus both between parties and between branches of government.

This model of the policy process has four im-portant consequences for the formulation of na-tional strategy. First, the American system is a dynamic and ever-changing one, evolving over time and responding to changes in broader so-cial, cultural, and economic environments. Sec-ond, the political turmoil of the 1960s and 1970s yielded a very destabilized process,[19] featuring weakened forces of coordination and unity and punctuated by the lack of continuity and by ab-rupt changes in the course of public policy. Third, as a result of changes and reforms, there are more actors in the policy process. This greatly compounds the tasks of coalition build-ing and executive-legislative cooperation, two necessary system lubricants. Finally, amidst the political instability, many major policy problems have not been dealt with forcefully, but instead have been sidestepped and put on the shelf.

PRESIDENTIAL POWER: THE REAGAN REVOLUTION

The models of American politics discussed thus far highlight centrifugal forces in the American system, that is, those factors that fragment and disperse governmental authority. However, in order for a political system to function, there need to be definite and pronounced centripetal processes, that is, those factors that unify, cen-tralize, coordinate, and afford a comprehensive view. The founding framers recognized this. In making the case at the Constitutional Conven-tion for a strong chief executive, Alexander Hamilton spoke of the necessity for having "en-ergy in government." Throughout American history, the American system has functioned due to adroit exercise of presidential power, supportive party institutions, and congressional cooperation and deference during crises, emer-gencies, and in the aftermath of landslide pres-idential elections benefiting the president's party in Congress. Therefore, to comprehend fully the nature of domestic factors in American policymaking, one must also be aware of a model of presidential power.

The most important statement on presidential

power in the American system is Richard Neustadt's *Presidential Power*.[20] Originally published in 1960, it since has been reissued with updated chapters in 1968, 1976, and 1980. The book has assumed the stature of a classic for its value not only to academics but to practitioners as well. So struck was President Kennedy with the book that he made it required reading for his White House staffers. Many practitioners and academicians turned practitioners have stated that *Presidential Power* is the rare book that offers useful advice to decisionmakers.

In a nutshell, Neustadt's argument is that under the Constitution the president has little formal authority. Moreover, in this system of checks and balances, there are many competitors for the president's power, and there is a strong possibility that the president will be hemmed in by Congress, bureaucracy, press, and public. This is especially disconcerting since the presidency is most capable of handling difficult problems with a unified, quick response.

According to Neustadt, to survive in our system of divided authority, a president must have the "power to persuade" that allows him to add to his clerk responsibilities (authority) and become a leader with power, engendering "self-enforcing" degrees. Specific advice is given to presidents and other top-level executives for developing the power to persuade:

—Appreciate power, understanding that there are limited stocks.

—Tend to power stakes, realizing that how decisions are made today affects one's ability to influence events tomorrow.

—Be shrewd, cunning, and skillful in the exercise of executive power.

—Develop bargaining skills.

—Guard professional reputation, taking strong care to preserve the winner's image.

—Secure and maintain public support, approval, and standing.

There have been several noteworthy system-maintaining exercises of presidential power: Lincoln and the Civil War, Teddy Roosevelt and "trust-busting," Franklin Roosevelt and the great depression and World War II, and Lyndon Johnson and the great society. Each of these permitted the system to come to grips with major problems, the separation of power to be temporarily overridden, and the president to put his imprint on public policy. The most recent example is the Reagan revolution. The first Reagan administration exercised presidential power in such a significant manner that we were forced to refine, modify, and even jettison notions and concepts about American politics which were helpful and valid as late as the

1970s. As such, the Reagan revolution was an excellent illustration of presidential power, indicating a course very different from the destabilizing tendencies of the 1960s and 1970s.

The Reagan administration took office in the wake of four failed one-term presidencies: Johnson, cut short by Vietnam; Nixon, cut short by Watergate; Ford, denied his own election bid; and Carter denied reelection. In 1980, on the eve of the Reagan administration, the prevailing interpretation of American politics stressed the decline of party (both in its electoral role and, more importantly, as a powerful force in organizing and administering government) and the dissensus (the absence of consensus) in American public opinion. The dominant view of American institutions stressed a weakened, imperiled presidency that had become an "impossible," "no-win"[21] job. Congress was considered to be internally paralyzed and hopelessly stalemated with the president,[22] but strengthened in this struggle by the 1974 Congressional Budget Act. The policy process featured the incremental growth and expansion of domestic programs, the supremacy of subgovernments and issue networks, and nation-centered federalism. The potential for government spending was viewed as virtually unlimited, best thought of as a "variable-sum" (no losers) as opposed to a zero-sum (a winner's gain is someone's loss) game.

Although many took issue with the Reagan programs and many expressed apprehension about the economic future, almost all acknowledged by 1985 that the Reagan administration had succeeded in strengthening presidential institutions, changing the agenda of government, and reshaping the focus of public debate on the proper role of government. In 1985 we saw a "renewed" Republican party[23] and an emerging conservative consensus on government spending, instead of party decline and dissensus. Presidential institutions were described as experiencing a return to "normal" presidential leadership, following the aberrant Johnson-Nixon-Ford-Carter years during which presidents performed their duties by playing the role of Senate majority leader, foreign minister, House minority leader, and outsider, respectively.[24] Presidential-congressional relations were more workable, involving cooperation on some selected issues, a working Republican majority in the Senate (and sometimes the House), and presidential use of the congressional budget process to accomplish presidential objectives. The policy process was characterized by macrobudgeting (focus on the whole picture), decrementalism in social programs (not incrementalism), an era of consolidation and federal divestiture (not innovation,), presidential override of subgovernments through cabinet counsels (cabinet working groups),[25] state (not federal) oriented federalism, and an intense guns versus butter debate (not variable-sum policymaking).

The Reagan revolution can be attributed to many factors, some windfall and some the result of political savvy.[26] Those that can be credited to political aptitude include the following:

—a forceful transition that had the administration "hit the ground running";

—a 100-days strategy that had the administration take advantage of its "honeymoon" period with Congress;

—an approach that focused the president on major issues and policy themes;

—selective prioritizing that allowed the administration to tenaciously pursue only a few basic programs such as economic recovery and inflation control;

—a solid staff of pragmatic, knowledgeable, and experienced individuals;

—effective presidential use of the media in presenting himself and his programs with simple, popular messages and employing media management;

—skillful direct and indirect presidential lobbying of Congress;

—the use of "trial balloons" and "heat shields," having good news released by the White House and bad news announced by agencies and departments;

—a series of successive legislative victories on tax reduction, budget priorities, and sale of Airborne Warning and Control System (AWACS) to Saudi Arabia;

—credit for presidential commissions tackling social security, Latin America, and the MX missile.

Serendipitous events that added to the first Reagan administration's power to persuade include:

—a prolonged honeymoon period with Congress as the result of the attempt on the president's life;

—skillful congressional leadership from Senate Majority Leader Howard Baker and House Republican leader Robert Michel;

—what Democrats call "teflon luckiness" referring to the fact that in the first term problems of the economy and foreign affairs did not get politically "stuck" on the president;

—a divided Democratic party;

—the absence of energetic and resolute Soviet leadership.

The exercise of power by the Reagan administration is verification of the Neustadt thesis. For the study of defense policy, the Reagan administration showed that constraints on the US presidency and the disunifying effects of the system can be overcome by skillful exercise of presidential power. Even in its most vulnerable moments the Reagan administrations showed that the system can be made to work.

In summary, although each of these models has its origins in a different historical period, they serve as overlays to cumulatively pinpoint the major features of the system: deliberately divided authority, grass-roots input from mass publics, subsystems responding to interested and affected publics, ongoing transformation, and the centripetal influence of executive power. It is in this setting and against this backdrop that defense policy is made.

DEFENSE POLICY AS PUBLIC POLICY: THE CONSEQUENCES OF A PLURALISTIC SYSTEM

A political system with the characteristics described above yields a policy environment that is most frustrating, but also challenging and full of opportunity. American defense policy—like policy made in other sectors—is made and executed in a distinctive process. In all sectors, both decision process and policy content reflect this unique setting.

To be sure, defense policymaking has its unique aspects. There is not an easily identifiable affected public. Policy responds to international developments in a game of reaction.[27] Policy is a highly "indivisible" collective good, that is, most segments of society ostensibly benefit equally from national security.[28] But, a thorough understanding of defense policy is enhanced by a focus on the similarities of all policy produced in the American system. This section describes the commonalities of policy and identifies the problems that afflict professional public managers both in and out of the security community. This is followed by an overview of proposals for improving and streamlining the American policy process.

CHARACTERISTICS OF THE POLICY PROCESS IN A PLURALISTIC SYSTEM

In the American pluralistic system, defense policy shares five major characteristics of the policy process with policymaking in other sectors. First, all policy develops according to a *set sequence*. As Charles O. Jones notes, this process involves eight major stages: problem identification, problem definition, policy formulation, policy legitimation, policy funding, policy implementation, policy evaluation and feedback, and policy adjustment.[29] Each stage involves different actors, different considerations, and different patterns of influence. It is a predictable, political, patterned, and practically never-ending process.

Second, policy is made under *conditions of constraint*. Policymakers do not have a blank check. Certain constraining forces impinge on decision latitude, ruling out certain options and militating toward, even mandating others.[30] The major categories of constraints include:

—cognitive (imperfect knowledge and understanding, lack of information);

—temporal;

—fiscal;

—structural (divided authority, federalism, and the limited domain and scope of governmental authority);

—legal (certain courses of action denied policymakers by the constitution);

—international (options ruled out by the exigencies of alliance and international politics);

—bureaucratic (professional norms, standard operating procedures, and bureaucratic interests);

—political feasibility;

—dissensus (atomized, divided public opinion);

—sunk costs of existing programs.

Third, there is a distinctive political or leadership elite culture that encourages negotiation, bargaining, and accommodation as a way of solving disputes. Due to multiplicity of players and the many potential veto and chokepoints in the legislative process, policy is almost always made with *compromise* and mutual adjustment, since potential opponents who are strategically located must be appeased with concessions.

Fourth, to contend with conditions of constrained decisionmaking and the stricture of compromise, policymakers employ certain shortcut *decision rules* that simplify the tasks. The major such decision rules include:

—disjointed incrementalism;

—"muddling through";

—successive limited comparisons of policy options;[31]

—remediality and seriality;[32]

—mixed scanning (perfunctory scan of ongoing programs and in-depth scrutiny of proposed changes at the margins);[33]

—"satisficing" (i.e., moving away from problems of satisfying minimum standards instead of devising an optimum solution);[34]

—policy incubation (letting policies evolve gradually and adopting them only when the policy climate is favorable).[35]

To summarize, in a decision environment that presents manifold complex decisions on a vari-

ety of topics, decisionmakers cope by employing nonrational (not irrational) decision strategies that follow familiar paths and do not deviate drastically from current practices or long evolving positions. In other words, incentives in the system favor status quo policies. These compromise the rational decisionmaking.

Fifth, and finally, *policy content* reflects constrained, compromised, and incremental policymaking. Policies produced in such a process tend to be "safe." In such a system it is very difficult to develop bold, aggressive, radical, and comprehensive policy.

Table I depicts the interconnections of these five characteristics of policymaking in a pluralistic system. The argument is that the stages of the policy process unfold and are indelibly influenced by various structural characteristics and decision constraints. To facilitate business, a political culture of compromise has developed. The combined effect of constraints and the norm of compromise leads to certain decision rule. In turn, these decision rules produce an inevitable type of policy output.

POLICY PROBLEMS

US defense policy professionals, elected officials, members of the media, and the general public have long complained about the difficulties involved in developing a rational defense policy in the American system. Specifically, the pluralistic policy process presents twelve very frustrating problems for the national security manager. Although these policy processes can be found to some degree in the policy problems of authoritarian regimes or in "responsible party" systems, they are endemic in the US systems of "rampant pluralism."[36]

The first is *overlap, replication* and *duplication* of responsibilities and efforts in government. American bureaucracy features redundant activities and resources. For example, US intelligence assets include not only the Central Intelligence Agency (CIA) but also the National Security Agency (NSA), the Defense Intelligence Agency (DIA), separate Department of Defense (DOD) and service analysts, and substantive experts in the State Department, Congress, and the NSA. Such multiple channels are beneficial to be sure, but they pose enormous problems for coordination and integration of information and programs.

The second is *turf battling* between president and Congress, between House and Senate, and between presidential staff bureaucracies (NSC, OMB) and line bureaucracies. Such tugs-of-war cast enormous uncertainty over defense policy by involving it in contests of power. The sale of AWACS planes to Saudi Arabia, and the Panama Canal and SALT II treaties are excellent examples of defense policymaking being colored

Table A2

An Organization Theory of Policymaking in a Pluralistic System

Stages of Policymaking ⟶	Structural Characteristics and Constraints ⟶	Political Style ⟶	Decision Rules ⟶	Inevitable Policy Output ⟶
Problem identification	Fragmented author-	Compromise	Incrementalism	Lukewarm,
Problem definition	ity	Accommoda-	"Muddling through"	watered-down,
Policy formulation	Separation of pow-	tion	Successive lim-	compartmental-
Policy legitimation	ers	Bargaining	ited compari-	ize policy
Policy funding	Checks and bal-		sons	Avoidance of
Policy implementa-	ances		Remediality	bold, radical,
tion	Bicameralism		Seriality	aggressive,
Policy evaluation/	Congressional com-		Mixed-scanning	comprehensive
feedback	mittee system		Satisfycing	policy
Policy adjustment	Growing staffs		Incubation	
	Distractive authori-			
	zation/appropria-			
	tion processes			
	Issue subsystems			
	Fiscal, political, cog-			
	nitive constraints			

by political rivalry between president and Congress. In addition, such turf battling frequently leads to excessive micromanagement of DOD operations.

A third problem is the *necessity for long lead times* in the development, procurement, and acquisition of major weapons systems. The development and actual deployment of both the B-1 bomber and the MX strategic missile were drawn out over almost two decades. The reasons for this are changing political circumstances that politicize weapon systems and the requirement for multiple, annual authorization and appropriation votes in both House and Senate. As David Caputo argues, "the decisionmaking process is made all the more difficult and time-consuming by the larger number of participants involved . . . Decisions resulting from it are apt to be less than definitive."[37]

Problems of *inefficiency* (high overhead and operating costs) and *ineffectiveness* (unending problems and programs) are the fourth bundle of policy problems indigenous to a pluralistic system. These problems have become highlighted in recent years as budget battles, and issues of austerity have taken center stage. In a system of divided and redundant authority, efficiency and effectiveness are difficult and elusive goals.

The fifth problem involves the short-term *focus* of the system. American policymaking is the antithesis of central planning. Policy focuses on brush fires. It is reactive (responding to events) rather than proactive (anticipating events). An excellent example is the rapid deployment force, a concept pursued only after America's impotence was dramatized by the Iranian hostage crisis.

The sixth can be dubbed the *government of strangers* problem. Hugh Heclo has argued that a persistent problem in the American system is perennial turnover among high-level political executives (appointees).[38] This turnover—which occurs usually every four years, sometimes less, and no longer than every eight years—creates a government comprised of individuals who do not know each other, who are novices, and who often must learn both the issues and how to employ political resources. They also have to learn to work together. The result is frequently predictable policy and a maladroit use of executive power.

Piecemeal policymaking is the seventh problem. With the dispersal of authority, policy is dealt with in detail only in the subgovernment's domain. Subgovernments consider only the interests of the subsystem. A central focus on general considerations that affect several issue subsystems rarely occurs in bureaucratic or congressional subsystems, despite the fact that most major problems are inextricably intertwined with other problems.

Eighth, pluralistic policymaking is afflicted with *policy contradictions* as different agencies frequently pursue incompatible programs and objectives. The best explanation for this occurrence is Morton Grodzins's "Multiple Crack Theory." To Grodzins, policy conflicts occur because of the many cracks and fissures in the fragmented and redundant structure of American politics.[39] These cracks and fissures provide numerous access points for interest groups set on influencing the system or gaining government contracts. Generally, if an interest is determined and persistent enough it can get some kind of favorable concession from government.

The result is a government that has difficulty prioritizing, is immobilized by special interests, and is pursuing contradictory objectives, with departments and agencies working at cross purposes as one hand of government undoes what the other is doing.

The ninth problem is *goal displacement and suboptimization*—that is, agencies and individuals losing sight of primary purposes, goals, "what it is all about."[40] Norton Long has argued persuasively that because executive agencies lack a constitutional power base, they develop their own source of power and political support for security and survival.[41] Others have detailed the ways in which bureaucracies and agencies further their own political interests in contests of power in which they vie for such prizes as budgets, personnel slots, roles and missions, power, position, and prestige.[42] The point to be emphasized is that in pursuing bureaucratic interests among both external clientele (groups) and internal constituencies (professional norms), agencies will often overlook or neglect their primary responsibilities for the delivery or provision of public services.

The tenth identifiable problem is that of *negative externalities*, a term used by economists to refer to the unanticipated, unexpected, and undesired policy consequences of public programs.[43] All programs have secondary and tertiary impacts, usually unrelated to and unintended by the program. Some can be positive (side benefits), but most are negative (side penalties). A major problem of contemporary American policy is the prevalence of negative externalities in the environment, the deficit, social programs, arms control, etc. They occur because of the lack of central planning and comprehensive focus. Perhaps the best illustration is the national highway system. Ostensibly initiated in the aftermath of World War II for national security reasons, in order to connect America's east and west coasts, the highways were built by linking metropolitan areas. In metropolitan areas, the highways "looped" or "ringed" the central city so as not to disrupt inner city traffic. But—an unforeseen outcome—this gave city dwellers an opportunity to live great distances from their places of work. The unforeseen and unintended consequence was the suburbanization and strip development that have occurred at the expense of the central city.

The eleventh problem is *cosmetic problem solving*. Because of the imperative to satisfice—to move away from problems and not necessarily toward solutions—problems are frequently dealt with by treating symptoms, not causes, failing to go to the root of identified ills. Hard solutions are dodged and ducked; tough choices are avoided.

The twelfth and final problem is the *mobili-zation of bias* in policymaking. In the American system, not all interests are mobilized or represented equally in the marketplace of politics and policy deliberation. Thus, certain (especially centrist) interests and points of view are not always considered when decisions are made.

PROPOSALS FOR REMEDYING POLICY PROBLEMS AND SHORTCOMINGS OF PLURALISM

There appear to be seven major streams of thought concerning policy problems in the American system, all of which should loom large during debates in the late 1980s and early 1990s. Ordered according to the amount of change proposed, from least to most, the groupings are pluralists, structuralists, personnelists, legislativists, party reformists, presidentialists, and parliamentarists.

Pluralists are best thought of as neo-Madisonians. They acknowledge the problems and shortcomings of American policymaking but are apprehensive of tampering with the unknown, fearing that cures may be worse then the ailments. For pluralists, muddling through and incrementalism are preferable to other types of policymaking.

Structuralists turn to structural reform as an antidote to muddling-through policymaking. They encourage more lateral coordination in government, the use of budgetary reforms—such as systems analysis, cost-benefit (PPBS) analysis, management by objective (MBO), and zero-based budgeting (ZBB)—and the employment of collegial decision processes and multiple advocacy. These last terms[44] refer to proposed methods for widening the spectrum of information and advice decisionmakers receive, in an effort to counteract hasty consensus, mobilization of bias, sycophancy, decisionmaker isolation, groupthink,[45] limited options, closed systems, and vest-pocket decisionmaking. Multiple advocacy would structure the decision process to take advantage of redundancy and bureaucratic politics in government by insuring that all agencies and actors with an interest in a given decision would have the opportunity to make comment and input.[46]

Personnelists point to management problems that stem from a lack of professionalism in government service. They advocate that development of generalist perspectives in government and the pursuit of a professional career managerial cadre in government akin to the British civil service system[47] and the US Senior Executive Service (SES). Proposed reform of the Joint Chiefs of Staff and proposals for a general staff are good examples.

Legislativists are those, such as Theodore Lowi,[48] who urge that Congress delegate less power and force itself or be forced by the courts to deal more forthrightly with pressing issues. Lowi calls this "juridical democracy"—using the courts to declare excessive delegation of power by Congress to be unconstitutional in order to force Congress to prioritize and perform its function of policy deliberation.

Party reformists are those who would resuscitate and strengthen political party structures as a way of coping with policy problems. The most common proposals involve making candidates for office more beholden to their party, increasing the role and clout of party officials, and developing meaningful platform and policy councils.[49]

Presidentialists advocate a strengthening of presidential power and presidential institutions. Among the proposals are line item veto authority for the president, repeal of the twenty-second amendment (limiting presidents to two terms), and the extension of the president's term from four to six years, a position favored by most of our recent presidents.

Parliamentarists call for the most far-reaching change. They propose major alterations to promote a cabinet or parliamentary type of "responsible" government, with some fusion of executive and legislative authority and structure. At the forefront of this movement is Lloyd Cutler, former counselor to President Carter.[50] He is joined by Robert McNamara, former secretary of defense, and Douglas Dillon, former secretary of the treasury. Their objective is more consolidation, unity, and coherence in the American system.

In assessing various reform perspectives, the following questions[51] should be asked:

—Is the proposal relevant to the problems identified by its proponent?

—What are the underlying cause-and-effect assumptions, and are they supported by experience?

—Is the proposal politically feasible, i.e., is there a strong likelihood that it could be adopted?

—What are the tradeoffs? What unforeseen and unanticipated consequences might there be? On balance, is the proposal desirable?

—How will this affect the way in which the security community does business? How will it affect the spirit and values of the American system?

Such a line of inquiry can be most illuminating. For example, multiple advocacy although feasible and desirable from the standpoint of broadening the decision process may not be desirable when time or security constraints are paramount. Likewise, the parliamentary alternative, attractive as it is, may be somewhat incompatible with American cultural diversity

and political traditions.[52] In sum, only extensive study, sophisticated analysis, and good judgment can provide the basis for decisions about the wisdom of various proposed changes.

CONCLUSIONS AND RECOMMENDATION: WORKING EFFECTIVELY WITHIN THE SYSTEM AND MAKING THE SYSTEM WORK

When studying domestic factors and constraints (especially the erosion of authority and the fragmentation and individualization of policymaking power), it is tempting to throw up one's hand and declare: "In this untidy system of policymaking, with recent trends that exacerbate the problem, what's the use of trying? Why not just muddle through? As noted above, sometimes muddling through is all that the system allows policymakers. But sometimes strategists can work with apparent contraints and turn them into opportunities. Such are the challenges to the national security policymaker attempting to develop a coherent national strategy.

The specific steps that can be taken to enhance the participation of national security professionals in the pluralistic policy process include:

—Have a sophisticated knowledge and understanding of how the system works.

—Have an appreciation for the strengths and benefits of a pluralistic system.

—Develop sophisticated organizational and individual styles and strategies for participating in the pluralistic process and resolving policy problems.

—Develop an informed anticipation of political events and change.

To appreciate how the system works one must fully understand three points. First, there is little unity in the American system, and this lack of unified authority was intended by the founding framers. Second, much of what occurs in the American system involves "patterned role playing." Various officials do what they do because of the jobs they have. Remember the helpful dictum "Where you stand on an issue depends on where you sit."[53] People in Congress tend to see and take stands on things in very different ways than those in bureaucracy, due to differences in constituencies, timeframes, and policy roles.[54] Third, to use a Douglas Yates metaphor, the policy process is a "penny arcade," a shooting gallery where things are constantly popping up and dealt with only through the luck of a game of chance.[55]

To appreciate the strengths and benefits of the American system, one needs to know that, despite the frustrations, shortcomings, and blemishes of American policymaking, the system can and does work when it has to. Exercises of presidential power, supported by political parties, skillful congressional leaders, and public consensus permit the fragmented system to come to grips with crises and emergencies, such as war, depression, and energy shortages. Unlike closed systems, the more open American system facilitates political change and the identification of policy mistakes, and discourages the ratification of error. In the United States, policy will rarely strain or significantly deviate from a firm public consensus. The system does not pursue the folly of rigid five-, ten- or fifty-year plans. Most importantly, over time the system works out its defects. Two good examples of this are the cases of presidential impoundments and the legislative veto. Both threatened the delicate equilibrium of separation of powers. Impoundment—especially as used by President Nixon—is the procedure whereby the president refuses to spend congressionally appropriated funds. The legislative veto, as it evolved over several decades, gave Congress an opportunity to block executive action with a one- or two-house vote, depending on the relevant legislation. In both instances the problem was rectified—impoundments by the 1974 Congressional Budget Act and the legislative veto by the Supreme Court *Chadha* decision declaring such vetoes unconstitutional exercises of executive authority by the Congress.

To develop sophisticated strategies, national security managers need to know what succeeds in the policy process. Organizations must realize that, as many old-time legislative liaison types attest, success requires dogged coalition building at successive stages. It is not easy. Making policy involves the creation of an instantaneous government on a given issue at a given time. For individuals, success comes from following a useful list of aphorisms:[56]

—Understand and become comfortable with our systems of creative tension.

—Appreciate that, as Lyndon Johnson was fond of saying, half a loaf (or program) is better than no loaf (or program).

—Be professional. Develop a professional reputation.

—Be knowledgeable about your programs. If you are not the expert, who is? Give informed advice.

—Understand that Congress has a legitimate role to play in the policy process.

—Always level with authorities. Be truthful and creditable. Remember, you are only as effective as your word.

—Realize that bad news is made worse by lying.

—Don't be excessively parochial, for it hurts your cause.

—Serve missions and programs, not bureaucracies. Serve program goals.

—Keep the message brief, simple, and understandable.

—Do your best to minimize policy spillover problems. Do your part to identify and counteract negative externalities, inefficiency, and ineffectiveness. Do your best to attenuate the dysfunctions of rampant pluralism.

It is difficult to give recommendations for developing a sophisticated anticipation of the future. All that can be said is that the future is constantly changing, affected by events both predictable and unpredictable.

Defense planners, when formulating strategy, need to be aware of the vicissitudes of the American political environment. They need to have a "toleration for ambiguity,"[57] an acceptance for uncertainty, and an ability to cope with what Clausewitz dubbed "friction," in this case the friction in American national life. They need to understand that their professionalism is one of the things that pulls the system together and provides constancy and consistency.

The government of 1787 was not intended to be neat and orderly. It did not place a premium on tidiness, harmony, or efficiency. It was designed to be chaotic and disorderly for the purposes of advancing self-government, limited government, and conflict resolution. Although such a system at times is very frustrating, its values are precious. They are American, they are pluralistic (as opposed to dictatorial), and they are worth defending.

NOTES

1. Carl Von Clausewitz, *On War*, ed. and trans. Michael Howard and Peter Peret (Princeton: Princeton University Press, 1976), p. 146.

2. See *The Federalist Papers*, no. 10.

3. This term is used in Richard E. Neustadt, *Presidential Power* (New York: Wiley, 1980), p. 26.

4. Corwin, quoted in Cecil V. Crabb, Jr. and Pat M. Holt, *Invitation to Struggle: Congress, The President and Foreign Policy* (Washington, D.C.: CQ Press, 1980), p. 1.

5. Alexis de Toqueville, *Democracy in America* (New York: Alfred A. Knopf, 1945), p. 251.

6. See Samuel P. Huntington, *American Politics: The Promise of Disharmony* (Cambridge, Mass.: Belknap Press, 1981) and "Paradigms of American Politics: Beyond the One, the Two and the Many," *Political Science Quarterly* 89 (Mar. 1974): 1–26.

7. The major works that promote a subgovernment model are: Ernest S. Griffith, *Congress: Its Contemporary Role*, 3d. ed. (New York: New York University Press, 1961), pp. 50–51; Douglas Cater, *Power in Washington* (New York: Vintage Books, 1964), pp. 17–21; James L. Freeman, *The Political Process: Executive Bureau-Leglislative Committee Relations*, rev. ed. (New York: Random House, 1965), p. 6; and Theodore J. Lowi, *The End of Liberalism* (New York: W. W. Norton, 1969).

8. This term was coined by Ira Sharkarsky, *The*

Politics of Taxing and Spending (Indianapolis: Bobbs-Merrill, 1969), p. 38.

9. David Broder, *The Party's Over* (New York: Harper and Row, 1971).

10. Byron E. Shafer, *Quiet Revolution: The Struggle for the Democratic Party and the Shaping of Post-Reform Politics* (New York: Russell Sage Foundation, 1983).

11. See, for example, Walter Dean Burnham, "American Politics in the 1970s: Beyond Party?" in *The American Party System*, 2d ed., ed. by W. Chambers and W. Burnham (New York: Oxford University Press, 1975), pp. 308–57 and Gerald M. Pomper, "The Decline of the Party in American Politics," *Political Science Quarterly* 92 (Spring 1977): 21–41. See also William J. Crotty and Gary C. Jacobson, *American Parties in Decline* (Boston: Little, Brown, 1980) and Jeane J. Kirkpatrick, *Dismantling the Parties: Reflections on Party Reform and Party Decomposition* (Washington: AEI, 1978).

12. Richard E. Neustadt, *Presidential Power* (New York: Wiley, 1960), p. 26.

13. Ralph K. Huitt, "Democratic Party Leadership in the Senate," *American Political Science Review* 55 (June 1961): 335.

14. James L. Sundquist, "Congress, the President and the Crisis of Competence in Government," in *Congress Reconsidered*, ed. by L. Dodd and B. Oppenheimer (Washington, D.C.: CQ Press, 1981), p. 358.

15. Anthony King, "The American Polity in the Late 1970s: Building Coalitions in the Sand," in *The New American Political Systems*, ed. by Anthony King (Washington, D.C.: AEI, 1979), pp. 371–95.

16. Theodore H. White, *America in Search of Itself* (New York: Warren Books, 1982).

17. For elaboration of this point see the case study on the B-1 Bomber in Norman J. Ornstein and Shirley Elder, *Interest Groups, Lobbying and Policymaking* (Washington, D.C.: CQ Press, 1978), pp. 187–220; and Edward J. Laurance, "The Changing Role of Congress in Defense Policymaking," *Journal of Conflict Resolution* 20 (1976): 213–55.

18. See: Norman J. Ornstein, "The House and the Senate in a New Congress," in *The New Congress*, ed. by T. Mann and N. Ornstein (Washington, D.C.: AEI, 1981), pp. 363–86.

19. For a strong elaboration of the argument see Nelson W. Polsby, *Consequences of Party Reform* (Cambridge: Oxford University Press, 1983).

20. Neustadt, *Presidential Power*.

21. For a summary of this line of reasoning see Harold M. Barger, *The Impossible Presidency: Illusions and Reality of Executive Power* (Glenview, Ill.: Scott, Foresman, 1984); and Paul C. Light, *The President's Agenda: Domestic Policy Choices from Kennedy to Carter with Notes on Reagan* (Baltimore: Johns Hopkins University Press, 1983) and "Presidents as Domestic Policymakers" in *Rethinking the Presidency*, ed. by T. Cronin (Boston: Little, Brown, 1982), pp. 368–70.

22. For this line of reasoning see: Laurence C. Dodd, "Congress, the Constitution, and the Crisis of Legitimation," in *Congress Reconsidered*, 3d ed., ed. by L. C. Dodd and B. Oppenheimer (Washington, D.C.: CQ Press, 1981), pp. 390–420.

23. For a statement about the reinvigoration of party see: David E. Price, *Bringing Back the Parties* (Washington, D.C.: CQ Press, 1984).

24. For amplification on this point of view see:

Charles O. Jones, "Presidential Negotiations with Congress," *Both Ends of the Avenue*, ed. by A. King (Washington, D.C.: AEI, 1983), pp. 96–130.

25. For an overview of the cabinet council system see: Edwin Meese III, "The Institutional Presidency: A View From the White House," *Presidential Studies Quarterly* 13 (Spring 1983): 191–97.

26. The following analysis relies heavily on discussions I've had with Ken Duberstein, David Gergen, and Earl Walker.

27. As used in Thomas R. Dye, *Understanding Public Policy*, 5th ed. (Englewood Cliffs: Prentice-Hall, 1984), chap. 9.

28. As used in Robert L. Lineberry, *American Public Policy* (New York: Harper and Row, 1977), p. 37.

29. See Charles O. Jones, *An Introduction to the Study of Public Policy*, 3d ed. (Monterey, Calif.: Brooks/Cole, 1984).

30. For an amplification of the concept of constrained decision-making see the discussion of bounded rationality in James G. March and Herbert A. Simon, *Organizations* (New York: Wiley, 1958), pp. 169–71.

31. An elaboration of these terms can be found in Charles E. Lindblom, "The Science of Muddling Through," *Public Administration Review* 19 (Spring 1959): 79–88.

32. For an elaboration of these terms see Charles E. Lindblom, *The Policy-Making Process* (Englewood Cliffs, N.J.: Prentice-Hall, 1968), pp. 24–26.

33. This concept is from Amatai Etzioni, "Mixed Scanning: A 'Third' Approach to Decisionmaking," *Public Administration Review* 27 (1967): 385–92.

34. This concept is from Herbert A. Simon, "A Behavioral Model of Rational Choice," *Quarterly Journal of Economics*, 69 (Feb. 1955): 99–118.

35. For an elaboration of this concept see Nelson W. Polsby, "Policy Analysis and Congress," *Public Policy* 17 (Fall 1969): 61–74; James L. Sundquist, *Politics and Policy* (Washington, D.C.: Brookings Institution, 1968); and Nelson W. Polsby, *Political Innovation in America: The Politics of Policy Initiation* (New Haven: Yale University Press, 1984).

36. This term is used by James W. Fesler in "Politics, Policy, and Bureaucracy at the Top," a paper delivered at the fall conference of The Center for the Study of the Presidency, Minneapolis, 1982. Also, see James W. Fesler, "Politics, Policy, and Bureaucracy at the Top," *The Annals of the American Academy of Political and Social Science* (Mar. 1983): 23–41.

37. David A. Caputo, *The Politics of Policymaking in America: Five Case Studies* (San Francisco: Freeman, 1977), p. 170.

38. Hugh Heclo, *A Government of Strangers: Executive Politics in Washington* (Washington, D.C.: Brookings Institution, 1977) and "Political Executives and the Washington Bureaucracy," *Political Science Quarterly* 92 (Fall 1977): 395–424.

39. Morton Grodzins, *The American System*, ed. by Daniel J. Elazar (Chicago: Rand McNally, 1966), pp. 274–76.

40. This term is used in Felix A. Nigro and Lloyd G. Nigro, *Modern Public Administration*, 5th ed. (New York: Harper and Row, 1980), pp. 76–78.

41. Norton E. Long, "Power and Administration," *Public Administration Review* 9 (Autumn 1949): 257–64.

42. See Graham T. Allison, "Conceptual Models and the Cuban Missile Crisis," *The American Political*

Science Review 63 (Sept. 1969): 689–718; George Appleby, *Politics and Administration* (Tuscaloosa, Ala.: University of Alabama Press, 1949); Morton H. Halperin and Arnold Kanter, "The Bureaucratic Perspective: A Preliminary Framework," in *Readings in American Foreign Policy*, ed. by Halperin and Kanter (Boston: Little, Brown, 1973), pp. 1–42; B. Guy Peters, *The Politics of Bureaucracy*, 2d ed. (New York: Longman, 1984); Frances E. Rourke, *Bureaucracy, Politics and Public Policy*, 3d ed. (Boston: Little, Brown, 1984); and Harold Seidman, *Politics, Position, and Power: The Dynamics of Federal Organization*, 2d ed. (New York: Oxford University Press, 1976).

43. This term is best defined in Lineberry, *American Public Policy*, p. 104.

44. For a good overview of such proposals see: Stephen Hass, *Organizing the Presidency* (Washington, D.C.: Brookings Institution, 1976) and Alexander L. George, *Presidential Decisionmaking in Foreign Policy: The Effective Use of Information and Advice* (Boulder, Colo.: Westview Press, 1980).

45. For a definition of this term see Irving L. Janis, *Groupthink*, 2d ed. (Boston: Houghton Mifflin, 1982).

46. See Alexander L. George, "The Case for Multiple Advocacy in Making Foreign Policy," *American Political Science Review* 66 (Sept. 1972): 751–85 and Roger B. Porter, *Presidential Decisionmaking* (London: Cambridge University Press, 1980), pp. 213–28.

47. Robert C. Wood, "When Government Works," *The Public Interest* 18 (Winter 1970): 39–51.

48. See Theodore J. Lowi, *The End of Liberalism* and *The Politics of Disorder* (New York: W. W. Norton, 1971).

49. See Price, *Bringing Back the Parties* and

James L. Sundquist, "Party Decay and the Capacity to Govern," in *The Future of American Political Parties*, ed. by Joel Fleishman (Englewood Cliffs, N.J.: Prentice-Hall, 1982), pp. 42–69.

50. The best statement on behalf of a parliamentary proposal is Lloyd N. Cutler, "To Form A Government," *Foreign Affairs* 59 (Fall 1980): 126–44.

51. For a detailed elaboration of this evaluational methodology see: David C. Kozak, "The Multiple Advocacy Prescription for Presidential Decisionmaking: An Explication and Analysis," in *The American Presidency: A Policy Perspective From Readings and Documents*, ed. D. Kozak and K. Ciboski (Chicago: Nelson-Hall, 1985), pp. 542–62.

52. For this reservation see William J. Keefe, *Parties, Politics and Public Policy*, 2d ed. (Hinsdale, Ill.: Dryden, 1976), pp. 163–67.

53. This term is used by Graham T. Allison, *Essence of Decision: Explaining the Cuban Missile Crisis* (Boston: Little, Brown, 1971), p. 176.

54. Stanley Heginbotham, "Dateline Washington: The Rules of the Game," *Foreign Policy* 53 (Winter, 1983–84): 157–72.

55. Douglas Yates, "Urban Government as a Policy-making System" in *The New Urban Politics*, ed. by L. Masotti and R. Lineberry (Cambridge, Mass.: Ballinger, 1976), pp. 247–49.

56. These aphorisms came from advice and recommendations of policy professionals too numerous to list in their nonattribution talks to classes at the National War College.

57. This concept is amplified in Harland Cleveland, *The Future Executive: A Guide for Tomorrow's Managers* (New York: Harper and Row, 1972).

THE POLITICAL DIMENSION OF MILITARY PROFESSIONALISM

JOHN H. GARRISON

Since the mid-1960s the military profession in America has been faced with a crisis of identity and purpose. At first dimly perceived, but coming into sharper focus since 1968, this crisis centers upon two grossly general questions about the nature of the profession: "What are we to be?" and "What are we to do?" The first of these two questions concerns itself with the substance of military professionalism—the attributes, characteristics, and attitudes of the professional soldier. What are they? What should they be? The second question—"What are we to do?"—concerns itself with the functions of the military profession—the missions assigned, the tasks performed, the actions undertaken. What are they and what should they be?

The very crisis nature of the issue indicates that there is less than unanimous agreement, both within and outside of the military, about the

answers to these crucial questions. This has led to much soul-searching and a large amount of self-criticism within the profession itself. In rather general terms, opinion within the military seems to have crystallized around two basically different positions. One is a traditionalist view which sees the profession as having quite narrow, well-defined boundaries which restrict it to purely military qualities, attitudes and functions. The second is a more contemporary or modernist view which defines the profession in terms of much wider, less well-defined boundaries encompassing a variety of different attributes, concerns, and activities beyond the purely military. These schools of thought are known by many names: "Traditionalist/Modernist," "Separatist/Fusionist," "Professional Soldier/Soldier-Statesman," etc. No matter how they are labeled, these two positions are quite

Printed with minor revisions and by permission of the author.

contradictory in their views on many issues and aspects of military professionalism. And none is more contradictory than their interpretation of the proper political dimension of the profession.

TRADITIONAL VIEW

The traditional view of the American military profession is perhaps most clearly stated in Samuel Huntington's classic work *The Soldier and the State*. Written in 1956, this book was the first modern scholarly examination of the profession of arms in America, and as such it became something of a bench mark against which more recent studies have been compared.

One of Huntington's central theses is the assertion that the true American professional military ethic is a peculiar product of the years of extreme military isolation and rejection from American society between the Civil War and World War I:

The very isolation and rejection which reduced the size of the services and hampered technological advance made these same years the most fertile, creative, and formative in the history of the American armed forces . . . Withdrawing into its own hard shell, the officer corps was able and permitted to develop a distinctive military character. The American military profession, its institutions and its ideals, is fundamentally a product of these years.[1]

A basic tenet of the traditional professional military ethic that emerged from this period is the belief that the professional soldier is, and should be, apolitical—above politics. Politics and officership do not mix; there exists between military affairs and politics a sharp, clear line that must be observed and maintained! (It might be added that the traditionalist belief in a sharp, distinct division between war and diplomacy is a logical corollary to this apolitical view.)

This attitude had as its founder and strongest proponent General William T. Sherman, Commanding General of the Army from 1869 to 1883. Sherman was adamant in stressing the complete separation of the military from politics; he is quoted by Huntington as stating that the principle should be to "keep the Army and Navy as free from politics as possible," and that "no Army officer should form or express an opinion" on party politics. A contemporary of Sherman's, Rear Admiral Stephen B. Luce, fostered the same apolitical tradition in the Navy during the immediate post–Civil War years.[2]

In summarizing the traditional view of the relationship between the military profession and politics, it is useful to once again examine *The Soldier and the State*. Describing various attributes of the professional military ethic, Huntington states that:

Politics deal with the goals of state policy. Competence in this field consists in having a broad awareness of the elements and interests entering into a decision and in possessing the legitimate authority to make such a decision. Politics is beyond the scope of military competence, and the participation of military officers in politics undermines their professionalism, curtailing their professional competence, dividing the profession against itself, and substituting extraneous values for professional values. The military officer must remain neutral politically.[3]

TOWARD POLITICAL INVOLVEMENT

This belief that professional officers should be apolitical or above politics was almost universally accepted within the profession from Sherman's day up until World War II. However, the situation began to change under the pressures of waging a truly global conflict. The leading wartime military commanders—Marshall, Leahy, Arnold, King, Eisenhower, MacArthur—were drawn into extensive political involvement, including the making of the highest national policies relative to war production, surrender terms, zones of occupation, and the governing of occupied territories.

This trend continued into the post–World War II period as America was confronted with a drastically changed international environment, with a revolutionary and frightening threat posed by nuclear weapons and long-range delivery vehicles, and ultimately with the cold war. American foreign policy became very concerned with national security considerations, and as Professor Sam Sarkesian of Loyola University (and a retired army officer) states, "The preoccupation with national security and a proper defense posture stimulated the growth of a vast defense establishment and concomitant political power and involvement in the political process."[4] In the years after the end of World War II, military men bore much of the responsibility for the reconstruction of Germany and Japan; administered many US economic and social programs overseas; moved into key positions in Washington's extensive foreign policy establishment; began representing the United States and dealing with foreign powers as members of military assistance missions; defended administration defense budget requests before Congress; and generally began actively to seek public support for a strong defense establishment.

Most of these activities have continued, and expanded, from the late 1940s on into the 1970s. And as this political involvement by military men persisted, a different view of the relationship between the military profession and politics gradually emerged to challenge the traditional apolitical standard. Huntington identified this new attitude as early as 1956, calling it "political-military fusion," the idea that military considerations could no longer be realistically separated from political considerations.[5] However, it has only been in the last ten years or so that this view has come to compete seriously and openly (within the military) with the long-standing apolitical image.

This more recent and drastically different approach has been given many labels—"political soldiers," "soldier-statesmen," and "military policy advisers," to name just a few. They all stem, however, from Huntington's concept of "fusionism." This theme has appeared with increasing regularity in recent works, authored by civilians and military men alike, discussing civil-military relations and contemporary military professionalism. It has become quite common to see assertions that "military and nonmilitary factors in national security policy are so closely interrelated that they may be thought of as inseparable," "the traditional distinction between civil and military affairs has become quite tenuous," and "there are no purely political or military solutions to our national problems."[6] One of the clearest statements of this "fusionist" view of the relationship between the military profession and politics has been offered by Sarkesian:

The nature of professionalism in the U.S. military is undergoing significant change under the impact of changing domestic and international forces . . . To the [traditional] professional concept of manager-technician, and the diminished heroic leader role, must now be added the political dimension, since the pressures of the new forces require political skills and reponses . . .

The military in the United States, although steeped in professionalism, has a political dimension which needs to be recognized if the military is to serve national security goals and if effective civilian control is to be maintained.[7]

The key to understanding this approach, however, lies not only in recognizing that the political dimension exists, but also in accepting this dimension as a proper one for the military profession. Advocates of the "political soldier" school enthusiastically welcome this interpretation as both necessary and fitting. Professor Sarkesian's assessment is indicative of this attitude: "Professionalism must now incorporate considerations of political skills as part of the individual role, and political effectiveness as part of institutional patterns. In other words, a new military ethos is developing which is a manifestation of an emergent military whose institutional role and professional status are tempered by political consciousness."[8] Rather naturally, this also leads to a belief by such advocates that professional officers must attain greater political sophistication and training in order that they may better understand and participate in the political as well as the military aspects of national defense. Colonel Don Bletz, US Army, represents a sizable number of mid-dle-grade officers from all services who adhere to this view. Writing in *Military Review* in 1971, Colonel Bletz summarized this thinking quite clearly:

Military professionals must fully accept the fact that military force can only be a means to an end and that end is, in the final analysis, the eventual resolution of some political problem . . .

It is not suggested that the military officer develop the same type of expertise as his civilian colleague [in the Foreign Service] . . . However, [military professionals] must be fully capable of participating in the formulation and implementation of viable politico-military policies by appropriate weighting of the dynamic politico-military equation.[9]

Although many military professionals have accepted the idea of "political-military fusion," it must be said that there is less than complete agreement among all interested parties, military and civilian, as to the desirability of this political dimension of the profession. There is a great deal of controversy and disagreement between followers of the traditional, apolitical approach and supporters of the "political soldier" school. Most who think it is undesirable for the professional officer to become practiced and involved in politics do so for one (or both) of two reasons: there is concern that such activity will dilute the unique *military* expertise and qualities of the professional, thereby degrading the essential *military* capabilities of the profession, and, second, there is fear that such activity will threaten civilian control of the military by making it possible for military men to have excessive influence or control over defense and foreign policy.

Many supporters of the traditional view, desirous of an apolitical military, proceed from the belief that there are certain qualities, skills, and attitudes which are essentially military, and are, in fact, necessary for the effective accomplishment of the purely military tasks required of the armed forces. These qualities, skills, and attitudes are not only essential, they are quite difficult and demanding, focusing as they do upon armed combat. Maintaining these attributes and capabilities in a high state of readiness is a constant and time consuming job. From this point, the argument leads to the conclusion that the time and effort spent by professional officers preparing for, or actively involved in, other less-or nonessential functions (such as political activity) reduce the readiness and effectiveness of the profession in its primary mission area—combat.

The second line of reasoning, the fear of the usurpation of civilian control, is somewhat less rigorous, and more emotional, in its development. It really stems from the long-standing fear of excessive and unchecked military power in American society, dating from colonial times.

The principle of firm and constant civilian control of the military, generally understood to mean ultimate and complete subordination of the armed forces to elected and appointed federal government officials, has been upheld as the necessary safeguard against such an eventuality. Critics of the "political soldier" school of thought believe that this essential civilian control can be (will be?, is being?) circumvented or subverted by the greater involvement of the military in political activities. According to this argument, such involvement can lead (will lead?, is leading?) to the military, rather than their civilian masters, determining how, where, when, and with what resources the armed forces of the United States will be used—to the potential (possible?, actual?) detriment or danger of the rest of society.

THE MEANING OF "POLITICS"

Any useful assessment of these two approaches—traditional and fusionist—to the military profession and its relationship with politics must really include some discussion of the meaning of that term "politics" as it applies to military professionalism. In general, as understood and used by the traditional school, "politics" refers to partisanship. As has already been mentioned, being apolitical means, in General Sherman's words, that "no Army officer should form or express an opinion" on party politics. Or, as Professor Huntington put it, the professional officer must maintain strict political neutrality, refusing to become embroiled in partisan questions of who will be elected or what party's proposals will be supported.[10] In his landmark book *The Professional Soldier*, Morris Janowitz concisely supports this interpretation, explaining that, "Under democratic theory, the 'above politics' formula requires that, in domestic politics, generals, and admirals do not attach themselves to political parties or overtly display partisanship." He goes on to say that, "In practice, with only isolated exceptions, regulations and traditions have worked to enforce an essential absence of political partisanship."[11]

This assessment of the effectiveness of the "above politics" formula is widely held to be true. The military profession and its professionals in America have remained remarkably free from partisan party politics. Since World War II, the relatively few instances where individual professional officers, almost invariably retired, have sought political office (the most prominent examples being MacArthur, Eisenhower, Edwin Walker, LeMay, and now Zumwalt) have been exceptions rather than the rule. And in no instance, whether involving a military man or a civilian, has the military profession as an organization involved itself in a partisan contest for political office.

However, partisanship is not the only meaning of the term "politics" applicable to the military. The noted political scientist, David Easton, provided a definition of politics more than twenty years ago that has come to be one of the most widely accepted in the field of social science. Easton said that politics is "the authoritative allocation of values,"[12] where values are simply anything valued by society, such as resources (money, raw materials, manpower, etc.), power, or prestige. The process of determining how such things of value will be allocated among competing segments of society, and doing so in a way that is accepted as legitimate (though not necessarily desirable) by society as a whole, is the process of politics. Every day, decisions are made by government officials concerning policy choices, budget expenditures, and program implementation which are all, in fact, authoritative value allocations. Put very simply, then, politics is nothing more than the process of determining who in society gets what!

If the term "politics" is understood in this light, it is quite obvious that the military is involved. The military profession, as an organization, is most certainly involved in the process of determining how the federal government and American society will allocate values. Military professionals are participants in this process as advisers to political decisionmakers, as advocates of particular policy alternatives, as implementors of final political decisions, and even as decisionmakers themselves. Every time a military leader determines, from the military point of view, that a particular weapon system (or force level, or foreign commitment, etc.) is advisable or required for the nation's security, he is helping to determine how the nation's financial, material, and manpower resources will be allocated. It is not really important that this military leader is not often the final decisionmaker on the issue. Even as adviser, or as nothing more than implementor of the decision, he influences the outcome—and that is politics!

INEVITABILITY OF POLITICAL INVOLVEMENT

Based upon these two, distinct meanings of the term "politics," it is clear that the military profession has not been and is not now involved in politics of the partisan party politics variety. It is also quite clear that the military profession is deeply involved in politics in terms of decisionmaking or value allocation politics.

This type of involvement was probably inevitable! At least, it was probably inevitable given the combination of two circumstances—the constitutional provisions for civilian control of the military that were established more than 185 years ago, and the massive growth in the size, responsibilities, and functions of the federal government since the 1940s, particularly in the field of international relations. These two factors have combined to establish patterns of political interaction and pressures for involvement that in all likelihood made military participation in the political decisionmaking process inevitable.

As has already been mentioned, the concept of civilian control of the military is deeply rooted in American tradition, stemming from colonial experience and solidified into basic principles of law incorporated into the Constitution. This concept was really the product of a vital concern that our Founding Fathers had with efforts to "overcome and contain the exercise of arbitrary military power. England's experience pointed out the dangers of embodying the powers both to declare and to wage war in one office" (i.e., the King).[13] From the very founding of our republic, Americans have been particularly worried about the "exercise of arbitrary military power." It was, after all, the arbitrary use of British military power in the North American colonies which was considered by many to be a factor contributing to the Revolution.

If this were the only military concern of the Founding Fathers, a simple answer was available—do away with the military entirely! However, there was another concern on their minds; they clearly recognized the need to have a strong, capable military arm to protect the new nation against the British, the French, the Indians, etc. In the words of Professor Elmer Mahoney of the US Naval Academy, this led the Founding Fathers to a "dichotomy of the country's need for military strength coupled with fear that the power given and the institutions created [by this military strength] could be inherently destructive if arbitrarily used."[14]

The solution to this dichotomy involved the incorporation of two basic principles into the Constitution. The first was the concept of civilian control; to guard against the basic distrust of the military, reliance was put in the principle of strictly subordinating the military arm of the government to the civilian. Beyond this, there was still the danger of vesting the powers of both declaring and waging war in the same office, making it possible for one civilian in government to arbitrarily use the military arm. To prevent this, reliance was placed on the second basic principle, a principle used by the framers throughout the Constitution to guard against the excessive concentration of power in any one official or element of the government. This was the basic principle of separation of powers and checks and balances. Specifically, civilian control of the military was divided between the federal government and the states, on the one hand, and between branches within the federal government on the other. The US Congress was

given the power to declare war, and to raise and support military forces (Article I. Section 8). The president was given the power actually to conduct military operations (i.e., wage war) as commander in chief (Article II, Section 2). Control over the state militias was divided between the state governors and the federal government (Article I, Sections 8 and 10; Article II, Section 2). The result of all of this was the establishment of a system of divided civilian control of the military.

This arrangement has had some interesting implications for the resulting operation and effectiveness of civilian control, and, collaterally, for the involvement of the military in politics. To begin with, this system of divided civilian control has not, according to some students of civil-military relations, resulted in effective military subordination to the civilian voice in government. Samuel Huntington believes that:

The very aspects of the Constitution which are frequently cited as establishing civilian control are those which make it difficult to achieve. Civilian control would be maximized if the military were limited in scope and relegated to a subordinate position in a pyramid of authority culminating in a single civilian head. The military clauses of the Constitution, however, provide for almost exactly the opposite. They divide civilian responsibility for military affairs and foster the direct access of the military authorities to the highest levels of government.[15]

The reason for this lies in an understanding of the intentions of the Founding Fathers in establishing divided civilian control. Huntington believes that: "The Framers' concept of civilian control was to control the uses to which *civilians* might put military force rather than to control the military themselves. *They were more afraid of military power in the hands of political officials* than of political power in the hands of military officers. (Italics added.)"[16]

Whether or not Huntington is correct in assuming that the *primary* concern was to limit *civilian* use of military force, it is fairly evident that it was a *concern* of the framers. And as long as it was a concern, a problem existed. "The simplest way of minimizing military power" would be the "maximizing of the power of civilian groups in relation to the military." However, "the maximizing of civilian power always means the maximizing of the power of some particular civilian group."[17] Consequently, control was divided among multiple civilian groups so that no one group or individual would have absolute authority over the military.

The result of this arrangement has been a struggle among the various groups sharing divided control, primarily the president and Congress, for dominance. This struggle is not related solely to the issue of control of the armed forces; rather, it is part of a more general power struggle between the executive and congressional branches resulting from the broad principle of separation of powers and checks and balances. What makes the civilian control issue such an important one is the fact that federal government activities in the field of international relations have become so extensive since the 1940s, and that the military has played such a major role in those activities. This, of course, has had great influence on the size of the military establishment and its impact on the nation's economy and society. The struggle itself is such that, in Huntington's view, "The Chief Executive identifies civilian control with presidential control" while "Congress . . . identifies civilian control with congressional control." However, "both Congress and the president are fundamentally concerned with the distribution of power between executive and legislative rather than between civilian and military."[18]

Huntington's belief in the inherent failings of divided civilian control is not universally accepted. Others, such as Professor Mahoney, claim that military subordination to civilian control is ineffective not because the designed system is faulty, but rather because the appropriate civilian officials fail to exercise the constitutional powers given them! "There is not that much wrong with the present system that faithful adherence to the letter and spirit of civil supremacy will not correct. The power to control is there, but the civilians have failed to use it, and the electorate has not held them accountable for their omission."[19] And, there are still others who contend that civilian control is not ineffective; that it is functioning as it should.

Whatever the point of view, the congressional-executive power struggle aspect of divided civilian control is a fact, and it is this very struggle for power between civilian branches of the government that has made military involvement in politics inevitable. In the years since World War II, the executive branch has gained the upper hand in this tug of war with Congress, and every effort is made by the president and Defense Department officials to maintain the power over the military that has been acquired, as well as to acquire still more. Congress, on the other hand, has resisted this accretion of power by the executive, and, more recently, has made vigorous attempts to restrict the exercise of some of that power (witness the 1973 War Powers Resolution) and to reassert greater congressional control over the military.

As both parties to this struggle have maneuvered for an advantage, the military itself has been drawn into the controversy. Each side seeks to enlist the support and assistance of military leaders, the services, and even the entire profession. Executive branch leaders call upon the Joint Chiefs of Staff and senior commanders to support administration proposals for reorganizing the Defense Department, while congressional committees call upon the service chiefs and their deputies to oppose Defense Department efforts to increase centralized control over the services by the Office of the Secretary of Defense. As a result, the military is compelled to become involved in what is essentially a political power struggle—a power struggle that takes the form of disputes over policy decisions. In Huntington's words, "The separation of powers" between Congress and the president "is a perpetual invitation, if not an irresistible force, drawing military leaders into political conflicts."[20]

NECESSITY OF POLITICAL INVOLVEMENT

Beyond the probable inevitability of political involvement by the military, a strong case can be made for its necessity. Many articles have been written by proponents of the "political-military fusion" school in which multiple and detailed arguments are offered in support of the necessity of this involvement.[21] Rather than catalog the various points made in those papers, it might be useful to focus attention on two fairly broad reasons which demonstrate why involvement by military professionals in decisionmaking or value allocation politics is necessary.

The first reason has to do with decisionmaking itself—the process by which alternatives are examined and assessed, and one or more are chosen as policy to be pursued. To begin with, in the area of national security virtually all policy alternatives available to political decisionmakers today relate in some way to the use, the threatened use, or the condition of military forces. During the decisionmaking process, each of the various competing policy alternatives must be analyzed and appraised to determine suitability (will the alternative accomplish the desired objectives?), feasibility (is the alternative compatible with available resources?), and acceptability (what are the relative costs of the alternative, and are they acceptable?).

This process of analysis and appraisal must, if it is to be meaningful and helpful to the decisionmaker, be accomplished by experts having the knowledge and experience to make realistic judgments of relative merit from among the alternatives being considered. Since virtually all such alternatives in the national security policy field involve some aspect of military force, it follows that experts in the development and application of such force—military professionals—*must* be involved in the decisionmaking process. If they are not included, serious overcommitments of available resources and/or serious policy shortcomings are possible. Therefore, military professionals must be involved in political decisionmaking since, in the words of Lieutenant Colonel William Simons, "military officials are uniquely qualified to describe the resources and costs demanded by available strategic alternatives."[22]

The second reason is closely related to the first. The availability of resources (money, matérial, manpower) has a major, even determining, impact on the national security policies that are selected and implemented by the government. Policy alternatives for which there are inadequate resources available are not really feasible and should not be selected. In effect, they are not even available as options.

In recent years, two trends have developed which greatly affect the resources that are available to the defense establishment. To begin with, many vital resources have become quite scarce (and or costly); the oil embargo and subsequent energy crisis are a prime example. In addition, the competition for these scarce resources has increased dramatically—among nations, among segments of American society, and among the agencies of government. Greater competition for scarcer resources means that it will be harder and harder for the defense establishment to obtain the resources necessary to provide viable policy alternatives and to insure the ability to accomplish assigned missions.

All this means that expert advice must be made available to those officials in both the executive and legislative branches responsible for determining the allocation of resources, and that this advice must be effectively supported and defended. Once again, since military professionals are the only real source of a particular type of expert advice needed—that concerning the development and application of military force—it stands to reason that military professionals must be involved in the political process of resource allocation. In the words of Professor Sarkesian:

The military will find itself in a position in which it will have to defend its policies and programs to a suspicious Congress . . .

Politically astute and technically knowledgeable officers will be a *sine qua non* for the effective representation of the military in external political activities . . . By necessity, the military will have to engage in the "kinds" of politics, and on a scale and intensity, characteristic of politically active interest groups.[23]

Therefore, the question of the necessity of professional military involvement in the political decisionmaking process should be answered with a firm "yes." The two reasons that have just been discussed amply demonstrate the need for military participation in the analysis and appraisal of security policy alternatives, and in the resource allocation process. If professional officers do not, or cannot, participate in such political decisionmaking processes, the national security policy that *is* formulated and implemented runs the risk of being less appropriate and less effective than it should, or could, be.

DEVELOPING POLITICO-MILITARY EXPERTISE TO INSURE "PROPER AND EFFECTIVE" POLITICAL INVOLVEMENT

This, then, brings us back to the two questions asked at the beginning of this paper concerning the nature of the military profession: "What are we to be?" and "What are we to do?" A partial answer to the second question must be that one function of the military profession is participation in the political decisionmaking process as it relates to national security. This is, in all likelihood, an inevitable function, but it is also a necessary and proper function.

What, then, of the first question: "What are we to be?" In answering this question, full recognition must be made of the profession's involvement in the political process. Part of what officers must be as military professionals must necessarily relate to that political involvement. The military must accept, both individually as professionals and institutionally as a profession, the new political dimension of military professionalism. Professional officers must become aware of the proper political role that the military should play—a very definite and limited role, but a legitimate one nonetheless. But much more than this, professional officers must also become knowledgeable in the purposes and procedures of the political process in which they participate. They must know how and why the process operates if they are truly to understand it; and they must understand it if their involvement in the political process is to be effective. The objective of this process is the security of the nation, and it is vital that the military be effective in performing its role toward achieving that objective.

NOTES

At the time of this writing, Colonel Garrison was a student at the Armed Forces Staff College. Printed by permission. This paper represents the views of the author and does not necessarily reflect the official opinion of the Armed Forces Staff College.

1. Samuel P. Huntington, *The Soldier and the State* (New York: Vintage Books, 1964), p. 229.
2. Ibid., pp. 230–32.
3. Ibid., p. 71.
4. Sam C. Sarkesian, "Political Soldiers: Perspectives on Professionalism in the U.S. Military," *Midwest Journal of Political Science* 16, no. 2 (1972): 240.
5. Huntington, *Soldier and State*, pp. 350–51.
6. Sources of quotations are, in order: Rocco M. Paone, "Civil-Military Relations and the Formulation of U.S. Foreign Policy," in Charles L. Cochran, ed.,

Civil-Military Relations (New York: Free Press, 1974), p. 83; Lt. Col. Jack L. Miles, "The Fusion of Military and Political Considerations," *Marine Corps Gazette* 52, no. 8 (1968): 28; Colonel Donald F. Bletz, "Military Professionalism," *Military Review* 51, no. 5 (1971): 12.

7. Sarkesian, "Political Soldiers," p. 239.

8. Ibid., p. 258.

9. Bletz, "Military Professionalism," pp. 16–17.

10. Huntington, *Soldier and State*, pp. 71, 231–32.

11. Morris Janowitz, *The Professional Soldier* (New York: Free Press, 1971), p. 233.

12. David Easton, *The Political System* (New York: Knopf, 1953), p. 134.

13. Elmer J. Mahoney, "The Constitutional Framework of Civil-Military Relations," in Cochran, ed., *Civil-Military Relations*, p. 34.

14. Ibid., p. 35.

15. Huntington, *Soldier and State*, p. 163.

16. Ibid., p. 168.

17. Ibid., p. 80.

18. Ibid., p. 81.

19. Mahoney, "Constitutional Framework," p. 38.

20. Huntington, *Soldier and State*, p. 177.

21. Among some of the better ones are: Lt. Col. William E. Simons, "Military Professionals as Policy Advisers," *Air University Review* 20, no. 3 (1969); Charles Wolf, Jr., "Is United States Foreign Policy Being Militarized?" *Orbis* 14, no. 4 (1971); and the Sarkesian, Bletz, and Miles articles cited above.

22. Simons, "Military Professionals," p. 5.

23. Sarkesian, "Political Soldiers," p. 255.

A NEW LOOK AT THE MILITARY PROFESSION

ZEB B. BRADFORD, JR., and JAMES R. MURPHY

The military establishment today occupies a very prominent and controversial place in American national life. The prominence is readily understandable—billions of the federal budget, more than two million Americans in the armed forces, worldwide defense commitments, and concern for the future. These facts, of course, generate considerable public controversy over national policies, priorities, costs, and alternatives. But more fundamental is the controversy in the minds of many soldiers and the thoughtful public over the very nature and purpose of their profession in America. Is there some truth in the emotion-charged allegation of "a huge, powerful and somewhat autonomous military establishment whose influence reaches into almost every aspect of our national life"? What is the military profession's reason for existing? What is the character of personal military professionalism?

Most of the general assumptions and concepts which define the role of the soldier in our society are inaccurate and misleading. Simply stated, our prevailing notions concerning the military profession rest upon a narrow concept of specialized function. A unique expertise is asserted as the distinguishing characteristic which justifies a place for the professional officer in the councils of government, and gives him a claim to the support and respect of our society. This explanation tends to narrow the perspective of our career officers and fails to provide a basis from which American society can gain a rational understanding of its military establishment.

The officer corps must accept most of the responsibility for these faulty conceptions that dominate the thinking about its basic character, for it has failed to question its own assumptions or to state its own case. The military has been too willing to leave theorizing about the profession of arms to civilian intellectuals, who, although often talented, have failed to grasp its essentials, simply because their viewpoint from outside the military prevents sufficient insight. The result is an artificial conceptualization in terms of conventional social theory which distorts the perspective of both the military and the public, creating confusion and obstruction.

Many Americans find it difficult to explain a large professional military establishment in a democratic society. Most social theory asserts an anomaly created by a group specializing in the techniques of violence and serving an essentially liberal society which deplores force as an affront to man's assumed rational nature. The American historical experience of long periods of refraining from involvement in international power politics, interrupted only by sporadic forays, has reinforced the assumption that the military is at best a necessary but unnatural evil. Indeed, the standing army of yesteryear was a cadre, and postwar demobilizations insured that it remained out of sight and mind. This condition helps to explain an unending search on the part of the American military for self-justification and its willingness to embrace theses of legitimacy advanced by civilian intellectuals in terms of a highly specialized and

Reprinted with minor revision from Army *(February 1969): 58–64. Copyright ©1969 by the Association of the US Army and reproduced by permission.*

unique purpose. This quest for meaning involves establishing a respectable conceptual basis for making claims upon society—in part for resources, but primarily for acceptance.

To deal with the military as an institution, we may start with the definition of the profession given by Samuel P. Huntington in *The Soldier and the State*—perhaps the best-known, most widely accepted, and certainly the most methodically developed conceptualization.

Dr. Huntington states that the military is a profession because it possesses three characteristics common to all generally acknowledged professions and essential to professional status: expertise, responsibility, and corporateness. For the unique expertise of the military, he adopts from Harold Lasswell the "management of violence." This is distinct from the mere application of violence, such as physically firing weapons, for the latter ability gives only technical competence or tradesman status. All activities conducted within the military establishment, Dr. Huntington says, are related to the management of violence. This peculiar expertise is the hallmark of the profession as a whole and distinguishes the professional officer. Furthermore, the military holds a monopoly on this particular expertise. No one else may both possess and apply it.

The second characteristic Dr. Huntington cites is social responsibility. The nature of military expertise imposes an obligation upon the military to execute its function not for selfish ends but only in the service of society. The military profession does not exist for self-interest, profit, or personal motives.

Corporateness, the final characteristic, means that there is a shared sense of organic unity and group consciousness which manifests itself in a particular professional organization. The organization formalizes, applies, and enforces the standards of professional competence. For the individual, membership in the organization is a criterion of professional status: laymen are excluded. In the case of the military, Dr. Huntington designates the officer corps as the professional organization. Not all officers are considered professionals in his view, however, since some lack functional competence in the peculiar military expertise of management of violence. Those only temporarily serving, with no thought of a military career, are only amateurs. Enlisted men, as a group, are considered tradesmen and are outside the professional corps, although many career soldiers may qualify for the higher status—most frequently those in the upper noncommissioned ranks.

Dr. Huntington's model is attractive in its consistency and logic, and it is true that the military does share in some measure the characteristics of other professional groups such as law and medicine. But the analogy is insufficient to describe the military as a profession. The error is due to the attempt to find characteristics in the military which allow it to fall within a conventional definition of profession which is better fitted to other recognized groups. Rather than being defined by its own distinguishing characteristics, the military is interpreted in accordance with a socially standardized definition. This approach leads to the search for a particular expertise upon which the military can peg its professional status.

There are two basic objections to this approach. First, "management of violence" (or similar formulations for the same thing, such as "the ordered application of force to the resolution of a social problem") is insufficient to describe what is actually required of the American military establishment in our contemporary global security commitments. Second, the military profession cannot be defined sufficiently in terms of any functional expertise.

Military expertise is not a constant; it is contingent and relative. Military expertise will vary according to whatever is required of the profession to support the policies of the state. The range of possibilities includes "management of violence," "peacekeeping," "deterrence," "nation-building," "revolutionary development," "civic action," or "pacification." There are many examples of military establishments being required to do all sorts of "nonmilitary" tasks. To name only a few in the American experience, we can point to the construction of the Panama Canal, the building of railroads to the West, the rehabilitation of domestic social groups, conservation of natural resources, work projects for the unemployed, and even polar and space exploration.

Within our defense establishment, the past decade has seen a great transformation in the skills required of the military. The widespread use of systems analysis and our mushrooming technology have created whole new dimensions of military expertise necessary for national security. Quite obviously, any attempt to decide who is a professional, based upon the relationship of his occupational skill to management of violence or combat role, is arbitrary and too restricted. It also can be self-denying, in terms of doing what is required by the country, if the military does not comprehend a broader role and develop the necessary skills as part of professional expertise. Combat expertise, of course, is the single most vital skill of the soldier, and one which is uniquely his to develop and use. Hopefully, however, this would not be to the exclusion of all other—perhaps more desirable or socially productive—employment of the military establishment. Indeed, a military created or existing solely for the purpose of war may be dangerous to the values and goals of a democracy. Furthermore, the facts simply contradict such a narrow concept of the function of the military profession. In times of national security crises, the historical American approach has been to augment a small permanent military cadre with a vast mobilization of "citizen soldiers." The resulting successful management of violence has not seemed to depend upon qualities of expertise exclusive to the "military professionals."

As Morris Janowitz points out in *The Professional Soldier*, there is a narrowing skill differential between military and civilian elites. No particular skill, including management of violence, is restricted to the military profession. To take an extreme case, are not the secretary of defense and his top aides "managers of violence"? Of course they are; indeed, these civilian leaders brought the managerial revolution to the Pentagon, but that does not make them a part of the military profession, as they themselves would be the first to insist. We could say that those outside the uniformed officer corps fail to qualify as professionals, but this reduces professionalism simply to being a member of a group. Yet there are officers who are not considered professionals, and there are those in uniform who are considered professionals but who contribute only indirectly to "managing violence." The military profession is more than a uniformed structure incorporating a functional expertise.

The other two characteristics which Huntington cites—responsibility and corporateness—do exist in the military, and are important. But they lack real meaning in the way he relates them to expertise. In explaining the responsibility of officership, Dr. Huntington says that "the expertise of the officer imposes upon him a special social responsibility." Here again we see the crucial importance placed upon expertise as the essential qualification of the profession. Responsibility is said to be a function of the peculiar military expertise. This is putting the cart before the horse, for responsibility must come first. A more accurate wording would turn the phrase around. "The special social responsibility of the officer requires of him an expertise." As explained earlier, the varying demands of state service may prescribe that the officer possess skills unrelated to "management of violence." Responsibility is more than a means of insuring that the military exercise its expertise in the service of the state. Far more essential to military professionalism is an internalized *sense* of responsibility, of allegiance to duly constituted authority.

The third characteristic—corporateness—is, as Dr. Huntington states, a mark of the military profession and helps to describe it. Yet the significance of corporateness is missed, due to two errors. One is a faulty interpretation of the role

of management of violence in defining the rank structure. The other is a failure to emphasize the primary role of the professional organization as that of institutionalizing essential values and formalizing these legally.

The first point is expressed in this way by Dr. Huntington: "The larger and more complex the organization of violence which an officer is capable of directing . . . the higher is his professional competence." We come full circle and are contradicted once more by the example of the secretary of defense, who is *not* a member of the military profession. More disturbing is the focus on the size and complexity of the organization directed as the measure of professional competence. Degrees of competence, of course, are reflected and rewarded by rank; and the higher the rank, the higher the position in any corporate hierarchy. But the forms of competence are so diverse that the corporate structure becomes a way of organizing large numbers of people with many different functions for a common purpose, as in any other bureaucracy. Beyond certain minimum standards of competence, in the military service, seniority and career-development patterns largely determine the rise through echelons of command. Certainly the size of the organization does not define personal professional competence.

It is evident that the concept of the military profession described by Dr. Huntington rests fundamentally upon the attribution of a particular expertise, and that other characteristics, such as responsibility and corporateness, are defined in terms of this expertise. Without a unique skill the edifice falls.

Huntington's model has been discussed at some length because it has come to dominate thinking about the military and provides a widely used vocabulary. The military itself has attempted to rationalize its position and to solicit support in Dr. Huntington's terms—terms which are inadequate. For example, Brig. General Robert N. Ginsburgh wrote in *Foreign Affairs*:

The maintenance of a high degree of military professionalism is essential to the preservation of our nation's security without a sacrifice of basic American values. The challenge to military professionalism is reflected in each of what Samuel P. Huntington calls the essential characteristics of a profession: corporateness, responsibility and—especially—expertise . . .

The challenge to military expertise is the most important aspect of the challenge to military professionalism, because expertise is, after all, the very basis of any profession.

General Ginsburgh then went on to state that in order to salvage the profession we must achieve "the abandonment of the fusionist theory whereby military and nonmilitary factors are so entwined that a separate expertise in the military aspects of national security is simply impossible."

This is a self-defeating split. If the military insists on such an artificial compartmentalization of security policy issues to defend some hypothetical military preserve, the profession does not deserve to be taken seriously. One thing has been clearly established in recent years: there is no "purely military" sphere in security policy. The military has a responsibility for participation in both the formulation and the execution of policy. At the same time, there is a continuing effort to integrate at all levels, in both formulation and execution, every aspect of policy—economic, political, and military. If he is to be termed a professional and entitled to a place in the councils of government, the soldier's horizons must be broad enough to encompass all of these factors as they apply to national security policy.

The narrow concept of the profession described by Dr. Huntington and exemplified by General Ginsburgh fails to encourage the career officer to develop his knowledge in fields such as economics and politics, which give meaning and purpose to the use of military power. This restricted approach is illustrated by the statement of retired Brig. General Lynn D. Smith in *Army Digest*:

As a professional soldier, you must understand the difference between national strategy and military strategy. Military strategy is defined in the *Dictionary for Joint Usage* as: "The art and science of employing the armed forces of a nation to secure the objectives of national policy by the application of force, or the threat of force."

If you master this art and science, you will earn all the stars and decorations your country can bestow. You will be so occupied that you will not have time to concern yourself with the debates on the fine points of the political, economic, and psychological aspects of national strategy.

If the military's own conception of the profession leads to the kind of narrowness evidenced above, we will forfeit our responsibility. National security policy is directed at protecting the essential values of our nation. If the US military is to contribute fully to that purpose, it must have the ability and the inclination to relate the objectives of military power and alternative means of achieving these objectives to other aspects of public policy.

If the analysis is correct thus far, it is clear that Dr. Huntington has failed to describe the truly unique and distinguishing characteristics of the military profession. The military is not a profession in the way that certain other groups are, such as law or medicine. The term has to mean something different in the military case. Expertise, corporateness, responsibility—all are applicable in describing the functioning and organization of the military institution, but they do not of themselves define the distinctive quality of professionalism in the military context. We may continue to refer to the military as a profession only if it can be established that the military is a unique social group, and that its uniqueness gives content to the meaning of "profession."

The military profession can be properly defined only in terms of *both* its purpose and the conditions placed upon the fulfillment of that purpose. The military exists only for the service of the state, regardless of the skills required or the functions performed. As a profession, the military does not condition this commitment, for, in the words of Lt. General Sir John Hackett, a distinguished British soldier-scholar, the contract for service includes an "unlimited liability clause."

The military's obligation of unconditional service to the lawful authority of the state is unique. From time to time, changes in the nature of expertise are required for this service. There may even be changes in the meaning of national security itself when viewed in terms of policies and programs. But these do not alter the basic character of the military profession. Many people outside the profession may have a self-imposed commitment to unconditional service to the state, but only the military possesses the obligation *collectively* as a defining characteristic. Certainly, in this respect, it is far different from any other profession.

A military establishment cannot confer upon itself professional status. This is the prerogative of the state. The status of the profession is bound irrevocably to that of the governing authority which it serves. When, for any reason, a military group challenges the governing authority of the state, it loses its professional character and becomes an armed political force. As history illustrates through numerous examples, a professional military organization goes through a severe crisis when, for one reason or another, it cannot identify either an effective governing authority or one to which it can concede legitimacy.

Crucial to the character of a professional military group is the existence of a lawful and effective state authority to which the military owes allegiance. A professional military is impossible without such authority. It is pointless to attempt to define abstractly the conditions under which a military group should renounce its professional status. One's own values will determine how the legitimacy of authority should be judged. As a *profession*, however, if that term is to be grounded in reality, a national military establishment must share the destiny and moral stature of its governing master. No state can deny the morality of its own policies or its own legitimacy. The military profession, therefore, can be defined only in terms of its unique, unconditional obligation to serve the lawful authority of the state. It will develop whatever expertise is required to fulfill its unlimited contract for public service.

Up to this point our analysis has dealt with the meaning of the military establishment as a profession, and has deferred a discussion of the meaning of professionalism in individual terms. Obviously, the line between the two is not distinct. The nature of the profession dictates the basis of personal professionalism. If the profession is to survive on the basis defined above, it will do so because of the personal commitment of the professional soldiers. The basis for this commitment must now be studied.

What *is* military professionalism? Unfortunately, here also a substantial mass of undergrowth must be cleared away before the forest becomes visible. It is common but inaccurate to conceive of military professionalism in terms of beliefs which a person must hold in order to be a professional soldier. Dr. Huntington offers an example of this tendency. In his view, the "military ethic" is "a constant standard by which it is possible to judge the professionalization of any officer corps anywhere, any time." Even if a particular ethic is not a prerequisite to professional status, it is nevertheless assumed that the typical officer will have a predictable set of social and political concepts, thus stereotyping the "military man."

The professional ethic, according to Dr. Huntington, "emphasizes the permanence, irrationality, weakness and evil in human nature . . . is pessimistic, collectivist, historically inclined, power oriented, nationalistic, pacifistic and instrumentalist in its view of the military profession. It is in brief, realistic and conservative." This sort of standard ideology is incorrect at best and can be dangerous in its support of stereotyped thinking by and about the military. Yet the possession of this view of man and society is widely assumed to be both true and necessary. The root of the problem here lies in setting down as reality what would seem in the abstract to be a compatible outlook for a warrior, and one which would theoretically support the role of a fighting group as a social organism. This error is in part caused by the incorrect explanation of the profession as resting on functions of violence. But neither careful study nor a walk through the Pentagon supports an argument that this set of norms is required for military professionalism.

In the first place, this set of beliefs simply does not pervade the officer corps with anything resembling universality. Morris Janowitz, in his sociological study, *The Professional Soldier*, concludes that, actually, "the political beliefs of the military are not distinct from those that operate in civilian society." This reflects the diverse origins and representative nature of the officer corps. The US officer corps has neither social class nor dynastic orgins. Coming randomly from across the range of a pluralistic society, it has a pluralism in attitudes within its ranks.

Even if the argument against a "professional ethic" is accepted, it is generally assumed that the military professional must be unique on one point at least: his acceptance of the use of force in the pursuit of national objectives makes him an inherently different type of person from his average fellow citizen.

This is incorrect. If the soldier is apparently willing to solve problems by military means, it is only when so ordered by competent authority. He has been taught to see the military implications of problems and is charged with the responsibility of providing a military instrument to civilian policymakers. But, even given this, there is no sharp line dividing the soldier from society at large. There are "hawks" and "doves" on all branches of the social tree. As history and contemporary affairs amply illustrate, both the soldier and his civilian fellow share a willingness, although sometimes reluctant, to bear and use arms in common cause. There are higher values which even the dedicated liberal holds which justify to him the use of force. Civilians and professional soldiers alike have found themselves in the service of the state in all of our wars. While the soldier may find it easier to resort to force than some of his civilian fellows, in most cases it is only a matter of degree. It is not a matter of values, but one of acquired and required perspective. As a product and dedicated servant of his society, the professional officer shares the core values of the nation. He is not a mercenary.

It must be realized that one can be a professional officer while holding any number of political and social beliefs, some of which contradict "conservative realism." As the tasks of the profession are dictated by the state, so are the imperatives of professionalism. The officer may believe what he likes and may view the world as he chooses, so long as he can find it within himself to serve the state on its terms. And there are many diverse motivations. One may serve because he wants glory, loves patriotism, seeks social advancement, or needs the work. One may fight the enemy with bloodthirsty joy or sadness in his heart. All that matters is that he do it well enough to fulfill his obligations. Liberal and conservative, idealist and realist, can and do share the profession of arms. A fixed and elaborate set of social beliefs is a false basis upon which to define professionalism. Archetypes are abstractions: soldiers are not.

Much effort has been expended thus far in attempting to demolish false notions about a military ethic. There is a hard core of truth, however, which does serve as a common denominator. If a particular ethic is not basic to professionalism, there must be something else. What is basic is related to the unlimited obligation of the military as a corporate professional community and functional organism. What is it? In individual terms it has often been termed personal commitment. But a better way of phrasing it is a "sense of duty."

Commitment implies less than duty. Whereas commitment may indicate what one must do in terms of a consciously made obligation, duty has personal moral value content. A sense of duty is a feeling of what one *ought* to do and must do in terms of one's values. Robert E. Lee called duty "the sublimest word in the English language." Here we come again to a point made earlier concerning the nature of a profession: that a sense of belonging and a professional organization are required. It was argued that a corporate unconditional obligation distinguishes the military. In more subjective terms, it can be called a "collective sense of duty."

The officer cannot be a respected member of his profession without subscribing to the operating norms of his professional community. These norms are in fact a necessity for the success of the group in fulfilling its tasks. Without a collective sense of duty the military could not function and certainly could not be trusted. Military professionals must share a sense of duty to the nation. The professional officer must be an unconditional servant of state policy; he must have a deep normative sense of duty to do this. The rigorous demands made upon the profession by this sense of duty, and the tasks required of it, explain the premium placed upon other "soldierly" qualities. One cannot do his duty unless he has courage, selflessness, and integrity. The military profession must have these group values as a functional necessity.

A sense of duty is necessary, but not sufficient, for professional status. The person must also have competence to perform the service required to fulfill his obligations. As described earlier, this may require one or more of a number of skills. Finally, he must be a member of the officer corps of the armed forces. By joining the officer corps he makes his professional commitment and adopts the community values as his own.

Apart from belonging to the officer corps, professionalism is, then, a status determined jointly by the officer and his government. Neither the state nor the officer corps will grant professional standing to the man who lacks the necessary competence or who will not agree to make a commitment to duty which on the part of the state is assumed to be unconditional. The unconditional quality of this commitment is sig-

nified by the career length and a life of selfless sacrifice, ranging from Melville Goodwin's "genteel poverty" to the Gettysburg "last full measure of devotion."

Professionalism thus has both *objective* and *subjective* content. It is objective in that professional status is granted by the state if certain performance criteria are met by the officer. It is subjective in that the officer must feel a sense of duty to serve the lawful government "for the full distance," even at the risk of his life. Mentally, he does not condition this obligation.

This analysis has attempted to deal with the difficult problems of the ideas which provide a frame of reference for the military professional in our society. Narrow concepts of the meaning of professionalism, especially those which are dependent on an exclusive expertise, have been rejected as inadequate to describe the uniqueness of the US military profession. We have merely stated what should be obvious: that the military profession is an unconditional servant of lawful state authority and that its collective sense of duty makes the role possible. A particular skill is not basic. Also rejected is any rigid pattern of social and political beliefs as being necessary for a professional ethic for the individual. Only a sense of duty which somehow justifies sacrifice to the officer himself is required.

Some officers may feel that the denial of a professional expertise akin to that of law or medicine is a self-inflicted wound. We would urge, on the contrary, that acceptance of what we have tried to argue as the real meaning of professionalism will enable officers to accept a more demanding role in national security policy. As a profession of unlimited service, no considerations are taboo in formulating military advice, and no skill which is necessary to the nation conflicts with the professional status of the officer.

This analysis should not be misinterpreted as a suggestion for an expanded mandate for the military. What is suggested is that we recognize the profession for what it is—a profession of unconditional service to the nation, but one engaged in a multitude of tasks. For more than two decades the military has been required to participate in the formulation of national security policy and to assist in executing it. This has required broad knowledge and competence in many diverse branches of public affairs. False notions concerning the meaning of military professionalism must not be allowed to prevent the officer corps from equipping itself mentally for its required duty. The military professional has no vested interest save that of the health and security of the nation itself. We must not forfeit this trust by adopting a self-conception which unwittingly betrays our duty.

EPILOGUE
Forthcoming Challenges in American Defense Policy

DOUGLAS J. MURRAY

The task of this concluding chapter is to consider the future, not for the purpose of prediction, but for identification. The goal is to identify the major challenges that American defense policy decisionmakers must address as this nation approaches the twenty-first century. Specifically in the next few pages we look at challenges arising from: the changes in the international system and in the threat to US interests; the strategy developed to meet these changes; the US alliance structure; arms control regimes; technology and man-machine relationships; resource constraints; and decisionmaking organizations and process.

In the first chapter of this book, the editors point to the reality that Samuel Huntington highlighted many years ago, namely, that defense policy is Janus-like in that it exists in both the worlds of international politics and domestic politics. The international system is both the principal source of opportunity for the nation-state to achieve its national objectives and of the threat to which it must respond. The threat to a nation's security is frequently generated in the international political-economic environment and responded to in the domestic or national one. The basis for national security arises, after all, from the security dilemma in which all states find themselves and which occurs because of the absence of any authority—or authoritative decisionmaking body—over the states in order to secure order among the states. The defense policy of any state—including the United States—is thus set both within the international system of interacting states and within the domestic, political, economic, and social systems of each individual nation-state.

This conceptualization of defense policy has utility for organizing and discussing the challenges listed above. In the first section of this chapter, the focus is on issues within the international dimension of defense policy, and in the second, the focus falls on domestic considerations.

THE INTERNATIONAL DIMENSION

THE NATURE OF THE SYSTEM

As analyzed in Part I of this book, the international environment provides part of the context of American defense policy. Changes in that environment pose the first set of challenges to American defense policymakers. These changes are likely to occur in one of five areas: the boundaries of the current international political system; the nature of the actors in the system; the structure of the system; the level of interaction among the states; and the rules of the system.[1]

During the twentieth century, the international political system has taken on a global character in which states and regional subsystems have become increasingly interdependent. Nation-states have been little influenced, however, by events beyond the earth. Nonetheless, the dawn of the space age beginning in the 1950s and 1960s has ushered in a period of transition in which the boundaries of the system are changing. By the next century, the boundaries of the current international system will extend beyond this planet, and the use of space will provide both a challenge and an opportunity for the United States. It is not clear at this point what the opportunities and risks of space are, but certainly the challenges will be of a greater magnitude than those which this nation faced as it encountered its territorial frontiers in the eighteenth and nineteenth centuries. Already space has become no longer simply a place to explore but one in which to work. But for what purpose and in what way? In a generation raised with a polemic involving "Star Wars" and the potential militarization of space, one forgets that the exploration of the New World five centuries ago provided a critical release valve for nationalistic competition, even as it provided a basis for national competition that often took a violent form. Perhaps expanding the existing boundaries of the current international political system to space offers the same opportunity.

Current official US space policy is based upon the principle of free access to and the peaceful use and development of space. A vital challenge for American defense policy, of course, is to secure these goals. The difficulty is that we have not decided how to do that, in part because we have failed to think of space in terms other than as a mere extension of the earth. As former Secretary of Defense Weinberger pointed out, "Space is recognized as being a medium within which the conduct of military operations in support of our national security can take place, just as on land, at sea, and in the atmosphere, and similarly from which military space functions of space support, force enhancement, space control, and force application can be performed."[2] Certainly in the years ahead, the nation must be concerned with how it will address these needs. Current defense plans call for a new space launch recovery program encompassing expendable and manned launch vehicles such as the National Aerospace Plane, a new emphasis on the commercialization of space, new technology initiatives such as the Strategic Defense Initiative, and an antisatellite system. But that is not enough.

If this nation is to meet the security challenges associated with an expanding international political system to include space, it must develop an all-encompassing strategy for space. The challenge to our service academies and universities is to produce great strategists that will develop for space what Mackinder did for land, Mahon for the sea and Douhet and Mitchell did for the air. Associated with this new way to see, study, and use space is the need to develop new organizational structures, both at the international and national level. The creation of a unified space command in 1985 is but the first step at the national level. The challenge is to internationalize it. Today just a few nations are in space. In the next century, few will not be there. Thus one of the challenges posed by the changing boundaries of the international political system will be to structure a regime that encompasses the international use of space.

Similar challenges result from changes in the nature of the actors in this system. During the first 200 years of America's existence, the major, if not only, actor in the international system has been the nation-state. It alone could best provide for the security and well-being of the people who had organized themselves within it. While the nation-state has retained its dominance in the post-1945 era, significant changes in both the number and type of international actors have created challenges for American security.

Owing largely to decolonization, the number of states has grown substantially in the past four decades. The US has been faced with a growing challenge as these new states claim a position in the international system. This challenge posed by the so-called Third World is reflected in the issues associated with the North-South dialogue and catalogued in an agenda of items called the New International Economic Order. These nations demand political and economic independence and a right to the commerce, trade, and wealth of the world. They demand a voice in the direction of the international environment and their role in it. The challenge for US policy makers is to address these needs, while ensuring the viability of the existing world economy. Failure to do so will result in continued and greater international instability, as evidenced by the fact that most current armed conflicts involve these Third World or developing nations. One need only think of US involvement in Central America and the Persian Gulf to understand the nature of the issue that faces the nation in the future. The proclaimed root causes of international terrorism are certainly found in the Third World.

Accompanying an increase in the number of states, there has also been an increase in the types of actors in the international system. On the one hand, these include international governmental organizations composed of state members like the United Nations or regional organizations like OPEC or NATO. These have been formed by states to achieve ends which a state, acting alone, could not achieve. Examples include regulation of the seas, air pollution, control of air traffic, and health issues. In addition, the role of nongovernmental or non-state actors has also increased. Also significant has been the growth of multinational corporations whose political as well as economic power rivals many of the Third World states discussed above. To these must be added a proliferation of other non-state and private organizations, most significant of which, from a security perspective, are international terrorist organizations. This proliferation of actors makes the international system more complex, more difficult to manage, and more risky. The challenge is not that these new types of actors will result in the devolution of the state, but that state governments must accommodate the demands of these various actors. The impact of the oil embargo by the OPEC nations in the 1970s or of terrorists in the 1980s demonstrates both what these actors can do and why their interests must be addressed.

The current structure of the international political system is an intricate and multitiered web of power relationships dominated, particularly in military matters, by the United States and the Soviet Union, but with significant roles, particularly in economic matters, played by the

People's Republic of China, Western Europe, and Japan. In early 1988, the private commission on Integrated Long Term Strategy pointed out, "By 2010, China and Japan will have the economic capacity to act as major world powers. Unless 'restructuring' produces startling new gains, the Soviet Union's share of the world economy will shrink."[3] While not all analysts would agree with this prediction, they do agree that the structure—the power relationships— in the next century will be different than they are today. Certainly Japan will not only be the nation with the highest GNP; it may also have a greater military role. China's place is less certain, but its role will be more significant. Similarly, the major alterations in the political and economic systems of Eastern Europe in 1989 and 1990, and changes within the Warsaw Pact, suggest a new role for a possibly restructured Europe in the international system. Further, beyond these major power centers, there are numerous and diverse other smaller centers of power, regional and economic-based. About these the Commission states, "The GNPs of middle regional powers like India and Korea are likely to grow substantially relative to those of Western Europe."[4]

Some political scientists refer to this power structure as multipolar or polycentric. The relationship is held together by growing political, economic, and social interdependence in which states interact on every level of human endeavor. It is unlikely that these trends will be reversed. As a result, American defense policy decisionmakers will have to deal with an increasingly more complex set of interrelationships, each of which has the potential to disrupt international stability and threaten US security and other national interests. The continuing challenge for United States policy will be to deal with these changing power relationships as they affect the pursuit of US policy goals.

One particularly difficult problem that could alter significantly the current structure of the international political system is nuclear proliferation and what the Commission on Integrated Long Term Strategy called "the worldwide diffusion of advanced weapons." The commission report states:

The relationship between the major and minor powers will change by the early 21st century. Today the United States and the Soviet Union can often decisively influence the military postures of smaller states by making weaponry available or denying it. In the years ahead, weapons production will be much more widely diffused, and the superpowers . . . will have less control over transfers of advanced systems. Many lesser powers will have sizeable arsenals.[5]

Despite the Nuclear Nonproliferation Treaty of 1968, the number of nations that have or will have the potential to develop nuclear weapons is increasing. It is estimated that by the next century more than 40 nations, including some that are at present relatively poor and underdeveloped will have the ability to build nuclear weapons.[6] Many argue that it is only a matter of time before some international terrorist group possesses the bomb. Some analysts maintain that a proliferation of nuclear weapons does not guarantee their use; others, in fact, argue that such a situation would enhance global stability. Nonetheless, most believe that a world in which nuclear weapons are owned by many nations is inherently unstable and dangerous.[7] The challenge is how to control such proliferation. Certainly one nation alone cannot do it. A possible solution is the creation of an arms control regime that will enforce a set of rules and regulations to govern the development and use of nuclear energy. As discussed earlier in this work, past efforts to create such a regime have been incomplete.

Constructing such a regime—or any of the rules that govern the international political system—must still deal with the reality of national sovereignty. As noted in chapter 1, the international system is anarchic, but this does not mean that it lacks order. International law, public and private, is an effective constraint on the system, at least as long as the parties to a dispute agree to its jurisdiction. The likelihood is that this need will not change. In fact, as states become increasingly interdependent, the requirement for more laws and regulations to govern their conduct for their own mutual benefit will increase. The future holds the possibility of creating more comprehensive regimes like the existing ones dealing with the sea, space, and Antarctica. The US should encourage the development of such regimes, since, to the extent that they establish a system of rules and conduct, they reduce uncertainty and the potential for disputes. The challenge for US decisionmakers, however, is to mold such regimes so that US national security interests are not compromised. No place is this more important than in the area of arms control.

The last 30 years have witnessed a maturation in the way nations deal with arms control. From a utopian fixation with total disarmament in the periods immediately following both world wars, governments have realized that arms control must be a realistic and integral part of a nation's national security policy. To quote from the 1987 official US Strategy Statement:

Arms control is not an end in itself, but an integral part of our overall National Security Strategy. It must be viewed as only one of several tools to enhance our national security and to promote our fundamental national interest in the survival of the United States as a free and independent nation. Our arms control objectives are fully integrated with our defense and foreign policies to enhance deterrence, reduce risk, support alliance relationships, and ensure the Soviets do not gain significant advantage over the United States.[8]

The question is not whether this nation should pursue arms control, but, as Henry Kissinger asks, "to what kind of arms control should the Administration commit itself?"[9] The challenge for the US, then, is to develop an arms control regime with the Soviet Union that realizes the US objectives listed above while meeting the fourfold purpose of any effective arms control agreement.[10] Developments in strategy, weapons technology, procurement, force posture, and arms control must not only be coordinated with one another, but must be considered as a single, integrated entity. Decisionmakers must be able to deal with such questions as the impact of a ban on all ICBMs on deterrence strategy, on US strategic force posture, and on conventional force levels and strategy. Further, while the focus in arms control originally was on nuclear weapons, the initiatives coming out of the Conventional Armed Forces in Europe (CFE) talks in Vienna offer promise to create new conventional arms control regimes and a significant alteration of the security equation in Europe. The challenge is to examine more effectively the relationship between nuclear and conventional deterrence and how arms control agreements in one domain impact on the other. As Henry Kissinger concluded, "We face two related tasks. First, arms control requires not so much a new proposal as a fresh concept. Second, it must become an organic part of defence policy."[11]

THE THREAT

Each of the changes in the nature of the international system outlined above will continue to pose significant challenges to US decisionmakers in the years ahead. They will also have an important impact on the nature of the threat to US security. In particular, it is this second set of challenges arising from the international environment to which US defense policy must directly respond.

The presidential statement on national security strategy in 1987 stated:

The most significant threat to U.S. security and national interests is the global challenge posed by the Soviet Union . . . Motivated by the demands of a political system held together and dominated by Marxist-Leninist ideology and the political party which represents it, Moscow seeks to alter the existing international system and establish Soviet global hegemony. These long-range Soviet objectives constitute the overall concerted framework of Soviet foreign and defense policy.[12]

Whether the reforms and arms control initiatives of Mikhail Gorbachev's and the significant changes in Eastern Europe which he initiated will significantly alter, or even negate, this threat assessment is yet unclear. Certainly one of the challenges facing US policymakers is how to respond to these Soviet initiatives. What seems more certain is that the form of the threat will change. For example, renewed independence by the nations of Eastern Europe may reawaken old nationalistic tendencies and animosities that could disrupt 40 years of stability in Europe. Similarly, changes in the structure of the international system, changes in Soviet strategy, or new technologies will impact on the US threat assessment and how to respond to it. Two manifestations are the change in the nature of the Soviet threat to the North American continent as the result of new technologies and, second, the challenges to US interests in the Third World as the result of state-supported terrorism.

In the four decades since the end of World War II, the Soviet threat to the North American continent has taken many different forms. As that threat has changed, so has the US response. Originally, throughout most of the 1950s, the threat was a thermonuclear attack delivered by Soviet long-range aviation bombers. It was in response to this threat that the current North American defense architecture, reflected in the North American Aerospace Defense Command and its assets, was constructed in the 1950s. This entire structure was created to protect Americans from nuclear gravity bombs delivered by Soviet long-range aviation bombers. In effect, NORAD existed to protect US retaliatory forces on which the US deterrent was based. In the 1960s, however, the Soviet threat changed. The introduction of ICBMs and SLBMs in the 1960s revolutionized deterrence and, with it, the nature of the US response in terms of the NORAD defense system. While there had been a defense against bombers, there was no effective defense against the ICBM. At best, the US would have enough warning time in order to launch its own missiles. Nevertheless, such an early warning capability enhanced missile and aircraft survivability and maintained the credibility of the US nuclear deterrent. The US thus undertook the development of an extensive network of missile warning and surveillance systems to include the Ballistic Missile Early Warning System (BMEMS), a Satellite Early Warning System operating Defense Support Program (DSP) satellites in synchronous orbit, and a space Detection and Tracking System (SPADATS).

Today, ballistic missiles still pose the principal nuclear threat to North American security. While this will likely remain the case for many years, two evolving characteristics of the Soviet strategic challenge will complicate the US defense problem. The first change is an entirely new type of threat, based on the Soviet use of space for both defensive and offensive purposes. The second change is the return of an earlier threat—an air-breathing threat—involving both bombers and cruise missiles.

The Soviet space challenge has two aspects: the use of space as a medium to defend against ballistic missiles and as a medium for other military operations. In both areas, the Soviets have made significant progress. While modernizing and improving the only operational antiballistic missile (ABM) system in the world, they have conducted extensive ABM research that rivals if not surpasses the United States' SDI efforts. It is estimated that the Soviets are spending over one billion dollars a year on laser, particle beam, and kinetic energy weapons to destroy ballistic missiles. The Soviet Union also has the only operational antisatellite system (ASAT). While the Soviet ASAT has limited capabilities, its development clearly indicates the direction in which the Soviets are moving with respect to the military uses of space. The Soviets have a robust space program launching 4 to 5 times as many satellites as the United States, and more than 80 percent of these are military in nature. They are developing three new space boosters, a heavy-lift space booster, and their version of the American shuttle caled the *Snowstorm*. Soviet manned space operations dwarf those of the United States. Since 1977, Soviet cosmonauts have logged more than eight man-years compared to the United States' one man year. Soviet cosmonauts have developed an impressive on-orbit repair capability—restoring a *SALYUT* space station that the United States had earlier declared nonoperational.

The Soviet space program, while perhaps posing no significant present threat to North American security, may, if successfully developed and deployed, put at risk the US offensive nuclear forces, jeopardizing the stability of the strategic nuclear relationship. As a former NORAD commander noted, the Soviet space program clearly poses new challenges for US defense policy:

I remind you, and you must remind others, that our country is involved in space for research, exploration, and the peaceful missions of surveillance, navigation, and communications. But we cannot turn this clock back nor put the genie of Soviet space or terrestrial military capabilities back into a bottle. Our peaceful use of space is increasingly threatened by Soviet military actions. Our security demands that we take the steps necessary to counter this potential threat. Remember also that this is a long-lead business. We must act early to meet future needs.[13]

A Soviet analyst was more explicit when he said, "Whoever can seize control of space—that main arena of future wars—will be able to change the correlation of forces so decisively that it will be tantamount to establishing world supremacy."[14]

The second change in the Soviet threat is the revival of the air-breathing challenge to North American security. This threat takes two forms: the development of new Soviet long-range bombers and cruise missiles. The Soviets are presently engaged in upgrading their long-range bomber force with the deployment of a totally new version of the Bear bomber—the mainstay of the bomber fleet. The *Bear H* is the first new production model of this aircraft in 15 years and is capable of carrying the Soviet AS-15 long-range air cruise missile (ALCM). Equipped with ALCM, the *Bear H* gives the Soviets a standoff capability to hit targets throughout North America. The Soviets are also configuring some of their older Bear aircraft (designated Bear Gs) to carry a new, supersonic air-to-surface missile, the AS-4. In addition, the Soviets continue to flight-test the *Blackjack*, a supersonic bomber similar to but larger than the US B-1B. The *Blackjack* will carry gravity bombs, cruise missiles, and a mix of both. The Soviets also continue to develop an air-refueling capability.

The Soviet AS-15 cruise missile is a subsonic, low-altitude missile with a range of about 3000 kilometers, not unlike the United States' *Tomahawk*. In addition the Soviets have a sea-launched version (SLCM), the SS-NX-21, that can be launched from Soviet submarines submerged off North American coasts. The Soviet Union is also developing a larger cruise missile with a range of 2000 nautical miles, called the SS-NX-24, that is planned to be used in both sea-launched and land-launched configurations.

The impact of Soviet bomber and cruise missiles enhancements on North American security is uncertain. Some would argue that these advancements give the Soviet Union the capability to conduct a "precursor strike." In such an attack, the Soviets would use both Bear H and Blackjack bombers equipped with advanced air-to-surface missiles and ALCM, in conjunction with SLCMs launched from Soviet submarines stationed off the mainland, to strike at the "eyes and ears" of the NORAD warning system. Such an attack could be either nuclear or conventional.

The fact is that, today, the United States does not have the capability to detect such a launch. In fact, we probably would not know about the existence of a low- and slow-flying SLCM coming in from the Atlantic Ocean until it detonated on target. The question, however, is whether the Soviet Union would risk such an attack, since such an attack would likely precipitate a US launch of its missiles, effectively pinning Soviet forces. Others argue that it is unrealistic to assume that the Soviet Union would opt for the risk of using its bombers, ALCMs and SLCMs when it could effectively use its superiority in ballistic missiles to destroy our retaliatory capability.

At present, no one knows how the Soviets

intend to use their bombers and cruise missiles and some would argue that the Gorbachev reforms discussed earlier make the question academic. However, we do know that they now have a capability that they did not have prior to 1984 and a capability that certainly adds to the flexibility of their threat and compounds US defense efforts. If a START agreement succeeds in reducing significantly both US and Soviet ballistic missile arsenals, it seems clear that such air-breathing capabilities will take on increasing strategic importance.

At virtually the opposite end of the conflict spectrum is the problem posed by the rise of terrorism and the role of military force as an appropriate response. Former Secretary of Defense Weinberger has called terrorism "one of the most vexing problems facing Americans and the world community today."[15] In a 12-year period, 1973 to 1985, there were more than 6500 terrorist incidents of which 150 were aimed at Americans resulting in 405 deaths. Terrorist incidents have risen at the rate of 12 to 15 percent a year.[16] One need only remember terrorist attacks on the US embassy and Marine barracks in Lebanon to characterize the nature of the problem. The Long commission, convened to look into the Marine barracks bombing, concluded: "The international terrorist acts endemic to the Middle East are indicative of an alarming worldwide phenomenon posing an increasing threat to US personnel and facilities."[17] State-sponsored terrorism directly challenges US interests worldwide, but the question of how to deal with such acts remains unresolved. The Long commission could only conclude that "an adequate response to this increasing threat requires an active national policy that seeks to deter attack or reduce its effectiveness."[18]

At issue is whether military action, the use of military forces such as in the Libyan raid on 14 April 1986, is an appropriate and effective response to terrorism. In this respect, Weinberger concluded, "Political and economic actions are all the more effective when the terrorist state understands clearly that behind these other measures stands effective military power, capable of an appropriate and timely response."[19] To do this, however, the US must have an effective defense organization to carry out the antiterrorist mission. At question, as described earlier in chapter 8, is whether the United States does. The US record to date has been at best mixed.

This section has surveyed changes in the international environment, both with respect to the nature of the international political system and the threat to US security interests. Clearly, new weapons technologies or new power arrangements or even new actors, such as terrorists, in the international system alter the nature of the threat to US security. American defense policy must be prepared to respond appropriately and effectively to these threats. How it does so—in terms of national response—is the subject of the next section.

THE NATIONAL/DOMESTIC DIMENSION

Having briefly looked at a set of issues arising out of the international environment, the focus now shifts to the domestic face of defense policy. This section examines a series of issues arising within US and other domestic political and economic systems that also pose challenges for American defense policy in the years ahead. These issues include: shifts in the US strategy designed to meet the threats discussed above; alterations in America's alliance posture; new technologies and man-machine relationships; and the structure, organization, and process of defense decisionmaking.

A former secretary of the Air Force used to comment that, while the old adage in football states that the best defense is a good offense, in terms of nuclear weapons, the only defense is a good offense. Since the beginning of the nuclear era, US nuclear strategy has been based on deterrence—reliance on nuclear weapons to deter war rather than fight war. Over the past few decades, that deterrent strategy has been manifested in a variety of defense policies, such as massive retaliation, flexible response, assured destruction, limited response options, realistic deterrence, mutual assured destruction, and the countervailing strategy. As discussed earlier in this book, the formation of these policies has reflected a debate about whether nuclear weapons should be seen as highly destructive, punitive weapons ("deterrence by punishment") or instruments which can be used in a controlled and perhaps limited way to fight a war ("deterrence by denial").[20] Regardless of this debate, however, nuclear deterrence has typically been viewed as a question of offensive retaliatory capability. Both schools have focused on the need to protect that retaliatory or second-strike capability. While they differ on *what* should be targeted—and therefore the kinds of weapons one should have—both schools had accepted the notion that deterrence depended on credible threats of offensive use rather than credible defensive capabilities.

As a practical matter, US targeting policy has reflected both schools of thought. Strategy has changed over the years, but in an evolutionary rather than revolutionary way, exhibiting what this author calles the "mausoleum" approach. We never reject a strategy; we keep it, alter it, and add to it, but never completely discard it. In part this evolution reflects changing threat perceptions (and projections). It also depends on what technology will allow. Over the past several years, there has been a tendency to strengthen the US capability to wage war with nuclear weapons, largely because of increasing weapons accuracies, improved delivery systems, and better command, control, communications, and intelligence (C³I). At the same time, attention has focused on the possibilities of strategic defense, and the traditional debate between "deterrence by punishment" and "deterrence by denial" has again been joined over the utility of strategic defenses.

In that respect, the Strategic Defense Initiative (SDI) has the potential of revolutionizing strategic policy, particularly if it proves possible to construct a stable deterrent relationship based on defensive rather than offensive capabilties. As President Reagan put in in 1983 when he announced the SDI program:

What if free people could live secure in the knowledge that their security did not rest upon the threat of instant US retaliation to deter a Soviet attack, that we could intercept and destroy strategic ballistic missiles before they reached our own soil or that of our allies?[21]

Carnes Lord has written, "The geopolitical position, political culture, and military tradition of the United States have together shaped a national strategic outlook which is fundamentally defensive. Obvious as this may seem, it is too often forgotten in contemporary discussions, particularly of nuclear strategy."[22] Lord was suggesting that there is such a thing as an "American strategic culture" and that this is oriented more toward the defense than the offense. If so, then the Strategic Defense and Air Defense Initiatives will return US policy to that orientation. The impact of that reorientation on US defense policy, however, will be significant. The importance of nuclear weapons themselves may change, especially as new "smart" conventional weapons are able to do some of the same warfighting jobs better.

Such a change in strategic orientation will require more than the effective adaptation of new technology. The challenge for US decisionmakers will be to integrate all four of the dimensions of strategy discussed by Michael Howard in Part I of this book—technological, operational, logistical, and social. As the Commission on Integrated Long-Term Strategy highlighted in early 1988:

Our strategy must be designed for the long term, to guide force development, weapons procurement, and arms negotiations. Armaments the Pentagon chooses today will serve our forces well into the next century. Arms agreements take years to negotiate and remain in force for decades.

Our strategy must also be integrated. We should not decide in isolation questions about new technology, force structure, mobility and bases, conventional and nuclear arms, extreme threats and Third World con-

flicts. We need to fit together our plans and forces for a wide range of conflicts, from the lowest intensity and highest probability to the most apocalyptic and least likely.[23]

The strategic challenge at hand is not only to maintain a credible and stable deterrent against major aggression, whether nuclear or conventional, against the United States and its allies. The US must also be able to deal with the recurring and lower-intensity conflicts in the Third World and develop a viable strategy for the use of military fora in such conflicts. Our strategy needs to define the proper role of military forces. Former Secretary of Defense Weinberger has set down a series of conditions for such involvement, and the Commission on Integrated Long-Range Strategy also specifies a series of propositions upon which to build such a strategy.[24] These conditions are a start, but only that. They need to form the basis of a consistent US policy that enjoys support both throughout the executive branch and within the Congress on a bipartisan basis.

ALLIANCE RELATIONSHIPS

An important aspect of US strategy has always been our alliance relationships. World War II ended a tradition of US isolationism and, with the Rio Treaty in 1947 and NATO in 1949, the United States began to enter into a series of peacetime collective security agreements on which our security—as well as that of our allies—relies. These alliances, both bilateral and multilateral, have served us well. However, now the US alliance relationship is in transition both with respect to NATO and with allies outside of the NATO framework. Within NATO, the nature of the consensus that created the organization and sustained it for nearly 40 years may be changing. Among our other allies, the level of support and commitment appears to be changing if not eroding. The lack of allied support for US operations in Grenada and Libya, and the demands to shut down US bases from Spain to the Philippines to Central America—bases vital to maintaining US and allied security interests—reflect these trends. The obvious challenge is how to alter both trends, particularly in light of the fact that "In the future, even more than in the last forty years, the United States will need its allies to share the risks and burdens of the common defense."[25]

In NATO the change is manifested in three areas: military, economic, and political. Militarily, the critical issues involve the viability of the traditional strategy, the military balance, arms control, and out-of-area issues. As discussed at length in Part II of this book, since 1967 the NATO strategy has been one of flexible response calling for a balanced mix of military capabilities across the spectrum of conflict. The strategy has depended upon the NATO triad of strategic nuclear, theater nuclear, and conventional forces deployed for the forward defense of allied territory against a Warsaw Pact attack. However, Europeans and Americans are not of the same mind as to which one of the three legs of this NATO triad is most important, or the extent to which the relationship among these three elements of the triad will change. The elimination of all land-based INF missiles over 500-kilometer range has reenergized this long-standing debate within the Alliance. At issue is the best way to maintain a stable deterrent in Europe, especially as one element of NATO's nuclear response capability is to be eliminated by superpower agreement. Improved conventional weapons technology may enable new capabilities once obtainable only with nuclear weapons, but conventional deterrence is not a fully credible substitute for nuclear deterrence. There are practical limits as well on a decreased reliance on nuclear deterrence. Conventional forces require money and manpower. Neither is likely to increase in the years ahead.

Along with declining defense budgets throughout NATO, many allies are having difficulty maintaining the size of their armed forces due to a decline in the number of draft-eligible men. This is especially true of West Germany which has provided over half of NATO's central front land forces and which has had an active force of 495,000. Estimates suggest that the available manpower in West Germany might be cut in half by the end of this century. Already the Bundeswehr has decreased to 478,000.[26] The challenge, therefore, is how to address the conventional military imbalance in Europe between NATO and the Warsaw Pact in light of these economic and demographic realities. Possible alternatives include new conventional technologies. The traditional US response is more quality to meet quantity, but such technologies are expensive to acquire. Another alternative is an increased role for women, but there is a significant social cost in this alternative as well. Another possibility is arms control, as being developed in the CFE talks, but this is not a total solution.

A changing political environment in Europe also complicates the future of the NATO alliance relationship. Allies do not always share US perceptions of the Soviet threat, particularly in light of the Gorbachev reforms and arms control proposals, as well as the political and economic changes throughout Eastern Europe. Many have a regional rather than a global threat perspective which results in a reluctance by the European NATO nations to support US interests outside the region, particularly in the Middle East, Central America, and Southeast Asia. Yet, if the interests of the West are to be protected through the end of this century, Euro-

pean cooperation will be needed. Increasing recognition by many European allies that their own interests are often at stake in the Third World—especially the Persian Gulf—is encouraging.

At the same time, there is a growing movement among the left-wing parties throughout Europe to reject the traditional NATO doctrine and to challenge standing NATO consensus. This trend is reflected in the call for alternative defense policies that would reject offensive weapons and rely more on internal defense of the nation only after it had been invaded. Similarly, the British Labour party has called for the "denuclearization" of Great Britain, while West German Social Democrats argue for "nuclear-free zones" in Central Europe. In short, there is a sense—again—that the NATO Alliance is "in transition." Political change, economic pressures, and debates about military posture provide the recipe for uncertainty. That uncertainty extends to the form US leadership should take and, indeed, the nature of the Alliance in the years to come. Yet to be factored in is the impact of the dissolution of Soviet hegemony in Eastern Europe and potential changes in the nature of NATO and the Warsaw Pact.

Even if NATO continues much as it is today, there are a number of issues that will need to be resolved if the Alliance is to be effective. These include strategy, force posture, and modernization; interoperability and standardization of systems and procedures; and a panoply of defense economic questions involving burden sharing and a cooperative development of major weapons systems. Henry Kissinger has written:

This state of affairs has deeper causes than particular policies on either side. The present NATO structure is simply not working either in defining the threat or in finding methods to meet it.

Existing arrangements are unbalanced. When one country dominates the Alliance on all major issues— when that one country chooses weapons and decides deployments, conducts the arms control negotiations, sets the tone for East-West diplomacy and creates the framework for relations with the Third World—little incentive remains for a serious joint effort to redefine the requirements of security or to coordinate foreign policies.[27]

Unfortunately, this lack of allied commitment and solidarity is not limited to Europe. As already mentioned, it entails many of our allies, whether it be calls by Canadian opposition parties to remove Canada from NATO and NORAD or the refusal of New Zealand—a key member of the ANZUS alliance—to permit US ships carrying nuclear weapons from using her ports. At stake is nothing less than our worldwide forward-deployment strategy and the overseas bases upon which it depends. In this respect,

Secretary of State George Shultz has usefully highlighted the fundamental questions which encompass all of the US alliance relationships:

Our differences with New Zealand are specific and immediate, yet they raise the most basic questions about alliances and about alliance responsibilities in the modern world: What is the purpose of our alliances? What qualities are unique to an alliance of democracies? How do we manage our alliances in a new era in furtherance of our common purpose?[28]

THE CHALLENGE OF GOOD STEWARDSHIP—DEFENSE ORGANIZATION AND PROCESS

Formulating sound military strategy and managing alliance relationships in a changing political environment are but two aspects of the challenge facing US decisionmakers. Equally important is the requirement to confront the demands of a national economy faced with, among other things, record deficits and global indebtedness resulting in a declining defense budget. The substantial growth in defense expenditures during the first Reagan administration has been reversed. From a peak in 1985 of $323 billion, total budget authority for defense has declined consecutively since then, and substantial cuts in force structure in all services are already planned. Prophetically, in 1987 the Commission on Integrated Long-Term Strategy concluded, "The resources available for defense will probably be constrained more than in the past, principally by concern over the national debt and pressures for social spending."[29] The meaning is clear: in the 1990s the Department of Defense will have to do with less. In fact, in 1989 Secretary of Defense Cheney called for a $180 billion cut in defense spending to be realized within the first part of this decade. We can expect, therefore, the size of the military forces in the next century to be significantly smaller with larger numbers being in the reserves. The obvious challenge is how to continue to do the job with shrinking forces and budgets.

The answer will in a large part be found in a more responsible, accountable and effective defense organization. In recent years the Department of Defense has come under increased criticism for fraud, waste, and abuse, symbolized most dramatically in paying exorbitant amounts for hammers and other common pieces of equipment. The resulting loss of confidence by the American public has undermined the efforts by the Department of Defense to justify continued levels of increased defense spending. In short, if confidence is to be restored and the Department of Defense is to meet the challenge faced by shrinking budgets, it will have to become a better steward of public trust and resources.

In 1986, the Packard Commission on Defense Management concluded:

Today there is no rational system whereby the Executive Branch and the Congress reach coherent and enduring agreement on national military strategy, the forces to carry it out and the funding that should be provided—in light of the overall economy and competing claims on national resources. The absence of such a system contributes substantially to the instability and uncertainty that plague our defense program. These cause imbalances in our military forces and capabilities, and increase the cost of procuring military equipment.[30]

If these shortcomings are to be addressed, reforms in organization and process are needed in two areas: the national security decision-making process both within and outside the Department of Defense, and the acquisition management system within DoD that is responsible for the multibillion-dollar defense procurement program. In recent years these two areas have come under extensive review by agencies inside and outside of government to include the US Senate Armed Services Committee under the leadership of senators Goldwater and Nunn, the Georgetown University Center for Strategic and International Studies, and the Packard Commission. The recommendations of those panels and the resulting legislative reforms are examined in detail in Part IV of this book. None of these problems will be easily solved, and the issues raised by these and other panels will form the agenda of organizational issues which defense decisionmakers will have to continue to address in the years ahead.

In the area of national security management, the principal issues include:[31]

—Uncertainty about the lines of responsibility between the National Security Council, the executive branch of government, and the Congress.

—An inadequate budget review process that involves Congressional duplication and a fixation either with minute details or gross dollar allocation at the expense of consideration of strategy, missions, and overall defense policy.

—Concern with daily and immediate issues at the expense of long-range planning in both the executive and legislative branches.

—Inefficiency and parochialism within the Department of Defense as a result of bureaucratization and interservice rivalry.

—Shortcomings in the quality of advice provided by the services and by the chairman of the Joint Chiefs of Staff.

—Inadequate inputs into the defense decision system by the operational commanders.

—Lack of effective planning and execution of

operations to deal with low-level conflicts and terrorism.

In addition to these broader issues of managing the development of national security policy, studies have repeatedly highlighted the following deficiencies in the way weapon systems are developed and acquired:

—A weapon systems acquisition process that is too slow, too costly, and often results in systems that do not perform up to specifications.

—Fragmentation of the acquisition process with no single responsible official in the Office of the Secretary of Defense.

—Needless duplication of effort and lack of uniformity among the services, failure to coordinate, and frequent overstatement of new weapon systems requirements.

—Failure to take advantage of economies of scale as a result of unstable funding allocations and lack of follow-through in program development, thereby discouraging long-term investment by contractors.

—Complex federal laws regulating defense procurement.

At issue is nothing less than the way our government organizes itself to formulate and implement defense policy and to manage its resources to that end. Many reform initiatives are already under way to correct these deficiencies, including efforts to strengthen the role of the chairman of the Joint Chiefs of Staff and the operational commanders in the field, to emphasize planning for low-intensity conflict, to create a more comprehensive defense planning and budgetary process, to centralize the weapons acquisition process under the secretary of defense, and to generate greater stability and accountability in the contractual relationship between government and private industry.

Organizational and legislative remedies are only part of the solution. Attitudes will also have to change. We will have to rethink the role and relationship of the individual military services as warfare gets more complex and as technology increasingly blurs the distinction between various dimensions of warfare. We will have to manage the tendency toward increasing centralization lest we lose flexibility and responsiveness in the process. In short, as the Tower Commission pointed out in its analysis of the Iran-contra affair, an integral part of the problem is the political pressure placed on policy-making institutions and the need to maintain integrity, objectivity, and coherence of vision in the midst of those pressures:

If but one of the major policy mistakes we examined had been avoided, the nation's history would bear one less scar, one less embarrassment, one less opportunity for opponents to reverse the principles this nation seeks to preserve and advance in the world.

As a collection, these recommendations are offered to those who will find themselves in situations similar to the ones we reviewed: under stress, with high stakes, given little time, using incomplete information, and troubled by premature disclosures. In such a state, modest improvements may yield surprising gains. This is our hope.[32]

This is also our challenge.

No less a challenge, and one so new that its nature is still uncertain, concerns how to restructure the defense organization, redefine the defense role, and redesign the force posture in light of both international factors (read changes in threat) and domestic imperatives (read budget constraints). The December 1989 military operation in Panama to seize General Noriega and bring him to trial in the United States on drug-related charges is perhaps indicative of the type of role and mission for US forces in this decade.

THE PROMISE OF TECHNOLOGY

Technology impacts on the nature of the threat and on the tools with which a nation responds to that threat. It can offer the promise of greater security; it can also undermine a stable relationship and create uncertainty and insecurity. Changes in military technology alter the man-machine relationship and influence strategy and the nature of a nation's defense policy. New technologies offer the promise of finding alternatives to nuclear war, of reducing casualties in conventional war, and enhancing the reliability, maintainability, performance, and effectiveness of existing weapons. At the same time, significant enhancements in range, levels of destruction, and accuracy associated with precision-guided weapons and other high-technology weapons will expand the geography, time, and destructiveness of war. New conventional weapons technologies have the potential of radically altering the military relationship between NATO and the Warsaw Pact. Less labor-intensive weapons can enable the West to maintain defensive strength despite reduced manpower levels. Yet they must be maintainable, survivable, and affordable.

In exploiting the opportunities provided by new defense technologies, the principal challenge is to anticipate their impact on military strategy, doctrine, and force posture. Selecting technologies for development will involve complex tradeoffs with substantial uncertainty in the decision process. Beyond that, however, is the need to adapt those technologies to our defense needs—and to adapt our defense structures to those technologies.

Among the most promising new defense technologies are those which deal with low observables, or stealth—"smart" weapons, precision-guided munitions that combine accuracy and long range; ballistic missile defense; and space capabilities for wartime operations.[33] Such technologies have the potential of revolutionizing warfare and, indeed, the nature of and requirements for deterrence. One effort—the Balanced Technology Initiative (BTI)—seeks to ensure that resources for research and development of advanced technologies are applied so as to maximize benefits for all services and missions. Their potential application can improve substantially both the US strategic deterrent and US and allied defense in areas such as NATO where the Warsaw Pact retains a substantial quantitative superiority.[34]

Perhaps one of the most significant areas of research includes the development of unmanned weapon systems or remotely piloted vehicles (RPVs). These systems are designed to carry out complicated missions involving inflight decisions and changes in targets. Not only are these systems significantly less expensive and labor-intensive, they reduce the exposure of military personnel to hostile environments. It is conceivable that future Air Force squadrons will be made up solely of RPVs. Projects like Pave Tiger—testing and developing unmanned minidrones to conduct such military operations as defense suppression, jamming, and attack— offer the potential to remove man from his fighting machines. The Air Force is also testing unmanned bombers, and similar studies are under way on land and sea systems. These projects are controversial and have yet to be proven. Nevertheless, the fact that major technological advances are being achieved suggests that defense decisionmakers are going to have to rethink not only the nature of conflict in the future but also the role of man in it.

While technology has the potential to resolve a number of the problems that defense decisionmakers will face in the future, it will also generate new ones. Technology cannot overcome all military problems. As Michael Howard and others have pointed out, the US has tended to stress technology often at the expense of the other three dimensions of classical strategy— operational, logistical, and social. Vietnam is a dramatic example of what can happen when technology dominates the decision process. In taking advantage of the promise of technology, we will have to ensure that it is applied within a coherent strategic framework. Technology can be adapted to strategy, but there is a limit to which strategy can be adapted to technology.

CONCLUSION

This brief essay has been an indirect look at the future. The goal has not been to predict the state of American security nor to forecast American defense policy in the next century. Rather the effort has been to identify the major signposts—the critical issues—which will delineate the future course of American defense policy.

The list has not been intended to be all-inclusive. Not mentioned have been a variety of issues dealing with personnel matters, the role of women, changes in lifestyles, civil-military relations, the viability of an all-volunteer force, and many others. Nevertheless, all of these problems will pose challenges to future generations of defense decisionmakers. If the right decisions are to be made, those decisionmakers will have to possess the knowledge, integrity, vision, and commitment necessary to address those challenges. This book is dedicated to preparing individuals both in and out of uniform to answer these challenges. In this sense, the task of education is perhaps the most significant challenge facing all of us in the years ahead.

NOTES

1. For a detailed explanation of the framework and how it applies to the current international political system, see K. J. Holsti, *International Politics: A Framework for Analysis* (Englewood Cliffs, N.J.: Prentice-Hall, 1983).

2. *Air Force Times*, 23 Mar. 1987.

3. Commission on Integrated Long-Term Strategy, *Discriminate Deterrence* (Washington, D.C.: G.P.O., Jan. 1988), p. 7.

4. Ibid.

5. Ibid., p. 9.

6. Ibid., p. 10.

7. For a discussion of both sides of the issue, see Harvard Nuclear Study Group, Carnesale et al., *Living with Nuclear Weapons* (New York: Bantam Books, 1983), pp. 215–31.

8. White House, *National Security Strategy of the United States*, Jan. 1987, p. 23.

9. Henry Kissinger, "A New Approach to Arms Control," *Time*, 21 Mar. 1984. Taken from *Observations* by Henry Kissinger (Boston: Little, Brown, 1985), pp. 151–65.

10. The four purposes of an effective arms control agreement are 1) to achieve and maintain strategic and crisis stability, 2) to reduce weapons, 3) to reduce damage if war should occur, and 4) to reduce defense expenditures.

11. Kissinger, "New Approach," p. 156.

12. White House, Jan. 1987, p. 6.

13. General Robert T. Herres, "Soviet Space Objectives," speech delivered to the Institute of Real Estate Management, Denver, Colo., 16 Oct. 1986, pp. 3–4.

14. Ibid.

15. H. Lawrence Ganett II, "Terrorism and the Use of Military Force," *Defense/87*, p. 26.

16. Ibid.

17. Ibid.

18. Ibid., p. 27.

19. James M. Dorsey, "Terrorism May Be Met by Force—Weinberger," *Washington Times*, 22 Jan. 1987, p. 7C.

20. For a detailed discussion of this distinction, see Fritz W. Ermath, "Contrasts in American and Soviet Strategic Thought," *International Security* 3 no. 2 (1978): 138–55.

21. From President Reagan's 23 Mar. 1983 speech.

22. Carnes Lord, "American Strategic Culture," *Comparative Strategy* 5, no. 1 (1985): 275.

23. Commission on Integrated Long-Term Strategy, p. 1.

24. Weinberger's conditions included "six major tests to be applied when we are weighing the use of US combat forces abroad. Elements to be identified by these tests were: vital to our national interest, 'the clear intention of winning,' well-defined political and military interests, a continual willingness to reassess the 'size, composition and disposition' of the forces involved, the recognition that the use of US troops was a 'last resort' "—and public and congressional support. See "What *Are* the Lessons of Vietnam?" by David Fromkin and James Chace, *Foreign Affairs* (Spring 1986): 730. The commission called for a strategy built on six propositions. These included: US forces will not in general be combatants; the United States should support anti-Communist insurgencies; security assistance requires new legislation and more resources; the United States needs to work with its Third World allies at developing "cooperative forces"; in the Third World, no less than in developed countires, US strategy should seek to maximize our technological advantage; and the United States must develop alternatives to overseas bases. See Commission on Integrated Long Term Strategy, pp. 16–22.

25. Commission on Integrated Long Term Strategy, p. 3.

26. The International Institute for Strategic Studies (IISS), *The Military Balance, 1983–84* (London, 1983), p. 145.

27. Henry Kissinger, "A Plan to Reshape NATO," *Time*, 5 Mar. 1984.

28. George P. Shultz, "The Goal of Our Alliances: A Collective Security System," *Vital Speeches of the Day*, vol. 51, no. 21, 15 Aug. 1985, p. 642.

39. Commission on Integrated Long Term Strategy, p. 57.

30. *A Quest for Excellence*, Final Report to the President by the President's Blue Ribbon Commission on Defense Management. Referred to as the Packard Commission Report, June 1986, p. xvii.

31. The list is derived from the reports and studies mentioned to include the *Packard Commission Report*, *The Tower Commission Report*, and the statement of General David C. Jones, USAF (ret.), former chairman of the Joint Chiefs of Staff.

32. Tower Commission Report, pp. v–7.

33. For a detailed discussion of these four, see Commission on Integrated Long Term Strategy, pp. 49–61.

34. For further analysis of these technologies, see Debra Polsky, "Initiative Explores Conventional Technology," *Army Times*, 4 Jan. 1988, p. 29.

ABOUT THE CONTRIBUTORS

WILLIAM E. BERRY, JR., Colonel, USAF, holds a Ph.D. from Cornell University and served in the USAF Academy Department of Political Science from 1978 to 1980 and from 1980 to 1984 when he was associate professor and director of Comparative and Area Studies. He subsequently served with US Forces Korea and as deputy director of National Security Studies at the National War College. He is currently the US air attache to Malaysia.

BRENT D. BRANDON, Captain, USAF, graduated from the USAF Academy in 1984 and received his M.P.P. from the John F. Kennedy School of Government, Harvard University, in 1984. He is currently an EF-111 navigator at Mountain Home AFB, Idaho.

CHARLES E. COSTANZO, Captain, USAF, was an assistant professor in the Department of Political Science, USAF Academy, from 1984 to 1988. He received his M.A. from the University of Pittsburgh with a certificate in international relations. He now serves as the military assistant to the Special Adviser for international and governmental affairs, Strategic Defense Initiative Organization, Washington, D.C.

R. JOSEPH DESUTTER, Lieutenant Colonel, USAF, received his Ph.D. from the University of Southern California. He served as an associate professor and director of American and Policy Studies in the Department of Political Science, USAF Academy, 1977–1979 and 1982–1985. After a tour on the Air Staff dealing with arms control policy, he served as military assistant to the director of the Office of Science and Technology Policy in the White House.

MARILYN H. GARCIA, Captain, USAF, graduated from the USAF Academy in 1985 and attended pilot training at Laughlin AFB, Texas. She flies C-130s at Andersen AFB, Guam.

DAVID L. GIDDENS, Colonel, USAF, received his Ph.D. from the University of Denver and was an assistant professor of Russian language in the Department of Foreign Languages, USAF Academy, 1980–1984. Colonel Giddens was an assistant air attache in the Soviet Union and has since served as special assistant to the assistant director of the US Arms Control and Disarmament Agency in Washington, D.C.

ROBERT P. HAFFA, JR., Colonel, USAF Retired, received his Ph.D. from the Massachu-

setts Institute of Technology. Colonel Haffa was a command navigator and a graduate of the National War College. He was professor and acting head of the Department of Political Science, USAF Academy, from 1982 to 1984. He subsequently served as the director of Long Range Plans on the Air Staff and then chief of the USAF chief of staff's staff group. He is the author of *The Half War: Planning U.S. Rapid Deployment Forces to Meet a Limited Contingency* and *Planning U.S. Forces*.

HENRY O. JOHNSON III received his M.A. from the University of Virginia and was an assistant professor in the Department of Political Science, USAF Academy, 1983–1986. He is currently an executive with IBM in Portland, Oregon.

DAVID C. KOZAK is professor of public policy and director of the Institute for Policy and Leadership Studies at Gannon University. A retired USAF officer, Dr. Kozak received his Ph.D. from the University of Pittsburgh and was an American Political Science Association Congressional Fellow. He was an associate professor in the Department of Political Science, USAF Academy, from 1977 to 1981 and then served as professor of public policy at the National War College. He is coauthor of *Congress and Public Policy: A Source Book of Documents and Readings* (with John D. Macartney), *The American Presidency: A Policy Perspective from Readings and Documents* (with Kenneth N. Ciboski), and *Bureaucratic Politics and National Security: Theory and Practice* (with James M. Keagle).

JAY L. LORENZEN, Major, USAF, is a doctoral candidate at the University of Denver. He received his M.A.L.D. degree from the Fletcher School of Law and Diplomacy after graduate studies as an Olmsted Fellow at the University of Tuebingen in Germany. A senior navigator, Major Lorenzen was an assistant professor in the Department of Political Science, USAF Academy, from 1984 to 1987 and subsequently served as the deputy political adviser to the commander-in-chief, US Space Command.

E. DOUGLAS MENARCHIK, Colonel, USAF, received his Ph.D. from George Washington University. A command pilot, Colonel Menarchik was an assistant professor in the Department of Political Science, USAF Academy, 1979–1982. Following a tour on the Air Staff, he served as military assistant to Vice President Bush. He is a graduate of the NATO Defense College in Rome and is currently the deputy director of operations for a C-5 wing at Altus AFB, Oklahoma.

DOUGLAS J. MURRAY, Colonel, USAF, has been professor and head of the Department of Political Science, USAF Academy, since 1984. He holds a Ph.D. from the University of Texas and is a graduate of the National War College. Following an earlier tour as associate professor and director of Comparative and Area Studies at the USAF Academy, Colonel Murray served as the deputy chief, Air Force Secretary's Staff Group in Washington, D.C. He is coeditor (with Paul R. Viotti) of *The Defense Policies of Nations* and the author of *The Theory and Practice of Canadian Defense* and *U.S.-Canadian Defense Relations: An Assessment for the 1980's*.

GREGORY J. RATTRAY, Captain, USAF, graduated from the Air Force Academy in 1984 and received his M.P.P. from the John F. Kennedy School of Government, Harvard University. Captain Rattray has since served as an intelligence officer at Headquarters, Strategic Air Command.

CARL W. REDDEL, Colonel, USAF, is professor and head, Department of History, USAF Academy. Colonel Reddel received his Ph.D. in Russian history from Indiana University and has served as a research associate at the University of Edinburgh. He also serves as a team chief for the US On-Site Inspection Agency in monitoring Soviet compliance with the Intermediate Nuclear Forces Treaty.

ERVIN J. ROKKE, Brigadier General, USAF, holds a Ph.D. from Harvard University and is associate deputy director for operations, National Security Agency, Ft. Meade, Maryland. General Rokke was dean of faculty at the Air Force Academy from 1984 to 1986 and, from 1977, professor and head, Department of Polit-

ical Science. General Rokke also served in Moscow as the US defense attache to the Soviet Union (1987–1989) and in London as the US air attache to the United Kingdom (1980–1982). He was coeditor, with Richard G. Head, of the third edition of *American Defense Policy*.

FRANK L. ROSA, Lieutenant Colonel, USAF, received his Ph.D. from Notre Dame and, until 1989, served as associate professor and director of International Relations and Defense Studies in the Department of Political Science, USAF Academy. A command navigator, Colonel Rosa has served in numerous operational assignments and is currently a politico-military affairs officer at US Atlantic Command.

RONALD J. SULLIVAN, Lieutenant Colonel, USAF, is currently politico-military affairs adviser to the US military representative to the NATO Military Committee in Brussels. A command pilot, Colonel Sullivan is a doctoral candidate with an M.A.L.D. degree from the Fletcher School of Law and Diplomacy. Colonel Sullivan was assistant professor and director of International Relations and Defense Studies in the Deparment of Political Science, USAF Academy, from 1981 to 1985. He subsequently served as the deputy political adviser to the command-in-chief, United States Air Forces, Europe.

PAUL R. VIOTTI, Colonel, USAF, is a senior tenure professor and deputy head, Department of Political Science, USAF Academy. Colonel Viotti received his Ph.D. from the University of California, Berkeley, and previously served in the Political Science Department as director of International and Defense Studies. From 1981 to 1983, he was deputy political adviser to the commander-in-chief, US European Command. He is coeditor (with Douglas J. Murray) of *The Defense Policies of Nations*, editor of *Conflict and Arms Control*, and coauthor (with Mark V. Kauppi) of *International Relations Theory: Realism, Pluralism, and Globalism*.

ABOUT THE EDITORS

SCHUYLER FOERSTER, Lieutenant Colonel, USAF, is a tenure associate professor and director of International Relations and Defense Studies, Department of Political Science, USAF Academy. He holds a D.Phil. in politics from Oxford University and masters degrees in international relations and public administration from the Fletcher School of Law and Diplomacy and the American University. A graduate of the USAF Academy, he has served as an intelligence officer in Southeast Asia and Washington. Most recently, he served as a politico-military affairs officer in the Office of the Defense Advisor, US Mission to NATO, and as a National Security Fellow at the John F. Kennedy School of Government at Harvard University. He has published several articles on strategy and arms control and is coauthor of *Defining Stability: Conventional Arms Control in a Changing Europe*.

EDWARD N. WRIGHT, Lieutenant Colonel, USAF, is a tenure associate professor and director of American and Policy Studies, Department of Political Science, USAF Academy. He holds a Ph.D in American government from Georgetown University and a masters degree in international relations from the University of Southern Mississippi. He has served in the Office of the Secretary of Defense and was a research associate at the Carnegie Endowment for International Peace where he was the executive director of a major study of national security issues and arms control. Most recently, he served as the special assistant for national security affairs to the Attorney General of the United States. He has written several articles on the American presidency and defense issues and is authoring a book entitled *Presidential Responsibility for Emergency Preparedness: the Reagan Response*.

INDEX